3rd International Conference on Environment and Sustainable Development of Territories: Ecological Challenges of the 21st Century

IOP Conference Series: Earth and Environmental Science
Volume 107

Kazan, Russia
27 – 29 September 2017

ISBN: 978-1-5108-5773-5
ISSN: 1755-1307

Printed from e-media with permission by:

Curran Associates, Inc.
57 Morehouse Lane
Red Hook, NY 12571

Some format issues inherent in the e-media version may also appear in this print version.

Printed by Curran Associates, Inc. (2018)

For permission requests, please contact the Institute of Physics at the address below.

Institute of Physics
Dirac House, Temple Back
Bristol BS1 6BE UK

Phone: 44 1 17 929 7481
Fax: 44 1 17 920 0979

techtracking@iop.org

Additional copies of this publication are available from:

Curran Associates, Inc.
57 Morehouse Lane
Red Hook, NY 12571 USA
Phone: 845-758-0400
Fax: 845-758-2633
Email: curran@proceedings.com
Web: www.proceedings.com

Table of contents

Volume 107

3rd International Conference Environment and Sustainable Development of Territories: Ecological Challenges of the 21st Century

27–29 September 2017, Kazan, Russian Federation

Accepted papers received: 18 December 2017
Published online: 31 January 2018

Preface

Papers

1. Geoecology and Exodynamic processes

4. Strategies for Biological Resources Protection and Optimization

6. Soil and Land Resources, their Conservation and Efficient Management, and Food Safety

7. Social, Economic and Ecological Aspects of Urban Territories Development Analysis and Management

IOP Conf. Series: Earth and Environmental Science **107** (2017) 011001 doi:10.1088/1755-1315/107/1/011001

Preface

The 3rd International Conference *Environment and Sustainable Development of Territories: Ecological Challenges of the 21st Century* was successfully held in Kazan, Russia 27–29 September 2017.

Currently, there are significant changes in the environment: the depletion of the planet's resources, environmental pollution, reduced biodiversity and frequent natural and man-made disasters which are bringing huge damage to the natural and socio-economic systems. All these problems resonate especially in 2017 – announced in Russia as the "Year of Environmental Protection". Under consideration during the conference were the fundamental basis of a comprehensive analysis of the state of the environmental components, the interdisciplinary aspects of the strategy for human survival and the development of scientific bases and technologies for the control and regulation of the quality of the environment and environmental technologies. The conference was attended by over 300 researchers from different countries, who discussed the results of their work within the framework of six main areas: Global Pollution, Environmental Impact Management, Ecological Risks and Sustainable Development; Geoecology and Exodynamic Processes; Climate Change, Macro-circulation Processes and the Ecology of Atmosphere; Strategies for Biological Resource Protection and Optimization; Mathematical Methods and GIS Technologies in Ecology and Management of Natural Resources; Soil and Land Resources, their Conservation and Efficient Management, and Food Safety; Social, Economic and Ecological Aspects of Urban Territory Development Analysis and Management.

The Conference became the platform for the coordination of efforts and developments in the area of environmental study, revealing the strategic areas of research, the ways of communication with state authorities and organizations who are working in the real economy.

More than 400 papers were submitted from universities and research institutes to the "*Environment and Sustainable Development of Territories: Ecological Challenges of the 21st Century*" Conference. All of the papers were subjected to peer-review by conference Programme Committee members and international reviewers. This proceeding included papers which fall within the scope of the IOP Conference Series on Earth and Environmental Sciences. The papers were selected on the base of their originality and relevancy to the conference.

On behalf of the organizers we would like to thank all the authors who have contributed to our conference, reviewers, chairpersons and sponsors.

We believe that the next conference will attract the attention of more and more researchers and become a significant event for our colleagues all over the world.

Professor
Svetlana Selivanovskaya
Kazan federal university, Russia

IOP Conf. Series: Earth and Environmental Science **107** (2017) 011002 doi:10.1088/1755-1315/107/1/011002

Peer review statement

All papers published in this volume of *IOP Conference Series: Earth and Environmental Science* have been peer reviewed through processes administered by the proceedings Editors. Reviews were conducted by expert referees to the professional and scientific standards expected of a proceedings journal published by IOP Publishing.

IOP Conf. Series: Earth and Environmental Science **107** (2017) 012001 doi:10.1088/1755-1315/107/1/012001

Hydroecological characteristic of coal-mining regions with crucial anthropogenic load (in the case study of the Yaiva river basin)

O A Berezina, N G Maksimovich and S V Pyankov

Perm State University, Institute of the Natural Science, 15, Bukireva str., 614990, Perm, Russia

E-mail: berezina.olga16@gmail.com

Abstract. Coal mines closure is usually associated with a list of environmental problems and one of them is their negative impact on a region's river system. Hydroecological features of the Yaiva river basin located on the Kizel coal basin territory have been assessed. This river basin is typical for the Kizel coal basin territory that is why it has been taken as a model to verify the research algorithms. Further these algorithms will be applied for the investigation of other rivers running in the Kizel coal basin territory. The main negative consequences of anthropogenic impact on the Yaiva river basin watercourse have been revealed.

1. Introduction

The Kizel coal basin (the Western Urals, Russia) occupies area of 200 km^2 and is located within West Urals folding zone adjacent to the pre-Ural boundary deflection. Folds have meridional and close to meridional orientation, they are elongated for tens of kilometers and complicated by numerous disjunctive dislocations [1]. Mining in the Kizel coal basin (KCB) had been carried out since 1796. The Kizel coal basin territory is within the drain area of the West Ural rivers that belongs to the Kama river basin (the Kama reservoir). All rivers are greatly influenced by the KCB. Coal mining makes ecological situation worse that is determined by lithologic-and-geochemical characteristics of coal-bearing formations. More than 50 elements have been found in coal, 12 of which have 10–1,000 times higher concentration in coal than in background strata [2].

The KCB is intensively karsted. The karst of the region is considered to be bare and covered that determines the dependence of the karst water regime of active circulation zone on atmospheric precipitation circulation regime [3, 4]. As a result of thickness diversity, great mining depth (up to 1,000 m) and overlying limestone being karsted the coal mining conditions are characterized as difficult. The increment of water in to the mines was 2,500 m^3/h in karst zones.

These two factors caused intensive environment pollution during coal mining. Water containing 4% of pyrite becomes acid (pH of 2–3) and is enriched with sulphate because of the pyrite oxidation. During the basin operation up to 100 million m^3 of mine waters were discharged into the local hydrological system without being purified [5].

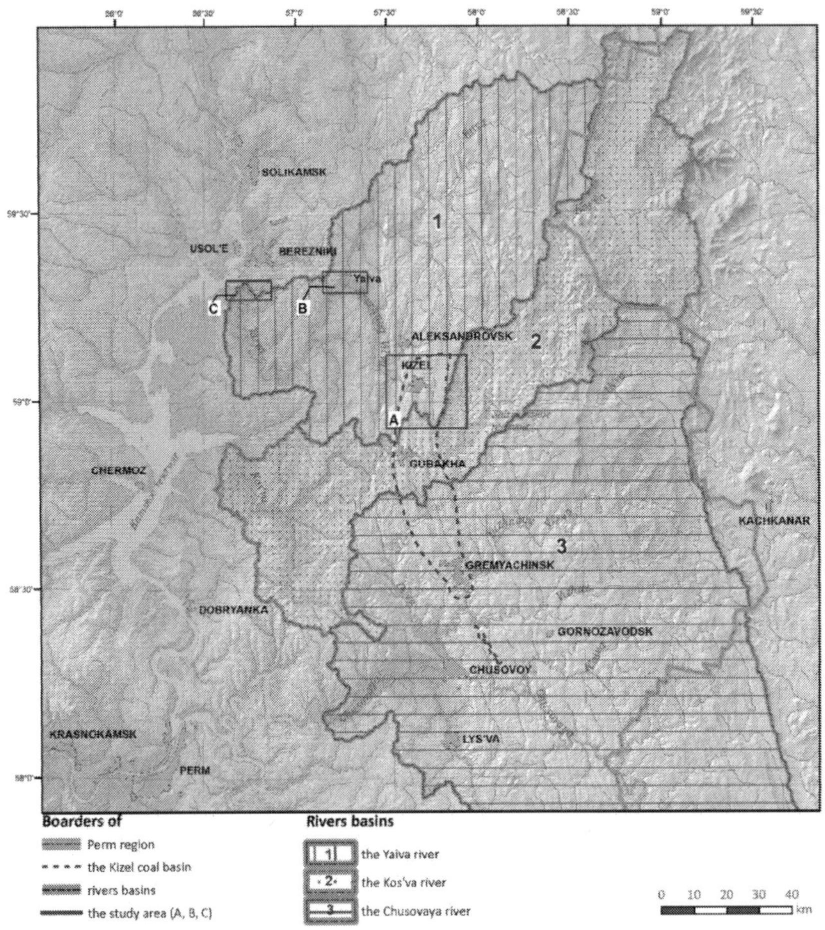

Figure 1. The Kizel coal basin surface waters.

The mines closure in the 1990s did not resolve the environmental problems [6]. After a gradual restoration of underground water level the mine water discharged during several years and according to several scientists the quality of the mine water discharge may worsen or become stable during 20-40 years [7]. When acid mine and drainage waters enter natural waterways, changes in pH and the formation of ochreous precipitates can have devastating effects on aquatic ecosystems [8, 9, 10, 11, 12].

The basins of three large rivers – the Yaiva, the Kos'va and the Chusovaya rivers were significantly effected by the anthropogenic load after the coal mines closure (figure 1). The Yaiva river basin is typical for the Kizel coal basin territory that is why it has been taken as a model to verify the research algorithms. Further these algorithms will be applied for the investigation of other rivers running the Kizel coal basin territory.

2. The problem statement

The Yaiva river starts in the North Ural mountains (the Kvarkush mountain range), 879 km from its mouth the river flows into the Kama reservoir making the Yaivinskii bay. The Yaiva basin is mostly located outside the boarders of the Kizel coal basin territory only the south-eastern part of the Yaiva catchment area is within its territory. The Severnaya Vil'va, being the first-order tributary of the Yaiva, is in the immediate region of the early developed coal field. Almost all pollution sources connected to coal production are located on the catchment area of the right tributary of the Severnaya Vil'va- the Bol'shoi Kizel river. The Bol'shoi Kizel and its tributaries are polluted as a result of the long-term

IOP Conf. Series: Earth and Environmental Science **107** (2017) 012001 doi:10.1088/1755-1315/107/1/012001

influence of mine waters (figure 2A). There are no pollution sources among the Bol'shoi Kizel's tributaries except the Severnyi Kizel catchment – the reason why the hydrochemical composition of the Severnyi Kizel's waters were not observed. The Bol'shoi Kizel when flowing into the Severnaya Vil'va has a negative impact on it.

There are three main pollution sources of surface water at the Yaiva catchment area: mine water discharge, polluted springs and tailingspiles' drainage water. There 7 discharges of mine water, 13 springs (7 of them are polluted) and 24 tailingspiles were discovered on the Yaiva river basin.

The monitoring of the surface water of the Kizel coal basin that is conducted by the Ural Centre for social-ecological monitoring of coal-mining territories [13], suggests regular water composition observations of the Yaiva river throughout its length. The observation results indicate a negative impact of the Kizel coal basin on the Yaiva river and its tributaries till it flows into the Kama reservoir. To analyse ecological condition of surface water objects standards for fishery water objects- the maximum permissible concentration (MPC_{fwo}) were used. The specific combinatorial index of water pollution (SCI WP) was also calculated according to the operating management directive RD 52.24.643-2002 [13] that also confirms a high anthropogenic load on the basin hydrological network. Based on the SCI WP values the surface water is divided into 5 classes according to its pollution (table 1). The management directive regulating the quality of surface and underground water for other purposes GN 2.1.5.1315-03 (MPC_{OP}) was used to assess the pollution of underground water (springs) [14], the composition of mine water discharge and of tailingspiles' drainage water.

Table 1. Classes of pollution and hydroecological assessment.

Class	Rank	Description	Hydroecological assessment [15]
1		Conditionally pure	normal
2		Slightly polluted	risk
3	a	Polluted	
	b	Significantly polluted	crisis
4	a	Polluted	
	b	Polluted	
	c	Significantly polluted	disaster
	d	Significantly polluted	
5		Extremely polluted	

3. Results and discussion

Water pollution in the monitoring section ranged from "polluted" (3 class "a") to "polluted" (4 class "b") based on the SC IWP calculation results by 14 components over 2013. According to the SCI WP the total hydroecolocical condition of watercourses was ranged as being normal, in risk, in crisis and in disaster (figure 2) [15]. Compounds of iron, aluminium, beryllium, lithium, manganese mostly contribute to water quality and this fact makes them critical indices of water pollution of these water objects. The schema (figure 3) provides the highest possible excesses of maximum permissible concentration (MPC) in the monitoring sections. The highest possible exceedance of MPC was observed at the Poludennyi Kizel river (below spring No 030): Fe – 9450, Mn – 1118, Al – 127, Be – 56.

Considering the above mentioned data we may conclude that surface water is contaminated along the rivers chain: the Poludennyi Kizel, the Bol'shoi Kizel (figure 3A), the Severnaya Vil'va (figure 3B), the Yaiva (figure 3C) rivers right up to the Kama reservoir. Special attention should be paid to the Bol'shoi Kizel river basin as most polluters are located there.

The Bol'shoi Kizel river (or the Kizel river) starts from the confluence of two rivers: the Poludennyi Kizel and Vostochnyi Kizel (figure 3A). The Bol'shoi Kizel stresses about 20 km from the Poludennyi Kizel's confluence to its mouth. Seven small rivers flow into the Bol'shoi Kizel and only one of them (the Severnyi Kizel) is not polluted by the Kizel coal basin mining. The tributaries bring a lot of pollutants into the Bol'shoi Kizel. Moreover the Bol'shoi Kizel is impacted by acid tailingspiles' drainage water and mine water discharge (figure 4). The Bol'shoi Kizel riverbed is stony and silted with hydrate Fe sediments.

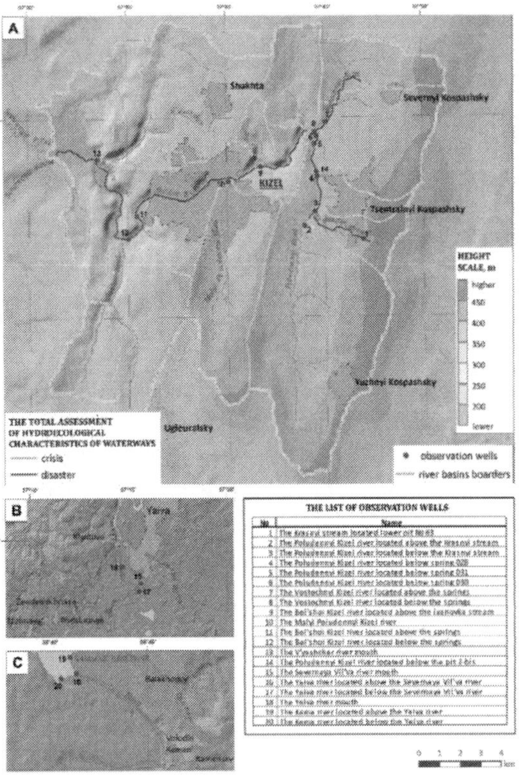

Figure 2. Total hydroecological assessment of watercourses and the measuring sections.

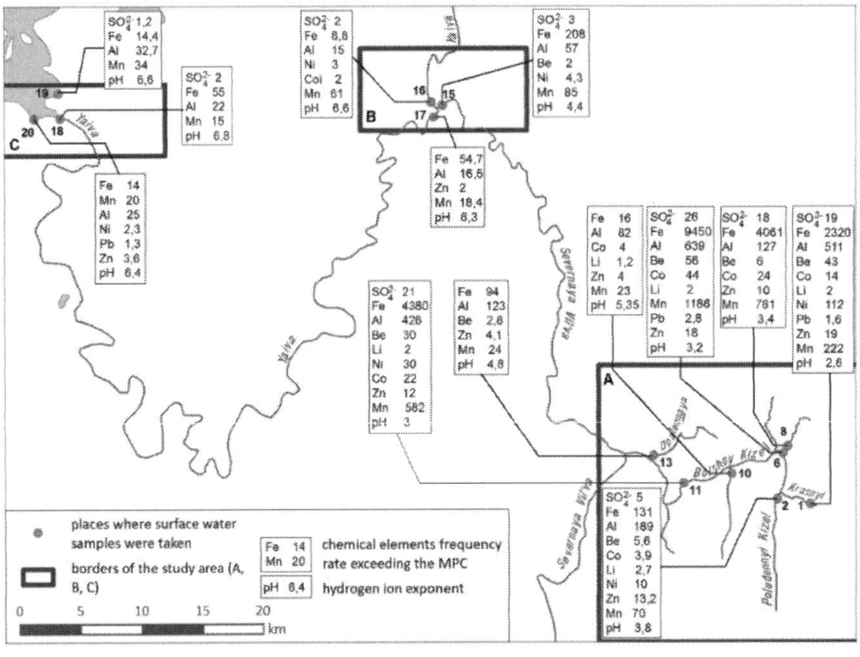

Figure 3. The Schema of monitoring sections location after the mines closure in the Yaiva basin.

Figure 4. Mine water discharge.

To characterize pollution sources that influence the Bol'shoi Kizel and its tributaries we have analyzed the mine water discharge and tailingspiles' drainage water data (table 2).

Table 2. Volume of mine water discharge into the Bol'shoi Kizel and its tributaries

The location of mine water discharge	River	Average discharge m^3/h	Maximum discharge, m^3/h	Minimum discharge, m^3/h
Subshaft of the Lenin Mine	The Bol'shoi Kizel	21.2	86.0	5.0
Shaft № 8 of the Lenin Mine	The Bol'shoi Kizel	268.0	587.0	94.0
The Volodarskii Mine adit	The Bol'shoi Kizel	13.4	40.0	1.0
Pit of 2-bis Kospashskaya Mine	The Poludennyi Kizel river and the Krasnyi stream	165.5	337.0	42
Pit № 63 of the Belyi Spoi Mine	The Poludennyi Kizel river and the Krasnyi stream	59.7	147.0	22.0
Total		527.8	1197	164

The overall rates of mine water that discharges immediately into the Bol'shoi Kizel (we do not consider the tributaries catchment) is on average 302 m^3/h, at that 89% of polluted mine water is discharged from shaft No 8 of the Lenin Mine. The maximum recorded discharge of mine water from this mine is a significant number– 587 m^3/h. The mine water discharge is acid (pH of 3-4), sulphate and has a high concentration of particular elements, first of all iron, aluminium, manganese, berylliu-m. The concentration of these elements significantly exceeds MPC_{OP}.

Annually on average 15,300 tons of sulfates, 6,000 tons of iron, 400 tons of aluminium and 57 tons of manganese come into the Bol'shoi Kizel directly (we do not consider the polluted tributaries); the amount of other pollutants is significantly low.

Tailingspiles' drainage water composition and volume surveys were made at 15 measuring sections. The maximum range of tailingspiles' drainage water was observed at the Vilodarskii Mine and at the Kospashskaya Mine – 20 and 22 m^3/h correspondingly.

Tailingspiles' drainage water has a high concentration of pollutants and is acid (pH < 3). High concentration of iron, aluminium, beryllium and manganese comes into water as a result of atmospheric precipitation infiltration. The concentration of these elements in tailingspiles' drainage water 100-1,000 times exceeds MPC_{OP} that leads to landscapes degradation. Heavy metals (cadmium, cobalt, nickel) were discovered practically everywhere but in low quantity. From time to time lead and zinc were revealed

IOP Conf. Series: Earth and Environmental Science **107** (2017) 012001 doi:10.1088/1755-1315/107/1/012001

in tailingspiles' drainage water samples. The concentration of these elements 2-70 times exceeds MPC_{OP}.

Mostly polluted is the tailingspiles' drainage water from a cone-shaped and flat tailingspiles from the Kospashkaya Mine. Concentration of main pollutants of mine waters is enormous: iron – 2083 - 6157 MPC_{OP}, aluminium – 5245 – 10955 MPC_{OP}, manganese – 250 – 880 MPC_{OP}, pH constantly < 3.0.

A pond at the Kizel river is of a particular interest. Its area is 30 hectares and it has been existing for a long time. Currently the pond is drained, however it has accumulated a lot of sediments (about 1.2 million m^3).

There a 2-km part of the Bol'shoi Kizel channel passes through a huge amount of sediments that are the pollution sources for the environment (figure 5).

Figure 5. Degraded landscape and bottom sediments at the Bol'shoi Kizel river's pond.

Thus, the Yaiva rive is greatly polluted by the sources located on its catchment area although they are tens of kilometres away from the river and are limited with watersheds. The Severnaya Vil'va river being the Yaiva river tributary and polluted by the Bol'shoi Kizel greatly impacts the Yaiva river. When the Bol'shoi Kizel flows into the Severnaya Vil'va river, the water of the latter changes the colour that is observed at the satellite image (figure 6).

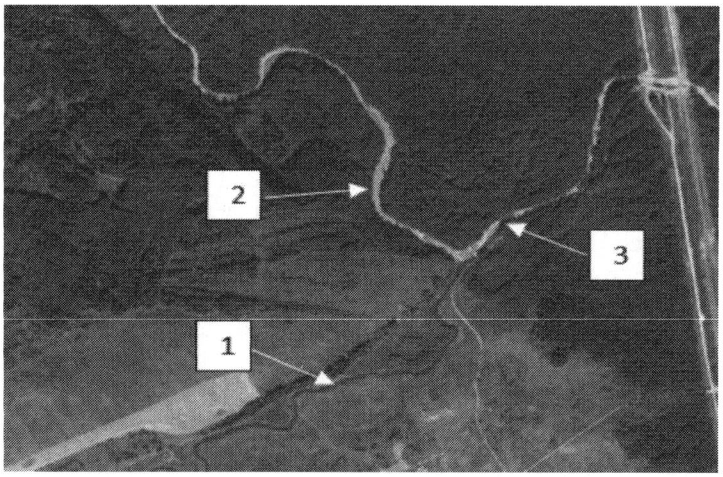

Figure 6. Place where the Bol'shoi Kizel flows into the Severnaya Vil'va.
(1 – the Severnaya Vil'va river; 2 – the Severnaya Vil'va after the Bol'shoi Kizel has flown into it; 3 – the Bol'shoi Kizel river)

The concentration of iron, aluminium and manganese naturally increases below the place where the Bol'shoi Kizel flows into the Severnaya Vil'va. These elements amount decreases in the Yaiva river mouth but unfortunately does not achieve the background level of measuring sections (above the Severnaya Vil'va mouth) and exceeds MPS_{FWO}. Long-term observations of the dynamics of iron, aluminium and manganese concentration do not demonstrate the improvement of the Yaiva water quality.

Thus, the quality and the quantity characteristics of the water objects running at the KCB territory significantly change:

- Chemical composition of rivers changes (the hydrochemical phase changes from hydrocarbonate-calcium-natrium to sulfate, chalybeate-aluminium, the mineralization increases from 90-159 mg/L to 640-6,000 mg/L and close to neutral hydrogen ion exponent becomes acid;
- Mine water discharges supply rivers, small ones in particular;
- Bottom sediments of many rivers and ponds are anthropogenic, acid and enriched with iron, aluminium, sulfates and heavy metals, as a result they are the sources of secondary pollution.

Under these circumstances the amount and quality of biota in rivers is low at some places. We argue, that rivers lose their natural characteristics and are considered to be new anthropogenic formations.

To assess the general hydroecological conditions of watercourses is necessary to determine the maximum anthropogenic load on the whole hydrological network of the Kizel coal basin. Further, a complex approach to a river system recovery will include the construction of artificial geochemical barriers [16] with the determination of their optimal location. Also the complex approach will include the development of an environment monitoring system based on both remote sensing images and field observations.

Acknowledgments
This work was supported by a grand of Russian Foundation for Basic Research No. 17-45-590793 p_a.

References

[1] Khayrulina E, Khmurchik V and Maksimovich N 2016 *Proc.of IMWA2016 Ann.Conf. Mining Meets Water – Conflicts and Solutions, Germany* pp 766-771
[2] Orem W and Finkelman R 2003 *Treatise on Geochemistry* **7** 191-222
[3] Maksimovich N and Gorbunova K 1990 *6th Int Congr of Int Association of Eng Geol* (CRC Press) pp 1457–61
[4] Sergeev V, Shimko T, Kuleshova M and Maximovich N 1996 *Water Sci. Tech.* **34**(7–8) 383–387
[5] Maksimovich N 2011 *Inzhenernaya geologiya (Engineering geology)* **3** 46-51(In Russian)
[6] Maksimovich N, Pyankov S and Khayrulina E 2017 *Mine Water and Circular Economy, IMWA* (Lappeenranta, Finland) pp 212-217
[7] Demchak J, Skosen J and McDonald L 2004 *Enviorn. Qual.* **33** 656-668
[8] Furrer G, Phillips B L, Ulrich K-U, Pöthig R and Casey W H 2002 *Science* **297**(5590) 2245–47
[9] Gray N F 1998 *Water Res* **32**(7) 2122–34
[10] Nordstrom D K, Alpers C N, Ptacek C J and Blowes D W 2000 *Environ Sci Technol* **34** 254–258
[11] Sivakumar M, Singh R N and Morton S G S 1994 *Mine Water Environ* **13**(1) 27–39
[12] Tiwary R K and Dhar B B 1994 *Mine Water Environ* **13**(3-4) 1–9
[13] RD 52.24.643-2002 *The method of integrated assessment of the contamination degree of surface waters by hydrochemical indicators* (ratified and enforced by Rosgidromet 03.12.2002)
[14] GN 2.1.5.1315-03. *Maximum permissible concentrations of chemicals in water bodies of drinking and cultural and community water supplies*
[15] Erina O N, Efimova L E and Zaslavskaya M B 2017 *Proceedings of the 3d International conference "Environment and region sustainable development: ecological challenges of the XXI century"* (AN RT Publ.) pp 353-355 (In Russian)
[16] Maximovich N and Khayrulina E 2014 Environmental Earth Sciences **72**(6) 1915–24

Consequences of the river valley bottom transformation after extreme flood (on the example of the Niida River, Japan)

D Botavin[1], V Golosov[1], A Konoplev[2] and Y Wakiyama[2]

[1] Faculty of Geography, Lomonosov Moscow State University, 1, Leninskie Gory, 119991, Moscow, Russia
[2] Institute of Environmental Radioactivity, Fukushima University, Fukushima, Japan

E-mail: dmitry.botavin@gmail.com; gollossov@gmail.com; alexeikonoplev@gmail.com; wakiyama@ipc.fukushima-u.ac.jp

Abstract. Detailed study of different sections of floodplain was undertaken in the Niida River basin (Fukushima Prefecture) after an extreme flood event which occurred in the middle of September 2015. The upstream part of the basin is located in the area with very high level of radionuclide contamination after the accident at Fukushima Dai-ichi NPP. Field and GIS methods were used, including direct measurement of the depth of fresh sediment and its area, with soil descriptions for the typical floodplain sections, measurement of dose rates, interpretation of space images for a few time intervals (before and after flood event) with the following evaluation of spatial changes in deposition for different floodplain sections. In addition, results of quantitative assessment of sedimentation rates and soil radionuclide contamination were applied for understanding the effect of extreme flood on alluvial soils of the different sections. It was established that the maximum sedimentation rates (20-50 cm/event) occurred in the middle part of the lower reach of the Niida River and in some locations of the upper reaches. Dose rates had reduced considerably for all the areas with high sedimentation because the top soil layers with high radionuclide contamination were buried under fresh sediments produced mostly due to bank erosion and mass movements.

1. Introduction

Extreme flood events are very typical of small mountain river basins in different parts of tropical and sub-tropical landscape zones [1, 2, 3, 4]. They usually result in both intensive erosion and sediment deposition on the river valley bottom [5, 6, 7, 8, 9]. Also, sedimentation rates considerably influence alluvial soil formation and contamination, depending on the sources of sediment [10, 11]. Detailed study of different sections of floodplain was undertaken in the Niida River basin (Fukushima Prefecture) after an extreme flood event which occurred in the middle of September 2015, when 385-456 mm of precipitation fell during 6 days with maximum 200 mm/day at the end of typhoon. The objectives of the study are: 1) to evaluate the erosion/deposition processes and their spatial changes in different sections of the river floodplain, with particular attention to the low reaches of the Niida River, and 2) to assess changes in alluvial soil contamination in comparison with the level of contamination before the extreme flood.

2. Material and methods

The Niida River basin has the total area of 248 km^2 and is located on the eastern part of the Fukushima prefecture. Level of initial radionuclide deposition after the FDNPP accident varies considerably along the Niida River with maximum contamination in its southwestern part (the Hiso River basin), with decrease in both northern and eastern directions. All previously used agricultural lands of the headwater uplands both in depressions and on hillslopes have been abandoned since spring 2011 immediately after the accident. Intensive decontamination work (complete removal of the contaminated upper topsoil layer for the long-term storage) is underway now starting from the northwestern part of the basin. Coastal lowlands were not seriously affected by the radionuclide fallout. Agricultural lands in this part of the basin are still in use. The Niida basin is characterized by wet monsoon climate with highly variable total annual precipitation (1300-1500 mm). Less than 5-7% of precipitation falls as snow. The alluvial sandy and loamy sandy soils are typical of the river floodplain. They are characterized by high moisture content and stratification with interlayers of light loam. More detailed description of the Niida river basin can be found elsewhere [12, 13].

Field and GIS methods were used, including direct field measurements of the depth of fresh sediment and its area with soil descriptions for the typical floodplain sections, measurement of dose rates; interpretation of space images of the river valley for a few time intervals (before and after flood event) with the following evaluation of spatial transformation of the river bottom, including river channel and different floodplain levels. Finally, area of three types of valley bottom modifications were distinguished. The first type is erosion sections, which appeared because of bank erosion. The second type is the deposition section which was associated with formation of new floodplain area (point bars). The third type is sedimentation area on the pre-exiting floodplain. Only areas with high rate of sedimentation rates (>5 cm per event) were included in this group.

3. Results and discussion

Results of quantitative assessment of sedimentation rates were used for understanding the effect of extreme flood on alluvial soils of the different reaches. The sedimentation rates in the upper part of the Niida River basin within intermountain depressions depended on the type of the channel. They are in the range 10-15 cm (one order of magnitude higher than the mean annual) for the artificially canalized channels with concrete embankments and levees, which are the dominant type for main rivers draining the basin headwaters upland depressions. However, sedimentation rates are only 1-3 cm for the uppermost section of the artificially canalized channels. Some valley reaches within uplands depression have wider channels, but unconfined by levees or embankments (though there can be levees on one or both sides but remote from the channel along the higher floodplain or low terrace level). Here in some section within the floodplain there was extremely high deposition of sand and pebble material with an average depth of 30-40 cm and a total weight of 400-500 tons which, in fact, amounts to one tenth of the total measured flow volume of suspended sediments.

The most detailed study was undertaken for the lower reach of the Niida River located within the coastal lowlands. It was possible to divide the lower reach into three sections according to decreasing of the channel gradients towards the river mouth. Evaluation of areas which had undergone by bank erosion, in-channel deposition and active overbank sedimentation was undertaken for each section and presented for the different types of channel, determined according to the river channel classification [14, 15] (table 1).

It was found that bank erosion is particularly high at the section 1, because of the high energy of the flow after the exit of the Niida river from the mountains to the coastal plain in particular in the river reaches with straight type of channel (table 1, figure 1). Generally, erosion prevails the in-channel deposition for the all sections of the lower reach of the Niida river except meandering channel type at the section 3 (table 1). However, the meandering type of channel is characterized by the relative equilibrium relationship of erosion and in-channel deposition even at the section 1 with the most active valley bottom transformation including overbank sedimentation (figure 2; table 1).

IOP Conf. Series: Earth and Environmental Science **107** (2017) 012002 doi:10.1088/1755-1315/107/1/012002

Table 1. Mean areas of the lower reach of the Niida river valley bottom affected by bank erosion, in-channel deposition and intensive overbank sedimentation

Type of channel	Bank erosion, m^2	In channel deposition, m^2	Floodplain sedimentation, m^2
		Section 1 *	
Meandering	16732	7178	32970
Braided	13228	608	21615
Straight	14983	1196	7894
		Section 2	
Meandering	36431	13144	79499
Braided	10691	2290	19523
		Section 3	
Meandering	14562	19713	71510
Braided	7473	1675	18776
Straight	3232	369	644

* Section of the river valley bottom of the lower reach were separated according to the decrease of the river channel gradient towards the mouth.

Figure 1. The erosion (1), deposition (2) and sedimentation (3) zones appeared after extreme flood at the section 1 (semi-mountain sub-reach) within the straight type of the Niida River channel.

Figure 2. The erosion (1), deposition (2) and sedimentation (3) zones appeared after extreme flood at the section 1 (semi-mountain sub-reach) within the meandering type of the Niida River channel.

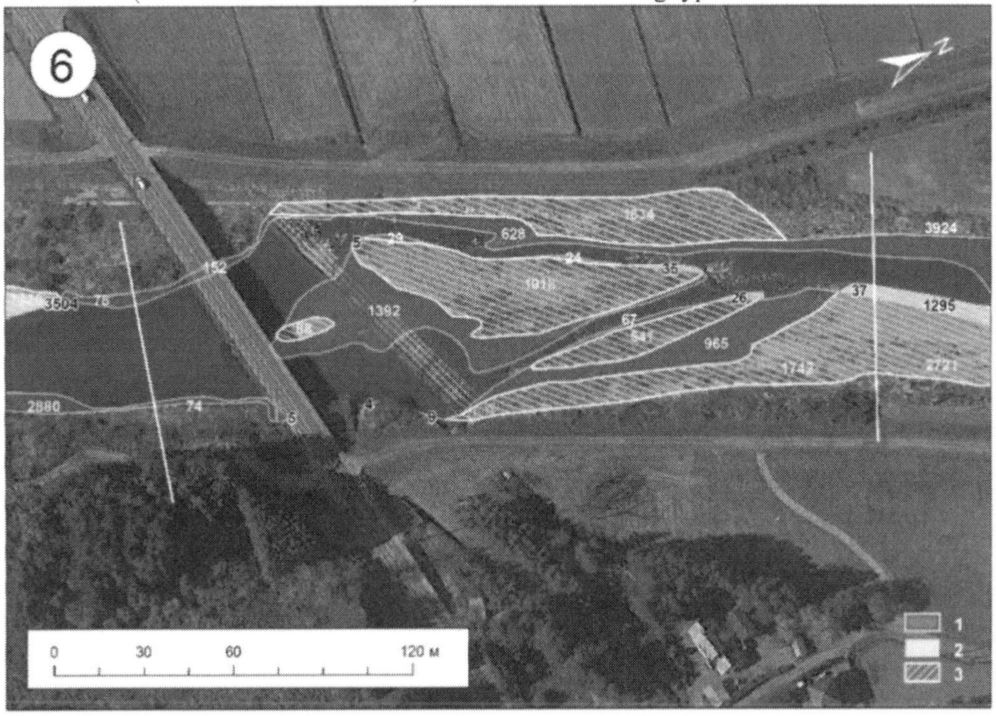

Figure 3. The erosion (1), deposition (2) and sedimentation (3) zones appeared after extreme flood at the section 1 (semi-mountain sub-reach) within the braiding type of the Niida River channel.

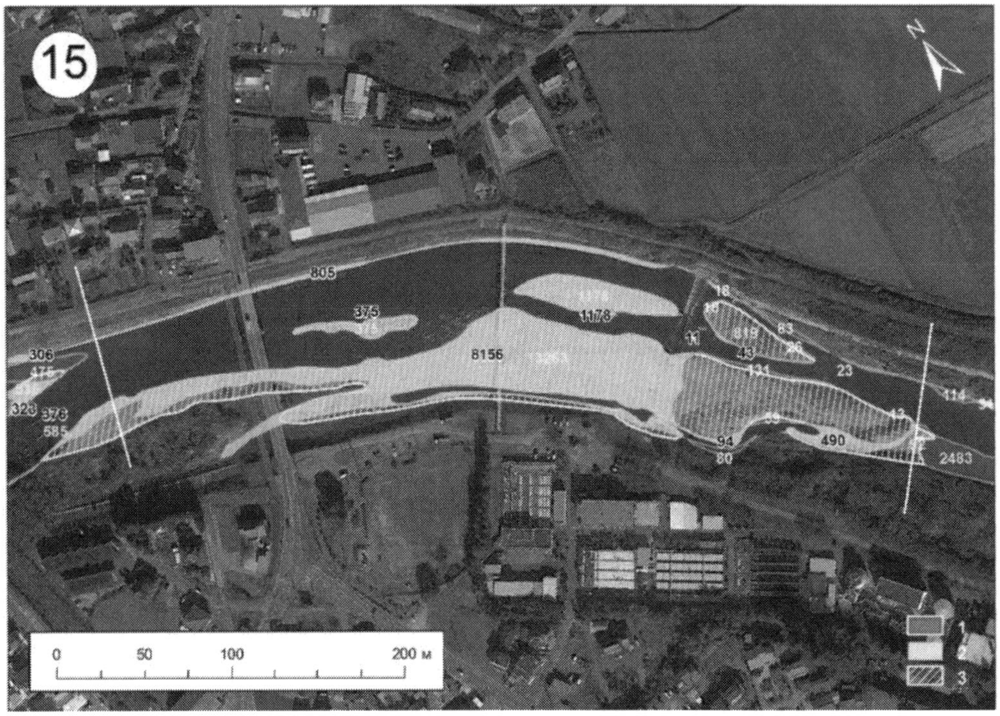

Figure 4. The erosion (1), deposition (2) and sedimentation (3) zones formed after extreme flood at the section 3 (plain sub-reach) within the braided type of the Niida River channel.

The braided type of the river channel is characterized by the different modification of the river valley bottom depending on the river channel gradient. Active erosion was identified for the braided channel at the section 1 (figure 3), while intensive in-channel deposition was observed at the section 3 (figure 4). It is necessary to stress that overbank sedimentation at the lower reach of the Niida River leads to the reduction of the dose rate, even though the dose rate was low initially because of sand-gravel grain size of the sediment. Dose rates were reduced by 3-5 times for all the floodplain sections in both upper and lower reaches of the Niida River with high sedimentation rate because the top soil layers with high radionuclide contamination were buried under fresh sediments produced mostly due to bank erosion and mass movements.

4. Conclusions

The detailed assessment of geomorphological and radioecological consequences of the extreme flood in the Niida River basin allows to conclude that different exogenic processes (landslides, scree, bank erosion etc.) were the main sediment sources for the river sediment discharge. Maximum channel deformations were observed within the lower reach of the Niida River because of extremely high energy of the stream with maximum water and sediment discharges. However, sedimentation rates on the floodplain of the upper reaches of the Niida River were also one order of magnitude higher than the mean annual rates. Finally, extreme event contributed to the serious reduction of the dose rate of the river bottom except for some high floodplain levels where sedimentation rates were negligible.

Acknowledgements
Work was funded by RFBR project no. 16-05-00815.

References

[1] Magilligan F J, Phillips J D, James L A and Gomez B 1998 *Journal of Geology* **106** 87–95
[2] Marchi L, Borga M, Preciso E and Gaume E 2010 *J. Hydrol.* **394**(1-2) 118–133
[3] Borga M, Stoffel M, Marchi L, Marra F and Jakob M 2014 *J. Hydrol.* **518**(B) 194–205
[4] Thompson C and Croke J 2013 *Geomorphology* **197** 156–169
[5] Baker V R 1977 *Geological Society of America Bulletin* **88** 1057–71
[6] Eaton B C and Lapointe M F 2001 *Geomorphology* **40** 291–309
[7] Gaume E and Borga M 2008 *J. Flood Risk Manag.* **1**(4) 175–189
[8] Costa J E 1988 *Rheologic, Geomorphic, and Sedimentologic Differentiation of Water Floods, Hyperconcentrated Flows, and Debris Flows, in Flood Geomorphology* ed V.R. Baker et al (Chichester: John Wiley & Sons) pp. 113–122
[9] Milan D J 2012 *Geomorphology* **138**(1) 319–328
[10] Vanacker V, Vanderschaeghe M, Govers G, Willems E, Poesen J, Deckers J and De Bievre B 2003 *Geomorphology* **52** 299–315
[11] Johnson R M and Warburton J 2002 *Earth Surface Processes and Landforms* **27**(9) 945–969
[12] Konoplev A V, Golosov V N, Yoschenko V I, Nanba K, Onda Y, Takase T and Wakiyama Y 2016 *Eurasian Soil Sciences* **49**(5) 570–580
[13] Konoplev A, Golosov V, Laptev G, Nanba K, Onda Y, Takase T, Wakiyama Y and Yoshimura K 2016 *J. Environ. Radioact.* **151** 568–578
[14] Alabyan A M and Chalov R S 1998 *Earth Surface Processes and Landforms* **23** 467–474
[15] Leopold L B and Wolman L B 1957 *US Geological Survey Professional Paper* **282-B** (US Government Printing Office, Washington, DC) 39–85

The Approach to Study the Kama Reservoir Basin Deformation in the Zone of a Variable Backwater

S A Dvinskikh, A B Kitaev and A A Shaydulina

Perm State University, Institute of the Natural Science, 15, Bukireva str., 614990, Perm, Russia

E-mail: hydrology@psu.ru

Abstract. A reservoir floor starts to change since it has been filled up to a normal headwater level (NHL) under the impact of hydrosphere and lithosphere interactions as well as under the impact of chemical and biological processes that occur in its water masses. At that complicated and often contradictory "relations" between features of geo- and hydrodynamic processes are created. The consequences of these relations are the alterations of values of morphometric indices of the reservoir water surface, depth and volume. We observe two processes that are oppositely directed. They are accumulation and erosion. They are more complex at the upper area of the reservoir – the zone of a variable backwater. The basin deformation observed there is lop-sided and relatively quiet, but with time these deformations make difficulties for water users. To provide good navigation and to reduce harmful effect of waters on other water consumption objects, it is necessary to study and to forecast constantly the basin transformation processes that occur at this zone.

1. Introduction

Reservoirs are artificial reservoirs, the function of which is to regulate the flow. They are relatively young and developing systems. Reformation processes of different parts of reservoirs are a subject of multiple researches. They are studied by scientists of Russia [1-7], Europe [8-10], Asia [11], the USA [12, 13] etc.

The study of individual aspects of channel processes in variable backwater region was handled by N.I. Makkaveev, V.A. Kritsky and V.I. Litvinsky, V.V. Lysenko, R.D. Frolov, N.B. Baryshnikov, R.B. Uralsky, S.L. Vendrov, V.F. Nikolaev et al. [2-4]. However, there is no complete picture of the processes taking place in this region.

The zone of a variable backwater (which length is 145 km.) of the Kama reservoir is located at the upper area of the Kama reservoir. The zone starts near Ust'-Pozhva village and finishes near Kerchevskii urban village (Figure 1), but even so the backwater zone some years stressed 46 km. higher than the Vishera river estuary. The beds of the Kama and Vishera rivers within the studied area are cutting in and straightforward. In the upper area the backwater zone is within the original Kama riverbed. A partial flooding of the floodplain of 0.5 to 1 meter depth is observed in the southern part of the area. The water-surface area is 108 km^{-2} and water masses capacity is 0.31 km^3 there. Point bar channel processes with the elements of mid-channel bars predominate at the Kama river. Sand and gravel are the main bed loads, sand and clay are the main floodplain sediments. Complex interrelation between the reservoir filling regime and spring tides periods at the main tributaries causes multiversion locations of boarders in a pinch zone in a variable backwater zone.

Channel deformation forms also have some peculiarities: a channel regime is characterized by a regression accumulation. We observe a self-regulated channel deformation in accordance to the altered erosion basis and the riverbed adapts to its new conditions. The consequences and the forms of this process are:

1. The marks of point bars and mid-channel bars rise and the size of both bars also increases.
2. Channel forms get larger because of the passing away of small point bars and the merger of adjacent point bars. As the result small single riffles get larger and more complex.
3. When the reservoir was built the channel cutting-in process increased at the territory between the river Vishera estuary and Tyul'kino urban village. Relatively fast increase of rivers confluence index (50-60 m/ year) and the angle decrease of their matching have led to the wash-out of the left bank estuary point bar of the Vishera river. In its turn it has caused the displacement of the dynamical axis of the Kama river water-course to its left bank and a left bank unnavigable channel has appeared between Tyul'kino islands [14].

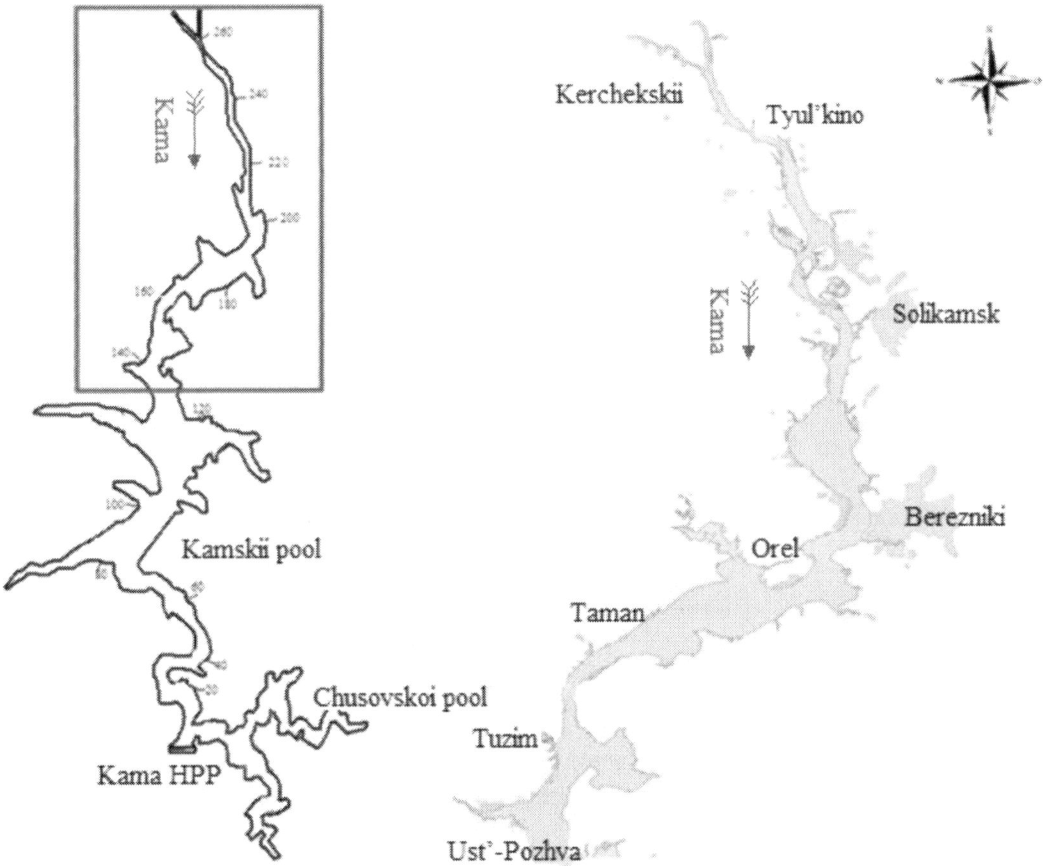

Figure 1. The zone of a variable backwater of the Kama reservoir

2. Material and methods

To determine the formation intensity of sediments accumulation and denudation zones during one year the channel surveys of several riffles (Kerchevskii, Ust'-Moshevskii, Malakhovskii, Verkhnii and Nizhne-Solomenskii, Teterinskii) have been matched.

The Kerchekskii riffle is located at 2575-2572 km. along the navigable pass. There is an intensive growth of bottom sediments at the left bank and the channel erosion at the right bank. Ridges with 175 km length may appear during a year.

The Verkhnii and Nizhne-Solomenskii riffles are located at 2.568-2.562 km along the navigable pass. There is a ridge increase in the middle course. Sediment accumulation zones are located randomly: at the right and left banks and in the middle course. As the result the number of shoals, point bars and riffles increases. The erosion zones are also located randomly and are observed everywhere. Significant basin depression deformations are observed within 2.563-2.562 km along the navigable pass: the left bank is characterized by intensive erosion, and the right one - by alluvion. The navigable pass shifts to the left bank.

The matching of the channel surveys of the Ust'-Moshevskii and Malakhovskii riffles has revealed how the Kama reservoir basin has changed during a year: the reservoir floor has been divided into two zones: a sediment accumulation zone has formed at the left bank and the right bank and bottom have washed out. As the result the navigable pass has shifted to the right bank.

There is an additional local erosion base level – the Glotikha and Usolka rivers at the Teterinskii riffle. As the result the storage capacity of the main channel has been almost exhausted. There point bars increase intensively especially along the left bank and as a result the bank stretches with the speed of 80 m per year and has already prevented the passes to the docks of the petroleum storage depot and an industrial port. In 5-7 years this will seriously prevent the "Sil'vinit" Ltd. water supply point activity. The right bank channel erosion behind the Teterinskii island is also possible. The shoaling of the left distributary, that increased since the channel dredging has stopped, can accelerate this process l leading to unpredicted challenges and consequences for both navigation and the operation of water consumption objects located close to the riffle. Currently the uncontrolled deformation process of the basin increases. One reason for that is that the activity to straighten the channel and to clean navigable passes from sediments has stopped, another one is an early reservoir drawdown. These reasons affect the hydrodynamic regime of the reservoir.

Hydrodynamic regime plays a significant role in basin transformation as it determines a process (erosion and accumulation) trend.

Process trend may be assessed by means of the dimensionless parameter η:

$$\eta = \frac{v}{Vp} \tag{1}$$

Where v is actual or calculated flow velocity (m/sec) and Vp is eroding velocity (m/sec). If $\eta > 1$, then we observe erosion, $\eta = 1$- stabilization, $\eta < 1$ – accumulation.

Actual flow velocities were obtained by observation data generalization. The data were taken from 1964 to 2016. Totally 600 velocity values were used. The calculated near-bottom velocities of the predetermined exceedance probability were calculated according to SP 33 101 2003 [15]. Three ratings of exceedance probability – 25%, 50% and 75% were used as the quantitative assessment of the flow velocity in a variable backwater zone.

An eroding velocity is one of the crucial flow velocities (together with a non-eroding velocity) and it is a qualitative expression of erosion-preventive soil stability [16] that has been calculated according to the formula:

$$V_p = lg\frac{8.8H}{d}\sqrt{\frac{2g(\rho_1-\rho)d}{1.75\rho}} \tag{2}$$

where H - vertical depth, m; d - sediments diameter, m (at that an average diameter is under the root, a maximum diameter is in the denominator); ρ_1 - sediments density, 2650 kg/m^3; ρ - water density, 1000 kg/m^3. Eroding disruptive velocity (V_p) corresponds to the beginning of mass particles displacement. If actual flow velocities exceed the average eroding velocity, then basin transformation, sediments erosion and re-deposition take place. The actual flow velocities are determined by a level regime that is why predominant transformation processes of the reservoir floor should be determined.

3. Results and discussion

The studies devoted to the level regime peculiarities [17] that were previously made have distinguished three typical sites in the zone of variable backwater. They are the area between Kerchevskii and Tyul'kino urban villages (where river conditions prevail); the area between Tyul'kino urban village and Berezniki city (where river and reservoir conditions are equal); and the area between Berezniki city and Ust'-Pozhva village (where reservoir conditions are observed during a whole year). The basin transformation process trend has been determined within these sites with η parameter.

The η parameter calculation has revealed that erosion processes prevail under all marked exceedance probabilities and levels in Tyul'kino urban village hydrometric section. In all the rest hydrometric sections (except Ust'-Pozhva village) erosion prevails under all considered levels when exceedance probability is 25% and lower. Stabilization and accumulation processes take place if exceedance probability is 25% and higher in the hydrometric section of Ust'-Pozhva village. The η parameter distribution in different months of a navigation process in the variable backwater zone of the Kama reservoir (Table 1) demonstrates that there is erosion under 107 and 108.5 m absolute levels in the area between Tuyl'kino urban village and Berezniki city in May and June, the stabilization is observed in August and the sediment accumulation - in July and September. The washing-out processes are observed only in May in the area between Berezniki city and Ust'-Pozhva village. In all other months (June-September) accumulation processes prevail (Table 1) and accumulative forms of a bottom relief develop.

Table 1.The η parameter in different months of navigation period in variable backwater zones of the Kama reservoir

Month	Tyul'kino urban village-Berezniki city		Berezniki city- Ust'Pozhva village	
	107 m above the sea level	108.5 m above the sea level	107 m above the sea level	108.5 m above the sea level
May	**1.97**	**1.94**	**1.20**	**1.17**
June	**1.48**	**1.45**	0.86	0.83
July	0.71	0.70	0.46	0.44
August	***0.92***	***0.91***	0.51	0.50
September	0.80	0.79	0.29	0.28

[a] Values that illustrate washing out/erosion processes are printed with a bold type and values that demonstrate stabilization are printed in bold italics

Underwater ridges are the examples of accumulative forms. Their values parameters depend on average turbidity in the course channel and hydraulic parameters (water level and discharge, flow velocity and sectional area). We have made nomograms for Tyul'kino urban village that are based on long-term research data and recommendations provided in the Departmental building codes (DBC) [18]. The hydrometric section selection is explained by the fact that there are differentiations of river and reservoir conditions. Nomogram feature is that it is necessary to have only turbidity value that is easily obtained under field conditions to determine ridge parameters (length and height). The ridge parameters are obtained by stages:

1. First we determine water discharge that corresponds to the present average turbidity value. The turbidity dependence on the water discharge constructed on long-term observation data for the hydrometric section in Tyul'kino has two point clouds (Figure 2). The first point cloud is presented by a direct linear dependence, the second one - by an inverse linear dependence. The second point cloud is formed from the beginning to the middle of May during a spring tides peak when the reservoir is filled and water level rises to a normal headwater level (NHL). So a backwater appears: water masses go upstream. At this period water mass "faces" the river water masses. In this connection the low point clouds correspond to river conditions and the upper point clouds correspond to the reservoir conditions.

2. The dependence between water level and flow velocity is made and a pair correlation coefficient is calculated on the basis of long-term observation data. In accordance with the obtained equation a flow velocity for level values with a 40 cm-step under different depths (a 5 cm-step) is calculated. We have obtained 3.240 values that have been used to build nomograms.

3. We calculate a ridges length l_r (m) under the determined water movement regime according to the dependence: $l_r = H^3\sqrt{C^2\ /g}$, where C is Chezy's coefficient at the designed vertical line that takes an average incline value into account; H is a vertical flow depth; g is gravity acceleration. The obtained value is used to make the ridge length dependence on a flow velocity (Figure 2).

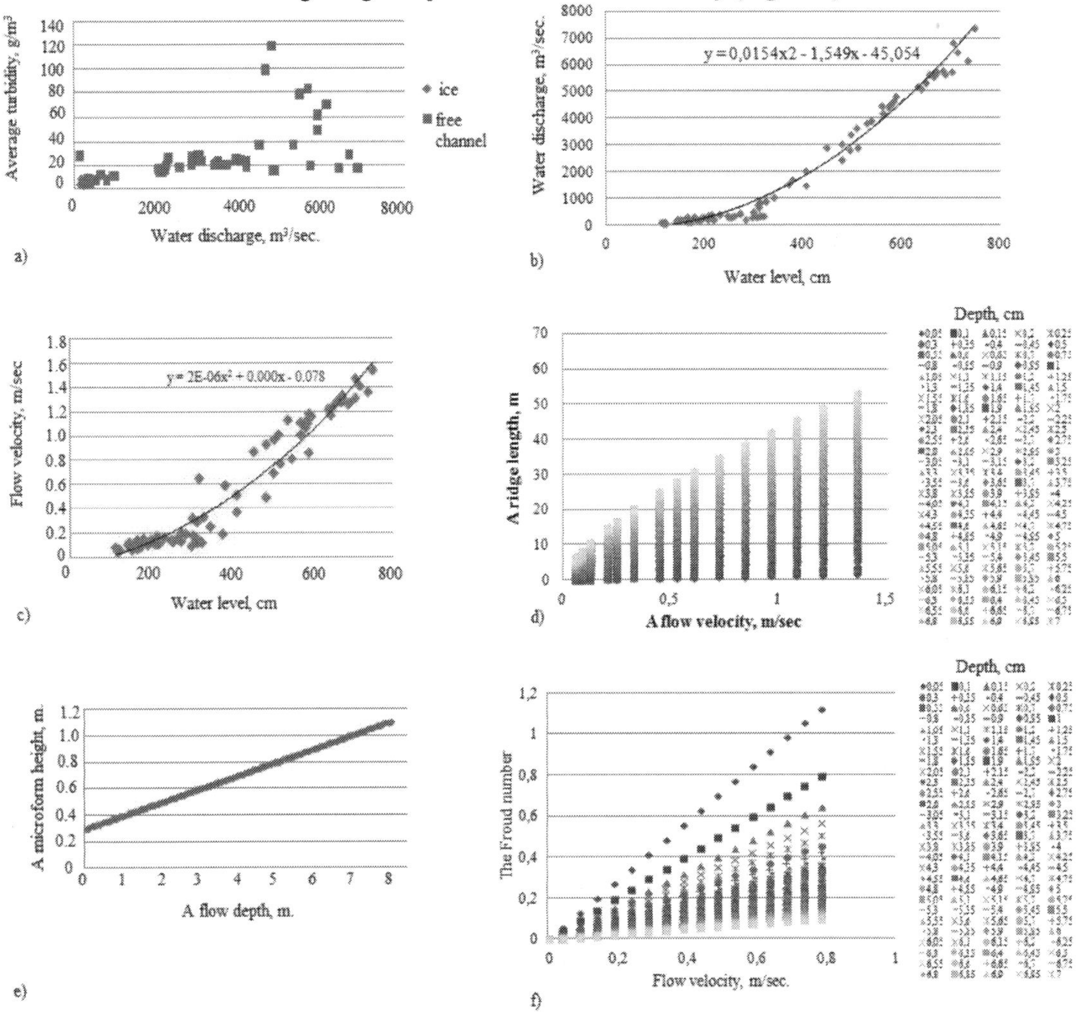

Figure 2. Nomograms to determine ridge parameters in a variable backwater zone of the Kama reservoir: a). Water discharge dependence on an average turbidity; b). Q = f(H); c). V = f(H); d). Nomogram to calculate a ridge length dependence on a flow velocity; e). Nomogram to calculate a ridge height dependence on a flow depth; f). Nomogram to calculate the Froud number dependence on flow velocity

4. We calculate a ridges height h_r (m): $h_r = 0.25H$ if H<1 m or $h_r = 0.20 + 0.10H$ if H>1 m (Figure 2). As we can see from the nomogram, the ridge height-flow depth function is linear. In the case study the correlation between values is strong, the correlation coefficient is 1. Ridges shift velocity (C_r, m/sec) is calculated by the formula: $C_r = 0.019vFr3$, where υ is an average flow velocity above the

place where a ridge is being determined, m/sec; Fr is the Froud number. If Fr > 1, water flow is considered to be rough, if the dependence is inverse the flow is quiet. At the case study area rough water masses finish at the 50 cm-depth and starting from the 55 cm-depth the flow becomes absolutely "quiet".

We calculate ridges shift period of the particular profile during a day using the formula: $\tau_r = l_r / C_r$. The largest observed ridge shifting occurs at low depth (up to 45 cm) and the 0.54 m/sec flow velocity.

The possibility to use the obtained nomograms has been proved in the case study of the hydrometric section located near Kerchevo village (0.5 km lower the Kerchevka river estuary). The turbidity in this area is 60 g/m^3. First we calculate water discharge corresponding to the present average water turbidity according to the dependence. If average water turbidity is 60 g/m^3, the water discharge will be 190 m^3/sec. Then using the dependence of water discharge on a water level, we calculate the level value (using the equation or the dependence curve), which is 157 cm. At that using a flow velocity dependence on a water level, we see that the velocity is 0.33 m/sec. Then using the obtained flow velocity, we calculate the ridge length (2.54 m), its height (using the flow depth) is 0.05 m, the Froud number is 0.108 and the ridge shifting velocity is 0.000056 m/sec. We have revealed that if the flow velocity during a year was constantly 0.25 m/sec, then we would not observe the evident movement.

Thus, the obtained dependences allow calculate possible parameters of the change of a ridge- a base bottom relief form, if there is an absence of field observation data.

During our research the following conclusions were made:

1. The reservoir construction has led to the formation of channel deformation processes. The reason for that is a lop-sided character of a regression accumulation that leads to a self-regulated reservoir basin transformation in a variable backwater zone. The η parameter is an index of a basin deformation trend.
2. Development of underwater ridges obviously demonstrates the process of a basin transformation. The forecast of the ridges development is based on the data of average turbidity and hydrological parameters (water level, water discharge and flow velocity).
3. Ridge development is characterized by its length and height. We suggest using nomograms to calculate these parameters in variable backwater zone of the Kama reservoir. The nomograms are based on dependences: a). the water discharge on an average turbidity and water level; b). the water level value on the flow velocity; c). the ridge length on the flow velocity and the ridge height on the flow depth. To calculate ridge parameters it is necessary to know only a turbidity value that is easily obtained under field conditions.

4. Acknowledgements

The research has been conducted with the financial support of RFBR grant № 16-45-590546 p_a.

References

[1] Matarzin Y U M 2003 *Gidrologiya vodohranilishch* (Hydrology of reservoirs) (Perm: Russia, Perm State University Press) p 296 (in Russian)

[2] Makkaveev N I, Belinovich I V and Khmeleva N V 1958 *Ruslovye protsessy: sbornik statei* (Channel processes: collection of articles) (Moscow: Academy of Sciences of the USSRPubl.) pp 318–337 (in Russian)

[3] Makkaveev N I 1972 *Inzhenerno-geograficheskie problemy proektirovaniya i ekspluatatsii krupnykh ravninnykh vodokhranilishch* (Engineering and geographical problems of design and operation of large flat reservoirs) (Moscow:Science Publ) pp 36–42 (in Russian)

[4] Berkovich K M 2012 *Ruslovye protsessy na rekakh v sfere vliyaniya vodokhranilishch* (Channel processes on the rivers in a sphere of influence of reservoirs) (Moscow: Geographical faculty of MSU Publ) p 163 (in Russian)

[5] Baryshnikov N B 2007 *Dinamika ruslovykh potokov* (Stream dynamics) (St. Petersburg: RSHU Publ) p 314 (in Russian)

[6] Baryshnikov N B *Ruslovye protsessy* (Channel processes) (St. Petersburg: RSHU Publ) p 439 (in Russian)

[7] Berkovich K M and Vinogradova N N 1975 *Vodnye resursy* (Water resources) **6** 81–87 (in Russian)
[8] Babiński Z 1992 *Współczesne procesy korytowe dolnej Wisły* (Contemporary fluvial processes of the lower Vistula (dolna Wisła)) (Prace Geograficzne, IGiPZ PAN) p. 157
[9] Babiński Z 2002 *Influences of barrages on processes of koritove aluvialnickh rivers with special taking into account of water «degree of Włocławek»* (Bydgoszcz: publishing House of Academy of Bydgoskiej) p 185
[10] Banach M 1993 *Czas. Geogr.* **3-4** 285–306
[11] Fu K D, He D M, and Lu X X 2008 *Quaternary International* **186**(1) 91–99
[12] Williams G P and Wolman M G 1984 *U.S. Geological Survey Professional Paper* **1286** 83
[13] Rathburn S, Finley J, Klein S and Whitman B 2004 *Assessing reservoir sedimentation using bathymetric comparison and sediment loading measurements* (Geological Society of America, Abstracts with Programs) vol 366, No 5 p 12
[14] Shaydulina A A, Kalyuzhnyi V K, Dvinskikh S A and Vasil'ev A V 2017 *Sovremennye problemy vodokhranilishch i ikh vodosborov. Trudy VI Mezhdunarodnoi nauchno-prakt. konf.* (Current challenges of reservoirs and their catchment areas. Proceedings of the VI International Scientific Practical Conference) (Perm: Russia, Perm State University Press) vol 1 pp 162–169 (in Russian)
[15] *Svod pravil po proektirovaniyu i stroitel'stvu: SP 33-101-2003. Opredelenie osnovnykh raschetnykh gidrologicheskikh kharakteristik: normativno-tekhnicheskii material* (Design and Building instructions: SP СП 33-101-2003. Determination of fundamental calculation hydrological characteristics: reference information) 2004 (Moscow: PNIIIS Gosstroya Rossii Publ) p 72 (in Russian)
[16] Goncharov V N 1962 *Dinamika ruslovykh potokov* (Dynamics of channel flows) (Leningrad: Gidrometeoizdat Publ) p 374 (in Russian)
[17] Shaydulina A A 2016 *Geographicheskiy Vestnik* (Geographical Bulletin) **4**(39) 44–56 (in Russian)
[18] *Vedomstvennye stroitel'nye normy. Uchet deformatsii rechnykh rusel i beregov vodoemov v zone podvodnykh perekhodov magistral'nykh truboprovodov (neftegazoprovodov)BCH 163 83* (Departmental building codes. Consideration of deformation of river channels and banks of reservoirs in zones of underwater trunk pipelines (oil-gas pipelines) DBC 163 83) 1985 (Leningrad: Gidrometeoizdat Publ) p 144 (in Russian)

Engaging expert communities in development of content of Russia`s regional geoportals (case study: "River basins in European Russia" geoportal)

P Ermolaeva[1,2]

[1]Department of Sociology, Kazan Federal University, 18, Kremlevskay Str., 420008, Kazan, Russia
[2]Center of Advanced Economic Research in the Academy of Sciences of the Republic of Tatarstan, 21, Ostrovskogo str., 420111, Kazan, Russia

E-mail: polina.ermolaeva@gmail.com

Abstract. The purpose of this study is to attract expert communities`intellectual resources to the process of developing a geoportal entitled "River Basins in European Russia". The results of a survey of experts (n=100) have shown that more than half of respondents had used a variety of geoportals` data in their professional life. Data on digital relief models, streamflows and landscape maps are of greatest interest. In order to obtain a comprehensive social and ecological analysis of the territory, experts have expressed interest in placing data on population in the basins and its density, the volume of used natural resources, and recreational zones on the designed geoportal. In practical sense, our study can be viewed as a fruitful ground for the development of upward vertical communication (from citizens to the government) and partially horizontal communication among citizens via their engagement in the environmental decision-making process.

1. Introduction

Nowadays the new virtual network resources and services are arising. They are aimed at enabling work of users with geoinformation – cards, atlases, satellite images, digital models and other types of information. Improvement and development of these resources and services implement mainly by development of specialized geoportals.

The concept 'geoportal' employs not only in the professional circle of geographers, but also in a wide public media discourse [1, 2]. Two key definitions of this concept are recognized. In professional discourse the geoportal conceptualizes as 'the portal that allows to organize search in network or in separate network resources based on binding properties of geo-information" [3, p 89]. In this arena, the geoportal is understood as a uniform point of access to geo-information of various sources and types. The functions of geoportals include search, visualization, loading of metadata and also downloading and the publication of spatial data (according to access rights and a type of the license for use of materials).

In general sense geoportals are understood as systems of directory type with maps and plans of settlements on which elements of infrastructure and other social objects are displayed. In this meaning geoportals can provide coordinates of interesting objects, a possibility of calculation of their areas or perimeters, laying of routes, printing of maps, etc. However their tasks do not conclude neither the search of internal nor external distributed resources of spatial data; a search as a part of functions of visualizers is limited to search of "offline" objects [4]. Geoportals in this understanding are a subject of discussion under the concept of 'neogeography' which is understood as set of methods of cartography and geographic information systems [5].

Today practically all industrial countries of Europe, America and Asia have national geoportals such as Geospatial One-Stop (USA), the Australian catalog of spatial data of ASDD, the geoportal of the European Union INSPIRE, etc. These geoportals are full-function portals which contain hundreds of thousands of sets of spatial metadata with good functions of data visualization. It is important that international geoportals support search of the data on their metadata, it is the main criterion by which we judge whether the geoportal is 'real' or represents only means of cartographical data visualization and even the website or the portal with the loud name 'geoportal', 'GIS-portal' or 'geoinformation portal' [4]. In Russia there is no federal geoportal with uniform bank of metadata. All works in the field of creation of geoportals are conducted in the conditions of lack of regulatory framework and normative and technical standards.

This article is the result of a joint interdisciplinary project on development of the geoportal "River Basins in European Russia" on the basis of the Institute of Environmental Sciences of the Kazan Federal University [6].Today, in the process of creation of geoportals and their further maintaining a great attention is paid to crowdsourcing, engagement of stakeholders in activities within geoportals.This process coincides with the new paradigm of actors` interaction changes – citizens become not only consumers of information but also its active creators [7, 8, 9, 10].

The interaction between citizens and public authorities changes radically. Civil participation in environmental decision-making processes in the "digital era" is expanding because of the proliferation of laptops, smartphones with Internet access, cameras, availability of geolocation function and specialised mobile applications that keep the public posted about urban issues in a real time mode [11, 12, 13, 14].

However, the analysis of literature has shown that in Russia a practice of involving concerned population groups into development of geoportals is not widespread. As a result, most of the data released by government are of no practical interest to users of geoportals [15, 16]. To fill this gap when developing the geoportal "River Basins in European Russia" methods of sociological science have been applied to obtain feedback on metadata that are of interest to professional groups of ecologists with further addition of these layers to the geoportal.

The novelty of the research project is that it identifies the intellectual resources of expert communities contributing to raising the profile of regional geoportals in experts` professional life. In addition, according to the concept of H. Lefebvre [17], city`s decision-making processes should be reoriented from the government level to the level of micro-groups, residents, and urban communities. Moreover, the right refers not only to the right to housing, work, and education, but, in a broader sense, the right to be a part of the city, to dwell in it, and to change it.

Thus, the involvement of different groups of citizens in recommendation development (during a sociological survey, within the framework of a research and practice workshop, and via the project`s web page) will increase their role in urban policy, what in turn, will reduce the degree of skepticism, paternalism and apathy inherent in the Russian population on ecological issues [18, 19, 20]. In the Russian Federation no similar study has previously been conducted.

2. Research methods

The objective of the study is to analyse the qualitative and quantitative characteristics of geoportals, in particular, the geoportal "River Basins in European Russia" such as the frequency and nature of the use of data of international and Russia`s geoportals in professional activities, study the content of interest depending on the target audience, suggest measures for optimization of the work of the designed geoportal based on professional interests of experts.

To achieve the objective, at the first stage, we have conducted semi-structured expert interviews with faculty and research fellows (ecologists, geographers, social ecologists, sociologists, and etc.), representatives of project and environmental organizations, representatives of the public sector, mass media, and environmental NPOs. A total of 100 experts from different regions in Russia have been interviewed. The survey and data analysis were carried out via the online questionnaire builder *Testograf*. In future, it is planned to attract broad segments of population to the development of geoportal.

3. Results and discussion

Most experts have used the data of geoportals in their practice. The most popular resources are Google Earth, Google Maps, Yandex Maps, Open Street Map, 2GIS, Wikimapia, Earth Explorer (USGS), Roskosmos geoportal, and private portals of Ulyanovsk, Belgorod, and Samara Regions, the Republic of Tatarstan, and etc. Thus, more than half of the experts surveyed had used a variety of data from geoportals in their professional life. Data on digital relief models, streamflows, and landscape maps are of the greatest interest.

In their work experts have turned to various data sources: land use type maps, geological maps, thematic maps of nature and images of Earth landscapes, morphometry of rivers and river basins, wildfire maps, property boundaries of cadastral plots of land, mobile signal strength, commercial objects` density, unauthorized waste disposal, location of buildings, information on land plots, their cost, number and proprietors, maps of characteristics of agriculturally used lands and others.

Experts have noted that the most useful data for placing on the geoportal are first and foremost digital terrain models; then, data on streamflows, a landscape map, a soil map and data on watersheds at river gauging stations (figure 1).

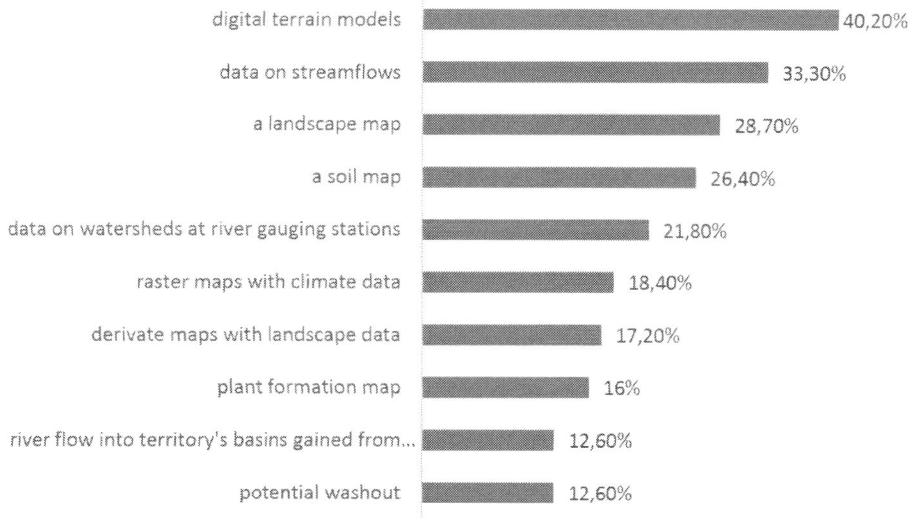

Figure 1. 'Which geoecological data are you interested in to visualize on the boundaries of the basins at the given geoportal?', %

Experts have also pointed out the importance of presenting social and environmental data that are direct and indirect factors for a full-scale analysis of the territory. Experts are interested in the following data in descending order: data on population and its density in the basins, the volume of natural resources used, recreational zones, the type of land use, the density of industrial objects in the studied area (including accidents at industrial enterprises), volumes of production and consumption wastes, transportation infrastructure, incidence rate, utility system and telecommunications infrastructure (figure 2).

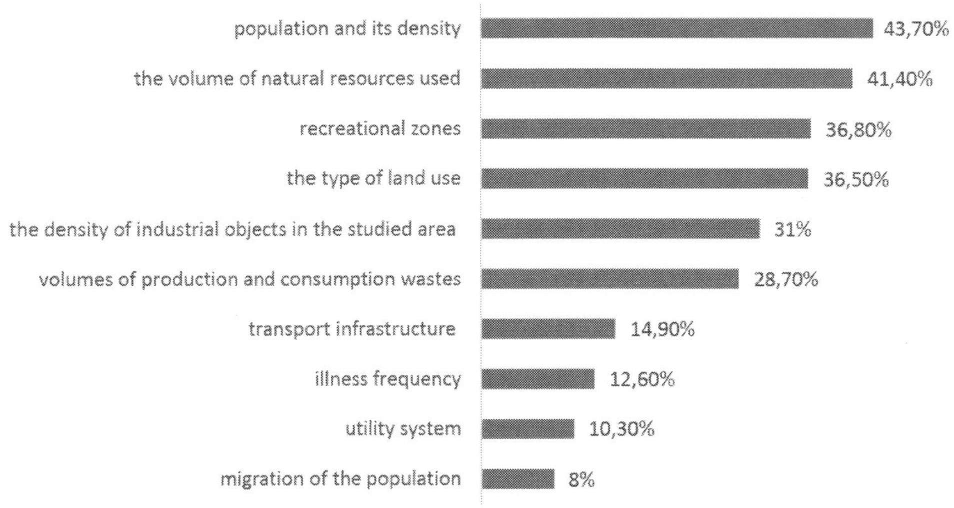

Figure 2. 'Which social and environmental data will you be interested to see at the given geoportal?', %

In order to obtain a comprehensive social and ecological analysis of the territory, experts have expressed interest in placing data on population and its density in the basins, the volume of natural resources used and recreational zones on the designed geoportal. In practical sense, our study can be viewed as a fruitful ground for the development of upward vertical communication (from citizens to the government) and in part, horizontal communication among citizens via their engagement in the environmental decision-making process

There are limitations that must be taken into account when considering the findings presented here. In this study, the ethnographic research approach is adopted because of its capacity to provide detailed, "thick" description of data to produce key research hypotheses that can be tested and further reinterpreted by employing the qualitative methods to build a theory. This means that no generalisation of results can be made at this stage of the research.

Acknowledgements
This work is funded by the Russian Science Foundation, Project No. 15–17–10008.

References
[1] Resch B and Zimmer B 2013 *Journ of Geo-Inform* **2** 1015–37
[2] Budhathoki N, Bruce B and Nedovic-Budic Z 2008 *GeoJournal* **72**(3) 149–160
[3] Gruzinov V 2014 *Geod and aeroph* **1** 89-94
[4] Koshkarev A, Tekynov V and TimoninS 2009 *Geo data* **3** 33–45
[5] Mitchell P 2017 *Routl Handb of Lit and Sp. Routl, Abing NY* 85–94
[6] Ermolaev O, Mal'tsev K and Ivanov M 2014 *Geog and Nat Res* **35**(3) 222–228
[7] Ermolov N, Tereshina M and Morozova E 2014 Ekologicheskaya politika v sovremennoy Rossii (Ecological policy in modern Russia) (Krasnodar: Perspec. of Educ.) p 188 (in Russian)
[8] Usacheva O 2014 *Sociol Res* **53**(2) 3–17
[9] Johnson G and Kulpa A 2007 *CyberPsyc and Beh* **10**(6) 773–779
[10] Kelleher T 2009 *Jour of Commun* **59** 172–188
[11] Castells M 2009 *The Rise of the Network Society: The Information Age: Economy, Society, and Culture* vol I (Wiley-Blackwell) p 656
[12] Grübler A 1998 *Technology and Global Change* (Camb. Uni. Press) p 462

[13] Efremenko D 2002 *Vvedeniye v otsenku tekhniki (Introduction to the assessment of technology)* (Mos MNEPU) p 250
[14] *World Economic Forum* 2013
[15] Ermolaeva P and Dyomkina E 2016 *E-econom Herald* **4** 74–78
[16] Ermolaeva P 2012 *Sociol Res* **12** 80–88
[17] Lefebvre H 1991 *(The Production of Space)* (Blackwell) p 464
[18] Ermolaeva P 2013 *Environ Asses and ProjApprais* **32** 66–71
[19] Yanitsky O 2016 *Mir Rossii* **3** 37–55
[20] Yanitsky O 2016 *New Developments in Environmental Sociology* (Official portal of the IS RAS) http://www.isras.ru/publ.html?id=4308 p 11

IOP Conf. Series: Earth and Environmental Science **107** (2017) 012005 doi:10.1088/1755-1315/107/1/012005

Small catchments DEM creation using Unmanned Aerial Vehicles

A M Gafurov

Institute of Environmental Sciences, Kazan Federal University, 18, Kremlevskaya str., 420008, Kazan, Russia

Email: gafurov.kfu@gmail.com

Abstract: Digital elevation models (DEM) are an important source of information on the terrain, allowing researchers to evaluate various exogenous processes. The higher the accuracy of DEM the better the level of the work possible. An important source of data for the construction of DEMs are point clouds obtained with terrestrial laser scanning (TLS) and unmanned aerial vehicles (UAV). In this paper, we present the results of constructing a DEM on small catchments using UAVs. Estimation of the UAV DEM showed comparable accuracy with the TLS if real time kinematic Global Positioning System (RTK-GPS) ground control points (GCPs) and check points (CPs) were used. In this case, the main source of errors in the construction of DEMs are the errors in the referencing of survey results.

1. Introduction

In the last decade, there was a trend to increase using of high-performance and high-precision technologies for monitoring erosion processes. In contrast to classical methods of recording erosion and accumulation such as, for example, the runoff method, the benchmark method, etc., new geodetic methods allow calculating the qualitative and quantitative changes in the microrelief by subtracting the extra-high resolution digital elevation models (DEM) from each other. For a long time, such high-performance geodetic instruments were terrestrial laser scanners (TLS), performance of which reached one million points per second [1]. However, despite the high performance and accuracy of scanning systems surveying large areas is very time-consuming. In addition, because of the specifics of taking measurements with ground-based laser scanners the surface scan can contain "shadows" – areas without fixed coordinates, resulting from a lack of direct visibility of the surface by the scanner.

Such limitations of TLS can be overcome using airborne laser scanners, however, the price of this equipment negates the possibility of its use on local catchments [2]. In addition, altitude errors limit the calculation of changes in the relief for different time intervals.

The development and cheapening of the technology of small flying systems coupled with the development of computer vision algorithms, has stepped up the number of works on the construction of digital terrain models using unmanned aerial systems. Such works first appeared at the end of the millennium in 1999 [3]. However, the massive use of this approach began in the second decade of the 21st century when the use of reference points (RP) allowed achieving submetric accuracy of the referencing [4, 5]. Already in 2013, the accuracy of the model reference to the control points obtained with the RTK-GPS receiver (real time kinematic global positioning system) was 0.29 m which, according to the authors [6], is equivalent to the accuracy obtained by scanning systems. Thus, the primary issue in surveying using a UAV is the organization of a precise referencing. In 2015 the achievable accuracy of the model referencing has increased to 0.02 m [7], in 2017 the tendency to increase the accuracy preserved [8].

Also, researchers are interested how accurate the DEM is, for which the DEM obtained with the UAV is often compared with the DEM obtained using the TLS. Recent work shows an average error not exceeding 5 cm [9, 10]. However, such errors are unacceptable when using this approach to study rill and sheet erosion which are the dominant exogenous process on almost the entire area of the arable slopes [11]. The purpose of this work is to develop a methodology for constructing high resolution DEMs using UAVs on the example of small catchments of different landscape zones in the European part of Russia.

2. Study area

The development and approbation of the technique of surveying in small catchments using UAV began in autumn 2016. As the first test site, a small catchment of the gully was located on the right slope of the basin of the river Temev Ruchey (right tributary of the Mesha River). The catchment area is a 500 m long slope composed of deluvial loams on which gray forest soils have formed.

In 2017, the works were carried out on the sites located in the landscape zones (from the forest to the steppe zone) of the basins of the rivers Temev Ruchey (Republic of Tatarstan), Veduga (Voronezh Region) and Medveditsa (Saratov Region). As objects, small catchments of dry valleys with an area of 1-2 km^2 were chosen. The length of the bottoms of the valleys varies from 400 to 1600 m. Also in the basin of the river Veduga of the Voronezh Region with the aim of conducting repeated surveys of rill erosion, which could arise as a result of summer-autumn rains, a catchment area of an agrogenic gully with an area of 0.1 km^2 was chosen. The coordinates of the objects are given in table 1.

Table 1. Study areas coordinates

Object	Longitude	Latitude
Gully catchment, Temev Ruchey River basin	49°39'01.88"E	55°39'17.69"N
Dry valley catchment, Temev Ruchey River basin	49°39'03.44"E	55°38'10.76"N
Gully catchment, Veduga River basin	38°37'17.35"E	51°41'45.50"N
Dry valley catchment, Veduga River basin	38°35'43.82"E	51°45'37.16"N
Dry valley catchment, Medveditsa River basin	45°24'36.95"E	51°50'21.51"N

3. Materials and methods

The work was carried out using an DJI Phantom 4 unmanned aerial vehicle. The UAV in the quadrocopter form factor is equipped with a 12mpx precalibrated camera with a 20 mm focal length in 35 mm equivalent as well as a three-axis active stabilization system. In contrast to the UAV with fixed wings, the form factor of the copter coupled with the stabilization system avoids fuzzy images as well as artifacts.

The survey was carried out in an automatic mode by loading a flight plan into the UAV controller, thus eliminating the human factor as a source of errors. For gully catchments in the basins of the river Temev Ruchey and Veduga, the flight was carried out at a 50 m from the take-off point altitude with 70% longitudinal and transverse overlapping of the images. For the dry valleys catchment the survey was carried out from a 100 m altitude with 70% longitudinal and transverse overlapping of images.

To provide the geodetic control special marks were used representing a 70x70 cm black and white chess pattern [12]. To register the coordinates of the marks a Trimble Geoexplorer 6000 GNSS receiver was used capable of operating in RTK (real time kinematic) mode. As reference, virtual base stations VRNZ and ATKS of the HIVE network respectively located in Voronezh and Atkarsk, as well as the base station of the IGS network, located at the Astronomical Observatory of Kazan Federal University, were used. Removing of the base station from the study area did not exceed 35 km. The data received by the receiver was processed in the RTKLib software where the data was adjusted from the rover and RINEX files of the base station as well as the removal of false points [13].

IOP Conf. Series: Earth and Environmental Science **107** (2017) 012005 doi:10.1088/1755-1315/107/1/012005

4. Results and discussion

The use of photogrammetry makes it possible to achieve the highest performance when working with UAVs. Thus, the average density of the resulting point clouds was 20 points/m^2. At the same time, the coverage area in the catchment area of the Temev Ruchey river basin was 2.33 km^2 (22 million points), 1.6 km^2 at the catchment area of the dry valley (20 million points) and 0.18 km^2 for the of the gully catchment area of the Veduga river basin (11 million points), and 2.77 km^2 for the dry valley catchment of the Medveditsa river basin (60 million points).

To obtain accurate referencing parameters of the model, reference points were used that made it possible to achieve the accuracy of the model referencing not exceeding 2 cm in vertical and 1 cm in the horizontal. Root mean square errors (RMSE) of referencing of control point by the example of the catchment of a dry valley in the basin of the river Veduga are presented in table 2.

Table 2. RMSE of control points

Control points	X (m)	Y (m)	Z (m)
1	-0.0043	0.0063	-0.0029
2	-0.0004	-0.007	0.003
3	0.0023	-0.007	0.0004
4	0.0047	0.003	-0.0002
5	-0.0038	0.0122	-0.017
Total	0.0035	0.007	0.0185

To assess the correctness of the obtained models, the DEM of small watersheds of a gully and dry valley in the basin of the river Temev Ruchey was compared with reference DEM. For a small catchment of the dry valley a DEM obtained from a topographic map with a scale of 1:10000 with an isoline interval of 1 m was used; a model obtained with the Trimble VX ground-based laser scanner (TLS) was used to estimate the model of the gully catchment area. Since the binding of the models obtained with the help of UAVs and with the help of TLS was carried out in the same coordinate system using the same control points, it becomes possible to estimate the quality of DEM by direct subtraction of the tested model from the reference one. As parameters for the assessment we used RMSE systematic error, maximum and minimum error, and the standard mean error as indicators. The results are presented in table 3.

Table 3. Results of UAV and TLS DEMs comparison

Number of points	577528
Systematic error, m	-0.01
RMSE, m	0.03
Minimum error, m	-0.12
Maximum error, m	0.1
Standart mean error, m	0.0002

Thus, using UAVs makes it possible to obtain a DEM comparable in accuracy to TLS DEM. To estimate the DEM obtained by using UAVs on the catchment areas of the dry valley to minimize errors associated with different coordinate systems and altitudes on the topographic map and in the production of DEM, a different approach was applied than that comparing DEMs obtained using UAV and TLS. The residual of slope angles and average lengths of slopes were taken as evaluation criteria. The results are presented in table 4.

Table 4. Results of UAV DEM and DEM obtained from a topographic map comparison

	Slope angles residuals	Average lengths from UAV DEM, m	Average lengths from topographic map DEM, m
Mean	0.097	1378.74	1373.61
Maximum	4	2757.49	2747.22
RMSE.	1.45	669.83	693.86
90% quantile	1.86	2481.74	2472.50
95% quantile	2.91	2619.61	2609.86

Analysis of the statistical indicators allows to state that there are no significant differences between the models obtained using UAV and DEM constructed on a 1:10000 scale topographic map, and all the differences fit the permissible 5%, and for the slope lengths do not exceed 1%. The results of the comparison of the models are confirmed in the studies of other scientists. For example, the same order accuracy (0.08 m) was reached when studying the relief in the basin of the Elbow river (Canada) [14]. The authors conclude that the UAV can achieve such accuracy and resolution which is difficult to achieve with other methods of remote sensing. Even more impressive results were achieved by researchers from Turkey [7]. The accuracy of the model was 0.06 m with a minimum, maximum deviation of altitudes was 0.0081 m and 0.085 m, respectively. Such results look impressive, however, such accuracy was achieved in comparison with GCP when both the real accuracy of the model as well as the estimation of available artifacts of model points is possible only with a continuous comparison of surfaces, which was done by us.

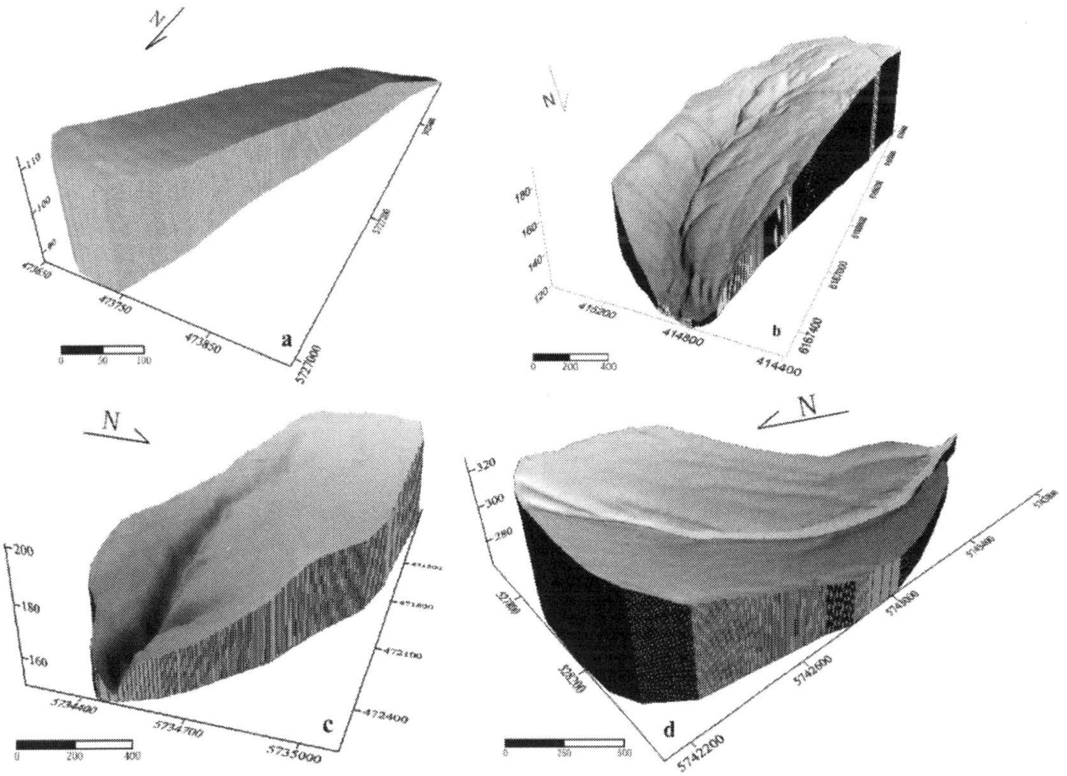

Figure 1. DEM for gully catchment in the Veduga river basin (a) and dry valleys in the basins of the river Temev Ruchey (b), Veduga (c), the river Medveditsa (d)

The evaluation of the accuracy of DEMs obtained using UAVs allows us to state that the models can be successfully used to calculate erosion losses in different parts of the temporary network. The comparison of the DEM and the UAV of gullies proved the applicability of the method for quantitative and qualitative assessment of gully erosion. DEM on small watersheds allows calculation of erosion losses in the rill and microrill using erosion models as well as direct subtraction of different-time DEMs from each other.

5. Conclusions

The construction of DEM for small catchments (figure 1) using a UAV demonstrates the potential of technology as a source of high resolution data obtained in a short time period. For the first time, data were obtained on the accuracy of the models obtained throughout the entire study area and not at individual points. In addition to the source of relief data, which can later be used to evaluate its changes after erosion processes, DEM data obtained with UAVs can be used to directly quantify the amount of material being moved after repeated surveys. However, in this case, as experience has shown in the use of TLS, the accuracy of model referencing comes to the fore, as well as the need to take into account the humidity and density of soil and ground, since when they swell / dry, surface changes reach the first centimeters. Despite the perspective of GNSS technology, obtaining high-precision data in the field is a very laborious task and it is recommended to make a geodetic control using tacheometric survey to obtain correct data about changes in the relief.

Acknowledgments

The research was carried out with the financial support of the Russian Scientific Foundation (project No. 15-17-20006).

References

[1] Gafurov A M, Yermolaev O P and Usmanov B M 2016 *Int. J. Pharm. Technol.* **8** 14822–32
[2] Prokop A, Schön P, Singer F, Pulfer G, Naaim M, Thibert E and Soruco A 2015 *Cold Reg. Sci. Technol.* **110** 223–230
[3] Asseline J, Noni G D and Chaume R 1999 *Photo Interprétation* **37** 3–9
[4] D'Oleire-Oltmanns S, Marzolff I, Peter K D, Ries J B and Hssaïne A A 2011 *Proc. 1st World Sustain. Forum* pp 1–13
[5] Vaaja M, Hyyppä J, Kukko A, Kaartinen H, Hyyppä H and Alho P 2011 *Remote Sens.* **3** 587–600
[6] Hugenholtz C H, Whitehead K, Brown O W, Barchyn T E, Moorman B J, LeClair A, Riddell K and Hamilton T 2013 *Geomorphology* **194** 16–24
[7] Uysal M, Toprak A S and Polat N 2015 *Measurement* **73** 539–543
[8] Agüera-Vega F, Carvajal-Ramírez F and Martínez-Carricondo P 2017 *Measurement* **98** 221–227
[9] Altyntsev M A, Arbuzov S A, Popov R A, Tsoi G V. and Gromov M O 2016 *ISPRS - Int. Arch. Photogramm. Remote Sens. Spat. Inf. Sci.* **XLI-B6** 155–159
[10] Neugirg F, Stark M, Kaiser A, Vlacilova M, Della Seta M, Vergari F, Schmidt J, Becht M and Haas F 2016 *Geomorphology* **269** 8–22
[11] Usmanov B, Yermolaev O and Gafurov A 2015 *Proc. Int. Assoc. Hydrol. Sci.* **367** 59–65
[12] Takasu T and Yasuda A 2009 *Int. Symp. GPS/GNSS* pp 4–6
[13] Mancini F, Dubbini M, Gattelli M, Stecchi F, Fabbri S and Gabbianelli G 2013 *Remote Sens.* **5** 6880–6985
[14] Tamminga A, Hugenholtz C, Eaton B and Lapointe M 2015 *River Res. Appl.* **31** 379–391

IOP Conf. Series: Earth and Environmental Science **107** (2017) 012006 doi:10.1088/1755-1315/107/1/012006

Study of medieval fortified settlements destruction under natural and anthropogenic factors using remote sensing data

I I Gainullin[1], P V Khomyakov[2] and B M Usmanov[1]

[1] Khalikov Institute of Archaeology, 30, Butlerova str., 420012, Kazan, Russia
[2] Institute of Environmental Sciences, Kazan Federal University, 18, Kremlevskaya str., 420008, Kazan, Russia

E-mail: busmanof@kpfu.ru

Abstract. Archaeological monuments are an essential part of the cultural landscape. Modern condition of archaeological monuments of the Republic of Tatarstan is discussed in this article. Fortified settlements, with the system of defensive fortifications were selected as the objects of study, as they are easily identified by remote sensing data. Due to the fact, that most of monuments are located on the small rivers banks, the first task of our study was to assess the risk of their destruction by natural processes. The second objective was to evaluate the role of the human factor in archaeological sites destruction. One of the main used methods is archival and modern remote sensing data analysis that also made able to correct the form of study settlements in comparison with existing plans as well as their size and location in the landscape. The results of research will help to identify trends in monuments state and to quantify the risks of their destruction.

1. Introduction

Worldwide, development of cultural heritage preservation strategy based on the analysis of the current monument state, forecasting and assessment of risks for archeological sites using advanced methods (analysis of remote sensing data, GNSS technologies and GIS) is an essential part of modern archaeological research [1]. For example, on the base of multi-method research it is possible to identify and document landscape changes for improving understanding, protection and management of cultural heritage at all scales from single monuments to entire landscapes [2]. Here, various methods of aerial photographs and satellite image processing are widely used: comparative analysis of changes in the state of the monument [3], planning of targeted archaeological protection and rescue works [4], reconstruction of ancient landscapes and initial appearance of archaeological objects of different historical epochs [5]. Thus, remote sensing data solves the following tasks in archaeology [6]: 1) best documentation and managing of rapidly disappearing ancient landscapes; 2) understanding of landscape formation processes; 3) identification and interpretation of economic, environmental and social influences that result in long-term settlement and land use patterns; 4) recognition and contextualization of the interplay between environment and human agency in evolution of ancient economies and transformations in socio-organizational complexity. In addition, the use and analysis of remote sensing data is an integral part of the non-destructive methods used in the study of archeological monuments along with geophysical research aimed at objects identification and minimization of destruction caused by archaeological excavations.

Over the last years, multi-rotor unmanned aerial vehicles (UAVs) were used in modern archaeological research due to their low price and ease of operation [7]. Now they are widely used in various fields where it is necessary to obtain remote sensing data quickly and inexpensively from a short distance. UAV used to obtain highly detailed aerial photographs which enable to produce orthophoto of study area, digital elevation models (DEMs) for detection and reconstruction of archaeological objects [8, 9], as well as the monitoring of their current state [10]. The use of UAVs also allows data combining for documentation and 3D visualization [11, 12].

The use of high-precision geodetic equipment is now becoming common and essential for archaeological works. Earlier it was enough to build topographic plans in conventional coordinate system using traditional instruments, but now archaeological research involves both digital geodetic equipment and high-precision GNSS receivers. Currently, GNSS technology is primarily used for archaeological excavations allowing archaeologists to get results in electronic form, helping to create an excavation grid and manage the overall organization process. In addition, satellite methods help to determine exact location of the find, its exact length and spatial orientation. When using UAVs, satellite methods are especially useful to justify the coordinates of ground reference points for more accurate positioning and aligning of aerial images [13]. Also GNSS technologies are effective for determining and studying the dynamics of dangerous exogenous processes that threaten the monument [14]. Thus, the use of GNSS technologies provides accurate data collection, helps to build complex multi-layer maps and to form geodatabases on the object under study.

This work is continuation of the research [15, 16] aimed at developing a system for analyzing of disturbance of cultural heritage site (archeological monument) territory using both archaeological research methods and geomorphological and geoecological research practices. Methodology for risk assessment of cultural heritage objects destruction is being developed on the basis of modern instrumental and cartographic-geoinformation approaches within the territory of Predvolzhye of the Republic of Tatarstan (RT).

2. Methods

Remote sensing data of maximum possible time spectrum over the past 60 years were selected. Search for fortified settlements on aerial photographs was carried out with the help of descriptions based on the results of field survey of past years. The next step was to estimate the fortified settlements condition.

The study of remote sensing data does not always make it possible to determine the actual situation on the sites under study, therefore, fieldwork is necessary to obtain operational data on the current state of settlements to determine the degree of their susceptibility to different types of impact and to justify the need for rescue activities.

Field surveys included:

- specification of location and visual features of settlement;
- photofixation of current monument state;
- GNSS survey of reference points;
- aerial survey of archaeological object under study.

Aerial photography was performed using a multi-rotor UAV DJI Phantom 4 (figure 1 a). The survey was performed with a 12-megapixel camera mounted on a quadrocopter; the gimbal provides a stable camera position during shooting. The UAV was controlled from a smartphone using Pix4D Capture software (figure 2) which allows to create a flight task and to configure the shooting parameters.

Aerial photography was performed with the following parameters:

- height – 50-100 m;
- pictures overlapping – 60-80%;
- camera position – 90°;
- coverage – area survey;
- meteorological conditions – no precipitation, wind no more than 15 m/s.

IOP Conf. Series: Earth and Environmental Science **107** (2017) 012006 doi:10.1088/1755-1315/107/1/012006

(a) (b)

Figure 1. Quadrocopter DJI Phantom 4 (a), special mark georeferencing by GNSS receiver (b).

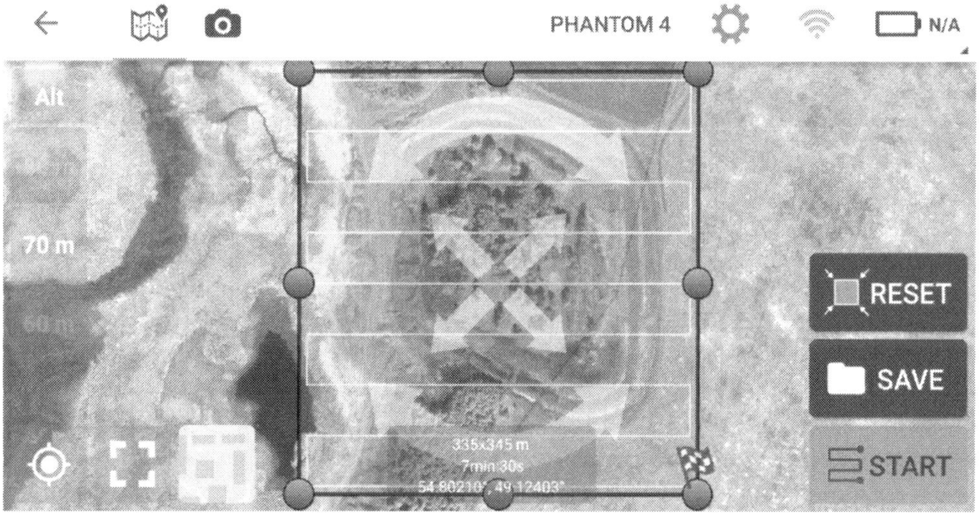

Figure 2. Flight task for Tankeevsky hillfort survey.

Coordinates in plan and height during aerial survey are determined by the GPS built in the UAV. This accuracy (4-5 m) is not sufficient for the construction of a topographic plan, digital terrain model (DTM) and digital elevation model (DEM) of an archaeological heritage object. To increase the accuracy of the survey before the flight a network of control points was formed on the territory.

For this purpose, the special marks were printed on the banner fabric with holes in the corners for fixing on the ground, and in the center – for a geodetic pole with antenna (figure 1 b). For high precise coordinates and height detection, GCPs centers were measured with a double frequency GNSS receiver with external antenna. Survey was performed in RTK regime in WGS84 with real-time corrections by satellite reference stations.

Coordinates of the GCPs centers were used to improve UAV data processing accuracy. Photogrammetric works were performed in Agisoft Photoscan in the following sequence:

- images inspection;
- pictures alignment;
- cloud of points construction;
- digital elevation model and orthophoto construction.

DEM analysis, mapping and calculations were conducted in Golden Software Surfer 13 software. A digital terrain model (DTM) with a step of 0.5 m showing the altitude characteristics of the settlement territory was generated by the point cloud. Profiles, inclination and aspect maps were built to describe the morphometric characteristics of settlement relief.

Ultra-high resolution orthophotos with a pixel size of 5 cm were also created. They were used to create electronic layers for topographic plans construction in MapInfo Professional 12.0 software.

3. Results

The mail goal of our investigation is a study of negative impact on cultural heritage objects with the help of modern technologies. In this context, to our opinion, the first priority of remote sensing is to provide both historical data and operational information about the current state of archaeological sites in the areas of intensive processes of coast transformation (sea, large water reservoir) and dangerous exogenous processes (landslides, debris, etc.) where the threat of destruction is maximal. As always the main criteria for the selection of favorable place to settle was the closeness of the water and the presence of natural fortifications, most of the archaeological sites are usually located on the banks and terraces of small rivers where influence of fluvial and exogenous processes is possible. Therefore, important task is geomorphological description of ancient settlements placement in order to assess the risk of their destruction by natural processes. Another factor that increases the danger for historical objects is widespread process of plowing, irrigation, active construction, etc. that started in the middle of the XX century. Intensive territory development led to a significant change in exterior of monuments and often – to their complete disappearance. Therefore, another objective is to evaluate the role of the human factor in archaeological sites destruction.

As a result of analysis of multi-temporal archival aerial and actual satellite imagery 12 hillforts of RT were selected for subsequent detail field survey aimed to describe their current state. This article presents the first results of UAV and GNSS survey on an example of Tankeevskoe fortified settlement I.

Tankeevskoe fortified settlement I is known since the XVII century. It is located in the Republic of Tatarstan in Spassky district 1.7 km south from the southern outskirts of the Polyanka village and 0.35 km west from the Bolgar-Ulyanovsk highway. An ancient settlement is surrounded from the north, east and south by a system of ramparts and ditches. From the west it is bounded by terrace cusp 6-7 m high, beyond which there is a narrow (25-35 m) floodplain of the Yasachka river (currently the bay of the Kuibyshev reservoir). The settlement area is flat and softly decreases to the western border by 2-2.5 m. Pine and birch plantations are located on the site of the settlement territory at the time of the field survey.

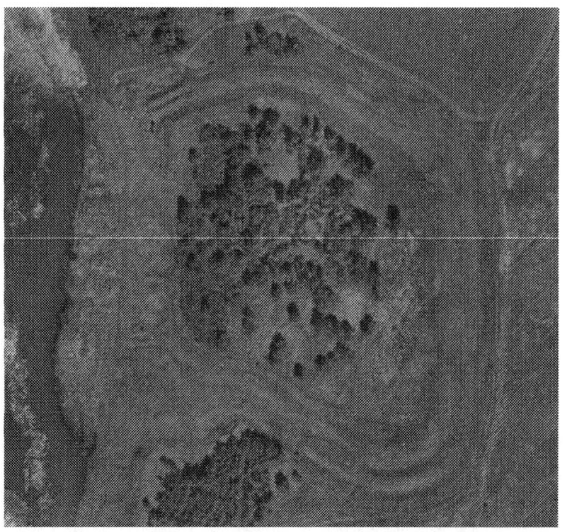

Figure 3. Orthophoto on the territory of Tankevskoe fortified settlement I.

Defensive structures are a triple system of ramparts in varying disturbance degrees. The extant outer (third) rampart (1-1.5 m) is found in the northeastern, eastern and southern parts of the settlement. The depth of ditches ranges from 2 to 4 m. In the northern part the outer rampart is not preserved practically due to the plowing of the adjacent territory. In the southwestern part defensive structures smoothly descend into a shallow ravine the mouth part of which is on the floodplain.

Based on the UAV survey, a dense cloud of points and a 3D model of the Tankeevskoe fortified settlement I were constructed required for DEM and orthophoto creation. The orthophoto resolution is 0.05 m (figure 3). With the use of an orthophoto, roads, vegetation, water objects, defensive structures were drawn for a topographical plan of the settlement.

Based on the digital model, hypsometric curves were constructed and preliminary morphometric analysis was performed (figure 4). Maps of territory inclination and aspects (figure 5) were also rendered to obtain characteristics for assessing the danger of modern exogenous processes development.

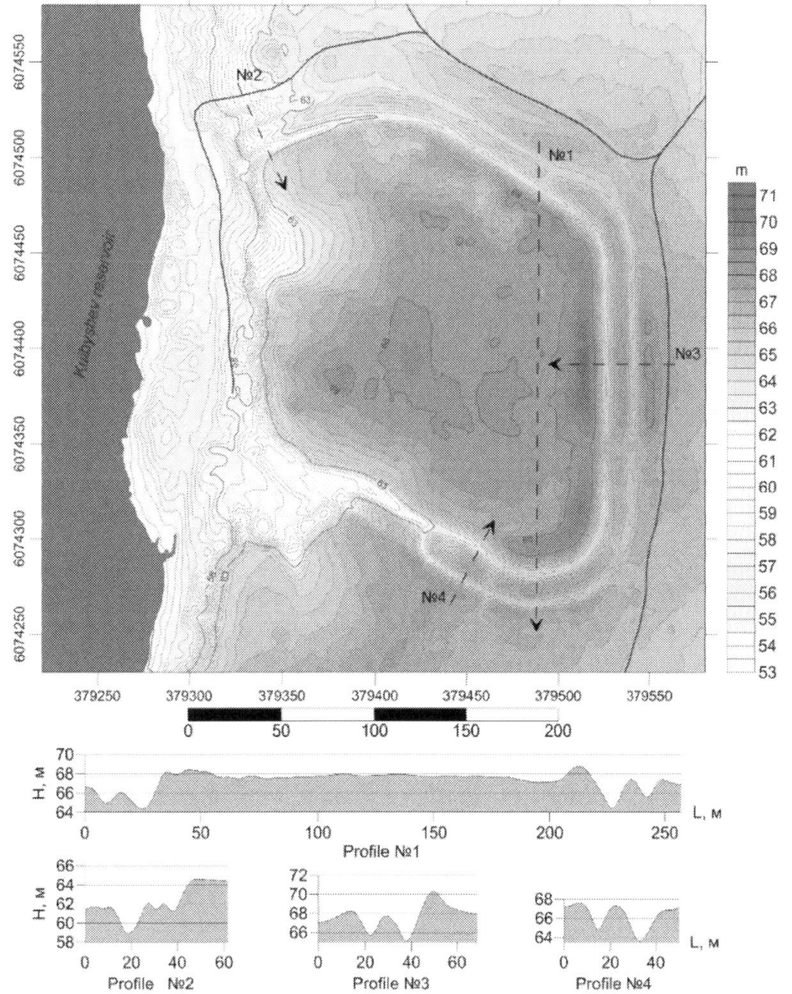

Figure 4. Digital elevation model of the Tankevskoe fortified settlement I. Elevation profiles: through the settlement – 1; trough defensive structures – 2-4.

IOP Conf. Series: Earth and Environmental Science **107** (2017) 012006 doi:10.1088/1755-1315/107/1/012006

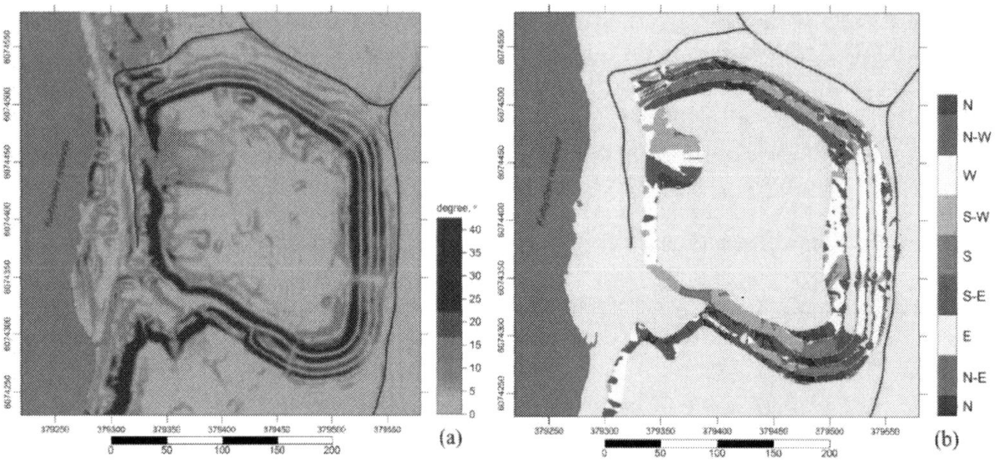

Figure 5. Inclination map (a) and aspects map (b) on the territory of the Tankevskoe fortified settlement I.

The results of UAV survey revealed that the Tankeevskoe settlement I is under significant exogenous and anthropogenic impact. There are furrowing traces from reforestation works on the settlement area. In the eastern part the ramparts are strongly deformed under the influence of slope processes. As a result the passes through defensive structures are practically not recognized. The plowing partially destroyed the third (external) system of fortifications. As a result of exogenous processes and anthropogenic impact, ramparts are destroyed in the northwestern part of the site. In the eastern and north-eastern parts, the remains of previously existing field roads are marked on the fortification. On the ridge of the north-western tip of the middle rampart a longitudinal excavation (possibly traces of archaeological research) with length of about 25 m and depth – 0.5-0.8 m, is noted.

It should be underlined that the negative impact on the territory of the Tankeevskoe settlement I was noted in the works of researchers of the past. According to SM. Shpilevsky in the XVII-XVIII centuries saltpeter was produced here. In the 1870s. the territory of ancient settlement "was plowed up under crops". At the time of the last survey of the ancient settlement in 1961, the fortified settlement area was also plowed up.

4. Conclusions

Thus, for the first time for the territory of Republic of Tatarstan, works were carried out to get actual data on fortified settlements transformation, dynamics of monuments destruction with the use of modern technologies and methods. The results of research will help to identify trends in the monuments state and to quantify the risks of their destruction. During the field works an original method for risk assessment of archaeological monuments destruction will be developed using unmanned aerial vehicles, photogrammetry, 3D-modeling, GNSS technologies.

References

[1] Wang J-J 2015 *Journal of Cultural Heritage* **16**(2) 210–220
[2] Risbol O, Briese C, Doneus M and Nesbakken A 2015 *Journal of Cultural Heritage* **16**(2) 202–209
[3] Lasaponara R, Masini N, Holmgren R and Forsberg Y B 2012 *Journal Of Geophysics And Engineering* **9**(4) S26-S39
[4] Banerjee R and Srivastava P K 2013 *Land Use Policy* **34** 193–203
[5] Reinhold S, Belinskiy A and Korobov D 2016 *Quaternary International* **402** 46–60
[6] Hritz C 2014 *Archaeol. Res.* **22**(3) 229–276

[7] Nikolakopoulos K G, Soura K, Koukouvelas I K and Argyropoulos N G 2017 *Journal of Archaeological Science: Reports* **14** 758–773

[8] Stek T D 2016 *Journal of Cultural Heritage* **22** 1066–1071

[9] Fernandez-Hernandez J, Gonzalez-Aguilera D, Rodriguez-Gonzalvez P and Mancera-Maboada J 2015 *Archaeometry* **57**(1) 128–145

[10] Asăndulesei A 2017 *Remote Sensing* **1**(17) 41

[11] Balletti C, Guerra F, Scocca V and Gottardi C 2015 *The International Archives of the Photogrammetry, Remote Sensing and Spatial Information Sciences* **XL-5**(W4) 215–222

[12] Tscharfa A, Rumplerb M, Fraundorferb F, Mayera G and Bischof H 2015 *ISPRS Annals of the Photogrammetry, Remote Sensing and Spatial Information Sciences* **II-1**(W1) 15–22

[13] Kršák B, Blišt P, Pauliková A, Puškárová P, Kovanic L, Palková J and Zeliznaková V et al 2016 *Measurement* **91** 276–287

[14] Gaynullin I, Sitdikov A and Usmanov B 2014 *International Multidisciplinary Scientific Conference on Social Sciences and Arts, SGEM2014 Conf. Proc.* book 3 vol 1 pp 339–346

[15] Gainullin I I, Khomyakov P V, Sitdikov A G and Usmanov B M 2016 *Proc. SPIE Fourth International Conference on Remote Sensing and Geoinformation of the Environment (RSCy2016)* vol 9688 96880Z

[16] Gainullin I I, Khomyakov P V, Sitdikov A G and Usmanov B M 2017 *Proc. SPIE Fifth International Conference on Remote Sensing and Geoinformation of the Environment (RSCy2017)* vol 10444 104440X

IOP Conf. Series: Earth and Environmental Science **107** (2017) 012007 doi:10.1088/1755-1315/107/1/012007

Evaluation of heavy metal pollution in bogs of Tomsk region on change in biogeochemical activity of ericaceous shrubs

L P Gaskova[1,2]

[1] Siberian Research Institute of Agriculture and Peat, 3, Gagarina str., 634050, Tomsk, Russia
[2] Tomsk State University, 36, Lenina str.,634050, Tomsk, Russia

E-mail: gashkova-lp@rambler.ru

Abstract. The article discusses the change in biogeochemical activity of plant species in bogs under the influence of various types of human impact (roads, cities, drainage of mires, fire). It has been established that ericaceous shrubs, depending on the species, react with varying degrees of intensity to anthropogenic influences. The biogeochemical activity of species increased by 2.5 to 4.8 times in polluted sites.

1. Introduction

Bogs are unique ecosystems that are the habitats of highly specialized species. The supply of mineral nutrition to plants in such bogs is very low; the main source of elements is atmospheric precipitation [1, 2]. Currently, the three main sources of metals in soils are industrialization, urbanization and agricultural practices. Heavy metals in the soil from anthropogenic sources tend to be more mobile and hence bioavailable than those from lithogenic soil formation [3]. The absorption and accumulation of heavy metals by plants depends on the chemical form of metal compounds, soil factors and characteristics of the plant [4, 5, 6]. Mires have very low environmental buffer capacity and therefore their response to atmospheric pollution is more pronounced than in many other ecosystems [7, 8]. It is necessary to know the levels of metals in peat and plants to assess potential environmental damage and develop mitigation strategies [9]. Large areas of bogs are characteristic of the Tomsk region, an area which is swamped by different sources, from 39.5% to 50% [10]. Significant areas of mires are affected by pollution as a result of the influence of the Tomsk–Seversk industrial agglomeration, roads, drainage, fires, peat extraction, oil and gas. Mires located near settlements are often used to discharge sewage.

2. Method and data

Taking into account the listed anthropogenic factors, 19 mire sites which are affected by one or more of these factors were selected. As a background, 13 bog sites significantly removed from pollution sources were selected. Previously conducted studies have shown that the content of heavy metals in plants of the mires is within or below background concentrations for the study region [11]. The research was conducted from 2014 to 2016, in mires located in Tomsk, Shegarsky, Bakcharsky, Kozhevnikovsky, Kolpashevsky, Kargasoksky and Parabelsk districts of the Tomsk region (figure 1). To study the effect of heavy metal contamination four kinds of shrub widespread in bogs of the Tomsk region, were selected: *Andromeda polifolia* L., *Chamaedaphne calyculata* (L.) Moench, *Ledum palustre* L. and *Vaccinium uliginosum* L. For all the investigated sites, plant and peat samples were selected in July 2015 for the determination of Zn, Cd, Pb and Cu. In plants of each of the species studied, a mixed sample of shoots of the current year was selected from an area of 10 × 10 m. Samples of peat were selected in the root layer. Determination of the content of heavy metals in plants and peat was conducted by inversion voltammetry in the laboratory analytical centre of the Siberian Research Institute of Agriculture and Peat. Statistical analysis was performed on the basis of StatSoft Statistica for Windows 6.0 and Excel 7.0. Comparison and significant differences between the samples were set using the nonparametric Mann–Whitney test. To determine the influence of pollution on the accumulation of heavy metals the coefficient of biological absorption (CBA), which reflects the ratio of the element content in plant ash to its content in peat [12], and the biogeochemical activity of the species (BAS) which is the sum of the CBA of the elements under consideration were calculated [13].

3. Results

When comparing the absolute values of the heavy metal content in plants it was found that not all of the elements investigated had a significantly increased concentration in plants from sites prone to anthropogenic impact. In *A. polifolia* an increase in the content of Zn and Cu was observed, in *V. uliginosum* the concentration of Zn, Cu and Cd increased, and the content of Zn, Pb and Cu increased in *C. calyculata*. Only in *L. palustre* was the concentration of all the elements studied increased from contaminated sites: Zn, Cd, Pb and Cu, when compared with similar areas of bogs remote from pollution sources.

Absolute values of the concentration of elements in plants do not take into account the influence of the content of heavy metals in peat. However, when comparing not absolute but relative indicators, for example the CBA, there were differences in all elements between background and contaminated sites ($U = 5$; $p < 0.05$).

The BAS level also significantly increased in sites prone to anthropogenic impact ($U = 85$, $p < 0.001$). In addition, the spread of values was much larger for the disturbed sites (figure 1).

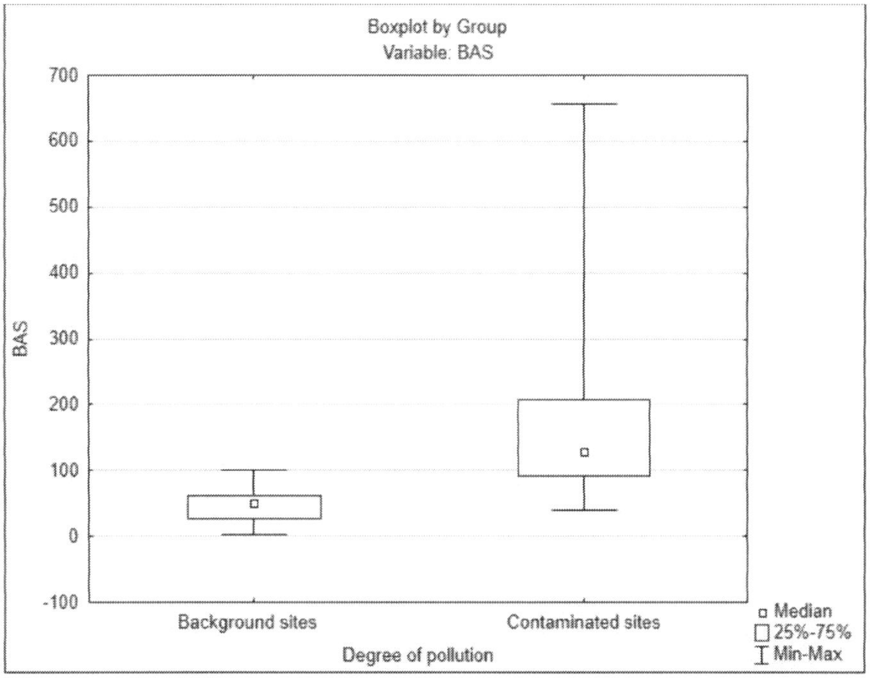

Figure 1. Differences in the distribution of BAS values for background and contaminated sites.

The biogeochemical activity of all the species studied significantly increased in the disturbed mire sites. In the background sites the biogeochemical activity was the same for different species of shrub. In areas subject to anthropogenic effects the equilibrium is violated so the biogeochemical activity of various species has increased unevenly. For *C. calyculata* and *L. palustre* biogeochemical activity of the species increased approximately 3-fold, while *A. polifolia* and *V. uliginosum* biogeochemical activity of the species increased 7-fold (figure 2).

Thus, *A. polifolia* and *V. uliginosum* were species most sensitive to changes in the microelemental composition of the environment.

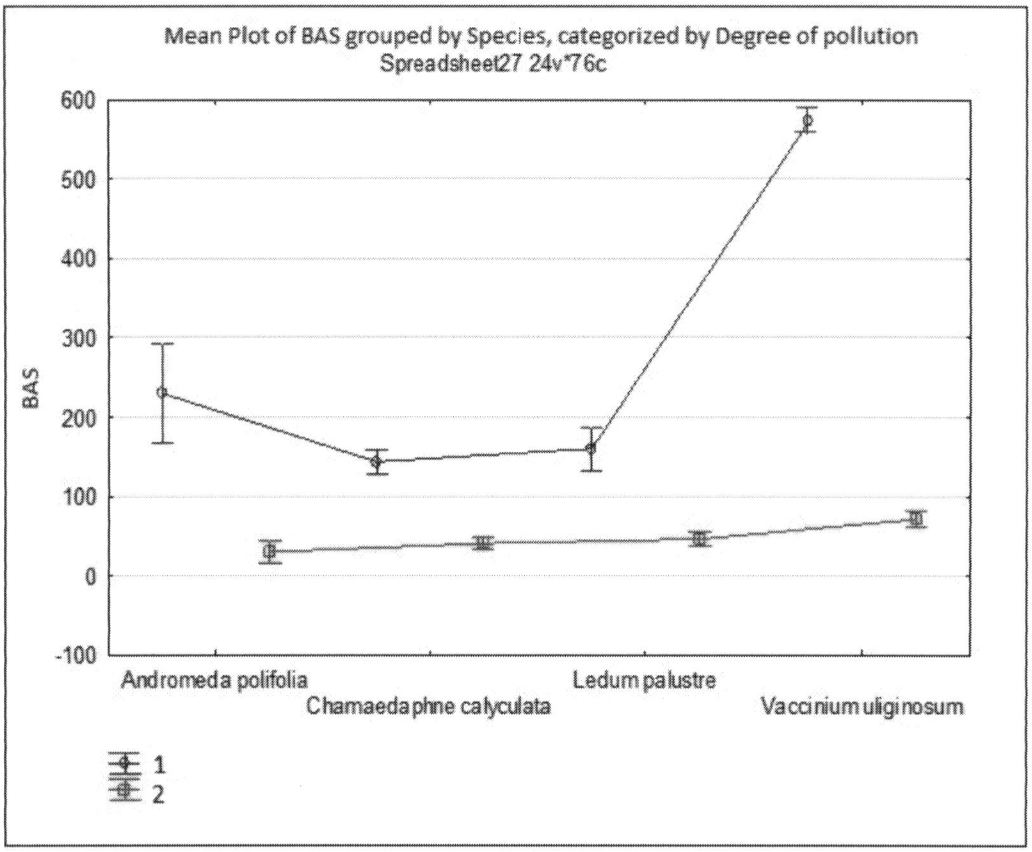

Figure 2. Comparison of biogeochemical activity values of species in sections 1 – subject to anthropogenic impact – and 2 – remote from pollution sources.

The biogeochemical activity of the species data obtained also showed a dependence on the type of pollution. There was an increase in biogeochemical activity of the species of approximately 2.5 times in areas affected by roads compared with the background values. In dry areas of mires, biogeochemical activity of the species increased by 4.3 times and in the zone of city influence by 4.6 times in comparison with uncontaminated areas. In mires exposed to fire, biogeochemical activity of the species increased on average by 4.8 times (figure 3).

The differences in the increase in the types of biogeochemical activity are due to different levels of contamination and varying availability of trace elements for plants. On mires affected by the city and roads the main source of pollution is atmospheric transport of elements from vehicles and industrial plants. On the drained bogs the degree of peat decomposition increases; as a result of this the availability of trace elements for plants is increased. After a fire in the mires the amount of mobile trace elements in peat also increases.

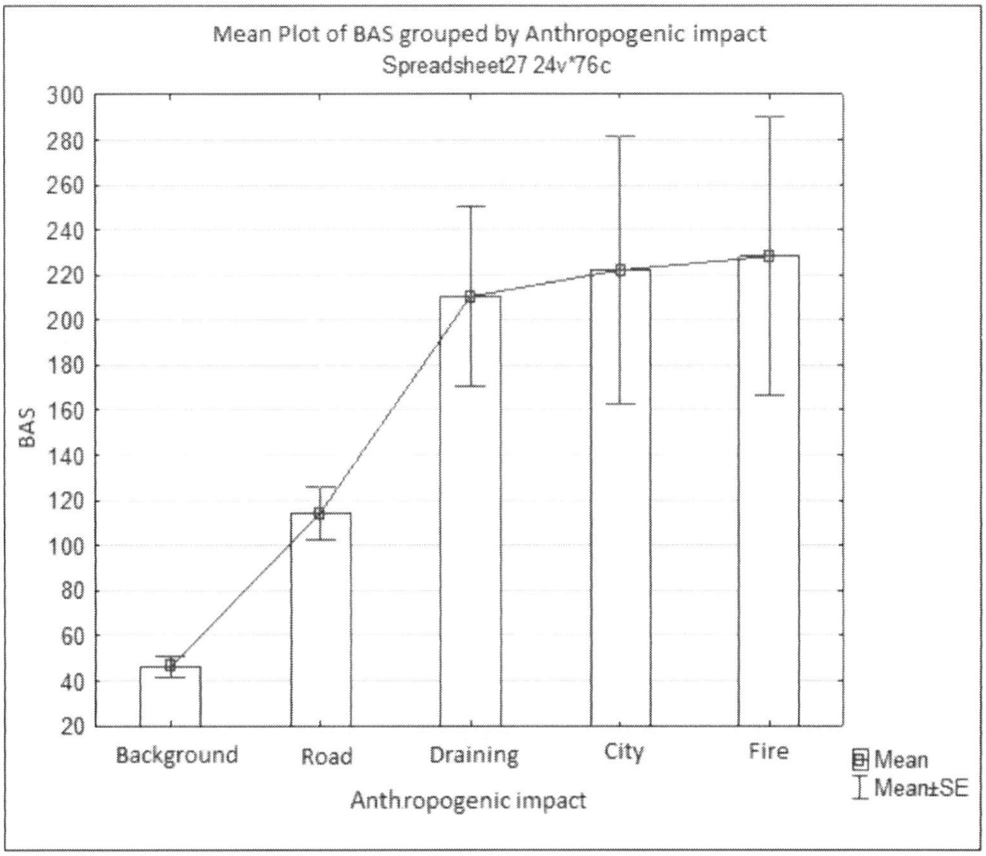

Figure 3. Influence of different types of anthropogenic impact on the change in biogeochemical activity of the species.

4. Conclusions

As a result of the study, it can be concluded that ericaceous shrubs can serve as indicators of pollution of mires with heavy metals. The most sensitive species to contamination with heavy metals from the environment were *A. polifolia* and *V. uliginosum*. Despite a slight increase in heavy metal content in contaminated sites bog plants had a sharply increased intensity of biological absorption under anthropogenic influence. Therefore, in determining the degree of pollution, instead of comparing the absolute values of the content of elements in plants and peat, it is more indicative to compare the biogeochemical activity of the species.

Acknowledgments

The study was carried out with the financial support of the Russian Foundation for Basic Research in the framework of the scientific project no. 16-45-700418-p_a.

References

[1] Glooschenko W, Holloway L and Arafat N 1986 *Aquatic Botany* **25** 179–190
[2] Gotellil N J, Mouser P J, Hudman S P, Morales S E, Ross D S and Ellison A M 2008 USA *Wetlands* **28(3)** 827–840
[3] Klos A, Rajfur M and Waclawek M 2011 *Ecol. Chem. Eng. S.* **18(2)** 171–183
[4] Kaczorek D, Brümmer G W and Sommer M. 2009 *J Environ Qual* **38(3)** 1109–19
[5] Van der Ent A, Baker A J M, Reeves R D, Pollard A J and Schat H 2013 *Plant Soil* **362** 319–334
[6] Chenery S R, Izquierdo M, Marzouk E, Klinck B, Palumbo-Roe B and Tye A M 2012 *Science of*

the Total Environment **433** 547–560
[7] Holdena J et al 2007 Environmental change in moorland landscapes *Earth-Science Reviews* **82** 75–100
[8] Wojtuń B, Samecka-Cymerman A, Kolon K, Klink A and Kempers A J 2013 *J. Environ. Sci. Health* A **48(7)** 686–693
[9] Balabanova B, Stafilov T, Baceva K and Sajn R 2010 *J. Environ. Sci. Health* A **45** 1504–18
[10] Dyukarev A G, Lvov Yu A and Khmelev V A 1991 *Natural Resources of the Tomsk Region* (Novosibirsk Science) (in Russian)
[11] Gashkova L P 2016 *The Bulletin of Ryazan State University named for S A Yesenin* **4(53)** 146–158 (in Russian)
[12] Ayvazyan A D 1974 *Geochemical Features of the Flora of the Landscapes of the South-Western Altai* (Moscow MGU) (in Russian)
[13] Perelman A I and Kasimov N S 1999 *Geochemistry of the landscape* (Moscow MGU) (in Russian)

Trend of Soil Erosion Processes within the Southern Half of the Russian Plain for the Last Decades

V N Golosov[1,2] , O P Yermolaev[1], G R Safina[1] , K A Maltsev[1], A V Gusarov[1] and I I Rysin[1,3]

[1] Institute of Environmental Sciences, Kazan Federal University, 18, Kremlevskaya str., 420008, Kazan, Russia
[2] Faculty of Geography, Lomonosov Moscow State University, 1, Leninskie Gory, 119991, Moscow, Russia
[3] Institute of Natural Sciences, Department of Ecology and Natural Resources Management, Udmurt State University, 1, Universitetskaya str., 426034, Izhevsk, Russia

E-mail: gollossov@gmail.com

Abstract. Complex approach is applied for assessment of recent trends of sheet, rill and gully erosion in different landscape zones of study area. Investigation is undertaken in 6 selected sectors (area of each transect is about 6-10 thousand km2), uniformly distributed over the area of the Russian Plain. Changes of the different factors, including some meteorological and hydrological parameters, land use change, USLE C-factor, were determined for the period 1980-2015. A set of field methods was used for quantification of sediment redistribution rates for the key small catchments. It was found that erosion rate decreased in forest and forest-steppe zone. Gully density decreases considerably in all landscape zones. The reduction of surface runoff from cultivated slope during snow-melting is the main reason of decreasing of sheet, rill and gully erosion rates in the forest, forest steppe and the north of steppe landscape zones. Increasing the proportion of perennial grasses in crop-rotation is the other factor of serious reduction of erosion processes in the forest zone.

1. Introduction

The most part of croplands is located in the Southern part of Russian Plain within south part of forest, forest-steppe and steppe landscape zone. It is area with relatively short history of intensive agriculture with duration from 4-5 centuries in the center part around Moscow in forest zone up to about 120-150 years on the south and south-east within the steppe zones [1]. The highest area of cropland in the most productive forest-steppe zone was reached at the end of 19th century after land reform when very steep slopes were cultivated, and maximum of gully erosion rates was reached immediately [2]. During the 20-th century cropland area was relatively stable within the forest and forest-steppe zone while it enlarged in the steppe zone because of virgin lands cultivation during 1950th.

Latest quantitative assessment of erosion rates for the entire European Russia was undertaken at the beginning of 1980th based on application of modified version of USLE and State Hydrological Institute model for evaluation of erosion rate during rain-fall season and snow-melting respectively [3]. Results of model calculations were verified based on the monitoring data of soil losses during snow-melting and evaluation of total soil losses in the small field ponds [4, 5]. It was shown that model calculations are in a good agreement with actual soil losses.

More than 35 years passed since the last quantitative assessment of soil erosion losses from croplands of the Russian Plain. Considerable economic changes had occurred after USSR collapse in 1991 with following land use and crop-rotation changes in some landscape zones of European part of Russia. In addition, global warming is affecting on dynamic of meteorological characteristics in the different parts of the world since the middle of 1970th. It is very likely that both land use and climate changes influenced on the soil erosion rates on the croplands of the Russian Plain. The objectives of this paper are evaluation of the trend of different erosion processes based on application of erosion models, interpretation of aerial photographs and satellite images and application of ^{137}Cs techniques for the dating of sediment redeposited in the valley bottoms of the first- order agricultural catchments located in the different landscape zones of the Russian Plain.

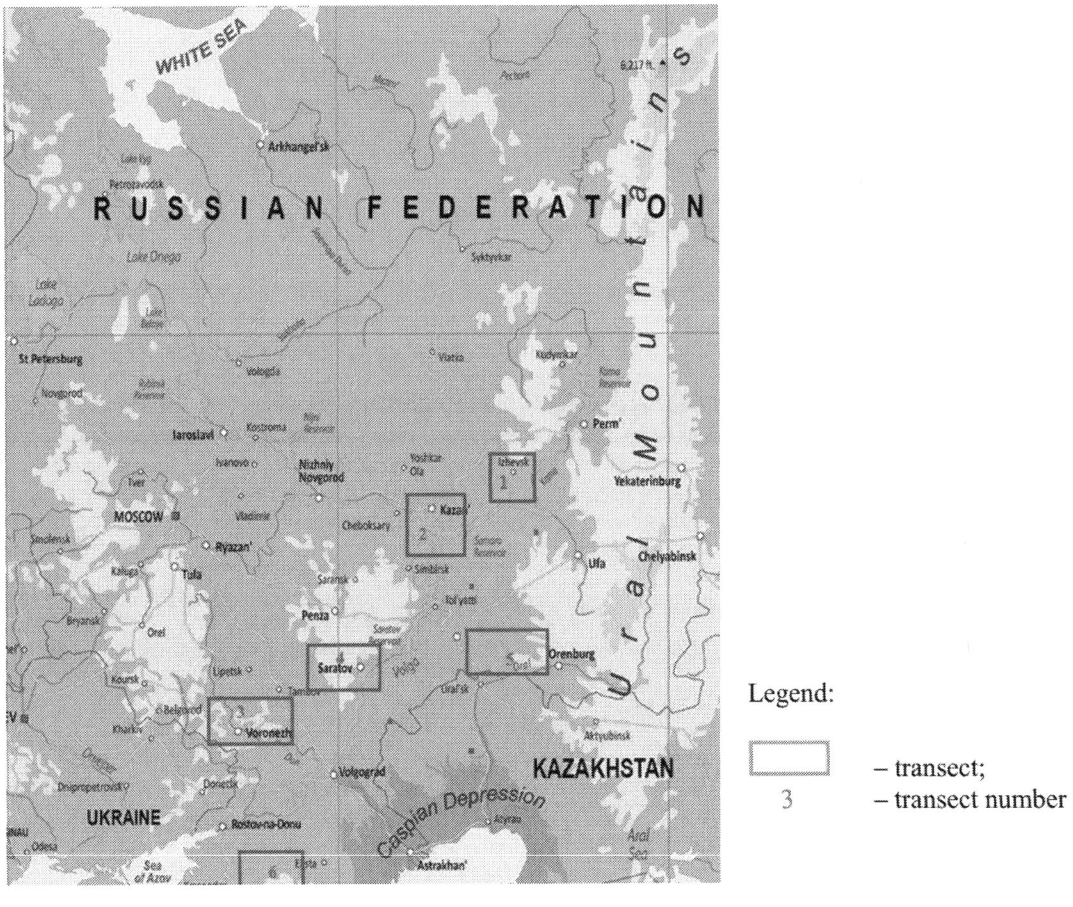

Figure 1. Map of the European part of Russia with location of studied transects.

2. Material and methods

2.1. Study area

The Russian Plain occupies area from the Caucasus mountain foothills and the Caspian Sea on the south, the Ural Mountains on the east, Arctic Ocean and White Sea on the north. Western boundary of the study area is similar with the western border of the Russian Federation (figure 1). Despite of quantitative assessment of soil losses done for the entire Russian Plain, the most attention was given to the southern half of studied territory to the south from the line Bryansk- Moscow – Nizhniy Novgorod – Kazan –

Perm because the most part of croplands is located within the given territory. The Northern half of the Russian plain is located in tundra, taiga and mixed deciduous-coniferous forests landscape zones with mostly moraine relief, poor soils with low content of organic matter, proportion of croplands from 0% to 20% (of the total area) and the significant proportion perennial grasses in crop rotations.

The southern half of the Russian Plain is located mostly in forest-steppe and steppe landscape zones where soils formed on the loess with high organic content dominate. So the proportion of cropland changes in the range 35-85%, and row crops (corn, sunflower, sugar beet, buckwheat etc.) are the significant components of the crop-rotation. A temperate continental climate with relatively cold winter and warm summer is typical for the most part of the Russian Plain. Total (snow + rain) annual precipitation reduces from the north-west to the south-east territory from 650 to 350 mm with increasing of evaporation in the same direction. So droughts are observed very often in particular in the south-eastern part of the steppe zone. During the most part of the 20th century soil erosion during snow-melting is observed within the most part of the Russian Plain with maximum on the north of the forest-steppe zone while soil erosion during rain storms is widespread mostly in steppe and south part of forest-steppe zone. Sometimes extremely heavy rains lead to severe erosion even in the forest zone [5].

2.2. Methods

Six transects located in different landscape zones of the southern part of Russian Plain were selected for the more detail assessment of possible trend of erosion rates during last decades (figure 1). The evaluation of erosion process trend was undertaken for the key river basins located within the each transect and for the small agricultural catchment located within each river basin. Complex approach was applied for evaluation of contemporary trend of the sheet, rill and gully erosion within each transect.

Field methods and techniques are used for determination of gully headcut retreat rate and sediment redistribution rates within small first-order catchments, including: large-scale geomorphological mapping, 3-D scanning, detail topographical survey, ^{137}Cs technique, soil-morphological method and others. A set of high-resolution satellite images including IKONOS, GeoEye-1, Spot, World-View was used for evaluation of land-use transformations, gully head density, length of active gully within the key river basins selected within each transect. In addition, long-term monitoring data of water discharges in the key studied basins were collected. Particular attention was given to analysis of water discharges during spring floods, as an indicator of the surface runoff from cultivated slopes in this period of year.

Calculation of erosion rates and total soil losses from croplands for different landscape zones of the Russian Plain and for the key river basins within each transect was done using modified version of USLE (rain-fall period) and State Hydrological Institute model (period of snow-melting). Rainfall erosivity factor was re-calculated for the 2012 based on linear relationship between layer and rainfall erosivity of individual rain [6]. Crop rotation coefficients for rainfall and snow-melting periods were calculated using statistical data about areas under different crops and method of cultivation and rainfall erosivity index for different months for each region. Changes of arable lands area were taken into consideration. Information about land-use changes and land area under different crops for different regions of Russia for period since USSR collapse was collected from Russian Federal State Statistics Service. Particular attention was given to evaluation of croplands dynamics in different landscapes for the period since 1991 [7]. Additional independent sources of information were used for correction of possible mistake in official statistics [8]. LS factor was calculated separately on 1985 and 2015 for the cultivated fields taking land use changes into consideration [9]. All other parameters were taken from data base constructed in a period of erosion rate calculation for the Russian Plain in 1980.

3. Results and discussion

Changes of the climate and land-use led to different changes of erosion rates and the total soil losses in different landscape zones of the Russian Plain. According to previous assessment the most considerable reduction of both erosion rate and the total soil losses was found for the different parts of the forest zone [10] (table 1).

Table 1. Changes of mean annual soil erosion rate and total soil losses in 1980 and in 2012 calculated using modified version of USLE (rain-storm erosion) and State Hydrological Institute (erosion during snow-melting) models [11].

Landscape zone of European part of Russia (without mountain area)	Mean annual soil erosion rate in 1980, t ha^{-1} year^{-1}	Mean annual soil erosion rate in 2012, t ha^{-1} year^{-1}	Change of mean annual soil erosion rate[a], (+/-)%	Total soil losses in 1980, 10^3 t	Total soil losses in 2012, 10^3 t	Change of total soil losses* (+/-)%
Northern and middle taiga zone	6.5	4.0	-38.4	6131.5	1808.8	-70.5
Southern taiga zone	7.3	4.1	-44.0	145031.9	35965.9	-75.2
Forest zone, total	7.3	4.1	-43.8	151163.4	37790.9	-75.0
Forest-steppe zone	4.1	3.3	-19.4	136449.7	79277.3	-41.9
Steppe zone	3.9	4.63	18.7	148617.8	127662.7	-14.1
Total European Russia	4.7	4.0	-15.0	436230.9	244830.9	-43.9

[a] Relationship between 1980 and 2012

There are several reasons of such trend, including reduction of arable land area, changes in crop rotation and serious decreasing of surface runoff during snow-melting. Also some reduction of erosion rates was determined for the forest-steppe zone with more serious decreasing of the total soil losses mainly because of reduction of arable land area. Also it is more likely that possible increasing of soil losses during rain-storm season does not compensate the decreasing of soil losses during snow-melting. However, soil erosion rate increased in the steppe zone according to results of the model calculation because of increasing of frequency of the rain-storms with layer of 40-50 mm. But the total soil losses in steppe zone also decrease because of reduction of arable land area. It is necessary to emphasize that area of arable lands partially restored since 2012 due to the plowing of previously abandoned lands since 2012.

Table 2. Changes of mean annual soil erosion rate for the key river basins in 1985 and 2015.

River basin	Basin area, km^2	Landscape zone, transect	Mean annual erosion rate, t ha^{-1} yr^{-1}		Reduction of erosion rate, %
			1985 year	2015 year	
Izh	2508	Forest, 1*	4.6	4.4	4.0
Mesha	4354	Forest ; forest-steppe, 2	7.8	6.8	12.8
Ulema	881	Forest-steppe, 2	5.8	5.0	13.8
Veduga	1193	Forest-steppe 3	5.6	5.6	0
Medveditsa	3611	Forest-steppe ; steppe, 4	2.6	1.6	38.5
Samara	2874	Steppe , 5	2.7	2.3	14.8

Here and further – transect number, see figure 1.

Results of calculation of the soil erosion rates show the trend of reduction of soil losses for the most river basins (table 2). The main reason of soil erosion rates decrease is serious decline of surface runoff from the cultivated slopes during snow-melting period. So the soil losses during snow-melting in particular in forest-steppe zone and in the northern part of steppe zone became very close to zero. In addition, some input into reduction of soil erosion rates was associated with decrease of LS factor due to abandonment of cultivated lands. Only the Veduga River basin, located in the western part of the forest-steppe zone, is characterized by the lack of any trend. This territory is characterized by increasing frequency of strong rainfalls [11], which should lead to increase of the rainfall erosion.

Results of the detailed assessments of sediment redistribution rates for periods of 1963-1986 and 1986-2015 at the key catchments, located in the Veduga and the Mesha River basins, were used for verification of the results of soil erosion rate calculations within the studied river basins. Both key catchments are characterized by a high proportion of cultivated land, which, according to the available evidence (topographic maps, aerial photographs, satellite imagery and data archives), has not changed appreciably since the 1950s. Both bomb- and Chernobyl-derived ^{137}Cs were used as chronological markers for documenting changes in soil erosion rates on croplands during the last 50-55 years in key catchments. The sediment accumulation rates in the dry valley bottoms reflect the intensity of soil erosion on the slopes of the dry valley catchments [12], [13]. Eight ^{137}Cs depth profiles located in representative cross-sections along the dry valley bottoms have been used to document sediment accumulation rates at key catchments. It was established that the mean annual sedimentation rates had reduced 4-5.4 times in both key catchments (table 3). In case of Mesha River basin key catchment is located in area with slightly higher erosion rates as compared to mean values for the entire river basin (table 2).

The calculated mean annual erosion rates had reduced from 8.4 to 7.1 t ha^{-1} (the Mesha key catchment) and from 5.5 to 5.0 t ha^{-1} (Veduga key catchment) in 1963-1986 and 1986-2015 respectively. It is very likely that soil losses were underestimated for 1963-1986 due to incomplete estimation of erosion losses during snow melt. The ephemeral gully erosion usually had formed in the hollow bottoms due to water concentration, and this type of erosion was not calculated by the SHI model. Soil losses from cultivated hollow catchments are almost one order magnitude higher as compared to soil losses from slope without hollows according to field monitoring data obtained as a result of 15- years' observations during snow-melt in the southern part of the forest zone of the European Russia [4].

Table 3. Estimates of mean annual sedimentation rates in dry valley bottoms at the key catchments for different periods based on interpretation of ^{137}Cs depth profiles.

Dry valley location	Landscape zone, transect	Co-ordinates	Mean annual sedimentation rates, cm yr^{-1}, for indicated time interval		Reduction of sedimentation rate
			1963-1986	1986 – 2015	
Veduga River basin	Forest and forest-steppe, 3	51.760539N 38.602139E	2.0 ± 0.2	0.5 ± 0.05	x 4
Mesha River basin	North of forest-steppe, 2	55.645670N 49.646451E	1.4 ± 0.15	0.26 ± 0.05	x 5.4

Table 4. Gully density in some key river basins of different landscape zones of the European Russia in 1990 and 2015 [14].

River basin	Landscape zone, transect	Gully density, km km^{-2}	
		1990 year	2015 year
Mesha	South of forest, forest-steppe, 2	0.56	0.04
Ulema	North forest-steppe,2	0.8	0.09
Veduga, Devitsa 1, Devitsa 2	Forest-steppe, 3	0.8	0.02
Medveditsa	North steppe, 4	0.43	0.01
Samara	Steppe,5	0.26	0.02

Quantitative assessment of the gully density dynamics for the key river basins indicates the sharp reduction of gulling (Table 4). It is another confirmation of the reduction of erosion processes intensity mostly due to the considerable decrease of the surface runoff during snow-melting in the last decades. It is well-known that the main factor of gully head retreat rates for the most part of landscape zone of Russian Plain (except for the south of steppe zone) is intensity of the surface runoff during the snow-melt [15]. The similar trend of gully erosion rates decrease was found as a result of long-term monitoring of gully head (about 150) in different parts of Vyatsko-Kamsckoe interfluve area for per 1978-2015 (Rysin et al., 2017).

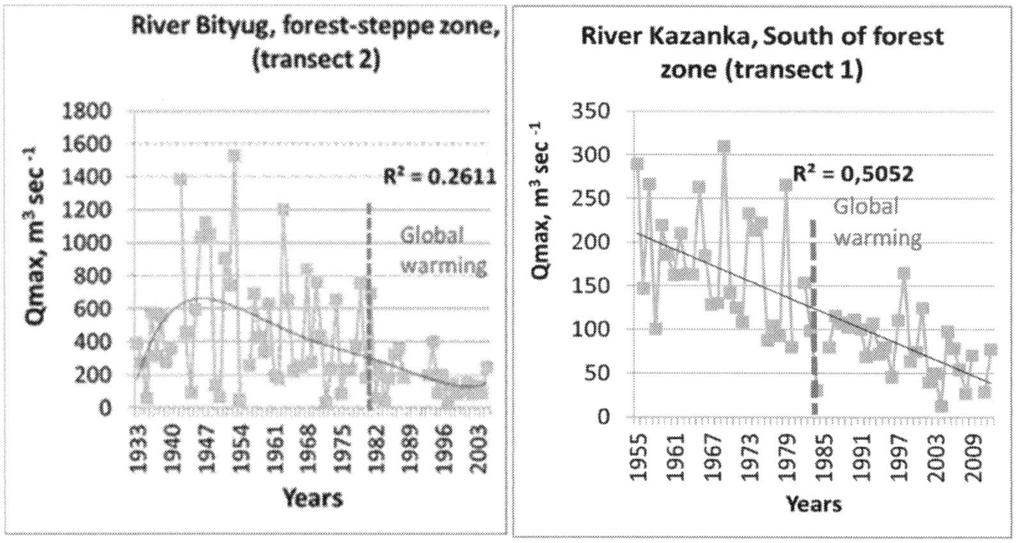

Figure 2. Trends of maximum water discharges during a period of spring snow-melting in the typical small rivers of the southern part of forest zone and forest-steppe zone.

The other confirmation of serious reduction of surface runoff from the cultivated slopes during last decade is the decrease of the maximum water discharges of small rivers located within transects 1-5 since the beginning of the 1980s when air temperature began to increase due to climate change (Figure 2). At the same time minimum water discharges observed during summer season increased during the same decades [16]. The reason is increasing ground water table due to growth of water infiltration during spring snow-melting and winter thaws.

At least, the results of monitoring the surface runoff from cultivated lands undertaken at the Novosil experimental station (Orel region, the north of forest-steppe zone) show that the coefficient of water surface runoff decreased from 0.5 for period from 1955-1980 to <0.1 since the late 1990th due to serious reduction of the mean depth of the frozen soil [17].

However, results of the model calculations for the entire steppe zone [11], [10] contradict evaluation of the erosion rates trend for the key river basins (Table 1 and 2). This contradiction may be due to the spatial irregularity of the soil erosion during rainfall season which prevails in the steppe zone. In particular, this assumption is confirmed by the results of estimating the trends of rain frequency with different precipitation layers. However, the results of field investigation within small catchments and the evaluation of the gully density dynamics in the Medveditsa and the Samara River basins indicate that clear decrease trend of the soil erosion rates is observed as compared to the two time windows (1963-1986 and 1986-2016). It is necessary to say that results of the ^{137}Cs analysis of samples taken within small catchment in the Kalaus River basin are not ready yet.

4. Conclusions

Application of erosion models enables to identify the erosion rates trends for the typical river basins located within the different landscape zones of the southern half of the Russian Plain. The reduction of the erosion rates was determined for the most river basins with the exception of the Veduga River basin, located in the western part of the forest-steppe zone. It is more likely that serious reduction of the soil losses during snow-melting due to decrease of the surface runoff from the cultivated slope is the main reason of negative trend of the mean annual erosion rates. This assumption is in good agreement with the results of monitoring observations of the river runoff which indicate a sharp drop in maximum water discharge during floods and an increase in minimum water flow in summer low water. Field verification of the model calculations on the small agricultural catchments based of the evaluation of sedimentation rates for the two time windows (1963-1986 and 1986-2015) confirm the identified trend for the previously mentioned landscape zones. The lack of changes of erosion rates in the Veduga River basin can be explained by some increase of soil losses during rainfall season in the last decades due to increase of number of heavy rains. Some uncertainty about the trend of the erosion rates in the southern and western parts of the steppe zone still exists.

Acknowledgements
The work was supported by the Russian Science Foundation (project No. 15-17-20006).

References

[1] Sidorchuk A Yu and Golosov V N 2003 *Hydrological processes* **17**(16) 3347–3358
[2] Veretennikova M V, Zorina E F, Kovalev S N, Lyubimov B P, Nikol'skaya I I and Prokhorova S D 2006 *Geography of gully erosion* (Izd-vo MSU, Moscow) p 324
[3] Larionov G A 1993 *Water and wind erosion: the main principles and quantitative estimates.* (Mosk.Univ. Publ. House: Moscow) (in Russian)
[4] Litvin L F 2002 *Geography of soil erosion on agricultural lands of Russia* (IKC Akademkniga, Moscow) p 255 (in Russian)
[5] Golosov V N 2006 *Erosion and deposition processes in river basins of cultivated plains* (GEOS, Moscow) p 296 (in Russian)
[6] Kanatieva N P, Krasnov S F and Litvin L F 2010 *Eroziya pochv i ruslovye processy* **17** 14–27 (in Russian)
[7] Rosstat 2015 *Russian Federal Service of State Statistics* (Moscow, Russia) Available from: http://www.gks.ru
[8] Lyuri D I, Goryachkin S V, Karavaeva N A, Denisenko E A, Nefedova T G 2010 *Dynamic of Agricultural lands in Russia in XX century and postagrogenic restoration of vegetation and soils* (GEOS, Moscow) p 416 (in Russian)
[9] Ivanov M A 2017 Changes of cropland area in the river basins of the European part of Russia for the period 1985-2015 years, as a factor of soil erosion dynamics. This volume
[10] Golosov V, Gusarov A, Litvin L, Yermolaev O, Chizhikova N, Safina G and Kiryukhina Z 2017 *Integrating monitoring and modelling for sediment dynamics Okehampton, UK, 11–15 July 2016, Proc. IAHS* (Copernicus Publications) eds A Collins, M Stone et al vol 375 23–27
[11] Litvin L F, Kiryukhina Z P, Krasnov S F and Dobrovolskaya N G 2017 *Pochvovedenie* **11** 1390–1400
[12] Golosov V N 1998 *Geomorphologie Relief, Processus, Environnement* **1** 53–64
[13] Golosov V N, Walling D E, Konoplev A V, Ivanov M M and Sharifullin AG 2017 Application of bomb- and Chernobyl-derived radiocaesium for reconstructing changes in erosion rates and sediment fluxes from croplands in areas of European Russia with different levels of Chernobyl fallout *J. Environment Radioactivity* (in press, Available online 18 August 2017)
[14] Medvedeva R A, Golosov V N and Yermolaev O P 2017 Spatial-temporal assessment of gully erosion in zone of intensive agriculture, European part of Russia *Geography and Natural Resources* (in press)
[15] Butakov G P, Zorina E F, Nikol'skaya I I, Rysin I I, Serebrennikova I A and Yusupova V V 2000 *Erozionnye I ruslovye processy* **3** (Moscow, MSU) 52–62
[16] Kireeva M B, Frolova N L, Winde F, Dzhamalov R G, Rets E P, Povalishnikova E S and Pahomova O M 2016 *Geography, environment, sustainability* **9**(4) 33–47.
[17] Petelko A I, Golosov V N and Belyaev V R 2007 *Proceedings of the 10-th international symposium on river sedimentation* (Moscow) vol 1 pp 311–316

IOP Conf. Series: Earth and Environmental Science **107** (2017) 012009 doi:10.1088/1755-1315/107/1/012009

Capability of applying morphometric parameters of relief in river basins for geomorphological zoning of a territory

M A Ivanov and O P Yermolaev

Institute of Environmental Sciences, Kazan Federal University, 18, Kremlevskaya str., 420008, Kazan, Russia

E-mail: maximko-87@mail.ru

Abstract. Information about morphometric characteristics of relief is necessary for researches devoted to geographic characteristics of territory, its zoning, assessment of erosion processes, geoecological condition and others. For the Volga Federal District for the first time a spatial database of geomorphometric parameters 1: 200 000 scale was created, based on a river basin approach. Watersheds are used as a spatial units created by semi-automated method using the terrain and hydrological modeling techniques implemented in the TAS GIS and WhiteBox GIS. As input data DEMs SRTM and Aster GDEM and hydrographic network vectorized from topographic maps were used. Using DEM highlighted above for each river basin, basic morphometric relief characteristics such as mean height, slope steepness, slope length, height range, river network density and factor LS were calculated. Basins belonging to the geomorphological regions and landscape zones was determined, according to the map of geomorphological zoning and landscape map. Analysis of variance revealed a statistically significant relationship between these characteristics and geomorphological regions and landscape zones. Consequently, spatial trends of changes of analyzed morphometric characteristics were revealed.

1. Introduction

Information about relief morphometry is necessary for solving problems of the geographic characteristics of the territory, estimation of erosion processes and assessment of the geoecological condition, etc. [1, 2]. Also, importance of a quantitative analysis of the relief is due to its high informativeness in carrying out various types of local and complex geographical zoning. Relief is the basic characteristic when creating series of landscape maps.

Digital elevation models based on regular coordinate grids are widely used now by professionals working in various areas of geography and geomorphology [3, 4].

Hydrological and geomorphological studies were started in the mid-1970s by A.P. Dedkov and V.I. Mozzherin at the Department of Landscape Ecology of Kazan University. The main objective of these studies at this stage is the creation of specialized GIS at different level of generalization where river basin acts as an operational territorial unit. The basin approach in the geomorphometric analysis of the territory allows to characterize not only the relief of a single basin but the entire study territory acting as spatial unit. There are many databases for different regions of the Earth based on basins [4, 5]. The most well-known global base is HydroSHEDS produced by the World Wildlife Fund in 2006 [6]. At present, the authors registered a number of spatial databases of the morphometric characteristics of the relief on the territory of the Volga Federal District.

The main objective of the study is to assess the capability of using morphometric parameters of the river basin relief for geomorphological zoning.

Main goals: 1. Definition of river basins belonging to geomorphological regions and landscape zones; 2. Finding out the dependence of the relief geomorphological characteristics on the geomorphological region and the landscape zone using variance analysis; 3. Analysis of spatial variability of geomorphological characteristics.

2. Material and methods

2.1. Study area

The research area is the Volga Federal District of the Russian Federation, located in the east of the Russian Plain. It's total area is 1 034 000 km^2. This choice is due to several reasons. First, the size of the study area should be large and the scale should corresponds to the transregional and global generalization level. Secondly, the territory should cover different landscape zones and areas with various morphogenetic types of relief. This will allow to provide further spatial extrapolation of the research results to not observed territories. The study area is located in the upland and lowland landscapes of the temperate plains, extending from the subzone of the northern taiga in the North to typical steppes in the South.

2.2. Input data

As the main data, the spatial database of morphometric parameters of the relief of 68787 river basins of the 2nd order was used [7, 8].

To determine the patterns of spatial variability of the morphometric parameters of the relief the vectorized USSR geomorphological zoning scheme, edited by S.S. Voskresensky [9, 10], and landscape map of the USSR with 1: 2500000 scale were used [11].

2.3. Analysis methods

Initially, using ArcGis, codes of the corresponding geomorphological region and the landscape subtype were assigned to each basin (figure 1).

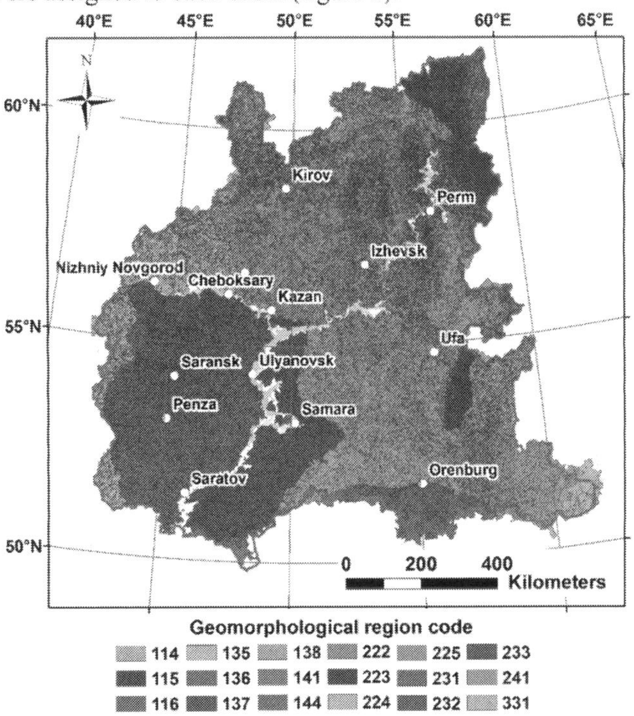

Figure 1. Map of belonging of the basins to geomorphological regions (Code description look in the Table 1).

IOP Conf. Series: Earth and Environmental Science **107** (2017) 012009 doi:10.1088/1755-1315/107/1/012009

According to the geomorphological zoning scheme the region is located within three geomorphological countries: Russian plain, Novozemelsky-Ural plain and Turanskaya plain, including 7 provinces and 18 regions (Table 1).

Table 1. Belonging of basins to geomorphological regions.

Country	Province	Region	Number of basins, pcs.	Area, km^2	Region code
Russian plain	Northern Russian province	Northern Dvina region	5769	42706.3	114
		Timan Ridge Area	849	6503.8	115
		Petchora Lowland area	10	175	116
	MidRussian province	Volga-Oka-Don Plain area	3993	73296.2	135
		Volga region Uplands and Ergeni	9847	185402.2	136
		Low Volga region	1760	80042.9	137
		Upper Volga region	35840	460142.3	138
	South Russian province	Caspian depression area	65	5073.9	141
		Sub Ural Plateau area	536	32316.2	144
Novozemelsky-Ural plain	The province of Urals axial zone	Northern Urals steeply sloping ridge-midlands	729	8326	222
		The area of ridge-remnant low mountains of the Middle Urals	394	5957.6	223
		The area of middle ridge of the Southern Urals	1795	21651.6	224
		The area of low mountains and the plateaus of the Southern Urals and Mugodzhar	1443	30356.8	225
	West Urals province	Parm area	2568	12691.8	231
		The Ufa-Chusovskaya area	1911	15350.8	232
		Sim Nugush ridge-remnant region	977	11904.7	233
	East Urals province	The area of the Ural-Tobolsk Plateau	240	25848.4	241
Turanskaya plain	Turgay Betpakdalinskaya province	Turgay Plateau area	61	10260	331

Due to a large number of regions the analysis was carried out separately for the Russian Plain country and Novozemelsky-Ural country. Besides, the basins belonging to Petchora lowland areas (total count is 10 pieces), the Caspian depression (total count is 65 pieces) and the basins belonging to the Turan plain (total count is 61 pieces) were excluded from analysis due to their insignificant amount.

To identify the dependence of the values of morphometric parameters on their belonging to a particular geomorphological area the analysis of variance was used where parameters acted as a dependent variable and specific geomorphological region acted as a factor.

The average length of slopes and the river network density parameters do not demonstrate the dependency on geomorphological regions. Under conditions of flat type of relief of the studied territory it was difficult to expect a different result. These parameters reflect indirectly the conditions of humidification. Consequently, their location is related to the landscape zones. In this regard, for the average length of the slopes and the density of the river network the presence of a connection with natural zones was analysed. For this aim, the analysis of variance was also carried out where the subtypes of landscapes were acted as factors (figure 2). To reduce the number of classes all mountain landscapes were joined into one type.

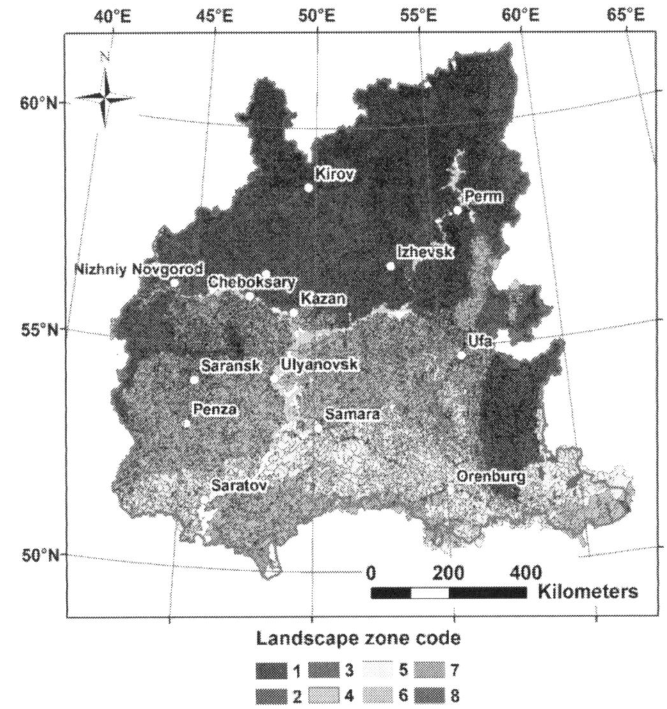

Figure 2. Landscape subtypes: 1 – Middle taiga; 2 – Southern taiga; 3 – Subtaiga (With broad-leaved and small-leaved-coniferous forests); 4 – Broadleaved forests and Forest-steppe; 5 – Typical steppes; 6 – Dry steppes; 7 – Semideserts; 8 – Mountains.

3. Results and discussion

According to the analysis of variance results it can be argued that during the transition from one geomorphological region to another such morphometric values as mean height, relative elevation, slope steepness, and LS factor [12] vary statistically significant. In all cases, the between-group dispersion is significantly larger than within-group and p-value is less than 0.05.

As can be seen from the graph (figure 3), the maximum heights within the Russian Plain are observed in the basins located in the regions corresponding to uplands of anticlises, and partially on syneclises with inversion relief. Basins with minimal values belong to downfolds.

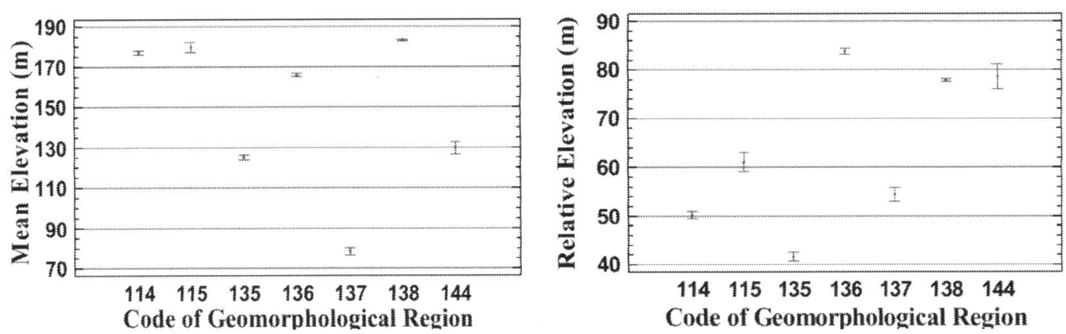

Figure 3. The dependence of average height and height range on the geomorphological region of Russian Plain.

Low values of relative elevation are observed in the basins located in the areas with an accumulative type of relief; it is increased in basins dedicated to the areas with erosion-denudation relief.

Average slope steepness values and, accordingly, LS factor behave in slightly different way (figure 4). In general, a pattern is the same for both factors but the minimum values of slope steepness are observed in basins located on Sub Ural plateau with a large depth of vertical dissection. This can be explained by two factors: the prevalence of such relief elements as plateau and low drainage density. Thus, the domination of the territory areas with the slopes close to zero degrees, provides low values at calculation of average slope steepness. On the contrary, the average steepness in the basins located in Northern Dvina region are higher than expected ones at low values of vertical dissection. This is explained by the high density of the river network and therefore by a large number of slope complexes.

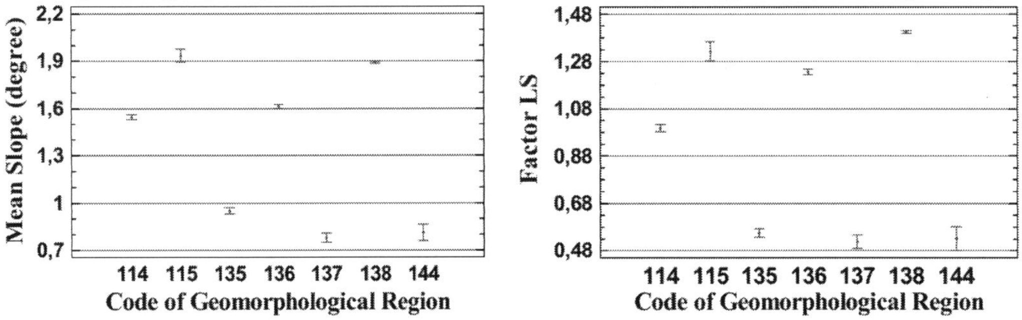

Figure 4. The dependence of slope steepness and LS factor on the geomorphological regions of the Russian Plain.

For basins located within the Novozemelsky-Ural country statistically significant patterns of spatial variability of the morphometric characteristics of the relief are also revealed.

Maximal values of the height (figure 5) are observed in the basins corresponding to the mid-mountains where the highest ridges and massifs are concentrated. Significantly smaller heights are observed in basins belonging to low-mountain areas. The maximum values of relative elevation correspond to the midland Province of the Urals axial zone. Minimum vertical dissection is noted in lowland regions and plateaus. Relatively high values in Sim Nugush ridge-remnant region is explained by the alternation of sharply outlined ridges with narrow and deep depressions.

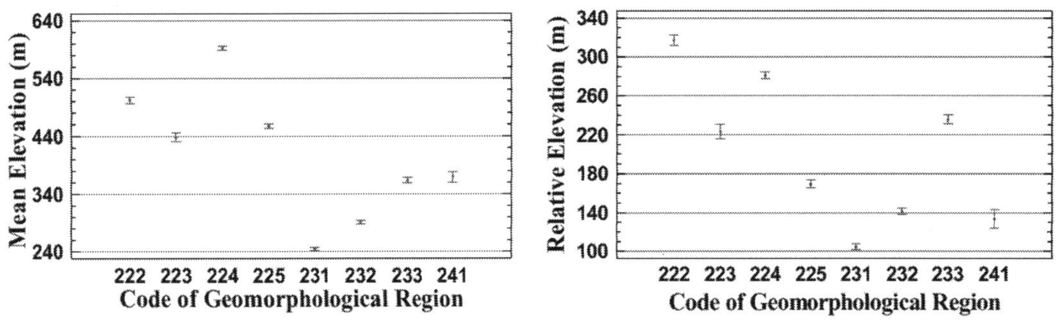

Figure 5. The dependence of average height and height range on the geomorphological area of Novozemelsky-Ural country.

This also explains the high values of average slope steepness (and consequently LS factor) in the basins which are located in this region (figure 6). The values of these factors in other geomorphological regions give quite a logical picture. Maximum deviations correspond to middle mountains, lower ones correspond to low mountains, and the minimal ones are observed in the Ural-Tobolsk Plateau characterized by aligned relief.

Such factors as average length of slopes and the river network density have an inverse dependence: the higher the density of horizontal dissection the shorter the slopes of river valleys. The density of the river network in the plains reaches a maximum in the taiga zone, and, is as it should be, decreasing to the North and South. The lowest river density is typical for the Caspian depression. With the shifting to the south (to the zone of forest-steppes and steppes), the density of the river network reduces and the number of temporary streams gradually grows. When the height goes up (on the hills, on the mountains) the density of the river network increases [13].

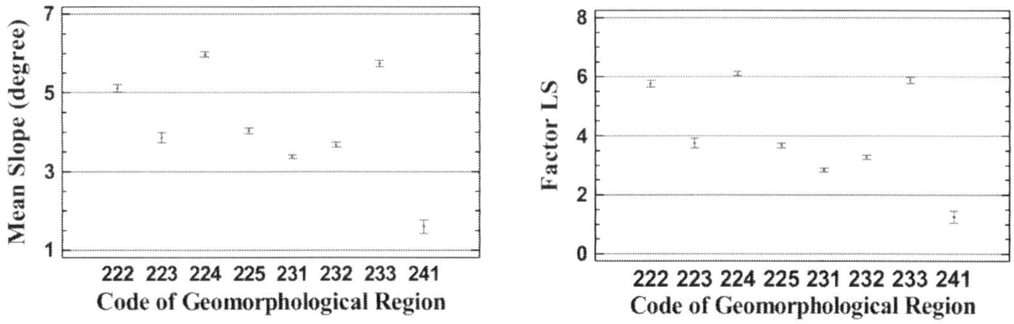

Figure 6. The dependence of slope steepness and LS factor on the geomorphological regions of Novozemelsky-Ural country.

As we can see on the graph (figure 7), the density of rivers is systematically reduced when moving from the zone of southern taiga and mixed forests to the zone of deciduous forests and forest-steppe, steppe and further to the semi-desert. It is primarily caused by climatic factors. In mountain landscapes high values of this factor are observed.

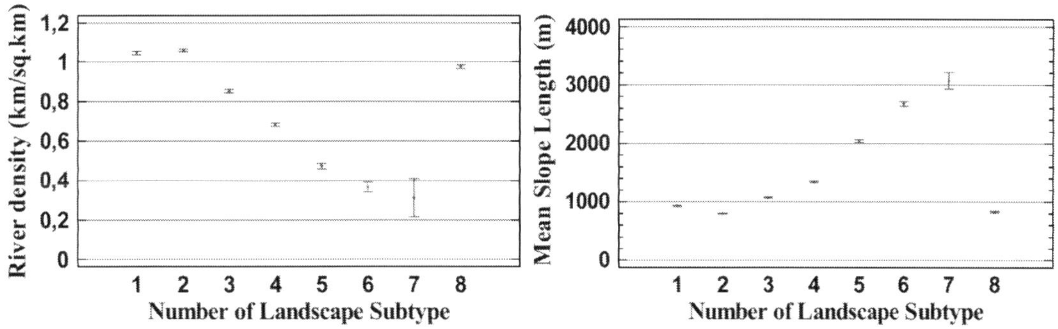

Figure 7. The dependence of mean slope length and river density on landscapes subtype.

The average length of the slopes, in contrast, increases in the row: southern taiga and mixed forests (mountain landscapes) - broad-leaved forests and forest-steppe - steppe - semi-desert.

The method of variance analysis revealed statistically significant relations of morphometric characteristics of relief with geomorphological regions and landscapes. Spatial variability trends of considered geomorphometrical characteristics, identified during the study on the basis of a basin approach, confirm the basic regularities described by other researchers.

Despite the fact that morphometry of the relief does not always explain its genesis, connection with geomorphological regions revealed for the four parameters, allows to assume that they can be used for geomorphologic zoning of the territory.

Acknowledgements
The work was supported by the Russian Science Foundation (project No. 15-17-10008)

References

[1] Yermolaev O 2004 *Sediment Transfer through the Fluvial System* **288** 60–66
[2] Yermolaev O 2017 *Eurasian Soil Science* **50**(1) 118-131
[3] Maltsev K, Yermolaev O and Mozzherin V 2015 *Proc. IAHS* **367** 326–332

[4] Yermolaev O and Maltsev K 2014 *Geomorphology* **1** 45–53
[5] Maltsev K, Yermolaev O and Mozzerin V 2012 *Proc. IAHS* **365** 251–258
[6] Lehner B, Verdin K and Jarvis A 2006 *HydroSHEDS Technical Documentation* (Washington: World Wildlife Fund US)
[7] Ermolaev O, Mal'tsev K and Ivanov M 2014 *Geography and Natural Resources* **35**(3) 222–228
[8] Ivanov M and Yermolaev O 2017 *Sovremennye problemy distantsionnogo zondirovaniya Zemli iz kosmosa* **14**(2) 98–109
[9] Voskresensky S S, Leont'ev O K, Spiridonov A I, Luk'yanova S A, Ul'yanova N S, Anan'ev G S, Andreeva T S, Varushchenko S I and Spasskaya I I 1980 *Geomorphological zoning of the USSR and adjacent seas* (Moscow: Vysshaya shkola) p 343
[10] Haliullina A and Samsonov T 2013 *XXXIII Plenary Meeting of the Geomorphological Committee RAS* **1** 106–110
[11] Gudilin I 1980 *Landscape map of USSR* (Moscow.: Ministerstvo geologii SSSR, Gidrospetsgeologiya)
[12] Moore I, Grayson R and Ladson A 1991 *Hydrological Processes* **5**(1) 3–30
[13] Rakovskaya E and Davydova M 2001 *Physical geography of Russia* (Moscow: Vlados) p 287

IOP Conf. Series: Earth and Environmental Science 107 (2017) 012010 doi:10.1088/1755-1315/107/1/012010

Changes of cropland area in the river basins of the European part of Russia for the period 1985-2015 years, as a factor of soil erosion dynamics

M A Ivanov

Institute of Environmental Sciences, Kazan Federal University, 18, Kremlevskaya str., 420008, Kazan, Russia

E-mail: maximko-87@mail.ru

Abstract. The work is devoted to the effect of the change of cropland area on the rate of soil erosion in sites of the European part of Russia located in different climatic, landscape and geomorphological conditions. Mapping of the croplands on the territory of 9 river basins for the two time slices (the middle of the 1980s, and the present-day period 2013-2015) was carried out by visual interpretation of multi-seasonal images Landsat 5 and Landsat 8. Using vector layers obtained as a result of digitization, the areas of croplands in the analyzed periods have been calculated, and changes which occurred in 30 years were estimated. The decrease of croplands area was revealed in all studied river basins. An assessment of the effect of cropland reduction on the soil loss rate was carried out. Using the SRMM DEM with a 30 m spatial resolution, the following morphometric characteristics of relief for cultivated and abandoned croplands were calculated: steepness of slopes, flowpath length, factor LS. Based on the results of calculations, the average values of the factor LS reduced from 1985 to 2015 on the croplands in all considered basins. The obtained data confirm that the reduction of the croplands area is one of the factors responsible for the decrease of modern soil loss rates observed in field studies.

1. Introduction

In numerous works devoted to the dynamics of land use and the estimation of the areas of abandoned croplands on the territory of the former USSR, the possible causes of these changes are discussed in detail [1, 2, 3, 4]. However, information about changes in the structure of agricultural lands for assessing the ecological condition of the territories isn't less important, especially for a territory with a significant share of cropland. The features of land use change largely determine the rate of erosion-accumulative and a number of other exogenous processes. Within the plain territories the reduction of croplands leads to a decrease of erosion processes rates, since the main loss is formed on the cultivated land. In the mountainous areas where the cultivated land is mainly located on terraced slopes, erosion-denudation processes on the abandoned cropland are activated [5].

Changes of cropland area on territories with a high percentage of cultivated land determine the volume of material moved by erosion processes and, to a certain extent, affect the change in loss intensity which depends on the relief parameters [6, 7]. Water soil erosion is one of the most important agents of soil mantle degradation and, at the same time, the main mechanism of the transport of pollutants sorbed by soil particles [8, 9] as well as carbon into permanent watercourses and reservoirs [10, 11]. Thus, the intensity of erosion processes and the spatial coverage of the territory on which they occur, largely determines the ecological pressure in the regions with a high share of cropland [12, 13] and is an important part in the global carbon cycle [14, 15, 16]. The river basin is the most convenient territorial unit for assessment of erosion processes rate trends and revealing their impact on the quality of surface waters [17, 18]. Equally, the optimization of land use in order to reduce the rate of erosion processes and their negative impact on the ecological state of surface waters and reservoirs, is advisable to be carried out using river basins as spatial units. In this case the interests of various land and water users can be taken into account, and ecological risks will be minimized [19].

The structure of agriculture lands is the most dynamic factor in comparison with such conservative conditions as lithology, relief, soil cover and climate, the changes of which are more gradual. Unfortunately, official statistics, for example, data about sown areas do not allow to estimate spatial changes in land use due to its representing using administrative region boundaries. Also, quantitative indicators of land use change vary widely in different sources. For example, in state ecological and agricultural reports land areas are given on the basis of their legal belonging to a particular category rather than real land use. Therefore, a large number of croplands which have not been used for several years, can still formally belong to this category. Thus, the reduction of croplands area according to official data is often greatly underestimated when compared with the actual value [20].

Satellite monitoring plays an important role in assessing the land use dynamics and vegetation cover condition [21]. Remote sensing data can be used as a source for the detailed agricultural land use change mapping [22, 23]. Multi-temporal multispectral medium resolution (10-30m) imagery (for example free Landsat data) is suitable for creation and updating of spatial databases of land use at the regional level [24, 25].

The main objectives of this study is to assess the change of cultivated croplands area in the river basins of the European territory of Russia located in different climatic, landscape and geomorphological conditions from 1985 to 2015, and to estimate affect of these changes on the soil erosion rate.

Main goals of the research: 1. Mapping of the croplands on the territory of 9 river basins for the two time slices (the middle of the 1980s, and the present-day period 2013-2015) by visual interpretation of multi-seasonal images Landsat 5 TM and Landsat 8 OLI. 2. Accuracy assessment. 3. Quantitative analysis of the croplands area change based on the resulting vector layers. 4. Calculation of a number of morphometric parameters of relief on cultivated and abandoned croplands using SRTM DEM with a spatial resolution of 1 arcsecond. 5. Analysis of changes of the given above parameters aimed to determine their effect on the soil erosion rate.

2. Material and methods

2.1. Study area

The studied territory is the southern half of the East European Plain which is the main agricultural region of Russia. Research was carried out within the 9 river basins located in different agroclimatic zones (figure1).

Selected sites are covering the territories with various landscape and climatic conditions and the farming features are specified by these conditions. According to the landscape map of the USSR, edited by I.S. Gudilin, scale of 1: 2500000 [26], studied basins within three landscape zones are located in: forest, forest-steppe and steppe. Izh river basin and north-eastern half of the Mesha basin are lying in the sub-taiga landscapes. Basins of the Ulema, Sviyaga, Veduga and the northern part of the Medveditsa basin are located in the forest-steppe landscapes. All other sites are corresponding to typical steppe landscapes.

2.2. Input data

As the input data for mapping, Level-1 standard data products Landsat 5 TM and Landsat 8 OLI were used. For each studied basin multi-seasonal images in a snowless period in two time slices: middle of 1980s and 2013-2016 years period were selected. For visual interpretation the 5, 4, 3 bands RGB combination for Landsat 5 and 6, 5, 4 bands for Landsat 8 were used. This combination provides the user with a great amount of information and color contrast. Healthy vegetation is bright green and soils are mauve. This combination is widely used for agricultural land dynamics analysis [27]. As the reference data we used high resolution images presented in Google Earth. Calculation of the relief morphometric characteristics was carried out using SRTM DEM with 1 arcsecond (~ 30 m) spatial resolution.

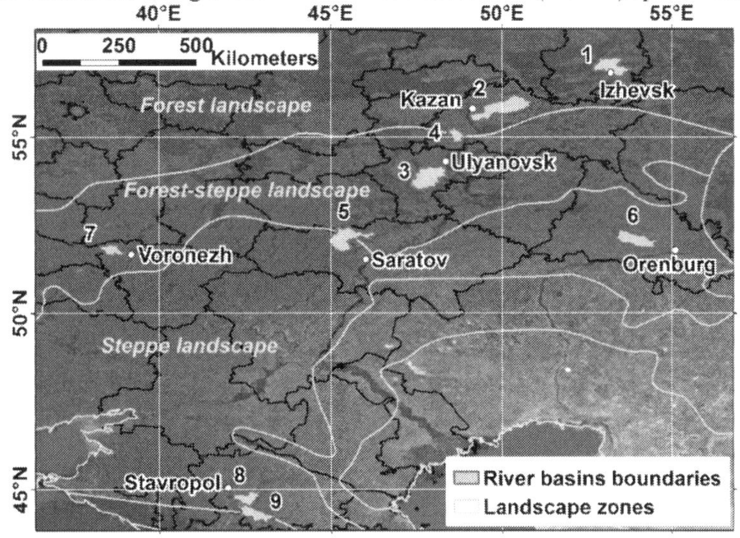

Figure 1. Studied basins: 1 - Izh (Udmurtian Republic), 2 - Mesha (Republic of Tatarstan), 3 - Sviyaga (Ulyanovsk Region), 4 - Ulema (Republic of Tatarstan), 5- Medveditsa (Saratov Region), 6 - Samara (Orenburg Region), 7 - Veduga (Voronezh Region), 8 - Kalaus, 9 - Kuma (Stavropol Territory)

2.3. Mapping technique

We were using mapping technique realized in the CORINE Land Cover 2000 project (CLC2000) [28, 29], adjusted to the regional features and purposes of our study. The main method is visual interpretation and further manual digitization.

Croplands, in general, are easy to be separated from land use classes such as water bodies, urban areas and woodlands (except for undergrowth, young plantations). Separation of cultivated croplands from abandoned cropland and grassland is problematic but usage of multi-seasonal images allows to divide used and unused agricultural lands more accurately. In addition to the standard interpretation keys of cropland, such as orthogonality of borders, fragmented pattern and color, it is possible to distinguish a number of classification characteristics that will depend on the territory such as: kind of use of particular field and season of survey, including tone and texture [30].

At the first stage, croplands boundaries were manually digitized according to the images of 1980s. Not only particular fields were digitized, but the common boundaries of the cultivated croplands as well. Previously abandoned fields were not assigned as a separate class. Resulting polygons were overlaid on the images of the second time slice. Using keys described above, abandoned croplands were determined and the boundaries were corrected accordingly. Such approach allows to preserve topology of unchanged objects on the layers of different periods. However, the position of boundaries of the same field on images for different years in some cases was shifted. This could be explained by accuracy of the Landsat spatial reference and actual differences in the boundaries of plowing in different years. Deviation of no more than 3 pixels of Landsat TM (less than 100 m) in width was acceptable, which corresponds to the geometry accuracy specified in the CORINE CLC project. In addition, 100 m is the minimum boundary displacement which is mapping as a land use change [31]. Result of digitization are 2 vector layers of cultivated croplands. Overlaying these layers, the areas of cropland reduction and increase over the period under review were obtained. Area of used and abandoned croplands was calculated in hectares.

2.4. Accuracy assessment

Validation was carried out according to the methodology used to assess the accuracy of the CLC in Portugal [32]. Reference information was used to compare results of the interpretation with reality and should have a higher level of accuracy than the information used for production of maps. Sources of such information are: aerial imagery; satellite images with better resolution than those used in the maps production; field research [33]. Congalton and Biging think that only field observations could be used to assess the accuracy and reliability of interpretation, but there is a number of limitations: territory inaccessibility, human and material resources, cost and time. The reference information should be close to date of the data used in maps production to avoid the impact of landscape changes [34] and should be independent [35, 36].

Since there are no high-resolution imagery for the 1980s, we estimated the accuracy of the croplands mapping for 2015. For this aim, high-resolution images, partially covering the territory of three sites (Mesha, Medveditsa and Kalaus), located in different landscape zones were used. Each selected image covers more than 15% of the basin area to provide representativeness (table 1).

Table 1. High-resolution images used for validation.

River basin	Satellite	Date of survey	Coverage, ha	% of basin area	Digital Globe Catalog ID
Mesha	GeoEye-1	12.07.14	67485	15.5	1050410010D5FD00
Medveditsa	WorldView-2	16.09.15	83153	23.03	1030010047C35A00
	WorldView-2	16.09.15			1030010049158C00
Kalaus	GeoEye-1	09.03.15	52806	29.85	1050410012519C00
	WorldView-2	16.12.14	17413	9.84	103001003B1BA200

Polygons, which are the results of croplands mapping, were overlaid on the above images. If the polygon boundaries were shifted by 100 m or more from the boundaries, recognized on high resolution images, they were corrected. Polygons adjusted using high resolution images were overlaid on the croplands 2015 layer obtained from Landsat images, and agreement and disagreement regions were extracted.

The areas of these regions were calculated; the error matrix was created and the producer's accuracy and user's accuracy were obtained. Results demonstrate that accuracy of interpretation in the examined areas is more than 98% (tables 2 and 3).

Table 2. Error matrix for accuracy assessment of mapping of the croplands 2015.

		Cropland area (ha) calculated using high-resolution satellite images		
Cropland area (ha) calculated using Landsat images		Mesha		
		Cropland	Other land use	Total
	Cropland	32596	426	33022
	Other land use	308	34154	34463
	Total	32905	34580	67485
		Medveditsa		
		Cropland	Other land use	Total
	Cropland	49291	425	49715
	Other land use	467	32970	33438
	Total	49758	33395	83153
		Kalaus		
		Cropland	Other land use	Total
	Cropland	25815	198	26012
	Other land use	208	43998	44206
	Total	26022	44196	70219

Table 3. Errors and accuracy.

	Mesha	Medveditsa	Kalaus
User's accuracy, %	98.7	99.2	99.2
Producer's accuracy, %	99.1	99.1	99.2
Commission error, %	1.3	0.8	0.8
Omission error, %	0.9	0.9	0.8

2.5. Calculation of morphometric parameters of relief of the used and abandoned croplands

Using DEM, rasters of slopes steepness and flow path lengths (determining relief factors for erosion processes rate estimation) were calculated for territory of the studied basins. As an integral factor, the LS-factor was calculated using the "Sediment Transport Index" tool in WhiteBox GAT using the following equation:

$$LS = (m+1)\times\left(\frac{A_s}{22.13}\right)^m \times \sin\left(\frac{B}{0.0896}\right)^n, \quad (1)$$

where A_s – is the specific catchment area, B - is the local slope gradient in degrees, m - the contributing area exponent usually set to 0.4, n - the slope exponent usually set to 1.3 [37]. Using the vector layers of cultivated and abandoned croplands as masks the average values of the above morphometric parameters for each basin were calculated.

3. Results and discussion

Based on the mapping results, the croplands areas in the studied time periods were calculated (table 4) and the changes that occurred over 30 years were estimated (table 5). Croplands is the dominant category of land use in almost all regions. The minimum percentage of croplands (25.6%) in the Izh river basin has been observed, which is located in the forest landscape zone, in the northernmost and therefore the coldest agroclimatic zone.

Table 4. Croplands areas.

River basin	Basin area, ha	Cropland area 1985 year, ha	Cropland area 1985 year, %	Cropland area 2015 year, ha	Cropland area 2015 year, %
Izh	250819	64228	25.6	40229	16
Mesha	435421	271257	62.3	243918	56
Sviyaga	343250	195150	56.9	141598	41.3
Ulema	88145	60753	68.9	56782	64.4
Medveditsa	361064	247290	68.5	223538	61.9
Samara	287416	167817	58.4	144412	50.3
Veduga	119347	84068	70.4	76610	64.2
Kalaus	176887	81544	46.1	70763	40
Kuma	246675	153988	62.4	127281	51.6

Table 5. Changes of croplands over the considered interval.

River basin	Decrease of cropland, ha	Decrease of cropland, %	Increase of cropland, ha	Increase of cropland, %	Total change, ha	Total change, %
Izh	24205	37.7	206	0.3	-23999	-37.4
Mesha	30106	11.1	2767	1	-27340	-10.1
Sviyaga	54468	27.9	916	0.5	-53552	-27.4
Ulema	4328	7.1	356	0.6	-3971	-6.5
Medveditsa	25748	10.4	1995	0.8	-23753	-9.6
Samara	26078	15.5	2673	1.6	-23405	-14
Veduga	8073	9.6	615	0.7	-7457	-8.9
Kalaus	15907	19.5	5125	6.3	-10781	-13.2
Kuma	28602	18.6	1894	1.2	-26707	-17.3

Due the different basins area, for an objective assessment of the croplands dynamics the changes were calculated as a percentage of the croplands area in the 1980s. The areas of abandoned croplands, newly plowed land plots, as well as the total change were calculated (table 5).

The decrease of croplands was established for the all studied river basins. Analyzing the obtained results, some regularities connected with the natural conditions of the basins location could be revealed, which in turn can cause some socio-economic reasons for the croplands reduction. The largest decrease of croplands (37.7%) is observed in the Izh river basin located in the forest landscape zone, namely in sub-taiga landscapes. The value is comparable to the percentage of abandoned land in the period 1990-2000 obtained for the number of regions of the Russian Federation using the automated classification method: 46% in Smolensk region, 30% in Kaluga region, 26% in Tula region, 28% in Ryazan region and 27% in Vladimir region [38]. It is worth to note, that these regions also belong to subtaiga and southern taiga landscapes, and the southern parts of Ryazan and Tula regions - to the forest-steppe.

Significantly smaller reduction of croplands (10%) observed in the basins belonging to the forest-steppe landscape zone: Mesha, Ulema, Veduga, Medveditsa. Here, the greatest plowing of the territory is observed - from 62 to 70%. The exception is the territory of Sviyaga basin where the croplands area decreased by 27.4%. If we consider natural conditions, this can be related with the location of the basin in the cold-moderate agroclimatic zone with the sum of active temperatures up to 2200 °C, and also with low-quality soils common to the basin.

In the basins located in the steppe zone (Kalaus, Kuma, Samara) the reduction of croplands is in the range 10-20%. There are territories with maximum values of the absolute heights range. Basins of Kalaus and Kuma rivers are characterized by wide slope complexes. In addition, solonetzic soils are widespread here. In contrast to all the sites, in Kalaus basin, with a general cropland reduction, a relatively large percentage of newly cultivated land is observed when compared with 1985.

Abandoned sites are mainly located on the edges of cropland massifs and border with river valleys, gullies and settlements. Colonization of abandoned croplands by grass, by bushes and forest vegetation is observed. It should be noted that croplands reduction is also associated with the transfer of land to another category, most often in woodlands and settlement. Such sites can not be interpreted as abandoned croplands.

Based on the results of calculations of the morphometric characteristics of the relief (Table 6), the mean values of LS factor reduced from 1985 to 2015 on cultivated croplands in all studied basins. At the same time for the abandoned croplands the maximum values of this factor were obtained, which indicate that areas vulnerable to erosion were abandoned.

Table 6. Morphometric parameters of relief of cultivated and abandoned croplands

River basin	Landscape zone	S_{1985}	S_{2015}	S_{Ab}	L_{1985}	L_{2015}	L_{Ab}	LS_{1985}	LS_{2015}	LS_{Ab}
Izh	South of the forest	2.59	2.60	2.59	120.24	116.78	126.29	1.50	1.47	1.56
Mesha	North of the forest-steppe	1.79	1.77	2.07	115.67	115.02	120.76	1.04	1.01	1.53
Sviyaga	North of the forest-steppe	1.52	1.48	1.62	115.98	119.02	109.82	0.87	0.84	0.98
Ulema	North of the forest-steppe	1.89	1.88	2.07	129.36	128.76	144.34	1.03	1.02	1.34
Medveditsa	North of the steppe	1.53	1.49	1.80	98.55	96.74	116.24	0.80	0.77	1.06
Samara	East of the steppe	2.01	1.92	2.54	129.48	129.44	129.71	1.19	1.12	1.63
Veduga	West of the forest-steppe	2.30	2.23	3.42	91.42	90.14	120.33	1.34	1.27	2.52
Kalaus	South of the steppe	2.57	2.43	3.23	121.44	118.53	144.39	1.48	1.38	2.04
Kuma	South of the steppe	2.84	2.79	3.22	113.05	110.25	132.79	1.75	1.67	2.26

S – mean slope, degrees; L – mean flow path length, meters; LS – mean LS-factor
Indices: 1985, 2015 – on the croplands of the corresponding year; Ab – on the abandoned croplands

The decrease of the LS-factor in different basins depends on changes of various morphometric parameters. In the basins of the rivers Medveditsa, Veduga, Kalaus, Kuma, there is a decrease of mean values of both slope steepness and the flowpath lengths. In the Izh and Ulema river basins at constant values of slopes, the decrease of the LS-factor was due to the reduction of flow path lengths. For the Mesha, Samara and Sviyaga rivers, the reverse picture is observed where the slope steepness was the primary factor in reducing the LS-factor at equal mean flow path lengths (and in the case of the Sviyaga River at an increased value).

Previously, for a number of regions of the European Russia, the croplands reduction in studied period was largely explained by socio-economic drivers [3]. The obtained results suggest that within the studied river basins, the most inconvenient land, susceptible to erosion and located at the lower parts of the convex slopes was abandoned. This trend is more pronounced in the river basins of the steppe and western part of the forest-steppe zone and in a less degree in the south of the forest and northern forest-steppe zones. Thus, it can be concluded that a reduction of croplands area provided decrease of the average annual erosion rate on the remaining croplands.

Acknowledgements
The work was supported by the Russian Science Foundation (project No. 15-17-20006)

References

[1] Baumann M, Kuemmerle T, Elbakidze M, Ozdogan M, Radeloff V C, Keuler N S, Prishchepov A V, Kruhlov I and Hostert P 2011 *Land Use Policy* **28**(3) 552-562
[2] Hostert P, Kuemmerle T, Prishchepov A V, Sieber A, Lambin E F and Radeloff V C 2011 *Environmental Research Letters*. **6**(4) 045201
[3] Prishchepov A V, Müller D, Dubinin M, Baumann M and Radeloff V C 2013 *Land use policy* **30**(1) 873-884
[4] Prishchepov A V, Müller D, Baumann M, Kuemmerle T, Alcantara C and Radeloff V C 2016 *Land-Cover and Land-Use Changes in Eastern Europe after the Collapse of the Soviet Union in 1991* pp 91-117
[5] Jose M G R 2010 *Catena* **81**(1) 1-11
[6] Mitasova H, Hofierka J, Zlocha M and Iverson R L 1996 *International J. of Geographical Information Sci.* **10** 629–641
[7] Van Oost K, Govers G and Desmet P J J 2000 *Landscape Ecology* **15** 579–591
[8] Morgan R P C 2005 *Soil Erosion and Conservation* (Oxford: Blackwell Publishing)
[9] Golosov V N, Gennadiyev A N, Olson K R, Markelov M V, Zhidkin A P, Chendev Y G and Kovach R G 2011 *Eurasian Soil Science* **44**(7) 794-801
[10] Lal R 1998 *Critical Reviews in Plant Sciences* **17**(4) 319–464
[11] Olson K R, Al-Kaisi M A, Lal R and Cihacek L J 2016 *Journal of Soil and Water Conservation* **71** 61-67
[12] Maetens W, Vanmaercke M, Poesen J, Jankauskas B, Jankauskien G and Ionita I 2012 *Progress in Physical Geography* **36**(5) 597–651
[13] Golosov V, Ivanova N, and Kurbanova G 2017 *Int J. of Sediment Res* **32**(1) 60-72
[14] Van Oost K et al. 2007 *Science* **318** 626–629
[15] Lal R 2003 *Environment International* **29** 437-450
[16] Quinton J N, Govers G, Van Oost K and Bardgett R D 2010 *Nature Geoscience* **3** 311–314
[17] Walling D E 1999 *Hydrobiologia* **410** 223–240
[18] de Vente J and Poesen J 2005 *Earth-Science Reviews* **71** 95–125
[19] Apitz S E and White S 2003 *Journal of Soils* and Sediments **3** 132–138
[20] Lyuri D I, Goryachkin S V, Karavaeva N A, Denisenko E A and Nefedova T T 2010 *Dynamics of Russian agricultural land in the XX century and postagrogenic restoration of vegetation and soil* (Moscow: GEOS) p 416
[21] Cohen W B and Goward S N 2004 *BioScience* **54** 535-545
[22] Kraemer R, Prishchepov A V, Müller D, Kuemmerle T, Radeloff V C, Dara A, Terekhov A and Frühauf M 2015 *Environmental Research Letters* **10** 054012
[23] Bartalev S A, Plotnikov D E and Loupian E A 2016 *Remote Sensing Letters* **7** 269–278
[24] Steven M D and Clark J A 1990 *Applications of remote sensing in agriculture* (London:

Butterworths) p 427
[25] Hansen M C and Loveland T R 2012 *Remote Sensing of Environment* **12** 66-74
[26] Gudilin I 1980 *Landscape map of USSR* (Moscow.: Ministerstvo geologii SSSR, Gidrospetsgeologiya)
[27] Quinn J W 2001 *Band Combination* (http://web.pdx.edu/%7Eemch/ip1/bandcombinations.html)
[28] Bossard M, Feranec J and Otahel J 2000 *CORINE Land Cover Technical Guide – Addendum. Technical report No 40* (Copenhagen (EEA)
[29] Buttner G, Feranec J, Jaffrain G, Mari L, Maucha G and Soukup T 2004 *EARSeL eProceeding* **3**(3) 331–346
[30] Jensen J R 2005 *Introductory digital image processing: A Remote Sensing Perspective, 3rd ed, Prentice Hall series in geographic information science* (Pearson, Upper Saddle River, NJ)
[31] Feranec J, Hazeu G, Christensen S and Jaffrain G 2007 *Land Use Policy* **24**(1) 234–247
[32] Caetano M, Mata F and Freire S 2006 *Global Developments in Environmental Earth Observation from Space* 2006 pp 459–467
[33] Congalton R and Biging G 1992 *Photogrammetric Engeering & Remote Sensing* **58** 1669-1671
[34] Congalton R and Green K 1999 *Assessing the accuracy of remotely sensed data: principles and practices* (Danvers: CRC Press) p 200
[35] Hammond T and Verbyla D 1996 *Int. J. of Remote Sensing* **17** 1261-1266
[36] Stehman S 1999 *International Journal of Remote Sensing* **20** 2423-2441
[37] Moore I D, Grayson R B and Ladson A R 1991 *Hydrological Processes* **5**(1) 3-30
[38] Prishchepov A V, Radeloff V C, Baumann M, Kuemmerle T and Müller D 2012 *Environmental Research Letters* **7** 024021

IOP Conf. Series: Earth and Environmental Science **107** (2017) 012011 doi:10.1088/1755-1315/107/1/012011

The elevation and its distribution in geomorphological regions of the European Russia

S V Kharchenko[1,2], O P Ermolaev[1] and S S Mukharamova [1]

[1] Institute of Environmental Sciences, Kazan Federal University, 18, Kremlevskay str., 420008, Kazan, Russia
[2] Institute of Geography, RAS, 29/4, Staromonetnyj per., 119017, Moscow, Russia

E-mail: xar4enkkoff@rambler.ru

Abstract. Spatial differences of elevation were analysed by side of view of geomorphological boundaries on the European Russia territory. Geomorphological pattern of the studied territory was taken from Geomorphological Map of the USSR at scale of 1: 2 500 000. There 2401 fragments for combinations of 58 types of structural landforms and 22 types of sculptural landforms were allocated. The elevation values computed by digital elevation model (cell size – 200 m, number of cells – 322M) based on SRTM (south of 60 nl.) and GDEM 2010 (north of 60 nl.) resampled data. It was founded that some types of structural (16 types) and sculptural (6 types) landforms located in the relatively thin intervals of elevation. Using of elevation above sea level is needed for effective automatic recognizing these landform regions.

1. Introduction

Geomorphometric data can give possibility to classification of landforms by characteristics of the Earth surface geometry. There is a lot of articles illustrating this approach [1, 2, 3, 4, 5]. Some regions with different landform origin have non-equivalent values of elevation and its distributions. Not all differences of landforms can be recognized by single parameter. However, there are little publications with assessment of representativeness of even simple geomorphometric parameters such as elevation. This paper is aimed at to identification of landform types subdivided using averages and ranges (standard deviation) of elevation. Back in the late 19th century Lapparent used statistical distribution of elevation as indicator of geological structure [6]. Then some papers were written on the problem of analysis and interpretation of the statistical distributions of terrain [7, 8, 9, 10].

There are premises for using other geomorphometric parameters (for example, some focal statistics [11, 12]) for rising demarcation effectiveness, but in the paper we will limit ourselves only by this weightiest parameter.

2. The problem statement

Soviet geographers I.P. Gerasimov and J.A. Mescherikov in the 1946 offered new way for hierarchically landform classification. They recognize three levels of the Earth surface organization which Gerasimov named by terms: "geotextures", "morphostructures" and "morphosculptures" [13]. These levels were actively used as units for geomorphological mapping in the USSR and Russia in the second half of the 20th century and at present. The base of this approach is dividing of the Earth surface by leading geomorphological processes on each hierarchical level. For geotextures they are planetary and geotectonic processes, for morphostructures they are mainly structurally-initiated endogenic processes. And for morphosculptures they are almost exclusively, exogenic processes.

In the study there was used geomorphological pattern of the European Russia from famous soviet Geomorphological Map of the USSR at the scale of 1: 2 500 000 [14] completed in 1981. The whole area of the Soviet Union included sites of 149 types of morphostructures and 32 types of morphosculptures. The territory of the European Russia included sites of 58 types of morphostructures and 22 types of morphosculptures. This scheme is mainly given in the tables 1 and 2.

Table 1. Typology of structural landforms on the European Russia territory from Geomorphological Map of the USSR (fragment).

Geotectonic region	Subregion		General condition of landform development	Type of structural landforms
1. Platform plains	1.1. Socle plains, tablelands and mountain massifs of Pre- Paleozoic shields with strong recent upwarp and block uplift		1.1.2. Structural-denudational	Semiburied step plains
	1.2. Socle plains, island low-mountain relief, massifs and ranges of platform basement protrusion		1.2.1. Structural-denudational	Semiburied plains and chains of hills
			1.2.2. Denudational	Plains and chains of hills
	1.3. Plains and plateaus Pre- Paleozoic and Phanerozoic plates was subjected of resiliency deformation with uplift and subsidence	1.3.1. Layered and denudation plains and plateaus with predomination of recent uplift	1.3.1.1. Structural-denudational	Table and step plateaus built by sediment deposits
				Stepped plains built by sediment deposits
				Sloped plains built by sediment deposits
				Sloped, partially stepped plains built by sediment deposits
			1.3.1.2. Denudational	Layer plateaus built by sediment deposits
				Weak sloped plains and plateaus built by sediment deposits
		1.3.2. Layered and accumulation plains generally created by recent subsidence built by loose Neogene and quarternary rocks	1.3.2.1. Layered and accumulation	Sloped plains
			1.3.2.2. Accumulation	Stepped plains
3. Orogens	3.1. Mountain structures updated by recent block uplift built by consolidated rocks of Pre- Paleozoic and Paleozoic folding		3.1.1. Structural-denudational mountains	Socle sloped plains
	3.2. Mountain structures created by recent socle and block uplift and fold deformation built by different consolidated sediments of Alpine folding		3.2.1. Structural-denudational mountains	Fold monoclinal ranges and plateaus
			3.2.2. Intramontane lowlands	Grabens
			3.2.3. Intermontane lowlands	Sloped plains
	3.3. Mountain structures created by recent uplift and volcanism	3.3.1. Areas of Cenozoic volcanism of Alpine folding	3.3.1.1. Mountains and intermontane basins	Volcanic mountains and high plateaus

Table 2. Typology of sculptural landforms on the European Russia territory from Geomorphological Map of the USSR.

Condition of development		Forms created by processes of...	Type of sculptural landforms	Area, tous. sq. km
1. Azonal	1.1. controlled by structural condition	1.1.1. erosion and complex denudation	Hill lands	4.9
			Ouval lands	661.2
			Hill and ouval lands	85.3
			Ridge lands	11.9
			Ridge and ouval lands	121.2
			Table and step lands	101.7
			Flat and weak wavy lands	390.9
			Mountain (incised) lands	238.5
			Plateaus (incised) lands	68.5
		1.1.2. complex denudation	Sloping relief (morphologically diverse)	54.0
		1.1.3. accumulation of different genesis sediments	Flat or weak wavy lands covered by quaternary layers of different origin	1030.5
2. Zonal	2.1. inherited-developing and modern	2.1.1. complex arid denudation	Flat or weak wavy lands	48.7
			Hills, ridges and ouvals (hummocky terrain)	12.9
			Table and step lands	50.0
	2.2. relict (generally quaternary)	2.2.1. exaration	Rolling and sloping hill lands, smoothed	50.2
			Ridge and hill lands	178.8
		2.2.2. cryogenic and glacial morphogenesis	Mountain relief with cryogenic and glacial moderation	59.1
		2.2.3. glacial and fluvioglacial accumulation	Rolling and sloping hill lands of moraine plains	625.3
			Ridge and hill lands of marginal and radial postglacial areas	102.9
		2.2.4. glacial accumulation and further erosion	Hill and ouval lands	94.4
			Ouval lands	112.5
		2.2.5. fluvioglacial accumulation and further erosion	Sloping ouval lands	178.0

3. Results and discussion

Structure. The localisation of structural regions in thin elevation interval was estimated as quotient of standard deviation and median of elevations. Only one region has the quotient less than 0.1 – it is "table and step plateaus built by sediment deposits" (subdivision 1.3.1.1). The median elevation here is 264 m, but standard deviation is 13 m only. This region is strongly localised in the elevation interval. The region is represented by three sites in the buffer ring of the studied area – in the Kostanay Region of Kazakhstan.

The six regions have the quotient more than 0.1 but less than 0.2: they are "semiburied step plain" (subdivision 1.1.2), "sloped plain built by sediment deposits" and "sloped, partially stepped plain built by sediment deposits" (subdivision 1.3.1.1.), "layer plateau built by sediment deposits" (subdivision 1.3.1.2.), "sloped plain" (subdivision 1.3.2.1.) and "fold monoclinal ranges and plateau" (subdivision 3.2.1.). The first region is represented by only one site placed in the Velyka Novosilka Region of Donetsk Oblast in the Ukraine. Median elevation of the semiburied step plains is 159 m and standard deviation is around 17 m. The second region is placed in the Central Russian Upland and its southeastern part known as the Kalach Upland, in the northern part of the Volga Upland and part of Bugulma-Belebey Upland. Median elevation of region is 196 m and standard deviation is 37 m. The third region is placed in Smolensk and Moscow Upland, some sites are in the Cis-Ural territory. Median elevation of region is 190 m and standard deviation is 34 m. Last defined regions are similar to their origin. The region of layer plateaus is located only in the Aktobe and Kostanay Regions of Kazakhstan with median elevation 242 m and standard deviation 29 m. The two regions have strong differences with previous divisions. Sloped plains from subdivision of 1.3.2.1 locates exclusively on the eastern flank of Donetsk coal basin (median elevation – 157 m, standard deviation – 25 m). And the fold ranges and plateaus are located in the far south of the studied area – in the Lesser Caucasus territory (median elevation – 358 m, standard deviation – 47 m). All six regions cover relatively small areas and haven't any morphometric diversity.

The nine regions have the quotient more than 0.2 but less than 0.3. Two of them are plains and chains of hills (included semiburied and opened landforms). The regions are represented by territories of Donetsk and Timan Ridges and also by some near-Ural areas. Median elevation of semiburied plains is 174 m and standard deviation is 39. The same parameters of opened plains are 305 m and 90 m.

The largest median elevation value (more than 2900 m) typical for structural region of volcanic mountains and high plateaus includes two sites of Keli (Qeli) Highland and Mount Kazbek. However, the region has quotient 0.22 at the standard deviation of elevation of 640 m. High-mountain grabens (for example, in the Great Caucasus) on the studied area locates on 2400 m above sea level at the elevation standard deviation of 599 m. The sloped plains with the sediment cover or without it (socle plains) are placed in piedmonts of Lesser Caucasus and Trans-Ural. They have equivalent elevations. The stepped plains of Pechora river basin have median elevation of 118 m and standard deviation of 32 m. And the last region is "sloped plains built by sediment deposits" (subdivision 1.3.1.1.). The region is represented by Valdai Hills with the median elevation of 187 m and standard deviation of 53 m.

The regions with the quotient more than 0.3 are considered to be not representative of using the elevation intervals as diagnostic attribute. Short data on characterized sixteen regions is given in the table 3. The geographical distribution of described structural regions is shown on the figure 1 (a).

Sculpture. We used the same technique for assessment of localization of morphosculptural regions in the elevation intervals as in the case of morphostructures. The threshold value of quotient was taken as 0.3 again. It is important that the values of quotient less than 0.1 were not found. It may mean that some structural regions fix more finely in the elevation intervals than sculptural regions, and elevation field was set by structural conditions in general.

The least deviations of elevation characterize areas created by processes of complex semiarid and arid denudation. Authors using geomorphological map of the USSR, singled out hummocky and flat (or weak wavy) terrains. The first have median elevation of 312 m and standard deviation of 37 m. Flat or weak wavy lands have surprisingly bigger value of standard deviation — 48 m (at the median value of elevation — 295 m).

Four morphosculptural regions relate with the areas of Pleistocene glaciations. Region of hill and ouval lands is located on the territory of old middle-Pleistocene glaciations (Moscow ice age, synchronically to Riss ice age in theWest Europe). Hilly (concentric) landforms are common in the band of marginal zone of the glacier. Ouvals and intermediate landforms from hills to ouvals are common outside glacier zone (periglacial band). It has better erosional moderation and large topographical dissection. Band of hilly lands has values of median elevation of 191 m and standard deviation of 36 m. Intermediate zone of sloped ouvals is characterized by values of 165 m and 32 m, respectively. Zone of typical ouvals is characterized by values of 182 m and 53 m respectively (essential erosional dissection).

One morphosculptural region is located in the area of the Valdai ice age glaciers ("ridge and hill lands of marginal and radial postglacial areas"). This territory is replete with such landforms as drumlins and eskers. Median elevation here is 161 m and standard deviation is 48 m. Short data on characterized six regions are given in the table 4.

Table 3. Morphostructural regions of European Russia located in the thin elevation intervals.

No	Region	Median elevation	St. deviation of elevation	Quotient	Geographical location of examples
1	1.3.1.1. table and step plateaus built by sediment	264	13	0.05	Kostanay Region of the Kazakhstan
2	1.1.2. semiburied step plains	159	17	0.11	Donetsk Oblast in the Ukraine
3	1.3.1.2. layer plateaus built by sediment deposits	242	29	0.12	Aktobe and Kostanay Regions of the Kazakhstan
4	3.2.1.2. fold monoclinal ranges and plateaus	358	47	0.13	Lesser Caucasus
5	1.3.2.1. sloped plains	157	25	0.16	eastern flank of Donetsk coal basin
6	1.3.1.1. sloped, partially stepped plains built by sediment deposits	190	34	0.18	Smolensk and Moscow Upland and some Cis-Ural Uplands
7	1.3.1.1. stepped plains built by sediment deposits	196	37	0.19	Central Russian, Volga and Bugulma-Belebey Upland
8	3.3.1.1. volcanic mountains and high plateaus	2906	640	0,22	Keli (Qeli) Highland and Mount Kazbek
9	1.2.1. semiburied plains and chains of hills	174	39	0,22	East of Donets Ridge
10	3.2.2. grabens	2408	599	0,25	Valleys on North Slope of Greater Caucasus
11	3.2.3.2. sloped plains	237	62	0,26	Lesser Caucasus
12	3.1.1.1. socle sloped plains	238	63	0,26	Central and North Trans-Urals
13	1.3.1.2. slightly sloped plains and plateaus built by sediment deposits	156	42	0,27	Uppers of North Dvina river basin
14	1.3.2.2. stepped plains	118	32	0,27	Pechora river basin
15	1.3.1.1. sloped plains built by sediment deposits	187	53	0,28	Valdai Hills
16	1.2.2. plains and chains of hills	305	90	0,30	South Trans-Urals, North Cis-Ural and part of Timan Ridge

Table 4. Morphosculptural regions of European Russia located in the thin elevation intervals.

No	Region	Median elevation	St. deviation of elevation	**Quotient**	Geographical location of examples
1	2.1.1. Hills, ridges and ouvals (hummocky terrain)	312	37	0,12	Southeast Ural
2	2.1.1. Flat or weak wavy lands	295	48	0,16	Southeast Ural
3	2.2.4. Hill and ouval lands	191	36	0,19	Line Smolensk - Syktyvkar
4	2.2.5. Sloping ouval lands	165	32	0,19	South of line Smolensk - Syktyvkar
5	2.2.4. Ouval lands	182	53	0.29	South of line Smolensk - Syktyvkar and Ural piedmonts
6	2.2.3. Ridge and hill lands of marginal and radial postglacial areas	161	48	0.30	North-West of European Russia

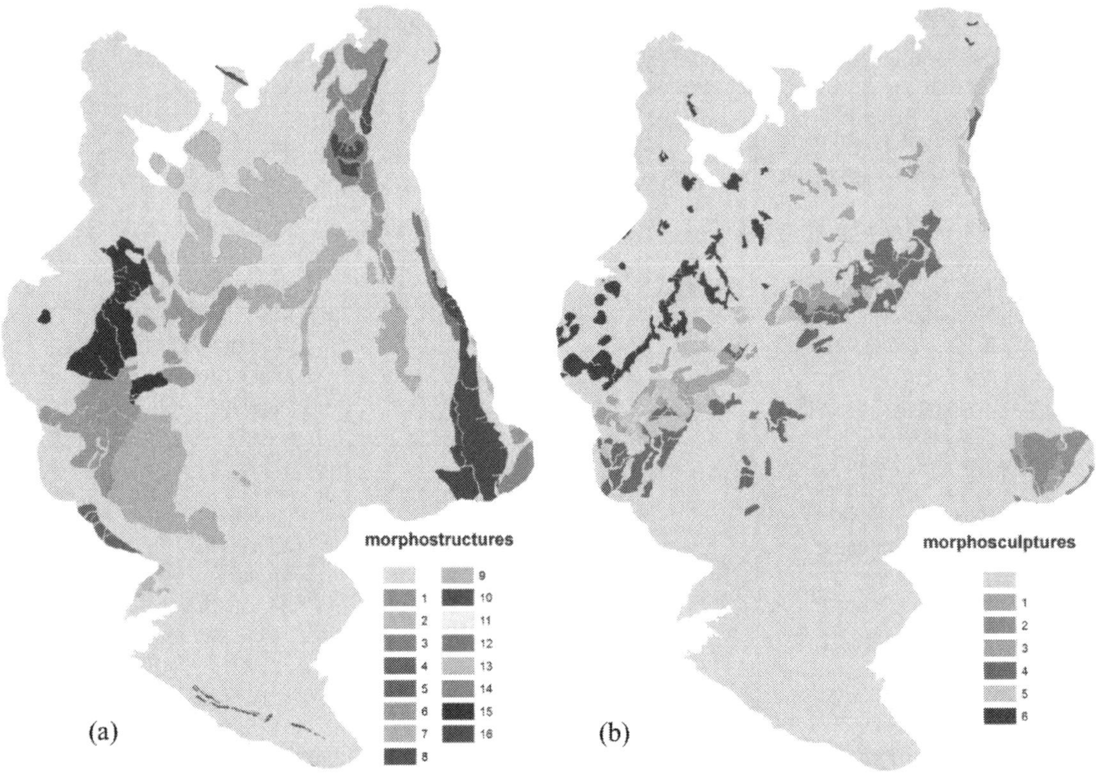

Figure 1. Types of morphostructures (a) and morphosculptures (b) of the European part of Russia localized in relatively thin intervals of elevation.

The geographical distribution of described sculpturtal regions is show on the figure 1 (b).

Thus we found out that:

1. Location of some types of morphostructures and morphosculptures are characterized by relatively thin intervals of elevation (quotient of standard deviation and median of elevations less than 0.3 in the article). Sixteen types of structural landform regions from fifty eight types on whole studied area have corresponding values of the coefficient. And six types of sculptural landform regions from twenty two types on whole studied area have corresponding values of the coefficient;
2. Structural landforms have mainly lower values of quotients than sculptural landforms. Consequently, the elevation field is due to the structural plan of the territory in general;
3. Relatively low dispersion of elevation corresponds to stepped and sloped plains and plateaus and, in a less degree, to some mountain structures such as grabens and high volcanic plateaus. In the case of sculptural landforms, low dispersion of elevation corresponds to the arid climate planation surfaces and to areas of typically post-glacial landforms;
4. The parameter of elevation can be used for automated classification of DEM and recognizing of described landform types.

Acknowledgements

The work is performed according to the Russian Science Foundation (15–17–10008). Besides, authors are grateful to Sergey Seminozhenko (Institute of Geography, RAS) for scan copies of Geomorphological Map of the USSR.

References

[1] Giles P T and Franklin S E 1998 *Geomorphology* **21**(3-4) 251–264
[2] Ventura S J and Irvin B J 2000 *Terrain analysis: principles and applications* (Hoboken: John Wiley & Sons) 267–294
[3] Drăguţ L and Blaschke T 2006 *Geomorphology* **81** 330–344
[4] Saadat H, Bonnell R, Sharifi F, Mehuys G, Namdar M and Ale-Ebrahim S 2008 *Geomorphology* **100**(3) 453–464
[5] Drăguţ L and Eisank C 2012 *Geomorphology* **141** 21–33
[6] Lapparent A 1885 *Traite de géologie* (Paris: Librairie F. Savy)
[7] Tanner W F 1959 *Am. J. Sci.* **257**(6) 458–460
[8] Sharapov IP 1964 *Rel'ef Zemli i matematica* (Moscow: Misl') pp 72–79
[9] Pike R J and Wilson S E 1971 *Geological Society of America Bulletin* **82**(4) 1079–84
[10] Harlin J M 1978 *Mathematical Geology* **10**(1) 59–72
[11] Kharchenko S V 2017 *Geomorphologiya* **2** 14–24
[12] Kharchenko S V 2017 *17th International Multidisciplinary Scientific GeoConference SGEM 2017 (Albena, Bulgaria)* pp 431–437
[13] Gerasimov I P and Mescherikov J A 1968 *Morphostructure Encyclopedia of Geomorphology ed RW Fairbridge* (New York: Reinhold Book Corporation) pp 731-2
[14] *Geomorphological Map of USSR at the scale of 1:2 500 000* ed I P Gerasimov and A A Aseev on 16 sheets

IOP Conf. Series: Earth and Environmental Science **107** (2017) 012012 doi:10.1088/1755-1315/107/1/012012

The genetic structure of the chloride ion runoff on the example of karst and non-karst geosystems of Arkhangelsk oblast

D N Khayrullina and A A Kurzhanova

Institute of Environmental Sciences, Kazan Federal University, 18, Kremlevskaya str., 420008, Kazan, Russia

E-mail: dinara-hi@yandex.ru

Abstract. This paper deals with the estimate the structure of the chloride ion runoff from the karst (on the example of the Sula river basin) and non-karst (on the example of the Vaga river basin) geosystems of Arkhangelsk oblast. The contribution of the surface component predominates in the structure of the chloride ion runoff.For example, the input of surface ion runoff is 49% (for the Sula river basin), 55% (for the Vaga river basin). In time aspect the highest values of variability of the components of the chloride ion runoff are noted for karst geosystems and vary from 38.5% to 55.4% and from 24.7% to 42.9% – for non-karst geosystems.Finally, there is prevalence of the local factors influence because the atmospheric component decreases while ion runoff increases.

1. Introduction

Quantitative indicators of natural-anthropogenic geosystems can reflect the anthropogenic intensification of geochemical cycles of elements.

So, the ion runoff reflects the geosystems state. Moreover, the assessment of its genetic components helps to assess the factors of its formation.The chloride ions (in particular, highly mobile water migrants) are the most indicative ions.

Entirely, the investigated rivers are located in different natural and anthropogenic conditions.

So, the Sula river basin has an area of 8500 km^2. As a rule, the river basin is located in the closest proximity to the Barents Sea (100 km) within the tundra zone. Groundwater of the river basin belongs to the fractured waters of Canino-Timan basin. On the one hand the middle annual precipitation is less than 400 mm, on the other hand 300 mm evaporates. The middle annual air temperature is 1.5°C, and the relative humidity is 85%.The middle height of the river basin is 97 m, the middle slope is 0.94‰.The forest cover is 35%, bogging is 1%.

The Vaga river basin has an area of 1410 km^2. The river basin is the most remote from the Onega Bay, the distance is 400 km. The river basin is located within the Onego-Vagsky protrusion of the northern part of the Moscow syneclise. Moreover, river basin drains the Permian rocks of the Sukhona horizon represented by dolomites, marls, clays. The middle height of the river basin is 172 m. 550 mm of precipitation falls out every year, 350 mm evaporates.The middle annual air temperature is about +2 °C, relative humidity is 70%. The forest cover is 84%, bogging is 5% [1].

Actually, river basins are characterized by a low proportion of groundwater feeding (to 27%) as a result of the influence of both climatic (the presence of permafrost rocks and taliks in the Sula river basin) and geological factors [2, 3].

The aim of research is to estimate the space-time variability of the structure of chloride ion runoff in the basins of the Sula and the Vaga rivers.

There is a number of problems:

1. Estimation of the atmospheric, underground and surface component of chloride ion runoff of the Sula and the Vaga river basins.
2. Analysis of space-time variability of the structure of chloride ion runoff.

2. Material and methods

The paper based on the materials of the Northern Department for Hydrometeorology and Environmental Monitoring of meteorological stations in the cities of Belozersk and Naryan-Mar and hydrological posts on the Vaga river (Gluboretskaya village) and the Sula river (Kotkina village) for the period from 1978 to 2007.

The methodology is based on the balance equation (1), proposed by V.P. Zverev (1971):

$$W_{total} = W_{i.atm} + W_{i.underground} + W_{i.surface} + W_{i.accum}, \quad (1)$$

where W_{total} – the total ion runoff; $W_{i.atm}$ – the ions of atmospheric precipitation supply; $W_{i.surface}$ – the surface ion runoff; $W_{i.underground}$ – the underground ion runoff; $W_{i.accum}$ – the ion accumulation in the surface horizons of the drainless regions (according to V.P. Zverev (1971) for the river basins of the Barents Sea and the White Sea this index is zero) [4].

The ions of atmospheric precipitation supply ($W_{i.atm}$) are calculated using the coefficient of runoff calculated by formula (2):

$$W_{i.atm} = n \cdot \frac{C \cdot S}{10^3}, \quad (2)$$

where S – the amount of atmospheric precipitation, mm, C – the ion concentration of atmospheric precipitation, mg/l, n – the runoff coefficient calculated by formula (3):

$$n = \frac{h}{S}, \quad (3)$$

where h – the layer of water flow from the territory of the river basin for the hydrological year, mm; S – the amount of atmospheric precipitation falling on the territory of the river basin for a hydrological year, mm [5-9].

The underground ion runoff was calculated by the formula (4):

$$W_{i.underground} = \frac{k \cdot a \cdot C_{winter} \cdot W_{water} \cdot 1000}{F}, \quad (4)$$

where k – the proportion of underground water runoff [4, 5]; a – a correction coefficient defined as the ratio of the middle long-time water flow for hydrological year to the water flow for particular year; C_{winter} – the ion concentration during the winter low water period when the minimum water flow rates are known, mg/l; W_{water} – total water flow in this hydrological year, km^3; F – the area of the river basin above the hydrological post, km^2 [10-14].

The total ion runoff was calculated by the formula (5):

$$W_{total} = \frac{\frac{\sum_{i=1}^{n}(Q_i \cdot C_i)}{\sum_{i=1}^{n} Q_i} \cdot a \cdot W_{water} \cdot 10^3}{F}, \quad (5)$$

where C_i, Q_i – the ion concentration (mg/l) and discharge (m^3/s) on the day of taking a water sample; n – number of samples per year [15-17].

Surface ion runoff was calculated using formula (6):

$$W_{i.surface} = W_{total} - (W_{i.underground} + W_{i.atm}), \quad (6)$$

The variability of ion concentrations in water and water discharge depends on an exponential law (figure 1a).

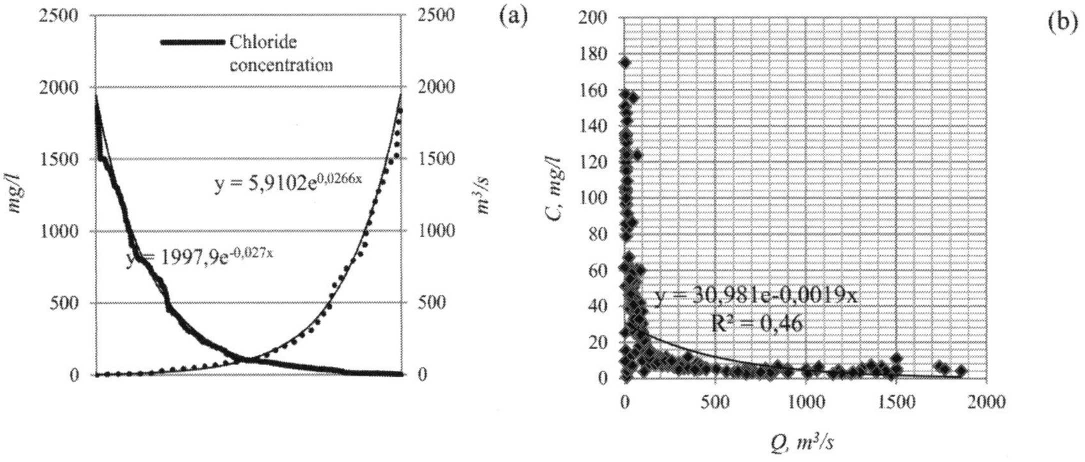

Figure 1. Exponential variability of the chloride ion concentration and water flow on example of the Sula river – Kotkina village (1978 - 2007) (a); the graph of the dependence of the chloride ion concentration on the discharge on example of the Sula river – Kotkina village (1978 - 2007) (b).

The determination the ion concentration C_{winter} when the discharges of deep winter low period are known was realized by constructing the exponential dependence C = f (Q) on the basis of known concentrations and discharges for the all period (figure 1b).

3. Results and discussion

There is the domination of the surface chloride ion runoff formed as a result of elution of ions by thaw and storm water from soil horizons (figure 2)

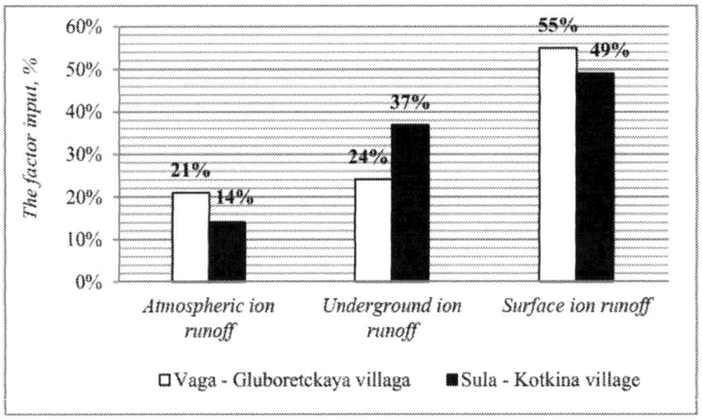

Figure 2. Genetic structure of chloride ion runoff within the basins of the Vaga river and the Sula river (1978-2007).

So, the input of surface chloride ion runoff to chloride ion runoff of the Vaga river is 55% (or 700 kg/km^2), and of the Sula river – 49% (or 4253 kg/km^2) for the middle long-term period.

The Sula river basin is karst therefore the underground component of chloride ion runoff is high (37% or 2936 kg/km^2). Conversely, the atmospheric component of chloride ion runoff is 14% (1121 kg/km^2).

Moreover, the underground component of chloride ion runoff of Vaga river basin is 24% (297 kg/km^2) because the river basin is composed of aquitard morainic loams (figure 2).

The atmospheric component is almost equal to the underground component of chloride ion runoff and is 21% (266 kg/km^2).

Primarily, the variability of the components of the chloride ion runoff within the karst geosystem is higher. Statistically, the coefficient of variation fluctuates from 38.5% to 55.4% within the Sula river basin (table 1).

Table 1. Coefficient variation (in the numerator) and coefficient *k* of the linear trend (in the denominator) of time variability of structure of chloride ion runoff

River basin	Components of chloride ion runoff			Ion runoff
	Atmospheric	Underground	Surface	
Sula – Gluboretckaya village	55.4% -40.35	46.9% -6.53	42.5% 42.16	38.5% 28.09
Vaga – Kotkina village	24.7% -33.94	42.9% 3.97	31.4% -131.58	35.1% 16.68

The maximum values are noted for the atmospheric component. Conversely, the minimum values are fixed for the surface component.

The coefficient of variation varies from 24.7% to 42.9% within the Vaga river basin. There is the minimum variability of the atmospheric component of chloride ion runoff, the maximum is observed for underground component.

Primarily, the most fixed increase in the multi-year variability of the components of the chloride ion runoff of the Sula river basin is noted for the surface component (*k* of the trend line is 42.16) (figure 3, table 1).

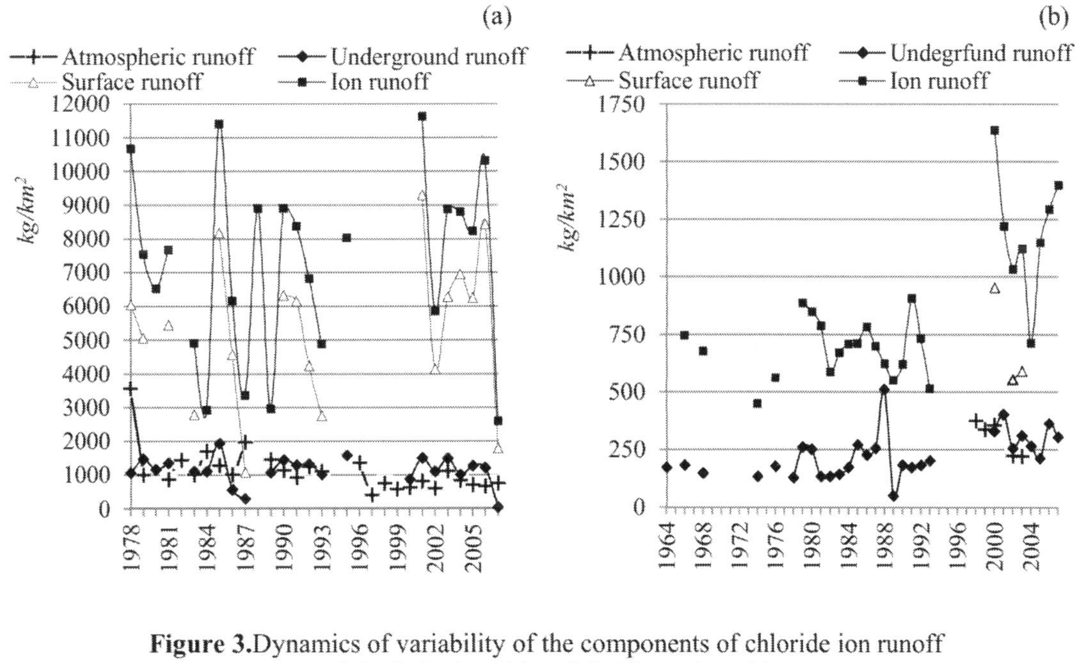

Figure 3. Dynamics of variability of the components of chloride ion runoff of the Sula river (a) and the Vaga river (b)

Conversely, there is significant decrease for the atmospheric component (*k* = -40.35). A small decrease is also fixed for the underground component (*k* = -6.53). This fact is connected with the comparative isolation of groundwater from the external factors. Broadly speaking, in the multi-year aspect the increase is recorded for the total ionic runoff (*k* = 28.09). It indicates a significant influence of the surface component of ion runoff.

Secondly, there is an increase (k = 16.68) for the total chloride ion runoff for the Vaga river. There is also a small increase (k = 3.97) for the underground component.

The other components are characterized by a significant decrease. So, k = -33.94 for the atmospheric component, k = -131.58 for the surface component because there is a short observation period.

In time aspect the highest values of variability of the components of the chloride ion runoff are noted for karst geosystems. There is prevalence of the local factors influence because the atmospheric component decreases while ion runoff increases.

Finally, the ion runoff consists of atmospheric, underground and surface components. Surface (soil) component of chloride ion runoff is the main factor of chloride ion runoff formation. In time aspect, there is the maximum variability of values of components of chloride ion runoff for karst river basins.

Acknowledgements
Authors thank the doctor of geographical sciences, professor N.P. Torsuev, associate professor of the department of landscape ecology K(P)FU V.A. Fedorova and V.A. Belonogov for help in finding and improving the methods of work.

References
[1] Hydrogeology of the USSR 1970 *Nedra* XLII p 288 (In Russian)
[2] Filenko R A 1966 *Waters of Vologda oblast* (Leningrad Publish House) p 132 (In Russian)
[3] Filenko R A 1974 *Hydrological zoning of the North of European part of the USSR* (Leningrad Publish House) p 223 (In Russian)
[4] Zverev V P 1971 *Hydrochem Mat* **56** 11–18 (In Russian)
[5] Kopoteva T N and Fedorova V A 2011 *Modern problems of geochem.* 169–171 (In Russian)
[6] Reynolds B, Fowler D, Smith RI and Hall JR 1997 *Journal of Hydrology* **194**(1–4) 305–329
[7] Khayrullina D D 2014 *GeoConference SGEM* 2014 pp 267–272
[8] Khayrullina D N and Torsuev N P 2012 *Gazette of Udmurt University* **2** 19–23 (In Russian)
[9] Reid J M, MacLeod D A and Cresser M S 1981 *Journal of Hydrology* **50** 129–145
[10] Wang H, Jiang X-W, Wan L, Han G and Guo H 2015 *Journal of Hydrology* **527** 433–441
[11] Rumsey C A, Miller M P, Susong D D, Tillman F D and Anning D W 2015 *Journal of Hydrology* **4**(B) 91–107
[12] Stephen U N, Celestine O O and Solomon O O 2017 *Journal of African Earth Sciences* **131** 145–155
[13] Ebrahim G Y and Villholth K G 2016 *Journal of Hydrology* **541**(B) 754–765
[14] Jódar J, Cabrera JA, Martos-Rosillo S, Ruiz-Constán A, González-Ramón A, Lambán LJ, Herrera C and Custodio E 2017 *Science of The Total Environment* **593-594** 760–772
[15] Belonogov V A and Torsuev N P 1997 *The News of RGS* **6** 58–68 (In Russian)
[16] Khayrullina D N and Fedorova V A 2014 *Advances in Environmental Biology* **8**(4) 1015–1020
[17] Khayrullina D N and Kurzhanova A A 2016 *Research Journal of Pharmaceutical, Biological and Chemical Sciences* **7**(5) 1530–1537

The main relief-forming factors of asymmetry of river valleys within the central part of East European Plain

A A Kourjanova and E V Petrova

Institute of Environmental Sciences, Kazan Federal University, 18, Kremlyovskaya Street, Kazan, 420008, Russia

E-mail: akurjan@mail.ru, helengeo@mail.ru

Abstract. The article examines the spatial distributions of asymmetry types in the slopes of the river valleys in the central part of the East European Plain. Five types of asymmetry are found in the studied region: planetary, climatic, hydrodynamical, structural (tectonic) and topographic.

Keywords: asymmetry, valleys, relief, East European Plain.

1. Introduction

Since the 1950s, both Soviet and foreign researchers have proposed a whole number of methods for quantifying the degree of asymmetry on the slopes of river valleys. The first quantitative indicator was proposed by A. Gloriad, J. Tricart [1]. Significant studies in the development of a quantitative determination of the asymmetry of river valleys slopes were carried out by V.N. Sementovskii [2], H. Karrasch [3], G.P. Butakov [4]. Later on studies were continued by H. French [5], C.B. Crampton [6], T. Czudek [7], H.Y. Akerman [8].

The asymmetry of the river valley in the center of the East European Plain was considered earlier except the territory of the Central Russian Upland [9, 10, 11, 12].

This paper shows the results of studying the asymmetry of the slopes of the river valleys in the central part of the East European Plain.

The considered territory extends in meridional direction from the Volga river head in the North to the Donetskaya ridge in the South. The Eastern border of this region is the Don valley, the Western border – the left tributaries of the Dnepr river. The landscape of the territory is flat, heightened to the West and lowered to the East.

2. Materials and methods

The degree of manifestation of the asymmetry of river valleys slopes was determined by the methods proposed by H. Karrasch [3] and G.P. Butakov [4].

For field researches the formula of H. Karrasch was used:

$$A'' = 1 - Agentle / Asteep \quad (1)$$

where Agentle, Asteep are the angles of inclination of the gentle and steep slopes in degrees.

At cameral works there was used the following formula:

$$A = 1 - Lsteep / Lgentle \quad (2)$$

where Lsteep / Lgentle are the lengths of the horizontal location of the steep and gentle slopes. In this case, the asymmetry coefficient will vary from 0 in symmetrical valleys to values close to 1.0 in sharply asymmetric ones. The coefficient is limited on both sides.

In the course of the work full study of large-scale topographic maps was conducted. The results of cameral work were monitored by field observations in selected key areas.

3. Results

The asymmetry appeared in all valleys of studied region. The average coefficient of the asymmetry for the whole region is equal to 0.45. From the North to the South the following regularity occurred: in the North region on the Valdai Hills the coefficient of asymmetry A consists of 0.46, in the South in the basin of the Moskva river it increases to 0.57, the maximal values (0.50-0.72) belong to the cold points and the minimal values (0.29-0.50) - to the warm ones. Moving from the North to the South the coefficient A decreases to 0.40 for the basin of the Oka river and to 0.47 for the basins of the Desna and Don rivers. In the Southern part of the region the coefficient of the asymmetry again increases to 0.50 in the basins of the Seim river and to 0.54 in basins of the Severskii Donets river. Here the maximal values of the coefficient of the asymmetry (0.47-0.87) belong to the warm points, but the minimal (0.19-0.40) - to the cold ones.

Five types of asymmetry are found in the studied region: planetary, climatic, hydrodynamical, structural (tectonic) and topographic.

The planetary type occurred in the valleys of all large rivers (Moskva, Protva, Ugra, Upa, Desna, Jizdra, Oka, Zusha, Seim, Severskii Donets, Oskol) excluding the basin of the upper Volga within Valdai Hills. The coefficient of the planetary type for the territory changes from 0.33 to 0.60 and the average is 0.49 (table 1, 2).

The minimal coefficient of the planetary asymmetry in the basin of the Jizdra river consists of 0.33. This region gives the most part of the symmetric sections (33%) and the lowest percent of the sharply asymmetric valleys (11%).

Table 1. The distribution of the planetary type of asymmetry

№	River valley	Average coefficient A	Symmetric valleys, %	Asymmetric valleys, %	Maximal expense of river, m/s
1	Upper Volga	0.64	11.6	53.5	1124
2	Moskva river	0.57	20.8	40.7	1095
3	Oka river	0.40	31.4	16.1	845
4	Protva river	0.59	-	82.2	1150
5	Ugra river	0.45	5.3	52.6	748
6	Upa river	0.42	7.7	38.5	705
7	Desna river	0.56	-	75.0	1110
8	Jizdra river	0.33	33.3	11.1	810
9	Zusha river	0.53	-	60.0	1040
10	Seim river	0.52	-	55.5	1185
11	Sosna river	0.55	4.8	61.8	1160
12	Oskol river	0.74	3.3	93.4	1200
13	Don river	0.69	3.7	80.5	1175

Table 2. The average coefficients of the different types of the asymmetry

Region	A common	Types of the asymmetry				
		Planetary	Hydrodynamical	Structural	Topographic	Climatic
Valdai Hills	0.46	–	0.49	–	0.47	0.51
The Moskva river	0.57	0.61	0.57	–	0.40	0.49
The Dnepr river	0.50	0.52	0.40	0.43	–	0.44
The Oka river	0.40	0.49	0.41	0.41	–	0.39
The Don river	0.47	0.69	0.51	0.54	–	0.44
The Severskii Donets river	0.54	0.59	0.53	0.44	–	0.52
Average	0.45	0.56	0.49	0.45	0.43	0.43

The possible explanation is the following:

1. The river flows in the latitudinal direction and its left bank has the southern exposition. Here passes the superposition of two types of the asymmetry: planetary and climatic. The climatic factor sufficiently decreases the difference between two slopes.
2. The strong twistiness of the river. It promotes the left bank to be usually steep (hydrodynamical asymmetry).

The maximal coefficients of the asymmetry of the planetary type are found in the Don valley (0.69) and in the low Oskol (0.82) e.g. in the largest rivers of the given region. For this valleys the most part of the sharply asymmetric sections of valleys and the minor part of the symmetric ones are accordingly observed (table 1).

The high correlation of the values of the coefficient of asymmetry and the average maximal expenses are calculated (table 1). The part of the whole length of the valleys due to the planetary type of the asymmetry consists of 10.9% (table 3).

Table 3. The part (%) of the types of asymmetry on the whole quantity of the asymmetric sections.

Region	Types of asymmetry				
	Planetary	Structural	Hydrodynamical	Topographic	Climatic
Valdai Hills	–	–	2.2	1.1	96.7
The Moskva river	20.2	–	15.7	9.6	54.5
The Dnepr river	15.1	0.8	3.1	–	81.0
The Oka river	10.3	3.5	4.3	–	81.9
The Don river	8.4	5.1	8.1	–	78.4
Average	10.9	1.8	6.6	2.1	78.6

The hydrodynamical type of the asymmetry is well developed in the river valley of the given region. For example: rivers Ugra, Upa, Jizdra, Sosna, Nepriadva tributary of the Oskol river and others. Passing as the planetary type the river valleys on the sections of the left meander have the steep left bank often directed to the South-East, the South and South-West. In addition the "right" asymmetry alternates by the "left" one (for the main rivers) in the confluences of the large right tributaries: the Oka and the Cherepue, the Upa and the Vatitsa, the Ugra and the Dolinka, the Don and the Devitsa, the Potudue and the Sosna, the Krasivaya Mecha and the Vyazoevka, etc. In 5-7 km downstream of the tributary mouth the normal asymmetry is reconstructed. A lot of occasions of the hydrodynamical asymmetry are found in the sections of the Volga valley, Tsna, Bol. and Mal. Koshy, Osuga rivers. The average coefficient of the hydrodynamical type for the region consists of 0.49 (table 2).

The part of the asymmetry of such type consists of 6.6% of the whole length of all valleys (table 3).

The hydrodynamical type is observed on the whole territory and the regularity of its distribution is not found.

The structural and tectonic types of the asymmetry are connected with the peculiarities of the formation of mountain rocks and with the manifestations of the tectonic structures in the landscape. On the territory of the investigated region a lot of structures effecting the formation of the river valley are revealed. Unfortunately, the connection between the construction of the land surface and the bed of the river valley was revealed only for several sections due to the absence of the detailed structural maps and to the fact that structures II and El are not well revealed.

We have only some examples where such connections are well observed: the Tuskar and the Rat rivers (basin of the Seim river) placed in the zone of the Kursk raising (II category) are distributed to the North-West. The vault is more raised in the Northern-Eastern part and it is lowered to the South-West. It characterizes the common asymmetry of the Kursk raising which can be called monoclinal. The asymmetry of the raising is the main reason of the stable right asymmetry of this rivers. The riverhead of Krasivaya Mecha river rounds from the South of the Ptanskoe raising, consequently the river valley has the curve and its right bank becomes steep directed to the North and North-West. Such picture is observed in the Sosna river where the river rounds Kozminskoe raising; here the steep right bank directs to the North-West. In this case the enlargement of the planetary type of the asymmetry occurs. At whole region the average coefficient of the asymmetry consists of 0.45 (table 2). The part of the structural type is 1.8% (table 3). In the Southern part of the region this type of the asymmetry is not revealed.

The level of climatic asymmetry development within the considered territory is showed in the table 4. The analysis of the spatial distribution of the coefficients of the climatic asymmetry suggests the regularity of the changes in the territory. The analysis of the levels of the expression for the asymmetry takes into account different mountain rocks in various regions. Using the correction coefficients to lithology for the level of river valleys slopes asymmetry in one case decreases the value A (the upper Seim river from 0.47 to 0.42) in other case increases it (basins of rivers Don, Oskol etc.) (table 4).

Table 4. The quantitative characteristics of the climatic asymmetry

№	Region	A	A correct	Steep slope, %	
				Warm points	Right bank
1	Valdai Hills	0.45	0.51	44.0	54.0
2	The Moskva river	0.49	0.49	38.1	60.0
3	The Dnepr river	0.41	0.44	56.2	49.0
4	The Seim river	0.47	0.42	72.6	56.0
5	The Oka river	0.40	0.39	68.2	49.0
6	The Don river	0.39	0.44	69.6	43.0
7	The Oskol river	0.48	0.51	82.5	62.0
8	The Severskii Donets river	0.49	0.52	72.5	54.0
	Whole	0.44	0.43	64.8	53.0

The slowly decreasing A is observed from 0.51 in the North within Valdai Hills to 0.35-0.44 in the center (in the basins of Oka river, upper Don river, middle section of Dnepr river). In the South the value A increases again to 0.5 and more in the basins of the Oskol and Severskii Donets rivers. The maximal values of the coefficients A 0.5-0.54 are revealed in the basins of the Severskii Donets river and Aidar in the South and also in the basins of the Shlyny and Tsna rivers in the North of the studied region.

The minimal coefficients A regard for the basin the upper Oka river. The average coefficient of the climatic asymmetry in this region is equal to 0.43. The part of the sharply asymmetric valleys in this basin consist of 16.1%. This is the lowest part of the sharply asymmetric valleys. The weak asymmetry and the small percent of the sharp asymmetric valleys are possibly explained by the picture of the hydrological net, the small development of the river valleys, the superposition of the several types of the asymmetry and, as a consequence, the effect of the climatic factors decreases. One of the reasons of the weak expression of asymmetry [2] is the presence of the pediments in this territory which are formed in the slopes of the warm points, they are smoothed and as a result the asymmetry of the slopes decreases.

For the Northern regions (the Valdai Hills and the basin of the Moskva river) the steep slopes dominate (directed to the cold points, N, NE, NW) (Table 4). In the upper Volga river only 38% of cases for steep slopes are directed to the warm points (SE, S, SW, W), but in the basin of the Moskva river accordingly – 19-40% and to the cold points – 62-70%. It shows the development of the northern type of the climatic asymmetry here. The border between the Northern and Southern types passes along the watershed of the Volga and Dnepr rivers, Moskva and Protva rivers, Moskva and Oka rivers.

In the upper Dnepr and the basin of the Protva river the part of the steep slopes directed to the warm points increases to 53-69% but in the basin of the Oskol river this part increases to 82%. The regular change of the part for the steep slopes directed to the warm points repeats the variation of the values for the coefficient A.

The connection of the steep slope with right or left slopes is not revealed. For several basins the part of the steep slope varies within 43%-60%, on the average consists of approximately 50% (table 4), that shows the independence of climatic asymmetry of the river valleys slopes in the direction of the river stream.

In nature, the asymmetry of the slopes of river valleys, only one type is observed rarely, as a rule we observe the superposition of two and more types. In any cases this superposition strengthens or decreases the level of the asymmetry.

4. Conclusions

The planetary type of asymmetry occurs in the valleys of all major rivers. The planetary type of asymmetry coefficient for the territory varies from 0.33 to 0.60. The hydrodynamic type is observed throughout the territory and the regularity of its distribution is not found. The average coefficient of asymmetry of hydrodynamic type for the region is 0.49. The structural and tectonic types are the main cause of stable right asymmetry in the rivers of the studied area. The climatic type of asymmetry, along with the planetary one is the main type in the studied region. It is typical for the most of small and medium rivers. The coefficient of this type of asymmetry can reach more than 95%.

References

[1] Gloriad A and Tricart J 1952 *Revue de Géomorphologie Dynamique* **3** 88-98
[2] Sementovsky V N 1963 *Regularities in the morphology of platform relief* (Publishing house of the Kazan University) p 170
[3] Karrasch H 1970 *Götting geogr. Abh.* **50** 299
[4] Butakov G P, Badanov B V, Mozzherin V I , Aleksandrova A 1977 *Landscape Research in the Territory of the Volga Region* **10** 53-79
[5] French H 1972 *Int. Geogr.* **1** 82–84
[6] Crampton C B 1977 *Larth. Surf. Process.* **2** 427–429
[7] Czudek T 1983 *Cas. Slezsk. musea* **A32** 150–180
[8] Akerman H Y 1984 *Geogr.ann.* **66** 267–284
[9] Kozmenko A S 1954 *Osnovyi protivoerozionnoi melioracii (Fundamentals of erosion melioration)* (Moskva: Selhozgiz) p 420
[10] Hrutskii S N and Koscova E V 1981 *Geomorfologiya* **3** 92–96
[11] Aleksentseva A I and Butakov E P 1978 *Klimaticheskii factor rel'efoobrazovaniya (Climatic factor of relief formation.)* (Publishing house of the Kazan University) pp 41–43
[12] Kourjanova A A and Butakov G P 1987 *Kolichestvennyii analiz ekzogennogo refefoobrazovaniya (Quantitative analysis of exogenous relief formation)* (Publishing house of the Kazan University) pp 24–30

IOP Conf. Series: Earth and Environmental Science **107** (2017) 012014 doi:10.1088/1755-1315/107/1/012014

Erosion Losses of Soils on Arable Land in the European part of Russia

K A Maltsev and O P Yermolaev

Institute of Environmental Sciences, Kazan Federal University, 18, Kremlevskaya str., 420008, Kazan, Russia

E-mail:, mlcvkirill@mail.ru

Abstract. The quantitative assessment of potential soil losses in arable lands of the European part of Russia is carried out in the article. The assessment was carried out using a mathematical model based on the mathematical dependencies of the universal soil loss equation and the mathematical dependencies of the State Hydrological Institute of Russia. Assessment of potential soil losses was performed using calculations in a geographic information system. To perform the calculations the database was created containing information on: the relief; properties of soils; climate and land use. The raster model of data organization was used to create the database and subsequent calculations. The assessment shows that the average amount of soil loss in the plowed land of the European territory of Russia is 11 t/ha per year. At the same time, about half of the territories are located in conditions where the soil loss value does not exceed 0.5 t/ha per year. The potential loss of soil taking into account the soil protection role of vegetation is 3.3 tons/ha per year. In addition, a spatial analysis of the distribution of soil loss by landscape zones shows that there is a consistent reduction in the potential loss of soil from the forest zone (20.92 t/ha per year) to the forest-steppe (10.84 t / ha per year), steppe (8.13 t/ha per year) and semi-desert (4.7 tons/ha per year) zone.

1. Introduction

Water erosion is the leading process of denudation within the land and simultaneously carrying out migration of pollutants transported together with sediments [1, 2, 3]. The issues of quantitative assessment of the intensity of erosion and spatial distribution within objects of different scale from small catchment to the continent, on the basis of erosion modeling, are given great attention [4].The European territory of Russia has more than once become the object of a quantitative assessment of erosion processes and soil erosion mapping. For example, S.S. Sobolev and I.F. Sadovnikov published "Soil-erosion map of the USSR" with scale of 1: 5,000,000 [3]. A small-scale map of agricultural lands erosion of a scale of 1: 2,500,000 was compiled for the same territory in the 1990s, by soil Institute named after V.V. Dokuchaev. The map data were obtained using cartographic generalization of large-scale soil-erosion survey data obtained by soil-morphological method.

In addition, in the late 80's and early 90's of the twentieth century on the basis of the modernized Universal Soil Erosion Equation and the National Hydrology Institute's model the folowing maps were developed at the Moscow State University: "Erosion hazardous lands of Russia" with a scale of 1: 1,500,000; "Erosion hazardous at the agricultural lands of Russia" with a scale of 1: 5,000,000, and map of "Soil Erosion" at a scale of 1: 10,000,000. These maps were developed for the Ecological Atlas of Russia. The "Erosion-hazardous lands of Russia" map shows the average annual intensity (in t/ha per year) of the total modern soil erosion in cultivated lands and natural forage lands. Estimation of erosion potential of the relief was carried out by a point-statistical method using maps of scale 1: 25,000 [3].

At present time, geoinformation systems are used to calculate erosion losses of soils using modern models.Many assessments of potential soil losses for large areas have been carried out recently for the territory of Europe [5, 6, 7] as well as in other areas - in the Eastern Basin in China [8] with the use of this approach.

2. Study area

The European territory of Russia (ETR) is located within the East European plain and is bounded from the west by the state border of the Russian Federation from the east by the Ural mountains, from the north by the coastal line of the Arctic Ocean, from the south by the Caucasian mountains and by the coastal line of the Black and Caspian seas. The total area of the territory is about 4 million km^2. This territory is located within several landscape areas: tundra, forest-tundra, forest, steppe, forest-steppe, semi-desert. The total area of cropland in this area is about 650,000 km^2. The population of this territory is about 95 million people.

3. Methodology

Mathematical modeling was used based on the methodology developed by the soil erosion laboratory of the Moscow State University to assess soil erosion. The essence of the methodology for estimating soil erosion losses proposed by the Moscow State University is set out in several monographs [9]. Therefore, here we will not elaborate on the technique in detail, but will present it in abbreviated form.

So, the soil erosion is estimated as the sum of erosion losses from rainwater runoff (Wd) and the flow of meltwater (Wt) in the proposed method. In turn, each of these indicators depends on a number of factors: average annual erosion potential of rainfall (EPR), conventional units; erosibility (soil erodibility), t/ha per EPR. Relief factor, which is a function of slope, slope length and shape; erosion index of the crop or crop rotation; H – a layer of surface slope runoff, depending on the water supply in the snow, precipitation during snow melting and the drain factor, mm. This technique has already been tested to calculate the potential loss of soil within the territories located on the European territory of Russia [10], but for the whole territory the technique is used for the first time.

4. Original data

Several raster models have been created to prepare the database for calculations: slopes, slope lengths, soil erodibility, erosion potential of rainfall (EPR), water reserves in snow, intra-annual redistribution of rainfall, land use.

4.1. Land Use Model

All calculations were carried out within the limits of arable land ETR. The situation of arable land was determined using the land-use model "TerraNorte RLC v.3" developed by the Space Research Institute of the Russian Academy of Sciences. This model is updated annually and we used the current version of the model for 2014. A full description of the methodology for developing this model is given in the article "Estimation of the area of fires on the basis of the integration of satellite data of different spatial resolution MODIS and Landsat-TM/ETM +" [11], while arable land was allocated on the basis of locally adaptive classification and perennial signs of decoding [12].

4.2. Terrain parameters

The raster patterns of slopes and slope length were prepared based of the global relief model GMTED2010. A detailed description can be found in the technical report of the creators of the model [13]. The model of the relief is distributed in parts (tiles), 12 parts were downloaded from the site https://lta.cr.usgs.gov/GMTED2010 for execution on the ETP. This model of relief has already been used to construct the boundaries of the catchments of the European territory of Russia [14], but for the first time it is used to calculate the potential soil loss. Based on the obtained DEM in the GIS ARCGIS 9.2, such characteristics of the relief as the slope and length of the slopes were calculated.

4.3. Soils

The "Unified State Register of Soil Resources of Russia (UGRSR)" was used, which is available at http://egrpr.esoil.ru/ for the preparation of data on the properties of the soil. A detailed description of the content of register is presented in the collective monograph [15]. A detailed description of the creation of a geo-information database of this register is presented in several articles of the employees of the Dokuchaev's Institute [16, 17].

4.4. Climate

There has been an increase in annual precipitation in both summer [18] and winter [19] in recent years. For the warm period of the year from 1960 to 2015 there is a tendency for an increase in precipitation capable of causing soil erosion (intensity> 10 mm / day) [20, 21]. A significant reduction in the snow melt slope flow is observed in study area despite the increase in the amount of precipitation in the cold period [22, 23].

Therefore, for calculations we decided to create our own cartographic model of the spatial distribution of the erosive potential of rainfall on the EPR. The EPR model was created on the basis of daily observations of precipitation at 239 climate stations that were conducted from 1965 to 2015, and function 1 [24]. Data on precipitation were downloaded from the official site of the Russian national climate agency "RIHMI-WDC" (meteo.ru)

$$R30 = 3.19e\,0.006P \quad (1)$$

where P is the mean annual sum of the precipitation bed of daily precipitation exceeding 10 mm. in the warm season (the warm season is determined by the average daily temperature above 0 degrees). Raster models of water reserves in snow and a raster model of intra-annual distribution of erosion potential of rainfall were created on the basis of maps developed in MSU [9].

5. Results and discussion

The erosion on the territory of the ETR varies widely and can reach thousands of t/ha per year (table 1). At the same time, the average amount of soil loss on the arable land is 11 t/ha per year. About half of the territories are located in conditions where the erosion value does not exceed 0.5 t/ha per year (figure 1) and among the remaining territories the largest area is occupied by territories characterized by losses from 10 to 15 t/ha per year. In general, we can say that the distribution of erosion histogram is close to the normal distribution (if the first range is considered separately). If we take into account the soil protection coefficients of vegetation which for some basins of the considered landscape zones vary from 0.2 to 0.4 for rain erosion and from 0.57 to 0.84 for snow melt erosion, and also taking into account the fact that the arable land ER is located within the climatic zones where at present moment losses from rain erosion significantly exceed the loss of soils from snow melt erosion, then the average annual losses on the ETR taking into account the vegetation's coefficients can be estimated at 3.3 t/ha per year. The data obtained can be to compare with the data of recent studies in the European Union [6, 25]. The average soil losses in European Union according different researches range from 2.6 to 2.76 t/ha per year. The soil losses in European Union were estimated for the all land use, but in our case the studies were performed only within arable land.

Table 1. Statistics of the soil erosion loss on the arable land.

	Minimum	Maximum	Mean	Standart Deviation
European Russia	0	3574.66	11.03	24.14

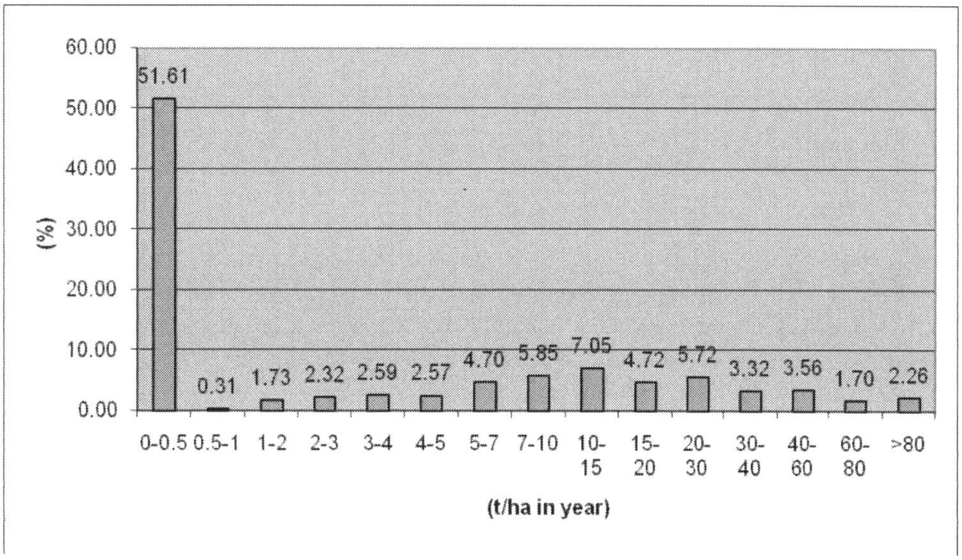

Figure 1. Frequency histogram of the soil erosion loss at the arable land of European Russia.

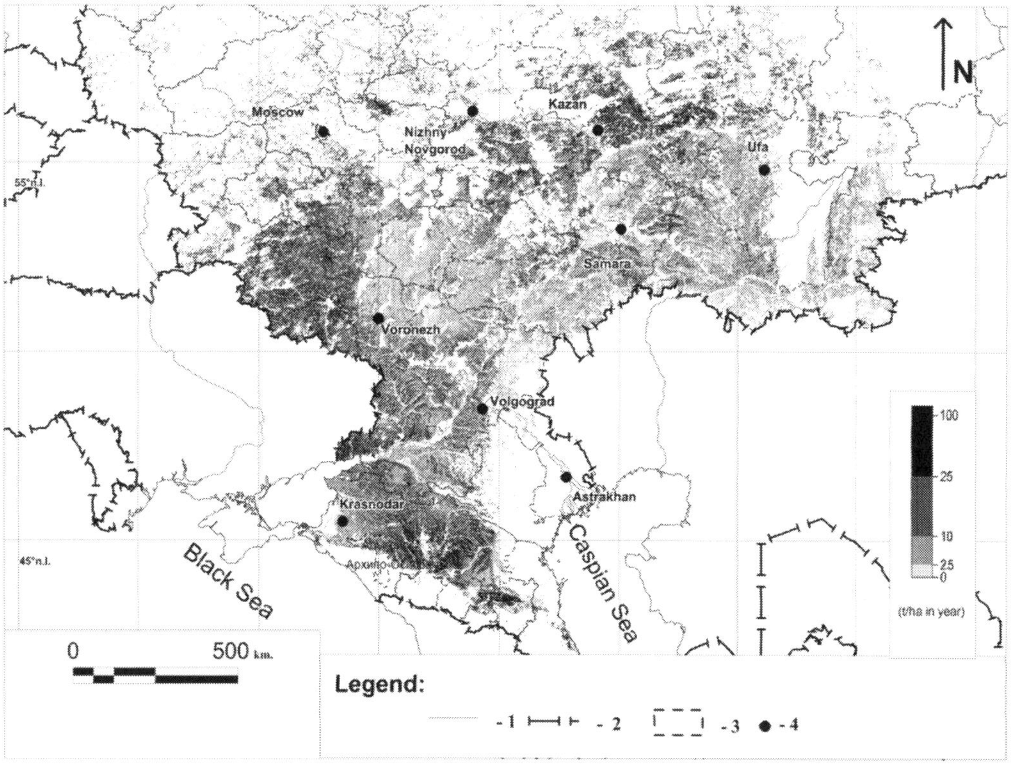

Figure 2. Spatial distribution of the magnitude of soil erosion within the European part of Russia (1-river, 2 state borders, 3- subjects borders of the Russian Federation, 4-populated areas).

Table 2. Statistics of the soil erosion loss within landscape zones

Landscape zones	Mean	Standard deviation	Maximum
Forest	20.92	34.93	1435.26
Forest-steppe	10.84	22.73	3127.04
Steppe	8.13	17.69	880.55
Semi-desert	4.7	9.51	373.24

Analyzing the spatial distribution of potential soil losses, according to the administrative units of the Russian Federation traditionally having large areas of arable land (figure 2), it can be said that the smallest erosion values characterize such subjects of federation as Tambov; Saratov; Volgograd; Voronezh regions. The Republic of Tatarstan, Tula, Oryol, Kursk, Belgorod regions, as well as Stavropol and Krasnodar Krai have high values of erosion soil losses. Also high and very high values of soil losses are on the North Caucasian subjects of the federation, as well as subjects of the federation located mainly in the forest landscape zone.

In general, analyzing the potential loss of soil from planar soil erosion, we can say that it decreases when moving from the forest landscape to the semi-desert zone (table 2). This, in our opinion, can be explained by the contraction of the average annual layer of surface runoff with a shift to the south, and by the predominance of soils in the south of the study area that are well cemented with a high content of humus and have a structure that is well resistant to erosion.

Acknowledgments
The work was supported by the Russian Science Foundation (project No. 15-17-10008).

References
[1] Walling D E 1983 *Journal of Hydrology* **65** 209-237.
[2] Lal R 2001 *Land Degradation & Development* **12** 519-276
[3] Litvin L F 2002 *Geography of soil erosion in agricultural land of Russia (*Moscow, Akademkniga) p 255
[4] Karydas C G, Panagos P and Gitas I Z 2014 *Int. J. of Digital Earth* **7** 229–250
[5] Boardman J and Poesen J. 2006 *Soil Erosion in Europe* (John Wiley and Sons, UK) p 855
[6] Panagos P, Borrelli P, Poesen J, Ballabio C, Lugato E, Meusburger K, Montanarella L, Alewell C 2015 *Environmental Science & Policy* **54** 438-447
[7] Ballabio C et al 2017 *Science of The Total Environment* **579** 1298–1315
[8] Wu Y and Chen J 2012 *Science of The Total Environment* **441** 159-168
[9] Larionov G A 1993 *Erosion and deflation of soils* (Moscow, Moscow State University Press) p 200
[10] Maltsev K A, Yermolaev O P and Mozzherin V V 2014 *AHS-AISH Proceedings and Reports* **367** 326–332
[11] Bartalev S A, Egorov V A, Efremov V Yu, Loupian E A, Stytsenko F V and Flitman E V 2012 *Sovremennye problemy distantsionnogo zondirovaniya Zemli iz kosmosa* **2** 9-27
[12] Bartalev S A, Egorov V A, Loupian E A, Plotnikov D E and Uvarov I A 2011 *Komp'yuternaya optika.* (Samara RAS Press) **1** 103-116.
[13] Danielson J J and Gesch D B 2011 *Global multi-resolution terrain elevation data 2010 (GMTED2010)* (U.S. Geological Survey Open-File Report) pp 26
[14] Ermolaev O P, Mal'tsev K A, Mukharamova S S, Kharchenko S V and Vedeneeva E A 2017 *Geography and Natural Resources* **38**(2) 131–138
[15] *Unified State Register of Soil Resources of Russia* 2014 (Moscow, Soil Institute V.V. Dokuchaeva) p 768
[16] Rukhovich D I, Koroleva P V, Kalinina N V, Vil'chevskaya E V, Simakova M S, Dolinina E A and Rukhovich S V 2013 *Eurasian Soil Science* **3** 251–267
[17] Rukhovich D I, Wagner V B, Vil'chevskaya E V, Kalinina N V and Koroleva P V 2011 *Eurasian*

Soil Science **9** 1043–55
[18] Groisman PYa, Knight R W, Easterling D R, Karl T R, Hegerl G C and Razuvaev V N 2005 *J. Clim.* **18** 1326-50
[19] Shiklomanov I A and Georgievsky V Yu 2002 *The impact of anthropogenic climate change on the hydrological regime and water resources: Climate change and its consequences* (St. Petersburg, "Nauka Press").
[20] Zolina O 2011 *Transactions of RAS* **5** 690-695
[21] Chizhikova N A 2016 *Erozionnyye, ruslovyye i ust'yevyye protsessy (Erosion, river bed and estuarine processes)* (Nizhny Novgorod) pp 251-258 (in Russian)
[22] Litvin L F, Golosov V N, Dobrovolskaya N G, Ivanova N N, Kiryukhina Z P and Krasnov S F 1998 *Soil erosion and channel processes* **11** 57–76
[23] Petelko A I, Golosov V N and Belyaev V R. 2007 *Experimence of design of system of counter-erosion measures* (Moscow, MSU Press)
[24] Kanatyeva N P, Krasnov S F and Litvin L F *Modern changes of climatic factors of erosion in the northern adventure* (Moscow, MSU Press) **17** 14-28
[25] Bosco C, de Rigo D, Dewitte O, Poesen J and Panagos P 2015 *Nat. Hazards Earth Syst. Sci.* **15**(2) 225–245

IOP Conf. Series: Earth and Environmental Science **107** (2017) 012015 doi:10.1088/1755-1315/107/1/012015

Soil evaluation for land use optimizing

O A Marinina

Institute of Environmental Sciences, Kazan Federal University, 18, Kremlevskaya str., 420008, Kazan, Russia

E-mail: marininao@mail.ru

Abstract. The article presents the method of soil classification proposed in the course of the study in which the list of indicators proposed by the existing recommendations is optimized. On the example of one of the river basins within the boundaries of the Belgorod region zoning of the territory was carried out. With this approach, the boundaries of the territorial zones are projected along the natural boundaries of natural objects and the productivity of soils is determined as the main criterion for zoning. To assess the territory by soil properties, the features of the soil cover of the river basin were studied and vectorization of the soil variety boundaries was carried out. In the land evaluation essential and useful for the growth of crops macro- and minor-nutrient elements necessary for the growth of crops were included. To compare the soils each of the indicators was translated into relative units. The final score of soil quality is calculated as the mean geometric value of scores from 0 to 100 points for the selected diagnostic features. Through the imposition of results of soil classification and proposed by the concept of basin nature management – land management activities, five zones were identified according to the degree of suitability for use in agriculture.

1. Introduction

For introduction of basin and ecoregional approaches [1] involving optimization of land resources use, the large-scale projecting of the territories of specific agricultural enterprises is necessary. For the purpose of more effective determination of crop rotations that is based on results of land evaluation, the possibility of using land for different types of agricultural land is revealed. Three components of the geo-planning are implemented when carrying out the planning of the territory by the main components: the formation of a natural framework for the environmental safety of the territory [2], the eco engineering and economic planning which allows the final justification for planning decisions through a combined analysis.

For the implementation of land management activities that are envisaged by the concept of basin nature management [3], the large-scale design of the territories of specific agricultural enterprises is necessary. For the purposes of more effective determination of crop rotations, it is proposed to use the results of soil surveys. The possibility of using land for different types of agricultural land (plowland, perennial plantations, hayfields, pastures, etc.) is revealed based on the results of the evaluation.

2. Material and methods

At the first stage, we have studied the features of the soil cover of the Toplinka river basin and vectorized the boundaries of soil varieties. It should be noted that it is expedient to determine the composition of soil varieties on the basis of data obtained as a result of land management, materials of large-scale soil survey of land use.

According to the Order of the Ministry of Agriculture of Russia "On approval of the State registration of indicators of the state of fertility of agricultural lands" in order to generate complete and reliable information on the status and dynamics of fertility of agricultural land by various organizations and institutions, subordinated to the Ministry of Agriculture of Russia, with a frequency of once every 5-15 years there is a collecting and generalization of data containing a full complex of soil, agrochemical, phytosanitary and ecologo-toxicological agricultural land use, the results of which are not used in the course of land evaluation works, but which can significantly increase the reliability of soil classification, which predetermines the determination of the normative crop yield of crops.

So, based on the data of a planned agrochemical survey of agricultural soils that was carried out by the Belgorod center of the agrochemical service, and using the data of the Red Book of Soils of the Belgorod Region (2007), the soils of the Toplinka river basin in the land evaluation were included are not only humus, the thickness of the humus horizon, the content of physical clay and the coefficients for negative properties of soils but also essential and useful for the growth of crops macro- and minor-nutrient elements [4]: the necessary macroelements: N, P, K , Mg, Ca, S, essential necessary trace elements: Mo, Mn, Fe, Ni, Cu, Zn, B, useful elements: Co, Al, Si. In order to compare the soils, each of the indicators was translated into relative units (points), the bonitet score on the basis of fertility factor was calculated by the formula [5]:

$$Y = A\left[1 - \left(1 - \frac{x-b}{a-b}\right)^3\right], \tag{1}$$

where Y is the (given) fertility score of the soil; X - parameter of a specific attribute of soil fertility; A - parameter of the fertility attribute of the soil-standard, b - the worst value of the same attribute of fertility; A - the volume of the scoring scale in points. The ratio x/a characterizes the saturation of the soil with a specific attribute, for example, the content of mobile phosphorus relative to its content at the potential fertility of the reference soil. The expression $(1 - \frac{x-b}{a-b})$ in the formula shows that the relative score (score) of one or another attribute of fertility is inversely proportional to the intensity of this attribute. The less the intensity of the attribute in the soil, the higher its score. At x=a, the actual parameter of the evaluation attribute of fertility is equal to the parameter of potential fertility, and the intensity of this attribute in the soil is zero and the score of the attribute (its parameter) is equal to the highest score, for example, over a closed scoring scale of 100. The exponent in the third degree indicates that the dependence of soil fertility on the parameters of its attributes is nonlinear, and the nature of the function has the form of a parabola, i.e. soil fertility is inversely proportional to the cube of tension of the limiting attribute.

Zoning of the territory was carried out on the example of one of the river basins within the boundaries of the Belgorod region. With this approach, the boundaries of the territorial zones are laid out along the natural boundaries of natural objects, and the productivity of soils is determined as the main criterion for zoning. To carry out the evaluation of soil properties and yields, we studied the features of the soil cover of the river basin on the first stage and made vectorization of the boundaries of soil varieties. The soil classification includes essential macro-and minor elements necessary for the growth of crops the content of which was determined during the planned agrochemical survey of soils by the center of the agrochemical service. To compare the soils each of the attributes was translated into relative units (points). Total score of the soil bonitet was calculated as the geometric mean value of the scores (0 to 100 points) by selected diagnostic attributes.

Five zones were identified according to the degree of suitability for use in agriculture through the imposition of results of soil classification and land management activities that are proposed by the project of basin nature management.

The environmental and economic results of the proposed zoning of the territory were assessed based on the results of the design. Taking into account existing developments for determination of the coefficients of land ecological properties [6, 7, 8] the coefficients of ecological stability and natural protection were calculated using formulas:

Coefficient of ecological stability:

$$K_{es} = \frac{\sum P_i \cdot B_i}{P}, \tag{2}$$

where B_i is a score corresponding to an area with a certain natural security [6] P is a total area of the study, P_i is an area of the site of the i-th species.

In case when the obtained value of K_{es} is <0.33, then the territory is environmentally unstable, with K_{es} = 0.34-0.50 is unstable stable, with K_{es} = 0.51-0.66 goes into a gradation of average stability, and at K_{es} > 0.67 is environmentally stable.

Coefficient of natural protection K_{np} was calculated by the formula [8]:

$$K_{np}=\frac{Psl}{Pdsl}, \tag{3}$$

where P_{sl} is an area of stabilizing lands; P_{dsl} is an area of destabilizing lands.

The higher the assessment of the natural protection of the territory (K_{np}) the greater is the stability of the landscape.

The scale of the coefficient of natural security [9]:

≤ 0.45 – Critical (instability is well pronounced);
0.46-0.55 – Stressed (unstable state);
0.56-0.59 – Satisfactory (condition is relatively stable);
0.60-0.70 – Relatively favorable (stability is sufficiently pronounced);
0.71 ≥ – Favorable (stability is sufficiently pronounced);

The coefficient of relative intensity of the organization of the territory is calculated by formula:

$$K_{ri}=\frac{Psl}{Pdsl}, \tag{4}$$

The following scale corresponds to the rating:

> 2.41 – Below optimal for more than 10%;
1.3-2.4 – Below optimal no more than 10%;
0.7-1.3 – Optimal;
0.39-0.7 – Upper optimal no more than 10%;
< 0.39 – Upper optimal for more than 10%.

3. Results and discussion

The territory of the selected river basin belongs to the forest-steppe zone with a significant spread of gray forest soils, dark gray forest soils, also available solonetzic and residual-carbonate black soils having individual, complexly directed types of vertical distribution of bonito characters, and in this connection the content and distribution of elements along the profile was calculated as the weighted average value for the proposed [10] horizons: 0-20, 21-40, 41-65, 66-100 cm.

According to the technical recommendations used in the determination of soil fertility in the cadastral evaluation of agricultural lands, the following attributes are used as the main ones: humus content in the plow layer, thickness of the humus horizon, content of physical clay in the plow layer and correction factors for the negative properties of soils. For the most reliable determination of the fertility of soils in agricultural lands, we consider it expedient to supplement the calculations with attributes characterizing the prosperity of soil by the most important for the growth of crops nutrition elements as well as by indicators characterizing certain agrophysical [11] and morphological properties of soils. Besides these indicators, in order to take into account the effect of agrogenic soil changes [12], the following indicators were included in the bonitet calculation: structural coefficient, mobile P and exchange K, ratio of humic acid carbon to the total carbon. The final score of soil bonitet is calculated as the mean geometric value of the partial estimates (from 0 to 100 points) for the selected diagnostic attributes. The weighted average bonitet score according to the proposed method for the area of the Toplinka river basin is 69 points, while according to the methodology used for cadastral valuation of agricultural land in the context of soil varieties it is 57 points. In this way the proposed methodology can facilitate more efficient and reliable performance of the fiscal function by the cadastre.

Thus, due to a more detailed set of diagnostic attributes that reflect more objectively the agrogenically conditioned physico-chemical transformations of soils, the technique suggested by the author can contribute to a more effective implementation of the fiscal function by the cadastre.

It should be noted that the specific indicator of cadastral value for the proposed methodology in some cases may be lower than the estimation obtained, basing on the use of three normatively recommended diagnostic attributes of soils. However, the integration of relatively conservative soil characteristics (thickness of the humus horizon, the content of physical clay) in the proposed methodology and the results of regularly updated (every 5 years) agricultural agrochemical survey data will contribute to a more objective and effective determination of the land tax, thereby reducing the cases of judicial contestation of its results.

Known in the world practice state mechanisms for monitoring of the use of soil fertility resources by land users should be more actively applied in Russian legislation both at the federal, regional and local levels. Russian law defines the rights and obligations of owners, users, including tenants, of land plots in the field of ensuring the fertility of agricultural land. In our opinion, introduction of regulatory and incentive measures controlling the reproduction of soils should be effective that is, besides all kinds of sanctions imposed on land users by wasteful use of land leading to land degradation, there should be established preferences and incentives (subsidies, concessional lending, reduction of the tax rate) for responsible land users, investing additional funds in ensuring soil fertility. Here, it is possible to apply the method of calculating the index of soil fertility used to determine the value of subsidies for rendering unrelated support to agricultural producers in the field of crop production in connection with the state of soil fertility.

Using the proposed method of soil evaluation, zoning of the territory was carried out on the example of the Toplinka river basin in the Belgorod Region. With this approach, the boundaries of the territorial zones are laid out along the natural boundaries of natural objects (the river basin) within the municipality. Scientific and legal support for land management activities in the organization of nature management on the basin principles is proved by the scientific studies of well-known russian scientists in this field of knowledge and adapted to the concept of basin based nature management in the Belgorod region [3].

Through the imposition of the results of soil evaluation and land management measures that were developed in the basin nature management project, five zones were identified according to the degree of suitability for use in agriculture, in analogy with earlier proposals but taking into account the natural boundaries of the design and management object - the river basin (figure 1).

The lands of the 1-3 zones can be considered as the main cultivatable fund, here for the transition to the biological path of agricultural development the following measures for rationalization of land can be offered:

- in areas with predominant slopes of 0-3° and soil bonitet of 86-100 and 61-85 points (1 zone), it is proposed to introduce field crop rotations where all types of land and any regional crop and land can be located;
- in areas with an inclination mainly of 3-5° and soil quality of at least 60 points (zone 2) it is advisable to place grain-crop rotations where the proportion of perennial grasses will be of the 50%order;
- plots on inclines of 5-7° with soil strength of 36-60 points (3 zones) - the most erosionally hazardous, here it is necessary to use grass and soil protection rotations, minimum tillage and cultivation of row crops should be limited.

Lands of the 4 zones on slopes 7-12° must be classified as transitional (low-profit) which can be used, for example, to accommodate natural forage lands or bee keeps.

The lands of the 5 zones that located on slopes above 15 ° with soil bonitet of 11-35 points and the lowest-yield (less than 10 points) are low-producing (unprofitable) territories which are expedient to withdraw into recurrent (irretrievable) conservation. They are possible to be restored with honey grass in case of proximity to bee keepers or transformation into forest plantations.

Thus in the course of the study, a methodology for soil classification was developed which expanded the list of indicators that are proposed by technical recommendations used in determination of soil fertility in the cadastral assessment of agricultural land. So, it is advisable to supplement the calculations with indicators characterizing soil availability with the most essential elements for the growth of crops as well as indicators characterizing individual agrophysical and morphological properties of soils in addition to the humus content in the plow layer, the thickness of the humus horizon, the content of physical clay and the correction factors for the negative properties of soils.

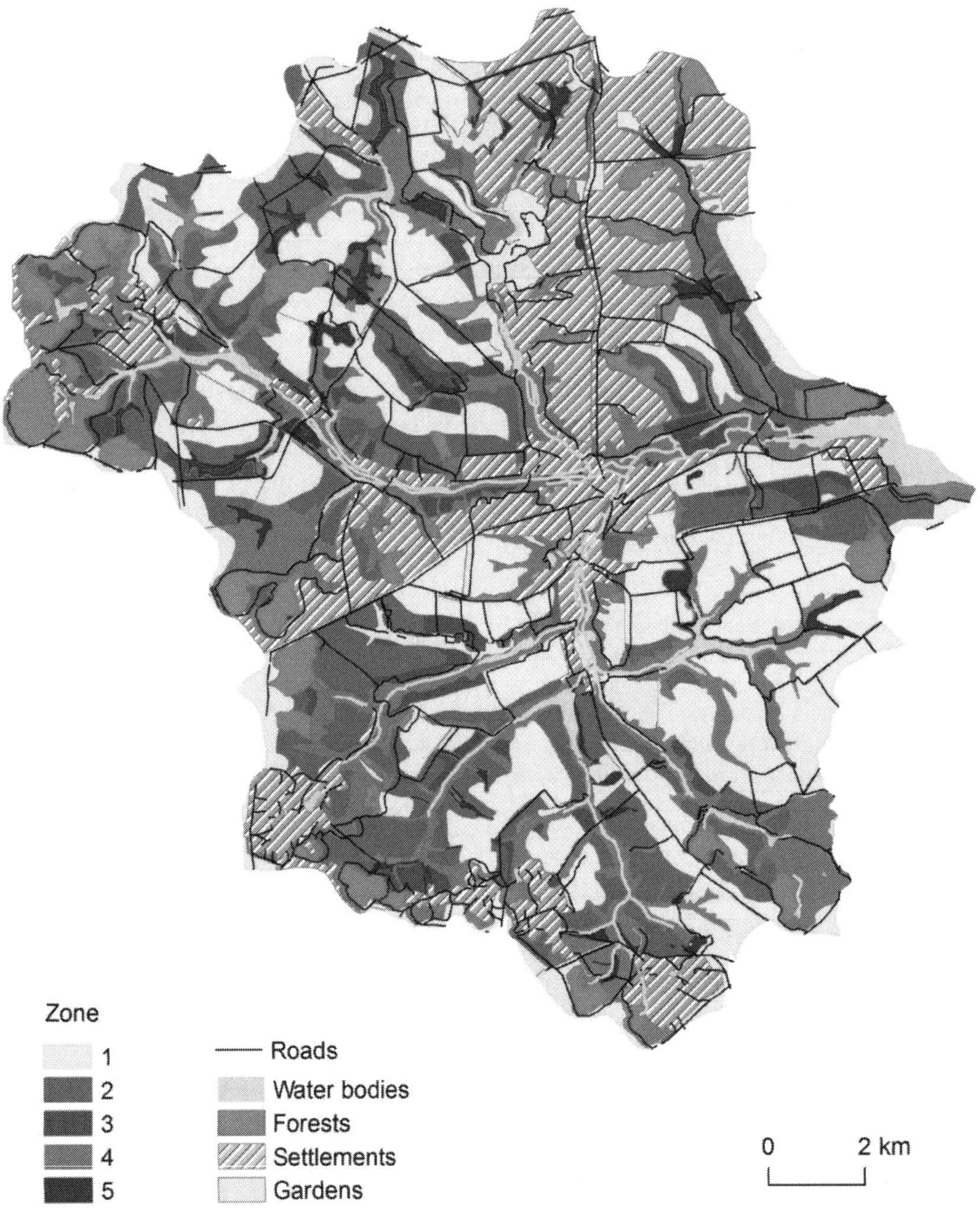

Figure 1. Scheme of territorial land zoning of the Toplinka river basin in the Belgorod Region (territorial zones 1-5 are explained in the text)

According to the requirements of the Ministry of Agriculture of Russia, with the frequency of once in every 5-15 years, there should be collected and generalized data, containing a full range of soil, agrochemical and environmental-toxicological surveys of agricultural land, the results of which are not used in land assessment but which can significantly increase reliability of land evaluation preceding the definition of normative yields of crops.

The results of the proposed method of land evaluation can serve as one of the main criteria for the allocation of territorial zones and zones with special conditions for the use of agricultural areas that are objects of land management.

Acknowledgements
The work is performed according to the Russian Science Foundation (15–17–10008).

References

[1] Yermolaev O P, Lisetskii F N, Marinina O A and Buryak Z A 2015 *Biosciences Biotechnology Research Asia* **12** 145-158
[2] Yang S F and An Y L 2016 *IOP Conference Series: Earth and Environmental Science* **39**(1) 012036
[3] Lisetskii F N, Zemlyakova A V, Terekhin E A, Naroznyaya A G, Pavlyuk Y V, Ukrainskii P A, Kirilenko Z A, Marinina O A and Samofalova O M 2014 *Advances in Environmental Biology* **8**(10) 536-539
[4] Bitjuckii N P 2011 *Mikroelementy vysshikh rasteniy (Microelements of higher plants)* (SPb: Izd-vo S.-Peterburgskogo un-ta) p 368 (in Russian)
[5] Lebedev P P 2001 *Teoriya i metody kadastrovogo kartografirovaniya s primeneniyem GIS (Theory and methods of cadastral mapping using GIS)* (Moscow: GUZ press) p 127 (in Russian)
[6] Volkov S N 2013 *Economy of agricultural and processing enterprises* **3** 23-26 (in Russian)
[7] Lopyrev M I 1999 *Design and implementation of ecological and landscape agrosystems in*
[8] *agricultural enterprises of the Voronezh region* (Voronezh: Istok Press.) p 186 (in Russian)
[9] *Agroecology* 2000 ed Chernikov V A, Chekeres A I (Moscow: Kolos Press) p 536 (in Russian)
[10] Kiryushin V I 2007 *Eurasian Soil Science* **40**(7) 785–91
[11] Marinina O A 2014 *Author's abstract* (Belgorod: Izd-vo BelGU Press) p 24 (in Russian)
[12] Vondráčková T, Kmec J, Čejka J, Bartuška L and Stopka O 2016 *IOP Conference Series: Earth and Environmental Science* **44**(2)
[13] Lisetskii F N, Stolba V F and Marinina O A 2015 *Geoderma* **239-240** 304–316

Trends of the gully erosion development in the territory of the Republic of Tatarstan

R A Medvedeva

Institute of Environmental Sciences, Kazan Federal University, 18, Kremlevskaya str., 420008, Kazan, Russia

E-mail: gregina8@mail.ru

Abstract. Gully erosion is one of the most active geomorphic processes and one of the major cause of land degradation worldwide. The aim of the study was identifying the dynamics of gully erosion development in the Republic of Tatarstan. The interpretation of satellite images were used for evaluation of the modern dynamics of gullies. Two key indicators of gully erosion (length density and gully head density) were determined. Maps of modern gully erosion for a part of the Republic of Tatarstan were constructed.

1. Introduction

Gully erosion is one of the most active geomorphological processes that determine the modern dismemberment of the relief, it causes significant soil and land degradation [1, 2]. Also, gullies affect the condition of the upper links of the hydrographic network [3]. Spatial-temporal assessment of gully erosion is an actual task due to the lack of data for the last 30 years for the study area and it is also important that during this period climate has changed [4, 5] and land-use change occurred [6, 7].

The most widely used method of gully erosion mapping is an interpretation of satellite images [8, 9, 10]. Nowadays, the availability of high-resolution and ultra-high resolution satellite images in the open access gives new possibilities for their processing and analysis in the process of the gully network visual interpretation.

The aim of the study was identifying the dynamics of gully erosion development in the Republic of Tatarstan. An assessment of the current density of the gully network was carried out within different physical-geographical regions of the Republic of Tatarstan: Prevolzhye and Prekamye (Western and Eastern).

2. Materials and methods

Prekamye occupies the northern part of the republic (21800 km^2). The valley of the Vyatka river divides it into two unequal parts: the western (Western Prekamye) – 17900 km^2 and the eastern (East Prekamye) – 3900 km^2. Prekamye is bounded from the southwest by the Volga River, from the south by the Kama River. Prevolzhye is limited from the north and east by the valley of the Volga River and occupies 14% of the area of the Republic of Tatarstan (9500 km^2). The study area is moderately dissected denudation plain lower plateau being denudation surface with heights of 140-160 m Neopleistocene age. There are similarities in the geological structure of the regions. Parent materials for local soil are loam, sandy loam, loam-like, sand, which thickness varies from the first meters to 10-12 m [11]. There are no perceptible differences in climatic conditions.

SAS.Planet program allows to select the satellite images with the best quality for a required area, digitize all gully forms and create a vectorized layer of the gully network. This base layer of gullies can be processed in any GIS.

The basic method of investigation was the visual identification of gully forms. To obtain data on modern gully erosion of Republic Tatarstan territory satellite images of 2010-2016 were used. A system of visual interpretation signs in the determination of active gully forms on satellite images was formed. Gullies of different types were distinguished: a) slope gullies; b) bottom gullies; c) coastal gullies.

All study area was divided into basins and sub-basin area displayed on the map 1:50,000 and 1: 100,000 scale. For each sub-basin the total number of gullies and the number of gullies of different types, the total length of the gully network, gully length density and gully head density were evaluated.

3. Results

Prevolzhye and Prekamye include 799 basins which areas vary from 5 km^2 to 353 km^2. Table 1 and figure 1 show the number of basins with maximum indices of gully length density and gully head density.

Table 1. Basins of Prevolzhye and Prekamye with maximum indices of gully length density and gully head density.

Physico-geographical region of the Republic of Tatarstan	Total number of basins	The number of basins with the maximum indices of gully length density (more than 0.1 km/km^2)		The number of basins with the maximum gully head density (more than 1 number of head/km^2)	
		Total	%	Total	%
Prevolzhye	285	14	5	19	7
Prekamye	514	44	9	49	10

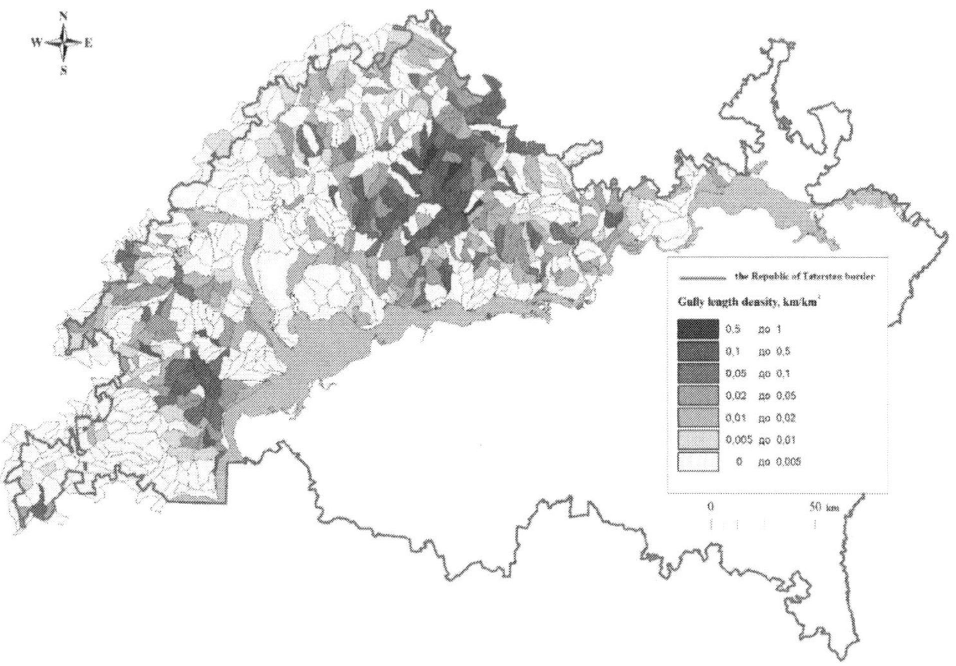

Figure 1. Gully length density in the territory of Prevolzhye and Prekamye of the Republic of Tatarstan

The mean value of the gully length density is 0.03 km/km^2, for individual basins it ranges from 0 to a maximum of 0.73 km/km^2. The maximum values are confined to the central parts of both regions. The minimum values of the gully length density are typical for most of the basins located in the southern part of Prevolzhye as well as in the basins adjacent to the Volga River upstream from the mouth of the Kama River.

In Prekamye the minimum gully length density values are mostly observed in the basins directly adjacent to the Volga and the Kama rivers and partly to the Vyatka River. The basins located in the central and north-eastern parts of Prekamye have a gully length density in the range 0.05-1 km /km^2.

The change in the gully head density is more uniform than the gully length density (on the average 0.3 number of head/km^2) (figure 2). The greatest values of the gully head density are in the basins with high gully length density.

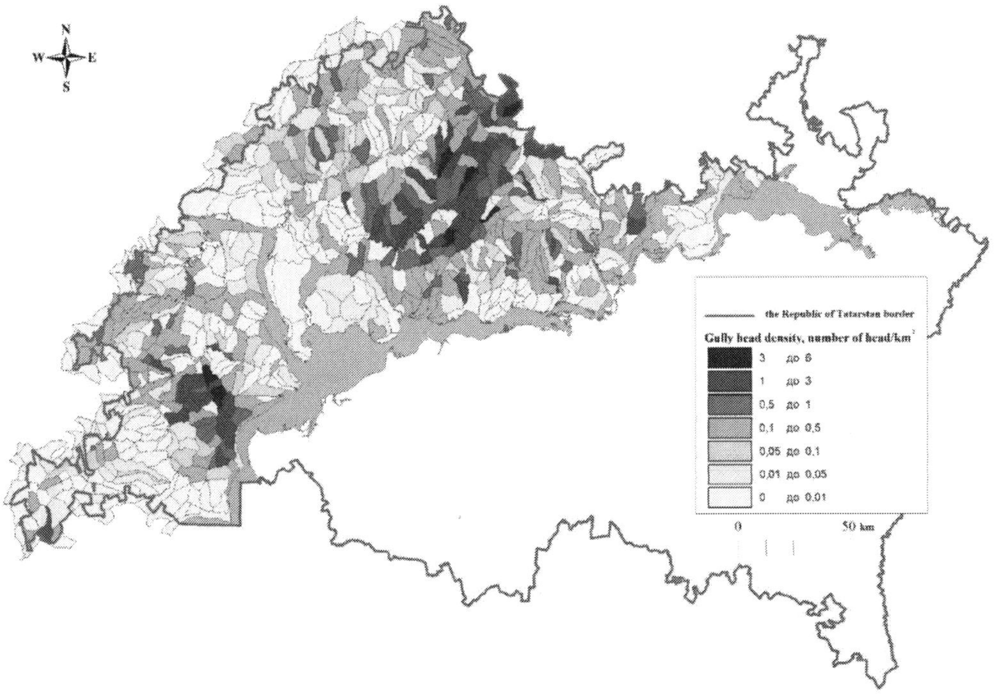

Figure 2. Gully head density on the territory of the of Prevolzhye and Prekamye of the Republic of Tatarstan

Gully head density on the territory of Prevolzhye and Prekamye increases from the periphery of regions to their central part as well as the gully length density considered above. The exception is the southern basin within Prevolzhye (from 1 to 3 number of head per km^2) and the north-eastern basin in the Prekamye (from 3 to 6 number of head per km^2).

The predominant type of gullies is slope gullies. In Prekamye and in Prevolzhye the same ratio of different types of gullies: slope gullies constitute 76% of the total, coastal gullies - 17%, bottom gullies – 7%.

If the study territory is considered by administrative units the highest indicators of gully density of the territory are observed in the following regions of the Republic of Tatarstan: Sabinsky, Kukmorsky, Tyulyachinsky, Pestrichinsky, Apastovsky and Tetyushsky.

4. Discussion

To study the dynamics of gully erosion within the Republic of Tatarstan, the results of the interpretation of satellite images were compared with the results obtained in 1985-1988 (the gullies were interpreted from aerial photographs of the 1960-1970s) (figure 3).

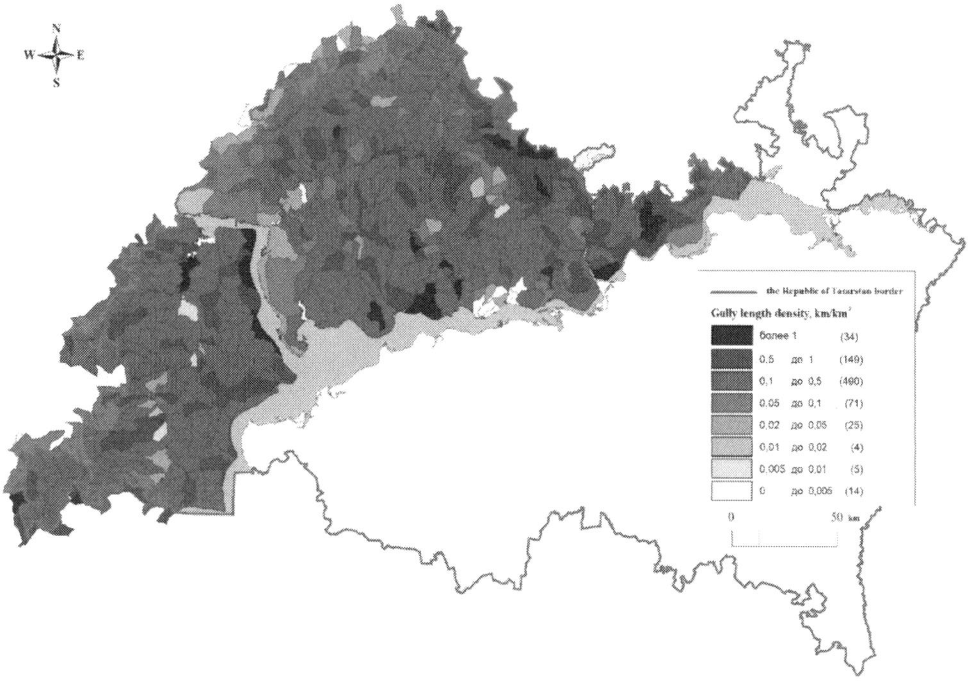

Figure 3. Gully length density according to the results of interpretation of aerial photographs of 1960-1970 [12, 13].

The results of the comparison of gully length density for two periods of time show its reduction. On average, for all basins the values of the gully length density were reduced by 0.35 km/km^2.

The distribution of the gully length density along the basins of both Prevolzhye and Prekamye in the 70s of the 20th century differs significantly from the current state.

In Prevolzhye the maximum values were in the basins that adjoin the Volga River upstream from the mouth of the Kama River. There was a very small number of basins with minimum values of the gully length density (from 0.005 to 0.05 km/km^2), most of them are located in the central part of Prevolzhye.

5. Conclusions

In the process of comparing aerial photographs of the 1960-1970 with satellite images of 2010-2106 allows to conclude that:

a) the quality of satellite imagery and aerial photographs allows to identify gully forms equally well in all stages of their development;

b) the most part of the active gullies in 1960-70 on aerial photographs with clear geometric outlines trasformed on the satellite images into valleys with completely overgrown slopes indicating their transition to the stabilization stage.

The process of reducing the area of arable land in connection with their abandonment and further overgrowing began in the mid-1980s. In the period 1985-2015, the reduction of arable lands in the Mesha River basin is 10%, in the basin of the Ulema River is 6.5%, located within the Western Prekamye and Prevolzhye of the Republic of Tatarstan, respectively [7].

The history of agricultural development is one of the leading reasons for the declining gully formation in the study area at present. Active plowing of land, a decrease in the area of meadows and deforestation in the XIX century led to an increase in the rate of erosion processes. The soil conservation measures conducted in the 1970s, the reduction of arable land and the achievement of a peak of development by many gullies promoted to declining trend of gulling. An additional factor in reducing the rate of gully formation was a sharp decrease in surface runoff from arable slopes during the spring snowmelt which occurred as a result of climate warming. The results of field monitoring of gullies growth in the Vyatka-Kama interfluve since 1978 to the present time confirm this conclusion [14, 15].

It is possible to conclude that most part of the gullies in the study area is currently inactive and massively passes into the balka stage.

Acknowledgments
The work was funded by the Russian Science Foundation (project №15-17-20006).

References

[1] Poesen J, Nachtergaele J, Verstraeten G, Valentin C 2003 *Catena* **50** 91–133
[2] Valentin C, Poesen J and Li Y 2005 *Catena* **63** 132–153
[3] Lyubimov B P, Nikolskaya I I and Prokhorova S D 2000 *Erosion of Soils and Channel Processes* **12** 297 (In Russian)
[4] Perevedencev J P, Shantalinskij K M and Vazhnova N A 2014 *Meteorology and Hydrology* **10** 19-31 (In Russian)
[5] Perevedentsev Yu P and Shantalinsky K M 2015 *Uchenye zapiski Kazanskogo Universiteta. Seriya Estestvennye Nauki.* **157** 8–19 (In Russian)
[6] Lury D I, Goryachkin S V and Karavaeva N A 2010 *Dynamics of agricultural lands of Russia in the 20th century and postagenic restoration of vegetation and soils* p 416 (In Russian)
[7] Ivanov M A 2016 *Theory and Methods of Modern Geomorphology* pp 362–369 (In Russian)
[8] Bocco G and Valenzuela C R 1993 *Remote Sensing Reviews.* **7** pp 233–240
[9] Zinck J A, Lopez J, Metternicht GI, Shrestha DP, Vazquez-Selem L 2001 *International Journal of Applied Earth Observation and Geoinformation* **3**(1) 43–53
[10] Desprats J F et al 2013 *Land Degradation & Development* **22** 24–32
[11] Butakov G P, Yermolaev O P and Muharamova S S 1993 *The Green Book of the Republic of Tatarstan* pp 221–231 (In Russian)
[12] Yermolaev O P, Rysin I I and Golosov V N 2017 *Geomorphology* **2** 38–51 (In Russian)
[13] Yermolaev O P 2017 *Eurasian Soil Science* **50** 118–131
[14] Rysin I I, Grigoriev I I, Zaitseva M Yu, Golosov V N, Sharifullin A G 2017 *Proc. IAHS* **375** 1–4
[15] Rysin I I, Golosov V N, Grigoriev I I, Zaitseva M Yu 2017 *Geomorphologiya* **1** 90–103 (in Russian)

Modern Approaches to Mathematical Modeling of River Runoff in the Territory of the European Part of Russia

S S Mukharamova, O P Yermolaev and E A Vedeneeva

Institute of Environmental Sciences, Kazan Federal University, 18, Kremlevskaya str., 420008, Kazan, Russia

E-mail: vedeneeva-evgeniya@mail.ru

Abstract. This article deals with the construction of mathematical models, which describe the dependence of the river runoff modulus in the territory of the European part of Russia on landscape-geographical conditions and anthropogenic load using modern regression-type statistical models. The obtained models reflect about 80% of the variability of the data and provide acceptable forecast accuracy. As well as these models reflect the main dependencies of river runoff on the conditions of its formation in the research scale. Spatial extrapolation of the river runoff values to the hydrologically unexplored areas of the European part of Russia is performed.

1. Introduction

The formation of river runoff is a complex multifactorial process. The study of dependencies between hydrological characteristics and different zonal and azonal factors is often too general, that under conditions of hydrometeorological monitoring network reduction can lead to serious errors in predicting runoff for unexplored river basins. The strengthening of anthropogenic load on basin geosystems and climate change necessitates the development of new approaches to establishing quantitative links between the characteristics of the river runoff and the totality of its controlling factors. One of such approaches can be mathematical modeling of river runoff.

General ideas about the mechanism of formation of river runoff were described in 1940-1950s in the works of such scientists as Alekseev G.A., Befani A.N., Velikanov M.A., Kalinin G.P., Kohler M.A., Linlesley R.K., Horton R.E. As a trend, mathematical modeling took shape in the 1970s, which was associated with the use of computers for processing large amounts of information [1, 2]. This period is characterized by the appearance of closed models that describe the entire hydrological cycle from the arrival of water to the catchment area to the formation of runoff in the main-stream station [3]. An important role in the development of the modeling of hydrological processes on the river basins in the Russian hydrology was played by the Kuchment's monograph, published in 1972. This work presented the foreign and domestic modeling experience accumulated by that time. At present, scientific groups under the leadership of Kuchment L.S., Burakov D.A., Harzman B.I., Gelfan A.N., Gusev E.M., Kondratiev S.A. and others have significant achievements in runoff modeling [4, 2].

Three main classes of models are distinguished depending on the nature of the description of the runoff formation processes and the extent of using theoretical and empirical information: "black box" type models, conceptual and physical-mathematical models [5]. This classification allows us to identify the most important principles that are the basis of the model. However, it is not unique. So, for example, three types of models are distinguished by the method of specifying a mathematical structure: non-linear multifactorial models ("neural networks"), multiple linear regression models and physico-mathematical models [6].

Physico-mathematical models use physical laws for quantitative description of the processes of runoff formation taking into account their spatial heterogeneity [5]. The universal deterministic modeling system "Stock-Erosion-Pollution" and the distributed deterministic hydrological model of the flow formation processes "Hydrograph" can be cited as examples of physico-mathematical models. Both models were developed under the leadership of Vinogradov Yu.B. [1, 7]. Other well-known physico-mathematical models (MIKE, SHE, TOPOG, KINEROS2) are oriented to rivers with a mild climate and, basically, describe the processes of rainwater flow formation [8]. The main direction of practical application of physico-mathematical models is connected with modeling situations in unobserved or in changed conditions, in particular, the pre-calculation of the risk and magnitude of possible hazardous floods, the assessment of the impact of human activities on river runoff [9].

The conceptual models are based on various concepts describing the physical processes of river runoff formation with preliminary spatial averaging of the input effects on the catchment area [10]. This approach to modeling involves the solution of systems of equations, which are based on various concepts describing the physical processes of river runoff formation. Examples of conceptual models are the model of river basin pre-flood cycle (Harzman), the model of spring flood hydrograph generation for plain rivers (Vinogradov), the SWMM model (Huber, Dickens) [1].

Along with models that take into account the spatial heterogeneity of input data and require a large amount of information about the properties of the catchment area, there is a number of models based on empirical relationships between the river runoff and the factors of its formation (the "black box" type model). Empirical relationships are identified on the basis of statistical methods, the theory of random processes or using artificial neural networks [11]. Despite the fact that such models practically do not use a priori information about the processes occurring in the river basin, they are successfully used in forecasting practice. In particular, artificial neural networks are well suited for predicting the structure and trends of data. Using the example of the Bolhovets River (Belgorod, Russia), neural network technologies were successfully applied to predict the dynamics of the water regime of small rivers depending on climatic factors [12].

This class of models includes the models of multiple linear regression, the mathematical structure of which is determined by the choice of the approximating function. The main goal of multiple regression is to construct a model with several factors and determine the impact of each factor separately, as well as their combined effect on the studying indicator under investigation. The advantages of this approach are clear interpretation and simplicity of the definition of parameters; limitations are the problem of choosing the essential factors and their independence, i.e. the absence of a linear connection between them. An example is the use of a linear regression model for the formation of the annual runoff of rivers in the Volga basin in a variety of meteorological factors [13].

The purpose of this research is to study the statistical dependencies of river runoff characteristics on landscape-geographical conditions and anthropogenic load using modern regression-type statistical models. The territory of the European part of Russia, where the main part of the population, industrial and agricultural potential of the country is concentrated, was selected as the study territory. The modulated element is the river runoff modulus. The use of river runoff modulus as a characteristic of the water runoff from a catchment area unit will allow, in the presence of a model describing the dependence of this indicator on landscape-geographical and anthropogenic conditions, extrapolate its value to the hydrologically unexplored areas of the territory of the European part of Russia.

2. Material and methods

Two multidimensional samples for the statistical study of the regularities of the river runoff formation were formed. The catchments referred to hydrological posts, which provided with river runoff data, as elements of the first sample, were made. The second sample is small river basins and their intervening areas (over 60.000), planar covering the entire territory of the European part of Russia. Both samples include a list of quantitative and categorical variables: the catchment area, 16 climate indicators, morphometric relief characteristics in the basin, interest forest coverage, ploughness, grassing, wetlands in the basin, the characteristics of the soil cover (the predominant type of soil and type of parent rock), and the predominant class of pre-Quaternary sediments . These variables {X} are regarded as explaining, i.e. describing the conditions of formation of river runoff. The first sample also includes a dependent variable Y - the river runoff modulus of catchments allocated relatively hydrological posts.

The basins of small rivers and their intervening areas of the European part of Russia were obtained by constructing in automatic mode with GIS technologies using the corrected model of the relief GMTED2010 and the raster model of the hydrographic network [14]. The boundaries of the basins, related to the hydrological posts, were constructed on the basis of the obtained relief model and a point layer representing hydrological posts.

Published materials of long-term regime observations at hydrological posts (Surface Water Resources of the USSR), data from open sources (http://caspi.ru/), as well as data from the «All Russian Research Institute of Hydrometeorological Information» were used as sources of information on river runoff. The source of the data on climate factors was the data of the Federal Service for Hydrometeorology and Environmental Monitoring weather stations (open data of the RIHMI-WDC). Source of data on soil cover characteristics: The Unified State Register of Soil Resources of Russia, presented as a vector layer, the objects of which are digitized contours of a soil map of the RSFSR with scale 1: 2,500,000. The model of the relief GMTED2010 with a spatial resolution of 250 m was used as a source of information on the relief of the study area. The map of land cover types, developed at the Space Research Institute for the Earth (Moscow-2015), served as a source of information on the types of earth cover and the degree of anthropogenic load. The geological conditions were recorded using the State geological map of pre-Quaternary deposits of scale 1: 1,000,000, represented as the sheets of maps of three generations in vector form and the form of scanned paper maps.

The construction of models was preceeded by an analysis of the structure of explanatory variables, the study of their mutual correlation with the subsequent choice of an independent subset {X}.

The modeling of the dependence of the river runoff modulus on the conditions of its formation was carried out from the data of the first sample using methods such as the generalized linear model (GLM) and the generalized additive model (GAM), separately for subsamples of basins located in plain and mountain areas. Calculations were carried out using programs written in the statistical environment R. The best subset of regressors in constructing models was selected by analyzing the statistical significance of the contribution of each regressor to the model and analyzing the VIF factor. In addition, the value of the information criterion AIC, which allows comparing models for a given dependent variable, was monitored. Standardized values of independent variables, which were used in the construction of models, allow considering the absolute values of the model coefficients as an estimate of the contributions of the relevant factors to the value of Y.

3. Results and discussion

Two models reflecting the main regularities of river runoff for mountainous and plain areas were obtained as a result of the calculations. The criterion for assessing the quality of the obtained models was the adjusted coefficient of determination. It should be noted that the complication of modeling methods in comparison with linear methods does not significantly improve the quality of the model.

Consider the model of the river runoff modulus constructed by linear methods for plain areas.

$$Log\,(Y) = -5.15 - 0.51 \cdot T + 0.21 \cdot log(R) + 0.08 \cdot log(S) + 0.10 \cdot F \qquad (1)$$

The presented model explains more than 80% of the variability of the data and well reflects the main regularities of river runoff in a given research scale.

The sum of the active temperatures (T), the annual precipitation (R), the average steepness of the slopes (S), and the percentage of forest coverage in the catchment area (F) are the most significant predictors of the model.

The obtained model is well interpreted in terms of the water balance equation, which is confirmed by the positive contribution of the sum of atmospheric precipitation and the negative contribution of the sum of the active temperatures. The positive effect of the average steepness of the slopes is due to its reverse effect on the amount of evaporation, in other words, with increasing steepness of the slopes the speed of water movement increases and evaporation losses decrease. The positive contribution of the percentage of forest coverage in the catchments is manifested in the following. The soil within forest areas is characterized by increased filtration capacity due to the presence of loose forest litter and less depth of soil freezing. This contributes to the replenishment of groundwater resources, which are drained by rivers during the low streamflow period, thereby increasing the annual runoff modulus.

The model built for mountain areas explains 72% of the variability of the data. The height of the catchment area (H) and the percentage of ploughness in the catchment area (A) were added to the model as the main predictors.

$$Log\,(Y) = -5.01 - 0.19 \cdot T + 0.29 \cdot log(R) + 0.72 \cdot log(S) - 0.20 \cdot A - 0.43 \cdot H \quad (2)$$

The negative impact of the ploughness percentage in catchment rate on the river runoff modulus is associated with the depletion of ground waters, which are drained by rivers during low streamflow period. This, in turn, is related to an increase in the agricultural development of catchments and, as a consequence, deterioration in the filtration properties of soils. The negative contribution of the average height of the catchments is due to the fact that the thickness of the aeration zone and the depth of occurrence of groundwater increase with increasing height. Deterioration of drainage conditions leads to a reduction in runoff.

Figure 1 shows diagrams illustrating the qualities of the constructed models - the observed values of the river runoff modulus are displayed against the model values.

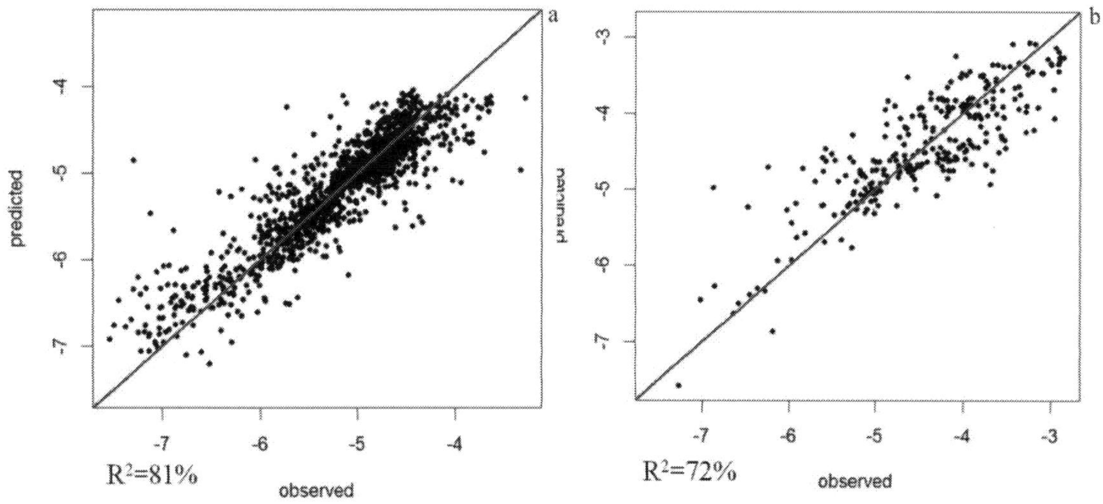

Figure 1. Diagrams illustrating the quality of the constructed models for the river runoff modulus: a). for plain territories, b). for mountain territories

The constructed models provide an acceptable forecast accuracy, which allows using the data of the second sample to calculate the predicted (model) values of the river runoff for river basins and intervening areas planar covering the study area. In this way, spatial extrapolation of the river runoff values (river runoff modulus and the annual runoff layer) to the hydrologically unexplored river basins of the European part of Russia was conducted. The adequacy of the obtained spatial models was verified by comparing them with previously constructed small-scale maps (National Atlas of Russia, etc.).

4. Acknowledgements

The work is performed according to the Russian Science Foundation (15–17–10008).

References

[1] Antokhina E 2012 Water regime of the rivers of the European territory of Russia and its study on the basis of the flow formation model(Moscow: Moscow State University Press) p 219
[2] Kuchment L Mathematical modeling of river flow (Gidrometeoizdat Press) p 191
[3] Koren' V Mathematical models in river flow forecasts(Gidrometeoizdat Press) p 200
[4] Gorshkova A, Urbanova O and Karimova A 2015 International Scientific and Research Journal 8 (39) 66-72
[5] Motovilov Yu 2013 Int. Conf. "The First Vinogradov Readings. The Future of Hydrology " (St. Petersburg) pp 25-27
[6] Ignatov A 2011 Geography and natural resources 4 114–118
[7] Vinogradov Yu and Vinogradova T 2010 Mathematical modeling in hydrology (Moscow: Publishing Center "Academy" Press) p 304
[8] Guide to hydrological practice 2012 (Geneva: World Meteorological Organization) p 324
[9] Kuchment L, Gelfan A, Kondratiev S and Lavrov S 2013 VII All-Russian Congress
[10] Geyman T 2015 South-Siberian Scientific Herald 4(12) 47-49
[11] Belyakova P 2015 Flood discharge of the Russian rivers of the Black Sea coast of the Caucasus(Moscow: Moscow State University Press) p 200
[12] Kuzmenko Ya, Lisetskiy F and Pichura V 2012 Modern problems of science and education 6
[13] Frolova N 1984 Geographical regularities of the spatio-temporal variability of the annual runoff of the Volga River basin (Moscow: Moscow State University Press) p 201
[14] Ermolaev O, Maltsev K, Muharamova S, Kharchenko S and Vedeneeva E 2017 Geography and natural resources 2 27-36

Constructing development and integrated coastal zone management in the conditions of the landslide slopes of Cheboksary water reservoir (Volga River)

I V Nikonorova

Chuvash State University named after I.N. Ulyanov, 428015, 15, Moskovsky Avenue, Cheboksary, Russia

E-mail: niko-inna@yandex.ru

Abstract. Uncontrolled construction and insufficient accounting of engineering-geological and hydro-geological conditions of the coastal zone, intensified technogenic impact on sloping surfaces and active urbanization led to the emergence of serious problems and emergency situations on the coasts of many Volga reservoirs, including the Cheboksary reservoir, within Cheboksary urban district and adjacent territories of Chuvashia. This article is devoted to substantiation of the possibility of rational construction development of landslide slopes of the Cheboksary water reservoir.

1. Introduction

The Volga River has a length of 127 km within the Chuvash Republic. The lower part of the Cheboksary reservoir and the upper section of the Kuibyshev reservoir are located here. Cheboksary hydroelectric power station (HPP) was built last in the cascade of the Volga hydrosystems. The hydraulic power plant is located in the zone where the water supply of the Kuibyshev reservoir is running low. Start of the first hydroelectric complex unit occurred in 1980. The first unit of the hydrosystem was commissioned in 1980. In the flood of 1981 the reservoir was filled to the level of 63 m. The reservoir has been operating at this level for 36 years. Forecasts for changes in the natural environment, including shores in the design and construction of the Cheboksary HPP were performed in terms of a water level mark of 68 m. As a result, all the forecast values for re-forming the coasts, siltation of the bottom and water pollution proved to be untenable. At the same time, the increased anthropogenic press on the slopes within the Cheboksary agglomeration leads to a decrease in the stability of the coasts which can cause dangerous abrasion and landslide processes. In addition to the intensive construction of residential buildings, industrial, recreational, agricultural and other facilities, excavation work on the right bank of the reservoir is actively being carried out(laying of electrical and telephone cables, repairing of heating mains and conduits, etc.). As a result, the groundwater regime is violated, which can also contribute to landslide. Already in 1973, during the construction of the pit of the Cheboksary hydropower plant, the landslide covered the southern shore of the pit and part of the indigenous slope of the Volga. The volume of the landslide mass of the soil was about 200 thousand m3. The water drainage system, power line and road were temporarily disrupted. 65 new landslides were counted at the beginning of operation of the Cheboksary hydroelectric power station in 1981 from the village of VasilSursk to the dam of the Cheboksary HPP.JSC "RusHydro", in charge of the operation of the Cheboksary water reservoir, has created a water protection zone for the preservation of coastal landscapes (at a distance of 1-3 km from the water edge). However, in violation, during the years of operation of the hydroelectric power station, dachas, garages and recreation centers were built here.

2. Material and methods

Landslide process is one of the most dangerous natural processes and many works in foreign and domestic scientific literature have been devoted to its study. They define the landslide process, development factors, classification and consider the mechanism of the process [1, 2, 3, 4, 5]. In conditions of shortage of land resources, people actively develop and build up lands with landslide slopes. In connection with this, applied engineering and geological scientific developments on risk reduction in landslide conditions become very relevant [6, 7, 8, 9]. Among the factors of development of landslide processes is the factor of creating reservoirs on the banks of rivers. So, the coastal zone becomes a zone of increased risk in economic development which requires the creation of an integrated management system. In the world practice there is such a management system for oceanic coasts [10, 11]. Obviously, the need has come for such documents for the coasts of reservoirs.

The main factors of landslide formation on the slopes of the Cheboksary water reservoir, in our opinion, are: 1) structural features of rocks of the sedimentary cover of the Russian plate (subhorizontal occurrence, their fracturing); 2) lithological structure of rocks (terrigenous, predominantly clayey, poorly argillized rocks of the Upper Permian Paleozoic, covert deluvial formations of river valleys, buried beams and ravines that extend to the reservoir; 3) the geomorphological features of the northern part of the Volga Upland (the mark of the Miocene alignment surface is about 210-215 m, and the main basis of erosion of Chuvashia is the Volga River with marked elevations of 53 m at the Kuibyshev Reservoir to 63 m in Cheboksary), erosion and landslide types of slopes with their limiting morphometric parameters (height and steepness); 4) climatic and hydrological conditions of the temperate zone, differently manifested on the slopes of different genetic types and exposures; 5) hydrogeological conditions of the slopes to the depth of the erosion cut, which are the consequence of their geological structure, drainage conditions, relief and climate. Hydrogeological and hydrological factors affect the stress field in the massif, the shear characteristics of soils, control the role of hydrodynamic and hydrostatic pressures on the stability of slopes. Our observations confirm the ever increasing role of human economic activity in landslide formation.

The combination of slope and wave geodynamic processes and lithology became the leading factors in the formation of landslide-dangerous engineering and geological conditions on the slopes of the Cheboksary water reservoir. Lithology is represented by interstratification of clayey strata with sandy-marly complexes of the Permian Paleozoic system. The abrasion of the Volga coasts is complicated by slope processes, and abrasion-landslide, abrasion-landslide, abrasion-scree, abrasion-accumulative types of reservoir banks are formed. The most widespread are landslides in places where the banks are composed of rocks of the Tatarian stage of Perm. As a result, landslide blocks successively displacing, capture significant parts of the slopes and lead to the formation of a stepped relief. In the Upper Permian rocks with relative heights of 30-50 m and steepness of the slopes 15-30°, types of landslides-block are formed. According to the degree of manifestation of landslide processes they belong to the II and III categories of danger. The dimensions of the displaced landslide blocks reach sizes 50-70 at 40-60 m. The heights of the walls of separation are up to 20-30 m. The length of the wall of detachment in front landslides reaches 130-170 m. Landslides form well-defined semicircles, frontal terraces and have such characteristic: separation walls, tongue parts and shafts of bulging. With the slope steepness of 20° mudflows-flows began to form due to the creation of the reservoir. In general, on the Volga right bank are landslides classified according to the displacement mechanism by types: extrusion, slip, flow and special. Based on the methodology of S.I. Bolisov and others, with our additions [12], the coasts of the Cheboksary reservoir were typified and their present state was determined with subsequent evaluation of the geological and geomorphological safety of the studied territory. The definition of complex geomorphological safety of the coastal zone of the Cheboksary reservoir included an analysis of engineering-geological conditions, intensity of terrain fragmentation, landscape differentiation, presence of dangerous exogenous processes, etc. The calculation of the coefficients of geological and geomorphological safety of the territory showed that the abrasion-landslide type of the coast is characterized by indicators from 15.2 to 17.3; Erosion type of coast - 17.4; Protected - 17.2-18.8; Abrasion-scree-shaped type - 15,8-17,3; Abrasion-accumulative and Accumulative type - 17,9-19,8. This gives grounds to refer them to the groups of banks with low and medium stability.

Depending on the engineering and geological conditions on the coast of the Cheboksary and Kuibyshev reservoirs, the following areas suitable for construction and economic use can be identified: 1) Sustainable and suitable for widespread development, represented by water dividing areas on the right bank of the Volga, accumulative surfaces of the left bank and flat river over-flooded terraces and deluvial slopes in the indigenous and Quaternary sediments. They can be suitable for agriculture. Possible problems here are: the issues of water supply, steep slopes and the level of groundwater. 2) Conditionally stable, economic development of which is difficult because of their relief. These are coastal slopes that have reached the equilibrium state without landslides and scree, slopes of valleys of large beams and tributaries. Possible uses are as gardens and meadows. Construction will require anti-landslide and anti-erosion measures here. 3) Unsuitable areas affected by landslide deformations.

As technogenic protection of the pre-river zone of the Cheboksary reservoir can be recommended: 1) Usage of polymeric materials to strengthen the beach surface from sand and pebbles. A negative factor is the weak resistance against wave action. In addition, the polymer coating will worsen the ecological situation of the coastal zone since in the river part of the beach this crust will be covered with microalgae mucus as well as boulder-stone blind in water. 2) Stone outline. Stone-block or tetra-pits are used; however this will completely remove the shore from the recreational nature management. This method was implemented directly at the structures of the Cheboksary hydroelectric complex. 3) Construction of interrupted breakwaters. On the relatively shallow left bank of the Cheboksary reservoir, this method of protection can give good results but on the deep right coasts, which are just prone to landslide deformations, this method is of little effect. And the construction of intermittent breakwaters on a steep landslide underwater slope is problematical. 4) Protection of the coast by groynes. With the deep underwater slope of the landslide type, the application of this option is also problematic. 5) Concrete counter-bank. It is the most effective method for socially significant territories implemented by Cheboksary, Novocheboksarsk, Sosnovka sites, etc.

In each case, the criteria for selecting the most effective technogenic protection of the Cheboksary water reservoir banks are: 1) the lower cost of the structure; 2) less material consumption; 3) higher degree of quenching of wave energy; 4) the best ecological properties.

As for the complex of anti-landslide measures (ALM), the following types of preventive measures have been proposed by builders and designers for many years: 1) Mechanical retention of landslide masses. 2) Cutting, unloading and terracing of the slopes. 3) Regulation and filling of beams, ravines. 4) Installation of networks of upland drainage. 5) Drainage of groundwater. 6) Agroforestry. All of them are implemented in one or another way in the practice of construction in Chuvashia.

Land-use planning to reduce losses from landslides includes active and passive methods, used both individually and in combination: 1) Active methods: transfer or modification of existing buildings; protection of existing buildings. 2) Passive methods: restriction of new construction; building regulation.

In the practice of prospecting for the construction in the areas of dangerous slope processes the relevance of the engineering-geological approach is beyond doubt. The essence of the engineering-geological approach is reduced to the following stages [13]: 1) Geological and geomorphological survey and compilation of engineering geological maps and sections based on survey data. 2) Compilation of correct calculated structural and functional models from the materials of this survey and sections both in the present state of the slope and, if necessary, in the past. 3) Selection of the optimal design formula and optimal design parameters for the stability of landslide slopes. 4) The calculation of the stability of landslide systems, at first based on laboratory data of the shear strength of soils. And then, if necessary, performing reverse calculations to clarify the shear strength of the soil of the bedrock (peak strength) and landslide accumulations after displacement (residual strength). 5) According to the data obtained, the stability forecast for the edge of a plateau not covered by a landslide. As a result, in the calculation schemes, the role and significance of each element and component will be taken into account in the functioning of the entire landslide geomorphosystem, in mechanisms and intensity of the manifestation of landslide deformations. Correct structural-functional models of landslide systems will reflect the mechanisms and structure of real landslides of the objects under study, which are more objective compared to simplified test models such as the circular cylindrical model of plane displacement, etc. These real models are compiled according to certain principles taking into account the genetic types of slopes, the types of landslides by the displacement mechanism, modern concepts of the structure and component-elemental composition of landslide systems. That is, the hierarchy of the landslide geomorphic system will be taken into account: types of landslide blocks, types of landslide tiers and landslide floors [13].

Monitoring of dangerous exogenous processes on the coast of the Cheboksary reservoir has been carried out since 1980 by various design and survey organizations ("RusHydro", "Engineering Shore Protection", "ChuvashTISIZ" and JSC "Institute Chuvashgiprovodkhoz"), as well as by scientists from the Chuvash State University. Observable sites are assigned to I and II significance categories. Geodetic measurements are carried out on profiles with frames. To protect valuable economic facilities on dangerous shores, concrete slopes and surge suppressors, artificial beaches, drainage systems and other engineering protection facilities have been created in the city Cheboksary, Novocheboksarsk, in the Yadrinskaya and Sosnovsky agricultural lowlands.

However, there are numerous new data on landslide phenomena of recent years that arose both under the influence of the Cheboksary hydropower plant regime and intensive urban construction (figure 1).

Figure 1. Construction of a private house on the Volzhskaya (former Rybatskaya) street in Cheboksary was stopped due to landslide deformations (2016).

One of the examples of construction on landslide slopes is the rapidly commissioned microdistrict VI "Center" in Cheboksary on the left bank of the river Cheboksarka (authors of the project for the construction of the microdistrict VI in the central part of the Cheboksary city – Ltd "Architectural Bureau "Classic", architect N.A. Rozhkova, designer M.A. Kitaev).

The conducted field studies made it possible to distinguish here three engineering geological (IG) areas: 1) Arrays of type A – leveling surfaces or plateaus (the upper tier of the relief), characterized by the greatest stability.2) Arrays of type B – slopes of various genetic types (middle tier of relief) characterized by medium and high degree of danger. 3) Arrays of type B – floodplain and bottom of beams (the lower tier of relief) characterized by an average and low degree of danger. The boundaries between the arrays A and

B are determined by the walls of the landslide system failures, the edges of the erosion slopes, and between the B and B - the rear seam of the floodplain of the river Cheboksarka [14].

Arrays of type A are almost entirely built up, genetically and structurally rather monotonous and are an object of engineering protection for the building of slopes. Between these tiers of relief are high-altitude slips landslide systems, subject to local sliding deformations, even with their temporary trimming for construction purposes. The building properties of ancient and long-slope slopes (type B) are determined by the structure and mechanism of these landslides, i.e. on which functional components of landslide geomorphic systems the designed buildings are located, and how the pattern of distribution of loads on the slope and the regime of underground and surface waters in connection with vertical planning and other design solutions will change. The structure of the entire slope polygon is complicated by the fact that the elements of landslides oriented almost along the strike of the Cheboksary valley are overlaid with an age-old buried and modern ravine-girder net, significantly complicating the construction properties of the landfill. Therefore, knowing the confinement of the project object to those or other elements of the landslide system, it is possible to predict in advance the character of its influence on the stability of the slope. The accumulative floodplain of the Cheboksarka River (type B), bounding by its rear seam the considered slope site, is for it a counterbank, a basis of erosion and a place for the accumulation of ravine prowl. As a result, recommendations were made on the construction conditions taking into account the identified and projected hazardous processes. Calculations of slope stability in the natural state taking into account the designed buildings as well as vertical planning and engineering preparation of the territory, communications and roads are performed. However, a local landslide in the playground area of the kindergarten (position 26) caused deformations in May 2014, which suspended the commissioning of the facility for 8 months. In addition, the formation of a deep landslide by cutting the slope with a foundation pit to position 24 began. There was a threat of landslides within position 25 due to over-wetting of the soil grounds. In an emergency order measures were taken to reduce the groundwater level to 5 m [14].

The presence of an object of unfinished hydraulic engineering construction – the Cheboksary HPP and the reservoir of the same name is the most important problem of economic development of the Middle Volga region. Therefore, the problem of functioning of the bank protection structures of the Cheboksary water reservoir remains the most urgent task of construction and territorial development. At the same time, the designer and the customer meet with a number of difficult problems due to imperfections in the regulatory framework at all stages of implementation of bank protection facilities.

First, these projects are relatively "obsolete" due to current changes of natural processes. Over time, planned and high-altitude deformations of the coastal slope occur, leading to the need for adjustments to the design estimates, the repetition of approval procedures and the passing of expertise. As a result of multi-stage (preparation of outgoing documentation, numerous approval and expertise of project, preparation for construction), the construction is delayed. And by the time the building permit is received, the project needs to be corrected. Therefore, it is necessary to simplify and speed up the procedure for approving state expertise of "Design and estimate documentation" (DED) for the construction of shore protection facilities. And then immediately proceed to the implementation of the facilities on the readiness of the DED for their construction.

Further, artificial beaches, etc. refer to deformable structures. And the regulatory requirements for the rules of their operation are imperfect. Therefore, it is advisable to take decisions on the need to develop a special section on the operation of these structures as part of the DED for the construction of deformable structures, section, including the need to monitor facilities, with the development of a program and separate estimates for its implementation, as well as additional workloads and accordingly estimates for annual repairs and even restoration of similar facilities.

Thirdly, the unique feature of anti-landslide and bank protection structures of reservoirs is the only or a small number of variants of their spatial location. However, in reality, the most optimal land plots under anti-landslide and bank protection structures have already been allocated to other facilities and without servitude compliance. This is especially true for water protection zones (WPZ). Therefore, when providing land plots in the WPZ, it is important to legislatively fix the obligatory introduction of an easement in the title documents or the right to withdraw the land plots needed for the landslide and bank protection structures.

Russian legislation on the one hand has many regulatory mechanisms for the protection and use of natural resources in the coastal zone. But in the Russian legislation there are no uniform statutes on integrated management of coastal zones, various elements of it regulate federal laws in the field of environmental management and environmental protection: Urban Planning Code, Land Code, Water Code, Forest Code; Federal laws: "On Environmental Protection", "On Subsoil", "On Ecological Expertise", "On Sanitary and Epidemiological Well-Being of the Population", "On Specially Protected Natural Territories", etc. As a result, a superficial opinion is formed that full-fledged environmental protection in the coastal zone is carried out. However, the narrowly sectoral character of legal norms is not very much combined with the integrated approach necessary to manage coastal and landslide risks. Territorial planning carried out within the framework of the Urban Development Code could be a good tool for sustainable development in coastal zones. The weak side is that the management of each of the elements of the natural environment is carried out by separate departments that are not formally subordinate to local government bodies. As a result, decisions on integrated environmental protection in the framework of territorial planning schemes are ignored. The lack of a single jurisdiction and mechanisms for resolving disputes leads to a rivalry between the subjects of land use in the coastal zone. The solution of problems should go with attraction of experts in the field of architecture and construction, territorial planning, environmental protection. Representatives of public environmental and hail protection organizations, local self-government bodies, enthusiasts of water sports and tourism, etc. should participate in the development of proposals. The search of the way of of coastal territory development in which the balance of interests of the population and investors is observed under condition of unconditional preservation of valuable objects of natural and cultural heritage remains the most urgent task.

Consideration of the social factor in the coastal zone, the prospects for the development of the territory and the escalating environmental problems are a task of great importance. The coasts of the reservoirs are the most important element in the economic development of the regions of the Volga region. Preventing or minimizing the negative consequences of coastal and landslide development is an important part of coastal zone management. Thus, the removal of sandy material for construction needs on the left bank of the Cheboksary reservoir can lead to a shortage of long-shore deposits and the replacement of the accumulative regime for abrasion, which will undoubtedly affect the formation of beaches and the functioning of the left-bank recreational zone. The problem of managing territorial development in the coastal zone is to find and maintain a dynamic balance between the processes of economic development and the processes that form the ecological state of the environment.

The next problem is the training of personnel with synthetic knowledge in the field of technical and natural sciences, hydraulic engineering, geotechnics, engineering geology and applied geomorphology. Previously, the departments on coastal protection in the design institutes were organized and successfully operated("Giprocommunstroy", "PNIIIS" in Moscow, the "Hydroproject" and systems of "Giprovodkhoz" in a number of the Volga cities).The construction of anti-landslide and bank protection structures was carried out by highly professional construction organizations, trusts, qualified personnel, appropriate equipment. At present, in this activity there are people who do not have the necessary experience, the quality of their work is doubtful. Anti-landslide protection and bank protection in their projects are reduced to the erection of retaining walls, wave walls and reinforced concrete buns without adequate drainage and other ALM. Over the past two decades a large number of design and production organizations have emerged that clearly having low professionalism and little knowledge of coastal and slope processes and the impact of erected bank protection structures on them. It is necessary to increase the requirements for qualification of workers in this field, in the likeness of tightening the requirements for cadastral engineers in the field of land management and cadastral activity by the Ministry of Economic Development in 2016.

This explains the complex approach in the scientific substantiation of the landslide risk management strategy in the coastal zone of the Volga reservoirs and in adjacent sloping areas. At the same time, the diversity of social, technical and natural factors must be taken into account. The complex scientific justification for anti-landslide and coastal protection measures should be based on reliable modern methods that allow assessing the effect of erected structures on the natural course of coastal and slope processes, on the ecology of the adjacent reservoir and the coastal section and also take into account social consequences. Recently much attention has been paid to the mathematical modelling of geo-control. However, without minimizing the importance of mathematical modelling, it should be noted that this method gives good results when applied as an accompanying one. Mathematical models should not be blindly copied; they should take into account regional engineering and geological features of slope, landslide and abrasion processes. So, underestimation of the features of the geodynamics of the coastal zone in any area of design leads to emergency consequences, as discussed above, and to financial costs. As a rule, only after the project's shortcomings are identified as a result of emergencies, developers are turning to scientists. A preventive scientific justification for protecting the coasts is necessary. For this purpose, in the universities, in the training of civil engineers, geotechnicians and designers focused on professional activities in the coastal zones it is necessary to introduce the teaching of the relevant disciplines of the variable cycle. It should be understood that there are no universal solutions in the field of landslide risk management, there are only basic methods of solution.

3. Results and discussion

Underestimation of the complex engineering and geological situation in the coastal zone of the Cheboksary water reservoir in urban development leads to a rise in the cost of creating various national economic objects. Often the construction process has to be suspended in connection with activating landslide deformations, which confirms the extreme need for thorough engineering and geological surveys at the design stages. Landslide-hazardous engineering and geological conditions on the slopes of the Cheboksary reservoir are caused by a combination of slope and wave geodynamic processes and lithology, represented by interlacing clay differences with sandy-marly complexes of the Permian Paleozoic system. Our observations confirm the growing role of human economic activity in landsliding. The variety of natural factors acting in the coastal zone, their influence on the constructed hydrotechnical and other capital structures and the feedback account predetermined high requirements for competent specialists in the complex interaction of knowledge in geotechnics and hydraulic engineering, design and construction of structures in the coastal zone and applied natural scientific disciplines. When resolving controversial issues of coastal nature management, it is necessary to adhere to the principles of decision-making on economic, social and environmental tasks at the state, regional and local levels in the light of the interests of both present and future generations of citizens; integrated approach to management of coastal and landslide risks; involvement in the discussion of wide circles of the public in the decision-making processes and in monitoring the management of landslide risks on the coasts of the Cheboksary and other Volga reservoirs.

References

[1] Cruden D M 1991 *Bulletin of the International Association of Engineering Geology* **43**(1) 27–29
[2] Cruden D M and Varnes D J 1996 *Landslides: Investigation and mitigation. Serial: Transportation Research Board Special Report* **247** 36–75
[3] *Hazardous exogenous processes* 1999 ed V I Osipov (Moscow: GEOS) p 290
[4] Sharpe C F S 1938 *Landslides and related phenomena: a study of mass-movements of soil and rock* (Columbia University Press, New York)
[5] Terzaghi K Mechanism of landslides 1950 *Application of Geology to Engineering Practice* S Paige (Geological Society of America, New York) pp 83–123
[6] Highland L and Bobrowsky P T 2008 *The Landslide Handbook – a guide to understanding landslides* (USGS Circular) 1325
[7] Jackson L E Jr, Bobrowsky P T and Bichler A 2012 *Identification, Maps and Mapping – Canadian Technical Guidelines and Best Practices related to Landslides: a national initiative for loss reduction* (Geological Survey of Canada) Open File 7059

[8] *Landslides. Analysis and Control.* 1978 ed R Schuster, R Krizek (National Academy of Sciences, Washington D C Springer-Verlag) Translation from English 1981 ed G S Zolotarev (Moscow: Peace) 1981 pp. 32–85 (in Russian)
[9] Porter M and Morgenstern N 2013 *Landslide Risk Evaluation – Canadian Technical Guidelines and Best Practices related to Landslides: a national initiative for loss reduction (*Geological Survey of Canada) Open File 7312 p 21
[10] Kay R and Alder J 1999 *Coastal Planning and Management* (London, E & FN Spon.)
[11] Ketchum B H 1972 The water's edge: critical problems of the coastal zone *Coastal Zone Workshop, 22 May-3 June 1972 Woods Hole, Massachusetts.* (Cambridge: MIT Press)
[12] Ilyin V N, Bespalova L A, Nikonorova I V and Sushko K S 2016 *Advances in modern natural science* **12**(2) 395–400
[13] Petrov N F, Nikonorova I V and Prokopieva N A 2016 *Proceedings of the 12th All-Russian Conference of Survey Organizations. December, 7-9, 2016* (St. Petersburg) pp 87–95
[14] Petrov N F, Nikonorova I V and Prokopieva N A 2017 *Sergeevskie readings* 19 *Geoecological safety of mining mineral deposits* (Moscow: Publishing House of the Peoples Friendship University) p 590

Investigation of the dynamics of ephemeral gully erosion on arable land of the forest-steppe and steppe zone of the East of the Russian Plain from remote sensing data

E V Platoncheva

Institute of Environmental Sciences, Kazan Federal University, 18, Kremlevskaya str., 420008, Kazan, Russia

E-mail: evgeniya689@mail.ru

Abstract. Spatio-temporal estimation of the erosion of arable soils is still an urgent task, in spite of the numerous methods of such assessments. Development of information technologies, the emergence of high and ultra-high resolution images allows reliable identification of linear forms of erosion to determine its dynamics on arable land. The study drew attention to the dynamics of the most active erosion unit – an ephemeral gully. The estimation of the dynamics was carried out on the basis of different space images for the maximum possible period (from 1986 to 2016). The cartographic method was used as the main research method. Identification of a belt of ephemeral gully erosion based on materials of multi-zone space surveys and GIS-technology of their processing was carried out. In the course of work with satellite imagery and subsequent verification of the received data on the ground, the main signs of deciphering the ephemeral gully network were determined. A methodology for geoinformation mapping of the dynamics of ephemeral gully erosion belt was developed and a system of indicators quantitatively characterizing its development on arable slopes was proposed. The evaluation of the current ephemeral gully network based on the interpretation of space images includes the definition of such indicators of ephemeral gully erosion as the density of the ephemeral gully net, the density of the ephemeral gullies, the area and linear dynamics of the ephemeral gully network. Preliminary results of the assessment of the dynamics of the belt erosion showed an increase in all quantitative indicators of ephemeral gully erosion for the observed period.

1. Introduction

Erosion is the leading process of soil cover degradation on agricultural lands. The intensively plowed territories are most susceptible to soil erosion and lead to a decrease in the fertility of the soil cover, the formation of gullies in contrast to the territory occupied by natural vegetation.

To identify qualitative changes in the structure of basin erosion, it is possible to use the so-called "erosion belts" on the slopes of the catchments. In the structure of basin erosion 6 erosion zones are distinguished reflecting patterns of surface runoff formation and transformation of the flushing mechanism: 1) raindrop destructing; 2) sheet and rill; 3) ephemeral gullies 4) gully; 5) the prevailing accumulation; 6) areas of no erosion. At the same time, up to 80-90% of the entire surface of the plowed slopes falls on the belt of sheet and ephemeral gully erosion. In this work, a belt of ephemeral gully erosion was chosen for the study, here there is a maximum washout of the soil and it is morphologically well expressed in space images, due to the linear pattern created by the ephemeral gully network.

The choice of materials for space surveys as a tool for mapping the dynamics of erosion is determined by three factors: the ability to determine the dynamics over different time periods, the high quality of displaying linear forms of erosion in space images and the possibility of research in different regions. The use of images distributed free of charge in digital form and at different time intervals allows determining various parameters of ephemeral gully erosion: the density of the ephemeral gullies (km/km^2), the density of the ephemeral gullies ($unit/km^2$), the area and linear dynamics of the ephemeral gully network.

The basic method of investigation was the visual identification of ephemeral gully erosion by visual analysis of the screen image of high-resolution space images.

2. Material and methods

For the study, sites were chosen in the forest, forest-steppe and steppe zones of the east of the Russian Plain within the Republic of Tatarstan and the Orenburg region. For the convenience of mapping, the entire study area was divided into basins (along first-order watercourses). For each pool, the total length of the ephemeral gully network, the density of the ephemeral gully net and the density of the ephemeral gullies were calculated.

Sites for the study were selected according to several criteria: 1) the time of shooting for the observed periods should be approximately the same; 2) the plots should be fully plowed, without any projective covering by agrocenoses; 3) the quality of the images (no clouds, etc.) should be good; 4) pictures on the sites should be in the public domain and have a high and ultra-high resolution; 5) for deciphering, interfluvial plowed spaces with a morphologically well-defined net of ephemeral gullies should be selected. In the course of the work it is established that the identification of the linear erosion belt is most reliably carried out in late spring, autumn and early summer images. In other seasons, interpretation of linear forms of erosion in the fields is accompanied by large errors due to their leveling by cultural vegetation.

As the initial data for deciphering is the current ephemeral gully network, the images downloaded from the Google Earth program have a spatial resolution of 0.5-1.5 m. These are modern images for the period 2009-2015 which allow covering the territory within 10 km^2. Landsat images for the territory of the Samara river basin (the Orenburg region) and the Mesh river basin (the Republic of Tatarstan) were also taken for mapping. Landsat images have a resolution of 30 m which allows you to explore more extensive territory and study the dynamics of the ephemeral gully network during a longer time period from 1984 to 2015. Both types of images have sufficient resolution for deciphering the ephemeral gully network.

Ephemeral gullies in the images are decoded reliably according to the characteristic linear pattern of the image. Unlike gullies, they are depicted in the form of thin, more often, dark bands and lines along slopes with expansion in the central part. Ephemeral gullies are depicted by meandering lines with a large variability of the phototone (color) from almost white to almost black (figures 1, 2). All the linear forms visually determined in the picture were decoded. Area, linearly stretched contours at the tops of ephemeral gully erosion in the hydrosphere areas with a tree-like pattern did not belong to the ephemeral gully network since these sections are formed by plumes of ephemeral gully erosion spatially strongly changing their planned location due to a number of reasons (cultivation of arable land, etc.).

Work with images is carried out in MapInfo. With the help of it the forms of linear erosion and the boundaries of arable lands were traced and translated into a vector format. Further, the area of arable land, the total length of the ephemeral gully network, the number of peaks of the ephemeral gully erosion, their density was calculated.

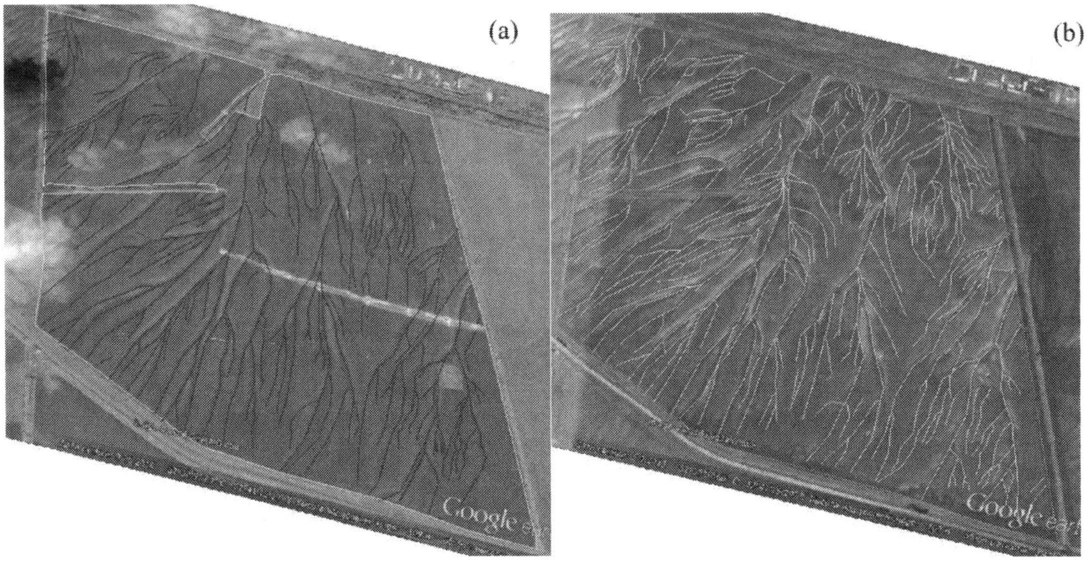

Figure 1. Structure of the ephemeral gullies in the belt of ephemeral gully erosion from the materials of interpretation of space images (Totsky village in the interfluve of the rivers Samara and Soroca): a) 2010; b) 2013

Figure 2. Structure of the ephemeral gully network in the belt of ephemeral gully erosion based on the materials of interpretation of the satellite images Landsat (Rodnikovoe Ozero village, Samara river basin: a) 1988; b) 2015

3. Results and discussion

On the basis of space images, the basin deciphering was carried out within the Republic of Tatarstan and the Orenburg region. In the course of studies to determine the dynamics of the ephemeral gully erosion belt a system of deciphering features (regional standards) for definition of ephemeral gully network has been developed, and field verification of the deciphering results has been carried out. A technique for geoinformation mapping of the dynamics of the ephemeral gully erosion belt has been developed, the main parameters quantitatively characterizing the dynamics of this belt have been obtained (tables 1, 2).

Table 1. Dynamics of a belt of ephemeral gully erosion on the example of a site near the village of Totskoye, Orenburg Region.

Indicators	25.09.2010	03.05.2013
Area of arable land (km^2)	6.36	6.36
Total length (km)	64.18	94.78
Number of vertices	144	204
Density of the ephemeral gullies	10.09	14.9
Density of the ephemeral gullys	22.6	32.07

Table 2. Figures of the belt of ephemeral gully erosion from Landsat images for 8 sites in the Samara river basin of the Orenburg Region.

Basin No.	Area of arable land, km^2		Total length, km		Number of vertices		Density of the ephemeral gullies, km / km^2		Density of the ephemeral gullies, unit / km^2	
	1988	2015	1988	2015	1988	2015	1988	2015	1988	2015
1	15.3	15.3	60.6	62.1	70	79	3.96	4.06	4.57	5.16
2	16.8	16.8	80.32	83.55	130	135	4.78	4.97	7.74	8.03
3	25.67	25.67	101.7	122.05	140	151	3.96	4.75	5.45	5.88
4	34.03	34.03	146.1	167.9	194	239	4.29	4.93	5.7	7.02
5	91.4	91.4	165.7	245.6	205	300	1.81	2.69	2.42	3.28
6	57.6	57.6	113.8	167.8	96	189	1.98	2.91	1.66	3.28
7	53.91	53.91	94.1	99.6	102	151	1.74	1.85	1.89	2.8
8	55.1	55.1	101.5	104.3	120	125	1.84	1.9	2.17	2.27

In general, conducted studies on 35 field sites in the region showed that there was an increase in all quantitative characteristics of the belt erosion from 1988 to 2016. In the Republic of Tatarstan, a significant increase in the total length of the ephemeral gullies is observed in the basin of the Mesha River near the village of Sokury (4% from 17.05.2012 to 27.04.2016) and in the Volga river basin near the village of Burtasy (37% from 12.04.2009 to 28.04.2016). However, it should be noted that it is quite difficult to estimate the overall trend of the dynamics of the modern ephemeral gully network of Google Earth images for 2009-2016, as it is not always possible to find pictures of good quality for only one season in one territory.

According to the images of the Landsat of the Samara River Basin, Orenburg Region it was possible to trace the dynamics of the ephemeral gully network for a sufficiently long time interval for one season, we considered two images - April 19, 1988 and April 29, 2016. The obtained data showed that in this territory there is an increase in all parameters of the ephemeral gully network. On the average, the density of the ephemeral gully network increased by 6%. The maximum increase in the total length of the ephemeral gullies corresponds to basin No. 5 along the Samara River (47%).

4. Conclusions

The use of high-resolution and ultra-high-resolution images makes it possible to reliably identify the belt of ephemeral gully erosion expressed by ephemeral gully network on the slopes of river basins by a direct and indirect set of deciphering features characteristic of this belt. At key sites approaches have been developed for geoinformation mapping of these linear forms of erosion as well as dynamics using high resolution satellite imagery. A system of quantitative indicators characterizing its development on arable slopes has been proposed for the belt of ephemeral gully erosion. The obtained results on the dynamics of the belt of ephemeral gully erosion allow us to reveal the time trend of the increase in its development from 1988 to 2016.

Acknowledgements
The work is performed according to the Russian Government Program of Competitive Growth of Kazan Federal University.

References
[1] Labutina I A 2004 *Deshifrirovanie aerokosmicheskih snimkov* (Aspect Press) p 184
[2] Desprats J F, Raclot D, Rousseau M, Cerdan O, Garcin M, Le BissonnaisY, Ben Slimane A, Fouche J and Monfort-Climent D 2013 *Land Degradation & Development* **22** 24–32
[3] Vrieling A 2006 *Catena* **22** 2–18
[4] Yermolaev O P 2002 *Erosiya v basseinovih geosystemah.* Unipress KGU p 265
[5] Yermolaev O P 2017 *Geoinformation Mapping of Soil Erosion in the Middle Volga Region.* Eurasian Soil Science **50**(1) 118–131
[6] Govers G 1991 Rill erosion on arable land in Central Belgium: Rates, controls and predictability. *Catena* **18** 133–155
[7] Mekuriaw A, Heinimann A, Zeleke G, Hurni H and Hurni K 2017 Journal of Geographical Sciences **27**(1) 79–94
[8] Saadat H, Adamowski J, Tayefi V, Namdar M, Sharifi F and Ale-Ebrahim S 2014 Catena **113** 306-313
[9] Van Lynden GWJ and Mantel S 2001 *International Journal of Applied Earth Observation and Geoinformation* **3**(1) 61–68
[10] Hoffmann E and Winde F 2010 Water SA **36**(1) 53–68
[11] Boardman J 2016 *Catena* **143** 123–127
[12] Casalí J, López J and Giráldez J 1999 *Catena* **36** 65–84
[13] Nachtergaele J and Poesen J 1999 *Earth Surface Processes and Landforms* **24**(8) 693–706
[14] Wang L, Huang J, Du Y, Hu Y and Han P 2013 *Remote Sens.* **5** 3826–3848
[15] Alatorre L. 2009 *Catena* **76** 182–190
[16] King C, Baghdadi N, Lecomte V and Cerdan O 2005 *Catena* **62** 79–93
[17] Floras S and Sgouras I D 1999 *International Journal of Applied Earth Observation and Geoinformation* **1**(1) 68–77
[18] Liberti M, Simoniello T, Carone M, Coppola R, D'Emilio M and Macchiato M 2009 *Geomorphology* **106** 333–343
[19] Zhang Y 2007 Soil and Tillage research **96** 28–41

IOP Conf. Series: Earth and Environmental Science **107** (2017) 012020 doi:10.1088/1755-1315/107/1/012020

Exploring the performance of the SEDD model to predict sediment yield in eucalyptus plantations. Long-term results from an experimental catchment in Southern Italy

P Porto[1, 2], **V Cogliandro**[1] **and G Callegari**[2]

[1]Dipartimento di Agraria, University Mediterranea of Reggio Calabria, Feo di Vito, 89122, Reggio Calabria, Italy
[2]Institute for Agricultural and Forest Systems in the Mediterranean (ISAFOM) - National Research Council (CNR), Via Cavour, 4/6, 87036, Rende (CS), Italy

E-mail: paolo.porto@unirc.it

Abstract. In this paper, long-term sediment yield data, collected in a small (1.38 ha) Calabrian catchment (W2), reafforested with eucalyptus trees (*Eucalyptus occidentalis* Engl.) are used to validate the performance of the SEdiment Delivery Distributed Model (SEDD) in areas with high erosion rates. At first step, the SEDD model was calibrated using field data collected in previous field campaigns undertaken during the period 1978-1994. This first phase allowed the model calibration parameter β to be calculated using direct measurements of rainfall, runoff, and sediment output. The model was then validated in its calibrated form for an independent period (2006-2016) for which new measurements of rainfall, runoff and sediment output are also available. The analysis, carried out at event and annual scale showed good agreement between measured and predicted values of sediment yield and suggested that the SEDD model can be seen as an appropriate means of evaluating erosion risk associated with manmade plantations in marginal areas. Further work is however required to test the performance of the SEDD model as a prediction tool in different geomorphic contexts.

1. Introduction

Soil erosion by water is a serious environmental problem in many areas of the world, and more particularly in the semiarid zones of Southern Italy, where rates of soil loss of ca. 100-150 t ha^{-1} year^{-1} have been documented during the last few decades. Problems of accelerated soil erosion and land degradation as well as the associated offsite impacts have emphasized the need for improved information on rates of soil loss and thus for reliable means of assessing soil erosion rates across a range of environments. Traditional approaches to soil erosion assessment, involving erosion plots and experimental catchments, have been used in many geographic contexts. It is well recognized that they have some limitations both in terms of cost and the representativeness of the results. For example, their operation for a sufficient time to produce reliable estimates of soil loss rates lies beyond the resources of most institutions and the scope of most short-term investigations. However, when long-term measurements are available, they represent a precious resource to calibrate and validate theoretical and empirical models able to predict soil erosion. The use of simulation models represents in fact an important tool to understand some of the key processes governing a catchment and also to extend the knowledge in areas where no measurements are available. Some of these models predict rates of soil loss and sediment yield using a distributed approach. However, in order to provide reliable results, it is important that such models are calibrated and validated using local parameters.

This paper presents some results based on the use of long-term sediment yield measurements obtained in the catchment W2 (1.38 ha in size) located in Calabria, southern Italy, aimed at exploring the performance of the SEDD (SEdiment Delivery Distributed) model which was originally developed in the same area [1]. The measurements of sediment yield derived from the suspended sediment loads at the catchment outlet and obtained for the period 1978-1994, have been used to calibrate the model specific parameter β of the sediment delivery component. Then, the SEDD model was successfully validated at the event and annual timescale using corresponding values of sediment yield related to the period 2006-2016.

2. The Sediment Delivery Distributed (SEDD) Model

The SEdiment Delivery Distributed (SEDD) model [1] has been extensively used to predict erosion rates and sediment yield in different areas of the world [1, 2, 3, 4, 5, 6, 7, 8]. The SEDD model requires a spatially distributed approach based on a subdivision of the catchment into morphological units. It consists of two components: the first component is RUSLE based and allows estimates of soil detachment from each morphological unit, i; the second component accounts for the sediment delivery ratio SDR_i.

For a given temporal scale t (event, annual, mean annual), and for areas with homogeneous soil and vegetation cover, the model provides estimates of sediment yield at the catchment outlet by the following equation:

$$Y_{c,t} = R_{c,t} K_c C_c P_c \sum_{i=1}^{N_u} L_i S_i SDR_i S_{u,i} \quad (1)$$

in which $R_{c,t}$ is the rainfall erosivity factor (t ha^{-1} per unit of K_c); K_c is the soil erodibility factor [9] (t h kg^{-1} m^{-2}); C_i (dimensionless) is the cover and management factor [10]; P_i (dimensionless) is the support practice factor; L_iS_i (dimensionless) is the topographic factor [11, 12]; and $S_{u,i}$ (ha) is the area of each morphological unit.

According to Ferro and Minacapilli (1995) [13], SDR_i represents the probability of the eroded particles, detached from the considered morphological unit i, to arrive to the nearest stream reach. It can be represented by the following expression:

$$SDR_i = \exp(-\beta t_{p,i}) = \exp\left(-\beta \frac{l_{p,i}}{\sqrt{s_{p,i}}}\right) = \exp\left[-\beta\left(\sum_{j=1}^{N_p} \frac{\lambda_{i,j}}{\sqrt{s_{i,j}}}\right)\right] \quad (2)$$

in which $t_{p,i}$ expresses the travel time of the soil particles of each morphological unit; $l_{p,i}$ and $s_{p,i}$ are, respectively, the length and the slope of the hydraulic path from the i-th morphological unit to the nearest stream reach; N_p is the number of morphological units localized along the hydraulic path j; $\lambda_{i,j}$ and $s_{i,j}$ are, respectively, length and slope of each morphological unit i localized along the hydraulic path j; β is the only calibration parameter. The latter can be estimated using the approach suggested by Ferro and Porto (2000) [1] that requires a recursive use of Eq. (1). In other words, if some independent measurements of sediment yield are available at the catchment outlet, these can be used to derive the value of β that provides the best agreement between measured and simulated values of sediment yield. Based on this assumption, Ferro and Porto (2000) [1] proposed two criteria to derive the 'best fitting' value. The first is based on a single, representative value corresponding to the median value of the theoretical distribution of β (namely, *MV* approach); the second, event-dependent is based on an empirical correlation existing between β and the runoff coefficient *RC* (namely, *RC* approach). Both the approaches require a calibration using an independent set of measurements. In this contribution, the measurements obtained from 1978 to 1994, corresponding to 52 single events, have been used to provide the empirical distribution of the β coefficient and the basic relationship between β and RC. Then, the measurements obtained for the period 2006-2016, consisting of 81 single events, served for validation.

3. The experimental catchment W2

The W2 (1.38 ha in size) is an ephemeral catchment within the larger Crepacuore basin which drains to the Ionian sea and is located near Crotone (35 m a.s.l., 39°09'02"N, 17°08'10"E) (figure 1). The catchment W2 was afforested in 1968 with Eucalyptus trees (*Eucalyptus occidentalis* E.) and was coppiced twice in 1978 and 1990 for wood production. The catchment was equipped in 1978 for measuring rainfall and runoff using respectively a mechanical raingauge located in the upper part of the catchment and an H-flume weir [14] coupled with a mechanical streamgauge located at the catchment outlet. Sediment yield records are also provided using a Coshocton wheel placed downstream the H-flume. Measures of rainfall, runoff and sediment yield are generally obtained at event scale. However, when short-time interval occurred between two or more consecutive events, only one measurement of sediment yield was possible, and the corresponding data of rainfall and runoff responsible for that sediment output were aggregated accordingly.

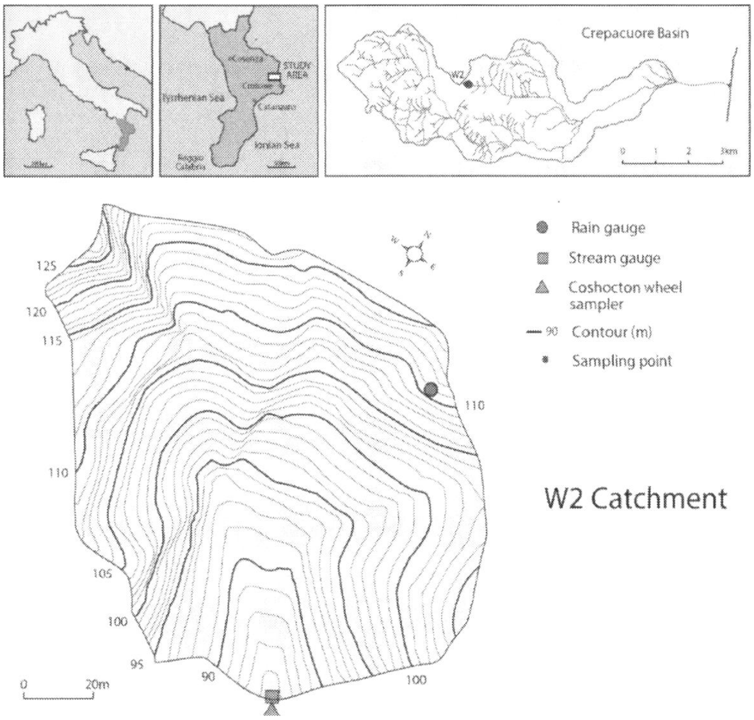

Figure 1. The study area and the experimental catchment W2.

4. Results and discussion

The model calibration – The dataset used for calibration relates to the period 1978-1994 during which 52 single events and the corresponding sediment yield values have been recorded. The model required, at first, the evaluation of the RUSLE parameters for the basic component. More specifically, the rainfall erosivity factor R_t for a given temporal scale t was obtained from rainfall data available for the raingauge station located in the catchment. A value of R_j from each single event, j, was calculated as the product of the storm rainfall kinetic energy E and the maximum rainfall intensity related to a 30-min time interval I_{30}. Because of the small size of the catchment, the values of the rainfall erosivity factor obtained at event scale were considered spatially invariant ($R_{c,t}$) over the catchment area. Similar assumptions were made for the soil erodibility factor K_c that was estimated using the properties of the soil samples collected within the catchment. The cover and management factor C_c, again assumed constant for the catchment was calculated using field data. Considering that no management works were carried out in the catchment, the support practice factor P_c was assumed equal to one. The topographic factor L_iS_i was calculated, for each morphological unit i, using the following equations coupled with a subroutine provided by the LIBERTYBASIC software.

$$L_iS_i = \left(\frac{\lambda_i}{22.13}\right)^{m_i}(10.8\sin\alpha_i + 0.03) \qquad \text{if } \tan\alpha_i < 0.09 \tag{3a}$$

$$L_iS_i = \left(\frac{\lambda_i}{22.13}\right)^{m_i}(16.8\sin\alpha_i - 0.5) \qquad \text{if } \tan\alpha_i \geq 0.09 \tag{3b}$$

in which λ_i = slope length of the i-th morphological area, α_i = slope angle, $m_i = f_i/(1 + f_i)$ and $f_i = \sin\alpha_i / 0.0896(3\sin^{0.8}\alpha_i + 0.56)$.

Once the input of the basic component was obtained, Eq. (1) was adopted to calculate 52 values of β (one for each event) using the recursive approach. The empirical frequency of β is shown in figure 2a together with the Gaussian distribution. Following the MV approach a median value $\beta = 0.0142$ to use, as a representative value in the validation phase, was obtained.

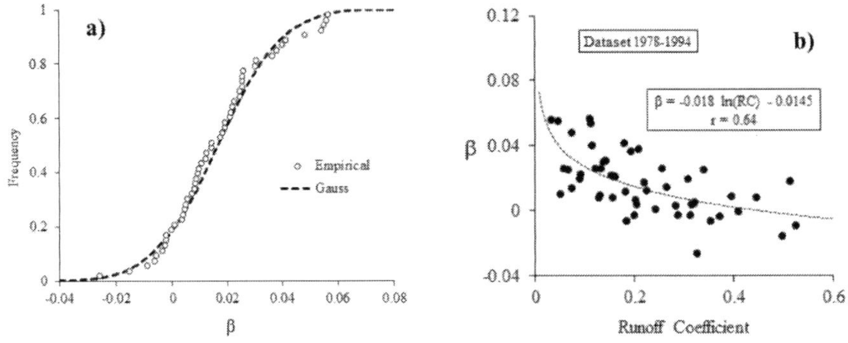

Figure 2. Comparison (a) between the empirical frequency and the Gaussian distribution of β and relationship (b) between β and RC for the 52 events.

The empirical relationship between β and RC, for the same 52 events, is provided in figure 2b and can be approximated by the formula $\beta = a$ lnRC – b, where a and b are parameters.

The model validation – The results provided in figure 2a,b afford a basis for validating the SEDD model at the event scale for the new dataset (81 events) obtained during the period 2006-2016. The validation has been conducted using both the *MV* approach and the *RC* approach. The results of this validation are presented in figure 3a,b.

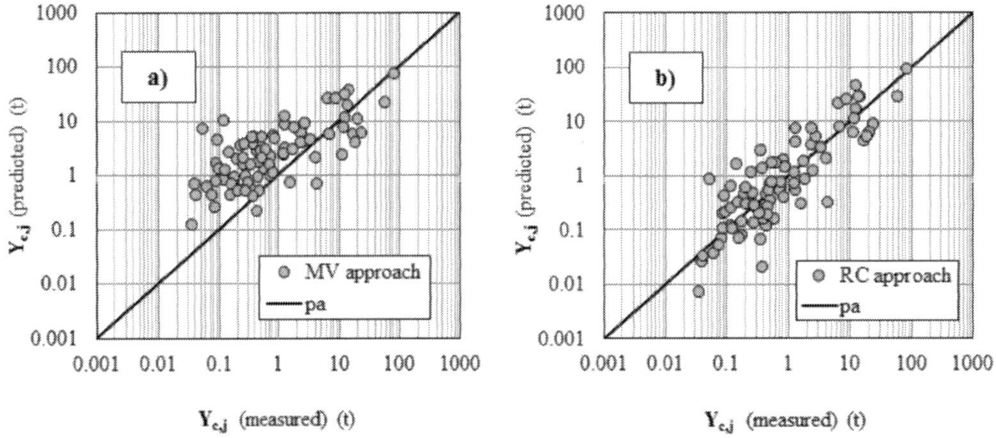

Figure 3. Comparison between the measured and predicted values of sediment yield for the validation period using (a) the MV approach and (b) the RC approach.

Clearly, perfect agreement between measured and predicted values of sediment yield cannot be expected because the SEDD model neglects any contribution due to channel erosion. However, looking at figure 3a, it seems evident that some overestimation for the small-medium values of sediment yield occurred. These results suggest that the β calibration parameter cannot be assumed time invariant for the entire period of measurements. This assumption is confirmed by a visual inspection of the graph presented in figure 3b where a better agreement with no systematic under- or overestimation between measured and predicted values of sediment yield can be observed. Figure 3b reports the results obtained using the *RC* approach which assumes that the β parameter has an inter-event variability and is dependent on the runoff coefficient. The use of runoff measurements to improve the prediction given by similar models was already emphasized by other authors. For example, Kinnel and Risse (1998) [15] and Bagarello et al (2011) [16] suggested to incorporate the runoff measurements to improve the prediction of the USLE model. Obviously, the recourse to runoff measurements means that one more variable must be taken into account when using the model, and the need to measure or predict event runoff, may limit the possibility to apply it. However, other investigations suggest that a reliable prediction of event runoff should be possible and this would simplify the applications [17, 18, 19, 20]. If the comparison is limited to the results of this study, the prediction shown in figure 3b provided a Nash and Sutcliff efficiency index of 0.69 that is higher than that obtained for the prediction presented in figure 3a (EI = 0.61). In other words, the reasonable agreement between the measured and predicted values of $Y_{c,t}$ suggests that the SEDD model provides meaningful estimates of event sediment yield for the study area, especially if the parameter β is calibrated using the runoff measurements.

In order to confirm this hypothesis, a further validation of the SEDD model, using the *RC* approach for estimating β, can be carried out at annual scale. The results of this validation are presented in figure 4, where the annual values $Y_{c,a}$ are obtained by summing the values of $Y_{c,t}$ for the events occurring during that year.

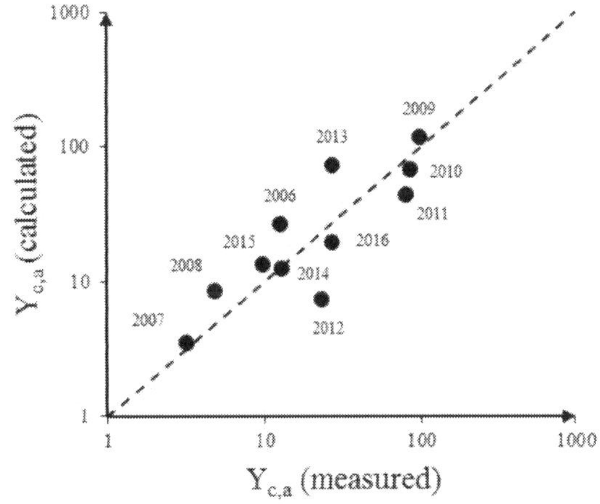

Figure 4. Comparison between the measured and predicted values of annual sediment yield (t) for the validation period using the RC approach.

Again, the model provides meaningful estimates of annual values of sediment yield with no systematic under- or overestimation between measured and predicted values.

5. Conclusions

The SEDD model that integrates a RUSLE input with a sediment delivery component through a distributed approach proved to be very effective in predicting sediment yield at catchment scale. In this context, if long-term measurements of sediment yield are available calibration and validation of this model is possible and its potential can be exploited further.

The study reported here involved the use of sediment yield data collected from a small catchment located in Calabria, southern Italy.

The results demonstrated that the sediment delivery component of the model and, more particularly, its parameter β, can be calculated using a first calibration dataset consisting of 52 erosive events occurred during the period 1978-1994. At event scale, the prediction capability of the model was tested by comparing measured sediment yield values and calculated ones for 81 independent events occurred during the period 2006-2016. The validation exercise established that the approach that allows calculation of β through the runoff coefficient gave the best performance and can be used if runoff data are available or can be easily estimated. This analysis is confirmed when the validation is extended at the annual scale where the performance of the model allowed to get reliable estimates of annual sediment yield.

However, further work is required to explore the potential for estimating the β parameter from alternative variables that can be easily measured and to establish the contribution of channel erosion in larger catchments.

Acknowledgements
The study reported in this paper was supported by a grant from MIUR PRIN 2010–2011.

References

[1] Ferro V and Porto P 2000 *Journal of Hydrologic Engineering* **5**(4) 411–422
[2] Bhattarai R and Dutta D 2007 *Water Resources Management* **21** 1635–47
[3] Bhattarai R and Dutta D 2008 *Hydrological Sciences Journal* **53** 1253–69
[4] Di Stefano C, Ferro V and Porto P 1999 *J. Agric. Engng. Res.* **74** 41–62

[5] Fu G, Chen S, and Mc Cool D K 2006 *Soil and Tillage Research* **85** 38–49
[6] Porto P and Walling D E 2015 *Journal of Hydrologic Engineering* **20**(6) 10.1061/(ASCE)HE.1943-5584.0001058.
[7] Taguas E V, Moral C, Ayuso J L, Pérez R, and Gómez J A 2011 *Geomorphology* **133** 47–56
[8] Vigiak O, Borselli L, Newham L T H, McInnes J and Roberts A M 2012 *Geomorphology* **138** 74–88
[9] Wischmeier W H, Johnson C and Cross B 1971 *J. Soil Water Cons.* **26**(3) 189–193
[10] Wischmeier W H and Smith D D 1965 *Predicting rainfall erosion losses from cropland East of the Rocky Mountains USDA Agr. Hand.* No 282 (Washington, USA)
[11] Renard K G, Foster G R, Yoder D C, and Mc Cool D K 1994 *Journal of Soil and Water Conservation* **49** 213–220
[12] Mc Cool D K, Brown L C, Foster G R, Mutchler C K and Meyer L D 1987 *Transaction of ASAE* **30**(5) 1387–96
[13] Ferro V and Minacapilli M 1995 *Hydrol. Sci. J.* **40**(6) 703–717
[14] Brakensiek D L, Osborn H B and Sheridan J M 1979 *Field Manual for Research in Agricultural Hydrology. Agriculture Handbook* no. 224 (US Department of Agriculture Science and Education Administration, Washington)
[15] Kinnell PIA and Risse LM 1998 *Soil Science Society of America Journal* **62** 1667–72
[16] Bagarello V, Di Stefano C, Ferro V, Kinnell P I A, Pampalone V, Porto P and Todisco F 2011 *Journal of Hydrology* **400** 267–273
[17] Yu B, Rose C W, Coughlan K J and Fentie B 1997 *Transactions of the ASAE* **40**(5) 1295–1303
[18] Vandervaere J P, Vauclin M, Haverkamp R, Peugeot C, Thony J L and Gilfedder M 1998 *Soil Science* **163** 9–21
[19] Gao G Y, Fu B J, Lü Y H, Liu Y, Wang S and Zhou J 2012 *Hydrology and Earth System Sciences* **16** 2347–64
[20] Mc Cool D K, Foster G R, Mutchler C K and Meyer L D 1989 *Transaction of ASAE* **32**(5) 1571–76

IOP Conf. Series: Earth and Environmental Science **107** (2017) 012021 doi:10.1088/1755-1315/107/1/012021

Gullies and dry-valley systems of Kazan as a territorial reserve for the development of the city

G R Safina, V A Fedorova and R A Medvedeva

Institute of Environmental Sciences, Kazan Federal University, 18, Kremlevskaya str., 420008, Kazan, Russia

E-mail: safina27@mail.ru

Abstract. The article considers possibilities of the use of gully systems of Kazan as a possible resource for territorial development. The city has extensively formed gully and dry valley networks located in different parts of the city. Multi-temporal analysis of cartographic material showed that currently there is an active development of gully systems especially in the central historical part of the city due to the technical capabilities and investment attractiveness. The nature of interaction of gully-dry valley systems and the city is of an overwhelming type.

1. Introduction

High levels of urbanization, urban population growth, inefficiencies and congestion in infrastructure and an increase in the number of vehicles are responsible for the shortage and scarcity of territories in urban systems. At present stage of many cities development, there is a need to revise and adjust some principles of urban planning in terms of more rational economic use of land resources and environmental safety [1].

There are several approaches to solve the problem of scarcity of land within city limits [2, 3]:

- Increasing the area of the city (new territory development);
- Seal buildings;
- Demolition (sanitation) with the subsequent construction on the site of more extensive and high-quality housing;
- Transfer of industrial enterprises out of city boundaries and development of the liberated territories;
- High-rise construction;
- Creation of artificial territories;
- Underground construction;
- Development of urban badlands.

According to the Explanatory Dictionary of the Russian language, the term "badlands" is defined as a terrain, inconvenient for ploughing and mechanized tillage [4]. In urban practice, unsuitable or inconvenient for development areas include land on which construction is complicated by adverse geological-engineering, hydrological, orographic, sanitation and hygiene [5, 6].

The urgency of the problem is due to the urban development situation when from the end of the twentieth century, the territories that were considered inconvenient by engineering reasons began to be developed again. In the past, limited technical capabilities excluded this kind of territories from urban-planning practice, and individual cases of their development were unique. Currently, the costs of their development and their urban-planning value equalize the inconvenient territories with other areas of the city. The inconveniences in the zone of historical center increase their investment attractiveness.

One of the criteria underlying the allocation of uncomfortable territories is the presence of gully network [7]. Gully erosion is one of the most important processes of the modern topography development. Exercising direct influence on morphology of the topography and morphometric characteristics of its forms the formation of gullies is a dangerous natural-anthropogenic process.

The functioning of gully systems within settlements plays a special role. Knowledge of geoecological regularities of interaction of natural objects with settlements within their limits consists of the study of individual development of gully systems and their interaction with the infrastructure of the city in direct contact with natural objects

This paper is focused on problems of gully and dry-valley networks territories (badlands) development in Kazan.

2. Material and methods

The original data was a multi-temporal cartographic material (plans of the city of Kazan) [8]. A great number of plans and maps were compiled for the city of Kazan over a thousand-year history, but not all material is suitable for work purposes since cartographic works have different loads, degree of detail of the image, etc.

The plans and maps selected for the job had to meet the following criteria: to have different ages (the map of 1936 and the google satellite images of 2014 were analyzed), to provide visual observation and comparison of gully and dry-valley systems. The changes in the morphological and morphometric characteristics of the gully systems were defined using the program MapInfo. Determination of the types of interaction between the gully and dry-valley system and the city was done according to the classification proposed by Kovalev S.N. [9]:

a) dependent on relief (all types of buildings are located, depending on the type and landform of relief);

b) subordinated to relief (most of the buildings in the settlements are inserted in relief);

c) subordinating to relief (localities where during construction partial conversion of relief throughout the site or in its substantial part happens);

d) suppressing to relief (settlements or their parts, big industrial enterprises, where in the process of growth or on design stage features of relief are not taken into account and construction is carried out with complete conversion of the territory depending on the needs).

3. Results and discussion

Gullies in the city of Kazan have mainly two negative side effects: on the one hand – the gullies "eat" usable area; on the other hand – they lead to deformation and destruction of structures of different purpose [10, 11].

At present, gully erosion has much more difficult engineering-geological conditions in Vahitovsky, Sovietsky, Privolzhsky, Moscovsky and Aviastroitelny parts of the city. Under a threat there are the objects of civil and industrial purpose. Gully formation in Kazan is mainly due to geomorphological and geological conditions. The territory of the town located within the complex alluvial terraces of the Volga and Kazanka Rivers is deeply dissected not only by the valley of Kazanka River. However, in the low order rivers, the relief is characterized by great ruggedness and rock formation, represented by sandy-loam, sometimes sandy loess and differences between the Pleistocene and have high smearing.

In addition, rainy summer and autumn and, as a rule, abundantly snowy winters also contribute to the development of gully erosion.

Within the present territory of Kazan there are the following areas of gullies and dry valley systems: development Gorsko-Ometevskaya, Moscovskaya, Derbishkinskaya, Tsaritsynskaya, Karavaevskaya, Zaretchenskaya, Sukhoretchenskaya, Krutushkinskaya. In general, the territory of Kazan is characterized by a moderate obrigatoriamente – an averaged coefficient gully dissection is 0.44 km/km^2. The density of gully network is shown in figure 1 [12].

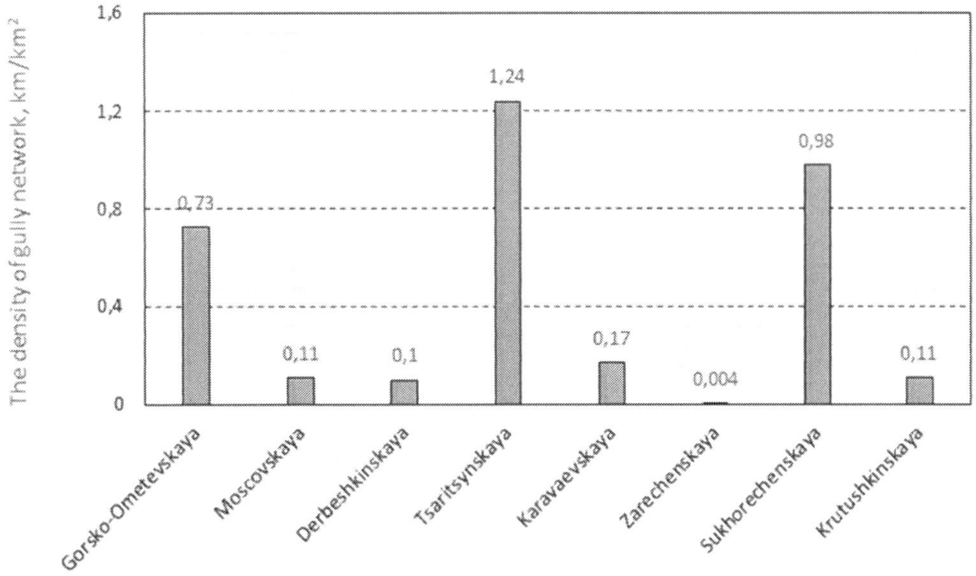

Figure 1. The density of gully network in Kazan.

The longest gully-dry valley system in the city of Kazan is the Gorsko-Ometevskaya which is easy to read on the map of 1936. It is seen that this gully system being sufficiently developed consists of three independent branches: "Nesterovsky", "Ometievsky" (in place of the modern settlement Ometevo) and "Zaometevsky". It should be noted that in 1936 the area was practically the outskirts of the city and the natural development of gullies is not obstructed.

The Gorsko-Ometevskaya system, well observed in 1936, has now significantly changed. The "Nesterovsky" gully, especially its upper reaches on the map are practically not recognized. Currently, most of the gullies are completely covered and are used for residential development. This is one of the most built-up areas. There is an intensive construction of the sides of the gully from the side of Vishnevskogo street and its surroundings.

The "Zaometievsky" gully also has almost disappeared. In its place there is the Ometevskaya highway, which is laid along the bottom of the gully. The fact that this is a former gully is reminiscent of even redesigned but rather steep slopes.

It should be noted that in Kazan gullies and dry valleys are being used for highways not for the first time. Some practical examples include: the transport artery of Pushkin Street which was formed at the bottom of a crooked ravine that ran from Lake Kaban. Likewise, the transport highway was formed at the bottom of the ravine along Tankovaya and Tolstoy streets.

The "Ometevsky" gully located in the area of the former settlement. Ometevo is the only gully that has changed a little. Preservation of this gully was facilitated, evidently, by the fact that up to the present time private low-rise buildings have occupied this territory. Steep layers of gullies are used for garden plots. In the same vein, the presence of some gullies are reflected in the names of the streets. For example, Vyatsky Val Street, Podgornaya Street are among others.

Analyzing the development of the entire Gorsko-Ometevsky gully stretch it can be noted that this gully system underwent significant changes during the observation period (figure 2). Determining the nature of the interaction between the gully system and the city we can state that:
- the type of interaction refers to the dependence on relief (gullies of "Nesterovsky" and "Zaometevsky");
- dry-valley systems of Ometevsky illustrate the type subordinate to relief.

The shortage of land within the city limits can be solved through various approaches. One of them is the use of badlands which include gulliesand dry valley system. Extensive gully systems in the city of Kazan are of considerable territorial reserves for development.

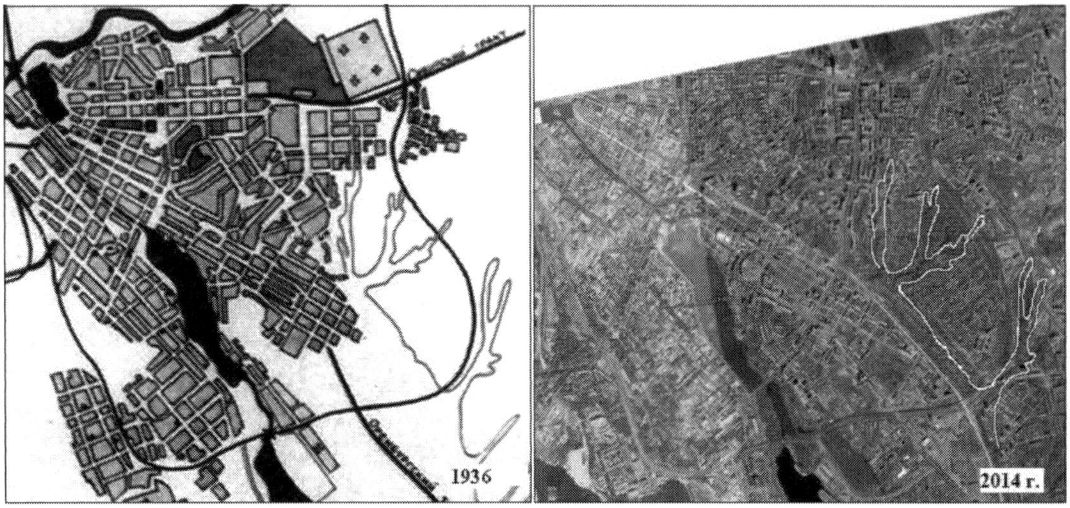

Figure 2. Gorsko-Ometevsky gully on the map of 1936 and on a satellite image of 2014 (Google).

Multi-temporal analysis of cartographic material showed that currently there is an active development of gully systems, especially in the central historical part of the city due to the technical capabilities and investment attractiveness. The nature of the interaction of gully systems and the city is overwhelming.

Acknowledgements
The publication was prepared with financial support of RFBR and the Government of the Republic of Tatarstan, project No. 17-13-16003, a (r).

References
[1] Safina G R, Fedorova V A, Sirotkin V V, Gasanov I M 2016 *International Journal of Pharmacy and Technology* **8**(3) 14864-71
[2] Safina G R and Fedorova V A 2012 *Geography at school* **5** 9–14
[3] Fedorova V A and Safina G R 2016 *Regional Geographical and Ecological studies: Current Problems* (Cheboksary) pp 448–456
[4] *Great Dictionary of the Russian language* 1998 (St. Petersburg: Norint) ed S A Kuznetsov p 1534
[5] Nayfeld L R and Tarasov N A 1968 *Development of Inconvenient Land for Urban Development* (Moscow) p 223
[6] Reznitskaya L M and Bergman O A 2017 *Electronic Scientific Journal Engineering Bulletin of Don* **2**
[7] Baymuratova S K 2005 *Dynamics Of Development Of Badland Territories In The structure Of A Large City, Case Study of Ufa.* Abstract of Cand. Diss. To The Soot. Degree of Cand. Architecture (Moscow) p 175
[8] *The Kazan Cartographer* (http://tat-map.ru/).
[9] Kovalev S N 2011 *Gully Systems in Cities* (Moscow: Print Coop Company) p 138
[10] Vanmaercke M et al 2016 *Earth-Science Reviews* **154** 336–355
[11] Golosov V, Gusarov A, Litvin L, Yermolaev O, Chizhikova N, Safina G and Kiryukhina Z 2017 *Integrating monitoring and modelling for sediment dynamics Okehampton, UK, 11–15 July 2016, Proc. IAHS* (Copernicus Publications) eds A Collins, M Stone et al vol 375 pp 23–27
[12] Zharkova N and Kupriyanov E V 2005 *Herald of the Tatarstan branch of the Russian Ecological Academy* (Kazan: Ekotsentr) №4 (26) pp 15–22

Preliminary estimating the contemporary sedimentation trend in dry valley bottoms of first-order catchments of different landscape zones of the Russian Plain using the ^{137}Cs as a chronomarker

A Sharifullin A Gusarov, A Gafurov and B Essuman-Quainoo

Institute of Environmental Sciences, Kazan Federal University, 18, Kremlyovskaya str., 420008, Kazan, Russia

E-mail: AGSharifullin@kpfu.ru, avgusarov@mail.ru

Abstract. A general trend of erosion processes over the last 50–60 years can be estimated by dating sediments washed off from arable lands and accumulated in the first-order dry valleys bottoms. Three small (first-order) catchments were chosen as objects of the study. They are located, respectively, in the southern part of the taiga zone, the zone of temperate broad-leaf forests and the forest-steppe zone of the Russian Plain. To date the sediments accumulated in the bottoms the radioactive caesium-137 (^{137}Cs) of global (since 1954) and Chernobyl origin (1986) had been used as a chronomarker. The average (for all the catchments) sedimentation rates during the global ^{137}Cs fallout period (1963(1954)–1986) are at least 0.88–2.71 cm per year.For the period that has passed since the Chernobyl accident (1986–2015(2016)) the average rates were 0.15–1.07 cm per year. The greatest reduction in the sedimentation rates is observed in the subzone of the southern taiga, the lowest one is in the forest-steppe zone of the Russian Plain. The main reason for such significant reduction in the rates of sedimentation of the soil erosion products in the dry valley bottoms was a reduction of surface runoff within the catchments during a snowmelt period, as well as crop-rotation changes there.

1. Introduction

The monitoring of erosion/accumulation processes within agricultural lands of Russia has not been implemented in recent decades due to high financial costs. Therefore, the actuality of estimating the processes increases year by year. This paper takes a look at the accumulation rates estimation of the soil/ground particles washed off from the catchment slopes, using the caesium-137 (^{137}Cs) as a chronomarker. It was made on the example of some catchments within the forest and forest-steppe zones of the Russian Plain. The approach has been applied by various researchers in many regions of the World [1–7*etc*]. The caesium-137 is an artificial origin isotope which appeared in the environment from the beginning of nuclear testing in 1954. There are ^{137}Cs of global fallout throughout the Earth with atmospheric precipitation during the nuclear tests (figure 1) in the atmosphere (with the extreme fallout in 1963), and the Chernobyl-derived ^{137}Cs associated with the accident at the Chernobyl nuclear power station (April–May, 1986), fell mainly in Eastern and Central Europe, Scandinavia.

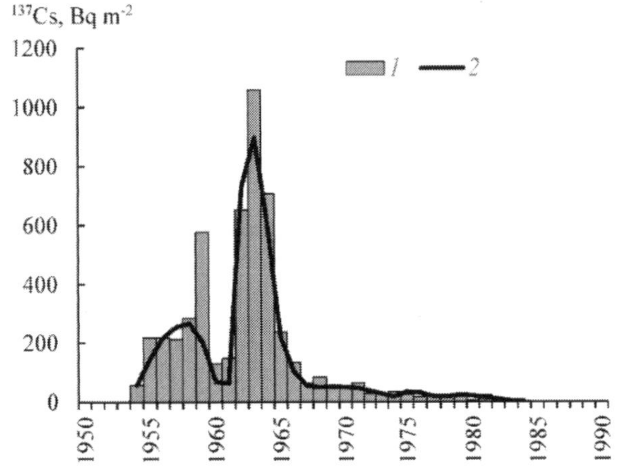

Figure 1. The ^{137}Cs fallout in the Northern hemisphere (*1*) [4] and the Leningrad Region of the former USSR (*2*) [8].

According to modern research [4], the extreme concentration of the Chernobyl-derived ^{137}Cs persists in a soil depth of 10–20 cm, and, depending on the granulometric composition of the soils, can migrate at different rates (from 0 to 0.35 cm per year) [9–13] during first years.

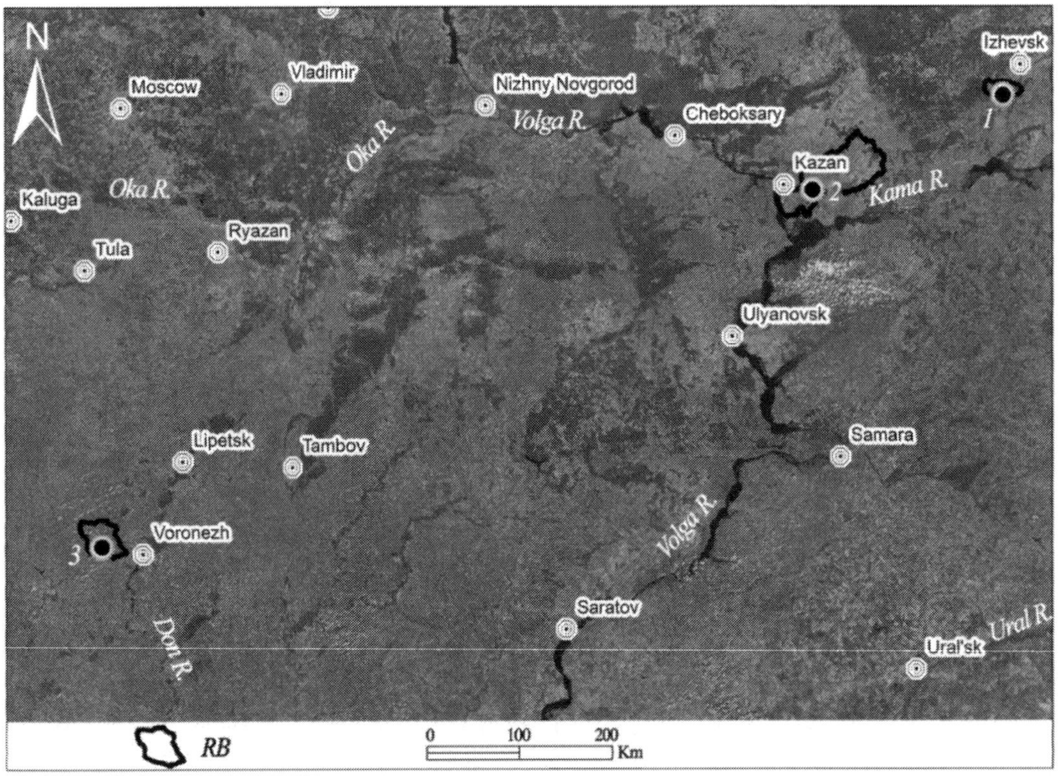

Figure 2. Location of the catchments under study in the Russian Plain.
The catchments: 1 – the Kuregovo (the Udmurt Republic), 2 – the Temeva Rechka (the Tatarstan Republic), 3 – the Veduga (the Voronezh Region); RB– the river basins.

Table 1. Some morphometric parameters of the catchments and their dry valleys.

The parameters	The catchments		
	Kuregovo	Temeva Rechka	Veduga
Catchment area, km^2	0.68	1.13	1.15
Average absolute height, m	167	161	177
Altitude amplitude, m	76	74	43
Dry valley length (with the hollows), m	400 (680)	1635 (2087)	1183 (1463)
Dry valley bottom area, km^2	0.006	0.024	0.020

2. Study area

To assess the erosion/sedimentation intensity trend, three key dry valley catchments located in different landscape zones of the Russian Plain, from the southern taiga to the forest-steppe zone, were selected: the Kuregovo catchment within the Agryzka River basin (the Udmurt Republic), the Temeva Rechka catchment within the Myosha River basin (the Republic of Tatarstan) and the Veduga catchment within the Veduga River basin (the Voronezh Region) (figure 2). The surfaces of the catchments under study are formed on quaternary deluvium/solifluction loams in the case of the Kuregovo and the Temeva Rechka, and on quaternary loess in the case of the Veduga.

The lengths of the dry valley bottoms of the catchments vary from 400 to 1635 m (table 1). Moreover, the Temeva Rechka catchment is dissected by a dense network of gentle hollows (up to 1.5 km long); the Veduga catchment has the same land forms only on its left side, and in the Kuregovo catchment the hollows are sporadic. There are several small erosion head-cuts (up to 0.5 m deep) in the dry valley bottoms of the catchments located in the Republic of Tatarstan and the Udmurt Republic. The average absolute heights in the key catchments vary from 161 to 177 m (table 1). The climate type is temperate continental with mean annual temperature from 2.7 °C in the Udmurt Republic (in the southern taiga subzone with coniferous and coniferous-deciduous forests) through 4.6 °C in the Tatarstan Republic (the temperate broad-leaf forests zone) to 7 °C in the Voronezh Region (the forest-steppe zone). The distribution of atmospheric precipitation over the territory of the Russian Plain depends on the western air masses transfer. The mean annual precipitation in all three key catchments varies from 522 mm (the catchment in the Udmurt Republic) through 553 mm (the Temeva Rechka catchment) up to 559 mm (the catchment in the Voronezh Region).

The most of the precipitation (66–74%) fall during the warm period (April–October). Over the past 30 years the precipitation increased in total amount in all the regions studied, as well as the number of rainfalls exceeding 10 mm, which can create surface runoff, increased.

Sod-podzolic soils (*Umbric Albeluvisols Abruptic* (WRB, 2006)) are predominant within the Kuregovo catchment, gray forest soils (*Greyic Phaeozems Albic* (WRB, 2006)) – within the Temeva Rechka catchment, and chernozems (*Haplic Chernozems (Aric, Pachic)* and *Luvic Chernozems* (WRB, 2006)) – within the Veduga catchment.

All the catchments studied are well ploughed (ploughed lands are more than 80% of total catchment area). Also, there are agricultural fields located very close to the dry valleysbrows. In the course of the last decades, even after the USSR collapse, according to Landsat space images, the ploughed areas were not abandoned.

3. Materials and methods

At each of the key catchments, the high-resolution aerial survey was carried out using unmanned aerial vehicle DJI Phantom 4 followed by the generation of digital elevation model (DEM) of the catchments (including their dry valleys). Various catchment elements such as the cropland/grassland borders, the hollows, the erosion cuts, furrows, runoff paths were extracted from the DEM. Based on obtained results, 3–4 soil (*Stratozems* (WRB, 2006)) sections (pits) were prepared in each of the dry valley bottoms (in different parts of the bottoms according to the peculiarities of sedimentation and sediment transit there).

In each section (pit), layer-by-layer soil samples from fixed area (15×15 cm, 10×10 cm) were vertically taken through 2, 3, 5 cm for the following ^{137}Cs-activity analysis.

The specific ^{137}Cs activity in the soil samples was determined using a coaxial germanium γ–spectrometer at the research laboratory of soil erosion and riverbed processes of the Lomonosov Moscow State University (Russia). Previously, all the soil samples were dried to an absolutely dry state at a temperature of 105°C and sieved through a sieve with a hole diameter of 1 mm. Based on results of the laboratory analysis, diagrams of the ^{137}Cs vertical distribution were constructed and the mean annual sedimentation rates for two time intervals (1963(1954)–1986 and 1986–2015 (2016)) were calculated.

Simultaneously, the meteorological data (for the period 1950–2015) have been collected for several weather stations located in the vicinities of the catchments under study in the cities of Kazan, Nolinsk, Sarapul, Cheboksary, Bakaly, Voronezh, Stary Oskol and Izhevsk to identify some regional features of the hydrometeorological characteristics changes (the materials from the All-Russian Research Institute of Hydrometeorological Information – World Data Center, ARRIHI–WDC; http://meteo.ru).

4. Results

In the course of detailed investigation of the dry valleys and description of the soil sections (pits) it was established that the bulk of the deposits accumulated in the dry valley bottoms comes from the ploughed lands due to sheet/rill erosion. Also, an additional source of the sediments was the material formed due to the receipt of the material from the gullies formed on the left side of the Veduga catchment and due to the erosion head-cuts within the catchments of Kuregovo and Temeva Rechka, mainly in the period until 1986.

The ^{137}Cs distribution in all the catchments under study is not the same. The greatest concentrations are observed in the Voronezh Region and the Udmurt Republic (figure 3), since in these areas, despite their different distance from the Chernobyl nuclear power station, a greater amount of the ^{137}Cs fell out in May 1986 immediately after the accident.

The Chernobyl-derived ^{137}Cs peak (1986) is quite clearly distinguished in the bottom sediments of all the dry valleys under study. The global ^{137}Cs peak (1963) is not defined so clearly on the depth distribution curves except the Temeva Rechka catchment (figure 3). The main reason for these peaks absence is a linear erosion of the sediments in the bottoms of the dry valleys and subsequent sediment redeposition occurred before 1986, and also, in some cases, insufficient depth of sampling for the ^{137}Cs content determination. That is why the accuracy of determining the sedimentation rates during the period 1963–1986 in the catchments of Kuregovo and Veduga was lower than for the period 1986–2015(2016). It was estimated taking into account the vertical ^{137}Cs migration which is approximately 0.35 cm per year during first 3–4 years with following reduction to 0.1 cm per year [5, 7], and the lower horizons are conventionally dated as potentially close to 1954, that is, in fact, the sedimentation rates were somewhat underestimated. In the Kuregovo catchment located in the subzone of the southern taiga it is established that the mean annual sedimentation rates were at least 1.8–2.5 cm per year and 0.15–0.75 cm per year for periods 1954(1963)–1986 and 1986–2015(6) respectively. In the zone of temperate broad-leaf forests (the Temeva Rechka catchment) the rates were 0.92–1.81 and 0.17–0.50 cm per year respectively; in the forest-steppe zone (the Veduga catchment) the rates were 0.88–2.71 and 0.47–1.07 cm per year, respectively. So, the greatest reduction in the average sedimentation rates is observed in the subzone of the southern taiga, the lowest one is in the forest-steppe zone of the Russian Plain.

This sedimentation trend was primarily determined by the climate change in the region, as well as changes in crop rotation. The climate change has affected the reduction of surface runoff from the slopes during the snowmelt period, caused by an increase in air temperature and a decrease in a soil freezing depth in winter [14]. An agricultural activity impact on the reduction of the sedimentation (erosion) rates in all the dry valley bottoms (catchments) under study was not so significant, with the exception of the Kuregovo catchment, where the percentage of perennial grasses incrop-rotation structure has been increasing since the 1990s.

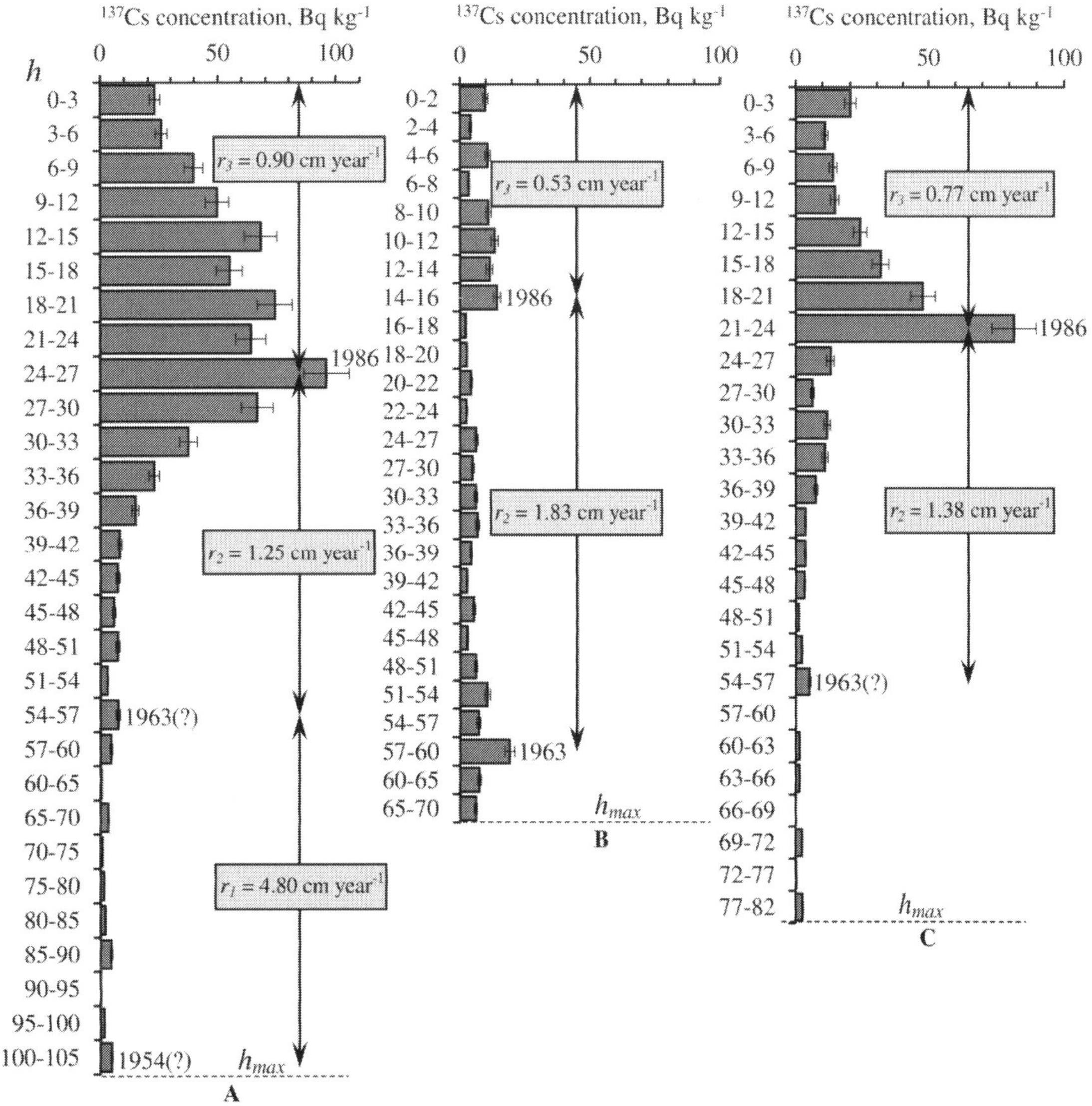

Figure 3. The vertical distribution of the ^{137}Cs concentration and sedimentation rates at the soil (sediment) sections (pits) within the dry valleys of the catchments under study.
The catchments: **A** – the Veduga (the Voronezh Region), **B** – the Temeva Rechka (the Tatarstan Republic), **C** – the Kuregovo (the Udmurt Republic); the average sedimentation rates for: r_1 – 1954(?)–1963, r_2 – 1963–1986, r_3 – 1986–2015(2016); h – the depths of soil (sediment) samples selection (h_{max} – the maximal depth of the selection)
Note. The figure shows only the upper pits in the dry valleys of the catchments.

5. Conclusions

The estimation of the sedimentation rates over two time intervals in the dry valley bottoms in the south of the forest zone (subzone of the southern taiga and the zone of the broad-leaf forests) and in the forest-steppe zone allows to speak about a significant decrease in the soil-erosion rates within the ploughed parts of the catchments over the last 30 years. The average rates of the soil material accumulation in the first-order valleys bottoms within all the catchments under study decreased at least 3–5 times. The main reason for such significant reduction in the rates of the erosive degradation of the ploughed soils was a decrease in the surface runoff from the catchment slopes during the snowmelt period caused by the general climatic warming in the Russian Plain. The agricultural activity influence on the indicated sedimentation (erosion) dynamics within the catchments under study was apparently insignificant, possibly with the exception of the subzone of southern taiga (the coniferous and coniferous-deciduous forests).

Acknowledgments
The reported study was funded by the Russian Science Foundation according to the research project no. 15–17–20006.

References

[1] Porto P, Walling D E and Callegari G 2011 *Hydrological Processes* **25** 886–900
[2] Golosov V N, Panin A V and Markelov M V 1999 *Physics and Chemistry of the Earth, Part A: Solid Earth and Geodesy* **24** 881–885
[3] Parfitt R L, Baisden W T, Ross C W, Rosser B J, Schipper L A and Barry B 2013 New Zealand *Geoderma* **192** 154–159
[4] Zapata F 2002 *Handbook for the assessment of soil erosion and sedimentation using environmental radionuclides* vol 219 (Springer)
[5] Golosov V N, Belyaev V R and Markelov M V 2013 *Hydrological Processes* **27** 781–794
[6] Golosov V N, Ivanova N N, Gusarov A V and Sharifullin A G 2017 *Eurasian Soil Science* **10** 1195–1208
[7] Zhang X and Walling D E 2005 *Journal of Environmental Quality* **34** 514–523
[8] Silant'ev A N and Shkuratova I G 1983 *Detecting Industrial Soil Contaminants and Atmospheric Fallout against the Global Contamination Background* (Leningrad: Gidrometeoizdat)
[9] Almgren S and Isaksson M 2006 *Journal of Environmental Radioactivity* **91** 90–102
[10] Karadeniz Ö, Çakir R and Karakurt H 2015 *Journal of Environmental Radioactivity* **146** 27–34
[11] Ciszewski D, Czajka A and Błażej S 2008 *Environmental Geology* **55** 1577–86
[12] Kirchner G, Strebl F, Bossew P, Ehlken S and Gerzabek M H 2009 *Journal of Environmental Radioactivity* **100** 716–720
[13] Korobova E M, Dogadkin N N, Shiryaev A A, Kolotov V P, Kononkova N N and Turkov V A 2014 *Journal of Geochemical Exploration* **142** 94–100
[14] Golosov V, Gusarov A, Litvin L, Yermolaev O, Chizhikova N, Safina G and Kiryukhina Z 2017 *Proceedings of the International Association of Hydrological Sciences* **375** 23–27

IOP Conf. Series: Earth and Environmental Science **107** (2017) 012023 doi:10.1088/1755-1315/107/1/012023

GIS-technologies application for calculation of potential soil loss of Marha River basin (Republic of Saha)

Y Shynbergenov, K Maltsev and N Sihanova

Institute of Environmental Sciences, Kazan Federal University, 18, Kremlevskaya str., 420008, Kazan, Russia

E-mail: shynbergenov.erlan@mail.ru

Abstract. In the article the presentation of estimation methods of potential soil loss in the conditions of Siberia with application of geographical information systems is resulted. For the reference area of the Marha river basin, which is a part of the Lena river catchment, there was created a specialized geographic information database of potential soil erosion, with scale of 1: 1,000,000. Digital elevation model "GMTED2010" and the hydroset layer corresponding to the scale of 1: 1,000,000 are taken to calculate the soil loss values. The formation of the geobase data is considered in detail being constructed on the basis of the multiplicative structure which reflects the main parameters of the relief (slope steepness, exposition, slope length, erosion potential of the relief), soil, climatic characteristics and modern types of land cover. At the quantitative level with sufficiently high degree of spatial detail results were obtained for calculating the potential erosion of soils. The average value of potential soil loss in the basin without taking into account the factor of land cover types, was 12.6 t/ha/yr. The calculations carried out, taking into account the types of land cover obtained from remote sensing data from outer space resulted in an appreciable reduction of the soil loss values (0.04 t/ha/yr.).

1. Introduction

The territorial development of Siberia and Far East, indicated in the plans of the Russian Federation Government, predetermines the scientific substantiation of projects on economic development and prediction of anthropogenic impact. Territorial analysis of geographic features and phenomena is most conveniently carried out on the basis of the basin approach [1, 2]. This also can be applied to such a process as erosion.

The works on calculating the potential loss of soils within the basins of different order by means of GIS technologies is widely spread both abroad and in Russia. The review of publications on river basins in Siberia showed the lack of systematization of works on this topic. Some works concern the problem of determination of long-term dynamics of water balance. Soil-erosion processes are taken only in the individual basins of the rivers Selenga, Tom', Ob, Irkut [3, 4, 5, 6, 7].

The purpose of this study is to quantify the potential soil erosion in the Marha River basin. The research tasks included the choice of the design model of potential soil loss, the selection of source materials, their transfer to the geo-information database and the calculation of the potential soil loss in the basin.

2. Study area

The object of study is the basin of the Markha river (Yakutia), flowing in conditions of lowlying relief, the catchment area of 89600 km^2 [8] which falls into the Vilyuy river – left tributary of the Lena.

3. Materials and methods

Calculations were carried out on the basis of methods related to the class of hydrophysical models. The soil loss model recommended by the Laboratory of Soil Erosion and Fluvial Processes of the Moscow State University was used which is a modified version of the USLE (Universal Soil Loss Equation) for Russia (rainfall runoff) and the RSHI (Russian State Hydrological Institute) loss formula (meltwater runoff) [9].

The main formulas used in determining the potential loss of soil are [10]:

Soil loss from rainfall runoff:

$$LR = EPO \cdot ES \cdot EPR \quad (1)$$

where LR – Soil loss from rainfall runoff, t/ha/yr., EPO – Eroding potential of rainfall runoff, ES - Soil erodibility, t/ha per unit EPO, EPR - Eroding potential of relief.

Soil loss from meltwater runoff:

$$LM = f \cdot (ES \cdot h \cdot L \cdot I) \quad (2)$$

where LM – Soil loss from meltwater runoff, t/ha/yr.,ES – Soil erodibility, h – Layer of surface runoff (reserves of water in snow, precipitation for during snowmelt, runoff coefficient), L – Slope length, m., I – Slope steepness.

Potential soil loss:

$$LP = LR + LM \quad (3)$$

The presented empirical dependencies, when performing calculations using GIS-tools are adapted for computer technology; a detailed description is contained in the monograph [9]. The loss model is based on the creation of a specialized geoinformation database using GIS tools composed of four blocks, i.e. multiplicative structure (table 1). All layers of thematic maps are presented in the form of point files [10].

Table 1. Input datasets used for the estimation of soil loss factors for Marha River catchment

Thematic factor	Parameter	Input datasets	Spatial resolution
Morphometric analysis of relief	Slope length Slope steepness Exposition	DEM GMTED2010	250x250 м
Morphological-genetic properties of soils	Soil number and name Humus content Class by structure Water permeability class Coefficient of stony Fraction of particles 0.1-0.05 mm from fraction content of 0.25-0.05 mm. Mechanical composition of soils	Soil maps in vector format; The Unified State Register of Soil Resources of Russia	1:2,500,000
Climatic characteristics	Erosion potential of rainfall (EPO) EPO Zone Reserves of water in snow	Data from Roshydromet meteorological stations FSBO "RIHMI-WDC"; Map-scheme "Regionalization of the North of Eurasia for the Interannual Distribution of the Erosion Potential of Precipitation"	250x250 м
Soil protection factor of cover and management	Types of land use	Map of land cover types	230x230 м

Morphometric analysis of relief. Digital relief models (DEM) can be divided into two categories: open access and distributed on a commercial basis. We used only open sources of data including DEM. Among them, the most qualitative global DEM for the region under study is GMTED2010 (Global Multi-Resolution Terrain Elevation Data 2010) [11]. The preparation of model of the relief is given in sufficient detail in a number of publications [11, 12, 13, 14]. On the basis of the used model of relief GMTED2010, raster layers of the morphometric characteristics of relief have been prepared: slope length and steepness, exposition.

Morphological-genetic properties of soils. As basic data the "Soil Map of Russia" which is in open access was attracted. It is created in a vector format, in scale of 1:2,500,000 by Soil institute of V.V. Dokuchayev. On this map, all polygonal objects are digitized contours of the RSFSR soil map [15, 16, 17]. The description of this map was obtained from the Unified State Register of Soil Resources of Russia (USRSR RF) [18]. Each polygon corresponds to the soil profile, to which data on the soil cover and soil-forming rocks are attached, there are altogether more than 240 soil parameters.

The database of potential soil loss contains the column "Soil Index" where to each index there corresponds the soil profile information which is filled in a separate form no. 2, presented in methodical instructions [10] and contains 13 parameters: soil name, humus content, class of a structure, waterpermeability, stoniness, granulometric composition, proportion of sand, losses norm and so forth. According to Soil map in Marha River basin there are 25 soils subtypes. Subsequently, a number of them were excluded from loss calculation due to the actual absence of erosion due to the spatial position in the basin. Areas relating not to soil formations are excluded from the soil map. These subtypes of soils and not soil formations subsequently were excluded from the map of Marha River basin.

Climatic characteristics. For definition of climatic indicators data of "RIHMI-WDC" are used: coordinates of meteorological stations [19]; air temperature and amount of precipitation (daily data) [20]. As additional materials for loss calculation the published cartographic materials were also attracted (erosive potential of rainfall, intra annual distribution (zones) EPO [9] and the manuscript map of water reserves in snow which was kindly provided by employees of Laboratory of Soil Erosion and Fluvial Processes. These cards have been digitized in the MapInfo software product with the subsequent rasterization, creation of dot files in the GIS-environment ArcMap.

Soil protection factor of cover. Types of a terrestrial cover need to be considered from a position of their influence on soil loss. Types of a vegetable cover allow to correct soil loss towards reduction, and such types as water objects, rocks, the urbanized territories etc., are excluded in general, during calculation.

The geodatabase on land cover types was created on the basis of materials of Space research Institute [21]. Values of factors vary from 0 (the protected soils) to 1 (open soils) [22]. Maps of erosive danger of lands are the results of modeling for the studied Lena River basin (figure 1). In our calculations, factor C (land use type) is used since factor P implies the consideration of an agrofon (borders of farms, pastures, etc.), which is practically absent in the basin. Details of the procedure are given in the work of Panagos et al [23].

4. Results

The created geoinformation database formed the basis of calculation of potential soil losses in a Lena river basin on earlier given formulas (1-3). Calculations are carried out in two steps: at first, potential soil loss is determined without a real picture of distribution of a vegetable cover, secondly – taking into account the types of land cover (table 2).

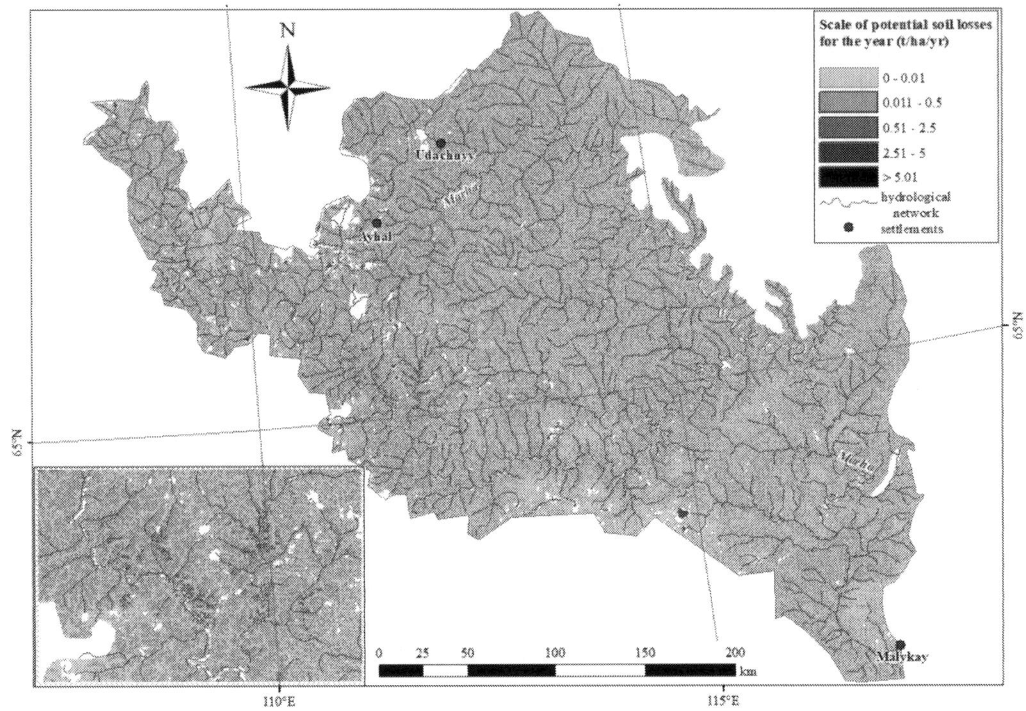

Figure 1. Lena River basin potential soil loss map.

Table 2. Lena river basin distribution by gradations of potential soil erosion.

Soil loss rates, t/ha/yr.	Soil loss (without considering the types of land cover)		Soil loss (taking into account the types of land cover)	
	Area, км^2	% of total area	Area, км^2	% of total area
0-0.01	36417	45.2	39053.4	49,9
0.01-0.5	109.06	0.1	38517.6	49,2
0.5-2.5	3018.4	3.7	733.8	0,9
2.5-5	4987.4	6.2	7.1	0 (0.01)
5-10	8668.3	10.8	0.6	0 (0.0008)
10-50	22596.3	28.1	-	-
50-100	3772.1	4.7	-	-
100-1000	944.4	1.2	-	-
>1000	0.6	0 (0.0007)	-	-
Total	80513.6	100	78312.5	100

5. Conclusions

Specialized cartographical geoinformation database of the Arctic catchment basins of an Asian part of Russia is created for the first time. As part of the database, the quantitative indexes of the parameters influencing soil erodibility are grouped according to thematic blocks (morphometric analysis of the relief, morphological and genetic properties of soils, climatic indices). Calculations of potential soil loss of the catchment basin of the Marha River were carried out based on the application of the model recommended by the Laboratory of Soil Erosion and Fluvial Processes. Based on the calculations performed, it was found that average soil loss rate per Marha river basin is 12.6 t/ha/yr, the greatest 1314.7 t/ha/yr., which corresponds to the maximum possible scenario of erosion. After calibrating the factors of model, taking into account the types of land cover, average soil loss rate per study area was

0.04 t/ha/yr., the maximum 7.8 t/ha/yr. Considering landscape and geographical features of the Lena river basin these values more correspond to realities.

Acknowledgments
The authors expresses their sincere gratitude to the employee of Laboratory of Soil Erosion and Fluvial Processes Mr. S.F. Krasnov for the given opportunity to use the soil loss calculating technology. The work is performed according to the Russian Science Foundation (15–17–10008).

References
[1] Lisetskii F N, Buryak J A, Zemlyakova A V and Pichura V I 2014 *Biogeosystem Technique* **2 (2)** 163–173
[2] Buryak Zh A 2014 *Nauchnye vedomosti Seriya Estestvennye nauki* ***23(194.29)*** 140–146 (in Russian)
[3] Ryzhov Yu V 2009 *Geographiya i prirodnye resursy* **3** 94–101 (in Russian)
[4] Haptuhaeva N N 2012 *Agrarnaya nauka Evro–Severo–Vostoka* **1** 24–26 (in Russian)
[5] Haptuhaeva N N and Tarmaev V A 2014 *Vest. SVNC DVO RAN* **3** 41–45 (in Russian)
[6] Golubev I A 2009 *Vest. Kras GAU, zemleustroystvo, kadastr i monitoring* **1** 80–83 (in Russian)
[7] Knaub R V 2006 *Geographical analysis of surface flushing factors and assessment of modern erosion on arable lands of Tom-Yai interfluve (within the Tomsk region)* Abstract of diss. cand. geo. sc. Tomsk p 11 (in Russian)
[8] Dedkov A P and Mozzherin V I 1984 *Erosion and sediment yield of Earth* (Kazan, Izd-vo KazSU) p 264 (in Russian)
[9] Larionov G A 1993 *Erosion and deflation of soils: basic patterns and quantitative estimates* (Moscow, Izd-vo MSU) p 200 (in Russian)
[10] Larionov G A et al 1996 *Methodological guidelines for the compilation of large-scale maps of erosion-hazardous lands for the justification of soil protection measures for on-farm land managemen*t (Moscow, Fondy Roszemproect) p 47 (in Russian)
[11] Danielson J J and Gesch D B 2011 *Global multi-resolution terrain elevation data 2010 (GMTED2010)* Open-File Report – 1073 (Reston U S Geological Survey) p 26
[12] Danielson J J and Gesch D B 2008 *International Archives of the Photogrammetry Remote Sensing and Spatial Information Sciences* **XXXVII (B4)** 1857–63
[13] Mal'tsev K A and Yermolayev O P 2014 *Geomorphology* **1** 45–53 (in Russian)
[14] Yermolayev O P, Mal'tsev K A and Ivanov M A 2014 *Geography and Natural Resources* **35**(3) 222–228 (in Russian)
[15] *Soil map of RSFSR, Scale 1:2 500 000* 1988 (Digitized original of the same name Soil map) ed Fridland V M (Moscow, *GUGK*) (in Russian)
[16] *The Unified State Register of Soil Resources of Russia* 2014 (Moscow, Soil institute of V V Dokuchaev) p 768 (in Russian)
[17] Rozhkov V A, Aliabina I O, Kolesnikova V M, Molchanov E N, Stolbovoy V S and Shoba S A 2010 *Pochvovedenie* **1** 3–6 (in Russian)
[18] *The Unified State Register of Soil Resources of Russia in Internet-resource of official support of the USRSR* [Electronic resource] http://egrpr.esoil.ru/index.php free Verified 18.06.2017 (in Russian)
[19] *Coordinates of meteorological station* (All-Russian Scientific Research Institute of HydroMeteorological Information – World Data Center (RIHMI-WDC) Federal Center for Hydromet. and Envir. Monit.) [Electronic resource] http://meteo.ru/data/155-meteostations free Verified 18.06.2017 (in Russian)
[20] *Air temperature and precipitation (daily data)* (All-Russian Scientific Research Institute of HydroMeteorological Information – World Data Center (RIHMI-WDC) Federal Center for Hydromet. and Envir. Monit.) [Electronic resource] http://meteo.ru/data/162-temperature-precipitation free Verified 18.05.2017 (in Russian)

[21] Bartalev S A, Egorov V A, Efremov V Yu, Lupian E A, Stitsenko F V and Flitman E V 2012 *Sovremennye problemy distantsionnogo zondirovania Zemli iz kosmosa* **9(2)** 9–27 (in Russian)
[22] Renard K et al 1997 *Agric. Handb.* **703** 404
[23] Panagos P et al 2015 *Environ. Sci. Policy Elsevier Ltd.* **54** 438–447

Spectroscopic methods for determining of zonal soils erosion (Chuvash Republic, Russia)

V V Sirotkin[1], S V Vasyukov[2], B M Usmanov[1] and G F Gainutdinova[1]

[1] Institute of Environmental Sciences, Kazan Federal University, 18, Kremlevskaya str., 420008, Kazan, Russia
[2] Federal Service for State Registration, Cadastre and Cartography, 56, Karl Marx str., 428003, Cheboksary, Russia

E-mail: busmanof@kpfu.ru

Abstract. Results of spectrographic study of zonal soils with different erosion degree on the territory of the Chuvash Republic are given. In the points of spectrographic research, the soil specimens were taken and agrochemically tested. Changes in received spectrograms were analyzed in soils subtypes context with subsequent comparison with agrochemical indicators. The obtained data will help to develop new rapid methods for determining soil fertility parameters for agricultural lands based on spectroscopic physical principles.

1. Introduction

There are many classifications of eroded soils but all variety of methods for determining erosion can be reduced to four large groups:

1. Based on the morphological differences between eroded and non-eroded soils, with taking into account the nature of the relief.
2. Based on the change in humus content.
3. Based on other chemical, biological or physico-chemical changes occurring in eroded soils.
4. Using several of the above listed indicators.

According to the classification of eroded soils, published in "All-union instruction on soil surveys and large-scale soil maps of land use composition", USSR standard, soil washout in comparison with the unwashed soils was taken as a basis of soils washout classification degree.

An exception was the classification of eroded sod-podzolic and light gray forest soils, where the degree of erosion is determined for agricultural soils by complex of factors: what horizons are under the arable layer, dependence on the slope inclination, and so on [1, 2]. In some classifications [3, 4, 5], indicators characterizing the relief (steepness, length, exposure of slopes, degree of slope dissection), soil-forming rocks, and erosion resistance of soils are used [6]. Currently, several common field methods for soil erosion determination are used for Chuvash Republic territory. Visually-morphological classification developed by S.S. Sobolev [7, 8] and adapted to the conditions of Chuvashia by S.I. Andreev and F.Y. Mikhailov [9, 10] is the most well-known. S.I. Andreev used a set of indicators as the basis for eroded agricultural soils classification and assessment: slope inclination, degree of humus horizon washout, decrease in genetic horizons thickness, plowing layer color, presence and quantity of rills, humus content (for loamy sod-podzolic and light gray forest soils), growth and development of agricultural crops. F.Y. Mikhailov used these diagnostic features of eroded soils in the context of soil types and degree of erosion [10]. It should be noted that the last Chuvash Republic soils descriptions received at 1980s, are still used in cadastral evaluation of agricultural lands as the basis for the tax accruals, calculation of land plots redemption price determining the size of penalties. Therefore, it should be realized that this information on the cadastral value of agricultural lands is at least unreliable. In addition, the given criteria of soil cover used to determine a significant reduction of agricultural land fertility and can be the probable cause for compulsory removal of agricultural land plots from the owners in accordance with the Russian legislation [11, 12].

In contrast to the geomorphological indicators that are easily described using GIS technology and external features of erosion that can be identified by analysis of remote sensing data, agrochemical parameters of soils (humus content, acidity, phosphorus and potassium content, etc.) are very dynamic and require direct field definition. As the network of representative sampling points should be sufficiently dense, development of accessible, low-cost and fast methods for determining agrochemical attributes of soil changes is required [13, 14, 15, 16]. In this article, the team of authors proposes the development of an express method for determining of soil erosion for agricultural lands based on spectroscopic physical principles. Infrared spectroscopy methods based on the interaction of molecules with electromagnetic energy in the infrared spectrum is recognized to be one of the most promising methods for soils study [17]. Technological advances in the application of the near and middle infrared spectrum over the past decades have made this band very popular in soil research and recognized as one of the most promising non-contact methods for determining soil parameters for various purposes [18, 19, 20]. Thus, the current state in the field of soils spectrometric analysis is contradictory. Apparently, there is a need for standardization of sampling and soil analyzes as well as geographical localization for specific types and subtypes of zonal soils [21, 22].

This paper presents the results of spectrographic studies of zonal soils on the territory of the Chuvash Republic.

2. Methods

In this work long-term observation points of the Federal Organization "State Center of Agrochemical Service "Chuvashskiy" organized in the well-studied soil areas with different soil varieties have been used.

The sample includes chernozem and gray forest soil varieties:

1. Dark-grey forest, light washout (9 samples).
2. Dark-grey forest, medium washout (40 samples).
3. Dark-grey forest, typical gley (light clay texture of surface soil layer) (10 samples).
4. Chernozems leached, medium thickness, medium-humic (7 samples).
5. Chernozems leached, light washout, medium loamy (9 samples).
6. Chernozems leached, light washout, (light clay) (11 samples).
7. Chernozems leached, light washout, (medium clay) (8 samples)
8. Chernozems leached, medium thickness, medium-humic (light clay) (7 samples).

To obtain the spectral characteristics of soils spectroradiometer HandHeld2 ASD was used. Besides, the main agrochemical indicators were obtained for all the samples as the criterion for agricultural soils rational use: Ph, humus, P2O5, K2O, total absorbed bases and hydrolytic acidity.

3. Results

Analysis and determination of spectrograms main types for the studied soils was carried out. As a result, five main types of spectrographic curves were identified for the studied soil varieties.

First type of the spectrographic curve (figure 1) includes all the studied varieties of soils with the exception of №8 – chernozems leached, medium thickness, medium-humic (light clay).

The second type of spectrographic curve (figure 2) includes all studied varieties of soils with the exception of №4 – chernozems leached, medium thickness, medium-humic.

The third type (figure 3) is typical for all studied varieties of soils, except №4 – chernozems leached, medium thickness, medium-humic and №5 – chernozems leached, light washout, medium loamy.

The fourth type of spectrographic curve (figure 4) was noted for №8 – chernozems leached, medium thickness, medium-humic (light clay) and №2 – dark-grey forest, medium washout and the fifth type (figure 5) of spectrographic curve is peculiar to dark gray forest medium-washed soil.

IOP Conf. Series: Earth and Environmental Science **107** (2017) 012024 doi:10.1088/1755-1315/107/1/012024

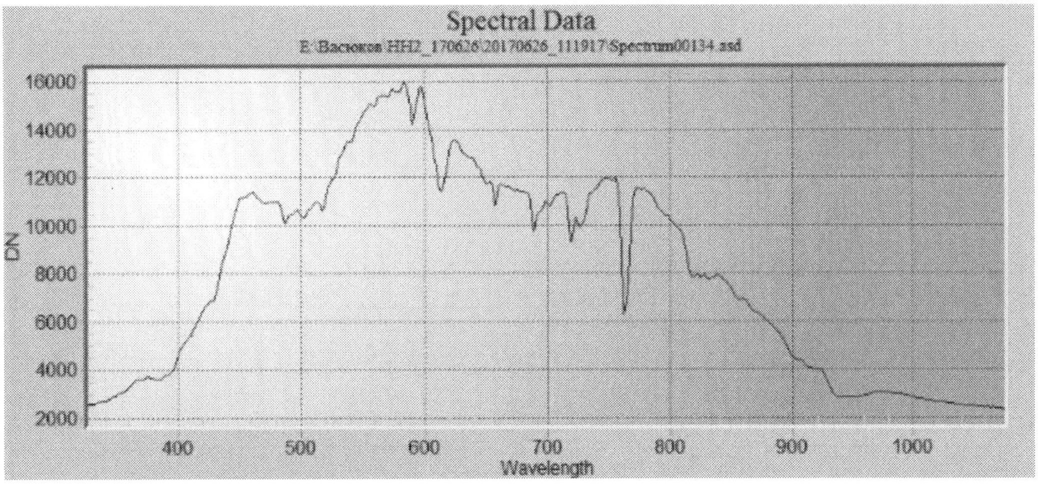

Figure 1. The first type of the spectrographic curve.

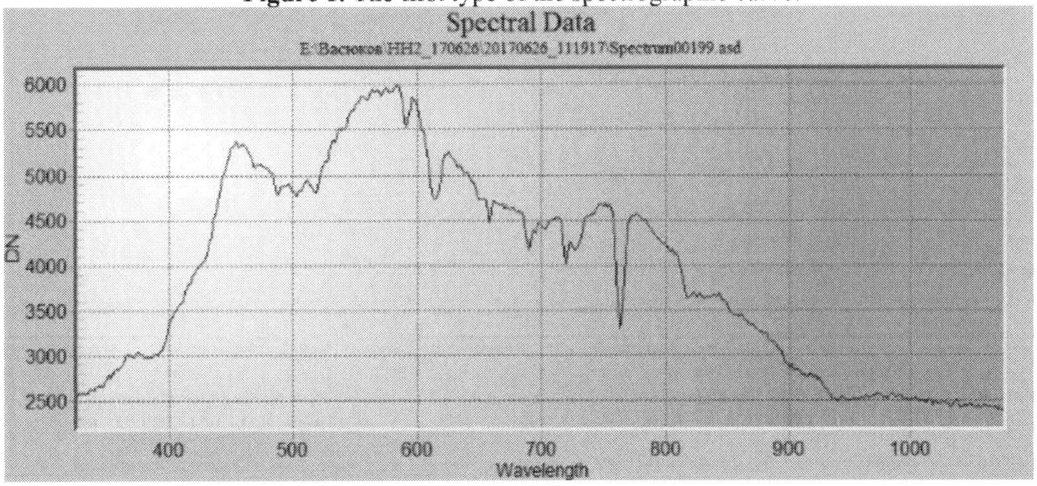

Figure 2. The second type of the spectrographic curve.

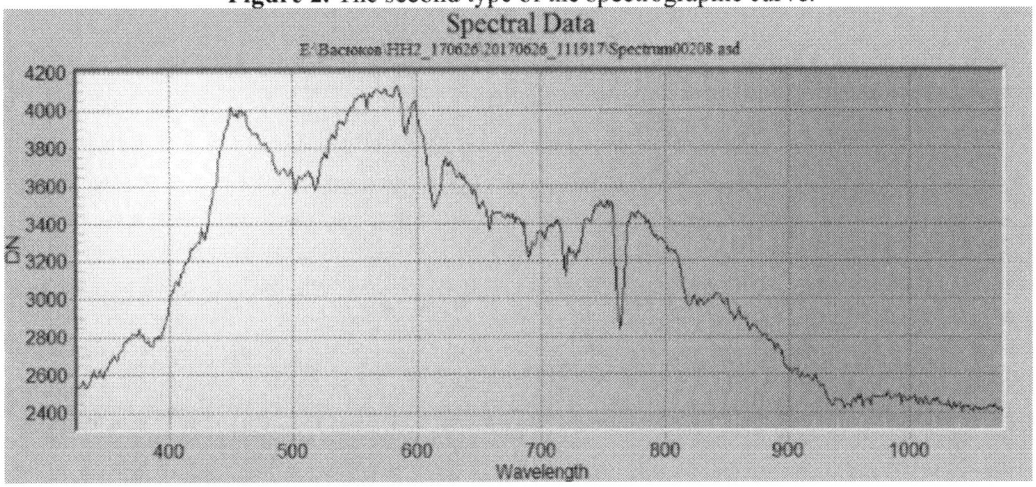

Figure 3. The third type of the spectrographic curve.

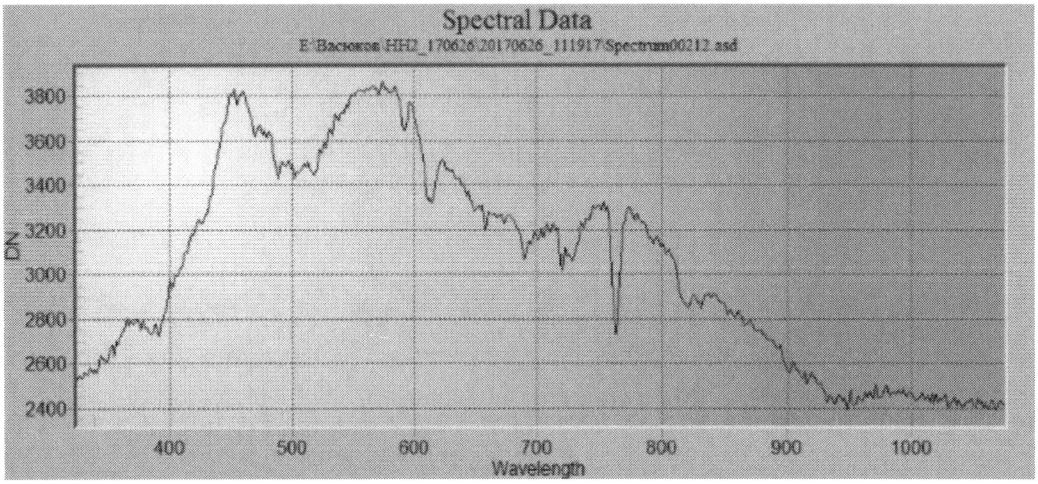

Figure 4. The forth type of the spectrographic curve.

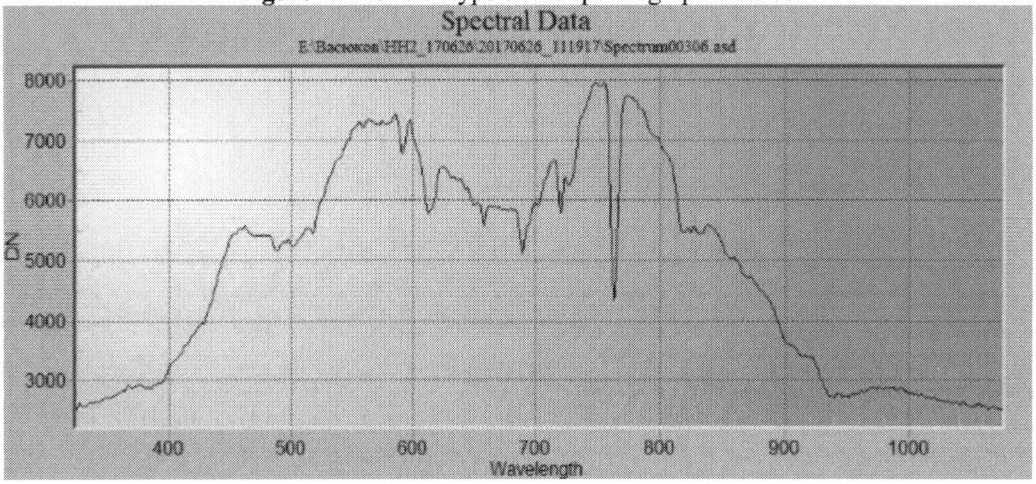

Figure 5. The fifth type of the spectrographic curve.

Further, the relationship between the spectrographic curves types and agrochemical indicators of soils was analyzed. The following results were obtained:

1. The first type of the spectrographic curve has maximum values of humus and K_2O;
2. The second type of the spectrographic curve has lower values of humus and K_2O;
3. Humus content and exchange potassium decreases from third to fifth type of the spectrographic curve.

Correlation analysis was performed to determine the relationships between agrochemical parameters and spectrographic curves types. Obtained values of correlation coefficients show statistically significant connections between spectrographic curve type and four agrochemical indicators. For the humus content and exchange potassium, an inverse correlation is observed – -0.73 and -0.66, respectively. Between the total absorbed bases and hydrolytic acidity the relationship is direct – 0.66 and 0.98, respectively.

Thus, spectrographic curve nature clearly reflects the changes in agrochemical indicators mentioned above. In this case, the marking indicator of humus content in different soil types is 550 and 750 nm peaks ratio. For well-humified soils the 550 nm peak is above the 750 nm peak. We can see the same picture with an increase in soils erosion degree within a single subtype. There is also a clearly traceable dependence of humus content and the 450 nm peak, especially within one type of soil. The more it extends within one soil type with unchanged other peaks, the lower is the humus content.

The obtained intermediate results make it possible to state about the possibility of using spectrographic methods to determine and to map soil erosion degree. However, for reliable interpretation of spectrographic data it is necessary to perform control measurements of soils with known humus content and erosion values. These can be pre-defined soil areas or control measurements of agrochemical indicators. In combination with a hypsometric basis (slopes inclination and length) the spectrographic data will allow to obtain maps of the soil cover erosion and determine the degree of soil humus content.

Thus, the results of the studies can be used as the basis for rapid mapping of soil erosion since spectrographic data takes time and material expenses on several orders lower than "classical" methods of studying and mapping of soil cover.

References

[1] *Klassifikatsiya i diagnostika pochv SSSR (Classification and diagnostics of soils of the USSR)* 1977 (Moscow: Kolos) p 257 (in Russian)

[2] *Obshchesoyuznaya instruktsiya po pochvennym issledovaniyam i sostavleniyu krupnomasshtabnykh pochvennykh kart zemlepol'zovaniya (All-union instruction on soil surveys and large-scale soil maps of land use composition)* 1973 (Moscow: Kolos) p 244 (in Russian)

[3] Kostyuchenko P A 1937 *Eurasian Soil Science* **8** (in Russian)

[4] Kozmenko A S 1948 *Agrolesomelioratsiya* (Moscow) p 156 (in Russian)

[5] Lidov V P and Didurenko N F 1955 *Eurasian Soil Science* **11** 15–21 (in Russian)

[6] Litvin L F 1998 *Geomorphology* **2** 13–22 (in Russian)

[7] Sobolev S S 1960 *Development of erosion processes in the European part of the USSR and their control* (Moscow) vol II (in Russian)

[8] Sobolev S S 1961 *Protection of soils from erosion and increase of their fertility* (Moscow: Selkhozizdat) p 231 (in Russian)

[9] Andreev S I 1971 *Soils of the Chuvash ASSR* vol 1. The history of development of soils of the republic and human impacts on them (Cheboksary: Chuvash publ. house) p 358 (in Russian)

[10] Mikhailov F Y 1973 *Reliable protection for our soils Proc.* (Cheboksary: Chuvashgiz) (in Russian)

[11] Federal Law № 101-FL *Concerning agricultural lands turnover* 24.07.2002

[12] Government Decree № 612 *Concerning approval of criteria of essential decrease of agricultural lands fertility* 22.07.2011

[13] Linker R 2011 *Fourier transforms – new analytical approaches and FTIR strategies* ed. prof. Goran Nikolic pp 385–404 Published online 01 April 2011

[14] Stenberg B, Viscarra Rossel R A, Mouazen A M and Wetterlind J 2010 *Advances in Agronomy* (Burlington: Academic Press) **107** 163–215

[15] Campbell D, Shiley D and Curtiss B 2013 *Measurement of soil mineralogy and CEC using near-infrared reflectance spectroscopy* (ASD Inc., a PANalytical company Boulder, Colorado, 80301, USA)

[16] McBratney A B, Minasny B and Viscarra Rossel R 2006 *Geoderma* **136**(1-2) 272–278

[17] Cécillon L, Barthès B G, Gomez C, Ertlen D, Genot V, Hedde M, Stevens A and Brun J J 2009 *E.J. of Soil Sci.* **60**(5) 770–784

[18] Rossel R A V, Walvoort D J J, McBratney A B, Janik L J and Skjemstad J O 2006 *Geoderma* **131**(1) 59–75

[19] Borenstein A, Linker R, Shmulevich I and Shaviv A 2006 *Applied Spectroscopy* **60**(11) 1267–1272

[20] Canasveras J C, Barron V, del Campillo M C, Torrent J and Gomez J A 2010 *Geoderma* **158**(1-2) 78–84

[21] Sirotkin V, Vasyukov S and Usmanov B 2016 *Water resources, forest, marine and ocean ecosystems conference Proc. SGEM 2016* book 3 vol II pp 357–362

[22] Sirotkin V, Vasyukov S and Usmanov B 2017 *Water resources, forest, marine and ocean ecosystems conference Proc. SGEM 2017* vol 17 issue 32 pp 639–646

Peculiarities of the thermal regime of the Russian plain depending on tidal oscillation Earth rotation speed

L M Akimov

Voronezh State University, University sq., 1, Voronezh, 394018, Russia
E-mail: akl63@bk.ru

Abstract. Typification of fields of anomaly of temperature in the central part of East European Plain depending on the main phases of the Moon taking into account these tidal fluctuations of speed of rotation of Earth is presented. The main regularities of spatial distribution of anomaly of temperature in December are revealed. The opposite dependence of distribution of anomaly of temperature on antiphases of the Moon is established.

1. Introduction

Despite numerous publications, the emergence of general reports on the problems of changing the contemporary climate [1], interest in this topic is not lost. This is due both to the instability of the modern climate, and to the possible negative and positive consequences of climate change in the natural and socio-economic spheres. There are still different views on the causes of the warming that is occurring [2-7].

The thermal regime of the territory is the main climatic indicator determining the development of economic activity. The increasing instability of weather conditions associated with global warming, currently observed, contributes to the development of long-term forecasting techniques involving various predictors. One of such directions is the use of data of tidal oscillations of the Earth's rotation speed due to the sun-lunar tides established by N S Sidorenkov [8].The thermal regime of the territory in this case is a reflection of the reaction of the change in the atmospheric circulation regime to oscillation in the rotation speed of the Earth.

The purpose of this article is to conduct studies of the temperature anomaly fields in the central part of the Russian Plain, taking into account the data of tidal oscillation of the rotation speed of the Earth. To realize this task, the temperature fields were typified according to the main phases of the moon in the territory under consideration.

2. Materials and methods

The moon is a natural satellite of the Earth, the closest celestial body to the Earth. The phases of the Moon are the periodically changing states of the illumination of the Moon by the Sun. There are four main phases of the moon, which gradually change into one another in the following sequence: new moon (phase I), first quarter (phase II), full moon (phase III), last quarter (phase IV). New Moon (Phase I) - The moon is in conjunction with the Sun; First Quarter (II phase) - The Moon forms the right quadrature with the Sun; Full Moon (III phase) - Moon in opposition to the Sun; The last quarter (phase IV) - the Moon forms a left quadrature with the Sun. This cycle is divided into four phases of approximately 7.5 days each. The moon passes its cycle from the new moon to the next new moon in 29.6 days.

Newton established that the motion of all celestial bodies is governed by the force of attraction. For example, the Earth moves in its orbit under the influence of the Sun's attraction, and the Moon - under the influence of the attraction of the Earth. On different days, the moon turns over different parts of the Earth. The moon draws its gravity a little to the atmosphere of the Earth to its side, making the atmospheric column higher, as, respectively, and atmospheric pressure [9]. This relationship between the phases of the moon and atmospheric pressure was first noticed in 1847 [10].

The Earth and the Moon revolve around the common center of gravity (barycenter) of the Earth-Moon system with a sidereal (relative to the stars) period of 27.3 days (day). The Earth presents an orbit that is a mirror image of the Moon's orbit, but its dimensions are 81 times smaller than the lunar orbit. The barycenter is always located inside the Earth, at a distance of about 4670 km from its center. The Earth's body turns without rotation (translationally) around the "stationary" (in the Earth-Moon system) barycenter. As a result of such a monthly rotation of the Earth on all terrestrial particles, exactly such a centrifugal force acts as at the center of the masses of the Earth. The sum of the vectors of the centrifugal force and the force of the Moon's attraction is called the tidal power of the moon. The tidal force of the Sun is determined similarly [8-10].

Tidal oscillations in the Earth's rotation speed are based on the theory of the moon-solar tidal potential. At present, 62 harmonics of the zone tide with periods from 5 days to 18.6 years are used to calculate tidal oscillations of the Earth's rotation speed [11]. The most significant of these are fluctuations with a half-monthly, monthly and semi-annual periods. The speed of the Earth's rotation is characterized by a relative value and is determined by the formula:

$$\Omega = \frac{2\pi}{86400} \qquad (1)$$

$П_з$ - is the duration of the earth's day; T - duration of standard (atomic or ephemeris) days, which is equal to 86400 s; и Rad·s^{-1} - angular velocities corresponding to the earth and standard days. Since ω varies only in the ninth to eighth digit, the values of ν are of the order of $10^{-9} – 10^{-8}$ [8].

The database that was used for the thermal regime of the central part of the Russian Plain consists of the daily actual meteorological parameters of observations on the state of atmospheric temperature (for December), at the main observation dates, from observations between 1948 and 2016, was placed on the official website of Roshydromet (Meteorological Center - URL: http://meteocenter.net/), at 11 meteorological stations. In the beginning, the total sample size for each station, for the month of December, was 16864 cases. Later, based on the temperature values in the main terms, average daily values were calculated, thereby the sample size decreased to 2108 cases. This sample (2108 cases) was divided into 4 groups corresponding to a certain phase of the Moon. The sample size for each phase of the Moon ranged from 486 to 534 cases. Such a large number of observations indicate a high reliability of the results of the study. For each station, the statistical significance of the results was analyzed at a significance level of p = 0.001. The reliability level for Arkhangelsk station with a sample length of 486 cases was 1.333084, which is significantly lower than the critical values of Student's t-test (P = 0.001) = 3.319 for a given sample length, which indicates high-value results [11, 12]. Similar results were obtained for all stations.

3. Results and their discussion

The spatial distribution of the temperature anomalies in December in the central part of the Russian Plain at the 1st phase (new moon), indicates an insignificant negative temperature anomaly throughout the territory, fluctuating from -0.1 °C Ufa to -0.4 °C Kursk.

IOP Conf. Series: Earth and Environmental Science **107** (2017) 012025 doi:10.1088/1755-1315/107/1/012025

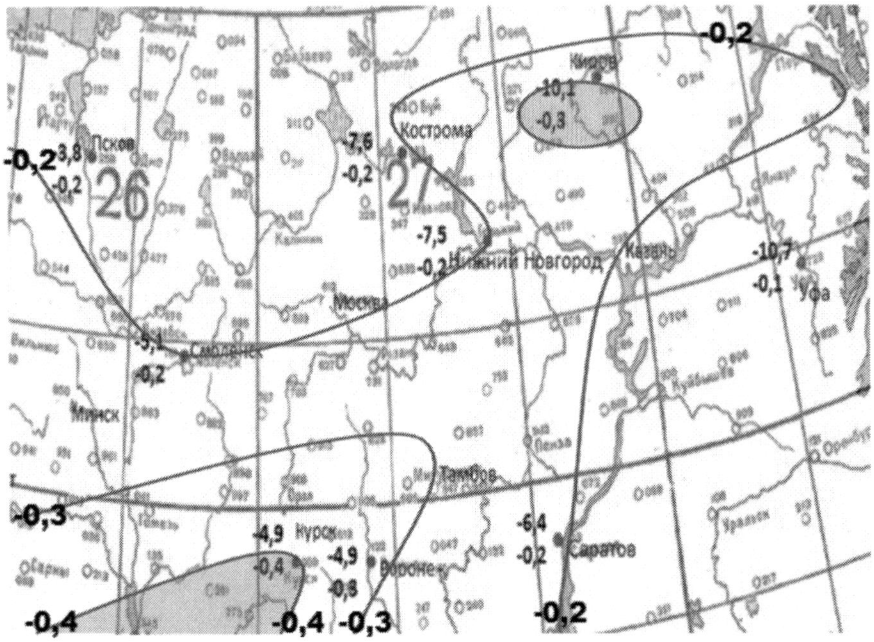

Figure 1. Spatial distribution of the temperature anomaly in December at the new moon.

The mean values of the temperature anomalies are within -0.2 °C, with insignificant increases in the central part of the territory under consideration and oriented in the meridional direction along the line Kursk (-0.4 °C), Tambov (-0.3 °C), Kirov (-0.3 °C).

The picture of the distribution of temperature anomalies for phase 2 (the first quarter) in December (figure 2) is more variegated, with negative values in the northwest (Pskov -0.3 °C) and positive foci in the southwest (Kursk 0.3 °C, Voronezh 0,3 °C), and also in the north-east (Kirov 0,3 °C, Ufa 0,2 °C).

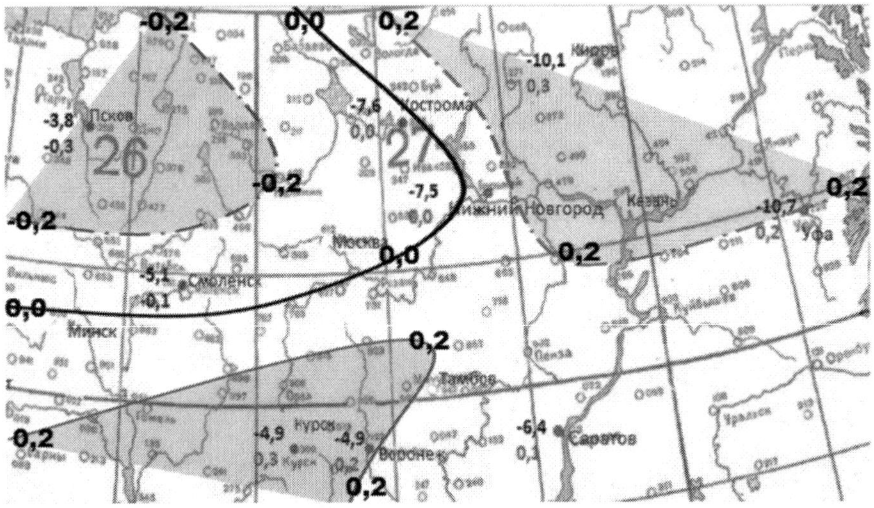

Figure 2. Temperature anomalies in December at the first quarter of the moon.

The nature of the distribution of temperature at the full moon (3 phase of the Moon), presented in figure 3, has positive values throughout the field reaching values of 0.6 °C Smolensk, Kursk, Saratov. Not significant positive anomalies are observed in Pskov and Ufa + 0.2 °C.

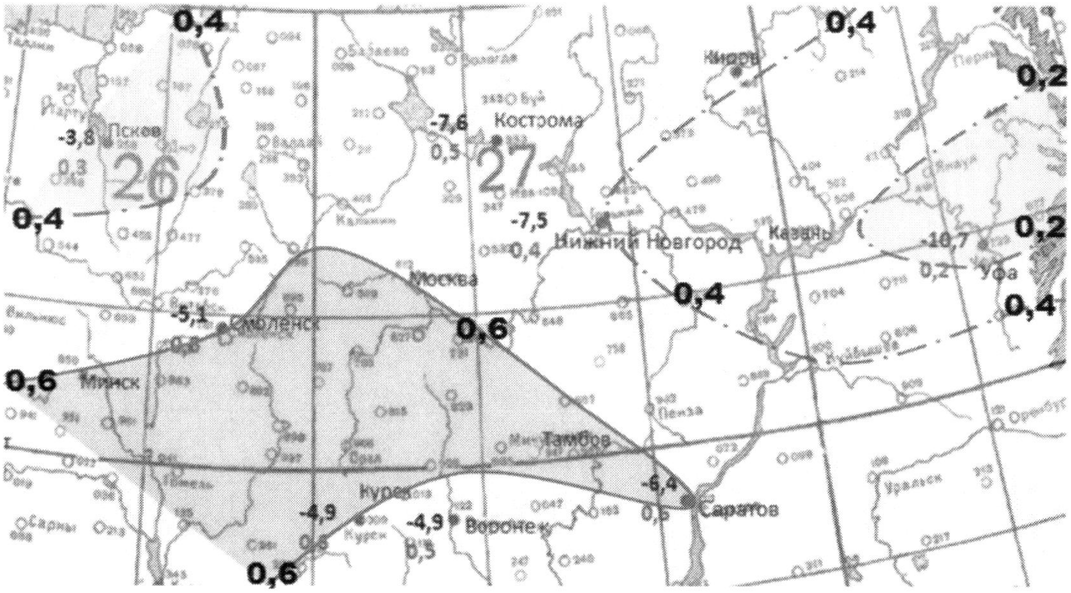

Figure 3. Temperature anomalies in December at full moon.

A distinctive feature of the 4-phase (last quarter) in December (figure 4) is the prevailing large negative temperature anomaly values located in the side altitudes from -0.3 °C (Kostroma, N. Novgorod) to -0.5 °C, Saratov, -0.7 °C Ufa. Negative temperature anomalies, during the 4th phase of the Moon, cover almost 3/4 of the territory of the Russian Plain. A positive focus of the temperature anomaly with a value of 0.3 °C is observed in the northwest of the territory in Pskov area.

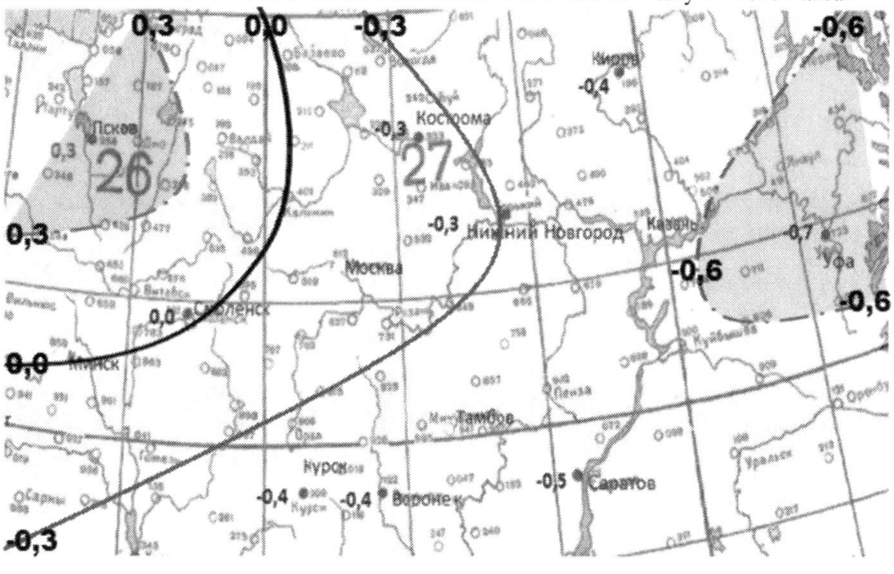

Figure 4. Temperature anomalies in December in the last quarter of the Moon.

4. Conclusions

The analysis of the thermal regime of the territory of the Russian Plain carried out in this way made it possible to reveal the main regularities of the spatial distribution of the temperature anomaly during different phases in the month of December. An opposite dependence of the distribution of the temperature anomaly on the antiphase of the Moon is established. For example, if negative temperature anomalies predominate in the Russian plain during the new moon (phase 1 of the Moon), then at full moon (phase 3) - positive. At the first quarter (the 2nd phase of the Moon) and in the last quarter of the Moon (4 phase), the signs of the temperature anomaly change to the opposite ones at a constant location of the foci of formation.

References

[1] IPCC 2013 *Climate Change 2013: The Physical Science Basis. Contribution of Working Group I to the Fifth Assessment Report of the Intergovernmental Panel on Climate* (Cambridge, UK and NY, USA: Cambridge University Press) 1535

[2] Akimov L M and Zadorozhnaya T N 2015 *Bulletin of VSU. Ser. Geography. Geoecology* **2** 15-20

[3] Akimov L M and Zadorozhnaya T N 2014 *Bulletin of VSU Ser. Geography. Geoecology* **3** 35-39

[4] Akimov L M and Zadorozhnaya T N 2013 *Bulletin of VSU. Ser. Geography. Geoecology* **1** 67-74

[5] Berry B L 2008 *Kholodok* **1** 72-84

[6] Perevedentsev Y P, Shantalinsky K M A and Auhadeev T R 2016 *Proceedings of the international scientific conference study of climate change using methods of classification of atmospheric circulation regimes* 22-26

[7] Wild G 1882 *On the air temperature in the Russian Empire* (St. Petersburg: Russian geographical society) **2** 359

[8] Sidorenkov N S and Sidorenkov P N 2002 *Patent RF no. 2182344*

[9] Sidorenkov N S 2002 *Atmospheric processes and the rotation of the Earth* (St. Petersburg: Gidrometeoizdat) 366

[10] Sidorenkov N S *On the importance of observations of the effects of tidal vibrations of the Earth's rotation speed* http://meteoweb.ru/articles/sidorenkov.pdf

[11] Hays W 1994 *Statistics. Cengage Learning*

[12] Nasledov A D 2006 *Mathematical methods of psychological research. Analysis and interpretation of data* (St. Petersburg: Speech) 392

IOP Conf. Series: Earth and Environmental Science 107 (2017) 012026 doi:10.1088/1755-1315/107/1/012026

On the forecast of runoff based on the harmonic analysis of time series of precipitation in the catchment area

A V Cherednichenko[1], A V Cherednichenko[2], V S Cherednichenko[1]

[1] SRI Of Biology and Biotechnology Problems al-Farabi Kazakh National university, al-Farabi Ave., 71, Almaty, 050040, Republic of Kazakhstan
[2] Zhasyl Damy Joint stock company Seifullina str., 597, Almaty, 050000, Republic of Kazakhstan
E-mail: geliograf@mail.ru, alexey3969@mail.ru

Abstract. It is shown that a significant connection exists between the most important harmonics, extracted in the process of harmonic analysis of time series of precipitation in the catchment area of rivers and the amount of runoff. This allowed us to predict the size of the flow for a period of up to 20 years, assuming that the main parameters of the harmonics are preserved at the predicted time interval. The results of such a forecast for three river basins of Kazakhstan are presented.

1. Introduction

Almost the whole territory of Kazakhstan is in a zone of insufficient runoff. Lack of drinking water or its unsatisfactory quality is one of the main problems hampering the development not only of agriculture, but also of a number of other branches of economic activity. The main source of water in each region is the local surface water-rivers, lakes and reservoirs. Therefore, the task of long-term forecasting of changes in runoff of surface waters is very important [1-3, etc.].

In [3], a harmonic analysis of the time series of the runoff of the rivers of Semipalatinsk and Almaty regions was carried out. As a result, it can be stated that in the time series of the runoff values there are harmonics in 3, 6, 9, 19 and 45 years in Almaty and close in the Semipalatinsk region, where, in addition, there is still a harmonica in 24 year. The author's forecast of the runoff for ten years ahead of 2010 showed its high accuracy. In this case, the forecast for one main half-century harmonic in both cases gave better results than the forecast for the combination of the three harmonics.

Since the runoff is the result of the amount of precipitation falling into the catchment area, we attempted to establish such a connection. The expediency of finding such a connection is because precipitation is the primary factor determining surface runoff. Consequently, the relationship between the fundamental harmonics in the rows of precipitation and runoff must be physically stronger than simply extrapolating the main harmonic or harmonics of the time series of the runoff to the perspective.

2. Methodology

We performed an analysis of the time series of annual precipitation for the period from the 1920s to the present, in conjunction with the runoff for three basins (regions): the Urals, Semipalatinsk and Almaty regions. We thus have an opportunity to compare the results for the regions of Semipalatinsk and Almaty with the results obtained in [3].

Rate the periodicity in observed data. Let there be n consecutive values of a time series of observations of the magnitude of Q. It will be approximated by a function of the form:

$$R = Q_0 + \frac{\partial Q}{2} sin(\omega t + \varphi) = Q_0 + bsin\omega t + ccos\omega t, \quad (1)$$

Here Q_0 is some constant for the accepted period of the approximation value, which oscillates the value of interest; ∂Q, ω, φ -accordingly amplitude, frequency and phase of these oscillations.

It is known that the parameters b and c are connected to the amplitude and phase of the oscillations in accordance with the rules of the addition of periodic values with the same frequency in the following way:

$$\partial Q = 2\sqrt{b^2 + c^2}, \quad (2)$$

$$tg\varphi = \frac{b}{c}, \quad (3)$$

Approximation of a time series will be carried out using the sum of squared differences between the series values and the approximating function S_Q.

$$S_Q = \sum_1^n (Q_i - R)^2 = \sum_1^n (Q_i - Q_0 - bsin\omega t_i - ccos\omega t_i)^2, \quad (4)$$

where Q- the study value taken from the data of observations in year t, i- is the number of years of observations; n is the duration of the series. The expression (1) approximates ranges of values of Q well, if the sum is the smallest [4, 5].

To determine the best approximating a sine wave with the specified period has been equal to Q and linked to derivatives of the expression (4) in the parameters Q, b and c:

$$\frac{\partial S_Q}{\partial b} = -2\sum_1^n (Q_i - Q_0 - bsin\omega t_i - ccos\omega t_i) = 0, \quad (5)$$

$$\frac{\partial S_Q}{\partial b} = -2\sum_1^n ((Q_i - Q_0 - bsin\omega t_i - ccos\omega t_i)sin\omega t_i) = 0, \quad (6)$$

$$\frac{\partial S_Q}{\partial c} = -2\sum_1^n ((Q_i - Q_0 - bsin\omega t_i - ccos\omega t_i)cos\omega t_i) = 0, \quad (7)$$

The solution of the system of equations (5) - (7) allows to obtain formulas for the calculation of amplitude and phase, a constant value Q_0, which varies around the approximating function for any period of approximation and consistency of periods. The sum of the squared differences between the approximating sine wave and the values of a series can be calculated by the formula (4). The results of the calculations are presented below in graphs. Thus to give an opportunity to assess the internal structure of the row, we retain all first harmonics.

The time series of precipitation were approximated by a sixth-degree polynomial to detect climatic fluctuations in them, and a harmonic analysis of the series was performed. Then the variance of the main (first) harmonics was estimated. As the rule, the first three harmonics choose 85-90% of the dispersion, so in the following analysis only the first three harmonics were used. A feature of the distribution of the dispersion between the first harmonics in the rows of precipitation is that, in contrast to the harmonics in the series of temperatures, they converge slowly. As a result, the first three harmonics contain significant values of the dispersion.

3. Results and discussion

Let's consider the results we obtained for the selected basins.

Figure 1 shows the time series of precipitation for Almaty. The performed harmonic analysis of this series makes it possible to identify three main harmonics with periods of 135, 42 and 21 years, which account for 60, 30 and 20 mm precipitation dispersion, respectively. In sum, these harmonics select more than 80% of the total precipitation dispersion (figure 2). Thus, in the time series of precipitation there is a harmonica with a period of 135 years, not typical for the series of precipitation in the territory of Kazakhstan. In the rows of precipitation, the first harmonics are, as a rule, half-century (Brickner) harmonics.

The maximum of this harmonic took place around 1975. However, at the same time, there was a minimum of the second harmonic (42 years), which significantly weakened the influence of the first harmonic. As a result, the maximum precipitation shifted in the sixties, when the second harmonic was at maximum, and the first - close to it. As a result, the climatic maximum of precipitation (figure 1) and the climatic maximum of the runoff (figure 3) were formed in the sixties. Figure 3 shows the time series of runoff values for Almaty and Semipalatinsk regions according to [3].

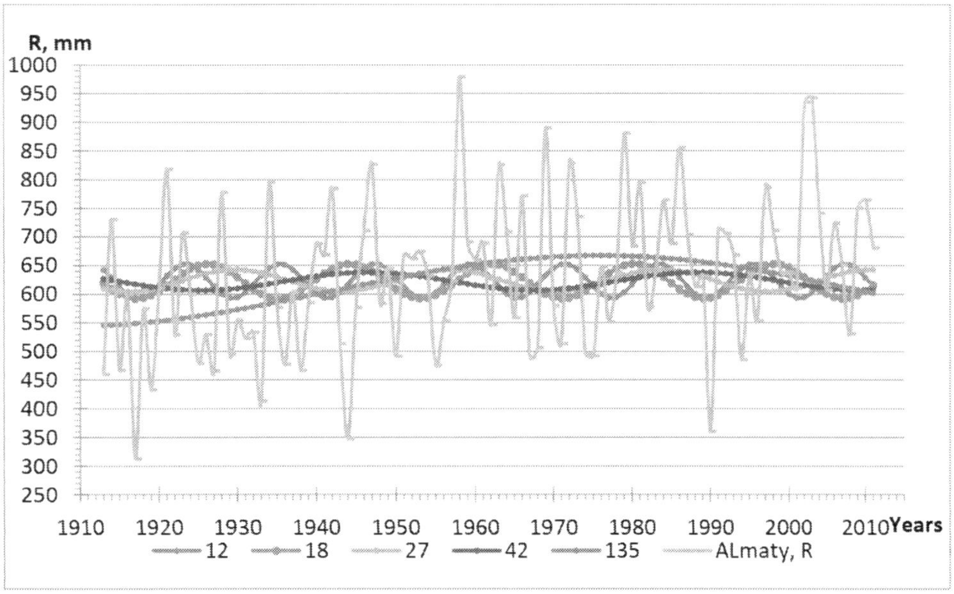

Figure 1. Temporal course of annual precipitation amount Almaty station.

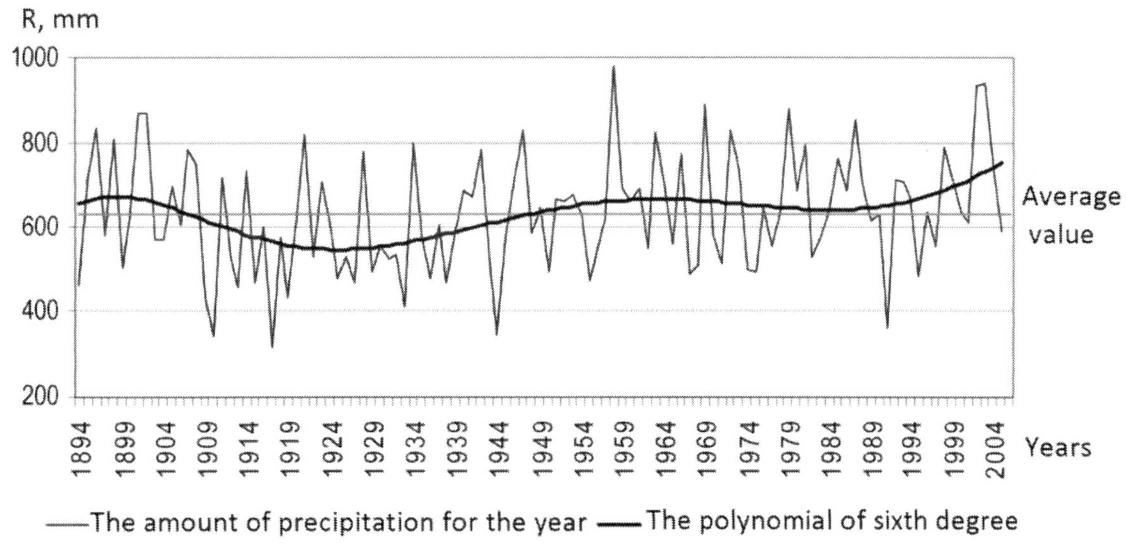

Figure 2. Harmonic analysis of the time series of precipitation. Almaty Station.

The author [3], approximating the runoff time series for Almaty region, received that the harmonic at 45 years is the most effective (figure 3a). The time series of precipitation, approximated by a polynomial of the sixth degree, taken from [2], is in good agreement with the runoff by the location of the extremes (figsures 1 and 3a). In the time interval under consideration, the maximum precipitation from the approximated curve falls on 1959, and the minimums fall for 1984 and 1926. The time of the last minimum of runoff thus coincides with the minimum of precipitation.

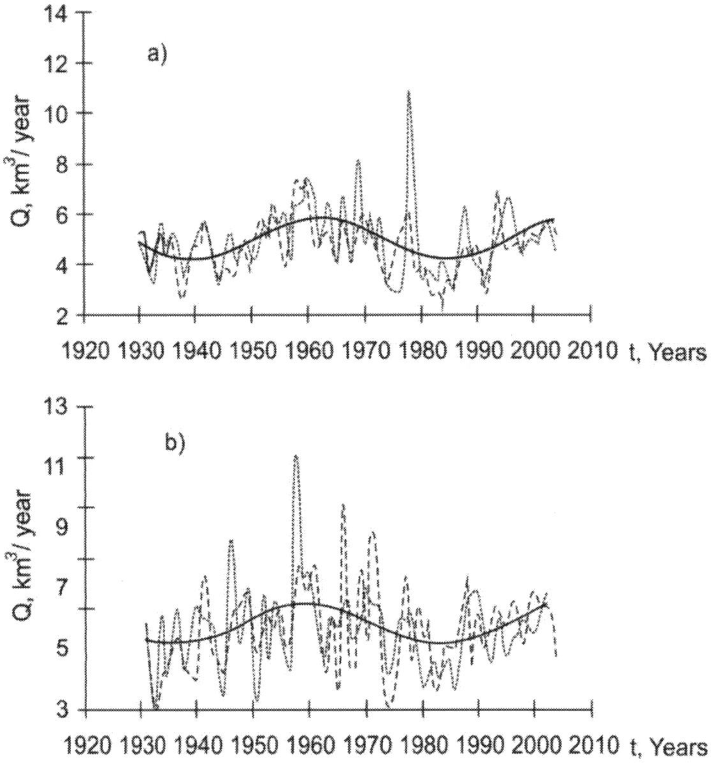

Figure 3. Dynamics of local flow fluctuations in Almaty (a) and Semipalatinsk (b) regions [3].

As for the minimum in the thirties, the minimum of precipitation took place in 1925-27, while the minimum of runoff, according to [3], was in 1940-41. We explain this difference by two factors. The minimum runoff in those years was much stretched, and the time series of runoff, taken for analysis in the cited work, began in 1931. This could affect the result of spectral and harmonic analysis. As it is known from theory, the harmonic "fits" so that it includes the maximum amount of dispersion of the entire analyzed series. And in this is the value of harmonic analysis, which makes it possible to build a prediction on the assumption of the preservation of the fundamental harmonics, since the characteristics of the harmonic depend little on the magnitudes of the series at its edges. With polynomial approximation, the approximating line "fits" so that the standard deviation is minimal in each of the time series sections.

The nature of the approximating curve at the edge of the series does not allow to confidently judge the trend concerning its new (trend) changes, to build a forecast; nevertheless, within the series we get an optimal approximation, even better than the summation of the fundamental harmonics in harmonic analysis [4, 5]. In our case, we are analyzing data not for the edges of the time series and the results of polynomial approximation should be believed.

Consequently, in the time series of precipitation and local runoff for Almaty, there is a good coincidence of extremes, allowing the use of the first harmonics and results of polynomial approximation in the series of precipitation, for the climatic forecast of surface runoff.

Let us now consider the temporal course of precipitation in Samarka, which most objectively characterizes the distribution of precipitation in the region, including Semipalatinsk (figures 3b, 4).

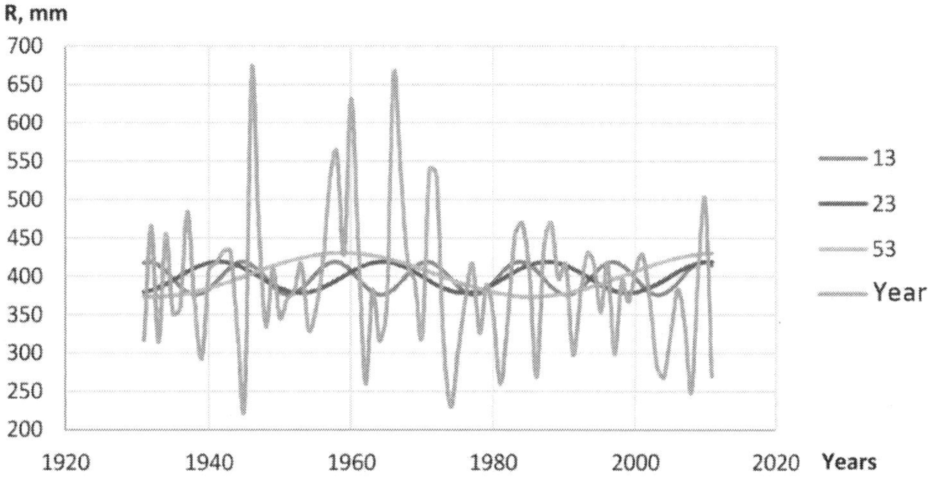

Figure 4. Time course of the annual amount of precipitation at the MC Samarka.

It can be seen that during the period under review, a maximum of precipitation took place in the sixties, before this time the amount of precipitation gradually increased, and after the decade it gradually decreases. In the time series, harmonics in 53, 23 and 13 years with fractions of 50, 30 and 25 mm, respectively, are distinguished. It can be seen that the fraction of the first harmonic in the total dispersion of precipitation is large, 50 mm. There is no century-old or close harmonic, which is typical for time series of precipitation in the territory of Kazakhstan. The curve of the approximation of the time series of precipitation by a polynomial of the sixth degree largely repeats the position of the first harmonic and therefore we do not give it. The maximum of the Brikner's harmonics, first allocated by us, falls on the sixties and the minimum at the eighties of the last century. The extreme of the first harmonic coincide with the extreme of the runoff according to [3]. This allowed the author [3] to make a forecast of the dynamics of the climatic runoff for only one harmonic.

The maximum of a half-century harmonic was observed in 1960. With a subsequent minimum in 1986, after which a new growth begins, and in recent years a new decreasing. In the sixties, i.e. in the years of maximum precipitation, there were at the maximum all three harmonics - 13, 23 and 53 years. In 1975-78, the half-century harmonica was on the decline, and 13 and 23-year-olds in the minimum. At this time, precipitation was at the minimum, too. By examining figures 4 and 3, one can find a good consistency between the amount of precipitation and the amplitudes of the harmonics and for other time intervals.

The maximum runoff of the rivers of Semipalatinsk region according to figure 3b coincide with the period of the greatest amount of precipitation in the sixties. The minima of the runoff almost coincide with the minimum of the amount of precipitation in the thirties and eighties. Thus, in the long run, the runoff values in this region are in good agreement with the temporal course of the amount of precipitation, although, as is known [1, 2, etc.], the yearly precipitation and runoff values should not be expected to coincide year after year.

Consider now the basin of the Ural River (Zhayik). As it is known, this river has two equivalent watersheds: one is located to the west of the South Urals and the Sakmara River flows into the Urals, and the second catchment is the catchment area of the Ural River itself, which is located to the east of the Southern Urals. In [2, etc.] it is shown that the precipitation regime in each of the catchments reflects well the sediments of the meteorological stations of Uralsk and Kustanai located to the west and east of the mountains of the Southern Urals.

Figure 5 shows the discharge values of the Ural River in the alignment of the Kushum post in the territory of Kazakhstan

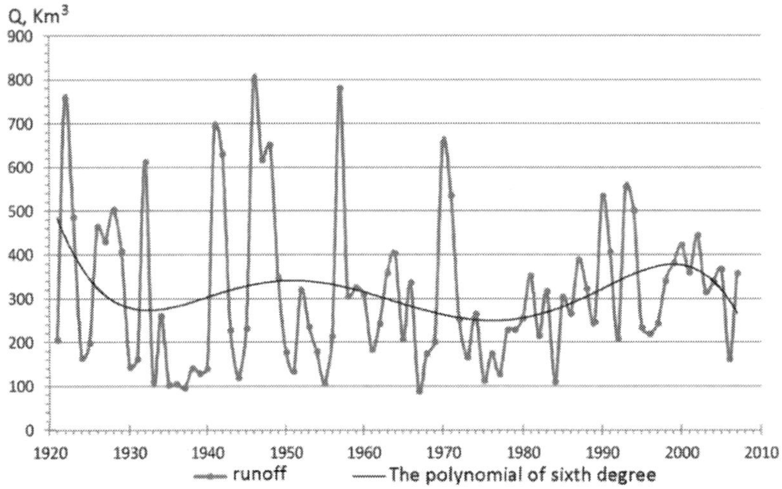

Figure 5. Time course of the Ural River runoff in the Kuchum post area.

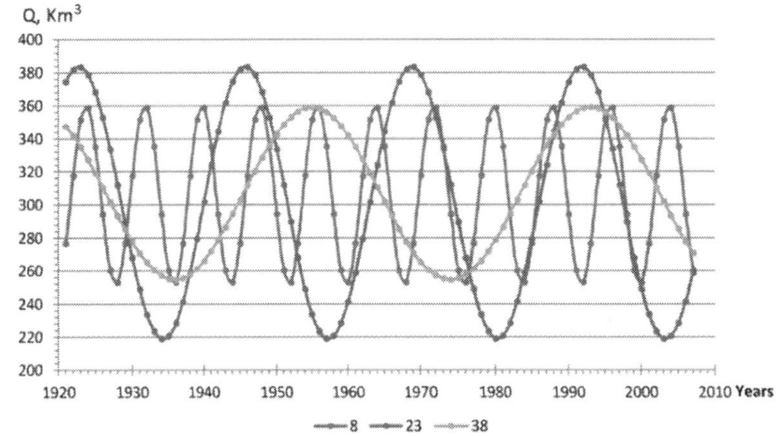

Figure 6. Results of the harmonic analysis of the discharge values. The Ural River.

It can be seen that harmonics 8, 23 and 38 years are distinguished. Checking the harmonics in the runoff for stability by elongating and shortening the series showed that all three harmonics of 8, 23 and 38 years are stable. The amplitude of all harmonics is significant. In this case, the amplitude of the harmonic in 23 is the largest, more than 160 m^3, harmonics of 38 and 8 years, more than 100 m^3. The more harmonics and the greater their contribution to the overall variance, the more difficult it is to predict the flow using them.

Let's see what harmonics are contained in the time series of precipitation at the stations of Uralsk and Kustanai. In the time series of the Uralsk station precipitation, harmonics in 8, 18 and 43 have been identified, and they are all of approximately the same amplitude. In the precipitation of Kustanai, harmonics of 8, 23, and 38 years are distinguished. The verification of the harmonics in the rows of sediments of these stations to stability illustrates that in Kustanai and Uralsk a harmonic appeared at 8 years, which should be considered unstable. In the total sediments of the two stations, harmonics of 8, 18, and 43 are observed, i.e. the same harmonics as in the series of precipitation in Uralsk. It is noteworthy, however, that if the amplitude of these harmonics in the precipitation of Uralsk is 60 mm each, then in the total precipitation of Uralsk and Kustanay 80, 100 and 100 mm, respectively, i.e. the amplitude of each harmonic increased due to the precipitation of Kustanai, although none of the harmonics taking place in Kustanai (except for the eight-year old) appeared. At the same time, in a time series of sums of precipitation at two stations, there is not a single harmonic, which would coincide with the harmonics present in the river discharge of the Ural River. The harmonic values separated from the time series exactly coincide in length with the harmonics contained in the time series of precipitation of the station of Kustanai. The most significant are only two harmonics: 38 and 23 year.

A joint analysis of the harmonics in the time series of the precipitation of each station and in the series of total precipitation on the one hand and the harmonics in the drain on the other showed that there is a connection, but it is rather complicated and sometimes ambiguous. For this reason, we further analyzed the harmonics in the Ural River discharge on one side and the harmonics in the macro circulation indices on the other. The initial data are consistent with [6-8].

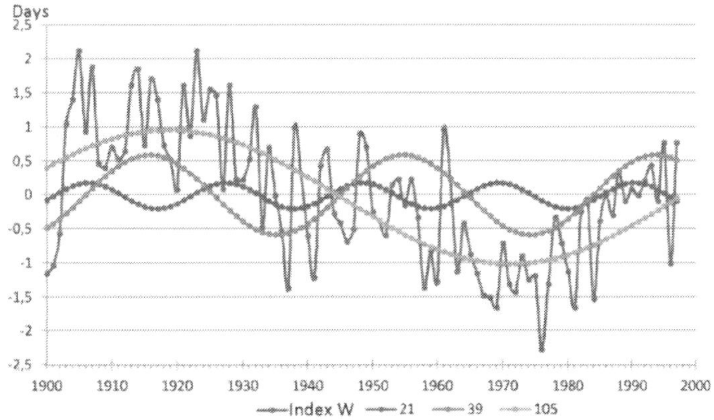

Figure 7. Harmonic analysis of the W index.

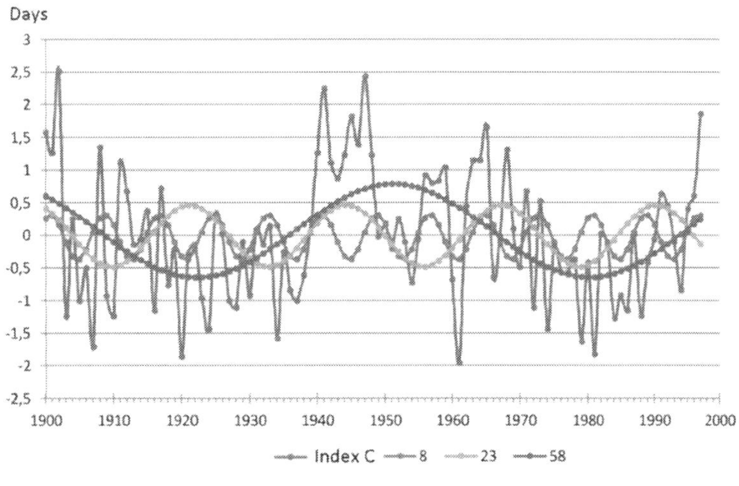

Figure 8. Harmonic analysis of the C index.

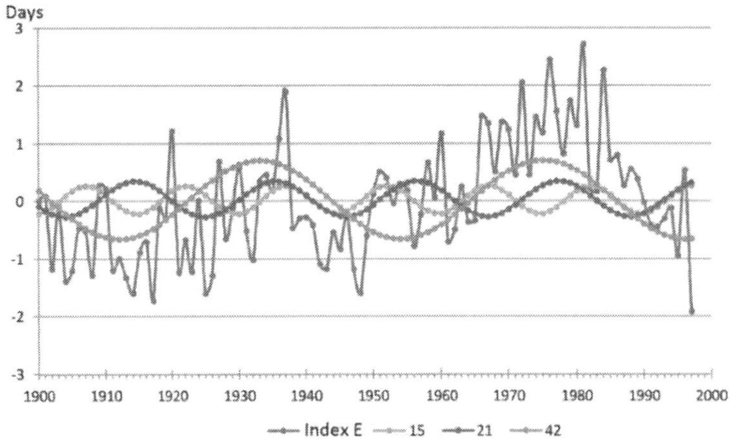

Figure 9. Harmonic analysis of the E index.

Among the harmonics of the W index, the century-old harmonica (103 - 105 years) determined the main extreme in the drain - a maximum in the twenties and 2000s and minimums in the sixties and seventies. The time series of the runoff is shorter by 20 years of the time series of the W index.

The 38-year-old harmonics in the W index with amplitude of about 28 days, which sometimes turns into a 29-year-old harmonic when testing stability, is synchronous to a 38-year harmonics in the drain.

Consequently, the two main harmonics of 108 and 38 years, the index W, have a response in the Ural River discharge in the form of the same harmonic duration with amplitudes of 110 and 80 m3, respectively.

From the harmonics of the index C, a harmonic of 23 y is expressed, with amplitude of more than 160 m^3, which is synchronous with the harmonics of 23 y in the runoff of the Ural River. The amplitudes of these harmonics completely coincide. This harmonic in amplitude exceeds the amplitude of the harmonics of the secular and 38 years in the runoff caused by the corresponding harmonics of the index W, i.e. it is most significant.

From the harmonics of the index E, we note the harmonic 42 y, which in the time interval under consideration is in oppose with a harmonic of 38 years of index W. The influence of other harmonics of the index E is weak.

Thus, the harmonics in the Ural River discharge are in good agreement with the harmonics of the secular and 38 years of the W index and the harmonics of the 23rd index of C. Extremes in the run take place when the amplitudes of the named harmonics coincide. Thus, a minimum of flow around the middle of the thirties of the last century was observed with a decrease in the amplitude of the secular harmonic of the index W, and the harmonics of the harmonics of 38, 23 and 8 years coincided. In the mid-seventies - at a minimum of harmonics of 38 and 8 years and a harmonics of 23 y, which was close to a minimum. The maximum runoff took place in the middle of the forties with a maximum of harmonics of 23 and 8 years and a harmonics close to the maximum of 38 years. At last the maximum of the drain in the nineties - at a maximum of harmonics of 38 and 23y on a rise close to the maximum of the secular harmonic of index W.

The sensitivity of the rivers in the droughty region to climate change is very high. The drain sensitively reacts to very slight climatic temperature variations and precipitation on the territory of formation of a drain. Anthropogenic impacts on components of a surrounding medium in a river basin can make on it the most adverse effect.

In [9], the authors, studying the climatic changes in air temperature and precipitation on the territory of the Middle and Southern Urals in the twentieth century, concluded "the ambiguity of trends of changes of climatic characteristics in the twentieth century in the Urals. Installed features rhythmic changes in air temperature and precipitation". To the authors the results similar to our own, we can only add that on the territory of Kazakhstan, adjacent to the watershed of the Ural River, and in the Kazakh part of the basin up to the Caspian Sea, we also recorded only fluctuations of climate. Because the methods of analysis we have are different, we fixed the "rhythms" of different durations: a century, of close to half a century (Brickner's), and a shorter one. The age-old cycle that had the largest amplitude close to the maximum, however, due to the shorter harmonics and here, the total climatic temperature trend has become negative.

In [10] methods of model operation influence of global climate change on the hydrological system of the Aral Sea located, in fact, in the center of Eurasia was studied. In spite of the fact that the field of formation of a drain is apart about one and a half thousand km from the sea, and the system of accounting of the consumed water on this site is unreliable, authors nevertheless concluded that the influence of climate change on the processes in the basin of the Aral Sea is significant. Reservoirs of an arid zone are sensitive to the slightest climatic temperature variations and rainfall.

In [11] authors studied the influence of climate change on an underground delivery of the rivers in Great Britain. And, though Great Britain is out of an arid zone, authors assume that such dependence exists. It indicates quite strong dependence of the surface drain from climatic temperature variations and rainfall irrespective of a natural zone. The existing balance between precipitation and temperature on the one hand and the surface drain on the other, being rather steady on a temporary interval of one year to several, it is very sensitive to climatic changes, i.e. changes on a time span up to 10 years.

In [12] not only temporary, but also space changes of an amount of precipitation at a size of a drain of the river of Yangtze (Yangtzy) during 40 years are analyzed. The received results confirm that the sensitivity of water systems to climatic changes is high even outside an arid zone. There is a number of other works devoted to this problem, but with similar regularities.

We can see that our results for the river basin of the Urals in general confirm the common conclusions received for other river basins. However, in view of existence of a number of features except aridity of climate, such as the existence of two pools of formation of a drain divided by mountains, different conditions of formation of extreme rainfall in each of them, etc., we received the particular communications and bonds and the expected scenarios of change of a drain in prospect.

4. Conclusion

As a result of the analysis of time series of precipitation and surface runoff, we have shown that the climatic forecast of runoff can be built on the expected climatic changes in the amount of precipitation in the basin. The climate forecast should be preceded by a joint in-depth analysis of the time series of runoff and precipitation in the region (the area of flow formation). Based on this analysis, a long-term runoff forecast is constructed. Analyzing time series of surface runoff, it is firstly necessary to make sure that earlier there was a general coincidence of extremes in them with the extremes of precipitation in the basin. The presence of such a coincidence indicates the prospect of finding links. Due to this and the joint analysis of the series, the most appropriate variant for the forecast is selected.

Depending on the regional and circulatory features, it may be sufficient to use only one first harmonic in the precipitation series if it contains 50% of the variance and more (as in the case of the Semipalatinsk region) for the forecast of runoff. For Almaty region, it was useful to use all three harmonics in the analysis with the use of results of polynomial extrapolation.

In complicated cases, when the conditions for the formation of surface runoff in different parts of the basin are different, it is advisable to use the drainage relationship with the results of a harmonic analysis of the forms of large-scale atmospheric circulation, since precipitation is associated with quite definite forms. Unfortunately, the widely used Wangengeim-Girs typification does not have a quantitative characterization of the intensity of large-scale processes. This reduces the effectiveness of its application in similar tasks.

References

[1] Shklyayev A S 1961 *Uchenyye zapiski PGU* **112** 108
[2] Cherednichenko A V, Davletgaliyev S K, Kozhakhmetova E P, Cherednichenko V S and

IOP Conf. Series: Earth and Environmental Science **107** (2017) 012026 doi:10.1088/1755-1315/107/1/012026

Baykhonova T A 2010 *Vestnik KazNU. Ser. Geograf.* **1** (30) 17-24
[3] Babkin A V 2008 *Materials of International scientific-practical conference dedicated to the 70th anniversary of the Institute of geography of RK* 153 – 158
[4] Sneyers R 1990 *Technical note* **143** 192
[5] Kendal M and Styuart A 1973 *Statisticheskiye vyvody i svyazi* (M.: Nauka) 900
[6] 1982 *Solar-terrestrial connections, weather and climate* ed. Mak-Norman B and Selige T (M.: izd "Mir") 384
[7] Girs A A 1971 *Many years fluctuations of atmospheric circulation and long-term hydrometeorological horecast* (L: Gidrometeoizdat) 280
[8] Dmitriyev A A and Belyazo V A 2006 *Space, planetary climatic variability and atmosphere of the polar regions* (St. Petersburg.: Gidrometeoizdat) 358
[9] Shklyayev A S and Shklyayeva L S 2011 *Vestnik ChSU* **5** 61-69
[10] Beek A T, Vos F and Florke M 2011 *Physics and Chemistry of the Earth* **36** 684-695
[11] Herra P M and Hiscok K M 2008 *Hydrological processes* **22** 73-86
[12] Jiang T, Su B D and Hartman H 2007 *Geomorphology* **85** 143-154

IOP Conf. Series: Earth and Environmental Science **107** (2017) 012027 doi:10.1088/1755-1315/107/1/012027

Variability of the snow cover and its characteristics on the territory of the Republic of Bashkortostan

R G Galimova

Department of Hydrometeorology and Geoecology, Bashkir State University, Zaki Validi str., 32, Ufa, 450076, Russia
E-mail: galim-rita@yandex.ru

Abstract. This article examines the variability of snow cover characteristics in the territory of the Republic of Bashkortostan (Russia). For analysis, the author used a long-term series of observations of meteorological stations for the period 1961-2010. The main tendencies of variability of snow cover height, water reserve, dates of snow cover establishment and destruction, as well as the duration of occurrence were revealed.

1. Introduction

Snow cover is one of the most important factors of the winter season. Having distinctive properties, the snow cover forms its own "snowy" climate. The greater height of the snow cover increases the temperature of the soil and reduces the depth of its freezing. These factors in a complex increase the absorbing capacity of frozen soils and, thereby, improve the conditions of wintering of plants. Atmospheric precipitation, accumulated during the winter in the snow cover, feeds surface and groundwater [1, 2].

The formation of snow cover is possible under conditions of negative radiation balance and fall of solid precipitation. The territory of Bashkortostan belongs to the region with the annual formation of snow cover throughout the whole winter.

2. Material and methods

The occurrence of the snow cover and further redistribution depends mainly on orography and differences in the hypsometric levels of the surface. On average, snow fall can be observed in October, but with variations in air temperature in the region of 0° C, the snow cover may repeatedly melt throughout the late autumn and pre-winter.

In a changing climate, the snow cover and its characteristics undergo spatiotemporal changes.

To examine the variability of snow cover characteristics, the author uses the average and maximum values of height for winter, snow reserves, dates of establishment and destruction, duration of occurrence.

As data for analysis, the database of the All-Russian Scientific Research Institute of Hydrometeorological Information - World Data Center was used [3].

3. Results and discussion

Throughout the territory of Bashkortostan there is an increase in the average and maximum height of the snow cover. This reflects a direct relationship between the increase in precipitation in the cold period and the thickness of the snow cover [4, 5].

IOP Conf. Series: Earth and Environmental Science **107** (2017) 012027 doi:10.1088/1755-1315/107/1/012027

The greatest increase in snow height is recorded along the western periphery of the forest-steppe zone within Bashkortostan. By meteorological stations Sterlitamak, Raevsky, Verkhnejarkevo, the increment for each decade is 5.5, 3.7, 3.4 cm, respectively. The lowest power increase is observed at the meteorological station Aksakovo and Bakaly (figure 1, 2).

Considering the overall trend line in absolute values, the increase is 6-10 cm across the territory. An exception is the meteorological station of Ufa-Dema, where a decrease in the height of the snow cover is noted. This is due to the fact that this meteorological station was built on the outskirts of the city, whereas at the moment it is experiencing the impact of urban infrastructure (because of the intensive urban development).

Aksakovo

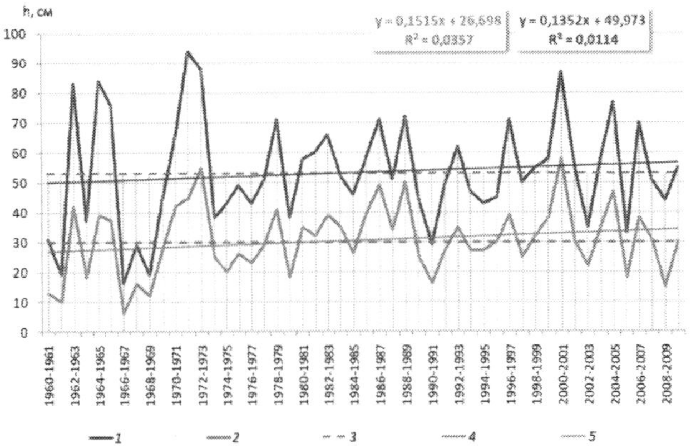

Sterlitamak

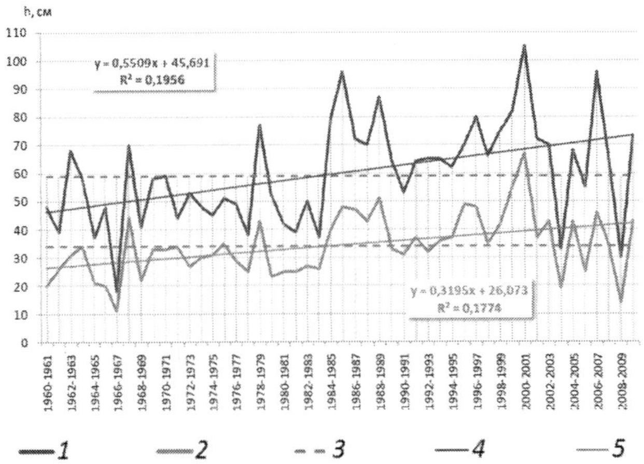

Figure 1. Dynamics of maximum and average height of the snow cover. 1 - maximum height of the snow cover; 2 - average height of the snow cover during the winter; 3 - average multi-year norm; 4 - trend line of maximum height; 5 - trend line of medium height.

The increase in the thickness of the snow cover during the winter is caused by snow falls. The main causes of snowfall is the frontal precipitation of solid sediments during the passage of cyclones.

Thus, during the passage of Atlantic cyclones in December 1935 and 1947, a record precipitation value for the winter period - 43 and 42 mm per day, respectively. The norm for December is 45 mm per month. At the same time, considering more broadly the entire synoptic situation on December 28, 1935, it should be noted that two days before this date, 30 mm were recorded per day and a powerful thaw was recorded (the air temperature increased by day to + 5 °C). More such catastrophic cases were not recorded [6].

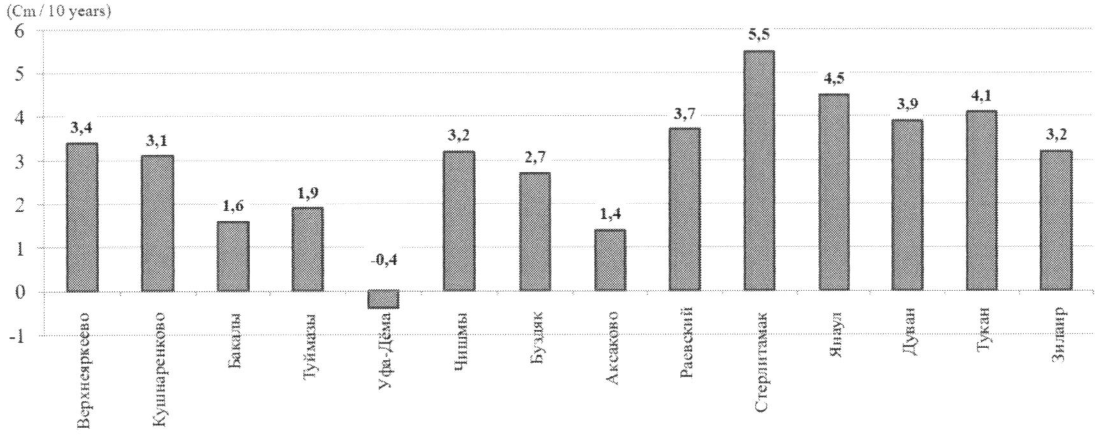

Figure 2. Coefficients of slope of the linear trend of the maximum height of the snow cover (cm·(10 years)$^{-1}$).

In the cold period, the highest correlation in atmospheric precipitation is observed with the North Atlantic Oscillation Index. Some authors note that the variability of the pressure field in the winter half-year is related to this index by about 40%. In the work [7] the correlation analysis showed that the greatest relationship was revealed in the cold period in the western forms of circulation. The same authors noted that with the increase in the share of the W circulation form, in accordance with the increase in the index of the North Atlantic Oscillation, the transfer of moist air from the Atlantic increases, which leads to the increase in precipitation during this period in the region under consideration.

Figure 3 shows the dynamics of the forms W, C, E for the period 1900-2009, which shows a significant tendency to increase the form of circulation W. In the work [8] based on the analysis of the constructed trends, the rates of variability in the forms of circulation are described. Thus, the forms of the western circulation W tend to increase 2.5 days for every 10 years in the winter and slightly during the summer period, the eastern circulation of E decreases by 3 days·(10 years)$^{-1}$ in the winter and increases by 3.9 days·(10 years)$^{-1}$ in the summer period, meridional circulation C increases slightly in the cold period and decreases significantly in the warm period - by 4.5 days·(10 years)$^{-1}$.

[9] noted that there is an increase in the tendency of the North Atlantic Oscillation Index, especially in the winter period. This increases the number of days with a positive NAO phase in this season (figure 4).

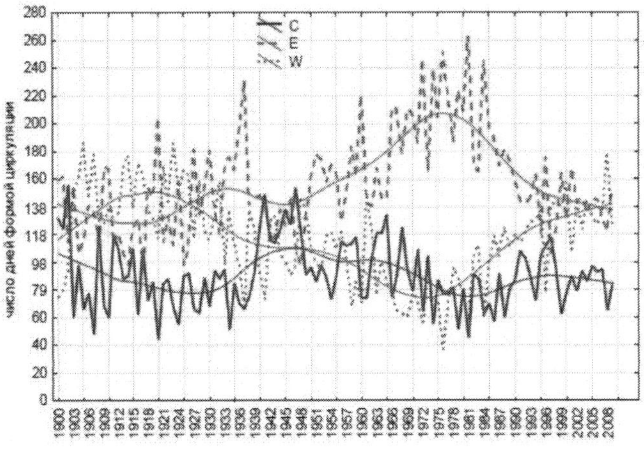

Figure 3. Dynamics of the forms of circulation W, C, E during the period of instrumental changes [10].

Figure 4. Number of days with index NAO > 1 in the cold period [9].

An increase in the snow mass entails an increase in snow accumulation and, accordingly, water reserves in the snow. In the greater part of Bashkortostan, there is a tendency for an increase in water reserves in the snow (figure 5).

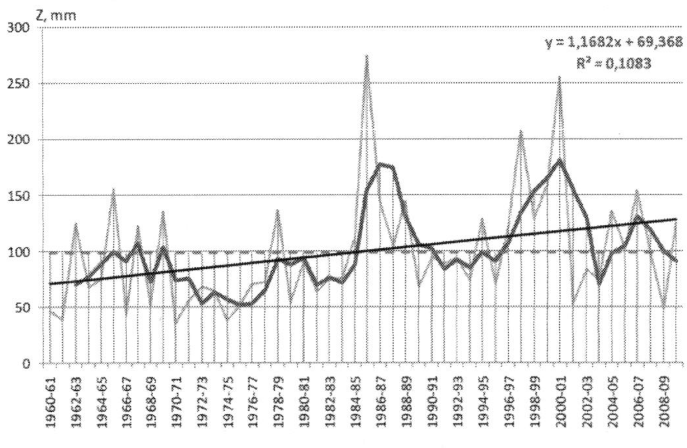

Figure 5. Dynamics of water reserves in the snow cover.

The melting of a larger mass of snow in the spring period, therefore, entails enhanced run-off processes with large snow reserves, this will lead to more intense surface erosion processes and ravine formation.

In addition to the above characteristics of the snow cover, the dates of the establishment and destruction of the snow cover are also important. From these dates will directly depend on the duration of snow occurrence.

A permanent snow cover begins to form with a stable transition of the average daily air temperature below 0 °C (the average multi-year date is November 1-4), but sometimes there are cases of exceptions. So, for example, in 2006 and 2008, the establishment of snow cover in the city of Ufa was noted in mid-December (which, apparently, was the result of the impact of a large city and its infrastructure). In other words, the establishment dates have a very wide spread in time.

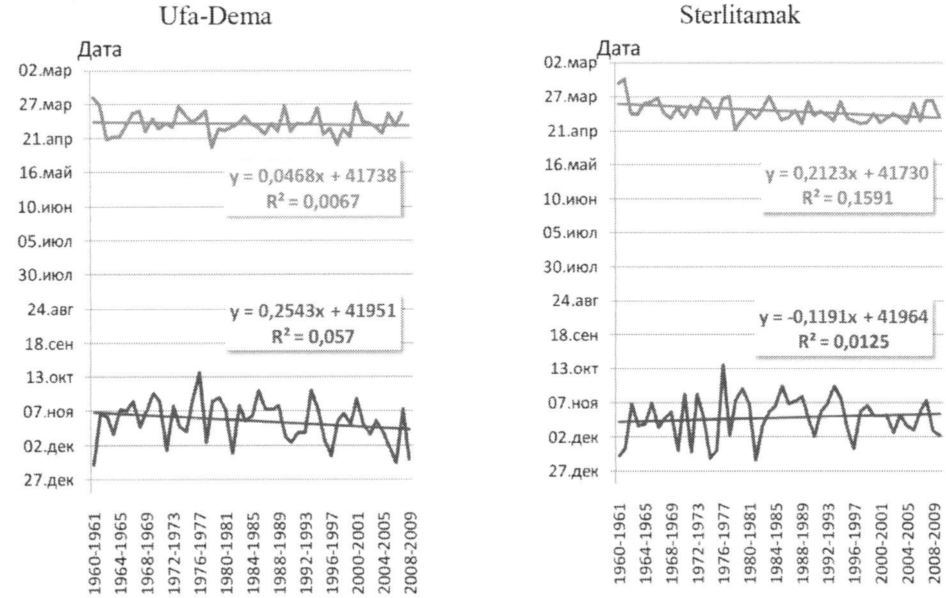

Figure 6. The dynamics of the establishment dates (blue) and the destruction (red color) of the snow cover.

Considering the dates of establishment in the multi-year profile (50 years), characteristic for this indicator is a wide spread of the timing of the beginning of the occurrence of snow - from October 13 to December 20 (more than two months). The extremely unstable temperature regime and the weather instability during this transitional season (fluctuations in the vicinity of 0 ° C and advection of warm air masses) lead to significant variability in the date of snow cover almost every year (figure 6).

Nevertheless, analyzing the graphs of long-term observations, there is a practically homogeneous tendency to shift the dates of snow cover to later dates. If earlier the establishment was observed in the first decade of November, then recently - since the middle of this month (figure 6). The meteorological station Aksakovo is characterized by the greatest variability, where every decade an average snow cover is established later for more than three days.

Regarding the dates of the destruction of the snow cover, it can be noted that they also shift to a later date, but not so much.

The destruction of the snow cover begins even at negative average daily temperatures, but when positive temperatures are observed during the day. The date of destruction of the snow cover in the study area can fluctuate from the third decade of March to the third decade of April. On the basis of the constructed graphs, the displayed displacement of dates is insignificant - on average 1-2 days. The greatest change was in the date at the meteorological station Sterlitamak - for every decade there was a shift of more than two days.

The tendency described above seems to be due to the increase in power, since the largest mass of snow will continue to melt. However, due to the overall increase in average winter and spring temperatures, the shift in the date of total destruction is not so significant.

Almost everywhere there is a tendency to decrease the duration of snow cover. On average, the reduction was 3-6 days. At the Aksakovo meteorological station, there occured the most significant decrease in duration (for 5-6 days). Exception is the meteorological station Sterlitamak: there is an increase in duration - on average for every 10 years 3 days. So, if at the beginning of the analyzed period the duration of days with snow was about 132, then by the end - along the trend line about 150 (figure 7).

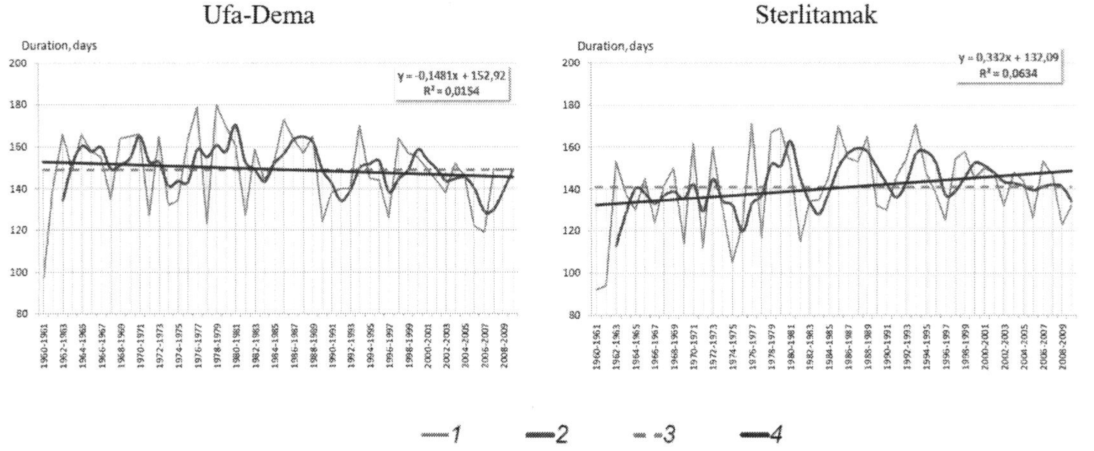

Figure 7. Dynamics of the duration of snow cover. 1 - duration of occurrence (days); 2 - moving average duration; 3 - medium; 4 – trend.

Based on the above mentioned results, it can be stated that at present the winter-spring season is shifting to a later date. This is mainly determined by several reasons:

1) the establishment of a stable snow cover occurs later (approximately from the middle of November);

2) the height of the snow cover noticeably increases both the average and maximum height (by 2-6 cm and up to 10 cm, respectively).

Thus, the shift in climatic seasons that has taken place will affect not only the characteristics of the snow cover, but also the factors that depend on it: surface and underground runoff, the temperature regime of the winter-spring period, humidification, the depth of freezing of the soil and other conditions.

References

[1] Richter D G 1946 *Nature* **4** 20-28
[2] Pearce R P 2001 *Meteorology at the Millennium* (NY, London: Academic Press) 346
[3] *All-Russian Scientific Research Institute of Hydrometeorological Information - World Data Center.* URL: http://meteo.ru/
[4] Perevedentsev Yu P, Malkhazova S M, Aukhadeev T R and Shantalinskiy K M 2016 *Geography, Environment, Sustainability* **9** 63-76
[5] Galeeva E M and Galimova R G 2014 *Bulletin of Buryat State University* **4** (2) 12-19
[6] Galimova R G 2015 *Materials of the interuniversity scientific-practical conference* 25-29
[7] Bezuglova N N and Zinchenko H S 2009 *Geography and natural resources* **3** 83-87
[8] Kurtov D Long-term variability of precipitation in the forest-steppe and steppe landscapes of the

Novosibirsk Region URL: http://iesen.nspu.ru/sbor13/t3/ch3.html
[9] Nesterov E S 2013 *North Atlantic Oscillation: Atmosphere and Ocean* (M.: Triada) 144
[10] Bokov V N and Vorobyov V N 2010 *Zapiski RSHU* **13** 83-88
[11] Galimova R G 2011 *Kazan science* **1** 454-456
[12] Naaim M, Naaim-Bouvet F, Faug T and Bouchet A 2004. *Cold Reg. Sci. Technol.* **39** 193-204

IOP Conf. Series: Earth and Environmental Science **107** (2017) 012028 doi:10.1088/1755-1315/107/1/012028

Water runoff vs modern climatic warming in mountainous cryolithic zone in North-East Russia

V E Glotov, L P Glotova

North-East Interdisciplinary Science Research Institute, Port str., 16, Magadan, 685000, Russia
Russian Academy of Sciences Far East Branch, Svetlanskaya str., 50, Vladivostok, 690990, Russia
E-mail: geoecol@neisri.ru; glotova@neisri.ru

Abstract. The article presents the results of studying the effects of current climatic warming for both surface and subsurface water runoffs in North-East Russia, where the Main Watershed of the Earth separates it into the Arctic and Pacific continental slopes. The process of climatic warming is testified by continuous weather records during 80-100 years and longer periods. Over the Arctic slope and in the northern areas of the Pacific slope, climatic warming results in a decline in a total runoff of rivers whereas the ground-water recharge becomes greater in winter low-level conditions. In the southern Pacific slope and in the Sea of Okhotsk basin, the effect of climatic warming is an overall increase in total runoff including its subsurface constituents. We believe these peculiar characters of river runoff there to be related to the cryolithic zone environments. Over the Arctic slope and the northern Pacific slope, where cryolithic zone is continuous, the total runoff has its subsurface constituent as basically resulting from discharge of ground waters hosted in seasonally thawing rocks. Warmer climatic conditions favor growth of vegetation that needs more water for the processes of evapotranspiration and evaporation from rocky surfaces in summer seasons. In the Sea of Okhotsk basin, where the cryolithic zone is discontinuous, not only ground waters in seasonally thawing layers, but also continuous taliks and subpermafrost waters participate in processes of river recharges. As a result, a greater biological productivity of vegetation cover does not have any effect on ground-water supply and river recharge processes. If a steady climate warming is provided, a continuous cryolithic zone can presumably degrade into a discontinuous and then into an island-type permafrost layer. Under such a scenario, there will be a general increase in the total runoff and its subsurface constituent. From geoecological viewpoints, a greater runoff will have quite positive effects, whereas some minor negative consequences of it can be successfully prevented.

1. Introduction

The territory of North-East Russia (NER) is geographically peculiar as it is washed by seas of the Arctic and Pacific oceans, which differ by their hydrothermal regimes (figure 1). The Arctic Ocean is a natural continuation of the Atlantic Ocean, and their drainage divide is a fragment of the Main Watershed of the Earth (MWE). Over the Arctic slope of the MWE, climatic conditions are formed at account of heat and water brought there by air masses forming over oceanic areas, the largest part of which is covered by ice even in warm seasons. As a result, the air over the land is cold and its water content is low. Climate over the Pacific slope depends on submeridional cyclones bringing heat and water from temperate warm zones of the Pacific. Such cyclones can reach high northern latitudes and in some cases can even attack the Arctic slope raising the air temperature and increasing precipitations in the Chukotian Sea basin. Different runoff-forming factors produce their effects on the total water runoff of different oceanic slopes [1].

The study purpose is to elucidate regular relationships between changes in MWE slope runoffs and modern changes in runoff-forming climatic factors.

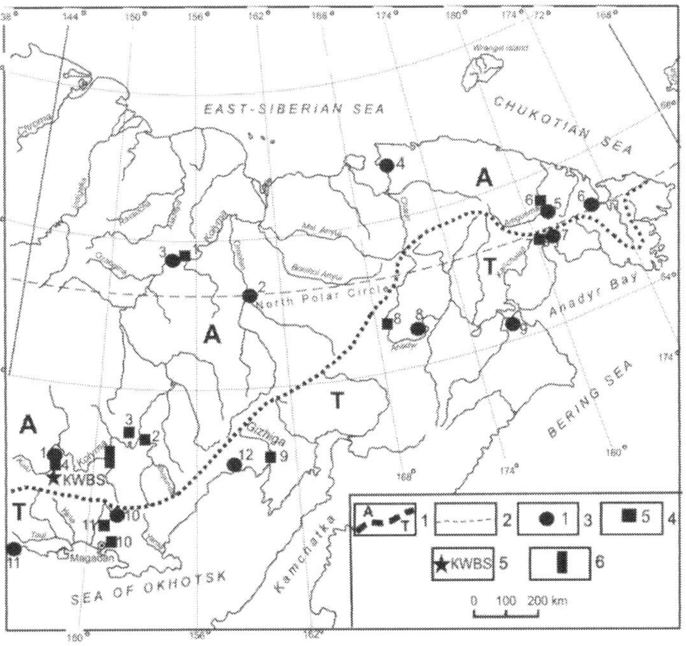

Figure 1. Schematized map of North-East Russia: **1** – the Main Watershed of the Earth (A – Arctic slope, T – Pacific slope); **2** – the Northern Polar Circle; **3** – weather monitoring stations, by their numbers in the text including: 1 – Seimchan, 2 – Omolon, 3 – Srednekolymsk, 4 – Pevek, 5 – Amguema, 6 – Uelen, 7 – Egvekinot, 8 – Markovo, 9 – Anadyr, 10 – Palatka, 11 – Okhotsk, 12 – Nayakhan; **4** – hydrologic stations, by their numbers in the text including: 1 – the Kolyma River, Srednekolymsk, 2 – the Kolyma River, Ust-Srednekan, 3 – the Taskan River, Taskan, 4 – the Kulu River, Kulu, 5 – the Detrin River, the Vakhanka River mouth, 6 – the Amguema River, bridge at 1744 km mark, 7 – the Izyskatelsky Creek, 2.5. km from its mouth, 8 – the Anadyr River, Novy Eropol, 9 – the Gizhiga River, 20 km from its mouth, 10 – the Khasyn River, Khasyn, 11 – the Dukcha River mouth; **5** – KWBS (the Kolyma Water Balance Station); **6** – water-power plant.

The study objects are runoff-forming climatic conditions, the study subjects are water runoffs over MWE slopes.

The research and actual significance of this study is presupposed by an important role of water runoff as an agent participating in fluvial denudation and deposition processes, and formation of placer deposits. Understanding the total runoff pattern and its changes is a necessary prerequisite for rational use of fresh water resources and solution of geoecological problems.

Research methods include analysis of published records of the Kolyma Water Balance Station (KWBS), which embrace periods since 1930s-1940s through the late 1980s. As a result of sociopolitical disturbances, which followed after that period in Russia, the majority of weather monitoring stations in NE Russia ceased their activities, the observations became irregular, so observation data available with us are only till 1988, and we have just some scarce data for later period.

The subsurface constituent of the water runoff is established by virtue of genetic subdivision of runoff hydrographs in consideration of ground-water recharge pattern in the cryolithic zone. According to S.F. Fotiev [2], the basic water exchange components within the continuous cryolithic zone include water hosted in seasonally thawing rocks (STR), superpermafrost and local continuous taliks. Within discontinuous cryolithic zones, besides STR and talik ground waters, continuous band- or ribbon-shaped taliks forming beneath stream channels of the 2nd and greater orders also become important in river recharge processes. With perennially frozen rocks (PFR) having thickness less than 100-120 m, subpermafrost fractured aquifers become important components of the active water exchange zone [3].

In compliance with hydrologic budget data for mountainous cryolithic areas [4, 5] supplied by the Kolymian station for the period 1947-2002, the largest monthly average discharge in September was 550 $l \cdot s^{-1}$ at the closing range of the Kontaktovy-Nizhny Creek in 1975. It is in September, that rains and ground waters are the only recharge sources for all streams there. In order to establish the participation of ground waters, the runoff hydrograph is created (figure 2) for period of September 1-20, 1975 prior to colder season and ice forming over river banks [6].

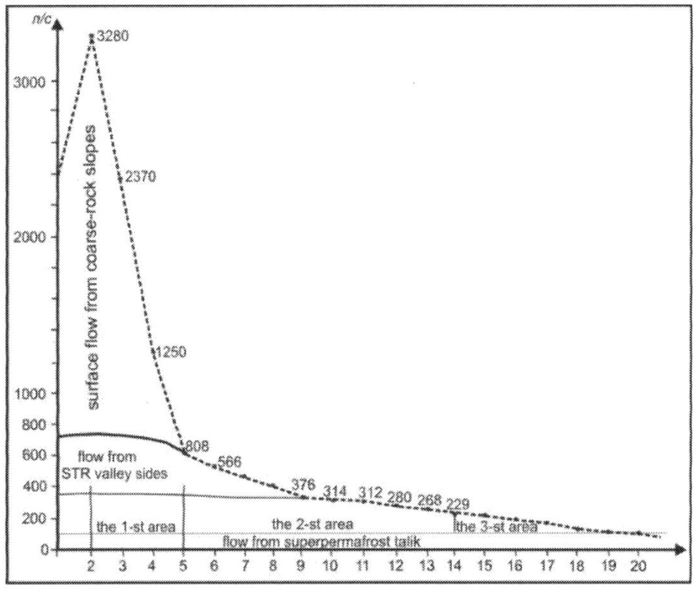

Figure 2. The Kontaktovy-Nizhny Cr. runoff hydrograph, September 1-20, 1975, the catchment area 21.2 km^2.

The hydrograph shows the peak runoff of 3280 $l \cdot s^{-1}$ on September 2, and till September 5, the following daily average runoff is represented as an almost straight declining graph. During three days, the discharge amount became 75% less of its daily average maximum. In the next 5 days, the discharge lessened to 314 $l \cdot s^{-1}$ that made about 10% of the peak value. We believe that such a drastic decrease in water discharge intensity is due both to ceasing water flow from coarse STR characterized by complete water yield and also to commencing drainage of less-yielding rock debris with small- and fine-grained infilling. Such a runoff type can be undoubtedly identified as a subsurface one. In that case, on September 5, the river was replenished at account of ground waters with daily average discharge of 808 $l \cdot s^{-1}$, which was much higher than the average monthly value. Proceeding from these data, we can conclude that throughout NE Russia in September the monthly average runoff is provided by subsurface recharge sources dominated by water flows from STR. In November, when STR layer is frozen, subsurface recharge is mainly provided by superpermafrost subchannel taliks. According to the Kolyma Water Station data, the ground-water table in mid-December comes down to depth about 5 m in superpermafrost talik [7]. Over the Arctic slope of MWE since January through April, and over the Pacific slope since January through March, the runoff is supplied by water flows from superpermafrost and continuous taliks, and there is also subpermafrost water over the Pacific slope [8 9].

The study objects description. The most representative climatic data reported during as long observation periods as 60 years and more are summarized in table 1, which represents the basic runoff-forming climatic factors including the annual average air temperatures and precipitations over NE Russia. Since climatic warming has been reported during the last 50-60 years [10], the table is distinguished into two observation periods: before 1960 and later on.

Table 1. The annual average air temperatures and precipitations (according to the Kolymian Hydrometeorologic Department).

№	Weather station; absolute elevation, m; co-ordinates	Observation periods	Annual average values during the observation periods	
			Air temperature, °C	Precipitations, mm
		Arctic Slope		
		The East Siberian Sea runoff basin		
1	Seimchan, 205 m; 62°55' N	1934-1959	-11.9	286
	152°25" E	1934-2010	-10.3	307
2	Omolon, 260;	1944-1959	-13.2	256
	65°44' N 160°32' E	1944-2010	-11.6	264
3	Srednekolymsk, 29; 67°27'	1887-1959	-12.6	195
	N 158°43' E	1887-2010	-11.2	234
4	Pevek, the sea shore, 8;	1940-1959	-10.4	136
	69°42' N 150°56 E	1940-2010	-9.5	190
		The Chukotian Sea runoff basin		
5	Amguema, 62;	1948-1959	11.9	422
	64°47' N 177°34' E	1948-1987	11.5	494
6	Uelen, 5;	1928-1959	8.2	386
	66°10' N 169°50 W	1928-2010	7.1	432
		Pacific Slope		
		The Bering Sea runoff basin		
7	Egvekinot, 29;	1946-1959	7.1	432
	66°15' N 179°09 E	1946-2010	6.9	557
8	Markovo, 28;	1895-1959	9.1	334
	64°41' N 170°25' E	1895-2010	7.2	432
9	Anadyr, 62;	1898-1959	7.7	312
	64°47' N 170°25' E	1898-2010	6.9	382
		The Okhotian Sea runoff basin		
10	Palatka, 341;	1940-1959	-6.6	383
	60°06' N 150°56' E	1940-2010	-5.7	411
11	Okhotsk, 12;	1844-1959	4.8	322.5
	59°22' N 147°12' E	1844-2010	-3.8	499
12	Nayakhan, 23;	1914-1959	-4.9	411
	61°55' N 158°59' E	1914-2010	-4.5	463

As this table shows, the processes of climatic warming have manifested themselves over the entire territory under consideration and are the most obvious over the Arctic slope of MWE and on the Pacific coasts. This warming phenomenon is everywhere associated with greater precipitations, the increase in which ranges from 3 % to 55 % with respect to preceding periods. Precipitations increased to the largest extent over the Pacific slope, and over the Arctic slope the increase makes up 3-7 % in intermountain areas and is as high as 20-40% in cyclone-accessible areas as Pevek, Srednekolymsk, Amguema.

2. The Study Results

In order to establish the relationships between the water runoff parameters and climatic factors, the data available on streams with all-year-round runoff regimes were examined with reference to atmospheric changes reported during long-term periods of observations, with just few exclusions of lacking data. It should be noted here, that due to socioeconomic disturbances in Russia at the end of 1980s, the majority of weather stations in NER were abandoned and hydrometric studies were interrupted. It was during that very decade that the Kolymian Hydro came into its full capacity operation, the result of which was a significant alteration of both subsurface and surface runoffs in the Kolyma River valley [11]. The runoff calculations for representative rivers are given in table 2.

Table 2. The representative runoff data for NER [12 - 16].

№	Rivers, closing range; catchment area, km^2	Observation periods	Annual average discharge, m^3·s^{-1}	Winter low-level subsurface runoff, month, m^3·s^{-1}			
				IX	XI	I	IV (III)*
		Arctic Slope					
1	Kolyma R., Srednekolymsk; 361 000	1927-1960	2234.1	3076	280.5	111.9	59.1
		1961-1980	2156	3256	254.4	111.3	58.1
		1981-1999	2030	-	-	-	-
2	Kolyma R., Ust-Srednekan; 99 400	1933-1960	724	960	76.8	17.4	6.02
		1961-1980	720.9	961.5	69.7	17.9	6.1
3	Taskan R., Taskan; 9 970	1938-1960	73.8	80.4	6.12	3.67	2.94
		1961-1999	71.6	86.1	6	3.24	2.5
4	Kulu R., Kulu; 10 300	1942-1960	92	132	12.1	2.78	1.02
		1961-1988	95.2	136.2	13	2.83	1.3
5	Detrin R., Vakhanka R. mouth; 5 630	1938-1960	52.3	89.9	7.17	0.82	0.21
		1961-1988	51.1	80.4	6.25	0.92	0.26
6	Amguema R., 1744 km; 26 700	1944-1960	280	280	18.9	0.28	0.006
		1961-1988	271	283	`7.6	0.25	0.005
		Pacific Slope					
7	Izyskatelsky Cr., 1.5 km from the mouth; 13.2	1947-1960	0.36	0.28	0	0	0
		1961-1980	0.31	0.38	0	0	0
8	Anadyr R., Novy Eropol; 43 300	1958-1970	502	443	50.4	16.3	10.7
		1971-1988	431.4	499	50.9	17.5	12.3
9	Gizhiga R., 20 km up the mouth; 11 700	1951-1960	148	132	24.5	10.1	7.27
		1961-1988	156	161	27.1	11.2	7.98
10	Khasyn R., Khasyn; 682	1941-1960	9.04	15.6	1.73	0.33	0.09
		1961-1999	9.19	13.9	1.72	0.02	0.01
11	Dukcha R., the mouth; 330	1961-1970	5.27	7.08	2.69	1.2	0.7
		1971-1988	5.66	7.28	2.78	1.25	0.8

* The minimum winter runoff is observed in April over the Arctic slope and in March over the Pacific slope.

IOP Conf. Series: Earth and Environmental Science **107** (2017) 012028 doi:10.1088/1755-1315/107/1/012028

As this table shows, the general climatic trend optimizing, the total annual runoff of the majority of rivers over the Arctic slope of MWE becomes lower. Due to effects of warmer climatic conditions, rivers become to a greater extent recharged with ground waters hosted in STR, superpermafrost and continuous taliks. The exclusions are just some streams (the Kulu River), the runoff of which becomes greater including the subsurface constituent under warmer climate, which is due to different tectonic and neotectonic settings [17].

Over the Pacific slope, the similar trend of total runoff's decrease associated with its growing subsurface constituent is reported from the Anadyr Bay of the Bering Sea. Since 1960, all rivers within the Sea of Okhotsk basin have increased their runoffs. This increase dynamics is exemplified by the Dukcha River with its total runoff having increased as high as 0.35 $m^3 \cdot s^{-1}$ during 17 years (since 1971) relative to period of 1961-1971. The ground-water share of the river recharge in winter low-level period has obviously increased.

3. Discussing the Results

In current warmer climate conditions, different formational trends of water runoffs over the oceanic slopes are explained as due to cryolithic zone characteristics [18]. A continuous cryolithic zone having thickness greater than hypergenous fracturing depth (100-150 m) is distributed over the Arctic slope of MWE and in the northern areas of the Pacific slope. In our opinion, a decrease in the total water runoff under such geocryological settings has been the result of a whole set of natural and industrial factors.

Natural effects caused by climatic warming include higher annual average temperatures of STR, their greater thickness and a longer duration of the warm season. The results are the growing biological productivity of herbaceous vegetation, shrubby and arboreal plants, and formation of mountain-tundra soils of mountain-pine belt at divides with elevations about 1300-1400 m [19]. Losses of water used for evapotranspiration and photosynthesis become greater. An increase in air temperature in warm seasons enhances water evaporation losses over rocky slopes and mountain tundra of any type.

Besides natural impacts, industrial factors as well negatively affect the total runoff of rivers over the Arctic slope of MWE and are the most significant in mountain valleys of minor and medium rivers having catchment areas about 5000 – 6000 km^2, where gold placers are mined by opencut methods. Thus, the bottoms of valleys of the Taskan, Detrin, Bererlyok and other river tributaries, and the lower portions of mountain slopes there are totally reworked by opencut activities from rivers headwaters down to mouth areas. Vegetation cover was destroyed and removed with gravel and pebble rocks outside the placer area. Alluvial deposits are washed off silt and clay and disposed in dumps. River channels were repeatedly displaced and superpermafrost taliks destroyed, and many artificial ponds were made with their surface water evaporation exceeding the amount of precipitations in warm seasons [20]. Industrial activities were accompanied by slope landslide processes destroying STR layers. These industrial activity effects combined with long-term natural alterations result in decrease tendencies of annual average water discharge and subsurface recharge of rivers in winter low-level period. The examples of such industrially altered rivers, as the Taskan and Detrin rivers, are presented in table 2.

Over the Pacific slope of MWE, within the Sea of Okhotsk basin, cryolithic zone becomes intermittent and continuous band-shaped taliks occur beneath channels of rivers of the 3rd and 2nd orders with catchment areas of 5 – 10 km^2. Taliks are frequent over mountain slopes and in divide areas beneath bottoms of lakes of glacial or tectonic origins. All these factors contribute to both the reduction of the PFR layer's thickness to 100-150 m and formation of subpermafrost water-bearing fractured rocks. The result is that the river recharge in winter low-level period is here many times greater than over the Arctic slope. A growth of total runoff amounts is reported here from everywhere even for rivers with industrially impacted valleys, such as the Dukcha and Khasyn rivers, due to a rise in subsurface replenishment, the share of which can be as high as 70% of the total recharge. From here it follows, that in discontinuous cryolithic zone conditions, precipitations to a great extent replenish the ground-water reserves within the active water-exchange zone, at account of which the surface streams with small catchment areas are recharged and have their runoffs all year round.

Current forecasts and predictions for expected climatic changes and possible responses of the cryolithic zone and hydrogeosphere are diverse and contradictory. The majority of researchers suggest a continuous rise in air temperature and precipitations amount. Presumably, the climate will become as warm as in the late Pleistocene, when the annual average temperature of air in NER territory was 4 – 4.5 °C higher than at present. So, over the coasts of the Anadyr Bay of the Bering Sea and at the Chukotian Sea the air temperatures will be 4-3 °C below zero and over the coasts of the East Siberian Sea temperatures will be 7-5 °C below zero. Within the Sea of Okhotsk basin, it can range from 2-3 °C below zero in continental areas to 0 °C over the coasts. In such temperature conditions, a continuous cryolithic zone becomes intermittent and then it transforms into the island-like or spot type. Over the Arctic slope of MWE, in compliance with the current situation described in this paper, a growing discontinuity of the cryolithic zone will be associated with its thickness reduction to 100-150 m in divide areas and even less over river valley slopes. A region-wide subpermafrost zone of hypergenous fracturing can form with its aquifer capacities generated by ice freezing and melting processes in fractures; such zone can be reasonable defined as the "cryohypergenous fracturing zone". In climatic conditions, which will be similar to those existing now in mountain-taiga areas of the northern Amur River basin, atmospheric precipitations will be almost completely absorbed by soils [21]. Thus, the climatic conditions incessantly becoming warmer, the total river runoff is expected to increase both over the Arctic and Pacific slopes of MWE within the NER territory. Such an increase will be provided by active water-exchange zones developing everywhere at account of greater areas of thawing rocks, smaller cryolithic thicknesses, and formation of subpermafrost aquifer zone of regional cryohypergenous fracturing. On the whole, this prediction is geoecologically favorable as it raises the industrial attractiveness of this region including its agricultural sector. However, the negative consequences are as well possible due to intense denudational and thermoerosional processes. In order to control this situation, long-term monitoring stations shall be established as soon as possible in areas of crucial climate-monitoring significance.

4. Conclusion

The results obtained from conducted studies serve for us as a basis to make the following conclusions:

- Long-term meteorological observations, as long as 160 – 50 years periods, over the NER territory, indicate unambiguously a progressive warming of climate that has become even more intense since the 1960s;
- Warmer climates and greater precipitations in areas of continuous cryolithic zone having thicknesses more than 150-180 m over the Arctic slope of MWE and in the northern part of the Pacific slope, that is the drainage area of rivers flowing into the Anadyr Bay of the Bering Sea, result in a decrease of the total runoff there, whereas the share of subsurface replenishment of streams is increased to a greater extent at account of STR water, and to a lesser extent – of subchannel taliks;
- Modern climatic changes over the Pacific slope of MWE, within the Sea of Okhotsk basin, where cryolithic zone is discontinuous, are associated both with an increase in total runoff of rivers and a considerable share of their subsurface recharge including subpermafrost aquifer zone of regional cryohypergenous fracturing and long subchannel continuous taliks;
- The runoff balance responses to modern climatic changes are influenced by their different ground-water recharge regimes in continuous and discontinuous cryolithic zone conditions;
- In the first case, the STR layer is the basic constituent of the active water exchange zone; it has a small thickness and its ground waters are placed close to the day surface, so the STR water is mainly used to provide the processes of evapotranspiration and plants photosynthesis, and evaporation of summer precipitations from rocky slopes; the losses of water become naturally greater due to higher air temperatures and also as a result of such industrial activities as opencut mining of gold placers in river valleys;

- In discontinuous cryolithic zone conditions, meteoric water is absorbed not only by STR layer, but also, via continuous taliks, by subpermafrost aquifer zone of cryohypergenous fracturing; therefore, warmer climate conditions favor both infiltration processes of atmospheric water and its participation in rivers recharges;

- The expected warming shall cause transformation of continuous cryolithic zone into discontinuous type over the Arctic slope of MWE and in the northern areas of the Pacific slope, and its subsequent transformation into the island-shaped and spot types within the Sea of Okhotsk basin of the Pacific slope; such transformations of the cryolithic zone shall favor an increase in the total runoff everywhere, including to the largest extent an increase in its subsurface constituent; from geoecological viewpoints, the expected warming and presumed responses to it of the cryolithic zone and hydrogeosphere shall have positive effects for people, however, a strict monitoring of these processes is obligatory.

References

[1] Glotov V E 2002 *The factors forming total runoff of small mountain rivers in the Subarctic* (Magadan: SVKNII DVO RAN) 182-201
[2] Fotiev S M 2013 *Eatrh's Cryosphere* **17** (2) 41-59
[3] Glotov V E and Glotova L P 2011 *Russ. J. of Pacific Geology* **30** (5) 93-104
[4] Bojcov Ju A, and Kuznecov A S 1979 *Kolyma* **10** 28-30
[5] Slaugter C, Glotov V E and Michailov W M 1995 *Water Air and Soil Pollution* 351-361
[6] 1977 *Observational data of Kolyma podnebesova station in 1975* (Magadan: KUGMS) **16** 155
[7] Glotov V E 2002 *The factors forming total runoff of small mountain rivers in the Subarctic* (Magadan: SVKNII DVO RAN) 102-141
[8] Glotov V E 2016 *Bulletin of the North-East Scientific Center, Russia Academy of Sciences Far East Branch* **3** 36-46
[9] Glotov V E and Glotova L P 2012 *Eatrh's Cryosphere* **16** 57-66
[10] Gavrilova M K 2003 *Review of the status and trends in the climate of Yakutia* (Jakutsk: JaF SO RAN) 12-18
[11] Glotov V E, Glotova L P and Ushakov M V 2014 *Environmental geoscience* **1** 20-27
[12] 1988 *Annual data on the regime and resources of surface land waters* **1**100
[13] 1985 *Annual data on the regime and resources of surface land waters* **1** 430
[14] 1966 *The main hydrological characteristics. North-East of the USSR* (L.: Gidrometeoizdat) **19** 379
[15] 1974 *The main hydrological characteristics. North-East of the USSR* (L.: Gidrometeoizdat) **19** 230
[16] 1978 *The main hydrological characteristics. North-East of the USSR* (L.: Gidrometeoizdat) **19** 226
[17] Glotov V E and Glotova L P 2010 *Bulletin of the North-East Scientific Center, Russia Academy of Sciences Far East Branch* **1** 89-98
[18] Woo M-K 1990 *Northern Hydrology: Canadian Perspectives* ed. Prowse T D **1** 63–76
[19] Pugachev A A 2002 *The factors forming total runoff of small mountain rivers in the Subarctic* (Magadan: SVKNII DVO RAN) 141-166
[20] Sushhanskij S I 2002 *The factors forming total runoff of small mountain rivers in the Subarctic* (Magadan: SVKNII DVO RAN) 18-35
[21] Kirjuhin V P 1981 *Zonality of underground waters of platform and homoscedastic areas* (L.: GO SSSR) 12-25

IOP Conf. Series: Earth and Environmental Science **107** (2017) 012029 doi:10.1088/1755-1315/107/1/012029

The changes in the composition of Cladocera community in bottom sediments of Lake Maloye Shibrozero (Zaonezhsky Peninsula) as a consequence of shifts of environmental and climatic conditions

A G Ibragimova[1], L A Frolova[1], D A Subetto [2,3], N A Belkina[2], M S Potakhin [2]

[1] Institute of Geology and Petroleum Technologies, Kazan Federal University, Kremlevskaya str., 4/5, Kazan, 420008, Russia;
[2] Northern Water Problems Institute, Karelian Research Centre of Russian Academy of Sciences, Alexander Nevsky prosp., 50 Petrozavodsk, 185003, Russia
[3] Herzen State Pedagogical University of Russia, Malaya Posadskaya str., 26, Saint Petersburg, 197046, Russia
E-mail: AGIbragimova@kpfu.ru

Abstract. The study aims to explore the evolution of lakes of the boreal zone during the late- and postglacial time on the south-eastern periphery of the Fennoscandian crystalline shield since the last deglaciation. In order to reconstruct the past for virgin territories of the Zaonezhsky Peninsula current investigation on bottom sediments of Lake Maloye Shibrozero was conducted. Analyzes were performed using the new paleoindicator – subfossil remains of Cladocera (Cladocera, Branchiopoda). The 28 samples of bottom sediments were analyzed. It has been determined that discovered Cladocera remains belong to representatives of 6 families and 38 taxa. Species inhabiting Palaearctic zone are predominant in lake deposits; most of the identified subfossil remains are related to the pelagic species inhabiting the open part of the lake. According to the Lubarsky scale the dominant of Cladocera community is *Bosmina (Eubosmina)* cf. *longispina*. Secondary taxa are *Chydorus sphaericus*, *Bosmina coregoni*, *Alonella nana*, *Alona guadrangularis*, *A. affinis*, *Chydorus gibbus*. At a depth of 650-653 cm, a partial replacement of *Bosmina (Eubosmina)* cf. *longispina* by *Bosmina coregoni* takes place with a simultaneous increase in the significance of *Chydorus sphaericus*, which is used to be an indicator of eutrophication and increasing trophic status of the reservoir. *Changes in Cladocera community could be attributed to decreasing the level of periglacial lake, as a result of which the Lake Maloye Shibrozero became a small isolated lake with the trend to* trophic status *increasing*. Cold-water species were replaced by thermophilic ones with a further return to a cold-water fauna. In the upper layers of the column an increase of the number of phytophilous species is noted.

1. Introduction

The problem of climate change is of particular relevance in view of the real threat not only for the economic sector, but for the life and wellbeing of the population, preservation of the biological diversity and ecological balance. Numerous studies on climate and ecological conditions changes during the past, which were and are being conducted worldwide, allow us to create models of future climate change, therefore database of regional environmental reconstructions is to be constantly added and updated.

Bottom sediments of lakes are the archives, which layers contain the information about the environmental conditions of the past on regional and global scale with a resolution from the millennia and centuries to the year [1]. One of the methods of paleolimnology is paleoindication based on the analysis of qualitative and quantitative content of remains of animal (Chironomidae, Ostracoda, Cladocera) and plant organisms (spores and pollen, diatoms) of bottom sediments.

Zooplankton, such as Cladocera, is a key component of aquatic ecosystems [2, 3]. Its distribution in the lake is strongly related to abiotic and biotic factors [4]. Cladocera analysis, as the most recent paleoindicator, has already had time to prove to be a reliable and successful way to reconstruct past conditions [3, 5]. For studies related to global change biology, fossil assemblages and paleolimnological methods provide a unique and invaluable source of information through the long-term perspective of ecosystem dynamics that cannot be resolved by monitoring data, which usually consist of only few years or decades of monitoring [6-8]. Chitin structure of Cladocera exoskeleton (headshields, carapacse, postabdomens, postabdomenal claws, etc.) are saved in bottom sediments of lakes, generally in a good state of preservation that allows to identify them to species and on the basis of obtained data reconstruct changing conditions of the past [3, 9].

Regions that are remote from the direct anthropogenic influence are indicator-regions, that carry reliable information of the changes of the environmental and climatic conditions. The category of weakly disturbed areas include the Zaonezhsky Peninsula which is situated on the northern coast of Lake Onega.

Using the method of Cladocera paleoindication the study of Maloe Shibrozero Lake (figure 1) on the Zaonezhsky Peninsula was carried out. According to the glacial theory during the Quaternary period the territory of the Fennoscandian shield area of 1.7 million km^2 repeatedly was exposed to strong covering glaciations and was the center of the European continental glaciation with the ice thickness of 3.5-4 km [10]. The transition from cold and dry climate of the late Pleistocene to the warm and humid conditions of the Holocene led to the changes in the natural environment both in the catchment areas of lakes and aquatic ecosystems [1].

Figure 1. Location of Maloe Shibrozero Lake (The Republic of Karelia, Russia).

The first fragmentary information about the rivers and lakes of the Karelia Republic was obtained in the 1920s and 1940s. In the early 1960-ies, the hydrographic studies of Zaonezhsky Peninsula was obtained by the water problem unit of the Karelian Branch of the USSR Academy of Sciences (now Northern Water Problems Institute, Karelian Research Centre of Russian Academy of Sciences) [11]. Introduction of a new method in paleoenvironmental and paleogeographic reconstructions provided additional information about the development of the aquatic ecosystems in post-glacial time and ecological and climatic situation of the region as a whole.

2. Material and methods

The Zaonezhsky Peninsula is located in the middle taiga subzone of the Karelia Republic, in the Northern Prionezhskiy geobotanical district, the vegetation of which is characterized by a great originality. Spruce forests take precedence over the pine [12].

The topography of the Peninsula is characterized by the presence of alternating long, narrow ridges and long, narrow depressions, presently occupied by lakes [13]. The western part of the Zaonezhye, where is Maloe Shibrozero Lake is located, is a typical area of selka tectonic denudation topography with its Quaternary period ice sheet structure. There is a system of ridges, oriented in the north-western direction, with a relative excess over the level of the lake up to 120 m [12].

Most of lake depressions of the Zaonezhye are formed by tectonic subsidence, glacial tectonics or other glacial processes. They differ considerably in their morphometric and hydrological indices. Water area of lakes varies from less than 0.01 to 30 km^2. Commonly the basins are highly elongated, sometimes dozens of kilometers long and no more than several hundred metres wide [11]. Extremely elongated lakes are relatively uncommon, occurring only in northern Scotland, northwestern England (Camberland Plateau) and northwestern Central Siberia (Putoran Plateau) [14].

The climate of Karelia has its own characteristics that distinguish it from other ecologically valuable areas of the Republic of Karelia. The average annual temperature in Zaonezhye is positive and amounts to 2.3 °C. The coldest month is February (average temperature is -10.2 °C - -11.0 °C), the warmest is July (average temperature is 15.8 °C -16.7 °C) [15].

On April of 2015, within the field works on the Zaonezhsky Peninsula, 6,15 m long of Maloe Shibrozero Lake sediment cores (N 62° 22.346', E 35° 12.660', 56.5m a.s.l., the area is 0.076 km^2, depth – 3.1 m) were recovered (figure 1), which once was a former part of Lake Onego. The expedition was carried out jointly with Northern Water Problems Institute, Karelian Research Centre of Russian Academy of Sciences (Petrozavodsk, Russia), Kazan (Volga region) Federal University (Kazan, Russia), Herzen State Pedagogical University of Russia (St.Petersburg, Russia) and University of Tartu (Estonia, Tartu).

Lithostratigraphic analysis of bottom sediments, analysis of ionic composition and gas mode were conducted. Samples for spore-pollen, diatoms, Chironomidae and Cladocera analysis were collected.

Water samples were taken from the surface, bottom and supernatant horizons. The component ratio of the ionic composition of the water classifies it as a hydrocarbonate group of magnesium, calcium. Supernatant water is characterized by a slight increase in the content of calcium, bicarbonates and pH (average in the lake - 6.81, in supernatant water - 7.45). According to these performances, the reservoir corresponds to moderate alkaline weak-acid neutral waters, supernatant water is classified as a moderate alkaline neutral weak-alkaline water. According to the results of the analysis, rising of the concentration of carbon dioxide, organic substances, ammonium ions, nitrates, iron, manganese and silicon from the surface to the bottom was observed. The content of nutrients indicates a mesotrophic status of the reservoir. There is a decline in the oxygen content with depth: 0.5 m -8.23 $mg \cdot L^{-1}$, 1 m - $\frac{7}{26}$ $mg \cdot L^{-1}$, 2 m - $\frac{5}{7}$ $mg \cdot L^{-1}$, 2.5 m - 4.1 $mg \cdot L^{-1}$.

According to the lithostratigraphical description the colour of brown organic homogeneous gyttja at a depth of 6.50 m from the surface of the lake turns greyer, indicating an increase of the content of mineral component in the sediments. At a depth of 6.75 the transition interval is noted - layered grey-brown silt with inclusion of plant remains (mosses), the thickness of the layers is different. Greybrownish clay gyttja occurring at a depth of 7.05 m, at a depth of 7.26 m turns to the dark grey silty-clay gyttja, and then at a depth of 7.50 m to grey layered sandy gyttja. Transitional zone, represented by streaky grey silt with black organic layers, is observed at a depth of 7.78 m. Sandy grey streaky clay with interlayers of hydrotroilite presented at depths of 7.82 – 8.01 m become more sandy moving down the column of bottom sediments– sandy layers reaches 4 cm in thickness. At the same depths the silt layers up to 4 cm are noted. At a depth of 9.40 m, the clay layers become lighter, layers become thinner. At depths of 9.70 – 9.95 m thin-bedded layer of pink clay is observed, which turns to the grey well-laminated clay.

Sand band between grey clays and overlying brown silts indicates the presence of a very sharp boundary between grey glaciolacustrine sediments and lacustrine Holocene sediments, that is the result of interruption of sedimentation due to a catastrophic drop in the water level of the Baltic Ice Lake at the boundary between the late Pleistocene and Holocene. A decrease in the level of lakes that are located within the catchment basin of the Baltic was caused by these changes [16].

The age of a pink banded clay horizon ("pink horizon"), which is used to be a stratigraphic marker of the Onego periglacial lake sediments and formed by diagenetic changes of sediments during the sharp fall of the Onega lake level, is about 11 300 years [17]. "Pink horizon" in Maloe Shibrozero Lake is noted at a depth below 970 cm from the lake surface.

The sample preparation and analysis of samples were carried out in the laboratory of "Paleoclimatology, paleoecology, paleomagnetism" of the Kazan Federal University. Samples were investigated under a light microscope Axiostar Plus Carl Zeiss at magnification x100-400. In each sample were identified at least 100 remains. Identification was carried out using specialized determinants of subfossil [18] and modern Cladocera [19]. Two half shells of the carapace were counted as an integral organism.

Analysis of changes in diversity of biotic groups is made using indices that determine the degree of species richness, diversity and dominance of Cladocera communities: index of Shannon-Weaver, index of uniformity of ecological groups of Pielou. The investigation included the values of saprobity index of Pantle and Buck. In order to identify classes of the dominance the scale of Lubarsky was used. Statistical and stratigraphic analyses implemented in the program C2 of S. Jaggens.

3. Results and discussion

According to the results of subfossil Cladocera analysis 3297 chitinous remains of Cladocera exoskeleton were detected and identified to generic assignment (*Eubosmina sp.*, *Eurycercus sp.*, *Pleuroxus sp.*, *Simocephalus sp.*, *Ceriodaphnia sp.*) and species. It was established that remains discovered in 28 bottom sediment samples belong to the representatives of 38 taxa and 6 families (Bosminidae, Chydoridae, Daphnidae, Polyphemidae, and Sididae, Leptodoridae).

Qualitatively species inhabiting the Palearctic (57.6 %) zoogeographic zones are dominated in Maloe Shibrozero Lake, the occurrence of species with cosmopolitan (22.17 %) and Holarctic (20.2 %) distribution is roughly equal. The majority of the identified remains are pelagic species inhabiting the open part of the reservoir (54.8 %). However, the high taxonomic diversity comes from species preferring overgrown, silt-covered or sandy areas.

According to the Lubarsky scale the dominant of Cladocera community is *Bosmina* (*Eubosmina*) cf. *longispina* (43.8 %). Among the secondary taxa are the following: *Chydorus sphaericus* (11.67 %), *Bosmina coregoni* (8.92 %), *Alonella nana* (5.7 %), *Alona guadrangularis* (5.61 %), *A. affinis* (5.06 %), *Chydorus gibbus* (3.88 %) (figure 2).

Bosmina (*Eubosmina*) cf. *longispina* might be present in the littoral zone, however, its volume in the pelagic zone, where *Bosmina* forms a massive group, far exceeds the littoral zone, that taking the tanatocenosis of *Bosmina* for a pelagic group is fully justified. *Bosmina* have the age-related changes in the valves and the headshields form, which raises questions regarding their taxonomic status. Systematics of living *Bosmina* group remains far from clear and needs of major overhaul [20]. Fennoscandian *Bosmina* group was previously studied by Nillssen and Larsson [21]. They believed that the most common Fennoscandian *Bosmina* with the North Holarctic habitat should be called

Bosmina longispina (= *B. Obtusirostris* Sars, 1862, *B. lacustris* Sars, 1862). In their opinion it was the taxa which first moved to the postglacial water bodies. According to the early studies, the dominance of *Bosmina sp* has been noted in Lake Onega since the formation of thanatocenosis [20].

Bosmina (*Eubosmina*) cf. *longispina* is totally dominating in the lower layers of the column, reaching 82.76 %, that indicates a well-developed open part of the reservoir. There is a high species diversity with the dominance of the Northern species. At depths of 796-686 cm remains of *Camptocercus liljeborji* (*Schoedler, 1862*) are detected. The species is an inhabitant of the shallow ponds, small lakes, rivers, river basins littoral zone in which the species mostly occurs in areas covered with aquatic vegetation. *C. liljeborji* is found in Germany, Romania, Slovenia, Sweden and Northern part of European Russia [22]. The occurrence of the species suggests a decrease in the water temperature of the lake [23]. The remains of the *C. liljeborji* are not detected up to the column, above the depth of 686 cm, the species is completely replaced by *Camptocercus rectirostris*, associated with warm water [24, 25] and occurring in Maloe Shibrozero Lake at all stages of formation.

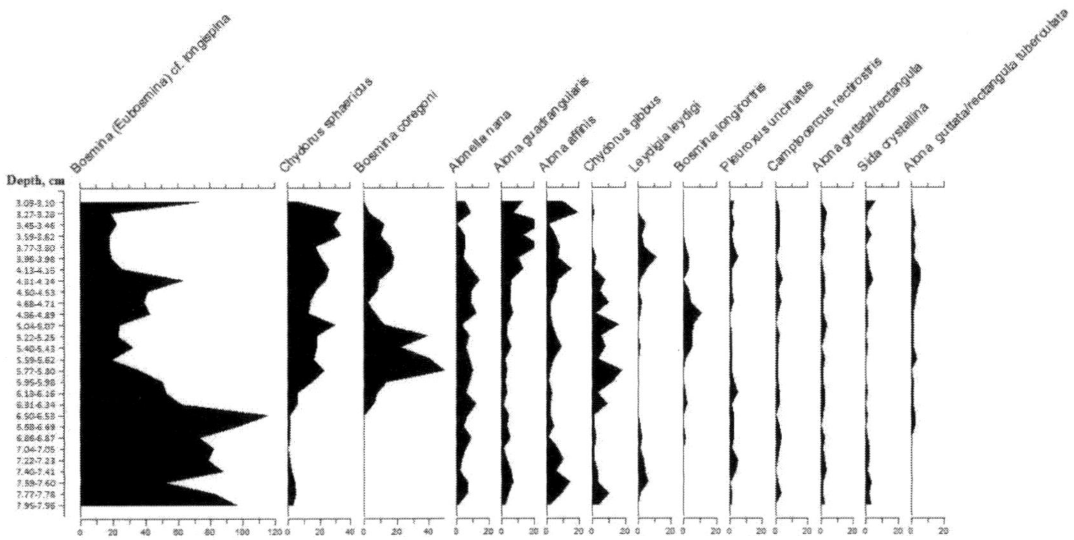

Figure 2. Relative abundances of the most common Cladocera taxa in the Maloe Shibrozero Lake.

The highest numbers of *Alonopsis elongata* remains are noted in the lower layers of the column. This species is typical for Northern and primarily Scandinavian countries [26]. The suggestion of cool climate was confirmed by the presence of the littoral species *Alona affinis*, *Alonella nana*, *Oxyurella tenuicaudis*, *Monospilus dispar* which can tolerate low temperatures [20, 27]. An increase in the numbers of individuals from the *Eubosmina*, along with simultaneous low percentage share of *B. longirostris* and littoral species, indicates low water fertility and temperature during this time [23].

The occurrence at this time of *Daphnia* species, particularly those from the *D. pulex* group, and *B. (E.) longispina*, typical for Northern (Scandinavian and Greenland) and mountain lakes, suggests that the lake was quite deep, and its waters were rather cool, with low trophic status [28, 29].

The representatives of the genus Pleuroxus (*P. uncinatus*, *P. trigonellus*, *P. laevis*) are found throughout the lake evolution, their relative abundance is 1.9%. The most numerous remains of the genus belong to *P. uncinatus* – inhabitant of the open littoral zone and silt-covered sand [20].

According to the results of the Cladocera analysis, the development of Alonella nana, which is one of the first invaders of the lake, occurred without sudden fluctuations in abundance. It is known that *Alonella nana* inhabits different types of Northern waters from oligo- dystrophic to eutrophic [20, 30, 31]. Phytophilous cold-water *Alona guttata/ rectangula* was also constantly present in the lake [27]. It is interesting to note that *Alona guttata/ rectangula* was found in all samples, whereas the closely related form *Alona guttata/ rectangula tuberculata* was found in samples above the depth of 686 cm.

The composition of subfossil Cladocera assemblages could have been the result of cool climate, with concurrent high water level of the lake.

Moving up the column, at the depths of 650-653 cm, a partial replacement of *Bosmina (Eubosmina)* cf. *longispina* by the larger taxa *Bosmina coregoni* takes place, at the same time an existence of another representative of the genus – *Bosmina longirostris*- is noted (figure 2). In accordance with Patalas [32] "subspecies" of *Bosmina coregoni* are common for lakes of different trophic status and can be present in oligotrophic and eutrophic water bodies [20]. Thus, in a number of lakes of Germany - Grosser Ploner See, Segeberger, and Schosee *Bosmina longispina* was replaced by *B. coregoni kessleri*, and then by *B. coregoni coregoni* with the eutrophication [33]. The decrease in the content of nutritious substances in the water is documented by a decrease in the frequency of *B. longirostris*, a species usually inhabiting fertile, shallow zones of lakes [34].

This observation is confirmed by the fact of simultaneous increase of importance of *Chydorus sphaericus*, which is an indicator of eutrophication, and cosmopolitan *Chydorus gibbus*- inhabitant of sandy-silty soils in the inshore zone of large water bodies found in sphagnum bogs [15]. The importance of phytophilous species (*Alona guadrangularis*, *A. guttata/ rectangula*, *A. affinis*, *A. costata*, *Alonella nana*, *A. exigua*, *A. excisa*) is increasing. According to a research conducted on Lake Charzykowskie [22], *Bosmina longirostris* prefers warm temperature and reducing its biomass under adverse conditions. Appearance of *B. longirostris* at the certain stage of the lake development may be considered as an improving of the environmental conditions and air temperatures rising.

Reducing the number of *Alonopsis elongata* above 686 cm depth may be associated with climate change in favor of higher temperatures and increasing the concentrations of Ca^{2+}. It is known that the species prefers acidic waters up to pH of 4.45 and with low Ca^{2+} concentrations of 0.4 and 2.2 mg dm^{-3} [35].

Above the depths of 413-416 cm, an increase in the number of *Alona affinis*, *Alona guadrangularis*, *Leydigia leydigi*, *L. acanthocercoides* remains is marked. Increasing biomass of *Alona affinis* may be associated with a decrease of temperature and increase of the overgrown part of the reservoir (the book Internet links, etc). It is known that members of the genus *Leydigia* appear or increase in numbers in biocenoses with eutrophication [3]. At the same time, the number of remains of *Bosmina coregoni* is decreasing above the depth of 468 cm, and the proportion of *Bosmina longirostris* is getting smaller to the point of extinction. This indirect evidence may indicate a decrease in the pelagic part of the lake and a cooling of the climate in the study region.

The average Shannon index in the Maloe Shibrozero Lake is 2.6 + 0.1, that allows to classify the lake as moderately polluted, which is consistent with the results of the hydrochemical analysis. The index values are in the range from 1.2 to 3.6. Index Pielou changes in the range from 0.23 to 0.69, with an average of 0.5, which characterizes the structure of Cladocera community as not sufficiently aligned. The saprobity index of the lake is 1.54, which determines the trophic status of the lake as β-mesosaprobic.

4. Conclusion

The study of the column of Maloe Shibrozero Lake bottom sediments using subfossil Cladocera analysis characterized changes that took place in the catchment area of the lake and in the region. It was established that the dominant of Cladocera community is *Bosmina (Eubosmina)* cf. *longispina*. The secondary species are *Chydorus sphaericus*, *Bosmina coregoni*, *Alonella nana*, *Alona guadrangularis*, *A. affinis*, *Chydorus gibbus*. The majority of the identified remains are pelagic species, but the high taxonomic richness of diversity is achieved by the taxa associated with specific substrates, primarily -overgrown areas. According to the results of Cladocera analysis the lake at the start of the foundation was a cold-water reservoir with a well-developed pelagic zone and was characterized by the presence of shallow areas and zones overgrown by macrophytes. With the development of Maloe Shibrozero the biomass of pelagic *Bosmina* group has decreased, trophic status has increased, as confirmed by the increase of the biomass of *Chydorus sphaericus*. *Changes in Cladocera community could be attributed to decreasing the level of periglacial lake, as a result of which the Lake Maloye Shibrozero became a small isolated lake with its own regime and processes.* These *factors can also be a consequence* of eutrophication. Moving up the column the presence of *Bosmina longirostris* is noted, which may be connected with the increase of air temperatures. Analysis of the upper part of the column of bottom sediments indicated an increasing of cold-water species that allows us to say that after some warming the weather got colder. The lake is moderately polluted and β-mesosaprobic, the structure of Cladocera community is not sufficiently aligned.

Acknowledgement
This work was funded by the subsidy allocated to Kazan Federal University for the state assignment in the sphere of scientific activities, by grants RFBR (research projects № 15-05-04442 and № 16-35-50067) and RSF № 14-17-00766, as well as at the expense of Russian Government Program of Competitive Growth of Kazan Federal University.

References

[1] Subetto D A 2009 *Bottom sediments of lakes: paleolimnological reconstructions* (St. Petersburg.: The Herzen state pedagogical University) 309
[2] Einarsson Á and Örnólfsdóttir E B 2004 *J. Aquat Ecol* **38** 253–262
[3] Frolova L A 2011 *Methodological approaches to use biological indicators in paleoecology: Cladocera in paleoecology* ed. Nazarova L B (Kazan: Kazan state University) 280
[4] Leibold M A 1990 *J. Limnol Oceanogr.* **35** 938–944
[5] Frolova L A, Ibragimova A G and Fedorova I V *Proc. 16th Int. Mult.Scient. Geo-Conf. SGEM 2016, Energy and Clean Technologies* **4** (2) 579-587
[6] Smol J P 1991 *J. Hydrobiologia* **214** 201–206
[7] Luoto T P and Nevalainen L 2013 *J. Aquatic Biology* **18** 47–58 doi: 10.3354/ab00487
[8] Frolova L A 2016 Subfossil Cladocera (Branchiopoda, Crustacea) in climatic and palaeoenvironmental investigations in Eastern Siberia *Proc. Int. Mult. Scient.GeoConf. SGEM 2016, Energy and Clean Technologies* **4** (2) 601-607
[9] Ibragimova A G, Frolova L A and Grekov I M 2016 *Research Journal of Pharmaceutical, Biological and Chemical Sciences* **7** (6) 3201-3206
[10] Chuvardinsky V G 2014 *Was there really the continental glaciations of Europe? Myths and relality* (Saarbryukken: Lambert Academic Publishing) 275
[11] Litvinenko A V and Bogdanova M S 2014 *Biogeography, landscapes, ecosystems and species of Zaonezhsky Peninsula in Lake Onega, Russian Karelia. Reports of the Finnish Environment Institute,* ed. Lindholm T, Jakovlev J and Kravchenko A (Helsinki: Finnish Environment Institute, Natural Environment Centre) **40** 41-52
[12] Filimonova L V and Lavrova N B 2015 *Proc. of Kar. RC RAS* **4** 30–47 doi 10.17076/bg22
[13] Golubev A, Rychanchik D, Romashkin A and Polin A 2014 *Biogeography, landscapes, ecosystems and species of Zaonezhsky Peninsula in Lake Onega, Russian Karelia. Reports of the Finnish environment institute,* ed. Lindholm T, Jakovlev J and Kravchenko A (Helsinki: Finnish Environment Institute, Natural Environment Centre) **40** 17-34
[14] Semenov V N 1993 *Kizhi Vestnik* **2** 53–59
[15] 2013 *Materials of comprehensive environmental survey underpinning the organization of the State natural Park "Zaonezhye"* 108
[16] Subetto D A, Nazarova L B, Pestryakova L A, Syrykh L S, Andronikov A V, Biskaborn B, Diekmann B, Kuznetsov D D, Sapelko T V and Grekov I M 2017 *Siberian Ecological Journal* **4** 369–380
[17] Demidov I N 2005 *Geology and minerals of Karelia* (Petrozavodsk: Karelian Research Centre of Russian Academy of Sciences) **8** 134-142
[18] Szeroczyńska K and Sarmaja-Korjonen K 2007 *Atlas of Subfossil Cladocera from Central and Northern Europe* (Świecie: Friends of the Lower Vistula Society) 84
[19] Kotov A A, Sinev A Ju, Glagolev S M and Smirnov N N 2010 *Cladocera in identification Key*

of zooplankton and zoobenthos of European Russia freshwater. Zooplankton, ed. Alekseeva V R, Calolihina S Ja (Moscow: Partnership of scientific publications KMK) **1** 151-276
[20] Smirnov N N 2010 *Historical ecology of the freshwater zoocenoses* (Moscow: Partnership of scientific publications KMK) 225
[21] Nilssen J P, Larsson P 1980 *J. of Zoological Systematics and Evolutionary Research* **18** 62–68. doi:10.1111/j.1439-0469.1980.tb00727.x
[22] Smirnov N N 1998 *J. Hydrobiologia* **386** 63–83
[23] Szeroczyńska K and Zawisza E 2011 *Quaternary International* **233** 185-193
[24] Milan M, Bigler C, Tolotti M and Szeroczyńska K 2017 *J Paleolimnol* 1-17 doi: 10.1007/s10933-017-9981-z
[25] Kosareva L R *et al* 2017 *ARPN Journal of Engineering and Applied Sciences* **12** (7) 1-15
[26] Hessen D O and Walseng B 2008 *Freshwater Biology* **53** 2026-2035
[27] Guilizzoni and Oldfield 1996 *Palaeoenvironmental analysis of Italian crater lake and Adriatic sediments* (Verbania Pallanza: Istituto Italiano di Idrobiologia) 357
[28] Frey D G 1991 *Journal of Paleolimnology* **6** 193-197
[29] Ulrich M, Wetterich S, Rudaya N, Frolova L, Schmidt J, Siegert C, Fedorov A N and Zielhofer C 2017 *The Holocene OnlineFirst* doi:10.1177/0959683617708454
[30] Chengalath R, Bruce W J, Scruton D A 1984 *Verh. Internat. Verein. Limnol.* **22** 419-430
[31] Nevalainen L *Sexual reproduction in chydorid cladocerans (Anomopoda, Chydoridae) in southern Finland - implication for paleolimnology* (Helsinki: University of Helsinki) 2008 54
[32] Patalas K 1971 *Trans. Amer. Microsc. Soc.* **90** (1) 117-118
[33] Hofmann W 1978 *Polskie Archiwum Hydrobiologii* **25** 167–176
[34] Szeroczyńska K 1998 *Studia Geologica Polonica* **112** 123-165
[35] Bledzki L A and Rybak J I 2016 *Freshwater Crustacean Zooplankton of Europe: Cladocera & Copepoda (Calanoida, Cyclopoida)* 923

IOP Conf. Series: Earth and Environmental Science **107** (2017) 012030 doi:10.1088/1755-1315/107/1/012030

Multiparameter models in the management of the development of territories, taking into account the influence of hydrometeorological factors

E P Istomin, N N Popov, A G Sokolov, A A Fokicheva

Russian State Hydrometeorological University, Malookhtinsky pr., 98, Saint-Petersburg, 195196, Russia

E-mail: biom@rshu.ru

Abstract: The article considers the geoinformation management of the territory as a way to manage the organizational and technical systems and territories distributed in space. The article describes the main factors for the development and implementation of management decisions, requirements for the territorial management system and the structure of knowledge and data. Mathematical one-parameter and multiparameter models of risk assessment of management decisions applied to the natural and climatic potential of the development of the territory were considered.

Content management as a focused type of human activity from a methodological point of view involves the use of certain approaches [1]. Known approaches in management are system, program-target, project, marketing and others. At the same time, modern socio-economic systems are complex spatially distributed systems and territories. This indicates the need to develop and apply a geo-information approach to management.

The phenomenon of geoinformation management in modern conditions leads to the need to consider and implement the object management system as a multidimensional spatial phenomenon that includes spatially distributed factors, resources and conditions. Space in this case should be regarded as a form that serves as a medium for the existence of other forms and objects. Objects as constructions of space are points characterized by a set of properties.

Dependence of the process of development and implementation of management decisions on the scale of systems and territories allows us to consider the problem of managing sustainable development of the latter as a strategic target for the socio-economic development of systems and territories of various levels and scales [1, 2]. The development strategy of the territory is based on:

Objective factors - resource-spatial and natural-climatic potential of the territory, infrastructure and labor opportunities for locating objects of economic and economic activity;

Subjective factors - the current level of scientific and methodological management, organizational decisions of the subjects of management, as well as the readiness of leading cadres to the complex processes of developing and implementing effective strategic solutions.

The territory as a control object includes several subsystems [3]:

Population as a subject and object of management in terms of democratic governance principles;

Natural resource subsystem;

Production subsystem - realizing the supply of a tangible product: industry, construction, agriculture and forestry etc.;

Non-productive subsystem - rendering of market and non-market services: transport and communication, markets, science and education, health, art, etc.

At the heart of the management principles of any system are two main system properties:

- purposefulness: compliance of the purpose of territory management with the objectives of subjects of management of the objects of the territory;
- sustainability: it presupposes the ability of the territory to resist the process of destroying the foundations of its functioning and to maintain the chosen mode of activity for a certain time.

In accordance with these principles, the organization of management of the development of organizational and technical systems and territory should be:

- applicable to various situations, including emergency situations, when communication with subjects and management objects may be lost for some time;
- flexible, which implies a change in the forms of management depending on the situation and changes in long-term trends;
- operational and continuous in time, i.e. Capable of timely responding to changes in the external and internal environment;
- effective, which means the ability of elected management decisions and the entire management process to ensure the achievement of the goal at a certain level of performance criteria (economic, social, etc.).

Space at the level of perception is intuitively understood as a place, volume or territory in which motion is possible, different positions and mutual locations of objects, as well as characterizing the diversity of objects from each other. It can be characterized by relations of proximity-range, the concept of direction, as an arena of events and actions, universally containing all places and accommodating objects and structures, sometimes as a specific place that largely determines the essence of events occurring in it.

With reference to the problems of geoinformation management, it is worth noting the content, coordinate and structural features of components of organizational and technical systems, in accordance with which it will be possible to isolate and explore the corresponding types of spaces: geographic: two- and three-dimensional, event space and object existence space or attribute space. Spatial factors in their content, specificity of the components of the territory can be divided into compliance with certain characteristics: natural-resource, institutional, socio-demographic, economic, managerial, geographical (coordinate), etc.

The analysis of the characteristics of the territory is a basic stage in the development of strategy development [4-7]. Methodically, this stage involves identifying entities that form an attractive competitive image of the territory, its natural-resource and economic potential, as well as entities and objects that impede the positive dynamics of development. Allowing to assess the advantages and weaknesses of the territory, the results of the analysis can be used as a basis for strategic management of the development of the territory.

Attributes of the basic components of the territory allow:

Assess the impact of the elements (territories) of the territory considered on the development prospects in accordance with the chosen system of criteria;

Assess existing limitations;

To build a decision tree as the basis for a management strategy for the development of the territory.

To solve these problems, it is possible to propose a typical structure of knowledge and data characterizing the territory presented in the table 1.

Natural, technogenic and social processes of territory development unfold in time and in space and are characterized by a set of random parameters. We will show the possibility of formalizing the characteristics of the territory using the example of the natural and climatic potential [8-11].

The components of the natural and climatic potential are random non-stationary processes, the study of the regularities of which leads to the necessity of forming, based on the measurement results, multifactorial models. An example is the model for assessing the risk of flooding, when the meteorological parameters such as the direction and force (speed) of the wind are acting as factors, and finding them within certain limits (in the tolerance zone) can lead to a rise in the water level above the established level, which is regarded as a flood. Different approaches can be applied to the solution of this problem. On the one hand, having a series of observations of the water level, one can construct a one-dimensional model, treating the level as a random process (a random function of time). On the other hand, one can consider the influence of meteorological factors that lead to a change in the water level. In this case we are dealing with a two-parameter model. In the end, both models describe one event - a flood. For the management of the territories, the knowledge of the occurrence of a flood is important, but not enough. It is necessary to take into account not only the fact of flooding, but also be able to predict the likely damage from flooding (risk).One-parameter models of risk assessment are well described and can be used for different cases [3, 8, 9]. However, more interesting are two-parameter and multi-parameter models, which can describe more complex processes.

Table 1. Typical structure of knowledge and data for the analysis of the territory.

Knowledge	***Data***	***Indicators***
Characteristics of the economic environment of the territory	The volume of production in the base and current periods	The level of gross domestic product (GDP)
	The price of the "market basket" in the reporting year and the base period	The rate of inflation
	The number of employed / unemployed, the number of economically active population, the total population	Employment level
	The amount of resources in the current period by types of resources, Production volumes	Resources, including energy resources
	Volume of exports and imports in cash	Trade balance (positive / negative)
	The number and size of firms on the market, barriers to entry, uniqueness of the supply chain, the status and significance of buyers and others	Level of competition
Characteristics of the political environment of the territory	Changes in laws, regulations affecting the operation of facilities	Current legislation and its possible changes
	Methods of regulating local markets: administrative, economic	Forms of state regulation and control over the sectoral markets
	City twin-cities, agreements and agreements on cooperation and so on, organizations operating in the foreign market	International Relations and Relations
	Territorial / state management system	State (political arrangement)
Characteristics of the social environment of the territory	Number of population, number of working population, unemployed, economically active population, number of employees on the territory and outside it	Demography
	Statistical data on the size and structure of household incomes	Structure of incomes and expenditures of the population
	Elements of culture	Traditions and habits Ethnic and religious characteristics
Characteristics of technological development of the territory	The amount of advanced technology used, the amount of R & D costs, the volume of production in monetary terms, the number of patent applications, the average life of fixed assets and the average life of their operation	Indicators of available and implemented technologies
	Total number of enterprises, number of enterprises using the Internet	Development of information technologies, communications

Knowledge	*Data*	*Indicators*
	Total distance of all roads, communication lines	Level of development of transport and communications
Spatial characteristics of the territory	Geographic coordinates of the boundaries of the territory	Regulations on climatic and meteorological zones
	Geographical coordinates of the boundaries of the territory and the coordinates of development centers, industrial zones and other	Regulations on development centers
	Geographical coordinates of the boundaries of the territory, neighboring territorial units	Regulations on other regions, territories
	Geographical coordinates of the boundaries of the territory and the coordinates of hazardous objects and areas of their influence	The situation regarding the dangerous zones and high-risk factors of neighboring territories
Characteristics of a territory management system	Geographical coordinates of the largest enterprises within the territory, special economic zones (industrial, zones of economic development, etc.)	Scheme of location of especially significant objects in the territory and conditions of their functioning
	Data on public services, administrative units, job descriptions, provisions for departments, communications between subjects of public administration	Organization of the management of the territorial system
	Job descriptions, regulations on departments of administrations of municipalities	Organization and implementation of strategic and operational management
Characteristics of the territory in the external environment	Weighted average estimates of economic, political factors, social, technological	Investment climate Image of the territory competitive advantages Opportunities for development Threats of development

Formulation of the problem:

1. The critical level of the characteristic of an extreme natural phenomenon is determined by two main factors and can be considered as a two-dimensional stochastic process.
2. We assume that we can monitor (regularly measure) the characteristics of an extreme natural phenomenon, and present the dependence as a one-dimensional random process $h(t)$.
3. It is necessary to determine (predict) the waiting time for the onset of an extreme natural phenomenon, provided that the values of the characteristics of the main factors at the time of the forecast are known. Determine the mathematical expectation of the non-occurrence of a two-dimensional random process over a specified tolerance field. We need to estimate a functional of the form $T(x_1, x_2)$.

In this formulation, the problem acquires a classical form, for the solution of which it will be necessary to adopt a number of restrictions related to the type of random processes.

Let there be given a vector random process $X_t = (X_{t1}, X_{t2})$. We assume that each of the one-dimensional stochastic processes is described by a stochastic differential equation of the form:

$$dX_{ti} = b_i(t)dt + a_i(t)d\eta_t, \quad (1)$$

In this formulation, the problem has a solution under the following conditions of unique solutions:

$$\begin{aligned} \frac{\varphi_2(u_2)-x_2}{u_2} - \rho\frac{\varphi_1(u_1)-x_1}{u_1} \geq c_* \\ \frac{\varphi_1(u_1)-x_1}{u_1} - \rho\frac{\varphi_2(u_2)-x_2}{u_2} \geq c_* \end{aligned} \tag{2}$$

where: $x_i \equiv M_{\Psi i}(t)$, $u_i(t) \equiv \sqrt[2]{\int_0^t a_1^2(t)dt} \leq \beta_i(t)$, $\varphi_i(u_i)$ – the auxiliary function for finding the extremum, $\rho\ (t) = \rho$ is a cross-correlation function.

In the end, we obtain estimates for the desired functional $T(x_1, x_2)$ - the mathematical expectation of the time of the process in the given region - the waiting time of the event.

Suppose that a two-dimensional random process can be described as a normal process, then, when condition (2) is satisfied, and known values of the mathematical expectation and variance, solving problem (1) we obtain a lower bound for the desired functional:

$$T_*(x_1, x_2) = \int_0^\infty P_2(\mathrm{t})\Phi(Z)\,dt \tag{3}$$

where: $\Phi\ (\cdot)$ is the Laplace function,

$$\Phi(Z) = \Phi\left(\frac{\Phi^{-1}\,(P_1(t))\cdot\left(\dot{u}_1^2+\sigma_1^2\right)^{1/2}+m_1-x_1-\rho\cdot P_1}{\sqrt{1-\rho^2}}\right)dt,$$

$$P_i(\mathrm{t}) = \Phi\left(\frac{\varphi_i(u_i,t)-m_i}{\sqrt{u_i^2(t)-\sigma_i^2}}\right),\ i=1,2.$$

In solving many problems, an upper bound is required for the given functional, which can be obtained and the problem has a solution.

This method allows, for a different class of tasks, to obtain estimates of the waiting time for an event to occur in an explicit form.

From here follows the management, which is formulated for the solution of each specific problem.

Application of the proposed model requires isolation of extreme natural phenomena and corresponding pairs of basic hydrometeorological factors. The transition from two-parameter to multi-parametric models is possible by combining pairs of main factors in accordance with their significance for the process of development of the natural phenomenon.

Conclusion

The analysis of natural, technogenic and social processes of territory development demonstrates the need to develop multi-parameter models of risk assessment. The proposed model is based on the approximation of complex multiparametric estimations by forming paired characteristics of unfolding in time and space. As a random factor, the characteristics of the territory can be proposed. To clarify the forecast solutions, it is necessary to construct a parametric description of specific events and development processes of the territory. The proposed approach to risk assessment has been tested based on the results of the evaluation of flooding in St. Petersburg and the forecast of emergency situations in the management system of housing and communal services.

References

[1] Istomin E P, Slesareva L C, Sokolov A G, Timofeeva A G and Fokicheva A A 2016 *Adoption of managerial decisions Series: "System of Public Administration"* (St. Petersburg: OOO "Andreevsky Publishing House") 454

[2] Istomin E P and Sokolov A G 2015 *Information technologies and systems: management, economics, transport, law.* **2** (16) 82-88

[3] Istomin E P, Sokolov A G, Abramov V M, Gogoberidze G G and Popov N N 2015 *International Multidisciplinary Scientific GeoConference Surveying Geology and Mining Ecology Management, SGEM, Bulgaria* **1** (2) 607-614

[4] Istomin E P, Sokolov A G, Abramov V M, Fokicheva A A and Popov N 2016 *International Multidisciplinary Scientific GeoConference Surveying Geology and Mining Ecology Management, SGEM, Bulgaria* **1** (2) 601-608

[5] Beskid P P *et al* 2016 *Risk management of sustainable development of objects and territories in a spatial aspect*: ed. Istomin E P (St. Petersburg: RSHU) 250

[6] Beskid P P *et al* 2016 *Development of the fundamentals of the methodology of geoinformation management of objects and territories*: ed. Istomin E P (St. Petersburg: RSHU) 286

[7] Showalter P S and Lu Y 2009 Geospatial techniques in urban hazard and disaster analysis *Springer* 452

[8] Istomin E P, Sokolov A G, Fokicheva A A, Slesareva L C and Popov N N 2016 *Managing risks to coastal regions and communities in a changing world* 110

[9] Istomin E P, Sokolov A G and Fokicheva A A 2015 *Analysis, forecast and management of natural risks in the contemporary world (Georgian-2015) Materials of the 9th International Scientific and Practical Conference* 170-176

[10] Istomin E P, Sokolov A G, Abramov V M, Gogoberidze G G and Fokicheva A A 2015 *International Multidisciplinary Scientific GeoConference Surveying Geology and Mining Ecology Management, SGEM Bulgaria* **1** (2) 729-736

[11] Istomin E P, Fokicheva A A, Korshunov A A and Slesareva L S 2016 *Scientific notes of the Russian State Hydrometeorological University* **44** 219-224

[12] Istomin E P, Popov N N, Sokolov A G, Abramov V M, Fokicheva A A and Slesareva L C 2016 *Managing risks to coastal regions and communities in a changing world* 111

IOP Conf. Series: Earth and Environmental Science **107** (2017) 012031 doi:10.1088/1755-1315/107/1/012031

Inter-annual variability of variations of ground and integrated atmospheric water vapor content in Europe

O G Khutorova, V E Khutorov, V V Kalinnikov and G M Teptin

Institute of Physics, Kazan Federal University, Kremlevskaya str., 18, Kazan, 420008, Russia

E-mail: olga.khutorova@kpfu.ru

Abstract. This paper presenting results of the seasonal, interannual mesoscale and synoptic variations of the surface and atmosphere integral moisture content variability analysis. More than 50 years of the meteostation network data and data from satellite navigation system receivers on Europe were analysed. The surface moisture content field in Europe decreases with increasing latitude and further from the coast. The intensity of variations in the partial pressure of water vapour, on the contrary, increases from northwest to southeast. It is shown, that the maximum contribution to the moisture content variability is given by seasonal variations. The predominant process of near ground and integrated atmospheric water content is year and half-year harmonics. In near ground and integrated atmospheric water content spectra the interannual variations were discovered. It is shown that the average annual values of the integral humidity, amplitude and phase of the annual and semiannual harmonics vary from year to year, as well as the intensity of the synoptic variations.

Last decades have shown a strong influence of climate variability on the main risks for the economy and ecology of the regions [1]. Temperature trends in different degrees are manifested in the regions of the Northern Hemisphere [2]. With increasing temperature, moisture content variability is often associated [3]. Water vapor plays a key role in the radiation balance of the Earth's atmosphere, as a result of which the water vapor content temporal and spatial variations at the interannual, synoptic and mesoscale level are closely interrelated with changes in the other meteorological fields and geophysical elements. In this case, one of the most important tasks is to search for the interannual regularities in the evolution of atmospheric moisture content variations.

In this work, long series of the atmosphere integral moisture content were constructed, calculated according to the GNSS receiver's every-second observations from the Kazan network [4] world service IGS GNSS in years of 2000-2016 [5]. The surface moisture content was estimated from the meteorological parameters [6] for years of 1966-2015. From the measurements of relative humidity and temperature, the partial pressure of water vapor was calculated e. The series of surface parameters are longer and allow the use of spectral analysis to highlight interannual variations. However, recently the development of the GPS – meteorology, which uses signals from global navigation satellite systems (GNSS) has been developed, which allows determining the integral moisture content (IWV) with high spatial-temporal resolution [4]. In Kazan, there is a network of receivers and it is shown that, according to GNSS receivers, it is possible to study temporal variations in atmospheric parameters over a wide range of time scales [7, 8].

Figure 1 shows the spatial cross sections of mean annual values of the partial pressure of water vapor and its standard deviation. Obviously, on the average, the surface moisture content field in

Europe decreases with increasing latitude and away from the coast. The intensity of variations in the partial pressure of water vapor, on the contrary, increases from northwest to southeast. The contribution to variance of variations in the partial pressure of water vapor yields all the processes from mesoscale to interannual.

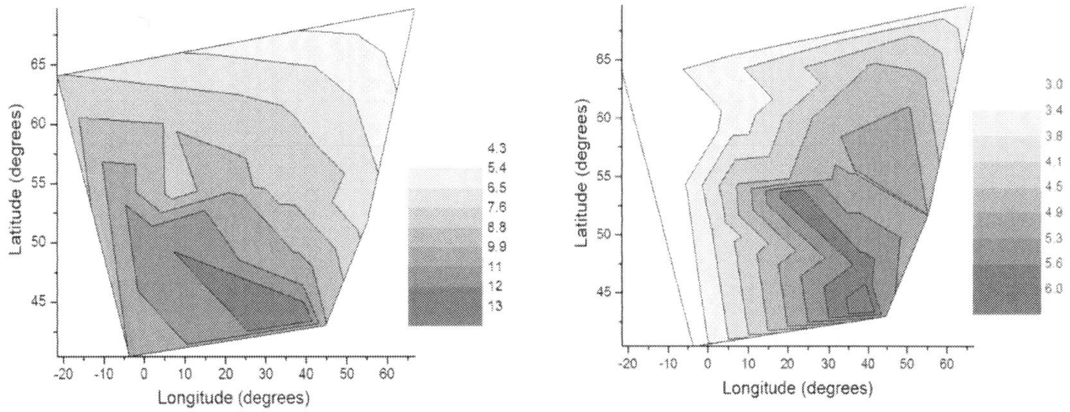

a) b)

Figure 1. a) Spatial dependence of mean annual values of the partial pressure of water vapor e, mbar; b) The spatial dependence of the value of the standard deviation of the partial pressure of water vapor σ_e, mbar.

The strongest in the spectra of both the partial pressure of water vapor and the integral moisture content are seasonal variations. Their amplitudes for several points are given in table 1.

Table 1. Periods and amplitudes of significant periodicities in the series of surface partial pressure of water vapor.

Location	Murmansk 69.0 N 33.0 E	Moscow 55.8 N 37.6 E	Kaliningrad 54.7 N 20.5 E	Sochi 44.6 N 39.7 E
Mean value, mbar	5.8	8.3	9.0	12.8
12 month, amplitude mbar	1.0	2.9	1.2	1.6
6 month, amplitude mbar	0.3	0.5	0.2	0.3

Figure 2a shows the spatial dependence of the amplitude of the annual harmonic of surface water content in Europe. As can be seen from table 1 and figure 2a, the mean annual values of the partial pressure of water vapor have a strong latitudinal dependence, related both to the inflow of solar radiation and to the influence of moist air masses. Spatial variability in Europe is from 2 to 8 mbar on average over 50 years.

Figure 2b shows the spatial dependence of the maximum time of the annual harmonic of surface water content in Europe. The time lag of the maximum from the maximum of the solar radiation influx (170-180 days) characterizes the influence of the ocean, its spatial variability in Europe from 180 to 220 days on average over 50 years.

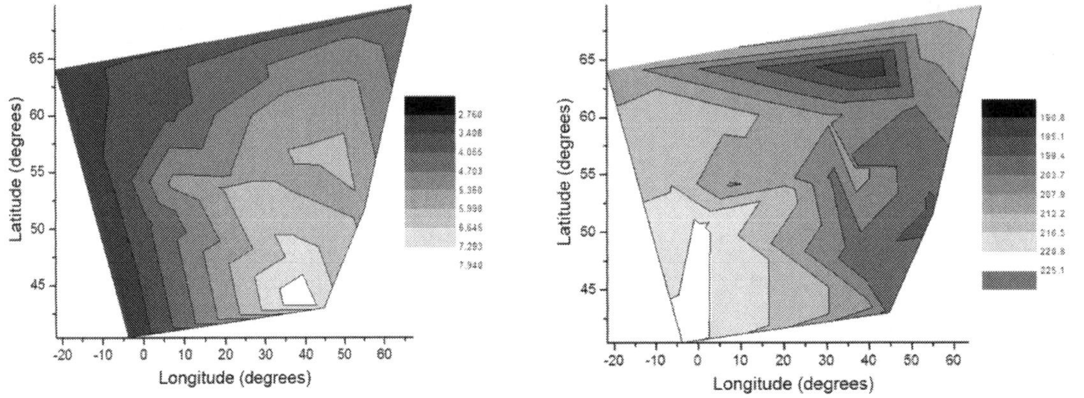

a) b)

Figure 2. a) Spatial dependence of the amplitude of the annual harmonic of the surface moisture content, mbar; b) Spatial dependence of the maxima time of the annual harmonic of the surface moisture content, day.

Harmonic analysis of time series of surface moisture content in a sliding time window showed that both the mean annual values of the integral moisture content, so the amplitudes and phases of the annual and semiannual harmonics vary from year to year.

Figure 3a shows the interannual variations of the annual harmonic amplitudes for several points in the European territory of Russia and their linear trends. It can be seen that the amplitudes of the annual harmonic increase on average with time, but the angle of inclination of the trend is varies with the continentality index of the each site.

However, a detailed analysis shows that, since 2000, there has been no trend towards an increase in the amplitudes. The same effect can be seen in figure 3b, which shows the interannual variations of the mean annual values of the integral moisture content and the amplitude of the annual harmonic for the city of Potsdam (Germany).

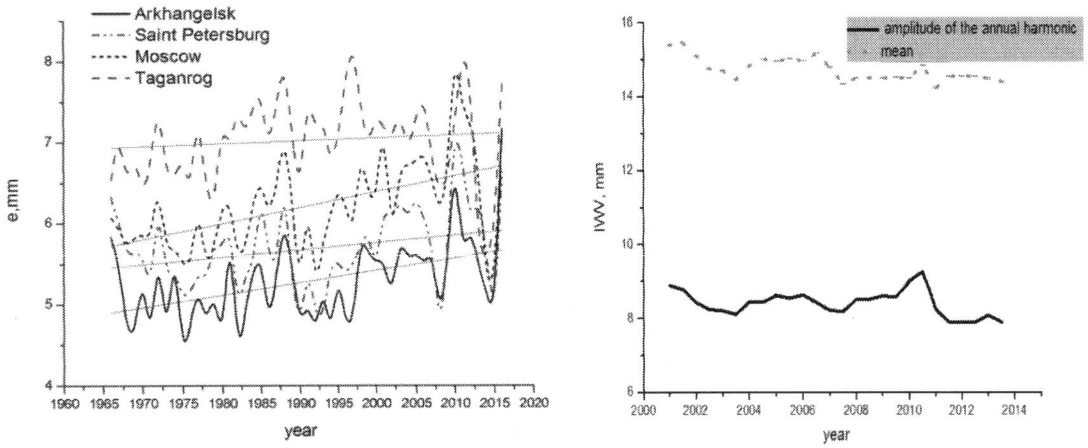

a) b)

Figure 3. a) Interannual variations of the annual harmonic amplitudes for several points in the European territory of Russia and their linear trends. b) Interannual variations of the integral moisture content mean annual values, the amplitude of the annual harmonic for Potsdam (Germany).

To study the interannual variations in the atmospheric moisture content, Fourier analysis [7]. Fourier analysis of 50-year series of surface water content with preliminarily excluded linear trends revealed that in the spectra there are quasiperiodic processes with time scales of 4-5 years, 10 years, 20 years. Similar variations were observed in the series of surface temperature. Figure 4 shows the spectra of partial pressure series of water vapor and surface temperature in the city of Orenburg, Russia. For various sites of Europe periodicity in the spectra reveals similar facts, but the ratio of amplitudes for temperature and water vapor varies at different points.

Due to the fact that the length of the integral moisture content is less than 17 years, the slower variability was analyzed with the help of sliding smoothing of the time series within the annual time window with a step in half a year. Figure 4 shows the interannual variations of the integral moisture content (IWV) series in Olsztyn (Poland) and the variations in the duration of the day (dLOD) with excluded trends. Figure 5 shows that the interannual dynamics of the integral moisture content of the atmosphere and the parameters of the Earth's rotation change in antiphase, that is, the processes are connected. The communication mechanism requires additional research. Perhaps the advection of water vapor and the speed of Earth's rotation depend on the interhemispheric momentum exchange [10] Perhaps for each parameter a harmonica with a period of 18.6 years is allocated associated with the nutation of the earth's axis under the influence of the Moon [11]. Interannual variations and a strong seasonal course of moisture content are also noted from satellite data [12]. Spectra of the integral moisture content there are also harmonics with periods of 4-5 years and about 10 years.

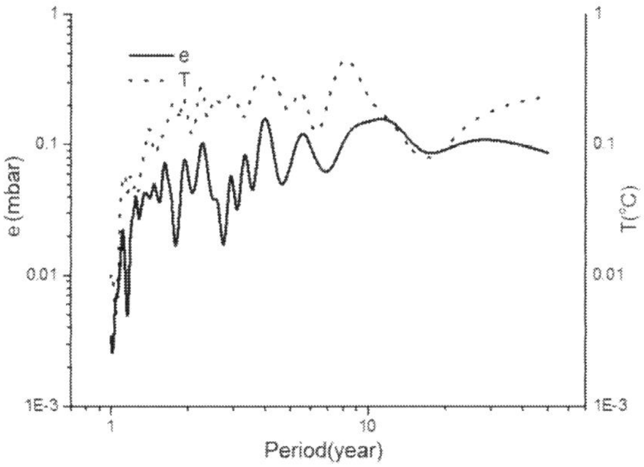

Figure 4. Spectra of interannual variations of partial pressure series of water vapor and surface temperature in Orenburg.

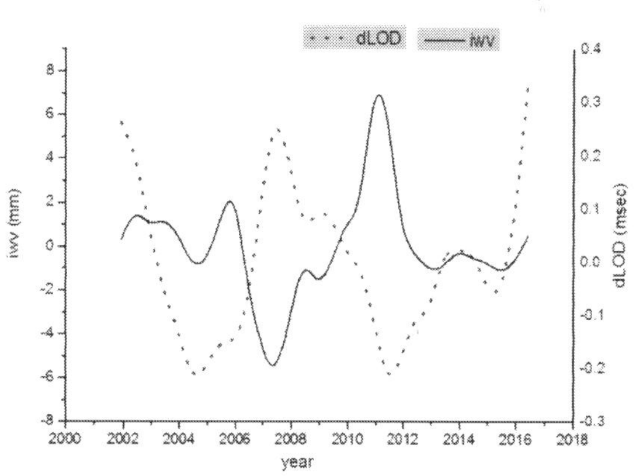

Figure 5. Interannual variations of the series of integral moisture content (IWV) in Olsztyn city and variations in the duration of the day (dLOD) with excluded trends.

Interannual variations are also present in the variability of intensity of synoptic processes in atmospheric moisture content.

Figure 6 presents a fragment of the atmosphere integral moisture content and its wavelet spectrum for Kazan city (Russia) for 2009-2015 years. It illustrates that in the spectra of synoptic variations there are significant modes 3-4, 7-10, 11-14, 20-30 и 40-45 days.

According to figure 6, it is definite that the intensity of the synoptic variations for all the investigated parameters is modulated by the harmonics of the annual variation. It is determined that in the time series under study, wave variations predominate with periods of 3 to 10 days, which are characteristic of the Rossby waves. However, variations with periods of 15 to 45 days are more powerful, although less frequent. Amplitudes of the variations of integral water vapor during the summer grow from 1 to sometimes 4 mm of precipitated water.

It is shown that the maximum contribution to the variability of moisture content is given by seasonal variations. For Kazan city, with an average value of IWV equal to 13.4 mm of precipitated water, the annual variation makes the greatest contribution to the overall dispersion; its share is 68%. The second for the contribution to the overall variance are the synoptic variations; their contribution in the dispersion is 20%. The average contribution of mesoscale variations to the overall variability of IWV in Kazan is 12%. Similar ratios of the contributions of different processes to the overall dispersion of variations are also characteristic of other series of integral moisture content in Europe.

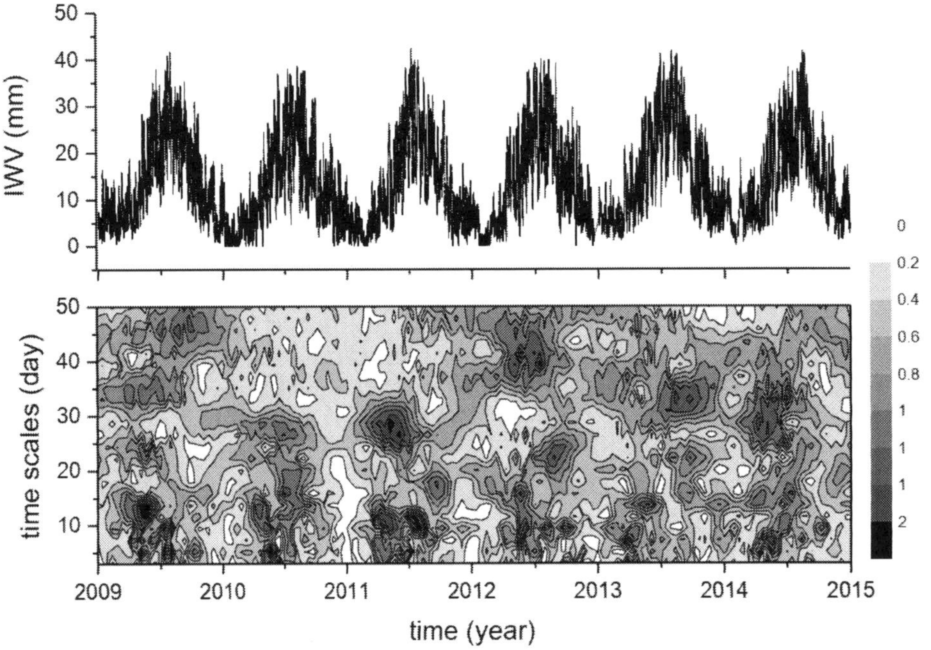

Figure 6. Integral water content of the atmosphere and its wavelet spectrum by measurements of GNSS receivers in Kazan (Russia) 2009-2015.

Long series of surface and integral water vapor content of the atmosphere in Europe are analyzed. It is shown, that predominant process of near ground and integrated atmospheric water content is year harmonic. In near ground and integrated atmospheric water content mean value the interannual variations were found. Also synoptic and mesoscale processes make important contributions to variability of atmospheric moisture content. It is shown that the average annual values of the integral humidity, amplitude and phase of the annual and semiannual harmonics vary from year to year, as well as the intensity of the synoptic variations.

Acknowledgments
The work was supported by the Russian Foundation for Fundamental Research (grant № 17-05-00863).

References
[1] Guiot J and Cramer W 2016 *Science* **354** 465–468
[2] Perevedentsev Yu P, Gogol F V, Naumov E P and Shantalinskiy K M 2007 *Probl. Anal. Riska* **4** 73–80
[3] Alshawaf F, Dick G, Heise S, Simeonov T, Vey S, Schmidt T and Wickert J 2016 *Atmos. Meas. Tech. Discuss.* doi:10.5194/amt-2016-151
[4] Khutorova O G, Kalinnikov V V and Kurbangaliev T R 2012 *Atmospheric and Oceanic Optics* **25(6)** 529-533
[5] *Internatioanal GNSS* URL: http//igscb.jpl.nasa.gov/

[6] Bulygina O N, Korshunova N N, Razuvaev V Nl *Description of the array of urgent data on the main meteorological parameters at Russian stations. Certificate of state registration of the database* No. 2014620549 URL: http://meteo.ru/english/index.php
[7] Khutorova O G, Vasilyev A A and Khutorova V E 2010 *Atmospheric and Oceanic Optics* **23(6)** 510-514
[8] Kalinnikov V V, Khutorova O G and Teptin G M 2012 *Bulletin of the Russian Academy of Sciences: Atmospheric and Oceanic Physics* **48** 631–638
[9] Jenkins G and Watts D 1969 *Spectral analysis and its attachments* **1,2** (San Francisco: Holden-Day)
[10] Richard D and Salstein D A 1983 *J. of Geo. Res.* **88** 5451-5470
[11] Sidorenkov N 2004 *Bulletin of the russian academy of sciences* **74** 701-715
[12] Astafeva N M, Raev M D and Sharkov E A 2007 *Modern problems of remote sensing of the Earth from space* **2** 20–26

The temperature characteristics of biological active period of the peat soils of Bakchar swamp

M V Kiselev[1], E A Dyukarev[1], N N Voropay[1,2]

[1]Institute of Monitoring of Climatic and Ecological System of the Siberian Branch Russian Academy of Sciences, Akademicheskii str., 10/3, Tomsk, 634055, Russia
[2]V B Sochava Institute of Geography of the Siberian Branch Russian Academy of Sciences, St. Ulan-Bator, 1, Irkutsk, 664033, Russia
E-mail: xplutox@yandex.ru, dekot@mail.ru, voropay_nn@mail.ru

Abstract. The results of the study of the peculiarities of the temperature regime in the five basic ecosystems of oligotrophic bogs in the south taiga zone of Western Siberia in 2011-2016 are presented. The soil temperature regime was studied using the atmospheric-soil measuring complex at different depths from surface to 240 cm. All sites were divided into two groups according the bog water level: flooded sites (hollow and open fen) and drained sites (ridge, tall and low ryam). Waterlogged sites are better warmed in the summer period, and slowly freeze in the winter period. The analysis of the annual cycle of temperature showed that the maximum surface temperature is observed in July. The minimum temperature on the surface observed in February or January. The greatest temperature gradient was recorded in the upper 2 cm layer. The gradient at the open fen was -2 $°C \cdot cm^{-1}$ in February and 1.1 $°C \cdot cm^{-1}$ in October. The peak of formation of the seasonally frozen layer occurs at the end of autumn, beginning of winter. The degradation of the seasonally frozen layer was observed both from top and bottom, but degradation from the top is faster.

1. Introduction

Bog complexes occupy a small area of 3-5% of the land surface, but play a significant role in regulating the gaseous composition of the atmosphere, the water balance of the biosphere, biological diversity on Earth [1, 2].

Thermal and water regimes of peat and mineral soils were significantly different. Peat accumulation is a complex organic system, which has specific properties: high water content and porosity, content of a large number of little decomposed organic matter [3, 4].

The temperature of peat soils affects the growth of ground vegetation (mosses, shrubs, etc.), the formation of climate of the territory is a key factor in controlling many biotic and abiotic processes that occur in soils: decomposition and mineralization of soil organic matter, emissions of greenhouse gases (CO_2, CH_4, N_2O), excretion of dissolved organic carbon [5-10].

In Western Siberia are located the main storages of peat. Peatland ecosystems of West Siberia contain up to 70 billion tons of carbon [11]. Such a large reservoir potentially may go into the atmosphere as CO_2 or CH_4 as the result of climate change or anthropogenic impacts and make a significant contribution to the carbon balance of the atmosphere.

2. Methods

Studies of the temperature regime of soils was carried out in the geophysical field station "the Vasyuganje" Institute of monitoring of climatic and ecological systems SB RAS, at the Bakcharskoe bog (area of about 1400 km^2), located in the interfluves of the rivers Iksa and Bakchar in Bakcharsky district of Tomsk region [12].

For the study we used 5 sites with different levels of bog waters (LBW), which were divided into two groups. Flooded areas – hollow on the ridge-hollow complex (RHC), Osaka-sphagnum bog (mean level of bog water 5 cm). Low-water sites – pine-dwarf shrub sphagnum phytocenosis with a normal forest is 18 m (high ryam) and oppressed the stand is 2-3 m (low ryam), the range on the MMC (mean level of bog water 25 cm) [13].

In the study sites for studying temperature regime was measured the soil temperature at 14 depths from the surface peat soil up to 320 cm and air temperature at a height of 2 m, using atmospheric and soil measuring complex [14]. The measurements were carried out in the period 1 April 2011 to 30 April 2016 (1858 days) with a time step of 15 minutes.

3. Results and discussions

Analysis of annual temperature of peat soil in the profile 0-240 cm can identify a number of differences between the observation sites. Despite the fact that the sites are located in the 300-1000 m from each other, density and moisture content of the peat at the sites varies little, all sites can be divided into two groups. Sites of the first group during the warm period are warmed up better than sites of the second group. The first group includes sites with water level located at 3-7 cm from the surface (hollow at RHC and open fen). Water level at the sites from the second group (tall and low ryam, ridge at RHC) decreases to 20-30 cm in the end of summer. While maximum annual surface temperature in July at hollow and open fen was 18.1 and 17.4 °C, respectively; for low ryam, tall ryam and ridge of RHC it was 17.3, 14.1 and 17.7 °C, respectively.

The soil temperature in the warm period is associated not only with the bog water table level, but also with the characteristics of the vegetation cover [7]. Low and tall ryams in the upper 60 cm layer have a lower temperature than the ridge at RHC, where the wood layer is very tenuous. Ryams due to the dense vegetation cover receive less solar radiation, and in conjunction with a low level of bog waters become colder than ridge at RHC.

During the study period, maximum average daily temperature was noted in the top 20 cm layer in 2012 due to the anomaly hot weather in June. The soil temperature ranged from 23.5 (ridge at RHC) to 24.9 °C (low ryam) on the surface and from 16.2 (tall ryam) to 19.5 °C (open fen) at a depth of 20 cm. Deeper than 30 cm the maximum temperature of the peat deposits ranged from 12.6 (low ryam) to 16.9 °C (open fen) at 40 cm and 5.5 (low ryam) to 7.9 °C (open fen) at 240 cm, respectively. The maximum temperature of the upper layers (0-60 cm) is strongly influenced by the air temperature and the features of the peat deposits (peat density, water content, vegetation cover, etc.).

The temperature gradient in the soil has the highest variations in layer 0-2 cm. Its value changes from -2.0 $°C \cdot cm^{-1}$ in February 2012 to 1.1 $°C \cdot cm^{-1}$ in October 2014 at the open fen, and from -1.1$°C \cdot cm^{-1}$ in October 2012 to 0.8 $°C \cdot cm^{-1}$ in July 2012 at tall ryam. In the warm period the greatest fluctuations in soil temperature gradient was observed in the layer 2-60 cm. In July they vary in the range from 0.05 $°C \cdot cm^{-1}$ in open areas (open fen, hollow at RHC) to 0.33-0.36 $°C \cdot cm^{-1}$ in the forested bogs (tall and low ryam). The value of the gradient decreases with increasing depth. At 240 cm depth the value of gradient at all sites over the entire observation period does not exceed 0.1 $°C \cdot cm^{-1}$. In the cold period, the greatest fluctuations of temperature gradients in the soil are observed in the layer 2-40 cm. In February, they vary from -0.01 $°C \cdot cm^{-1}$ in open areas (open fen, hollow at RHC) to 0.15-0.21 $°C \cdot cm^{-1}$ for forested bogs (ridge at RHC, tall and low ryams).

During the propagation of thermal wave from the surface into the soil there is a lag of this wave. The timing of the maximum temperature depends not only on the type of site (open, forested), but also on weather conditions of a particular year [1]. The lag of the daily average temperature wave in the layer 0-15 cm is almost absent. In some cases, the maximum temperature in the layer of 0-5 cm can be

IOP Conf. Series: Earth and Environmental Science **107** (2017) 012032 doi:10.1088/1755-1315/107/1/012032

recorded later than in the underlying layer. The greatest variation in the lag of the temperature wave is observed at depths of 20-80 cm and varies from 0 to 50 days. With increasing depth there is slowing of the propagation of a thermal wave and at a depth of 240 cm the maximum is reached on average in 100 days. With increasing depth in addition to the lag in occurrence of maxima decreases the maximum temperature itself. On the surface it is an average of 24-27 °C and at a depth of 240 cm – 5-8 °C, depending on the site.

To estimate biologically active period for the flora and fauna of a peat deposit was identified periods with temperatures above 10 °C. The biologically active period was defined as a time when the peat provides a maximum activity of bacteria, growth of vegetation (mosses, sedges, shrubs) [15]. The activity of the bacteria increases the rate of decomposition of organic materials within the peat, which leads to some increase in the formation of carbon dioxide and methane in peat deposits and higher emissions.

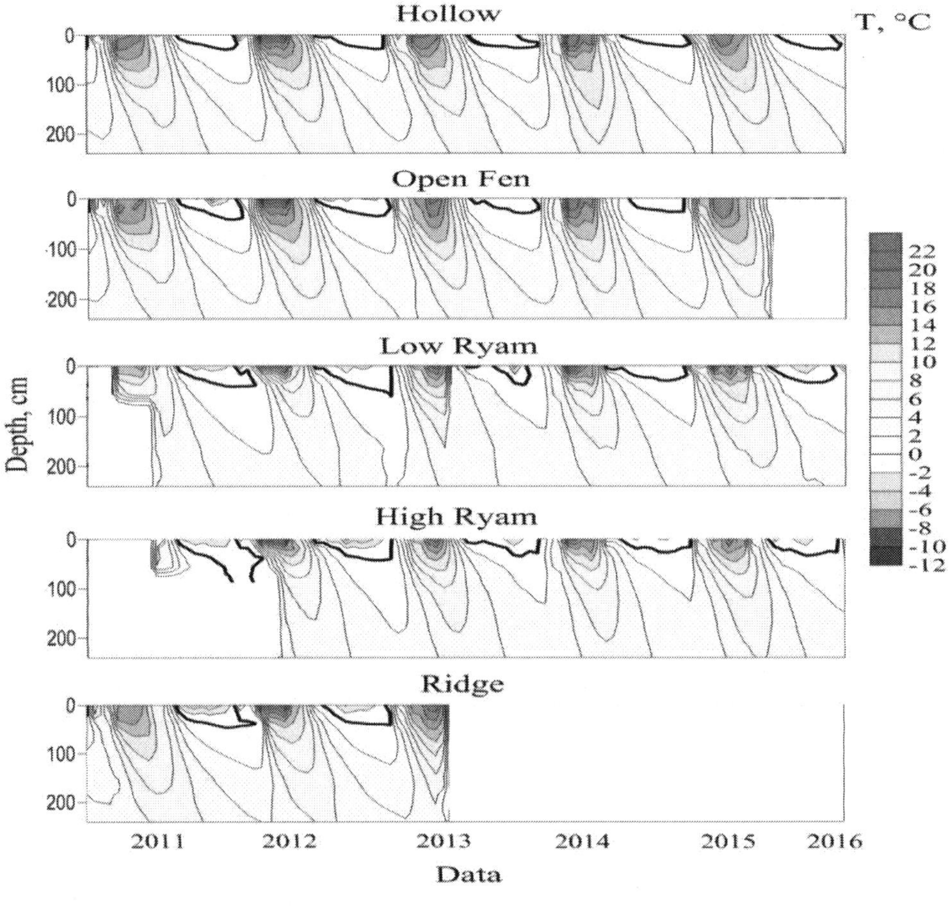

Figure 1. Distribution of daily average soil temperature in the studied peat deposits.

The duration of the frost-free period (temperature above 0 °C) at the studied sites varies at depths ranging from surface to 60-80 cm (figure 1). The temperature of the deeper peat layers is always positive. The average duration of the frost-free period for all sites is 180-230 days throughout the whole peat column. The transition of soil temperature above 0 °C in the direction to positive temperatures occurs in April – May on the surface and with increasing depth shifts towards the summer months. At 40-80 cm the transition to the frost-free period occurs in June – July. Transition into the negative temperature occurs in October-November on the surface and with depth shifted to December-January.

The depth where the biological active period was observed varies from 60 cm at forested sites (high and low ryam, ridge at RHC) up to 120 cm at open areas (hollow at RHC and open fen). The duration of the biological active period is less than frost-free period by 2-3 times. On open sites at surface and depth up to 60 cm the duration is 110-120 days, deeper (to 120 cm) – 85-95 days. At the same time in 2014 at the open fen and in 2013 at hollow the duration of the biologically active period in the top 120 cm layer was about the same – an average of 90 days, the reason for that was the slow warming of peat deposits at the beginning of the warm period.

At low-water sites at depths from the surface to 15 cm the average length of biologically active period was 85-90 days, at 20-60 cm depth – 70-75 days. It should be noted that in some years the duration of this period at depths of 40-60 cm can be reduced to 40-50 days.

On flooded sites, the transition temperature of the peat deposit from +10 °C to higher temperatures is observed mainly in May and June at depths from the surface to 80 cm, deeper than 80 cm - in July. The end of biologically active period at these sites is mainly observed in September and early October at depths and deeper than 80 cm in October. On low-water sites the biologically active period is shorter and occurs from June to September at depths of 0-40 cm and deeper than 40 cm from July to end of August – beginning of September.

The sum of active temperatures (above 10 °C) on low-water sites changes from 2000 to 1500 °C in the layer 0-20 cm, and then it decreases with depth up to 800 °C for 60 cm. Flooded sites are warmed up better and the sum of active temperatures on them varies from 2500 to 1900 °C in the layer of 0-20 cm, then it decreases to 1000-500 °C at a depth of 80-120 cm, respectively.

4. Conclusions

The sites with high water level (hollow at RHC, open fen) are warm up better in the summer period and have a higher temperature on the surface than at forested bog sites with lowered water level (ridge at RHC, high and tall ryam). While maximum annual surface temperature in July on hollow and open fen it is 18.1 and 17.4 °C, respectively; for low ryam, tall ryam and ridge of RHC it was 17.3, 14.1 and 17.7 °C, respectively.The full heating of the peat deposits at depths of 2 m is ending in October due to large lag in the heat wave propagation.

The start of warm period is a march on surface and August-September on 60 cm warm period for all sites is 180-230 days throughout the peat column. The biological active period have maximum depth 120 cm for flooded sites and 80 cm for low-water sites with different duration. Duration on that period more on flooded sites – average 110-120 days from 0 to 60 cm and about 90 deeper. The sum of active temperature more too on flooded sites and ii more about 500-300 °C.

Acknowledgments

This study was supported by Russian Foundation for Basic Researches (grant № 16-07-01205 and 16-45-700562) and Russian Academy of Sciences (project № VIII.77.1.1).

References

[1] Golovatskaya E A and Dyukarev E A 2011 *Russian Meteorology and Hydrology* **36** (6) 84-93
[2] Voropay N N, Maksyutova E V and Balybina A S 2011 *Environ. Res. Lett.* **6** 9
[3] Golovatskaya E A and Nikonova L G 2017 *Eurasian Soil Science* **50** (5) 580-588
[4] Romanov V V 1961 *Hydrophysics of bogs* (Leningrad: Gidrometeoizdat) 359
[5] Golovatskaya E A, Dyukarev E A, Ippolitov I I and Kabanov M V 2008 *Reports of the Russian Academy of Sciences* **4** 1-4
[6] Golovatskaya E A and Dyukarev E A 2012 *Eurasian Soil Science* **45** (6) 588-597
[7] Trofimova I E and Balybina A S 2015 *Geography and Natural Resources* **36** (3) 234-244
[8] Trofimova I E and Shekhovtsov A I 2011 *Geography and Natural Resources* **32** (4) 363-369
[9] Zang Y, Chen W, Smith S L, Riseborough D W and Cihlar J 2005 *Geophys. Res.* **110**
[10] Maksyutova E V and Gustokashina N N 2009 *Geography and Natural Resources* **4** 87–92
[11] Ciais P *et al* 2013 *Carbon and Other Biogeochemical Cycles. In: Climate Change 2013: The*

Physical Science Basis. Contribution of Working Group I to the Fifth Assessment Report of the Intergovernmental Panel on Climate Change ed. Stocker T F *et al* (Cambridge: Cambridge University Press, UK and NY, USA) 570

[12] Dyukarev E A, Golovatskaya E A, Duchkov A D and Kazantsev S A 2009 *Russian Geology and Geophysics* **50** (6) 579-586

[13] Kiselev M V, Voropai N N and Dyukarev E A 2016 *Reports of higher educational institutions. Physics* **7-2** 93-97

[14] Kurakov S A *Sensors and systems* **4** 29-32

[15] Borodina I A, Kizhner L I, Voropay N N, Bogoslovskiy N N and Erin S I 2016 *Proceedings of SPIE - The International Society for Optical Engineering* **10035** 1-5

[16] Alekseeva M N, Preis Y I and Dyukarev E A 2015 *Bulletin of the Tomsk Polytechnic University, Geo Assets Engineering* **32**(4) 143-146

[17] Tesař M, Šír M, Krejča M and Váchal J 2008 *IOP Conf. Series: Earth and Environmental Science* **4** doi: 10.1088/1755-1307/4/1/012029

IOP Conf. Series: Earth and Environmental Science **107** (2017) 012033 doi:10.1088/1755-1315/107/1/012033

Formaldehyde content in atmospheric air of Saratov

N V Korotkova, N V Semenova

Faculty of geography, Saratov State University, Astrahanskaya str., 83, Saratov, 410012, Russia

E-mail: fonadia@yandex.ru

Abstract. The analysis and estimation of the sources of formaldehyde emissions into the atmospheric air in Saratov are presented. Average monthly, average annual and average long-term concentrations of formaldehyde are calculated. The dynamics of the content of formaldehyde in the atmospheric air of the city is characterized. Seasonal variability of formaldehyde emissions is shown. The spatial distribution of formaldehyde concentrations in the city is given.

1. Introduction

The quality of air is formed as a result of a complex interaction of natural and anthropogenic factors. The state of the air environment is a very important element of urban ecosystem and has a great influence on all aspects of economic and industrial activities of the city's inhabitants. Saratov as a large industrial center which especially needs a system of observations, assessment and forecast in monitoring the state of the environment.

In terms of its industrial potential, Saratov belongs to the largest cities of Russia with a diversified industry, among which, first of all, are the oil refining, chemical and construction industries, enterprises for production and distribution of electricity, gas and water, transport and communications. At the enterprises of the electric power industry emissions are discarded into the atmosphere without purification. Enterprises are located mainly in the southern, north-western and north-eastern districts of the city. An important role in the formation of air pollution in the surface layer of the atmosphere is played by the exhaust gases of cars that pollute the atmosphere on the level of medium height of man and pose a greater danger to public health in comparison with emissions from industrial sources [1, 2].

Observations of atmospheric air in the Saratov region are carried out by the Saratov Center of Hydrometeorology and Environmental Monitoring every day (except on public holidays), with a frequency of six days a week, three times a day. The monitoring network for air pollution monitoring in Saratov includes six main posts. Wherein, measurements of formaldehyde concentrations are measured at all monitoring posts (MP's) of the city. This allows territorially differentiating the results of measurements and conducting a spatiotemporal analysis. Pollution of atmospheric air is determined from the values of impurity concentrations. In this regard, the level of air pollution in Saratov is defined as high. The most common atmospheric pollutants are sulfur dioxide, nitrogen oxides, carbon monoxide, chlorine, formaldehyde, phenol, hydrogen sulphide, ammonia, benz (a) pyrene, dust. Formaldehyde has a significant impact on the pollution of the city atmosphere among other pollutants [3, 4].

2. The problem statement

Formaldehyde refers to small gas impurities; however, it is important in atmospheric chemistry and it is one of the main gas components-air pollutants in cities. It is also considered as priority dangerous compound.

Formaldehyde is classified as a toxic and dangerous substance of the 1st hazard class in accordance with Russian regulatory documents regulating the content of pollutants in the atmospheric air of populated areas, (until 2014 formaldehyde had 2nd class of danger). Formaldehyde hygienic standards in the atmospheric air are: TLV –one time maximum = 0,05 $mg \cdot (m^3)^{-1}$, TLV -average daily = 0,01 $mg \cdot (m^3)^{-1}$. It should be noted that from July 1, 2014 in Russia, the TLV value of formaldehyde and its hazard class have been changed. TLV -average daily changed from 0.003 $mg \cdot (m^3)^{-1}$ to 0.01 $mg \cdot (m^3)^{-1}$ (change No. 10 in 2.1.6.1338-03 "TLV of pollutants in the atmospheric air of populated areas", registered in the Russian Ministry of Justice in April 11, 2014 No. 31909) [4].

At elevated concentrations, formaldehyde has diverse toxic effect: irritates the mucous membranes of the upper respiratory tract, throat, eyes, causes queasiness and headache [5, 6]. In addition to general toxicity of this compound it can also be a carcinogen [7, 8].

In this regard, valuation of formaldehyde release into the environment is the most important condition for the development of a strategy for reducing air pollution in cities.

Incoming sources of formaldehyde into the atmosphere are varied. They can be divided into groups: industrial and other facilities that use formaldehyde in their activities; stationary burning of fuel and waste; mobile sources; Formaldehyde containing materials; city fires, landfill and industrial waste; natural sources [2, 9, 10].

The source of formaldehyde emissions from vehicles are automobile exhaust gases. The content of formaldehyde in exhaust gases is increased in comparison with stationary fuel combustion systems. This is due to the fact that unlike stationary fuel combustion systems, the combustion duration in internal combustion engine is limited to fractions of a second, and the cold walls of the engine chamber prevent the complete combustion of fuel, which leads to the releasing of products of incomplete combustion [2, 11, 12].

3. Results and discussion

The monitoring network for air pollution monitoring in Saratov includes six main posts. Wherein, measurements of formaldehyde concentrations are measured at all monitoring posts of the city. This allow to differentiate the results of measurements territorially and to conduct a spatiotemporal analysis [13].

The average concentration of formaldehyde in Saratov is 0.019 $mg \cdot (m^3)^{-1}$ or 1,9TLV (according to Russian standards).

The change in formaldehyde concentration over the past five years is shown in table 1. Figure 1 it shows the dynamic change in average annual concentrations of formaldehyde.

Table 1. Average monthly, average annual and Average long-term concentrations of formaldehyde in Saratov for the period 2011-2015.

Year	Mouth												Avg-ann
	I	II	III	IV	V	VI	VII	VIII	IX	X	XI	XII	
2011	0.008	0.008	0.009	0.009	0.015	0.018	0.025	0023	0.016	0.018	0.016	0.013	0.015
2012	0.010	0.014	0.019	0.017	0.015	0.010	0.017	0.013	0.012	0.018	0.016	0.015	0.015
2013	0.015	0.018	0.020	0.021	0.021	0.023	0.024	0.028	0.024	0.021	0.019	0.019	0.021
2014	0.020	0.020	0.020	0.021	0.023	0.027	0.034	0.034	0.030	0.026	0.023	0.020	0.025
2015	0.020	0.017	0.016	0.017	0.025	0.028	0.023	0.016	0.019	0.014	0.011	0.009	0.018
Averege	0.015	0.015	0.017	0.017	0.020	0.021	0.025	0.023	0.020	0.019	0.017	0.015	0.019

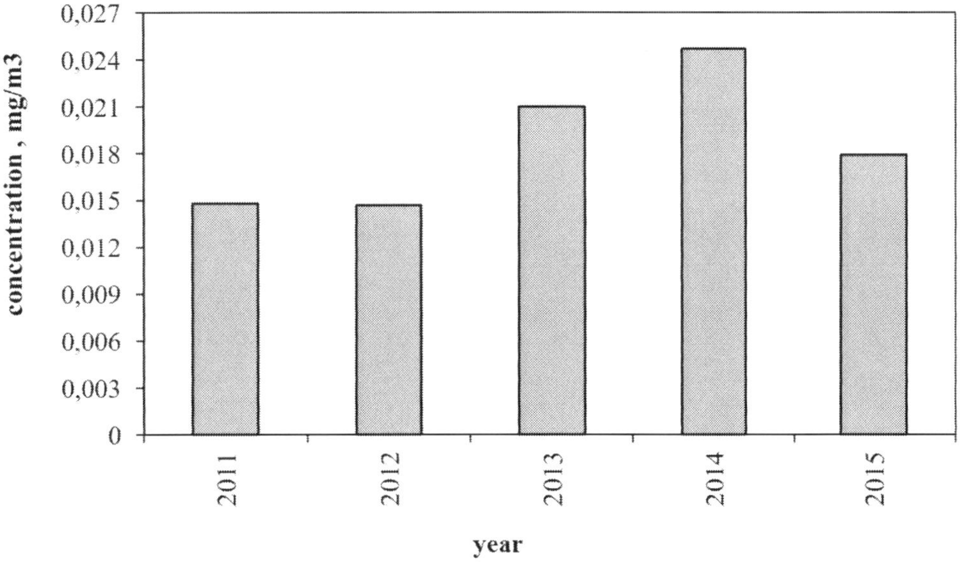

Figure 1. Dynamics of changes in average annual concentrations of formaldehyde.

The highest concentrations of formaldehyde were observed in 2014. The average annual value was 2,5TLV (according to Russian standards). The maximum values were recorded in July and amounted to 3,4TLV (according to Russian standards). The minimum was in 2012. The average annual concentration value is 1.5TLV (according to Russian standards).

During annual course (figure 2) it is possible to note an increase in the concentrations of formaldehyde from winter to summer (from July to a maximum of 2.5TLV) and the reduction from summer to winter (with a minimum January, February and December 1.5TLV).

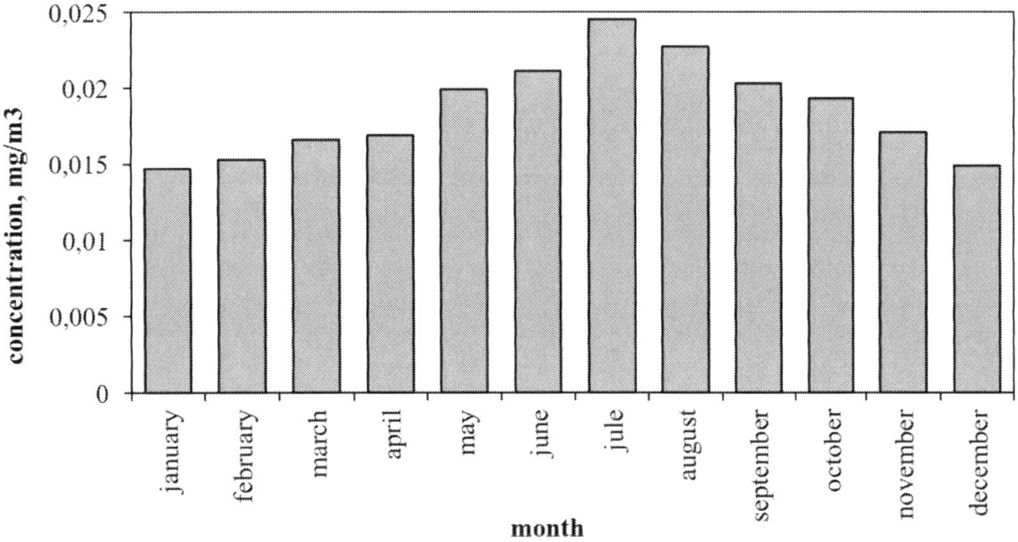

Figure 2. Dynamics of changes in the average monthly concentration of formaldehyde in Saratov.

According to figure 3, in the seasonal course, there is a obvious increase in formaldehyde concentrations in summer (average 2.3TLV). The lowest concentration of formaldehyde in the seasonal course is observed in winter. Spring and fall concentrations of formaldehyde range almost on the same level 0.018-0.019 mg·$(m^3)^{-1}$.

Formaldehyde is a pollutant which concentration values most often exceed daily average threshold limit value.

Analysis of complex air pollution index (API) (table 2) for the last five years has shown that there is a tendency to reduce pollution levels. In 2015 The API decreased almost 2 times in comparison with 2013, but this is largely due not to a decrease in the amount of emissions, but due to law changes in 2014 for threshold limit vales for formaldehyde.

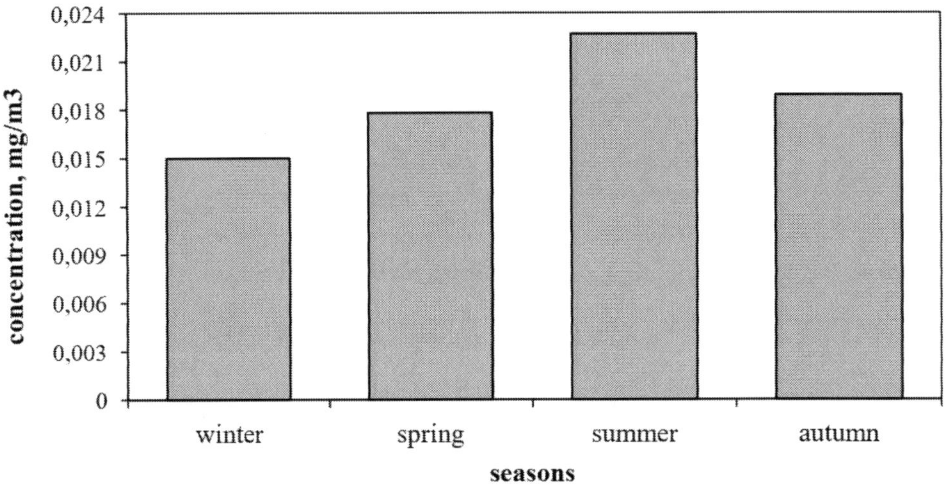

Figure 3. Seasonal dynamics of the concentration of formaldehyde in Saratov.

Table 2. API values for the period from 2011 to 2015 In Saratov.

Year	2011	2012	2013	2014	2015	Avg. annual
API	11.03	11.29	16.39	13.89	6.15	11.75
Level	high	high	very high	high	increased	high

In this way, the analysis of the structure of pollutant emissions has shown that formaldehyde contributes significantly to the formation of high API values. This ratio is maintained for MP's located near highways, and for MP's located inside the residential zone (figure 4).

The most polluted by this impurity is atmospheric air near MP-8, which is located near a highway with intensive traffic. The average annual impurity concentration here was 2.1TLV. Also quite high values are observed at the point MP-6 (2.01TLV), which is located near the JSC "Saratovstroysteklo" plant, which is the source of formaldehyde emission. The high value of the concentration on MP-6 is determined by the proximity of the highway, so the main source of formaldehyde here, as well as on MP-8 are vehicles.

The table 3 shows some statistical characteristics of the formaldehyde concentration during the study period: mean monthly (q), standard deviation (σ), standard deviation error (σ_σ), confidence interval of average concentrations ($I_{0.95}(q)$). The calculated statistical characteristics for formaldehyde have a small variability. The error standard deviation is only about 0.001, i.e. 0.1%.

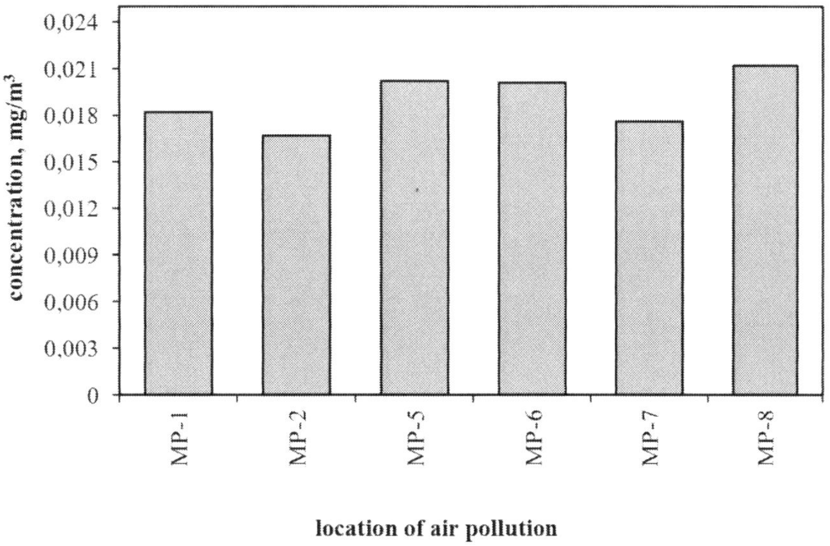

Figure 4 - Concentration of formaldehyde on various MP's of Saratov.

Table 3. The statistical characteristics of the formaldehyde concentration during the study period.

Parameter	Seasons				Year
	winter	spring	summer	autumn	
q, $mg \cdot (m^3)^{-1}$	0.015	0.018	0.023	0.019	0.019
σ	0.004	0.005	0.008	0.006	0.005
σ_σ	0.001	0.001	0.001	0.001	0.001
$I_{0.95}(q)$	0.013-0.017	0.016-0.020	0.020-0.026	0.017-0.021	0.017-0.020

4. Conclusions

From the analysis of the formaldehyde content in the atmosphere of Saratov following conclusions can be drawn:

- Formaldehyde makes a significant contribution to the pollution of the atmosphere of the city;
- Sources of formaldehyde are both emissions from industrial facilities and vehicles;
- The amount of formaldehyde is seasonal;
- Formaldehyde concentrations often exceed the TLV;
- A high level of formaldehyde can adversely affect the health of the city's population.
- The calculated statistical characteristics for formaldehyde have a small variability. The error standard deviation is only about 0.001, i.e. 0.1%.

References

[1] 2013 *About situation and protection of the environment of the Saratov region in 2012* State report (Saratov) 250

[2] 2014 *About situation and protection of the environment of the Saratov region in 2013* State report (Saratov) 240

[3] 2015 *About situation and protection of the environment of the Saratov region in 2014* State report (Saratov) 235

[4] 2016 *About situation and protection of the environment of the Saratov region in 2015* State report (Saratov) 247

[5] Kakareka S V 2012 *Nature management* **21** 75-82
[6] *AP-42. Compilation of Air Pollutants Emission Factors* 1995 Stationary Point and Area Sources **1** 95
[7] 1992 *Final Report on the Identification of Formaldehyde as a Toxic Air Contaminant* 16
[8] 1995 *Wood Dust and Formaldehyde. International Agency for Research on Cancer (IARC) Monographs on the Evaluation of Carcinogenic Risks to Humans* **62** 405
[9] Kakareka S V and Malchikhina A V 2011 *Natural Res.* **1** 107-115
[10] Skubnevskaya G I 1994 *Pollution of the atmosphere with formaldehyde: analytic review* 150
[11] Ksandopulo G I 1980 *Chemistry* 256
[12] 1989 *Environmental Health Criteria Formaldehyde* (Geneva: WHO)
[13] Semenova N V and Korotkova N V 2017 *Proc. Int. Conf. Environmental problems of geographical regions of Russia, dedicated to the 110th anniversary of the birth of the candidate of geographical sciences, associate professor, head of the Department of Geology and Geography, Dean of Faculty of Natural Sciences of the Kuibyshev Pedagogical Institute T.A. Aleksandrova* (Samara: Samara State Social and Pedagogical University) 214-219

IOP Conf. Series: Earth and Environmental Science **107** (2017) 012034 doi:10.1088/1755-1315/107/1/012034

Interannual variability of anticyclone activity and temperature extremes in the Black sea region

O Yu Kovalenko, E N Voskresenskaya

Institute of natural and technical systems, Lenin st., 28, Sevastopol, 299011, Russia
E-mail: kovalenko_olga89@mail.ru

Abstract. Manifestations of North Atlantic oscillation (NAO) during its negative and positive phases and during different El Niño (EN) types in the anticyclone parameters and air temperature extreme thresholds in the Black sea region are studied. It is estimated quantitative characteristics of anticyclone frequency anomalies and temperature extremes associated with the NAO phases and the EN types.

1. Introduction

The cyclone and anticyclone activity have a significant impact on the structure of hydrometeorological fields. The most illustrative parameters are average and extreme air temperature. The abnormally hot weather conditions in summer 2010 and 2017 in Europe are perfect examples. They were caused by the prolonged existence of anticyclone [1–3]. The recent decade's studies of climate changes of cyclone parameters in different regions have been given sufficient attention to, in contrast with anticyclones. The anticyclone variability in the Black sea region can be estimated by the papers where the only large-scale region is analyzed. In particular, the anticyclone activity weakening in Eurasia from 1990's to the 2000's and a significant intensification of anticyclone activity afterwards is found in [4]. The authors suggest that the character of anticyclone changes is accompanied by a decrease of mid-latitude surface air temperature over Eurasia and an increase in the frequency of extremely cold weather events. In [5] it is shown that the anticyclone frequency in the Black Sea region was increased in late 1960's - late 1980's during all seasons, especially in winter. At the same time, the anticyclone depth in that region increased in spring, but decreased in summer. However, the authors of this work do not consider the relationship with temperature, and especially with its extreme values. Such study is presented by [6], where the relationship between semi-permanent anticyclones and temperature anomalies in the Atlantic-European region, including the Black sea region, for the period 1952–2000 is analyzed. It is noted that the anticyclone frequency and average temperature during the last decades of the 20th century increased, but changes of extreme high temperature frequency were not detected. In accordance with the findings of [6] it is necessary to cogitate about the changes in anticyclone activity and extreme temperature in local regions, and also together with changes associated with circulation regimes.

It is now known that the interannual climate processes in the ocean-atmosphere system, such as the North Atlantic Oscillation (NAO) and El Niño (EN), are responsible for the changes of global circulation of atmosphere in the Northern Hemisphere, and, as a consequence, the changes of different hydro-meteorological characteristics [e.g., 7]. The study of their role in the generation of anomalies in the Atlantic-European hydrometeorological fields is one of the main tasks of the largest international scientific communities programs (e.g., EGU, EMS and others). The NAO manifestations in the sea level pressure (SLP) field are well described on the example of anticyclone blockings. The positive

NAO phase during winter is accompanied by low blocking activity in Europe [8, 9]. Changes of winter anticyclone characteristics during positive and negative NAO phases were examined by [10] for the Atlantic-European region. In [5] the relationship of anticyclones and NAO during all seasons only for the Black Sea is studied. These authors demonstrate that the NAO intensification in 1960-1990 is accompanied by more frequent winter anticyclones over the Southern Europe, including the Black Sea region.

Recent studies of EN events show the individual differences of various events in intensity, duration, and time of the beginning of ocean anomaly. There are successful attempts to classify them [e.g., 11–15]. So, the regional hydrometeorological manifestations of different EN types have to be different as well [11, 14]. This conclusion agrees with that reported by [12] for the Black Sea region. These authors have shown that the air temperature anomalies during two different EN types are different in sign. At the same time, the relationship between interannual climate processes and anticyclone activity, and also extreme temperature in the Black sea region is shown fragmentarily.

The aim of this study is to analyze the changes of anticyclone activity and extreme temperature over the Black Sea region, associated with the NAO phases and different EN types in 1950-2012.

2. Data and methods

The 1000-hPa geopotential height fields within 20-80 °N latitudinal belts for 1951–2012 were extracted from NCEP/NCAR [16]. Its temporal resolution is four time intervals (00, 06, 12, 18 UTC) and spatial resolution–2.5° latitude × 2.5° longitude. There are some publications, in which this data array was compared with observation data and other reanalyses [e.g., 17, 18]. The Black sea region is limited by 37–50 °N, 27–45 °E.

A technique by [10] is applied to separate the anticyclone eddies and their frequencies are calculated. The selected anticyclone parameter is reduced to the area of 1 km^2, dividing frequency by the area of a region. The Black sea anticyclone case snapshot, received by the technique and NCEP/NCAR data, is shown in figure 1a (24 March 1976, 0 UTC).

Daily data of air temperature observations from the European Climate Assessment & Dataset project [19] at 13 stations in the Black sea region for 1951–2012 were used. These Black sea stations are Anapa (44.9 °N, 37.33 °E), Armavir (44.98 °N, 41.12 °E), Feodosiya (45.03 °N, 35.38 °E), Gelendzhik (44.56 °N, 38.08 °E), Henichesk (46.17 °N, 34.82 °E), Izmail (45.37 °N, 28.85 °E), Kerch (45.4 °N, 36.42 °E), Krasnodar (45.03 °N, 38.98 °E), Odessa (46.43 °N, 30.77 °E), Rostov-on-Don (47.25 °N, 39.75 °E), Simferopol (44.96 °N, 34.12 °E), Sochi (43.6 °N, 39.72 °E), Tuapse (44.1 °N, 39.07 °E). The time series of air temperature on selected stations do not contain more than 25% of the date void for 1950–2012. The map where stations are marked is shown in figure 1b.

Using the temperature data set, the values of the 5th (P5) percentile of the intraseasonal daily temperature anomalies (i.e. threshold value of the extreme low temperature), calculated from the smoothed annual cycle, were obtained [21].

The NAO phases and different EN types manifestation on analyzed climate characteristics was studied using composite analysis. The monthly NAO index for 1951–2012 was used from website [20]. To select the NAO phases, about 15% of the total sample size (~30 months) were selected from either end of the ranked sample, i.e., for each phase, positive or negative, with maximum absolute values of the NAO index. This technique is described in detail in [10]. The EN classification by [12] is attracted to estimate the EN influence on anticyclone parameter and extreme temperature. Composite significance was estimated by the Student's t-test (significance level $p < 0.1$).

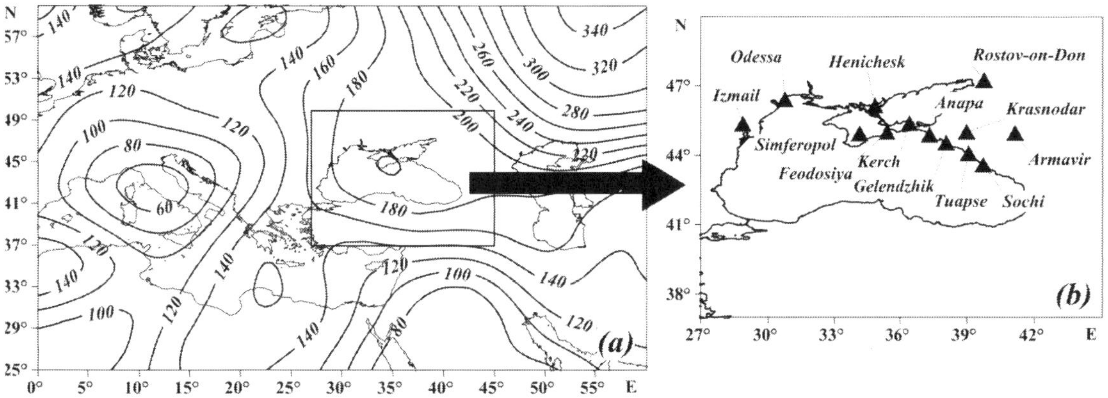

Figure 1. Geopotential height at 1000 hPa for 24 March 1976, 0 UTC (a) and location of stations (b) in the Black sea region.

3. Results and discussion

3.1. Manifestation of NAO

It is known that the NAO has a strong influence on the structure of hydrometeorological fields over the Atlantic-European region during all seasons [10, 22]. The phases of this mode are characterized by spatial shift of the Iceland Low and Azores High. It is accompanied by increase/decrease in the SLP gradient between these action centers of atmosphere, strengthening/weakening of the zonal flow [10, 22, 23]. As a result, the cyclone and anticyclone tracks are displaced, characteristics of these eddies are changed [18, 24].

Composite analysis shows significant changes of anticyclone and extreme temperature parameters during different NAO phases only in the winter. It is found out that winter positive NAO phase (NAO+), relative to the negative phase (NAO-), is accompanied by about 50% increasing of anticyclone frequency in the Black Sea region. This result coincides with study of [25]. The winter threshold value of the extreme low temperature (P5) in the Black Sea region during NAO+, relative to the opposite phase, increases by ~ 60%. This demonstrates a smaller absolute threshold value during NAO+. For example, the P5 value at the Krasnodar station for NAO+ is -2.9 °C, and it for NAO- is -10.7 °C.

Previous studies [5, 10, 26] shows that the positive NAO phase is associated with the strong anticyclone frequencies increase in the subtropical gyre between 30° and 40 °N and displacement of anticyclone tracks to south Europe. As a result, it was to cause the increasing of extreme winter daily temperature at the end of the 20th century. However, here we present quantitative measures of both anticyclone frequencies and thresholds of extreme low temperature for the NAO phases over the Black sea region from the end of the 20th century to the beginning of the 21st century.

3.2. Manifestation of different El Nino types

To analyze the EN manifestations on anticyclone and extreme temperature parameters during different types, the classification, more detailed in [14], is attracted. According to this classification, the EN events are separated into three types: spring (SPR), summer-autumn short-lived (SAS) and summer-autumn long-lived (SAL) types. Their typical characteristics are given in table 1. During the 62-yr analyzed period, the first type was observed 8 times, the second type - 7 times, and the third type - only twice. To obtain statistically significant estimates the lengths of time series (i.e. 7–8 or 2 months·year^{-1}) are not enough, but it is possible to demonstrate the different character of regional manifestations of the EN types. In this investigation the analysis of manifestations of different EN types on studied climate parameters is given for only the SPR EN type and SAS EN type.

The seasonal average changes of anticyclone parameters in the Black Sea region during SPR EN type, relative to SAS EN type, are insignificant. However, the two EN types are accompanied by anticyclone frequency anomalies of the opposite signs during certain months, especially in April and September, and by positive frequency anomalies in cold months, when the EN event is intensive (figure 2). At the same time the SPR EN type is characterized by more intensive extreme cold temperatures than SAS type (figure 3). Such temperature distribution is consistent by [14].

Table 1. Typical properties of different El-Nino types by [14].

Characteristic	Spring type	Summer-autumn short-lived type	Summer-autumn long-lived type
Beginning	April	July	September
Maximum SST anomaly in Nino 3.4 region, °C	+1.85	+1.24	+1.26
Duration, months	12	9	20

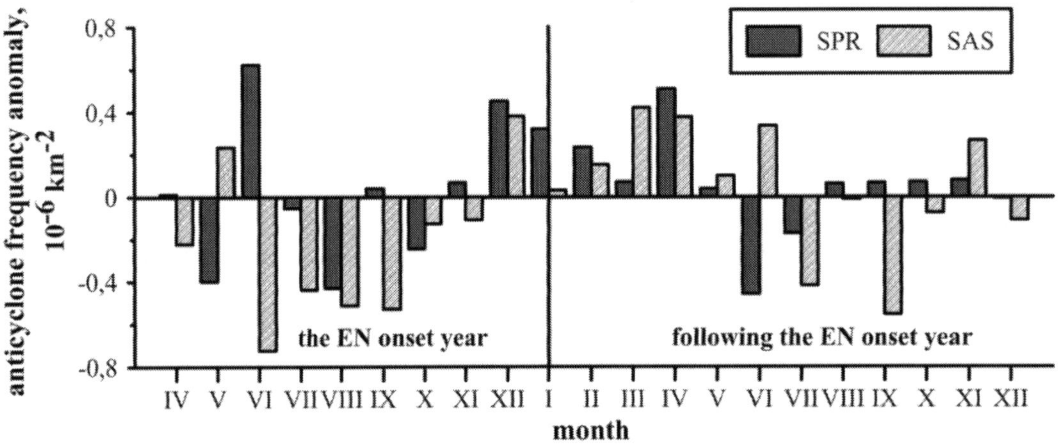

Figure 2. Composites of anticyclone frequency anomalies during SPR and SAS EN types in the Black sea region.

4. Discussion and conclusions

Thus, the changes of anticyclone activity and extreme temperature over the Black Sea region, associated with the NAO phases and the EN types have been studied. It is shown that the most intense NAO manifestations in the anticyclone parameters and extreme temperature anomalies are in winter. The positive NAO phase, relative to the negative one, is accompanied by a statistically significant increasing of anticyclone frequency and decreasing of the threshold of extreme cold temperature in the Black Sea region during winter.

The mechanism of NAO influence on anticyclone activity over Europe has been specified . It is estimated that anticyclone frequency in the Black Sea region in winter positive NAO phase is about 50% higher than in negative one. At the same time it is shown that winter threshold value of the extreme low temperature (P5) in the Black Sea region during NAO positive phase increases by ~ 60% relative to the opposite phase.

The interannual changes of anticyclone frequency and extreme low temperature are examined in association with the EN taking into account its classification. It is shown that the manifestations of EN types are more intense in the autumn-winter. The SPR EN type and SAS type are characterized by anticyclone frequency anomalies of the opposite signs in the Black Sea region. The absolute threshold extreme of cold temperature values in the region under consideration are higher during SPR EN than in SAS type.

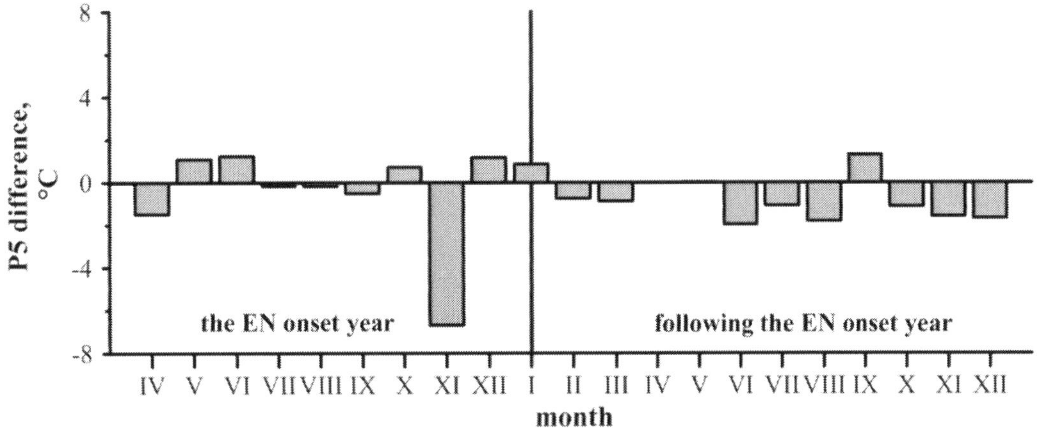

Figure 3. The difference of P5 thresholds during SPR and SAS EN types in Krasnodar station.

Acknowledgement
The reported study was partially supported by RFBR according to the research project № 16-05-00231.

References

[1] Shakina N P and Ivanova A R 2010 *Russian Meteorology and Hydrology* **35** (11) 721–730
[2] Trenberth K E and Fasullo J T 2012 *J Geophys Res.* **117** (17) D17103
[3] Global Climate Report. June 2017 (Electronic Materials) – URL: https://www.ncdc.noaa.gov/sotc/global/201706
[4] Xiangdong Z, Chuhan L and Zhaoyong G 2012 *Environ Res Lett* **7** 044044 doi:10.1088/1748-9326/7/4/044044
[5] Polonsky A B, Bardin M Y and Voskresenskaya E N 2007 *Physical Oceanography* **17** (6) 348–359
[6] Bardin M Y 2007 R*ussian Meteorology and Hydrology* **32** (2) 75–84
[7] Hartmann D L *et al* 2013 *Observations: Atmosphere and Surface. In: Climate Change 2013: The Physical Science Basis. Contribution of Working Group I to the Fifth Assessment Report of the Intergovernmental Panel on Climate Change* ed Stocker T F *et al* (Cambridge, United Kingdom and New York, USA: Cambridge University Press) 223–232
[8] Rogers J C 1984 *Mon Weather Rev* **112** 1999–2015
[9] Scherrer S C, Croci-Maspoli M, Schwierz C and Appenzeller C 2006 *Int. J. Climatol* **26** 233–249
[10] Bardin M Y and Polonsky A B 2005 *Izvestiya. Atmospheric and Oceanic Physics* **41** (2) 127–136
[11] Horii T A and Hanawa K 2004 *Geophys Res Lett* **31** L06304
[12] Voskresenskaya E N and Mikhailova N V 2010 *Dopov Nac akad nauk Ukr* **3** 120–124
[13] Kao H Y and Yu J Y 2009 *J Climate* **22** 615–632
[14] Kug J S, Jin F F and An S I 2009 *J Climate* **22** 1499–1515

[15] Zheleznova I Vand Gushchina D Y 2015 *Russian Meteorology and Hydrology* **40** (3) 170–179
[16] Kalnay E *et al* 1996 *Bull Am Meteorol Soc* **77** 437–471
[17] Akperov M G and Mokhov I I 2010 *Izvestiya Atmospheric and Oceanic Physics* **46** (5) 574–90
[18] Voskresenskaya E N and Maslova V N 2011 *Advances in Science and Research* **6** 237–43
[19] *European Climate Assessment & Dataset project* (Electronic Materials) – URL: http://eca.knmi.nl/
[20] *North Atlantic Oscillation* (Electronic Materials) – URL: http://www.cpc.ncep.noaa.gov
[21] Bardin M Y 2014 *Extremeness of climate. In: Second Roshydromet assessment report on climate change and its consequences in Russian Federation General summary* ed Yasukevich V V *et al* (Moscow: Roshydromet) 14–16
[22] Nesterov E S 2003 *Russian Meteorology and Hydrology* **1** 47–55
[23] Quiroz R S 1984 *Mon Weather Rev* **112** (9) 1894–1912
[24] Lau N C 1988 *J Atmos Sci* **45** (19) 2718–2743
[25] Voskresenskaya E, Bardin M Yu and Kovalenko O 2016 *Quaternary International* **409** 70–74
[26] Brown S J, Caesar J and Ferro C A T 2008 *J. Geophys. Res* **113** D05115

IOP Conf. Series: Earth and Environmental Science **107** (2017) 012035 doi:10.1088/1755-1315/107/1/012035

Latent fluctuation periods and long-term forecasting of the level of Markakol lake

A S Madibekov[1], A V Babkin[2], A Musakulkyzy[1], A V Cherednichenko[3]

[1] Institute of Geography, Pushkin str., 99, Almaty, 050000, Kazakhstan
[2] Department of Atmospheric Dynamics and Space Geosciences, Russian State Hydrometeorological University, Maloohtynskay str., 97, Saint - Petersburg, 195196, Russia
[3] Zhasyl Damu JSC, Seifullin str., 597, Almaty, 050022, Kazakhstan
E-mail: madibekov@mail.ru, abav@mail.ru, mus_ain@mail.ru, geliograf@mail.ru

Abstract. The analysis of time series of the level of Markakol Lake by the method of "Periodicities" reveals in its variations the harmonics with the periods of 12 and 14 years, respectively. The verification forecasts of the lake level by the trend tendency and by its combination with these sinusoids were computed with the lead time of 5 and 10 years. The estimation of the forecast results by the new independent data permitted to conclude that forecasts by the combination of the sinusoids and trend tendency are better than by the trend tendency only. They are no worse than the mean value prediction.

1. Introduction

The lake levels are very visible indicators of climate change, water balance and human influence [1-3]. The issues of studying long-term forecasting of water levels in lakes have been given a lot of scientific attention in works [4-6]. In this study, using the example of a climatically conditioned lake, a valid method is proposed that is applicable for all lakes based on the detection of periods and provides the possibility of predicting the water level with a high degree of reliability.

The water level of Markakol lake experiences significant fluctuations over time. Its rises and falls impact on various sectors of modern production associated with the utilization of its water resources. Long-term forecasting of the lake level is necessary for the socioeconomic development of its basin according to the objectives of the State Water Resources Management Program of Kazakhstan (Decree of the President of the Republic of Kazakhstan dated 04.04.2014 №786) and "Water Security of the Republic of Kazakhstan" National program, under which "Water Resources of Kazakhstan and their Utilization" Geospatial Information System is being developed and the lakes in the Republic are being studied. Forecasting of the lake level would be expedient taking into account the latent times in its vibrations.

This study presents an analysis of fluctuations in the annual values of the lake level and, in view of the identified periods, makes its verification forecasts for the short term. The time series was analyzed from 1942 - the beginning of regular observations, till 2003. Over the next 10 years its values have been used to estimate the forecasts' results based on a new independent material. Evaluation of forecasting results has shown that for the combination of the identified sinusoids and trend tendencies, the forecasts turned out to be better than for the trend tendencies and no worse than for the mean value.

Markakol Lake is located in Kazakhstan in the eponymous bolson edged with the mountain ridges of Kurchum and Azutau. In the north-east it adjoins the Bobrovskaya depression, and in the south-west along the Kalzhyr river it has access to the Zaisan valley (figure. 1). The lake is tectonic. It is the second largest lake in the basin of the Ertis river. The lake is located at an altitude of 1419.3m above sea level and has a prolate oval shape. It extends from the northeast to the southwest. The length of the lake is 38 km, and its width is 19.0 km, its catchment area is 1180 km^2, its water surface is 455 km^2, legislation of its shoreline is 106 km, its mean depth is 14.3 m, the maximum depth is 24 m. The lake basin contains 6.5 km^3 of water. By the nature of water exchange, Markakol Lake is a drainage one. There is a large number of rivers and currents of water running into the lake, including the Topolevka, Mate-Bai, Urunhayka, and other rivers, and the Kalzhyr River taking off from the lake [7].

Observations of the water level and temperature near the lake shore were started in 1937, and meteorological observations at the Markakol-Zapovednik station - in 1982. The climate in the lake basin is sharply continental: the average yearly temperature is -2.4 °C, the average temperature in January, which is the coldest month, is -21.9 °C, and in July, being the warmest month, it is 14.4 °C.

Figure 1. Aerial view of Markakol Lake (Source - Google Earth).

A steady 0°C crossing of air temperature in the spring usually occurs in the middle of April, and in the second decade of October in the fall.

2. The problem statement or Material and methods

The time series of the lake level is analyzed using "Periodicities" method [8, 9]. This method is based on the approximation of the time series of hydrometeorological characteristics by sinusoidal functions in sequence with a step-by-step change in the period. The amplitude and phase of the best approximating sinusoid, as well as the additional addend around which it oscillates, and the sum of its quadratic differences with the values in the series are calculated for each period using the least squares adjustment [10].

In the analysis of the relation of the sum of quadratic differences of the approximating sinusoid to approximation period, the local minima of this sum should be observed for individual periods. Such a minimum of the sum of the quadratic differences of approximating sinusoid and the values in the series may be a sign of the presence of periodicity here.

When adding or combining the identified sinusoids, the correlation of their sum with the time series should consistently increase, reflecting peculiarities of its oscillations. The combination of sinusoids, which can also take into account the linear trend of the series, essentially represents its pattern, which can be used as a forecast expression for calculating future values of the lake level.

The results of the long-term forecast of the lake level can be estimated based on two different criteria. The forecast of the lake level should be considered verified if the difference between its predicted and actual values is less than its permissible error Δ, equal to 0.674 of mean-square deviation of its time series σ [11]. The predicted results for a multi-year interval are estimated by the number of verified annual forecasts and by the sum of squares of their errors S_f. Its actual error δr, which can be compared with the time series values of the lake level and its mean square deviation is calculated taking into account the sum of squares of the forecast errors,

$$\delta r = 0{,}674\sqrt{\frac{S_f}{l}} \tag{1}$$

where l is the verification forecast interval, 5 and 10 years, respectively. If it is required to compare the results of forecasting of different lakes levels, it is necessary to compare relative errors $\frac{\delta r}{\Delta}$, rather than their actual errors. The relation of the actual forecast error to its permissible error has no dimensionality and is calculated in fractions of 1.

The long-term forecast of the lake level on the specified interval is the better, the larger is the number of verified forecasts and the smaller is its relative error. A successful long-term lake level forecast should be no worse than a prediction based on the mean value of its time series.

3. Results and discussion

The results of approximation of the time series of the lake level are presented in table 1. The first column of this table indicates approximation periods T, the second column indicates the constant value around which the best approximating sinusoid of this H_0 period oscillates; the third and fourth columns show its amplitude $\frac{\delta H}{2}$ and phase φ_H, respectively. The fifth column calculates its sum of quadratic differences with S_H time series of the lake level.

For certain periods, there are minima of the sums of quadratic differences of approximating sinusoids and the series values observed. Characteristics of sinusoids with the minima of sums of quadratic differences with series values are in italics. The minima of these sums are registered for the periods of 3, 6, 8, 12, 14 and 25 years respectively.

In the bottom rows of table 1, the identified periods in lake level fluctuations are presented in the decreasing order of η correlation of their sinusoids with a time series. The highest correlation with a number of levels is observed in periods of 14 and 12 years. It was 0.390 and 0.365, respectively. We are going to use these two sinusoids with the greatest correlation with a time series exceeding 0.35 to calculate the verification forecast.

Figure 2 shows fluctuations in the Markakol Lake level [12], the identified sinusoids with the periods of 12 and 14 years, the trend tendency and a combination of these sinusoids and trend tendencies. Sequences of increased lake level values with time series maxima in 1946, 1958 and 1969 and its decreased values with the minima in 1951 and 1963, the sinusoidal extremums with the periods of 12 and 14 years are closely spaced. The sinusoidal extremum with a period of 14 years is somewhat ahead of the corresponding extremum of the 12-year-old sinusoid at the maximum lake level in 1946 and its minimum in 1951; at the maximum lake level in 1958 these extremums occur almost simultaneously.

In the set of decreased values with a minimum lake level in 1963 and increased values with a maximum in 1969, the corresponding sinusoidal extremums with a period of 14 years lag behind the extremums of harmonics with a 12-year period. The time difference of sinusoidal extremums with the periods of 14 and 12 years is becoming greater in the region of decreased level values with the minimum values in 1974 and 1982 and in the future, these sinusoids virtually oscillate in their opposite phases.

In the first half of the analyzed interval, at which the corresponding sinusoidal extremums with the periods of 12 and 14 years are closely spaced, their combination with the trend tendency well describes the groups of years of increased lake level values with the maxima in 1946, 1958 and 1969, and its decreased values with the lows in 1951 and 1963. The decreased level values time interval with the minimum values in 1974 and 1982 is described by a smaller-amplitude basin in this combination, since the time of sinusoidal extremums with the periods of 12 and 14 years significantly diverges there.

IOP Conf. Series: Earth and Environmental Science **107** (2017) 012035 doi:10.1088/1755-1315/107/1/012035

Relatively small amplitude fluctuations in the lake level after 1989 are reflected by a smooth line formed by a combination of sinusoids with the periods of 12 and 14 years, oscillating in different phases, and a trend tendency. The derived forecast equation fails to describe the peak lake level in 1988. Its correlation with the time series equals 0.516.

Table 1. Sinusoidal Approximation of the time series of water level in Markakol Lake (1942 – 2003).

T, years	H_0, m	$\frac{\delta H}{2}$, m	φ_H, radian	S_H, m^2
3.0	*153.2775*	*2.2854*	*4.1756*	*4890.386*
4.0	153.3246	1.1432	-0.1532	5010.476
5.0	153.3127	0.1485	1.1332	5050.287
6.0	*153.3028*	*3.1911*	*4.6408*	*4740.345*
7.0	153.2792	2.0819	2.4486	4917.954
8.0	*153.2682*	*3.3387*	*0.0314*	*4700.315*
9.0	153.3240	1.4267	2.7949	4987.716
10.0	153.3714	2.4240	2.5629	4867.161
11.0	153.1765	4.4659	1.7497	4428.943
12.0	*153.3080*	*4.7237*	*0.7937*	*4378.584*
13.0	153.1685	4.5847	4.0762	4399.329
14.0	*152.9807*	*4.9656*	*2.6031*	*4284.018*
15.0	153.1674	4.5188	-1.4637	4400.974
16.0	153.4240	3.6297	-0.1930	4655.557
17.0	153.4840	2.2330	1.4720	4903.723
18.0	153.3913	1.1720	-1.3958	5008.690
19.0	153.3115	1.4674	4.2768	4987.528
20.0	153.2946	2.1111	-0.5672	4916.667
21.0	153.3206	2.5358	3.8733	4848.937
22.0	153.3576	2.7782	-0.7113	4801.404
23.0	153.3822	2.9314	-1.3789	4771.890
24.0	153.3883	3.0499	2.1825	4754.858
25.0	*153.3830*	*3.1417*	*3.9539*	*4747.649*
26.0	153.3749	3.1844	4.1530	4750.168
27.0	153.3676	3.1554	2.9593	4762.074
28.0	153.3605	3.0549	0.5248	4780.852
29.0	153.3508	2.9093	3.2639	4802.134
30.0	153.3349	2.7582	-1.2763	4821.370
31.0	153.3092	2.6382	-0.4325	4835.208
32.0	153.2713	2.5742	-0.4075	4841.865
	T, years			η
	14.0			0.3897
	12.0			0.3649
	8.0			0.2635
	6.0			0.2480
	25.0			0.2451
	3.0			0.1783

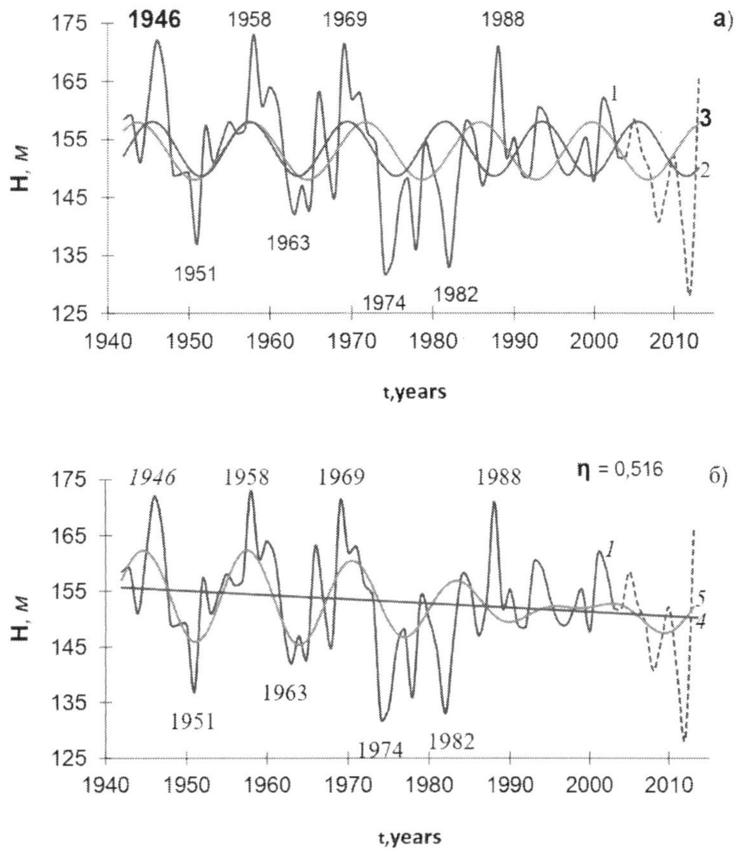

Figure 2. Fluctuations in the water level of Markokol Lake – Urunhai vil.: 1 - observation data (dotted line is the verification forecast interval for 2004-2013), 2 - approximating sinusoid with a period of 12 years, 3 - approximating sinusoid with a period of 14 years, 4 - linear trend tendency (its equation is presented in the lower left corner (b)), 5 - a combination of sinusoids with the periods of 12 and 14 years and trend tendency, η - its correlation with the time series.

4. Verification Forecasts of the Level of Markakol Lake

Based on the forecast equation, combinations of sinusoids with the periods of 12 and 14 years and a trend tendency - calculations were made for the lake level verification forecasts with a lead time of 5 and 10 years respectively. The forecasting results and their evaluation are presented in table 2.

The first column of this table indicates years when verification forecasts had been calculated, the second column shows the actual values of the lake level and the third column shows the difference of the mean and the actual lake level values for the analyzed period (1942-2003). The fourth and the sixth columns show lake level computation results based on the trend tendency and its combination with the sinusoids with the periods of 12 and 14 years, the fifth and the seventh columns indicate their differences with the actual level value.

The average value of the lake level time series for 1942 – 2003 was 153.31 m, with mean-square deviation of 9.03 m. The admissible forecast error was 6.08m. The forecast for the year is thus verified if the difference in the columns 3, 5 and 7, respectively, is less than this value.

The mean value forecast however was not verified for 2008-2009 and for 2011-2013 but proved to be right for all other years. Therefore, the mean value forecast was verified 4 times for the first five years and 5 times for the entire verification period. The relative mean value forecast error calculated factored in the differences presented in column 3, was 0.703 for the first five years, and 1.206 for the entire ten-year verification period.

Table 2. Results of Verification Forecasts of the level of Markakol Lake for 2004 – 2013

t, years	H_{actual}, m	$H_{mean} - H_{actual}$, m	H_{trend}, M	*trend* – H_{actual}, m	H_{comb}, m	$H_{comb} - H_{actual}$, m
1	2	3	4	5	6	7
2004	151.9	1.41	150.92	-0.98	152.46	0.56
2005	158.4	-5.09	150.84	-7.56	151.74	-6.66
2006	152.8	0.51	150.77	-2.03	150.64	-2.16
2007	149.2	4.11	150.69	1.49	149.36	0.16
2008	140.8	12.51	150.62	9.82	148.22	7.42
2009	146.0	7.31	150.54	4.54	147.55	1.55
2010	152.0	1.31	150.46	-1.54	147.63	-4.37
2011	140.0	13.31	150.39	10.39	148.54	8.54
2012	129.0	24.31	150.31	21.31	150.17	21.17
2013	166.0	-12.69	150.24	-15.76	152.19	-13.81
N_5		4		3		3
$\frac{\delta r}{\Delta_5}$		0.703		0.628		0.506
N_{10}		5		5		5
$\frac{\delta r}{\Delta_{10}}$		1.206		1.105		1.014

Forecasts by trend tendency and its combination with sinusoids with the periods of 12 and 14 years did not prove to be true in 2005, 2008, 2011, 2012 and 2013. According to these schemes, 3 forecasts were verified for the period of 2004 - 2008, and 5 forecasts had proven to be true for the entire verification forecast interval of 2004-2013. The relative error of the forecast by trend tendency and its combination with the sinusoids with periods of 12 and 14 years with a lead time of five years was 0.628 and 0.506, respectively, and with a lead time of 10 years – 1.105 and 1.014.

5. Conclusions

Analysis of the Markakol Lake level time series for 1942 – 2003 using "Periodicity" method indicates the presence of harmonics with the periods of 12 and 14 years in its fluctuations. The combination of the trend tendency and these sinusoids describes well the fluctuations in the values of this series. The trend tendency and its combination with these sinusoids were used to compute verification forecasts of the lake level with a lead time of 5 and 10 years. The forecasting results were evaluated based on a new independent material and compared with the forecasts for the mean value of the time series.

Since by one of the results assessment criterion forecast was shown to be better by mean value while by another criterion it was better by trend tendency and its combination with the identified sinusoids, we can conclude that forecasting results based on mean value, trend tendency and its combination with sinusoids are of approximately the same quality. Nevertheless, the forecasts by a combination of trend tendency and sinusoids has proven to be better than those by trend tendency: with the same number of correct forecasts, the relative error is smaller in forecasts based on combination of sinusoids and the trend tendency.

When forecasting for the entire verification period of 2004 – 2013, the same number of forecasts was verified in all three schemes. Therefore, the forecast, which has a smaller relative error is a better one. Thus, the best forecast was obtained by the combination of sinusoids and trend tendency, and then by trend tendency, and the worst forecast was the one based on the mean value of the time series.

To make long-term lake level forecasts, it is necessary to analyze the time series of the levels of all lakes in the Republic of Kazakhstan, where there are ongoing continuous series. Reliable harmonics, which can be used in forecasting, should be identified for multiple, rather than one series. At the same time, the fluctuations phase for the same period should change gradually depending on the geographical location of the lakes.

For a more reliable forecasting of lake levels for the next 5-10 years, it is necessary to analyze their series for shorter time intervals, for example, for seasonal or monthly intervals.

References

[1] Van der Kamp G, Keir D and Evans M S 2008 *Canadian Water Resources Journal*, **33** (1) 23-38
[2] Zoljoodi M and Didevarasl A 2014 *Atmospheric and Climate Sciences* **4** 358-368
[3] Demargne J, Wu L, Seo D J and Schaake J 2007 *Proceeding of the Symposium HS2004 at IUGG2007, Perugia, July 2007*. IAHS 313
[4] Trolle D, Hamilton D P, Pilditch C A, Duggan I C and Jeppesen E 2011 *Environmental Modelling & Software* **26** (4) 354–370
[5] Garcia Molinos J, Viana M, Brennan M and Donohue I 2015 *J. PLoS ONE* **10** (3) e0119253 doi:10.1371/journal.pone.0119253
[6] Liu L, Xu Z X, Reynard N S, Hu C W and Jones R G 2013 *J. of Flood Risk Management* **6** (1) 14–22
[7] Filonets P P and Omarov T R 1974 *Lakes of the Northern, Western and Eastern Kazakhstan* (L.: Gidrometeoizdat) 138
[8] Babkin A V 2005 *Meteorology and Hydrology* **11** 63-73
[9] Babkin A V and Semeikin N I 2014 *News of the Russian Geographical Society* **1** 43-48
[10] Linnik Yu V 1962 *Least Square Method* (Moscow: Nauka) 350
[11] Apollov B A, Kalinin G P and Komarov V D 1974 *Hydrological Forecasting Course* (L.: Gidrometeoizdat) 419
[12] Musakulkyzy A and Madibekov A S 2016 *International Research and Practice Conference on "Water resources of Central Asia and their Utilization", on progress review for the UN "Water for Life" decade, Almaty*, 360-365

IOP Conf. Series: Earth and Environmental Science **107** (2017) 012036 doi:10.1088/1755-1315/107/1/012036

Transmissivity of the atmosphere above the Russian territory: observed climatic changes

A N Makhotkin[1], E L Makhotkina[1], I N Plakhina[2]

[1]Voeikov Main Geophysical Observatory, Karbyshev st., 7, Saint-Peterburg, 194021, Russia
[2]A.M.Obukhov Institute of Atmospheric Physics RAS, Pyzhevsky, 3, Moscow, 119017, Russia
E-mail: alex.makhotkin@gmail.com, inna@ifaran.ru

Abstract. We report the systematic investigation of spatial-temporal changes of integral and aerosol atmosphere turbidity above the territory of Russia in the period 1976-2016. The data referring both to the whole territory of Russia and some certain regions are discussed.

1. Introduction

Ambient monitoring is a complex observing system; it includes environmental conditions observation, estimations and predictions of changes of ambient conditions affected by natural and anthropogenous factors. Solar irradiation is one of the basic climate forcing factors. Solar irradiation measurements have been in the focus of researchers' interests during last decades [1–8]. Transmissivity of the atmosphere, solar angle and cloudiness determine solar irradiation income at a ground surface, therefore being essential characteristics of irradiation regime. The most prominent and long-term changes in the transmissivity of the atmosphere occur under influence of volcanic eruptions and anthropogenous activity; relatively short-range changes are the result of smoke haze caused by peat-bog and forest fires.

In the mid- 20th century the researchers' attention was centered on the investigation of integral transmissivity of the atmosphere. Last decades the issue of systematic and regular investigation of long-term changes of aerosol turbidity of the atmosphere (aerosol optical thickness, AOT) has come to the fore. In the 60's 33 stations (11 stations at European part of USSR, ETC, and 22 – at Asian, ATC) have been chosen for determination of atmosphere transmissivity. Further data from 228 stations were used for investigation of the dependencies of coefficient transmissivity change.

Nowadays data from different observing systems underlie the atmosphere transmissivity monitoring. The system includes land-based actinometrical network of RosGidroMet (116 stations that register direct solar irradiation), international network AERONET (Aerosol Robotic Network – 9 stations in Russian Federation) and satellite monitoring system, implemented by satellites Terra and Aqua.

Data, obtained from Roshydromet actinometrical network, provide information about integral and aerosol transmissivity of the atmosphere (from the mid- 20th century). AERONET data and satellite measurements have provided information about AOT of the atmosphere only since 2000 year. Nevertheless, these data are used when analyzing the tendency of AOT change in different regions [9–13]. AERONET data alongside with satellite information allow characterization of the atmosphere transmissivity at vast areas.

The aim of the work is to systematize and generalize data on spatial-temporal changes of integral and aerosol transmissivity of the atmosphere both at the Russian territory as a whole and at certain regions for the period from 1976 to 2016.

2. Material and methods

Used data base

The objects of investigation were series of monthly and annual values of turbidity factors T2 and AOT over the period of 40 years (1976–2016) for the basic regions of Russia. Three regions in European Russia were considered: North, South and Center. Six regions from Asian Russia were chosen: Ural, West Siberia, North-East, Central Siberia, Baikal region, Transbaikal, Far East. Calculation methodology used for integral turbidity characterization and general characteristics of the considered regions are given in [14, 15]. The series of transmissivity characteristics for a number of stations were formed in the period from 1957 to 1960. The said series are regularly amplified with the current data.

A tricenarian period (1976–2005) was chosen as a base period for determination of normal values of T_2 and AOT.

The changes of monthly and annual values of T_2 and AOT were analyzed for two periods: 1976–2016 and 1994–2016. The second mentioned period is featured by the absence of powerful volcanic eruptions and, therefore, the increase of income solar irradiation and growth of the atmosphere transmissivity. Sometimes it is called «brightening» — the period of aerosol translucence of the atmosphere [3].

The series of monthly and annual values of T_2 and AOT are described by linear equations. The velocity of change of T_2 and AOT is estimated by a coefficient in a regression equation ($a \times 10$), for 10 years.

The results of the analysis are represented for three regions: Ural, the South of European part and the North-East of Asian Russia. This choice provides a glimpse of observable changes in regions with different aerosol load.

3. Results and discussion

The atmosphere turbidity over the territory of Russia increases in the direction from North-East to South-West. The highest transmissivity of the atmosphere is observed in the north-west of Asia. The South of European part of Russia is characterized by high turbidity [9–13]. The European Russia is featured by a growth of turbidity in the direction from north to south; the same tendency is valid for Asian part, though it is less prominent.

The time dependencies of T_2 and AOT that have been registered within 55 years can be divided into three periods: 1) 1960-1981 years – stable growth of atmosphere turbidity; 2) 1983–1992 years – slight decrease of atmosphere turbidity; 3) from 1994 till present – significant decrease of atmosphere turbidity. Noticeable growth of average annual values of T_2 and AOT is caused by powerful eruptions of volcano El Chichón and volcano Pinatubo.

Figure 2 shows the alteration of the annual courses of T_2 and AOT in regions with different aerosol load. The turbidity turndown along the Russian territory is determined by the curves for the North-East of Asian Russia and the South of European Russia.

The World Meteorological Organization (WMO) recommended to choose the period 1961-1990 as the basic period for monitoring and estimating current climate changes. Taking into consideration the duration of available series of normal values of T_2 and AOT, we have chosen a tricenarian period 1981–2010, which should be applied (according to WMO recommendation) when performing climatic service works. The comparison of average values within different periods was completed using data from several stations. The averages within the period 1961-1990 are just over than the averages within other periods 1981–2010 and 1976–2005; averages within the period 1976-2005 are the closest to the base averages WMO.

The analogous dissimilarities for the long-term annual averages for individual months fluctuate in a wider range: in winter-time these differences are more noticeable than in summer.

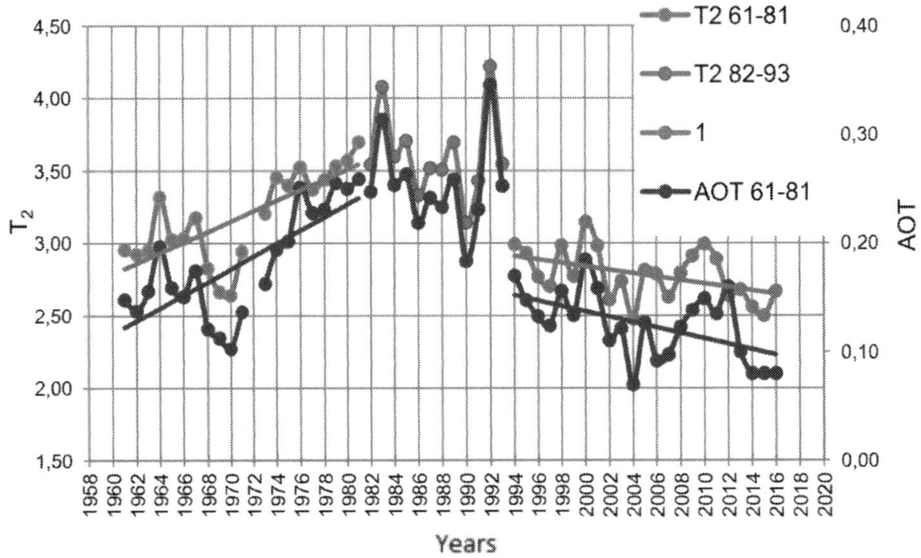

Figure 1. T_2 (1) and AOT (2) courses at the station "Tsimlyansk", period 1961-2016.

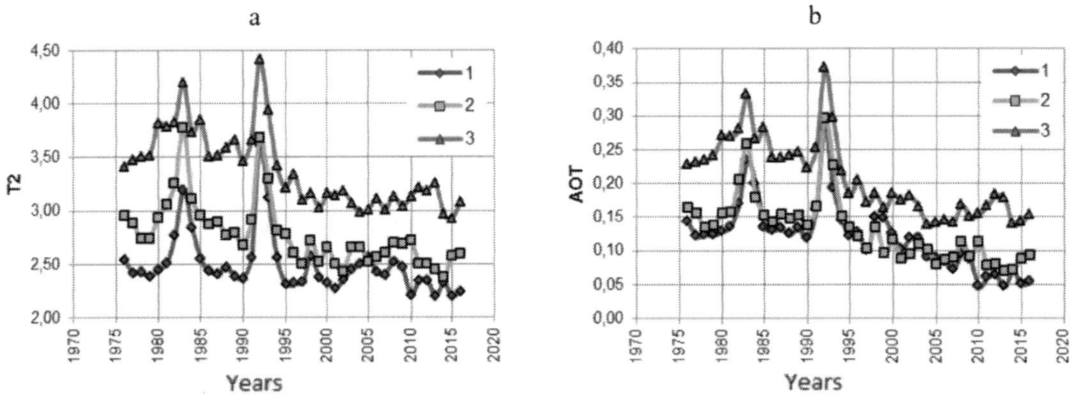

Figure 2. T_2 (a) and AOT (b) courses in the period 1976-2016. 1 – North-East of Asian Russia, 2 – Ural, 3 – South of European Russia.

Table 1. Comparison of average annual T_2 values.

Station	T_2 av 1961-1990	Δ T_2 av 1981-2010	1976-2005	1955-2016	1994-2016
Umba	2.74	–0.08	–0.05	–0.08	–0.27
Valday	2.90	–0.15	–0.09	–0.13	–0.39
Kamennaya Ctep	3.25	–0.15	–0.03	–0.25	–0.46
Tsimlyansk	3.30	–0.13	–0.02	–0.19	–0.50
Verhnee Dubrovo	2.76	–0.03	–0.05	–0.02	–0.11
Khabarovsk	2.91	–0.06	–0.04	–0.10	–0.31

The average values of T_2 and AOT for the regions are represented in figure 3. It clearly demonstrates the diversity of atmosphere aerosol load. The highest transmissivity is typical for the North-East of Asian Russia; the region with the highest turbidity is the South of European Russia. The atmosphere transmissivity remains stably high over recent years.

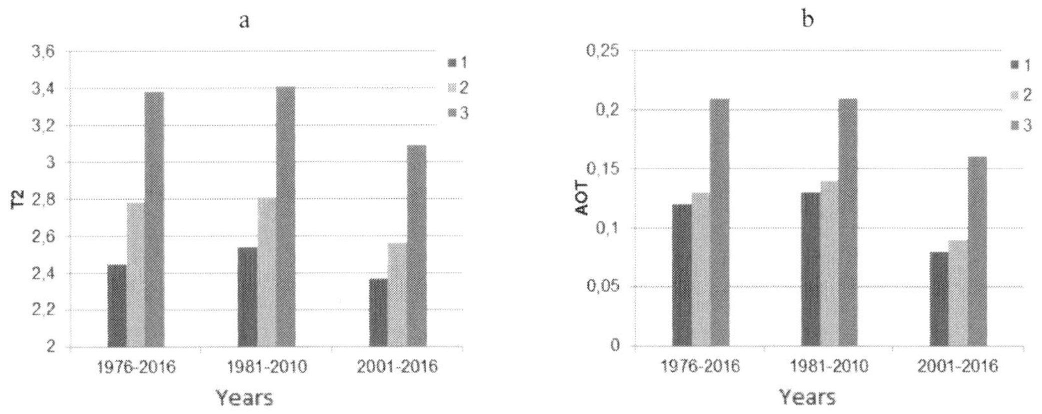

Figure 3. Average long-term T_2 (a) andAOT (b) in different periods.
1 – North-East of Asian Russia, 2 – Ural, 3 – South of European Russia

The evaluation of the annual course of T_2 and AOT shows the decrease of the average monthly values for all the regions; the most drastic reduction is observed in the first part of the year. However, the annual course trends of T_2 remain intact – the extremes refer to the same months. The diminishing of the average monthly values at Asian territory resulted in decrease (in some cases even in vanishing) of spring maximum of AOT (in April) [16].

The linear approximation of series of monthly values of T_2 and AOT allowed to estimate changes of the said characteristics (figure 4).

The period from 1994 till present is characterized by the absence of powerful volcanic eruptions. Only short-term (not more than 2 months) and local atmospheric disturbances have been registered during this period. A noticeable decrease of average long-term annual and monthly values of integral (T_2) and aerosol (AOT) atmosphere turbidity arised in period the of "aerosol transmissivity" (see table 2).

We have analyzed the structure of deviations of monthly turbidity characteristics from its average long-term values. Fractional histograms of T_2 and AOT anomalies frequency (%) have been constructed and analyzed; the step being 0.5 σ. The anomalies for the periods 1976-1993 and 1994-2015 have been considered. The results of the analysis for the two regions are represented in figure 5.

The anomalies of AOT and T_2 that have been registered during the period 1976-1993 stay within the + σ with a probability of 65-80 %. No reversed-polarized anomalies exceeding -2σ have been observed within the period under review. Normal-polarized anomalies exceeding 2 σ and, sometimes, 3 σ are dedicated by perturbations caused by volcanic eruptions. The possibility of occurrence of such high anomalies does not exceed 8%. Maximal frequency of AOT and T_2 usually lies in the range $0.0 \div -0.49\ \sigma$. West Siberia and South of European Russia are exceptions, because maximal frequency for the said regions lies in the range $0.01-0.50\ \sigma$.

The anomalies of monthly AOT and T_2 values within the "aerosol transmissivity" period fit into the range $+\sigma$ (in most cases), but the structure of the anomalies' values distribution is different – histogram shifts towards negative values. The sift is the more significant, the more pronounced is the difference between the average values of the considered periods. There are practically no normal-polarized anomalies exceeding 2σ within the period 1997-2015 while there is a noticeable increase of

IOP Conf. Series: Earth and Environmental Science **107** (2017) 012036 doi:10.1088/1755-1315/107/1/012036

AOT negative deviation frequency in the range $-2\sigma \div -\sigma$. The AOT maximal frequency is -0.99 до -0σ and T_2 - $-0.99 \div -0.50\ \sigma$.

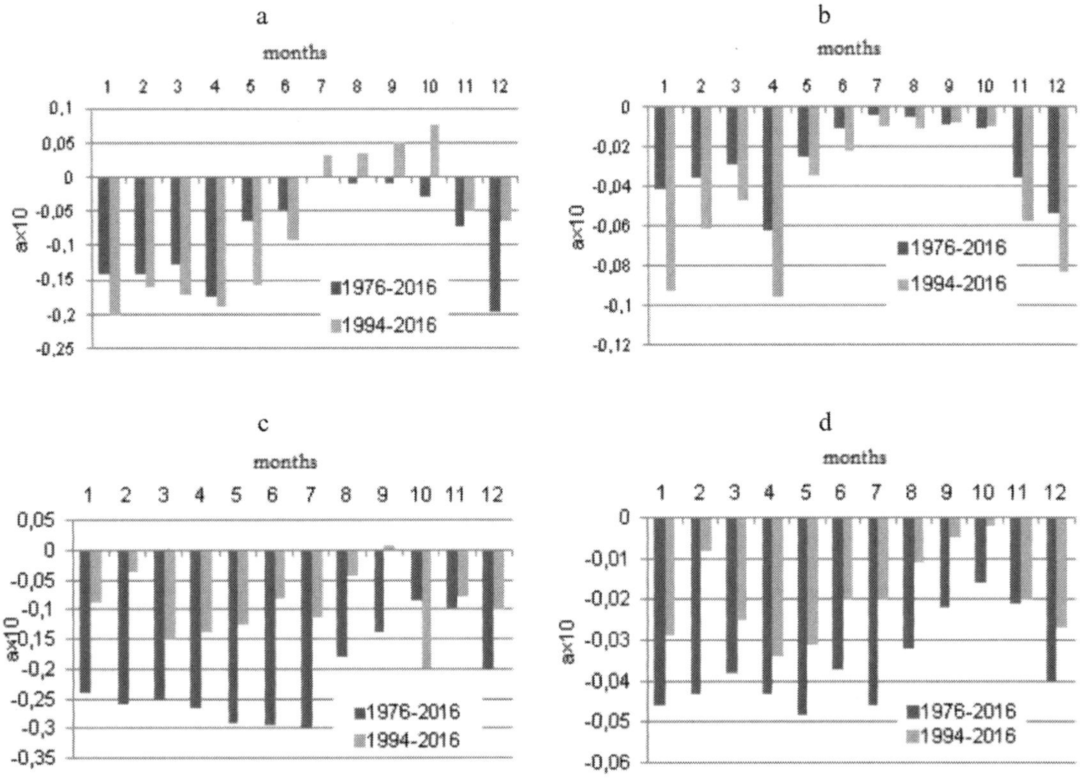

Figure 4. Annual course of coefficient of a linear trend of T_2 (a, c) and AOT (b, d) in 10 years: a, b – North-East of Asian Russia, c, d – South of European Russia.

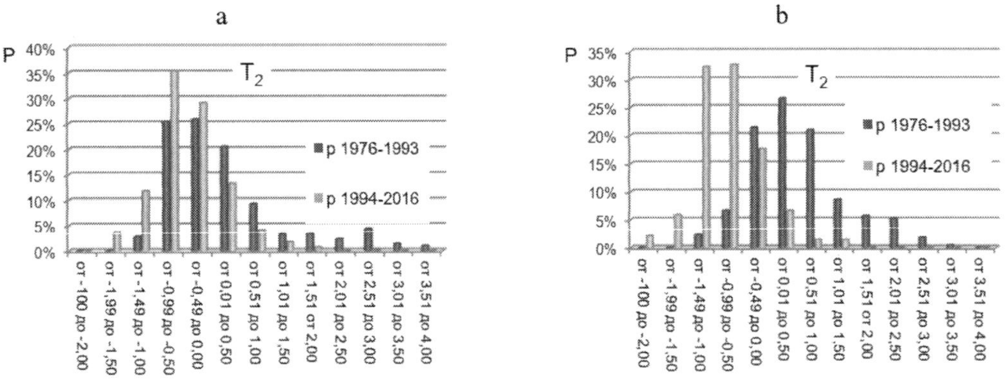

Figure 5. Histograms of T_2 anomalies frequencies (P) for the regions: North-East of Asian Russia (a) and South of European Russia (b).

The annual and monthly values of AOT and T_2 within the territory of West Siberia and the North-East of Asian Russia have less changed than in other regions within the "aerosol transmissivity" period. The histograms of the T_2 anomaly frequency for both considered periods coincide with each other. Histograms for AOT are also similar, though there is a slight shift of the anomalies towards negative values.

The histograms of T_2 and AOT anomalies for the South of European Russia differ qualitatively. The most prominent transmissivity changes (for the period 1976–2015) have happened in this region. Anomalies have maximal frequency in the range –0.99 ÷ –0.50 σ. AOT and T_2 anomalies have probability of 82–85 % and stay within the limit 0 ÷ −1.49σ.

Table 2. Divergence of average annual values of T_2 and AOT.

Region	T_2		ΔT_2	AOT		ΔAOT
	1976–1993	1994–2015		1976–1993	1994–2015	
North of European Russia	2.84	2.56	−10.3	0.17	0.10	−47
Center of European Russia	3.33	2.77	−17.8	–0.21	0.12	−50
South of European Russia	3.71	3.13	−16.6	0.26	0.17	−39
Ural	3.02	2.59	−15.1	0.17	0.10	−47
West Siberia	2.91	2.62	−10.3	0.17	0.11	−40
North-East of Asian Russia	2.63	2.38	−9.8	0.15	0.10	−36
Central Siberia	2.84	2.53	−11.5	0.18	0.11	−47
South of Asian Russia	2.88	2.57	−11.3	0.19	0.12	−41
Far East	3.01	2.68	−11.4	0.19	0.13	−35

Comments: $\Delta T_2 = (T_{2\,1976-1993} - T_{2\,1994-2015}) / T_{2cp}$; ΔAOT = $(AOT_{1976-1993} – AOT_{1994-2015}) / AOT_{cp}$. $\mathbf{T_2}$, AOTav. – mean for the period 1976–2005 .

4. Conclusion

Monitoring of atmosphere transmissivity is based on the data, available by different observing systems: land-based actinometrical network, AERONET and satellite monitoring system. Data available by the above-mentioned systems located in different parts of the Earth are coinciding.

The decreasing tendency of the integral and aerosol turbidity of the atmosphere has been observed within the territory of Russia during last decades; substantial interannual variability of monthly and annual T_2 and AOT values is in evidence. Northern regions of Russia (both Asian and European) are defined by the highest transmissivity of the atmosphere.

Aerosol cleaning of the atmosphere above Russia in the period 1994–2017 is characterized by a stable decrease of atmosphere turbidity. There are some peculiarities in the course of T_2 and AOT anomalies in several regions. Practically no normal-polarized anomalies of T_2 have been registered during the period 1994-2010 in the South of European Russia though recently they have been noticed in the South of Asian Russia. Periods with normal-polarized anomalies are typical for the Far East.

Acknowledgement

This work was supported by RFBR grant № 10-05-01086.

References

[1] Alpert P *et al* 2005 *Geophysical Research Letters* **32** L17802
[2] Wild M 2009 *Journal of Geophysical Research-Atmospheres* **114** D00d16
[3] Wild M 2012 *Bulletin of the American Meteorological Society* **93** (1) 27–37
[4] Kishcha P *et al* 2009 *Journal of Geophysical Research* **114** D00D14. doi:0.1029/2008JD010975.

[5] Khlebnikova E I and Sall I A 2014 *Proceedings of MGO* **570** 34–49
[6] Plakhina I N, Makhotkina E L and Pankratova N V 2009 *Izvestiya, Atmospheric and Oceanic Physics.* **45** (4) 489–500
[7] Plakhina I N, Makhotkina E Land Pankratova N V 2011 *Izvestiya, Atmospheric and Oceanic Physics.* **47** (6) 1–6
[8] Makhotkina E L and Plakhina I N 2014 *Proceedings MGO* **572** 57–87
[9] De Meij A, Pozzer A and Lelieveld J 2012 *Atm. Env.* **51** 75–85
[10] Poddubnyi V A *et al* 2013 *Izvestiya, Atmospheric and Oceanic Physics.* **49** (3) 314–322
[11] Sakerin S M, Andreev S Yu, Bedareva T V and Kabanov P M 2012 *Atmospheric and Oceanic Optics* **25** (6) 484–490
[12] Sakerin S M *et al* 2012 *Izvestiya, Atmospheric and Oceanic Optics* **25** (11) 958–962
[13] Makhotkina E L, Plakhina I N and Makhotkin A N2015 *Proceedings MGO* **579** 162–177
[14] Makhotkina E L, Plakhina I N and Lukin A B 2005 *Russian Meteorology and Hydrology* **1** 28–36
[15] Plakhina I N, Makhotkina E L and Pankratova N. V. 2007 *Russian Meteorology and Hydrology* **2** 19–21

IOP Conf. Series: Earth and Environmental Science **107** (2017) 012037 doi:10.1088/1755-1315/107/1/012037

Variability of the circulation processes in the Lower Volga Region on the background of global climate trends

S V Morozova[1], E A Polyanskaya[1], G F Ivanova[1], N G Levitskaya[2], K E Denisov[3] and N P Molchanova[3]

[1] Saratov State University, Astrakhanskaya Str., 83, Saratov, 410012, Russia
[2] Research Institute of Agriculture of the South-East, Tulikova str., 7, Saratov, 410010, Russia
[3] Saratov State Vavilov Agrarian University, Teatralnaya sq., 1, Saratov, 410012, Russia
E-mail: swetwl@yandex.ru

Abstract. In this article we consider variability of the character of atmospheric circulation during different periods of climatic variability in the Lower Volga region as part of the Atlantic-Eurasian hemisphere sector. Variability of the circulation regime is analyzed with the involvement of the regional typification of synoptic processes by V. L. Arkhangelsky - E. A. Polyanskaya. It is shown that during the phase of the second wave of global warming compared to the period of stabilization in the South-East of the Russian Plain we actually see the decrease of the polarfront cyclones occurrence frequency, the increase of cyclones frequency on the Arctic front and the intrusion of Arctic anticyclones. The revealed character of synoptic processes changes is reflected in the reconstructions of geopotential fields at the middle level of the troposphere. The climatically meaningful conclusion was also made about the stabilizing role of the atmospheric circulation and the formation of the specific features of the climatic variability that does not have analogues in the period of regular meteorological observations.

1. Introduction

Climate changes occurring on the Earth generally and in certain regions are obvious. However, the question about the reasons these changes stays debatable. Today except anthropogenic ones, natural reasons are also indicated, such as change in the flow of solar radiation, volcanic eruptions, phenomena and processes in the ocean [1-4]. Undoubtedly, in short-time intervals (within several decades) climate variability primarily is defined by regime fluctuation of general circulation of atmosphere (GCA). A great number of publications is devoted to studying of the regime of general circulation, curriculum vitae of which is are in IPCC. In earlier researches, carried out by the authors of the current article, the variability of circulating processes in different climatic periods was studied by the nature of the pressure field on the average level of troposphere , since climate ridges and troughs on AT-500 are integral indicators out of all the variety of synoptic processes ongoing above the Earth's surface [5-8].

2. Material and methods.

The most striking indicator of modern climate tendencies are changes in air temperatures. According to the data of (http://www.cru.uea.ac.uk/cru/data/temperature/#datdow), the authors of the web-site made a chart of temporary motion of anomalies of average annual air temperature of the Northern hemisphere (figure 1). In this graph the time intervals clearly state, where average near-surface temperature has the same variability tendency:

1) Little ice age in Europe, shown in the graph by interval since the middle of the 19th century until the end of 1900s.
2) The first wave of global warming was observed since the middle of 1840s until the middle of 1905 until 1940s.
3) The period of stabilization (relative cold snap) that came out in the 1950s-1960s.
4) The second wave of global warming started in the middle of 1970s and continued until the present moment with an essential slowdown.

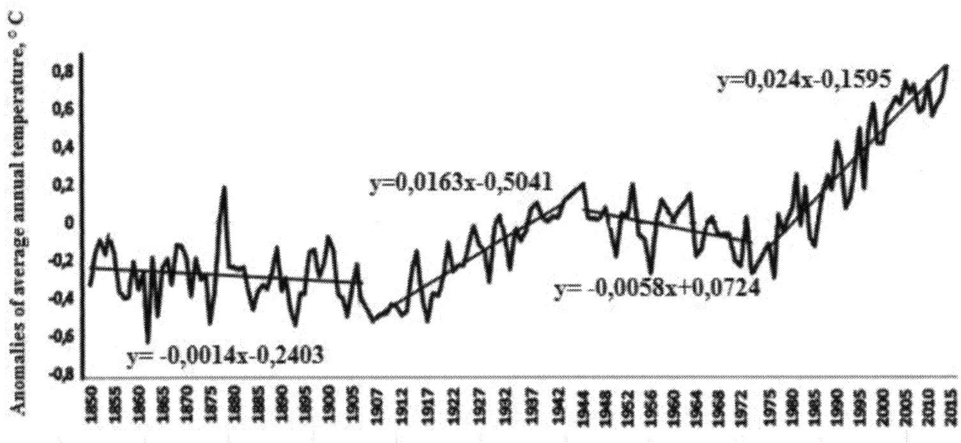

Figure 1. Variability of the average hemispheric air temperature.

Variability of the average hemispheric air temperature in each of the four aforementioned intervals you can see the trend lines. As the graph shows, the lines have different incline that value of linear coefficients α. They differ not only in value but in sign as well. To define if the noticed change is climatically significant, we checked those with the synoptic one. We used the confidence interval methods with applying to Student`s statistics and 95% level of significance by the formula:

$$\bar{x} \pm t\gamma \frac{s}{\sqrt{n}},$$

in which $\bar{x}$ - average value; $t\gamma$ - Student's statistics, γ - Significance level indicator (95 %);. s - mean square; n - Number of members of the series.

Analysis of synoptic processes near the earth's surface was carried out with the help of regional typification of Arkhangel'skiy [10] and Polyanskaya [11], Morozova [12]. According to this typification, 7 types of synoptic processes are distinguished in the Lower Volga region.

I. The cyclonic activity on the Arctic front.
II. The impact of the Arctic anticyclone.
III. The impact of the winter Asian anticyclone.
IV. The impact of the subtropical anticyclone.
V. Flat pressure gradient of high and low pressure.
VI. The cyclonic activity on the Polar front.
VII. Field of deformation.

3. Results and discussion

Using the formula above we calculated the borders of confidence intervals, which are represented in table 1. Except for this, in the table shows several statistical characteristics for each of the time period.

Table 1. The statistical significance estimation of anomalies change of the average annual North hemisphere`s temperature.

Periods, years	Statistical characteristics of the significance of changes.			Confidence Intervals
	α	$\bar{x}$	σ	
1) Small glacial age (1850-1907)	-0.0014	-0.28	0.147	[-0.508; -0.315]
2) First wave (1908 – 1943)	0.0163	-0.19	0.194	[-0.240; -0.130]
3) Stabilization (1944 – 1974)	-0.0058	-0.02	0.124	[-0.078; 0.038]
4) Second wave (1975 – 2016)	0.0240	0.357	0.320	[0.273; 0.439]

According to the average quadratic deviation (σ), characterizing the range of random value concerning its average value, we can see that the greatest variability is typical for the second wave of global warming. It is noteworthy that during the little ice age and period of stabilization, the temperature range was less than in relative warm ones. Maximum variability of the temperature regime in the second wave of global warming compared to all other periods shows the increase in climatic instability.

Estimating the confidence intervals (table 1) and making sure they do not overlap, we can conclude the statistical significance of changes. Since the changes were statistically significant, the null hypothesis about the homogeneity of a series is rejected, and significant climate changes are seen. The conclusion about the statistical significance of climate change allowed the author call natural climatic periods of the earth's climate system (ECS) selected periods of time, although, undoubtedly, natural changes have an anthropogenic component.

Variability in circulation was studied in three locations: Samara, Volgograd, Astrakhan. In each of them the annual frequency of SYNOPTIC processes (average number of days) was calculated within a period 1949 – 1969 years, corresponding to stabilization period and 1998-2007 years which is also called the second wave of global warming (table 2).

You can see from the exact analysis of the represented data (table 2) that the stabilization period in all three areas, with the exception of Samara, the process of cyclonic activity on the polar front prevailed over all others. In Samara repeatable processes, cyclonic activity at polar and Arctic fronts are equally probable.

According to the data in table 2 we can conclude that during the second studied period (the second wave of global warming) compared to the first (stabilization) is the frequency of occurrence of cyclonic activity on the Arctic front, and, as a result, an increasing number of Arctic incursions into the region. At the same time there is reduction of the frequency of occurrence of polarfront cyclones. Also there are much less prominent cases of the influence of subtropical highs in the Lower Volga Region. The number of days with the influence of the Asian winter anticyclone has decreased slightly (from 37 to 33 days per year). Simultaneously with it, low-gradient fields of high and low pressure increases in frequency from the stabilization period to the second wave of global warming. Moreover, these trends are typical for all locations.

Climatically significant effects of the analysis can be considered as a conclusion about the stabilizing role of the General atmospheric circulation. The increase in the invasion region of the Arctic anticyclone contributes to more frequent penetration of the relatively warm areas of cold air masses, which somewhat moderates the growth of near-surface air temperature. The increase in the frequency of low-gradient fields may indicate a greater uncertainty about the future character of development of the atmospheric circulation. Thus, the atmospheric circulation can be regarded as a stabilizing factor, tending to the preservation of climatic and ecological balance on the Earth's climate system. This conclusion does not contradict the role of the General circulation of the atmosphere as climate-forming factor, smoothing temperature contrasts between the Pole and the equator.

IOP Conf. Series: Earth and Environmental Science **107** (2017) 012037 doi:10.1088/1755-1315/107/1/012037

Table 2. The average number of days with synoptic processes in the Lower Volga region (average annual).

Type of process	Periods, years	The average number of days with the process			
		Samara	Volgograd	Astrakhan	Average
I	1949-1969	84.2	56.5	42.1	60.9
	1998-2007	95.4	63.5	43.6	67.4
II	1949-1969	73.2	61.9	58.8	64.6
	1998-2007	105.6	103.0	100.0	102.9
III	1949-1969	40.9	33.8	35.6	36.8
	1998-2007	36.2	30.5	32.1	32.9
IV	1949-1969	38.0	56.5	62.4	52.9
	1998-2007	21.0	32.5	35.9	29.8
V	1949-1969	23.7	25.9	30.6	26.7
	1998-2007	51.4	64.4	87.3	67.7
VI	1949-1969	84.5	111.4	95.4	97.1
	1998-2007	34.3	51.1	42.5	42.6
VII	1949-1969	17.8	19.2	28.8	21.9
	1998-2007	21.0	20.2	23.2	21.5

We can note that synoptic processes in different seasons of the year determine different weather conditions. For instance, the cyclonic activity softens weather conditions: in the winter- causing relative warmth, in the summer – reducing hot temperature background. Arctic anticyclones in the winter determine the dramatic intense cold waves in the summer. Under their influence a particularly severe drought was formed. Therefore, the increase of the frequency of occurrence of the invasion of Arctic anticyclones is affected by another characteristic of the climate degree of continentalism. We note that one of the features of warming is to decrease the degree of continentality due to the increase in winter temperatures. The invasion of Arctic cores increases the seasonal temperature differences, increasing annual amplitude of temperature and, consequently, the degree of continentality of the climate. This aspect of the participation of the General circulation of the atmosphere in climate-forming processes can also be considered as a mechanism preventing irreversible processes in the earth's climate system, which can occur under the influence of anthropogenic factors, primarily caused by the emission of carbon dioxide.

We considered the repeatability of synoptic processes in a differentiated way, in summer and winter (table 3).

Analyzing the seasonal differences of the synoptic processes occurrence frequency in the region (table 3), we note that in winter and summer from the first to the second period the number of impacts on a region of cyclonic activity on the Arctic front increases. It is interesting, that the increase of the invasion of the Arctic cores frequency in the second analyzed period is typical only for summer season. In winter, the frequency of their incursions in the phase of the second wave of global warming is a bit less than in the period of stabilization. The number of cases of the subtropical ridges impact becomes less in winter and summer, also during both seasons the frequency of cyclones, developed in the polar front is reduced. Flat pressure gradient and deformation fields from the first to the second period become larger in winter and summer.

The most interesting and important for the physical theory of climate factor is the summer increase in the frequency of invasion of the Arctic anticyclone. As already mentioned, the integral characteristic of earth-surface processes is the structure of the pressure field of the middle troposphere. In the figure presented the pressure field on the middle level of the troposphere over the same study periods for the Central month of the summer season is represented. It is constructed from data on the value of the geopotential at the nodes of a regular grid increments in latitude and longitude 5°. Data visualization was performed using MAPINFO software complex on the basis of GIS-technologies.

Table 3. The average number of days with synoptic processes in the Lower Volga Region.

Type of process	Periods, years	The average number of days with process					
		Samara		Volgograd		Astrakhan	
		winter	summer	winter	summer	winter	summer
I	1949-1969	25.1	14.4	19.3	7.6	15.3	5.0
	1998-2007	30.5	18.6	19.9	14.2	13.8	9.2
II	1949-1969	15.9	15.3	14.4	11.2	13.5	10.3
	1998-2007	12.3	30.0	12.9	27.3	12.6	26.2
III	1949-1969	22.3	-	17.5	-	18.8	-
	1998-2007	22.2	-	18.7	-	19.7	-
IV	1949-1969	2.7	16.9	5.0	23.5	5.9	26.2
	1998-2007	0.6	10.7	3.1	14.1	2.8	17.8
V	1949-1969	4.4	11.4	4.8	11.4	5.0	13.1
	1998-2007	10.6	20.6	13.9	22.9	21.0	28.7
VI	1949-1969	14.8	29.5	24.8	34.2	21.8	30.0
	1998-2007	8.6	9.0	12.4	11.5	11.5	7.5
VII	1949-1969	1.8	1.2	4.5	3.9	3.5	1.5
	1998-2007	5.3	4.8	4.7	3.9	4.9	4.9

As can be seen from the image, the pressure fields on the middle level of the troposphere differ significantly among themselves in two studied time intervals. Pressure field on the space of the first natural synoptic region (I n.s.r.) in the stabilization period, presented by ample narrows, especially distinct over the North Atlantic and the Black sea (figure 2a), the existence of which identify areas of negative values of the geopotential anomalies over these regions (figure 2b). In the same period, anticyclonic axis of the clamping forces, directed from the region of the Azores to the Arctic is well observed. It determines the background of high pressure over Western Europe (except the Northern territory), the European part of Russia and Western Siberia. Over Central Asia an extensive area of geopotential heights negative deviations stands out. The centerline (572 hPa) of the planetary high-altitude frontal zones (PHFZ) in this period is about 50 °N. The quasilatitudinal location PHFZ indicates poorly developed meridionality of the space I n.s.r. in that period. Note that in the period of stabilization the Lower Volga region was located in front of the high-altitude pressure narrow. The front part of the narrow is the area of active cyclogenesis. High frequency of the polarfront cyclones prevented the development of long and extensive drought.

In the phase of slow warming 1998-2007 (figure 2c) there were noticeable changes in the structure of the isohips field. It turns out to be quite perturbed with chaotic arrangement of ridges and narrows, reflecting the unstable nature of atmospheric processes. Over the regions small pressure waves are detected. Small-scale ridges are observed over the North Sea and the West of Scandinavia, and also over Ukraine, the Middle and Lower Volga region. Over Central Europe there is a small high-altitude narrow. The centerline PHFZ over the Atlantic and Western Europe decreased to 45 degrees latitude, and above the East moved 3° to the North. This curvature of the centerline PHFS is testified of increased meridionality of air currents.

Analysis of geopotential anomaly fields (figure 2d) shows that the area of low values of the geopotential (corresponding to the Icelandic low) has moved south and has occupied the whole investigated part of the Atlantic, with the result that the Azores ridge was over the Mediterranean Sea and united with the anticyclone located in the South-East ETR. Probably, this anticyclone emerged as a result of the interaction of the Azores and Arctic centers of atmosphere action. Thus, the center of positive geopotential deviations emerged over the Middle and Lower Volga, which means frequent penetration of the Arctic cores in this period in the South-East ETR that contributed to the formation of blocks and severe droughts. According to [2], during the second wave of global warming the frequency of droughts does not increase, but their intensity and geographical coverage.

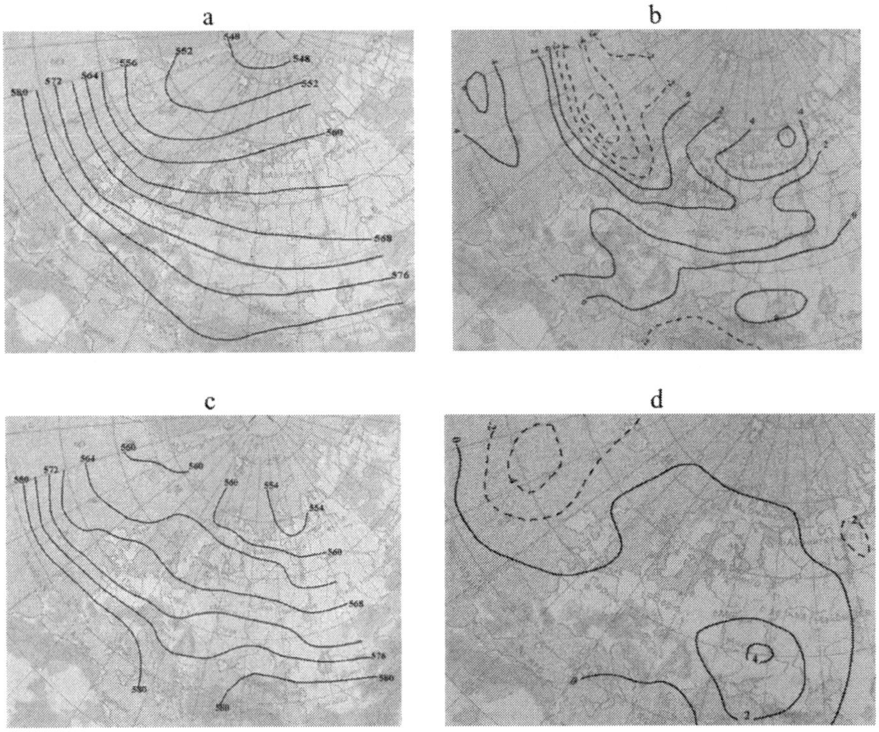

Figure 2. Average field isohips the surface of 500 hPa, July:a) 1949 – 1969, b) 1998 – 2007. Average field isanormal the geopotential surface 500 hPa, July: c) 1949 – 1969, d) 1998 – 2007.

Thus, over the Middle and Lower Volga Region was the center of positive deviations of the geopotentiality, which means frequent penetration in this period of time in the South-East ETR Arctic cores that contributed to the formation of blocking and severe droughts. According to [2] during the second wave of global warming increases the frequency of occurrence of droughts and their intensity and geographical coverage.

Thus, we can conclude that in the future with a possible increase in average global temperature, the contribution of the temperatures of summer seasons caused by increase in temperature due to anticyclone conditions, needs to become more meaningful. If in the second wave of the global warming temperature increase was determined mainly by increasing temperatures in the cold part of the year, in the future a significant contribution to this increase will be brought by summer seasons as well due to the increased meridionality of processes, and, as a consequence, the formation of positive temperature anomalies. Based on the data represented above, we can point out some trends of regional climate change, namely the increase in the frequency of dry phenomena in summer in the Volga region.

From the period of stabilization to a phase of slow warming in the Lower Volga region we see an increase in the number of days with the processes of cyclonic activity on the Arctic front and as a consequence an increase in the number of intrusions in the Lower Volga region of the Arctic cores. Subtropical anticyclones began to affect the Lower Volga region less also polar front activities decreased over the region. In this case above the Lower Volga region we observed flat pressure gradient more, generally, of high pressure. The preservation of the identified trends may cause a

IOP Conf. Series: Earth and Environmental Science **107** (2017) 012037 doi:10.1088/1755-1315/107/1/012037

further increase in surface temperature without changing the degree the climate continentality. This feature has no analogs in the era of instrumental meteorological observations.

References

[1] IPCC 2013 *Climate Change 2013: The Physical Science Basis. Contribution of Working Group I to the Fifth Assessment Report of the Intergovernmental Panel on Climate* (Cambridge, UK and NY, USA: Cambridge University Press) 1535

[2] 2014 *Roshydromet second assessment report on climate change and its consequences in the Russian Federation* (Moscow: The Federal Service for Hydrometeorology and Environmental Monitoring of Russia) 1001

[3] Perevedentsev Yu P 2009 *Climate theory* (Kazan: Kazan State University Press) 504

[4] Sherstyukov BG 2011 *Changes, variability and fluctuations in the climate* (Obninsk Publishing house Press) 294

[5] Morozova S V 2011 *Circulation of the atmosphere as a factor in the variability of the regional climate. Global and regional climate change* (Kiev: Nika-Center Press) 96-101

[6] Morozova S V 2012 *Bulletin of Saratov University. New series. Series: Earth Sciences* **12** (2) 34-37

[7] Morozova S V 2015 *Proceedings of Hydrometcentre of Russia* **358** 88-98

[8] Morozova S V and Polyanskaya E A 2017 *Klimate&Nature* **1** 33-48

[9] Gmurman V E 2003 *Probability Theory and Mathematical Statistics* (Moscow: Moscow Higher School Press) 480

[10] Arkhangel'skiy V L 1968 *Regional Synoptics of the Lower Volga Region* (Saratov: Saratov State Press) 208

[11] Polyanskaya E A 1986 *Synoptic processes and weather phenomena in the Lower Volga region* (Saratov: Saratov State University Press) 208

[12] Morozova S V 2013 *The research of synoptic processes by the method of "ethalon"* (Saratov: Saratovn State University Press) 164

Hydrobiological investigations of Kytalyk Wildlife Reserve polygonal ponds (North-Eastern Yakutia)

G Nigamatzyanova[1], L Frolova[1], L Pestryakova[2]

[1] Laboratory of Paleoclimatology, Paleoecology, Paleomagnetism, Institute of Geology and Petroleum Technologies, Kazan Federal University, Kremlevskaya str, 4/5, Kazan, 420111, Russia
[2] Institute of Natural Sciences, North-Eastern Federal University in Yakutsk, Belinsky str, 58, Yakutsk, 677027, Russia
E-mail: gulnaraniga@mail.ru

Abstract. In the following article there are introduced the first researching results of 27 water bodies of polygonal tundra in Kytalyk Wildlife Reserve in the summer 2011. The evaluation of physic-hydrochemical indexes of water bodies is given. The basic structure-forming characteristics of zooplankton communities are analyzed. The ecological state of the lakes is estimated.

1. Introduction

Researching of plankton communities of polygonal ponds typical for permafrost regions [1] is one important component of the modern state of cryolithozone lakes evaluation. Species composition and level of quantitative development of aquatic invertebrates are the high-sensitive indicators for the water pollution extent and the impact of climate change on northern ecosystems [2, 3]. Changes of the organisms' existence conditions are reflected in the species composition, quantitative indicators, the ratio of individual taxonomic groups, and the structure of the zooplankton organisms' population. Kytalyk resource wildlife reserve which is located in the basin of the river Chroma of the Yano-Indigirka interfluve (East Siberian Sea coast) was created with the support of the World Wildlife Fund (WWF) with an issue to protect the east population of Siberian Crane, as well as to conserve other rare and economically important animals and plants. Furthermore, the Paleolithic sites of primitive man and repositories of mammoth fauna are located on the territory of natural sanctuary [4, 5] (figure 1). The landscape of the reserve is diverse: arctic tundra and complex mires, larch-tundra woodlands and mountain tundra. Polygonal ponds, which are the habitat for numerous bird species, some of which are in the Red List of IUCN, Russia and the Republic of Sakha (Yakutia) [5], and containing unique tundra plankton communities, play a significant role in the reserve.

The altitude above the sea level of the investigated region varies between 0.4-71 m. The geographic coordinates of the wildlife reserve are 70°30'-72°35' N. and 143°00'-152°30' E. Square footage of the area is 1608,000 hectares [7].

2. Material and methods

Data for the current study was collected in the summer of 2011 in the frames of a joint Russian-German expedition to the lower course of the Indigirka River. The most part of study area is covered with Quaternary permafrost deposits. The landscape consists of a flood plain and terraces of the Berelekh River, its branches and tributaries, a yedoma hills of the Middle and Late Pleistocene ice complex, alas and thermokarst lakes formed in Lateglacial and Holocene. Flora is represented by tussock-sedge, dwarf shrub, moss tundra. According to meteorological data, the climate is characterized as sharply continental, the average air temperatures in July and January are +9.7 °C and -36.6 °C, respectively. The temperature of the frozen soil is -4 - -6 °C with a permafrost thickness of 200-300 mm [8].

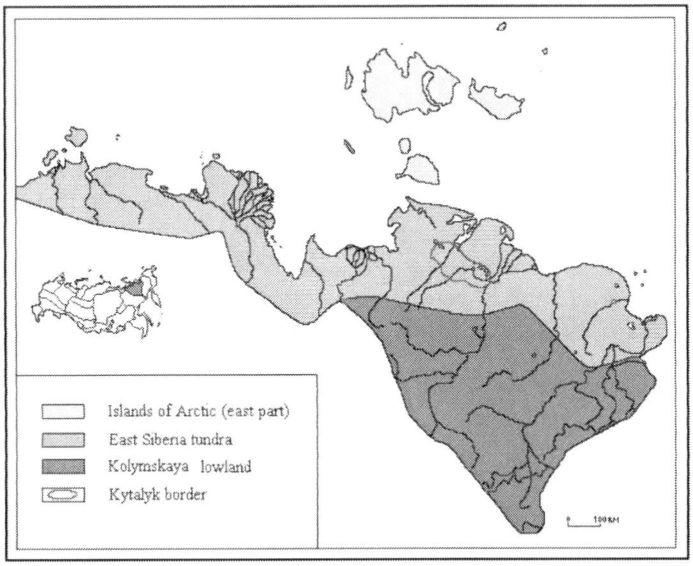

Figure 1. Map for Kytalyk Site in Republic Sakha (Yakutia) [6].

The study of modern polygonal reservoirs in certain key areas, covering different development stages, can serve as an informative natural model, making possible to trace the character and impact depth of the main ecological factors on the biota. Herewith the structural elements of the polygons provide a wide range of hydrological conditions from the water habitats (at the center of the polygons) to xeromorphic conditions (above the ridges). Selection of polygonal ponds and lakes for hydrological research was determined by the factor of formation and development stage, i.e. we chose the water reservoirs with a strongly marked polygonal shape, lakes, which are formed in result of confluence of two or more small lakes of the neighboring polygons or polygonal lakes in the overgrowing stage. During of the research for the first time samples of water and samples of zooplankton from 27 different water bodies of polygonal tundra were collected. Also, physical and hydrochemical studies of selected reservoirs were done, during which such indicators like water depth, temperature of water and air, pH, specific electrical conductivity of water, content of ions NH_4^+ and NO_3^-, phosphates, dissolved oxygen in water were determined. Sampling of zooplankton was conducted in the littoral part of water reservoirs with the help of a small conical network of Apstein (inlet diameter 25cm, size of the mesh is 100 μm). Fixation of organisms was done with 4% formaldehyde solution. In total during the research 36 samples of zooplankton were selected.

Main aims of current research of zooplankton communities were the following: analysis of zooplankton communities' species composition, comparison the qualitative characteristics of zooplankton communities in the water reservoirs, analysis of the zoogeographic structure of zooplankton, eco-faunistic description of dominant zooplankton species, analysis of morphological, hydrochemical and hydrobiological features of the researched lakes, evaluation of water quality. To achieve these aims we calculated the following indexes and indicators: Pantle-Buck's saprobic index in modification of Sládeĉek [9], based on regard of complex of physiological properties of the organism, conditioning its' ability to live in water with one or another content of organic substances; index of saprobity by Zelinka and Marvan, taking into account the saprobic valency of each indicator species [10]. Statistical processing of data was done using the package of the program of the integrated system STATISTICA 6.0. A non-normal distribution of values in the samples was determined using the Shapiro-Wilk's criterion and the Kolmogorov-Smirnov's criterion with Lillieforce's correction. The Spearman correlation analysis was made in the Nonparametric distribution module of Rotifera species depending on the physical and hydrochemical parameters of the water reservoirs, taking into account the correlation coefficient at significance level $p \leq 0.05$.

3. Results and discussion

According the results of the water samples, the studied polygonal ponds are distinguished by low mineralization (to 100 mg·l^{-1}), the pH values change from acidic to neutral, the average values were 6.3±0.08 (table 1), content of biogenic and organic substances were low. The studied water reservoirs of the polygonal tundra are small, the depth of the polygonal ponds ranges from 0.2 to 2.0 m, with an average depth of 45.1 cm, often overgrown. In most reservoirs, little or moderately decomposed plant detritus with a small amount of sandy silt was found. The average air temperature for the period of research was +11.9 °C, the water temperature - +11.5 °C. The average value of electrical conductivity was 35.2 μS·cm^{-1} that indicates a low degree of water mineralization of the studied reservoirs. Content of the dissolved oxygen in the reservoirs is sufficient; the average value was 8.7 mg·l^{-1}.

Table 1. Results of physical and hydrochemical analysis of water polygonal ponds.

	Min	M	Max	M±m
Depth (cm)	15.00	45.08	150.00	45.08±4.08
Tair (°C)	5.50	11.96	25.00	11.96±1.05
Twater (°C)	5.50	11.50	23.50	11.50±0.90
Conductivity (μS·m^{-1})	19.00	35.20	153.00	35.20±5.14
Mineralization (mg·l^{-1})	12.35	22.85	99.45	22.85±3.34
pH	5.50	6.30	7.10	6.30±0.08
NH^{4+} (mg·l^{-1})	0.002	0.03	0.13	0.03±0.01
NO_3^- (mg·l^{-1})	0.04	0.41	0.79	0.41±0.04
PO_4^{3-} (mg·l^{-1})	-0.01	0.03	0.29	0.03±0.01
Dissolved oxygen (mg·l^{-1})	5.40	8.68	11.80	8.68±0.31
Alkalinity (mmol·l^{-1})	0.20	0.51	1.60	0.51±0.06
Acidity (mmol·l^{-1})	0.40	0.42	0.60	0.42±0.01
Water hardness (°dH)	1.50	3.97	7.50	3.97±0.28

In zooplankton of water bodies of the wildlife reserve there were determined 71 species, belonging to 26 families. The main part of the list of species was presented by Rotifera (70.4%), crustaceans accounted for 19.7% and 9.9% of the copepods (Copepoda, type Arthropoda, Crustacea) and cladocerans (Cladocera, class Branchiopod, Crustacea) respectively.

Among Rotifera we found 50 species from 17 families. The most widely represented families were Brachionidae (8 species) and Trichocercidae (7 species), as well as Testudinellidae, Notommatidae, Lecanidae and Euchlanidae (4 species in each). Cosmopolitan and eurytropic *Rotaria rotatoria* (Pallas, 1766) dominated in the samples and was found in 91.7% of all samples. The cosmopolitan *Trichothria tetractis* (Ehrenberg, 1830) (occurrence frequency 80.6%) and *Lecane crenata* (Harring, 1913) (75%) became subdominants in our samples.

The copepods numbered 14 species, including 7 species of Cyclopoida, 5 species of Copepoda and 2 species of the family Camptocamptidae (order Harpacticoida). Order Cyclopoida was represented by 4 species of the subfamily Cyclopinae, 3 species of the subfamily Eucyclopinae, among which there were both cosmopolitan species and species specific to northern regions. Among order Copepoda 4 species of the family Diaptomidae, having mainly Palearctic area distribution, and *Heterocope appendiculata* (Sars, 1863) of the family Temoridae, were recognized. Copepodite development stages of the copepods were recorded in 91.7% of all samples, in 75% of the samples nauplius stages were registered.

Cladocera were represented in the zooplankton community by 7 species belonging to 5 families. By two of these species were attributed to the families Bosminidae and Chydoridae, each of the families Daphniidae, Polyphemidae and Euricercidae had one representative in the community. *Chydorus* cf *sphaericus* had a frequency index 91.7%, Holarctic *Daphnia pulex* (Leydig, 1860) - 77.8%, cosmopolitan *Alona guttata* (Sars, 1862) - 66.7%.

The species diversity of zooplankton in polygonal ponds varied in wide range. The lowest diversity of zooplankton was observed in water bodies marked KYT-15, KYT-16, KYT-18 and KYT-20, 5-6 representatives of zooplankton were recorded that may be explained by the small size of water bodies. Within communities of these small lakes, Rotifera and Cladocera were found in almost equal quantities and only one Copepoda species per lake, except for the reservoir marked KYT-8, where Copepoda species were not found.

The following lakes were characterized by relatively rich species diversity: KYT-2 - 22 species, KYT-7 - 19 species, KYT-8 and KYT-25 - 17 zooplankton species. Within the community of these reservoirs, only Rotifera (14-16 species) were among dominants, Cladocera were represented by 3-4 species. Copepoda were absent, except for a reservoir KYT-2, in which two representatives of copepods were identified.

The lake with the marking KYT-1 was notable for a maximal high diversity - 39 species in the community, which could be most likely explained with the presence of a large number of samples (10 samples), selected during monitoring from this reservoir, in contrast to other lakes of wildlife reserve, in each of them the samples were selected just once. In the zooplankton species community KYT-1 rotifers were dominated - 71.8% of all species. Cladocera and Copepoda accounted for 17.9% and 10.3% respectively. Species *Trichocerca cylindrica* (Imhof, 1891), *L. crenata* Rotifera, *Ch.* cf *sphaericus*, *Al. guttata* Cladocera, as well as immature development stages of Copepoda species had 100% occurrence in the samples of this water reservoir. There were not revealed changes in the zooplankton species composition during the monitoring.

According to the Pantle-Buck's index of saprobity in the Sladechek modification, most of the studied reservoirs refer to the β-mesosaprobic zone, 2.03 ± 0.4. Some water bodies belong to the oligosaprobic zone. The index value for the lake with the KYT-1 marking was 1.6 that also indicates the belonging to the β-mesosaprobic zone.

According to the index of saprobity, calculated by the method of Zelinka and Marvan, oligosaprobic reservoirs with a deviation to the β-mesosaprobic zone predominate in the investigated region. The value of this index for the KYT-1 water reservoir also indicates an oligosaprobic zone with a deviation into the β-mesosaprobic zone.

In consonance with zoogeographical zonation, the main amount of species of limnetic fauna was composed of organisms, which have a cosmopolitan distribution (68% of the total species composition). Such species like *R. rotatoria*, *Ch.* cf *sphaericus*, which are massive for most of the studied reservoirs, have a worldwide distribution, but among the often-recognized species in the samples there were also cold-water species, which have a limited northern distribution area. Thus, *D. pulex* is characterized as a Holarctic species, preferring temporary and small permanent water bodies. The proportion of species with the Holarctic and Palearctic distribution area accounted for 17% and 15% of the total species composition, respectively.

By biotopic preference the eurytopic organisms are dominated (36 species), other groups: littoral, planktonic and benthic - are presented in lower amounts.

According the analysis, there is a statistical significance between such parameters as species number of Rotifera and the content of NH_4^+ in lake water ($p = 0.02$). Therefore, the influence of the ammonium index in water on the distribution of species of rotifers in water bodies was identified.

In the studied water reservoirs of polygonal tundra, the indexes of species diversity and of zooplankton communities' abundance were high enough that is specific for tundra zone generally [11, 12]. Complex of dominant species is typical for small polygonal water bodies of Northern lakes ecosystems. Comparison of species composition of the tundra lakes' zooplankton, given in the literature for this region, namely for north-western part of Sakha Republic (Yakutia), and also in our results for Kytalyk wildlife reserve, showed the zonal similarity of zooplankton indicators [1]. Its composition and distribution within the water area of the researched reservoirs were driven by the physic-morphological features of the latter, namely, the shallow depths and development of coastal vegetation.

4. Conclusion

The water bodies of the polygonal tundra of the Kytalyk resource wildlife reserve were studied for the first time in the region of the lower course of the Indigirka River, physical and hydrochemical water indicators of lakes and zooplankton communities were analyzed. The water reservoirs are mostly overgrown and are characterized by small size and relatively small depths, the presence of plant detritus in the decomposition stage, low mineralization and low values of organic and biogenic substances, sufficient oxygen solution in the water and pH values close to neutral. A relatively rich species diversity of the zooplankton community with a predominance of Rotifera species was estimated. The species *Rotaria rotatoria* (Pallas, 1766), *Chydorus* cf *sphaericus*, *Trichothria tetractis* (Ehrenberg, 1830), *Daphnia pulex* (Leydig, 1860) and *Alona guttata* (Sars, 1862) belong to the structure-forming complex of zooplankton. By the taxonomic composition, a rotatory-cladocera complex is developed in the researched lakes. Species, which have a cosmopolitan distribution area by zoogeographic characteristics, and organisms, eurytropic by the biotope preferences, are dominated in the community. The influence of ammonium content on the number and distribution of Rotifera species in water bodies was recognized. Due to the results of the calculated indexes, on the basis of data of the zooplankton community, the studied water reservoirs of the polygonal tundra can be characterized as pure and moderately polluted, with low values of saprobity.

Acknowledgments
This work was funded by the subsidy allocated to Kazan Federal University for the state assignment in the sphere of scientific activities, by grants RFBR (research projects № 15-05-0444), and also based on program of competitiveness improvement of the Kazan Federal University.

References

[1] Nigamatzyanova G and Frolova L 2017 *17th international multidisciplinary scientific geoconference SGEM* **17** 891-899
[2] Frolova L A, Nazarova L B, Pestryakova L A and Herzschuh U 2013 *Siberian ecological j.* **1** 3-15
[3] Frolova L A 2016 *16th International Multidisciplinary Scientific GeoConference SGEM 2016 Proceedings, Energy and Clean Technologies* **4 (2)** 601-607
[4] Potapova N A, Nazyrova R I, Zabelina N M, Isaeva-Petrova L S, Korotkov V N and Ochagov D M 2006 *Svodnyy spisok osobo okhranyaemykh prirodnykh territoriy Rossiyskoy Federatsii (spravochnik)* Ch II (M: VNII prirody) 364
[5] 2011 *Atlas biologicheskogo raznoobraziya morey i poberezhiy rossiyskoy Arktiki* (M: WWF Rossii) 64
[6] 2005 Siberian crane wetland project Kytalyk Wetlands URL: www.scwp.info
[7] 2000 *Wetlands of Russia* ed. V G Krivenko (M: Wetlands International Global Series) **3** 490
[8] Tumskoy V and Schirrmeister L 2012 *Joint Russian-German Polygon Project East Siberia*

2011 - 2014: The expedition Kytalyk 2011 ed. L Schirrmeister (Bremerhaven: Alfred-Wegener-Institut für Polar und Meeresforschung) 5-10

[9] Sládeĉek V 1973 *Arch. Hydrobiol. Ergebnisse der Limnologie* **7** 218

[10] Zelinka M 1961 *Arch. Hydrobiol* **57 (3)** 71-81

[11] Kononova O N, Dubovskaya O P and Fefiliva E B 2014 *Journal of Siberian Federal University* **3 (7)** 303-327

[12] Rautio M, Dufresne F, Laurion I, Bonilla S, Vincent W F and Cristoffersen K 2011 *Ecoscience* **18 (3)** 204-222

Relationship between glacier melting and atmospheric circulation in the southeast Siberia

O P Osipova[1], **E Y Osipov**[2]

[1] V B Sochava Institute of Geography SB RAS, Ulan-Batorskaya st., 1, Irkutsk, 664033, Russia
[2] Limnological Institute SB RAS, Ulan-Batorskaya st., 3, Irkutsk, 664033, Russia
E-mail: olga@irigs.irk.ru, eduard@lin.irk.ru

Abstract. The interaction between climate and cryosphere is a key issue in recent years. Changes in surface mass balance of mountain glaciers closely correspond to differential changes in atmospheric circulation. Mountain glaciers in southeast Siberia located on East Sayan, Baikalsky and Kodar ridges have been continuously shrinking since the end of the Little Ice Age. In this study we used daily synoptic weather maps (Irkutsk Center of Hydrometeorology and Environmental Monitoring), 500 hPa, 700 hPa and 850 hPa geopotential height and air temperature data of NCEP/NCAR reanalysis to assess relationships between atmospheric circulation patterns and the sum of positive temperature (SPT), a predictor of summer ice/snow ablation. Results show that increased SPT (ablation) is generally associated with anticyclones and anticyclonic pressure fields (with cloudless weather conditions) and warm atmospheric fronts. Decreased SPT (ablation) is strongly correlated with cyclones and cyclonic type pressure fields, cold atmospheric fronts and air advections. Significant correlations have been found between ablation and cyclonic/anticyclonic activity. Revealed decreasing trends in the SPT in three glaciarized ridges at the beginning of the 21st century led to changes of air temperature and snow/ice melt climates.

1. Introduction

Atmospheric circulation is a basic climate factor and the problem of interaction between atmosphere and cryosphere at global and regional levels is still up to date. Mountain glaciers are good indicators of such interaction at global, regional and local scales. It should be noted that response of inland glaciers to climate change is studied insufficiently. Typical examples of such glaciers are mountain glaciers of the south of Eastern Siberia, located in high-mountain environments of East Sayan, Baikalsky and Kodat ridges [1]. The atmospheric circulation patterns controlling energy fluxes at glacier surfaces operate under combination of such factors as difficult orography and remoteness from oceanic atmosphere centers of action. The studies of relationships between the glacier mass-balance and atmospheric circulation were carried out in some glaciarized regions, such as Arctic, North Pacific and Atlantic [2-7]. For Siberian glaciers such studies are still lacking [8, 9]. In this study we assessed the relationship between summer processes of regional atmospheric circulation (in lower and middle troposphere) and glacier ablation in southeast Siberia ridges over the period of 2001−13. We analyzed a set of regional synoptic whether maps and identified combinations of atmospheric pressure field characteristics associated with high and low glacier ablation (melting). We studied three glaciarized mountain ridges located in south-eastern Siberia: the East Sayan Range, the Baikalsky ridge and the Kodar Range (figure 1). The East Sayan extends from north-west to south-east over ~ 700 km (the highest peak is Munku-Sardyk, 3491 m a.s.l.). The Baikalsky is submeridional ridge extending along the western coast of Lake Baikal (the highest peak is Chersky Mountain, 2588 m). The Kodar is the northern sublatitudinal ridge (the highest peak is Bam, 3072 m a.s.l.), bordered by the Vitim and Olyokma tributaries of the Lena River.

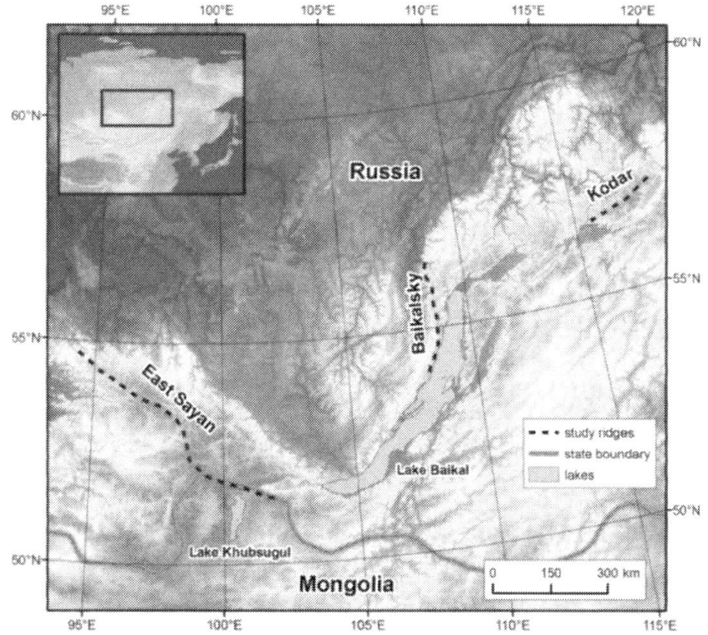

Figure 1. Studied glaciarized mountain ridges (East Sayan, Baikalsky and Kodar). SRTM model is used as a background. Upper left inset shows location of the studied region in Asia.

2. Material and methods

We used daily synoptic maps (for mean sea level, 850, 700, and 500 hPa) of the Irkutsk Center for Hydrometeorology and Environmental Monitoring for June–August 2001–13 to perform synoptic analysis of the regional atmospheric circulation patterns over our study area. The map data included such information as mean sea level pressure field, baric-topography maps (for levels 850, 700, and 500 hPa), air masses, atmospheric fronts, cyclones, anticyclones, and other forms of baric topography. The vertical extents of cyclones and anticyclones were estimated from the 500 hPa height field which corresponds to the middle troposphere level (about 5.6 km). The 700 hPa geopotential height (about 3 km a.s.l.) is close to the glacial zones of the study ridges and leading atmospheric air flows, as well as main pressure patterns were identified mainly based on this level. The atmospheric fronts, different types of advection, and the presence of moist air masses were determined on the weather maps of 850 hPa geopotential height.

Due to lack of mass-balance measurements on the glaciers in the study area we used the sum of positive temperatures (SPT) as a proxy of potential glacier ablation (melting) rate [10]. Daily temperatures over the period of 2001–13 were calculated between two atmospheric levels (700 and 850 hPa) using previously estimated lapse rates. We used the daily NCEP/NCAR reanalysis data as temperature and geopotential height records [11]. The SPT during the period of 2001–13 were calculated for mean elevation of glacier population for each ridge (see below).

3. Results

3.1. Sum of positive temperatures.
Statistical characteristics of the sum of positive temperatures (SPT) calculated in three studied glaciarized regions for the period 2001–13 are shown in table 1. The SPT show negative trend during the study period in all three regions with the trend for the Baikalsky Range is more pronounced (statistically significant). The lowest mean SPT is for Kodar Range and the highest on the Baikalsky Range, mainly, due to both mean glacier elevations (glaciers of the Baikalsky Range are the lowest) and with their latitude (glaciers of Kodar Range are the most northern). The mean duration of the SPT period ranges from 99 (Kodar) to 126 days (Eastern Sayan). Using the SPT thresholds we identified (based on deviations of the SPT values from the mean) years with maximal (SPT > [mean + st. dev.]), minimal (SPT < [mean − st. dev.]) and moderate ([mean − st. dev.] < SPT < [mean + st. dev.]) values of the SPT. For all three regions the years of 2001 and 2002 were characterized by the highest SPT. Abnormally low SPT observed in 2013 for the Eastern Sayan and the Baikalsky ranges and in 2003-2004 for the Kodar.

Table 1. Sum of positive temperatures (SPT) calculated for mean glacier elevations of three glaciarized regions for the period of 2001–13. Significant linear trend value ($p<0.05$) is in bold.

Parameter	East Sayan	Baykalsky	Kodar
Mean glacier elevation (m)	2260	1960	2240
Mean latitude (°)	53.3	55.1	56.9
Linear trend (°)	−176	−218	−35
Mean SPT (°C)	857	926	663
standard deviation of SPT (°C)	103	103	78
Max threshold (°C)	960	1030	740
Min threshold (°C)	754	823	585
Mean duration of SPT period (days)	126	123	99
Years with maximal SPT	2001, 2002, 2007	2001, 2002	2001, 2002
Years with minimal SPT	2009, 2013	2013	2003, 2004
Years with moderate SPT	2003–06, 2008, 2010–12	2003–12	2005–13

3.2. Summer pressure fields
During the summer months the cyclonic (30-34% of the SPT period), anticyclonic (20-28%) and low gradient (24-27%) pressure fields were prevailing over the three glaciarized regions (figure 2). Passing of cyclones over the study area with sharply continental climate, as a rule, contributed to reduce the air temperature. By increasing the influence of cyclonic pressure field the sum of positive temperature are reduced. Analysis of baric topography maps showed that the cyclones moved mainly from south-west, north-west, west and south and anticyclones from the north-west, north-east and south-west. High cyclones (i.e. visible on the 500 hPa surface) accounted for 20% of the total, and contributed significantly to the air temperature decreases and hence, a reduction in the snow/ice melting. Positive air temperature anomalies with cloudless weather conditions occurred in anticyclonic pressure field facilitated the increased ablation. Sum of positive temperature increased with increasing the period of anticyclonic type of pressure field (figure 2). In 95% of cases, the anticyclones were secondary (or local), with period of 1-2 days, whereas high slow-moving anticyclones accounted for no more than 5% of cases.

The SPT increased with increasing the period of anticyclonic pressure field influence (figure 2). In 2001–02 anticyclonic pressure was prevailing over all studied regions. Another maximum of anticyclonic pressure field activity occurred over the East Sayan in 2007. The largest number of days with cyclones (227) was observed over the Kodar. Enhanced cyclonic activity over the three regions was observed in 2003. Cyclonic pressure field patterns (with maximum in 2011–13) were dominated over the study period. Low gradient pressure pattern (25% of cases) is associated with a moderate glacier ablation summers. In general, the peaks of the SPT coincide with an increase in anticyclone and anticyclonic pressure field patterns, while the minimal SPT observed in summers with enhanced cyclone and cyclonic pressure field patterns.

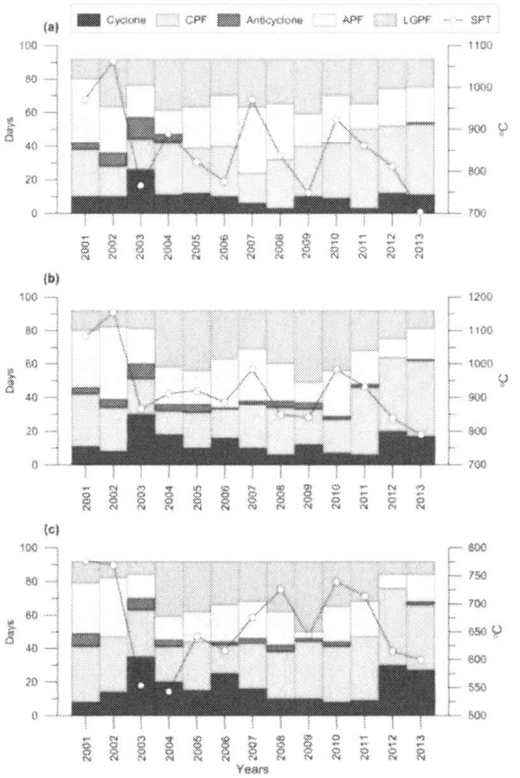

Figure 2. Frequency distribution (in days) of atmospheric pressure patterns over the East Sayan (a), Baikalsky (b) and Kodar (c) ridges during summer (June-August) periods of 2001–13. CPF is cyclonic pressure field, APF is anticyclonic pressure field and LGPF is low gradient pressure field.

Frontal activity occurred over the three study ridges both in low and high pressure fields. Warm atmospheric fronts were the most frequent and peaked in 2002 when the sum of positive temperature was the highest. Occluded fronts (complex fronts as a combination of warm and cold fronts) were more unusual. The largest amount of cold fronts with usual development of convective clouds and heavy rainfall moved over the Eastern Sayan and Baikalsky ridges in 2013. Maximal number of cold fronts over the Kodar Range was recorded in 2011. During the summer months the mid-latitudes continental air (relatively warm and humid) was dominated over the study area. The number of days with heat and cold advection are almost equal, but their temporal distribution is non-uniform. In 2002 the heat air advection occurred in about 80% of cases (days), while in 2013 the cold air advection was prevailing.

3.3. Relationship between the SPT and atmospheric circulation

The correlation coefficients between the sum of positive temperature and regional atmospheric circulation characteristics are presented in table 2. There are the links between the SPT and almost all types of pressure fields and atmospheric fronts. We found significant positive correlations between the SPT and anticyclonic fields, warm atmospheric fronts and warm air advections. A significant negative correlation was found between the SPT and cyclonic activity and cold air advections. The highest correlation coefficients were observed over the Baikalsky Range and the Eastern Sayan. The anticyclonic pressure fields, warm atmospheric fronts and heat air advections were associated with increased SPT and snow/ice ablation, whereas cyclones and cold air advections contributed to lower SPT and ablation. The SPT (and ice melting) on the Kodar increased significantly under influence of anticyclonic pressure field and warm and occluded atmospheric fronts, and decreased when passing the cyclones through the ridge.

Table 2. Correlation coefficients between the sum of positive temperatures calculated for mean elevation of glaciated zone of three ridges and frequency of some atmospheric circulation patterns. Significant values ($p < 0.05$) are in bold.

Atmospheric parameter (unit)	Ridge		
	East Sayan	Baikalsky	Kodar
Cyclones (days)	–0.24	–0.44	–0.78
Cyclones (count)	–0.49	–0.63	–0.60
Cyclonic pressure field (days)	–0.54	–0.20	0.23
Anticyclones (days)	0.22	0.19	–0.01
Anticyclones (count)	0.31	0.12	0.02
Anticyclonic pressure field (days)	0.56	0.88	0.70
Low gradient pressure field (days)	0.13	–0.28	–0.03
Cold fronts (days)	–0.10	0.10	0.28
Occluded fronts (days)	0.32	0.18	0.52
Warm fronts (days)	0.60	0.74	0.46
Cold air advections (days)	–0.65	–0.73	–0.15
Heat air advections (days)	0.65	0.73	0.15
Warm and moist air masses (days)	–0.16	–0.17	–0.16

4. Discussion

Previously we found that the area of southeast Siberia glaciers had decreased by ~59% between 1850 and 2011 corresponding to a significant positive temperature trend at least over the past 130 years [12, 13]. The period of 2001–05 was the warmest in the instrumental temperature record (since 1882) that contributed to increased glacier melting. However, during the second part of the first 21st century decade temperatures decreased across the study region. The most intensive area shrinkage of glaciers occurred between 2001 and 2011, probably, due to delayed response of glaciers to temperature increase from mid-1970s to mid-2000s (by ~2.5 °C). The results of this study suggest that the positive temperature anomaly recorded in 2001–2002 across the region occurred due to the peculiarities of summer atmospheric circulation, such as anticyclonic pressure field influence (cloudless weathers), warm front activity with warm air advections. A similar link between summer ablation and macro-circulation mechanisms in the atmosphere has been established from mass balance studies of glaciers in North-East Siberia [14]. The authors concluded that the positive anomalies of summer temperature (in the background anticyclonic type of pressure field) were caused by mainland (non-Arctic) anticyclone or low gradient pressure field.

The largest ice area reduction since ~1995 has occurred on Kodar glaciers [15, 16]. Probably this is due to high sensitivity of the glaciers both to climate change and to local topography influence. Also it was found that enhanced melting of Kodar glaciers often correlates with such atmospheric circulation patterns as lower pressure under influence of warm and humid air mass, and heat air advection at the altitudes of 2.5-3.0 km [12]. More moderate ice reduction the Baikalsky Range glaciers compared with the other two ranges is probably due to: 1) specific climate in the Lake Baikal surrounding, which is characterized by higher winter temperatures and lower summer ones; 2) enhanced influence of secondary Lake Baikal cyclones formed under the influence of mountain terrain and bringing significant portions of snowfall to the glaciers.

The geographic location of the glaciarized mountain ridges can also affect the circulation. East Sayan Range is an orographic barrier for cyclones to move from west to east. Some part of precipitation from western and southern cyclones falls in this area, while a significant number of cyclones, rounding the East Sayan move to the north-east of the study area (Baikalsky and Kodar ridges). Probably, ice melting on the East Sayan is reduced by frequent early summer snowfalls from southern and western cyclones. The predominance of air transformation processes in a low pressure field is mostly associated with the air masses of warm cyclone sector (mid-latitude continental or tropical air).

Decreasing SPT trends in 2001–13 suggest that summer ablation tends to decrease and glacier mass balance will gradually recover. During the same period (2001-2013) the SPT in the high-mountain zone (~3000 m a.s.l.) of the East Sayan ridge demonstrated a negative trend [9]. Shortening of duration of elementary atmospheric circulation mechanisms unfavorable for glacier mass-balance during the last decade was revealed in the North-East Siberia [14]. The authors also found a decrease in air temperature associated with either cyclonic activity (cold air advections, cloudiness, precipitation and heat losses through evaporation) or advections of cold arctic air. [17] revealed a similar negative trend in ablation on the Djankuat Glacier in Caucasus Mountains after 2000.

Acknowledgement
The present study supported by the Russian Foundation for Basic Research (grant № 15-05-04525), and by Siberian Branch of the Russian Academy of Sciences (project № 0345-2015-0033).

References
[1] Osipov E Y and Osipova O P 2014 *Ann Glaciol* **55** (66) 167–176
[2] McCabe G J and Fountain A G 1995 *Arctic and Alpine Res* **27** (3) 226–233
[3] Hannah D, Gurnell A and McGregor G 1999 *Interactions between the cryosphere, climate and greenhouse gases* Birmingham 155–163
[4] Hodge S, Trabant D, Krimmel R, Heinrichs T, March R and Josberger E 1998 *J Climate* **11** 2161–2179
[5] McCabe G, Fountain A and Dyurgerov M 2000 *Arctic Ant Alp Res* **32** (1) pp 64–72
[6] Washington R, Hodson A, Isaksson E and Macdonald O 2000 *Int J Climatol* **20** 473–487
[7] Watson E, Luckman B and Yu B 2006 *The Holocene* **16** (6) 783–790
[8] Osipova O P and Osipov E Y 2014 *Geography and Natural Resources* **35** (1) 77–81
[9] Osipova O P and Osipov E Y 2015 *Environ Earth Sci* **74** (3)1947–1956
[10] Ohmura A 2001 *J Appl Meteor* **40** 753–761
[11] Kalnay E et al 1996 *Bull Am Meteor Soc* **77** 437–471
[12] Osipov E Y, Ashmetiev A Y, Osipova O P and Klevtsov E V 2013 *Ice Snow* **3** (123) 45–54
[13] Osipov E Y and Osipova O P 2014 *Ann Glaciol* **55** (66) 167–176
[14] Ananicheva M D and Kononova N K 2007 *Data of glaciological studies* **103** 58–67
[15] Stokes C, Shahgedanova M, Evans I and Popovnin V 2013 *Global Planet Change* **101** 82–96
[16] Osipov E Y and Osipova O P 2015 *Environ Earth Sci* **74** (3) 1969–1984
[17] Shahgedanova M, Popovnin V, Petrakov D and Stokes C R 2007 *Ann Glaciol* **46** 355–361

IOP Conf. Series: Earth and Environmental Science **107** (2017) 012040 doi:10.1088/1755-1315/107/1/012040

Climatic changes in the troposphere, stratosphere and lower mesosphere in 1979-2016

Y P Perevedentsev, K M Shantalinskiy, V V Guryanov

Institute of Environmental Sciences, Kazan Federal University, Kremlevskaya st., 18, Kazan, 420008, Russia

E-mail: ypereved@kpfu.ru

Abstract. Changes in thermal characteristics in the atmospheric layer from 1000 to 0,1hPa are studied based on reanalysis data. It was demonstrated that during 1979-2016 temperature increased in the troposphere in January and July, while cooling was observed in the stratosphere, and air warming in lower mesosphere in summer. Most pronounced long-period cyclic changes were registered for temperature in the upper stratosphere and the lower mesosphere, and for ozone mixing ratio in the middle stratosphere.

Modern measurement and information technologies enable to carry out analysis of atmospheric processes even at high altitudes. The present paper studies thermal regime peculiarities of the troposphere, stratosphere and lower mesosphere. It is known that in atmospheric process troposphere air is warmed by the earth surface, while in stratosphere warming is caused by ultraviolet radiation absorption by ozone molecules. In the troposphere, air temperature drops with altitude, while in the stratosphere, on the contrary, it increases with altitude, and decreases again in the mesosphere. According to [1], the troposphere, stratosphere and mesosphere have fundamental differences from the standpoint of radiation conditions, vertical temperature distribution, and chemical composition. Stratosphere-troposphere exchange-related issues have been studied in [2]. It states that a continuous mass and energy exchange takes place through the tropopause, which exerts a strong influence on atmospheric chemistry, energy and the global climate. Vertical temperature gradient γ characterizing static stability of the atmosphere is an important parameter of the atmosphere and climate system as a whole. Requisite conditions for baroclinic instability and, correspondingly, cyclonic activity in the atmosphere depend on γ [3].In the latter paper it was noted that, under the conditions of anthropogenic climate warming, the troposphere gets warmer, while the stratosphere and mesosphere are cooled, especially near the mesopause (80-90 km). Using the monthly ERA-Interim global reanalysis data [4] about air temperature distribution and ozone mixing ratio in regular grid points 2.5 × 2.5° on 26 isobaric surfaces (from 1000 to 0.1 hPa) for 1979-2016, the present paper estimates a number of thermodynamic properties in January and July. All calculations are averaged over the entire Northern Hemisphere (NH), polar (90-65 °N), temperate (65-30 °N) and tropical 30-0 °N) latitudes. It gives an opportunity to characterize the atmosphere on the whole as a unified dynamic system controlled by internal and external factors, individual layers of which more or less actively interact with each other, physically differing one from another by the nature of thermal stratification and energy-related processes [5]. This interaction occurs through the wave mechanism, with the help of which meteorological disturbances propagate upward from the troposphere to the stratosphere, conditioned by atmosphere baroclinity, nonuniform heating of the underlying surface, and relief. In the previous articles the authors examined spatial-temporal variations in temperature, wind speed and atmospheric pressure fields in the troposphere [6, 7].

Table 1presents vertical distribution of long-term means for air temperature Av (°C) and linear trend slope A (°C·year^{-1}) on 26 isobaric surfaces. The data averaged for the NH both for January and July demonstrate that Av temperature drops with height in the troposphere, rises in the stratosphere and again decreases in the lower mesosphere. Temperature inversion in lower troposphere is observed in the polar zone in January. A marked annual cycle of temperature and rather notable zonal differences in temperature distribution are registered, as well. The process of temperature change with time is vertically inhomogeneous. Linear trend slope (A) supports the latter observation. As for the troposphere, especially intense heating takes place in winter in the polar zone, where A values reach 0.107 °C·year^{-1} near the earth surface. Cooling ($A < 0$), most pronounced in the temperate and tropical zones, is observed in the stratosphere, while in the lower mesosphere heating ($A > 0$) is observed over the whole hemisphere. Thus, at 3 hPa level in January, A for the NH reaches 0.118 °C·year^{-1}, while in the tropical zone A=-0.175 °C·year^{-1}. However, in the polar zone in the lower mesosphere, the temperature significantly decreases in January, in contrast to the tropical zone. In general, LTSC distribution confirms greenhouse origin of the contemporary global warming. Nevertheless, noticeable differences conditional on radiation and dynamic factors exist between the year seasons and latitudinal zones.

Specifically, cooling is more vividly expressed in the summer stratosphere, than in the winter stratosphere (figure 1). The latter testifies the weakening of vertical links between the layers through wave action and the absence of stratospheric warmings typical for winter, which contribute to fast heating of polar regions.

The analysis shows that processes proceed more homogeneously at high altitudes in summer, being influenced by the stratospheric ozone, thanks to which a vigorous heating of air masses takes place in the middle stratosphere. As figure 1 evidences, temperature trends change their sign vertically.

Vertical correlations between atmospheric layers have been studied for the entire Northern hemisphere and its zones in January and July. Analysis of the obtained results shows that in the polar zone in January close correlation exists at all the levels in the layer of 1000-0.1 hPa, the correlation coefficient r>0.9 in the layer of 1000-5 hPa, except for the layer 400-300 hPa, where r = 0.743. The correlation slightly decreases in the upper stratosphere – lower mesosphere. In July, the correlation coefficients are significantly lower than those of January, especially at stratosphere-mesosphere levels, which is the evidence of atmosphere stratification. Thus, temperature variations between the levels of 10 and 7 hPa (r = 0.268), 5 and 3 hPa (r = 0.076) are weakly correlated between each other in the upper stratosphere and the lower mesosphere. During this time the ozone screen is essential to thermal regime, while dynamic mixing is not as intense (weakening of vertical wave interaction). In the temperate latitudes in January, the correlation in the layers of 250-200 (r = 0.793), 150-100 (r = 0.671), 2-1 hPa (r = 0.716) slightly weakens; the correlation coefficients in the troposphere in July are high except for the layers of 200-150 hPa (r = 0.615) – tropopause; 0.29-0.1 hPa (r = 0.474) – stratopause. Consequently, the vertical correlations in the polar zone in July have a more complex structure, while in the temperate zone in July the correlations significantly weaken in the transition layers: in the tropopause and the stratopause. Analysis of the vertical correlation relationships in January across the Northern hemisphere indicates that correlations are strong in troposphere, their dramatic weakening is observed between 100 and 70 hPa (r = 0.040), which can be explained by processes taking place in the tropical zone only. Stronger correlations are observed in the upper stratosphere – lower mesosphere. In July, the correlations are strong in the troposphere, in the layer of 200-150 hPa r = 0.773; 150-100 hPa r = 0.496; 100-70 hPa r = 0.740, i.e. in the lower stratosphere correlation relations weaken. In the upper stratosphere, such correlations are stronger in the layers 2-1 hPa, where r = 0.417, and 0.29-0.1 hPa, where r = 0.575. In general, the correlations are rather strong, their weakening is conditioned by physical differences between layers and seasonal restructuring of meteorological fields.

IOP Conf. Series: Earth and Environmental Science **107** (2017) 012040 doi:10.1088/1755-1315/107/1/012040

Table 1. Long-term means for temperature Av (°C) and linear trend slope A (°C·year^{-1}) on 26 isobaric surfaces for 1979-2016.

P, hPa	January								July							
	Northern hemisphere		Polar zone of the NH		Temperate zone of the NH		Tropical zone of the NH		Northern hemisphere		Polar zone of the NH		Temperate zone of the NH		Tropical zone of the NH	
	Av, °C	A, °C·year^{-1}	Av, °C	A, °C·year^{-1}	Av, °C	A, °C·year^{-1}	Av, °C	A, °C·year^{-1}	Av, °C	A, °C·year^{-1}	Av, °C	A, °C·year^{-1}	Av, °C	A, °C·year^{-1}	Av, °C	A, °C·year^{-1}
1000	9.82	0.024	-20.52	0.107	-0.60	0.026	22.63	0.012	22.23	0.020	4.46	0.025	19.54	0.028	26.89	0.013
925	6.19	0.023	-19.23	0.078	-3.77	0.026	17.95	0.012	18.63	0.023	3.86	0.047	16.38	0.031	22.53	0.012
850	3.48	0.020	-19.65	0.059	-5.87	0.024	14.41	0.011	15.18	0.021	1.67	0.038	12.97	0.028	18.86	0.013
700	-3.20	0.020	-24.88	0.052	-12.33	0.021	7.37	0.014	6.63	0.018	-5.73	0.028	4.19	0.022	10.35	0.013
600	-9.93	0.016	-30.77	0.044	-18.98	0.018	0.45	0.011	-0.83	0.016	-12.22	0.021	-2.97	0.023	2.50	0.009
500	-18.32	0.015	-38.69	0.041	-27.59	0.017	-7.83	0.010	-9.12	0.013	-20.59	0.023	-11.63	0.023	-5.46	0.004
400	-29.28	0.019	-48.65	0.038	-38.53	0.016	-18.94	0.018	-20.05	0.020	-32.15	0.025	-23.00	0.022	-15.94	0.017
300	-42.65	0.023	-58.70	0.045	-50.60	0.018	-33.85	0.025	-34.99	0.027	-45.55	0.029	-37.97	0.027	-31.06	0.027
250	-49.36	0.022	-61.20	0.051	-54.98	0.018	-43.07	0.020	-43.54	0.024	-48.24	0.018	-45.90	0.024	-40.93	0.025
200	-55.19	0.014	-61.06	0.042	-56.16	0.014	-53.57	0.010	-51.59	0.011	-44.46	-0.013	-51.28	0.013	-52.82	0.012
150	-61.39	0.015	-61.70	0.052	-56.34	0.013	-65.54	0.011	-60.08	0.004	-44.02	-0.012	-55.05	0.005	-66.45	0.006
100	-69.23	0.019	-64.75	0.091	-59.22	0.005	-78.17	0.021	-66.52	-0.005	-44.91	-0.009	-58.29	-0.013	-76.33	0.003
70	-68.72	-0.016	-67.00	0.108	-60.19	-0.005	-76.06	-0.042	-62.61	-0.033	-44.35	-0.019	-56.86	-0.026	-69.89	-0.041
50	-64.40	-0.026	-68.48	0.120	-59.88	-0.009	-67.59	-0.060	-57.88	-0.040	-43.86	-0.026	-53.95	-0.032	-63.07	-0.048
30	-59.45	-0.018	-68.69	0.155	-58.49	-0.009	-58.98	-0.049	-52.07	-0.029	-42.95	-0.027	-49.62	-0.030	-55.36	-0.029
20	-55.81	-0.006	-66.23	0.155	-56.51	-0.002	-53.82	-0.032	-47.41	-0.017	-40.22	-0.023	-45.31	-0.025	-50.14	-0.010
10	-49.00	0.002	-59.14	0.147	-50.89	0.016	-46.03	-0.030	-39.62	-0.015	-31.90	-0.006	-37.14	-0.021	-42.74	-0.012
7	-43.24	-0.047	-52.45	0.066	-45.01	-0.030	-40.51	-0.077	-34.75	-0.006	-26.87	0.065	-32.02	0.016	-38.11	-0.034
5	-36.87	-0.103	-46.02	-0.013	-38.69	-0.073	-34.10	-0.140	-28.83	-0.029	-21.73	0.111	-26.24	0.029	-31.95	-0.096
3	-28.36	-0.118	-38.82	-0.089	-31.21	-0.054	-24.55	-0.175	-18.03	-0.094	-7.25	-0.028	-14.51	-0.063	-22.43	-0.128
2	-22.84	-0.077	-34.19	-0.140	-27.35	-0.046	-17.53	-0.094	-11.33	-0.071	4.38	-0.064	-7.38	-0.090	-16.77	-0.056
1	-16.76	0.014	-26.33	-0.098	-22.50	-0.023	-10.67	0.060	-6.85	0.054	9.67	-0.008	-3.98	0.046	-11.49	0.068
0.8	-15.54	0.037	-23.72	-0.074	-21.14	-0.010	-9.76	0.091	-7.04	0.082	8.13	0.006	-4.60	0.085	-11.15	0.090
0.51	-15.68	0.072	-19.15	-0.040	-19.47	0.023	-12.06	0.129	-12.18	0.113	-0.34	0.047	-10.39	0.113	-15.29	0.122
0.29	-21.93	0.087	-19.08	-0.029	-22.82	0.041	-21.59	0.142	-24.69	0.106	-17.70	0.088	-24.30	0.078	-25.97	0.131
0.1	-36.66	0.043	-26.68	-0.090	-34.44	-0.001	-39.87	0.098	-49.00	0.030	-55.90	0.002	-51.25	-0.032	-46.18	0.086

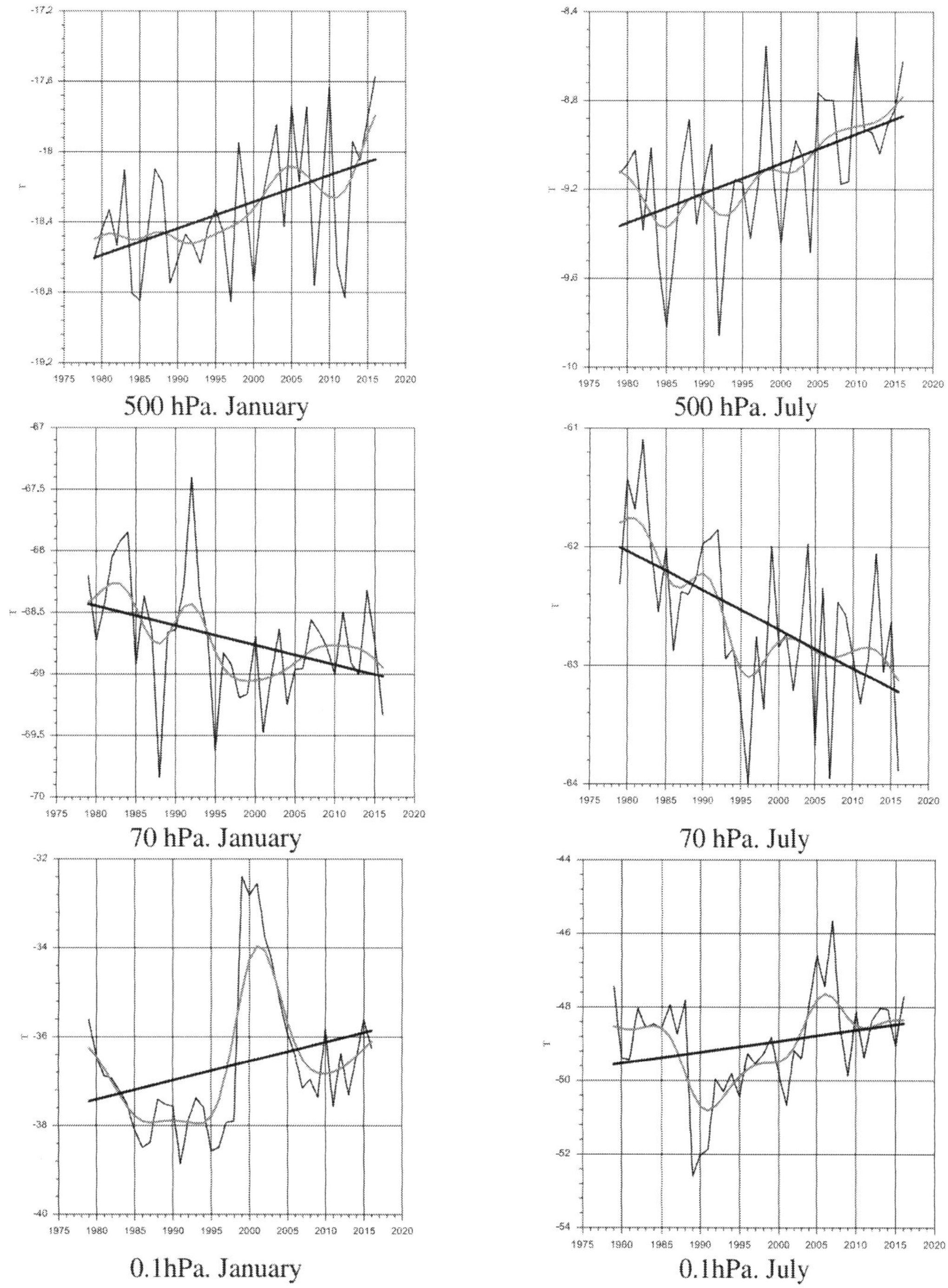

Figure 1. Air temperature, averaged over the whole Northern hemisphere at the levels of 500, 70 and 0.1hPa. Shown are the raw data, linear trend, and low-frequency components with the period > 10 years.

Vertical correlations between atmospheric layers have been studied for the entire Northern hemisphere and its zones in January and July. Analysis of the obtained results shows that in the polar zone in January close correlation exists at all the levels in the layer of 1000-0.1 hPa, the correlation coefficient $r > 0.9$ in the layer of 1000-5 hPa, except for the layer 400-300 hPa, where $r = 0.743$. The correlation slightly decreases in the upper stratosphere – lower mesosphere. In July, the correlation coefficients are significantly lower than those of January, especially at stratosphere-mesosphere levels, which is the evidence of atmosphere stratification. Thus, temperature variations between the levels of 10 and 7 hPa ($r = 0.268$), 5 and 3 hPa ($r = 0.076$) are weakly correlated between each other in the upper stratosphere and the lower mesosphere. During this time the ozone screen is essential to thermal regime, while dynamic mixing is not as intense (weakening of vertical wave interaction). In the temperate latitudes in January, the correlation in the layers of 250-200 ($r = 0.793$), 150-100 ($r = 0.671$), 2-1 hPa ($r = 0.716$) slightly weakens; the correlation coefficients in the troposphere in July are high except for the layers of 200-150 hPa ($r = 0{,}615$) – tropopause; 0.29-0.1 hPa ($r = 0{,}474$) – stratopause. Consequently, the vertical correlations in the polar zone in July have a more complex structure, while in the temperate zone in July the correlations significantly weaken in the transition layers: in the tropopause and the stratopause. Analysis of the vertical correlation relationships in January across the Northern hemisphere indicates that correlations are strong in troposphere, their dramatic weakening is observed between 100 and 70 hPa ($r = 0.040$), which can be explained by processes taking place in the tropical zone only. Stronger correlations are observed in the upper stratosphere – lower mesosphere. In July, the correlations are strong in the troposphere, in the layer of 200-150 hPa $r = 0.773$; 150-100 hPa $r = 0.496$; 100-70 hPa $r = 0.740$, i.e. in the lower stratosphere correlation relations weaken. In the upper stratosphere, such correlations are stronger in the layers 2-1 hPa, where $r = 0.417$, and 0.29-0.1 hPa, where $r = 0.575$. In general, the correlations are rather strong, their weakening is conditioned by physical differences between layers and seasonal restructuring of meteorological fields.

It should be emphasized that interaction between the layers was also studied by the authors earlier [8], but for a shorter period (1986 – 1999), with the spectrum-based approach. A wave structure of the zonal circulation in troposphere-stratosphere and upper mesosphere-lower thermosphere was investigated for a spectral range for time scales of planetary waves (2 – 30 days). To establish interrelations between layers, we got coherent wave structures, which depended on season, interannual variability, and altitude.

Assessment of the impact of individual circulation modes (North Atlantic Oscillation (NAO), Arctic Oscillation (AO), Scandinavian Oscillation (SO) and others) on near-surface temperature in extra-tropical latitudes of the Northern hemisphere, which was done with the help of composite analysis, permitted us to identify regions with a strong response. The warming effect of the NAO was most significant in winter over Central and Eastern Europe, as well as over Central Siberia; the Arctic Oscillation had its maximum manifestation in temperature in Eastern Europe and Siberia, while Scandinavian Oscillation led to considerable air temperature changes in January in Eastern Europe, West and Middle Siberia [9].

Analysis of altitude-time sections of first-order differences, which was conducted for low-frequency components (LFC) with the period exceeding 10 years for air temperature (AT) ($°C \cdot year^{-1}$) in January, July and for annual data was performed as well (figure 2). The most stable situation was observed in the troposphere. A sustainable (rather weak) AT growth had been observed since 1994 (0.1 $°C \cdot year^{-1}$). Admittedly, a weak AT drop was registered in January from 2006 to 2010 (-0.1 $°C \cdot year^{-1}$). In the stratosphere, trend coefficients changed their sign within certain time periods, and the negative trend coefficients achieved the value of -0.2 $°C \cdot year^{-1}$ in the layer 15-30 km in January. At altitudes higher than 30 km these differences grew in January, and starting from the altitude of 45 km both negative and positive sign change centres of LFC differences were observed. An intense warming was registered in the layer of 45 km and higher within the period from 199 5to 2001, when in January the value of the first-order LFC differences exceeded 1.2 $°C \cdot year^{-1}$. At the same time, the primary maximum was located within the altitudes of 45-65 km, and in the layer between 45 ad 57 km

(0.9 °C·year^{-1}) in July. We found a ~ 8-year cyclicity with cooling in 1980-1988 and in 2002-2008 and warming in 1994-2002. Therefore, the upper stratosphere and lower mesosphere demonstrated an unstable temperature behaviour in comparison with the troposphere [10-12].

a b

c

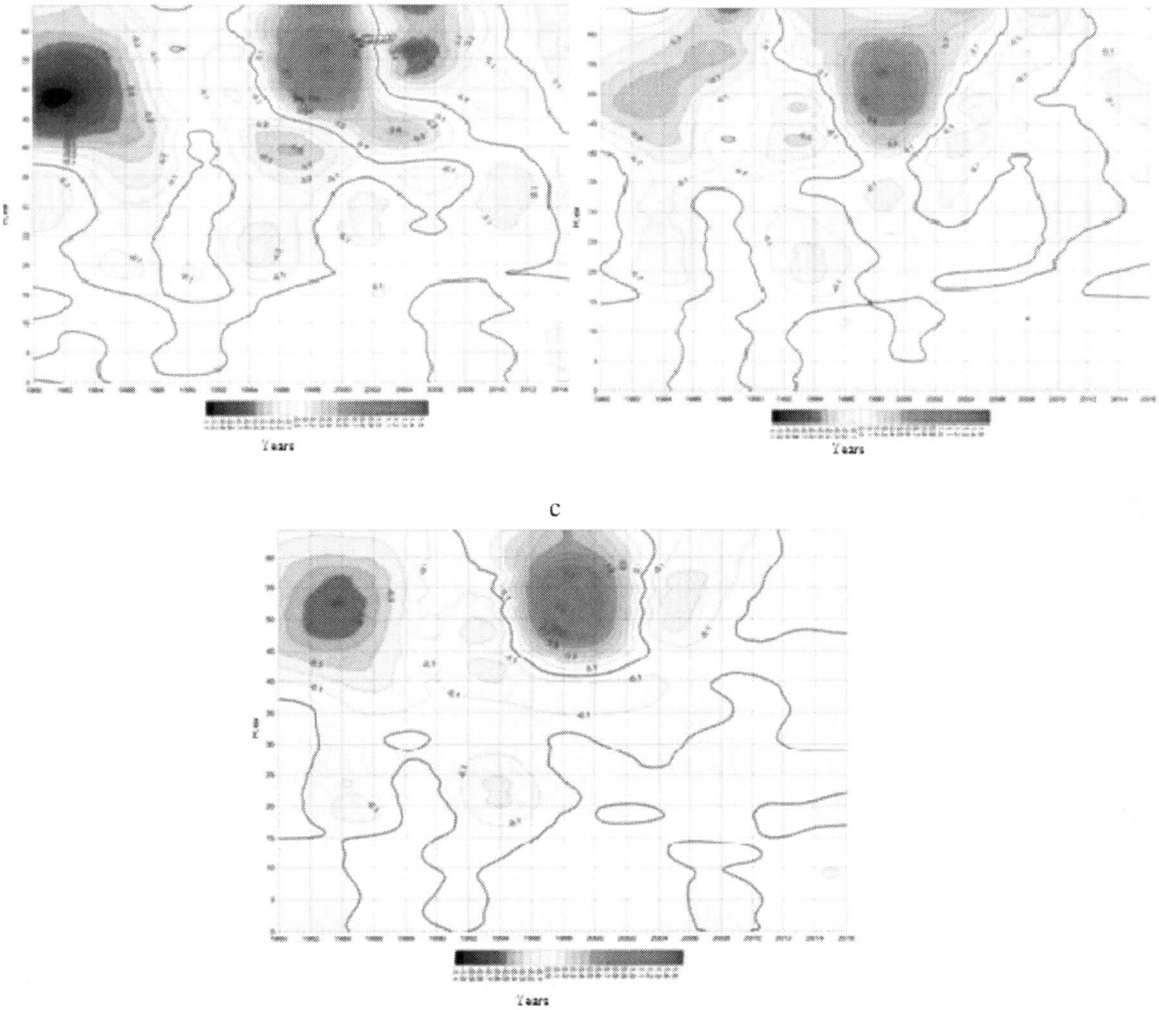

Figure 2. First-order differences of LFC with the period exceeding 10 years for air temperature in the Northern hemisphere (°C·year^{-1}): a – January, b – July, c–annual mean.

A similar work was done for ozone mixing. It was found that the major interannual changes of this value occurred in the layer of 20-45 km (figure 3). Besides, maximum dynamics was registered for air temperature in winter, while for ozone in summer. In the stratosphere heat is emitted when ozone absorbs ultraviolet radiation; furthermore, ozone absorbs the radiation coming from the earth surface.

Conclusions

1. ERA-Interim reanalysis data offer an unbiased reflection of fundamental patterns of dynamics in the key climatic indices in the atmospheric mass up to the altitude of 64 km.
2. The significant spatiotemporal inhomogeneity in thermal characteristic distribution, which results from differences in physical mechanisms regulating thermal regime in atmospheric layers, is discovered.

3. The sign of the vertical slope coefficient of air temperature linear trend confirms the greenhouse origin of the contemporary global warming.
4. Centres of temperature growth and decrease are observed in the upper stratosphere and lower mesosphere (45-60 km) with the cyclicity of 8 years. In the ozonosphere an analogous process occurs at smaller altitudes, in the layer of 25-40 km.
5. To get a fuller picture of changes in the thermal behaviour of the atmosphere one should also use data about circulation in its different layers.

a

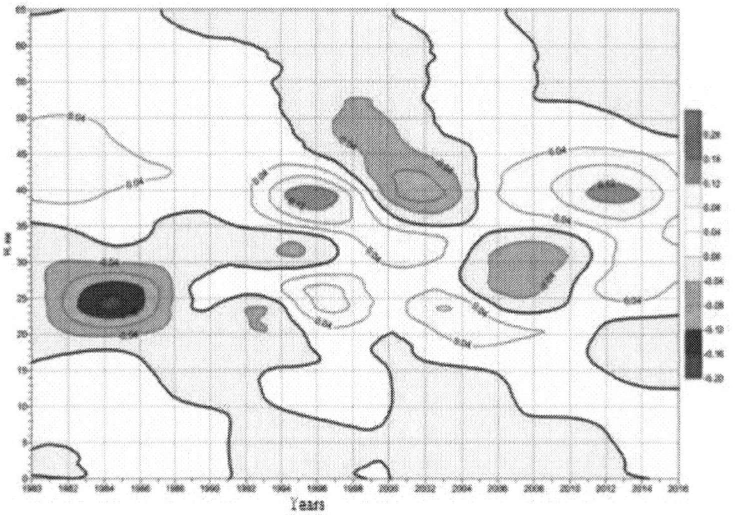

b

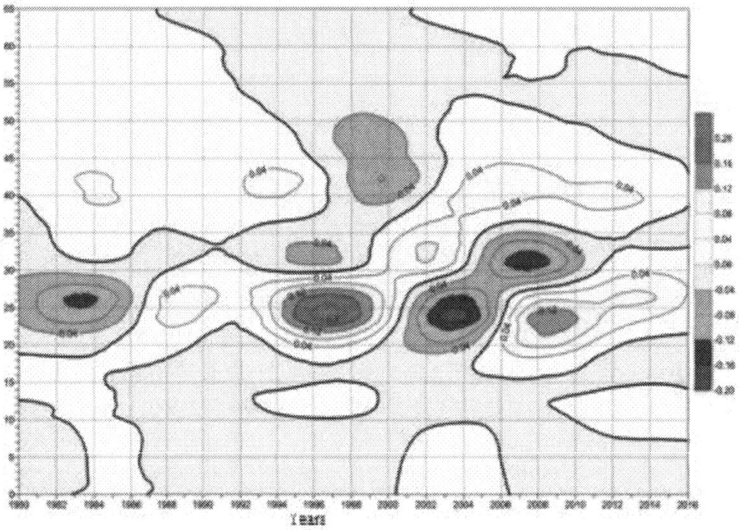

Figure 3. First-order differences of LFC with the period exceeding 10 years for ozone mixing ratio ($10^{-6}\cdot year^{-1}$) in the Northern hemisphere: a – January, b – July.

Acknowlegments
The work was supported by the Russian Foundation for Basic Research (grants № 15-05-06349, 15-05-06399, 17-45-160693).

References

[1] Mohanakumar K. 2008 *Stratosphere-troposphere Interactions* (Springer) 416
[2] Ivanova A R 2016 *Meteorology and Hydrology* **3** 22-45
[3] Mokhov I I and Akperov M G 2006 *Izvestiya, Atmospheric and Oceanic Physics.* **42 (4)** 467-475
[4] Dee D P et al 2011 *Q.J.R. Meteorol. Soc.* **137** 552-597
[5] Perevedentsev Y P 1984 *Circulation and energy-related processes in the middle atmosphere* (Kazan: Kazan University Publishing House) 167
[6] Perevedentsev Y P and Shantalinskiy K M 2014 *Russian Meteorology and Hydrology* **39** (10) 650-659
[7] Perevedentsev Y P, Vasilev A A, Shantalinskiy K M and Guryanov V V 2017 *Russian Meteorology and Hidrology* **42** (7) 461-470
[8] Fahrutdinova A N, Perevedentsev Y P, Guryanov V V and Kulikov V V 2001 *Adv. Space Res.* **27** (10) 1667-1672
[9] Perevedentsev Y P, Vilfand R M and Shantalinskiy K M 2016 *Proceedings of the Hydrometcentre of Russia* **360** 5-25
[10] Barnett J J and Chandra S 1990 *Advances in Space Research* **10** (12) 7-10
[11] Hamilton K, Wilson R J, Mahlman J D and Umschied L J 1995 *J. Atmos. Sci* **52** 5–43
[12] Holton J R and Qehbrein W M 1980 *Pure Appl. Geophys* **118** 284–306

IOP Conf. Series: Earth and Environmental Science **107** (2017) 012041 doi:10.1088/1755-1315/107/1/012041

Model-based calculations of surface mass balance of mountain glaciers for the purpose of water consumption planning: focus on Djankuat Glacier (Central Caucasus)

O O Rybak[1,2,3], **E A Rybak**[1,2]

[1] Scientific Research Center of the Russian Academy of Sciences, Theatralnaya str., 8-a, Sochi, 354000, Russia

[2] Branch of Institute of Natural and Technical Systems, Kurortny Av., 99/18, Sochi, 354024, Russia

[3] Earth System Sciences & Departement Geografie, Vrije Universiteit Brussel, Pleinlaan, 2, Brussels, B-1050, Belgium

E-mail: o.o.rybak@gmail.com

Abstract. Mountain glaciers act as regulators of run-off in the summer period, which is very crucial for economy especially in dynamically developing regions with rapidly growing population, such as Central Asia or the Northern Caucasus in Russia. In overall, glaciers stabilize water consumption in comparatively arid areas and provide conditions for sustainable development of the economy in mountainous regions and in the surrounding territories. A proper prediction of the glacial run-off is required to elaborate strategies of the regional development. This goal can be achieved by implementation of mathematical modeling methods into planning methodologies. In the paper, we consider one of the first steps in glacier dynamical modeling – surface mass balance simulation. We focus on the Djankuat Glacier in the Central Caucasus, where regular observations have been conducted during the last fifty years providing an exceptional opportunity to calibrate and to validate a mathematical model.

1. Introduction

Mountain glaciers play an exceptionally substantial role in functioning of the regional economies and in the everyday life of the local communities in mountain and piedmont areas. Indeed, they act as main regulators of water run-off. Fresh water accumulated by glaciers is gradually released during the melting season. For example, in summer season melting glaciers contribute up to 70-80% to the total river run-off [1]. Obviously, growing economy and population increase demand higher water consumption. That is why accurate prediction of future available water resources is a challenging problem for the scientific community.

Consequences caused by degradation of mountain glaciation under condition of changing climate [2, 3] are rather acute especially in developing countries, first of all in the countries experiencing sharp deficiency of fresh water. In some regions situation is aggravated by very uneven distribution of water resources, for instance, in Central Asia. Here, main rivers originate in the mountains of Kyrgyzstan and Tajikistan, but the main consumer is Uzbekistan with rapidly growing population and developing

IOP Conf. Series: Earth and Environmental Science **107** (2017) 012041 doi:10.1088/1755-1315/107/1/012041

economy. This situation provokes tensions between the countries, which may anytime in future evolve in a full-scale armed conflict.

Current changes of regional climate do not necessarily follow global warming trend [4]. This will probably also be the case in future. Nevertheless, elaboration of strategies of development of local economies and maintaining social infrastructure require scientifically based possible scenarios of availability of water resources. Therefore, accurate prediction of future glacial run-off is a key problem of sustainable development in the region where hydrological regime is highly dependent on glacial run-off. In turn, evaluation of future run-off requires dynamical modeling of mountain glaciers.

Prognostic estimates of the glacial run-off can be obtained in several steps, which can be schematically represented as follows:

1. Choice of a desired climatic scenario (either schematic or based on modeling).
2. Choice of a type of a mathematical model of a mountain glacier.
3. Determining so-called reference glaciers in the region with the maximum available information and direct measurements (on topography, flow velocity, mass balance etc).
4. Determining the way of extrapolation of model results on the rest of the glaciers in the study region
5. Model calibration – defining values of the key tuned model parameters [5, 6].
6. In case desired climatic scenario is based of prognostic modeling on a GCM, it is necessary to choose an optimum downscaling methodology [7]. Another approach is implementation of a kind of simplified scenario(s) by setting schematically evolving climatic forcing [8].
7. Model validation – reconstruction in numerical experiments past observations on the glacier (flow velocity, mass balance) by forcing the model with a set of the past climatic records.
8. Carrying out a series of prognostic numerical experiments and further interpretation of their results [9].

The main obstacle for accurate prognostic calculations is the lack of observations. It is explained, first of all, by bad accessibility of the mountain glaciers and by unfavorable working conditions. Of course, growing amount of data obtained with remote sensing partially solves the problem. Nevertheless, remote methods cannot totally replace direct instrumental measurements at the glaciers. Monitoring of all or even of only biggest glaciers is quite demanding in terms of regular financial investments. That is why it is expedient to perform regular observations on certain typical (reference) glaciers. Djankuat Glacier is such reference glacier for Central Caucasus. Djankuat is a valley glacier with morphometric characteristics typical for the glaciers in the region. In the years 2012-2013 its area and volume were equal to 2.45 km^2 and 0.077 km^3 respectfully [10]. Regular observations have been conducted here for a half of the century. This is the reason to choose Djankuat for calibration and validation of a mathematical model of mountain glacier dynamics. In the current paper we consider on of the necessary steps in prediction methodology - surface mass-balance modeling of Djankuat glacier and collate results with observations.

2. Methods and data

2.1. Model formulation

We employ an energy-balance model to calculate amount of energy E coming on the surface of a skin layer on the surface of a glacier (figure 1):

$$E = SW(1-\alpha) - E_{eff} + SHF + LHF, \qquad (1)$$

where SW is incoming short-wave radiation, α - surface albedo, E_{eff} – effective radiation (balance of long-wave radiation), SHF and LHF – sensible and latent turbulent heat fluxes. Using skin-layer approach supposes that heat exchange between the surface of snow/ice and rest of the glacier is neglected. Normally, ablation areas of the glaciers are covered with debris. Debris cover transforms essentially processes of heat exchange [11], therefore we take into account presence of this layer where necessary.

Annual mass balance on the surface of the glacier, *SMB*, is expressed as a difference between accumulation, *AC*, and run-off, *RO*:

$$SMB = \sum_{1}^{365} [ACC - RO], \tag{2}$$

Accumulation is calculated as the sum of solid and liquid precipitation *ACC=PS+PL*, from which we subtract evaporation (only in the zone of ablation), *SU*. The latetter is proportional to *LE*. We avoid explicit description of post-depositional redistribution of snow because it is rather individual quantity for the particular glacier. Instead, the latter is regarded as a tuned parameter. Ratio of *PS* and *PL* depends on daily average air temperature $\overline{T}_A$. Hourly values (daily cycle) are expressed as follows:

$$T_A = \bar{T}_A - \tilde{T}_A \cos\left(2\pi \frac{t}{24}\right), \tag{3}$$

where $\tilde{T}_A$ – is the daily amplitude; $t = 0, \ldots, 23$ – time (hours). $\tilde{T}_A$ is a key tuned parameter determining the area of surface melting. It has to be defined from available observations on particular glaciers. Actually, $\tilde{T}_A$ acts as a switcher between melting and no-melting conditions:

$$\begin{array}{ll} M = \max(E, 0)/L_m & T_A \geq T_0 \\ M = 0 & T_A < T_0 \end{array}. \tag{4}$$

Run-off is evaluated as a sum of melted water and liquid precipitation reduced by refrozen and retained water.

More details of the model formulation are given in [2, 12].

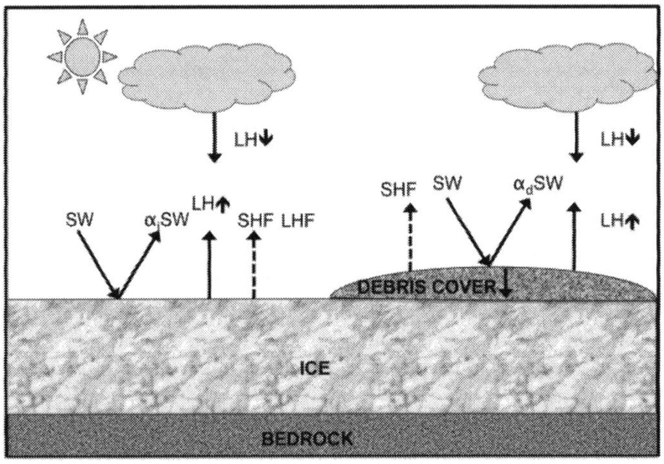

Figure 1. Heat fluxes on the surface of the mountain glacier taken into account in the mathematical model.

2.2. Input data and set up of a numerical experiment

To validate the model, in the current study we utilized data from two weather stations located close to the Djankuat Glacier – Terskol and Mestia. Both stations are located within several tens of kilometers to the glacier. Another reason of choice was completeness of data. Surface air temperature measured at Terskol (2144 m above sea level, a.s.l.) on the Northern macroslope of the Main Caucasus Chain can be easily

IOP Conf. Series: Earth and Environmental Science 107 (2017) 012041 doi:10.1088/1755-1315/107/1/012041

remapped to the area of the Djankuat Glacier. We carried out comparison of parallel observations at Terskol and at the automatic weather station (AWS) located directly on the glacier at the elevation of about 3100 m a.s.l. during several summer seasons (data obtained in summer of the year 2011 is shown in figure 2a) and derived a simple linear transformation expression $T_{Djankuat} = -2.91 + 0.48T_{Terskol}$ to recalculate air surface temperature observed at Terskol to the elevation of AWS Djankuat (figure 2b). Unfortunately, available precipitation record at Terskol is corrupted. This was the reason to utilize daily precipitation data from Mestia, located at the Southern macro-slope of Caucasus at the elevation of 1434 m a.s.l. Preliminary analysis of climatic averages proved that monthly and annual precipitation averages at both sites are quite similar. Unfortunately, precipitation was not measured directly on the glacier, therefore direct comparison could not be carried out. For model validation we have chosen a 10-year time segment with minimum gaps in the records: from October, 1, 1999 to September, 30, 2009 (ten conventional hydrological years).

Absorbed solar radiation, effective radiation and available melting energy are calculated every hour. Precipitation, refrozen/retained water run-off and mass balance are evaluated once a day.

We limit climatic forcing to only two records – daily temperature and precipitation rate. In view of using our modeling approach for evaluation of the future glacier dynamics, we find it reasonable to restrict prognostic forcing variables to only those chosen and to parameterize the others – cloudiness, absorbed radiation etc. For instance, cloudiness in our approach is set to 0.2 when daily precipitation is zero and to 1.0 when precipitation is not zero.

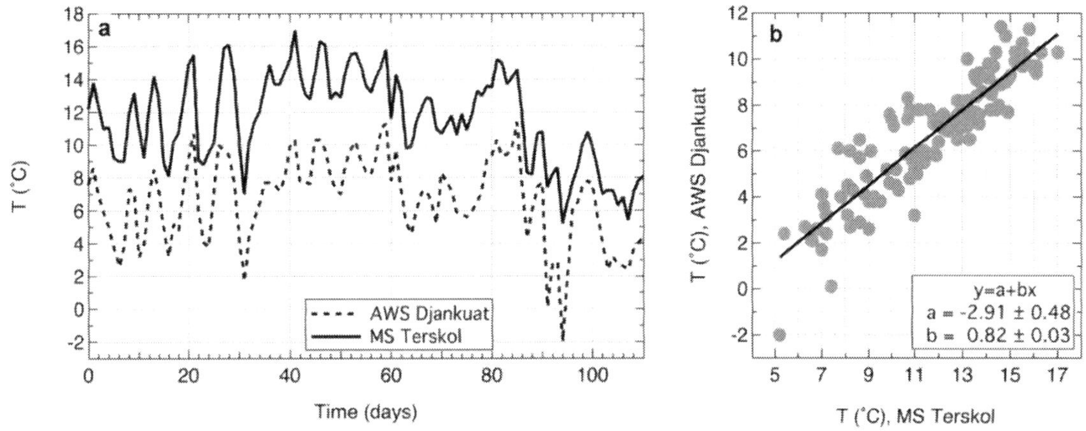

Figure 2. Parallel observations of air surface temperature at Terskol weather station and Djankuat weather station in summer of the year 2011 (a), linear relation between both records (b).

3. Results and discussion

Specific mass balance obtained during ten model years was compared to the similar characteristic evaluated from direct accumulation and ablation measurements (V V Popovnin, personal communication) (figure 3). Two solid curves indicating annual mass balances are quite similar. Model failed to reproduce positive mass balance in the year 2004. That is why cumulative mass balance in the end of the decade is 300 mm more negative in the modeled case. We established several crucial model parameterizations and assumptions affecting final results most of all. Among those are parameterization of albedo, ratio between direct and diffuse radiation, daily amplitude of surface air temperature. Comparison with fields of surface mass balance and ablating rate based on observations (V G Pastukhov, personal communication) with simulated results indicate that model fails to reproduce small details of observations. This mostly happens because albedo parameterization is rather simple and cannot describe its spatial and temporal distribution exactly. This is perhaps not necessary for predictions of glacial run-off since the overall result is reasonable. Therefore, we find it expedient to follow a rather simplified approach described in [12]. Another source of deviations between the model and reconstructions is possible not exact masks of debris covered areas.

We find that obtained results are promising, because we managed to elaborate a comparatively simple mathematical model, which can reproduce processes of heat exchange of a mountain glacier using only two input variables - surface air temperature and precipitation rate. In future, we plan to couple surface mass balance calculations and ice flow simulation using a model previously applied to one of the alpine glaciers [11]. This will enable to carry out prognostic calculations basing on the results of global climate modeling downscaled to a particular mountain glacier as discussed in [7].

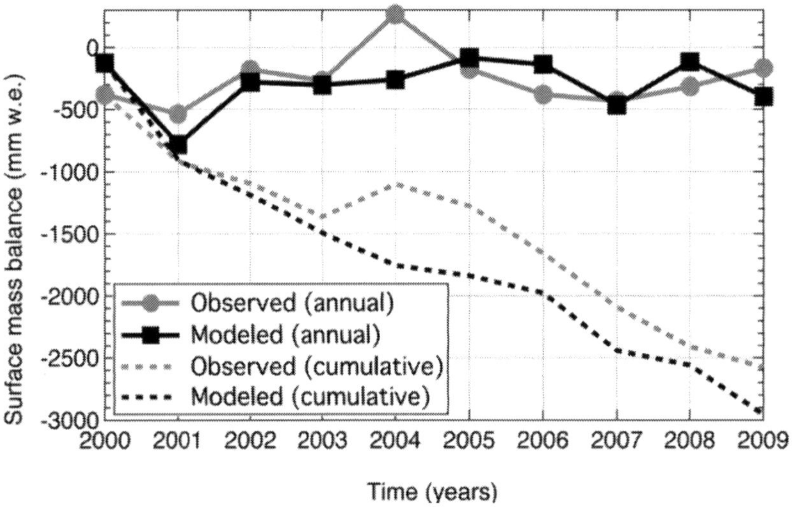

Figire 3. Observed and modeled specific surface mass balances (mm of water equivalent) of the Djankuat Glacier, annual 2000-2009 and cumulative.

Acknowledgements
This research was supported by Russian Foundation for Basic Research (grant № 15-05-00567). Authors are grateful to our colleagues, Dr Victor Popovnin, Mariia Kaminskaya, Valintin Pastukhov (Department of Geography, Lomonosov Moscow State University), Dr Stanislav Kutuzov and Dr Ivan Lavrentiev (Institute of Geography of RAS), for providing data of observations and for numerous preliminary discussions. We also appreciate multiyear efforts of Dr Victor Popovnin who has been organizing and sustaining mass balance measurements at Djankuat Glacier for the last several decades and who was an excellent guide for the authors when they visited Djankuat Glacier.

References
[1] Petrakov D A, Lavrentiev I I, Kovalenko N V and Usubaliev R A 2014 *Earth's Cryosphere ('Kriosfera Zemli')*, **18** (3) 83-91

[2] Shahgedanova M, Nosenko G, Kutuzov S, Rototaeva O and Khromova T 2014 *The Cryosphere* **8** 2367-2379
[3] Stokes C R, Gurney S D, Shahgedanova M and Popovnin V 2006 *J Glaciol* **52** (176) 99-109
[4] Shahgedanova M, Popovnin V, Aleynikov A, Petrakov D and Stokes CR 2007 *Ann Glaciol* **46** 355- 361
[5] Rybak O O, Rybak E A, Kutuzov S S, Lavrentiev I I and Morozova P A 2015. *Ice and Snow ('Liod ISneg')* **2** (130) 9-20
[6] Zekollari H, Huybrechts P, Fürst J J, Rybak O and Eisen O 2013 *Ann Glaciol* **54** 343-351
[7] Morozova P A and Rybak O O 2017 *Ice and Snow ('Liod I Sneg')* **4**, in press
[8] Rets E P, Frolova N L and Popovnin V V 2011 *Ice and Snow ('Liod I Sneg')* **4**(116) 116-124
[9] Zekollari H, Fürst J J and Huybrechts P 2014 *J Glaciol* **60** (224) 1155-1168
[10] Lavrentiev I I, Kutuzov S S, Petrakov D A, Popov G A and Popovnin V V 2014 *Ice and Snow ('Liod I Sneg')*,**4** (128) 7-19
[11] Nicholson L and Benn D I 2006 *J Glaciol* **52** (178) 463–468
[12] Nemec J, Huybrechts P, Rybak O and Oerlemans J 2009 *Ann Glaciol* **50** 126-134

Long-term changes in the hydroclimatic characteristics in the Baikal region

N N Voropay [1,2], **N V Kichigina** [1]

[1] V B Sochava Institute of geography of Siberian Branch of the Russian Academy of Science, Ulan-Bator str., 1, Irkutsk, 664033, Russia
[2] Institute of Monitoring of Climatic and Ecological Systems SB RAS, Academic av., 10/3, Tomsk, 634021, Russia
E-mail: voropay_nn@mail.ru, nkichigina@mail.ru

Abstract. Since the end of the 19th century, global air temperature has been increasing. The period after 1976 is called the period of the most intensive warming. In Russia, the average annual air temperature rises at a rate of + 0.43 ° C / 10 years. The change of precipitation over the last 50-60 years on average in Russia is not significant. In the Baikal region, precipitation increase during the warm period (10-11%) and decrease during the cold period (4%). It is reflected on hydrological regime and the factors of river flow formation. The regional features of the hydrological regime dynamics of the Baikal region against the background of climate change are considered. Groups of the rivers with similar alternations of low water and high-water periods are allocated. Trends in runoff are analyzed. The increase in air temperature leads to intra annual redistribution of river flow. The majority of statistically significant trends of river run off are observed during the cold period of year.

1. Introduction

According to the data of instrumental observations, the global temperature and temperature in Russia have been increasing since the late 1880s. The period after 1976 is called the period of the most intensive warming [1]. The average annual rate of warming is +0.43 $°C·(10\ years)^{-1}$. In this case, a statistically significant trend is not observed only in the winter. During the same period (1976-2012), the annual precipitation trend is positive (+0.8 $mm·month^{-1}·(10\ years)^{-1}$), on average in Russia. It describes 24% of the total interannual variability of the series (that is, the trend is small, but statistically significant even at the 0.5% level). Basically, spring precipitation changes, and autumn precipitation increases slightly. On average, precipitation for the period 1961-1990 is less than precipitation for the period 1981-2010. This change is less than 1 mm/month. That is, on the average, there have been no significant changes in the mode of precipitation in the last 50-60 years on the territory of the Russian Federation.

According to data of weather stations, the trend of air temperature for the period 1976-2012 is calculated for the territory of the Predbaikalie and Transbaikalie [1]. A statistically significant trend in air temperature is observed for the whole year (0.32 $°C·(10\ years)^{-1}$), in spring (0.53 $°C·(10\ years)^{-1}$) and in summer (0.52 $°C·(10\ years)^{-1}$). In winter and autumn, there are also positive trends (0.06 and 0.26 $°C·(10\ years)^{-1}$, respectively), but they are not statistically significant. For the whole region, a statistically significant increase in annual precipitation is observed (1.1 $mm·month^{-1}·(10\ years)^{-1}$). There is an increase in liquid and mixed precipitation (by 10-11%) and a decrease in solid precipitation (by 4%).

The instrumental observations (1961–2008) showed a persistent long-term tendency towards a rise in mean annual temperature (0.4 °C·(10 years)$^{-1}$), with the largest contribution coming from changes in the winter months [2]. The maximum air temperature changes were recorded in February (0.62–1.94 °C·(10 years)$^{-1}$). Minimum statistically significant trends were recorded in the summer months. The tendencies for a change of the monthly precipitation are insignificant, because they are 10–100 times lower than the interannual variability [3, 4]. At the same time, much research points to an increase in precipitation at the global (planetary) level, as well as in a large number of the individual regions of the Russia and neighbouring countries [5, 6].

Some researchers [7, 8] distinguish four gradatons of precipitation intensity (i, mm·h^{-1}). For these gradations, the duration of precipitation is determined by: weak precipitation ($i \leq 1.8$ mm·h^{-1}), moderate precipitation ($1.8 < i \leq 6$ mm·h^{-1}), strong precipitation ($6 < i \leq 60$ mm·h^{-1}) and very strong precipitation ($I > 60$ mm·h^{-1}). In the Predbaikalie and Transbaikalie in recent years, a decrease in weak precipitation by 14% (about 140 h) has been observed. The duration of moderate, strong and very strong precipitation increases by 5, 7 and 10%. This is 1.8, 0.9 and 0.02 hours respectively.

A fair amount of attention has recently been focused on the issue relating to climate fluctuation and its consequences [9, 10]. The studied impact of the consequences of the climatic changes on the hydrological regime of rivers there is mainly observed on the macro level. However, the regional manifestation of climate change has a number of intrinsic properties associated also with local physical–geographical conditions [11, 12].

Baikal region is located in the center of the Eurasian continent, at a great distance from the oceans, which is the huge inertial accumulator of solar energy. It largely determines the direction and speed of warm oceanic and air masses on Earth, which affect the climate of the planet. The features of physical–geographical position of the region and sharply continental climate have to define the features of climate change.

The study of regional features of Baikal region rivers flow, the laws of their long-term fluctuations and intra annual redistribution, extreme values, taking into account the local geographical features of the runoff formation in the background of global climate change gets us closer to solving the problems associated with both economic and environmental tasks.

2. Objects and data

The Baikal region includes three subjects of the Russian Federation, united by belonging to the basin of Lake Baikal, they are the Irkutsk Region, the Republic of Buryatia and the Chita region. In the hydrographic relation the Baikal region includes the Lake Baikal basin, nearly all the Angara River basin, the Upper Lena River basin (with the Vitim and Olekma River basins) and the upper parts of the Lower Tunguska.

The long-term series of monthly and daily precipitation at weather stations are the initial information. Weather stations are located on the territory of the considered basins. The data were analyzed on the territory of the Angara river basin (9 weather stations), the Upper Lena River basin (14 weather stations) and Baikal Lake basin (5 weather stations). Extreme characteristics of atmospheric precipitation are calculated.

The objects of our study were the run-off series in majority for rivers with a catchment area not greater than 100 km^2. Such river flows are formed in a more or less homogeneous physical-geographical condition within the same geographical zone. River run off cross-sections were selected in such a way as to reflect the full range of physical-geographical features of the territory.

We used information on run-off obtained from hydrological gauging stations to assess changes in the hydrological regime. We examined the data series for mean annual and monthly run-off. Besides we examined run off series of genetically homogeneous phase of water regime – spring snow melt and summer rainfall maxima, annual and summer minima. Analysis of run-off trends was undertaken mainly for an identical 55-year-long period of instrumental observation, from 1961 to 2014, except the Lower Tunguska, where the period from 1961 to 1999 was conceded (due to the lack of data). We used information from 15 gauging stations within the Lena basin, 10 - within the Angara basin, 12 -

within the Lake Baikal basin, and 3 gauging stations within the Lower Tunguska. We used long-term observational data from the Irkutsk and Ulan-Ude Administration for Hydrometeorology and Environmental Monitoring, and data from handbooks of the series Surface Water Resources of the USSR, the State Water Cadastre, as well as data from hydrological yearbooks.

3. Methods

The WMO Working Group on Climate Change Detection proposed a set of 27 extremal indexes for describing climatic extremes. Most of the indices determine the number of days when meteorological values go beyond a certain predetermined threshold. The maximum five-day precipitation (R5d), the daily precipitation intensity (SDII), the number of days with precipitation not less than the threshold set by the investigator (Rnnmm), the sum of strong precipitation above the specified percentile distribution (R95p and R99p) are the main indices in the analysis of extreme precipitation. Extremely large precipitation has a significant effect on the regime of rivers. A large amount of precipitation in winter and spring causes a high level of spring flood; in summer strong rains often cause catastrophic floods. According to the analysis of the indices, in the Baikal region the growth of heavy precipitation prevails throughout the year.

The assessment of long-term cyclic flow fluctuations with use of integral difference curves is executed. These curves reflect not only the annual fluctuations in flow, but also indicate long-term exhaustion (the descending curve sites) or accumulation (the ascending curve sites) of river flow. Allocation of water content phase is carried out on the main turning points of curves.

The coefficient of a linear trend determined by a method of the smallest squares is used as a measure of run off changes intensity. The statistical significance of the trends is estimated by the statistical test of the null hypothesis of no difference between the regression model and the experimental data from 95% confidence interval.

Only data series without gaps or with single gaps were used in processing. Missing values were replaced by an average value, calculated for the corresponding variable. Despite the low accuracy, this method is appropriate due to its simplicity and the small number of gaps. Values with a large deviation from the average value were estimated and also treated as missing values.

4. Results

Monthly precipitation for the period 1966-2015 was analyzed. In general, statistically insignificant trends predominate. A significant increase in precipitation over the cold period is observed at nine stations. Precipitation in November-March increases by 2-11 mm·(10 years)$^{-1}$, precipitation in October-April increases by 4-18 mm·(10 years)$^{-1}$. In the cold period, the maximum number of weather stations on which there is a significant increase in precipitation (6 stations) was recorded in October and December. The largest number of stations (5 in each month), where precipitation increases during the warm season, is in May and June. Moreover, in May, these are weather stations located in the basin of the Angara River, in June - in the basin of the Lena River. The trends in both cases are 4-7 mm·(10 years)$^{-1}$. Statistically significant positive trends in annual precipitation are observed at two stations in the Angara River basin and at five stations in the Lena River basin. The coefficients of the trend are equal to 19-24 mm·(10 years)$^{-1}$. The weather station Khamar-Daban (basin of Lake Baikal) deserves special attention. This weather station is most highly located in comparison with the other weather stations under consideration. Here the annual precipitation decreases (-24 mm·(10 years)$^{-1}$, R^2 = 2%), and the maximum decrease (-34 mm·(10 years)$^{-1}$, R^2 = 15%) is observed in July. A long-term change in the sum of atmospheric precipitation at weather stations located in the basins of Lake Baikal, the Angara River, and the Lena River is shown in figure 1, 2.

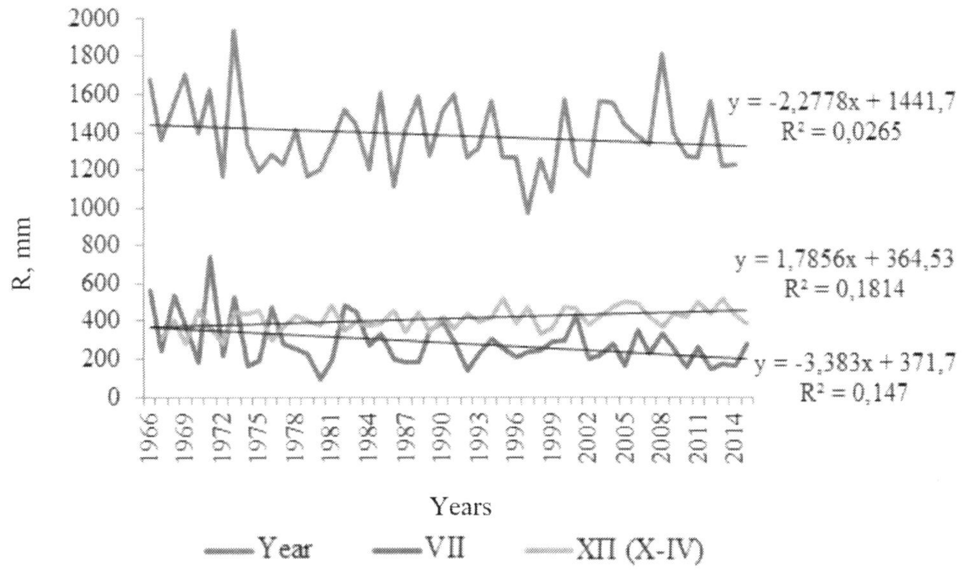

Figure 1. Long-term changes of precipitation at the weather station Khamar-Daban.

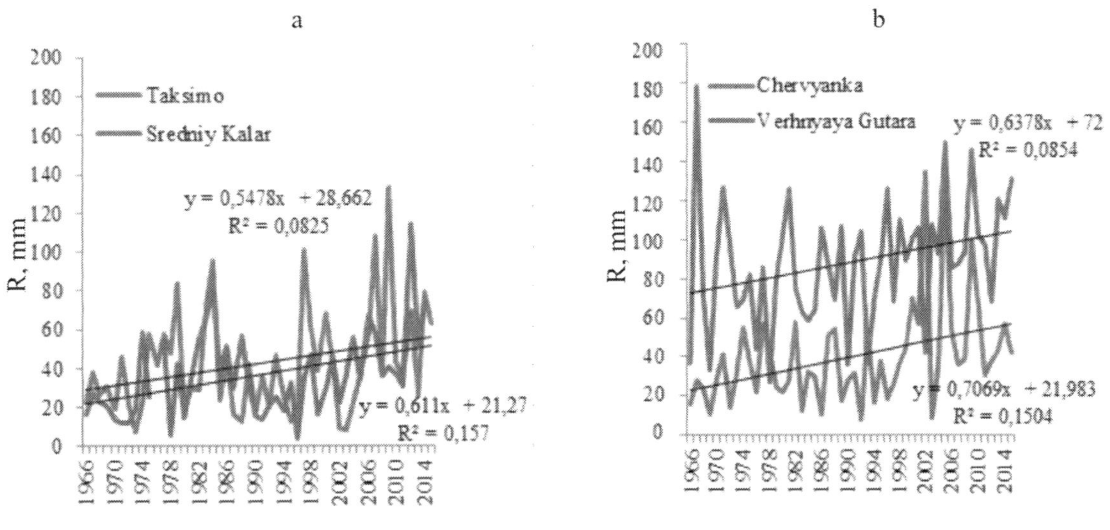

Figure 2. Long-term changes of monthly precipitation (May) at meteorological stations located in the basin of the river Angara (a), the river Lena (b).

Integral difference annual flow curves clearly illustrate the long periods of low and high water. They show the general tendency of long-term fluctuations of a flow in river basins. In the central and southern of Lake Baikal basin (figure 3, line 1) now is low-water period, which began in 1995. In the north of Lake Baikal basin is high-water period at the same time (line 2), which probably ended in 2012. The Angara basin is marked the small reduction of the flow on the number of rivers from 2008 (line 4). This low-water period replaced the high-water period, which was here from 1987 to 2006. In Upper Lena basin the high-water period ended which lasted from 2000 to 2012 (line 3).

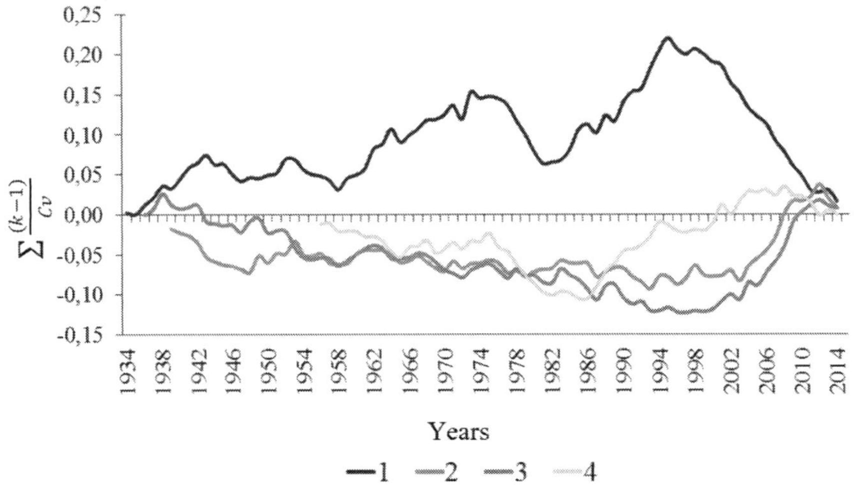

Figure 3. Integral difference curves of the river flow, where k – modular coefficient, Cv – variation coefficient: 1 - Selenga-Mostovoi (Lake Baikal basin), 2 - Verch.Angara-Zaimka (Lake Baikal basin), 3 - Lena-Zmeinova (Lena River basin), 4 - Irkut-Tibelti (Angara River basin).

Within the Angara and Upper Lena basins annual run-off changes were directed differently and were not generally statistically significant. A statistically significant positive tendency was observed here on most of the gauging stations considered in the cold season (from October to March–April). Within the Upper Lena river basin, a tendency for run-off reduction in May, Jun and July was observed. At the same time within the Vitim River basin a tendency for run-off increasing in May was observed. Within the Angara basins the statistically significant negative tendency of spring snow melt and summer rain fall maxima was predominate. At the same time within the Upper Lena basin the statistically significant positive tendency of spring snow melt and summer rain fall maxima was predominate in some rivers.

Within the Lake Baikal basin annual run-off changes were directed differently and were statistically significant in some cases. In the north of Lake Baikal basin a statistically significant positive tendency was observed in the cold season (from October to March–April). At the same time within central and southern Lake Baikal basin a tendency for run-off reduction was predominate for most of months. This is in accordance with the results obtained by the integral difference curves – now here marked low water period.

Floods repeatability as the number of excesses of the water levels corresponding to the beginning of flooding was calculated. On the majority of hydrological gauging stations the floods repeatability during 1981 - 2014 became less in comparison with earlier period. With the exception of some hydrological gauging stations, where there was a significant increase in floods repeatability, for example, Chuksha-Savelyevsky (from 10 to 18%), Vitim-Bodaibo (from 41 to 68%), Lena-Kirensk (from 55 to 74%), etc.

5. Conclusion

The increase in air temperature is mainly due to the winter months which leads to intra annual redistribution of river flow. The majority of statistically significant trends of river run off are observed during the cold period of year. Hence, there emerge more favorable (compared with the mid-20th century) heat balance conditions for flow during the winter and summer–autumn low water period. Winter run-off is determined first of all by underground power. It is possible to assume, that now in the Baykal region during the milder winters, the share of underground power increases because of the lower freezing and earlier thawing of the soil.

The local change in floods repeatability indicates rather its anthropogenic origin. Increased flood damage is associated with the active development of flood-prone areas in recent times.

Acknowledgements
The authors thank the Russian Science Foundation for the support of this work under the grant № 17-05-00400.

References

[1] Kattsov V M *et al* 2014 *Second Roshydromet assessment report on climate change and its consequences in Russian Federation* (Moscow: Federal service for hydrometeorology and environmental monitoring) 1009
[2] Voropay N N, Maksyutova E V and Balybina A S 2011 *Env.Res.Let* **6** 19-23
[3] Voropay N N, Gagarinova O V, Ilicheva E A, Kichigina N V, Maksyutova E V, Balybina A S and Osipova O P 2013 *Hydroclimatic studies of the Baikal natural territory* ed Korytnyy L M (Novosibirsk: GEO) 188
[4] Maksyutova E V, Kichigina N V, Voropai N N, Balybina A S and Osipova O P 2012 *Geography and natural resources* **4** 72-80
[5] IPCC 2012 *Summary for policymakers. In: Managing the Risks of Extreme Events and Disasters to Advance Climate Change Adaptation. A Special Report of Working Groups I and II of the Intergovernmental Panel on Climate Change* (Cambridge, UK and NY, USA: Cambridge University Press) 1-19
[6] Semenov V A 2007 *Surface Water Resources of the Mountains of Russia and Neighboring Territories* (Gorno-Altaisk: RIO Gorno-Altaisk. un-ta) 148
[7] Bogdanova E G 1998 *Year, Map in Atlas Nature and Resources of the Earth* (Moscow-Vienna ED HOLZER) **I** 57
[8] Bogdanova E G 2001 *Climate of Russia* ed Kobysheva N V (St. Petersburg: Gidrometeoizdat) 317-329
[9] Georgievsky V Y and Shiklomanov I 2003 *World Water Resources at the Beginning of the 21st Century* (Cambridge: Cambridge University Press) 390–413
[10] Bedritskiy A I *et al* 2008 *Assessment Report on Climate Change and its Consequences in the Russian Federation* (Moscow: Federal service for hydrometeorology and environmental monitoring) 1008
[11] Groisman P Y 2005 *J. of Climate* **18** 1326–1350
[12] Kundzewicz Z W 2006 *Bulletin of WMO July 2006*

IOP Conf. Series: Earth and Environmental Science 107 (2017) 012043 doi:10.1088/1755-1315/107/1/012043

Interannual climate anomalies in the Atlantic-European region associated with La-Nina types

E N Voskresenskaya, O V Marchukova, V N Maslova, A S Lubkov

Institute of Natural and Technical Systems, Lenin str., 28, Sevastopol, 299011, Russia
E-mail: elena_voskr@mail.ru

Abstract. La Nina manifestations in the meteorological fields of the Atlantic-European region are studied in the paper taking into account its classification based on spatial and temporal features. Monthly global data on sea surface temperature (SST) from COBE SST2 and HadISST data sets in 1900 – 2014, air temperature and sea level pressure from 20th Century Reanalysis V2c were used. Two La Nina types differing one from another in character of SST cold anomaly propagation over the equatorial Pacific region were classified. It is shown that interannual ocean-atmosphere anomalies manifest in the Atlantic-European region mostly in winter due to teleconnection mechanism. Additionally, the role of North Atlantic Oscillation and East Atlantic Oscillation in the level of La Nina types manifestation are analyzed too.

1. Introduction

El Nino — Southern Oscillation (ENSO) is a quasi-periodic (two to seven years) signal of the air-sea interaction in the equatorial Pacific [1]. It characterizes by two episodes – the warm (El Nino) and cold (La Nina) ones.

ENSO provokes great interest of world leading scientists and experts due to its global climate manifestations in the environment especially its negative consequences [2]. Investigation of the ENSO influence on the regional hydrometeorological anomalies around the Globe is one of the most important issues of the study within the principal world and regional climate projects since the late second half of XX century. ENSO manifestations in the Atlantic-European region is an important problem within lot of international programmes such as CLIVAR, EUROMED, etc. The El Nino and La Nina impacts on the atmospheric circulation in the Atlantic-European region was generally studied in [3-6]. The authors [4, 6] emphasized that atmospheric circulation variability in the Atlantic-European region is more prominent in La Nina years than in El Nino years.

As a rule, La Nina events are characterized by an extreme SST drop, growth of Southern Oscillation index and trade wind strengthening in the equatorial Pacific Ocean. Unlike El Nino, the La Nina events gained the scientific interest some later. For a long time La Nina was considered as El Nino feature [7]. Only in late 1990s after 1998 – 2000 La Nina events, which led to catastrophic consequences around the Globe, researchers began to investigate this event more carefully [8].

Nowadays, taking into account different approaches El Nino events are classified into two to three types [9-12]. However, the question of La Nina classification is still open for discussion. Some authors have shown that the zonal location of maximum cold SST anomalies in the central Pacific does not reflect obvious character of each individual La Nina event [11, 13, 14]. On the other hand, there are the publications which confirm the existence of two La Nina types [15, 16]. The results of papers [6, 16, 17] show that different La Nina types have different climatic responses in anomalies of precipitation, surface wind direction and surface air temperature in the lower atmosphere both in the tropical and non-tropical regions. Nevertheless, the mathematical confirmation of La Nina differences has not been obtained yet, thus there is no generally recognized classification at the moment. Therefore, the aim of this paper is to clarify the preliminary La Nina classification and then study the pattern of the each La Nina type manifestation in climatic anomalies in the Atlantic-European region in 1900 - 2014.

2. Data and methods

Monthly SST data sets from Met Office Hadley Centre (HadISST) [18] and Japan Meteorological Agency (COBE SST2) [19] of 1° space resolution from 1900 to 2014 are used in this study. Besides this, Southern Oscillation index (SOI) obtained for 1856 – 2013 in the University of East Anglia Climatic Research Unit is used too [20]. To remind, SOI index is calculated as a normalized difference of sea level pressure (SLP) between the Darwin and Tahiti in the Pacific. At the same time, Nino3.4 index was calculated as SST anomaly in Nino3.4 region (5 °N- 5 °S, 170 °W-120 °W) using HadISST and COBE SST2 data sets. The time series (SOI and Nino3.4) were detrended and tested for Gaussian distribution.

Sea level atmospheric pressure (SLP) and air temperature (T_a) from the 20th Century Reanalysis V2c with 1° space resolution in 1900 – 2014 [21] were analyzed as the main parameters to study La Nina manifestations in the Atlantic-European region. At the same time, North Atlantic Oscillation (NAO) index and Eastern Atlantic Oscillation (EA) indices were used too.

Seasonal variability was filtered from all data sets. In order to assess seasonal variability the average monthly values of $\tilde{F}(\varphi,\lambda,z_0,\tau))$ were calculated for all time series for the each month:

$$\tilde{F}(\varphi,\lambda,z_0,\tau) = \frac{1}{n}\sum_{i=0}^{n-1} F(\varphi,\lambda,z_0,\tau + 12i), \quad (1)$$

where φ is latitude, λ - longitude, z_0 - depth, τ - 1,2, ..., 12 months, and n – number of observation years (in our case 114). Then monthly anomalies $\delta\tilde{F}(\varphi,\lambda,z_0,t)$ of the obtained seasonal variability and linear trend

$$T(\varphi,\lambda,z_0,\tau) = a + bX\,, \quad (2)$$

were calculated: $\delta\tilde{F}(\varphi,\lambda,z_0,t) = F(\varphi,\lambda,z_0,\tau) - \tilde{F}(\varphi,\lambda,z_0,\tau) - T(\varphi,\lambda,z_0,\tau)$, where X is a serial value of SST, SLP or T_a time series, a and b are linear trend coefficients.

Thus anomalies for the each grid point were calculated relative to 1900-2014 period for monthly SST, SLP and T_a series after removing linear trend and seasonal variability.

La Nina classification was made on the basis of cluster analysis (hierarchical method). As a measure of relationship between clusters the Euclidean distance was chosen which is the most popular metric in cluster analysis representing a geometric distance in the multidimensional space. It is determined by the formula:

$$d_{E_y} = \sqrt{\sum_{k=1}^{m}\left(x_{ik} - x_{jk}\right)^2} \quad (3)$$

where x_{ik} and x_{ik} are values of the k variable at i and j distance, m is a number of characteristics. Two variables were applied for La Nina geographic coordinates and values of SSTA in October-December (La Nina events reach of the mature phase in these months in 80% cases).

Composite analysis [22] was used to study the patterns of La Nina type manifestations in the meteorological fields associated with the mature phase of event. Spatially averaged strongest negative monthly SST anomalies were chosen to define the La Nina mature phase and a 3 months period was a time criterion for composites [16]. Composites of SLP and T_a anomalies associated with different La Nina types were calculated for the each month from December of the La Nina onset («0») year to November of the next («+1») year for the 1900 – 2014 interval. Statistical significance of the composites was assessed by the Student's test.

Using the Atlas of Extratropical Storm Tracks for 1961-1998 from the NASA GISS web page (http://data.giss.nasa.gov/stormtracks/) composite schemes of winter storm tracks were obtained.

3. La Nina spatial classification

The initial task of this study is to identify La Nina events on the basis of specific criteria. The list of operational definitions and indices of La Nina and El Nino suggested by 26 member-countries of the World Meteorological Organization (WMO) is published [23]. And it is noted that the researcher is responsible for criteria choice to identify the cold and warm ENSO episodes.

Similar to El Nino, the La Nina is usually identified by Nino3.4 index (see above). In present paper Nino3.4 index was calculated for the period 1900-2014. Criterion chosen to identify La Nina events was a cold SST anomaly (SSTA) with -0.5°C threshold and minimum duration (D) of 5 months as follows:

$$\text{SSTA in Nino3.4} \leq -0.5°\text{C},$$
$$D \geq 5 \text{ months}.$$

However, one primary criterion is not enough to provide the complete selection of cold episodes of ENSO. Accordingly, we also involved the Nino1+2 index (SST anomaly of -0.75 °C in the region limited by coordinates 0°-10 °S and 90°-80 °W). This permitted us to make sure that the selected event is really La Niña, but not just a period with a fall of average SST in the Pacific during 5 months. The cold anomaly can progress in the Eastern equatorial Pacific (Nino1+2), but it stay weak in the Central equatorial Pacific (Nino3.4). To check the La Nina occurrence and clearly illustrate the intensity and spatial scales of anomalies, the maps of the cold SST anomalies during the mature La Nina phase were built.

Thus, using both HadISST and COBE SST2 data sets and the abovementioned criteria 21 La Nina events were selected within 114 years period: in 1909-1910, 1916-1917, 1924-1925, 1933-1934, 1938-1939, 1942-1943, 1949-1951, 1954-1956, 1964-1965, 1967-1968, 1970-1971, 1973-1974, 1975-1976, 1983-1984, 1984-1985, 1988-1989, 1995-1996, 1998-2000, 2007-2008, 2010-2011, 2011-2012. Long-lasting La Nina events, such as 1954-1956, 1949-1951 and 1998-2000, were considered as two different La Nina events in spatial classification because they have two significant SSTA maxima with a 12 month periodicity. Note, that such events were identified by some authors [6, 16, 17], however, using much shorter time series, in particular from 1950 to 2013.

The analysis of all monthly maps revealed the following. Some La Nina events are characterized by the occurrence and development of cold SST anomaly in the central equatorial Pacific, while the others – in its eastern part, near Peru and Ecuador coast. The similar feature is described in [6, 16, 17].

To get the objective La Nina classification, the hierarchical cluster analysis was applied to both HadISST and COBE SST2 data sets. The results shown in the form of a dendrogram in figure 1 indicate the presence of two clusters. Accordingly, 12 La Nina years with the lowest SSTA values in the east are in the first cluster named the East Pacific La Nina type (EP), and 12 La Nina years with the most negative SSTA values in the central equatorial Pacific belong to the second cluster named the Central Pacific type (CT).

Figure 2 illustrates a clear difference between the two La Nina types in location of the most negative SSTA during its mature phase in October-December. Maps were obtained for the each identified type of events on the basis of HadISST (figure 2 a, b) and COBE SST2 (figure 1 c, d) data sets. Two independent data sets produce identical results, and this additionally demonstrates the significance of our results. Composites of SSTA spatial distribution during the mature La Nina phase in October-November for the EP La Nina type are shown in figure 2 a and b and CP La Nina type – in figure 2b and d. However, it does not mean that cold anomalies occur in the same places. The main feature of the CP La Nina type is the location of the most negative SSTA in the equatorial Pacific during the mature phase, and accordingly as for the EP La Nina type it is in the East Pacific, near the western coast of South America. So the results of cluster analysis in figure 2 show the place of typical location of the most negative SST anomalies during mature phase of La Nina. It is important for the subsequent study of atmospheric and oceanic responses.

IOP Conf. Series: Earth and Environmental Science **107** (2017) 012043 doi:10.1088/1755-1315/107/1/012043

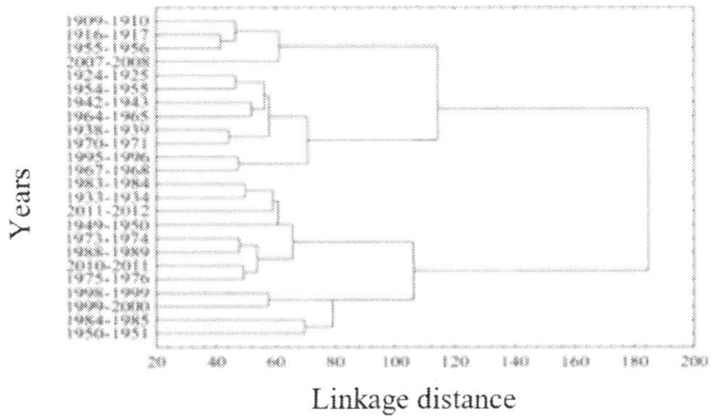

Figure 1. Dendrogram of cluster analysis of classification of La Nina events for 1910-2014 taking into account two criteria: geographic coordinates and values of SSTA in October-December during the mature phase of La Nina.

The Student's test (t) estimations of statistical significance of mean values of the composites ($\bar{x}$) were used to confirm the correctness of classification. To solve this task, two hypotheses were stated: $H_0: \bar{x}_1 \neq \bar{x}_2$ and $H_1: \bar{x}_1 = \bar{x}_2$. Then the empirical Student criterion (t) and critical criterion (t_{cr}) were estimated. H_0 was rejected if $t > t_{cr}$, therefore the difference between the average values of the elements in the same composite was too large and not statistically significant. H_0 was adopted if $t < t_{cr}$, therefore the difference between the average values of the elements in the composite was acceptable and statistically significant. The 95% statistically significant areas of the composites are dotted in figure 2.

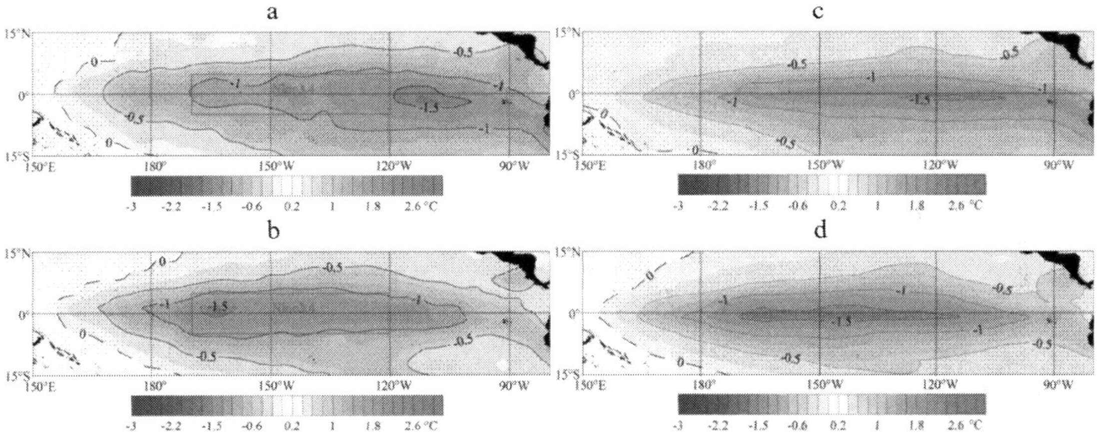

Figure 2. Composites of SSTA spatial distribution by HadISST (a, b) and COBE SST2(c, d) data sets during the mature La Nina phase in October-November in 1900 – 2014 for the EP La Nina type (a, c) and CP La Nina type (b, d). The areas with 95% statistical significance are dotted.

The average composite anomaly during the CP La Nina type is approximately -1.3 °C, while the most negative anomaly reached -2.22 °C in 1988-1989. The CP events usually start in summer, while the period of their mature phase is mostly in November-January. The typical maximum SOI index for 12 CP La Nina events is about 2.69. As for the EP La Nina type, the average anomaly is -1.10 °C which is less than during EP La Nina events. Some of them occur both in summer and autumn, but all of them reach their maxima in autumn-winter. Their mature phase falls on October-December at 0.9 probability level. The typical maximum value of the SOI index for EP La Nina events is about 1.71 which is about a unit less than the typical SOI value for the CP La Nina events. This considerable difference in the atmospheric pressure index allows to assume that the atmospheric processes play an important role in the EP La Nina development.

4. Manifestations of La Nina types

Two main meteorological characteristics – SLP and T_a in the Atlantic-European region were studied for each month from December «0» to November «+1» La Nina years for 12 EP and 12 CP events in 1900 – 2014. The strongest signal in the Atlantic-European region during La Nina years was noted in winter from December to March. The maximum La Nina response in the North Atlantic is usually in January.

As a rule, the mature phase of EP La Nina type is in October-December, while in the case of CP type it is in November-January. Due to lagged teleconnection La-Nina manifestations in the Atlantic-European region is observed 1-2 month later after its most negative SSTA in the Pacific Ocean. La Nina influence manifests in changes of the North Atlantic Oscillation (NAO) and the East Atlantic (EA) oscillation. CP type is followed by the positive NAO index in that winter, while the EP La Nina type – by the negative one. At the same time, the EA index is negative during both the EP and CP La Nina types, but it is more intense for the EP type.

The composite maps of SLP and T_a in January and February associated with EP La Nina type (12 events) and CP La Nina type (12 events) based on the 20th Century Reanalysis V2c data sets are shown in figure 3. Statistically significant negative SLP anomaly related to EP La Nina years (figure 3 a, b, e, f) is observed in December-February over subtropics and temperate latitudes in the North Atlantic. At the same time positive SLP anomaly is located in the North-East of the Atlantic-European region. This structure of the SLP field corresponds to the negative EA pattern. The analyses of all monthly composites shows that EP La Nina years are accompanied by EA negative phase strengthening, which causes cold winters in the Eastern Europe (figure 3 b, f) and late spring throughout the European region.

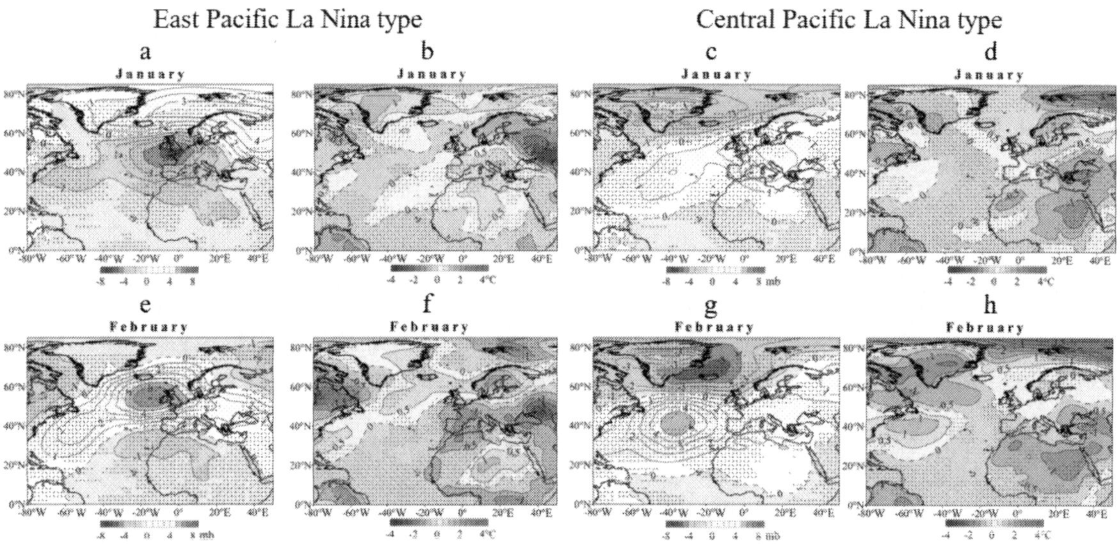

Figure 3. Composites of sea level pressure anomalies (a, e), air temperature anomalies (b, f) during the EP La Nina type (12 events) and composites of sea level pressure anomalies (c, g) and air temperature anomalies (b, d) during the CP La Nina type (12 events). The areas with 95% statistical significance are dotted.

CP La Nina type (figure 3 c, d, g, h) is characterized by the opposite spatial structure of anomalies relative to the EP type. In winter, a positive SLP anomaly is observed in the temperate latitudes (40-60 °N) over the North Atlantic. A negative SLP anomaly is in high latitudes and reaches -6 mb (figure 2 c, g). This structure corresponds to the positive phase of the North Atlantic Oscillation. In this case the trajectories of the North Atlantic cyclones shift about 200-400 km to the North and bring a positive temperature anomaly to the Western and Northern Europe (figure 3 d, h). The characteristic feature is that due to cyclonic activity a significant positive air temperature anomaly occurs in the Barents Sea region in December - March during the CP La Nina type.

Thus, the La Niña classification implemented in this paper has permitted to find the features of two La Nina types associated with the global processes in the ocean-atmosphere system and their manifestations in the fields of meteorological anomalies by the means of strengthening or weakening of the EA and NAO.

Composites of winter storm tracks during the mature phase (+1 year) of La Nina were obtained. In connection with the limitation of the data set 8 events were considered: 4 of them for are CT and 4 for the ET La Nina types. Composite schemes of storm tracks in January and February calculated for each type of La Nina (figure 4) agree with the composite structure of SLP and air temperature anomalies shown above. North Atlantic storm tracks are more spreaded toward the South during the EP La Nina type (figure 4 a, c) as well as for the negative EA oscillation phase and more concentrated northward for the CP type (figure 4 b, d) as well as for the positive NAO phase. In the result cyclonic activity over the Central Europe and Mediterranean basin intensifies during the EP La Nina type.

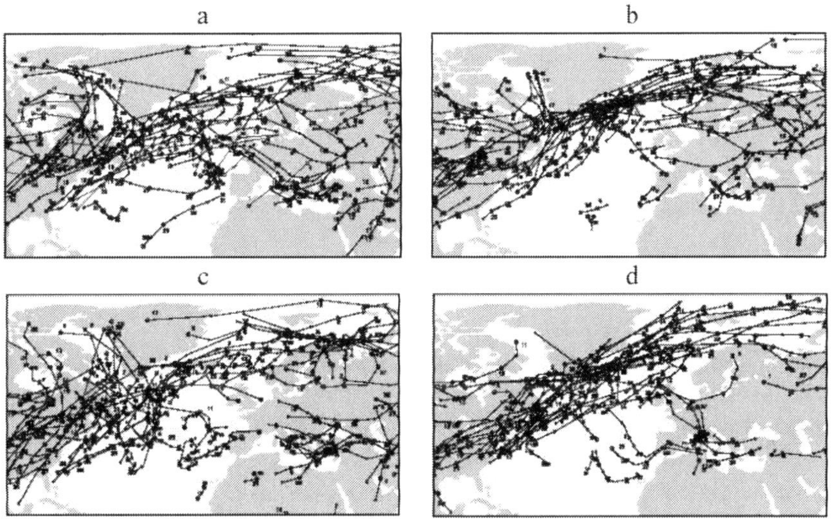

EP La Nina type **CP La Nina type**

Figure 4. Composite schemes of storm tracks in January (a, b) and February (c, d) in "+1" year of the (a, c) Eastern Pacific La Nina type (1965, 1968, 1971, 1996) and (b, d) Central Pacific La Nina type (1974, 1976, 1984, 1989).

5. Conclusions

On the basis of HadISST and COBE SST2 data sets in 1900 - 2014 using the criteria of -0.5 °C SST anomaly and its duration for no less than 5 months in the Nino3.4 region, the La-Nina events were selected and analyzed. It was found that the La Nina contribution in the total variance of the ENSO variability is 48.5%.

Two La Nina types (East Pacific and Central Pacific) were detected using the hierarchical cluster analysis method taking into account the most negative SSTA values and their spatial location in the mature La Nina phase. It was shown that the CP La Nina events are more intense than the EP La Nina events estimated by the SSTA and SOI values. This result confirms the important role of the atmospheric processes in formation of the CP type of La Nina.

The manifestations of the two La Nina types were shown in the meteorological fields of the Atlantic-European region. The EP La Nina years are characterized by the strengthening of the negative

EA phase causing the cold winters in eastern Europe, including the Azov-Black Sea region, with anomalies reaching -2.5 °C in January and February. The CP La Nina type manifests in the strengthening of the NAO positive phase which causes a positive air temperature anomaly in western and northern Europe. There is a significant positive air temperature anomaly in the Barents Sea region during the CP La Nina type from December to March due to the cyclonic activity over the region.

Acknowledgement
The work was carried out with the partial financial support of the Russian Foundation for Basic Research (RFBR) grant (project № 16-05-00231-A).

References

[1] Philander S G 1990 *El Nino, La Nina and the Southern Oscillation* (CA: Academic Press, San Diego) 304
[2] Rasmusson E M and Wallace J M 1983 *Science* **222** 1195-1202
[3] Dong B-W, Sutton R T, Jewson S, O'Neill A and Slingo J M 2000 *Geophys. Res. Let.* **27** 985-988
[4] Nesterov E S 2000 *Russian Meteorology and Hydrology* 8 74-83
[5] Mokhov I I and Smirnov D A 2006 *Izvestiya. Atmospheric and Oceanic Physics* **42** (5) 598–614
[6] Zhang W, Wang L, Xiang B, Qi L and He J 2014 *Climate Dynamics* **44** (5-6) 1351-1366
[7] Glantz M H 1998 *A La Nina Summit: A Review of the Causes and Consequences of Cold Events. Environmental and Societal Impacts Group* (Boulder: NCAR) 43
[8] Glantz M H 2002 *La Nina and its impacts: facts and speculations* (New York: The United Nations University) 313
[9] Wang B 1995 *J. Clim.* **8** 267-285
[10] Ashok K, Behera S K, Rao S A, Weng H and Yamagata T 2007 *J. Geophys. Res.* **112** C11007
[11] Kao H Y and Yu J Y 2009 *J. Clim.* **22** 615-632
[12] Yeh S W, Kug J S, Dewitte B, Kwon M H, Kirtman B P and Jin F F 2009 **461** 511-514
[13] Kug J S, Jin F F and An S I 2009 **22** 1499-1515
[14] Ren H L and Jin F F 2011 *Geophys. Res. Lett.* **38** L04704
[15] Shinoda T, Hurlburt H E and Metzger E J 2013 *J. Geophys. Res.* **116** C12001
[16] Yuan Y and Yan H M 2013 *Chin. Sci. Bull.* **58** (3) 406-415
[17] Singh A, Delcroix T and Cravatte S 2011 *J. Geophys. Res.* **116** C06016
[18] Rayner N A, Parker D E, Horton E B, Folland C K, Alexander L V, Rowell D P, Kent E C and Kaplan A 2003 *J. Geophys. Res.***108** (D14) 4407
[19] Hirahara S, Ishii M and Fukuda Y 2014 *J. of Climate* **27** 57-75
[20] Ropelewski C F and Jones P D 1987 *Monthly Weather Review* **115** 2161-2165
[21] Compo G P, Whitaker J S and Sardeshmukh P D 2006 *Bull. Amer. Met. Soc.* **87** 175-190
[22] Laken B and Calogovic J 2013 *J. Space Weather Space Clim.* **3** A29
[23] 2006 *Catalogue of Indices and Definitions of El Niño and La Niña in Operational Use by WMO Members / WMO Commission for Climatology; CCl-XIII (2005) Expert Team on El Niño and La Niña Definitions* (Geneva: World Meteorological Organization)

Changes of extreme precipitation in Southern Russia

E V Vyshkvarkova, E N Voskresenskaya

Institute of natural and technical systems, Lenin St., 28, Sevastopol, 299011, Russia
E-mail: aveiro_7@mail.ru

Abstract. Features of extreme precipitation and their long-term changes over Southern Russia were analysed in the paper using daily precipitation data from 21 hydrometeorological stations in 1950 – 2010. Two climate indices for extreme precipitation (RX1day and RX5day) and concentration index data series for the study period were calculated and analysed. It was defined that precipitation extreme changes are characterized by spatial and temporal variability. It was found the tendency to wetter in extreme anomalies, however, the trends are mainly negligible. Spatial distribution of precipitation concentrations demonstrates that the northern, southern and western regions are characterized by moderate values, while the coast of the Caspian Sea and the territory of Caspian Depression has shown high values of CI.

1. Introduction

The temperature and precipitation behavior has significantly changed during past centuries. It is confirmed by IPCC report – "It is certain that Global Mean Surface Temperature has increased since the late 19th century" and "averaged over the mid-latitude land areas of the Northern Hemisphere, precipitation has likely increased since 1901 (medium confidence before and high confidence after 1951)" [1]. This also applies to the extreme values of temperature and precipitation – "It is very likely that the numbers of cold days and nights have decreased and the number of warm days and nights have increased globally since about 1950" and "It is likely that since about 1950 the number of heavy precipitation events over land has increased in more regions than it has decreased" [1].

Events with extreme precipitation may lead to catastrophic consequences, e.g. drought, floods and others. A graphic example is the floods in Krasnodar Krai of Russia in early July 2012, when the equivalent of two-five months precipitation norm fell in short time [2, 3]. Droughts in many Russian regions and following them forest fires caused great disasters in 2010 and 2012. In this work we used daily data because only daily resolution can provide information about the frequency, number of rainy days, or gaps in precipitation series. The south of Russia is a large and world famous agricultural and recreational region. The territory is characterized by diverse climate and complex orthography (Crimean Mountains, Caucasus Mountains, deltas of large rivers, coastal zone of the Sea of Azov, the Black Sea and the Caspian Sea). These factors together with natural variability of meteorological parameters can lead to the above-mentioned catastrophic events.

Thus, the aim of the paper is to study the typical characteristics of extreme precipitation in Southern Russia in 1950 – 2010 and its temporal distribution.

2. Data and method

Daily precipitation totals data observed at 21 meteorological stations in period 1950 – 2010 were analyzed. These data were obtained from the KNMI Climate Explorer (http://climexp.knmi.nl) and National Oceanic and Atmospheric Administration (NOAA, http://www.ncdc.noaa.gov). Stations names and their locations are presented in table 1 and in figure 1.

IOP Conf. Series: Earth and Environmental Science 107 (2017) 012044 doi:10.1088/1755-1315/107/1/012044

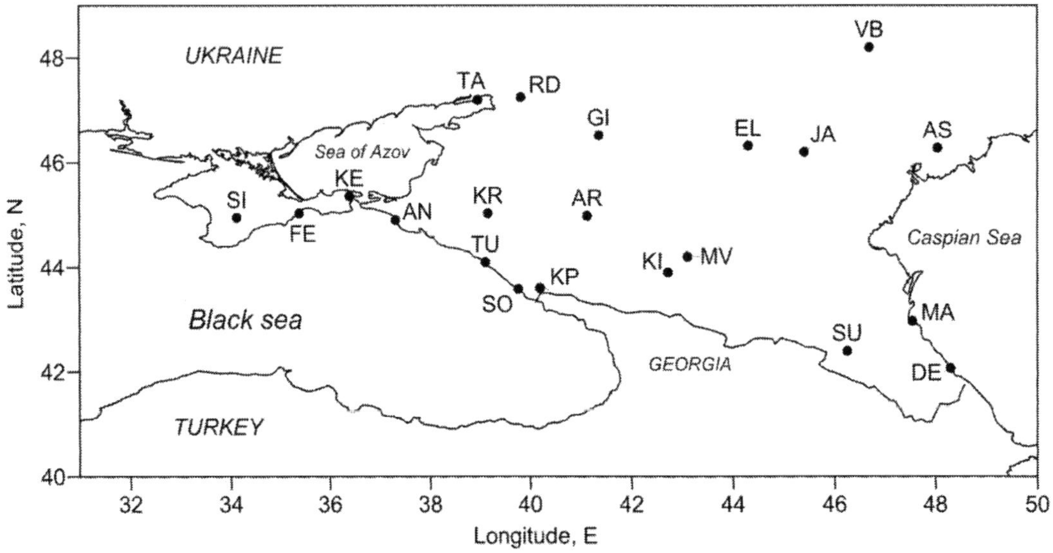

Figure 1. Study region. The black points mark stations location.

Table 1. List of hydrometeorological stations, their WMO (World Meteorological Organization) number, latitude, longitude and altitude (m).

Station name	WMO no.	Latitude (N)	Longitude (E)	Altitude (m)
Anapa (AN)	37001	44.90	37.30	30.0
Armavir (AR)	37031	44.98	41.12	159.0
Astrakhan (AS)	34880	46.28	48.05	-23.0
Derbent (DE)	37470	42.07	48.30	-19.0
Elista (EL)	34861	46.32	44.30	151.0
Feodosia (FE)	33976	45.03	35.38	22.0
Gigant (GI)	34740	46.52	41.35	79.0
Jaskul (JA)	34866	46.20	45.40	-7.0
Kerch (KE)	33983	45.36	36.39	49.0
Kislovodsk (KI)	37123	43.90	42.72	890.0
Krasnaja poljana (KP)	37107	43.60	40.20	566.0
Krasnodar (KR)	34927	45.03	39.15	29.0
Makhachkala (MA)	37472	42.97	47.55	28.0
Mineralnye vody (MV)	37054	44.20	43.10	316.0
Rostov-na-Donu (RD)	34730	47.25	39.82	66.0
Simferopol (SI)	33946	44.95	34.12	181.0
Sotchi (SO)	37099	43.58	39.77	57.0
Sulak (SU)	37461	42.40	46.25	2927.0
Taganrog (TA)	34720	47.20	38.95	30.0
Tuapse (TU)	37018	44.10	39.10	41.0
Verkhnij Baskuncak (VB)	34579	48.20	46.70	35.0

Two methods to analyze extreme precipitation were used. First of them is climate indices. The extremes precipitation indices were selected from the list of climate indices recommended by the World Meteorological Organization–Commission for Climatology (WMO–CCL) and the Research Programme on Climate Variability and Predictability (CLIVAR) Expert Team on Climate Change Detection, Monitoring and Indices (ETCCDMI) [4, 5]. The main function of these indices is to assess the changes in the intensity, frequency and duration of precipitation events [6]. Climate extreme indices that were used in the paper are presented in table 2. Selected indices represent frequency of extreme precipitation. Data quality control and extreme climate indices calculation were done using standardized software RClimDex (freely available at http://etccdi.pacificclimate.org /software.shtml), developed by ETCCDMI.

Table 2. List of ETCCDI climate indices used in the work.

Index	Definition	Units
Rx1day	Maximum 1-day precipitation total	mm
Rx5day	Maximum 5-day precipitation total	mm

Annual and monthly time series of selected climate indices for each station were calculated. Annual indices distribution, their first, second and third quartiles, maximum and minimum values were analysed using box-and-whiskers plots approach [7, 8].

Next, the concentration index (CI) method was applied. This index allows evaluating the contribution of very rainy days to the total amount, because a few daily events produce a high percentage of total monthly, seasonal or annual precipitation. The methodology proposed by Spanish climatologist J. Martin-Vide and described in [9]. Later this methodology was applied to investigation of precipitation concentration in many regions – for the territory of Europe [10], in China [11], in northern Africa [12], in USA [13] and around the world [14].

3. Results

To show the features of extreme precipitation at the first step of study we consider the spatial distribution of typical precipitation extremes over the Southern Russia territory in period 1950 – 2010 and their long-term changes using selected indices.

3.1. Climate indices

A graphical representation of the annual RX1day and RX5day indices distribution on the basis of box-and-whiskers plots are demonstrated in figure 2. The bottom and top of the box are the first and third quartiles; the band inside the box is the second quartile (the median). The dashes at the end of whiskers show maximum and minimum values, the dots show outliers

The box plot of extreme precipitation events in short periods of time (RX1day) is presented in figure 2a. The index shows monthly maximum 1-day precipitation. Minimum values of RX1day range from 8.3 mm·day^{-1} in eastern part of region to 48.5 mm·day^{-1} for the Black sea coastal stations. Maximum values vary from 50.5 mm·day^{-1} in Jaskul to 179.5 mm·day^{-1} in Tuapse. Annual median values of RX1day index ranges from 22.5 mm·day^{-1} in the northeastern part to 82 mm·day^{-}1 on the Black Sea coast. The biggest values of index RX1day are typical for the Black sea coastal stations.

Analysis of the wettest consecutive five days (RX5day) is presented in figure 2b. Distribution of the annual maximum five-day precipitation accumulation is similar to distribution of RX1day index. Minimum values range from 10.8 to 106.2 mm for the eastern part of region and the Black sea coastal zone respectively. Mean values range from 32.3 mm in the northeast of territory to 162.5 mm for the Black Sea coastal zone. The maximum values of RX5day index is typical for the Black Sea coastal stations and reached 285.1 mm during the studied period. The smallest change of RX5day index is typical for Elista station (continental climate) and the largest for Tuapse. Summarizing the results of precipitation extremes analysis it can be concluded that the maximum values of indices are typical for the Black Sea coast and the minimum for the northeastern part of the region.

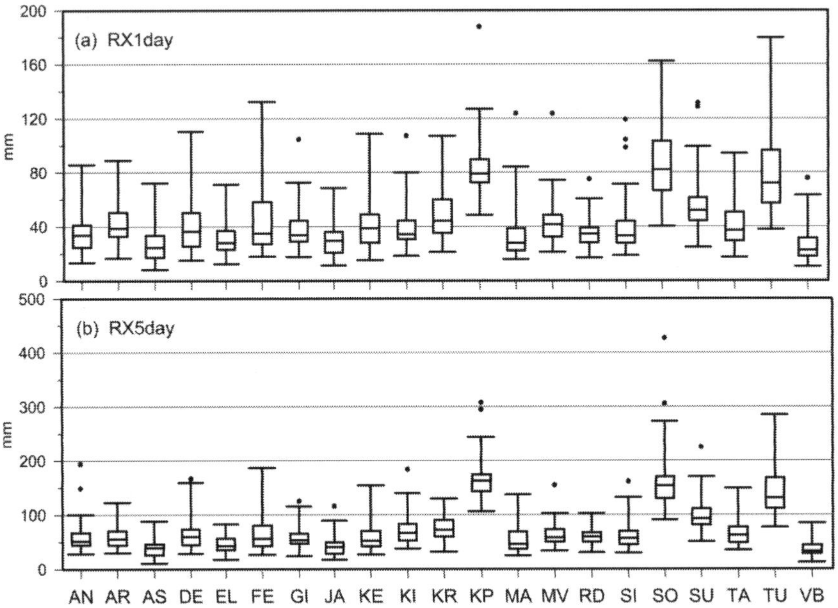

Figure 2. The box plots of the RX1day (a) and RX5day (b) indices for the period 1950 – 2010.

Distribution of extreme precipitation trends is characterized by strong spatial variability. Spatial distribution of trends RX1day index is shown in figure 3a. Linear trends are predominantly positive and range from 0.1 mm·(10 years)$^{-1}$ (the southeastern part of studied region) to 3.5 mm·(10 years)$^{-1}$ (the coastal zone of Black Sea). But there are only several significant ($p < 0.05$) trends. Negative trends are typical of the north-western part of the studied region, coastal zone of the Sea of Azov (continental part of Russia). Trend coefficient varies between 0.04 and 1.8 mm·(10 years)$^{-1}$, but all of them are not significant. Analysis of long-term changes of the wettest consecutive five days (RX5day) are presented in figure 3b. Upward changes are typical for the region under consideration. Positive trends ranges have the values between 0.5 and 5.4 mm·(10 years)$^{-1}$. Negative trends are few in number and not significant. So, the analysis of long-term precipitation changes generally demonstrates insignificant increase to wetter conditions.

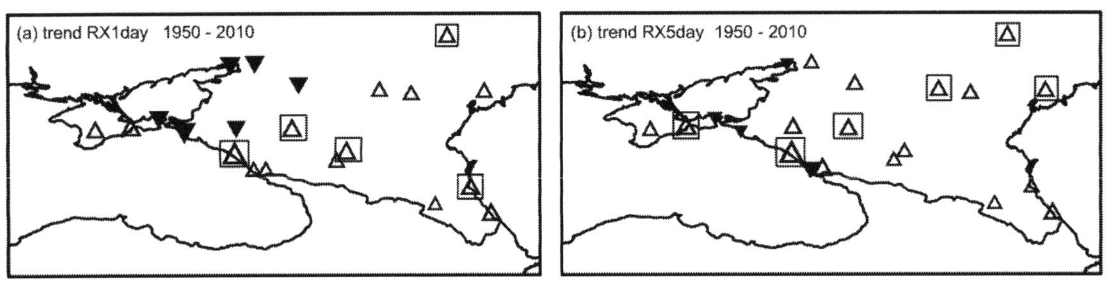

Figure 3. Annual trends for the (a) RX1day (mm/10 years) and (b) RX5day (mm/10 years) indices for the period 1950 – 2010. White upward triangles mean positive linear trend, black downward triangles mean negative linear trend. Significant precipitation trends ($p < 0.05$) are indicated by external square. The sizes of triangles are proportional to the trend values.

3.2. Precipitation concentration

The concentration index was calculated for 21 meteorological stations for the period 1950 – 2010. According to the previous studies $CI > 0.61$ is high, $0.55 < CI < 0.61$ is moderate and $CI < 0.55$ is low.

The minimum value of CI is typical for north-eastern part of the region and for Anapa station. High values of CI are found in the Caspian Sea coast and the Caspian Depression.

The northern part of the region is characterized by sufficiently high values of the CI (0.58 – 0.59). The climate of this region is moderate-continental, with a short, cold, snowy winter and a long, hot, dry summer. In Ciscaucasia and the northern part of the Greater Caucasus index values vary within 0.58 to 0.6. This is explained by the uniform distribution of rainy days throughout the year. As it was said above high values of concentration index are observed on the coast of the Caspian Sea and in the territory of the Caspian Depression. It can be explained by geographical location and the atmospheric circulation over this region. In winter and summer precipitation regime is irregular, which leads to high values of the CI. High value of CI was also found in the eastern part of the Crimean peninsula.

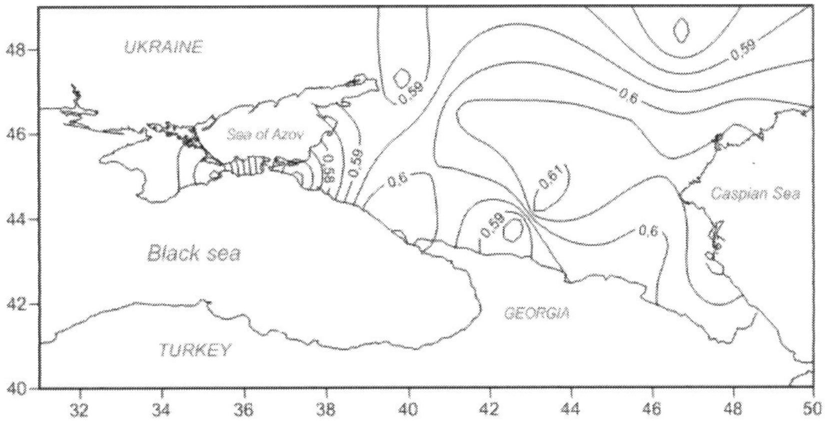

Figure 4. Spatial distribution of concentration index values.

4. Conclusions

The study of precipitation extremes distribution in Southern Russia and their long-term changes in 1950 – 2010 showed the following results. The maximum values of climate indices (RX1day and RX5day) are typical for the Black Sea coastal zone. Linear trends of precipitation indices are predominantly positive, but only a part of them is significant. Analysis of precipitation concentrations made it possible to identify several regions with different regimes. The eastern part of the region showed the highest concentration of precipitation, whilst the northern, southern and western parts exhibit relatively regular precipitation. The highest CI values were found on the coast of the Caspian Sea and in the territory of the Caspian Depression.

Acknowledgments
The reported study was funded partly by RFBR according to the research project № 16-05-00231.

References

[1] IPCC 2013 Climate *Change 2013: The Physical Science Basis. Contribution of Working Group I to the Fifth Assessment Report of the Intergovernmental Panel on Climate* (Cambridge, UK and NY, USA: Cambridge University Press) 1535

[2] Meredith E P, Semenov V A, Maraun D, Park W and Chernokulsky A V 2015 *Nature Geoscience* **8** 615–19

[3] Kotlyakov V M, Desinov L V, Dolgov S V, Koronkevich N I, Likhacheva E A, Makkaveev A N, Medvedev A A and Rudakov V A 2013 *Regional Research of Russia* **3** (1) 32–39

[4] Frich P, Alexander L V, Della-Marta P, Gleason B, Haylock M, Klein Tank A M G and Peterson T 2002 *Clim. Res.* **19** 193–212

[5] Peterson T C 2005 *Bulletin WMO* **54** 83–6

IOP Conf. Series: Earth and Environmental Science **107** (2017) 012044 doi:10.1088/1755-1315/107/1/012044

[6] Alexander L V *et al* 2006 *J. Geoph. Res.* **111** D05109
[7] McGill R, Tukey J W and Larsen W A 1978 *The American Statistician* **32** (1) 12–6
[8] Benjamini Y 1988 *The American Statistician* **42** (4) 257–62
[9] Martin-Vide 2004 *Int. J. Clim.* **24** 959–71
[10] Cortesi N, González-Hidalgo J C, Brunetti M and Martin-Vide J 2012 *Nat. Hazards Earth Syst. Sci.* **12** 2799-2810
[11] Li X, Jiang F, Li L and Wang G 2011 *Int. J. Climatol.* **31** 1679-1693
[12] Benhamrouche A, Boucherf D, Hamadache R, Bendahmane L, Martin-Vide J and Teixeira Nery J 2015 *Nat. Hazards Earth Syst. Sci.* **15** 617-625
[13] Roye D and Martin-Vide J 2017 *Atmospheric Research* **196** 237-247
[14] Monjo R and Martin-Vide J 2016 *Int. J. Climatol.* **36** 3828–3838

IOP Conf. Series: Earth and Environmental Science **107** (2017) 012045 doi:10.1088/1755-1315/107/1/012045

Impact of climate changes on population vital activities in Russia in the early 21st century

A N Zolotokrylin, V V Vinogradova, T B Titkova, E A Cherenkova, D D Bokuchava, I A Sokolov, A V Vinogradov, E D Babina

Institute of Geography, Russian Academy of Sciences, Staromonetnyy per., 29, Moscow, 119017, Russia
E-mail: vvvinog@yandex.ru

Abstract. The study substantiates the approach to the assessment of impact of climate change on vital activities of population in Russia in the face of increasing climate extremes. The obtained results reveal the occurrence of the essential climate extreme events over the period 1991–2013 in Russia that are vital for population activities. Annual amounts of interdiurnal temperature differences and pressure were calculated. Propagation of heat and cold waves, trends and frequencies of daily precipitation extremes were evaluated. The map "Zoning the territory of the Russian Federation by natural living conditions of the population" adapted for modern climate (2001–2010), illustrates the climate changes in the early 21st century. The modern warming of climate has led to a significant easing of discomfort in the territory of Russia. The steady decline of the absolutely unfavorable zone resulted from the expansion of less unfavorable areas is observed, especially in the Northern and Arctic regions. In the south the boundary of unfavorable territories shifts toward the north. It results in the expansion of the conditionally unfavorable area in West Siberia and in the south of East Siberia. In European Russia the favorable area expands and shifts far to the northern regions.

1. Introduction

The continuous global warming and growing climate extremes have an ambiguous impact on various areas of population life. Climate change is a significant threat for the population vital activities and especially for population health. Since the mid -1970s the average surface air temperature in Russia has been increasing at an average rate of 0.43 °C per decade that excesses the global warming rate more than twofold [1]. The most significant climate changes are observed in the Arctic regions and subarctic areas of permafrost. According to the observation data the tendency to slowdown in the climate warming has not yet been traced (except for winter). The main contribution to the observed temperature rise in Russia is caused by changes in the concentration of greenhouse gases. The surface pressure in the significant part of Russia decreased in the second half of 20th century. At the same time almost everywhere in the territory the wind speed became less [1].

This study presents new results indicating the geographical distribution of extreme climate events with a significant impact on the population vital activity in Russia over the period 1991–2013. For the first time, the annual amounts of the average daily temperature differences and pressure were calculated. There were also obtained the evaluations of heat and cold wave propagation, trends and the frequency of daily precipitation extremes.

2. Material and methods

The available observations over the average daily maximum and minimum temperatures and daily precipitation total from the VNIIGMI (All–Russian Research Institute of Hydrometeorological Information – Worldwide data centre) [2] (http://meteo.ru/) for the whole territory of Russia (458 weather stations) and surface pressure values from the NCEP/NCAR reanalysis archive with a spatial resolution of 2.5° × 2.5° were used in this study [3]. The study covered the peculiarities of the geographical distribution of heat and cold waves with characteristics significant for population activities, the maximum precipitation totals and the average daily temperature and pressure differences (swings) in the late 20th – early 21st century. The maximum (or minimum) daily temperatures exceeding the 95th percentile value (or below the 5th percentile) of maximum (or minimum) temperature distribution over the base period were considered as a threshold value to identify the heat and cold waves. Trends in daily total precipitation and frequency of precipitation extremes for the various periods of the year were calculated. The assessment was performed for the whole territory of Russia and for the most sensitive regions with a high density of population where precipitation extremes lead to the negative outcome for environment and human life activities. The results of numerous foreign and domestic studies indicate the impact of the average daily temperature and pressure differences on the amount of hospitalization and mortality caused by circulatory system diseases [4–6]. In this regard, the changes in average annual number of events with critical temperature and pressure differences were analyzed. The annual total of the critical average daily temperature and pressure differences exceeding 6 °C and 8 hPa, respectively was calculated. The significance of changes was estimated using the Student's t test.

3. Results and discussion

3.1. Weather climatic anomalies in Russia

In the early 21st century, the number of days with extremely hot weather increased non–uniformly both over time and through territories [7–11]. In the late 20th century, the growth of positive temperature anomalies was observed mainly in the south of European Russia, in the south of Eastern Siberia and in the Far East. In the north of European Russia, in Western and Eastern Siberia the number of such days decreased. The number of heat waves with a duration exceeding three days increased and the temperature growth of heat waves reached 4–8 °C. The most significant duration of lengthening of heat waves was registered in European Russia, in Transbaikalia and in the northeast of Russia. The growth of the average maximum temperature in the heat wave speeded up, especially in European Russia, where it reached 10–12 °C in eastern and southern areas. The maximal heat wave duration increased almost throughout Russia, with the exception of the Kola Peninsula, Taimyr and the south of Western Siberia. Within the European territory, southward of 60° North latitude it increased by 10–15 days, and in Central Black Soil Regions – by 20 days.

The number of days with extremely low temperatures went down almost throughout Russia except the far northeast regions and Chukotka. The maximum decrease of low temperature days was observed in the southern Urals, in the Middle Volga and in the western regions of Russia. The duration of cold waves decreased significantly in many regions of Russia as well except the Kola Peninsula, north of European Russia and the centre of Siberia. But in the early 21st century the number of cold days and the maximal duration of cold waves began to grow in the southern and central regions of European Russia and in the south of Siberia. Absolute minimum temperatures and average minimum temperatures in the cold wave grew and the areas with extremely low temperatures decreased significantly.

The maximum number of the average daily temperature differences (exceeding 30 events per year) was observed in Siberia. On the Arctic coast, it went down to 25 events per year. Geographically the interdiurnal pressure differences in the European Russia territory have the area–based distribution: the maximum totals are observed on the Arctic coast (more than 40 events per year), and the minimum ones – in the south of the territory (less than 10 events per year). The number of pressure differences in Eastern Siberia increased up to 40 events per year, toward east and south it went down to 10–20 events per year. It should be noted that the maximum number of temperature and pressure differences was observed in the low-populated regions with severe climatic conditions and the minimum ones – in the highly populated regions with mild climatic conditions.

It was established that the frequency of extreme precipitation in winter and spring in the period 1991–2013 increased significantly by an average of 20–40% compared with the climatic norm of 1961–1990 across much of Russia. Against the background of winter and spring positive trends, changes in daily precipitation extremes increase the risks of spring catastrophic floods. In the summer, strong positive trends in precipitation extreme changes were observed in the southern part of the Far Eastern District in the Amur River basin, increasing flood risks in this region.

The consolidation of the results obtained is presented on the integral map of the extreme natural phenomena distribution across the territory of Russia (figure 1).

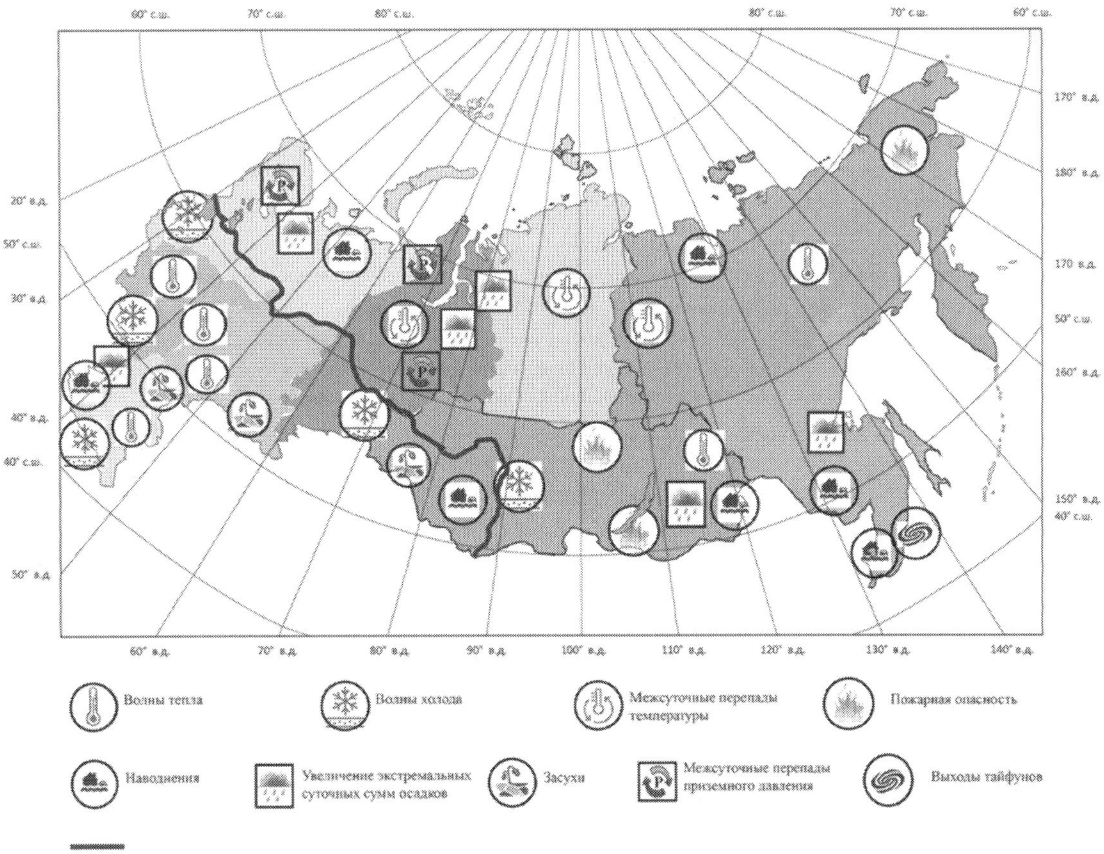

Figure 1. Extreme natural phenomena in Russia for the period 1991–2013. The solid line shows the southern borders between the Far North and the Arctic and the terrains equated to the Far North.

The map analysis shows that the number of the heat waves increased in European Russia, in the Irkutsk region and Yakutia. The growing heat wave frequency and the decreasing daily precipitation extreme amount caused the higher fire hazard in the forest area of Eastern Siberia. Cold waves were observed in European Russia and in the south of Western and Middle Siberia [5, 9]. Areas of distribution of the maximum daily temperature and pressure differentials are located in the northern part of European Russia and in Western Siberia. The areas with maximum average daily temperature and pressure differences were located in the northern part of European Russia and in Western Siberia. The growing amount of daily precipitation extremes was observed in the Krasnodar Territory, in the north of European Russia and in Western Siberia. Destructive floods occurred in Krasnodar, Altai, Khabarovsk and Primorye Territories [12]. Severe droughts became more frequent in the Volga region and in the south of Cis-Ural region [13, 14]. Taking into consideration the facts mentioned above, the climate extremes negative for population vital activities have increased almost on the whole territory of the country.

This fact allowed to determine the aspects of adverse environment impact and especially the impact of climate changes on the living conditions of the population and to establish the areas where this impact is stronger.

3.2. Environmental living conditions of population in Russia in the early 21st century

The map "Zoning of the territory of the Russian Federation by natural living conditions of the population" was adapted to the modern day climate conditions (2001–2010) on the base of the meteorological network data [2]. At this stage, the most dangerous and rapidly changing climatic factors, such as thermal, cold and wind, were analyzed.

A brief description of the areas with unfavorable environment conditions emphasized on the map for the modern climate is given below.

I – The area with absolutely unfavorable environment conditions occupies the northern part of Asian Russia from Taimyr to Chukotka. It includes the Taimyr Peninsula where it expands up to 65 parallel north in the Putorana Plateau region and the Arctic ocean coast, mostly northward from 70 parallel north. Territory landscapes are presented by arctic and partly sub–arctic types and by mountains in Subarctic and in the north boreal forest (taiga). This zone is featured by severe winters with little snow in Eastern Siberia. The summer is very cool on the whole arctic coast. With increasing distance from the coast toward the south the summer becomes cold. It is excessively humid in the entire zone territory. According to medical and geographical characteristics an optimal period of residence for the newcomers in this zone is 1–2 years on the plains and up to 1 year in mountainous areas. This zone is not recommended to create residential areas population.

II – The very unfavorable zone covers tundra and forest–tundra landscapes of European Russia and coastal areas of cold seas. It is characterized by the extremely high natural impact on the population with a critical stress of adaptation systems of the incomers and with trends to the health failure decompensation. The prevailing pathology in these areas is determined mainly by climatic conditions – meteorological stresses, cardiovascular pathology, cold poly neuroses, cold injuries, traumatism and decreases in immunity and disorders in biological rhythms. The very unfavorable living conditions take place in the highlands of the Altai Mountains that is characterized by an excess of ultraviolet radiation, drought and extremely high seismic intensity. This territory features extremely intensive impacts on human health, with a critical stress of adaptation systems of the incomers and with trends to the health failure decompensation. With the same prevailing pathologies as in the first zone, the optimal period of residence is slightly longer: 2–3 on the plains and 1–2 years in mountainous areas. The territory is not suitable for mass populating as well.

III – The unfavorable zone stretches as a narrow strip from the Kola Peninsula to Yakutia. Then it becomes wider due to high seismic ever frozen mountain systems of Eastern Siberia. In the mountainous areas of Tuva and Buryatia the unfavorable zone is placed along the state boundary. Unfavorable living conditions are observed in the west of Kamchatka and on the northern islands of the Kuril Ridge. This territory is presented mostly by north taiga and partly by middle taiga woodlands, as well as by mountain forest lands in the middle taiga in Eastern Siberia. The unfavorable zone is featured by slightly severe and snowy winter in European Russia, severe and high-snowy winter in Western Siberia and very severe and less snowy winter in Eastern Siberia. Cool summer is observed in European Russia and on mountain regions, except of the middle taiga landscapes in Western Siberia, and Central Yakutia, as well as in the basins of Tuva and Buryatia where it is mild. This zone has an intensive environment impact on human health. It manifests itself in very high stress of adaptation systems of incomers' organisms and hindered compensation. Concerning the environmental conditions the territory is suitable for local populating.

IV – The conditionally unfavorable zone extends considerably and is traced in the middle taiga landscapes of the European Russia plains in Western and Eastern Siberia, as well as in Transbaikalia, the Amur basin and in Primorye. Its natural conditions are specific for Sakhalin and south islands of the Kuril Ridge. In addition, they are specific for the droughty lands. The zone is characterized by semi-severe and snowy winter in European Russia, severe and dry winter in Siberia and semi severe and snowy in the Far East region. Summer over the most zone territory is not quite warm and in the Far East it is cool. There is a high risk of natural stresses in this zone and additional investments are required to maintain normal living conditions.

V –The conditionally favorable zone narrows slightly in European Russia, and spreads out to the southern half of Western Siberia and Krasnoyarsk Territory. This zone is presented as well in the southern part of the Amur basin and in the western part of Primorye. The following climatic conditions are typical for this area: a moderately mild snowy winter in European Russia, a slightly severe snowy winter in Western Siberia and a slightly severe low-snowed winter on the Far East. Summers are usually moderately warm. The conditionally favorable zone is characterized by a slight natural impact on human health with a progressive compensation. Permanent residence of the incomers is possible in these areas, but the costs to maintain vital activities are higher compared to the central part of European Russia.

VI – The favorable zone includes Central European Russia and the adjacent black soil areas. In the West of the European territory it spreads out to the north up to 63–65th parallel north. The environment conditions here are moderately favorable for the population. A winter is moderately mild and a summer is warm. Extreme environment impacts on vital activities are very rare. As a rule, these effects manifest themselves as very cold winters and severe droughts, the frequency of which is increasing at the present time.

VII – The most favorable zone for life activity covers the southern part of European Russia, with the exception of conditionally unfavorable dry-lands, as well as the highlands of the North Caucasus that belong to the conditionally favorable zone. This territory features very mild winters and very warm summers. Extreme natural factors effecting the population living (such as a very cold winter, very severe drought, or very severe flooding, or earthquake) may occur here once in 50 years. It should be noted that there is an increased risk that strong and destructive tornadoes may cause landfall from the Black Sea to the Russian coast of the Caucasus with a frequency of one tornado per decade. At the present time, the risk of extreme precipitation and flooding increases in this zone.

A comparison between the map "Zoning of the territory of the Russian Federation by natural living conditions of the population" for the early 21st century (2001–2010) and the similar map compiled for the long-term average annual conditions (1961–1990) [15, 16] shows that the modern climate warming has led to a significant easing of discomfort in the territory of Russia. It is most noticeable in the northern and Arctic regions where the territories with absolutely unfavorable conditions are significantly collapsed due to expansion of very unfavorable zones (a softer discomfort gradation). The southern boundary of unfavorable territories shifts toward the north. It results in the expansion of the conditionally unfavorable area in West Siberia and in the south of East Siberia. In European Russia the favorable area has expanded and shifts far to the northern regions (approximately to 60˚ North latitude). The steady decrease of the absolutely unfavorable zone resulted from the expansion of less unfavorable areas is observed. By the beginning of this century the area of absolutely unfavorable zones collapsed by 8% and the area of conditionally unfavorable zones increased by 8%. The unfavorable zone area decreased by 5% and the favorable zone has slightly expanded – by 3%, mainly in the early 21st century.

Changes in the total score of discomfort under the conditions of the modern warming are mainly caused by heat and cold factors. The changes are related to the decrease in the sum of negative temperatures and in the number of days with extremely low temperatures as well as to the rise of sum of active temperatures and the prolongation of the frost-free period.

Acknowledgement
The research was carried out at the expense of the grant from the Russian Science Foundation (Project № 16–17–10236).

References

[1] 2014 *Roshydromet second assessment report on climate change and its consequences in the Russian Federation* (Moscow: The Federal Service for Hydrometeorology and Environmental Monitoring of Russia) 1001
[2] Archive of VNIIGMI (All–Russian Research Institute of Hydrometeorological Information – Worldwide data centre) http://meteo.ru/
[3] https://www.esrl.noaa.gov/psd/data/gridded/data.ncep.reanalysis
[4] Kysely J, Pokorna L, Kyncl J and Kriz B 2009 *BMC Public Health* **9** 19. doi:10.1186/1471-2458-9-19.
[5] Revich B A and Shaposhnikov D A 2016 *Problems of forecasting* **2** 125–131
[6] Saez M, Sunyer J, Tobias A, Ballester F and Antó J M 2000 *European Journal of Public Health* **10** 58–63
[7] Blazejczyk K 2000 *Instytut Geografii Uniwersytetu Jagiellońskiego, Prace Geograficzne* **108** 101–108
[8] Revich B A 2011 Heat *Human Ecology* **7** 3–9
[9] Revich B A and Shaposhnikov D A 2012 *Problems of forecasting* **2** 122–139
[10] Robinson P J 2001 *J App Meteor* **40** 762–775
[11] Vinogradova V V 2017 *Bulletin of the Russian Academy of Sciences. Geographic ser.* **4** 68–77
[12] http://www.meteorf.ru
[13] Zolotokrylin A N, Titkova T B, Cherenkova E A and Vinogradova V V 2015 *Current problems in remote sensing of the earth from space* **12** 155–161
[14] Zolotokrylin A N, Titkova T B, Cherenkova E A and Vinogradova V V 2016 *J. Arid Ecosystems* **6** 5–10
[15] Zolotokrylin A N, Krenke A N and Vinogradova V V 2012 *Zoning the territory of Russia on the natural conditions of life* (Moscow: GEOS) 156
[16] Vinogradova V V, Zolotokrylin A N and Krenke A N 2008 *Bulletin of the Russian Academy of Sciences. Geographic series* **5** 106–117

IOP Conf. Series: Earth and Environmental Science **107** (2017) 012046 doi:10.1088/1755-1315/107/1/012046

An integrated risk assessment of the environmental hazards influence on Kazakhstan section (from Khorgas and Dostyk dry ports to Aktau seaport) of the international transport corridor

F Zh Akiyanova, Z K Arykbayeva, A M Atalikhova, B A Dauilbayev, N B Zinabdin, A B Kubeyev and K A Tkach

Institute of Geography Astana branch, Astana, Kabanbai batyr ave.8, 010000, Republic of Kazakhstan

e-mail: af_ingeo@mail.ru

Abstract. The article outlines research results on the assessment of natural hazards impact risk on the international transport corridors' Kazakhstan section (from Khorgas and Dostyk dry ports to the seaport of Aktau) functioning. Based on the component-by-stage analysis of physical and geographical conditions with the use of qualimetric approach, the areas with different risk levels of natural disasters were identified. To minimize the risk of natural problems exposure, a set of environmental recommendations has been developed.

1. Introduction

In the context of modern globalization, a noticeable increase in the trade and ridership has been noted on the Eurasian continent [1]. Kazakhstan's advantageous geographical position as a land bridge between the main macroeconomic markets, such as China, Russia, the countries of European Union and Asia-Pacific region, significantly shortens cargo delivery time and increases the potential for transit opportunities and foreign trade balance of all the participants (figure 1). The implementation of the transit potential is one of the priority directions of the economic policy of Kazakhstan, which is crossed by six international transport corridors: Northern TARM, Central TAR, South TAR, North-South, Chongqing-Duisburg, TRACECA. Natural characteristics of the territories near transport corridors play an important role in the efficient and stable operation of the transit [2]. They can have both direct and indirect influence on the operation of the corridor. In order to provide possible convenient conditions for the international cargo transportation and trade and economic relations, on the assumption of the need to ensure a reliable and stable year-round communication, it was important to study the environmental risks that could affect international transport corridors' capacity [3]. In this regard, it is important to study and assess the risks, identify areas of hazardous natural processes and develop recommendations for their minimization and safe transportation.

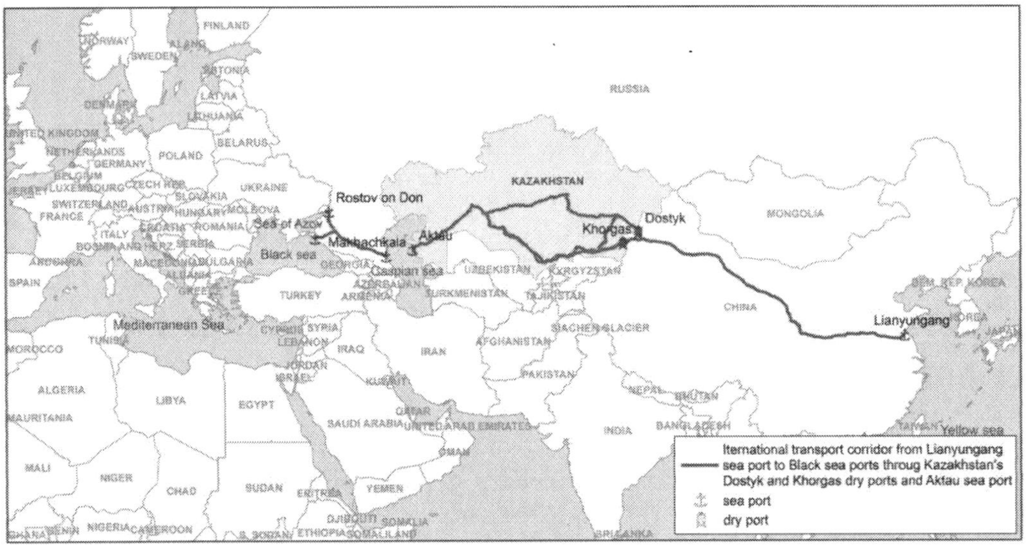

Figure 1. International transport corridor's route from Lianyungang seaport to Black Sea ports through Kazakhstan.

2. Assessment

The Kazakhstan section of the multimodal transport corridors from Khorgas and Dostyk to the Aktau port was studied, it includes sections of various types of transport - rail, road, pipeline, and their combination. International transport corridor's total length within the territory of Kazakhstan is around 5424 km. The transport corridor passes through the territory of 8 administrative regions of Kazakhstan, including 47 administrative districts and 17 city akimats. The area of the investigated zone is approximately 522.8 thousand km^2. As of 01.01.2015, about 44% of the population of Kazakhstan or 7.8 million people live here [4].

The natural characteristics of the territories along which the transport corridor goes are quite diverse. The greater part of the corridor lies along the flat territory, the rest along the more dismembered foothill regions. The corridor is located within two natural zones - semi-deserts and deserts. The risk has been assessed as the hazardous natural phenomena and processes influence probability on the operation and the construction of the Kazakhstan section of the multimodal transport corridor [5]. To assess the risk, an exploratory activity, as well as a comprehensive analysis of the natural conditions including relief, climate, natural waters, soil and vegetation, were carried out.

To assess the risk of impact, main components of the natural environment were determined with indicators and quantitative parameters. To assess the relief impact on the functioning of the transport corridor, there were used the indicator of compartmentalization and exogenous processes exposure risk [6]. The climate is determined by the risk of extremely high and low air temperatures exposure, strong fogs, heavy rainfall and snowfalls, the risk of ice and frost, strong wind, heavy snowstorms and dust storms. For soils - the risk of soil degradation and the content of mobile metals. The indicators and quantitative parameters for risk levels of surface and ground waters, vegetation cover, are given as an example in tables 1, 2 [7, 8].

The risk levels (high, medium high, medium, medium low, low) of each of the components impact on the Kazakhstan section of the multimodal transport corridor were assessed based on the parameters of the natural environment analysis and mapped in ArcGIS10.2. A qualimetric approach was applied to the impact integrated assessment, which is related to the differences between components of the natural environment, because of their nature, scale and influence on each other. Using the method of analyzing the hierarchies (Analytic hierarchy process) of Thomas Saati the weight coefficients of each type of risk were defined (table 3) [9].

Table 1. Parameters for risk levels of surface and ground waters.

Hazard type	Parameter	Low risk level	Medium risk level	Medium high risk level	High risk level
Flooding	the repeatability of water level excess, %	10-50	5-10	2-5	< 2
	maximum rising level, м	< 1	1-1.5	1.5-2	> 2
Groundwater flooding	depth, м	> 5	3-5	2-3	< 2
	relief compartmentalization, м	> 50	10-50	5-10	< 2
	drainage conditions	heavily drained	slightly drained	not drained	coastal zones
Rivers drying out	The frequency of years with the rivers drying out, %	wasn't observed	below 30	30-70	over 70

Table 2. Parameters for risk levels of vegetation.

Hazard type	Parameter	Low risk level	Medium risk level	Medium high risk level	High risk level
Productivity reduction	Productivity, % from average long-term values	<80	50-80	10-50	>10
Foliage cover reduction	Foliage cover, %	50-80	15-30	10-15	>10
Vegetation degradation from overgrazing	Average livestock density, livestock unit per km^2	0.3-0.68	6.8-20.4	20.4-36.1	36.1-109.6
Vegetation loss from fire	Fire frequencies	rare fires	rare spotting fires	frequent fires, regeneration with a change of dominants	burnout of all layers

Table 3. Weight coefficients of each type of natural risk.

Risks	Climate	Waters	Relief	Soils	Vegetation
Climate	0	2	7	0	1
Waters	9	0	7	2	1
Relief	3	8	0	1	4
Soils	8	6	8	0	7
Vegetation	9	8	7	9	0
Total	*29*	*24*	*29*	*12*	*13*
Proportion, %	*100*	*83*	*100*	*41*	*45*

Taking the obtained results into account risk levels were recalculated and on their basis the integrated index that reflects the general risk of natural hazards impact was determined. They were calculated as the sum of scores of all types of risk: $\text{II} = \sum(\text{G} + \text{M} + \text{H} + \text{S} + \text{P}) / \text{n}$ where, II –

integrated index; G – geomorphological risk level; M – meteorological risk level; H – hydrological risk level; S – soil risk level; P – vegetation risk level; n – the number of natural risks types

The integrated index is represented by 5 categories of risk of natural hazards impact on transit potential: from low to high. Based on the grading of the risk of natural hazards impact, a map of integrated risk was drawn up (figure 2).

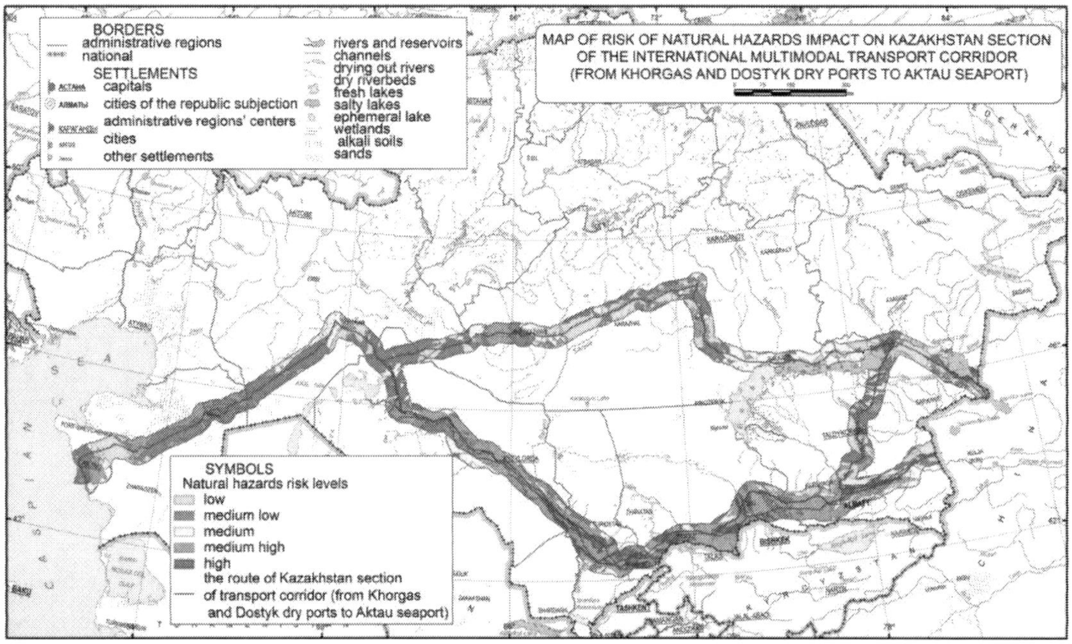

Figure 2. Map of integrated risk of natural hazards impact on Kazakhstan section of the international multimodal transport corridor (from Khorgas and Dostyk dry ports to Aktau seaport).

3. Conclusion

Thus, the analysis showed that 45% of the investigated zone along the Kazakhstan section of the multimodal transport corridor is characterized by a low and a medium low risk of natural hazards. 17% of the transport corridor area is subjected to put at medium risk levels and 38% to medium high and high levels (figure 3). It should also be noted that their distribution across the corridor is uneven.

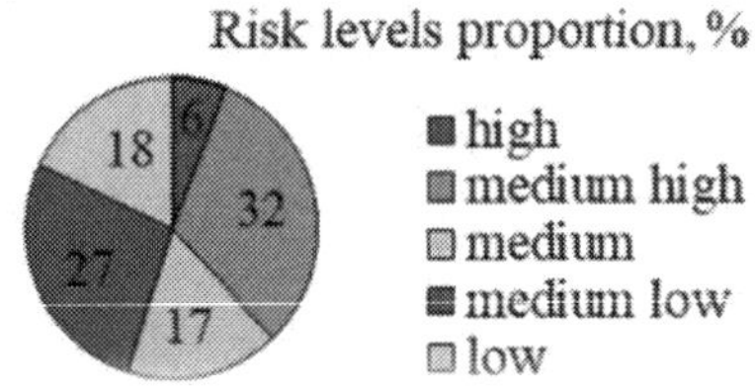

Figure 3. The diagram of risk levels proportion of natural hazards impact on the functioning of the Kazakh corridor section

The greatest dangers to the functioning of the transport corridor are the territories characterized by a dissected relief and a dense hydrographic system, where the risk of hydrological hazards is possible. This combination of natural conditions is typical for the foothill zones of southern and southeastern Kazakhstan (figure 2). These areas can be considered as ecotonic ones – with high biological diversity, so they are also characterized by a high level of vegetation degradation risk [10].

Risk for the sections of the transport corridor located in the immediate vicinity of large water bodies or a watercourse is increased, too. Significant danger is represented by the transport corridor sections that run along the floodplain and alluvial-deltaic plain of the Syrdaria river, which is associated with a high risk of flooding.

The level of risk is significantly reduced for sparsely populated desert and semi-desert areas of the transport corridor with a predominantly flat terrain and a poor hydrographic system. Here there are only some areas that can be obtained due to the manifestations of climatic hazards: dust storms and heavy snowstorms (near the cities of Shalkar and Zhezkazgan), ice-frost phenomena (southern part of the Ulytau mount), strong fogs (near Lake Shalkar) [7].

Thus, a detailed assessment and mapping of the risk of possible natural hazards impact on multimodal transport corridor within the territory of Kazakhstan allowed the developing of a set of georeferenced and risk level depended recommendations to minimize the influence on the functioning of the transport corridor. The recommendations will contribute to increasing the security of cargo and passenger traffic and strengthening the transport potential of Kazakhstan from its western (seaport of Aktau) to eastern (dry port of Khorgas) "gateways".

References

[1] Acar A Z and Pınar Gürol 2016 An Innovative Solution for Transportation among Caspian Region *Procedia - Social and Behavioral Sciences* **229** 78-87

[2] Dindar S, Kaewunruen S, An M and Osman M H B 2016 Natural Hazard Risks on Railway Turnout Systems *Procedia Engineering* **161** 10.1016/j.proeng.2016.08.561

[3] Newman J P, Maier H R, Riddell G A, Zecchin A C, Daniell J E, Schaefer A M, van Delden H Khazai B, O'Flaherty M J and Newland C P 2017 Review of literature on decision support systems for natural hazard risk reduction: Current status and future research directions *Environmental Modelling & Software* **96** doi.org/10.1016/j.envsoft.2017.06.042

[4] 2016 *Demographic yearbook of Kazakhstan Statistical book* (Astana: Agency on statistics of the Republic of Kazakhstan) p 100

[5] De Ville de Goyet C, Zapata-Marti R and Osorio C 2006 Natural Disaster Mitigation and Relief *Disease Control Priorities in Developing Countries* (New-York: Oxford University Press) pp 1147-62

[6] Muralikrishna I V and Manickam V 2017 Environmental Risk Assessment *Environmental Management Science and Engineering for Industry* 135-53

[7] Akiyanova F.Zh. Blagoveshchenskiy V.P., Kunaev M.S. Malkovskiy I.M., Medeu A.R. 2010 *Atlas of natural and technological hazards and risks of emergencies in Kazakhstan* Ed.-in-chief A R Medeu (Almaty: Republican Cartographic Factory)

[8] Akiyanova F.Zh., Iskakov N.A., Kunaev M.S. Malkovskiy I.M., Medeu A.R., Nurmambetov E.I 2010 *National atlas of the Republic of Kazakhstan* (Almaty: Republican Cartographic Factory)

[9] Kousalya P, Reddy G M, Supraja S and Prasad S V 2012 Analytical Hierarchy Process approach *An application of engineering education* (*Mathematica Aeterna)* **2** 861–78

[10] Mander Ü 1993 *Energy, Environment and Natural Resources Management in the Baltic Sea Region* (Copenhagen: Nordic Council of Ministers)

Ecological-geomorphological assessment of the suburban area of Astana

F Zh Akiyanova, N B Zinabdin, A Zh Kenzhebayeva, F G Adilbekova, A T Ilyassova and E M Karakulov

The Astana Branch of Institute of Geography, Kabanbay batyr ave, 8, 010000, The Republic of Kazakhstan
e-mail: af_ingeo@mail.ru

Abstract. The results of ecological-geomorphological assessment of the suburban zone of Astana is presented in the paper. Climatic and hydrological factors, which are the agents of pollutants' transport and caused the development of exogenous processes in the suburban area of Astana were studied and mapped. On the base of the geoinformation technologies and field studies the geomorphologic structure and morphogenetic processes were studied. The analysis of the data complex led to assess ecological-geomorphological conditions of the suburban area of Astana.

1. Introduction

One of the problems of modern society development is, a significant deterioration of the ecological condition of urban settlements and urbanized territories. The uncontrolled growth of large cities and their suburban areas, high rates of urban ecosystems pollution, the impact of hazardous exogenous processes lead to increased environmental tension, degradation of the natural resource potential of the adjacent territories, affect the safety of local population life [1], and also cause finacial losses [2].

This problem is, also relevant for the suburban area of the capital region of Kazakhstan, which is located approximately in the 50-km zone around the capital. The suburban area, whose total area is 12.118 km^2, includes the part of three administrative districts of the Akmola region: Tselinogradsky, Arshalynsky and Shortandinsky. The growth rates of the population of these regions within the hourly availability from Astana far exceed the projected ones for the last two decades. Within the zone, occupying 8% of the territory of the Akmola region, at the beginning of 2017, 17% of the region population lived. Over the last 15 years, the territory of construction of settlements close to the city has increased almost tenfold, which was accompanied by the expansion and construction of roads and engineering support networks. The riverbeds, floodplains and low terraces within territory of the city and suburban area were significantly amended, that is related as to town-planning organization of territory, as to uncontrolled building materials mining. The dynamic development entailed significant changes in the natural environment, which affected the stability of the basic component of the natural environment – relief and processes shaping it.

2. The problem statement, materials and methods

The aim of the research was an ecological-geomorphological assessment of the impact of modern exogenous processes on the organization of the territory of the actively developing suburban zone of Astana. To achieve the goal, the geomorphological mapping of the suburban zone of Astana was accomplished with the definition of the hazardous geomorphological processes development degree, the sources of the natural environment pollution were studied and the main factors of pollutants' transport and accumulation were determined [3,4]. Geomophological mapping is important for geochronology of the newest stage, study of natural resources, assessment of natural hazards [5]. Based on these data, the ecological-geomorphological assessment of suburban zone of Astana was accomplished.

Methods of geoinformation analysis, digital satellite images interpretation, traditional methods of relief studying in cameral and field conditions, digital terrain model creating, ecological-geomorphological mapping and modeling of exogenous processes were used [6,7].

The main natural factors, conductive to the development of exogenous processes were studied. Climatic features of suburban area were studied on the base of data on meteorological stations of Astana and Arshaly village, that located within the study area. In addition to climatic factors analysis the data on 16 meteorological stations of Akmolinsk and Karagandy regions, which have continuous, long-term monitoring raw from 1981 (RSE «Kazhydromet») were used. Through the application of the method of spatial analysis Spatial Analyst ArcGIS 10.2 the climatic inventory and estimated maps were made.

Hydrological factors analysis also was one of the integral parts of ecological-geomorphological assessment. The analysis of hydrological rates was made using data of monitoring on hydrological station of RSE «Kazhydromet». The data on the Esil River were used on the stations Priishimskoe, Turgen, Volgodonovka villages and Astana for the period from 2010 to 2015 years. On the stations Telman, Novoishimka villages and Koktal district for the period of 2015 year, the data on the Nura River were used on the station R. Koshkarbayev village for the period from 2010 to 2015 year. For the spatial analysis of hydrological data, the methods of geoinformation technologies were used. With the use of geoinformation methods of spatial analysis, the complex of inventory and estimated maps of hydrological and soil factors was made.

In addition, the assessment and mapping of the environment pollution sources were accomplished; the assessment of ambient air, soil and water quality was given. On the basis of data on the pollution sources and excess of pollutants limits (MPC), taking into account climatic and hydrological parameters, the pollutants spread halo in «Era» software (1.7 version), that was developed by «Logos-Plus» company (Novosibirsk). The complex of conducted researches, together with the data about exogenous processes activity, let assess ecological-geomorphological situation of suburban zone of Astana.

3. Results and discussion

The research area is located within the northern part of the Central-Kazakhstan hummocky topography, composed by dislocated rocks of the Dopaleozoic and Paleozoic. The up-to-date relief was formed as a result of a long-term continental regime of development under the influence of a complex of denudational processes, which contributed to the formation of a thick weathering crust. A characteristic type of relief is, the flat and slightly undulating denudation socle plains at absolute elevations of 340-392 m with prevailing surface slopes of up to 3-7°. They are developed along the complexly sedimentary, effusive, sometimes intrusive rocks of the Paleozoic. The accumulative plains within the suburban zone are genetically represented by deluvial-proluvial, proluvial, lacustrine-alluvial and alluvial types.

The results of studies were also the assessment of ecological and geomorphological conditions of the territory. As a result of studies of climatic and hydrological factors, mean annual air temperature, atmospheric precipitation, wind direction and velocity were studied as the main climatic factors contributing to the development of erosion and deflation processes. With the application of Spatial Analyst module of ArcGIS 10, the areas most vulnerable to erosion and deflation during the year were determined. The analysis showed that, soils of the investigated territory are at risk of erosion from May to August considering the average monthly air temperature and the total number of hours with precipitation exceeding 30 mm for a warm period. Due to the fact that, the development of deflation processes is facilitated by wind speeds exceeding 4 meters per second in the warm period of the year, it is revealed that the high risk of deflation is observed from April to May and in October. Despite the decrease of wind speed from June to September, the area of research is still subject to a low risk of removal of finely dispersed material because of the significant increase in air temperature. It is determined that, there is a risk of soil deflation processes almost throughout the whole warm period, as from 42% to 53% of all winds from April to October have a speed higher than 4 meters per second.

Thus, there is two periods of more intensive development of erosion and deflation processes in the study area in the warm season of the year. The risk of deflation increases in April, May and October for the southern and eastern parts of suburban area. In the south, deflation processes are activated within the accumulative alluvial-lacustrine plains, which are extended from the southwest to the northeast and framed by raised wavy and sloping-wavy deluvial-proluvial plains. In the southeastern and eastern parts of the suburban zone, such activation occurs within the widely developed sag-and-swell socle plain with small areas of the rocky hilly and ridge watershed hummocky terrain.

The risk of erosion processes rises from May to August and in September, geographically increasing from south to north. The risk of erosion processes is increased in the northern part of the suburban area within the basin of the river Koluton, which is the right-bank tributary of the Esil river and the upper reaches of the Sileti river basin. The valleys of these rivers were formed within the undulating denuded socle plains with parts of the hummocky terrain.

The integral assessment of climatic factors, which contribute to the development of erosion and deflation processes, has shown that, the eastern part of the investigated zone from the Prirechnoye village to the Nikolaevka village is at high risk of negative processes complex development.

The increased risk zone for the development of erosion and deflation processes is observed to the north and east of Astana. The rest of the territory is subject to a medium risk of erosion and deflation processes that depend on climatic factors.

The modern river network within the suburban area of Astana is underdeveloped due to aridity, the predominance of flat relief with plenty of lake depressions, structural features of the soil and vegetation cover. For the territory, the presence of temporary streams, disappearing in summer, is characterized. River network density does not exceed 0.05-0.10 $km \cdot km^{-2}$, while the density of gully network reaches 0.2-0.3 $km \cdot km^{-2}$ [8].

Area's flat relief causes mild severity watershed, at the same time within the suburban area of Astana. Three of the eight water basins watersheds of Kazakhstan are located: Esil (Esil River and its tributaries), Nura-Sarysu (Nura River) and Ertis (Silety River). Rivers are typical plain, valleys are wide and flat, trapezoidal shape. There are areas of riverbeds bifurcation and runoff intercept on Nura-Esil entre rios.

The nature of the underlying surface of the river valleys determines their vulnerability to eroding and accumulating work of river. The Esil River valley consists mostly of loose deposits (sandy loam and loam, underlanded by sand, gravel and pebbles) and the output of rock is only observed in the upper, which causes the prevailing of the processes of erosion and transfer substances.

Particular importance in the analysis of river hydrodynamics has the hydrological regime, especially intra-annual distribution of runoff. The Esil River is a typical river of Kazakh type with exceptionally snow-fed, that defines the sharp intra-uneven flow distribution. The predominant phase of water regime is a spring high tide (90-95% of annual runoff) extending over a short period (April - May) and marked a sharp increase in water consumption. Thus, according to data of hydrological post the Esil River – Volgodonovka village in 2015, the water consumption from the start of the flood (the 10th of April) rose to its peak of up to 4 days in 5 times, from 138 $m^3 \cdot s^{-1}$ up to 682 $m^3 \cdot s^{-1}$ (figure 1a). Also, increased water consumption is observed during floods, caused by heavy rainfall. Thus, during the spring flood active work of river water flow creates favorable conditions for rocks' erosive washout and loose deposits removal. At the low water period eroding capacity decreases and significant changes in the riverbed and the shoreline are not observed, the accumulation processes are activated.

The average altitude of the Esil River basin is about 350 m, at the site to Astana it is higher and reaches 460 m [8]. The Esil river slope along at length is also changing: from the source of the Esil

River to confluence of his right inflow Kargala slope is 2.8 ‰ at the absolute altitude of 560-450 m. Further, to Volgodonovka village (450-369 m) slope is reduced to 0.7 ‰. Downstream slope decreases to 0,4 ‰ with an average altitude of 344 m. Flow rate depends on to the channel slope as well as the degree of river flow regulation. The average maximum rate of flow for the period 2011-2015 years during the passage of water flood is 1.08-1.75 $m{\cdot}s^{-1}$ in the upper reaches of the river Esil before the Astana reservoir. Below Astana reservoir it is reduced to 1.02 $m{\cdot}s^{-1}$, below in the range of the city suburban area it is already 0.72 $m{\cdot}s^{-1}$. It is known that even on plain rivers at a flow rate of 1.6 $m{\cdot}s^{-1}$ ridges of 0.3-1 m height can be formed.

Sinuosity factor of the Esil river bed within the suburban area to hydrological post in Astana (park) is 1.69, at the downstream it is reduced slightly to 1.61. Indicators show the type of meandering riverbed Esil River within the suburban area. The meandering type of the channel sections the are subject to a greater erosive effect of the water flow, since with a greater bending radius of the flow the effect of the centrifugal force increases, which creates conditions for the development of lateral erosion. This type is characterized by meandering within inundated levels, creating multiple flow path, and reaches alternating sills, erosion and the accumulation of deposits.

Intra-annual distribution of the Nura river is caused to the Kazakh type of water regime and snow-fed character of river. According to the hydropost Nura River - R. Koshkarbayev village in 2015 water discharge was increased by 3 times, from 317 to 882 $m^3{\cdot}s^{-1}$ for 8 days (figure 1b). The Nura river is also characterized by increased runoff during the water floods. The slope of the Nura river bed within the suburban area is 0.1-0.2 ‰ at average altitude of 345-365 m. The average long-term maximum flow velocity of the Nura river is 1.06 $m{\cdot}s^{-1}$. Loose deposits, in which the Nura River valley within the study area is developed, contribute to their erosion, transport, and redeposition. Sinuosity of the Nura river bed within suburban areas of Astana is 1.48, the channel has a moderate-winding type.

A characteristic feature of the Esil and Nura rivers, whose main phase is an strongly pronounced spring water tide, contributes to the high risk of flooding caused by extreme water level rises during a high and significant rain flood. It should be noted that, nowadays the risk of flooding in rivers is reduced due to the flow regulation. At the same time, a high risk of flooding and inundation of low river valleys surfaces in high-water years and flooding of low areas surrounding plains during the spring snowmelt remain. It should also be noted that, over the past decades Esil and Nura rivers bed within Astana and suburban area have undergone significant anthropogenic transformation, modification of morphometric characteristics, and consequently, ecological-geomorphological conditions.

Moreover, the active anthropogenic conversion of bed, plains and low terraces, especially within the city and urban management zone, overlap on changes of the natural processes of hydrological and hydrodynamic conditions. Anthropogenic transformation of the riverbed within the suburban area and Astana was held with different purposes, from the formation of water-regulating and spillway (the creation of the Astana reservoir, Nura - Esil channel) to the strengthening and improvement of the coastal zone. Within the city at artificialisation rectification and expansion of the riverbed Esil is conducted, and in some areas covering with ground. The analysis carried out on the basis of satellite images for the period 1975-2015 years the Esil river bed bias on 3 areas from 403 to 907 m. Regulation of Esil river bed, concreting coast within the city causes too low flow velocity – 0.02 $m{\cdot}s^{-1}$, which even in the period of maximum flood reach only 0.19 $m{\cdot}s^{-1}$. Such low rates contribute to the formation of an oxygen-free environment, overgrown by aquatic vegetation. Currently, arrangements for the reconstruction of Esil river bed within the city and suburban areas continues at multiple sites. They are aimed at ensuring the sustainability of existing hydraulic structures and protection of Astana from the floodwaters.

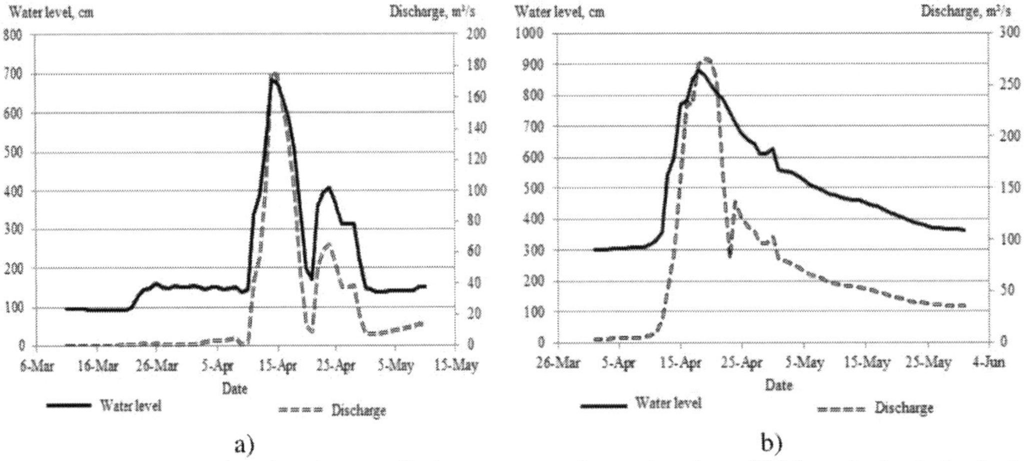

Figure 1. The water level and water discharge course for spring time, 2015 on the hydrological stations: a) the Esil River - Volgodonovka village, b) the Nura river - R. Koshkarbayev village.

Assessment and mapping of the environmental pollution sources within the suburban area and the city of Astana are based on published official data. Sources of pollution are mainly represented by the enterprises producing construction materials (asphalt, concrete, cement, bricks), housing and communal services (Cogeneration plant-1 and Cogeneration plant-2) and landfills for municipal solid waste (MSW). Other 15% of industrial enterprises are related to the mining sector, the main activity of which is the mining of construction materials, that became a reason of geomorphological hazards development [9].

The dynamics of air pollution within the territory of Astana for 2010-2016 is studied on the basis of atmospheric pollution index [10]. In 2010 higher degree of air pollution was indicated, which increased the risk to the health of the city's population. During the period of 2011 – 2015 the index decreased and corresponded to a safe level. Then, in 2016, the index increased to 7 what corresponds to a high degree of air pollution. In general, air pollution was traced in the eastern and south-eastern parts of the suburban area. Average value of the excess of MPC was 6.7, while maximum excess of MPC was 8.4 and was observed in the eastern part of the city.

To assess the ecological state of soils, concentration coefficients for 14 polluting components of I, II and III classes toxicological hazard were calculated using data provided by the analysis of the soil samples taken along the roads, railways and river valleys.

The elements of I hazard class are presented by lead, zinc, cadmium, fluorine, arsenic and mercury. In 28% of the samples there is observed excess of MPC from 1.2 to 4.3 times on lead and in 14.8% of the samples there is excess 1.008 - 1.945 times on zinc.

The highest number of excess of MPC is observed for polluting elements of II hazard class. In 98% of the samples there is recorded excess of MPC from 1.1 to 8.2 times on copper, in 78.7% of the samples – from 1.02 to 4.1 times on nickel and in 54.8% of the samples – from 1.2 to 5.5 times on boron. Concerning pollutants of III hazard class, there is an excess of vanadium from 1.01 to 1.33 times. Meanwhile, an excess of MPC from 1.33 to 2.18 times on manganese were recorded in two samples. According to the soil pollution indices for assessing contamination, the territory of the suburban zone of Astana city primarily refers to the permissible degree of soil contamination. Only within two sections of the industrial zone close to the villages of Saryadyr and R. Koshkarbayev and one area inside of the city of Astana moderately hazardous category of soil contamination is recorded. Maximum values of MPC excess vary from 17.2 to 22.7.

Based on the data provided by the Akimats of the city of Astana, an analysis of the ecological state of water bodies was carried out for 11 monitoring points of surface water and groundwater. Samples of water of Esil river close to Koktal district (complex water pollution index – 7.86) and groundwater near the storage lake Taldykol (8.42) are classified as very dirty. Sample of water taken from Esil river close to the village of Karaotkel (5.31) is recorded as dirty, while water samples of Shenet lake (3.29; 3.16) as well as Silety river close to the village of Prirechnoe are classified as contaminated. Water samples of the Astana reservoir (0.42), groundwater sources near the village of Shenet (0.94), siding 41 (0.63), the village of Arnasay (0.53) and siding 42 (0.20).

Analysis of data on the quality of water of Nura river has shown environmental state change over several years – from contaminated in 2011 to a high level of pollution in 2015 (3.14) and a moderate level of pollution in 2016 (2.3). Thus, within the suburban zone, the cleanest waters are observed along the Esil river to the city of Astana, the most polluted waters are characterized by river sections below the city of Astana due to the city's influence, the large volume of produced pollutants and sewage. There is also some correlation between the changes in the complex index of water pollution and the water content of the years.

The analysis results of climate, water and geomorphology factors, and data of sources and quantitative parameters of pollution allowed managing the ecological and geomorphological zoning. The territories were identified by the prevailing degree of development of erosive and deflationary processes, from mild to severe. The areas of flat and slightly-undulating watersheds and slighty-dissected flat areas of river terraces refer to the low-hazard zone in terms of the erosion on the basis of typing of relief and analysis of morphometric parameters. The medium-dissected watershed surfaces (denudation plains), valley territories refer to relatively erosion-dangerous areas [11]. Erosion-prone areas are the most dissected areas with a dense network of gullies and ravines. Erosive and deflationary processes are observed as a whole at 3.5% of farmland. The total area of eroded soils of Arshalynskii, Shortandinskii and Tselinogradskii districts is 9.4%. Total deflated soil area of these districts is less than 1%.

Within the framework of ecological and geomorphological evaluation of suburban areas of Astana the relief resistance to the external influences was also studied. It was given a particular focus on human impact on the transformation of relief. Changing the morphometric characteristics, morphological relief appearance and the lithogenic base lead to reduced stability of the relief, a change in direction and activization of dangerous exogenous processes [12]. It is determined that most of the area refers to the relatively stable areas with erosive potential of 0.3 to 0.5 ton·km^{-2}. The areas with an unstable terrain, characterized by sufficiently high rate of erosion and the potential reaching 7.1m·km^{-2} are limited.

Conclusions

It is determined that, the sheet and linear erosion and deflation processes are the largest development processes in the territory of suburban area of Astana in accordance with the ecological and geomorphological features of the terrain. The modern geomorphological processes of natural and man-made and only man-made genesis were identified in addition to the processes of the natural genesis. Analysis of the spatial development of these processes in relation to the zones of environmental pollution, residential and industrial facilities, engineering and transport networks, and agriculture allowed determining the degree of danger to the life of the population. The research results will form the basis of recommendations development aimed at reducing the impact of dangerous exogenous processes on the territory suburban area of Astana.

References

[1] Kurolap C A and Fedotov V I 2000 *Vestnik Voronezhskogo Gosudarstvennogo universiteta, Seriya: Geografiya. Geoekologiya.* **1** 120 – 3

[2] Graves A R, Moris J, Deeks L K, Rickson R J, Kibblewhite M G, Harris J A, Farewell T S and Truckle I 2015 The total costs of soil degradation in England and Wales *Ecological Economics* **119** 399 – 413

[3] Valiullin R R 2015 Theoretical aspects of geomorphological mapping *Proc. Conf. Geographical sciences and education* (*Geograficheskiye nauki i obrazovaniye*) (Astrakhan: Astrakhan State University) 121 – 3 (in Russian)

[4] Abdybachaev U, Moldobekov B and Ormukov Ch 2017 Modern Map of Landslide Hazard for Sulukta Town and Its Agglomeration Area, South-West Mountainous Margins of the Fergana Basin *Workshop on World Landslide Forum:Advancing Culture of Living with Landslides* 779 – 84

[5] Bishop M P, James L A, Shroder Jr J F and S J Walsh 2012 Geospatial technologies and digital geomorphological mapping: Concepts, issues and research *Geomorphology* **137** 5 – 26

[6] Likhacheva E A, Timofeyev D A, Borunok A K, Koshkarev A V and Chesnokova I V 2002 The principles of natural and natural-technogenic hazards and geomorphological risks mapping *The relief of the human environment (ecological geomorphology*) Eds Likhacheva, D.A. Timofeev (Moskow: Media-Press) p 213 – 22 (in Russian)

[7] Makkaveev A N, Makhorina E I 2009 *Valleys, buried under technogenic soil layer as one of the active components of large cities morpholithosystems Essays on the geomorphology of the urban area* Eds E A Likhacheva and D A Timofeev (Moskow: Media-PRESS) pp 291 –313 (in Russian)

[8] 1958 *The resources of surface water bodies of virgin and fallow lands Issue 1. Akmola region of the Kazakh SSR* Eds V A Uryvaev et al (Leningrad: Gidrometeoizdat) p 790

[9] Tarolli P and Sofia G 2016 Human topographic signatures and derived geomorphic processes across landscapes *Geomorphology* **255** 140 – 61

[10] 2017 *Environmental monitoring bulletin of the Republic of Kazakhstan 2010 – 2016* (RSE «Kazhydromet», National Hydrometeorological Service of Kazakhstan, Ministry on Energy of the Republic of Kazakhstan) (in Russian) kazhydromet.kz/ru

[11] Akiyanova F Zh and Vasilchenko N I 2015 *Ecosystems of Central Asia under current conditions of socio-economic development: Proc. Int. Conf.* **1** (Ulaanbaatar, Mongolia) 372 – 376 (in Russian)

[12] James L A, Harden C P and Clague J J 2013 Geomorphology of human disturbances, climate change, and hazards *Treatise on Geomorphology* (CA, San Diego: Academic Press) **13** 1–13 10.1016/B978-0-12-374739-6.00339-0

IOP Conf. Series: Earth and Environmental Science **107** (2017) 012048 doi:10.1088/1755-1315/107/1/012048

Specifics of the methodological approach to the study of nanoparticle impact on human health in the production of non-metallic nanomaterials for construction purposes

A M Ayzenshtadt[1], M A Frolova[1], T A Makhova[1], V E Danilov[1], Piyush K Gupta[2] and Rama S Verma[2]

[1]Higher Engineering school, Northern Arctic Federal University named after M V Lomonosov, Severnaya Dvina Emb 17, Arkhangelsk, 163002, Russia
[2]Stem cell and molecular biology lab, Bhupat and Jyoti Mehta school of Biosciences, Department of Biotechnology, IIT-Madras, Chennai, India
e-mail: aizenmaria@gmail.com

Abstract. Minerals samples of mixed-genesis rocks in a finely dispersed state were obtained and studied, namely sand deposit (Kholmogory district) and basalt (Myandukha deposit, Plesetsk district) in Arkhangelsk region. The paper provides the chemical composition data used to calculate the specific mass atomization energy of rocks. The energy parameters of the micro and nano systems of the rock samples – free surface energy and surface activity – were calculated. For toxicological evaluation of the materials obtained, next-generation sequencing (NGS) was used to perform metagenomic analysis which allowed determining the species diversity of microorganisms in the samples under study. It was shown that the sequencing method and metagenomic analysis are applicable and provide good reproducibility for the analysis of the toxicological properties of selected rock samples. The correlation of the surface activity of finely dispersed rock systems and the species diversity of cultivated microorganisms on the raw material was observed.

1. Introduction

The ability of inert materials virtually harmless to the human body to become dangerous in a finely-disperse state is a known fact not associated with the appearance of nanotechnologies. This is primarily due to the inability of the lungs to process or remove nanoparticles, thus causing a reflex narrowing the bronchioles and alveoli, which lowers the blood oxygen level [1, 2]. Therefore, the issues of predicting the nanoparticle impact on human health and the environment are relevant for nanomaterial creation and application processes.

Currently, there are three main causes of the harmful effect of nanoparticles on human health [3]. The first one is that, the main substance of nanoparticles can be toxic. The second one is that, some nanoparticles not exhibiting toxic properties associated with their chemical compositions can act as catalysts for toxic substances. The third one is a specific action on the body associated with dimensional effects. At the same time, it should be noted that, due to the developed surface the systems consisting of finely dispersed materials can be the centers on which various microorganisms are sorbed.

The building material industry is based on the use of mixed-genesis rocks as main raw material resources. This material is also basic for the production of construction nanocomposites (nano additives to concrete, dry mortars, reinforcing nanofibers, etc.). At the same time, however, there is currently no information on the toxicity index of rock nanoparticles and the properties of these particles as the carriers of various biologically active forms.

Free (unbound) nanoparticles used for the production of nanomaterials are quite widespread – some of them are of natural origin (sand suspension in desert regions of the world, products of volcano emissions, smoke particles from forest fires, sea salt nanocrystals, viruses); some of them are of anthropogenic origin (soot, exhaust gases, volatile particles of paints contain nanoparticles of anthropogenic origin), and some of them are of industrial origin (particles of titanium oxides, silicon, metals used in pharmacology, cosmetic industry, production of building composites) [4]. It is believed that there are three main ways of nanomaterials entering the human body, namely inhalation, dermal contact and ingestion. Perhaps there are other ways, such as through the olfactory nerve directly into the brain. The main ways of nanoparticles entering the environment are described in detail in the paper [5].

At the same time, there are general properties of finely dispersed systems associated with dimensional effects and determined by the activity of the particle surface. Our studies [6, 7] introduce a novel criterion – surface activity (k_S) – to characterize the activity of nanoparticles in transformations. It shows the proportion of free surface energy relative to the total energy reserve stored in the rock in the process of genesis obtained as a result of mechanical raw material activation. This criterion is calculated based on the chemical composition of the material (reflected in its specific mass energy of atomization (E_m) [8]), specific surface of the system (S_{sp}) and the free surface energy of the unit surface area numerically equal to surface tension (σ_k). Moreover, one of the methods that allow determining this parameter is the OWRK method [9]. The paper [10] demonstrates the applicability of the material surface activity for improving the composite nanostructured binder composition to obtain highly efficient fine concrete. However, the species diversity of microorganisms on the surface of such systems is still an open issue. The solution of this problem, in our opinion, is constrained primarily by the lack of reliable methods to determine biologically active materials directly on the surface of finely dispersed (micro and nano) particles. Therefore, the aim of this paper was to assess the possibility of obtaining this information for mixed-genesis rocks using next-generation sequencing (NGS) based on metagenomic analysis [11, 12].

2. Material and methods

Mortar sands and basalts classified as building stones are of great practical and scientific interest as the objects of research taking into account their significant reserves in the Arkhangelsk region and genetic classification of rocks. Therefore, polymineral sands from the Kholmogory district and basalt from the Myandukha deposit (all deposits are in the Arkhangelsk region) were used as raw material.

Previously, prior to experiments, rock raw materials were washed from clay inclusions and brought to constant weight at a temperature of 40±2°C. Experimental finely dispersed samples were obtained by mechanically grinding the raw materials on the Retsch PM100 planetary ball mill. Optimal grinding conditions were selected for each raw material type to ensure the maximum degree of dispersion and good reproducibility of results.

The dimensional characteristics of the obtained finely dispersed rock samples were identified using photon-correlation spectroscopy at the Delsa Nano Zeta Potential and Particle Size Analyzer.

Chemical analysis of the obtained finely dispersed fractions was carried out by X-ray fluorescence spectroscopy using the Shimadzu EDX-800 HS spectrometer.

The specific mass energy of atomization (E_m) of the analyzed samples was calculated on the basis of the standard heats of chemical compound formation to determine the mineral composition of the raw material taking into account the true density of the material (ρ). The true densities of test samples were determined experimentally by the pycnometric method.

The OWRK method was used to calculate the surface tension of finely dispersed systems (σ_k) based on the experimental determination of a contact angle of wetting the test sample surfaces with reference fluids – decane, ethylene glycol, glycerin and distilled water.

Specific surface area (S_{sp}) was measured on the PSH-10a device by the Kozeny-Karman gas permeability method.

The surface activity of finely dispersed systems of the rocks under study was calculated using the formula (1):

$$k_s = \frac{\sigma_k S_{sp}}{E_m} \quad (1)$$

Next-generation sequencing (NGS) was used to evaluate the microbial surface contamination of the obtained material samples. Metagenomic analysis was performed for this purpose which allowed determining the species diversity of microorganisms on the test object surfaces without isolation and cultivation of microorganisms.

NGS is a complex high-performance diagnostic and prognostic molecular genetic method that can be used to determine the bacteriological (toxicological) component of various natural samples of organic origin. High-performance sequencing technology was implemented on the commercial platform NextSeq 500 (Illumina), and included the following stages: preparation of libraries, sequencing and analysis of the data obtained. To analyze the genetic variants the software provided by Illumina was used.

3. Results and discussion

Mechanical dispersion of rock raw materials produced the following fractions with a satisfactory degree of polydispersity: mass median particle size was 96±4 nm for polymineral sand, and 135±15 nm for basalt.

Table 1 shows the chemical composition (in terms of oxides) of the main components of the analyzed systems. Table 2 presents the calculated energy characteristics of finely dispersed systems of the rocks under study.

The obtained data show that the material under study has a developed active surface that can be a favorable factor for the adsorption of various forms of bacteria and microorganisms.

Table 1. Chemical composition (in terms of oxides) of the mineral raw materials.

Oxide content, wt. %							
SiO_2	MgO	Al_2O_3	Na_2O	Fe_2O_3	CaO	K_2O	Loss on ignition
Sand							
77.48	1.8	12.13	2.76	1.61	1.45	2.09	0.68
Basalt							
47.75	14.60	13.80	0.89	13.09	8.29	0.32	1.26

Table 2. Energy characteristics of finely dispersed systems of the rocks.

Samples	$E_m \cdot 10^{-3}$, kJ kg^{-1}	S_{sp}, m^2 kg^{-1}	$\sigma_k \cdot 10^3$, J m^{-2}	$k_s \cdot 10^5$
Sand	67.16	2138.00	12.00	3.8
Basalt	26.56	1243.10	12.00	5.6

For the toxicological characteristics of the sand from the Kholmogory district and basalt, their metagenomic analysis (microbial diversity analysis) was carried out in the research laboratory of stem cells and molecular biology of the Madras Institute of Technology (India). The results of the studies are presented in figure 1.

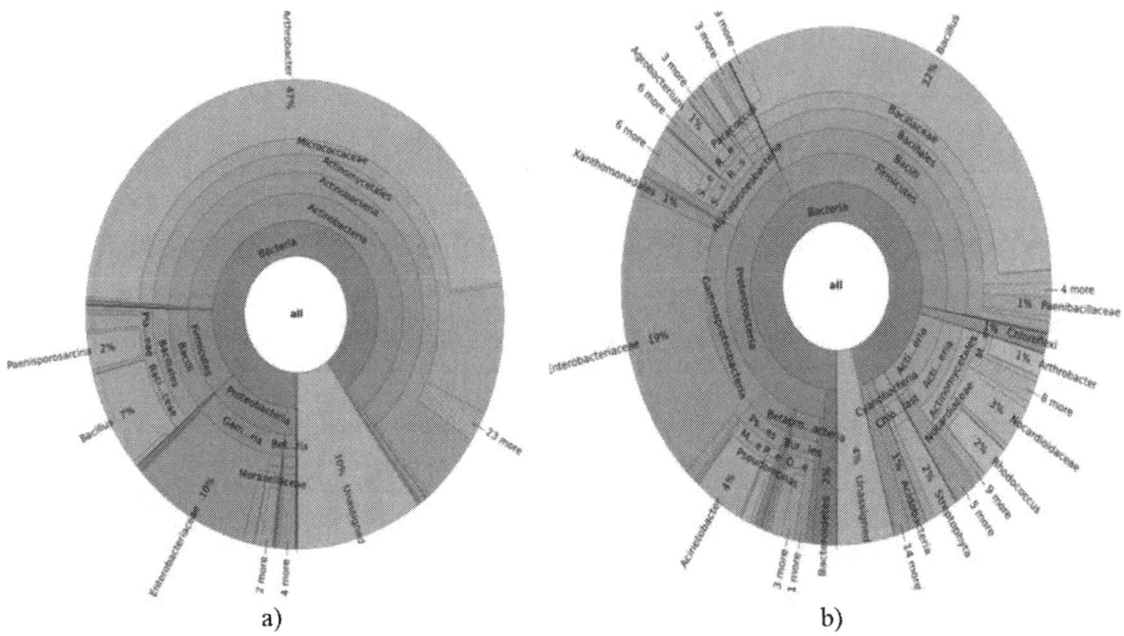

Figure 1. Metagenomic analysis: a) of the sand samples; b) of the basalt samples.

Sequencing data and metagenomic analysis have shown that rock samples have significant differences in the content and types of microorganisms that can imply variable degrees of toxicity. The basalt sample is richer in the content of various groups of microorganisms (family Bacillaceae and family Enterobacteriaceae) (and therefore potentially more toxic), while the sand sample contains predominantly arthrobacteria (genus Arthrobacter) typical for soils. The obtained results are well comparable with the surface activity of finely dispersed rock systems. Thus, it is noted that, for basalt this parameter has a value of about twice as high as for polymineral sand.

Thus, the following provisions can be noted as the summary of research results:

1. Next-generation sequencing and metagenomic analysis are applicable and provide good reproducibility for the toxicological analysis of rock samples.
2. The correlation of the surface activity of finely dispersed rock systems and the species diversity of cultivated microorganisms on the raw material was observed.

Acknowledgement
The reported study was partially supported by RFBR, research project № 16-53-45019 IND_a "Research of patterns and mechanisms of influence of the nanoparticles on human health in the industrial production and application of nonmetallic nanomaterials for construction".

References

[1] Karkishchenko N N 2009 Nanosafety: new approaches to risk assessment and toxicity of nanomaterials *Biomedicine* **1** 5 – 27
[2] Shang L, Nienhaus K and Nienhaus GU 2014 Engeneering nanoparticles interacting with cells: size matters *J Nanobiotechnology* **12** 5
[3] Godymchuk A Yu, Saveliev G G and Zykov A.P.2012 *Ecology of Nanomaterials: Textbook* ed L N Patrikeeva and AA The Revil. (Moskow: Beanom, Laboratory of Knowledge) p 272
[4] Antsiferova I V and Makarova E N 2013 *Vestnik PNIPU* **15** 59 – 67
[5] Antsiferova I V 2012 *Vestnik PNIPU* **14** 55 – 6
[6] Veshniakova L A, Ayzenstadt A M and Frolova M A 2015 *Physics and Chemistry of Materials Processing* **2** 68 – 72

[7] Ayzenstadt A, Frolova M, Mahova T, Veshniakova L and Rama S Verma 2016 *Int. Multidisciplinary Scientific GeoConference-SGEM: Conf. Proc.* (Albena) **3** 53 – 60
[8] Abramovskaya I R, Ayzenstadt A, Lesovik V, Veshniakova L, Frolova M and Kazlitin S 2012 *Industrial and Civil Engineering* **10** 23 – 5.
[9] Gayda J, Ayzenshtadt A, Tutygin A and Frolova M 2016 *Procedia Engineering* **143** 90 –7
[10] Veshniakova L, Drozdyuk T, Ayzenstadt A, Frolova M and Tutygin A 2016 *Materialovedenie* **5** 45 – 8
[11] Ansorge W J 2009 *Biotechnol.* **25** 195 – 203
[12] Berglund E C, Kiialainen A and Syvänen A C 2011 Next-generation sequencing technologies and applications for human genetic history and forensics *Investig. Genet.* **2** 23

Occurrence of Hydrocarbon Degrading Genes in the Soils of the Republic of Tatarstan (Russia)

L R Biktasheva[1], R P Shalyamova, U A Guseva[1] and P Yu Galitskaya[1]

[1]Institute of Environmental Sciences, Kazan Federal University, Kremlevskay str.,18, 420008, Russia
e-mail: biktasheval@mail.ru

Abstract
Oil pollution is one of the most serious environmental problems nowadays. The ability of soils for self-restoration is important, when choosing the strategy of pollution control. This ability depends on the pull of microbes able to decompose hydrocarbons that were present in the non-polluted soil prior to pollution. In this study, the occurrence of alkane degrading genes in the soils of the Republic of Tatarstan being one of the oil processing regions in Russia, was investigated. It was found that alkane degrading genes belonging to group I were present in 20 of the 25 soil samples, and their abundances ranged between 0.01 and 0.07%. Alkane degrading genes belonging to group II were not detected in the samples investigated, and those belonging to group III were present in all the samples, and their abundances ranged between 0.06 and 7.25%. No correlation between the alkane degrading gene copy numbers and pH and organic carbon content in soils was revealed.

1. Introduction

Taking into account the importance of soils for agriculture and food supply, their pollution with oil products is a serious threat [1,2]. When the soil is polluted, natural degradation of oil hydrocarbons is carried out by the microorganisms from the soil community present in it before the pollution. They are called hydrocarbon oxidizing microorganisms, capable of synthesizing enzymes to degrade alkanes and aromatic hydrocarbons – monooxygenases and dioxygenases [3].

There are microorganisms capable of synthesizing several enzymes to degrade hydrocarbons simultaneously, but usually only one enzyme can be synthesized by such microorganisms [4,5]. Alkane-monooxygenases complex consists of alkane-monooxygenase, rubredoxin-reductase and rubredoxin. Together with cytochrome P450-hydroxylase it is responsible for the degradation of alkanes [3,6–9]. The most widespread enzymic complex is alkane-monooxygenases enzymic complex, so the most researches are devoted to it. Gene *alk* responsible for encoding enzyme synthesis has an alternating structure, due to which enzymes possess different properties: for instant, *alkB* gene from *Pseudomonas* sp. is responsible for the synthesis of enzyme degrading short-chain alkanes C_6-C_{12}, *alkM* genes from *Acinetobacter* sp. encode the enzymes to decompose alkanes C_{10}-C_{40}, genes *alkB1* and *alkB2* from *Rhodococcus* sp. synthesize enzymes degrading alkanes C_8-C_{32} [7,9–13]. Owing to the growth of the number of such genes, when the soil is polluted, the assessment of their number can be used to monitor oil pollution, both new and old [14,15].

In pollution-free soils alkane-degrading microorganisms can be found as well, which is connected with the presence of alkane-like substances in the environment [3]. Such soil characteristics as organic matter content, mineral and granulometric composition, acidity, humidity, etc. determine the structure of microbial coens [5,16,17]. The degradation of hydrocarbons is carried out by some representatives of soil microbial community (hydrocarbon degrading microbes), which can depend on soil properties [11]. Thus, studying the occurrence of genes in pollution-free soils can reveal the relation of this group of microorganisms and soil characteristics, and it can also show the soil of what type is more or less capable of self-restoration.

The objective of our investigation was to monitor the occurrence of different groups of *alk* genes (*alkB*, *alkM* and *alkB1*) in different soil types in the Republic of Tatarstan. To reveal the relation between the occurrence of genes and soil properties we have analyzed the following agrochemical characteristics: acidity and organic carbon content.

2. Materials and methods

2.1 Soil samples

Twenty-five pristine (uncontaminated) soils were sampled from various sites in republic of Tatarstan, far from residential areas, trails and oil production sites (Table 1). Prior to sampling, plants were removed and soil was sampled to a depth of 10 cm. Five replicates were sampled, and one representative sample was prepared. For chemical and physical analysis samples were air-dried, sieved (2 mm mesh size), and stored at 4 °C in the dark for further use. Subsamples for molecular biological analysis were collected with sterile spatulas, transported in coolers to the laboratory, and stored at -80°C until they were analyzed.

Table 1. Sampling Regions.

Sample	Sampling Region	Soil Type
P1	Sovetsky District, Kazan	
P2	Sovetsky District, Kazan	
P3	Pestrychinsky District	
P4	Aviastroitelny District, Kazan	Eutric Podzoluvisols
P5	Laishevsky District	
P6	Zelenodolsky District	
P7	Zelenodolsky District	
P8	Arsky District	
G1	Kamsko-Ustinsky District	
G2	Kamsko-Ustinsky District	
G3	Laishevsky District	
G4	Laishevsky District	
G5	Laishevsky District	Haplic greyzem
G6	Laishevsky District	
G7	Kamsko-Ustinsky District	
G8	Kamsko-Ustinsky District	
G9	Kamsko-Ustinsky District	
Ch1	Nurlatsky District	
Ch2	Nurlatsky District	
Ch3	Buinsky District	Haplic Chernozems
Ch4	Buinsky District	
Ch5	Menzelinsky District	
Ph1	Apastovsky District	
Ph2	Apastovsky District	Luvic Phaeozems
Ph3	Apastovsky District	

2.2 Chemical and biological analyses

Hydrocarbon content was assessed using IR-spectrometry with an AN-2 analyser (LLC Neftehimavtomatika-SPb, Russia). The total organic carbon content in waste samples was estimated according to ISO 14235:1998, pH according to ISO 10390:2005 [18,19].

2.3 Molecular analysis
Total genomic DNA was extracted from the samples using the FastDNA®SPIN Kit for Soil (Bio101, Qbiogene, Heidelberg, Germany) according to the manufacturer's instructions. The DNA samples were stored at -20 °C or analyzed immediately.

The copy number of the *alk* genes in each sample was quantified with qPCR analysis. Real-time PCR was performed using a 0.1 U µl-1 SynTaq Polymerase, 1x Buffer with SYBR Green, 2.5 mM $MgCl_2$, 200 µM dNTPs each, 0.2 µM primer each and 1 µl of DNA template in the CFX96 Touch Real-Time PCR Detection System (Bio-Rad, Munich, Germany). qPCR assays were performed with the group-specific *alk* primers listed in Table 2 [10,20]. PCR amplification was conducted using the following protocol: 15 min at 95℃, followed by 39 cycles at 95℃ for 30 s, 30 s at 50°C, and 30 s at 72℃. Cycle thresholds were determined by comparison with standard curves constructed using several concentrations of a positive clone. The plasmid DNA concentration was determined on a Qubit 2.0 Fluorometer (Invitrogen, Thermo Fisher Scientific, Waltham, Massachusetts, USA), and the copy number of the target gene was calculated directly from the concentration of the extracted plasmid DNA. Ten-fold serial dilutions of a known copy number of plasmid DNA were subjected to real-time PCR in five replicates to generate an external standard curve.

To evaluate the abundance of n-alkane-degrading bacteria relative to total bacteria, the percentages of *alk* genes in proportion to 16S rRNA were also calculated. Bacterial 16S rRNA genes were quantified using 341f /534r primers [21,22]. The qPCR program consisted of initial denaturation at 95 ^{0}C for 5 min followed by 39 three-step cycles of 95^{0}C for 10 s, 55^{0}C for 30 s, and 72^{0}C for 30 s. The standard curves were generated for bacteria using serial DNA dilutions of DNA of *Pseudomonas putida*.

Three replicates for each sample were used for qPCR analyzes. All of qPCR assays performed with efficiency of more than 94%, R2 values greater than 0.99.

Table 2. Oligonucleotide sequences of primers used in this study.

Primers	Sequence	Phylogenetic affiliation
alk If alk Ir	5'-CAT AAT AAA GGG CAT CAC CGT-3' 5'-GAT TTC ATT CTC GAA ACT CCA AAC-3'	*Ps. putida (alkB)*, *Stenotrophomonas* spp. *(alkB)*
alk IIf alk IIr	5'-GAG ACA AAT CGT CTA AAA CGT AA-3' 5'-TTG TTA TTA TTC CAA CTA TGC TC-3'	*Acinetobacter* spp. *(alkB)*
alk IIIf alk IIIr	5'-TCG AGC ACA TCC GCG GCC ACC A-3' 5'-CCG TAG TGC TCG ACG TAG TT-3'	*Ps. fluorescens (alkB)*, *Ps. aeruginosa (alkB1)*, *Rhodococcus* spp. *(alkB1)*, *Burkholderia* spp. *(alkB)*, *Amycolatopsis* spp. *(alkB)*

2.4 Statistical analysis
Sampling and chemical analyses were carried out in triplicate and biological analyses in quintuplicate, and all results were expressed on an air-dried sample basis. Statistical analyses were performed using Origin 8.0 (OriginLab, Northampton, USA) and R Statistical Software (R 3.0.0, R Foundation for Statistical Computing Version, Vienna, Austria) [23] packages. The confidence of data generated in the present investigations has been analyzed by standard statistical methods to determine the mean values and standard errors (SEs). The means were compared using Fisher's least significant difference at α=0.05.

3. Results and Discussion
To test the hypothesis of relation between the distribution of genes responsible for alkane-degrading enzymes we have realized the analysis of agrochemical characteristics of soils under investigation.

Previous investigations showed that pH and organic carbon content were the main factors influencing the structure of soil microbial communities [3,11,24]. As a result of this, we have chosen the same parameters. The data obtained are presented in Table 3.

Table 3. Physical and Chemical Properties of the Soils.

Sample	pH	Organic C, %
P1	5.45	0.76
P2	5.99	0.72
P3	5.85	0.92
P4	5.05	2.17
P5	5.88	1.66
P6	5.84	1.18
P7	5.82	2.53
P8	6.23	1.77
G1	6.40	1.76
G2	6.10	1.08
G3	5.49	1.07
G4	6.40	1.27
G5	5.70	1.08
G6	5.64	1.99
G7	6.64	2.97
G8	6.78	1.92
G9	6.60	2.11
Ch1	7.29	4.54
Ch2	6.21	3.49
Ch3	6.60	3.21
Ch4	6.82	4.51
Ch5	7.25	3.10
Ph1	7.15	3.90
Ph2	7.30	4.52
Ph3	7.20	3.00

As we can see in Table 3, soils belonging to the type of eutric podzoluvisols were characterized with lower pH values varying between 5.05 and 6.23, haplic greyzem – between 5.49 and 6.78, values close to neutral ones were detected in soils belonging to the types of haplic chernozems and luvic phaeozems. These results in whole coincide with the typical parameters of such soil types [25].

Organic carbon content also corresponds to the typical values for such types of soils. The lowest organic carbon content was found in samples P1 and P3 belonging to the type of eutric podzoluvisols (0.72 to 0.92%), and the highest – in samples Ph2 and Ch1 belonging to the type of luvic phaeozems and haplic chernozems (4.52 to 4.54%).

In all the samples the relation of *alk*-genes and the number of 16S rRNA gene copies was carried out using the qPCR method (Table 4).

Table 4. Percentages of *alkB* genes copy numbers in proportion to 16S rRNA gene copy numbers.

Sample	alkI/ 16S rRNA ratio (%)	alkII/ 16S rRNA ratio (%)	alkIII/ 16S rRNA ratio (%)
P1	0.01	N/D[a]	0.80
P2	0.02	N/D	0.47
P3	N/D	N/D	0.88
P4	0.02	N/D	4.84
P5	0.05	N/D	7.25
P6	0.05	N/D	7.01
P7	0.04	N/D	7.04
P8	0.04	N/D	1.61
G1	0.01	N/D	0.06
G2	N/D	N/D	1.29
G3	N/D	N/D	0.72
G4	0.01	N/D	1.49
G5	0.01	N/D	1.16
G6	0.02	N/D	1.38
G7	0.01	N/D	0.26
G8	0.04	N/D	3.15
G9	0.04	N/D	4.88
Ch1	N/D	N/D	0.92
Ch2	0.01	N/D	0.68
Ch3	N/D	N/D	0.93
Ch4	0.02	N/D	0.75
Ch5	0.05	N/D	4.10
Ph1	0.07	N/D	0.92
Ph2	0.02	N/D	2.06
Ph3	0.06	N/D	6.58

[a] – not detected

Relative abundance of genes belonging to group alk I varied in the range of 0.01 to 0.07% in samples P1 and Ph1, correspondingly. In samples P3, G2, G3, Ch1 and Ch3 genes of the group haven't been detected.

Some authors note that the number of *alk*-genes grows, when the amount of organic matter grows, and the maximum variety of the community is observed in soils, where pH is close to neutral values [16,26]. Others note the wide-spread occurrence of alkane degrading bacteria in acid soils with low organic matter content [11]. However, in our case, the influence of agrochemical properties of soil on the occurrence of *alk*-genes belonging to group one (alk I), associated with *alkB* gene from *Pseudomonas putida* and *Stenotrophomonas* spp. hasn't been detected (R^2=0.23). Absence of significant relation can be connected with a narrow range of differences in agrochemical characteristics of the soils sampled.

Genes belonging to group two (alk II) being the gene of *alkM* and the genus of *Acinetobacter* spp. haven't been detected in the pollution-free samples under analysis. Similar results were obtained in the research realized by Luz (2004), where *alkM* genes haven't been found in pure soils, either [13].

AlkB and *alkB1* genes responsible for the synthesis of alkane-monooxygenases enzyme in group three (alk III) are more often found in the representatives of *Pseudomonas* spp. (*Ps. fluorescens, Ps. aeruginosa*), *Rhodococcus* spp., *Burkholderi*a spp. and *Amycolatopsis* spp. genus. This group of genes was the most highly represented in all the samples, being 10 to 100 times higher than the presence of alk I group. Relative abundance of *alk* genes varied from 0.06 to 7.25% in samples G1 and P5, correspondingly. The highest presence of genes belonging to this group was registered for the soil type of eutric podzoluvisols, and the lowest – for the soil type of haplic greyzem.

The analysis of the data received hasn't revealed any connection between the agrochemical characteristics of soil and the spread of that group of genes (R^2=0.21). The abundance of genes of the group connected with the presence of certain representatives of *Pseudomonas* spp., *Rhodococcus* spp., *Burkholderia* spp., and *Amycolatopsis* spp. genus can be explained by the variety of the group presented and the wide occurrence of the carriers of such genes in soil [20,27,28]. For example, in many investigations it is registered that the representatives of *Pseudomonas* spp. and *Rhodococcus* spp. genus are widely spread ones in soil communities in cold and moderate climate zones [27–29].

It can be pointed out that the most numerous group is that of alk III genes, genes belonging to alk I group were found in 20 samples, and gene carriers of *alkM* (alk II group) haven't been detected. Our results coincide with those received by Margesin et al. (2003) [20]. The authors point out that pure soil is characterized by low amounts of representatives of *alk*-genes from *Pseudomonas* spp. and *Acinetobacter* spp., while *alkB1* and *alkB2* from *Rhodococcus* spp. are found in large quantities in pollution-free soils.

Conclusions

Thus, it has been discovered that the greatest amount of genes detected belongs to alk III group (*alkB and alkB1*), genes belonging to alk I group (*alkB*) were also found. Those genes belong to *Pseudomonas* spp., *Rhodococcus* spp., *Burkholderia* spp. and *Amycolatopsis* spp. genus. *AlkM* genes from *Acinetobacter* spp. have been detected in none of the 25 samples. No relation between the characteristics of soil (acidity and organic carbon content) and the distribution of genes responsible for alkane degradation was revealed either.

Acknowledgement

The work is performed with the financial support of Russian Scientific Foundation, grant No 17-74-20183.

References

[1] Lamy E, Tran TC, Mottelet S, Pauss A and Schoefs O 2013 *Int Biodeterior Biodegrad.* **83** 85–91
[2] Galitskaya P, Biktasheva L, Saveliev A, Ratering S, Schnell S and Selivanovskaya S 2015 *Biogeosciences* **12** 3681–3693
[3] Van Beilen JB and Funhoff EG 2007 *Applied Microbiology and Biotechnology* **74** 13–21
[4] Harayama S, Kasai Y and Hara A 2004 *Current Opinion in Biotechnology* **15** 205–214
[5] Powell SM, Bowman JP, Ferguson SH and Snape I 2010 *Soil Biol Biochem.* **42** 2012–2021
[6] Liu YC, Li LZ, Wu Y, Tian W, Zhang LP and Xu L 2010 *Bioresour Technol.* **101** 310–316
[7] Yousaf S, Afzal M, Reichenauer TG, Brady CL and Sessitsch A *Environ Pollut.* 2011 **159** 2675–2683
[8] Tanase AM, Ionescu R, Chiciudean I, Vassu T and Stoica I *Int Biodeterior Biodegrad.* 2013 **84** 150–154
[9] Das D, Baruah R, Sarma Roy A, Singh AK, Deka Boruah HP and Kalita J *Genomics* 2015 **105** 182–190
[10] Kohno T, Sugimoto Y, Sei K and Mori K 2002 *Microbes Environ.* **17** 114–121
[11] Jurelevicius D, Cotta SR, Peixoto R, Rosado AS and Seldin L 2012 *Eur J Soil Biol.* **51** 37–44
[12] Hassanshahian M, Ahmadinejad M, Tebyanian H and Kariminik A 2013 *Mar Pollut Bull.* **73** 300–305
[13] Luz AP, Pellizari VH, Whyte LG and Greer CW 2004 *Can J Microbiol.* **50** 323–333
[14] Widada J, Nojiri H and Omori T 2003 *Applied Microbiology and Biotechnology* 45–59
[15] Liu Q, Tang J, Bai Z, Hecker M and Giesy JP 2015 *Sci Rep.* **5** 11068
[16] Fierer N and Jackson RB 2006 *Proc Natl Acad Sci USA* **103** 626–631
[17] Jones RT, Robeson MS, Lauber CL, Hamady M, Knight R and Fierer N 2009 *ISME J.* **3** 442–

453
[18] ISO 14235 1998 *International Organization for Standardization* 5
[19] ISO 10390 2005 *International Organization for Standardization* 5
[20] Margesin R, Labbé D, Schinner F, Greer CW and Whyte LG 2003 *Appl Environ Microbiol.* **69** 3085–3092
[21] Muyzer G, De Waal EC and Uitterlinden AG 1993 *Appl Environ Microbiol.* **59** 695–700
[22] Smits THM, Devenoges C, Szynalski K, Maillard J and Holliger C 2004 *J Microbiol Methods* **57** 369–378
[23] Team R. 2013 *R A Lang Environ Stat Comput.* **55** 275–286
[24] Saul DJ, Aislabie JM, Brown CE, Harris L and Foght JM 2005 *FEMS Microbiology Ecology* **53** 141–155
[25] Unified State Register of Soil Resources in Russia [Internet]. Available: http://atlas.mcx.ru/materials/egrpr/content/intro.html
[26] Delille D, Coulon F and Pelletier E 2004 *Microb Ecol.* **47** 407-415
[27] Aislabie J, McLeod M and Fraser R. 1998 *Appl Microbiol Biotechnol.* **49** 210–214
[28] Bej AK, Saul D and Aislabie J 2000 *Polar Biol.* **23** 100–105
[29] Whyte LG, Greer CW and Inniss WE 1996 *Can J Microbiol.* **46** 99–106

Influence of Biopreparations on the Bacterial Community of Oily Waste

L R Biktasheva, P Yu Galitskaya and S Yu Selivanovskaya

Institute of Environmental Sciences, Kazan Federal University, Kremlevskay str.,18, 420008, Russia
e-mail: biktasheval@mail.ru

Abstract. Oil pollution is reported to be one the most serious environmental problems nowadays. Therefore, methods of remediation of oily polluted soils and oily wastes are of great importance. Bioremediation being a perspective method of sanitation of oil pollutions, includes biostimulation of the polluted sites' indigenous microflora, and in some cases additional introduction of active strains able to decompose hydrocarbon. The efficacy of introducing such biopreparations depends on the interactions between the introduced microbes and the indigenous ones. In this study, the influence of bacterial consortium (*Rhodococcus jialingiae, Stenotrophomonas rhizophila* and *Pseudomonas gessardii*) introduction on the bioremediation of an oily waste sampled from a refinery situated in the Mari El region (Russia) was estimated. Single and multiple inoculations of the consortium in addition to moistening and aeration were compared with a control sample, which included only aeration and moistening of the waste. It was shown, that two of the three introduced strains (*Rh. jialingiae* and *Ps.gessardii*) gene copy numbers were higher in the inoculated variants than in the control sample and with their initial counts, which meant that these strains survived and included into the bacterial community of the wastes. At the same time, bacterial counts were significantly lower, and the physiological profile of waste microflora slightly altered in the inoculated remediation variants as compared with the control sample. Interestingly, no difference in the degradation rates of hydrocarbons was revealed in the inoculated remediation variants and the control sample.

1. Introduction

Oily wastes generated during the processes of oil extraction, transportation and refinery present a serious problem and are characterized by a complex multicomponent composition including oil products, mechanical impurities and water [1–4]. Disposal of oily wastes in the environment can lead to different toxic effects induced by hydrocarbons, radioactive elements and heavy metals [5–9].

Among many methods of oil waste disposal biological ones attract more and more attention. One of biological methods of oil waste disposal is bioaugmentation, which involves the introduction of microorganisms into the polluted object for bioremediation [10–12].

Single strains with high ability to decompose hydrocarbons as well as microbial consortia may be used for bioaugmentation. Usually for bioaugmentation used hydrocarbon oxidizing microorganisms of the following taxonomic groups: *Mycobacterium, Sphingomonas, Pseudomonas, Acromobacter, Acinetobacter, Alcalogenes, Arthrobacter, Bacillus, Brevibacterium, Cornybacterium, Flavobacterium, Nocardia, Pseudomonas, Dietzia, Methylobacterium, Rhodococcus, Vibrio* for bacteria, and *Chrysosporium, Bjerkander, Irpex, Agrocybe, Lentinus, Trametes, Pleurotus, Botryosphaeria* for fungi [13–18]. Wu et al. (2016) notice that the introduction of the strain *Acinetobacter* SZ-1 allowed to 34% reduce of total hydrocarbon content in contaminated soils [13]. Hassanshahian et al. (2014) achieved a 95% oil reduction in polluted water, using the strain *Alcanivorax borkumensis* SK2T [14]. Covino et al. (2016) used for bioaugmentation the following species of fungi - *Pleurotus ostreatus u Botryosphaeria rhodina* [15]. However, only the introduction of *P. ostreatus*, which reduced the total hydrocarbon content in the soil on 86.8%, was effective.

Some authors suppose that, the introduction of a consortium is more effective, since each strain of microorganisms will consume every component of oil [12,16,19,20]. Cerqueira et al. (2014) used the for bioremediation group of strains (*Stenotrophomonas acidaminiphila, Bacillus megaterium, Bacillus cibi, Pseudomonas aeruginosa, Bacillus cereus*) and their consortium [16]. Authors came to the conclusion that the most effective was the use of the consortium.

Pacwa-Plociniczak et al. (2016) described that a consortium of *Bacillus subtilis T0-1* and *Pseudomonas sp. P-1* is three times more effective than their individual application for bioremediation of contaminated soils, and Dellagnezze et al. (2016) found that the efficiency of using a consortium of bacteria reaches 99% for oil-polluted seawater [17,21] .

The efficacy of bioaugmentation process directly depends on the influence of indigenous community on the survivability of the introduced strains [22,23]. The analysis of changes in taxonomic and functional diversity of the microbial community can help to estimate the response of the microbial community to the introduction of microorganisms, and thus to help to control bioaugmentation process [24,25].

The objective of our study is to estimate the changes of a man-made formation microbial community in response to the introduction of a bacterial consortium. To estimate the influence of the introduction oily waste inoculation process was carried out with the help of hydrocarbon-oxidizing strains. Bioaugmentation process was carried out one and three times, a sample with no inoculation was used as a control sample. In the course of the experiment the following parameters were estimated: oil product content in the samples, survivability of the introduced strains and the changes in the total number of bacteria using qPCR method, as well as the changes in the functional diversity of the communities using BIOLOG® EcoPlates.

2. Materials and methods

2.1 Experimental Design

Oily waste containing mechanical impurities and oil products was chosen as the research target. The oily waste used contained 177.8 g kg-1 of total petroleum hydrocarbons (TPH). The oily waste was collected from a refinery, Mari El, Russia (latitude: 54°50′26″ N, longitude: 52°27′08″ E).

Three variants of the experiment were carried out: a sample with no introduction of microorganisms (I), a sample with a single introduction of microorganisms (II), and a sample with triple introduction of microorganisms (III). The introduction of the bacterial consortium was carried out on day 1 (II), and on days 1, 28 and 56 (III) of the experiment in the amount necessary to achieve a final concentration in the sample equal 10^8 CFU g^{-1}. The experiment was carried out under the same conditions: constant room temperature, moistening and aeration.

Three hydrocarbon-oxidizing strains belonging to the species of *Rhodococcus jialingiae, Stenotrophomonas rhizophila* and *Pseudomonas gessardii* previously extracted from the oily waste under analysis were chosen for the introduction.

2.2 Chemical and biological analyses

Hydrocarbon content was assessed using IR-spectrometry with an AN-2 analyser (LLC Neftehimavtomatika-SPb, Russia).

Community level physiological profiles (CLPP) were assessed by the BIOLOG® ECOplates (BIOLOG, Hayward, CA). Each 96-well plate consists of three replicates of 31 sole carbon substrates and a water blank. The plates were incubated at 25 °C for 168 h, and the color development in each well was recorded as optical density (OD) at 590 nm with a plate reader (Thermo Scientific Multiskan MK3, Shanghai, China) at regular 24 h intervals.

Microbial activity in each microplate, expressed as average well-color development (AWCD) was determined as follows: $AWCD = \sum OD_i/31$, where OD_i is the optical density value from each well. The Shannon–Weaver index (H) was calculated using an was calculated using Eq. $H = - \sum p_i \operatorname{Ln} P_i$, where pi is the ratio of the activity on each substrate (OD_i) to the sum of activities on all substrates ($\sum OD_i$).

2.3 Molecular analysis
Total genomic DNA was extracted from the samples using the FastDNA®SPIN Kit for Soil (Bio101, Qbiogene, Heidelberg, Germany) according to the manufacturer's instructions. The DNA samples were stored at -20 °C or analyzed immediately.

Quantitative polymerase chain reaction (qPCR) was conducted using 16S 984f /1378r primers for bacteria and species-specific primers listed in Table 1 [26–28]. Species-specific primers were designed based on the available draft genomes using Primer— BLAST tool from NCBI and assessed by Oligo 6 software. PCR reactions were conducted with a 0.1 U μl-1 SynTaq Polymerase, 1x Buffer with SYBR Green, 2.5 mM $MgCl_2$, 200 μM dNTPs each, 0.2 μM primer each and 1 μl of DNA template in the CFX96 Touch Real-Time PCR Detection System (Bio-Rad, Munich, Germany). The qPCR program for total bacteria consisted of initial denaturation at 95 °C for 5 min followed by 39 three-step cycles of 62–60 °C for 45 s, 95 °C for 15 s, and 72 °C for 30 s. PCR amplification for introduced species was conducted using the following protocol: 15 min at 95 ℃, followed by 39 cycles at 95 ℃ for 30 s, 30 s at T_m shown in Table 1, and 30 s at 72 ℃. The standard curves were generated using serial DNA dilutions of DNA of *Bacillus pumilus, Rh. jialingiae, S. rhizophila, Ps. gessardii.* The concentration of DNA was measured on a Qubit 2.0 Fluorometer (Invitrogen, Thermo Fisher Scientific, Waltham, Massachusetts, USA) using the Picogreen ds DNA reagent (Invitrogen Ltd., Paisley, UK). Three replicates for each sample were used for qPCR analyzes. All of qPCR assays performed with efficiency of more than 94%, R2 values greater than 0.99.

Table 1. Species-specific primers used in this study.

Primers	Sequence (5'-3')	Tm (°C)	Phylogenetic affiliation
JiF JiR	TAATACGTAGGGTGCAAGCGTT ATTTCACCGCTACACCAGGAAT	60	*Rhodococcus* spp.
PGeF PGeR	ATACAGAGGGTGCAAGCGTT GCATTTCACCGCTACACAGG	60	*Pseudomonas* spp.
StF StR	CGTGCGTAGGTGGTTGTTTAAG ATTTCACTGCTACACCGGGAAT	60	*Stenotrophomonas* spp.

2.4 Statistical analysis
Sampling and chemical analyses were carried out in triplicate and biological analyses in quintuplicate, and all results were expressed on an air-dried sample basis. Statistical analyses were performed using Origin 8.0 (OriginLab, Northampton, USA) and R Statistical Software (R 3.0.0, R Foundation for Statistical Computing Version, Vienna, Austria) [29] packages. The confidence of data generated in the present investigations has been analyzed by standard statistical methods to determine the mean values and standard errors (SEs). The means were compared using Fisher's least significant difference at α=0.05. The values in figures were expressed as mean ±SE of the corresponding replicates.

3. Results and discussion
The efficacy of bioaugmentation is evaluated first of all according to the rate of the contaminant decomposition. In the two inoculated samples as well as in the control sample, similar hydrocarbon content dynamics was observed – it decreased from 60 g kg^{-1} to 25-27 g kg^{-1} during 63 days, and further remained stable. Single and triple introduction of the hydrocarbon-oxidizing consortium did not influence significantly on the consumption of oil products in the samples. It can be explained with poor survivability of the species introduced or with competitive suppression by the indigenous community [23]. However, using qPCR analysis it was revealed that two of the three introduced strains did survive and even grew in the remediated waste (Figure 1 a-c).

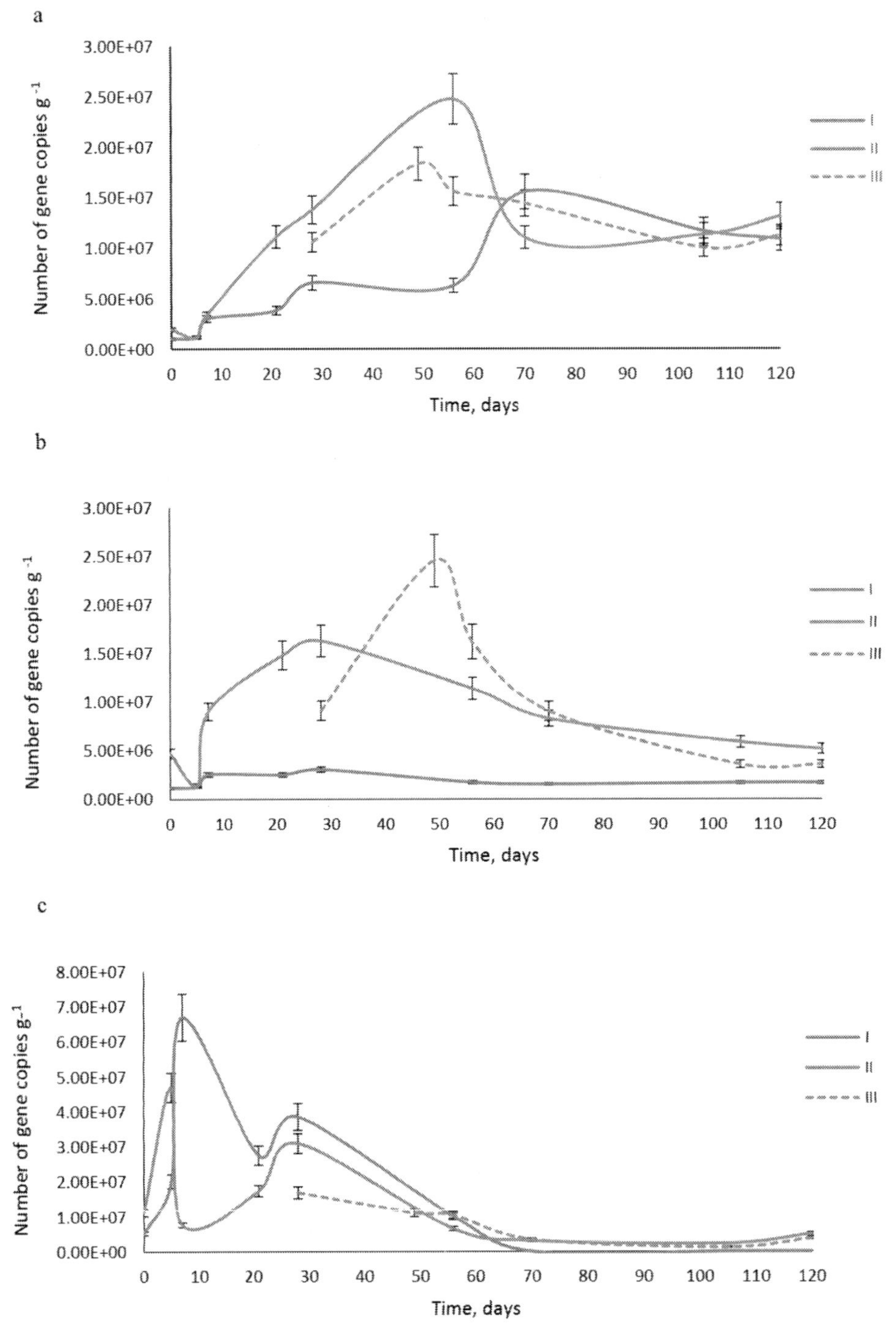

Figure 1. Specific Gene Copy Numbers of the Introduced Strains in the Process of Bioremediation (I, II, III – Variants of Bioremediation Methods; a – *Rh. jialingia;* b – *Ps. gessardii;* c – *S. rhizophila*).

Thus, the number of microorganisms belonging to the *Rh. jialingia* group after the first inoculation in sample II grew considerably from 1.94 x 10^6 gene copy number g^{-1} on the first day and reaching the maximum on day 56 – 2.47 x 10^7 gene copy number g^{-1}. The second and the third inoculations (sample III) did not help to increase the number of the species. After 70 days the number of *Rh. jialingia* species representatives became equal for all the three samples. The number of representatives of *Ps. gessardii* in the sample with no introduction changed considerably during the remediation and remained within the limits of 1.17 x 10^6 to 3.05 x 10^6 gene copy number g^{-1}. The first and the second inoculations induced the growth of the introduced strain as compared to the control sample. Thus, in sample II after the first inoculation the number of *Ps. gessardii* representatives increases from 4.71 x 10^6 gene copy number g^{-1} on the first day up to 1.63 x 10^7 gene copy number g^{-1} on day 28 of the experiment. The second inoculation in sample III induced the growth of the number from 9.09 x 10^6 gene copy number g^{-1} on day 28 up to 2.45 x 10^7 gene copy number g^{-1} on day 49, whereas the third inoculation did not have any significant effect. After 70 days, when the content of oil products used as a substrate decreases, the number of *Ps. gessardii* representatives decreases in all the samples. The reduction of the number of the introduced strains after 70 days can be explained with the reduction of oil product concentration being the main source of carbon in oily wastes [14,17]. As for the third strain, *S. rhizophila,* the introduction did not lead to the increase of its copy numbers in the inoculated variants as compared with the control sample which can be explained by competitive exclusion of this strain by the species of the indigenous community.

Absence of hydrocarbon decomposition rate on the background of introduced decomposers growth may be a sign that the indigenous microbial community was interrupted. Indeed, the maximum bacterial gene copy number was observed in the non-inoculated waste (4.06 x 10^9 gene copy number g^{-1}), while in the inoculated samples it ranged between 1.11 x 10^8 and 2.35 x 10^9 gene copy number g^{-1}. This can be explained by the fact that the introduced strains, despite their low relative abundance in the whole waste microbial community, destabilized it and inhibited indigenous species [30].

Then, the influence of the introduced biopreparation on the functional diversity of the wastes' microflora was estimated. Community level physiological profiling reflecting a number of carbon substrates that are decomposed by the heterotrophic members of the community was analyzed by means of BIOLOG® EcoPlates method [31–33]. Levels of AWCD are presented in Figure 2.

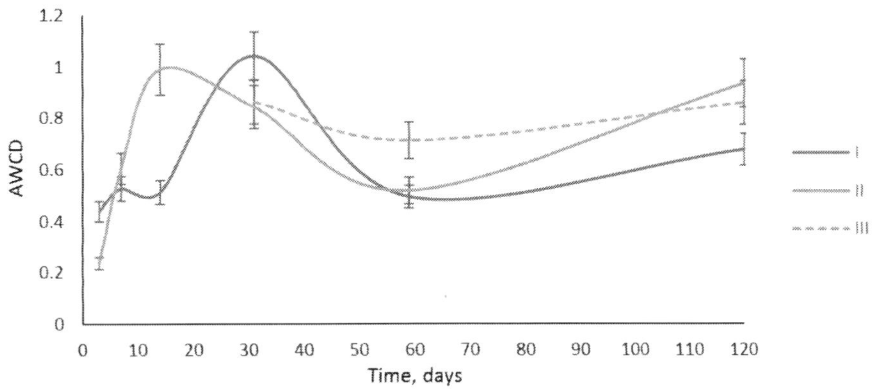

Figure 2. AWCD Values Estimated For Microbial Communities of the Remediated Waste Samples (I, II, III – Variants of Bioremediation Methods).

It can be concluded, that single introduction on day 1 lead to a certain reduction of AWCD value, however, on days 7 and 14 AWCD in sample II was higher than in the control sample. The second and the third inoculation also help to increase AWCD value, but it is not significant as compared with the control sample (I) and the sample with single inoculation (II). A similar regularity was witnessed by Mansur et al. (2016) in bioaugmented samples [34].

Functional diversity of microbial communities calculated based on Shannon index is presented in Table 2.

Table 2. Dynamics of Functional Diversity of the Microbial Communities in Wastes

Method of Bioremediation	Shannon Index on Different Days of Bioremediation					
	Day 3	Day 7	Day 14	Day 31	Day 59	Day 120
I	3.15	3.14	3.12	3.22	3.11	3.02
II	2.92	3.01	3.08	3.09	3.17	3.11
III				2.86	3.10	3.17

It can be observed that the introduction of destruction preparations in both variants (II and III) led to short term declination of Shennon index as compared with the control sample. As in the case of total bacterial copy numbers, it may be explained by the destabilization of the indigenous community of the wastes by the introduced microbes, and partial inhibition of the indigenous species. Interestingly, in the literature we found a description of an opposite effect – Dejonghe with co-authors (2001) demonstrated the deviation of Shennon index in the bioaugmented samples [35]. Indeed, the effects of biopreparation inoculations may vary strongly depending on many factors such as indigenous community structure, physical and chemical characteristics of the remediated waste, survivability of the introduced strains, etc. Therefore, these effects should be estimated in each specific case of bioremediation. However, the effect observed in our study lasted only for a short time – for several days after the introductions, Shennon index increased in the inoculated samples and later remained on the level of the control sample.

The analysis of using substrate groups has shown that in all the three samples carbons and carboxylic acids recover the most effectively, and substrates belonging to phenol compounds and amine group recover the least effectively, which was also observed in the study by Alisi et al. (2009). It was found out that the lowest number of oxidized substrates contained in the control sample (I) [36]. It was proved by other authors as well, thus, Teng et al. (2010) notice that the control samples are characterized by a lower number of oxidized substrates than the bioaugmented ones [37].

Conclusions

It can be concluded, that despite the high survival rate of two of the three introduced strains – *Rh. jialingiae* and *Ps. gessardii*, the introduction of the biopreparation did not lead to the improvement of petroleum hydrocarbon degradation in the remediated oily waste. This may be due to the deviation of total bacterial counts and short term alteration of physiological community profile as compared with the control sample. The poor efficacy of bioaugmentation may be also connected with relatively low abundance of the introduced strains as compared with total bacterial counts in the remediated waste.

Acknowledgement
The work is performed according to the Russian Government Program of Competitive Growth of Kazan Federal University.

References

[1] Xu L, Ravnskov S, Larsen J, Nilsson RH, and Nicolaisen M 2012 *Soil Biol Biochem.* **46** 26–32
[2] Mrayyan B, Battikhi MN. 2005 *J Hazard Mater.* **120** 127–134
[3] O'Rourke D and Connolly S 2003 *Annu Rev Environ Resour.* **28** 587–617
[4] Elektorowicz M 1994 *Environ Technol (United Kingdom).* **15** 373–380
[5] Swoboda-Colberg N 1995 *Microbial Transformations and Degradation of Toxic Organic Chemicals. New York: Wiley-Liss Inc.* Young L, Cerniglia C, editors. 27–74.
[6] Propst TL, Lochmiller RL, Qualls CW and McBee K. 1999 *Chemosphere.* **38** 1049–1067
[7] Zappi ME, Rogers BA, Teeter CL, Gunnison D and Bajpai R 1996 *J Hazard Mater.* **46** 1–12
[8] Galitskaya P, Gumerova R, Ratering S, Schnell S, Blagodatskaya E and Selivanovskaya S 2015 *J Plant Nutr Soil Sci.* **178** 825–833

[9] Galitskaya P, Biktasheva L, Saveliev A, Ratering S, Schnell S and Selivanovskaya S 2015 *Biogeosciences* **12** 3681–3693
[10] Jacques RJS, Okeke BC, Bento FM, Teixeira AS, Peralba MCR and Camargo FAO 2008 *Bioresour Technol.* **99** 2637–2643
[11] da Silva LJ, Alves FC and de França FP 2012 *Waste Manag Res.* **30** 1016–1030
[12] Galitskaya P, Akhmetzyanova L and Selivanovskaya S 2016 *Biogeosciences* **13** 5739–5752
[13] Wu M, Dick WA, Li W, Wang X, Yang Q and Wang T 2016 *Int Biodeterior Biodegrad.* **107** 158–164
[14] Hassanshahian M, Bayat Z, Cappello S, Smedile F and Yakimov M 2016 *J Environ Sci (China).* **43** 136–146
[15] Lladó S, Covino S, Solanas AM, Viñas M, Petruccioli M and D'annibale A 2013 *J Hazard Mater.* **248-249** 407–414
[16] Cerqueira VS, Peralba M do CR, Camargo FAO and Bento FM 2014 *Int Biodeterior Biodegrad.* **95** 338–345
[17] Pacwa-Płociniczak M, Płaza GA and Piotrowska-Seget Z 2016 *Appl Soil Ecol.* **105** 76–85
[18] Cerqueira VS, Hollenbach EB, Maboni F, Vainstein MH, Camargo FAO and Peralba M do CR 2011 *Bioresour Technol.* **102** 11003–11010
[19] Das K and Mukherjee AK 2007 *Bioresour Technol.* **98** 1339–1345
[20] Mukred AM, Hamid AA, Hamzah A and Yusoff WMW 2008 *Online J Biol Sci.* **8** 73–79
[21] Dellagnezze BM, Vasconcellos SP, Angelim AL, Melo VMM, Santisi S and Cappello S, 2016. *Mar Pollut Bull.* **107** 107–117
[22] Andreoni V, Cavalca L, Rao MA, Nocerino G, Bernasconi S and Dell'Amico E, 2004 *Chemosphere* **57** 401–412
[23] Mao J, Luo Y, Teng Y and Li Z 2012 *Int Biodeterior Biodegrad.* **70** 141–147
[24] Kruglov Y V 2016 Microbial community of the soil: physiological diversity and methods of research. *Agric Biol.* **51** 46–59 (IN RUSSIAN).
[25] Yakushev AV 2015 *Pedology.* **4** 429–446 (IN RUSSIAN).
[26] Heuer H, Krsek M, Baker P, Smalla K and Wellington EMH 1997 *Appl Environ Microbiol.* **63** 3233–3241
[27] Nübel U, Engelen B, Felsre A, Snaidr J, Wieshuber A and Amann RI, 1996 *J Bacteriol.* **178** 5636–5643
[28] White TJ, Bruns S, Lee S and Taylor J. 1990 *PCR Protocols: A Guide to Methods and Applications.* 315–322
[29] Team R. R Development Core Team. 2013 *R A Lang Environ Stat Comput.* **55** 275–286.
[30] Cunliffe M and Kertesz MA 2006 *Environ Pollut.* **144** 228–237
[31] Choi KH and Dobbs FC. 1999 *J Microbiol Methods.* **36** 203–213
[32] Garland JL 1997 *FEMS Microbiol Ecol.* 289–300
[33] Insam H and Goberna M 2004 *Mol Microb Ecol Manual.* 853–860
[34] Mansur AA, Taha M, Shahsavari E, Haleyur N, Adetutu EM and Ball AS 2016 *Int Biodeterior Biodegrad.* **115** 179–185
[35] Dejonghe W, Boon N, Seghers D, Top EM and Verstraete W 2001 *Environ Microbiol.* **3** 649–657
[36] Alisi C, Musella R, Tasso F, Ubaldi C, Manzo S and Cremisini C 2009 *Sci Total Environ.* **407** 3024–3032
[37] Teng Y, Luo Y, Sun M, Liu Z, Li Z and Christie P 2010 *Bioresour Technol.* **101** 3437–3443

Transformation of environmental conditions in large former Soviet countries: regional analysis

V R Bityukova and M S Borovikov

Lomonosov Moscow State University, Faculty of Geography, The Department of Economic and Social Geography of Russia, Moscow, Russia
e-mail: bituykova@yandex.ru; maximborovikov@gmail.com

Abstract. The article studies changes in the structure of environmental conditions of regions in the large former Soviet countries (case study of Russia and Kazakhstan) that have formed considerable contrasts in the placement of industrial complex and population settlement during the previous development stages. The changes related to the transition to market economy have led to essential transformation of environmental conditions. A complex index allowing to assess changes at the regional level in Kazakhstan and Russia and to reveal main similarities and differences between those changes is applied to studying the transformation of regional and industry structure. The article examines both industry-specific and spatial patterns forming environmental conditions at the regional level.

1. Introduction

The collapse of the Soviet Union has significantly changed the structure and scales of anthropogenic impact in the former Soviet republics: despite the decrease in absolute measures, relative measures remain too high to state that the environment in the countries is in a satisfactory state [1-5].

Transition to market economy in the former Soviet countries has transformed the scales and geographical structure of anthropogenic impact: a number of industries have been completely reorganized (both in terms of environmental and economic efficiency and in terms of location), some industries have disappeared, while others have appeared only after 1991; millions of people have moved, changing the structure of settlement; new borders have disturbed some economic ties and created new ones; and active state and private investment into new hubs of economic growth has led to new sources of impact.

At the same time the environmental conditions in the industrially developed former Soviet countries rich in natural resources, in particular, Russia and Kazakhstan, still bear the features inherited from the period of the Soviet industrialization [6, 7]. Their role can be revealed through a comparative study of the two countries, the industrial complex of which was formed within a single space.

Integral index of anthropogenic impact has been developed in order to assess changes in environmental conditions in the regions of the Russian Federation and the Republic of Kazakhstan. The scale and method of the research were chosen based on availability of statistical information for various levels and time periods, comparability of environmental indicators in Russia and Kazakhstan and adequacy of the available statistics [8, 9].

Integral indices of environmental conditions started to appear almost simultaneously in the USSR and in the West. But, differences in development conditions of this research method and in its purposes, impossibility of sharing practices predetermined separate ways already at the initial stage of formation of complex indices. Generalizing the experience of foreign scientists, governmental and international commissions available in this field, it is possible to identify a number of features. First, the use of basic indicators (GDP, gross domestic savings, net domestic product) normalized based on the cost estimate of depletion of natural resources, ecological damage, natural capital [10, 11]. Second, integral indicators of sustainable development (ecological sustainability, real progress, etc.) including both indicators of quality of life and ecological indicators [12 - 15]. The third important feature is the initial orientation towards the search of indicators reflecting a contribution of the countries to global problems, i.e., ecological footprint, living planet index [16]. And finally, proper environmental indices based on either indicators of anthropogenic impact (air emissions, waste water) or "environmental efficiency" (impact of an ecological factor on human health and ecosystems) [17, 18].

Large former Soviet countries, such as Kazakhstan and Russia, require a multiscale environmental research: country assessments lacking a regional component are unacceptable for countries as vast as they are. Subregional assessments are even more promising, however, the ecological statistics at the level of districts and municipalities is too scarce for a comprehensive complex assessment of anthropogenic impact over time. Thus, two levels have been chosen for the research: country and regional, allowing assessment of industry-specific and territorial shifts of areas of anthropogenic impact over time.

Key problems faced by researchers while elaborating a technique for impact assessment at the regional level are a weak statistical base, frequent change of calculation procedures and incomparability of some indicators within intercountry comparisons due to various approaches to measurements and calculations. For example, a key group of indicators highly relevant for the assessment of anthropogenic impact in the Russian regions connected with the impact on forest resources is less relevant for Kazakhstan. Taking into account a common group for conventional forests and haloxylon desert woodlands in the Kazakhstan statistics, the use of indicators of impact on forest resources demands a careful approach and expert adjustment.

The main purpose of this research is to know how anthropogenic impact in Russia and Kazakhstan has changed by using the integral index.

2. Research method

Creation of the integral index included four main stages:

Stage 1. Selection of priority components. In this case, ecological indicators of almost all significant sources of impact were taken into account: production sector, power industry, motor transport, agriculture and forestry, population as a source of impact (through household waste and indirectly through other indicators), indicators of radiative effects and others.

Stage 2. Selection of indicators for each group was based on the principles of systematicity, reliability, availability of statistics, pronounced territorial differentiation, clearly interpreted dynamics. Each type of anthropogenic impact has a set of absolute and relative (specified in brackets) indicators: A_1 – density of air emissions of the production sector and motor transport per the acreage of the cities, production sector, transport; A_2 – coefficient of emission toxicity calculated as the ratio of emission volume normalized to a single toxicity to the gross air emission volume [19]; B_1 and B_2 – respectively, share of water consumption and wastewater discharge in surface-water supply; C_1 – solid waste density in view of their hazard class; C_2 – share of disturbed industrial lands; D_1 and D_2 – share of cultivated and reclaimed lands respectively; D_3, D_4 and D_5 – mineral, organic fertilizers and pesticides respectively per 1 hectare of arable lands; D_6 – cattle livestock per the area of pastures; E_1 – share of total standing volume of cut timber; E_2 – the area of felling per calculated cut; E_3 – share of unauthorized felling; F_1 – share of inhabitants in radioactive contamination zones; F_2 and F_3 – soil pollution density with ^{137}Cs radionuclides of the forests and the area respectively; F_4 – share of lands of the cities, production sector, motor transport polluted by radionuclides; F_5 – share of lands of former nuclear test sites and major accidents related to radioactive pollution.

Stage 3. Translation of initial value matrix for assessment indicators into values normalized for addition of different-sized indicators. The following criteria can be used when selecting normalization principles: the nature of asymmetry of separate indicators, their contribution to the final assessment and interpretation of the totals. While searching for an optimum normalization principle, three most widespread methods were applied: ranking, linear scaling and logarithmic scaling. Logarithmic scaling proved to be the most acceptable method for the integral index. For the first time suggested for assessment of territorial differences of anthropogenic impact in the Russian regions, this method not only removes excessive differences of indicators but also adequately reflects the increasing pollution level parallel to the growing integral index (equation (1))[20].

$$X = \frac{\log_{X_{min}} X_i}{\log_{X_{min}} X_{max}} \tag{1}$$

x_i – index value in i region.

Stage 4. Integral index of anthropogenic impact (IAI) is calculated through the aggregation of group average values for the above-mentioned normalized indicators according to the formula (2):

$$IAI = (\frac{A_1 + A_2}{2}) + \left(\frac{B_1 + B_2}{2}\right) + \left(\frac{C_1 + C_2}{2}\right) + \left(\frac{D_1 + D_2 + D_3 + D_4 + D_5 + D_6}{6}\right) + \left(\frac{E_1 + E_2 + E_3}{3}\right) + \left(\frac{F_1 + F_2 + F_3 + F_4 + F_5}{5}\right) \tag{2}$$

In order to make the assessment more objective, two calculation options were applied to the integral index: the index of the power of anthropogenic influence (PAI) based on absolute measures and the index of the intensity of anthropogenic influence (IAI) based on relative measures. Use of both absolute and relative measures is especially important taking into account large regions in the studied countries. Application of absolute or relative measures depends on the research objective. If it is focused on the impact on the natural landscape or population health, absolute values are more justified as the whole volume of pollutants caused by a source spreads in the landscape and influences the population.

3. Results

Integral IAIs were calculated for four time periods: 1990 – the beginning of the transition period, 1998 – the end of crisis, 2003 – the end of the first stage of compensatory growth, 2014 – the most recent statistics available. In general, the results have demonstrated the adequacy of the developed technique for identification of territorial differentiation of environmental conditions in large countries.

The total index has shown a high, though reducing degree of interregional distinctions in terms of the intensity of environmental pressure on the territory of the two countries (from 3.6 to 1.8 times during 1990-2014) (figure 1). Various changing trends in individual factors have also determined the dynamics of the complex regional index after the collapse of the Soviet Union.

Decreasing AI has become an underlying trend, and internal fluctuations corresponded to the dynamics of economic indicators. At the same time in the early 2000s, during the maximum increase in oil prices, pollution volumes moved to the extracting regions with low population density. As the result, an export-oriented raw zone Taimyr – Yamal – Urals that concentrates over 55% of industrial production in the country and 70% of air emissions is distinguished based on the environmental pollution (especially, air emissions related to the combustion of associated petroleum gas). Unlike the majority of old developed regions where pollution is localized in the cities, the impact is dispersed across the territory in newly developed extracting regions.

Use of pollution density indices (as compared to absolute values of volumes) moves three types of regions down in rankings, both in Russia and in Kazakhstan. First, large regions with an intense but rather localized AI (Krasnoyarsk Krai, Tyumen, Karaganda, Arkhangelsk, Rostov, Vologda, Irkutsk). Second, regions with large rivers under serious anthropogenic pressure but with low density of impact (Novosibirsk, Omsk, Perm). Third, strong agro-industrial regions (Stavropol Krai, Altai Krai,

Kostanay) have seen an advancing production decline in the leading agricultural regions with the most intensive agriculture since the 1990s. On the contrary, industrial regions with localized impact (Tula, Lipetsk, Belgorod) and especially federal cities of the Russian Federation and cities of republican significance rise in ranking. Agricultural pressure exacerbates ecological stress in the Black Earth and southern regions of the Russian Federation, as well as in traditionally agrarian regions of Kazakhstan – North Kazakhstan and Akmola.

However, centers of ecological stress, unevenly spread across the territory of both Russia and Kazakhstan, are the same for any assessment method. These are the Chelyabinsk, Sverdlovsk, Kemerovo, Moscow and Bryansk regions. All Russian macro-regions have regions with serious anthropogenic impact that distinguishes them from others. Old developed regions are characterized with a more evenly distributed impact density which is more localized beyond the Urals. In general, regional differences are rather balanced.

In Kazakhstan, the center of ecological stress, Pavlodar-Karagandy-East Kazakhstan area, has actually begun to smear during the considered period due to decreasing impact in the East Kazakhstan region and new centers of ecological stress: oil-extracting West and densely populated South of Kazakhstan.

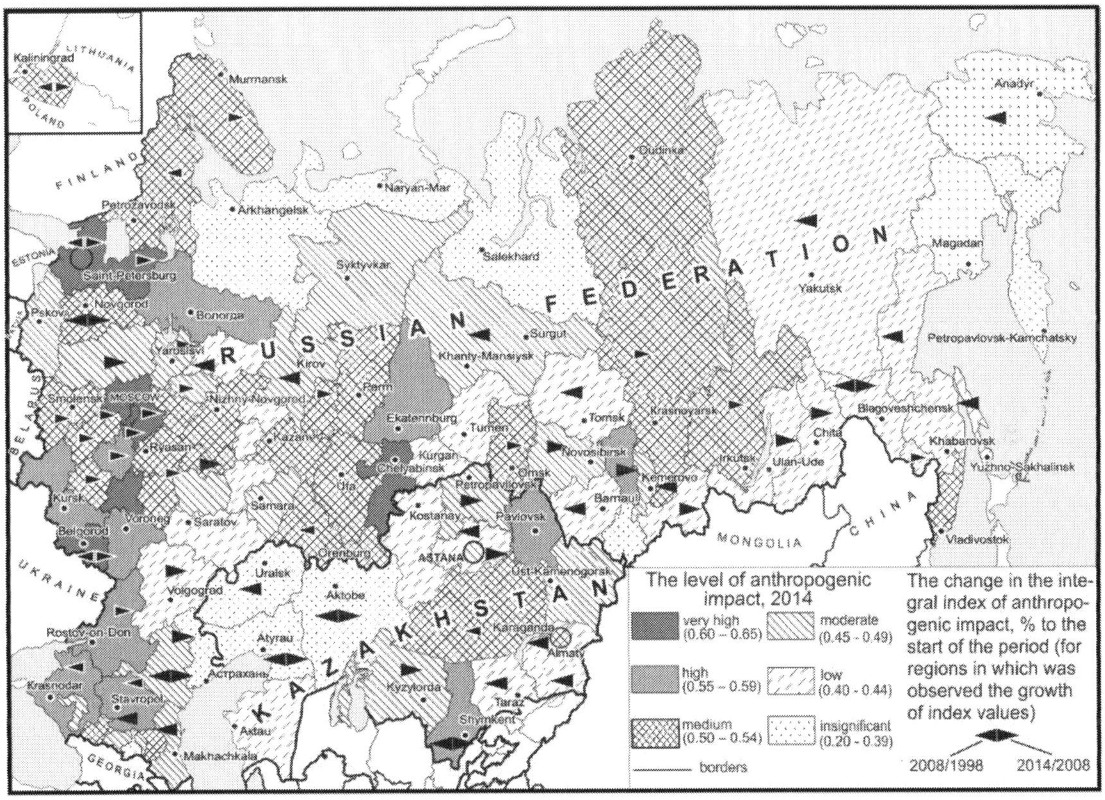

Figure 1. Changes in the integral index of anthropogenic impact in 1998 – 2014.

Despite considerable economic shifts and transition to a new type of economy, the inherited factors still play a crucial role in the ranking of regions. Almost all regions with a very high level of anthropogenic impact have large industries since the USSR, that determine specialization of the region. The Chelyabinsk and Sverdlovsk regions and metallurgical and industrial centers of the European part of Russia, i.e., the Lipetsk, Vologda, Tula regions, are the center of regions with the greatest intensity of anthropogenic impact. Capital regions are an important part of this center: the

Moscow region suffers additional pressure due to an inflow of population and sources of impact, while the Leningrad region has high rates of IAI mostly owing to the industrial development. Radioactive pollution, though having reduced considerably, is still noticeable in the zones of accidents of the previous period (Bryansk and other regions affected by the Chernobyl accident and the Chelyabinsk region that still bears the footprint of the accident on the Mayak Production Association).

Environmental rankings of regions based on the complex index are an efficient way of determining the priorities of environmental policy. The integral index provides new insights into the complexity and versatility of the environmental conditions. Nevertheless, a clear understanding of calculation specifics of the complex indicator is necessary to make the ranking a convenient decision-making tool in terms of environmental control and environmental management. The selected AI measurement technique considerably affects the outcome. It is important to use AI density for large countries with isolated intense sources of impact and extensive unaffected territories. Moreover, Kazakhstan that has experienced essential changes of agricultural areas (arable lands, pastures have reduced in the regions by 30-50%) and urban lands during the land reform, has the territorial structure of the intensity integral index considerably changed towards increase since 2003, while the indices for 1990 and 1998 prove to be underestimated.

Key factors maintaining territorial structure of AI areas in Russia and Kazakhstan are as follows:

1. Inherited development. The largest AI centers in the territorial structure of Kazakhstan and Russia are still the regions where the industrial base was developed and formed in the Soviet period: the Urals and industrial regions of the European part, Pavlodar and Karaganda regions. Despite low efficiency and insignificant economic feasibility related to production and extraction of some minerals, a number of large enterprises seriously harming the environment continue to exist. It is all the fault of the existing structure of population settlement (miner's towns) and impossibility to move people from depressive monotowns. This factor is more typical of Russia than of Kazakhstan.

2. Institutional factor. Despite the stated models of the free market and general privatization, only few foreign companies could establish full control over large industrial enterprises in Russia that has contributed to the stability of the territorial structure of AI areas.

Factors influencing transformation of AI territorial structure:

1. Becoming part of the global resource market. Soaring hydrocarbon prices have caused active development of oil and gas fields in the East of Siberia and the Western regions of Kazakhstan. It created new AI centers in the Western Kazakhstan: increase in oil production of the Atyrau region from 2.5 million tons in 1990 up to 31.9 million tons in 2014, development of the Karachaganak oil and gas field in the West Kazakhstan region caused increase in oil production from 4 million tons in 1990 to 13 million tons in 2014.

2. Institutional factor. Emergence of the western companies in mining and manufacturing industry of Kazakhstan has reduced the level of environmental impact in terms of specific indicators, for example, Kazakhstan has almost half as much emissions than Russia per unit of oil produced. It influences the territorial structure of pollution areas: new industrial centers in Kazakhstan are less significant on a national scale than in Russia (based on relative measures).

3. State environmental policy. A weak state policy in relation to the oil companies has made Khanty-Mansi Autonomous Okrug a "leader" in air pollution for the first time in 2004-2007. Having doubled air emissions, the district "has outrun" Krasnoyarsk Krai. After signing the resolution "On measures to encourage the reduction of air pollution products flaring gas in flares" in 2009 which sets the target indicator of combustion of associated gas on flares of no more than 5% of the extracted associated gas and provides for an increased payment for its combustion above the limit, large-scale investment has been made. As the result of the utilization coefficient of associated petroleum gas growing from 75.5% in 2011 up to 91.7% in 2013 and 93.2% in 2014 there was a reduction of its combustion in flares and a decrease of pollutant emissions by 21.4%.

4. Reaching the limit of ecological capacity in some regions. Most of regions in Kazakhstan are located within semi-arid and arid zones where ecological limits of water sources, pastures and arable lands are almost reached. When using relative measures, the Mangystau and South Kazakhstan regions

demonstrate very high specific rates of anthropogenic impact on limited water sources and land resources. Therefore, any minimum increase of anthropogenic impact in regions with ecological capacity close to its limits leads to a serious degradation of the environment.

5. Demographic changes as a factor transforming the territorial structure of anthropogenic impact are typical first of all of Kazakhstan that has AI increasing in the South due to population growth and falling in the North due to depopulation.

4. Conclusion

Integral assessment of environmental conditions in the regions of Russia and Kazakhstan has revealed both similarities caused by the consequences of an accelerated industrialization during the Soviet period and differences in the formation of environmental situation caused by different natural and social and economic conditions. The impossibility of blaming decisions made 50 years ago for all problems becomes clear, the reasons for environmental problems are much deeper. The regions of advancing development that possess competitive advantages have high AI level. Rich resource extracting regions see AI strengthening throughout the whole natural complex, faster than the economy develops. The largest city agglomerations experience the effect of the economy of scale that attracts investment and labor, creating greater transport pressure on infrastructure, waste disposal problems, deforestation, environment degradation, depletion of water resources. The state is responsible for the development of the whole territory, measures of regional environmental policy gradually improve the situation.

References

[1] Bityukova V R, Kasimov N S and Vlasov D V 2011 *Ecology and Industry of Russia* **4** 6 – 18 (in Russian)
[2] Bobylev S N, Minakov V S, Soloviova S V and Tretiakov V V 2012 *Ecology-economical index of Russian cities* (Moscow: WWF Russia) p 124 (in Russian)
[3] Klyuev N N 2007 *Izvestiya of RAS Geographical Series* **1** 13 – 23 (in Russian)
[4] 2011 *Mercer Health and Sanitation Index* (London: Mercer) p 11
[5] 2017 *OECD Urban Policy Reviews: Kazakhstan* p 232
[6] Bezuglaya E Yu and Smirnova I V 2008 *Urban Air and its Changes* (Saint-Petersburg: Asterion) p 254 (in Russian)
[7] Bityukova V R 2011 *Regional Research of Russia (RRR)* **1** 394 – 98
[8] Bityukova V R and Kasimov N S 2012 *Geofizika (Geophysics)* **29** 53 – 67
[9] Klyuev N N 2011 *Izvestiya of RAS Geographical Series* **5** 13 – 27 (in Russian)
[10] Dixon J and Jian X 2007 *Promoting Market-oriented Ecological Compensation Mechanisms: Payment for Ecosystem Services in China* (China: World Bank) p 28 42177
[11] 2010 *Green Growth Strategy Interim Report: Implementing Our Commitment for a Sustainable Future* (Paris: OECD) p 94
[12] Batten J and Edwards C 2016 *Sustainable Cities Index 2015 Balancing the economic, social and environmental needs of the world's leading cities* (Arcadis) p 40
[13] 2005 *Environmental Sustainability Index* (Yale: Center for Environmental Law and Policy Yale University, Center for International Earth Science Information Network Columbia University) p 63
[14] Ewing B, Goldfinger S and Wackernagel M 2008 *The Ecological Footprint Atlas* p 85
[15] 2010 *National Human Development Report 2009 From Exclusion to Equality: Realising the rights of persons with disabilities in Kazakhstan* (Kazakhstan: UNDP) p 161
[16] 2009 *Sino-Singapore Tianjin Eco-City: A Case Study of an Emerging Eco-City in China (Technical Assistance (TA) Report)* (The World Bank, AusAID) p 168
[17] 2013 Scandinavian Destination. Sustainability Index. Assessing and reporting on the sustainability performance of Scandinavian major meetings destinations *Scandinavian*

Chapter of the International Congress and Convention Association (ICCA) (MCI Sustainable Services) p 10
[18] 2012 *The Green City Index. A research project conducted by the Economist Intelligence Unit* (Germany, Munich: Siemens AG Corporate Communications and Government Affairs) p 2
[19] Boubel R W, Fox D L, Turner D B and Stern A C 1994 *Fundamentals of air pollution. Third edition* (Academic press) p 574
[20] Bityukova V R, Kirillov P L 2011 *Regional research* **31** 130 – 56 (in Russian)

IOP Conf. Series: Earth and Environmental Science **107** (2017) 012052 doi:10.1088/1755-1315/107/1/012052

Influence of Vladivostok coastal waters pollution on a microflora of mussel *Crenomytilus grayanus*

E A Bogatyrenko[1], T I Dunkai[1], L S Buzoleva[1,2] and A V Kim[1]

[1] Far Eastern Federal University, Vladivostok, Sukhanova str., 8, 690091, Russia
[2]Somov Research Institute of Epidemiology and Microbiology, Vladivostok, Sel'skaya str, 1, 690087, Russia
e-mail: kim-sandra@mail.ru

Abstract. Taxonomic structure of bacterial community for mussel *Crenomytilus grayanus* digestive system from coastal waters of Vladivostok city, which is characterized by considerable anthropogenic impact was studied. Specimens of *Micrococcaceae* family predominated in the mollusc microscopic flora by the quantity of selected strains (62%). The order of *Actinomycetales* (namely genera of *Actinomyces and Pimelobacter)* were large (11%). In addition, there were numerous representatives of Enterobacteriaceae family (9%), *Vibrio* genus *(9%), and Paracoccus* genus (9%) selected. There were found such pathogenic and opportunistic microorganisms, as *Escherichia coli*, *Listeria monocytogenes*, and *Klebsiella spp* in microbial communities of mussels. Even though pathogenic microflora does not predominate in the biotic community, it nevertheless shifts the balance to increase extrinsic microflora within the hydrobiont. Detection of such potentially harmful bacteria indicates the insufficient sanitary and epidemiological state of the water area researched, the bay persistently polluted with municipal domestic drain waters.

1. Introduction

The Sea of Japan is a favourable site for fishery and cultivation of various hydrobionts aggregations due to its environmental and geographical factors [1]. However, increasing pollution of water areas considerably affects natural and cultivated populations of marine animals [2, 3].

Vladivostok is a seaport with over 600000 people, and it is situated on the Muraviev-Amursky peninsula and the islands in the Peter the Great Gulf of the Sea of Japan. Vladivostok inshore waters are rich with marine animals. Species like herrings, smelts, navagas, flatfish, greenlings, rudds, so-iuy mullets, mussels, trepangs, scallops, octopuses, crabs inhabit them [4]. Despite natural original conditions, the water areas, which are adjacent to the city, have been tremendously influenced by anthropogenic stress since the mid-nineties of the last century. The main sources of seawater pollution close to the city are industrial wastes and domestic sewage from Vladivostok city, and other residential areas and settlements, as well as polluted waters of rivers running into the sea nearby. During the period of preparation for the APEC Vladivostok Summit 2012, there were three complexes of treatment facilities constructed, and the fourth complex was redeveloped, that was set to improve ecological situation in this region [5].

Crenomytilus grayanus is prevalent in the waters of southern Primorye, including many bays of Vladivostok islands. This surf type is popular among local people, however, systematic monitoring of microbiological safety for these objects is lacking. While mussels let water through mantle cavity, nutritious matter, as well as various organic and inorganic pollutants, also microorganisms which are agents of infectious diseases, can get into animal bodies with seawater and be accumulated inside [2]. Thus, industrial waste and sewage waters in the big city can produce a negative effect on natural microbial communities in the marine environment, thereby degrading the sanitary-epidemiological state of water areas and opening the risk of human population infections through contaminating seafoods [2, 6]. Therefore, the research objective was to estimate the influence of the anthropogenic pollution on the microflora composition of *Crenomytilus grayanus* and its habitat in Vladivostok coastal water areas.

2. Material and methods

2.1. Sampling

The Ajax bay was chosen as the research area, the bay is situated in northeastern part of the Russian Island within the Vladivostok city (Primorsky krai of the Russian Federation). The seafloor of the bay is substantially stony, and sand-stony. Previously, the technogenic pollution of the bay with oil hydrocarbons, phenols, and heavy metals was observed [7].

Adults of *Crenomytilus grayanus* of the same size, water and sea ground samples from the species microhabitat for experiments. Samples were taken at a depth of 5 - 6 metres in the Ajax Bay during the late summer-early autumn periods during 2014-2015. In the laboratory environment, the mussel gastrointestinal tracts were extracted with a scalpel, and they were homogenized. After serial dilution the tissue homogenate, water and sediment were plated on solid modified marine broth (MMB) medium, thereafter they were cultivated at the temperature of 25ºC inside the thermostat during two days [8]. The obtained single colonies were separated and replated for isolated bacterial strains could further be taxonomically determined.

2.2. Identification of obtained strains of cultivated heterotrophic bacteria

Identification of obtained bacteria strains was carried out on a basis of the analyzed nucleotide sequences in the 16S rRNA gene fragment. DNA extraction was made using the improved Echt method [9-11]. Two primers were chosen for the operation: 926F 5'-AACTCAAAGGAATTGACGG-3' and 1522R 5'-AAGGAGGTGATCCARCCGCA-3'[12]. A positive control (earlier separated bacterial DNA) was employed for the PCR efficiency evaluation. A negative control (no DNA adding) was used to prove unavailable contamination of foreign DNA reagents.

PCR was performed with 25 mcL reaction mixture, containing 1×TaqMan buffer B, 2.5 mM of $MgCl_2$, 250 mcM of dNTP, 2 activity unit of Taq DNA Polymerase (Sileks, Russia), 15 ng of DNA, and 0.25 mcM of each primer (Syntol, Russia). Initial denaturation was performed at 95 °C (2 minutes) with 37 following denaturating cycles at 95°C during 15 seconds, annealing at 56°C during 10 seconds, and elongation at 72°C during 40 seconds, with final elongation at 72°C during 1 minute.

All PCR-products were visualized through horizontal electrophoresis in 1.5% agarose gel with the Tris/Borate/EDTA buffer (TBE buffer), containing ethidium bromide (10 mg l^{-1}). Sequencing reaction was carried out with 10 mcL reaction mixture under the following protocol: 6 mcL of distilled water, 2 mcL of quintuply sequencing buffer, 0.6 mcL of Big Dye Terminator Cycle Sequencing kit v.3.1 (Thermo Fisher Scientific, the USA), 0.4 mcL of primers, and 1 mcL of PCR-products obtained. Sequencing reaction with forward and reverse primers was performed in different test tubes. Reaction was processed in the Bio-Rad T 100 amplifier (the USA) under the following protocol: initial denaturation at 96°C (1 minute) with 30 further denaturating cycles at 95°C during 10 seconds, annealing at 56°C during 9 seconds, and elongation at 60 °C during 50 seconds, with final elongation at 60°C during 60 seconds. Reaction products were purified with the use of ethanol. 8 mcL of distilled water and 32 mcL of 96% ethyl alcohol were added to a sample, which was thereafter centrifuged at maximum rpm within 15 minutes. After that, the previous alcohol was decanted, and 120 mcL of 70% ethyl alcohol was infused to centrifuge the sample for within 10 minutes, then the liquid was promptly discharged. Dried sample with addition of formamide (18 mcL) was denatured at 95°C during 2 minutes and sequenced in ABI 3130 Genetic Analyser (Perking-Elmer Biosystems, the USA).

The assembling and alignment of sequences were done with Pregap 4 v.1.4 software (England), and the analysis of DNA sequences was fulfilled via BLAST resource (Basic Local Alignment Search Tool) (NCBI, the USA).

3. Results and discussion

While making the research 181 strains of cultivated heterotrophic bacteria were isolated, there were 67 bacteria strains segregated out of water. 61 ones out of sediment, 53 ones out of mussels (table 1).

Table 1. Taxonomical diversity of bacterial communities from Ajax bay.

Taxon	Quantity of strains		
	water	sediment	mussel
order Actinomycetales:	**10**	**5**	**6**
genus *Actinomyces*	4	5	3
genus *Corynebacterium*	3	0	0
genus *Pimelobacter*	0	0	3
genus *Rhodococcus*	3	0	0
family Micrococcaceae:	**8**	**15**	**14**
genus *Arthrobacter*	2	8	6
genus *Micrococcus*	6	7	4
genus *Kocuria*	0	0	4
family Enterobacteriaceae:	**8**	**7**	**5**
genus *Escherichia*	6	6	3
genus *Citrobacter*	1	0	0
genus *Klebsiella*	0	1	1
genus *Proteus*	0	0	1
genus *Yersinia*	1	0	0
family Clostridiaceae:	**2**	**5**	**4**
genus *Clostridium*	0	3	2
genus *Sarcina*	2	2	2
family Flavobacteriaceae:	**4**	**1**	**0**
genus *Flavobacterium*	2	1	0
genus *Olleya*	2	0	0
family Halomonadaceae:	**5**	**2**	**0**
genus *Cobetia*	3	1	0
genus *Deleya*	2	1	0
family Moraxellaceae:	**9**	**5**	**2**
genus *Acinetobacter*	4	3	0
genus *Moraxella*	5	2	2
Other families:	**19**	**17**	**19**
genus *Acetobacter*	0	1	0
genus *Aerococcus*	1	0	4
genus *Bacillus*	2	4	3
genus *Cytophaga*	0	1	0
genus *Enterococcus*	1	4	0
genus *Listeria*	1	1	1
genus *Microbacterium*	2	0	0
genus *Paracoccus*	1	2	5
genus *Phyllobacterium*	1	0	0
genus *Pseudomonas*	4	5	1
genus *Pseudoalteromonas*	0	0	1
genus *Staphylococcus*	3	2	2
genus *Streptococcus*	1	0	0
genus *Vibrio*	4	1	5
Total:	**67**	**61**	**53**

All investigated microorganism strains were indentified and assigned to the 34 genera of bacteria. Representatives of the family *Micrococcaceae* predominated in the mollusc microflora as 26% of the segregated strains. Bacteria of the *Actinomycetales* order, namely genera *Actinomyces* and *Pimelobacter,* appeared also numerous and made up 11%. In addition, many specimens of *Enterobacteriaceae* family (9%), *Vibrio* genus *(9%), and Paracoccus* genus (9%) were selected.

The presence of pathogenic and opportunistic microorganisms, such as *Escherihia coli*, *Listeria monocytogenes*, and *Klebsiella spp* in the microflora of mussels in the Ajax Bay turned out to be important feature of the microflora. Detection of similar potentially hazardous bacteria proves the insufficient sanitary and epidemiological state of the water area researched, which is caused by persistent pollution with municipal domestic drain waters, and also by significant recreational pressure in summer time. As a strong factor, which has an influence to the ecological conditions and the microbal variety in the Ajax bay waters, a heavy contamination with waste oil products from private small-size vessels can appear. It is interesting to note that, only in the Ajax bay waters 2 bacteria strains belonging to the *Olleya* genus were determined, some of the specimens can be found in sea waters while extensive oil spills occur [13], as literature data report, which play the role of decomposers for polycyclic oil hydrocarbons [14], including high-toxic benzo[a]pyrene [15].

When analyzing hydrobionts microflora composition, one cannot fail to take into account their environmental conditions, as well as the peculiar microbiota of this environment [16]. Though, water and sediment microbiota have a substantial impact on the researched animals microflora (table 1), the certain groups of microorganisms, detected in the environmental objects of a habitat, did not occur among mussel symbionts. Thus, for example, there were not any *Corynebacterium, Rhodococcus, Flavobacterium, Olleya, Cobetia, Deleya, Acinetobacter, Enterococcus*, and *Microbacterium* bacteria genera in mollusc associates that indicates, a definite existing selectivity during the development of *Crenomytilus grayanus* bacteria communities. It is worth mentioning that in order to exclude probable desultory data only the microorganism groups, selected in two or more strains from the habitat, were accounted.

In addition, our data demonstrated, that in the mussel microflora of the Ajax bay there were some bacteria found, which presence in the water and ground samples of this area was not discovered: the representatives of genera of *Kocuria* (4 strains), *Pimelobacter* (3 strains), and *Pseudoalteromonas* (1 strain).

Other authors received similar results at *Crenomytilus grayanus* microflora investigations [17]. Thus, *Streptomyces, Enterobacteria* and yeasts occur only as mollusc associates, and were not determined in the waters of mollusc natural habitat [17]. Mediterranean oysters example demonstrates that *Cytophaga, Flavobacterium, α-Proteobacteria, Bacillus, Shewanella* bacteria are not medium-selected, but only in mollusc microflora representative [18]. During the study on *Mytilus galloprovincialis* microflora Cavallo et al [19] pointed out that the exceptional luminescent bacteria for mussels became the following genera: *Hotobacterium* and *Lucibacterium*, as well as *Vibrio, Aeromonas, Pseudomonas, Acinetobacter/ Moraxella, Flavobacterium, Micrococcus,* also bacteria of *Enterobacteriaceae* family. *Moraxella, Cytophaga, Flavobacterium, Bacteroides, Bacillus, Micrococcus* and *Streptomyces* were discovered only in mussels, and not in water, as indicated in the paper about *Mytilus trossulus* [6].

Such facts could be related with the rare specimens of inhabiting microorganisms accumulated inside mollusc body. Penetrating hydrobiont organism, they can colonize its gastrointestinal tract to possess thereby an advantage over free-living microorganisms. Therefore, also pathogenic bacteria, potentially hazardous for human beings, can take a preferential development under specific circumstances that indicates the necessity of a permanent microbiologic monitoring.

As the conducted research found, the anthropogenic contamination of Vladivostok water areas has a considerable impact on microbial communities residing in water, sediments and gastrointestinal tract of *Crenomytilus grayanus* in the Ajax bay. Although opportunistic microflora does not prevail in the biocenosis, it still upsets the balance toward the increasing quantity of the non-typical for hydrobiont microflora. Meanwhile, the ability of mussels to accumulate certain microorganism groups exposes us to higher risks of contaminated products.

Acknowledgments
The research was fulfilled with financial support of the Russian Scientific Fund (Agreement No. 14-50-00034)

References

[1] Nemtseva N V 1998 *The microbiological characteristic of the biocenotic relationship of hydrobionts and its value for sanitary assessment of reservoirs* (Chelyabinsk) p 42
[2] Lazhentceva L Yu 2012 *Research of water biological resources of Kamchatka and northwest part of the Pacific Ocean* **26** 33 – 52
[3] Khristoforova N K, Tsygankov V Y, Lukyanova O N and Boyarova O N 2016 *Environ Pollut* **213** 727 – 31
[4] Yakimenko L V and Ivanenko N V 2016 *Territory of new opportunities Bulletin of the Vladivostok state university of economy and service* **8** 256 – 65
[5] Chernyaev A P and Nigmatulina L V 2013 *Izvestiya TINRO* **173** 230 – 38
[6] Beleneva I A, Zhukova N V and Maslennikova E F 2003 *Microbiology* **72** 472 – 77
[7] Moshchenko A V and Shaihlislamova L E 2010 *Izvestiya TINRO* **161** 199 – 211
[8] Youchimizu M and Kimura T 1976 *Fish Pathol* **10** 243 – 59
[9] Echt C S, Erdahl L A and McCoy T J 1992 *Genome* **35** 84 – 87
[10] Kiselev K V, Tyunin A P and Karetin Y A 2013 *Acta physiologiae plantarum* **35** 1843 – 51
[11] Kiselev K V, Dubrovina A S and Tyunin A P 2015 *Journal of plant physiology* **175** 59 – 67
[12] Lane D J 1991 16S/23S rRNA sequencing *Nucleic acid techniques in bacterial systematic* ed E Stackebrandt and M Goodfellow (Chichester, United Kingdom: John Wiley and Sons) 115 – 75
[13] Gutierrez T, Singleton D R, Berry D, Yang T, Aitken M D and Teske A 2013 *The ISME journal* **7** 2091 – 2104
[14] Isaac P, Sánchez L A, Bourguignon N, Cabral M E and Ferrero M A 2013 *Int Biodeterioration & Biodegradation* **82** 207 – 14
[15] Okai M, Watanabe A, Ishida M and Urano N 2015 *Genome announcements* **3** 10.1128/genomeA.01328-15
[16] Bogatyrenko E A and Buzoleva L S 2016 Characterization of the gut bacterial community of the Japanese Sea cucumber Apostichopus japonicus *Microbiology* **85** 116 – 123
[17] Ivanova E P and Mikhailov V V 1992 *Microbiology* (Moscow) **61** 367 – 72
[18] Pujalte M J, Ortigosa M and Macián M C 1999 *Internatl microbial* **2** 259 – 66
[19] Cavallo R A, Acquaviva M I and Stabili L *Environ Monit and Ass* **149** 465 – 75

Biofilm formation as a method of survival of *Escherichia coli* and *Pantoea spp* in the marine environment

L S Buzoleva[1,2], Y S Golozubova[1], A I Eskova[2], A V Kim[1] and E A Bogatyrenko[1]

[1]Institute of Natural Sciences, Far Eastern Federal University, Suhanova str., 8, 690091, Russia
[2] Somov Institute of Epidemiology and microbiology Environmental Studies, Selskaya str.,1, 690000 Russia
e-mail: buzoleva@mail.ru

Abstract. The article shows the formation of biofilms of bacteria Escherichia and Pantoea, which were isolated from sea water, both in monoculture and in associations with marine heterotrophs. It studied the influence of the nutrient medium and temperature on the biofilm-forming properties of marine strains. The highest biofilm formation properties were found in monoculture in family enterobacteria compared to saprophytic marine bacteria, regardless of the medium and the culture temperature. In association with saprophytes, *Pantoea spp*. possess more pronounced biofilm-forming properties at 37 °C compared to the control than at 22 °C and 5 °C irrespective of the culture medium. *Escherichia coli*, in association with saprophytes, have less pronounced biofilm formation properties than monoculture, regardless of the temperature and culture medium.

1. Introduction

The coastal zone of the sea, which is the place of the predominant concentration of human activity and taking a large amount of waste. Waste requires constant monitoring of the content of pollutants and changes in the main parameters of the environment caused by anthropogenic impacts. For coastal sea waters, actively used in the economic activity of the coastal regions, the highest pollution is observed simultaneously with different pollutants (oil hydrocarbons, phenols, heavy metals, etc.) [1,2]. The active use of coastal zones for recreational purposes causes the emergence of a large number of sanitary-indicatoryn microorganisms in the marine environment that affect not only marine microbial communities, but also pose a threat to human health. Microbiological monitoring in ecologically unfavorable areas is promising for assessing the state of the environment, as well as ecological mapping of pollution of coastal zones. It is known that the growth of bacteria in the form of biofilms is a form of response to a variety of stresses, forcing them to adapt to the negative effects of the environment (and all possible variants of this effect). Organic pollution of the marine environment by pathogenic microorganisms is an important environmental factor affecting the vital activity of biological systems at different levels of their organization [3]. But, the marine environment is not optimal for the life of opportunistic and pathogenic bacteria, although, according to the literature, a number of authors note their long existence under these conditions [4]. One of the possible mechanisms for survival of enterobacteria in sea water may be their ability to biofilm formation with marine saprophytic microorganisms [5-7]. Previously, it was shown that the joint incubation of such and enterobacteria as *Esherichia coli* [8-16]. Based on this, the purpose of our investigation was to study the ability of *Esherichia coli and Pantoea spp.* to form biofilms with marine saprophytic bacteria.

2. Methods

Microorganisms were isolated from sea water b. Golden Horn (Primorsky Krai, Russian Federation), characterized high microbial contamination followed by identification with the RapiD 20 E test kit (Biomerieux) and PCR confirmation with appropriate primers. The formation of biofilms was studied using a "tablet method" [17] by growing strains in 150 μl. Manupalition was on beef-extract broth and marine microrganisms medium (in a 96-well polystyrene plate for 72 hours (biofilm maturation stage according to the bacterial growth indications) at 5°C, 22°C, 37°C. Removal From the wells of plankton cells, staining of biofilms with 1% crystal violet solution, rinsing the wells with distilled water, adding 96% of ethanol to them. It was measured the optical density (OD) of the supernatant at a wavelength (540 nm) using a Multiscan ascent photometer (Thermo Electron Co., China).

3. Results and discussion

In total, it was allocated 9 bacterial isolates from seawater that have been investigated for the formation of biofilms in monoculture at different temperatures in two kinds of culture media. It was shown experimentally that, properties of biofilms studied microorganisms isolated from sea water, depending on the species. High properties of biofilms was observed, 2 isolates belonging to family Enterobacteriaceae identified as a strain of 16 M *Escherichia coli* and 42M *Pantoea spp*. And 7 strains of other marine microorganisms (table 1). *Escherichia coli* is very stable in the environment for a long time are stored in the soil, water, feaces, can cause cholera, dizenteria disease and enterits. Pantoea can cause infections of the kidneys and urinary tract infections (acute pyelonephritis, acute exacerbation of chronic prostatitis), genitals, the respiratory system [18].

Table 1. Influence of temperature and culture medium on the biofilm-forming properties of bacteria isolated from sea water[a].

	Optical density (OD)					
	Beef-extract broth			Marine microrganisms medium		
Strains	5° C	22° C	37° C	5° C	22° C	37° C
Pantoea spp.	0.843±0.01	1.064±0.01	0.467±0.01	0.723±0.01	1.162±0.01	0.301±0.01
Escherichia coli	2.012±0.01	2.354±0.01	1.091±0.01	2.001±0.01	2.353±0.01	1.762±0.01
Rhodococcus spp.	0.623±0.01	0.838±0.01	0.701±0.01	0.655±0.01	0.923±0.01	0.776±0.01
Halomonas spp.	0.634±0.01	1.102±0.01	0.823±0.01	0.553±0.01	0.987±0.01	0.745±0.01
Micrococcus spp.	0.560±0.01	0.921±0.01	0.733±0.01	0.736±0.01	1.030±0.01	0.899±0.01
Vibrio spp.	0.455±0.01	0.853±0.01	0.800±0.01	0.544±0.01	0.765±0.01	0.664±0.01
Arthrobacter spp.	0.645±0.01	0.811±0.01	0.766±0.01	0.685±0.01	0.844±0.01	0.738±0.01
Pseudomonas spp.	0.789±0.01	0.923±0.01	0.856±0.01	0.663±0.01	0.976±0.01	0.765±0.01
Acinetobacter spp.	0.564±0.01	0.782±0.01	0.633±0.01	0.544±0.01	0.773±0.01	0.667±0.01
Medium(control)	0.10	0.10	0.10	0.10	0.10	0.10

[a]the indicators M ± m; The differences are statistically significant with respect to control ($p < 0.05$)

The highest values of optical density were obtained in the monoculture for enterobacteria in comparison with saprophytic marine bacteria, regardless of the medium and the culture temperature. By the degree of biofilm formation in monoculture at 22°C, microorganisms can be arranged in the following order: *Escherichia coli> Pantoea spp. > Halomonas spp. ≥ Micrococcus spp. ≥ Pseudomonas spp. ≥ Rhodococcus spp. ≥ Vibrio spp. ≥ Acinetobacter spp.* It is shown that, the factor influencing the biofilm formation in all the strains studied is the temperature, in contrast to the nutrient medium. The highest optical density values in monoculture were observed at 22°C, compared to other temperatures, for all strains under study, which is consistent with literature data [19]. Perhaps this is due to the fact that the strains studied were isolated from sea water, where the average temperature in summer does not exceed 22°C-24°C. Enterobacteria possessing the most pronounced biofilm-forming properties in monoculture have been investigated for the possibility of biofilm formation in associations with bacteria isolated from sea water. It was shown that for bacteria Pantoea spp. The most optimal for biofilm formation in associations with marine microorganisms was the temperature of 37°C, regardless of the culture medium (table 2).

Table 2. Effect of temperature and culture medium on the biofilm-forming properties of *Pantoea spp.* in association with microorganisms isolated from sea water[a]

Pantoea spp and heterotrophs	Optical density (OD)					
	Beef-extract broth			Marine microrganisms medium		
	5° C	22° C	37° C	5° C	22° C	37° C
Rhodococcus spp.	0.622±0.01	0.728±0.01	0.689±0.01	0.876±0.01	1.123±0.01	1.100±0.01
Halomonas spp.	0.734±0.01	1.211±0.01	1.011±0.01	1.008±0.01	1.435±0.01	1.324±0.01
Micrococcus spp.	0.823±0.01	1.157±0.01	1.145±0.01	0.876±0.01	1.165±0.01	1.109±0.01
Vibrio spp.	0.945±0.01	1.157±0.01	1.099±0.01	0.768±0.01	1.020±0.01	0.896±0.01
Arthrobacter spp.	0.554±0.01	0.789±0.01	0.625±0.01	0.622±0.01	0.823±0.01	0.738±0.01
Pseudomonas spp.	0.574±0.01	1.183±0.01	0.442±0.01	1.461±0.01	0.763±0.01	0.564±0.01
Acinetobacter spp.	0.813±0.01	1.276±0.01	0.901±0.01	0.943±0.01	1.192±0.01	0.781±0.01
Pantoea spp. в (monoculture)	0.843±0.01	1.064±0.01	0.467±0.01	0.721±0.01	1.164±0.01	0.301±0.01

[a]the indicators M ± m; The differences are statistically significant with respect to control ($p < 0.05$)

Optical density in associations *Pantoea spp.* and marine bacteria is higher than in monoculture (except *Pseudomonas spp* at 37°C. There is no significant differences in the results on the studied media, both in monoculture and in associations at 22°C,. The same data were obtained at a temperature of 5°C (except for *Halomonas spp., Pseudomonas spp,* where optical density is higher in associations than in monoculture). In the case of the association of *Escherichia coli* with marine heterotrophs, it was shown that they were more active in the formation of biofilms in monoculture than in association with marine bacteria (table 3).

Table 3. Influence of temperature and culture medium on the biofilm-forming properties of *Escherichia coli* in association with microorganisms isolated from sea water.[a]

Escherichia coli and heterotrophs	Optical density (OD)					
	Beef-extract broth			Marine microrganisms medium		
	5° C	22° C	37° C	5° C	22° C	37° C
Rhodococcus spp.	0.328±0.01	1.016±0.01	1.154±0.01	1.103±0.01	1.431±0.01	1.231±0.01
Halomonas spp.	0.227±0.01	1.140±0.01	1.701±0.01	1.364±0.01	0.773±0.01	1.003±0.01
Micrococcus spp.	0.404±0.01	1.506±0.01	0.463±0.01	0.472±0.01	0.884±0.01	0.552±0.01
Vibrio spp.	0.276±0.01	0.678±0.01	0.601±0.01	0.652±0.01	0.491±0.01	0.333±0.01
Arthrobacter spp.	0.154±0.01	1.585±0.01	1.553±0.01	2.193±0.01	1.693±0.01	1.552±0.01
Pseudomonas spp.	1.008±0.01	1.096±0.01	1.090±0.01	0.435±0.01	0.626±0.01	0.634±0.01
Acinetobacter spp.	0.564±0.01	0.787±0.01	0.657±0.01	0.539±0.01	0.692±0.01	0.593±0.01
Pantoea spp.	0.878±0.01	1.185±0.01	1.034±0.01	1.077±0.01	1.218±0.01	1.118±0.01
Escherichia coli (monoculture)	2.012±0.01	2.354±0.01	1.094±0.01	2.001±0.01	2.351±0.01	1.763±0.01

[a]the indicators M ± m; The differences are statistically significant with respect to control ($p < 0.05$)

Thus, pathogenic and conditionally pathogenic bacteria are able to form biofilms both in monoculture and in association with other bacteria that are present in sea water, which allows them to remain permanently in an unfavorable environmental conditions and is one of the mechanisms of adaptation.

4. Conclusion

The main factor that influences biofilm formation of microorganisms isolated from the marine environment is temperature, but not the nutrient medium. The highest biofilm formation in monoculture shown at 22°C for all studied strains. It is shown that the biofilm forming properties of the microorganisms studied, isolated from sea water, depend on the species. Thus, the highest values were obtained in the monoculture for *Escherichia coli and Pantoea* spp in comparison with other marine bacteria irrespective of the medium and the culture temperature. By the degree of decrease in the capacity for biofilm formation, the species can be arranged in the following order: *Escherichia coli> Pantoea spp. > Halomonas spp ≥Micrococcus spp ≥Pseudomonas spp. ≥ Rhodococcus spp. ≥Vibrio spp. ≥ Acinetobacter spp.* In association with saprophytes, *Pantoea* spp. possess more pronounced biofilm-forming properties at 37°C compared to the control than at 22°C and 5°C irrespective of the culture medium. *Escherichia coli,* in association with saprophytes, have less pronounced biofilm formation properties than monoculture, regardless of the temperature and culture medium.

Acknowledgments
The research was fulfilled with financial support of the Russian Scientific Fund (Agreement No. 14-50-00034)

References

[1] Degtyareva V A and Buzoleva L S 2010 Assessment of the sanitary-microbiological state of coastal waters in Vladivostok *Izvestia Tinro* **163** 349 – 54
[2] Karpova V A, Kovalchuka Y L, Kharchenkov U V and Beleneva I A 2012 The effect of microfouling on marine corrosion of metals and destruction of protective coatings *Chemistry of Surfaces* 7 803 – 09
[3] Buzoleva L S, Kalitina E G, Bezverbnaya I P and Krivosheeva A M 2008 Microbial communities in the coastal surface waters of Zolotoi Rog bay under the conditions of strong anthropogenic pollution *Oceanologia* **6** 882 – 88
[4] Somov G P and Buzoleva L S 2004 *Adaptation of pathogenic bacteria to abiotic factors of the environment* (Vladivostok: Primpigraph Combine) p167
[5] Dickschat J S 2010 Quorum sensing and bacterial biofilms *Natural Product Reports* **27** 343 – 69
[6] Ushkareva O A, Anganova E V, Dukhanina A V and Verkhozina E V 2014 Ability of water isolates to form biofilm communities under different cultivation conditions *Siberian Medical Journal* 7 102 – 04
[7] Rendueles O, Travier L, Lambert P L, FontaineT, Magnus J, Denamur E and Ghigo J M 2011 Screening *Escherichia coli* species biodiversity reveals new biofilm associated antiadhesion polysaccharides *mBio* **3** e00043-11 10.1128/mBio.00043-11
[8] Cergole-Novella M C, Pignatari A C and Beatriz E C 2015 Guth Adhesion, biofilm and genotypic characteristics of antimicrobial resistant Escherichia coli isolates *Brazilian Journal of Microbiology* **46** 167 – 71
[9] Donné J, Kerckhoven M V, Maes L, Cos P and Dewilde S 2016 The role of the globin-coupled sensor YddV in a mature E. coli biofilm population *Biochimica et Biophysica Acta* 835–9
[10] Garankina A Y, Kapustina O A and Kartashova O L 2011 Regulation of the ability of opportunistic microorganisms to biofilm formation *Veterinary Journal* **59** 61 – 63
[11] Gritsenko A V, Zhurlov O S and Andreichev V V 2012 Analysis of the relationship between seroresistance and physical and chemical properties of Escherichia coli with the ability to biofilm formation *Bulletin of the OSU* **4** 201 – 05
[12] Grudniak A M, Włodkowska J, Wolska K I and Chaperone Dna J 2015 Influences the Formation of Biofilm by *Escherichia coli Polish Journal of Microbiology* **64** 279 – 83
[13] Fasolato L, Vignaroli C and Luna G M 2015 Understanding the association of Escherichia coli

with diverse macroalgae in the lagoon of Venice *Sci Rep* dx.doi.org/10.1038/srep10969

[14] Literak I, Dolejska M, Janoszowska D, Hrusakova J, Meissner W, Rzyska H, Bzoma S and Cizek A 2010 Antibiotic-resistant Escherichia coli bacteria, including strains with genes encoding the extended-spectrum beta-lactamase and QnrS, in waterbirds on the Baltic sea coast of Poland *Applied and environmental microbiology* **24** 8123-34 10.1128/AEM.01446-10

[15] Thakura P, Chawlaa R, Tanwara A, Chakotiyaa A S, Narulab A, Goela R, Arorac R and Sharmaa R K 2016 Attenuation of Adhesion, Quorum Sensing and Biofilm mediated virulence of Carbapenem Resistant Escherichia coli by selected Natural Plant Products *Microbial Pathogenesis* **96** 76 – 85 doi.org/10.1016/j.micpath.2016.01.001

[16] Re D S, Valle J, Charbonnel N, Beloin C, Latour-Lambert P 2013 Identification of commensal *Escherichia coli* genes involved in biofilm resistance to pathogen colonization *PLOS ONE* 10.1371/journal.pone.0061628

[17] Christensen G D, Simpson W A and Younger J J 1985 Adherence of coagulase-negative staphylococci to plastic tissue culture plates: a quantitative model for the adherence of staphylococci to medical devices *Journal of Clinical Microbiology* **22** 996 – 1006

[18] Pozdeev O K and Fedorov P V 2007 *Enterobacteria: a guide for doctors* (GEOTAR-Media) p 720

[19] Anganova E V, Savilov E D, Ushkareva O A, Ablov A M and Dukhanina A V 2014 Ability of pathogenic and conditionally pathogenic enterobacteria to form biofilms *Bulletin VSNTS SB RAMS* **5** 34 – 7

Antibiotic resistance of microorganisms in agricultural soils in Russia

N V Danilova, P Yu Galitskaya, S Yu Selivanovskaya

Institute of Environmental Sciences, Kazan Federal University, Kremlevskay str.,18, 420008, Russia
e-mail: natasha-danilova91@mail.ru

Abstract. Antibiotics are medicines that are widely used in livestock production not only for the prevention and treatment of infectious diseases, but also for accelerating the growth of animals. The application of manure for fertilizing agricultural soils leads to the entry into the soil ecosystem not only of the antibiotics themselves, but also the resistance genes to them. In this study, 30 samples of arable soils were tested for the presence of the *tet*(X) gene, which encodes bacterial resistance to antibiotics of the tetracycline group. Using real-time PCR, it was found that 27 out of 30 soil samples contained *tet*(X). 52% of these samples were heavily contaminated, 34% had a medium level of contamination and 14% were slightly contaminated by the resistance gene *tet*(X).

1. Introduction

Almost immediately after the active use of antibiotics for medical purposes in the treatment of human diseases they were also used on a large scale in agriculture [1–3]. The use of antibiotics in animal husbandry aims to achieve the following objectives: disease prevention and treatment and in addition, in sub therapeutic doses they are used as growth stimulators for animals [4–6]. From the literature it is known that the use of antibiotics in agriculture is quantitatively superior to their use in medicine [1,7].

Antibiotics are substances that have weak adsorption in the body of the animal, so up to 90% of the used dose of the drug is excreted unchanged or as metabolites in the excrement and urine [8–10]. Antibiotics in large quantities therefore enter into the environment when fertilizing soils with manure from farm animals [4]. Antibiotics and their metabolites are able to remain bioactive in soils for several months [1,11]. The mobility and duration of antibiotic residues in the soil depends not only on the physico–chemical properties of the compounds themselves but also on the type of soil, its pH, climatic conditions [7,11]. Many studies indicate the adverse effects of antibiotics on soil microbial communities: the activity of microorganisms is inhibited, changes occur in the ratio of the bacteria and fungi counts, the natural cycle of the elements is disrupted [8,12,13]. All this can ultimately affect the quality and fertility of agricultural soils and harm agricultural production.

In addition to antibiotic themselves, manure contains intestinal microorganisms that are carriers of antibiotic–resistant genes[14]. Therefore, another danger of using manure from animals treated with antibiotics is the spread of resistance genes from the intestinal microflora of livestock to the microorganisms of soil ecosystems [15,16]. Often antibiotic resistance genes are stored in a bacterial cell in the so–called mobile genetic elements – plasmids, transposons, integrons [17]. In the process of horizontal transfer antibiotic-resistant genes can be rapidly transferred from the donor bacteria to the recipient bacteria, thereby leading to an increase in antibiotic-resistant soil microbial communities [18]. Interestingly, antibiotic resistance genes persist in soils for many years, whereas antibacterial drugs themselves are destroyed within a few months [19].

Recently, the spread of antibiotics and resistance genes in agricultural soils has been of great scientific interest, since it concerns the issue of public health. Genes resistant to antibiotics are able to enter the human body through the food chain (unwashed root crops) and in future they could subsequently hamper the treatment of infectious diseases [20]. Data about agricultural soil pollution by antibiotic resistance genes in Russia is currently not available.

In the present study, 30 samples of agricultural soils selected around the city of Kazan (Republic of Tatarstan) were examined for contamination by the *tet*(X) resistance gene, which is responsible for the destruction of tetracyclines in the bacterial cell. Antibiotics of the tetracycline group are the most commonly used drugs in the animal husbandry of Russia. The Republic of Tatarstan occupies one of the leading places in the agriculture of the Russian Federation, therefore it is important to assess the extent of the spread of antibiotic resistant genes in the arable lands of this region.

2. Material and methods

30 soil samples were taken from cropland around the city of Kazan (N 55°47′, E 49°06′) (Russia). The sampling plots were situated in areas with high agricultural activity (table 1). Soil samples were taken from a depth of 0–20 cm, and cleaned of roots. Immediately after delivery to the laboratory the level of respiratory activity, microbial biomass, soluble organic carbon content and the particle size distribution were estimated as initial parameters of soil samples. To enable further DNA extraction, samples were stored at – 80°C.

The respiratory activity (RA) was assessed according to ISO 14240–1 [21]. The level of microbial biomass (MB) was determined using the method of substrate–induced respiration (ISO 14240–1) [21].The dissolved organic carbon (DOC) content was determined according to ISO 14235:1998 [22] by a spectrometric method. The particle size distribution of the soil samples was measured on a Microtrack Blue Wave Sample Dell Delivery Controller laser particle analyzer using the Microtrack Blue Wave according to ISO 13320:2009 [23]. Prior to measurement, soil samples were dried and ground to a fine particle state. Extraction of DNA from soil samples was carried out using the FastDNA Spin Kit for Soil kit (MP Bio, Germany) according to the manufacturer's instructions. DNA purification was carried out using the QIAquick PCR Purification Kit (Quiagen, Germany) according to the manufacturer's instructions. Detection of the tetracycline *tet*(X) resistance gene was carried out on the real–time cycler using the following primers: *tet*(X)–F GAAAGAGACAACGACCGAGAG and *tet*(X)–R ACACCCATTGGTAAGGCTAAG [24]. The MasterMix reaction mixture contained the following components: DNA template – 1 μl, forward and reverse primers (10 μM) – 0.5 μl each, dNTPs (10 μM) – 2.5 μl, 10x Buffer with SYBR Green – 2.5 μl, $MgCl_2$ (25 μM) – 2.5 μl, Syn Taq polymerase (5 U μl-1) – 0.2 μl and ddH_2O – 15.3 μl. Amplification was performed on a BioRadCFX-96 cycler (BioRad, Munich, Germany) using the following temperature program: primary denaturation at 94°C for 3 minutes, then 39 three–step cycles at 94°C for 45 seconds, at 55°C for 45 seconds, and at 72°C for 45 seconds. The level of contamination of soil samples was determined based on the number of cycles of C_T necessary for fluorescence to reach the threshold level.

Table 1. Soil samples investigated

Sample number	Coordinates of the soil selection site (N, E)	Information on manure treatment of the soil
1	55.880665, 48.630881	cow and chicken manure
2	55.885069, 48.595461	cow and chicken manure
3	55.867224, 48.620585	chicken manure
4	55.857325, 48.625194	chicken manure
5	55.863052, 48.626243	chicken manure
6	55.863786, 48.585559	–
7	55.869621, 48.697672	chicken manure
8	55.867390, 48.670235	chicken manure
9	55.886093, 48.665341	–
10	55.937169, 48.749259	chicken manure
11	55.941670, 48.776962	chicken manure
12	55.865529, 48.817045	–
13	55.900259, 48.932387	chicken and turkey manure
14	55.893049, 48.866680	chicken and turkey manure
15	55.966047, 49.163583	–
16	55.895389, 49.288864	cow manure
17	55.798431, 49.384178	goose manure
18	55.820842, 49.459933	chicken and cow manure
19	55.815556, 49.520443	chicken and cow manure
20	55.802975, 49.580727	chicken manure
21	55.774133, 49.655554	chicken manure
22	55.791594, 49.614571	chicken manure
23	55.792456, 49.586793	chicken manure
24	55.559365, 49.276804	chicken manure
25	55.618271, 49.130810	chicken manure
26	55.553450, 49.203348	chicken manure
27	55.627570, 49.996571	–
28	55.062182, 50.952912	cow manure
29	55.175664, 50.781207	cow manure
30	55.021046, 51.038493	cow manure

3. Results and discussion

The mobility and bioavailability of antibiotics in soil depends on conditions that prevail in the soil (soil type, pH, temperature) [1]. So these factors may influence the development of antibiotic resistance in soil microbial communities under the antibiotic pressure. Furthermore, the fate of antibiotics in the soil may depends on the level of activity of microorganisms, since high activity of microorganisms promotes faster biodegradation, and hence less risk of genes [17]. In addition, the length of time that introduced organisms carrying antibiotic resistance can persist in the soil varies with temperature, moisture, pH, and the indigenous community present [17].

Therefore on the first stage of the study, the agrochemical and microbiological parameters of the investigated soils were analyzed. The measurement results are shown in table 2.

Table 2. Agrochemical and biological parameters of the soils samples

Sample number	DOC, mg g^{-1}	Soil type	RA, CO_2–C mg g^{-1}*h	MB, mg kg^{-1}
1	0.07	Silty loam	1.27	281.52
2	0.09	Silty loam	1.92	264.82
3	0.11	Silty loam	1.73	263.51
4	0.44	Silty loam	1.23	240.4
5	0.05	Silty loam	1.19	382.2
6	0.14	Silty loam	0.80	298.06
7	0.45	Silty loam	0.96	280.97
8	0.43	Sandy loam	2.48	442.16
9	0.30	Sandy loam	2.39	479.68
10	0.11	Silty loam	2.90	578.88
11	0.12	Silty loam	2.09	557.48
12	0.09	Silty loam	1.25	373.33
13	0.04	Sandy loam	2.29	323.67
14	0.06	Sandy loam	1.17	314.52
15	0.24	Silty loam	1.88	392.00
16	0.13	Sandy loam	2.32	334.96
17	0.27	Silty loam	2.51	450.09
18	0.25	Silty loam	2.07	368.70
19	0.02	Silty loam	2.02	466.97
20	0.09	Silty loam	1.70	476.78
21	0.21	Silty loam	1.77	361.86
22	0.04	Silty loam	2.57	367.99
23	0.08	Silty loam	1.96	402.28
24	0.04	Silty loam	2.43	642.09
25	0.11	Sandy loam	2.06	706.22
26	0.09	Silty loam	2.66	635.13
27	0.15	Silty loam	3.13	608.00
28	0.06	Silty loam	4.61	706.84
29	0.11	Silty loam	3.49	732.85
30	0.11	Silty loam	3.54	935.77

Estimation of dissolved organic carbon (DOC) content in soils showed that 13 of the 30 samples were of low DOC level (samples 1, 2, 5, 12–14, 19, 20, 22–24, 26, 28), 9 samples were of medium DOC level (samples 3, 6, 10, 11, 16, 25, 27, 29, 30) and 8 – of high DOC level (4, 7–9, 15, 17, 18, 21).

The results of the granulometric composition determination of the soils tested showed that the particle size type distribution of 20% of the soil samples was estimated to be sandy loam and 80% of the soil samples – to be silty loam. It is known that in clay soils the antibiotic binds and is inaccessible to microorganisms [3,7]. Studies have shown that under a broad range of environmental conditions tetracyclines (tetracycline, chlortetracycline, and oxytetracycline) can adsorb strongly to clay particles [18]. So among the arable soils investigated soils that were closest to the «sandy loam» type (samples 8, 9, 13, 14, 16, 25) were the most predisposed to dissemination of antibiotic resistance.

As can be seen from the table 2 the arable soils tested characterized by different levels of RA and MB. Thus, these parameters ranged between 0.80 and 5.28 CO_2–C mg g^{-1} h^{-1} and between 263.51 and 935.77 μg kg^{-1}, respectively. Since the high microbial activity affects the rate of antibiotic biodegradation, soils with a low level of microbial activity (samples 1–7, 12, 14, 20, 21) may be susceptible to the spread of resistant genes [18].

Many studies indicate a link between the application of manure contaminated by antibiotics and the presence of antibiotic resistance in treated arable soils [25–27]. So, in the next stage the contamination of the soils by the *tet*(X) gene was analyzed by real–time PCR. The results are shown in table 3. To assign a level of contamination to arable lands by the *tet*(X) resistance gene, the CT values obtained by real–time PCR reaction were ranked in three categories. Score 0 was assigned to the samples without contamination by *tet*(X) gene, scores 1, 2 and 3 were given to samples with low, middle and high contamination, respectively. As can be seen from the table 3, 27 of the 30 agricultural soils tested contained the tetracycline resistant gene *tet*(X). Out of 27 contaminated agricultural soils, 52% were heavily polluted, 34% showed medium level pollution and 14% were slightly polluted by *tet*(X) gene. According to the literature a high abundance of *tet*(X) gene has been found in manure and manure-treated soils by other authors [28–31].

Table 3. Contamination of soil samples by *tet*(X) gene

Sample number	C_T value	Contamination level by *tet*(X) gene	Sample number	C_T value	Contamination level by *tet*(X) gene
1	25.78	3^d	16	25.65	3
2	28.38	0^a	17	25.42	3
3	25.29	3	18	26.45	2
4	25.73	3	19	27.59	1
5	25.51	3	20	26.55	2
6	25.87	3	21	26.61	2
7	26.86	2^c	22	26.08	2
8	28.38	0	23	27.42	1
9	24.61	3	24	26.33	2
10	23.64	3	25	32.11	0
11	25.39	3	26	26.91	2
12	25.61	3	27	26.02	2
13	27.11	1^b	28	26.19	2
14	27.64	1	29	25.45	3
15	25.55	3	30	24.84	3

Contamination of soil samples by *tet*(X) gene:
a – no contamination
b– low contamination
c – middle contamination
d – high contamination

Despite the absence of any kind of manure treatments for soil samples 6, 9, 12, 15, 27 for at least the last 10 years, these soils had middle and high levels of pollution by resistance gene *tet*(X). Probably the presence of antibiotic resistance in these soils can be associated with the previous practice of fertilizing by manure from tetracycline medicated animals. Many studies also point to the fact that antibiotic resistant genes are persist in agricultural soil for a long time even when there is no continued application of manure [32–34].

In terms of the manure type that leads to the greatest antibiotic resistance gene contamination, chicken manure was the most problematic. Previously, we demonstrated that chicken manure is more polluted by the *tet*(X) resistance gene compared with cow, pig, goat and rabbit manures [35].

Conclusion

This study has effectively demonstrated that the agricultural soils sampled around Kazan are at potential risk of antibiotic resistance dissemination because of their agrochemical and biological properties. Among the soils investigated, more than 90% were contaminated by the tetracycline resistant gene *tet*(X). The results obtained confirm the data that antibiotic resistant genes spread in the soil through the introduction of manure and remain in the soil despite the termination of the practice of annual manure spreading.

Acknowledgement

The work is performed according to the Russian Government Program of Competitive Growth of Kazan Federal University. This work was supported by a subsidy allocated to Kazan Federal University from Russian Foundation for Basic Research, project No 16–04–00443.

References

[1] Sarmah A, Meyer M and Boxall A 2006 A global perspective on the use, sales, exposure pathways, occurrence, fate and effects of veterinary antibiotics (VAs) in the environment *Chemosphere* **65** 725–759

[2] Kemper N. 2008 Veterinary antibiotics in the aquatic and terrestrial environment *Ecol. Indic.* **8** 1–13

[3] Thiele-Bruhn S. 2003 Pharmaceutical antibiotic compounds in soils - A review *J. Plant Nutr. Soil Sci.* **166** 145–167

[4] Du L and Liu W. 2012 Occurrence , fate , and ecotoxicity of antibiotics in agro-ecosystems . A review *Agron. Sustain. Dev.* **32** 309–327

[5] Jechalke S, Heuer H, Siemens J, Amelung W and Smalla K 2014 Fate and effects of veterinary antibiotics in soil *Trends Microbiol.* **22** 536–545

[6] Caniça M, Manageiro V, Jones-Dias D, Clemente L, Gomes-Neves E, Poeta P, Dias E and Ferreira E 2015 Current perspectives on the dynamics of antibiotic resistance in different reservoirs *Res. Microbiol.* **166** 594–600

[7] Halling-Sorensen B, Nielsen SN, Lanzky PF, Ingerslev F, Holten Lutzhoft HC and Jorgensen SE 1998 Occurence, fate and effects of pharmaceuticals substance in the environment - A review *Chemosphere* **36** 357–393

[8] Thiele-Bruhn S and Beck IC 2005 Effects of sulfonamide and tetracycline antibiotics on soil microbial activity and microbial biomass *Chemosphere* **59** 457–465

[9] Song L, Li L, Yang S, Lan J, He H, McElmurry SP and Zhao Y 2016 Sulfamethoxazole, tetracycline and oxytetracycline and related antibiotic resistance genes in a large-scale landfill, China *Sci. Total Environ.* **551–552**, 9–15

[10] Gao M, Song W, Zhou Q, Ma X and Chen X 2013 Interactive effect of oxytetracycline and lead on soil enzymatic activity and microbial biomass *Environ. Toxicol. Pharmacol.* **36** 667–674

[11] Selvam A, Xu D, Zhao Z and Wong J 2012 Fate of tetracycline , sulfonamide and fluoroquinolone resistance genes and the changes in bacterial diversity during composting of swine manure *Bioresour. Technol.* **126** 383–390

[12] Kotzerke A, Sharma S, Schauss K, Heuer H, Thiele-Bruhn S, Smalla K, Wilke BM and Schloter M 2008 Alterations in soil microbial activity and N-transformation processes due to sulfadiazine loads in pig-manure *Environ. Pollut.* **153** 315–322

[13] Böhme L, Langer U and Böhme F 2005 Microbial biomass, enzyme activities and microbial community structure in two European long-term field experiments *Agric. Ecosyst. Environ.* **109** 141–152

[14] Aydin S, Ince B and Ince O 2015 Development of antibiotic resistance genes in microbial communities during long-term operation of anaerobic reactors in the treatment of pharmaceutical wastewater *Water Res.* **83** 337–344

[15] Tang X et al 2015 Effects of long-term manure applications on the occurrence of antibiotics

and antibiotic resistance genes (ARGs) in paddy soils : Evidence from four fi eld experiments in south of China *Soil Biol. Biochem.* **90** 179–187
[16] Alonso A, Sánchez P and Martínez JL 2001 Environmental selection of antibiotic resistance genes *Env. Microbiol.* **3** 1–9
[17] Chee-Sanford JC, Mackie RI, Koike S, Krapac IG, Lin Y, Yannarell AC, Maxwell S and Aminov, RI 2009 Fate and Transport of Antibiotic Residues and Antibiotic Resistance Genes *J. Environ. Qual.* **38** 1086–1108
[18] Heuer H, Schmitt H and Smalla K 2011 Antibiotic resistance gene spread due to manure application on agricultural fields *Curr. Opin. Microbiol.* **14** 236–243
[19] Fang H, Han Y, Yuanming Y, Xiong P and Yunlong Y 2014 Variations in dissipation rate, microbial function and antibiotic resistance due to repeated introductions of manure containing sulfadiazine and chlortetracycline to soil *Chemosphere* **96** 51–56
[20] Wang F, Qiao M, Chen Z, Su J and Zhu Y.2015 Antibiotic resistance genes in manure-amended soil and vegetables at harvest *J. Hazard. Mater.* **299** 215–221
[21] ISO 14240-1:1997 1997 Soil quality–Determination of soil microbial biomass – Part 1: Substrate– induced respiration method
[22] ISO 14235:1998 1998 Soil quality–Determination of organic carbon by sulfochromic oxidation
[23] ISO 13320:2009 2009 Particle size analysis – Laser diffraction methods
[24] Zhang W, Huang M, Qi F, Sun P and Van Ginkel SW 2013 Effect of trace tetracycline concentrations on the structure of a microbial community and the development of tetracycline resistance genes in sequencing batch reactors *Bioresour. Technol.***150** 9–14
[25] Sharma VK, Johnson N, Cizmas L, Mcdonald TJ and Ki H 2016 A review of the influence of treatment strategies on antibiotic resistant bacteria and antibiotic resistance genes *Chemosphere* **150** 702–714
[26] Yeom J, Yoon S and Kim C 2017 Quantification of residual antibiotics in cow manure being spread over agricultural land and assessment of their behavioral effects on antibiotic resistant bacteria *Chemosphere* **182** 771–780
[27] Tien Y, Li B, Zhan T, Scott A, Murray R, Sabourin L, Marti R and Topp E 2017 Impact of dairy manure pre-application treatment on manure composition , soil dynamics of antibiotic resistance genes and abundance of antibiotic-resistance genes on vegetables at harvest *Sci. Total Environ.* **581–582** 32–39
[28] Agersø Y and Jensen L 2002 The identification of a tetracycline resistance gene tet (M), on a Tn916-like transposon, in the Bacillus cereus group *FEMS Microbiol. Lett* **214** 251–256
[29] Asai T, Kojima A, Harada K and Ishihara K 2005 Correlation between the Usage Volume of Veterinary Therapeutic Antimicrobials and Resistance in Escherichia coli Isolated from the Feces of Food-Producing Animals in Japan *Jpn. J. Infect. Dis.* **58** 369–372
[30] Srinivasan V, Nam H and Oliver SP 2008 Distribution of Tetracycline and Streptomycin Resistance Genes and Class 1 Integrons in Enterobacteriaceae Isolated from Dairy and Nondairy Farm Soils *Microb Ecol.* **55** 184–193
[31] Wu N, Qiao M, Zhang B, Cheng WD and Zhu YG 2010 Abundance and diversity of tetracycline resistance genes in soils adjacent to representative swine feedlots in China *Environ. Sci.Technol.* **44** 6933–6939
[32] Peng S, Feng Y, Wang Y, Guo X, Chu H and Lin X 2017 Prevalence of antibiotic resistance genes in soils after continually applied with different manure for 30 years *J. Hazard. Mater.* **340** 16–25
[33] Zhao X, Wang J, Zhu L, Ge W and Wang J 2017 Environmental analysis of typical antibiotic-resistant bacteria and ARGs in farmland soil chronically fertilized with chicken manure *Sci. Total Environ.***593–594** 10–17
[34] Cheng W, Li J, Wu Y, Xu L, Su C, Qian Y, Zhu YG and Chen H 2016 Behavior of antibiotics and antibiotic resistance genes in eco-agricultural system: A case study *J. Hazard. Mater.* **304** 18–25

IOP Conf. Series: Earth and Environmental Science **107** (2017) 012054 doi:10.1088/1755-1315/107/1/012054

[35] Danilova N, Galitskaya P and Selivanovskaya S 2016 Multiresistance of bacte-ria to veterinary antibiotics in dung and manure samples of farm animals. *Uchenye Zapiski Kazanskogo Universiteta. Seriya Estestvennye Nauk.* **158** 507–516

IOP Conf. Series: Earth and Environmental Science **107** (2017) 012055 doi:10.1088/1755-1315/107/1/012055

System methodology application to make water resources management plan for unstudied rivers

S A Dvinskikh and O V Larchenko

Perm State University, Perm, Bukireva street, 15, 614990, Russia
e-mail hydrology@psu.ru

Abstract. Current public monitoring network is not able to involve in and to control water chemical composition of a rivers basin, and there is no coasts monitoring of water objects. As a result, the complete comprehension of rivers use and pollution is impossible. Due to this fact, a new conception of water resources management has been worked out. The conception is based on new approaches to define parameters that characterise usage potentialities and range.

1. Introduction

Currently, to keep balance between economic progress and geosystems possibilities is extremely important. This fact has been mentioned in the sustainable development program. However, not all parts of the geosystem play an equal role to keep their sustainability. Water is one of the significant parts [1, 2]. Its sustainable use and protection must be implemented within river basins. It may be explained by the fact that, any activity, even if water is not included directly in industrial processes, influences water resources quantity and quality. The lack or not sufficient river basins management is the reason for that. Due to this, an integrated water resources management (IWRM) has been worked out in the European Union (EU) [3, 4]. In the Russian Federation, the water code determines the management tasks for water objects use, rehabilitation and protection. These tasks are explained in the complex water objects use and protection scheme (CWOUPS) [5, 6]. The purpose of the scheme is to register water consumers within basins of large rivers. Currently complex water objects use and protection schemes have been worked out for main river basins. In addition, the next step is to apply the results for water resources management. However, there are huge challenges there. Also, we have to pay attention to the following facts:

1. Average and small basins as key points of sustainable development are not considered in these schemes.
2. CWOUPS includes water quality analysis based on monitoring data but the monitoring network is slightly developed or is absent. In this case, water quality characteristic (and water forecast in particular) may not be objective, and the management plan for its protection may be incorrect.
3. Co-ordination order and realization of target documents have not been worked out.
4. The scheme does not have "a direct action", it is not a plan but rather a foundation for working out and acceptance of plans at federal and regional levels; target indices are not standards but rather are aims to improve river basin condition. The scheme does not determine the permissible load but rather is the fundamental for its determination. Water resources management is conducted from local authorities to water users, so called from "the top to the bottom" authority level.

2. Material and methods

Efficient water consumption suggests such impoundments use when negative consequences of anthropogenic impact do not have a place. In this case, management issues become significantly important. These issues are connected to the regulation of an object (environment) and a subject (a human) interactions and subject-object relations that occur as the result of these interactions [7]. They are presented in a form of blocks in figure 1. The first block includes objects that are key environment components: leading (geological composition and relief; climate, surface and underground waters) and supporting (soils, fauna and flora). The second block includes subjects and their possible effects on objects included in the first block. The third block includes a geosystem that has changed as the result of its functioning (the change of meteorological indices, hydrodynamic regime features, etc.). This block, being the result of subject-object relations, is not passive: its components influence the object (feedback) and make it function within the conditions that have already changed (figure 1).

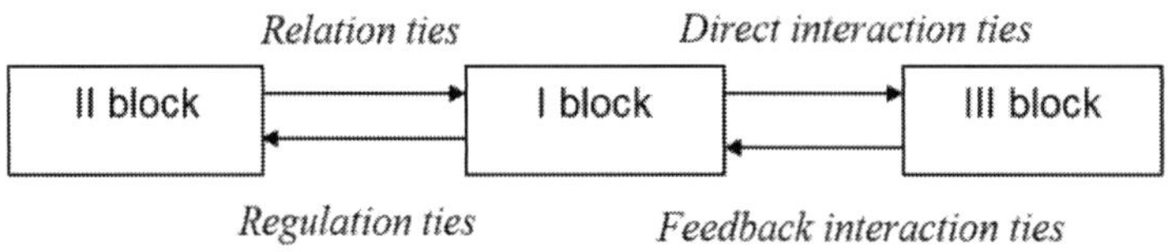

Figure 1. Geosystem functioning scheme [1]

These blocks may be considered as an aggregate of basis (I block is an object) and factors (II and III blocks are a subject and subdominant components correspondingly. These components are connected with the subject by relations, interactions and regulation. The role of the second and third blocks will depend on the effect type on the subject. If a subject includes both natural and anthropogenic components and the influence of the latter increases the impact of the natural one, then the geosystem becomes opened and slackens subject-object ties (figure 2). An object changes in the result of transformation quantitative changes into qualitative ones.

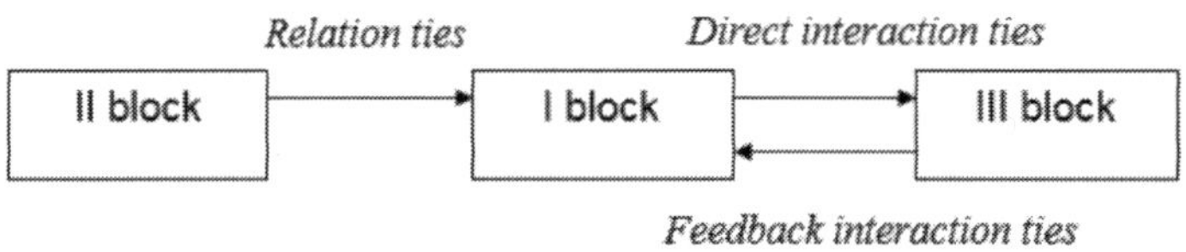

Figure 2. Geosystem functioning scheme with strong anthropogenic influence [1]

As the result of these changes, the geosystem changes the initial development trend and looses ecological balance. The more an object changes, the closer the geosystem to an ecological crisis is (challenges of the Aral and the Sevan lakes, etc). To liquidate crisis another type of ties- between I and II blocks- is created. This type of ties is called adaptation ties (figure 3).

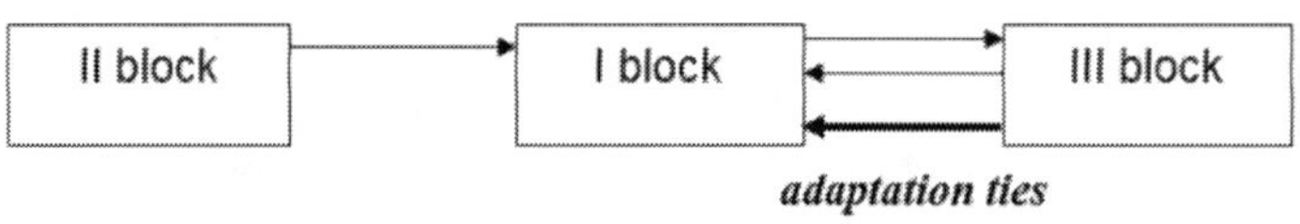

Figure 3. Geosystem functioning scheme with adaptation ties [1]

Most current activities (diversion of runoff, building of waste disposal plans, etc) allow only to adapt the changed conditions (III block) to the geosystem (I block). However, adaption ties do not always guarantee the liquidation of ecological hazards. That is why it is necessary to create management ties that close the geosystem. These ties should regulate the relations between a human (II block) and environment (I block) to prevent ecological balance destruction caused by the functioning of the created system (figure 4).

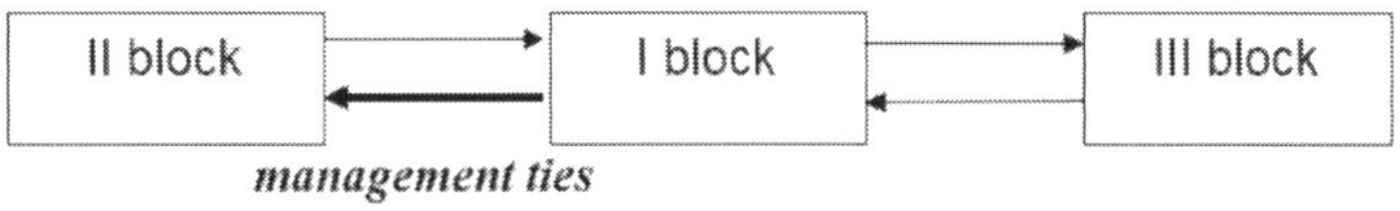

Figure 4. Geosystem functioning scheme with management ties [1].

3. Results and discussion

Are management ties mentioned in CWOUPS? To reveal this, two projects statements (table 1) were compared.

Table 1. The comparison of the CWOUPS and IWRM statements.

	CWOUPS [10,11]	IWRM [8,9]
Purpose	To determine the permissible anthropogenic water objects load; to determine water resources demand in future; to provide water objects protection; to determine main activity trends to prevent negative impact of waters	To provide coordinated development of water, land and resources associated with them to optimize ecological and social well-being without ecosystem sustainability risk
Main task	To make management-decision tools to achieve target indices of water quality of water objects of the case study river basin; the target indices are determined by the CWOUPS; to reduce negative consequences caused by flooding or by other types of water negative impact.	To preserve water for people and food production; to create working places; to protect vitally important ecosystems; risks management; to make public awareness; to strengthen political will to act; to provide international and inter-industry cooperation.
The data are provided	For planning and implementation of water protection and water-management activities in the frameworks of the federal investment program and other target programs; for the development of water protection and water-management activities; for making offers concerning payment for water objects usage; water use management.	For national organizations; to develop management strategies from local authorities to water users and vice versa from water users to local public authorities; for management at a state/province/ region levels; for river basin management structure.
When being developed the following aspects should be considered	The social-economic forecasts, based on industry branches and on regions; territorial planning schemes; current regional and local documents concerning the territory development planning, protected natural territories; plans and programs concerning water-management and water protection activities; international agreement provisions in the field of the joint use and protection of transboundary water objects (for transboundary water objects mentioned in the Scheme).	Consciousness of workers involved in payment collection and data assessment about basin water resources, involved in making plans and offers about investment and payment for water use; budget for organization maintenance and investment in water infrastructure; international role of basin organizations; local authorities importance; participation of public authorities and communities.

Thus, from table 1 despite the differences among these programs, still there are some similarities; they are both devoted to multi-purpose decrease of negative influence on water resources with the help of object-oriented programs. The main approach is basin, but IWRM is a philosophy, policy and guide in the field of water resources management, whereas CWOUPS is only a part of IWRM connected to complex water resources use and gives recommendations only.

Nowadays, a new system to control water-management activity has been created in the multipurpose water-resources scheme of Russia [10, 11]. However, current legislation system and legal acts do not always correspond to ecological, social and political situation. The key challenge is the lack of a scientifically valid methodology. Moreover, a system-methodological approach (SMA) may become such a methodology [12]. The application of this approach needs a clearly defined social-eco-political development strategy of a region. The SMA has been used to study ecological situation in large towns and cities of Perm Krai (Berizniki, Solikamsk, Perm), as well as to study the role of water objects in the creation of ecological surroundings in Perm Krai [13].

The present research presents a water resources management conception that suggests the selection of fundamentally new approaches to determine indices that characterise possibilities and ranges of usage. To obtain impartial data about a main river, solving management tasks from the lower (local) levels of authorities to upper (regional) ones, i.e. from particular objects and issues to larger ones is strongly suggested. The system methodology that is being used suggests studying "reason-result" relations between different components of the geosystem, such as *driving force* (ecological changes), *loads* (on the environment), *condition* (of the environment), *impact* (on the society, economy, ecosystem) and *response* (of the society). The initial point is the pilot analysis of the geosystem. This analysis determines the geosystem borders, demonstrates obvious disadvantages of the obtained data (e.g. not enough data about water chemical composition), identifies main loads (settlements, intensive agricultural activity, industrial agglomerations, etc.) and specifies subsystems that have been potentially affected (ploughed up land, lawn-and-gardens, recreational zones, etc.). The geosystem is spatially divided into river systems of different range (subsystems). Their size is determined by an actual river basin length. The minimum basin range that is used a river study unit is determined by a hydrological regime, potential impacts distribution, their "impact range" and available information (i.e. initial data volume and quantity). The key purpose of researches is determined during the discussion with privies. The privies are water users who are directly or indirectly involved in river system usage. They are invited, for example, to participate in a consulting council.

The data that have been obtained are used to make management decisions. The Nytva river basin has been chosen as a pattern as it is a typical case study basin of medium and small rivers. There are some average industrial and agricultural enterprises and settlements on the basin territory. Nowadays, drinking water shortage is an acute problem for the Nytva population. This challenge is closely connected to the basin siltation and eutrophication.

The procedure to work out management decisions (in the case study of the Nytva river basin) is divided into three parts: project domains, schedule of works and the expected results [14]. In this article, only the main project domains and expected results (tables 2-3) are considered.

Table 2. Procedure for management decision making in the field of river basins water resources (in the case study of the Nytva river basin): project domains.

Challenges	Project domains
Practical:	
Ecological situation in the river basin and the reservoir eutrophication	1. To characterize water objects pollution 2. To study technogeneous loads 3. To distinguish reasons of eutrophication
Scientific:	
To work out schemes for water resources management of the basins of unstudied rivers that suffer from anthropogenic impact	1. To develop methodology for "the Nytva river basin" geosystem study 2. To study the current management system 3. To access the main factors that influence the runoff, chemical composition, technogeneous loads and their roles in the basin ecological situation 4. To make a simulation model of chemical composition and eutrophication formation 5. To make schemes for water resources management of the basins of unstudied rivers that suffer from anthropogenic impact
Methodical:	
The selection and development of calculation and research methods	1. To develop a calculation method for runoff and hydrological regime features when there are no observational data 2. To use a balance method to access chemical composition and hydrodynamic features of the reservoir 3. To prove the array pitch of technogeneous load intensity distribution in the basin
Field and laboratory researches	1. Reconnaissance method (selection of sites and sample sites, water, soil, sediments and snow samples for chemical analysis) 2. To measure flow speed with a subsequent water flow determination 3. Morphological description of soils 4. Laboratory tests of chemical and physical-chemical indices in the selected samples 5. To access the amount of phosphates and nitrates in soils, their mobility, water migration and amount in the basin water objects
Statistical treatment of data and results mapping	1. Data collection, analysis and statistical treatment 2. Mapping of technogeneous loads sources and the basin zoning (subsystems separation) according to their type and impact intensity 3. The basin territory zoning according to phosphates and nitrates amount (for 95% and 5% flow probabilities and lag time)
Modern computer technology use	Development of the Nytva river basin GIS that contains different layers

Table 3. Procedure for decision making in the field of river basin water resources management (in the case study of the Nytva river): the expected results.

Research theme	Expected results
Development of the research methodology	Methodology of the integrated assessment, management and forecast of the basin water resources condition
Study of the current management system of the basin water resources	Development of the current management system: its advantages and disadvantages
Assessment of the main factors that determine the runoff, chemical composition and their impact on the basin ecological situation	Formation of the surface runoff, chemical composition and suspended matters flows. Equations of water and chemical balance
Technogeneous loads and their influence on the basin ecological situation	Types and intensity of technogeneous loads on the basin
Study the reasons of the Nytva reservoir eutrophication	To characterise watershed factors that contribute to eutrophication, their spatial distribution, nitrogen and phosphorus content
Assessment of the basin ecological situation	The basin zoning according to anthropogenic loads type and amount and according to ecological situation destruction
Field and laboratory researches	Data about runoff, chemical composition, hydro-biological indices
Development of the Nytva river basin GIS that contains different thematic layers	Databases and digital maps collections
Development of the simulation model of chemical composition and eutrophication formation	A simulation model
Development of a management scheme of water resources of the basins of previously unstudied rivers that suffer from anthropogenic impact	A management scheme of water resources of the basins of previously unstudied rivers that suffer from anthropogenic impact

The result of the present research is the development of a multi-functional procedure to create a plan for water resources management in previously unstudied average basins for the administrative support of the decision making of their sustainable development. The study believes that, this plan implementation to manage a basin together with CWOUPS will reduce efforts to achieve the specified water quality characteristics and will minimize ecological risks.

References

[1] Newson M 1998 *Land, Water and Development: Sustainable Management of River Basin Systems* (London, New York: Routledge) p 423

[2] Danilov-Danilyan V I 2007 Voda – strategicheskiy faktor razvitiya ekonomiki Rossii [Water is a strategic factor in the development of the Russian economy] *J. Bulletin of the RAS* **77** 108 – 44

[3] 2000 Directive 2000/60/EC of The European Parliament and of The Council *J. Of The European Communities* (Luxembourg) p 72
[4] 2017 *Competition for Water Resources: Experiences and Management Approaches in the US and Europe* ed Jadwiga Ziolkowska and Jeffrey Peterson (Amsterdam: Elsevier) p 478
[5] Bogacheva N Yu, Khranovich I L and Chen Zi 2000 Obosnovanie strategiy ratsionalnogo ispolzovaniya vodnyih resursov v usloviyah riska [Rationale for strategies for the rational use of water resources under risk] *J. Engineering ecology* **6** 2 – 21
[6] Danilov-Danilyan V I and Bolgov M V 2009 O vodnoy strategii Rossiyskoy Federatsii na period do 2020 goda [On the water strategy of the Russian Federation for the period until 2020] *Proc. Int. Conf. on Water problems of large river basins and ways to solve them* (Barnaul: Russia) 59 – 81
[7] Dvinskikh S A 2007 Metodologicheskie aspekty formirovaniya riska na vodnykh ob"ektakh [Methodological aspects of risk formation at water objects] *Proc. Int. Conf. on Current problems of water resources and their catchment areas* **1** (Perm: Russia/Perm State University) 8 – 17
[8] 2012 *Handbook for integrated water resources management in transboundary basins of rivers, lakes and aquifers* (Stockholm: Sweden /International Network of Basin Organizations) p 124
[9] 2009 *Handbook for integrated water resources management in basins* 2009 (Stockholm: Sweden / International Network of Basin Organizations) p 104
[10] Zajceva I S 2003 Nekotoryie regionalnyie osobennosti ispolzovaniya vodnyih resursov v sovremennoy Rossii [Some regional features of water resources use in modern Russia] *J. Regional Research of Russia* **4** 77 – 85
[11] Agasandyan G A, Gasanov I I and Ereshko F I 2003 Novyie podhodyi v probleme kompleksnogo upravleniya vodnyimi resursami [New approaches to the problem of integrated water resources management] *J. Computing Center of the RAS* **54**
[12] Devyatkova T P 2009 Sistemno-metodologicheskii podkhod k planirovaniyu prirodookhrannoi deyatel'nosti [System methodological approach to plan nature protection activity] *Proc. Conf. on hydrology and hydroecology of the Ural* (Perm: Russia/Perm State University) 101–121
[13] Dvinskikh S A, Devyatkova T P and Larchenko O V 2015 Opyt ispol'zovaniya sistemnogo podkhoda v gidrologicheskikh issledovaniyakh [Experience of using a system approach in hydrological research] *J. Geographical Bulletin* **1** 44 – 51
[14] Dvinskikh S A and Larchenko O V 2015 Novaya osnova dlya upravlencheskikh reshenii. Primenenie mnogofunktsional'nogo metoda pri sozdanii plana upravleniya vodnymi resursami neizuchennykh rek [A new basis for management decisions. Multi-functional method application to make a management plan for water resources of unstudied rivers] *J. Water MAGAZIN* **2** 46 – 50

Monitoring of Soil-Borne Pathogens in the Agricultural Soils of the Pestrechinsky District (Tatarstan, Russia)

K.O. Dzhabarova, P.A. Kuryntseva, P.Y. Galitskaya, S.Y. Selivanovskaya

Institute of Environmental Sciences, Kazan Federal University, Kremlevskay str.,18, 420008, Russia
E-mail: polinazwerewa@yandex.ru

Abstract. A recent agricultural trend is aimed to develop organic farming technologies. Organic farming means no mineral fertilizers, pesticides, antibiotics and other chemical substances not characteristic of natural conditions should be used in farm production. When choosing the regions, where this technology can be successfully realized, it is important to evaluate not only the physical and chemical qualities of soils, but also the degree of their infestation with phytopathogens. The Pestrechinsky District of the Republic of Tatarstan, where transfer to organic farming is being planned, was chosen as such a region. Agricultural lands were marked at the map of the administrative region, 100 sampling site were generated using GIS Technologies. It was found out that soil microbial community was characterized by a typical ratio and count of yeast fungi ($3.4 \cdot 10^5$ – $1.6 \cdot 10^6$ CFU·g^{-1}), mold fungi ($1.0 \cdot 10^4$ – $1.7 \cdot 10^5$ CFU•g-1) and bacteria ($1.6 \cdot 10^6$ – $3.1 \cdot 10^7$ CFU·g^{-1}). In all the selected soil samples plant pathogenic fungi of the *Fusarium* genus were found (26 to 250 CFU·g^{-1}), and as for another genus of plant pathogenic fungi, *Alternaria,* their count was rather low (0 to 9 CFU·g^{-1}), herewith in 46 samples out of 100 they were absent.

1. Introduction

A recent agricultural trend is aimed to develop organic farming technologies. Organic farming means no pesticides, toxic chemicals and mineral fertilizers having negative effect both on the soil covering and the soil microbial community should be used in farm production. All the measures listed above are used in so called conventional farming that, on the one hand, increases crop yields, but on the other hand caused soil degradation, soil and water pollution [1, 2, 3]. However, before organic farming can be implemented into practice, it is necessary to estimate the level of soil infestation. The Republic of Tatarstan is one of the leading subjects of the Russian Federation in the area of agriculture. Total cultivated area of the Republic is 3.7 mln ha, 52% out of which are under cereals (summer and winter wheat, rye, barley, oats, and pea), 37 % – feed crop (maize, roots, and grass), 7 % – industrial crops (sugar beet), 3 % – vegetables and potatoes. But regions with the high level of industrial production are not suitable for the introduction of organic farming. Out of 43 municipal areas of the Republic about 15 municipal areas correspond to the necessary requirements. As the Pestrechinsky District is located close to the million city of Kazan and still have relatively favorable ecological situation, the government of the District decided to organize organic farming in it. Before the implementation of this technology into practice it is necessary to estimate the level of agricultural land infestation with phytopathogenic organisms in the region. This study is aimed to monitor the level of soil contamination with *Fusarium* genus and *Alternaria* genus in the Pestrechinsky District of the Republic of Tatarstan. Many *Fusarium* species are pathogenic microorganisms that may cause serious diseases in agricultural crops, garden or forest plants. Thus, some species of *Fusarium oxysporum* are responsible for root rot or wilt in over a hundred species of plants [4, 5]. Some species, such as *F. graminearum, F. culmorum* or *F. avenaceum* lead to the destruction of grain crops, for example they cause *Fusarium* head blight in wheat. Being scientifically and economically important *F. graminearum* and *F. oxysporum* were included into the "Top 10" list of the most dangerous pathogenic fungi [6]. The species of *Fusarium* are also capable of producing mycotoxins, which contaminate agricultural crops, mostly cereals, and the food produced out of them. In wheat *F. graminearum* usually produce such trichothecenes as deoxynivalenol, and in maize *Fusarium verticillioides* produce fumonisin. These substances are classified as mycotoxins and they are harmful for the health of humans and animals [7]. The damage caused by *Alternaria* genus can be preharvest or postharvest. Cereals, vegetables and fruits may suffer [8]. Some *Alternaria* species not only spoil many articles of food directly, but also generate metabolites, which belong to the group of phytotoxins and mycotoxins, thus, *Alternaria* is harmful for plants, for animals and for people.

2. Material and methods

The present study was conducted on soils collected in the Pestrechinsky District located in the Republic of Tatarstan, Russia (N 55°45′, E 49°39′). A total number of 100 soil samples was selected in the agricultural lands to evaluate soil phytopathogenic properties. Sampling was carried out according to GOST 17.4.4.02-84 [9]. Sampling sites were chosen with the help of GIS Technologies. Landsat-8 satellite images dated 07/15/2013 were taken. The images were synthesized in pseudocolors (synthesis of 6, 5, and 4 channels); the procedure of synergism of the colored synthesized image with the resolution of 30 m and panchromatic channel with the resolution of 15 m was carried out. Thus, colored images were received with the spatial resolution of 15 m, which corresponds to the scale 1:50 000. The borders of agricultural lands of the Pestrechinsky Municipal District were digitized in EasyTrace software according to certain deciphering signs, such as the angularity of the borders of fields, the diversity of colors in the image, and distinctive colors. Using the vector layer received QGIS software generated 100 sampling sites in a random, and a rectangular buffer 100 m on a side was built for each sample (Figure 1).

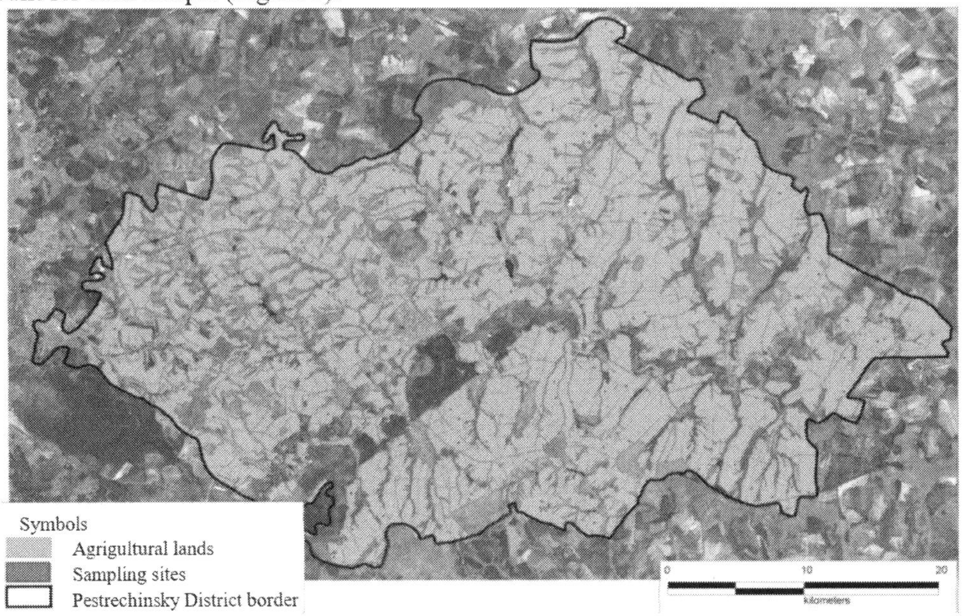

Figure 1. Map of the Pestrechinsky District, the Republic of Tatarstan, Showing Sampling Sites to Estimate Phytopathogenic Soil Characteristics

Total microbial count was carried out using the method of serial dilutions in semisolid agar with further inoculation of a sample drop on the surface of a membrane filter. Determination of the count of yeast fungi and mold fungi was carried out using the method of inoculation of serial dilutions into Sabouraud's medium, *Fusarium spp.* and *Alternaria spp.* plant pathogenic fungi count was determined by means of inoculation into agarized Czapek's medium and further microscopic examination [10].

3. Results and Discussion

Microfungi and bacteria form the main component of soil microbial biomass, while as oozoa and archean comprise no more than 3% of it [11]. According to the ratio of fungi and bacteria in soil its condition can be determined [12]. The selected soil samples were characterized by typical cultivated bacteria content [10, 13, 14]; the data are presented in Figure 2. The lowest total microbial count was found in sample No 47 and was $1.6 \cdot 10^6$ CFU$\cdot$g^{-1}, and the highest – in sample No 13 and was $3.1 \cdot 10^7$ CFU$\cdot$g^{-1}. On average, $5 \cdot 10^6$ to $1 \cdot 10^7$ CFU$\cdot$g^{-1}bacteria were present in the examined agricultural soils of the Pestrechinsky District, 68 out of the 100 soil samples were characterized with such content of cultured bacteria.

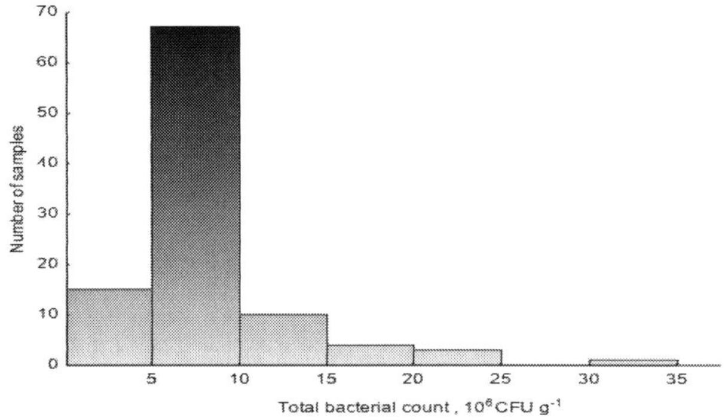

Figure 2. Distribution of Samples According to the Total Microbial Count

Then the content of yeast fungi in soil samples was analyzed (Figure 3). The count of fungi in the samples varied from $3.4 \cdot 10^5$ CFU$\cdot$g^{-1} to $1.6 \cdot 10^6$ CFU$\cdot$g^{-1}, herewith, in a big number of samples the count of yeast fungi was $6 \cdot 10^5$ CFU$\cdot$g^{-1} to $1 \cdot 10^6$ CFU$\cdot$g^{-1}. Evaluating soils, Semenov et al. [15] found out that the typical ratio of fungi and bacteria in soil is 4-12, which corresponds to our results.

The count of mold fungi varied from $1.0 \cdot 10^4$ to $1.7 \cdot 10^5$ CFU$\cdot$g^{-1} (Figure 4). The most typical concentration of mold fungi in agricultural soils of the Pestrechinsky District was $0.6 \cdot 10^5$ CFU$\cdot$g^{-1} to $1 \cdot 10^5$ CFU$\cdot$g^{-1} (57 samples out of 100).

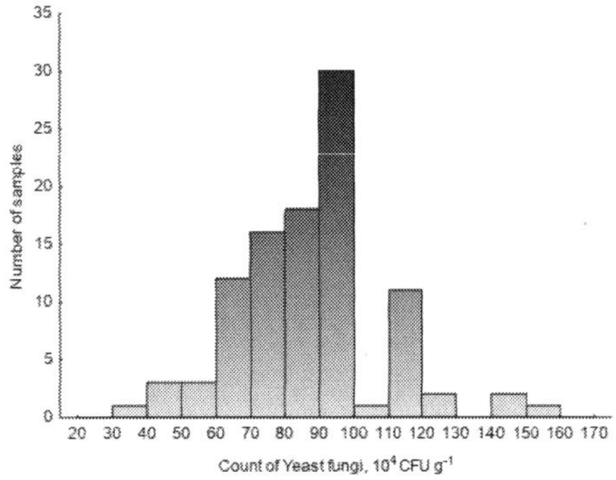

Figure 3. Distribution of Samples According to the Count of Yeast Fungi

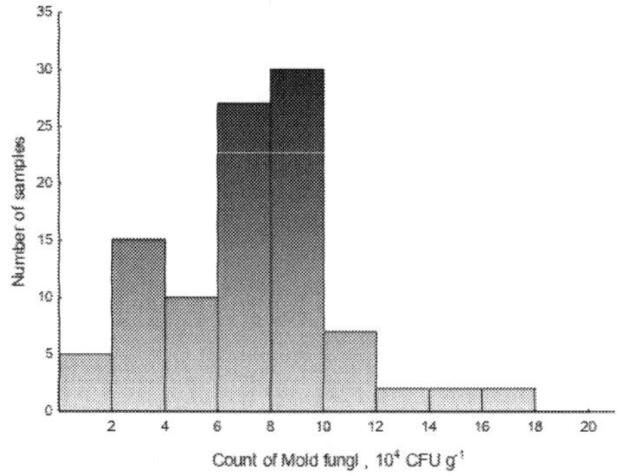

Figure 4. Distribution of Samples According to the Count of Mold Fungi

Fusarium is considered to be one of the most destructive soil-borne diseases [16]. Wide spread occurrence of Fusarium diseases is connected both with the use of contaminated seed material, and with planting nursery transplants taken from contaminated greenhouse soils. The count of plant pathogenic fungi of the *Fusarium* genus in the samples analyzed varied from 26 to 250 CFU·g^{-1} (Figure 5), whereby for 61 samples out of 100 the count of mold fungi was 60 CFU·g^{-1} to 100 CFUg^{-1}. It is worthwhile mentioning that no soil samples free from contamination with plant pathogenic fungi of the *Fusarium* genus were found. However, contamination of agricultural soils in the Pestrechinsky District cannot be considered high, as in the investigations carried out by Yao et al. [16] the total count of *Fusarium oxysporum* was high in the initial soil, reaching 17336 ± 2461 CFU·g^{-1} dry soil.

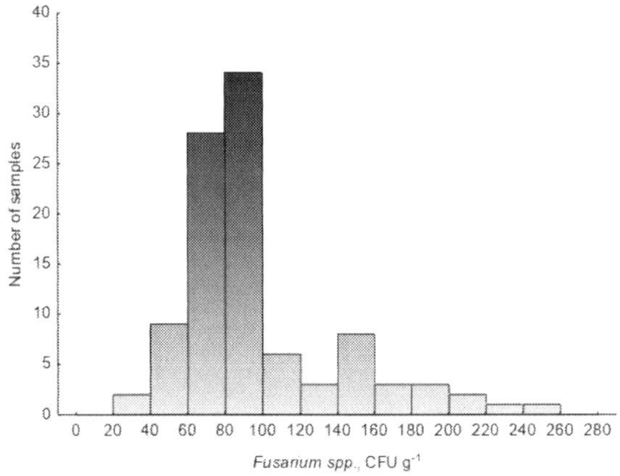

Figure 5. Distribution of Samples According to the Count of Plant Pathogenic Fungi of the *Fusarium* genus

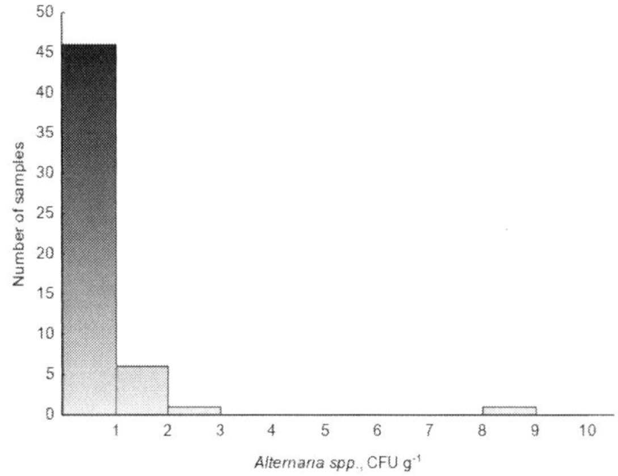

Figure 6. Distribution of Samples According to the Count of Plant Pathogenic Fungi of the *Alternaria* genus

Alternaria is a universal genus of plant pathogenic fungi including endophytic, saprobic and pathogenic species, which are able to survive on a wide range of hosts [17]. Some spaces of the genus are able to cause significant losses in many agricultural crops. Some taxa of *Alternaria* genus are characterized as postharvest pathogens dangerous for immuno-compromised patients, as they may cause phaeohyphomycosis in them. Besides, they produce airborn allergens. As *Alternaria* genus have serious negative effects on human health, it is very important for researches and other specialists to identify *Alternaria* species promptly. According to the results of our investigations, presence of *Alternaria* genus plant pathogenic fungi in the agricultural lands of the Pestrechinsky District is mostly not typical (Figure 6). The count of *Alternaria* genus plant pathogenic fungi was rather low (0 to 9 CFU·g^{-1}), herewith in 46 samples out of 100 they were absent.

Conclusions

In whole, the amount and the ratio of bacteria, yeast fungi and mold fungi in the tillable lands of the Pestrechinsky District are typical for agricultural soils. However, all the samples under analysis are contaminated with fungi of the *Fusarium* species. The count of plant pathogenic fungi of *Alternaria* species is low, or there are no fungi of these species. Thus, it is reasonable to realize organic farming on the tillable lands of the Pestrechinsky District in combination with biological methods of plant protection against *Fusarium* blight.

Acknowledgements

The work is performed according to the Russian Government Program of Competitive Growth of Kazan Federal University

References

[1] Reganold J., Elliott L. and Unger Y. Long-term effects of organic and conventional farming on soil erosion 1987 *Nature* **330** 370-372

[2] Springael D., Top E. Horizontal gene transfer and microbial adaptation to xenobiotics: new types of mobile genetic elements and lessons from ecological studies 2004 *Trends Microbiol.* **12** 53-58

[3] E. Peruzzi, I.H. Franke-Whittle, M. Kelderer, C. Ciavatta and H. Insam Microbial indication of soil health in apple orchards affected by replant disease 2017 *Applied* Soil *Ecology* **119** 115-127

[4] Armstrong G.M., Armstrong J.K. Formae speciales and races of *Fusarium oxysporum* causing wilt diseases P.E. Nelson, T.A. Toussoun, R.J. Cook (Eds.), *Fusarium*: Disease, Biology, and Taxonomy 1981 Pennsylvania State University Press, University Park 391-399

[5] Kistler H.C. et al. Systematic numbering of vegetative compatibility groups in the plant pathogenic fungus *Fusarium oxysporum* 1998 *Phytopathology* **88** 30-32

[6] Dean R. et al. The Top 10 fungal pathogens in molecular plant pathology 2012 *Mol. Plant Pathol.* **13** 414-430

[7] Edel-Hermann V., Gautheron N., Mounier A. and Steinberg C. Fusarium diversity in soil using a specific molecular approach and a cultural approach 2015 *Journal of Microbiological Methods* **111** 64-71

[8] Patriarca A., Vaamonde G. and Pinto V.F. Reference Module in Food 2014 *Science Encyclopedia of Food Microbiology (Second Edition)* 54-60

[9] GOST 17.4.4.02-84 Protection of nature. Soil. Methods of selection and preparation of samples for chemical, bacteriological, helminthological analysis. Moscow: USSR State Committee for Standards 1984

[10] Zenova G.M., Stepanov A.L., Likhacheva A.A. and Manucharova A.A. Workshop on Soil Microbiology. Moscow: Moscow State University 2002 p.120

[11] Gattinger A., Ruser R., Schloter M. and Munch J.C. Microbial community structure varies in different soil zones of a potato field 2002 *Journal of plant nutrition and soil science* **165** 421-428.

[12] Bardgett R.D. The biology of soil: a community and ecosystem approach 2005 Oxford University Press 242

[13] Panov A.V., Sokolov S.L., Kosheleva I.A. and Boronin A.M: Effect of soil contamination on the composition of the microbial community in 2013 *Microbiology* **82** 239-246

[14] Labutova N.M., Shcherba A.V., Galova A.V. and Orlova E.E. Peculiarities of the state of bacterial and fungal microbiocenosis of oil-contaminated sod-podzolic soil in laboratory experiments. "Anthropogenic transformation of the natural environment". 2010 Materials of the International Conference **1** 9-14

[15] Semenov M.V., Stolnikova E.V., Ananieva N.D. and Ivashchenko K.V. Structure of the microbial community of the soil of the catena of the right bank of the river 2013 *OCI 2013: Proceedings of the Russian Academy of Sciences Biological series* **3** 299-308.

[16] Yaoa Y., Xuea Z., Honga C., Xiang Zhua F., Chena X., Wanga W., Caib Z., Huanga N. and Yang X.. Efficiency of different solarization-based ecological soil treatments on the control of Fusarium wilt and their impacts on the soil microbial community 2016 *Applied Soil Ecology* **108** 341-351

[17] Woudenberg J.H.C., Groenewald J.Z., Binder M. and Crous P.W. Alternaria redefined 2013 *Studies in Mycology* **75** 171-212

Monitoring Soil Bacteria with Community-Level Physiological Profiles Using Biolog™ ECO-Plates in the Republic of Tatarstan (Russia)

G Sh Galieva, I M Gilmutdinova, V P Fomin, S Yu Selivanovskaya and P Yu Galitskaya

Institute of Environmental Sciences, Kazan Federal University, Kremlevskaya str., 18, 420008, Russia
e-mail: goolnaz0708@gmail.com

Abstract. Conservation of soil fertility is one of the most important tasks of the present time. As microorganisms are among the key factors in forming soil fertility, monitoring their state in natural and anthropogenically changed soils is an important component of compulsory environmental monitoring. Modern methods make it possible to evaluate the diversity and the functions of soil microorganisms, however, unfortunately, not all the soils are analyzed with their help up to the present moment. The present investigation is aimed to evaluate the functional diversity of five natural soil samples in the Republic of Tatarstan (belonging to sod-podzol, sod-carbonate, alluvial, and gray types) using the method of Biolog EcoPlate according to the index of average well color development, alpha-biodiversiry Shannon index (H), amount of substrates consumed ®, and strategy of consumption of various carbon substrate groups. It was shown that the highest AWCD index was found in sample No 3 – alluvial soil type (3.159±0.460), the lowest one – in sample No 5 – gray soil type (0.572±0.230). Correlation of biological activity of microorganisms with organic matter content in soil was shown.

1. Introduction

One of the main tasks of modern environmental management is the preservation of soil diversity. To make scientific researches in the area of ecologization and anthropogenic landscape recovery effective it is necessary to examine natural soils that have been preserved [1]. Microorganisms play the most important part in soil formation, they are the basic agents of the circulation of elements [2, 3]. The scale of soil microbiological processes influence on the biosphere can be compared only with photosynthesis [4]. In soil science physiological groups of microorganisms being certain indicators of biochemical processes in soil, which are important for its fertility and the productivity of agricultural crops, are of great interest [5]. In 1991 J.L. Garland and A.L. Mills [6] offered to use Biolog system to evaluate functional diversity and developed the main approaches to the interpretation of carbon-bearing substrate consumption ranges received (SCR) using the methods of multivariate statistics. These methods have been widely recognized as an instrument of soil content monitoring [7, 8] to compare the functional diversity of microbial communities both in polluted and unpolluted soils [9, 10]. Thus, the method of Biolog Eco Plate was used in examining soils in the Netherlands, in Europe [8], in Hungary [11], in Poland [12], and in People's Republic of China [13].

In the territory of Russia the method of multisubstrate testing «Ecolog», the composition of which is similar to Biolog Ecoplate, was used to estimate the physiological profiles of bacterial communities in the territories of taiga forests in the Komi Republic, and the Republic of Bashkortostan [14, 15]. However, the information on the SCR for soil microorganisms in the Republic of Tatarstan is very poor.

The objective of the present investigation was to estimate the quality of soils typical for the Republic of Tatarstan (belonging to sod-podzol, sod-carbonate, alluvial, and gray types) with the help of Biolog EcoPlate method.

2. Materials and Methods

Soil samples were selected from the topsoil (0 to 20 cm) in the following sites: No 1 sod-podzol soil (55°42'32.58"N, 49°23'5.28"E), sample No 2 sod-carbonate soil (55°45'15.08"N, 49°18'45.83"E), sample No 3 alluvial soil (55°54'55.8"N, 49°12'49.3"E), sample No 4 alluvial soil (55°91.77'.28.7"N, 49°21'98"E), sample No 5 gray soil (55°75'39.85"N, 49°86.48'61"E).

Initial characteristics of the investigated soils were determined according to standard methods: electrical conductivity and pH as per ISO 26423–85 [16], total nitrogen content (N_{total}) as per ISO 11261:1995 [17], humus content – using the translation method of organic carbon concentration, determined as per ISO 14235:1998 [18]. Grain size measurement was carried out using the method of laser diffraction with the help of Microtrac Bluewave device, in accord with ISO 13320:2009 [19].

To estimate the functional diversity of soil microbial communities Biolog system (Biolog, Int) was used. For this soil extract was previously prepared (1:40) for 30 min. Soil extract was dug into the holes of Biolog Ecoplate plates containing one of low-molecular carbon-bearing substrates (Table 1) in triplicate and tetrazolium salt as an indicator of the substrate consumption.

Table 1. Carbon substrates in Biolog ECO MicroPlates (Biolog, Ing).

Groups of Substrates	Substrates
Polymers	a-cyclodextrin
	glycogen
	Tween 40
	Tween 80
Carbohydrates	D-cellobiose
	i-erythritol
	D-galactonic acid g-lactone
	N-acetyl-D-glucosamine
	glucose-1-phosphate
	b-methyl-D-glucoside
	D,L-a-glycerol phosphate
	a-D-lactose
	D-mannitol
	D-xylose
Carboxylic acids	g-hydroxybutyric acid
	a- ketobutyric acid
	D-galacturonic acid
	D-glucosaminic acid
	itaconic acid
	D-malic acid
	pyruvatic acid methyl ester
Amino acids	L-arginine
	L-asparagine
	glycyl-L-glutamic acid
	L-phenylalanine
	L-serine
	L-threonine
Amines	phenyl ethylamine
	putrescine
Phenolic compounds	2-hydroxybenzoic acid

4-hydroxybenzoic acid

Substrate consumption rate was estimated with the help of a plan-table photometer Thermo Scientific Multiskan FC at 595 nm after the incubation in darkness at 25°C during 24 and 168 hours.

With the help of the results received the following parameters were calculated: ACWD (average well color development), alpha-biodiversiry Shannon index, amount of substrates consumed by microorganisms (R), types of the substrates consumed. The formulae used to calculate the indices are presented in Table 2.

Table 2. Formulae for the calculation of average well color development (AWCD), shannon diversity (H), and index of intensity of utilization of substrates (R).

Index	Formula	Definitions
The average well color development	AWCD $=\sum \frac{Ai-A0}{31}$	Ai – is the optical density within each well; A_0 – is the absorbance value of the control well
Shannon diversity	H=$-\sum pi * ln(pi)$ $pi = \frac{Ai}{\sum A}$	pi - proportional color development of the well over total color development of all wells of a plate
Index of intensity of utilization of substrates	R=1, if is the optical density i wells > 0.25 R=0, if is the optical density i wells < 0.25	R - Index of intensity of utilization of substrates

Statistical processing was carried out using the statistical analysis procedures of Microsoft Excel application program package.

3. Results and discussion

According to Soil Atlas of Tatarstan the following soil types are typical for the Republic: sod-podzol soils (17.0%), gray forest soils (32.4%), sod-carbonate soils (3.1%), black soils (39.7%), black meadow soils (2.5%), brown and gray soils (7.1%), bog soils (1.0%), floodplain soils (4.1%), alkali soils (0.1%), and alkaline soils (0.1%). The first three types, as well as two azonal alluvial soil types were analyzed in this study. To achieve homogeneous data when determining the functional profile of microbial communities sampling was carried out in equal temporary and weather conditions. The characteristics of soil samples selected are presented in Table 3.

Table 3. Characteristics of soil samples used in the study.

Soil Sample	pH	Humus, %	Ntot, mg g^{-1}	Sand (%)	Silt (%)	Clay (%)
1	6.02±0.02	1.1±10.2	0.77±0.02	17.30±0.05	81.39±0.05	1.31±0.05
2	8.33±0.03	2.0±8.9	1.35±0.01	22.50±0.05	73.27±0.05	4.23±0.05
3	6.56±0.01	3.6±11.3	0.91±0.02	25.41±0.05	72.87±0.05	1.72±0.05
4	6.64±0.04	2.4±14.1	0.40±0.01	22.34±0.05	73.61±0.05	4.05±0.05
5	6.51±0.03	1.4±10.1	0.59±0.01	26.64±0.05	70.08±0.05	3.28±0.05

Table 3 shows that pH of water extracts of soil samples No 1, 3, 4, and 5 is neutral, and pH of water extract of soil sample No 2 is weakly alkaline [20]. The analysis revealed differences in humus and total nitrogen content: the highest humus content was found in sample No 3 (3.6±11.3), the lowest – in sample No 1 (1.1±10.2), the highest total nitrogen content was found in sample No 2 (1.35±0.01), the lowest – in sample No 4 (0.40±0.01). Grain-size analysis of the samples selected has shown practical absence of differences in fraction content in the samples. In all the samples selected the fraction of dust prevails varying from (70.08±0.05) for sample No 5 to (81.39±0.05) for sample No 1, which classifies all the soil samples selected as heavy clays according to their grain-size distribution [20].

At the following step soil extracts were prepared and analyzed using Biolog EcoPlate method. It was shown that during the first 7 days of the investigation average well color development (AWCD) has grown in all the samples selected – by 1.4 on average. The growth of average well color development index shows that microbial community metabolic activity in relation to carbon substrates under analysis is high. The highest AWCD index was found in sample No 3 (3.159±0.460), the lowest one – in sample No 5 (0.572±0.230). The values received correspond with the data obtained by other authors [21, 22]. Likewise, the highest R value showing the number of substrates consumed by the community microorganisms was determined in sample No 3 (31), the lowest one – in sample No 5 [20]. The correlation between the indices of ACWD and R was found by other authors as well [21, 22]. The highest values of ACWD and R in sample No 3 are most likely connected with high humus content in it as compared to the other samples. Thus, in sample No 3 humus content was 3.6%, in the other samples it varied from 0.4 % to 2.4 % (Figure 1).

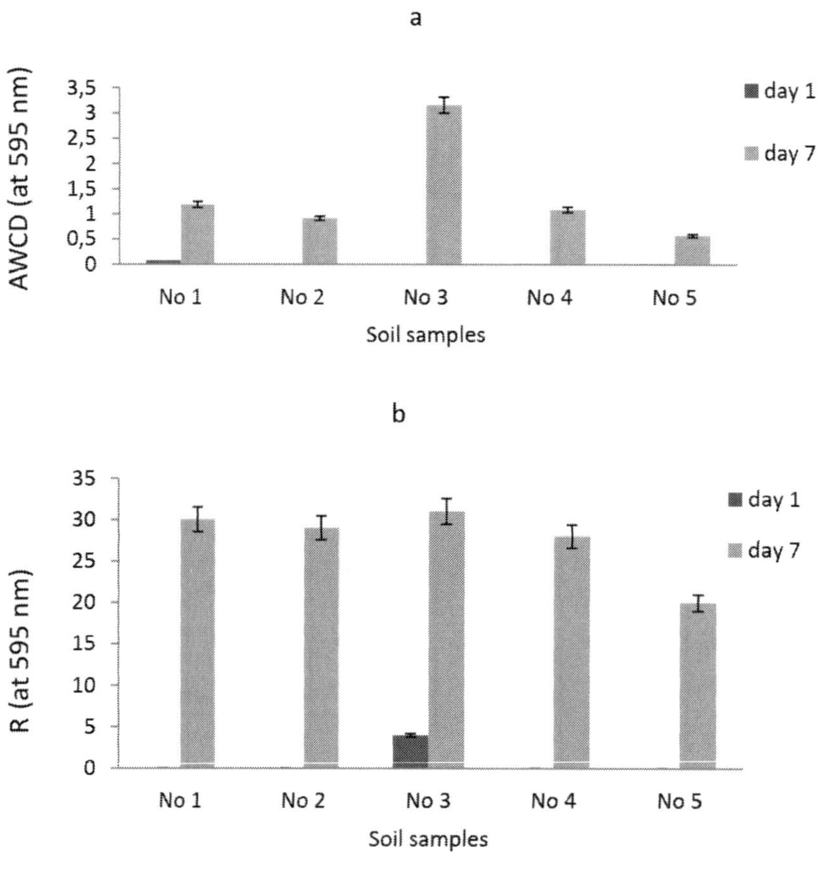

Figure 1. Average well color development AWCD (a) and index of intensity of utilization of substrates R (b) of five soil samples selected in Tatarstan, estimated on day 1 and day 7 of incubation.

High microbial counts and diversity are usually attributed to rich soils. Thus, in the investigations carried out by T.G. Dobrovolskaya et al. [23] and V.A. Kovda et al. [24] tight direct correlation between the biological activity of microorganisms and organic matter content in soil was shown. In the study carried out by high metabolic activity of soil microflora is connected with the influence of the environment, such as the type of litter and the availability of organic matters [13].

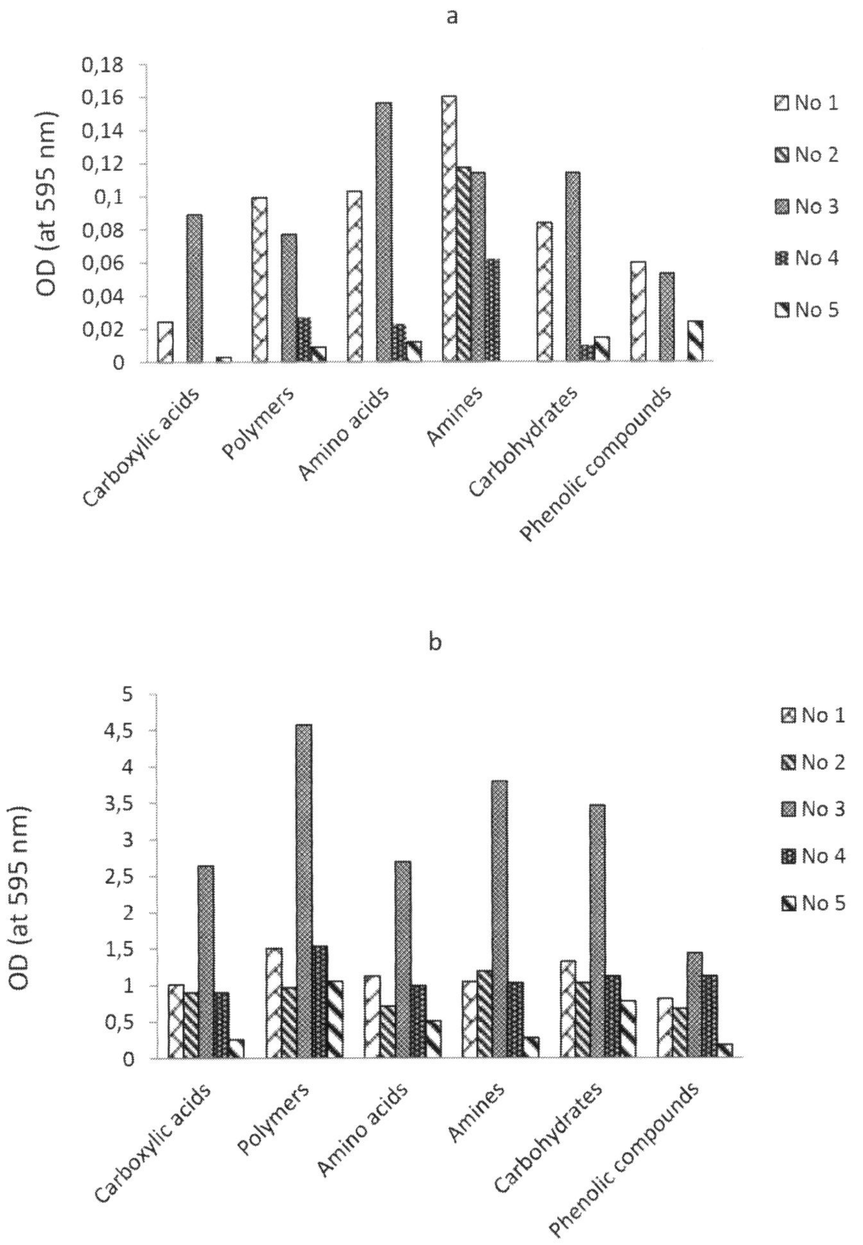

Figure 2. Level of consumption of substrates by microbial communities of soils, sampled in the republic of Tatarstan, estimated on day 1 (a) and day 7 (b) of incubation.

To compare the strategy of substrate consumption by the microbiomes contained in the soils under analysis, the matters contained in a Biolog EcoPlate plate were classified into 6 main groups: amino acids, carbohydrates, carboxylic acids, polymers, amines and amides and phenol compounds. The analysis of intensity of utilization of substrates belonging to the indicated groups helped to find out that the microorganisms in sample No 3 actively utilize all the types of substrates, the microorganisms in samples No 1, 2 and 4 show average intensity of utilization in respect to 5 types of substrates (except for phenol compounds), and microorganisms in sample No 5 show the lowest level of utilization intensity (Figure 2). The low level of phenol substrate consumption is probably connected with the complexity of phenol substrate chemical structure and they require more time for decomposition [11].

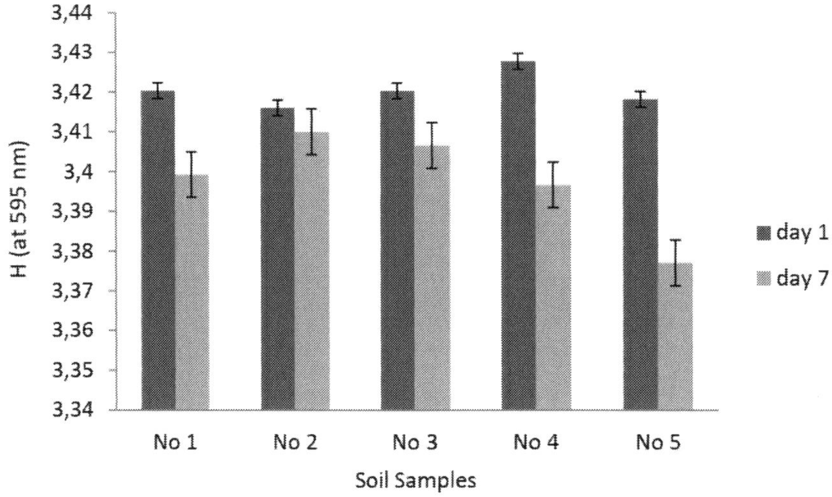

Figure 3. The Shannon index of five soil samples selected in Tatarstan, evaluated on the 1st day and the 7th day of incubation.

Shannon index showing the level of alpha-diversity in the community was analyzed on day 1 and day 7 of the experiment (Figure 3). In all the samples it practically had not changed during the time indicated, which probably shows that all the species in the community were active at the moment of sampling, there were no dormant species [25]. Such state of the community is most likely determined by favorable abiotic factor conditions – sampling was carried out in July 2017 at the average temperatures of 23°C to 25°C and the humidity pf 75% to 80%. It is interesting to note that in samples 1 to 4 Shannon index was practically the same – 3.41 points. In sample No 5 it was slightly lower (3.38 points), which is probably connected with relatively low humus content in the sample. In whole, the data correlate with the data obtained by other authors [11, 26].

Conclusions

In the present investigation physiological profiles of natural soils sampled in the Republic of Tatarstan were determined for the first time. Average well color development index, alpha-biodiversiry Shannon index, substrate consumption index, strategies of consumption for different carbon substrate groups determined using Biolog Ecoplate method, show that the biological activity of microbial communities is higher in soil sample No 3 out of the 5 soil samples selected. The interrelation of biological activity and organic matter content in soil was also revealed – the highest humus content was found in sample No 3.

Acknowledgement

The work is performed according to the Russian Government Program of Competitive Growth of Kazan Federal University.

References

[1] Alexandrova A, Bereghnaya N and Grigorian B 2012 *Red soil book of the Republic of Tatarstan* (Kazan, PH: Foliant) 192 (In Russian)
[2] Galitskaya P, Biktasheva L, Saveliev A, Ratering S, Schnell S and Selivanovskaya S 2015 Response of soil microorganisms to radioactive oil waste: Results from a leaching experiment *Biogeosciences* **12** 3681-3693
[3] Selivanovskaya S and Galitskaya P 2011 Ecotoxicological assessment of soil using the Bacillus pumilus contact test *European Journal of Soil Biology* **47** 165-168
[4] Volkova I 2013 *Soil ecology. Textbook* (Yaroslavl, PH:YaGU) (In Russian)
[5] Klimentev A 2003 *Ecological, scientific, legal aspects of the Red Book of Soils // The role of soil in the formation of landscapes* (Kazan, PH: Fen) 140-143 (In Russian)
[6] Garland J and Mills A 1991Classification and characterization of heterotrophic microbial communities on the basis of patterns of community level sole-carbon-source utilization *Appl. Environ. Microbiol* **57** 2351-2359
[7] Kruglov Y 2016 Microbial community of soil: Physiologycal. Diversity *Agricultural Biology* **51** 46-59 (In Russian)
[8] Rutgers M, Wouterse M Drost *S,* Breurea *A,* Mulder C, Stone D, Creamer R, Windinge A and Bloem J 2016 Monitoring soil bacteria with community-level physiological profiles using Biolog ECO-plates in the Netherlands and Europe *Applied Soil Ecology* **97** 23-35
[9] Gryta A, Frac M and Oszust K 2014 The Application of the Biolog EcoPlate Approach in Ecotoxicological Evaluation of Dairy Sewage Sludge *Applied Biochemistry and Biotechnology* **174** 1434–1443
[10] Preston-Mafham J, Boddy L and Randerson P 2002 Analysis of microbial functional diversity using sole-carbon-source utilization profiles - a critique *FEMS Microbiol. Ecol* **42** 1-14
[11] Feigl V, Ujaczki E, Vaszitat E and Molnal M 2017 Influence of red mud on soil microbial communities: Application and comprehensive evaluation of the Biolog EcoPlate approach as a tool in soil microbiological studies *Science of the Total Environment* **595** 903-911
[12] Wolinska A, Frac M, Szafranek-Nakonieczna A, Zielenkiewicz U and Stępniewska Z 2017 Microbial biodiversity of meadows under different modes of land use: catabolic and genetic fingerprinting *World J Microbiol Biotechnol* **33** 154-164
[13] Tian J, McCormack L, Wang J, Guo D, Wang O, Zhang X, Yu G, Blagodatskaya E and Kuzyakov Y 2015 Linkages between the soil organic matter fractions and the microbial metabolic functional diversity within a broad-leaved Korean pine forest European *Journal of Soil Biology* **66** 57-64
[14] Lapteva E, Vinogradova Yu and Perminova E 2016 The use of multisubstrate testing to evaluate the ecological state of taiga soils in the Russian European Northeast *International Symposium BIODIAGNOSTICS-2016* (In Russian)
[15] Semenova I, Zulkarnaev A, Ilbulova G and Suyndukov Y 2011 Monitoring microbial soil societies in a vicinity of sibai concentration plant by multisubstrate *Fundamental research* **9** 139-141 (In Russian)
[16] ISO 26423-85 Soil. Methods for determining the specific electric conductivity, pH and dense residue of soil water extract 5
[17] ISO 11261:1995 Soil quality – determination of total nitrogen – modified Kjeldahl method 4
[18] ISO 14235:1998 Soil quality – determination of organic carbon by sulfochromic oxidation 5
[19] ISO 13320:2009 Particle size analysis – laser diffraction methods 51
[20] Ganghara N, Borisov B and Baybekov R 2002 *Workshop on soil science* (Moscow, Agroconsalt) 280 (In Russian)
[21] Gilmullina A, Galitskaya P, Saveliev A, Kuzyakov Y and Selivanovskaya S 2016 Changes in mineralization activity of microbial communities depending on physico-chemical properties of soils and cadmium contamination *Uchenye zapiski Kazanskogo universiteta Seriya estestvennye nauki* **158** 440-454

IOP Conf. Series: Earth and Environmental Science **107** (2017) 012057 doi:10.1088/1755-1315/107/1/012057

[22] Boshoff M, de Jonge M, Dardenne F, Blust R and Bervoets L 2014 The impact of metal pollution on soil faunal and microbial activity in two grassland ecosystems *Environmental Research* **134** 169-180
[23] Dobrovolskaya TG 2002 *The structure of bacterial communities of soils* (Moscow, PH: Academkniga) 285 (In Russian)
[24] Kovda V 1988 *Soil and soil formation* (Moscow: Higher School) 400 (In Russian)
[25] Blagodatskaya E and Kuzyakov Y 2008 Mechanisms of real and apparent priming effects and their dependence on soil microbial biomass and community structure: critical review *Biol Fertil Soils* **45** 115-131
[26] Frac M Oszust K and Lipiec J 2012 Community level physiological profiles (CLPP), characterization and microbial activity of soil amended with dairy sewage sludge *Sensor (Basle)* **12** 3253-3268

Changes in enzyme activity and functional diversity in soil induced by Cd and glucose addition

A R Gilmullina, P Yu Galitskaya and S Yu Selivanovskaya

Institute of Environmental Sciences, Kazan Federal University, Kremlevskaya str., 18, 420008, Russia
e-mail: gilmullinaar@mail.ru

Abstract. Toxic heavy metal (HM) contamination is a major global issue as it may have an indirect effect on the health of soil, plants, animals and, consequently, on human health. Agricultural soils' fertilization is one of the reported sources of HM pollution in the world. In this case simultaneous input of stimulating and inhibiting agents into soil takes place, and effects of the combined influence of these agents is hardly predictable. In this study, a simultaneous inhibiting and stimulating effect of Cd and glucose on soil microbes was studied in a model experiment. Enzyme activities (phosphatase, ß-glucosidase and cellobiohydrolase) and functional diversity (*BIOLOG® EcoPlates ™*) were assessed as a test functions. Cd (300 µg Cd g^{-1}) amendment had a negative effect only on phosphatase activity. Glucose (3 mg C g^{-1}) addition inhibited ß-glucosidase activity and stimulated functional diversity. In joint addition of Cd and Glucose the leading effect belonged to that agent which had the greatest effect in case when it was added separately.

1. Introduction

Toxic heavy metal (HM) contamination is a major global issue as it may have an indirect effect on the health of soil, plants, animals and, consequently, on human health [1 - 3]. Cadmium (Cd) is one of the most toxic metals because it can be easily transferred through food web [4]. Generally, Cd is distributed at concentration between 0.1 and 1.0 mg kg^{-1} in different soils. The physicochemical properties of soils can influence the concentration of Cd [1]. For example, acid pH can improve Cd mobility [5] and clay particles decrease Cd mobility by colloids formation [6].

Several studies investigated that HM contamination may occur from different kinds of anthropogenic activity such as industrial wastes disposal, car emissions and agricultural treatments [7 - 9]. The main HM contamination sources occurring due to agricultural treatment are the use of composted waste and sewage sludge as fertilizers [10] These fertilizers contain available organic matter and the amendment is known to be enough good to improve soil fertility. However, such treatment may also contribute HM contamination inhibiting soil microbial community [11].

Soil biological activity reflects the status of soil microbial community. Soil microorganisms take part in nutrient cycling and the decomposition of organic matter. Any soil disturbance may cause the changes in diversity and activity of soil microbial population [12]. Activity of soil microbial population may be estimated by different parameters such as respiration, soil enzymes, community level physiological profiling etc. Soil enzymes are microbial metabolites, therefore, enzyme activity measurement allows to indirectly track the changes in carbon (C), nitrogen (N) and phosphorous (P) cycle in soil. Biolog Ecoplate contains 31 different C-containing sources (aminoacids, carbohydrates and carboxylic acids). This technique is based on measuring oxidation ability of the C substrates to create patter on potential sole C source utilisation [13]. So, application of this method allows to estimate soil microbial community involved in C cycling [14].

HM have different effect on soil microbial activity and the effect mainly depends on its concentration and soil parameters. There are several studies which have already demonstrated various effects of HM on phosphatase, urease, dehydrogenase activity and functional diversity of microbial communities [7, 14, 15]. Also investigations concerning the effects of organic compounds on soil microbial communities are presented [1, 11]. However, the effect of simultaneous inhibition and stimulation remains unclear.

Thereby, present study aimed to investigate combined effect of HM and organic matter on enzyme activity and functional diversity of microbial communities in the soil with different physical properties taking into account the individual effects of inhibition and stimulation agents.

2. Material and methods

Soil samples were sampled from three different depths (0-20 cm; 20-40 cm; 40-60 cm) in temperate mixed forest located at Laishevo district, Republic of Tatarstan (Russia). Soil was classified as sandy loamy luvisol. Soil initial parameters are shown on Table 1.

Table 1. Soil physiochemical properties.

Parameter	Soil sample		
	s0-20	s20-40	s40-60
Corg, %	2.9	0.7	0.2
DOC, mg g^{-1}	0.0162	0.0077	0.0016
pH	5.8	5	5.1
moisture, %	21.1	4.8	6.5
sand, %	54.5	81.3	85.1
silt, %	41.2	18.5	14.1
clay, %	4.2	0.1	0.7
Cmic, mg g^{-1}	0.301	0.169	0.011

Prior to experiment soil samples were moistened up to 40 % of WHC (water holding capacity) and pre-incubated at 23°C for two weeks. Then soil samples were treated with water solutions containing 3 mg C g^{-1} glucose, 300 μg Cd g^{-1} $CdCl_2$, mix of 3 mg C g^{-1} glucose and 300 μg Cd g^{-1} $CdCl_2$. Final addition of water solutions adjusted the moisture of soil samples to 70 % of WHC. Soil samples treated only with water were used as control samples. Treated soils were incubated for 14 days at 23C and samples were taken at the 1st, 3rd, 7th and 14th days for estimation of physiological activity of soil microbial community which was measured as enzyme activity and functional diversity. For these analyses soil samples were suspended in water (1:10).

Enzyme activity was estimated using 4-methylumbelliferone (MUF) containing substrates: MUF-ß-d-cellobioside MUF for cellobiohydrolase, MUF-ß-d-glucopyranoside for ß-glucosidase and 4-MUF-phosphate for phosphatase. 50 μl of soil extracts and 50 μl of 200 μM MUF-containing substrate were transferred into black microplates. The fluorescence intensity was measured at an excitation wavelength of 355nm and an emission wavelength of 460 nm by microplate reader (*Fluoroskan Ascent*^{mM}).

Functional diversity was determined using *BIOLOG*® *EcoPlates* ^{mM} . Soil extracts were diluted by 10^{-4} and then transferred to microplates. The plate contains 31 different organic substrates with tetrazolium redox dye. When the substrate is oxidised the colour of well is changed which is measured by spectrophotometer (*Multiskan FS, 595 nm*). Colour change was measured every day during 13 days of incubation.

The microbial response was estimated using the next indices: average well colour development (AWCD), functional richness(R) and Shannon index (H). AWCD was calculated as follows:

$$AWCD = \frac{1}{31}\sum(A_i - A_0), \quad (1)$$

where A_i is optical density of well i, A_0 is optical density of blank well.

Functional richness (R) was calculated as total number of wells with optical density higher than 0.25. Functional diversity was estimated by calculation of H according to:

$$H = -\sum p_i ln(p_i) \quad (2)$$

where p_i is the proportional colour change in each well with substrate to the sum of colour change of all wells with substrate.

3. Results and discussion

Soil enzymes are known to be sensitive indicators of soil health. The assessment of enzyme activity gives information about the intensity of soil processes. Enzyme activity was 10-50 times higher in topsoil (0-10 cm) compared to subsoil samples (20-40 cm and 40-60 cm) in all treatments and all sampling days (Figure 1). Besides, enzyme activity did not differ significantly in subsoil samples regardless of different treatments.

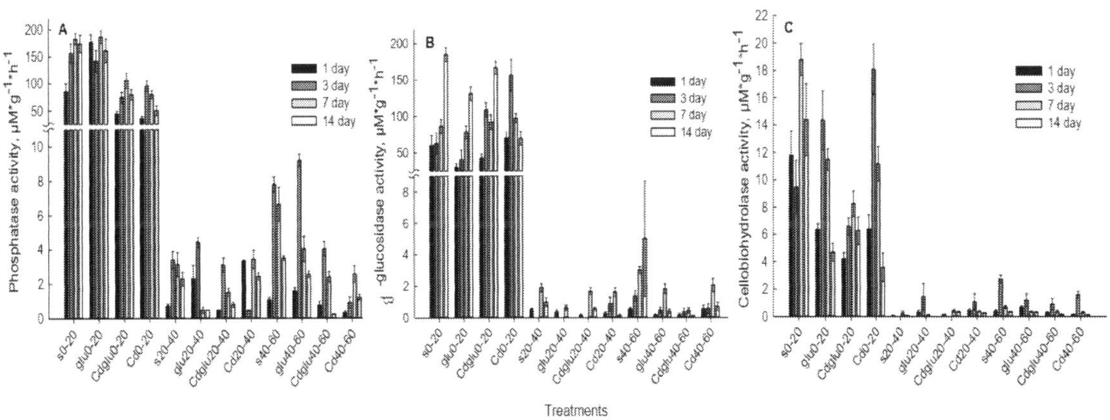

Figure 1. The effect of Cd and glucose on enzyme activity: (a) phosphatase, (b) ß-glucosidase and (c) cellobiohydrolase in different top- and subsoil layers during 2 weeks of incubation. Vertical bars demonstrate SE (n=3).

Phosphatase activity was the highest among other enzyme activities and was sensitive to Cd addition (Figure 1a). Cadmium addition inhibited phosphatase activity by 286-297 % in the both of Cd-contained topsoil treatments (+/- Glucose) at the 1st day of incubation. By the incubation time inhibiting effect of Cd slightly diminished but remained high (137-256 %).

C-cycle enzymes were presented by ß-glucosidase and cellobiohydrolase which are involved in cellulose degradation. ß-glucosidase activity was inhibited by 60-120 % in topsoil samples amended with glucose after the 1st day of incubation (Figure 1b). This might occur due to inhibition by product as ß-glucosidase hydrolyses cellobiose releasing glucose molecules. But then at the 3rd day of incubation ß-glucosidase activity rose to control value and remained at the same level up to the end of experiment. Cellobiohydrolase activity also decreased after glucose amendment at the 1st day of incubation (Figure 1c). However, cellobiohydrolase activity was several-fold lower in comparison to ß-glucosidase and phosphatase activity.

Biolog Ecoplates were used for assessment of functional diversity of soil microbial community. This microplate system contains 31 C substrates which are predominantly amino acids, carbohydrates and carboxylic acids. The technique based on the measurement of substrate oxidation which gives the picture of functional diversity, i.e. the pattern of sole C source oxidation. The data is usually presented by calculation of different indices.

Figure 2 presents the data of calculated indices: average level of C source utilisation (AWCD), number of oxidised substrates (R) and functional diversity based on variety of oxidised substrates (H). Physiological activity was higher in topsoil samples compared to subsoil samples. Besides, both of subsoil samples were almost similar to each other and there was no effect of both of agents (Glucose/Cd) on the calculated indices in subsoil.

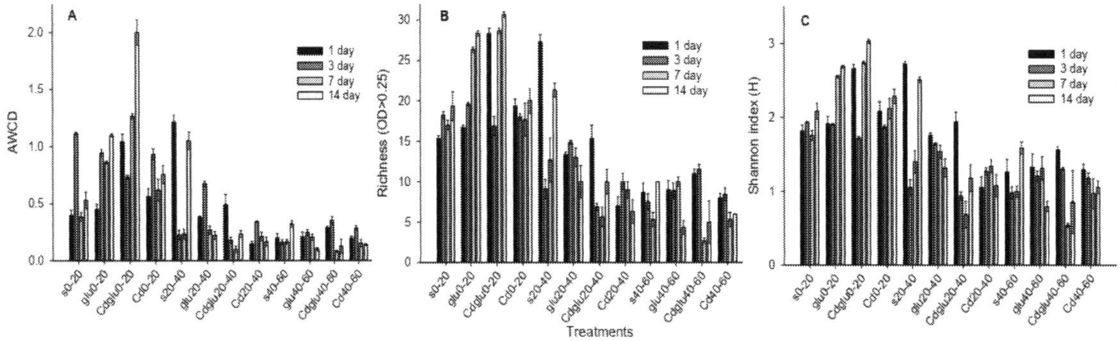

Figure 2. The effect of Cd and glucose on functional diversity: (a) average well colour development (AWCD), (b) Richness (R) and (c) Shannon index (H) in different top- and subsoil layers during 2 weeks of incubation. Vertical bars demonstrate SE (n=3).

AWCD, R and H were significantly higher in the treatments amended with glucose. Calculated indices (AWCD, R and H) started to increase only after the 3rd day of incubation in soil treated with glucose (glu-20). The interesting point is that samples treated with Glucose and Cd (Cdglu0-20) responded very quickly and were characterised by increase of AWCD (150 %), R (50 %) and H (80 %) even at the 1st day of incubation remaining at the same level for all incubation time. The highest values of calculated indices were observed for Cdglu0-20 at the 14th day of incubation.

In this study Cd amendment had a negative effect only on phosphatase activity which is in agreement with finding of other studies [1, 8]. But there was no effect of Cd on C-cycle enzymes and functional diversity. Contrary to our finding, some studies demonstrated the decrease of ß-glucosidase activity [1, 7] and functional diversity [13, 14] after HM treatment during long-term experiments. In our case the experiment was planned as short-term, so we suppose that these parameters respond to HM treatment with delay. Sole glucose affected on ß-glucosidase activity which can be explained by product inhibition. Glucose treatment also increased functional diversity and this is in agreement with studies which examined the effect of different stimulating agents (organic and inorganic fertilisers) on microbial activity [16, 17]. Additionally, it is necessary to note that the increase of physiological activity in sample treated with Cd and Glucose was quicker and higher than in sample treated only with glucose. This demonstrates that in case of sole glucose addition microorganisms competed for easily available C source at the same time regulating their population. In contrast, joint addition of Cd and glucose permitted to develop Cd-tolerant population without any competition.

Acknowledgement
This work was supported by RFBR №15-04-04520 grant, and published with support of the Russian Government Program of Competitive Growth of Kazan Federal University.

References

[1] Tejada M 2009 *Geoderma* **153** pp 254–268
[2] Mahar A, Wang P, Li R and Zhang Z 2015 *Pedosphere* **25** pp 555–568
[3] Renella G, Mench M, Landi L and Nannipieri P 2005 *Soil Biol. Biochem.* **37** pp 133–139
[4] Landi L, Renella G and Moreno L J 2000 *Biol. Fertil. Soils* **32** pp 8–16
[5] Giller K E, Witter E and McGrath S P 1998 *Soil Biol. Biochem.* **30** pp 1389–1414
[6] Bewley R J F and Stotzky G 1983 *Sci. Total Environ.* **31** pp 41–55
[7] Kızılkaya R, Aşkın T, Bayraklı B and Sağlam M 2004 *Eur. J. Soil Biol.* **40** pp 95–102
[8] Wang J, Lu Y, Ding H and Shen G 2007 *Environ. Geochem. Health* **29** pp 395–403
[9] Yao J et al *Journal of Thermal Analysis and Calorimetry* 2009 **95** 517–524
[10] Galitskaya P Y, Saveliev A A and Selivanovskaya S Y *Contemporary Problems of Ecology* **8** pp 780-788
[11] Moreno J L, Hernández T, Pérez A and García C 2002 *Appl. Soil Ecol.* **21** pp 149–158
[12] Wang F, Yao J, Si Y, Chen H, Russel M, Chen K, Qian Y, Zaray G, Bramanti E 2010 J. *Hazard. Mater.* 173 pp 510–516
[13] Muhammad A, Xu J, Li Z, Wang H and Yao H 2005 *Chemosphere* **60** pp 508–514
[14] Boshoff M, Jonge M De, Dardenne F, Blust R and Bervoets L 2014 *Environ. Res.* 134 pp 169–180
[15] Pan J and Yu L 2011 *Ecol. Eng.* **37** pp 1889–1894
[16] Liu B, Gumpertz M.L, Hu S and Rustaino J B *Soil Biol. Biochem.* 2007 **39** pp 2302–2316
[17] Sradnick A, Murugan R, Oltmanns M, Raupp J and Joergensen R G 2013 *Appl. Soil Ecol.* 63 pp 23–28

Assessing the Potential of Using Biochar as a Soil Conditioner

D M Glazunova, P A Kuryntseva, S Y Selivanovskaya and P Y Galitskaya

Institute of Environmental Sciences, Kazan Federal University, Kremlevskaya str., 18, 420008, Russia
E-mail: polinazwerewa@yandex.ru

Abstract. Biochar is a product of pyrolysis of biomass such as plant tissues, manures, sewage sludge, organic fraction of municipal solid wastes etc. Nowadays, biochar is being discussed as an alternative fertilizer that improves the air and water balance of the soil and provides soil microbiota with slow releasing biogenic elements. Many factors such as initial substrate properties, pyrolysis temperature and regime may influence biochar characteristics. In this study, characteristics of the two biochars prepared from chicken manure (ChM) and sewage sludge (SS) at 550 °C were analyzed in order to reveal their agricultural potential. It was found, that the ChM biochar had a pH value of 5.80±0.21, which was 1.6 lower than the pH of the SS sample. The electrical conductivity of the ChM sample was 6 times higher than that of the SS sample, being 6.42±0.30 mS cm-1 and 1.02±0.10 $mS \cdot cm^{-1}$, respectively. The cation exchange capacity was estimated to be 7.6±0.26 and 45±0.14 $cmol \cdot kg^{-1}$ in the ChM and SS samples, respectively. In the ChM sample total organic carbon content was 24.93±3.2%, which is nearly twice as large as that in the SS sample (12.36±4.1%), whereas total nitrogen content was estimated to be 0.33±0.03% and 0.10±0.01% for ChM and SS samples, respectively. Using scanning electronic microscopy and laser particle size distribution analysis, it was shown that the SS sample was more homogeneous in its structure and consisted of particles having a lower size of 1 to 200μm with particles of 10 to 100μm being the most frequent, while the ChM sample was nonhomogeneous and its particle size varied between 2 and 2000 μm. To observe the influence on plants, 1% of biochar was added to soil, and wheat seeds were planted. The germination index estimated for soil treated by SS biochar was estimated to be 97%, while that of soil treated by ChM biochar was lower at about 78%.

1. Introduction

Currently the world faces several serious problems: CO_2 emissions systematically increase, the amount of organic wastes grows, soil-carbon content decreases, and global climate changes – people are worried about global warming. In some regions, due to farming practices and intensive tillage soil organic carbon has decreased by 30 to 50%, as many fields have been cultivated for more than 100 years [1]. Herewith, both in big cities and in rural areas great amounts of organic wastes are generated, such as sewage sludges and different types of dung and manure. Thus, one of poultry plants in the city of Kazan (the Republic of Tatarstan, Russia) produces 50,000 tons of chicken manure every year. Another source of bulk wastes is sewage sludge forming in all wastewater treatment plants, where sewage is treated in aerotank at some stage. These types of organic wastes need additional treatment and cannot be used directly as organic fertilizers. The limiting factors of using them in plant growing are the content of toxic compounds (ammonium in chicken manure, organic matter decomposition products in SS), and possible content of pathogenic microflora. Biological methods of treating such wastes, such as anaerobic fermentation and composting are often expensive and long, and the quality of the end product is unstable. One of the effective methods of recycling such wastes is the pyrolysis, as a result of which biochar and liquid pyrolisis fuel are produced. Biochar is produced from a biomass heated without oxygen or with little oxygen [2–4], it is a solid porous substance rich in carbon with an odorous surface, and it contains various functional groups. The specific surface area and the stability of biochar are high, as well as the amounts of oxygen contained in functional groups. The distinctive properties of biochar make it a promising new method of dealing with global problems: climate change mitigation (reduction of greenhouse gas), waste treatment, energy production, soil remediation improvement, modification of physical and chemical properties of soil and the improvement of soil properties, gaseous N emissions reduction, alteration of soil nutrient availability, reduction of nutrient leaching and increase of crop yields [5–15]. The use of biochar for the above mentioned purposes is possible if it possesses certain physical and chemical properties, which in their turn strongly depend on the raw material and the parameters of the carbonization process [11]. Biochar can be produced from many types of raw material: such as woodchips, byproducts of timber industry, industry, sludge, organic wastes, plant residues, and chicken manure [16–18]. The factors influencing the pyrolysis process products are: the temperature of reaction, heating rate, and processing time [2,11,14,17,19]. Generally, the higher the temperatures are, the lower is the yield of biochar (and the more is the amount of liquid and gas), but the better is its quality: the content of nutrients like N, P, and K, porosity and the surface area [2,14,17,19]. When the pyrolysis temperature increases, the yield of biochar and the content of organic compounds in it decreases, and the percentage of mineral substances and the syngas yield increases [11,20]. It happens because the increasing temperature of pyrolysis activates the process of the biomass organic constituents (such as lignin, cellulose, and hemicelluloses) transformation and release as gases and volatiles while mineral compounds, like Mg, Ca and P, remain, and their concentration in biochar therefore increases [21]. Mohammad et al. [20] and Zhang et al. [22] discovered that biochar yield and the yield of acidic functional groups reduced, when the pyrolysis temperature increased, and the yield of the main functional groups, ash content, pH, and carbon stability became higher. Higher pH at high pyrolysis temperatures was explained by the decrease of organic functional groups, like – COOH and –OH. The highest bio-oil yield was observed at about 500°C, as at higher temperatures cracking occurs [11,20]. The usual pyrolysis temperature range is 200 to 800°C [23,24]. The processes of pyrolysis can be slow or rapid depending on the rate at which the temperature increases [11]. Slow pyrolysis is characterized by long residence of pyrolyzed vapors in the reactor, when the temperatures are low, and such long vapor-phase reactions enhance the char yield [11,20]. Inguanzo et al. [25] described the char obtained at the higher rate of temperature elevation (60°C/min compared to 5°C/min) as having low volatile substance content and a high ash yield (including fixed carbon). The researches came to the conclusion that the high rate of temperature elevation improves the quality of biochar. However, the rate of temperature elevation did not have such influence on the product quality at high pyrolysis temperatures [11,26]. That is why the usual aim of rapid pyrolysis is to produce the higher yield of a liquid product [11,27,28]. Short vapor residence time reduces the production of gas because of secondary cracking and liquid product yield remains higher, when the cooling is fast. Residence time in the process of pyrolysis influences the product composition [11]. Thus, biochar yield increases, when the time of pyrolysis increases [22]. Besides, residence time influences the specific surface area of biochar and its pore characteristics.

To assess the potential of using biochar as a soil improver it is necessary to evaluate the initial characteristics of biochar. We've chosen biochar produced from such organic wastes as chicken manure (ChM) and sewage sludge (SS) as the targets of research.

2. Material and Methods

We've chosen two types of biochar as the targets of research: i) biochar produced from chicken manure, and ii) biochar produced from sewage sludge. Both types of biochar were produced in an industrial pyrolysis plant in the city of Naberezhnye Chelny (the Republic of Tatarstan, Russia), using the method of rapid pyrolysis at the temperature of 550°C and the retention time of 8 min. The pH of the biochar was determined with the help of a pH-410 meter and its electrical conductivity (EC) – with the help of a conductmeter Anion 4100 – 10g of biochar in 100 ml of deionized water were suspended [29]. Moisture was measured according to the weight loss by drying wet samples at 105°C [30]. The CEC for the biochar samples was determined in this research using a modified barium chloride compulsive exchange method

(Skjemstad). Wet oxidation with potassium dichromate [31] was used to determine TOC and the TKN was determined by the Kjeldahl method [32]. Ash content was measured by means of burning a biochar sample in a muffle furnace at 550°C for 3 hours [33]. The extraction method in the solution of 0.1 M K_2SO_4 outlined by Chefetz et al. [34] was used to determine the DOC values of each sample. A 0.5-ml sample of solution was taken after extraction and it was analyzed immediately by dichromate oxidation in accord with the tube digestion procedure elaborated by Nelson and Sommers [35]. Total metal content, P, S, and B content was determined using a Schimadzu ICPE – 9000 (Japan) atomic absorption spectrophotometer after sample extraction by digestion with concentrated HNO_3 using the photometric method of PND F 16.1:2:3.11-98 [36].

Biochar particle size was determined using the Microtrack BlueWave laser particle analyzer in accord with ISO 13320-1 [37] . The size of pores and the form of particles were determined using the method of electronic microscopy. To realize this, the sample on a chuck was put into the chamber of Quorum Q 150T ES vacuum apparatus. Conductive layer apply by technique cathode sputtering using alloy Au/Pd by quantity proportion 80/20. The thickness of alloy is 15 nm. Measuring was carried out by using a Merlin Carl Zeiss field-emission high-resolution scanning electron microscope. Observation photo of morphology surface apply at accelerating voltage of incident electron 5 kV and current probe 300 pA in order to minimum modify sample. In the test with oats (*Avena sativa*) biochar was added to the soil in an amount of 10 % by weight. The germination test was carried out on gray forest soil (Haplic Greyzem). Each of 5 pots was sown with 10 oat seeds and after 10 days, the percentage germination, biomass and root length were determined. The soil without added biochar was used as the control. The germination index (GI) was evaluated in accord with Zucconi et al., 1981 using the following equation:

$$\mathrm{GI}\ (\%) = \frac{\text{Seed germination Sample } (\%) * \text{Root Length Sample (mm)}}{\text{Seed germination Control } (\%) * \text{Root Length Control (mm)}} \times 100$$

All measurements were realized in triplicate. All tables present the values and the standard deviations.

3. Results and Discussion

To assess the possibility of using biochar as soil improver it was necessary to evaluate its physical and chemical properties and its influence on plants. The data is presented in Table 1. One of very important parameters influencing the quality of biochar as soil improver is its pH. The biochar pH depends on the raw material and the production conditions. Most commonly biochar pH is greater than 7, so biochar is alkaline [2]. That is why using biochar as a soil improver can neutralize soils that are acidic [38]. Herewith, the lowest pH values are observed in biochars produced from dung and manure [2]. Our results coincide with those of other researches. Thus, the chicken manure biochar (ChM sample) has pH value of 5.80±0.21, which is 1.6 lower than the pH of the SS sample. The pH of biochar depends on the ash content – the higher ash content is the higher is pH [4,39]. Some researchers [2,40,41] have found a relationship between ash content and the value of pH: biochars with high ash content often have high pH. Our samples prove those results, as ash content in the SS sample is higher (54.85±0.06%), and its pH is higher as well (7.40±0.24). In whole, the data obtained coincide with those described by other researches, that is, the chicken manure biochar pH varies between 5.8 and 10.16 [42,43] , and sewage sludge biochar pH – between 7.7 and 9.54 [18]. In the study realized by Cely et al. [42] chicken manure biochar with the ash content of 20.24% was analyzed.

The ChM sample electrical conductivity was 6 times higher than that of the SS sample, being 6.42±0.30 mS cm^{-1} and 1.02±0.10 mS cm^{-1}, correspondingly. Being the product of pyrolysis, biochar is characterized by low moisture content (6.75±0.10% and 7.57±0.10% for ChM and SS samples, correspondingly). Thus, in the research carried out by Li et al. [44] moisture content for chicken manure biochar was determined as 10.2%, and for sewage sludge biochar – 3.59 to 9.53%.

It was found out that the biochar cation exchange capacity (CEC) affects the fertility of soil [2]. The biochar CEC also influences on the soil CEC, improving its physical and chemical properties, that is why it was important to assess the CEC of the samples we obtained. So, ChM sample CEC is 7.6±0.26 cmol kg^{-1}, and SS sample CEC is 45±0.14 cmol kg^{-1}. The CEC of the biochar produced from different types of biomass can differ greatly, depending on the type of pyrolysis (its temperature being the most important factor) and the initial substrate, and is 2 to 516 cmol·kg^{-1} [2,18]. Biochar samples we have analyzed were produced in the same conditions, thus, only the characteristics of the initial substrate influenced the difference in the CEC value. When biochar is introduced back into the soil, the functional groups on its surface are oxidized and superficial CEC increases [45,46]. The maximum CEC is observed in the biochar with the largest surface area [47]. The values of CEC for chicken manure biochar described by other researches vary between 9.14 and 66.8 cmol kg^{-1} [42,43] and the CEC values for sewage sludge biochar vary between 2.36 and 30 cmol kg^{-1} [18].

Table 1. Characteristics of biochar

Sample	ChM	SS
pH	5.80±0.21	7.40±0.24
EC, $mS \cdot cm^{-1}$	6.42±0.30	1.02±0.10
Moisture, %	6.75±0.10	7.57±0.10
CEC, $cmol \cdot kg^{-1}$	7.6±0.26	45±0.14
Ash, %	33.85±1.06	54.85±0.06
TOC, %	24.93±3.2	12.36±4.1
DOC, $mg \cdot g^{-1}$	0.30±0.02	0.20±0.01
TKN, %	0.33±0.03	0.10±0.01
P $mg \cdot kg^{-1}$	2125.00±15.59	5250.00±12.15
S $mg \cdot kg^{-1}$	4900.00±17.45	5583.33±14.12
B $mg \cdot kg^{-1}$	5.63±0.23	14.33±2.10
Al $mg \cdot kg^{-1}$	386.25±5.10	7350.00±24.10
Cd $mg \cdot kg^{-1}$	1.88±0.21	7.42±0.20
Ca $mg \cdot kg^{-1}$	5412.50±2.56	12541.67±52.45
Co $mg \cdot kg^{-1}$	0.53±0.10	51.83±0.92
K $mg \cdot kg^{-1}$	6525.00±13.24	2475.00±27.45
Mg $mg \cdot kg^{-1}$	2550.00±3.45	3241.67±1.58
Na $mg \cdot kg^{-1}$	675.00±1.22	199.58±1.21
Cr $mg \cdot kg^{-1}$	23.75±0.59	155.00±1.65
Cu $mg \cdot kg^{-1}$	73.38±0.12	90.00±0.22
Fe $mg \cdot kg^{-1}$	1762.50±1.23	13583.33±45.12
Mn $mg \cdot kg^{-1}$	312.50±3.10	433.33±2.87
Ni $mg \cdot kg^{-1}$	7.50±0.24	35.33±2.97
Pb $mg \cdot kg^{-1}$	11.25±0.59	31.25±1.64
Zn $mg \cdot kg^{-1}$	292.50±2.12	608.33±5.46
Mo $mg \cdot kg^{-1}$	0.46±0.02	0.71±0.01
GI,%	78	97

Being the product of organic substrate pyrolysis, biochar is the source of carbon, but unlike initial sewage sludge and chicken manure, biochar mostly contains slow-releasing carbon, which makes it possible to use it as a long-term fertilizer. That is why it was important to assess the content of dissolved organic carbon (DOC) easily assimilable by microorganisms and total organic carbon. In the ChM sample TOC content was 24.93±3.2%, which is nearly twice as large as TOC content in the SS sample (12.36±4.1%), for which high ash content is registered. According to the results of the investigation, sewage sludge biochar C content varies in the range of 15.92 to 66.7% depending on the characteristics of the initial sewage sludge and the type of pyrolysis [18,44]. High organic matter content of 213.52 g kg^{-1} [43] and high carbon content is particularly characteristic for biochar produced from dung and manure – 33.25-52.1% [18,43,44]. In both samples DOC content was low (0.2-0.3$mg \cdot g^{-1}$), and in this connection their use as soil improvers won't lead to carbon sequestration.

The biochar samples obtained were characterized by low nitrogen content of 0.33±0.03% and 0.10±0.01% for ChM and SS samples, correspondingly. In whole sewage sludge biochar can contain 0.91 to 9.3 % of nitrogen, depending on the initial nitrogen content in the sewage sludge and the type of pyrolysis, and the highest concentration of nitrogen was marked for low temperature pyrolysis [17,44]. According to the data obtained by other researchers, chicken manure biochar is characterized by the nitrogen content of 1.62 to 5.03% [17,42–44,48]. Herewith it is considered that the typical N content in biochar is less than 5 wt% [2,44]. Biochar samples are rich in phosphorus (2125.00±15.59 mg·kg-1 and 5250.00±12.15 mg·kg-1 in ChM and SS samples, correspondingly), which will help to use biochar as soil improver and to reduce the use of phosphate fertilizers.

Currently the Russian legislation has no legal framework in respect of the quality of biochar used as a soil improver. That is why we can compare the obtained product of recycling chicken manure and sewage sludge with GOST R 53765-2009 [49] regulating the use of poultry manure as organic fertilizer, with GOST R 54651-2011 [50] regulating the use of organic fertilizer produced from sewage and GN 2.1.7.2041-06 regulating maximum permissible concentrations (MPC) of chemical substances in soil. According to the data received, gross sulfur content in both samples of biochar exceeds the specified standard for soils (GN 2.1.7.2041-06). According to GOST R 54651-2011 [50] biochars obtained can be classified as fertilizers from group 2, as zinc ion content exceeds the specified standard in both samples. Compared to GOST R 53765-2009 [49] chicken manure biochar corresponds to all the standards, except for nitrogen content, which is connected with using the pyrolysis method to process the chicken manure.

The analysis of particle size was carried out using two methods: scanning electronic microscopy (Figure 1) and Microtrack Bluewave laser particle size analyzer.

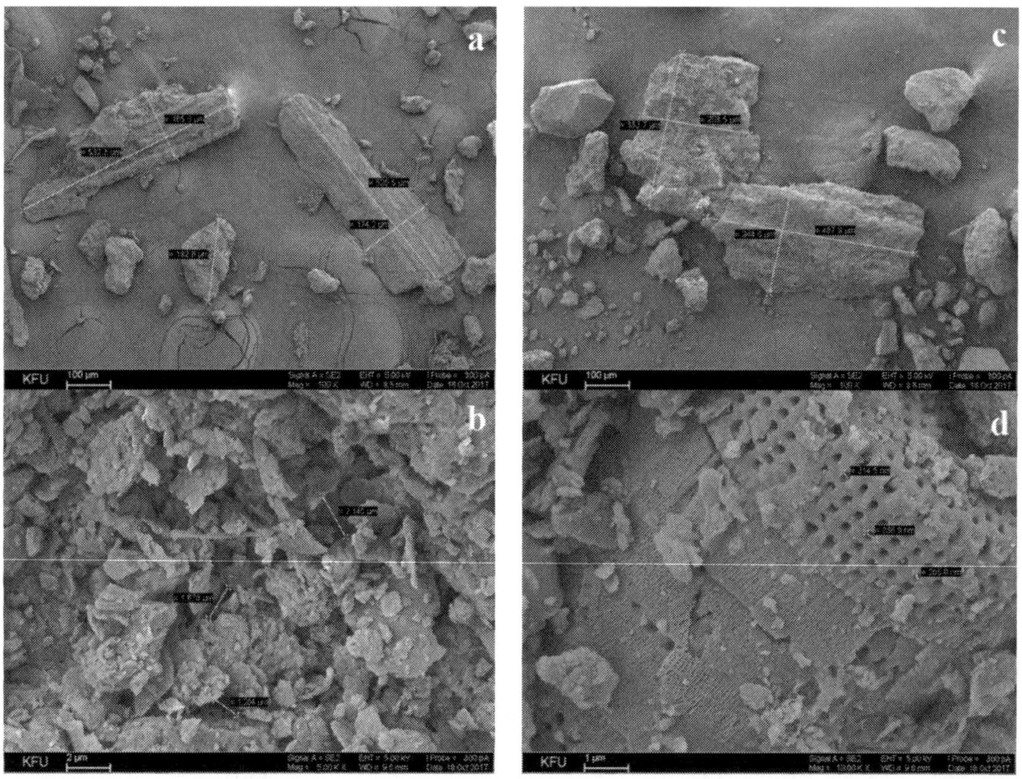

Figure 1. Scanning electron microscopic images of a) ChM (Bar=100μm), b) ChM (Bar=2μm), c) SS (Bar=100μm), d) SS (Bar=1μm)

Both methods showed that the SS sample is more homogeneous in its structure and consists of particles having lower size of 1 to 200μm, particles having the size of 10 to 100μm are more frequent. The ChM sample is nonhomogeneous and its particle size varies between 2 and 2000 μm. Particle size distribution in this sample is not normal.

It is worthwhile mentioning that the SS sample has a more prominent porous structure. The biochar specific surface area is of high importance as it increases its capacity to adsorb organic compounds and metal ions [18,48,51]. Generally, biochar that has a large specific surface area is characterized by a narrow micro-pore distribution, has good pore structure and possesses high adsorption capacity. Large specific surface area is the quality characteristic of biochar [2].

It is important to estimate the possibility of using biochar as soil improver taking into account its physical and chemical properties, but the characteristics received cannot provide an accurate forecast of the influence of biochar on the growth and development of plants. That is why the influence of the two types of biochar on the germination capacity, root and stalk length, and the biomass of wheat was assessed. According to the results received, the introduction of the SS sample into soil increased the length of above-ground sprouts by 23%; changes of germination capacity, root length and the biomass were unreliable. The germination index was 97%, which shows that there was no negative influence of biochar introduction. The introduction of ChM sample into soil lead to the shortening of root shoot; the changes of other parameters were unreliable. The germination index for the sample was 78%.

Conclusion

So, the two types of biochar obtained under the same pyrolysis conditions from different substrates not only differ in their chemical composition, but also in their physical structure. Such differences explain the fact that the introduction of sewage sludge biochar does not have negative influence on the growth and development of wheat, while as the introduction of chicken manure biochar into soil reduces the germination index of wheat.

Acknowledgments

The work is performed with support from Ministry of Education and Science of Russian Federation, project No 14.581.21.0024.

References

[1] Schlesinger W H 1986 Changes in Soil Carbon Storage and Associated Properties with Disturbance and Recovery *The Changing Carbon Cycle* (New York, NY: Springer New York) pp 194–220

[2] Tan Z, Lin C S K, Ji X and Rainey T J 2017 Returning biochar to fields: A review *Appl. Soil Ecol.* **116** 1–11

[3] Tan X, Liu Y, Gu Y, Xu Y, Zeng G, Hu X, Liu S, Wang X, Liu S and Li J 2016 Biochar-based nano-composites for the decontamination of wastewater: A review *Bioresour. Technol.* **212** 318–33

[4] Lehmann J 2007 A handful of carbon *Nature* **447** 143–4

[5] Lehmann J and Joseph S 2009 *Biochar for environmental management : science and technology* (Earthscan)

[6] Ladygina N and Rineau F 2013 *Biochar and soil biota* (Taylor & Francis)

[7] Deluca T H, Mackenzie M D, Gundale M J and Holben W E Wildfire-Produced Charcoal Directly Influences Nitrogen Cycling in Ponderosa Pine Forests

[8] Yanai Y, Toyota K and Okazaki M 2007 Effects of charcoal addition on N_2O emissions from soil resulting from rewetting air-dried soil in short-term laboratory experiments *Soil Sci. Plant Nutr.* **53** 181–8

[9] Chan K Y, Van Zwieten L, Meszaros I, Downie A and Joseph S 2007 Agronomic values of greenwaste biochar as a soil amendment *Aust. J. Soil Res.* **45** 629

[10] Major J, Rondon M, Molina D, Riha S J and Lehmann J 2010 Maize yield and nutrition during 4 years after biochar application to a Colombian savanna oxisol *Plant Soil* **333** 117–28

[11] Anon 2016 Production and utilization of biochar: A review *J. Ind. Eng. Chem.* **40** 1–15
[12] Gundale M J and DeLuca T H 2006 Charcoal effects on soil solution chemistry and growth of Koeleria macrantha in the ponderosa pine/Douglas-fir ecosystem *Biol. Fertil. Soils* **43** 303–11
[13] Anon 2007 Molecular signature and sources of biochemical recalcitrance of organic C in Amazonian Dark Earths *Geochim. Cosmochim. Acta* **71** 2285–98
[14] Lehmann J, Gaunt J and Rondon M Mitigation and Adaptation Strategies for Global Change (2006) 11: 403–427 Bio-char sequestration in terrestrial ecosystems – a review
[15] Lee J W, Hawkins B, Day D M and Reicosky D C 2010 Sustainability: the capacity of smokeless biomass pyrolysis for energy production, global carbon capture and sequestration *Energy Environ. Sci.* **3** 1695
[16] Kuppusamy S, Thavamani P, Megharaj M, Venkateswarlu K and Naidu R 2016 Agronomic and remedial benefits and risks of applying biochar to soil: Current knowledge and future research directions *Environ. Int.* **87** 1–12
[17] Xu X, Zhao Y, Sima J, Zhao L, Mašek O and Cao X 2017 Indispensable role of biochar-inherent mineral constituents in its environmental applications: A review *Bioresour. Technol.* **241** 887–99
[18] Ding Y, Liu Y, Liu S, Huang X, Li Z, Tan X, Zeng G and Zhou L 2017 Potential Benefits of Biochar in Agricultural Soils: A Review *Pedosphere* **27** 645–61
[19] Putun A E, Ozbay N, Onal E P and Putun E 2005 Fixed-bed pyrolysis of cotton stalk for liquid and solid products *Fuel Process. Technol.* **86** 1207–19
[20] Al-Wabel M I, Al-Omran A, El-Naggar A H, Nadeem M and Usman A R A 2013 Pyrolysis temperature induced changes in characteristics and chemical composition of biochar produced from conocarpus wastes *Bioresour. Technol.* **131** 374–9
[21] Zhang J and Wang Q 2016 Sustainable mechanisms of biochar derived from brewers' spent grain and sewage sludge for ammonia–nitrogen capture *J. Clean. Prod.* **112** 3927–34
[22] Zhang J, Liu J and Liu R 2015 Effects of pyrolysis temperature and heating time on biochar obtained from the pyrolysis of straw and lignosulfonate *Bioresour. Technol.* **176** 288–91
[23] Hossain M K, Strezova V, Chan K Y, Ziolkowski A and Nelson P F 2011 Influence of pyrolysis temperature on production and nutrient properties of wastewater sludge biochar *J. Environ. Manage.* **92** 223–8
[24] Song W and Guo M 2012 Quality variations of poultry litter biochar generated at different pyrolysis temperatures *J. Anal. Appl. Pyrolysis* **94** 138–45
[25] Inguanzo M, Menéndez J ., Fuente E and Pis J . 2001 Reactivity of pyrolyzed sewage sludge in air and CO2 *J. Anal. Appl. Pyrolysis* **58–59** 943–54
[26] Abu El-Rub Z A, Bramer E A and Brem G 2008 Experimental comparison of biomass chars with other catalysts for tar reduction *Fuel* **87** 2243–52
[27] Boucher M ., Chaala A, Pakdel H and Roy C 2000 Bio-oils obtained by vacuum pyrolysis of softwood bark as a liquid fuel for gas turbines. Part II: Stability and ageing of bio-oil and its blends with methanol and a pyrolytic aqueous phase *Biomass and Bioenergy* **19** 351–61
[28] Kopetz H, Weber T, Palz W, Chartier P and Ferrero G L 1998 Biomass for Energy and Industry: 10th European Conference and Technology Exhibition, proceedings of the International Conference *Biomass Energy Ind. 10th Eur. Conf. Technol. Exhib. Proc. Int. Conf.* 1829
[29] ISO 10390:2005 Soil quality -- Determination of pH
[30] ISO 11465:1993 Soil quality -- Determination of dry matter and water content on a mass basis -- Gravimetric method
[31] ISO 14235:1998 Soil quality - Determination of organic carbon by sulfochromic oxidation 5
[32] ISO 11261:1995 Soil quality — Determination of total nitrogen — Modified Kjeldahl method 4
[33] ISO 10694:1995 Preview Soil quality -- Determination of organic and total carbon after dry combustion (elementary analysis)
[34] Chefetz B, Chen Y, Hadar Y and Hatcher P G 1998 Characterization of Dissolved Organic Matter Extracted from Composted Municipal Solid Waste *Soil Sci. Soc. Am. J.* **62** 326
[35] Nelson D W and Sommers L E 1996 Total carbon, organic carbon and organic matter *Methods of Soil Analysis. Part 3. Chemical MethodsSoil* ed J M Bartels (Sci. Soc. Am.) pp 986–998

[36] PND F 16.1:2:3.11-98 Quantitative chemical analysis of soils. Method of conducting measurements of materials in solid objects by spectrometry with ICP
[37] ISO 13320:2009 Preview Particle size analysis -- Laser diffraction methods
[38] Yuan J H, Xu R K, Wang N and Li J Y 2011 Amendment of Acid Soils with Crop Residues and Biochars *Pedosphere* **21** 302–8
[39] Warnock D D, Lehmann J, Kuyper T W and Rillig M C 2007 Mycorrhizal responses to biochar in soil – concepts and mechanisms *Plant Soil* **300** 9–20
[40] Yuan J-H, Xu R-K and Zhang H 2011 The forms of alkalis in the biochar produced from crop residues at different temperatures *Bioresour. Technol.* **102** 3488–97
[41] Singh B, Singh B P and Cowie A L 2010 Characterisation and evaluation of biochars for their application as a soil amendment *Aust. J. Soil Res.* **48**
[42] Cely P, Gasco G, Paz-Ferreiro J and Méndez A 2015 Agronomic properties of biochars from different manure wastes *J. Anal. Appl. Pyrolysis* **111** 173–82
[43] Zhao B, Xu R, Ma F, Li Y and Wang L 2016 Effects of biochars derived from chicken manure and rape straw on speciation and phytoavailability of Cd to maize in artificially contaminated loess soil *J. Environ. Manage.* **184** 569–74
[44] Li D-C and Jiang H 2017 The thermochemical conversion of non-lignocellulosic biomass to form biochar: A review on characterizations and mechanism elucidation *Bioresour. Technol.*
[45] Steiner C, Glaser B, Geraldes Teixeira W, Lehmann J, Blum W E H and Zech W 2008 Nitrogen retention and plant uptake on a highly weathered central Amazonian Ferralsol amended with compost and charcoal *J. Plant Nutr. Soil Sci.* **171** 893–9
[46] Liang B, Lehmann J, Solomon D, Kinyangi J, Grossman J, O'Neill B, Skjemstad J O, Thies J, Luizão F J, Petersen J and Neves E G 2006 Black Carbon Increases Cation Exchange Capacity in Soils *Soil Sci. Soc. Am. J.* **70** 1719
[47] Gaskin J W, Speir A, Morris L M, Ogden L, Harris K, Lee D and Das K C 2007 Potential for Pyrolysis Char to Affect Soil Moisture and Nutrient Status of a Loamy Sand Soil
[48] Zhao X, Ouyang W, Hao F, Lin C, Wang F, Han S and Geng X 2013 Properties comparison of biochars from corn straw with different pretreatment and sorption behaviour of atrazine *Bioresour. Technol.* **147** 338–44
[49] GOST R 53765-2009 Poultry dung. Raw material for manufacture of organic fertilizers. Specifications
[50] GOST R 54651-2011 Organic fertilizers on basis of sewage sludge. Specifications
[51] Inyang M, Gao B, Pullammanappallil P, Ding W and Zimmerman A R 2010 Biochar from anaerobically digested sugarcane bagasse *Bioresour. Technol.* **101** 8868–72

IOP Conf. Series: Earth and Environmental Science **107** (2017) 012060 doi:10.1088/1755-1315/107/1/012060

Eco-organic tourism as an element of the sustainable development of territories

D V Ivanov[1] and I I Ziganshin[2]

[1] Deputy Director, Research Institute for Problems of Ecology and Mineral Wealth Use of Tatarstan Academy of Sciences, Kazan, Russia
[2] Senior Researcher, Department of biogeochemistry, Research Institute for Problems of Ecology and Mineral Wealth Use of Tatarstan Academy of Sciences, Kazan, Russia
e-mail: water-rf@mail.ru

Abstract. Organic agricultural production is an ecologically oriented alternative to traditional agriculture and is aimed at ensuring food and environmental safety of the population and territories. The development of ecological tourism on the basis of organic production farms is considered as a promising direction in the development of regional programs for the sustainable development of territories. Studies have shown that, farms engaged in the production of organic products in the Republic of Tatarstan have a significant potential for the development of eco-organic tourism using elements of the nature protection structure.

1. Introduction

The concept of Sustainable Development, adopted in 1992 at the United Nations conference on environment and development in Rio de Janeiro, affects all branches of the world economy, including agriculture, the basic basis of the economies of states and regions of the world. The concept is based on four major principles: satisfaction of the basic needs of all living people, equal standards of this satisfaction for the entire population of the planet; sensible, careful use of natural resources; preservation of opportunities for future generations to realize their basic needs [1].

Traditional agriculture is a significant source of anthropogenic pollution of the environment on a global and regional scale. The large-scale use of synthetic mineral fertilizers, pesticides, genetically modified organisms, antibiotics and growth hormones led to a deterioration in the quality of agricultural products, which is hazardous to human health and sharply raised the issue of the need for an alternative, environmentally oriented type of agricultural production that meets the principles of sustainable development. In many countries of the world, organic farming has been developing in recent decades, aimed at the production of natural foods that are not capable of causing negative reactions in the living body and produced in ecologically clean territory.

The Codex Alimentarius, developed by FAO-UNESCO and the WHO to establish harmonized international food standards to protect consumers' health and ensure fair trade in food, defines organic farming as "an integrated production management system that stimulates and enhances the well-being of the agrarian ecosystem, including biological diversity, biological cycles and soil biological activity, which is achieved by using all possible agronomic, biological and mechanical methods as opposed to the use of synthetic materials to perform specific functions within the system" [2].

The most active organic agriculture is developing in the countries of the European Union, where much has been done and is being done for its development, including the principles and rules of organic farming [3-5]. According to directives of the European Union No. 834/2007 of 28.06.2007, organic agriculture should fulfill its integrated social role, ensuring food security by producing high-quality healthy food, promoting the development of rural areas, protecting and preserving the natural environment [4].

In recent years in the Republic of Tatarstan, one of the most economically developed regions of the Russian Federation, having highly developed agriculture, specializing in the cultivation of grain crops, sugar beets and potatoes, and in the production of meat, milk and eggs, there has been a significant demand from the population for natural food. At the same time it should be noted that, the development of organic farming in Tatarstan, as in Russia as a whole, is in the making. In the local mass media, there are many proposals of purchasing "eco-friendly foods". However, their compliance with the requirements and standards of organic farming is almost impossible to establish, as the system of organic certification in Tatarstan is not yet established.

The law on organic production has not been adopted in Russia, unlike the United States or the European Union, where the national organic standards operate. In addition, the by-laws that regulate the rules and procedures for certification and accreditation of certifying bodies are not adopted.

State standard R 56508-2015 [6] on organic products, introduced in 2016, has not radically changed the situation, there are no accredited organizations in the country for certification under this standard. It is important to note that, certification requires not only an assessment of the conformity of agricultural technology in the field of crop production, livestock and food production to the requirements of organic agricultural production, but also an integral assessment of the environmental situation of the territory on which the agricultural enterprise is located.

2. Materials and methods

In 2015, a study was conducted on farms in Tatarstan for compliance with the requirements of organic farming. Based on its results, 39 farms in 21 municipal districts were allocated for producing products that can be certified as organic [7]. All farms are classified as small and medium-sized businesses, the most suitable for organizing various areas of environmentally oriented tourism, such as relaxing ones in ecologically clean places with beautiful landscapes, excursions to natural and cultural heritage sites, communication with animals, participation in harvesting and in various activities.

3. Results and discussions

Considering the location of organic farms across the territory of the Republic of Tatarstan, it should be noted that, most of the organic farms are concentrated in the areas adjacent to the capital of the republic of Kazan, which is largely due to their orientation towards a general consumer. All of the organic farms are located in ecologically safe areas with a large concentration of specially protected natural territories, which allows organizing excursions to these unique natural heritage sites (figure 1).

The regional system of specially protected natural territories includes 171 objects of various categories, status and protection regimes, including federal, regional and local importance and is an important part of the tourist and recreational potential of the Republic of Tatarstan [8]. However, the existing rich tourist and recreational potential of specially protected natural territories is not adequately reflected in the programs of tourist services. Studies have shown that, the key problem of development of tourist and recreational activities in specially protected natural territories is the lack of the tourist infrastructure [9]. In many respects, the reasons for its absence are related not only to the fact that they are usually located at a considerable distance from large settlements, but also with the fact that initially they did not have the task of receiving visitors. Transport distance, as well as poor development of road infrastructure exacerbates the situation with visiting unique natural sites, not allowing visiting them during a day, without the organization of spending a night.

An effective way to solve this problem can be the development of tourism on the basis of farms engaged or aimed at the production of organic products. They have an orientation toward rational use of natural, cultural, historical and other resources of the locality and its specifics in creating a tourist product that does not violate the integrity of ecosystems, involving local people in the tourist business.

At the same time, infrastructure problems are solved, including organization of reception of people with disabilities [10].

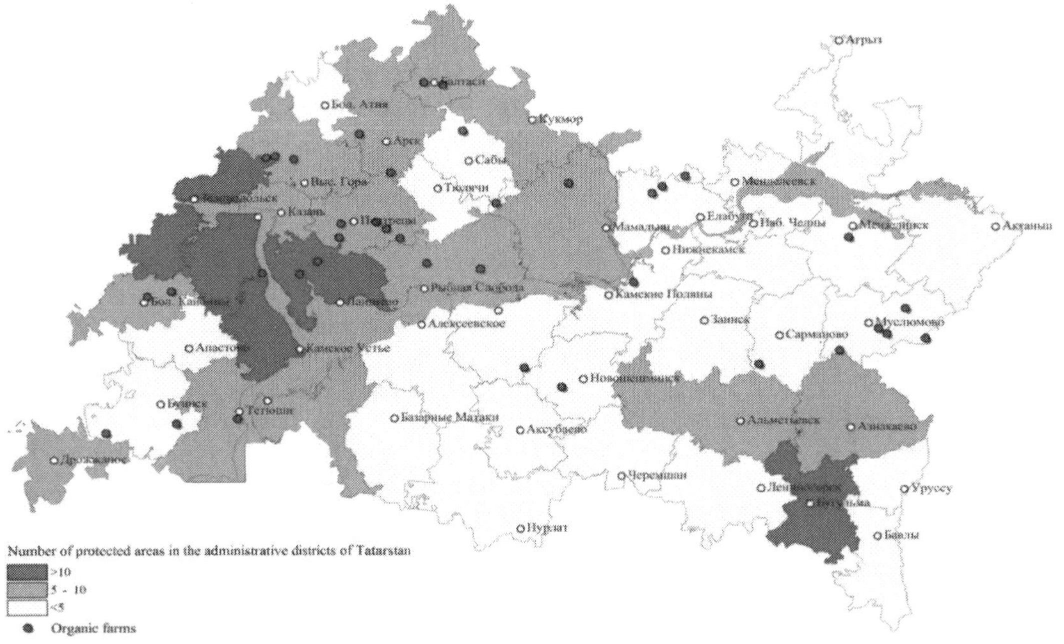

Figure 1. Spatial distribution of specially protected natural territories and organic farms across the territory of the Republic of Tatarstan.

Despite the undoubted advantages of eco-organic tourism for both holidaymakers and those who provide this service, only a small part (25%) of the investigated farms of organic farming are currently actively developing tourist services. The rest of the farms are ready to develop such an environmentally oriented business, subject of the support of the state.

To study the problems and possibilities of development of eco-organic tourism in the territory of the Republic of Tatarstan, a SWOT analysis was carried out (table 1). The analysis showed that, farms engaged in the production of organic products have a significant potential for the development of eco-organic tourism, including organization of excursion trips to specially protected natural territories. This is due to their location on the territory with a high concentration of unique natural objects and a growing demand for their products. Identified strengths and external opportunities can be considered as the basis for a competitive strategy for the development of eco-organic tourism based on organic farming and should be implemented through a system of activities aimed at their use and strengthening. It should be noted that, the complex problems present in the sphere of development of this type of tourism are quite acute and systemic. Eco-organic tourism can and should become an important element of sustainable development of territories. A scientific approach, based on a thorough study of the possibilities for receiving visitors to each individual farm of organic farming, as well as each natural facility is required.

Table 1. SWOT-analysis of the potential for the development of eco-organic tourism in the Republic of Tatarstan.

Strengths	Weaknesses
- rational use of energy and natural resources; - production of high-quality natural products; - suitable soil and climatic conditions for conducting organic farming; - high natural and recreational potential of the territory; - the presence of a developed system of specially protected natural territories; - availability of event events related to the history and culture of the peoples; - availability of infrastructure for accommodation and catering in farmsteads and farms; - good transport accessibility of organic farming farms; - favorable ecological situation in rural areas.	- poor quality of provided accommodation and nutrition services; - lack of specialists in the service of tourists; - insufficient information for supporting of organic farming; - seasonality of services related to climatic conditions; - lack of investment resources and financial support from the state; - the unwillingness of rural residents to host foreign people; - lack of developed ecological routes and observation platforms in specially protected natural areas; - high cost of products and services provided by organic farms.
Opportunities	**Threats**
- growth of popularity of ecological tourism among the population; - increasing the material well-being of the rural population and improving the socio-economic climate in the countryside; - receiving additional profits through the development of tourism; - an opportunity for the development of a variety of types of tourist services; - the possibility of free promotion of the image of farms of organic farming through social advertising; - replenishment of the republican and local budgets by increasing the taxable base.	- imperfection of the regulatory and legal framework for rural tourism; - weak entrepreneurial activity of the rural population; - high competition with outbound tourism; - deterioration of the ecological conditions of natural objects; - lack of a program to support organic farms engaged in rural tourism; - lack of calculations of the carrying capacity of natural areas; - deterioration of the ecological status of specially protected natural areas due to recreational impact; - a general decline in the living standards of the population in Russia.

Acknowledgement
The publication was carried out with the financial support of the Russian Humanitarian Science Foundation (RNF) and the Government of the Republic of Tatarstan in the framework of the scientific project №16-16-16012.

References

[1] 1993 The Rio Declaration on Environment and Development *UN document* A/CONF.151/26/Rev1(Vol I) (New York: UN) pp 3 – 7

[2] 1999 *Guidelines for the production, processing, marketing and labeling of organically produced foods* (Rome: The Codex Alimentarius Commission and the FAO/WHO Food Standards Programme) p 91

[3] European Union 2001 European Union Council Regulation No. 2092/91 of 24 June 1991 on organic production of agricultural products and indications referring thereto on agricultural products and foodstuffs *Official Journal of the European Union* **198** 101

[4] European Union 2007 European Union Council Regulation No. 834/2007 of 28 June 2007 on organic production and labeling of organic products and repealing Regulation No. 2092/91 *Official Journal of the European Union* **189** 23
[5] European Union 2008 European Union Council Regulation No.889/2008 of 5 September 2008 laying down detailed rules for the implementation of Council Regulation (EC) No. 834/2007 on organic production and labeling of organic products with regard to organic production, labelling and control *Official Journal of the European Union* **250** 84
[6] 2015 *State standard R 56508-2015 Organic production products – Rules of production, storage, transportation* (Moscow: Standardinform) p 71
[7] Grigoryan B R, Koltsova T G, Sungatullina L M and Sachabiev I A 2016 Conformity assessment of agricultural enterprises of the Republic of Tatarstan to the requirements of organic agricultural production *Russian J. of applied ecology* **4** 40 – 45
[8] 2007 *State Register of specially protected natural territories of the Republic of Tatarstan* (Kazan: Idel-Press) p 407
[9] Ziganshin I I and Ivanov D V 2016 Potential for accessible tourism development in specially protected natural territories of the Republic of Tatarstan *Russian J. of applied ecology* **2** 42 – 47
[10] Ziganshin I I and Ivanov D V 2016 *Imperatives of development of hospitality infrastructure in the region: Proc. Int. Sci. Conf.* (Volgograd) 90 – 93

IOP Conf. Series: Earth and Environmental Science **107** (2017) 012061 doi:10.1088/1755-1315/107/1/012061

Monitoring of oil hydrocarbons pollution in the Sea of Japan, based on detection of marker genes in microbial communities

A V Kim[1], L S Buzoleva[1,2], E A Bogatyrenko[1], T I Zemskaya[3] and E V Mamaeva[3]

[1]Far Eastern Federal University, Vladivostok, Sukhanova str., 8, 690091, Russia
[2]Somov Research Institute of Epidemiology and Microbiology, Vladivostok, Sel'skaya str, 1, 690087, Russia
[3]Limnological Institute, Siberian Branch, Russian Academy of Sciences, Irkutsk, Ulan-Batorskaya str, 3, 664033, Russia
e-mail: kim-sandra@mail.ru

Abstract. By means of molecular biology techniques, metabolic potential of microbial communities within the regions of inshore water areas in the Sea of Japan with various anthropogenic load was explored. Presence of functional genes, responsible for oil hydrocarbons destruction, for microbial communities within the regions of inshore water areas in the Sea of Japan was first researched. In total microbial DNA from water mass in the regions with chronic anthropogenic pollution, the genes, responsible for oxidation of broad range of *n*-alkanes and polycyclic aromatic hydrocarbons, were found. Detection of marker genes in the background water area (in the Vostok Bay) was ever indicating ecological deterioration within this territory. Thereby, it was demonstrated, that molecular genetic methods, aimed at marker gene detection in total bacterial DNA from environment objects, proved themselves to be more effective technique for identification of oil hydrocarbons water pollution, in comparison with trivial culturable methods.

1. Introduction

A unique diversity of locally inhabiting organisms belong to the ecosystem of the Sea of Japan, due to its distinctive regional physiographic features. Likewise, a strong anthropogenic impact, caused by population growth, coastal urbanization, industrialization, as well as tourist development in the Russian Far East, are therein observed [1]. Sources of pollution of Far Eastern seas are enterprises of pulp-and-paper, power, and oil-gas industry, housing and communal services, ship-building and ship-repairing enterprises, merchant marine and navy. Traditional higher contents of oil itself and oil products are noticed, as a result of untreated domestic waste in bays, gulfs, port areas, including lay-up bases for vessels in all regions of the Far East, also inside inshore water areas in the Sea of Japan close to Vladivostok [2].

Ecological importance of bacteria acting as petroleum hydrocarbons destructors is very great, because it is known, that a complete destruction of them can not be done by higher organisms. In a coastal zone with permanent pollution with oil and its products, specific communities of heterotrophic microorganisms emerge, able to oxidize wide range of hydrocarbons and products of their transformation [3].

Earlier within the inshore water areas in the Sea of Japan the research of water mass microbial community was via culturing and microscopy techniques conducted, data on enzymatic activity were received, total microorganisms count was performed [4]. Taking into account the need to develop an effective and modern monitoring system for marine environment pollution, by oil hydrocarbons polluted. At first, the research objective was set to carry out the molecular genetic analysis of water mass samples, taken from water areas with various anthropogenic load in the Sea of Japan, as well as to recognize marker genes in microbial communities, responsible for petroleum hydrocarbons destruction.

2. Material and methods

2.1. Researched regions

As the regions for the research water areas are undergoing different anthropogenic stress in the Sea of Japan, these regions were chosen: Zolotoi Rog Bay, the Nakhodka Bay, the Kievka Bay, the Stark Strait, and the Vostok Bay (figure1).

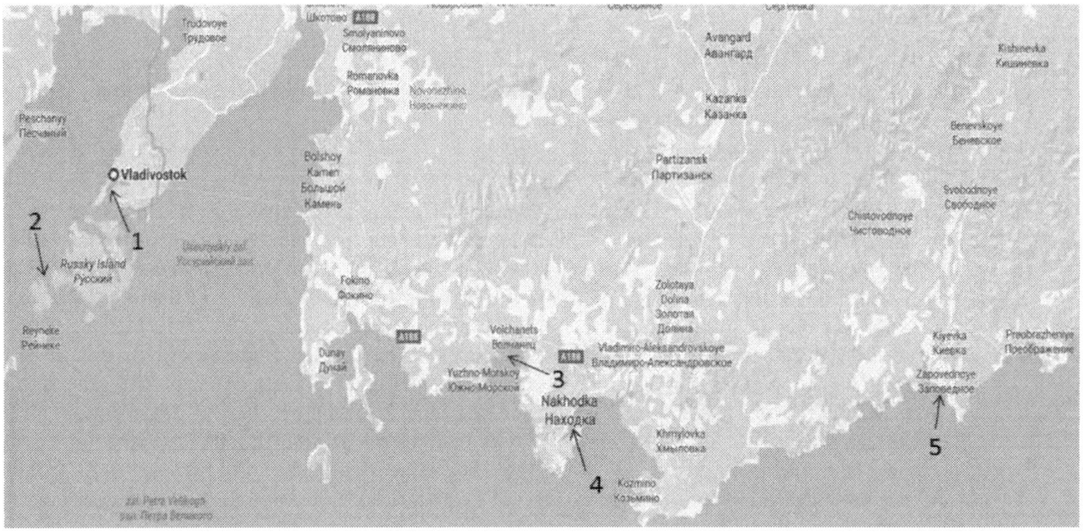

Figure 1. Place of sampling. 1 - Zolotoi Rog Bay; 2 - Stark strait; 3 - Vostok Bay; 4 - Nakhodka Bay; 5 - Kievka Bay.

Zolotoi Rog Bay, with its large industrial and port enterprises present, is considered to be one of the most troublesome districts. The heaviest pollution of chemical, petrolic, and thermal kind is traced in the bay, since treatment systems are not exploited [5].

The seaport, big in terms of cargo turnover in the Russian Far East, is situated along the Nakhodka Bay. The Partizanskaya river in southern Primorye, the second largest by river flow volume after the Razdolnaya river, falls into the head of the bay. Industrial and town public utilities, discharging sewage waters into the bay, vessels and port facilities, also river flows are the main sources of inshore waters pollution in the bay [6].

The Vostok Bay is located in the eastern part of the Peter the Great Gulf, near Nakhodka town. There are the Volchanka river flowing into the Vostok Bay in the northwest, and the Litovka river flowing into the Vostok Bay in the northeast. In 1989 the State Complex Marine Sanctuary was established, the Srednaya Bay is located under its authority [7]. Residents from all over the Far East spend their holiday time on the coast of the bay since July until the early September. There are no enterprises polluting the environment on the coast there.

The Kievka Bay is situated in the eastern part of Primorye coastline, it has latitudinal extension, open from the southern side, and classified as a type of unrestricted water exchange and substantial riverside flow. On the riverside one can find none of housing facilities, and little recreational load happens in summer time [7].

The Stark Strait divides The Russian and Popov islands, strait depth is about 3 till 4.5 meters near the shore. Depth in the northern part of the strait range from 15 till 21 meters, and in the south it is 7.7—14.6 meters. Heavy north-directed current is observed [7]. The area is characterized by minimal anthropogenic load [8].

2.2. Sampling and plating

Samples were taken out of water mass in 5 various spots of each researched area in July 2016. For the purpose of obtaining biomaterial, water samples of at least 5 liter volume were infiltrated through Milliporec filters of 25 mm in diameter (pore diameter – 0.22 mcm).

In order to receive pure bacteria cultures, which are able to biodegrade hydrocarbons, water samples were plated on Voroshilova-Dianova agar medium with addition of crude oil at 1% concentration, and incubated at room temperature during 7 days (24 hours each) [9]. Further work was performed on collection of obtained isolates to reveal functional genes, responsible for oil hydrocarbons degradation.

2.3. Research molecular methods

For extraction of total DNA each filter was minced and supplemented with 1 ml of TE buffer, 0.05 g of polyvinylpyrrolidone, and lysozyme (final concentration at 15 mg/ml), after that the mixture was incubated at 37°C within 1 h, stirring it every 10 minutes. Lysate and sodium dodecyl sulfate (SDS) at 10% concentration were combined to get final solution concentration of 1 %, next incubed during 15 munites at 37 °C temperature. Thereafter 3 successive freezing cycles in liquid nitrogen were started, and defrosting cycles at 56°C (10 min) were continued. The next stage was centrifugation at 13.4 thousand rmp within 15 minutes. Further total DNA separation was made using reagent kit of Lytech (Russia) according manufacturer's inscription from clause No.2.

For DNA separation out of pure culture strains a phenol-chloroform extraction method was applied [10, 11].

While conducting PCR-amplification, specific primers were operating on functional bacterial genes, which are responsible ones for decomposition of broad range of alkanes (*alk)* and polycyclic aromatic hydrocarbons (*nar/nah*) (table 1). PCR of total DNA and pure cultures was performed in 10 mcL amplification kit "AmpliSens" (Russia), containing 5 mcL of Redmix, 0.2 mcL of dNTP, proper primers 0.2 mcL each, and 3.4 mcl of H_2O. Amplifiers BIS (Russia) and «MyCycler» (Bio-Rad Laboratories Inc., USA) were enabled into amplification process.

Table 1. Oligonucleotide primers.

Primer	Sequence 5' – 3'	Reference
	Primers for *alk* genes	
ALK 2F	GAGACAAATCGTCTAAAACGTAA	[12]
ALK 2R	TTGTTATTATTCCAACTATGCTC	
ALK 3F	TCGAGCACATCCGCGGCCACCA	
ALK 3R	CCGTAGTGCTCGACGTAGTT	
Ps.putidaF Alk-B	TGGCCGGCTACTCCGATGATCGGAATCTGG	[13]
Ps. putidaR Alk-B	CGCGTGGTGATCCGAGTGCCGCTGAAGGTG	
	Primers for *nar/nah* genes	
NarAa-1010F	TACCTCGGCGACCTGAAGTTCTA	[14]
NarAa-1611R	AGTTCTCGGCGTCGTCCTGTTCGAA	
NarAb-2392R	GATTGTTGTCTGATCTAGCAGCA	
NarAb-2031F	GCACTCGTCACCGAGGATCTC	
NahAc-Ac1014r	CTCRGGCATGTCTTTTTC	[15]
NahAc-Ac149f	CCCYGGCGACTATGT	

For optimal DNA amplification with primers for *alk* genes (350-550 bp) the selection of conditions for amplification was made. The following programmes were chosen: 94 °C – 3 min (1 cycle), 94 °C – 60 sec, 44 ° C / 54 ° C – 60 sec (for pure cultures DNA/total DNA), 72 °C – 30 sec (30 cycles); 72°C – 3 min.

For optimal DNA amplification with primers for *nar/nah* genes (404-625 bp) also the selection of conditions for amplification was made.

The following programmes were chosen: 95°C – 3 min (1 cycle), 94 °C – 40 sec, 55 ° C – 60 sec, 72 °C – 60 sec (30 cycles); 72°C – 3 min.

Amplification products were divided through electrophoresis in 1 % agarose gel, coloured by ethidium bromide, and visualized in ultraviolet light. PCR-products purification was made through cutting fragments of expected length and gel extraction.

Sanger sequencing was by capillary instrument ABI 3130x1 Genetic Analyser conducted. Primary structure of obtained fragments was in the Center for collective use "Genomics" in Novosibirsk city identificated.

3. Results and discussion

In the research process, the total DNA, separated from bacterial filters, was tested for functional genes, responsible for destruction of *n*-alkanes and polycyclic aromatic hydrocarbons (PAHs). Similar genes were detected in 3 water areas out of 5 (table 2).

Table 2. Identification of the marker genes responsible for decomposition of alkanes and polycyclic aromatic hydrocarbons in total DNA from of water samples.

Area	Marker genes			
	ALKB	NarAa	NarAb	NahAc
Zolotoi Rog Bay	+	-	+	-
Nakhodka Bay	+	-	-	-
Stark Strait	-	-	-	-
Vostok Bay	+	+	-	+
Kievka Bay	-	-	-	-

Genes, determining alkane hydroxylase (*alk*) synthesis, providing degradation of short-chain and long-chain alkanes, were in DNA of microorganisms in the Zolotoi Rog Bay, the Nakhodka Bay, and the Vostok Bay detected. Genes, responsible for PAHs destruction, were in total DNA, isolated from water mass of the Zolotoi Rog Bay and the Vostok Bay, identified. In other samples genes, investigated with the help of chosen primers, were not found.

Identification of genes, responsible for oil hydrocarbons decomposition, in total DNA samples was the evidence of oil contamination with respect to oil-contaminated areas. Marker genes, detected in the Zolotoi Rog Bay and the Nakhodka bay, proved data on chronic pollution by oil products in these bays. Meanwhile, absence of marker genes in samples from background regions (the Kievka Bay and the Stark Strait) indicated the insignificant anthropogenic load within those water areas. Detection of marker genes in the background Vostok Bay, however, showed the environmental deterioration on that territory. One of possible reasons of current situation can be geographical proximity of the Vostok bay and the Nakhodka Bay. Polluted waters from the Nakhodka Bay might be transferred by means of a current to the Vostok Bay, undergoing minimal anthropogenic load.

Besides carrying out the research of total bacterial DNA, we studied metabolic peculiarities for separate bacterial strains, selected from water in specified regions. As a result of the work conducted in agarose medium, containing crude oil, 7 strains of bacteria were obtained: 6 ones from the Nakhodka Bay, and 1 from the Golden Horn Bay. From water samples, taken in other water areas, none of bacteria cultures, able to provide oil oxidation, were separated. It is important to mention, though in total DNA marker genes were in samples from the Vostok Bay determined, we did not succeeded in obtaining pure cultures of bacteria, which can grow in oil-containing medium. This can be testifying to both weak concentration of oil-oxidizing bacteria in territorial water, thus, impossible for us to isolate them as a separate culture, and possible ability for oil degradation in the Vostok Bay for uncultured bacteria.

Molecular genetic studies of obtained bacteria strains demonstrated, that all 7 strains contain 3B-group *alk* genes, responsible for degradation of broad range of *n*-alkanes, and do not contain 3B-group *alk* genes, responsible for PAHs decomposition (table 3). The similarity of bacteria comprising *alk* genes with gram-negative bacteria of *Acinetobacter* genus, which have decoded *alk* genes structure and participate inside various ecosystems in degradation of oil and oil derivatives, was via phylogenetic analysis of producted nucleotide sequences revealed, resembling bacteria of *Micrococcus*, *Bacillus*, and *Corynebacterium* genera too.

Table 3. The analysis of pure cultures strains of the bacteria isolated from the Sea of Japan on the basis of fragments of *alk*B of genes.

Area	Strain №	Length, bp	Similarity, %	Closest sequence of functional genes (NSBI)	Source of isolation of the closest relative
Nakhodka Bay	1NN2a/1	170	99%	CP007437 *Micrococcus luteus*	Collection of strains, Germany
	4N/3	205	95%	CP001628 *Micrococcus luteus*	Collection of strains, Germany
	23N/8	221	98%	DQ288004 Bacterium alkW92	Soil bacterium
			96%	AJ009584 *Acinetobacter calcoaceticus* (*alkM* gene)	Strain cultivated on alkanes (C_{12}, C_{16})
			95%	HG530362 *Acinetobacter guillouiae* (*alkB* gene)	Soil bacterium
	NGIL3/7	220	99%	CP000227 *Bacillus cereus*	Strain from industrial production
	NG1a/10	164	93%	CP007437 *Micrococcus luteus*	Collection of strains, Germany
	17N/11	222	82%	EU853413 *Corynebacterium variabile* (*alkB* gene)	Oil-oxidizing bacteria from coastal surface waters of Xiamen Island
Zolotoi Rog Bay	ZR32/2	220	99%	CP000227 *Bacillus cereus*	Strain from industrial production

For the first time, the presence of microbial functional genes, responsible for oil hydrocarbons destruction, in the inshore water areas in the Sea of Japan was researched. In total, microbial DNA from water mass in the regions with chronic anthropogenic pollution to the genes, responsible for oxidation of broad range of *n*-alkanes and polycyclic aromatic hydrocarbons, were found. Meanwhile, absence of marker genes in samples from background regions (the Kievka Bay and the Stark Strait) indicated the insignificant anthropogenic load within those water areas. Detection of marker genes in the background water area (in the Vostok Bay) was ever indicating ecological deterioration within this territory. Analysis of the data received gave the opportunity to specify molecular genetic techniques, intended for marker genes detection in total bacterial DNA from environment objects, as more effective technique for identification of petroleum hydrocarbons water pollution in comparison with trivial culturable methods.

Acknowledgments
The research was fulfilled with financial support of the Russian Scientific Fund (Agreement No. 14-50-00034)

References
[1] Chenje M and Johnson P 1996 *Water in Southern Africa (*Harare/Maseru: SADC, IUCN, SARDC)
[2] Vashchenko M A 2000 *Russian Journal of Marine Biology* **26** 155 – 66
[3] Mironov O G 2002 *Mor. Ekol. Zh* **1** 56 – 66
[4] Buzoleva LS, Bogatyrenko Ye A, Yes'kova A I, Kim A V, Dolmatova Ye S and Golozubova Yu S 2016 *Hydrobiological journal* **52** 66 – 74
[5] Koryakova M D, Nikitin V M, Zvyagintsev A Yu and Belogurova L S 2002 *Russian Journal of Marine Biology* **28** 138 – 42
[6] Naumov Yu A and Naidenko T H 1997 *Izvestiya TINRO* **122** 524 – 37
[7] 2005 *Lotsiya severo-zapadnogo berega Yaponskogo morya ot reki Tumannaya do mysa Belkina* (Sailing Directions for the Northwestern Coast of the Sea of Japan, from the Mouth of Tumannaya River to Cape Belkin) (St.-Petersburg: GUNiO) p 390
[8] Zhigula L D 2007 *Vologdinskie chteniya* **63** 41–3
[9] Buzoleva L S, Bogatyrenko E A, Repina M A and Belkova N L 2017 *Microbiology* **86** 338 – 45
[10] Marmur J 1961 *Moi. Biol.* **3** 208 – 18
[11] Rochelle PA 1992 *FEMS Microbiol. Lett.* **100** 59 – 66
[12] Sei K, Mori K, Kohno T and Maki H 2003 *Journal of Chemical Engineering of Japan* **36** 1185-93
[13] Whyte L G, Schultz A, van Beilen J B, Luz A P, Pellizari V, Labbe D and Greer CW 2002 *FEMS Microbiology Ecology* **41** 141 – 50
[14] Andreoni V, Bernasconi S, Colombo M, van Beilen JB and Cavalca L 2000 *Appl. Environ. Microbiol.* **2** 572 – 7
[15] Ferrero M, Llobet Brossa E and Lalucat J 2002 *Appl. Environ. Microbiol.* **68** 957 – 62

Atmospheric emission of $^{137}Cs_{82}$ from Beloyarsk nuclear power plant

G A Kolotkov

Institute of Atmospheric Optics SB RAS, Tomsk, Academician Zuev square, 1, 634021, Russia
e-mail: kolotkov@iao.ru

Abstract. Citing Beloyarsk nuclear power plant (Russia) as an example, the problem of remote detection of radioactivity in the atmospheric pollution is examined. The comparative analysis of injected radionuclides into the atmosphere from the nuclear power plant with advanced fast neutron reactor is carried out. The main radionuclides throw out into the atmosphere from the nuclear power plant are beta-radionuclides. The secondary and tertiary spectra of beta-electrons decay for artificial radionuclide $^{137}Cs_{82}$ is calculated, using Spencer-Fano's equation. The averaged parameters of initial beta – electrons generated by $^{137}Cs_{82}$ decay in the atmosphere is calculated.

1. Introduction

The actuality of this research concludes in accumulation of the huge amount of radioactive materials, only spent fuel more than 270000 tons, safety time defined by thousand years. Time decay of radionuclides and their mobility in the environment allow a component of radioactive waste to distribute widely in the natural food chain [1, 2]. Existing decisions could be divided on; building a new nuclear waste storage or recycle recovered fuel. For example, countries that do not possess the latest nuclear technologies are planning to build additional radioactive waste storage facilities, while in Russia this problem is solved by designing and launching nuclear power plants (NPPs) with fast neutron reactors (BN reactor). To implement the agreement on the disposition of plutonium, signed in 2000 between Russia and the United States, Russia is developing and commissioning fast neutron reactors for burning MOX fuel. The work of such NPPs is highly expected since they will increase the efficiency of the use of uranium, ^{238}U and ^{232}Th may be involved in the fuel cycle, which in nature is much larger than ^{235}U, the main fuel for thermal neutron reactors. The nuclear reactors work with great expectations on fast neutrons, as it would: increase the efficiency of the use of uranium, ^{238}U and ^{232}Th stocks may be involved in the fuel cycle, which of nature is much greater than ^{235}U - the main fuel for thermal reactors. Including can be used and so-called "depleted uranium" remaining after enrichment of nuclear fuel ^{235}U. It is interesting and relevant to consider the gas-aerosol radioactive emissions and theoretically justify the possibility of remote detection of increased radioactivity in air emissions from the Beloyarsk power plant with nuclear reactor BN (BAES) [3, 4].

2. Analysis of averaged gasoaerosol emission in the atmosphere by BAES working in the normal mode

Beloyarsk nuclear power plant is located in 42 km east of Yekaterinburg city and 3 km north of Zarechny town. In the event of hypothetic accident situation, it could be thrown from the tube out a bulk of radioactivity in the atmosphere and the radioactivity composition of which are not known. The purpose of the given work was the estimate of electron density formed in the emission plume from Beloyarsk nuclear power plant. Despite the main part of radioactivity in an emission plume are β-active radionuclides named in the paper radioactive inert gases (RIG) we consider separate isotope ^{137}Cs, ^{14}C, ^{89}Sr, ^{90}Sr from a plume of Beloyarsk nuclear power plant. In the example of calculation of electron density in the emission plume was chosen ^{137}Cs.

It is necessary to note that radioactive inert gases are Ar, Kr, Xe and etc. consists in an emission plume from enterprises of nuclear fuel cycle constantly. The nuanced analysis of radionuclides content in emission plume from enterprises of nuclear fuel cycle consists in [5].

The amount of radioactivity and the composition of radionuclides in emissions depend on the type of reactor. Radioactive noble gases (RBG) are manifested through the radioactive decay of products in fuel. Moreover, RBGs can appear as a result of neutron activation. Thus, ^{41}Ar can also be contained in the air. The main elements contained in the emissions of nuclear power plants with a BN reactor: ^{24}Na (33%), ^{137}Cs (33%), ^{134}Cs (16%). Some radionuclides with a small half-life are not included in the table 1.

Table 1. Annual estimation of the atmospheric emissions from various nuclear reactors.

Radionuclide	BWR, Bq·y^{-1} [6]	PWR, Bq·y^{-1} [6]	FBR, Bq·y^{-1} [7]	BAES, Bq·y^{-1} [8-10]
^{41}Ar	$9.25\cdot10^{11}$	$9.25\cdot10^{11}$	–	–
^{83m}Kr	$< 3.7\cdot10^{10}$	$3.7\cdot10^{10}$	–	–
^{85m}Kr	$5.55\cdot10^{12}$	$5.2\cdot10^{11}$	–	–
^{85}Kr	$1.073\cdot10^{13}$	$1.739\cdot10^{13}$	–	–
^{87}Kr	$7.4\cdot10^{12}$	$1.11\cdot10^{11}$	–	–
^{88}Kr	$8.88\cdot10^{12}$	$8.51\cdot10^{11}$	–	–
^{131m}Xe	$6.66\cdot10^{11}$	$3.034\cdot10^{12}$	–	–
^{133m}Xe	$< 3.7\cdot10^{10}$	$4.44\cdot10^{12}$	–	–
^{133}Xe	$1.184\cdot10^{14}$	$4.44\cdot10^{14}$	–	–
^{135m}Xe	$2.738\cdot10^{13}$	$< 3.7\cdot10^{10}$	–	–
^{135}Xe	$4.07\cdot10^{13}$	$3.182\cdot10^{12}$	–	–
^{138}Xe	$5.18\cdot10^{13}$	$< 3.7\cdot10^{10}$	–	–
^{131}I	$1.11\cdot10^{10}$	$9.25\cdot10^{8}$	$9\cdot10^{2}$	–
^{133}I	$9.07\cdot10^{10}$	$8.51\cdot10^{8}$	–	–
^{14}C	$3.515\cdot10^{11}$	$2.96\cdot10^{11}$	$1.2\cdot10^{5}$	$7.72\cdot10^{11}$
^{3}H	$1.591\cdot10^{12}$	$4.07\cdot10^{13}$	$9.7\cdot10^{7}$	
$^{89,90}Sr$	–	–	–	$3.28\cdot10^{6}$
^{60}Co	–	–	–	$4.04\cdot10^{6}$
^{137}Cs	–	–	–	$9.2\cdot10^{6} + 3.94\cdot10^{7}$
RIG	–	–	–	$4.39\cdot10^{12} + 3.14\cdot10^{14}$
Total activity	$2.7\cdot10^{14}$	$5.2\cdot10^{14}$	$1.5\cdot10^{8}$	$3.3\cdot10^{14}$

The Symbol "–" doesn't mean that aforementioned radionuclides are not completely emitted from a nuclear reactor with a fast breeder reactor, which means that we do not have sufficient information about radionuclide composition of the emissions. Main radionuclides release in the atmosphere by a nuclear power plant in a normal working regime. Taking into account the summary radioactive noble gases activity and the proven ability to remotely detect an increased concentration of radioactivity in the emissions of nuclear power plants and RCP, we can conclude that existing radiometric systems are able to detect emissions from a nuclear power plant with a BN reactor.

3. Calculation of spectrum radionuclide $^{137}Cs_{82}$

Table 1 data shows that significant proportion of emitted radionuclides are radioactive inert gases (RIG): Ar, Kr, Xe. In this paper, we will consider only ^{137}Cs because of its radiological hazard. The probability of beta-electron appearance in the result of decay radionuclide $^{137}Cs_{82}$ is shown below in the table 2.

Table 2. Spectrum of beta-decay radionuclide $^{137}Cs_{82}$ [11].

E, keV	$N_\beta(E)$	E, keV	$N_\beta(E)$	E, keV	$N_\beta(E)$
0	3.41^{-3}	391	$9.33\cdot10^{-4}$	782	$8.78\cdot10^{-6}$
20	3.37^{-3}	411	$7.30\cdot10^{-4}$	802	$7.39\cdot10^{-6}$
39	3.27^{-3}	430	$5.36\cdot10^{-4}$	821	$6.15\cdot10^{-6}$
59	3.18^{-3}	450	$3.56\cdot10^{-4}$	841	$5.06\cdot10^{-6}$
78	3.09^{-3}	469	$2.06\cdot10^{-4}$	860	$4.11\cdot10^{-6}$
98	3.00^{-3}	489	$9.77\cdot10^{-5}$	880	$3.30\cdot10^{-6}$
117	2.92^{-3}	508	$4.73\cdot10^{-5}$	899	$2.60\cdot10^{-6}$
137	2.83^{-3}	528	$4.25\cdot10^{-6}$	919	$2.01\cdot10^{-6}$
156	2.74^{-3}	547	$3.89\cdot10^{-6}$	939	$1.53\cdot10^{-6}$
176	2.65^{-3}	567	$3.54\cdot10^{-5}$	958	$1.14\cdot10^{-6}$
196	2.54^{-3}	587	$3.22\cdot10^{-5}$	978	$8.22\cdot10^{-7}$
215	2.43^{-3}	606	$2.90\cdot10^{-5}$	997	$5.78\cdot10^{-7}$
235	2.31^{-3}	626	$2.61\cdot10^{-5}$	1017	$3.92\cdot10^{-7}$
254	2.18^{-3}	645	$2.33\cdot10^{-5}$	1036	$2.56\cdot10^{-7}$
274	2.04^{-3}	665	$2.07\cdot10^{-5}$	1056	$1.59\cdot10^{-7}$
293	1.89^{-3}	684	$1.83\cdot10^{-5}$	1075	$9.30\cdot10^{-7}$
313	1.72^{-3}	704	$1.60\cdot10^{-5}$	1095	$5.04\cdot10^{-8}$
332	1.53^{-3}	723	$1.40\cdot10^{-5}$	1115	$2.44\cdot10^{-8}$
352	1.34^{-3}	743	$1.21\cdot10^{-5}$	1134	$9.67\cdot10^{-9}$
372	1.14^{-3}	763	$1.03\cdot10^{-5}$	1154	$2.28\cdot10^{-9}$

To calculate the averaged parameters of beta decay electrons and thermalized electrons we use Spencer-Fano's equation (1, 2):

$$\Sigma(E)\Phi(E)-\int_{E}^{E_0}\Sigma_S(E'\to E)\Phi(E')=S(E) \tag{1}$$

where Σ, Σ_S are the integral and differential energy transmission macroscopic cross section, derived by multiplying of appropriate microscopic cross section on atomic concentration in a homogeneous medium:

$$\Sigma_S(E'\to E)=\frac{d\sigma}{d\varepsilon}N \tag{2}$$

where $\varepsilon = E'-E$ is the energy lost by a particle in one act of interaction; $S(E)$ - the spectrum of beta, el·cm^{-3}·s; $\Phi(E)$ - the electron-flux density, having energy E (decision function), el·cm^{-2}·s.

The solution (1) was carried out with use of the numerical method on the following recurrence relations (3):

$$\Phi_0=\frac{S_0}{\Sigma_0},\quad \Phi_n=\frac{\left[S_n+\Delta E\left(\sum_{i=1}^{n-1}(\Sigma_{in}\Phi_i)+\frac{1}{2}\Sigma_{0n}\Phi_0\right)\right]}{\Sigma_n} \tag{3}$$

where n changes from 1 up to M; M is the amount of intervals in width ΔE on which the area of energy change $(E_{\min}, E_0)$ is broken. Energy E_0 corresponds to the peak energy of the beta –electrons spectrum; $E_{\min}$ is the minimal energy, when particle energy loss is missing (it insufficiently even for excitation of the first level of atoms). The coefficient n designates the left-hand boundary of the energy interval. In our case Φ_0 is the density of the electron stream, having energy E_0; S_0 is the density of electrons generation at the beta-decay, having energy E_0; Σ_0 is the macroscopic integral cross section of the energy transfer, providing transition of an electron from an energy interval with the left-hand boundary $E_2 = E_0 - 2\Delta E$ (and Σ_{in} at $i = n$ are equal 0).

Major factor effecting components of the air in the emission plume are high energy electrons formed in the result of beta-decay radionuclides. For the atmosphere $I \approx I_N\delta_N + I_O\delta_O = 95.5$ eV, where δ_N and δ_O are the parts of nitrogen and oxygen atoms in the air, respectively. Dissociation energies of molecules are for N_2 is 9.78 eV, for O_2 of 5.12 eV that is on the order less than average ionization potential of the air. Ionization potential of oxygen molecule is I_{O_2} =12.15 eV, nitrogen molecule is I_{N_2} =56 eV, oxygen atom is I_O =13.57 eV, nitrogen atom is I_N =14.6 eV. Concentration of oxygen atoms is of $11.4 \cdot 10^{18}$ atoms·cm^{-3}, nitrogen atoms is of $42.6 \cdot 10^{18}$ atoms·cm^{-3}. In table 3 averaged parameters describing the thermalization of fast electrons, formed by beta-decay are shown.

Table 3. The averaged parameters describing thermalization of electrons of beta-decay radionuclide $^{137}Cs_{82}$.

Radionuclide $^{137}Cs_{82}$	Initial	Secondary	Tertiary
The mean free path of electrons, cm	0.014	$3.592 \cdot 10^{-3}$	$1.703 \cdot 10^{-3}$
Time of run, between acts of energy transfer, s	$7.752 \cdot 10^{-13}$	$1.995 \cdot 10^{-13}$	$9.459 \cdot 10^{-14}$
Frequency of energy transfer, 1·s^{-1}	$1.29 \cdot 10^{12}$	$5.012 \cdot 10^{12}$	$1.057 \cdot 10^{13}$
The average kinetic energy of the electrons formed, keV	40.841	13.613	4.319

4. Results and discussion

Below presents the results of resolve of Spencer-Fano's equation.

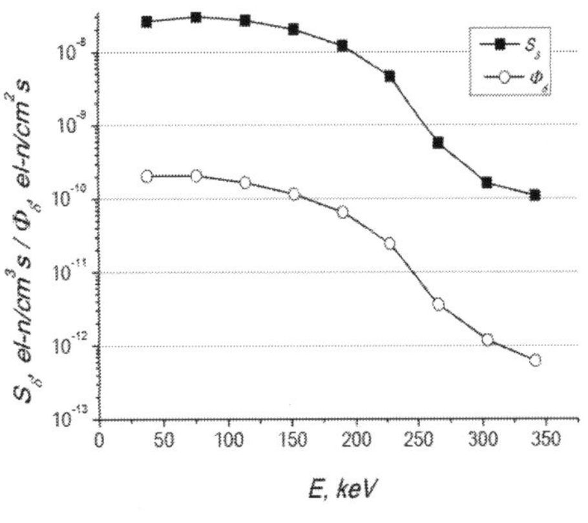

Figure 1. Spectrum of secondary electron of beta decay (S_δ) and flux density of beta secondary electron (Φ_δ) .

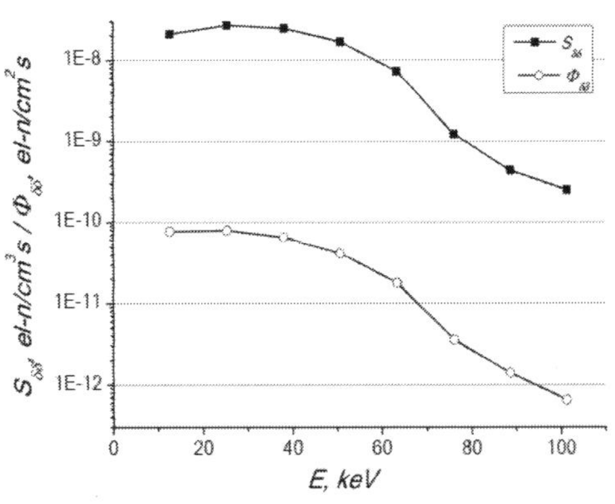

Figure 2. Spectrum of tertiary electron of beta decay ($S_{\delta\delta}$) and flux density of beta tertiary electron ($\Phi_{\delta\delta}$).

The averaged parameters of initial beta – electrons generated by $^{137}Cs_{82}$ decay in the atmosphere and results on the figure 1 and 2 show that averaged energy of primary, secondary and tertiary electrons of ^{137}Cs decay are exceeding considerably potential of the air ionization. In such manner, the process of ionization will be continued. Further research will be continued to solve the problem remote monitoring of nuclear power plant with fast neutron reactor.

Acknowledgement
The reported study was funded by RFBR, according to the research project No. 16-38-60115 mol_a_dk.

References

[1] Eckert V 2015 Moment of truth nears for nuclear waste time bomb *Reuters* *www.reuters.com/article/2015/06/19/us-nuclear-waste-idUSKBN0OZ1A020150619*
[2] Bridges O and Bridges J W 1995 *J. Radiol. Prot.* **15** 223
[3] Mitenkov F 2005 *Nauka I Zhizn* **3** 27 – 40
[4] Rybin A A and Momot O A 2017 *J. Phys.: Conf. Ser.* **781** 012013
[5] Jones A et al 2013 *J. Radiol. Prot.* **33** 163
[6] Kolotkov G A and Penin S T 2013 *Journal of Environmental Radioactivity* **115** 69 – 72
[7] Tykva R and Berg D 2004 *Man-Made and Natural Radioactivity in Environmental Pollution and Radiochronology* (Netherlands: Springer) **7** p 416
[8] 2009 Radiation situation on the territory of Russia and neighboring countries in 2008 year *Yearbook* (Obninsk RPA Typhoon) p 297
[9] 2016 Radiation situation on the territory of Russia and neighboring countries in 2015 year *Yearbook* (Obninsk RPA Typhoon) p 348
[10] 2017 Radiation situation on the territory of Russia and neighboring countries in 2016 year *Yearbook* (Obninsk RPA Typhoon) p 398
[11] Rubcov P M 1989 *Beta and antineutrino decay of radionuclides Handbook* (Moscow Energoatomizdat) p 800

IOP Conf. Series: Earth and Environmental Science **107** (2017) 012063 doi:10.1088/1755-1315/107/1/012063

Technogenic and natural radionuclides in the bottom sediments of the Sea of Azov: regularities of distribution and application to the study of pollutants accumulation chronology

A N Kuznetsov, Yu A Fedorov and V M Yaroslavtsev

Southern Federal University, Rostov-on-Don, Bolshaya Sadovaya, 105/42, 344090, Russia

e-mail: ankuznecov@sfedu.ru, fedorov@sfedu.ru, yaroslavtsev.vlad@gmail.com

Abstract. The study of pollutants vertical distribution in seabed sediments is of high interest as they conserve the information on the chronology of pollution level in the past. In the present paper, the results of layer by layer study of Cs-137, Am-241, Pb-210 specific activities as well as concentrations of petroleum components, lead and mercury in 48 sediment cores of the Sea of Azov, the Don River and the Kuban River are examined. In most sediment cores, two peaks of Cs-137 and Am-241 are detected. The upper of them was formed due to the Chernobyl accident in 1986 and the other is related to the global nuclear fallout of 1960s. The specific activity of naturally occurring atmospheric Lead-210 decreases exponentially with the sediment core depth. However, it is influenced by fluvial run-off, coastal erosion, Radium-226 and Radon-222 decay. The data on the radionuclides distribution in the seabed sediments is used to date them. According to the results of dating, most of petroleum components, lead and mercury quantities are concentrated in the upper sediment layer formed in the last 50 to 70 years i.e. in the period of the most important anthropogenic pressure.

1. Introduction

The study of pollutants distribution in the sediment cores is of particular interest as they conserve the information on the chronology of pollution level in the past and so may be considered as the "memory" of aquatic ecosystems. Therefore, the dating of different sediment layers is of high importance. For this purpose, the method of radioactive tracers such as ^{137}Cs and ^{210}Pb is largely applied [1-4].

Cesium-137 and Americium-241 are technogenic radionuclides, their presence in the environment and formation of specific activity peaks in sediment cores are related to the above-ground tests of thermonuclear weapons in 1950s and 1960s and to the accident at the Chernobyl Nuclear Power Plant in 1986. Unlike ^{137}Cs, ^{241}Am is formed directly in bottom sediments due to the decay of ^{241}Pu. So, over time the specific activity of ^{241}Am increases progressively while the specific activity of ^{137}Cs decreases. Lead-210 is a natural radionuclide. It is permanently generated in the atmosphere as a result of decay of radioactive gas Radon-222, and then it precipitates on the earth surface. In the sediment cores, the specific activity of Lead-210 decreases exponentially until the secular equilibrium with the parent radionuclide

Radium-226. As the half-life of Pb-210 is not very long (22.3 years), this method is suitable only for dating sediments formed during the last century [3, 4].

2. Materials and methods

Earlier in the uppermost layer of bottom sediments of the Sea of Azov and main rivers of its drainage basin the content and composition of technogenic material and the concentrations of different pollutants were determined [5]. The results of this study were very important; however, they are not sufficient for the localization of the sediment layer of anthropogenic impact and analysis of its formation chronology. The present paper considers the results of complex many-years (from 1995 to 2015) studies carried out in the Sea of Azov, the Strait of Kerch, the Don River and the Kuban River and based on the uniform common methodology [6-11]. During the period of this research, five scientific expeditions were carried out in summer and autumn, 48 sediment cores from 20 to 100 cm long were extracted by a 1 m long gravity corer and split longitudinally into 241 samples. The sampling point's locations are depicted in figure 1.

In the samples the concentrations of principal petroleum components (aliphatic, mono- and di-aromatic hydrocarbons (HC), polycyclic aromatic hydrocarbons (PAH) and asphaltic components (AC): resins and asphaltenes) as well as total lead and mercury were determined. The choice of the ingredients was not casual. For the considered region, petroleum components, lead and mercury are the most common and typical contaminants due to the functioning of large industrial centers of Russia and Ukraine, high concentration of population, municipal and transport facilities, port complexes and intense navigation. By now, the volumes of petroleum products and other cargos shipped through the Azov - Don waterway have increased 2 to 3 times compared with the late 1990s. This undoubtedly augments the risk of emergencies related to the oil spills such as the series of ship wreckages in November 2007 during the heavy storm in the Strait of Kerch accompanied by a spill of 1.3 thousand tons of heavy fuel oil. In order to date the studied samples, the specific activities of radionuclides ^{137}Cs, ^{241}Am and ^{210}Pb were measured there. In total, more than thousand determinations of the abovementioned ingredients were made up. The petroleum components were extracted and studied with the use of thin-layer chromatography, infrared and luminescence photometry [8, 9]. The presence of hydrocarbons of contemporary biological origin was also identified. The heavy metals were studied by the atomic absorption spectroscopy [10]. The specific activities of the radionuclides were measured by the method of direct gamma spectrometry [11].

3. Results and discussion

According to the results of the study, in the uppermost sediment layer the maximum specific activity of Cesium-137 (103.3 $Bq \cdot kg^{-1}$ of dry weight (d.w.)) was detected in the southern part of the Sea of Azov adjacent to the Strait of Kerch. This fact may be accounted for the isotope sorption by silty sediments from the near-bottom higher salinity waters entering from the Black Sea. The minimum values occurred in the mouth areas of the Don River and the Kuban River. In general, the sediment cores extracted in central and southeastern parts of the Sea of Azov were characterized by higher specific activities of Cesium-137 and Americium-241 than ones taken in the Golf of Taganrog. A well-defined upper peak of ^{137}Cs specific activity related to the Chernobyl accident was found in all the studied sediment cores. In the deeper parts of the Sea of Azov and of the Gulf of Taganrog, in the hydrodynamic conditions favorable for the sediment precipitation and accumulation, this peak was detected at 5-10 cm sediment core depths (51-116 $Bq \cdot kg^{-1}$). In the shallow eastern part of the Gulf of Taganrog the maximum ^{137}Cs specific activity was fixed in the superficial layer (35-79 $Bq \cdot kg^{-1}$). In the delta and in the near-delta part of the Don River the peak is located at 20-40 cm depth (14–43 $Bq \cdot kg^{-1}$) suggesting higher sedimentation rates. In most sediment cores the second peak of ^{137}Cs activity was also found in deeper layers. It may be associated with global fallout from aboveground testing of thermonuclear weapons in 1950s and 1960s. This peak is not as distinct as the first one due to radioactive decay and probably lateral and vertical migration.

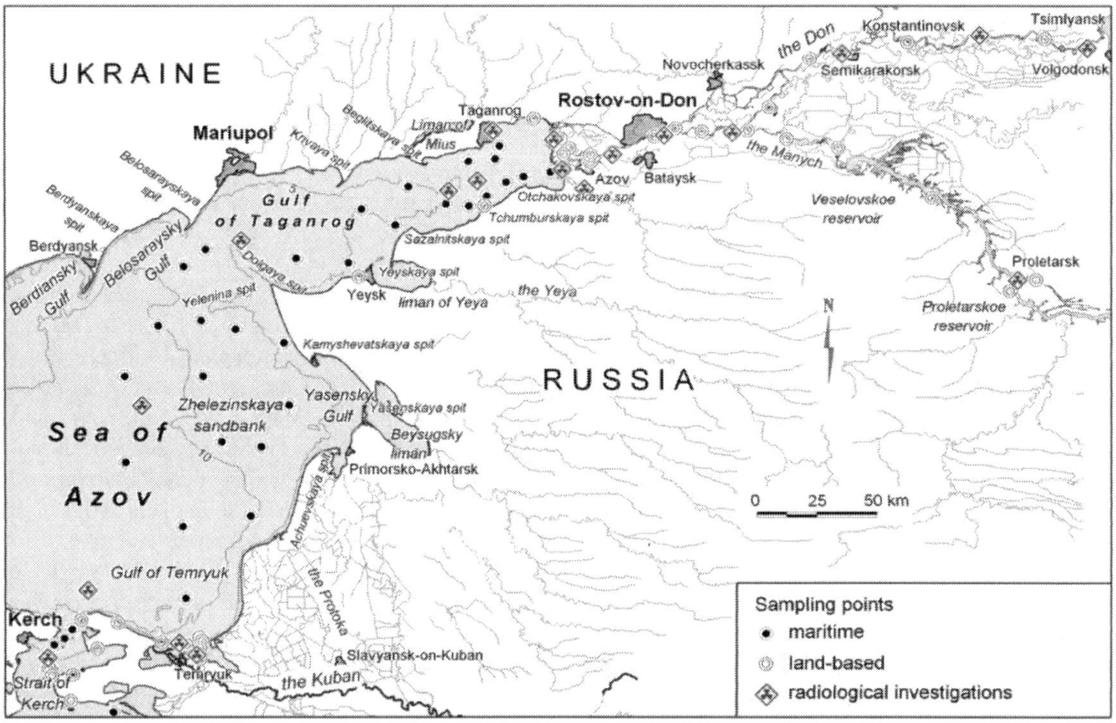

Figure 1. Sampling points in the Sea of Azov and in the Lower Don River basin.

The maximum specific activity of Americium-241 (18.0 Bq·kg^{-1}) was detected in the deeper part of the Sea of Azov. Significant activities of Am-241 are detected in most sediment cores. In some of them two peaks are found. Obviously, the upper peak corresponds to the Chernobyl fallout and the other located in a deeper sediment core layer corresponds to the global radioactive fallout of 1960s. Am-241 is a decay product of Pu-241 (half-life is about 14 years). It accumulates over time since the fallout of plutonium isotopes, as its half-life is 30 times longer than the half-life of Pu-241. The level of ^{241}Am pollution of bottom sediments is relatively high due to ^{241}Pu significant contribution to the global fallout and its almost total decay into ^{241}Am. Vertical distribution of Am-241 specific activity in the sediment cores is similar to that one of Cs-137. Conversely, its deeper peak corresponding to the period of global radioactive fallout of 1960-s is much more distinct and often even more important than the peak corresponding to the accident in Chernobyl.

The specific activity of Lead-210 decreases with the sediment core depth from 240–727 Bq·kg^{-1}in the superficial layer to 179–288 Bq·kg^{-1}at 50 cm depth. However, regular stirring-up of shallow sea sediments and multitude of terrigenous sources of Lead-210 complicate the use of this data for the dating purposes. Exponential character of decrease of the ^{210}Pb specific activity is not often evident because of distorting influence of the isotope import with the fluvial run-off, coastal erosion material, and in consequence of the ^{226}Ra and ^{222}Rn decay. In order to eliminate the influence of these factors, the method of multiple regression analysis was applied to the balance equation (1) that includes all the above-mentioned sources of ^{210}Pb isotope [9].

Secular equilibrium activity of ^{226}Ra — Secular equilibrium activity of ^{222}Rn

$$A(^{210}Pb)=A(^{210}Pb_a)e^{-\frac{\lambda x}{s}}+A(^{210}Pb_t)e^{-\frac{\lambda x}{s}}+\frac{A(^{226}Ra)}{\lambda}(1-e^{-\frac{\lambda x}{s}})+\frac{A(^{222}Rn)}{\lambda}(1-e^{-\frac{\lambda x}{s}}) \quad (1)$$

Total specific activity of ^{210}Pb in each sediment layer; Specific activity of atmospheric ^{210}Pb; Specific activity of terrigenous ^{210}Pb; Specific activity of ^{210}Pb generated in bottom sediments in consequence of ^{226}Ra decay; Specific activity of ^{210}Pb generated in bottom sediments in consequence of ^{222}Rn decay

In this equation, λ is the Pb-210 decay constant (0.0311 year^{-1}); x is the depth of considered layers in the sediment cores (cm); s is the average sedimentation rate (cm· year^{-1}).

According to the results of radiological analysis, the highest sedimentation rates are in the delta and near-delta part of the Don River (5 to 14 mm of wet matter per year). The lowest rates are in the shallow eastern part of the Gulf of Taganrog (1 to 2 mm of wet matter per year) in spite of considerable sediment runoff of the Don River. That is because of low depths (2 to 5 m) and high hydrodynamic activity of water column favoring regular stirring-up and carrying out of sediments. In the deeper central and southern parts of the Sea of Azov, the sedimentation rates rise up to 4 mm per year. The results of the sediment precipitation rates determination by the method of sediment traps prove these estimates for the eastern and northeastern parts of the Sea of Azov (2 to 3 mm of wet matter per year). However, in the shallow Gulf of Taganrog the rates of observed sediment precipitation are much higher than the sedimentation rates determined by the radiological methods. This may be accounted for the regular remobilization of sediments under storm conditions.

Now let us consider vertical distribution of common pollutants in the studied sediment cores. The total concentrations of petroleum components in the uppermost sediment layer vary from 0.05 to 1.21 mg g^{-1} d w. The highest values are detected in the Don River and the Kuban River deltas (on average 0.51 and 0.47 mg·g^{-1} d w respectively), in the southeastern and central parts of the Sea of Azov (on average 0.33 and 0.22 mg g^{-1} d w respectively).

Most of the petroleum components quantity is concentrated in the upper sediment layers formed in the last 50 to 70 years, i.e. in the period of high anthropogenic pressure. In the Gulf of Taganrog, the thickness of this layer is about 10 to 15 cm, in the dipper sectors of the sea it raises up to 20 cm and even reaches 30 to 50 cm in the lower parts and deltas of the Don River and the Kuban River. These differences are accounted not only for the pollution chronology but also and mostly for the different sedimentation rates. In the Don River and the Kuban River sediments, the resins and asphaltenes being the conservative indicators of chronic oil pollution prevail over hydrocarbons. On the contrary, in the Gulf of Taganrog and in the principal part of the Sea of Azov, except for the impact zones surrounding industrial centers and port terminals, the hydrocarbons dominate. A huge part of them consists of contemporary metabolic products of aquatic organisms. In most sediment, cores taken in the Gulf of Taganrog and the Gulf of Temryuk the highest concentrations of oil components are detected at 5 to 10 cm sediment core depths. This sediment layer is located under the peak of ^{137}Cs specific activity related to the Chernobyl accident and may be dated back to the period from 1955 to 1985. In the Don River sediments, such layer is also found, but in the reason of higher sedimentation rates its thickness rises up to 30 cm. In the central part of the Sea of Azov, the highest concentrations are detected in the uppermost sediment layer that has been formed during the last 15 years.

The concentrations of lead in the sediment cores vary from 12 to 43 μg g^{-1} d w, on average 18 μg g^{-1}. It is found out that in the Gulf of Taganrog, lead reacts actively with organic matter while in the principal part of the Sea of Azov their concentrations do not correlate. The highest concentrations of lead are detected in the upper sediment layers up to 20 cm sediment core depth. In the deeper layers, they usually droppe rapidly. A direct correlation (with coefficients ranging from 0.30 to 0.94) is found between the radioactive Lead-210 specific activity and the total lead concentration. This means that the decay of ^{222}Rn isn't the only source of ^{210}Pb precipitating from the atmosphere. Lead-210 is also present in the coal dust and coal ash transported with air masses and surface run-off.

The concentrations of mercury in the sediment cores varies from 0.025 to 0.28 μg·g^{-1} d w, on average 0.067 μg g^{-1}. In most cases, they decrease with sediment core depth. However, in some sediment cores, as it is also found out for the petroleum components, the maximum values were fixed in the deeper sediment layers formed between 1955 and 1985. In all the sediment cores, the content of mercury closely correlates with the content of organic carbon.

4. Conclusion

Therefore, the comparative analysis of the data on the vertical distribution of petroleum components, heavy metals, technogenic radionuclides Cs-137 and Am-241 and natural radionuclide Pb-210 in the sediment cores of the Sea of Azov, the Don River and the Kuban River allowed locating the layer of high anthropogenic impact and dating it. This superficial sediment layer contains the most of petroleum components, lead and mercury quantities. Its thickness varies from 15 to 50 cm. According to the results of radiological dating, these sediments were formed in the last 50 to 70 years, i.e. in the period of high anthropogenic pressure. Moreover, the most polluted sediment layers were formed in 1970s and 1980s.

Acknowledgement

The study was supported by the Russian Foundation for Basic Research (project no. 15-05-004977) and Russian Science Foundation (project no.17-17-01229).

References

[1] Pennington W, Cambray R S and Fisher E M 1973 Observation on lake sediments using fallout ^{137}Cs as a tracer *Nature* **242** 324 – 6

[2] Delaune R D, Patrick W H, Jr and Buresh R J 1978 Sedimentation rates determined by ^{137}Cs dating in a rapidly accreting salt marsh *Nature* **275** 532 – 3

[3] Faure G 1987 *Principles of Isotope Geology* (USA: John Wiley and Sons) p 589

[4] Kuptsov V M 1986 *Absolute geochronology of ocean and sea bottom sediments* (Moscow: Nauka) (In Russian)

[5] Khrustalev Yu P 1998 Geochemistry of the Sea of Azov Bottom Sediments *Geoecological investigations and subsurface mineral protection Scientific and technical information* **4** (Moscow: Geoinformmark) 3 – 14

[6] Fedorov Yu A et al 2007 Multidisciplinary Ecosystem Studies in the Russian Part of the Sea of Azov *Oceanology* **47** 294 – 7

[7] Fedorov Yu A, Dotsenko I V, Kuznetsov A N, Belov A A and Loginov E A 2009 Regularities of C_{org} Distribution in Bottom Sediments of the Russian Part of the Sea of Azov *Oceanology* **49** 211 – 17

[8] Fedorov Y A, Stradomskaya A G and Kuznetsov A N 2006 Regularities in the Transformation of Oil Pollution in Watercourses Based on Long-Term Observational Data *Water Resources* **33** 300 – 9

[9] Kuznetsov A N and Fedorov Yu A 2014 Oil Components in the Mouth Area of the Don R and in the Sea of Azov: Results of Many-Year Studies *Water Resources* **41** 55 – 64
[10] Fedorov Yu A, Khansivarova N M and Predeina L M 2003 Peculiarities of Mercury and Lead Distribution in the Bottom Sediments of the Gulf of Taganrog and the Sea of Azov *Water management* **5** 51 – 58
[11] Fedorov Yu A, Kuznetsov A N and Trofimov M E 2008 Sedimentation Rates in the Sea of Azov Inferred from Cs-137 and Am-241 Specific Activity *Doklady Earth Sciences* **423** 1333 – 34

The Issue of transporting pollutants with atmospheric precipitation

A Madibekov and L Kogutenko

Institute of Geography, Almaty, Kazakhstan
e-mail: madibekov@mail.ru

Abstract. A research of the pollution of atmospheric precipitation was conducted. The database of the chemical composition of atmospheric precipitation made by National Monitoring Network of the Republic of Kazakhstan for the period from 2000s to 2011 was generalized and analyzed. The research area covers the big territory of Ile-Balkhash river basin in the South-East Kazakhstan. The research shows that pollutants can be transported over long distances with atmospheric precipitation. Based on the results of the air masses tracking we identified that the main sources of emissions is located in the city of Balkhash.

1. Introduction

At the present stage of development of the Republic of Kazakhstan the positive trend of industry, energy, production of nonferrous and ferrous metals, increasing of resources consuming is observed. At the same time increasing of pollutant emissions to the atmosphere was detected. For example, emissions have increased by several times in comparison with the level of 1950s which leads to heavy pollution of the atmosphere, surface water and soil [1, 2].

In recent years another source of environmental pollution has appeared - atmospheric precipitation. This source of pollutants is not under control and it is quite difficult to take it into account. However, evaluation of its impact is required for analysis, forecast, regulation and quality management of natural environments. In this regard, the assessment of the influence of atmospheric precipitation on the territory of the Ile-Balkhash basin is relevant and has important economic significance.

Many studies have been devoted to the study of transport and deposition of pollutants in the atmosphere [3-10]. A number of works, which deal with spatial and temporal features of the distribution of pollutants in the atmospheric precipitation, is devoted to the research of atmospheric pollution, as one of the individual components of air pollution in the territory of the Republic of Kazakhstan (RK) [11, 12]. The distribution of atmospheric pollution has mostly uneven nature. The concentrations of heavy metals (HM) are located in industrialized advanced regions or in urbanized areas, in addition, and they have a certain downward trend [13-16]. Despite the reduction in the content of HM in atmospheric precipitation over the last decade, the level of air pollution still remains high, and the amount of research is inadequate. At the same time, the research of the chemical composition has other important value. Precipitation contribute to the cleansing of the atmosphere by washing various impurities out of the atmosphere, and their chemical composition is an integral characteristic of pollution of a certain layer of the atmosphere which they have been passed through. In addition, all substances, which have been washed out from the atmosphere, precipitate at various distances from the source of pollution, depending on weather conditions [17-19].

The purpose of the work is to study the qualitative state of the chemical composition of atmospheric precipitation according to data of the Weather Station (WS) located in the Ile-Balkhash basin.

2. Data and research methods

Currently according to the decision of the World Meteorological Organization (WMO) a network of stations to determine trends of global environmental pollution was created. The program of work includes studies of the chemical composition of atmospheric precipitation [20]. In the RK, these obligations are assigned to "KazHydromet", where the monitoring of the state of atmospheric precipitation is conducted, the results of this research are reflected in the newsletters (http://aarhus.kz/ru/1-10/). The chemical composition of atmospheric precipitation for the period of 2000-2011 was analyzed. WS Balkash, Aul-4, Kapshagai, Almaty, Mynzhylky located directly in the Ile-Balkhash basin were selected as the object of the research (figure 1).

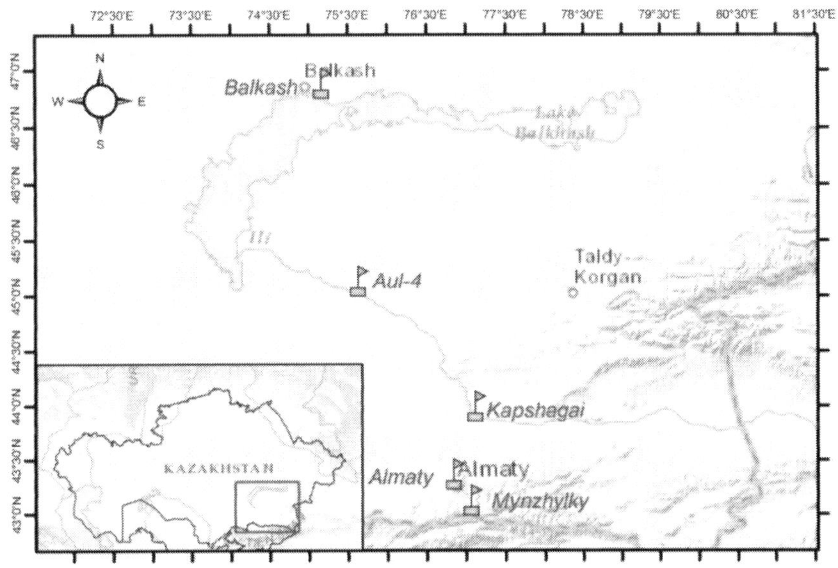

Figure 1. Research area and location of Meteorological station.

Manual on control of air pollution [21] was used to determine the chemical composition of precipitation samples, and The European Monitoring and Evaluation Programme (EMEP) manual - for quality control of the measurements [22]. The methods of estimation of atmospheric pollutions which was used in that research were fully described in [21, 23, 24]. The turbidimetric analysis was used to determine sulfates, titration analysis was used to estimate chloride and alkalinity, photocolorimetry was used to determine ammonium ion and nitrate. Potassium and magnesium ion content was evaluated by atomic absorption analysis, and sodium and potassium ion was determined by flame photometry. Hydrogen ion pH was estimated by using glass electrode; electrical conductivity was evaluated in the cell with electrodes from non-blackened platinum where temperature control at 25°C was used.

These methods of determining pollutants have a number of advantages over other methods. It is one of the easiest methods available for experimental material, the possibility to assess the ways, sources, speed and dynamics of entering of pollutants, to identify the places of their accumulation.

3. Results and discussion

The formation of the chemical composition is a complicated process, in which the important role is played by the weather conditions. Cloud droplets formed as a result of condensation and the process of coagulation, are determined in their chemical composition by the nature of the condensation nuclei. The life and development of individual droplet or crystal run in an environment characterized by various meteorological variables. The processes of evaporation or condensation of water vapor in the atmosphere can affect the concentration of impurities in the precipitation. Therefore, while studying the process of the formation of precipitation composition, it is necessary to take into account the whole range of weather conditions, from macro-synoptical, determining the transfer of condensation nuclei in the free atmosphere, to the microphysical, causing the onset and development of elements of precipitation [17].

Figure 2(a) shows the diagram of the annual atmospheric precipitation. Their amount varies widely, from 129 to 900 mm·year^{-1}. WS Balkash and WS Aul-4 have the least amount of precipitation (not more than 130 mm·year^{-1}). The biggest amount of precipitation falls at the WS Mynzhylki, more than 900 mm/year falls here. Large differences in the amounts of precipitation are caused by the climatic conditions.

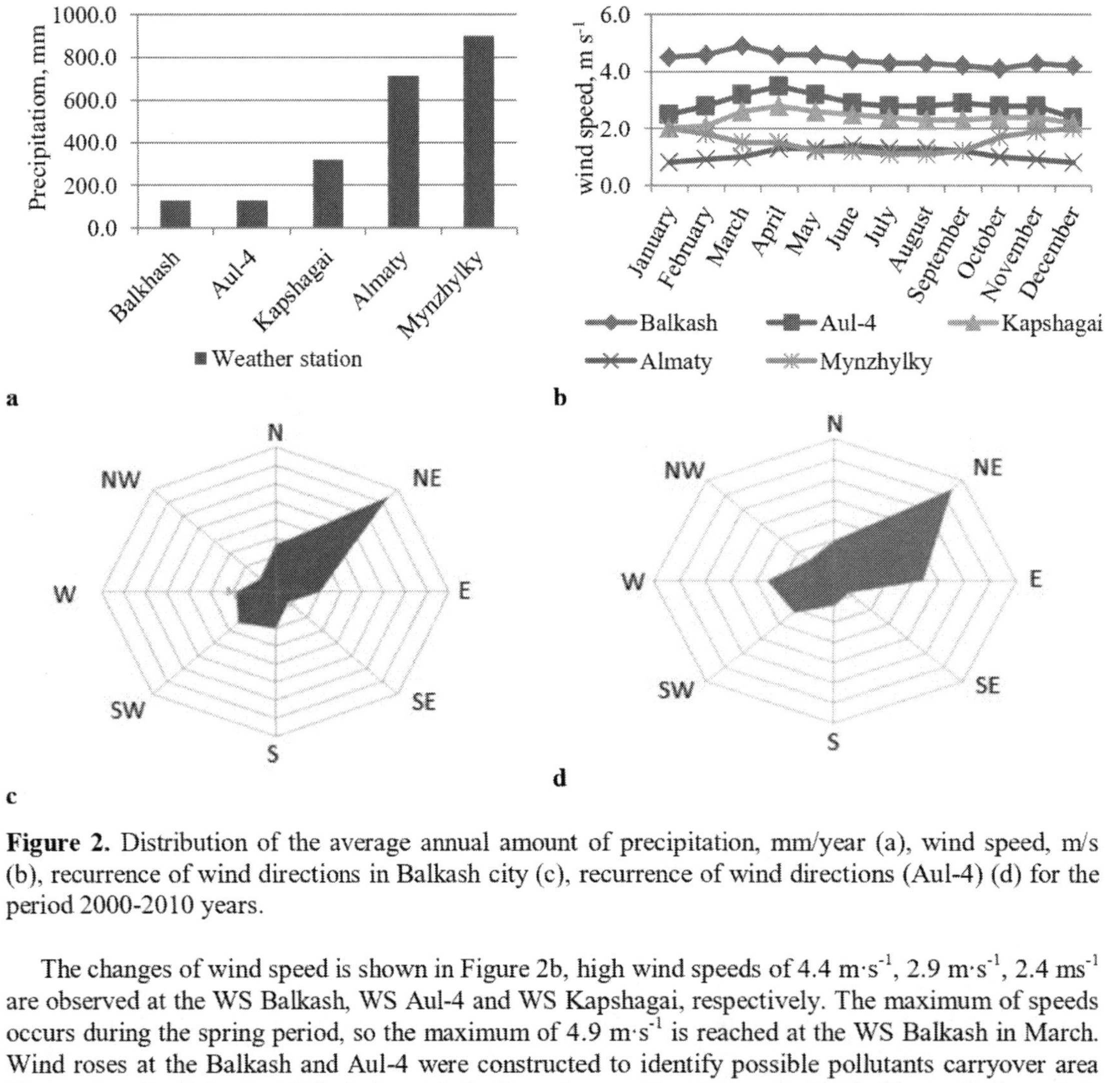

Figure 2. Distribution of the average annual amount of precipitation, mm/year (a), wind speed, m/s (b), recurrence of wind directions in Balkash city (c), recurrence of wind directions (Aul-4) (d) for the period 2000-2010 years.

The changes of wind speed is shown in Figure 2b, high wind speeds of 4.4 m·s^{-1}, 2.9 m·s^{-1}, 2.4 ms^{-1} are observed at the WS Balkash, WS Aul-4 and WS Kapshagai, respectively. The maximum of speeds occurs during the spring period, so the maximum of 4.9 m·s^{-1} is reached at the WS Balkash in March. Wind roses at the Balkash and Aul-4 were constructed to identify possible pollutants carryover area (figure 2c, d). It is established that wind of north-east direction was observed most often, their recurrence reaches up to 37%. Thus, based on the local weather conditions, i.e. deficit of precipitation, high wind speed, recurrence of wind, it can be assumed that the conditions for the transport of pollutants are favorable enough.

In this research [25] the author examined in detail the issues of the impact of pollution from the tailing dump of the Balkhash - Ore-dressing and processing plant, where up to 60 thousand tons of tailing [26] can be carried over in a single dust storm, at the same time researches [27-31] show that pollutants entering the atmosphere spread in it over long distances and the influence of a large industrial city is found at a distance of up to 100-150 km from it in the direction of the wind. Atmospheric precipitations, being a sensitive indicator of air pollution in this work, are used as one of the parameters of the study of the transport of pollutants at the distances, in this regard it is interesting to trace whether the fallout of pollutants is found in distant from the sources, in this case, from Balkhash city. The spatial distribution of pollutants in atmospheric precipitation is shown at the Figure 3. There is a marked increase in the concentration of basic anions over WS Aul-4, where the average values of sulfates can reach 32.29 mg·l^{-1}, of chlorides – 16.25 mg·l^{-1}, nitrates - up to 36.48 mg·l^{-1}. These values are several times smaller small in other observation points (figure 3a).

The distribution of the main cations is characterized by high variability (figure 3b). The prevailing cations are presented by ions of ammonium and sodium, there content amounts to 13.8 and 14.94 mg·l^{-1} at the WS Aul-4 and Mynzhilki, and to a lesser extent by ions of potassium, manganese and calcium with their average concentration of 0.32; 0.44; and 1.2 mg·l^{-1} respectively. Ions of ammonium and sodium are the main soluble component of the continental aerosol in areas remote from sources of pollution [32]. The predominance of this compound in the precipitation may mean the impact of long-range transport of impurities on the surface of Balkhash lake up to the delta of Ile river, where the Aul-4 station is located. The increase in cations at the Mynzhylki point may indicate the impact of city on air pollution. The polluting impurities penetrate in the vertical direction influencing the physical and chemical processes in the atmosphere.

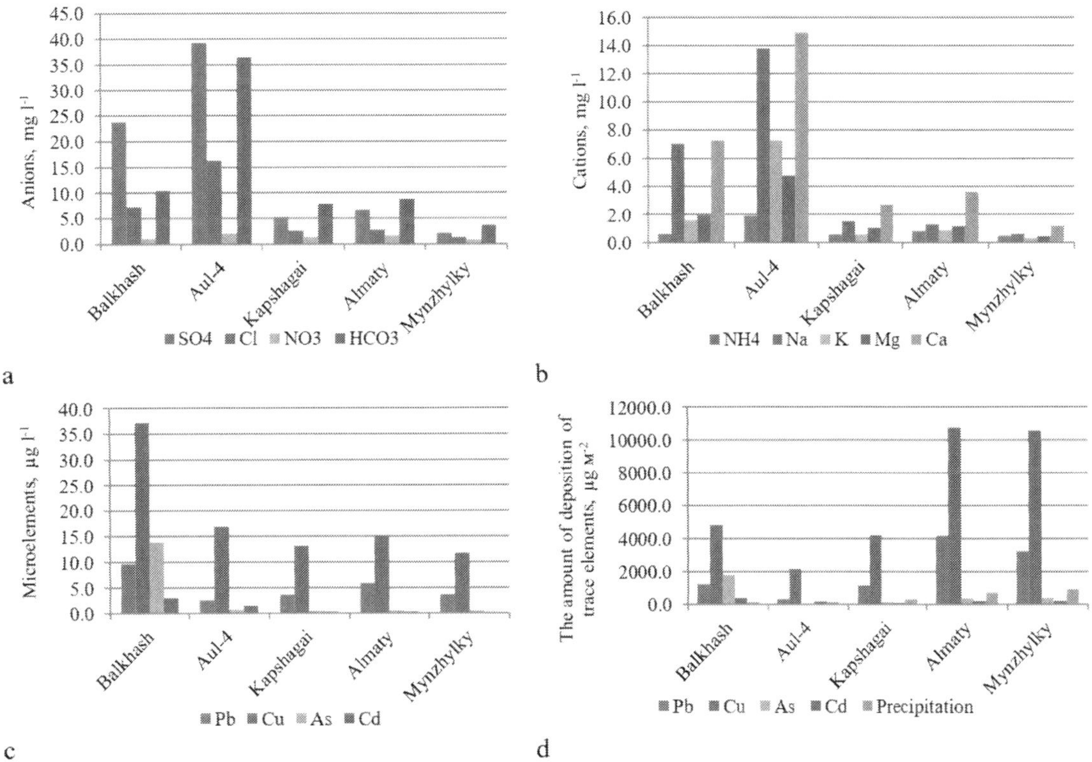

Figure 3. Distribution of long-term concentrations of pollutants: a) Anions; mg/l; b) Cations, mg/l; c) Microelements, µg·l^{-1}; d) Amount of fallout of microelements µg · m^{-2}.

An important indicator of the quality of atmospheric precipitation is the content of microelements. On the studied territory of Ile-Balkhash basin, precipitation is characterized as polluted with these compounds (figure 3c). The content of cadmium in the city of Balkash is especially marked. It concentration reaches 3.03 $\mu g \cdot l^{-1}$ in Balkash and 1.43 $\mu g \cdot l^{-1}$ in Aul-4, with the maximum permissible concentration (MPC) – 1 $\mu g \cdot l^{-1}$ [33]. That can worsen quality of soil by getting to the ground as well as natural waters by falling directly to the water body of due to surface runoff. The existing pollution confirms the existence the powerful source of entering pollutants to the atmosphere and transportation of these pollutants to the large distance up to 300 – 400 km.

It was also of scientific interest in the assessment of quantitative fallout of microelements on the underlying surface of the earth. It was revealed that the amount of their fallouts depends on the intensity of precipitation only for copper (correlation coefficient is 0.59). As it shown in Figure 3d copper and lead fall mostly with precipitation. In addition, the center of fallout is characteristic in the Almaty city, where the main factor of pollution is vehicles. Therefore it is not surprising that up to 10.76 $mg \cdot m^{-2}$ of copper and 4.1 $mg \cdot m^{-2}$ of lead fall there.

The interrelationship of fallout of heavy metals were calculated, so the correlation coefficients between the metal concentrations were as follows: 0.91 for arsenic and cadmium, 0.98 for copper and arsenic, 0.88 for copper and lead, 0.71 for lead and cadmium, 0.90 for lead and arsenic, which indicates the presence of a single source of incoming of these microelements. Negative correlation coefficients of the amount of precipitation were obtained for copper, lead, arsenic (-0.57,-0.23,-0.49, respectively) between their concentrations in atmospheric precipitations and the amount of precipitation, which may indicate the presence of a slight effect of "dilution and washout".

Conclusion

As a result of the study, it was found out that the considered meteorological characteristics, such as water scarcity, high wind speeds, recurrence of wind direction contribute to the transport of pollutants. It was revealed that the impact of pollution of the major metallurgical plant is found at a distance from it of up to 300-400 km, at the WS Aul-4. Also, it was identified that polluting impurities penetrate in the vertical direction to a height of 3 km, the height of location of the WS Mynzhylki. Heavy metals, in particular, cadmium, as a result of the transport, reaches the delta of Ile river, where its concentrations exceed the MPC almost in 1.5 times. It was revealed that the amount of fallout of heavy metals has a weak relationship to the intensity of precipitation; their high correlation dependences indicate the general source of pollutants entering the atmosphere.

Atmospheric precipitation should be considered as another source of pollution of river waters, so the assessment of the impact of atmospheric precipitation on the chemical composition of river waters is relevant and important in assessing the quality of river waters and planning water protection activities and runoff regulation.

References

[1] 2013 *The III-VI National Communication of the Republic of Kazakhstan to the UN Framework Convention on Climate Change (UNFCCC)* ed S Sakenov, V Kryukova, A Tonkobayeva and A Yesseneyeva (Astana: Forma Plus) p 265

[2] Dahl C and Kuralbayeva K 2001 Energy and the environment in Kazakhstan *Energy Policy* **29** 429 – 40 doi.org/10.1016/S0301-4215(00)00137-3

[3] Ghada F E, Suzanne D, Manal M and Abeer A M 2014 Sedimentology, geochemistry, pollution status and ecological risk assessment of some heavy metals in surficial sediments of an Egyptian lagoon connecting to the Mediterranean Sea *Journal of Environmental Science and Health*, A **49:9** 1029 – 44 10.1080/10934529.2014.894853

[4] Varol M 2011 Assessment of heavy metal contamination in sediments of the Tigris River (Turkey) using pollution indices and multivariate statistical techniques *Journal of Hazardous Materials* **195** 355 – 64

[5] Uluturhan E, Kontas A and Can E 2011 Sediment concentrations of heavy metals in the Homa Lagoon (Eastern Aegean Sea): Assessment of contamination and ecological risks *Marine Pollution Bulletin* **62** 1989 – 97
[6] Gao X and Chen Ch T A 2012 Heavy metal pollution status in surface sediments of the coastal Bohai Bay *Water Research* **46** 1901–11
[7] Galperin M, Sofiev M and Afinogenova O 1995 Long-term modelling of airborne pollution within the Northern Hemisphere *Water, Air, and Soil Pollution* **85** 2051–56
[8] Fujita S, Takahashi A, Weng J H, Huang L F, Kim H K, Li C K, Huang F T C and Jeng F T 2000 Precipitation chemistry in East Asia *Atmospheric Environment* **34** 525 – 37
[9] Li P, Feng X B, Qiu G L, Shang H and Li Z G 2009 Mercury pollution in Asia: A review of the contaminated sites *Journal of Hazardous Materials* **168** 591 – 601
[10] Aardenne J A, Carmichael G R, Levy H, Streets D and Hordijk L 1999 Anthropogenic NO_x emissions in Asia in the period 1990 *Atmospheric Environment* **33** 633 – 46 doi.org/10.1016/S1352-2310(98)00110-1
[11] Madibekov A S, Cherednichenko V S, Cherednichenko A V, Nyssanbaeva A S and Zhumalipov A R 2014 Heavy metal content in the snow cover in the Republic of Kazakhstan *Advances in Environmental Biology* **8** 1393 – 98
[12] Sultanova D M and Cherednichenko V S 2015 The chemistry of precipitation on the territory of the Republic of Kazakhstan *Proc. of the Fourth Int. Conf. of Young Scientists Indication of the state of the environment: theory, practice, education (Moskow)* (Moscow: Moscow State Pedagogical University) 140 – 43
[13] Madibekov A S 2011 Distribution of microelements in atmospheric precipitation over the South Kazakhstan *The Internet magazine of the National Certification Committee of the Kyrgyz Republic* **3** vak.kg/jurnalVAK/
[14] Kumar K S, Sajwan K S, Richardson J P and Kannan K 2008 Contamination profiles of Heavy metals, organochlorine pesticides, polycyclic aromatic hydrocarbons and alkylphenols in sediment and oyster collected from marsh/estuarine Savannah GA, USA *Marine Pollution Bulletin* **56** 136 – 49
[15] Hu X, Wang C and Zou L 2011 Characteristics of heavy metals and Pb isotopic signatures in sediment cores collected from typical urban shallow lakes in Nanjing China *Journal of Environmental Management* **92** 742 – 48
[16] Buccolieri A, Buccoilieri G, Cardellicchio N, Dell'Atti A, DiLeo A and Maci A 2006 Heavy metals in marine sediments of Taranto gulf (Ionian Sea, Southern Italy) *Marine Chemistry* **99** 227 – 35
[17] Petrenchuk OP 1979 *Experimental study of atmospheric aerosol* (Leningrad: Gidrometeoizdat) p 264
[18] Bai J, Cui B, Chen B, Zhang K, Deng W, Gao H and Xiao R 2011 Spatial distribution and ecological risk assessment of heavy metals in surface sediments from a typical plateau lake wetland *China. Ecol. Modell.* **222** 301–06
[19] Mingbiao L, Jianqiang Li, Weipeng C and Maolan W 2008 Study of heavy metal speciation in branch sediments of Poyang Lake. *Journal of Environmental Sciences* **20** 161–166
[20] WMO 1974 Operations manual for sampling and analysis techniques for chemical constituents in air and precipitation №299 (Geneva : Secretariat of the WMO*)* 2 – 56
[21] 1991 RD52.04.186-89 Manual on Atmospheric Pollution Control *Regulatory Document* (Moscow: Gidrometeoizdat) p 319
[22] *Manual for sampling and chemical analysis (EMEP/CCC-Report 1/95, Revision1/2001. O-7726.2001)* 2001 EMEP (Kjeller: Norwegian Institute for Air Research) p 176
[23] 1991 RD 52.04.186-89 Guide for Estimation of Precipitation Chemical Composition *Regulatory Document* (Moscow: Gidrometeoizdat) p 60

[24] An J, Ueda H, Wang Z, Matsuda K, Kajino M and Cheng X 2002 Simulations of monthly mean nitrate concentrations in precipitation over East Asia *Atmospheric Environment* **36** 4159-71
[25] Baimakova Y V 2002 Assessment of the impact of the tailing dumps to the Balkhash concentration plant on the environment *Newsletter of the KazNU. Geograph. series* **2** 48 –57
[26] Turchaninov I A 1981 Problems of fixing of dust-forming surface of active tailing dumps of Polar region *Engineering Geology* **5** 107 – 10
[27] Selezneva E S 1963 Spatial changes in the concentration of condensation nuclei according to data of horizontal flights over ETC *Proceedings of the MGO* **141** 3 – 18
[28] Selezneva E S 1964 On the distribution of condensation nuclei over the Ukrainian Meteorological polygon *Proceedings of the MGO* **154** 3 – 10
[29] Selezneva E S 1971 On the removal of contaminants from the atmosphere by clouds and precipitation *Meteorological aspects of air pollution* ed Selezneva E S and Petrenchuk O P (Leningrad: Gidrometeoizdat) 253 – 59
[30] Abdo M H 2005 Assessment of some heavy metals, major cations and organic matter in the recent sediments of Bardawil Lagoon *Egyptian Journal of Aquatic Research* **31** 214 – 28
[31] Rodhe H, Langner J, Gallardo L and Kjellstrom E 1995 Global scale transport of acidifying pollutants *Water, Air and Soil Pollution* **85** 37 – 50
[32] Brimblukumb P 1988 *The composition and chemistry of the atmosphere* (Moscow: Mir) p 351
[33] 2004 SanPiN 3.02.003-04 Sanitary and epidemiological requirements for the protection of water from pollution *Sanitary Rules and Norms* (Almaty: Ministry of Health) p 120

Screening of the state of urban ecosystem with the use of bioindication method (on the example of Kazan city)

E A Minakova[1], A P Shlychkov[2] and A V Arinina[1]

[1]Institute of Fundamental Medicine and Biology, Kazan Federal University, Kremlevskay str, 18, 420008, Kazan, Russia
[2] Institute of Ecology and Environmental Sciences of the Republic of Tatarstan, Kremlevskay str, 18, 420097, Kazan, Russia
e-mail: ekologyhel@mail.ru

Abstract. The urban environment is a complex of natural, natural-anthropogenic and socio-economic factors that exert a large and diverse impact on urban residents. In addition to traditional environmental monitoring, we propose to use a new bioindication method based on the evaluation of morphological changes in the leaves of *Betula pendula* Roth by fluctuating asymmetry (FA) to assess the quality of recreational areas. Such screening for the purpose of assessing of the environment state is very informative, since the bioindication assessment is an integral characteristic of the quality of the environment which is under the influence of all the abundance of chemical, physical and other factors. The two-sided symmetry of a leaf was calculated on the sites in the middle of the park zone, on the border of the park and on a roadside strip. The results of the study showed a connection between the FA values and the distance to the highway, and also revealed the absence of significant differences in FA indicators at the surveyed sites, which may indicate insufficient sizes of recreational areas and their insufficient potential to contribute to improving the quality of the environment.

1. Introduction

The urbanized environment, especially a comfortable urban environment, the developed infrastructure, social and economic benefits bear in themselves the mass of the environmental problems which are now extremely actual. Atmospheric air of the modern cities contains hundreds of organic and inorganic substances of various chemical classes arriving from numerous sources, as a rule, of anthropogenous character.

Biological monitoring allows to estimate integrally a state of environment and to carry out screening of anthropogenous impact on live objects. Perspective approach to biomonitoring of difficult natural and anthropogenous systems is the assessment of a condition of populations from an ontogenetic point of view on the basis of a morphogenetic monitoring method, or an assessment of the level of stability of live organisms development.

Intensive studying of the phenomena of asymmetry devoted to the use of the fluctuating asymmetry phenomenon at environmental monitoring of natural populations such as plant populations has begun relatively recently and is marked by works of both foreign and Russian researchers [1-8]. Rather large number of works is devoted to an assessment of stability of development of organisms in nature [9-11], as well as to exert influence on stability of development of climatic factors [12]. Other application of the approach is its use for monitoring of health of the environment in the protected natural territories or for regional monitoring [13]. In recent years, there have been many researches devoted to the use of a FA phenomenon in assessment of anthropogenous impact in urban, recreational environment and in vicinities of industrial facilities.

2. Material and methods

Multicomponent influence complicates an the assessment of anthropogenous impact, therefore for these purposes the bioindication methods give a chance of the integrated characterization of quality of the environment which is under the influence of all variety of physical, chemical and other factors. Biological monitoring allows to estimate integrally a state of environment and to carry out screening of the anthropogenous impact on living objects.

Perspective approach to difficult natural and anthropogenous systems biomonitoring is the assessment of the populations conditions from the ontogenetic point of view on the basis of the morphogenetic monitoring method or an assessment of living organisms development stability level.

The plant is an important object for characterizing the state of the natural environment. They are the main producers; plants are sensitive objects allowing to evaluate the whole complex of impacts characteristic of the given territory; they lead an attached way of life, the state of their organism reflects the state of a particular local habitat; the use of plants consists of the accessibility and simplicity of collecting material for research. Now the methods of bioindication which are most often put into practice consider morphological changes of the higher plants. The main approach to an assessment of morphological changes, owing to violations of a homeostasis of development, is morphogenetic which basis is the characteristic of stability of development, and one of the main morphological criteria is the size of the fluctuating asymmetry. The Fluctuating Asymmetry (FA) refers to small, not directed (random) deviations from bilateral symmetry of organisms or their parts. The basis of this research is the method of an environment quality assessment [14]. FA measures small non-directional deviations from the usually strict bilateral symmetry and characterizes the morphogenetic homeostasis or stability as a result of the imperfections in the growth processes. The value of FA gives the characteristic of the morphogenetic homeostasis or stability of development and is defined as a result of ontogenetic processes imperfection and represents the insignificant, not directed deviations from strict bilateral symmetry and can be characterized as one of the most usual and available display of development random variability. At the normal state of environment its level is minimal, at the increasing negative impact it increases that leads to an increase of asymmetry [15].

The birch tree (Betula pendula Roth.), the Birch family Betulaceae, was chosen as the research object for definition of development stability extent, the Birch family Betulaceae. The research data was collected in years 2014 – 2016. Seven platforms were put in place in the recreational territories of Kazan (figure 1). On each platform two sampling points were targeted:

- in a park zone
- on the border of a park and a roadside strip at a distance of 10 to 30 m from a zone of highways influence.

In total, 42 selections were collected and analysed during the research in recreational zones of Kazan from 2014 to 2016 yrs and 42000 measurements were executed.

For an assessment of an indicator of the fluctuating asymmetry they used a leaf as the body exhibiting bilateral symmetry. Leaves in quantity of 10 pieces from 10 trees in each point were gathered till a full stop of leaves growth (since the middle of August), from the short shoots evenly on all perimeter in the lower part of tree crown on the trees of generative age growing in identical light conditions. Parameters of the left and right half of a leaf plate were measured. To receive statistically reliable data 100 leaves were gathered on each test platform. For each leaf, on both left and right side, values of five measurements (figure 2) are as follows:

1 – width of the left and right halves of a leaf;
2 – length of a vein of the second order, the second from the leaf basis;
3 – distance between the bases of the first and second veins of the second order;
4 – distance between the ends of the same veins;
5 – corner between the midvein and the second from the leaf basis a vein of the second order

Figure 1. Platforms of selection of material in Kazan: 1. Gorky Park - Vakhitovsky district; 2. Kyrlay park - Moskovsky district; 3. Recreation Center Park of Chemists - Moskovsky district; 4. Uritsky Park - Moskovsky district; 5. Continent Park - Novo-Savinovsky district; 6. Krylja Sovetov Park - Aviastroitelny district; 7. Derbyshki Birchwood - Sovetsky district.

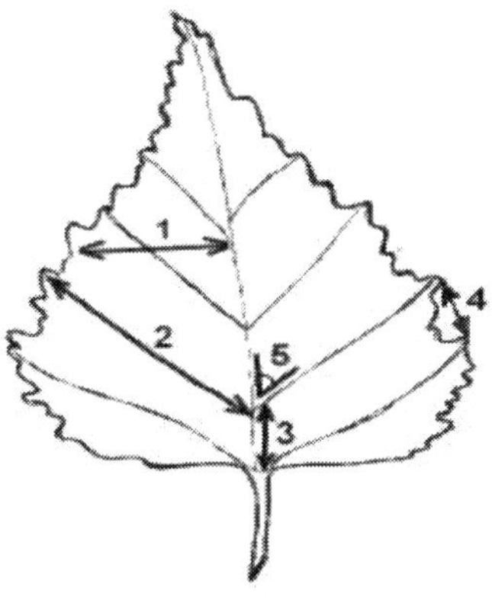

Figure 2. The scheme of measurements of a birch leaf used for an assessment of size of the fluctuating asymmetry.

The size of the fluctuating asymmetry of plants was estimated by means of an integrated indicator that is the value of average relative deviations per sign (a ratio of differences to the sum of measured values on the left side and on the right side of a leaf, referred to the number of signs) [16]:

$$FA = ABS(L - R)(L + R)^{-1}$$

where: FA – an integrated indicator of the fluctuating asymmetry;
ABS – an absolute value;
L – sign value on the left side of a leaf;
R – sign value on the right side of a leaf.

Statistical processing was carried out on the basis of the Microsoft Office Excel program. For detection of statistical distinctions between indicator sizes of the fluctuating asymmetry in paired comparison Student's t-criterion was used. Distinctions higher than 95% are considered statistically significant. In some cases, methods of nonparametric statistics, in particular Spearman's rank correlation coefficient, were applied.

The scale offered by V. M. Zakharov and coauthors for a birch was applied to an assessment of quality of the environment (table 1).

Table 1. Scale of an assessment of quality of the environment in terms of developmental stability indicators of a birch (*Betula pendula Roth*).

Score	Quality of the environment	Value of the developmental stability indicator
1	Conditionally normal	< 0.040
2	Initial (insignificant) deviations	0.040–0.044
3	Average level of abnormalities	0.045–0.049
4	Significant (significant) deviations	0.050–0.054
5	Critical condition	> 0.054

3. Results and discussion

The results stated in the real work are a part of complex territorial bioecological researches on the territory of the Republic of Tatarstan [17, 18] conducted at the Department of Bioecology, Hygiene and Public Health of the Kazan Federal University since 2012.

Results of the taken measurements and FA calculated values are shown in figure 3. On a rating scale of an assessment of quality of the environment, by the indicators of violation of development stability of Betula pendula Roth. on all relevant platforms, the FA level exceeding the size of conditional norm is shown (> 0.04) that testifies to a critical condition of the environment in the studied recreational territories of Kazan.

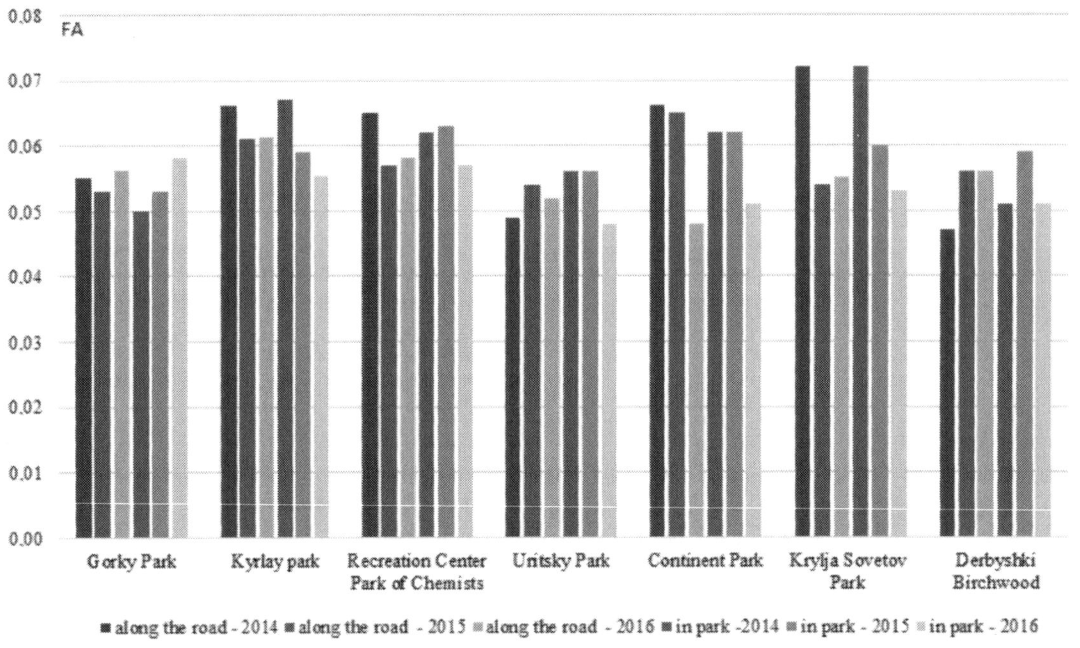

Figure 3. Dynamics of the FA index in parks of Kazan (2014 – 2016 years).

By comparing the FA indexes in inside-park and roadside platforms within one site found no significant differences in the FA indicators, which can testify to an insufficient sizes of Kazan city recreational zones and as a result of this it can testify to an inadequate capacity of the recreationarea to promote improvement of environmental quality.

The surveyed recreation areas of Kazan were divided into two groups in compliance with the scale of an assessment of environment (table 1, figure 4):

- territories with critical condition of environment (Park of Chemists, Continent Park, Kyrlay Park, Krylja Sovetov Park) – FA > 0.055;
- territories with essential deviations of quality of environment (Gorky Park, Uritsky Park, the Birchwood square) – FA < 0.055.

In general, uneven distribution of the FA index across the territory of the city was observed during research, and on the most part of its territory the environmental quality didn't meet conditional standard and was characterized as critical.

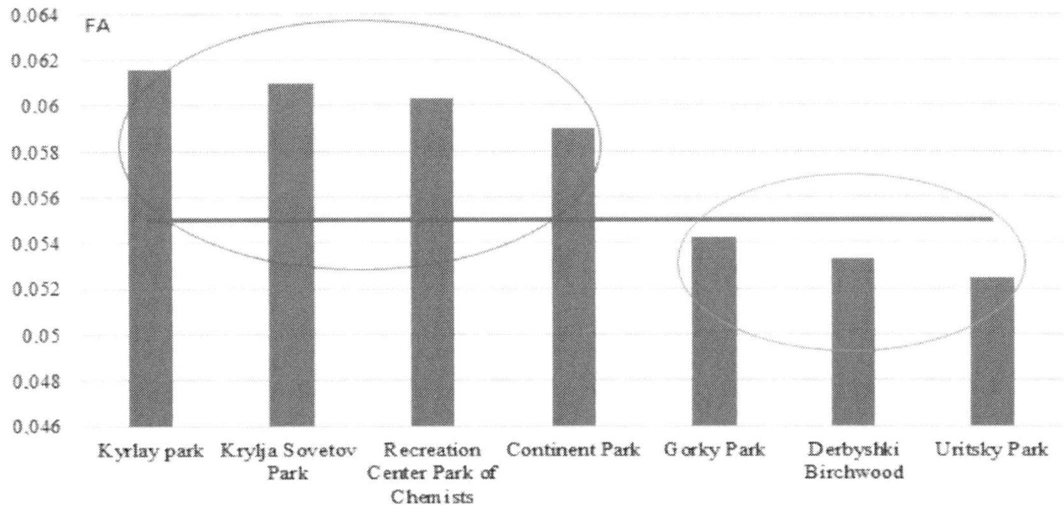

Figure 4. Average values of FA in the studied parks for 2014-2016 years.

During the conducted research we found a link between values of FA and a distance to the highway (figure 5). It shows that at a distance of 30 or more meters from the highway the quality of the environment is already characterized as conditionally normal.

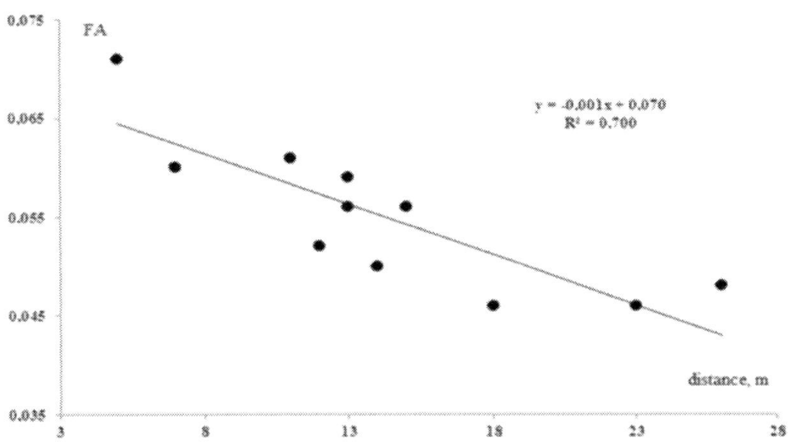

Figure 5. Dependence of size FA on the studied platforms from distance to the highway, m.

The coefficient of correlation is significant at the level α = 0.05. The factor of S defines 70% of dispersion of FA. For the analyzed $R^2 > 0.5$ it testifies to the noticeable force of communication between sizes FA and S.

Received average values of FA indicators on platforms for the entire period of supervision and the trend of change of the FA index (figure 6) is constructed. The analysis of dynamics of the average years given in a 2014 – 2016 retrospective on platforms in parks and on roadside sites allowed to reveal a trend of the FA index change towards improvement of a state of the environment (the FA index <0.055). It revealed that during the considered period of 2014 – 2016 yrs. the average values of the integral FA index in Kazan exceeded the value of the conventional norm (FA> 0.044).

During the follow-up period, the trend of decreasing FA values for Kazan city from V point (critical quality of the environment, FA > 0.054) to IV and III points (significant and average deviations from the norm) was observed on the scale of environmental quality assessment, which allows to speak about self-healing of birch pellucid (*Betula pendula Roth.*).

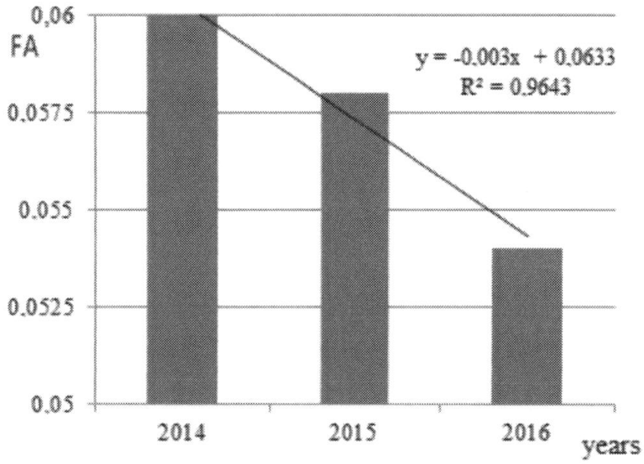

Figure 6. Average values of FA to Kazan (in parks and along roads).

Considering the reported results and materials of research on the size of FA of birch leavesin other cities of Russia showed that the FA values received in Kazan are comparable to the results obtainedin other cities of the Russian Federation: in Togliatti the indicator of FA varies from 0.035 to 0.056 [19], in Stavropol 0.025-0.056 [20], in Khanty-Mansiysk 0.053-0.057 [21], in Ufa 0.053 – 0.059 [22], in Barnaul 0.046 – 0.084 [23], in Yakutia in an urban environment 0.042-0.056 [24].

Acknowledgement
The group of authors thank V M Zakharov, the developer of the FA method (Koltzov Institute of Developmental Biology of Russian Academy of Sciences), for the advisory and methodical help.

References

[1] Leary R and Allendorf F 1983 Developmental stability and enzyme heterozygosity in rainbow trout *Nature* **301** 71 – 72
[2] Leary R and Allendor F 1989 Fluctuating asymmetry as an indicator stress. Chance use in protect nature *Acta Zool. Fenica* **4** 214 – 7
[3] Martel J and Lempa K 1999 Effect of Stress and Rapid Growth on Fluctuating Asymmetry and Insect Damage in Birch Leaves *OIKOS* **86** 208 – 16
[4] Moller A and Eriksson M 1995 Pollinator Preference for Symmetrical Flowers and Sexual Selection in Plants *OIKOS* **73** 15 – 22
[5] Parsons P 1990 Fluctuating asymmetry: an epigenetic measure of stress *Biol. Rev.* **65** 131 – 45

[6] Parsons P 1992 Fluctuating asymmetry: a biological monitor of environmental and genomic stress *Heridid* **68** 361 – 64
[7] Wilsey B and Saloniemi I 1999 Leaf Fluctuating Asymmetry in Tree-Line Mountain Birchs *OIKOS* **87** 34 – 45
[8] Zakharov V 1987 *Asimmetriya of animals* (Moscow: Science) p 216
[9] Bulatova E 2009 *Siberian News of the Samara scientific center of the Russian Academy of Sciences* **11** 363-368
[10] Gelashvili D 2001 *Bulletin of the Nizhny Novgorod University of N. I. Lobachevsky* **1** 64 – 72
[11] Zakharov V 2005 *Bulletin of the Nizhny Novgorod university of N. I. Lobachevsky* **1** 77 – 84
[12] Zakharov V 2011 Achievements of modern biology **131** 425 – 30
[13] Ushakova V 1999 Using of the method of fluctuating asymmetry in conducting environmental monitoring in Dzerzhinsk city *Mess. of the VI Congress of the Theriological Society* (Moscow) p 263
[14] Zakharov V 2000 *Health of the environment: assessment technique* (Moscow) p 68
[15] Palmer A and Strobeck 1992 *Acta Zool. Fenn.* **191** 57 – 72
[16] *Methodical recommendations about performance of an assessment of quality of the environment about a condition of living beings (assessment of stability of development of live organisms in the level of asymmetry of morphological structures)* 2003 (Moscow: Ministry of Natural Resources of the Russian Federation) p 24
[17] Minakova E 2013 *Mess. of the Tatarstan office of the Russian Ecological Academy* **3** 96 – 98.
[18] Minakova E 2015 *Messenger of the Kazan technolog. univ.* **18** 225 – 30
[19] Belyaeva Yu 2013 *Samara Luka: problems of regional and global ecology* **23** 167 – 74
[20] Tkachenko Y 2012 *Bulletin of agrarian and industrial complex of Stavropol Territory* **1** p 109
[21] Gurtyak A and Uglev V 2010 *News of Tomsk Polytechnical University* **317** 200 – 4
[22] Kulagin A 2015 *Forest plantings of the Ufa industrial center* p 195
[23] Davydova N 2014 *Conf. Problems of rational environmental management, ecology, a cadastral assessment and monitoring of lands* (Altay: Altay State University Press) 325 – 27
[24] SoldatovaV 2016 *Assessment of quality of the environment of the territory of Yakutsk on an indicator of violation of stability of development of a birch* (Yakutsk publishing house) p 112

New strains of oil-degrading microorganisms for treating contaminated soils and wastes

A Yu Muratova, L V Panchenko, D V Semina, S N Golubev and O V Turkovskaya

Institute of Biochemistry and Physiology of Plants and Microorganisms, Russian Academy of Sciences, Saratov, Prospekt Entuziastov, 13, 410049, Russia
e-mail: amuratova@yahoo.com

Abstract. Two new strains *Achromobacter marplatensis*101n and *Acinetobacter* sp. S-33, capable of degrading 49 and 46% of oil within 7 days were isolated, identified, and characterized. The application of *A. marplatensis* 101n in combination with ammonium nitrate (100 mg·kg^{-1}) for 30 days of cultivation resulted in the degradation of 49% of the initial total petroleum hydrocarbon content (274 g·kg^{-1}) in the original highly acid (pH 4.9) oil-contaminated waste. Up to 30% of oil sludge added to a liquid mineral medium at a concentration of 15% was degraded after 10 days of cultivation of *A. marplatensis* 101n. Application of yellow alfalfa (*Medicago falcata* L.) plants with *Acinetobacter* sp. S-33 for bioremediation of oil-sludge-contaminated soil improved the quality of cleanup in comparison with the bacterium- or plant-only treatment. Inoculation of *Acinetobacter* sp. S-33 increased the growth of both roots and shoots by more than 40%, and positively influenced the soil microflora. We conclude that the new oil-degrading strains, *Acinetobacter* sp. S-33 and *A. marplatensis* 101n, can serve as the basis for new bioremediation agents for the treatment of oil contaminated soils and waste.

1. Introduction

In the modern world, hydrocarbons are the most common and demanded raw material and energy source, and, as a result, the worst pollutants of the environment, especially in the regions where they are extracted, transported, and processed. As a result of the activities of oil and gas enterprises, there is a need to clean up soil and water bodies affected by oil pollution, as well as the problem of utilization of processing wastes containing significant amounts of hydrocarbon pollutants [1]. A modern approach to the purification of oil-contaminated objects is the application of bio- and phytoremediation technologies based on the use of vital activity of living organisms. Plant-microbial associations and symbiosis with a flexible metabolism and unique enzyme systems have great potential for survival under extreme conditions of oil-polluted ecosystems, and their survival is caused not only by increased tolerance to xenobiotics but also by the active removal of toxic substances from the environment [2, 3]. Microorganisms that utilize hydrocarbons as the sole carbon and energy source are the basis of such bioremediation technologies as biostimulation, bio-augmentation, and rhizoremediation [4-6]. In the past decades, numerous bioremediation agents have been developed that contain oil-degrading bacteria based on monocultures and microbial consortia, with fertilizers, structuring agents, and emulsifiers for improving the quality of remediation [7]. The use of microbial oil-degrading agents over the years has also revealed some problems, among which are the lack of equal effectiveness of existing bioremediation agents under different climatic conditions; gradual loss of specific activity of the microbial strains forming the basis of a particular bioremediation agent; and the frequent discrepancy between commercial products and the declared characteristics of bioremediation agents, not only in oil-degrading activity, but also in microbial composition. At the same time, it has been noted that the bioremediation agents based on natural hydrocarbon oxidizing microorganisms isolated in a specific climatic zone are the most effective, because the introduced microflora that is not characteristic of a particular ecosystem can be suppressed by indigenous microbial populations [8,9]. All the above-mentioned indicates that the search for new strains of oil-degrading microorganisms and the development of alternative bioremediation agents based on them remain relevant, because their application in environmental biotechnology is still in demand with environmental organizations.

New oil-degrading microbial strains isolated from oil-sludge and oil-contaminated rhizosphere soil were studied in this work for use in bio-and phytoremediation technologies.

2. Material and methods

2.1. Microorganisms and cultivation conditions

We used 80 oil-hydrocarbon-degrading microorganisms: 26 strains from the Collection of Rhizosphere Microorganisms of the IBPPM RAS and 54 new isolates from oil contaminated rhizosphere and nonrhizosphere soil samples.

Samples of oil sludge and oil-contaminated soil, and methods of direct plating and enrichment cultivation were used to isolate microorganisms able to degrade petroleum hydrocarbons. For the cultivation of microorganisms, beef-extract agar, LB broth and Bushnell-Haas's mineral medium for hydrocarbon-oxidizing microorganisms were used [10].

The total number of culturable heterotrophic microorganisms (THMs) in the samples was determined by the standard plating dilutions method. The number of hydrocarbon-oxidizing microorganisms (HOMs) was determined by the method of membrane filters and by cultivation of immobilized microbial cells on Bushnell-Haas's agar medium, as described previously [11]. Inoculated plates were incubated at 29 °C for 7 days, after which the CFU was calculated.

2.2. Study of oil-degrading activity of microbial isolates

The degradative activity of the isolated microbial strains was studied by culturing them in tubes or flasks with Bushnell-Haas's liquid medium containing 1% (w v^{-1}) of crude oil as the sole source of carbon and energy. Cultivation was carried out under aeration conditions on a rocking shaker (160 rpm) for 7 days. The initial optical density of the microbial suspension was $D = 0.5$ ($\lambda = 590$ nm). The degradative activity of the microorganisms was determined by the loss of oil from the medium after cultivation and expressed as a percentage.

2.3. Identification of microbial isolates

The taxonomic affiliation of the isolated microorganisms was determined on the basis of a study of the cultural-morphological and physiological-biochemical features, as well as 16S rRNA gene sequence analysis, by using the EzTaxon server (http://www.ezbiocloud.net/eztaxon) [12] and using Bergey's Manual of Systematic Bacteriology [13,14].

2.4. Analyses of oil content

The residual oil content in the medium after cultivation was determined by preliminary extraction with nonpolar solvents, gas and liquid chromatography, and gravimetric analysis. For the primary evaluation of the degradative activity of microorganisms, the content of oil remaining in the medium was determined by gravimetric analysis after preliminary trichloromethane extraction of oil hydrocarbons from the culture medium [15]. Gas and liquid chromatography with selective extraction of oil fractions was used to characterize the range of microbial degradation of hydrocarbons. For gas chromatographic analysis, the extraction of oil from the culture medium was carried out with carbon tetrachloride, and the extracts were analyzed according to [15] on a Shimadzu 2010 gas chromatograph fitted with an Equty-1 non-polar capillary column (Supelco, USA) with a flame ionization detector (40 ms), makeup gas, He; H2 flow, 40 ml·min^{-1}; air flow, 400 ml·min^{-1}; makeup gas flow 30 ml·min^{-1}; carrier gas, He. The sample volume was 1 µl in all cases. Fractional analysis of the contaminant remaining after the cultivation of microorganisms was carried out by separating the oil into fractions by liquid column chromatography, selective extraction of various hydrocarbon classes with different solvents, and a gravimetric method for determining their share [16]. Residual oil was divided into the following fractions: paraffins, naphthenes, mono- and bicyclic aromatic hydrocarbons (MBCAs), polycyclic aromatic hydrocarbons (PAHs) and alcohol-benzene tars (ABTs). In actual waste samples the concentration of total petroleum hydrocarbons was determined by gravimetric analysis by using the standard procedure [17]. All samples were analyzed in triplicate.

2.5. Bioremediation of original oil-contaminated soils and wastes

Two types of wastes were used to test the activity of selected oil-degrading strains: oil contaminated soil and oil sludge.

The oil-contaminated soil contained 24.6% of oil sludge and 2.1% of mechanical impurities; the pH value was 4.9. For studying the possibility of bioremediation, oil-contaminated soil was thoroughly mixed, a sample of soil (200 g) was taken for preliminary analyses, and 600-g portions were placed in 2-liter containers. There were four experimental treatments: 1) untreated soil (control); 2) treatment with a mineral nitrogen fertilizer; 3) treatment with oil-degrading microorganism; 4) joint use of mineral fertilizer and a microorganism. The soil was treated according to the following scheme: loosening, daily; moistening of the soil up to 40% of the total moisture capacity, once a week; introduction of the oil-degrading microorganism *A. marplatensis* 101n (up to a final cell concentration of not less than 10^7CFU·g^{-1} of soil), every 2 weeks; application of nitrogen fertilizer (NH_4NO_3, 100 mg·kg^{-1}), every 3 weeks. The bioremediation experiment lasted 30 days.

Oil sludge was a thick viscous liquid containing 8.2% of water, 1.6% of mechanical impurities, and 90.2% of oil hydrocarbons; the pH value was 6.4. Oil sludge was treated by cultivating *A. marplatensis* 101n in a flask with liquid Bushnell-Haas's medium containing 15% (w·v^{-1}) of oil sludge. The initial optical density of the microbial suspension was $D = 0.42$ ($\lambda = 590$ nm), which approximately corresponded to not less than 10^8 cells per ml. Cultivation was carried out under aeration conditions on a circular shaker at 140 rpm, at 28° C for 10 days. At the end of the cultivation, the residual oil content in the medium was analyzed.

To test the effectiveness of application of the oil-degrading microorganisms alone and in combination with plants, we used the original oil-sludge-contaminated soil (pH 8.3; petroleum hydrocarbons, 14.3 g·kg^{-1}; P_2O_5 150 mg·kg^{-1}; N-NO_3 10 mg·kg^{-1}; N-NH_4 - 25.9 mg·kg^{-1}), *Acinetobacter* sp. strain S-33 and yellow alfalfa plants (*Medicago falcata* L.). The soil experiment was carried out in 2-l pots containing 1 kg of soil in three replicates: 1) untreated soil (control), 2) treatment with *Acinetobacter* sp. S-33 to a final cell concentration in the soil of at least 10^7 CFU/g soil (bioaugmentation); 3) plant cultivation (phytoremediation); 4) cultivation of the plant and inoculation with *Acinetobacter* sp. S-33 (phytoremediation with bioaugmentation). To compensate for nitrogen deficiency, which can limit the biodegradation of oil in the soil, ammonium nitrate (NH_4NO_3) in concentrations of 200 mg·kg^{-1} was added every 3 weeks to 2-4 treatments of the experiment. Experimental conditions were: light period, 16/8 h (day·night^{-1}); humidity, 40%; watering by weight; temperature, 21-24 °C; duration, 2.5 months. At the end of the experiment, loss of oil hydrocarbons, microbial counts, and plant biomass were estimated.

2.6. Statistics

All experiments and analyzes were performed at least in triplicates. Data were subjected to statistical processing with Microsoft Office Excel 2003.

3. Results and discussion

3.1. Oil-degrading activity of selected microbial strains

As a result of screening of 80 strains of rhizosphere and non-rhizosphere microorganisms, two most active oil degraders were selected: strain S-33 was isolated from the rhizosphere of *Medicago falcata*, and strain 101n was isolated from an oil sludge sample. A study of the degradative activity of these microorganisms toward oil (1%) showed that strain S-33 was capable of destroying 49%, and strain 101n 46%, of oil for 7 days. As shown by gas chromatography, these microorganisms destroyed 78.4 and 85.2% of oil, respectively. Hydrocarbons with an average length of the carbon chain (C12-C-28) were the most degraded (figure 1).

Fractional analysis of residual oil indicated that all main fractions were subjected to microbial degradation: paraffins, naphthenes, mono-, bi- and polycyclic aromatic hydrocarbons, as well as alcohol-benzene tars (figure 2). Strain 101n better degraded aromatic fractions, whereas the rhizosphere strain S-33 degraded saturated oil hydrocarbons the most.

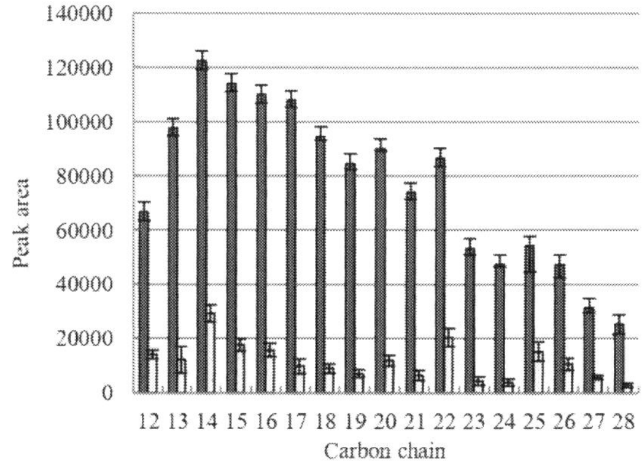

a

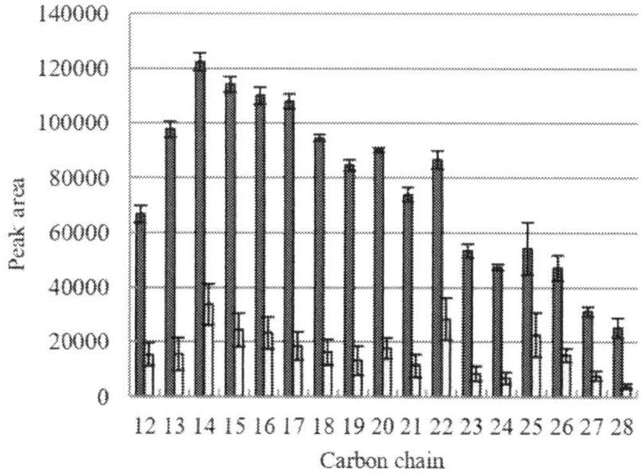

b

Figure 1. Degradation of oil hydrocarbons by strains 101n (*a*) and S-33 (*b*) according to gas chromatography data: ■, initial untreated control; □, after cultivation of microorganisms.

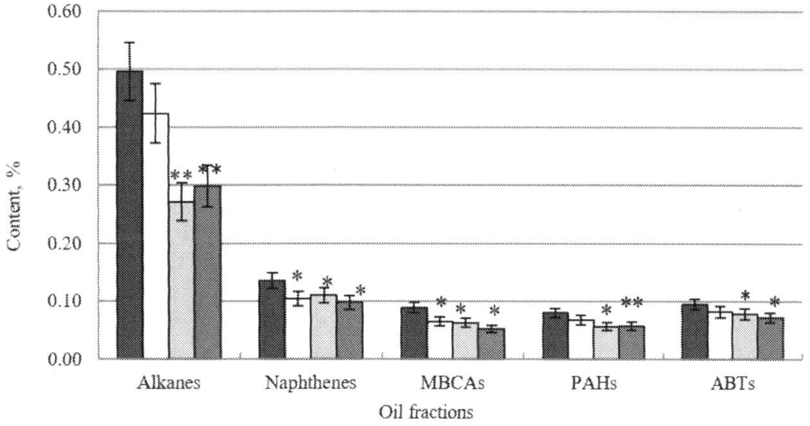

Figure 2. Degradation of oil fractions by strains S-33 and 101n grown in 1% ($v \cdot v^{-1}$) oil-containing mineral medium for 7 days: ■, initial untreated control; □, final untreated control; ▒, strain S-33 ▒, strain 101n; *, significant differences between treatments and initial control (P≤ 0.05); **, significant differences between treatments and untreated control (P≤0.05)

Comparison of the selected microorganisms with the previously described biological agents showed that the degradative activity of the new strains toward oil was in the range of activity reported for other microbial strains used, for example, in such biological agents as Ekoil [18] and Roder [19], other compositions [20].

On the basis of a study of cultural-morphological, physiological-biochemical and genetic characteristics, new oil-degrading strains S-33 and 101n were identified as *Acinetobacter* sp. and *Achromobacter marplatensis*, respectively.

3.2. Testing of oil-degrading activity of A. marplatensis 101n with original oil-containing wastes

The oil oxidizing activity of *A. marplatensis* strain 101n with the use of original oil-containing soil showed that the most successful bioremediation approach was the combined use of the oil-degrading microorganism and nitrogen fertilizer, in which the degradation of petroleum hydrocarbons was 49% (table 1).

Table1. Changes in characteristics of oil-contaminated soil during remediation

Experimental treatments	pH	Microbial count (CFU g^{-1})		Oil hydrocarbons	
		THMs	HOMs	Content ($g\ kg^{-1}$)	Degradation (%)
Oil-contaminated soil					
Initial(0 days)	4.9	5.4×10^6	1.2×10^4	274	
Untreated control (30 days)	3.5	1.0×10^6	0	260	5.2
Agrotechnics– loosening, watering, fertilization (30 days)	3.1	1.1×10^7	2.4×10^5	244	11.0
A. marplatensis 101n (30 days)	3.0	5.0×10^6	1.7×10^5	216	21.2
A. marplatensis 101n + Agrotechnics(30 days)	3.0	2.5×10^6	3.9×10^5	140	49.0
Oil sludge					
Initial	6.4	2.5×10^6	–[a]	858	
Abiotic control (0 days)	7.2	–	–	150	
A. marplatensis 101n (10 days)	7.1	$\geq 10^{8}$[b]		105	29.9

[a] HOMs were not detected by direct plating of the oil sludge sample, but they were detected as a result of enrichment culture.

[b] the initial seed dose of the microbial suspension of *A. marplatensis* 101n.

This is a very good result, with account taken of high acidity of the soil treated. The effectiveness of microbial degradation of hydrocarbons by *A. marplatensis* strain 101n was confirmed by cleaning the soil using a single microorganism by 21%. The stimulation of oil-oxidizing activity of the indigenous microflora was not effective enough for bioremediation of this soil – only 11%. The application of the microorganism with nitrogen fertilizer led to distinct changes in the soil as early as the first days after treatment. The soil became more friable, the lumps of pollutant were more easily destroyed during loosening, presumably due to the emulsifying activity of the introduced oil-degrading microorganism, which was clearly visible in the first laboratory experiments. However, the dynamics of such changes slowed down with time, which may be due to a decrease in soil pH and to inhibition of the strain's activity. By the end of the experiment, there was considerable acidification of the soil in all experimental treatments, which indicated its low buffer capacity. A decrease in pH to 3.0-3.5 inhibited the development of native bacterial populations but promoted the growth of fungi, which was observed in the untreated control, in which micromycetes did not compete with the bacteria introduced into other experimental treatments. In all variants of the experiment, with the exception of the untreated control, in comparison with the baseline, there was an increase in the number of HOMs, including micromycetes, which should also facilitate the cleanup.

A. marplatensis strain 101n successfully degraded oil hydrocarbons as part of oil sludge. As can be seen from Table 1, the degradation of oil sludge by *A. marplatensis* 101n was 30%. A comparison of the degradative activity of strain 101n tested for the treatment of oily wastes with previously reported data on other biological agents [21] allowed us to conclude that the microbial bioremediation agents developed on the basis of this microorganism can be promising for treating oil-containing wastes.

3.3. Application of Acinetobacter sp. S-33 and M. falcata for remediation of oil-contaminated soil

The prospects of using alfalfa for phytoremediation of oil-contaminated soil were shown previously [16, 22], and the additional application of a microbial inoculant can improve the remediation efficiency. As a result of a 2.5 months' soil experiment on bioremediation of the original oil-sludge-contaminated soil (14.3 $g \cdot kg^{-1}$) by using *Acinetobacter* sp. strain S-33 and yellow alfalfa, it was shown that the use of the oil-degrading microorganism for purification both as an monoculture bioagent and in combination with the plant significantly increased pollutant loss (Figure 3*a*). Soil cleanup in the control was 20%, and the application of yellow alfalfa yielded 34%. Application of *Acinetobacter* sp. S-33 only allowed cleaning of the soil by 35%; and the use of this microorganism in combination with alfalfa, by 39%. Fractional analysis of the residual contamination showed that the plant-microbial association was the most effective in remediation of soil from aromatic hydrocarbons, significantly degrading the MBCA and PAH fractions (figure 3*b*).

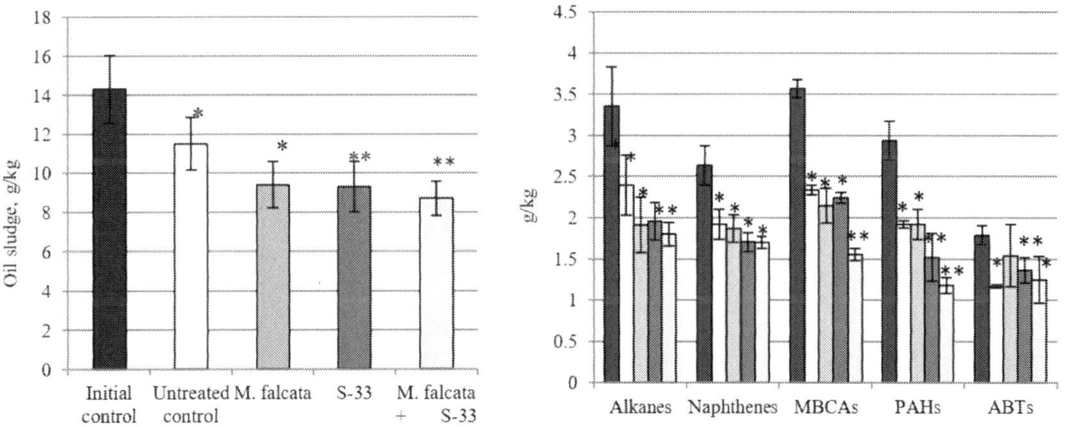

a *b*

Figure 3. Bioremediation of original oil-sludge-contaminated soil by different approaches: *a*, TPH content in soil; *b*, oil fraction content in soil; ■, initial oil-contaminated soil; □, untreated control; ▨, phytoremediation with *M. falcata*; ▩, bioaugmentation with *Acinetobacter* sp. S-33; □, phytoremediation (*M. falcata*) with bioaugmentation (*Acinetobacter* sp. S-33);
*, significant differences between treatments and initial control ($P \leq 0.05$);
**, significant differences between treatments and untreated control ($P \leq 0.05$).

Introduction of *Acinetobacter* sp. S-33 into soil had a significant effect on the soil microflora. It reduced the number of THMs by 46%, the planting of alfalfa increased eightfold (table 2). Additional application of *Acinetobacter* sp. S-33 in combination with alfalfa resulted in 10-fold increase in the THMs in comparison with the control and had a marked effect on the population of HOMs, reducing it by 20 and 24% in treatments without the plants and with alfalfa. Such quantitative changes in the oil-degrading microbial populations could reflect a competition between the introduced and the indigenous microorganisms.

Table 2. Effect of the introduced strain *Acinetobacter* sp. S-33 on the soil microflora.

Experimental treatments	Microbial count, $\times 10^7$(CFU g^{-1})	
	THMs	HOMs
Untreated control	11.67 ± 3.04	2.52 ± 0.18
Acinetobacter sp. S-33	6.30 ± 2.10	2.04 ± 0.41
M. falcata	88.66 ± 31.64	29.93 ± 1.86
M. falcata +*Acinetobacter* sp. S-33	111.22 ± 51.61	22.94 ± 6.26

Inoculation of alfalfa plants with *Acinetobacter* sp. S-33 increased the accumulation of shoot and root biomass by 44 and 47%, respectively. Thus, in our soil experiment, the rhizobacterium *Acinetobacter* sp. S-33 exhibited plant-growth-promoting activity and, thus, can be considered to be a promising agent for the phytoremediation of oil-contaminated soil.

Conclusion

From the results of our experiments, we conclude that the new oil-degrading microbial strains *Acinetobacter* sp.S-33 and *A. marplatensis*101n can serve as the basis for microbial bioremediation agents for the treatment of oil contaminated soils and wastes.

References

[1] Holliger C, Gaspard S, Gold G, Heijman C, Schumacher W, Schwarzenbach R P and Vazquez F 1997 *FEMS Microbiol Rev.* **20** 517 – 23
[2] Lin X, Li X, Li P, Li F, Zhang L and Zhou Q 2008 *Bull Environ Cont Toxicol.* **81** 19 – 24
[3] Fatima K, Afzal M, Imran A and Khan QM 2015 *Bull Environ Cont Toxicol.* **94** 314 – 20
[4] Das K and Mukherjee A K 2007 *Bioresour Technol.* **98** 1339 – 45
[5] Liu P W G, Chang T C, Whang L M, Kao C H, Pan P T and Cheng S S 2011 *Int Biodeterior Biodegrad.* **65** 1119 – 27
[6] Liu H et al 2014 *Int Biodeterior Biodegrad.* **87** 52 – 9
[7] Varjani S J and Upasani V N 2017 *Int Biodeterior Biodegrad.* **120** 71 – 83
[8] Andreeva I S, Emelianova E K, Zagrebelnyi S N, Ol'kin S E, Reznikova I K and Repin V E 2006 *Biotechnology in Russia* **1** 57 – 69
[9] Fedorenko V N, Serezhkin I N, Lamova Y A, Kniazyuk M K, Netrusov A I and Shestakov AI 2015 *Biotechnologiya* **6** 72 – 8
[10] Bushnell L D and Haas H F 1941 *J Bacteriol.***5** 653 – 74
[11] Muratova A, Hübner Th, Narula N, Wand H, Turkovskaya O, Kuschk P, Jahn R and Merbach W 2003 *Microbiol Res.***158** 151 – 61.
[12] Kim O S et al 2012 *Int J Syst Evol Microbiol.* **62** 716 – 21
[13] 2005 *Bergey's Manual of Systematic Bacteriology* vol 2: The Proteobacteria, Parts B Eds G Garrity et al (Springer US) p 658
[14] 2005 *Bergey's Manual of Systematic Bacteriology* vol 2: The Proteobacteria, Part C Eds G Garrity et al (Springer US) p 425
[15] Drugov Yu S and Podin A A 2007 *Ekologicheskie analizy pri razlivah nefti i nefterpoduktov (Ecological analyses for oil and oil product spills)* (Moscow: Binom, Laboratoriya znanyi) p 77
[16] Muratova A Yu, Dmitrieva T V, Panchenko L V and Turkovskaya O V 2008 *Int J Phytorem.* **10** 486 – 502
[17] 2010 PND F 16.1:2:2.2:2.3:3.64-10 (Environmental Regulation Document at the Federal Level No. 16.1:2:2.2:2.3:3.64-10) Methods of measurement of the mass fraction of oil in the samples of soils, sediments, sludges, sewage sludge, waste production and consumption by

IOP Conf. Series: Earth and Environmental Science **107** (2017) 012066 doi:10.1088/1755-1315/107/1/012066

gravimetric method) (in Russian) *PND F (Federal environmental regulatory documents)* (Moscow: The Federal Supervisory Natural Resources Management Service) p 18

[18] Ermolenko Z M, Kholodenko V P, Chugunov V A, Zhirkova N A and Rasulova G E 1997 *Microbiology* (*Mikrobiologiya*) **66** 542 – 45

[19] Zhukov D V, Murygina V P and Kalyuzhnyi S V 2007 *Appl Biochem Microbiol.* **43** 587 – 92

[20] Khabibullina F M, Shubakov A A, Archegova I B and Romanov G G 2002 *Biotekhnologiya* **6** 57 – 62

[21] Rogozina E A, Andreeva O A, Zharkova S I, Martynova V A and Orlova N A 2010 Comparative characteristic of native biopreparations proposed for cleanup of soils and grounds from pollution. *Neftegazov. geol., Teor. prakt.* **5** www.ngtp.ru/rub/7/37_2010.pdf

[22] Panchenko L, Muratova A and Turkovskaya O 2017 *Environ Sci Poll Res.* **24** 3117 – 30

Soil diagnosis of an urban settlement with low levels of anthropogenic pollution (Stepnoe, Saratov region)

C T Ngun, Ye V Pleshakova and M V Reshetnikov

Saratov State University, Saratov, Astrakhanskaya Street, 83, 410012, Russia
e-mail: clementngun@yahoo.com

Abstract. A soil diagnosis of an urban territory Stepnoe (Saratov region) was conducted within the framework of soil research monitoring of inhabited localities with low levels of anthropogenic impact using chemical and microbiological analysis. Excess over maximum permissible concentration (MPC) of mobile forms of Cr, Zn and Cd were not observed within the researched territory. A universal excess over MPC of mobile forms of Ni, Cu and Pb was established which is most likely connected with anthropogenic contamination. It was discovered that, at the territory of the Stepnoe settlement, mobile forms of heavy metals compounds (HM) in most cases formed paragenetic associations with high correlation coefficient and despite this, an excess over MPC was not significant. This point to a common mineralogical origin of the elements inherited from the parent rock. The values of the total index of chemical contamination were not above 16, which puts the researched samples in a category with permissible contamination. The indices of the total number of heterotrophic bacteria, iron-oxidizing and hydrocarbon-oxidizing bacteria in most samples corresponded to normal indices for chestnut solonetsous and saline soils. In some samples, a deviation from the normal indices was observed justifying the impact of specific contaminants on the soil.

1. Introduction

In the modern world, compounds of heavy metals (HM) have been known to be part of universal pollutants of soil cover [1]. They influence soil biological properties while inhibiting soil enzyme activities [2]. This is connected with repression of microbial biochemical activity and also a qualitative and quantitative reduction in their structure [3]. Problems of soil pollution with HM in districts of large cities have been well studied [4-6] but for settlements with a population less than 50 thousand with less industrial activities, this problem has practically not been studied.

Within the framework of soil research monitoring in settlements which serve as administrative centers in Saratov region we carried out soil diagnosis of an urban village Stepnoe using chemical and microbiological analysis. According to Aleksenko's classification [7] the researched territory falls under a group of small settlements, communities and small villages that represent a transition from agricultural, grassland and forest landscapes to city landscapes of inhabited locality.

In the course of this work, the content of mobile forms of copper, zinc, lead, cadmium, chromium and nickel was determined, the ecologically harmful heavy metals hazard ratio and the total index of chemical contamination Z_c by HM were calculated. Microbiological indices of the soil samples were also evaluated for the total number of heterotrophic microorganisms (HMs), number of iron-oxidizing bacteria (IOMs) and hydrocarbon-oxidizing bacteria (HOMs). Soils are often exposed to mixed pollution. Oil hydrocarbons and HM are known to be widely distributed and dangerous soil pollutants [8]. In connection with the above said, the aim of this research was to biogeochemically analyze soils of a territory with low levels of anthropogenic impact (for example, the settlement Stepnoy, Saratov region).

2. Material and methods

Objects of this research were soil samples obtained from the territory of the urban village Stepnoe (figure 1) with a size of 8 km^2 and a population of about 12.340 inhabitants (according to data obtained in 2015). This settlement is located within the Syrtian Plain, belonging to the East European Plain. The relief is gently sloping. Soils are chestnut solonetsous and saline soils.

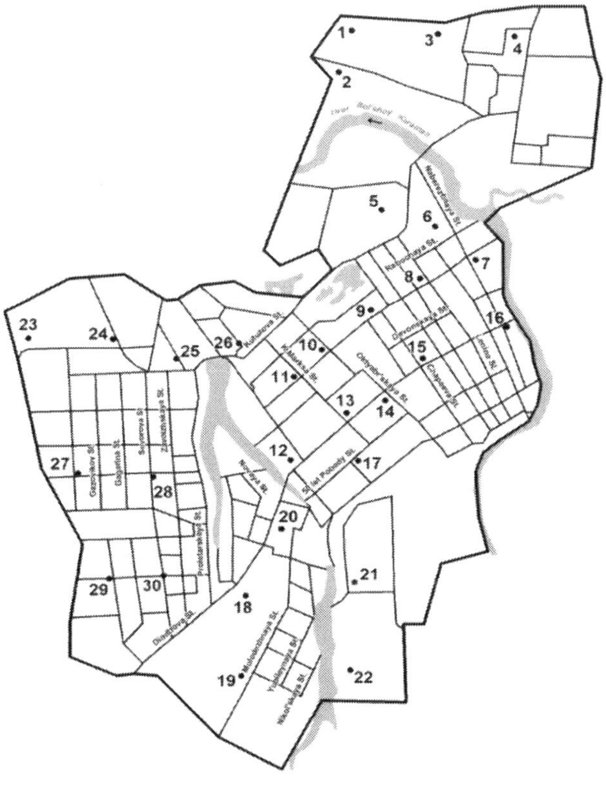

Figure 1. Map showing soil sample positions of the urban settlement territory Stepnoe.

Selection and preparation of samples was carried out in accordance with State Standards [9]. Soil sample points were distributed taking into consideration the wind rose, peculiarities of the micro-topography, distribution of the building plans and basic amenities. In accordance with State Standards, a test was conducted on the upper layers of the soil horizon «A» to the depth of 5 cm where basic pollutants from the atmosphere are usually accumulated. The size of the sample area varied from 2-3 to 10 м^2. Soil sampling was carried out using the envelope method – one sample obtained from the center and 4 samples from the corners of the sample point and also from 2-3 samples points around the top of the envelope. The weight of the combined samples varied in a scale of about 0.5-1.0 kg.

Determination of HM concentration in the soil samples was conducted using atom-adsorption spectrometric method (Kvant-2AT). The mobile acid-soluble form of the metals (Cu, Zn, Ni, Cd, Pb) was determined in 1 M HNO_3 extracts [10]. To determine the ecologically dangerous concentrations of HM in the soil cover, the actual concentration of each HM was compared with its maximum permissible concentration (MPC), while calculating the heavy metal hazard ratio K_o using the formula (1):

$$K_o = Ci \cdot MPC^{-1} \quad (1)$$

Where C_i – content of HM forms in the sample, [mg kg^{-1}];

MPC – maximum permissible concentration, [mg kg^{-1}].

The total index of chemical contamination Z_c was determined to help evaluate the level of geochemical transformation by the mobile forms of HM using the formula (2):

$$Zc = \sum Ko_n - (n-1) \quad (2)$$

where Z_c – total index of chemical contamination by HM in the sample;
n – number of determined elements;
Ko – heavy metal hazard ratio in the sample.

For calculating Z_c, an excess over of MPC was used (*Ko*).

The total number of culturable heterotrophic microorganisms in the soil (1 g, wet weight) was determined by plate dilution technique with a beef-extract agar medium. The number of hydrocarbon-oxidizing microorganisms was evaluated on Mills et al. [11] modified mineral agar medium, g l^{-1}: Na_2HPO_4 – 6.0; KH_2PO_4 – 3.0; NaCl – 0.5; NH_4Cl – 1.0 with vaseline oil (1%) as a sole carbon and energy source. An account of the number of neutrophilic iron-oxidizing microorganisms was carried out on a selective agar medium with the following composition, g l^{-1}: $FeSO_4 \times 7H_2O$ – 5.9; $(NH_4)_2SO_4$ – 0.5; $NaNO_3$ – 0.5; K_2HPO_4 – 0.5; $MgSO_4 \times 7H_2O$ – 0.5; lemon acid – 10.0, sucrose – 2.0, peptone – 1.0, pH 7.0 [12].

Plating on beef-extract agar and on selective media for an account of the number of hydrocarbon-oxidizing and iron-oxidizing bacteria was carried out from dilutions 10^{-2}, 10^{-3}, 10^{-4}, 10^{-5} in replicates. The plates with inoculants were incubated at 28-30°C for 2-5 days at a temperature of 28-30°C in an incubator and colony-forming units (CFU) were counted and calculated per g of dry soil [13].

Data were analyzed statistically using the software (Microsoft Excel 2010). The differences in probable errors $p<0.05$ (95% confidence interval) was considered the confidence. The interrelationships between the chemical and microbiological indices of soil were evaluated using pair correlation coefficients for $p<0.01$.

3. Results

The concentrations of mobile forms of Cu, Zn, Pb, Cd, Cr and Ni were determined in the 30 soil samples selected from the territory of the urban settlement Stepnoe (table 1). Concentrations of mobile forms of Cr observed in all soil samples were between 1.33 to 4.22 mg kg^{-1} which were less than the maximum permissible concentration (MPC_{Cr} = 6.0 mg kg^{-1}) [14]. Mobile forms of Zn were noticed in all soil samples in concentrations between 4.38 to 15.74 mg kg^{-1} and was also below the maximum permissible concentration (MPC_{Zn}= 23.0 mg kg^{-1}) [15].

Concentrations of mobile forms of Cd observed in all soil samples were between 0.08 to 0.11 mg kg^{-1} which were less than the maximum permissible concentration (MPC_{Cd} = 1.0 mg kg^{-1}) [14]. The heavy metal hazard ratio varied from 0.08 to 2.19 and the MPC was not exceeded. Mobile forms of Ni were noticed in all soil samples in concentrations between 7.66 to 14.20 mg kg^{-1}, MPC_{Ni} = 4.0 mg kg^{-1} [14]. The heavy metal ratio varied at a scale of 1.92 to 3.55. An excess over MPC was established in all the soil samples with the highest value observed in sample №3 (3.55 MPC). Concentrations of mobile forms Pb of observed in all soil samples were between 3.64 to 7.62 mg kg^{-1}, MPC_{Pb} = 6.0 mg kg^{-1} [14]. The heavy metal hazard ratio was at a range of 0.61 to 1.27. An excess over MPC was observed in 4 samples with sample №3 having a maximum value of 1.27 MPC. Mobile forms Cu of were noticed in all soil samples in concentrations between 3.05 to 7.82 mg kg^{-1}, MPC_{Cu} = 3.0 mg kg^{-1} [16]. The heavy metal ratio varied from 1.02 to 2.61. Excess over MPC were stated in all samples with sample №3 having a maximum value of 2.61 MPC.

A geochemical order for mobile forms of HMs in the territory of the urban settlement Stepnoe was as follows: Ni>Zn>Pb>Cu>Cr>Cd (based on average concentrations) and Ni>Cu>Pb>Cr>Zn>Cd (based on increments above normal indices). Analyzing the research results, the elements were distributed into 2 groups: Group 1 – Ni, Pb, Cu, soil samples in which an excess over MPC was observed; Group 2 – Cr, Zn and Cd, soil samples in which an excess over MPC was not noticed. A calculation of the total index of contamination showed the following – in the researched territory Z_c

varied at a scale from -0.8 to 3.92. All of the soil samples (100%) were put in a category with permissible levels of pollution (Z_c from 0 to 16).

Table 1. Results of soil chemical analysis of the settlement Stepnoe, Saratov region.

Soil Sample Identifier	Concentration of heavy metals, mg kg^{-1} of soil						Heavy metals hazard ratio (*Ko*)						*Zc*
	Cr	Ni	Pb	Zn	Cd	Cu	Cr	Ni	Pb	Zn	Cd	Cu	
1	2.89	10.90	5.69	9.24	0.09	5.74	0.48	2.73	0.95	0.40	0.09	1.91	1.56
2	3.60	11.60	6.41	10.49	0.09	6.86	0.60	2.90	1.07	0.46	0.09	2.29	2.40
3	4.22	14.20	7.62	15.74	0.11	7.82	0.70	3.55	1.27	0.68	0.11	2.61	3.92
4	3.44	13.41	6.14	10.34	0.11	5.63	0.57	3.35	1.02	0.45	0.11	1.88	2.39
5	3.70	13.13	6.49	12.08	0.11	6.19	0.62	3.28	1.08	0.53	0.11	2.06	2.68
6	2.80	11.08	5.82	9.59	0.10	5.58	0.47	2.77	0.97	0.42	0.10	1.86	1.58
7	1.49	9.68	4.30	9.43	0.10	3.85	0.25	2.42	0.72	0.41	0.10	1.28	0.17
8	1.83	9.25	4.64	7.57	0.09	4.34	0.31	2.31	0.77	0.33	0.09	1.45	0.26
9	1.33	8.58	3.99	6.34	0.09	3.63	0.22	2.15	0.67	0.28	0.09	1.21	-0.40
10	1.40	7.99	4.15	6.47	0.08	3.64	0.23	2.00	0.69	0.28	0.08	1.21	-0.50
11	1.60	8.14	4.82	5.64	0.08	4.55	0.27	2.04	0.80	0.25	0.08	1.52	-0.05
12	1.63	8.53	4.90	5.57	0.09	4.33	0.27	2.13	0.82	0.24	0.09	1.44	-0.01
13	1.81	8.99	5.50	5.87	0.09	4.40	0.30	2.25	0.92	0.26	0.09	1.47	0.28
14	1.94	8.77	5.10	5.98	0.08	5.16	0.32	2.19	0.85	0.26	0.08	1.72	0.43
15	1.51	7.66	4.66	6.46	0.08	3.94	0.25	1.92	0.78	0.28	0.08	1.31	-0.38
16	1.69	8.45	4.90	5.70	0.08	4.06	0.28	2.11	0.82	0.25	0.08	1.35	-0.11
17	1.58	9.10	4.74	5.73	0.10	4.12	0.26	2.27	0.79	0.25	0.10	1.37	0.05
18	1.64	8.42	4.95	6.48	0.09	4.22	0.27	2.10	0.82	0.28	0.09	1.41	-0.02
19	1.74	8.32	5.01	5.92	0.09	4.51	0.29	2.08	0.84	0.26	0.09	1.50	0.06
20	1.64	8.12	5.23	5.52	0.09	4.73	0.27	2.03	0.87	0.24	0.09	1.58	0.09
21	1.83	10.78	5.02	6.28	0.10	4.46	0.31	2.70	0.84	0.27	0.10	1.49	0.70
22	1.76	10.10	4.41	5.99	0.10	4.09	0.29	2.52	0.73	0.26	0.10	1.36	0.28
23	1.54	9.56	4.10	5.25	0.10	3.80	0.26	2.39	0.68	0.23	0.10	1.27	-0.07
24	1.67	10.36	4.20	5.45	0.10	3.57	0.28	2.59	0.70	0.24	0.10	1.19	0.10
25	1.55	8.83	4.78	6.00	0.10	4.30	0.26	2.21	0.80	0.26	0.10	1.43	0.05
26	1.42	8.28	3.64	4.38	0.08	3.05	0.24	2.07	0.61	0.19	0.08	1.02	-0.80
27	1.54	9.36	4.66	5.90	0.10	3.99	0.26	2.34	0.78	0.26	0.10	1.33	0.06
28	1.81	10.74	4.56	5.97	0.10	4.15	0.30	2.69	0.76	0.26	0.10	1.38	0.49
29	1.51	9.40	4.89	6.02	0.10	4.16	0.25	2.35	0.81	0.26	0.10	1.39	0.17
30	1.42	9.12	4.50	4.93	0.09	4.01	0.24	2.28	0.75	0.21	0.09	1.34	-0.09

Results of a correlation analysis conducted to identify possible paragenetical associations of elements showed the presence of strong correlation connections (values of r from ±0.7 to ±1.0) in most cases, this is typical of associations such as: Cr-Ni, Cr-Pb, Cr-Zn, Cr-Cu, Ni-Pb, Ni-Zn, Ni-Cd, Ni-Cu, Pb-Zn, Pb-Cu, Zn-Cu (table 2). Average correlation interconnections (values of r from ±0.3 to ±0.699) which is usual for the pairs: Cr-Cd, Pb-Cd, Zn-Cd, Cd-Cu. On the basis of these results and also considering the fact there was no excess over MPC, it can be proposed that, these elements most probably have the same mineralogical origin inherited from their parent rock.

Table 2. Correlation coefficients between the numbers of different microorganisms and chemical parameters of soil.

Parameters	C_{Cr}	C_{Ni}	C_{Pb}	C_{Zn}	C_{Cd}	C_{Cu}	Z_c	HMs	HOMs	IOMs
C_{Cr}	1	0.89*[a]	0.92*	0.91*	0.56	0.94*	0.98*	-0.24	-0.26	-0.41
C_{Ni}		1	0.75*	0.83*	0.82*	0.76*	0.93*	-0.14	0.02	-0.35
C_{Pb}			1	0.84*	0.46	0.96*	0.93*	-0.27	-0.38	-0.36
C_{Zn}				1	0.56	0.86*	0.92*	0.02	-0.12	-0.30
C_{Cd}					1	0.43	0.65*	-0.01	0.37	-0.27
C_{Cu}						1	0.95*	-0.24	-0.36	-0.36
Z_c							1	-0.19	-0.20	-0.04
HMs								1	0.50	0.63*
HOMs									1	-0.02
IOMs										1

[a]*Statistically significant differences for $p<0.01$

Abbreviations: HMs – heterotrophic microorganisms; HOMs – hydrocarbon-oxidizing microorganisms; IOMs – iron-oxidizing microorganisms

It is well known that the total number of heterotrophic microorganisms serves as a reliable sensitive monitoring indicator in soil analysis [17]. Iron-oxidizing soil microorganisms carry out the dissociation of iron compounds and iron-organic complexes [18], which can be as a consequence of technogenic soil contamination and in this case with HM. It is based on this the indices of the number of iron-oxidizing bacteria was chosen for the microbiological monitoring analysis of the soil samples. In most of the researched soil samples the total number of heterotrophic microorganisms was between 24.8 to 99.4×10^5 CFU g^{-1} of soil (figure 2).

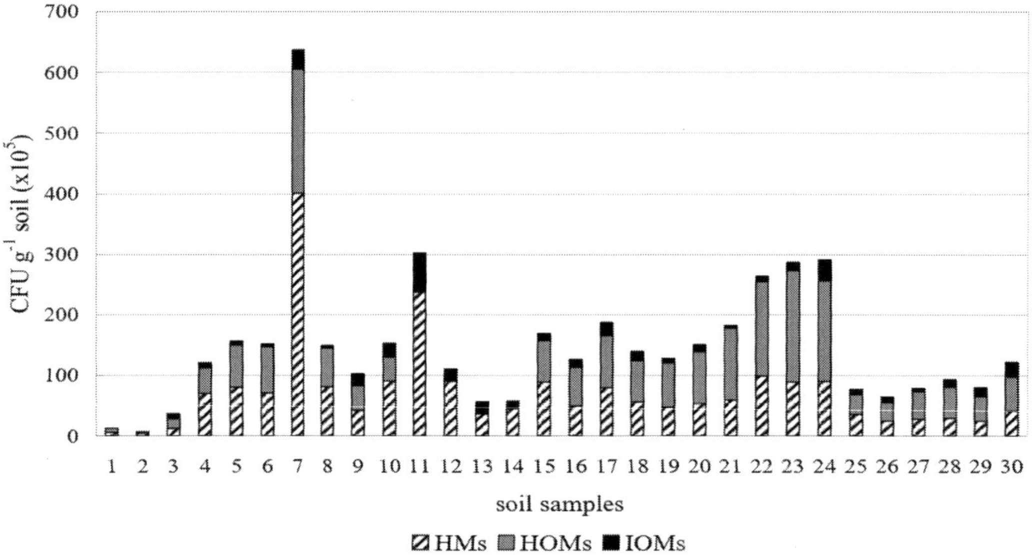

Figure 2. Number of microorganisms in soil samples of the urban settlement territory Stepnoe.

The number of iron-oxidizing microorganisms ranged from 4.6 to 34.7×10^5 CFU g^{-1} of soil. The number of hydrocarbon-oxidizing microorganisms in 18 samples was between 29.5 to 86.0×10^5 CFU g^{-1} of soil. These indices of the number of bacteria corresponded with the normal indices for chestnut solonetsous and saline soils. In three samples (№ 1, 2 and 3) obtained at the boarder of the researched territory, the total number of heterotrophic microorganisms and the number of hydrocarbon-oxidizing microorganisms was low, comprising of 4.3 of 12.4×10^5 and 2.7 to 15.8×10^5 CFU g^{-1} soil respectively. In samples №1 and 2 the number of iron-oxidizing microorganisms was also low (0.04 and 0.01×10^5 CFU g^{-1} soil). Sample № 11, 12, 13 and 14 differed with low values for the number of hydrocarbon-oxidizing microorganisms. Sample №7 was observed to possess a high number of heterotrophs, hydrocarbon-oxidizing and iron-oxidizing microorganisms, while in sample № 11 – the number of heterotrophs and iron-oxidizing microorganisms was very significant. This could be connected with the peculiarity of the contaminants and the level of soil pollution in that given sample point, which was located at the central part of the settlement. As a whole, a significantly positive correlation between the total number of heterotrophic microorganisms and the number of iron-oxidizing microorganisms was observed $r = 0.63$ ($p<0.01$) (table 2).

An increased hydrocarbon-oxidizing microbial content was observed in soil samples № 21, 22, 23 and 24. This can serve as a proof of fresh hydrocarbon contamination, as it is well known that the presence of fresh hydrocarbon soil pollution leads to an increase in the content of hydrocarbon-oxidizing microorganisms [19].

Between the amount of bacteria of the researched physiological group and the concentrations of HM in the soil samples from the settlement, strong correlation connections were not discovered. A positive correlation coefficient ($r =0.37$, $p<0.01$) was noticed between the number of hydrocarbon-oxidizing microorganisms and the concentration of Cd. It is a well known fact that Cd is quite often present in petroleum products [20, 21], which explains why when volatile hydrocarbons vanish from contaminated soils in the course of time or hydrocarbons undergo biodegradation, Cd which had been accumulated in soil continues to show an influence on soil biocenosis.

Conclusion

An examination of the geochemical peculiarities, in this case the presence of mobile forms of HM in the soils of the urban settlement Stepnoe brought about the following conclusions. A universal excess over MPC of mobile forms of the following elements was established within the boundaries of the researched territory – nickel, copper and lead. This increment most probably was due to anthropogenic contamination. An analysis of the correlation interconnections showed a significant correlation relationship practically between all the elements, which most likely indicates their belonging to one of the mineralogical associations.

Results of the soil microbiological analysis of the urban settlement Stepnoe showed: Three soil samples were distinct and showed a low soil microbial content with regards to the researched physiological groups, these samples were also characterized by high values of the total index of chemical contamination. A high microbial content of all the researched groups was observed in one of the samples (№ 7), while a maximum value for iron-oxidizing microorganisms and a high heterotrophic microbial content was noticed in sample № 11 justifying the influence of a specific contaminant on the soil.

In general, the identified geochemical anomalies (content of mobile forms of heavy metals) indicate favorable ecological and geochemical conditions in the territory of the urban settlement Stepnoe. Values of total index of chemical contamination were not above 16, which help place the 30 soil samples in a category with permissible contamination. According to the results of microbiological analysis, 25 soil samples can be assigned to the category with permissible contamination.

The absence of significant correlation between the number of heterotrophic, hydrocarbon-, and iron-oxidizing microorganisms and the indicators of chemical soil contamination also speaks well of the favorable ecological situation in the territory of the settlement Stepnoe.

Acknowledgement

This research was conducted with support of the Russian Science Foundation (research project No. 17-77-10040).

References

[1] Robert B 2010 *J Chem Ecol.* **36** 46 – 58
[2] Tuomela M, Steffen K and Kerko E 2005 *Microbial Ecol.* **53** 179 – 86
[3] Murata T, Kanao-Koshikawa M and Takamatsu T 2005 *Water Air Soil Poll.* **164** 103 – 18
[4] Salah E A M, Yassin K H and Abd-Alsalaam S 2015 *International Journal of Scientific Research in Science, Engineering and Technology (IJSRSET)* **6** 1646 – 52
[5] Su C, Jiang L and Zhang W 2014 *Environmental Skeptics and Critics.* **3** 24 – 38
[6] Angelovičová L, Bobuľská L and Fazekašová D 2015 *Carpath J Earth Env.* **10** 193 – 201
[7] Alekseenko V A and Alekseenko A V 2013 *Khimicheskie elementy v geokhimicheskikh sistemakh. Klarki pochv selitebnykh landshaftov Chemical Elements in Geochemical Systems. Soil Clusters in Residential Landscapes* (Rostov n/D: Izd. YuFU) (in Russian) p 383
[8] Gennadiev A N, Pikovskii Yu I, Tsibart A S and Smirnova M A 2015 *Eurasian Soil Sci.* **48** 1076 – 89
[9] 1984 GOST 17.4.4.02-84. Okhrana prirody. Pochvy. Metody otbora I podgotovki prob dlya khimicheskogo, bakteriologocheskogo, gelmintologicheskogo analiza (State Standard 17.4.4.02-84. The Nature Conservation. Soils. Methods of Selection and Preparation of Samples for Chemical, Bacteriological, Helminthological of Analysis) *Russian national standards (GOST) (in Russian)* (Moscow) p 8
[10] Dedina J and Tsalev D L 1995 *Hydride Generation Atomic Absorption Spectrometry* (New York: Wiley) p 526
[11] Mills A, Breuil Lu C and Colwell R R 1978 *Can J Microbiol.* **24** 552 – 7
[12] Granina L Z, Parfenova V V, Zemskaya T I, Zakharova Yu R and Golobokova L P 2003 *Berliner Palaobiologische Abhandlungen.* **4** 121– 8
[13] Foght J and Aislabie J 2005 Enumeration of soil microorganisms *Manual for Soil Analysis – Monitoring and Assessing Soil Bioremediation.* ed Margesin R and Schinner F (Berlin, Heidelberg, Germany: Springer-Verlag) 261–80
[14] 2006 GN 2.1.7.2041-06. Predel'no dopustimye koncentracii (PDK) i orientirovochno-dopustimye koncentracii (ODK) himicheskih veshhestv v pochve: Gigienicheskie normativy (Maximum Permissible Concentrations (MPC) of Chemicals in Soil. Hygienic Standards) *Hygienic Standards* (Moskow: Ministry of Health) (in Russian) p 15
[15] 2003 SANPIN 2.1.7.1287-03. Pochva, ochistka naselennyh mest, bytovye i promyshlennye othody, sanitarnaja ohrana pochvy. Sanitarno-jepidemiologicheskie trebovanija k kachestvu pochvy. Kontrol' (Soil, Cleaning of Populated Areas, Household and Industrial Waste, Sanitary Protection of Soil. Sanitary and Epidemiological Requirements to Soil Quality. Control, Moscow) *Sanitary Rules and Norms* (Moskow: Federal Service for Surveillance in Protection of Consumer Rights and Human Welfare) (in Russian) p 15
[16] 1992 Metodicheskie ukazanija po opredeleniju tjazhelyh metallov v pochvah sel'hozugodij i produkcii rastenievodstva (Methodological Guidelines on the Determination of Heavy Metals in Agricultural Soils and Crops) (Moscow: TsINAO) (in Russian) p 61
[17] Sumampouw O J and Risjani Y 2014 *Int J Ecosyst Ecol Sci.* **4** 251 – 8.
[18] Sabrina H, Michael S and Johnson B 2011 *Microbiology.* **157** 1551 – 64.
[19] Panov A V, Esikova T Z, Sokolov S L, Kosheleva I A and Boronin A M 2013 *Microbiology* **82** 241 – 8
[20] Mansfied C T, Barman B N, Thomas J V, Mehrotra A K and McCann J M 1999 *Anal Chem.* **71** 81 – 107
[21] Pekka Vähäoja 2006 *Acta Univ Oul* A **458** 76

IOP Conf. Series: Earth and Environmental Science **107** (2017) 012068 doi:10.1088/1755-1315/107/1/012068

Toxicity assessment of polluted sediments using swimming behavior alteration test with Daphnia magna

O V Nikitin, E I Nasyrova, V R Nuriakhmetova, N Yu Stepanova, N V Danilova and V Z Latypova

Institute of Environmental Sciences, Kazan Federal University, Kremlevskaya str., 18, 420008, Russia

e-mail: olnova@mail.ru

Abstract. Recently behavioral responses of organisms are increasingly used as a reliable and sensitive tool in aquatic toxicology. Behavior-related endpoints allow efficiently studying the effects of sub-lethal exposure to contaminants. At present behavioural parameters frequently are determined with the use of digital analysis of video recording by computer vision technology. However, most studies evaluate the toxicity of aqueous solutions. Due to methodological difficulties associated with sample preparation not a lot of examples of the studies related to the assessment of toxicity of other environmental objects (wastes, sewage sludges, soils, sediments etc.) by computer vision technology. This paper presents the results of assessment of the swimming behavior alterations of *Daphnia magna* in elutriates from both uncontaminated natural and artificially chromium-contaminated bottom sediments. It was shown, that in elutriate from chromium contaminated bottom sediments (chromium concentration 115±5.7 μg l^{-1}) the swimming speed of daphnids was decreases from 0.61 cm s^{-1} (median speed over the period) to 0.50 cm s^{-1} (median speed at the last minute of the experiment). The relocation of Daphnia from the culture medium to the extract from the non-polluted sediments does not essential changes the swimming activity.

1. Introduction

Various bioassay methods are actively used for assess the quality of natural waters and other environmental objects as well as the safety of existing or new artificially created chemical substances. In the tests, the evaluation is performed on the basis of observations and experiments with "living sensors" – organisms whose vital activity is disturbed in the toxic conditions [1–3]. Despite the fact that the range of organisms used as tests is very wide, there is no organism that is universally sensitive to all existing toxicants. Therefore, tend to use batteries of tests organisms [4, 5].

Cladocera (an order of minute branchiopod crustaceans) are one of the most frequently used organisms in ecotoxicological studies [6]. A feature of most of them is the filtration method of feeding, in which considerable amounts of water are passed over a specialized filtering structure. Cladocera are able to extract from the water not only the necessary food but also dissolved and suspended solids including the components of contaminants.

In most cases, as test objects used crustaceans *Daphnia magna*, *D. longispina*, *Ceriodaphnia dubia*, *C. affinis* and some others. The most universal test object for sensitivity and adequacy of response to different toxicants can be considered *D. magna*. Biotesting methods using *D. magna* are widely used for the purposes of environmental control, the laboratory culture of *D. magna* is one of the most commonly used in the practice of determining the acute and chronic toxic effects of individual chemicals and water samples with a complex chemical composition. Typically the toxicity is assessed by the EC50, LC50 end-points and also by a statistically significant change in the parameters of vital activity of daphnids under toxic and control conditions. Mortality of crustaceans is mainly used as a test reaction in acute tests; observations of changes in fertility and quality of offspring are carried out in the study of chronic toxic effects [7].

It is possible to significantly expand the list of recorded parameters by using additional information about the test organism based on its functional characteristics including behavioral responses [8,9] Moreover, analysis of the literature shows that behavioral reactions yielded significant responses at concentrations of toxicants well below reported LC50 values [10].

Behavioral responses most clearly manifested at the organism level, are between the biochemical and ecosystem levels of biomonitoring. However, at the same time behavioral alterations are directly based on biochemical processes, reflect the adaptation of an individual organism as well as the potential effects on the population level. Behavioral responses appear to compare favourably with biochemical and physiological responses in terms of sensitivity and efficiency. Furthermore, in addition to their integrative nature and ecological importance, the assessment of behavioral responses is a non-destructive method that makes possible continuous long-term environmental monitoring [11, 12].

Behavioral responses of aquatic organisms have been used for decades as methods for environmental monitoring but these types of studies have previously received much less attention than others focused on different subject such as developmental or reproductive toxicology [10]. Probably this was due to the problems in recording and quantification of the behavioral characteristics of test organisms. Many difficulties eventually were removed and at present, in addition to visual registration, a quantitative description of behavioral reactions is carried out by various methods: optical, ultrasonic, electrical, magnetic etc.

At the current stage of the development of technology, it is possible to sensitive determination of the presence of toxic substances in water by the behavioral characteristics of test objects, including such small organisms like Daphnia, using digital analysis of video recording [13–16]. To assess the toxicity used such daphnids behavioral characteristics as swimming time and speed, swimming trajectory, hopping frequency, vertical and horizontal distribution, distance travelled, number of turnings and turning angle, resting time, duration of quiescence, sinking rate, gravitaxis, swarming and spinning. Swimming mobility expressed by scalar quantity-speed and its vector quantity-velocity (usually expressed in millimeters per second) is one of the most reliable and widely used parameter of Daphnia behavioral activity [16].

Most studies have evaluated the toxicity of aqueous solutions of pure chemicals by the behavioral alterations of the test organisms. In addition to these studies, it is of interest to estimate the toxicity of other environmental objects (wastes, sewage sludges, soils, sediments etc.) using a digital analysis of the behavioral responses of test organisms exposed to water extracts of solid sample. In this case, there are methodological difficulties associated with sample preparation, such as staining and increasing the turbidity of the water solution. This can restrict the application of computer vision technology since this will make it impossible to recognize objects. The aim of this study was to assess the alterations in the swimming behavior of *D. magna* in elutriates from both uncontaminated natural and artificially chromium-contaminated bottom sediments.

2. Material and methods

Sediment samples for the experiment were collected in 2015 from the Almetyevsk reservoir (54°55'1" N 52°16'24" E). The reservoir is located in the northern part of Almetyevsk (Republic of Tatarstan, Russia). The reservoir exists within the riverbed and the low floodplain of the Steppe Zay River. The reservoir is shallow; the prevailing depth is about 0.9 m. Water area is 1.11 km^2, the maximum length and width is 1.78 and 1.0 km respectively.

Sediments were dried at the temperature of 105 °C and ground, than they were sieved through a 1 mm mesh. To simulate contaminated bottom sediments in the initial non-polluted sediments ("IS") potassium dichromate ($K_2Cr_2O_7$) was added in an amount of 1.41 grams per 100 grams of dry weight and mixed thoroughly ("CS"). Chromium was selected because of the high toxicity it represents to different zooplankton organisms. The resulting samples were filled with distilled water (Milli-Q, Millipore) at a ratio of "solid phase : liquid" – "1:10" in beakers. Then they were exposed on the platform shaker for 8 hours; the supernatant was allowed to settle for 12 hours and filtered ("IS-elutriate" and "CS-elutriate"). The sediments and extracts of the ones were analyzed for metals content by emission spectrometer ICPE-9000 (Shimadzu).

The behavioral experiments were performed with *D. magna* Straus, cultured in accordance with the Russian standard procedure [10]. For experiments we used parthenogenetically reproducing laboratory clones of Daphnia cultivated at the Department of Applied Ecology at the Kazan Federal University (Kazan, Russia). *D. magna* are cultivated at constant conditions of temperature (20±2°C), under 2000–2500 lux light intensity and 12/12 h light/dark photoperiod in a laboratory incubator.
Daphnids were fed daily with 1–2 mL *Chlorella vulgaris* suspension cultured in Tamiya medium. The daphnids used in this experiment originate from one single female to eliminate variation among clones. *D. magna* individuals were selected according to size (3.23±0.03 mm). Size measurements carried out using an eyepiece micrometer.

Measuring of swimming activity of the daphnids were determined and analyzed in real time by the hardware-software system "TrackTox". The scheme of experiment is presented in figure 1. Three daphnids were transferred from permanent culture into transparent polycarbonate test chamber (100 mm × 45 mm × 10 mm) with 25 ml of culture medium (1) and subsequently was placed into "TrackTox" (2) under static conditions (20±2 °C), without any water exchange during the experiment. The transfer of daphnids was carried out rapidly and carefully by micropipette in order to minimizing the stress to the test organisms. The next 30 min measuring of swimming activity (with rate 3 times s^{-1}) was implemented by "TrackTox" (3). In order to observe the behavioral response of the *D. magna* to elutriate from non-polluted and contaminated sediments daphnids consecutively were transferred from test chamber with culture medium ("CM") into test chambers with "IS-elutriate" (4) and "CS-elutriate" (5). In each case swimming activity was determined within 60 min by "TrackTox". During the experiment data are visualized on display (6). At the end of the experiment the data have been stored and exported to a statistical software package for further processing. Thus the total exposure time for each group of three was 150 (30+60+60) minutes. The acute toxicity of the same samples was additionally assessed (within 48 hours).

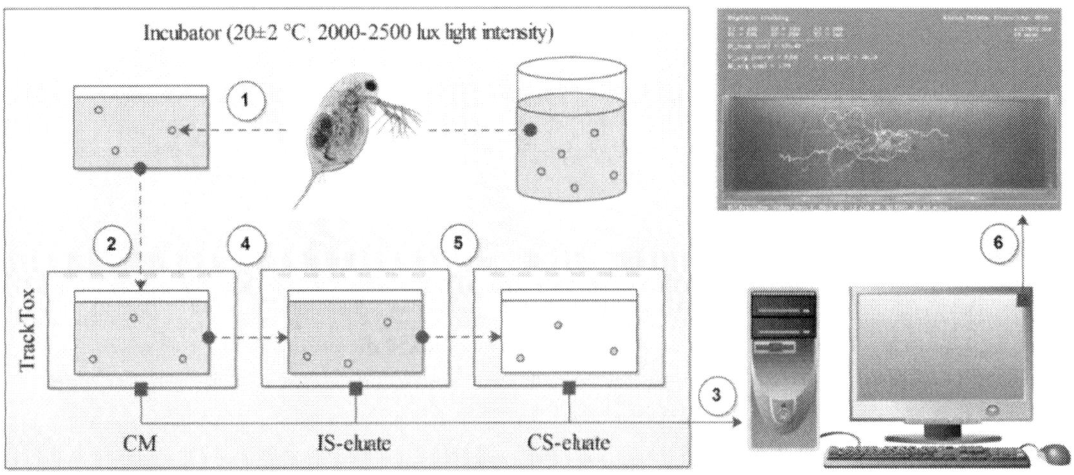

Figure 1. Scheme of Daphnia swimming activity measuring (dashed lines – movement of objects, solid lines – receiving the information). Explanations of experiment stages are given in the text above.

Normal distribution of data (Kolmogorov-Smirnov test) was previously verified. Differences were considered significant at values of $P < 0.05$. The full range of variation, arithmetic mean, standard deviation (SD), standard error of the mean (SE), median, and interquartile range were calculated. The swimming speeds were averaged for time intervals (every minute) as well as for the consolidated group of thirty daphnids (number of experiments is N=10). Statistical analysis was performed using STATISTICA 8.0 (StatSoft).

3. Results and discussion

A chemical analysis both of initial non-polluted sediments and water elutriates of them showed a low level of metals contamination (table 1). The chromium content in natural bottom sediments was 23±4.6 mg kg^{-1}, in artificially contaminated sediments – 9100±1820 mg kg^{-1}. The concentrations of chromium in "IS-elutriate" and "CS-elutriate" were 90±4.5 µg l^{-1} and 2300±115 µg l^{-1} respectively. In the latter case, difficulties were encountered with the recognition of daphnids by "TrackTox" (due to staining of the solution), so that the sample was further diluted to a concentration of 115±5.7 µg l^{-1} ("CS-elutriate$_d$").

Table 1. The concentration of metals in bottom sediments (mg kg^{-1}) and in eluents of them (µg l^{-1}).

	Cr	Al	Cd	Cu	Fe	Mn	Ni	Pb
IS	23±4.6	2025±526.5	<0.05	15±3	5000±1400	950±285	21±7.35	6.5±1.63
CS	9100±1820	1850±481.0	<0.05	18±3.6	4875±1365	900±270	19±6.65	4.9±1.23
IS-elutriate	24.4±1.22	52±3.12	0.3±0.03	6.5±0.39	81±4.05	20±1.4	3.9±0.23	9±0.72
CS-elutriate	2300±115.0	23±1.38	0.1±0.01	11±0.66	75±3.75	63±4.41	5.5±0.33	11±0.88

The summary results of the *D. magna* behavior alteration experiments in the following sequence: "CM" → "IS-elutriate" → "CS-elutriate$_d$" are shown in figure 2.

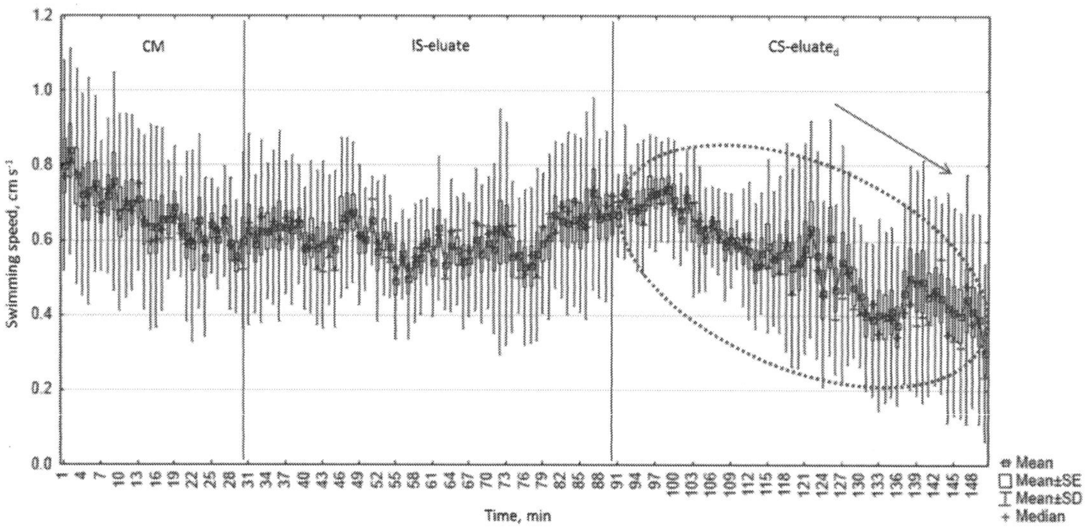

Figure 2. Summary results of the swimming activity of Daphnia in a series of experiments.

The average swimming speed of daphnids in culture medium was 0.67±0.01 cm s^{-1}, after transfer to the "IS-elutriate" it practically did not change and was 0.60±0.01 cm s^{-1}, after transfer to the "CS-elutriate$_d$" it it further decreased to 0.55±0.01 cm s^{-1} (table 2). More detailed information about swimming activity of Daphnia can give a per-minute analysis of the dynamics – in this case, the inhibiting effect of chromium can be traced particularly clearly.

Table 2. Parameters of Daphnia swimming activity (cm s^{-1}) in different mediums.

Swimming speed	Mean	Min	Max	Std.Dev.	Std.Err.	Q25	Median	Q75
Values in culture medium								
CM_p (whole period)	0.67	0.12	1.65	0.23	0.01	0.52	0.65	0.78
CM_{fm} (first minute)	0.80	0.44	1.43	0.28	0.07	0.55	0.77	1.04
CM_{lm} (last minute)	0.60	0.37	1.29	0.23	0.06	0.43	0.52	0.73
Values in IS-elutriate medium								
IS_p (whole period)	0.60	0.09	1.53	0.20	0.01	0.46	0.61	0.72
IS_{fm} (first minute)	0.63	0.13	1.03	0.25	0.07	0.42	0.65	0.86
IS_{lm} (last minute)	0.67	0.26	1.10	0.22	0.06	0.54	0.72	0.78
Values in CS-elutriate$_d$ medium								
CS_p (whole period)	0.55	0.02	1.38	0.24	0.01	0.37	0.58	0.71
CS_{fm} (first minute)	0.67	0.50	0.86	0.11	0.03	0.57	0.70	0.73
CS_{lm} (last minute)	0.30	0.04	0.81	0.24	0.06	0.10	0.24	0.50

An additional estimate is the toxicity index widely used in practice:

$$T,\% = 100 \times (V_{bc} - V_{ac}) / X_{bc} \quad (1)$$

V_{bc} – is the swimming speed in the baseline, V_{ac} – is the swimming speed in the analyzed conditions. It can be noted that the toxic effect in the series with chromium, despite a pronounced decrease in speed by the end of the experiment, is not revealed by this approach. The toxicity index for the average speed in this case is only 8.3%.

Probably this is a feature of data aggregation, when there is a significant variation in values of speed during the experiment. To eliminate the influence of this factor, it is possible to use for the calculation of the toxicity index not the integral value of the swimming speed for the entire observation period, but only the boundary values that characterize the final state of Daphnia (at the last minute), both in the control and in the experimental conditions (table 3).

Table 3. Variants of the toxicity indexes calculated by the different values of swimming speed.

Toxicity index	By mean values	By median values
$(CM_p - IS_p) / CM_p$	10.4%	6.2%
$(IS_p - CS_p) / IS_p$	8.3%	4.9%
$(CM_p - IS_{lm}) / CM_p$	0.0%	–10.8%
$(IS_p - CSl_m) / IS_p$	50.0%	60.7%
$(IS_{fm} - CS_{lm}) / IS_{fm}$	52.4%	63.1%
$(CM_{lm} - IS_{lm}) / CM_{lm}$	–11.7%	–38.5%
$(IS_{lm} - CS_{lm}) / IS_{lm}$	55.2%	66.7%

The toxicity index calculated by this way was 55 and 67% for the mean and median speed, respectively. Since the data distribution in the samples differs from the normal one, it is possible to recommend the use of a toxicity index calculated just by the median. The values obtained correspond more accurately to the actual picture of the swimming activity of Daphnia observed in the experiment.

Daphnia swimming behavior is complex, multiparametric and it may be characterized by several parameters reflecting changes induced by various compounds on sensitive (i.e. nervous and endocrine) systems [16]. A number of results indicate that various substances may alter swimming speed. In particular, our earlier studies showed that *D. magna* change their behavior under the exposure of potassium dichromate, zinc sulfate, esfenvalerate pesticide, microcystin-LR [17–19]. In the case of the exposure of metal salts, there was a statistically significant decrease, and in the case of a pesticide and cyanotoxin, an increase in the swimming speed of Daphnia with a short exposure time of 10–30 minutes.

Some authors also suggested that the swimming speed may be also altered by some metals. In particular, daphnids exposed to sublethal cadmium [20,21] or copper stress [22] tends to manifest inhibited swimming activity when compared to the unexposed, similarly-sized control organisms. The mechanisms by which metals decrease Daphnia speed have not been completely clarified, however it could be hypothesized that oxidative stress known to be induced by these substances in other animal species may also be responsible for the toxic effects on the behavior of crustaceans [16].

4. Conclusion

The conducted study showed the possibility of using a biotesting method based on computer vision to assess the quality of contaminated bottom sediments. In the elutriate from chromium contaminated bottom sediments, the Daphnia swimming speed decrease from 0.61 cm s^{-1} (median speed over the period) to 0.50 cm s^{-1} (median speed at the last minute of the experiment). The relocation of daphnids from the culture medium to the extract from the non-polluted sediments does not essential changes the swimming activity.

Acknowledgement

The work is performed according to the Russian Government Program of Competitive Growth of Kazan Federal University

References

[1] Stepanova N 2014 *Int. Multidisciplinary Scientific GeoConference Surveying Geology and Mining Ecology Management SGEM* **2** 215 – 22
[2] Stepanova N Y, Nikitin O V and Latypova V Z 2016 *International Journal of Pharmacy and Technology* **8** 25096–102
[3] Szczerbińska N and Gałczyńska M 2015 *Archives of Polish Fisheries* **23** 185–96
[4] Selivanovskaya S, Galitskaya P, Schnell S and Hung Y-T 2010 *International Journal of Environment and Waste Management* **6** 183–96
[5] Kokkali V and van Delft W 2014 *TrAC Trends in Analytical Chemistry* **61** 133–55
[6] 2017 EPA ECOTOX Knowledgebase URL: *https://cfpub.epa.gov/ecotox*
[7] Blaise C and Férard J-F 2005 *Small-scale freshwater toxicity investigations vol 1 – Toxicity test methods* (Dordrecht: Springer)
[8] Borcherding J 2006 *Hydrobiologia* **556** 417–26
[9] Olkova A S 2016 *Water and Ecology* **4** 75–95
[10] Melvin S D and Wilson S P 2016 *Chemosphere* **93** 2217–23
[11] Gerhardt A, Svensson E, Clostermann M and Fridlund B 1994 *Environment International* **20** 209–19
[12] Gerhardt A 2007 *Human and Ecological Risk Assessment: An International Journal* **13** 481–91
[13] Jeon J, Kim J H, Fee B C. and Kim S D 2008 *Science of The Total Environment* **389** 545–56
[14] Noss C, Dabrunz A, Rosenfeldt R R, Lorke A and Schulz R 2013 *PLoS One* **11** e80960
[15] Chevalier J, Grote M, Keller M, Pandard P and Cachot J *J Environ Anal Toxicol* **4** 246
[16] Bownik A 2017 *Science of the Total Environment* **601–602** 194–205
[17] Nikitin O and Latypova V 2014 *Int. Multidisciplinary Scientific GeoConference Surveying Geology and Mining Ecology Management SGEM* vol 2 (Sofia: STEF92 Technology) 85–92
[18] Nikitin O 2014 *Advances in Environmental Biology* **8** 74–8
[19] Nikitin O V, Petrova V M and Latypova V Z 2015 *Research Journal of Pharmaceutical, Biological and Chemical Sciences* **6** 1729–36
[20] Wolf G, Scheundersb P and Selens M 1998 *Comp. Biochem. Physiol.* **120** 99–105
[21] Baillieul M and Blust R 1999 *Aquat. Toxicol.* **44** 245–54
[22] Untersteiner H, Kahapka J and Kaiser H 2003 *Aquat. Toxicol.* **65** 435–42

IOP Conf. Series: Earth and Environmental Science **107** (2017) 012069 doi:10.1088/1755-1315/107/1/012069

Soil enzymes as biodiagnostics indicator of heavy metal pollution of urbanozem

E I Novosyolova, O O Volkova and R R Turyanova

Biological faculty, Bashkir State University, Street Zaki Validi, 32, 450076, Russia
e-mail:novoselova58@mail.ru

Abstract. The article presents a comparative analysis of the impact of the introduction of different doses of copper and cadmium on the activity of redox enzymes of urbanozem, collected from different territories of Ufa. The studies established the inverse relationship of the activity of catalase and polyphenol oxidase, and the direct one of the activity of peroxidase that depends on the doses of heavy metals, that allows to recommend their use as bioindicator of pollution of urbanozem with these metals. The reaction of the studied enzymes on the introduction of heavy metals is an indicator of their toxicity to living things at the molecular level. Comparative analysis of the impact of cadmium and copper in different doses on the activity of soil enzymes did not reveal a uniform regularity. Each of the metals showed their toxicity in different ways depending on the duration of their impact.

1. Introduction

Destructive anthropogenic pressure on soil cover in the cities led to the emergence of a new specific type of soils – urbanozem, it is characterized by the disruption of the natural arrangement of horizons, soil regimes, physico-chemical properties, biological activity, etc. [1]. A large concentration of industrial enterprises, motor vehicle emissions, the amount of litter in the territory of the city leads to penetration into the soil a huge number of pollutants, including heavy metals (HM).

One of the most sensitive impact of heavy metals on the soil cover is its biological activity, which reflects such indicators as the functional activity of microorganisms, enzyme activity, respiration rate, etc.

Soil enzymes react with a change of their activity even to minor anthropogenic impact [2] and can be used as bioindicators. To develop biodiagnostics scales, it is necessary to conduct studies in this direction on different types of soils.

2. Material and methods

In this regard, in a model laboratory experiment, the effect of copper and cadmium on the activity of soil enzymes and respiration intensity of urbanozem was studied. HM were introduced into the soil as water solution of sulphate of copper in doses below and above the values of roughly allowable concentration (RAC) (66 mg per kg) - dose 1 is 16.5, dose 2 is 33, dose 3 is 66 and dose 4 is 132 mg per kg of soil and cadmium acetate with RAC of 2 mg per kg – dose 1 is 5, dose 2 is 10, dose 3 is 20 and dose 4 is 40 mg per kg of soil. The activity of catalase, polyphenol oxidase and peroxidase was determined according to the methods described by F Kh Khaziev [3], respiration in A Sh Galstyan [4] on the 3-d, 90-th, 180-th and 360-th day from the beginning of the experiment.

3. Results and discussion

It is known that the enzymes play an important role in the formation of soil fertility and are one of its indicators [5]. Along with this, they have a high sensitivity to such negative environmental factors as heavy metals [6]. The group of redox enzymes include catalase, dehydrogenase, peroxidase, oxidase, etc., which are involved in the processes of transformation of organic matter in soil. Evaluation of the effect of different doses of cadmium and copper on catalase activity showed that doses of contaminants below the values of RAC have an inhibitory effect, which clearly reflects figure 1. This is the evidence of the high toxicity of the studied metals for the enzyme catalase, inhibition of which enhances the negative effect on living organisms because of violation of processes of decomposition of hydrogen peroxide carried out by this enzyme.

Comparative analysis of the influence of metals on catalase activity showed the greater toxicity of cadmium compared to copper on the 3-d, 90–th and 180–th day. Moreover, with increasing dose the inhibitory effect was increased: the percentage of reduction of catalase activity amounted : on the 3-d day after contamination of soil with cadmium to 24-41%, copper - to 12-35%; on the 90-th day – to 10-45% and 5-35%; on the 180-th day – to 13-43% and 12-30%, respectively. However, at the end of the experiment, it was revealed that, the intensification of the processes of decomposition of hydrogen peroxide amounted 11-41% with cadmium contamination, and 11 -63% with copper, which may be due to the reduction of their toxicity in connection with transformation of heavy metals in soil and transfer to an inactive status.

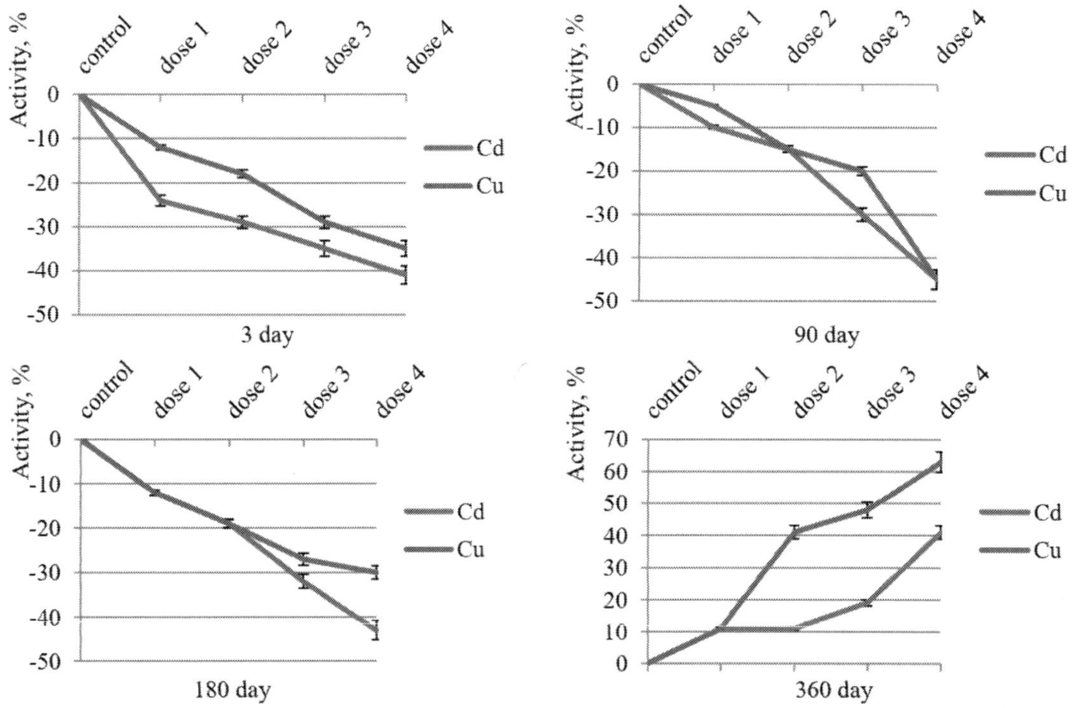

Figure 1. The percentage change of catalase activity in an urban soil contaminated with different doses of cadmium and copper.

An important role in the processes of humification belongs to the oxidase and peroxidase [3]. In contrast to catalase the activity of peroxidase increased with increasing doses of cadmium and copper in the soil during the first three observation times. On the 3-d and the 90-th day cadmium intensified the oxidation of organic substances, that took place with the participation of this enzyme, to a greater extent than copper: the growth of activity after introducing of cadmium during these periods amounted

IOP Conf. Series: Earth and Environmental Science **107** (2017) 012069 doi:10.1088/1755-1315/107/1/012069

to 6-28% and to 10-55%, in the experiment with copper this figure was equal to 2-23% and 1-6%, respectively. As can be seen from figure 2 on the 180-th day the picture has changed and in variants of experience with copper at doses numbered 1-3 the growth of activity of peroxidase was higher than in the experiment with cadmium and made up 26-31% and 19-29%, respectively. At the end of the experiment, the enzyme activity in the experimental variants with the metals were below control values: copper at doses numbered 1-3 was more toxic for the enzyme (decrease of activity at 6-25%) compared to cadmium (5-14%) (figure 2).

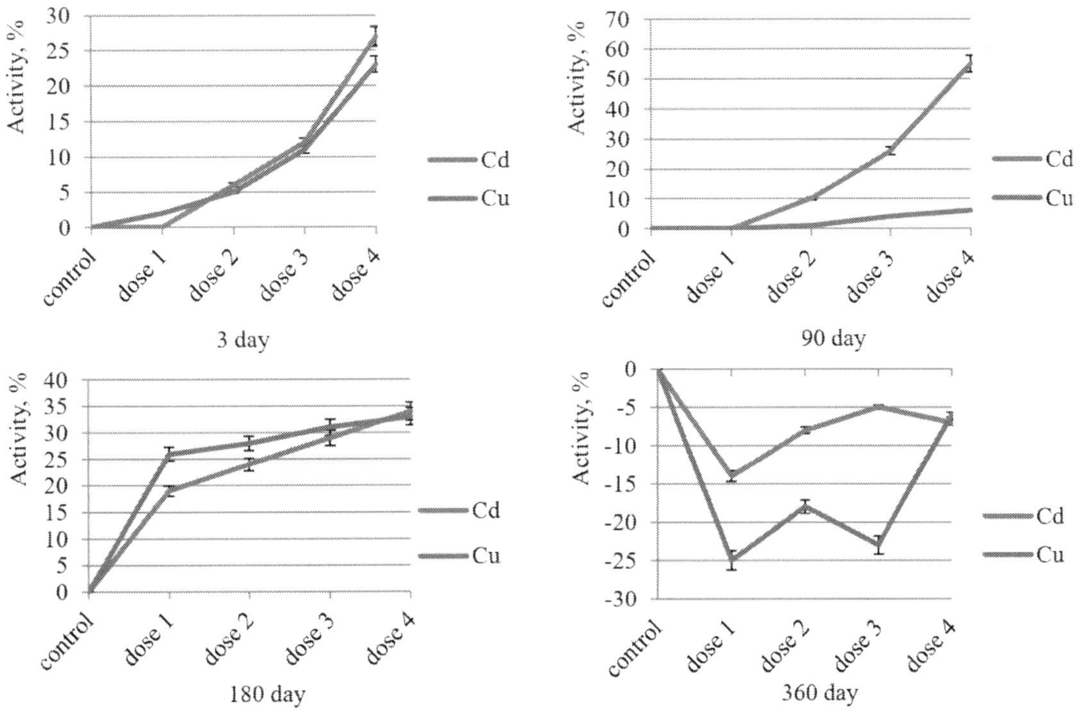

Figure 2. The percentage changes in the activity of peroxidase in an urban soil contaminated with different doses of cadmium and copper.

A comparative analysis of the charts in figure 1 and 2 revealed an antagonistic change in the activity of catalase and peroxidase. Similar results were obtained on other soil types [7] contaminated with cadmium, copper, zinc and lead. One of the possible causes of such an addiction, apparently, is participation of hydrogen peroxide in the reactions with these enzymes [3].

The studied metals suppressed the activity of oxidation processes of organic substances with the participation of polyphenol oxidase during the whole experience (figure 3). On the 3-d, 90-th and 360-th day cadmium in a number of variants of the experience was more toxic than copper. On the 180-th day an inverse relationship was observed. In comparison to control value a valid decrease of enzyme activity with cadmium and copper pollution, was installed in all variants of experience and was for cadmium 6-23% on the 3rd day, for copper respectively 6-21%, on the 90-th day it was 5-29% and 1-23%, on the 180-th day – 11-23% and 10-25%, and on the 360-th day – 1-30% and 3-19%.

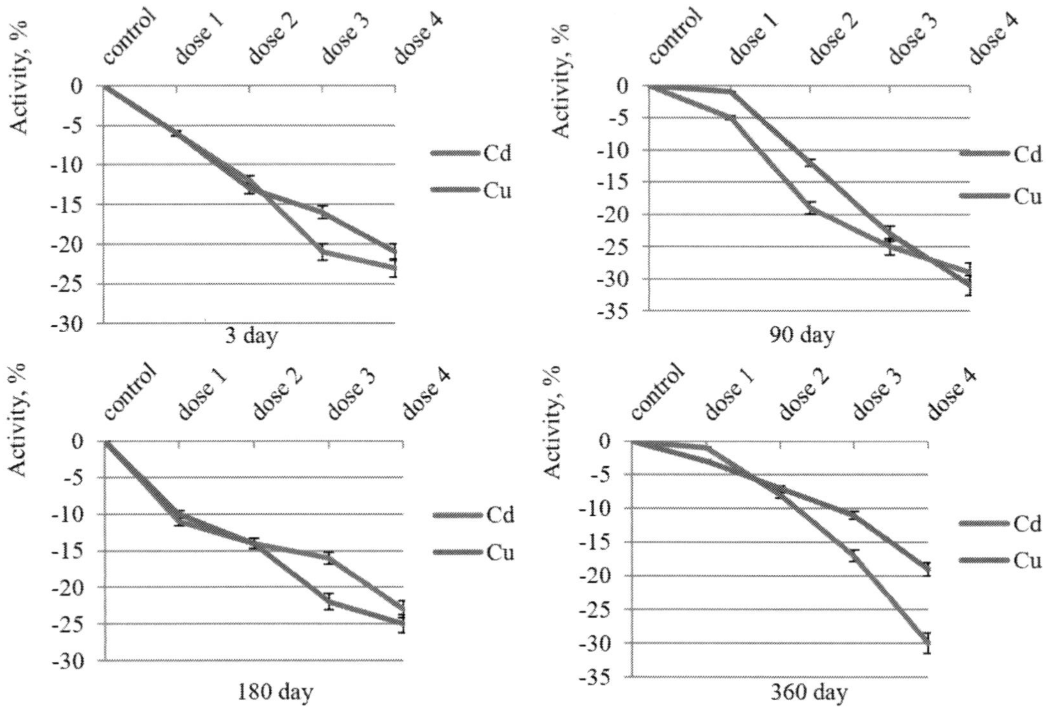

Figure 3. The percentage changes in the activity of polyphenol oxidase in an urban soil contaminated with different doses of cadmium and copper.

Along with the activity of enzymes heavy metals suppressed the respiration of the soil as well. The increase in the dose of cadmium in urbanozem reduced this figure to 4-52% on the 90-th day, and to 11-76% on the 180-th day, to 2-27% on the 360-th day (figure 4). The introduction of copper in doses of 16.5 and 33 mg per kg increased respiratory rate, respectively to 11- 22% by the 90-th day and by the 180-th day within doses of 16.5-66 mg per kg to 6-16% in comparison with the control value. Perhaps, this effect of copper is due to the fact that, it belongs to the group of trace elements and is involved in biological processes, intensifying them. Higher copper doses of 66 and 132 mg per kg in soil showed toxic effect in comparison with the control value. Directly proportional dependence of the intensity of respiration and depending on the dose of metal was evident on the 180–th day.

The ambiguous relationship between the dose of metal and respiration, probably, stems from the fact that, this indicator is integral and reflects the functional activity of soil biota.

Thus, the decrease in the activity of studied enzymes during the introduction of cadmium and copper on the background of increase of the respiration rate, indicates a high sensitivity of enzymes (molecular level). Affords you the opportunity to recommend this indicator for the diagnosis of early changes in biochemical transformations of organic matter in the soil when small doses of pollution occur. The respiration in these purposes can only be used when cadmium pollutes the soil.

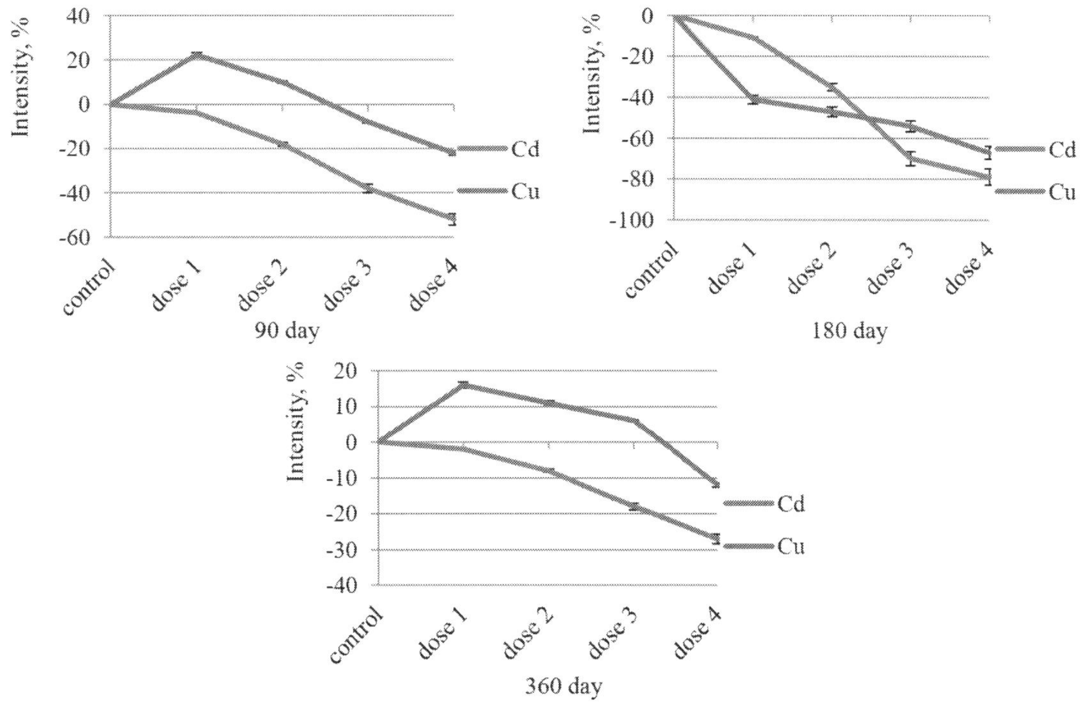

Figure 4. The percentage changes of the respiration rate in an urban soil contaminated with different doses of cadmium and copper.

The obtained dependences of the change in the activity of catalase, peroxidase and polyphenol oxidase can be used to compile an indicative biodiagnostics scales for urbanozem contamination of cadmium and copper after confirmation of the results in field studies.

References

[1] Ufimtseva M D, Teryokhina N V and Abakumov E V 2011 *Vestnik of Saint Petersburg State University* **4** 85 – 97
[2] Mikanova O 2006 *J Geochem Explor.* 220 – 23
[3] Khaziev F Kh 2005 *Methods of soil Enzymology* p 252
[4] Galstyan A Sh 1974 Enzymatic activity of soils of Armenia p 275
[5] Chunderova A I 1970 *Soil science* **7** 22 – 8
[6] Khan S, Cao Q, Hesham AEL, Xia Y and He J 2007 *J. Environ Sci.* 834 – 40
[7] Novoselova E I and Volkova O O 2015 *J International Research* **4** 19 – 21

IOP Conf. Series: Earth and Environmental Science **107** (2017) 012070 doi:10.1088/1755-1315/107/1/012070

Variation in the carbon cycle of the Sevastopol Bay (Black Sea)

N A Orekhova and S K Konovalov

Marine Hydrophysical Institute of RAS, Kapitanskaya str., 2, 299011, Russia
e-mail: natalia.orekhova@mhi-ras.ru

Abstract. Continuous increase in CO_2 inventory in the ocean results in dramatic changes in marine biogeochemistry, e.g. acidification. That is why temporal and spatial variabilities in atmospheric pCO_2 and dissolved inorganic carbon, including CO_2, pH and alkalinity in water, as well as organic and inorganic carbon in bottom sediments have to be studied together making possible to resolve the key features of the carbon cycle transformation. A 30% increase of pCO_2 in the Sevastopol Bay for 2008 – 2016 evidences changes in the DIC components ratios and a significant decrease in the ability to absorb atmospheric CO_2 by surface waters. High organic carbon content in the bottom sediments and predominance of organic carbon production in the biological pump at inner parts of the bay reveal ongoing transformation of the carbon cycle. This has negative consequences for recreation, social and economic potentials of the Sevastopol region.

1. Introduction

Increasing atmospheric CO_2 inventory, leading to climate changes and natural carbon cycle's transformations, has become the key scientific issue over the last decades. For last 120 years, the annual atmospheric CO_2 growth is about 0.4% due to global industrialization. Anthropogenic activities, such as cement production, fossil fuel burning, and forest slash, result in about $6 \cdot 10^{15}$ gC/year increase in the atmospheric CO_2 content. Nowadays, the global ocean CO_2 uptake from the atmosphere accounts for only ~ 30% of the annual production of CO_2, while it was ~ 50% in 1800 – 1994 [1], revealing a lowering ocean capability to compensate for anthropogenic CO_2 production. As a part of the carbon cycle, the carbonate system serves as the ocean buffer capacity and secures its ability to uptake CO_2 from the atmosphere. Increasing CO_2 inventory in the ocean results in acidification. As a result, dramatic changes occur in marine biogeochemistry. This has irreversible consequences, such as degradation of biota's population and diversity, dissolution of coral reefs and shells with carbonate skeletons, etc. As a result, recreation, social and economic potentials of a region deteriorate. That is why temporal and spatial variabilities in atmospheric pCO_2 and dissolved inorganic carbon, including CO_2, pH and alkalinity in water, as well as organic and inorganic carbon in bottom sediments have to be studied together making possible to resolve the key features of the carbon cycle transformation.

The carbon cycle includes atmosphere, hydrosphere and lithosphere compartments. Coastal areas account for only 7% of the World Ocean [2], but they are most valuable for all sorts of anthropogenic exercise and important for investigations. Coastal oceans are key locations, where terrestrial, oceanic and atmospheric parts of the global carbon cycle come together [3]. To make it more complicated, coastal ecosystems are under heavy anthropogenic pressures over the last century due to shipping and maritime activities, intensive land use, coast-based pollution, etc.

The Sevastopol Bay is a typical coastal marine system under heavy anthropogenic pressures. It is a maritime zone with seaports and docks. Domestic sewage are often loaded to the bay with limited treatment carrying nutrients and fresh organic matter. Protection piers between the outer part of the bay and open sea constructed in the middle 1970's restrict water exchange between the bay and sea by 40 – 70% [4]. The bay's ecosystem reveals oxygen deficit in the bottom waters and anaerobic/sulfidic conditions at and in bottom sediments, severe eutrophication, decline in biodiversity, dramatic changes in natural biogeochemical regimes and cycle's [5].

Dissolved inorganic carbon (DIC) is one of the most representative parameter of the ecosystem's stability and recoverability. For usual marine conditions, $[HCO_3^-] : [CO_3^{2-}] : [CO_2]$ equals 100:10:1 [6, 7]. Variations in one of the DIC components result in repartitioning of all dissolved inorganic carbon forms. An increase in the CO_2 flux, for example, results in acidification and increasing bicarbonate ion concentration (equation 1).

$$CO_2\,(g) \leftrightarrow CO_2\,(aq.) \leftrightarrow CO_2\,(aq.) + H_2O \leftrightarrow H^+ + HCO_3^- \leftrightarrow 2H^+ + CO_3^{2-} \quad (1)$$

The initial buffer capacity supports and varies due to absorption of atmospheric CO_2, but it compensates and limits variations in pH and CO_2 inventory in seawater. A doubling CO_2 concentration results in the DIC increase by only 10% [8]. Yet, any shift in DIC evidence changes in production and degradation processes, as well as in CO_2 fluxes. Variations in DIC, CO_2, pH, and alkalinity make possible to assess the ecosystem's stability.

The main purpose of our study is the investigation of the present state of the Sevastopol Bay's ecosystem and assessment of the carbon cycle transformation due to intensive anthropogenic pressures.

2. Material and methods

Monitoring of the Sevastopol Bay has been performed at Marine Hydrophysical Institute since 1998 [4]. Samples are taken from the surface and bottom waters with Niskin bottles at locations/stations presented at figure 1. Sediment columns are collected in plexiglass tubes by scuba diver.

Salinity and temperature are measured with a CTD probe, whereas pH and alkalinity are determined in a shore laboratory following standard analytical procedures [9]. DIC and pCO_2 values, as well as other carbonate system's parameters, are calculated as suggested in [9]. The dissociation constants of carbonic acid, recommended by the Department of Marine Sciences of UNESCO, have been used [9]. The boron content has been calculated from the salinity, as boron was assumed a conservative part of salinity [8]. The effect of dissociation of water, phosphoric acid, sulfuric acid, hydrofluoric acid and other acids have not been taken into account.

Data from the 1st station have not been taken into account (figure 1) for calculation of mean values, because fresh waters of the river Chernaya, rather than marine waters of the bay, have been usually traced there.

Organic and inorganic content have been analyzed in bottom sediments according to [10]. Voltammetric profiling with a glass Au-Hg microelectrode has been applied [11] to detect key redox species (oxygen, sulfides, Mn (II), Fe (II, III), FeS_{aq}) in porewaters.

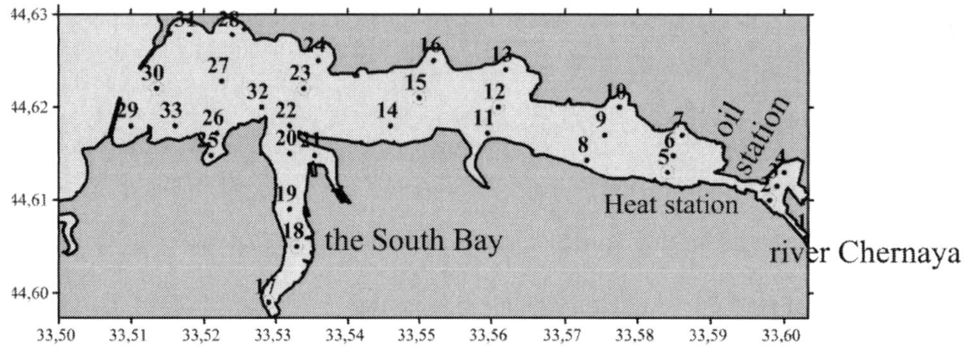

Figure 1. Sampling locations in the Sevastopol Bay (Red rings are for stations of bottom sediments sampling).

3. Results and discussion

The most important natural drivers of the ecosystem of the Sevastopol Bay include water exchange with the adjacent marine area, wind, seasonal variations in temperature, and fresh water load by the river Chernaya. Morphometry features support eddy dynamic structures in the bay [4]. All these and human activities result in specific hydrochemical water column features, including the carbonate systems' components.

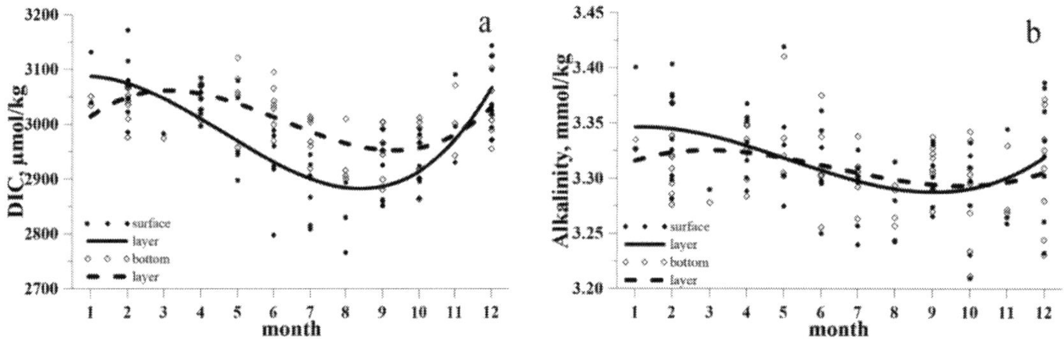

Figure 2. Monthly dynamics of DIC (a) and Alkalinity (b) in the surface and bottom waters of the Sevastopol bay (data for 1998 – 2016).

In cold time (from November to March), waters of the bay are vertically well mixed supporting equal values of DIC, pCO_2 and pH from surface to bottom (figure 2a). In warm months (from April to October) values of DIC decline from spring to summer and increase in fall. Yet, values of DIC in the surface layer decrease faster, as compared to those in the bottom layer, resulting in significantly different values of DIC. Unlike DIC, alkalinity reveals similar mean values in the surface and bottom layers of water throughout the year. Yet, intra-annual variations in alkalinity are similar to those for DIC: maximum values appear on wintertime and minimum values appear in summer. Intra-annual variability of these parameters depends on water dynamics and CO_2 fluxes, which is governed by physical, chemical and biological processes [6, 8].

Atmospheric CO_2 dissolves in seawater and supports photosynthetic (equation 2) organic carbon production in the surface layer. In the bottom layer, organic matter respiration (equation 3) results in oxygen utilization and CO_2 production.

$$6CO_2 + 6H_2O \leftrightarrow C_6H_{12}O_6 + 6O_2 \quad (2)$$

$$C_6H_{12}O_6 + 6O_2 \leftrightarrow 6CO_2 + 6H_2O \quad (3)$$

Other reactions governing the carbonate system are presented by equations 4 and 5. However, the contribution of these reactions to DIC concentration does not exceed 3% [12]. In the Sevastopol Bay, these reactions occur mainly in the bottom layer.

$$CO_2 + H_2O + CO_3^{2-} \leftrightarrow 2HCO_3^- \quad (4)$$

$$Ca^{2+} + 2HCO_3^- \leftrightarrow CaCO_3 + CO_2 + H_2O \quad (5)$$

Spring phytoplankton blooming results in decreasing DIC concentrations. The lowest values of DIC and a maximal difference between the surface and bottom layers are detected in August-September (2990 and 3014 $\mu mol \cdot kg^{-1}$ surface and bottom layers respectively). This is a result of several reasons; an increase in the photosynthetic activity, a decrease in gases solubility, and a strong summer water stratification, preventing CO_2 supply from bottom waters.

Intensive organic carbon accumulation in bottom sediments also results in depletion of DIC in waters. Currently, high organic carbon concentration ($\geq$4%) is observed in the bottom sediments of the Sevastopol Bay [5].

When the temperature decreases from summer to fall, solubility of gases increases (particularly oxygen) and bottom sediments serve as a source of CO_2 for bottom waters, while water mixing in late fall and winter results in equal DIC concentrations from surface to bottom.

Alkalinity does not depend on the CO_2 flux, but its value is determined by the presence of bicarbonates and carbonates. Thus, only a combination of water exchange in the bay and freshwater discharge explains intra-annual oscillations in the value of alkalinity (figure 2b).

Unlike DIC and alkalinity, pCO_2 is highly dynamic and rapidly reacting to any changes occurring in a system. It depends on temperature, CO_2 concentration and pH (6).

$$pCO_2 = \frac{[CO_2]}{K_0} \quad (6)$$

where $[CO_2]$ – equal concentration of CO_2 µmol/kg; K_0 – Henry's constant of carbon dioxide, µmol/(l atm).

Knowing the surface pCO_2 one could identify an ecosystem as a source of CO_2 to or sink from the atmosphere. When the value of pCO_2 in seawater is greater than its value in the atmosphere (currently it is about 405 µatm [1]), the water column is a source of CO_2. The Sevastopol Bay's waters absorb CO_2 from the atmosphere because pCO_2 is from 225 – 388 µatm, except for summer time, when it reaches 426 µatm. Such situation is typical for coastal organic carbon saturated marine environments [2, 3].

Our data show noticeable seasonal variations in surface pCO_2, because of variations in temperature, hydrodynamics, biological and chemical processes. Since the components of the carbonate systems are in equilibrium, an increase of the CO_2 flux results in lower values of pH. Seasonal variations in pH are as much as 0.09 pH units in the bottom layer, but they do not exceed 0.03 pH units for the surface layer. The surface layer is in contact with the atmosphere, supporting active exchange of CO_2, seasonal variations in pH are mostly the result of better dissociation of carbonic acid and increased solubility of CO_2 at higher summer temperatures. Unlike the surface layer, the bottom layer is isolated by the presence of a strong pycnocline and respiration of organic carbon results in active production of CO_2 and decrease in pH. Thus, the seasonal dynamic of the carbonate systems' components in the Sevastopol Bay indicates its dependence on biological and chemical processes of photosynthesis and organic matter degradation in summer. In cold time, the load of the river Chernaya and intensive water mixing are two major factors to control on the carbonate system.

Inter-annual dynamic of the carbonate systems' components for the last 20 years, from 1998 to 2016, reveals an increase in DIC, pCO_2, and acidification, especially in the surface waters (figure 3). Most significant changes take place for pCO_2 (up to 17%). According to NOAA data pCO_2 continuously increases in the atmosphere [1]. For this period, the atmospheric pCO_2 has increased by up to 11% providing an additional CO_2 source. Yet, pCO_2 has increased at a higher rate since 2008, as compared to pCO_2 in the atmosphere (figure 3b). From 2008 DIC concentration has increased by up to 5% and pCO_2 has increased by up to 30%. These considerable values indicate negatives changes, occurring in the ecosystem of the Sevastopol Bay.

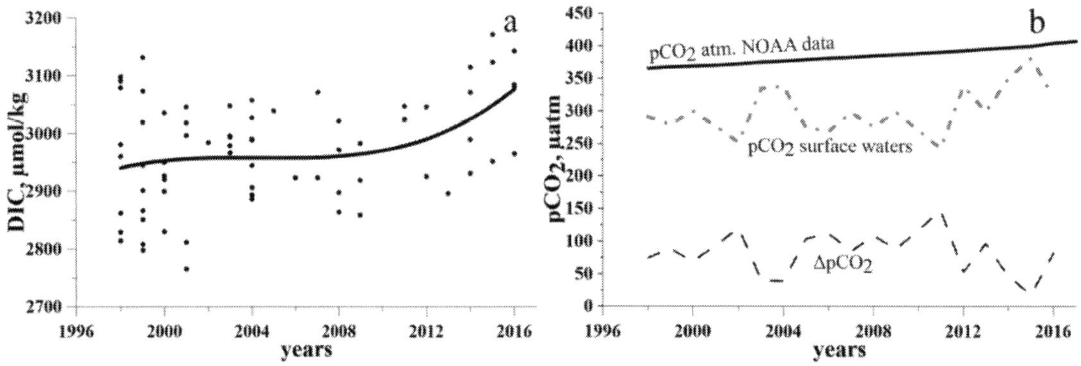

Figure 3. Inter-annual variations in DIC concentration surface waters (a) and pCO_2 and the difference between pCO_2 in the atmosphere and surface water (b) of the Sevastopol Bay

There are inter-annual oscillations in pCO_2 in the surface waters of the Sevastopol Bay, but the tendency suggests an uprising trend in excess of the trend of CO_2 in the atmosphere (figure 3b). Using annual pCO_2 data variation, we have estimated a pCO_2 difference between the atmosphere and the bay's surface water that reflects an ability of the Sevastopol Bay to absorb CO_2 from the atmosphere. This ability has dropped from 1998. Data from 2011 to 2015 suggested that the ecosystem of the Sevastopol Bay would exhaust its sorption ability by 2018 and would be a source of CO_2 for the atmosphere. But in 2016, pCO_2 dropped down suggesting that the Sevastopol Bay would remain a sink of CO_2 from the atmosphere.

Studies of the bottom sediments geochemistry and porewater chemistry in 2003 – 2016 reveal pronounced changes in the carbon cycle also. Bottom sediments are a steady system and any changes in this system occur at a longer time-scale comparing to the water column. Organic and inorganic carbon is main constitutions of the carbon cycle in the sediments. Currently, the organic carbon content in the sediments is ~3.3% with variations from 1.9 – 7.2%, over the bay's area. In the areas under heavy anthropogenic pressure (for example, the South Bay, where shipping docks are located, figure 1) the content of organic carbon exceeds 4%. The $CaCO_3$ content in bottom sediments is on average 31%. There are two types of the carbon pump: biological and physical [13]. In the biological carbon pump, main processes are production organic carbon or its respiration to CO_2. One of these processes can dominate in the ecosystem or they can be equal. A rain ratio parameter (γ) is used to identify the dominate process of carbon production in the sediments (7):

$$\gamma = \frac{CaCO_3}{(C_{org} + CaCO_3)}, \quad (7)$$

where $CaCO_3$ и C_{org} is molality of inorganic and organic carbon, respectively.

The rain ratio parameter can vary from zero to one. If this parameter is zero, than only organic carbon is produced.

The rain ratio parameter is on average 0.55 in the Sevastopol Bay for 2016 and it is 0.63 in the outer part of the bay. Thus, the inorganic pump is prevailed. But spatial variations in γ-parameter have revealed areas with prevailing organic carbon pump (0.39 – 0.44). These are areas under high anthropogenic pressure.

Organic carbon production and accumulation in sediments favor oxygen consumption and CO_2 production (equation 3). Voltammetric studies have revealed that the depth of penetration of oxygen in bottom sediments does not exceed 2 mm, but oxygen is often absent even at the surface of sediments. Thus, organic carbon rich sediments are a source of CO_2 for the water column.

4. Conclusion

The most important natural drivers of the ecosystem of the Sevastopol Bay include water exchange with the adjacent marine area, wind, seasonal variations in temperature, and fresh water load by the river Chernaya. Eutrophication in the form of organic carbon and nutrients load is the most prominent anthropogenic influence of the carbon cycle in the bay.

A 30% increase of pCO_2 for the period of 2008 – 2016 evidences changes in the DIC components ratios and a significant decrease in the ability to absorb atmospheric CO_2 by surface waters. High organic carbon content in the bottom sediments and predominance of organic carbon production in the biological pump (γ-parameter ~0.4) at inner parts of the bay reveal ongoing transformation of the carbon cycle. All these changes evidence negative shifts in the bay ecosystem lowering its self-recover possibility.

Acknowledgement

This work has been funded from the RFBR research project №16-35-60006 mol_a_dk "Long-term changes in the carbon cycle characteristics of the Sevastopol bay" and the state project №0827-2014-0010 "Fundamental Oceanography".

References

[1] Takahashi T, Sutherland S C and Kozyr A 2009 *Global Ocean Surface Water Partial Pressure of CO2 Database: Measurements Performed During 1968–2008 (Version 2008) ORNL/CDIAC-152, NDP-088r* (Tennessee, Oak Ridge, U.S. Department of Energy) p 12 10.3334/CDIAC/otg.ndp088r.

[2] Borges A V, Schiettecatte L S, Abril G, Delille B and Gazeau F 2006 Carbon dioxide in European coastal waters *Estuarine, Coastal and Shelf Sci* **70** 375 – 87

[3] Ribas-Ribas M, Anfuso E, Gomez-Parra A and Forja J M 2013 Tidal and seasonal carbon and nutrient dynamics of the Guadalquivir estuary and the Bay of Cadiz (SW Iberian Peninsula) *Biogeosciences* **10** 4481 – 91

[4] Ivanov V A, Ovsyany E I, Repetin L N, Romanov A S and Ignatueva OG 2006 *Hydrology and Hydrochemical Regime of the Sevastopol Bay and its Change under Climate and Anthropogenic influence* (Sevastopol: ECOSI-Hydrophysica) p 90 (in Russian)

[5] Moiseenko O G, Orekhova N A 2011 Investigation of the mechanism of the long-term evolution of the carbon cycle in the ecosystem of the Sevastopol bay *Physical Oceanography* **21** 142 – 52

[6] Zeebe R E and Wolf-Gladrow D 2001 *CO_2 in Seawater: Equilibrium, Kinetics, Isotopes* (Elsevier Oceanography Series) p 346

[7] Heinze C, Meyer S, Goris N, Anderson L, Steinfeldt R, Chang N, Le Quéré C and Bakker D C 2015 *Earth Syst. Dynam.* **6** 327 – 58

[8] Millero F J 2007 The marine inorganic carbon cycle *Chem Rev* **107** 308 – 41

[9] Thermodynamic of the carbon dioxide system in seawater 1987 *Unesco Technical Papers in Marine Science* **51** 3 – 21

[10] *Manual for the Geochemical Analyses of Marine Sediments and Suspended Particulate Matter* 1995 (Nairobi: UNEP/IOC/IAEA) p 74

[11] Orekhova N A and Konovalov S K 2009 Polarography of the bottom sediments in the Sevastopol Bay *Physical Oceanography* **19** 111 – 23

[12] Zeebe R E and Wolf-Gladrow D *Carbon dioxide, dissolved (Ocean)* ZeebeWolfEnclp07.pdf

[13] Rost B and Riebesell U 2004 Coccolithophores and the biological pump: responses to

environmental changes *Coccolithophores: from molecular processes to global impact* (Berlin: Springer) pp 99 − 125

Ligninolytic basidiomycetes as promising organisms for the mycoremediation of PAH-contaminated Environments

N N Pozdnyakova, S A Balandina, E V Dubrovskaya, C N Golubev and O V Turkovskaya

Institute of Biochemistry and Physiology of Plants and Microorganisms, Russian Academy of Sciences, 13 Prospekt Entuziastov, Saratov, 410049, Russia
e-mail: pozdnyakova_n@ibppm.ru

Abstract. Primary screening of ligninolytic fungi belonging to wood- and soil-inhabiting basidiomycetes revealed their ability to degrade three-ringed PAHs with formation of quinone metabolites at the first stage. The degradative activity was both species and strain specific, and some differences in the "chances" for the formed quinones were found. They were the main end metabolites in the degradation of PAHs by *Stropharia rugosoannulata* and *Agaricus bisporus*. During PAH degradation by strains of *Trametes versicolor*, *Pleurotus ostreatus*, *Schizophyllum commune*, and *Bjerkandera adusta* similar metabolites were detected during the cultivation, but they were utilized further. The results supported the hypothesis that the degree of PAH degradation may depend on the composition of the extracellular ligninolytic complex of the fungi: in the presence of a single ligninolytic enzyme, laccase, the accumulation of quinone metabolites takes place; their further utilization is possible with the participation of ligninolytic peroxidases. The data obtained showed the necessity not only to identify the metabolites formed, but also to study the activity of the basic ligninolytic enzymes. It is important for the correct selection of fungal strains for mycoremediation.

1. Introduction

Polycyclic aromatic hydrocarbons (PAHs) are a large group of organic compounds containing two and more fused aromatic rings. The widespread occurrence of PAHs is due to their generation from the incomplete combustion of pyrolysis of numerous organic materials, such as coal, oil, petroleum, and wood. PAHs exhibit the most structural variety in nature, as compared to other classes of non-halogenated molecules in the biosphere [1, 2]. The fate of PAHs in the environment includes volatilization, photooxidation, chemical oxidation adsorption on soil particles, and leaching [3]. They are difficult to degrade in natural matrices, and their persistence increases with their molecular weight. Because of their toxic, mutagenic, and carcinogenic properties, the US EPA lists 16 PAHs as priority pollutants [1-3].

Many reviews exist on the biodegradation of this class of pollutant. PAHs are degraded by bacteria, fungi, and algae [1-6]. Among the ways to address environmental contamination, biotechnological methods of remediation have serious advantages over physical and chemical methods owing to their ecological feasibility and relatively low cost. In the past decade, mycoremediation, a biotechnology that uses the degradative potential of fungi of various ecological-physiological and taxonomic affiliations, was developed [7, 8]. The main role in this technology belongs to a thorough study of fungal strains being able to effectively degrade pollutants.

Ligninolytic fungi are a physiological rather than a systematic group of fungi which are the most active degraders of lignin in nature. They have certain advantages over other microorganisms, including high growth rate and high activity of their extracellular enzymes, which catalyze the degradation of lignin [9]. *Phanerochaete chrysosporium, Pleurotus ostreatus, Bjerkandera adusta,* and *Trametes versicolor* are most commonly used for the degradation of such compounds owing to their production of ligninolytic enzymes such as lignin peroxidase, Mn-peroxidase, and laccase. The rate of biodegradation depends on culture conditions such as temperature, oxygen, pH, accessibility of nutrients, ionization potentials and concentrations of PAHs, and agitated or shallow culture [1, 10, 11].

Pollutants containing aromatic structures, such as PAHs, can be involved in the oxidation of lignin components of wood. This phenomenon, with species and/or strain specificity, is the reason to search for and study active strains of ligninolytic fungi which can be used for the development of ecological biotechnologies.

2. Material and methods

The main objects of study were 10 strains of *wood-inhabiting* (*Bjerkandera adusta* MUT 3398, *Schizophyllum commune* IBPPM541, *Pleurotus ostreatus* f. Florida IBPPM540, *Trametes versicolor* DSM11269, *Trametes versicolor* MUT3403, *Pleurotus ostreatus* 336, *Pleurotus ostreatus* MUT2977, and *Pleurotus ostreatus* D1 (Jacquin) P. Kummer) and *soil-inhabiting basidiomycetes* (*Stropharia rugosoannulata* DSM11372 and *Agaricus bisporus* F-8).

The fungi were grown at 24-26°C in a basidiomycetes rich medium [12] with our modifications ($g \cdot l^{-1}$): NH_4NO_3, 0.724; KH_2PO_4, 1.0; $MgSO_4 \times 7H_2O$, 1.0; KCl, 0.5; yeast extract, 0.5; $FeSO_4 \times 7H_2O$, 0.001; $ZnSO_4 \times 7H_2O$, 0.0028; $CaCl_2 \times 2H_2O$, 0.033; D-glucose, 10.0; peptone, 10.0. PAHs (anthracene, phenanthrene, or fluorene) were added on day 2 of growth as a chloroform stock solution to a final concentration of 0.05 $mg \cdot l^{-1}$. The control treatments contained 100 µl of chloroform. After 14 days of cultivation, residual PAHs were extracted from the flasks (without separation of the cultivation medium from mycelia) with chloroform (5 ml, three times). The resulting extracts were evaporated and were analyzed as described below. The mycelium was separated from the cultivation medium by filtration. The mycelium increment (mg of dry biomass) was determined by weighing.

The PAHs and their metabolites were analyzed by GC (Shimadzu 2010) with a flame photometric detector. Compounds were separated by using an HP5 column (Agilent) and helium as the carrier gas. The column temperature was kept at 200°C for 3 min, then programmed to increase to 270°C at a rate of 15°C min^{-1}, and finally kept at 270°C for 2 min. Before GC analysis, 2,2'-diphenic and phthalic acids were methylated with CH_3COCl. Commercial anthracene (RT=4.15 min), 9,10-anthraquinone (RT=5.39 min), phenanthrene (RT=4.08 min), 9-phenanthrol (RT=6.33 min), 9,10-phenanthrenequinone (RT=6.86 min), 2,2'-diphenic acid (RT=6.7 min), and phthalic acid (RT=8.99 min) were used as the marker compounds to identify the residual PAHs and metabolites. GC analysis was carried out at the Center for the Collective Use of Research Equipment in the Field of Physicochemical Biology and Nanobiotechnology, "Simbioz" (Institute of Biochemistry and Physiology of Plants and Microorganisms, Russian Academy of Sciences).

Laccase activity was measured by the oxidation rate for 2,2'-azino-bis(3-ethylbenzothiazoline-6-sulfonic acid)diammonium salt (ABTS) at 436 nm (ε=29300 M-1cm^{-1}), according to Niku-Paavola et al. [13]. Mn-peroxidase was measured by the oxidation rate for 2,6-dimethoxyphenol with H_2O_2 and Mn^{2+} at 468 nm (ε=14800 $M^{-1}cm^{-1}$), according to Heinfling et al. [14]. Peroxidase activity was calculated as the difference between the values for 2,6-dimethoxyphenol oxidation with and without H_2O_2. Lignin peroxidase was measured by the formation of the oxidation product of veratryl alcohol at 310 nm (ε=9300 $M^{-1}cm^{-1}$) [15]. One unit of enzyme activity ($U \cdot ml^{-1}$) is defined as the amount of enzyme that oxidizes 1 µM of substrate per min.

All experiments were run in triplicate; each has been repeated at least three times. Standard statistical methods were used to determine the mean values and standard deviations (SD). Data were processed with Microsoft Excel 2003.

3. Results and discussion

We studied 10 strains of ligninolytic fungi belonging to the genera *Agaricus*, *Bjerkandera*, *Pleurotus*, *Schizophyllum*, *Stropharia*, and *Trametes*. All the basidiomycetes actively degraded PAHs, as evidenced by decreases in their concentrations in the cultivation medium (Figure 1) and by the formation of the corresponding metabolites. The sorption of PAHs by the fungal mycelium was insignificant and did not exceed 5-7%. From the three tested compounds, anthracene was the least available to all fungi. The degradation of this compound varied from 39.8 to 100% for *Sch. commune* and *P. ostreatus*, respectively. The *Pleurotus* and *Trametes* strains metabolized them more intensely (up to 90%). Phenanthrene and fluorene were degraded more actively by all studied fungi. The decrease in their concentration with *P. ostreatus* D1 was about 100%.

According to their availability to the fungal cells, the PAHs were ranged in the following order: anthracene → phenanthrene → fluorene. Anthracene, with the lowest ionization potential (7.43 eV), has a low solubility (0.07 mg·l^{-1}). Perhaps this makes the molecules of anthracene poorly available for the fungal attack. The well-soluble phenanthrene (1.8 mg·l^{-1}) has a sufficiently high ionization potential (8.03 eV) and, as a consequence, is well accessible to the fungi. Fluorene has a "successful" combination of a low ionization potential (7.88 eV) and high solubility (1.98 mg·L^{-1}), which that makes it the best accessible PAH among these studied. Comparison of the obtained data with the physicochemical properties of these substances suggests that in this case, a coupling of two factors – solubility and ionization potentials – is essential.

The degradative activity was both species and strain specific. For example, among the *Pleurotus* strains tested, *P. ostreatus* D1 was more active, whereas *T. versicolor* MUT was more active than *T. versicolor* DSM.

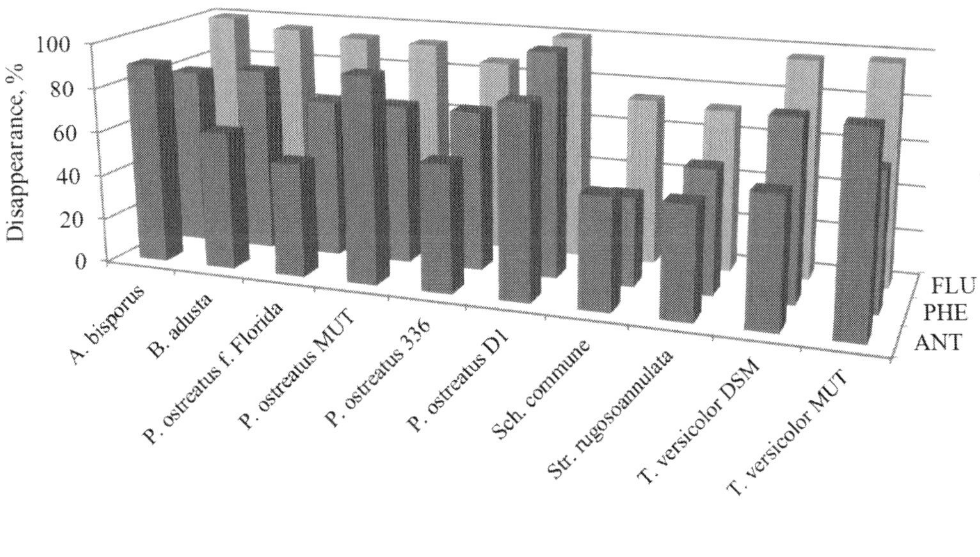

SD values do not exceed 5%

Figure 1. PAH elimination by the fungi studied: ANT – anthracene, PHE – phenanthrene, FLU – fluorene

It is well-known that the first and main metabolites of PAHs are the corresponding quinones [11, 16-23]. Two possible "chances" for the formed quinones are known. First, they can be the end metabolites of PAH degradation; their accumulation was found in *Phanerochaete chrysosporium* [16, 20], *B. adusta* [17, 20], and *Pycnoporus sanguineus* [18]. The second possibility is the subsequent utilization of quinone metabolites, as shown in *P. ostreatus* [17], *P. eringii* [19], *Ph. laevis* [21, 22], and *Polyporus* sp. S133 [23-24]. In this case, phthalic and benzoic acids were metabolites from "deep" degradation of fluorene [19] and anthracene [24]. The mechanism of the degradation of phenanthrene by *Polyporus* sp. S133 was determined through identification of several metabolites; 9,10-phenanthrenequinone, 2,2'-diphenic acid, salicylic acid, and catechol. 9,10-Oxidation and ring cleavage to give 9,10-phenanthrenequinone is the major fate of phenanthrene. The identification of 2,2'-diphenic acid in culture extracts indicates that phenanthrene was initially attacked through dioxygenation at C_9 and C_{10} to give cis-9,10-dihydrodiol. Dehydrogenation of phenanthrene-*cis*-9,10-dihydrodiol to produce the corresponding diol, followed by *ortho*-cleavage of the oxygenated ring, produced 2,2'-diphenic acid [23].

In our experiments, some differences were found also. The main end metabolites in the degradation of PAHs by *Str. rugosoannulata* and *A. bisporus* were their quinone derivatives: 9,10-anthraquinone, phenanthrene-9,10-quinone, and 9-fluorenone. During PAH degradation by both strains of *Trametes,* the traces of the corresponding quinones were found. With the *P. ostreatus* strains, *Sch. commune*, and *B. adusta*, similar metabolites were detected during the cultivation, but they disappeared when the cultivation time was extended. This indicated that they were utilized further. Indeed, GC-analysis revealed phthalic acid as the metabolite resulting from "deep" degradation of anthracene and fluorene by *P. ostreatus* strains. 2,2'-Diphenic and phthalic acids were produced by "deep" phenanthrene degradation by these fungi. These metabolites can be included in the basal metabolism of the fungi.

It is known that the ligninolytic enzyme system catalyzes the key stages of PAH degradation by fungi. The addition to the cultivation medium of some compounds can simultaneously enhance the degradation of PAHs and the activities of ligninolytic enzymes [10]. Examining of the activity of the main enzymes of the ligninolytic complex in the culture fungi's medium revealed the production of laccase, Mn-peroxidase, and lignin peroxidase. Differences in the activity and composition of the ligninolytic complexes of individual species were noted: *P. ostreatus* strains, *T. versicolor* strains, and *Sch. commune* produced laccase and Mn-peroxidase; *Str. rugosoannulata* and *A. bisporus* produced only laccase under the experimental conditions, and *B. adusta* produced lignin peroxidase and Mn-peroxidase (table 1).

Table 1. Activity of the fungal enzymes.

Fungus	Activity, U·ml^{-1}							
	Laccase				Mn-peroxidase			
	control	ANT[a]	PHE	FLU	control	ANT	PHE	FLU
A. bisporus	54.2	71.2	38.3	56.4	–	–	–	–
B. adusta	–	–	–	–	3.6	3.2	3.4	21.5
P. ostreatus f. Florida	49.9	7.3	46.2	16.9	242.5	38.4	78.8	14.2
P. ostreatus MUT	108.4	24.9	174.7	158.2	392.6	118.5	494.9	292.0
P. ostreatus 336	98.9	33.2	29.8	5.6	145.5	25.9	7.1	12.2
P. ostreatus D1	9.2	8.0	24.0	0.3	6.0	75.0	21.0	6.7
Sch. commune	26.8	11.1	13.2	15.9	86.3	3.0	27.5	3.2
Str. rugosoannulata	2.1	5.1	4.8	2.5	–	–	–	–
T. versicolor DSM	134.1	125.9	141.6	73.7	24.1	9.6	15.3	10.4
T. versicolor MUT	167.4	103.4	176.3	164.1	179.8	128.5	64.1	86.4

[a]ANT – anthracene, PHE – phenanthrene, FLU – fluorene

No correlation between enzyme activities and percentage of PAH degradation was found in all fungi studied. On the whole, the activity of Mn-peroxidase in the *P. ostreatus* strains was higher than that of laccase. But in the *T. versicolor* strains, laccase activity was higher than that of Mn-peroxidase. In any case, the presence of the PAHs and/or their degradation products decreased the activities of the ligninolytic enzymes (table 1). The activity of lignin peroxidase of *B. adusta was* unaffected by the

PAHs or their metabolites in the cultivation medium: 36.8 U·ml^{-1} (no PAHs), 42.3 U·ml^{-1} (with anthracene), 39.5 U·ml^{-1} (with phenanthrene), and 37.2 U·ml^{-1} (with fluorene).

It is known that PAHs are available for oxidation by all three ligninolytic enzymes; the corresponding quinones are the products of these reactions [25-27]. Correlation of our data on the metabolites formed during PAH degradation with those on the production of enzymes allows us to conclude that, ligninolytic enzymes catalyze the initial attack on the molecules of these pollutants. However, in the presence of a single ligninolytic enzyme, laccase, their accumulation takes place, and further utilization is possible with the participation of Mn-peroxidases. In the case of *B. adusta* the first attack on the molecules of PAHs can be catalyzed by both ligninolytic peroxidases, produced simultaneously.

Finally, despite some differences, the degradation of the PAHs followed the same scheme, forming quinone metabolites at the first stage: 9,10-anthraquinone in the case of anthracene, 9-fluorenone in the case of fluorene, and phenanthrene-9,10-quinone in the case of phenanthrene. The presence in the cultivation medium of only laccase (*Str. rugosoannulata* and *A. bisporus*) resulted in accumulation of the corresponding quinones in the cultivation medium. Successive production of laccase and Mn-peroxidase (*P. ostreatus*) resulted in the formation and subsequent utilization of these metabolites. Finally, if both enzymes were highly active (*T. versicolor* DSM), the quinone metabolites formed were degraded quickly.

The obtained data supported our hypothesis that, the degree of degradation of the PAHs can depend on the composition of the extracellular ligninolytic complex. Earlier, we showed that the completeness of degradation of fluorene, fluoranthene, pyrene, and chrysene by *P. ostreatus* D1 depends on the cultivation conditions. For example, in Kirk's medium about 65.6 % of the initial pyrene was metabolized after 3 weeks, with pyrene-4,5-dihydrodiol accumulating. This process was accompanied by laccase production only. In the basidiomycetes rich medium, *P. ostreatus* D1 metabolized up to 89.8% of pyrene within 3 weeks without pyrene-4,5-dihydrodiol accumulation. Phenanthric and phthalic acids were identified as the metabolites. Accumulation of phenanthric acid, with its subsequent disappearance, was observed. Pyrene metabolism in the basidiomycetes rich medium was accompanied first by laccase production and later by versatile peroxidase production. The data indicate that both enzymes (laccase and versatile peroxidase) are necessary for the complete degradation of pyrene [28]. This effect was confirmed for the other PAHs: chrysene, fluorene, and fluoranthene. Under the laccase production conditions, transformation of chrysene by *P. ostreatus* D1 occurs with accumulation of the quinone metabolite. Under conditions of production of both laccase and versatile peroxidases, chrysene degradation occurs, leading to phthalic acid formation and its further utilization [29]. Fluorene and fluoranthene were transformed in Kirk's medium (under conditions of laccase production), with the formation of a quinone metabolite and 9-fluorenone, when fluoranthene and fluorene were used as the substrates, respectively. More complete degradation with the formation of an intermediate metabolite, phthalic acid, which underwent subsequent utilization, occurred in the basidiomycetes rich medium (under the production of both laccase and versatile peroxidase) [30].

Therefore, the affiliation of the fungi with different ecophysiological groups and their cultivation conditions affect the composition and dynamics of production of the ligninolytic enzyme complex and, consequently, the completeness of PAH utilization. Comparison of the obtained data permits us to suppose that laccase can catalyze the initial attack on the PAH molecules, which results in quinone formation. It should be noted that the quinone derivatives are highly toxic, as we showed earlier with *Sorghum bicolor* L. Moench and *Medicago sativa* L. seedlings [31]. Perhaps peroxidases catalyze the following oxidation of these compounds, resulting in pollutant mineralization. In addition, it is necessary to investigate the role of different enzymes and factors in fungal degradation of the PAHs.

In this regard, for the correct selection of fungal strains for mycoremediation, it is necessary not only to identify the metabolites formed, but also to study the activity of the basic ligninolytic enzymes. This will allow the development of a technological process to avoid the accumulation of toxic substances in the treated objects. On the basis of their degradative properties and the composition of the ligninolytic enzyme system, the fungi *Trametes* and *Pleurotus* can be chosen for detailed study and for the development of technologies of mycoremediation of PAH-contaminated environments.

Acknowledgement
This research was supported by a grant from the Russian Science Foundation (project no. 16-14-00081). We thank Prof. G.C. Varese (Department of Life Sciences and Systems Biology, University of Turin) and Dr. D. Schlosser (Helmholtz Centre for Environmental Research – UFZ) for kindly providing us with fungal strains. We are grateful to Dr. M.P. Chernyshova for GC-analysis. We are grateful to Dmitry N. Tychinin for his assistance in preparation of the English text of this paper.

References
[1] Kadri T, Rouissi T, Kaur Brar S, Cledon M, Sarma S and Verma M 2017 *J. Envirom Sci.* **51** 52 – 74
[2] Abdel-Shafy H I and Mansour M S M 2016 *Egypt J Petrol.* **25** 107 – 23
[3] Haritash A K and Kaushik C P 2009 *J Hazard Mater.* **169** 1 – 15
[4] Adams G O, Fufeyin P T, Okoro S E and Ehinomen I 2015 *Int J Environ Biorem Biodegrad.* **3** 28 – 39
[5] Diaz M, Mora V, Pedrozo F, Nichela D and Baffico G 2014 *J Appl Phycol.* **27** 321 – 25
[6] Luo L, Wang P, Lin L, Luan T, Ke L and Tam T A H 2014 *Process Biochem.* **49** 1723 – 32
[7] Tortella G, Duran N, Rubilar O, Parada M and Diez M C 2015 *Crit Rev Biotechnol.* **35** 165 – 72
[8] Winquist E et al 2014 *Int Biodeterior Biodegrad* C **86** 238 – 47
[9] Wong D 2009 *Appl Biochem Biotechnol* **157** 174 – 209
[10] Ting WTE, Yuan SY, Wu SD and Chang BV 2011 *Int Biodeterior Biodegrad* **65** 238 – 42
[11] Han M-Oб Choi H-T and Song H-G 2004 *J. Microbiol.* **42** 94 – 98
[12] Bezalel L, Hadar Y and Cerniglia C 1997 *Appl Environ Microbiol.* **63** 2495 – 2501
[13] Niku-Paavola M, Karhunen E, Salola P and Raunio V 1988 *Biochem J.* **254** 877 – 84
[14] Heinfling A, Martinez M, Martinez A, Bergbauer M and Szewzyk U 1998 *FEMS Microbiol Lett.* **165** 43 – 50
[15] Tien M and Kirk K 1984 *Proc Nat Acad Sci USA* **81** 2280 – 84.
[16] Vyas BRM, Bakowski S, Sasek V and Matucha M 1994 *FEMS Microbiol Ecol.* **14** 65 – 70
[17] Schützendübel A, Majcherczyk A, Johannes C and Höttermann A 1999 *Intern Biodeterior Biodegrad.* **43** 93 – 100
[18] Li X, Wang Y, Wu S, Qiu L, Gu L, Li J, Zhang B and Zhong W 2014 *Biotechnol Appl Biochem.* **61** 549 – 54
[19] Hadibarata T and Kristanti RA 2014 *Fungal Biology.* **118** 222 – 27
[20] Field JA, ed Jong E, Feijoo Costa G, de Bont JAM 1992 *Appl Environ Microbiol.* **58** 2219 – 26
[21] Hammel KE, Green B and Gai WZ 1991 *Proc Natl Acad Sci USA.* **88** 10605 – 8
[22] Bogan BW and Lamar RT 1996 *Appl Environ Microbiol.* **62** 1597 – 603
[23] Hadibarata T, Tachibana S and Askari M 2011 *J Microbiol Biotechnol.* **21** 299 – 304
[24] Hadibarata T, Khudhair AB and Salim MR 2012 *Water Air Soil Pollut.* **223** 2201 – 08
[25] Torres E, Tinoco R and Vazquez-Duhalt R 1997 *Wat Sci Technol.* **36** 37 – 44
[26] Acevedo F, Pizzul L, Castillo M, Cuevas R and Diez M 2011 *J Hazard Mater.* **185** 212 – 19
[27] Pozdnyakova N, Rodakiewicz-Nowak J, Turkovskaya O and Haber J 2006 *J Molecular Catalysis B: Enzymatic.* **41** 8 – 15
[28] Pozdnyakova N, Nikiforova S, Makarov O, Chernyshova M, Pankin K and Turkovskaya O 2010 *World J Microbiol Biotechnol.* **26** 205 – 11
[29] Nikiforova S, Pozdnyakova N, Makarov O, Chernyshova M and Turkovskaya O 2010 *Microbiology (Moscow).* **79** 456 – 60
[30] Pozdnyakova N N, Chernyshova M P, Grinev V S, Landesman E O, Koroleva O V, and Turkovskaya O V 2016 *Appl Biochem Microbiol.* **52** 621 – 28

IOP Conf. Series: Earth and Environmental Science **107** (2017) 012071 doi:10.1088/1755-1315/107/1/012071

[31] Dubrovskaya E V, Pozdnyakova N N, Muratova A Yu and Turkovskaya O V 2016 *Rus J Plant Physiol.* **63** 172 – 79

IOP Conf. Series: Earth and Environmental Science **107** (2017) 012072 doi:10.1088/1755-1315/107/1/012072

Development of taxation system for oil production companies in Russia

S V Salmina[1], I M Sboeva[1], J I Selivanovskaya[1, 2], A R Khafizova[1], V P Fomin[1]

[1]Kazan Federal University, Kremlevskaya str., 18, Kazan, 420008, Russia

[2]Kazan National Research Technical University named after A.N. Tupolev (KAI), K. Marx str., 10, Kazan, 420111, Russia

e-mail: JISelivanovskaya@mail.ru

Abstract. The present article is devoted to the taxation system for oil production companies in Russia. The role of oil production companies in the realization of the fiscal function of the state is shown. Tax and due receipts at the consolidated budget of the Russian Federation from major economic sectors in the years 2013–2015 are presented and analysed. An investigation of oil production taxation peculiarities is carried out. In particular, mineral extraction tax analysis is made, the said tax being one of the basic taxes paid by oil production companies. The authors come to a conclusion that mineral extraction tax in Russia needs reforming. Based on the investigation realized possible ways of taxation system development in respect of oil production companies in Russia are proposed. Thus, taking into account the fact that oil industry is very important for budget revenue formation, initially it is planned to test the new taxation system principles in a limited number of deposits, so called 'pilot projects'. For highly profitable minefield deposits it is planned to introduce progressive and regressive index, varying depending on oil prices. Within the framework of the investigation the authors come to a conclusion that it is necessary to introduce gradually the taxation system based on the definition of surplus profit depending on the cost effectiveness and taking into account oil prices.

1. Introduction

At the present time oil companies in Russia play a very important part in budget revenue formation and provide a considerable part of receipts at the consolidated budget of the Russian Federation. Oil sector taxation mechanism in Russia began its formation in the 90's of the XX century. Oil production enterprises started paying resource taxes aimed for resource rent extraction together with standard taxes common for all the enterprises of the country. The taxation base for resource tax assessment was oil price, and not the financial result of operating activities. Tax assessment method for oil production companies formed in the 90's was reorganized in 2001. As a result, a new chapter was included into the Tax Code of the Russian Federation – Chapter 26 Mineral Extraction Tax, which replaced three previous payments: mineral replacement tax, royalty and excise tax.

2. The problem statement

In Russia mineral extraction tax is included into the resource rent. However, according to many modern scientists, current mechanism of mineral extraction tax assessment does not fully correspond to the modern oil-field development economic conditions, which prompts the necessity to substitute such mechanism by a tax assessment mechanism based on the financial result.

Oil production companies are major budget revenue generating enterprises in the Russian Federation, which is visualized by the data presented in Table 1.

Table 1. Tax and due receipts at the consolidated budget of the Russian Federation from major economic sectors in the years 2013–2015 [1].

	2013		2014		2015	
	Receipts, million roubles	Share, %	Receipts, million roubles	Share, %	Receipts, million roubles	Share, %
TOTAL budjet revenue, including:	11,322,650	100.00	12,606,293	100	13,707,087	100.00
Mineral extraction tax, including:	3,274,721	28.92	3,764,146	29.86	4,297,049	31.35
- crude oil and natural gas extraction; rendering of services in these areas	3,157,265	27.88	3,643,163	28.90	4,100,925	29.92
Manufacturing	2,137,088	18.87	2,283,541	18.11	2,474,250	18.05
Wholesale and retail trade; repair of vehicles, motorbikes, household goods and personal demand items	1,145,575	10.12	1,336,439	10.60	1,518,714	11.08
Real estate operations, lease and rendering of services	1,121,028	9.90	1,271,593	10.09	1,493,563	10.90
Transport and communications	706,159	6.24	783,621	6.22	801,370	5.85
Construction	586,862	5.18	607,080	4.82	601,091	4.39
Other types of activity	2,351,217	20.77	2,559,873	20.29	2,521,050	18.39

According to the data visualized in Table 1, the share of tax revenue at the consolidated budget of the Russian Federation from oil extraction companies in the years 2013–2015 averages 30 % and rises every year. Moreover, it should be noted that crude oil and natural gas extraction, as well as rendering of services in these areas, in the described period averages 96 % of all mineral extraction, which obviously evidences the importance of the industry for the state budget and the development of economy in whole.

With a view to investigate the peculiarities of taxation system for oil production companies in Russia we've examined mineral extraction tax (MET) receipts at the consolidated budget of the Russian Federation in 2013–2015 visualised in Table 2.

According to the data presented in Table 2 mineral extraction tax receipts at the consolidated budget of the Russian Federation in 2013–2015 average 75 %. In its turn crude oil and natural gas MET reach 98 % of all receipts, which yet again proves the importance of MET both for the development of the industry and the region. It is therefore necessary to pay special attention to existing problems of mineral extraction tax assessment and to find ways to solve such problems. Thus, as existing deposits deplete, the structure of reserves worsens and oil production costs increase, current tax assessment system fails to ensure the necessary cost-effectiveness of existing field production. As a result about 30–40 % of oil reserves registered in the State Balance and technologically extractable will not be developed because of total unprofitability of development in the context of current tax assessment system. Simultaneously, as the current tax assessment system is based on industry average costs, highly profitable oil deposits have relatively low tax burden, and the rent from such assets received by the state is not full.

This fact is evidenced by the practice of annual introduction or revision of single point tax deductions in respect of oil extraction activity; in particular, at the present moment there is preferential taxation for shelf fields, two export customs duty privileges, eleven regional tax allowances, and nine mineral extraction tax allowances for certain collector types and oil quality. Therewith allowances are often stated the way that their use and administration requires professional knowledge in geology and field development technology, which means that current conditions can't provide tax administration simplicity.

Table 2. MET receipts at the consolidated budget of the Russian Federation in 2013–2015 [1].

	2013		2014		2015	
	Receipts, million roubles	Share, %	Receipts, million roubles	Share, %	Receipts, million roubles	Share, %
Total taxes connected with mineral extraction, including:	3,274,721	100	3,764,146	100	4,297,049	100
Mineral Extraction Tax (MET), including:	2,481,561	76	2,824,460	75	3,179,625	74
- crude oil and natural gas MET; rendering of services in these areas	2,434,347	74.34	2,771,858	73.64	3,100,934	72.16

Peculiarities of existing mineral extraction tax allowances in the area of oil production in Russia are presented in Table 3.

Table 3. Conditions and peculiarities of mineral extraction tax allowances in Russia.

Reason for deduction (region, deposit characteristics)	Year, from which the deduction takes effect	Allowable extraction volume, million tons	Duration of deductions, years
Field depletion	2007	Depletion above 80 %	Without limitation
High-viscosity oil	2007	Without extraction volume limitation	Without limitation
Northern shelf within the Arctic Circle	2009	35	10
Size of reserves	2012	Reserves less than 5 million tons	Without limitation

As you can see in Table 3, beginning from 2007 mineral extraction tax rate in Russia depends on the depletion of reserves. Beginning from 2009 there are allowances for several new oil production territories, beginning from 2012 the list of oil production territories, where allowances are applicable, is enlarged. However, the character of innovations in the tax system doesn't correspond to the requirements of modern taxation system.

3. Results and discussion

One of the most important aspects of functioning of Russian taxation system is the problem of tax burden and its influence on an enterprise business activity. The main task of all tax reforms is to lower tax burden under the condition of compensation for lost income. The more oil and oil products the company sells, the more hydrocarbons are exported, and the more tax receipts there are, and vice versa, the more natural gas is sold, and the more of it remains in the domestic market, the less is the tax share in the income. This is quite logical, because the higher the selling price is, the greater is its difference from production and sales costs, and the seller can afford to pay more taxes. Therewith, increasing tax share in the income does not always mean increasing the seller's tax burden, but often shows such seller's high tax possibilities regarding such income. The dynamics of tax levies, tax burden and cost-effectiveness for oil companies is visualised in Table 4.

Table 4. Dynamics of tax levies, tax burden and cost-effectiveness for Russian oil companies

Companies	Volume of taxes paid, billion $			2015		
	2013	2014	2015	Specific value of taxes per barrel, $	Tax burden versus income, %	Profitability of sales, %
Lukoil	38.4	39.3	38.9	51.1	27.5	5.5
Rosneft	46.7	54	78.2	44.9	53	11.7
Gazpromneft	17.4	21.1	18.6	49.5	39.3	12.5
Surgutneftegas	22.1	26.5	26.3	50.8	63.5	21.3
Tatneft	10.9	10.3	10.1	41	50	12.4
Bashneft	6.8	7.5	5	21.9	28.2	8.2
RussNeft	4.4	3.5	1.7	21.9	27	6
Slavneft	2.9	3.3	3.1	23.8	50.8	23.5

As it is shown in Table 4 tax burden is very different for different companies. Thus, in Surgutneftegas it equals 63 %, and in Bashneft 28.2 %. Because of high tax share in the income, especially in such companies as Surgutneftegas, Rosneft, Tatneft and the higher amount of taxes journalists, economists and public servants working with oil production companies more and more often come to a conclusion that tax burden in oil and gas complex in Russia is excessive.

Thus, A V Novak insists that "the level of fiscal tax burden on oil industry in Russian is the highest compared to existing tax systems in other oil producing countries" and according to him equals 70 % of the income. The minister suggests lowering the tax share in the income to 65 % [2]. Therewith he doesn't find crucial the situation with depreciation effect for our companies, but, of course, the budget was seriously affected, because the income depends on oil prices.

In the cases, when 70 % tax burden on income is mentioned, it is meant that from the total volume of products and services offered by the company only one type of activity is taken into account – exported crude oil, in which customs duties and mineral extraction tax really make it about 70 %. Meanwhile it is not taken into account at all that in recent years; the state actively began to grant considerable mineral extraction tax export duty allowances for oil and gas production in the Northern Regions, in Eastern Siberia and in shelf field.

In this situation it is not taken into account that most companies are vertically integrated, and crude oil is not the only end product of their activity. Thus, for example, in Gazpromneft and Bashneft only 19 % and in Lukoil 25.5 % of the income is received from the sale of crude oil; and the price per barrel for all the hydrocarbons extracted and sold in 2013 was $78, while the world oil price was $108. The difference between the world oil price and specific average revenue per barrel in Russian companies of nearly $30 is connected with the fact that the companies not only sell crude oil outside CIS, but they also sell it to CIS countries and sometimes in the domestic market. And in each selling destination there exists its own level of prices and different tax payments are applicable.

In the world practice the problems of the taxation system for oil production companies were investigated in the articles of the following authors: T D Skolrud, G I Galinato [3], S Tappen, J Baek [4], J G Weber, Y Wang, M Chomas [5], E A Poltavtseva [6], B J A Willigers, K Hausken [7], S Bell, A Hindmoor [8]. Problems and peculiarities of the taxation system for oil companies in Russia were presented in the articles of the following authors: A R Khafizova, I A Fassakhov [9], L P Lunden, S Group [10], A Cherepovitcyn, N Smirnova [11], E Garden [12]. Factors influencing oil extraction were examined by R S Khisamov, V P Lavushchenko, L I Motina, [13], A Russell, R A Dawe [14].

Dramatic differences in the level of tax burden for companies calculated as the ratio of tax payments to the income rise two questions: are these tax burdens relevant and why are their levels so different for different companies? We think that tax burden calculated as the share of tax payments in the income can't be relevant, especially in oil and gas industry. First of all it is connected with the fact that pricing in this sector doesn't actually form under the influence of competitors, whose activity makes the prices stay at the level of average production costs and the minimum margin, but under the influence of other factors that provide a considerable difference between the market price and the production cost. Russian oil production companies have greater tax burden per extracted barrel than the biggest foreign oil production companies, whose average tax burden per extracted barrel equals USD29.24, while for Russian oil production companies it reaches USD43.22.

Average oil and oil product sale prices and export duties in Russia, including average sale prices within and outside CIS, the level of prices in the domestic market, and average export duties for the years 2013–2015 are visualized in Table 5.

Table 5. Average oil and oil product sale prices and export duties in Russia.

Aspects	2013		2014		2015	
	$/ton	%	$/ton	%	$/ton	%
Average oil sale price in countries outside CIS	834	100	807	100	789	100
Average oil sale price in CIS countries	559	67	392	49	377	48
Average oil sale price in Russian domestic market	541	65	327	41	346	44
Average oil export duty	321	38	395	49	396	50
Average oil product export duty	171	20	357	44	357	44
Average fuel oil product export duty	68	8.2	262	33	262	33

In 2015, when oil was sold in the external market, export duty for it was 50% of the price. If we add mineral extraction tax and some other taxes to the export duty, the total amount will be 70% of the income, as it has been mentioned already. In the biggest oil and gas companies of the USA world oil price was 10 times higher than extraction costs, in Brazil and China it was 13 to 14 times higher, in England – 17 times higher, and in Russian companies specialized in oil extraction it was 16 to 40 times higher.

4. Conclusion

Thus, the level of fiscal tax burden on oil industry in Russian is the highest compared to existing tax systems in other oil producing countries. It is necessary to point out that the higher the selling oil price is, the greater is its difference from production and sales costs, and the seller can afford to pay more taxes. Therewith, increasing tax share in the income does not always mean increasing the oil seller's tax burden, but often shows such seller's high tax possibilities regarding such income. In other words, an oil producing organization receives considerable surplus profit. That is why tax burden can only be assessed more adequately with the help of such criterion as profitability of oil production companies' costs.

The presence of natural resource rent, which is taken into account in the taxation system in many countries, predetermines the reforms of domestic taxation system in line with the world practice, where taxation is differentiated depending on certain environmental conditions, which will help to make the position of business entities working in different environmental conditions equal, and to make the development of depleted deposits cost-effective.

The process of formation of a flexible taxation system should be regarded as a very important part of the total system of oil industry state regulation, including the regulation of development and extraction processes, prices, and access to transportation infrastructure. This process should be carried out gradually, and each step should be synchronized with activities on the reformation of other elements of the regulation system, and with the development of oil product market. We find it appropriate to divide the strategy of transformation to a flexible taxation system for oil production industry into successive steps.

One of the steps of transformation to a rent taxation system is the introduction of differentiated mineral extraction tax rates in respect of oil. To achieve this it is necessary to form an appropriate system of accounting of operational, financial and economic performances in oil production companies. Later it will be necessary to adjust differentiated tax rates, simultaneously reforming the system of oil price regulation.

The following step supposes a transfer to differentiated ad valorem mineral extraction tax rates expressed as a percentage, which will give an opportunity to account domestic and external market prices more adequately. The necessary conditions for the realization of this step are: the introduction of a reference price system used to count the price of taxable products; extension of the limits of tax rate differentiation taking into account such factors as the size of the deposits and their depletion level.

The next step is the introduction of progressive and regressive excess profit tax, depending on the profitability and taking into account oil prices. We think it possible to build a strategy of transfer to a flexible taxation system for oil industry with its gradual transformation from production approach to economic approach. All this will create the necessary conditions to change mineral extraction tax for rental payments.

Transfer to the financial result taxation in oil industry will help to:

- seize additional surplus profit from the development of highly profitable oil deposits, which in current taxation system have a relatively low tax level;
- provide for the cost effectiveness of mature production field development, as well as the extraction of reserves difficult to recover and reserves with no infrastructure;
- take into account oil price changes in the world markets, as well as other macroeconomic aspects.

Taking into account the fact that oil industry is very important for budget revenue formation, initially it is planned to test the new taxation system principles in a limited number of deposits, so called 'pilot projects'.

For highly profitable minefield deposits it is suggested to introduce progressive and regressive index, varying depending on oil prices.

Within the framework of the investigation the authors come to a conclusion that it is necessary to introduce gradually the taxation system based on the definition of surplus profit depending on the cost effectiveness and taking into account oil prices.

Acknowledgement
The work is performed according to the Russian Government Program of Competitive Growth of Kazan Federal University.

References

[1] Summary reports for the Russian Federation in a whole and for subjects of the Russian Federation Available from: https://www.nalog.ru/rn77/related_activities/statistics_and_analytics/forms/
[2] Koliender A 2007 Russian companies don't have high income from high prices *The Wall Street Journal* November
[3] Skolrud T D, Galinato G I 2017 Welfare implications of the renewable fuel standard with an integrated tax-subsidy policy *Energy Economics* **62** p 291
[4] Tappen S, Baek J 2016 Can oil tax policy change exploration levels?: Empirical evidence from Alaska oil legislation *Energy Sources, Part B: Economics, Planning and Policy* **11** p 1055
[5] Weber J G, Wang Y, Chomas M 2016 A quantitative description of state-level taxation of oil and gas production in the continental U.S *Energy Policy* **96** p 289
[6] Poltavtseva E A 2013 Vietnam taxation of petroleum companies *NeftyanoeKhozyaistvo (Oil Industry)* **5** p 117
[7] Willigers B J A, Hausken K 2013 The strategic interaction between the government and international oil companies in the UK: An example of a country with dwindling hydrocarbon reserves *EnergyPolicy* **57** p 276
[8] Bell S, Hindmoor A 2014 The Structural Power of Business and the Power of Ideas: The Strange Case of the Australian Mining Tax *New Political Economy* **19** pp 470-486
[9] Khafizova A R, Fassakhov I A 2015 Development of the oil companies taxation system *Mediterranean Journal of Social Sciences* **6** p 21
[10] Lunden L P, Group S 2013 Russian tax and license policy - A hurdle for offshore development? Society of Petroleum Engineers - *SPE Arctic and Extreme Environments Conference and Exhibition, AEE 2013* **3** p 2691
[11] Cherepovitcyn A, Smirnova N 2013 Features of the Russian license and tax policy at offshore oil and gas fields development Society of Petroleum Engineers - *SPE Arctic and Extreme Environments Conference and Exhibition, AEE 2013* **3** p 2632
[12] Garden E Russia limits access to arctic *Offshore Engineer* **39** p 64
[13] KhisamovR S, Lavushchenko V P, Motina L I 2013 Economic factors contributing to sustainable oil production in the Republic of Tatarstan *Neftyanoe Khozyaistvo (Oil Industry)* **7** p 8
[14] Russell A, Dawe R A 2013 How the oil revenue is shared *Energy Sources, Part B: Economics, Planning and Policy* **8** p 346

Responsibility for the Violation of Ecological Safety Requirements

J I Selivanovskaya[1,2] and I Gilmutdinova[1]
[1]Kazan Federal University, Kremlevskaya str., 18, Kazan, 420008, Russia
[2]Kazan National Research Technical University named after A.N. Tupolev (KAI), K.Marx str., 10, Kazan, 420111, Russia
e-mail: JISelivanovskaya@mail.ru

Abstract. The article deals with the problems of responsibility for the violation of ecological safety requirements from the point of view of sustainable development of the state. Such types of responsibility as property, disciplinary, financial, administrative and criminal responsibility in the area are analysed. Suggestions on the improvement of legislation are put forward. Among other things it is suggested to introduce criminal sanctions against legal bodies (enterprises) for ecological crimes with punishments in the form of fines, suspension or discontinuation of activities.

1. Introduction

Environmental protection is one of the global problems of the present time. According to the Constitution of the Russian Federation every person has a right to live in favourable environment, to receive reliable information on its condition and to be compensated for the damage to his/her health or property caused by an environmental crime (art. 42) [1]. The Russian Federation is a party to a number of international cooperation agreements in the area (for example, the Convention on Biological Diversity, the Stockholm Convention on Persistent Organic Pollutants).

2. Methods

The framework of the investigation is formed by the Constitution of the Russian Federation, Civil Code of the Russian Federation, Labour Code of the Russian Federation, Administrative Violations Code of the Russian Federation, Criminal Code of the Russian Federation, Federal Law of January 10, 2002 No 7-FZ On Environmental Protection, and Administrative Violations Code of the Republic of Tatarstan. In the course of the study the following methods of investigation have been used: historical and legal method, formal and logical method, systematic and structural method, specific sociological method, statistical method and the method of comparative law.

3. Results

At the present time the matters of ecological safety become more and more important. According to Art. 1 of the Federal Law of January 10, 2002 No 7-FZ On Environmental Protection, ecological safety is the protection of natural environment and vital interests of a human being from possible negative influence of business or other types of activity, emergency situations of technogenic and natural origin and their consequences [2]. The content of the concept 'ecological safety' is wider than that of the concept 'environmental protection', as it includes not only a favourable environmental conditions, but also meets ecological interests and needs of a human being, society and the state [3]. Federal Law General Technical Regulations On Ecological Safety was drafted in Russia [4]. Technical Regulations should be adopted for the purpose of environmental protection, which is why they define general provisions of technical regulation in the sphere of ecological safety and specify general requirements to ecological safety of manufacturing, storage, transportation and disposal of products. As the Law has not been adopted yet, the matters of violation of ecological safety requirements are regulated by Federal Law On Environmental Protection.

Sustainable development of any region is only possible in favourable environmental conditions, which depend on the compliance with environmental legislation. Federal Law On Environmental Protection establishes property, disciplinary, financial, administrative and criminal responsibility for the violation of environmental legislation in Russia [2]. However, there is also financial responsibility. As financial responsibility supposes the relationship between an employee and an employer, it is regulated by labour legislation, but in the cases when financial responsibility results from an environmental crime, it can also be regarded as environmental responsibility. It is necessary to point out that only criminal and administrative responsibility for the violation of environmental legislation is mutually exclusive, which means that in one and the same case of violation one and the same person can only be brought to either administrative or criminal responsibility. All other forms of environmental responsibility can be simultaneous.

Civil (or property) responsibility for the violation of ecological safety requirements consists in the liability to recover environmental damage. In accordance with the legislation legal and physical bodies that have harmed the environment as a result of contamination, depletion, damage, destruction, irrational use of natural resources, degradation and destruction of natural ecological systems, natural complexes and natural landscape, as well as other violation of environmental legislation, shall compensate in full for such damage. Environmental damage caused by an entity involved in economic or other activities is compensated according to rates and methods for calculating the extent of damage caused to the environment approved in the established procedure, taking into account the losses including the missed profit. Thus, both actual damage and missed profit should be compensated. According to Federal Law On Environmental Protection there are two types of damages: environmental damage (Art. 77, 78) and damage to human health and property as a result of violation of environmental legislation (Art. 79) [2]. Therewith other classifications of damages, the compensation for which takes the form of property responsibility are described in literature. For example, economic harm is described as impairment of property interests as a result of destruction, damage, contamination and depletion of natural objects; and ecological damage is defined as impairment of the natural condition of natural objects, weakening of their ecological connections and functions, impairment of the ability of natural objects to clean themselves and to recover, weakening of protective properties of the nature, and a human being [5].

Disciplinary responsibility for the violation of ecological safety requirements is established in cases of unlawful, culpable non-performance or inadequate performance by an employee of his/her labour functions connected with the realization of plans and activities in the area of environmental protection and rational exploitation of natural resources, as well as for the violation of environmental legislation requirements within the scope of his/her official duty. According to Art. 192 of the Labour Code of the Russian Federation, a disciplinary offence is non-performance or inadequate performance by an employee of his/her employment duties [6]. It is necessary to point out that it is only possible to bring someone to disciplinary responsibility in case of carrying out professional activities connected with the use of natural resources or influencing environment. In this regard such professional (operational, scientific, etc.) activity is subject to special requirements concerning the standards of environmental legislation, environmental quality regulations, and supposes carrying out plans and activities in the area of environmental protection and rational utilization of natural resources. Besides, there is an important condition, that according to the regulations, or individual administration of law, or under contract the person has labour functions in operational activity of the enterprise/organization or the functions of an executive officer.

Law-enforcement offices have certain difficulties, connected with the fact that environmental legislation does not contain specific definitions of disciplinary cases. In every particular case it is the employer who has to assess the fact of violation of environmental legislation and to make a decision of disciplinary punishment. Thereby the definition of the disciplinary offence is formal and the guilty person is brought to responsibility for the violation of ecological rules, independent of consequences.

Administrative responsibility is provided for ecological offences defined in the administrative legislation. Administrative legislation comes under the joint competence of the Russian Federation and its Subjects. The matters of joint competence of the Russian Federation and its Subjects are regulated by federal laws and laws and other regulatory legal acts adopted by the Subjects of the Russian Federation according to such federal laws (Part 2, Art. 76 of the Constitution of the Russian Federation) [1]. At the federal level administrative responsibility in the sphere of environmental protection is regulated by Chapter 8 of the Administrative Violations Code of the Russian Federation – Administrative Offences in the Area of Environmental Protection and Natural Resource Management – containing both general and specific definitions of ecological offences. General cases include, for example, failure to meet ecological requirements in the process of planning, project feasibility description, engineering, siting, construction, reconstruction, commissioning, and operation of enterprises, constructions or other objects (Art. 8.1 of Administrative Violations Code of the Russian Federation), failure to meet ecological and sanitary and epidemiological requirements for the management of production and consumption wastes and other dangerous substances (Art. 8.2 of Administrative Violations Code of the Russian Federation), etc. [7]. Specific definitions of ecological offences include, for example, administrative violations in the spheres of land protection and use, mineral resources protection and use, aquatic resources aquatic resources, atmospheric air protection, etc. In case of an ecological offence within the scope of specific standards, such specific standards prevail, according to the rules of conflicts of general and specific norms. However, if an ecological offence committed is not included into specific standards of the Administrative Violations Code of the Russian Federation, general standard is used. The positive aspect of the situation is the fact that virtually no ecological offence will remain unpunished, even if environmental protection legislation changes, and administrative responsibility legislation fails to introduce corresponding changes in due time. On the contrary, there is a negative influence of general norms, present in the legislation: very often the executor of law prefers to use general standards and doesn't try to prove that certain ecological offences are specific, which lowers the efficiency of administrative responsibility in the area of environmental protection.

In the Republic of Tatarstan administrative responsibility in the sphere environmental protection is regulated by the Administrative Violations Code of the Republic of Tatarstan, that in fact contains only one definition of offence in this area – destruction of rare and endangered animals or plants (Art. 3.1 of Administrative Violations Code of the Republic of Tatarstan) [9].

Criminal responsibility is provided for environmental crimes. Environmental crimes are the most serious violations of ecological safety requirements, which, as a rule, lead to mass mortality of animals or plants, spread of epidemics and epizootics, harm to human health or other serious consequences [9].

Criminal Code of the Russian Federation includes Chapter 26 Ecological Crimes, which, as well as the Administrative Violations Code of the Russian Federation contains general and specific definitions of environmental crimes. In 2014 in Russia 9,479 people were convicted for ecological crimes defined in Chapter 26 of the Criminal Code of the Russian Federation, in 2015 the number of them was 8,960 people, and in 2016 – 8642 people. However, other chapters of the Criminal Code of the Russian Federation contain the definitions of crimes connected with ecological safety requirements. For example, a new definition of ecological crime of 'ecocide' was included into Section Crimes Against Peace and Human Safety (Art. 358) [10]. Ecosystem here is regarded as an additional object of crime. This crime contains mass destruction of plants or animals, poisoning the atmosphere or aquatic resources, as well as other actions leading to environmental catastrophes.

IOP Conf. Series: Earth and Environmental Science **107** (2017) 012073 doi:10.1088/1755-1315/107/1/012073

In the Russian Federation only physical bodies are subject to criminal responsibility, that is, if a legal body (an enterprise) harms environment and thus commits an ecological crime defined in the Criminal Code of the Russian Federation, it is not the enterprise itself, but the guilty person. This doesn't always correspond to the principle of justice, as the punishment of such a person will not correspond to the level of environmental damages caused by the enterprise [11]. Both Russian and foreign authors make suggestions to introduce criminal responsibility for enterprises, first of all for ecological crimes [12, 13]. Thus, B. V. Volzhenkin points out that criminal responsibility of a legal body can follow if the punishable offence was approved or took place with the knowledge, to the advantage, or in the interests of such a legal body (or its unit) [14]. In this case sanctions against legal bodies shall include fines, suspension or discontinuation of activities, and the criminal legislation should be changed correspondingly.

Establishment of different types of responsibility for the violation of environmental legislation is one of the methods of fighting environmental crimes, as it realizes the warning function of responsibility. Therewith, a necessary condition of fighting such crimes is constant improvement of the legislation by means of elimination of conflicts and gaps in the legal regulation of ecological responsibility matters.

4. Conclusions

1. Compliance with the requirements of ecological safety is the guarantee of sustainable development of the Russian Federation, and is achieved as well by the establishment of responsibility for violation.

2. The content of the concept 'ecological safety' is wider than that of the concept 'environmental protection', as it includes not only a favourable environmental conditions, but also meets ecological interests and needs of a human being, society and the state.

3. As at the present time Russia hasn't adopted yet a special purpose legislative act regulating relations in the area of ecological safety, the matters of violation of ecological safety requirements are regulated by Federal Law of January 10, 2002 No 7-FZ On Environmental Protection.

4. In Russia the violation of environmental legislation supposes property, disciplinary, financial, administrative and criminal responsibility.

5. Administrative Violations Code of the Russian Federation and Criminal Code of the Russian Federation, regulating the matters of administrative and criminal responsibility contain both general and specific definitions of ecological offences. In this situation virtually no ecological offence will remain unpunished, even if environmental protection legislation changes, and administrative responsibility legislation fails to introduce corresponding changes in due time. However, very often the executor of law prefers to use general standards and doesn't try to prove that certain ecological offences are specific, which lowers the efficiency of administrative and criminal responsibility in the area of environmental protection.

6. It is necessary to establish criminal responsibility for legal bodies (enterprises) for the violation of ecological safety requirements, because if only executive officers of such enterprises are brought to responsibility, the punishment of such a person will not correspond to the level of environmental damages caused by the enterprise, which is against the principle of justice. Criminal sanctions against legal bodies shall include fines, suspension or discontinuation of activities.

Acknowledgement

The work is performed according to the Russian Government Program of Competitive Growth of Kazan Federal University.

References

[1] Constitution of the Russian Federation (adopted by national referendum on December 12, 1993) Available from: http://www.consultant.ru/document/cons_doc_LAW_28399/
[2] Federal Law of the Russian Federation of January 10, 2002 No 7-FZ On Environmental Protection Available from: http://docs.pravo.ru/zakon-ob-ohrane-okruzhayushey-sredy/

[3] Dong S B, Sun J, Cao Z Y 2014 *Proc. Int. Conf. on Advances in Energy and Environmental Science (Guangzhou)* p 650
[4] Draft Federal Law General Technical Regulations On Ecological Safety Available from: http://www.rosteplo.ru/Npb_files/npb_shablon.php?id=987
[5] Gimadeev M *2000 Encyclopedic Dictionary of Ecology* (Kazan: Priroda) p 274
[6] Labour Code of the Russian Federation of December 30, 2001 No 197-FZ Available from: http://docs.pravo.ru/document/view/7/
[7] Administrative Violations Code of the Russian Federation of December 30, 2001 No 195-FZ Available from: http://docs.pravo.ru/document/view/9/
[8] Administrative Violations Code of the Republic of Tatarstan of December 19, 2006 No 80-ZRT Available from: http://docs.pravo.ru/document/view/3307861
[9] Knopp L 1994 Neues Umweltstrafrecht und betriebliche Praxis Betriebs-Berater 32 pp 2219-2224 (in German)
[10] Criminal Code of the Russian Federation of June 13, 1996 No 63-FZ Available from: http://docs.pravo.ru/kodeks-ugolovniy/
[11] Baizhomartova K A, Kegembayeva Z A, Baiseitova A T 2016 Measures of responsibility of juridical persons for ecological offenses *Ecology, Environment and Conservation* **22** p 2151
[12] Ebenroth C T 1997 Die strafrechtliche Verantwortung des Vorstandes für Umweltstraftaten und gesellschaftsrechtliche Vermeidungsstrategien *Betriebs-Berater* **28** p 1941 (in German)
[13] Grebennikov V V, Sangadzhiev B V, Vinogradova E V 2016 International environmental safety as object of crimes, leading to possibility of legal entities criminal liability *International Journal of Environmental and Science Education* **11** 13-15
[14] Volzhenkin B V 2002 *Crimes in the sphere of economic activity* (St. Petersburg) p 109

Bioconversion of oil sludge into biomass of lipid metabolites for use as a source of biofuel

T N Shchemelinina, N V Matistov, M Yu Markarova and E M Anchugova

Institute of Biology of Komi Scientific Centre of the Ural Branch of the Russian Academy of Sciences (IB Komi SC UB RAS), Syktyvkar, 28 Kommunisticheskaya str, 167982 Russia
e-mail: tatyanakomi@mail.ru

Abstract. The possibilities for the generation of biofuel from the results of the accumulation of lipids in oil-contaminated environments were studied. This type of accumulation occurs in the biomass of yeast strains *Rhodotorula* sp. VKM Y-2993D; in bacteria like *Pseudomonas libanensis* B-3041D and in consortia of microalgal strains such as *Acutodesmus obliquus* Syko-A Ch-055-12, *Chlorella* sp. SYKO A Ch-011-10, *Monoraphidium* sp., and *Anabaena* sp. The most promising of these for processing petroleum hydrocarbons into biofuels was found to be the consortium of microalgal strains, the content of palmitic acid of which reached 49.0 %, thereby achieving a mid-range cetane number.

1. Introduction

The accumulation of persistent pollutants which are almost indestructible in natural environments, or are destroyed only very slowly [1], both disrupts ecosystems and the connections in the biosphere established through long-term evolution, and undermines the capability of those ecosystems to self-regulate. Such persistent pollutants include oil sludge, as well as oil products (oils, diesel fuel, petrol, resins, etc.) that often have accumulated and contaminated the environment for years.

Legislation in many countries, including Russia, encourages the inclusion of industrial wastes into economic turnover, as sources of secondary raw materials. At present, numerous non-biological technologies are evolving for the processing oil wastes. These are designed to exploit the potential of the resource to generate various commercial products, such as road-building materials, secondary hydrocarbon crude, etc. Biological methods, however, are mainly aimed at enhancing the degradation of the residual oil contaminants with a loss of energy resource.

It is well-known that, among renewable energy sources, biomass has been widely used as the basis for generating biofuels (biodiesel, bioethanol) [2, 3]. Studies are ongoing on to evaluate a range of conditions and methods for the bioconversion of various substrates into microbial biomass. The accumulation of microorganism biomass in industrial wastes is a promising energy-related and environment-friendly solution to this problem [4]. We have proposed a fundamentally new approach to obtaining secondary energy raw materials from industrial wastes by way of the bioconversion of oil sludge into lipid metabolites with petroleum-oxidizing microorganisms, and their accumulation of fatty acids that can subsequently used for generating biodiesel. Through the accumulation of useful microorganism biomass, this has the advantage of helping to solve, not only the problems of waste utilization but also of bioremediation at sites of oil spillages.

Biodiesel is a multicomponent liquid fuel consisting of methyl and ethyl esters of higher unsaturated fatty acids [5]. Fatty acids and higher fatty alcohols accumulate in the cells of the microorganism in the form of droplets, and as the phospholipids in membranes, etc. [6]. Fatty acids obtained from microorganisms need to conform to CN standards in order to be used for biodiesel production. Diesel fuels with CN 45 to 55 ensure the best performance in up-to-date diesel engines. The higher the overall content of palmitic and stearic acids, the higher the quality of the biodiesel [7].

2. Material and methods

The purpose of this work is to identify features of the microbial bioconversion of petroleum hydrocarbons into lipid metabolites.

For our experiments, we used native microorganisms (strains of bacteria, yeast, microalgae) extracted from oil-contaminated soils of the Usinsk district. These were incubated in soil microcosms enriched with mineral oil (30%) at room temperature, under natural lighting, at 220 rpm over periods of 3, 30 and 90 days. The following microcosms were used: 1 – with an inoculum of yeasts and bacteria, 2 – with microalgae, 3 – with an alga-bacteria-yeast consortium. For the investigation we used *Rhodotorula* sp. VKM Y-2993D yeasts, *Pseudomonas libanensis* B-3041D bacteria and a consortium of microalgal strains: *Acutodesmus obliquus* Syko-A Ch-055-12, *Chlorella* sp. SYKO A Ch-011-10, *Monoraphidium* sp. and *Anabaena* sp., respectively. As a control, the concentration of fatty acids in an oil-contaminated microcosm without microorganism inoculation was determined.

Methyl esters of fatty acids were identified using the method of K. Sinyak-et al [8]. Saturated fatty acids contained in the lipid metabolites of the microbiota were determined using chromatography-mass-spectrometry [9].

3. Results and discussion

Analysis of the conditions needed for the bioconversion of oil hydrocarbons into fatty acids showed that the maximum accumulation of fatty acids occurred after 3 days of incubation. It is likely that, during their use of the easily oxidized oil products, the cells accumulated lipids as reserve materials, and that as the readily used resources disappeared and environmental conditions deteriorated the cells then accessed the reserves for continuation of their metabolic processes. As a result, after 30 and 90 days the fatty acids content had fallen again, almost to zero. The component composition of fatty acids in the system included carboxylic acids with chain lengths from C_{12} to C_{18} and having even numbers of carbon atoms. The strain specificity of the accumulation of stearic and palmitic acids is noteworthy. In the biomass of the bacteria, yeast and microalgal consortia, palmitic acid predominated (table 1). Table 1 shows that accumulation of palmitic acid $C_{16:0}$ occurs in all the variants compared with the control, but that the maximum accumulation, matching the CN standards was observed in microcosm 2 with the microalgal inclusion. Furthermore, this microcosm was also high in linoleic acid – an output of 22.9%, requiring separation of these acids: palmitic acid for biodiesel generation and linoleic acid for therapeutics and cosmetics.

Table 1. Content of fatty acids in different microcosms (bacteria, yeast, microalgal, consortium), %.

Variant	$C_{12:0}$[a]	$C_{14:0}$	$C_{14:1}$	$C_{16:0}$	$C_{16:1}$	$C_{18:0}$	$C_{18:1(9)}$	$C_{18:1(11)}$	$C_{18:2cis}$
C	0	5.5	2.39	33.6	13.56	4.72	10.1	2.08	28.0
1	0.5	3.6	2.3	44.2	24.2	1.7	1.8	7.8	14.0
2	1.3	4.9	2.7	49.0	4.3	3.3	9.0	2.5	22.9
3	2.0	6.4	6.0	34.4	14.0	1.0	7.1	6.2	22.8

[a]$C_{12:00}$ – lauric acid; $C_{14:0}$ – myristic acid; $C_{14:1}$ – myristoleic acid; $C_{16:0}$ – palmitic acid; $C_{16:1}$ – palmitoleic acid; $C_{18:0}$ – stearic acid; $C_{18:1(9)}$ – oleic acid; $C_{18:1(11)}$ – vaccenic acid; $C_{18:2}$ – linoleic acid.

It is interesting to compare the CNs of the resulting products with the existing standards for diesel, and bio-diesel fuels from oil seed rape and from Mucorales fungal lipids *Cunninghamella japonica* (table 2). The need to devote large areas of land for the generation of biodiesel by growing gramineous plants restricts such activity, particularly considering that, only the seed oil is used for production, and the majority of the plant biomass is not used. Furthermore, as oil-generating plants also require frequent crop rotation [10] conclusion can be drawn that, reforming oil wastes into biofuel with the help of microorganisms is a viable alternative.

Table 2. Comparative Analysis of Fuel.

Source	CN	Reference
Traditional diesel fuel	47	[11]
Biodiesel (EN14214)	>51	[12]
Plant raw material, *Brassica napus*	54.4 51.74	[11] [13]
Mucoral fungus biomass *C. japonica* F-1204 (-)	55.68	[14]
Biomass of a consortium of microalgal strains *A. obliquus, Chlorella* sp., *Monoraphidium* sp., *Anabaena* sp.	49.0	Proprietary data

However, as the overall content of palmitic and stearic acids is not very high (table 1), the resulting biodiesel cannot be considered to be of premium grade.

Thus, of the microorganisms tested the consortium of microalgal strains (*Acutodesmus obliquus, Chlorella* sp., *Monoraphidium* sp., *Anabaena* sp.) generating 49.0% palmitic acid and thereby providing an acceptable cetane number, was found to be most promising in respect of reprocessing waste oil products into biodiesel.

Presumably, attempts can be made to increase the yield of stearic acid, it will be necessary to include into the consortium additional microorganism capable of synthesizing this acid.

Acknowledgements
This study was performed with funding from the State Task "Technology of living systems" based on higher plants, incubated plant cells, microorganisms and enzymes for the production of biomedical products, biofuel and the protection of the environment" No. 115021210021 and "Bioekobalans – LLC".

References
[1] 1989 *Encyclopedic Dictionary of the Environment* (Chisinau: Chief Editorial Board of the Moldova Soviet Encyclopedia) p 408
[2] Yuzbasheva E Y, Yuzbasheva T V, Mostova E B, Perkovskaya N I and Sineokiy S P 2014 *Biotechnology* **2** 8 – 23
[3] Beopoulos A, Nicaud J M and Gaillardin C 2011 *Appl Microbiol Biotechnol* **90** 193 – 206
[4] Shchemelinina T N, Matistov N V, Kovaleva V A, Markarova M Yu, Gruzdev I V, Volodin V V and Anchugova E M 2017 *Appl Biochem Microbiol* **53** 387 – 94
[5] Gelfand E D 2012 *Tekhnologiya biotopliv (Technology of biofuels)* (Arkhangelsk: Northern (Arctic) Federal University named after MV Lomonosov) p 60
[6] Khotimsky B G, Akopian A I 1970 *Preobrazovanie neftey mikroorganizmami (Reforming of oils with the help of microorganisms)* (Leningrad: All-Russia Petroleum Research Exploration Institute) p 281
[7] Kovalenko G A, Perminova L V, Beklemishev A B, Yakovleva E Y and Pykhtina M B 2014 *Kataliz v promyshlennosti* **6** 71 – 9
[8] AS 542932, USSR, G N 01 1 / 28 Sinyak K M, Danilenko I I and Vasyurenko Z P 1977 *Mode of preparation of samples of lipids at Kiev Institute of Epidemiology, Microbiology and Parasitology № 2138675 Bull 2 (In Russian)* http://dfo-mpo.gc.ca/Library/113290.pdf

[9] Gabov D N, Beznosikov V A, Kondratenok B M and Gruzdev I V 2010 *Eurasian Soil Science* **43** 1102 – 8
[10] Yuzbasheva E Yu, Yuzbashev T V, Mostova E B, Perkovskaya N I and Sineokiy S P 2014 *Biotechnol* **2** 8 – 23
[11] Gunstone F, Harwood J and Dijkstra A 2007 *The Lipid Handbook* (CRC Press Book) p 1472
[12] Knothe G, Gerpen J V and Krahl J 2005 *The Biodiesel Handbook* (Eds Illinois: AOCS Press) 278 – 285
[13] Lunin V V, Sergeyeva Y E, Galanina L A, Mysyakina I S, Ivashechkin A A , Bogdan V I and Feofilova E P 2013 *Appl Biochem Microbiol* **49** 46 – 52
[14] Bogdan V I, Koklin A E, Krasovsky V G, Lunin V V, Sergeyeva Y E, Ivashechkin A A and Feofilova E P 2013 *Sverkhkriticheskie flyuidy: teoriya I praktika* **8** 46 – 52

IOP Conf. Series: Earth and Environmental Science **107** (2017) 012075 doi:10.1088/1755-1315/107/1/012075

The carbonaceous sorbent based on the secondary silica-containing material from oil extraction industry

I V Starostina[1], D V Stolyarov[1], Ya N Anichina[1] and E V Porozhnyuk[1]

[1]Belgorod State Technological University named after V G Shukhov, Kostyukova str, 46, Belgorod, 308012, Russia
e-mail: starostinairinav@yandex.ru

Abstract. The object of research in this work is the silica-containing waste of oil extraction industry – the waste kieselghur (diatomite) sludge from precoat filtering units, used for the purification of vegetable oils from organic impurities. As a result of the thermal modification of the sludge, which contains up to 70% of organic impurities, a finely-dispersed low-porous carbonaceous mineral sorption material is formed. The modification of the sludge particles surface causes the substantial alteration of its physical, chemical, adsorption and structural properties – the organic matter is charred, the particle size is reduced, and on the surface of diatomite particles a carbon layer is formed, which deposits in macropores and partially occludes them. The amount of mesopores is increased, along with the specific surface of the obtained product. The optimal temperature of sludge modification is 500°C. The synthesized carbonaceous material can be used as an adsorbing agent for the purification of wastewater from heavy metal ions. The sorption capacity of Cu^{2+} ions amounted to 14.2 $mg{\cdot}g^{-1}$ and for Ni^{2+} ions – 17.0 $mg{\cdot}g^{-1}$. The obtained values exceed the sorption capacity values of the initial kieselghur, used as a filtering charge, for the researched metal ions.

1. Introduction

In the present conditions of the intensive industrial production growth the problem of environmental protection against the industrial pollution becomes a more and more vital task. One of the particularly acute problems is, the generation and treatment of wastewaters, which get into the surface water bodies, pollute the subsoil and ground waters, deteriorate the water situation and complicate the water utilization [1].

The effluents of many chemical, petrochemical and machine-building industry enterprises are characterized as multicomponent ones, which contain the heavy metal compounds, oils, fats, suspended substances and petroleum products. Heavy metals are classified among the substances which have cumulative, additive, carcinogenic and mutagenic effect [2], so the waste waters, containing heavy metal compounds, are subject to fine purification. For this purpose a number of methods are used – the reagent treating methods [3], sorption methods [4], biochemical and electrochemical methods [5] and others [6]. The most preferred are sorption processes, which are highly efficient and allow using both natural [7] and artificial materials [8], as well as agricultural waste products [9, 10] and industrial waste [11].

This article presents the research findings of the use of waste kieselghur sludge (WKS) from oil extraction industry – diatomite as a feedstock for obtaining a carbonaceous sorption material.

At present, the spent kieselghur sludge is the waste belonging to hazard category 4 (low-hazardous substance) [12]. Of all the possible ways of its utilization, the most common is the above-ground storage - over 95% are dumped on waste deposits, producing the adverse effect on the environment [13]. Besides, the inhalation of its pulverized form has a negative effect on the immunity and respiratory system [14]. So, taking into account the growing scale of vegetable oils production in Russia, the problems of enlarging the range of spent powdered sorbents application are becoming more and more relevant.

In this regard, the objective of this work is evaluating the influence of the WKS thermal modification conditions on the textural and sorption properties of the obtained carbonaceous product.

2. Material and methods

As an object of research the waste kieselghur sludge, generated at the stage of refined oil winterization – i.e. its fractionating by cooling, at the OAO «EFKO», Alekseevka city, Belgorod region, Russia, was considered.

The mineral composition of kieselghur sludge was studied by means of X-ray phase analysis with a diffractometer «DRON-2.0».

The microscopic study of kieselghur sludge samples was carried out by using a high-resolution scanning electron microscope «Hitach -8-800».

The thermal modification of waste kieselghur sludge was carried out by heating an averaged sample of WKS without oxygen in the SNO 25/12 electric furnace in the temperature range from 450°C to 650°C within 1 hour. The obtained product was sifted through the sieve with openings diameter 0.1 mm.

The porosity and specific surface of the native kieselguhr and the waste kieselghur sludge, thermally modified at various temperatures, were determined with a low-temperature nitrogen adsorption method by using an automated sorption unit TriStarII 3020.

The particle size of the initial and thermally modified WKS was determined by using FRITSCH Analysette 22 NanoTec Plus laser particle analyzer.

The sorption properties of the native kieselghur, used as a filter, and the thermally modified WKS were determined by their ability to absorb heavy metal ions (Cu^{2+} and Ni^{2+}) in static conditions at the constant temperature (20°C) from salt test solutions. To prepare test solutions the following chemical reagents were used: $CuSO_4 \cdot 5H_2O$ qualified as «p.a.» (analytically pure), produced by Kyshtym copper anode plant, and $NiCl_2 \cdot 6H_2O$ qualified as «c.p.» (chemically pure), produced by Cherkasy chemical reagents plant. The adsorption amount for Cu^{2+} and Ni^{2+} ions were determined in the concentration range of heavy metal ions from 3 to 400 $mg \cdot l^{-1}$, the solutions volume was 50 ml, material consumption – 0.5 g, the mixing time – 24 hours. The pH value of the medium was kept within the range of 4.0 – 4.3 in order to prevent the hydrates formation process, the correction was made with HCl solution. Then the liquid was separated by centrifugation, and the concentrations of Cu^{2+} and Ni^{2+} ions were determined by spectrophotometrical method using sodium diethyl dithiocarbamate and dimethyl glyoxime, respectively, with a spectrophotometer «KFK-3» according to standard techniques [15].

3. Results and discussion

The winterization process in refined vegetable oils production implies the separation of organic impurities from the oil – the glutinous impurities and wax-like substances, which form a fine and sedimentation-stable suspension, increase the oil turbidity and make it difficult to provide the high energy and nutrient value of the product [16]. The wax-like compounds are mostly accumulated in the ground tissues of oil crops, and in the course of processing they pass to vegetable oils. The total content of organic impurities can amount to 70%, and among them the wax-like substances can make from 1.5 to 12 % [17], which depends on the species and breed of oil-bearing crops, and on the processing technology. The organic impurities are removed in precoat filtering units with the use of natural sorption materials – perlite, diatomite rocks, bleaching clay and others.

In the setting of OAO «EFKO», Alekseevka, Belgorod region, Russia, as a result of oil purification the waste kieselghur sludge (WKS) is generated, which is represented by a dingy white slow stock with 60 - 70% of fat-and-oil mixture. The mineral part of the sludge is represented by diatomite – a light sedimentary siliciferous rock, consisting of the skeletons of microphytic diatomic algae – diatoms and radiolarians of a wide range of forms and sizes - from 10-20 nm to 0.5 mm. Taking into account the size and ordering of the pores, diatomite can be characterized as a natural macroporous material (figure 2.3). According to [18], the pores with radius of 4 – 40 μm make about 15% of the total pore space.

As follows from the carried-out research, diatomite is mostly represented by siliceous hydrates of various water cut – varieties of opal of $mSiO_2 \cdot nH_2O$ type, quartz impurities as a crystalline component (figure 1), which is coherent with the results, presented in the work [19]. According to the researchers [19], the diatoms' shucks appear as alternating tridymite and cristobalite layers. The opal globules are formed as early as the diatomic ooze formation stage and are registered in the diffraction pattern as an amorphous phase, namely an opal-cristobalite-tridymite phase [20].

Diatomite, which is used in the food industry as a filtering material, is thermo-chemically modified to provide its chemical inertness by means of acid treatment with the subsequent calcination at temperature 800-1000°C. This process results in the multistage removal of crystallization and adsorption water from the opaline silica and clay minerals, contained in the mineral constituent of the initial diatomite. At heating diatomite to 1000 °C the opal globules transform to cristobalite, its crystallinity degree is increased, so at studying the mineral composition of the waste kieselghur sludge under study, only cristobalite and quartz reflections are registered in the diffraction pattern (figure 1). The clay minerals, mostly represented by kaolinite, are present in the amorphized state, which is indicated by the low-intensity halo in the angular range 2Θ 8-16°.

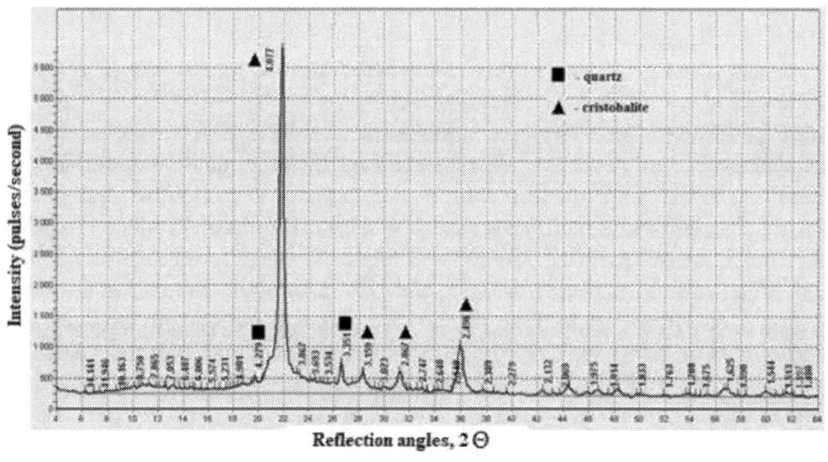

Figure 1. The X-ray pattern of the kieselguhr sludge mineral part.

To remove the organic impurities, which pass from the purified oil to the spent sludge, and to increase its sorption activity the waste kieselghur sludge was thermally modified in the temperature range from 450 to 650°C within 1 hour. At this, the organic matter is charred, a carbon layer is formed on the surface of diatomite particles and the particle size is reduced (figure 2); as a result, a finely-dispersed product is obtained. The obtained carbonaceous mineral material was further used as an adsorbing agent to purify test solutions from Cu^{2+} and Ni^{2+} ions.

As known, the adsorption rates of various pollutants depend on a number of parameters: the properties of an adsorbing agent – its surface area, particle size and porosity, the properties and concentration of pollutants, and the conditions of the adsorption process [21, 22]. The carbon particles and other products of the organic matter oxidation, formed as a result of the WKS thermal modification, can deposit in macropores and large mesopores and occlude them. As the physical adsorption processes of diatomite are mostly provided by shucks pores and interporous areas, such processes can affect the physical, mechanical and sorption properties of the finished product. So, the influence of the WKS thermal modification conditions on the textural and sorption properties of the obtained carbonaceous product has been studied.

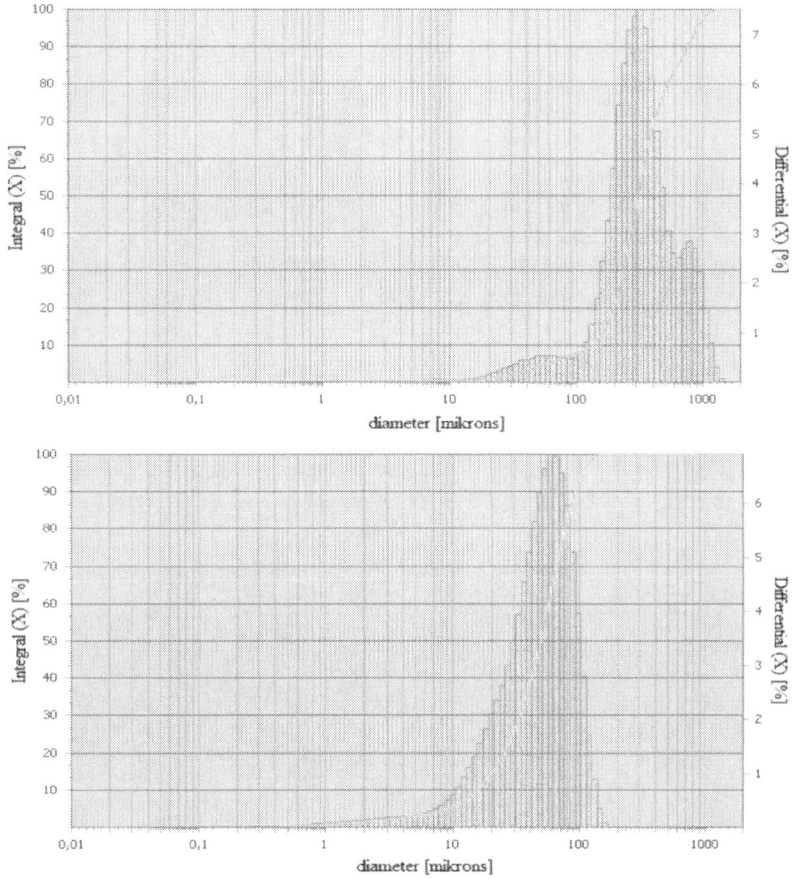

Figure 2. The differential and integral distribution of the WKS particles (a - initial, b – thermally modified at temperature 500°C).

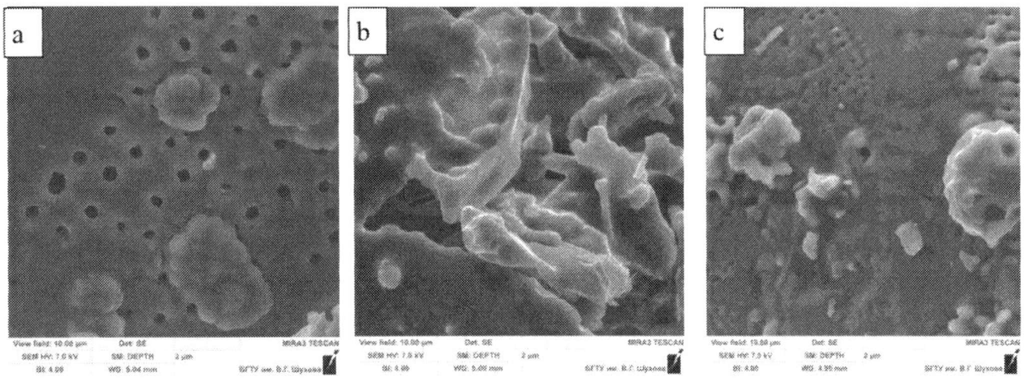

Figure 3. Results of the microscopic study of the particles: a – native (initial) kieselghur, used as a filtering material; waste kieselghur sludge, thermally modified at temperatures,°C: b – 470; c – 500.

At the low modification temperature (450-470°C) the organic matter charring process is incomplete, the products and semiproducts of the organic impurities decomposition deposit in macropores and large mesopores and «seal» or occlude them (figure 3). It results in the reduction of the obtained product specific surface - from 2.0 $m^2 \cdot g^{-1}$ (for the native diatomite) to 0.8 – 0.9 $m^2 \cdot g^{-1}$ (table 1).

The increase of the treatment temperature to 500°C provides the more complete organic matter oxidation, removal of semiproducts and the highest carbon particles formation. It results in the increase of the specific surface to 3.3 $m^2 \cdot g^{-1}$ and the formation of the primarily mesoporous structure with the prevailing pores of 15 – 79.6 nm (table 1). The obtained material appears as a black finely-dispersed powder with the prevailing particles of 55-60 μm.

The further increase of the WKS modification temperature causes the more intensive organic impurities charring up to their complete burn-off with the release of gaseous products.

Table 1. The influence of waste kieselghur sludge modification temperature on its textural characteristics

№	The WKS modification temperature, °C	The effective size, nm	The volume of pores less than 190 nm in diameter, $cm^3 \cdot g^{-1}$	The pore distribution relative to the total pore space, %	Specific surface, $m^2 \cdot g^{-1}$
1	Native kieselghur, used as a filtering material	15.0	0.005	20.8	2.2±0.4
		142.3 and more		79.2	
2	470	4.4	0.005	0.4	0.9±0.2
		8.4		0.2	
		15.0		19.4	
		51.4		34.0	
		79.6		0.9	
		142.3 and more		45.1	
3	500	15.0	0.005	25.4	3.3±0.2
		24.0		34.0	
		79.6		40.6	
4	530	79.6	0.003	2.6	3.9±0.2
		142.3 and more		97.4	

The volume of mesopores is reduced, the primarily macroporous structure of the obtained adsorbent is formed, and the total pore space is reduced. Thus, at temperature 530°C mostly the pores with the effective size over 140 nm are formed, though the specific surface of the obtained product remains high – 3.9 $m^2 \cdot g^{-1}$, probably due to loosening of the particles as a result of vapor and gas generation.

So, as a result of waste kieselghur sludge temperature modification the finely-dispersed low-porous carbonaceous mineral materials are formed.

The waste kieselghur sludge, thermally modified at various temperatures, has been used as an adsorption material to purify test solutions from heavy metal ions by an example of Cu^{2+} and Ni^{2+} ions.

The findings of studying the WKS sorption properties dependence on the modification temperature demonstrate (table 2) that the sorption of Cu^{2+} and Ni^{2+} ions with the surface of modified sludge alters, reaching its maximum level – 14.3 $mg \cdot g^{-1}$ and 17.0 $mg \cdot g^{-1}$, respectively, at temperature 500°C, which can be defined as optimal.

The findings of studying the dependence of WKS sorption properties on the modification temperature have shown that the sorption of Cu^{2+} and Ni^{2+} ions with the surface of the modified sludge alters, reaching its maximum level at the modification temperature 500°, which can be defined as optimal. The obtained values exceed the sorption capacity values of the initial kieselghur, used as a filtering charge.

Table 2. The influence of waste kieselghur sludge modification temperature on the sorption capacity of the obtained material for Ni (II) and Cu (II) ions, $mg \cdot g^{-1}$.

Extracted ions	Initial Kieselghur[a]	The temperature of WKS treatment, °C							
		450	470	500	530	550	570	600	630
Ni (II)	9.5	13.4	14.0	17.0	15.7	14.3	13.5	13.0	12.2
Cu (II)	9.12	12.2	12.8	14.3	12.2	13.2	11.8	10.9	10.4

[a]kieselghur, used as a filtering charge

So, the waste kieselghur sludge can be considered as a valuable organo-mineral raw stuff for obtaining carbonaceous sorption materials, which can be used in purification of wastewater from heavy metal ions.

4. Conclusions

As a result of temperature modification of waste kieselghur sludge, generated at vegetable oils refining stage, a finely-dispersed low-porous carbonaceous mineral material is formed. The modification of the WKS particles surface results in the considerable alterations of its physical, chemical, adsorption and structural properties: the organic matter is charred, and on the surface of diatomite particles a carbon layer is formed, which deposits in macropores and partially occludes them. The amount of mesopores is increased, and the specific surface of the obtained product increases as well.

The optimal modification temperature is 500°C. The synthesized carbonaceous material can be used as an adsorbing agent for purifying wastewater from heavy metal ions; the maximum sorption capacity for Cu^{2+} and Ni^{2+} ions is 14.2 $mg \cdot g^{-1}$ and 17.0 $mg \cdot g^{-1}$, respectively. The obtained values exceed the sorption capacity values of the initial kieselghur, used as a filtering charge, for the researched metal ions.

Ackowledgments

The work is realized in the framework of the Program of flagship university development on the base of the Belgorod State Technological University named after V G Shoukhov, using equipment of High Technology Center at BSTU named after V G Shoukhov.

References

[1] Baymakhanov M T et al 1983 *Purification and monitoring of non-ferrous metal industry wastewaters* (Moscow: Metallurgy) p 39
[2] Fomin G S 1995 *Water — Inspection of chemical, bacteriological and radiation safety according to international standards* (Moscow: Protector) p 624
[3] Shaydurova G I, Shevyakov Ya S and Golub L S 2015 Selection and efficiency evaluation of the treating agents for electroplating industry wastewaters neutralization *Ecology and industry in Russia* **1** 54 – 5 10.18412/1816-0395-2015-1-54-55
[4] Chaturvedi D and Sahu O 2014 Adsorption of heavy metal ions from wastewater *Global Journal of Environmental Science and Technology* **2** 20 – 8.
[5] Tran Th-K, Leu H-J, Chiu K-F and Lin Ch-Y 2015 Electrochemical treatment for wastewater contained heavy metal the removing of the COD and heavy metal ions *International journal of Engineering research & Science* **1** 96 – 101
[6] Gunatilake S K Methods of removing heavy metals from industrial wastewater *Journal of Multidisciplinary Engineering Science Studies* **1** 12 – 18.

[7] Vezentsev A I, Goldovskaya L F, Volovicheva N A and Korolkova S V 2008 Research of Cu (II) and Pb (II) ions sorption efficiency with the native forms of montmorillonite clays of the Belgorod region *Sorption and chromatographic processes* **8** 807 – 11.

[8] Samiey B, Cheng Ch-H and Wu J 2014 Organic-inorganic hybrid polymers as adsorbents for removal of heavy metal ions from solutions: a review *Materials* **7** 673-726.

[9] Stepanova S V and Shaykhiev I G 2013 The usage of agricultural waste for removing iron ions from model waters. *Ecology: education, science, industry and health: Proc. of the 5th Int. research and practice conf* (Belgorod: BSTU publishing office) 45 – 7.

[10] Emenike P C, Omole D O, Ngene B U and Tenebe I T 2016 Potentiality of agricultural adsorbent for the sequestering of metal ions from wastewater *Global Journal of Environmental Science and Management* **2** 411 – 42.

[11] Porozhnyuk L A and Chemerichenko E N 2013 Removal of Ni^{+2}, Cu^{+2} and CrO_4^{-2} with aluminium-containing technogenic waste *Bulletin of BSTU named after VG Shukhov* **4** 161 – 3

[12] 2007 GOST 12.1.007-76 Occupational safety standards system. Noxious substances. Classification and General Safety Requirements *Russian national standards* (Moscow: Standartinform) p 7

[13] Osepchuk D V, Chikov A E and Omelchenko N A 2014 *Actual problems of intensive livestock development* zoovet.info/vet-knigi/107-zyvotnovodstvo/problemy-chast1/7113-ispolzovanie-otkhoda-filtratsii-rastitelnogo-masla-v-kachestve-istochnika-lipidov-dlya-zhivotnykh-21-01-2013

[14] Spent kieselguhr http://www.akonit-ut.ru/utilizaciya-otxodov/pishhevyx/kizelgur-otrabotanniy/

[15] 2001 *State water quality control* (Moscow: Publishing House of Standards) p 687

[16] Prokofiev V Yu and Razgovorov P B 2010 Physical and chemical processes at adding kaolinite clays to vegetable oils *Plant materials chemistry* **2** 159 – 64

[17] Supyrev A V, Yarullin R N, Mustafin M T, Yarullin R R and Sultanov I Y 2013 Method of waste filtering material reclamation: Patent of the Russian Federation Ru 2488425 *Patent of the Russian Federation*

[18] Bakr H E G M M 2010 Diatomite: Its Characterization, Modifications and Applications *Asian Journal of Materials Science* **2** 121 – 36

[19] Ilyicheva O M, Naumkina N I and Lygina T 2011 Interpretation of X-ray diffraction data of opal-cristobalite-tridymite phase *Mineralogical perspectives: materials of the int. seminar (Syktyvkar)* (Russia, Syktyvkar: IG Komi NC UrB RAS) pp 51 – 3

[20] Ubaskina Yu A and Petrenko E V 2012 The production of bleach clays from diatomite: the calcination processing procedure Part 1. The properties of diatomite at calcination *New technologies* **2** 62 – 5

[21] 1983 *Chemistry of industrial wastewaters* (Moscow: Chemistry) p 360

[22] Smirnov A D 1982 *The sorption purification of water* (Leningrad: Chemistry) p 168

IOP Conf. Series: Earth and Environmental Science **107** (2017) 012076 doi:10.1088/1755-1315/107/1/012076

Priority chemical pollutants of drinking water in the city of Kazan: approach based on risk assessment

N V Stepanova, N S Arkhipova and S F Fomina

Institute of Fundamental Medicine and Biology, Kazan Federal University, K. Marx street, 74, 420008, Russia

e-mail: stepmed@mail.ru

Abstract. Assessment of non-carcinogenic risks from chemical substances ingested with drinking water included peroral, skin and inhalation routes of contact with water. The study was carried out for children aged 3-6 years living in 4 districts (zones) of the city of Kazan. Regional exposure factors (REF) at the median (Me) and the 95-th Percentile (95P) levels were identified according to the results of the questionnaire survey. The value of total hazard indices (THI) calculated with application of REF at the median (Me) and the 95-th Percentile (95P) levels made THI_{Me} = 14.2 and 17.1, and THI_{95perc} = 13.03 and 16.3 in zones with a combined type of water supply. The ingestion of chemical substances with drinking water in different zones of the city of Kazan implies, alert and high levels of non-carcinogenic health risk for the child population.

1. Introduction

The quality of drinking water exercises significant influence over the public health. According to the data of the State Report of Rospotrebnadzor (The Russian Federal Service for Surveillance on Consumer Rights Protection and Human Wellbeing) "On the Status of Sanitary and Epidemiological Wellbeing of the Population in Russian Federation" last year, the number of deaths associated with consumption of polluted drinking water increased by a factor of three and reached almost 18.9 thousand in comparison with 5.9 thousand in the year of 2015 [1]. Morbidity associated with a water factor increased as well: last year 1.486 million cases of genito-urinary diseases, digestive diseases, skin and subcutaneous tissue disorders, musculoskeletal disorders, endocrine diseases, circulatory diseases and others (1.455 million according to the results of the year of 2015) were registered. The major negative effect on the human health is exerted by chemical impurities contained in the tap water, both of natural origin, for example nitrates, and industrial origin – metal compounds (iron, manganese, strontium, aluminium) and other substances [2-4]. The presence of chlorine compounds in drinking water remains an urgent problem [5-7]. Analysis of the present stage of the drinking water quality indicates that, uniform requirements for composition and properties of the drinking water are not unified on a world-wide scale and should take into account and reflect the national peculiarities of public water supply domestically [8]. The World Health Organization (WHO) underlines in the 4th Guidelines on the Control of the Drinking Water Quality that, the approaches based on the health risk assessment should be used for justification of managerial decisions on provision of the drinking water safety [9]. The dose-effect and dose-response relationships in children under 6 years old due to differences from the older children and the adults in structural and functional characteristics are responsible for their high vulnerability on exposure to chemical substances [10-13].

2. Materials and methods

The study was carried out for children aged between 3-6 years living in 4 districts (zones) of the city of Kazan. The water supply source of selected zones (Kirovsky (the 1st zone) and Vakhitovsky districts (the 3rd zone)) is the "Volzhsky" water intake. The population of Sovetsky (the 2nd zone) and Privolzhsky districts of the city (the 4th zone) use drinking water of combined nature (the "Volzhsky" water intake and underground water sources). The assessment of non-carcinogenic risk was carried out according to the values of the upper limit of the 95% CI of the results of studies carried out on the basis of an accredited laboratory of the Federal State-Funded Healthcare Institution "The Center of Hygiene and Epidemiology in the Republic of Tatarstan" in keeping with Guidelines on the population health risk assessment (Guidelines P 2.1.10.1920-04) and the Environmental Protection Agency (US, EPA) [14 - 16]. The assessment of exposure to chemical substances via the ingestion, inhalation and dermal routes was carried out with application of the assessment of chronic daily intake (CDI) with application of standard formulas [14]. Information on maximum concentrations of pollutants corresponding to the upper limit of statistical confidence interval of the 95% probability was used in calculation due to the fact that, risk potentials, reference doses and concentrations used for assessing the "dose-effect" relationship are oriented on this particular criterion. The value of the absorbed dose calculated on the basis of a reference dose (RfDo) on peroral route of ingestion and absorption ratio in gastro-intestinal tract (GIABS) are used as an approximate measure of permissible skin effect (RfDd) of chemical substances [14]. The value of the absorbed dose on ingestion of chemical substances, when washing and taking a bath (a shower), is assessed in our study for children 3≥6 years old on skin exposure. EPA applies various approaches to assessment of DAevent (absorbed dose per event ($mg \cdot cm^{-2}$-event)) concerning non-organic and organic chemical substances [16] in current policy.

3. Results and discussion

In total, the risk assessment process consists of four steps: hazard identification, exposure assessment, dose-response assessment, and risk assessment. Regional exposure factors (REF) at the median (Me) and the 95-th Percentile (95P) levels were identified according to the results of the questionnaire survey [17]. Reference values for calculating the impact and risk assessments are given in table1.

Analysis of the total hazard indices (HI) calculated from the values of REF (Me and the 95th Perc) for the child population of the city of Kazan showed (apart from the 1st zone) the excess of the upper limit of the reference level (3.0) for three systems: the blood, the kidneys and the cardiovascular system. The major contribution to development of general toxic effects for the critical organs and systems on peroral ingestion is made by, oil products (from 29 to 54.7%) in the 2nd and the 4th zones; by chloroform (from 10 to 30.6%) and nitrates (from 12.8 to 35.9%) in all zones; by magnesium (up to 11.2%) in the 3rd zone; and by fluorides (from 13.7 to 14.3%) in the 1st and the 3rd zones. The peroral route of ingestion (65.4% - 83.3%) is the main route in all zones. Inhalation route makes from 16.01% to 33.2 % and is caused by chloroform (65.4%) in all zones, the proportion of the skin route being from 0.5% to 1.1% (table 2).

Ingestion of chemical substances with drinking water in different zones of the city of Kazan is indicative of alert and high levels of non-carcinogenic risk for the health of the child population living in the 2nd and the 4th zones. Quantitative assessment identified differences in regional exposure factors (Me and the 95th Perc) by a factor of 1.25, which are revealed on characterization of the exposure doses and hazard quotients of the chemical substances ingested in an integrated manner (simultaneously via several routes) with drinking water. The value of total hazard indices (THI) calculated with application of REF at the median (Me) and the 95th Perc levels made THI_{Me} = 14.2 and 17.8 and THI_{95perc} = 13.03 and 16.03 in the zones with combined type of water supply. The levels of non-carcinogenic risk in the 1st and the 3rd zones exceeded the upper limit of the reference level as well: at the Me level (6.1 and 9.1) and at the 95th Perc level (7.63 and 11.45).

Table 1. Input Parameters for Calculating Exposure and Intake.

Input parameters	Unit	Values (for children), Me	Values (for children), 95 Perc	Reference
Concentration in water (CW)	$\mu g \cdot L^{-1}$	$\mu g \cdot L^{-1}$	$\mu g \cdot L^{-1}$	This study [17, 18]
Ingestion rate (IR)	$L \cdot day^{-1}$	1.0	2.0	[17]
Concentration in air (C_{air})	$mg \cdot L^{-1}$	Calculation method	Calculation method	[14]
Ventilation rate (VR)	$m^3 \cdot hour^{-1}$	0.5 (children)	0.5 (children)	[14]
Absorption efficiency in alveoli (AE)	-			[14]
Water flow rate (QL)	$L \cdot hour^{-1}$	30	30	[14]
Dimensionless Henry's law constants (H)	-			IRIS
Water temperature (T)	44			This study
Air-water mass transfer efficiency (Theta)		Theta= $3000000/(2.5/Dw^{0.67})+[(R*T/(H*Da^{0.33})]$		[14]
Universal gas constant (R)		8.31		[14]
Skin surface area (SA)	cm^{-2}	5400	7050	[18]
Fraction of skin in contact with water (F)	Percent	100	100	[18]
Water diffusion coefficient (Dw)	$cm^2 \cdot c^{-1}$	For organic substances, $Dw=22*0.00001/MV^{0.67}$. For majority of non-organic substances, Dw is close to zero		[14]
Air diffusion coefficient (Da)	$cm^2 \cdot c^{-1}$	For organic substances, Da $=1.9/MV^{0.67}$. For majority of non-organic substances, Da is close to zero		[14]
Exposure time (ET)	minute $\cdot day^{-1}$	30	90	[17, 18]
Conversion factor (CF)	$L \cdot cm^{-3}$	0,001	0,001	[14]
Exposure duration (ED)	year	3	3	[15]
Exposure frequency (EF)	$day \cdot year^{-1}$	296	364	[17, 18]
Mean exposure time (AT)	day	ED·365	ED·365	[14, 15]
Body weight (BW)	kg	10.3	16.4	[17, 18]

Table 2. Total hazard indices of chemical substances ingested with drinking water via different routes, REF (Me).

Research zones	Peroral route	Skin route	Inhalation route	Total risk (THI)
Zone 1	4.018 (65.8%)	0.059 (0.96%)	2.027 (33.2 %)	6.104
Zone 2	11.838 (83.3%)	0.098 (0.69%)	2.275 (16.01 %)	14.211
Zone 3	6.256 (68.3%)	0.100 (1.1%)	2.811 (30.7 %)	9.166
Zone 4	10.768 (82,6%)	0.069 (0.53%)	2.199 (16.9 %)	13.036

Forming a list of priority monitored indices constituting a real danger for the population health, and its sensitive groups in particular, it is an integral part of elaborating effective measures on prevention or minimization of possible negative effect on the health. The important task in this context is scientific justification not only of the sufficiency of the list of controlled chemicals with the account of nature and peculiarities of their impact on the body, but their priority as far as the probability of health damage is concerned. The concept of the drinking water safety based on the requirements of the sanitary rules and regulations, as well as the results of the population health risk assessment, is the basis for selection of priority monitored indices characterizing chemical pollution of drinking water and subject to constant monitoring [19]. The safety of drinking water in chemical composition is determined by its conformity with standards in composite indices, the content of harmful chemical substances due to anthropogenic origin and global widespread, as well as the presence of harmful chemical substances, which enter and are formed in water in the process of its treatment in the water-supply system, and the substances in the territory of a certain region or city. The priority pollutants of drinking water in the city of Kazan, which determine from 62.6 % to 99. 0 % of the level of the total non-carcinogenic risk in all zones of Kazan, are oil products (in total), chloroform, nitrates (in NO3), magnesium and fluorides. Oil products are a complex mixture of certain hydrocarbons, and this fact complicates the identification of their toxicity in case of water pollution and implies that, traditional approaches to risk assessment for them (on inhalation and dermal routes) are considerably inappropriate. Currently, the content of oil products in drinking water (as well as the value of per oral reference level) is standardized only in RF, compared to the drinking water standards of EC, WHO and other countries [20]. There are data in the literature that, pre-chlorination results in a multiple increase in concentrations of chlorine-containing compounds when passing through the water purification stations, moreover, their content increases further on transportation in the water-supply system reaching the highest values in the outermost points of the system [21, 22].

The value of total hazard indices (THI) calculated with application of REF at the median and the 95-th Perc levels is indicative of alert and high risks on complex ingestion of chemical substances with drinking water. The obtained results imply the justification of managerial decisions on the health risk minimization for the child population of the Sovetsky (2nd zone) and Privolzhsky districts of the city (4th zone).

Acknowledgement
This work was funded by the subsidy allocated to Kazan Federal University for the state assignment in the sphere of scientific activities 19.9777.2017/8.9

References

[1] 2016 On the Status of the Sanitary and Epidemiological Welfare of the Population in the Russian Federation in the Year of 2015 *State report* (Moscow: Federal Service on Surveillance in the Sphere of Consumer Rights Protection and Human Welfare) p 200
[2] Zhang L et al 2015 *Sci. Total Environ* **5** 163 – 70
[3] AbRazak N H, Praveena S M, Aris A Z and Hashim Z 2015 *J. Epidemiol. Glob. Health* **5**. 10.1016/j.jegh.2015.04.003
[4] Andra S S and Makris K C 2014 *Water Health* **12** 81 – 93
[5] Chowdhury S and Hall K 2010 *Environmental International* **36** 453 – 60
[6] Gopal K, Tripathy S S, Bersillon J L and Dubey S P 2007 *J. Hazard Mater* **140** 1 – 6
[7] Whitaker H, Nieuwenhuusen M J, Best N, Fawell J, Gowers A and Elliot P 2003 *Journal of Exposure Analysis and Environmental Epidemiology* **13** 17 – 23
[8] Villanueva C M, Kogevinas M, Cordier S, Templeton M R, Vermeulen R Nuckols J R, Nieumenhuijsen M J and Levallois P 2014 *Environmental Health Perspectives* **122** 10.1289/ehp.1206229
[9] 2011 Guidelines for Drinking-Water Quality 4th edition *Guidelines World Health Organization (WHO)* (Gutenberg: WHO) p 564
[10] Firestone M 2010 *J. of Exposure Science and Environmental Epidemiology* **20** 227 – 8
[11] Margot T B and Foos B P 2009 *Hum. Ecol. Risk Assess* **15** 923 – 47
[12] 2011 *Summary of principles for evaluating health risks in children associated with exposure to chemicals* 2011 (Switzerland: WHO) p 50
[13] Brown M and Foos B 2009 *Hum. Ecol. Risk Assess* **15** 923 – 47
[14] Rakhmanin J A et al 2004 *Guidelines for health risk assessment for the population on exposure to chemical substances polluting the environment (P 2.1.10.1920-04).* (Moscow: Federal Center of the State Committee for Sanitary and Epidemiological Control) p 143

[15] U S EPA 2005 *Guidance on Selecting Age Groups for Monitoring and Assessing Childhood Exposures to Environmental Contaminants* (Washington: U.S. EPA) p 50
[16] U S EPA 2011 *Exposure Factors Handbook (Final Report)* (Washington: U.S. Environmental Protection Agency, DC) EPA/600/R-09/052F
[17] Stepanova N V, Valeeva E R and Fomina S F 2015 *Hygiene and sanitation* **5** 56 – 61
[18] Stepanova N V, Valeeva E R, Ziyatdinova A I and Fomina S F 2016 *Research J. of Pharmaceutical, Biological and Chemical Sciences* **7** 1677 – 81
[19] Masaev V T 2002 SanPin 2.1.4.1074-01. Drinking water Hygienic requirements for water quality of centralized drinking water supply systems *(instead SanPin 2.1.4.559-96)Sanitary rules and norms* (Moskow: Russian Ministry of Health) p 67
[20] 2008 WHO/SDE/WSH/05.08/123 Petroleum Products in Drinking-water. Background document for development of WHO *Guidelines for Drinking-water Quality* (Switzerland: WHO) p 20
[21] Karim Z, Mumtaz M and Kamal T 2011 *J of Chemical Society of Pakistan* **33** 215 – 9
[22] Basu M, Gupta S K, Singh G and Mukhopadhyay U 2011 *Environ Monit Assess* **178** 121– 34

Recovery of ostracod with known ages in differently textured sediments and comparison of toxicity of heavily contaminated sediments with ostracod *Heterocypris incongruens* and amphipod *Hyalella azteca*

N Yu Stepanova, O V Nikitin, V Z Latypova, I B Vybornova, G S Galieva and R V Okunev

Institute of Environmental Sciences, Kazan Federal University, Kremlevskay str., 18, Kazan, 420008, Russia
e-mail step090660@yandex.ru

Abstract. The recovery of 1-, 4-, 6,-, and 8-d-old ostracods (*Heterocypris incongruens*) from sediments with different texture has been evaluated. The recovery of ostracods at all ages has been in agreement with the acceptability criterion of 80% of survival for sediment tests. The recovery of ostracods has turned out to be equal to or more than 80% for sand and silt sediments, respectively. The comparison of survival rates between ostracods and amphipods has shown good convergence in the tests of heavily contaminated sediments (R^2=0.75, $p<0.05$). The sediment quality criteria (TEC) have been exceeded mostly for total petroleum hydrocarbons (100% samples), Cr (100%), Ni (87%), Cu (87%), Pb (47%), and Cd (53%). The content of acid volatile sulfides (AVS) has been significantly higher than that of simultaneously extracted metals (SEM). The obtained results have indicated that, metals (Cu, Zn, Cd, Ni, and Pb) are non-bioavailable. Only one sample has exceeded TEC for PAHs (dibenz[a,h]anthracene). It was observed that, no significant correlation between the effect of toxicity and the chemical content.

1. Introduction

Sediments, due to their capability to accumulate many organic and inorganic compounds, represent a powerful factor influencing self-purification of water bodies. At the same time, they can become a source of secondary pollution. Currently, bioassay tests have been widely used for assessment of water and sediment pollution [1, 2, 3]. There are numerous methods for routine toxicological assessment using freshwater organisms from different trophic levels and characterized by different feeding and habitat requirements. Some of these methods are comparatively less robust, and the protocols for such species (amphibians, oligochaetes, etc.) are considered only as guidelines [4]. Most bioassay species are used for sediment toxicity assessment depending on cultivation conditions, time, personnel labor costs, etc. Among all methods, whole sediment tests with crustaceans are very popular. Whole sediment testing with the amphipod crustacean *Hyalella azteca* has been used especially extensively in North America. It is extremely valuable for detection and quantification of the toxic hazard of contaminated sediments. Nowadays, a new technique called "microbiotest", which is independent of culturing and maintaining of live stocks of test species, has gained increasing popularity. The choice for a suitable test species has eventually fallen on the ostracod crustacean *Heterocypris incongruens*, whose biological cycle is known to shift to dormant egg (cysts) production under particular environmental conditions. *H. incongruens* is a cosmopolitan benthic bivalved microcrustacean living in various types of small water bodies within the temperate climate zone [5]. Previously, the researches [5-7] showed that the sensitivity of ostracods is similar to the sensitivity exhibited by the amphipod *H. azteca* and the chironomid *C. riparius*.

The *objective* of this study was to assess the recovery of ostracods with known ages in differently textured sediments, as well as to compare the toxicity of heavily contaminated sediments with ostracods (*Heterocypris incongruens*, microbiotest) and amphipods (*Hyalella azteca*, routine technique).

2. Materials and methods

The Admiralteiskii pond is situated in the center of the city of Kazan (Republic of Tatarstan, Russia). It was created during the construction of engineering protection structures of the city against flooding, when the Kuybyshev Reservoir was filled in 1957, and represents a number of reservoirs connected by channels. The pond has total length of about 3.5 km, average width of 30 m, and depth of 0.5-1.5 m. Being cut by two dikes, the pond serves as a pool for collection of rainwater and melted snow with their subsequent transfer to the Kuybyshev Reservoir, in the water protection zone of the water intake area of the Volga River. The pond has been used for a long time as a recipient of wastewaters from the industrial enterprises located along the pond bank.

Sediment samples were collected in June 2015 from 15 sites. Four core samples, with a total volume of 2 L, were used to take the upper 15 cm of sediment using a Van Veen grab to obtain 2 L of sediment. The sediment samples were composited and homogenized by stirring in a stainless steel bowl with a stainless steel spatula. A portion was removed for chemical analysis and toxicity testing. The samples were stored in the dark at 4^0C before toxicity testing or chemical analysis.

During the analysis of whole sediment samples, the grain size, total organic carbon (TOC), metals, polycyclic aromatic hydrocarbons (PAHs), and total petroleum hydrocarbons (TPHs) were investigated [10, 11, 12].

2.1. Whole sediment toxicity testing of ostracods

The whole sediment toxicity tests with ostracods were performed according to the Standard Operational Procedure [8]. Ostracods at the start of exposure were collected 52 h after the onset of cyst incubation, in the standard freshwater at 25°C, and under continuous illumination. Neonates were pre-fed with Spirulina microalgae for 4 h before their transfer to the testing cups.

The direct contact test of the whole sediment was performed in six replicates, simultaneously to that of the control sediment (sand from Toxkit) also in six replicates, using six-well microplates. Each well was supplemented with 1 g of the sediment and 2 mL of the water with algae as food. To be precise, 10 neonates were moved into each well.

Exposures were carried out for 6 days without water renewal or additional feeding.

Survival (S) was calculated as in equation (1):

$$S = \frac{n_{end}}{n_{start}} \cdot 100\% \quad (1)$$

where: n_{end}: the number of ostracods at the end

n_{start}: the number of ostracods at the start

Growth (G) was calculated as in equation (2):

$$G = (1 - \frac{\overline{L_c} - \overline{L_s}}{\overline{L_c}}) \cdot 100\% \quad (2)$$

where: $\overline{L_c}$ the mean length increment of ostracods in the control after 6 days from 6 replicates (i.e., their length at the end minus their length at the start, μm)

$\overline{L_s}$ the mean length increment of ostracods in the sample after 6 days from 6 replicates (i.e., their length at the end minus their length at the start, µm).

The length measurement was performed with the help of a special micrometer slip after ostracod fixation with drop of Lugol's solution. The length was not measured if mortality was more than 30% (in one sample only).

2.2. Whole sediment toxicity testing of amphipods

The whole sediment toxicity tests with amphipods were carried out according to the Standard Operational Procedure [9].

H. azteca of various ages were cultured at 23^{0}C in glass aquaria containing 20 L of water (hardness 220 mg $CaCO_3$/L) and with light intensity of about 500 lux. Amphipods were fed with Tetramin^{R}fish food and presoaked in maple leaves. Amphipods with known age were obtained from the mixed-aged culture using the sieve (500-µm mesh). Mature amphipods retained in the sieve were pipetted into another sieve (425-µm mesh) placed in a glass pan with water. The mature amphipods were left in the pan overnight to release neonate amphipods. The <24-h-old amphipods were held in the pan for 7 days and fed with maple leaves and Tetramin^{R}fish food. The water was changed two times.

On the day before testing, 100 mL of the sediment and 175 mL of overlying water were added to 300-mL beakers. The exposure was performed for 14 days at 23^{0}C and a 16 light: 8 dark photoperiod at the light intensity of about 200 lux. A total of 10 amphipods were exposed in each beaker. Four replicates were tested for each sediment sample. The overlying water was changed three times a week. Amphipods were fed with algae every day. On day 14 of the exposure, the sediments in each beaker were passed through a sieve (300-µm mesh). The survived amphipods were counted.

Survival (S) was calculated as in equation:

$$S = \frac{n_{end}}{n_{start}} \cdot 100\% \quad (3)$$

where: n_{end}:the number of amphipods at the end
n_{start}: the number of amphipods at the start

Statistical processing was performed using Statistica 8.0® (StatSoft, Tulsa, OK, USA) and the data were expressed as mean ± standard deviation for each toxicity endpoint from each treatment, for granulometric and chemical data. Since the data distribution was not normal, the revealed differences between the samples were assessed using the nonparametric Mann–Whitney U-test and Kruskal–Wallis analysis of variance (nonparametric ANOVA). Correlations between the parameters were assessed using Spearman's Rank Order Correlation Coefficient. The differences were considered statistically significant at $p \leq 0.05$. Toxicity criteria were used as less than 80% deviation of control for all tests.

3. Results

3.1. Recovery of the ostracod Heterocypris incongruens with known age from differently textured sediments

The largest difficulty of testing procedures involving small organisms (400-800 µm), such as ostracods, is their recovery at the end of experiments. The standard methods for performing sediment toxicity tests with *H. incongruens* require the minimum control survival of 80% at the end of the 6-d exposure [8]. However, the low "survival" of ostracods at the end of the sediment test may be a result of the low recovery of smaller organisms from sediments.

The sediment tests with *H. incongruens* were started with 1-d-old ostracods. The small size of *H. incongruens* can make it difficult to isolate organisms at the end of the sediment test. The length of 1-d-old ostracods was about 220± µm, whereas the length of ostracods after 4 days was 442± µm. The most rapid growth was observed between the first and fourth days of testing and did not change considerably in the following four days (figure 1).

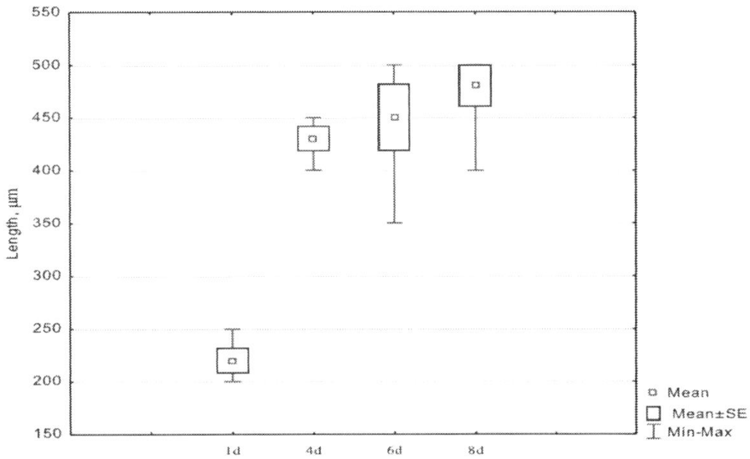

Figure 1. Changes in the length of ostracods during sediment toxicity test.

A special investigation was carried out to evaluate the recovery of ostracods with known age (1, 4, 6, and 8-d-old) in three differently textured clean sediments used as control (1 – sand from the Mariy Chodra National Park (Mari El Republic, Russia), 2 – sand from Ostracodtoxkit, 3 – silty sediment from the Spring River (bioassay practice in the US Geological Survey, Columbia Environmental Research Center).

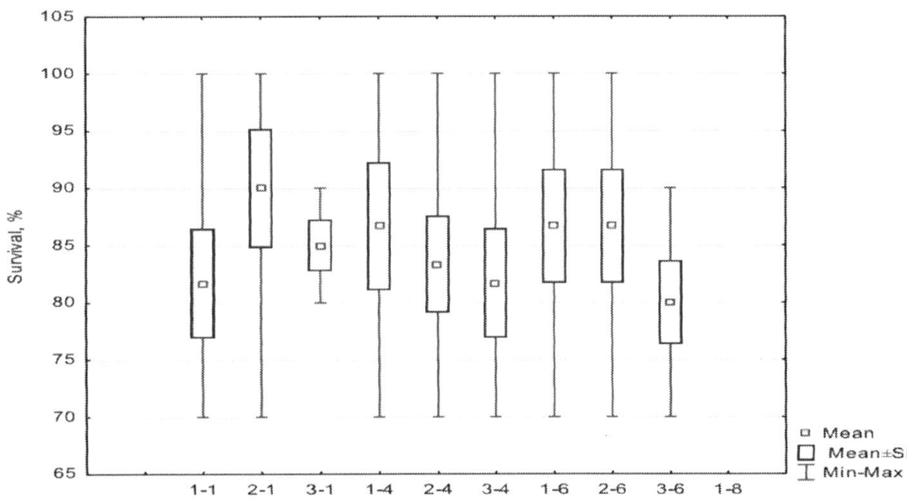

Figure 2. The recovery of ostracods with known age in different types of sediments (1-1 – sand (Mariy Chodra National Park, Russia) and 1-d age; 2-1 – sand (Ostracodtoxkit) and 1-d age; 3-1 – silt (Spring River, USA) and 1-d age; 1-4 – sand (Mariy Chodra National Park, Russia) and 4-d age; 2-4 – sand (Ostracodtoxkit) and 4-d age; 3-4 – silt (Spring River, USA) and 4-d age; 1-6 – sand (Mariy Chodra National Park, Russia) and 6-d age; 2-6 – sand (Ostracodtoxkit) and 6-d age; 3-6 – silt (Spring River, USA) and 6-d age; 1-8 – sand (Mariy Chodra National Park, Russia) and 8-d age; 2-8 – sand (Ostracodtoxkit) and 8-d age; 3-8 – silt (Spring River, USA) and 8-d age.

The recovery of ostracods was equal or higher than 80% in all variants (figure 2). These results indicated that neither ostracod age nor sediment texture influenced the recovery of ostracods.

3.2. Repeatability of results of toxicity measurements of sediments from the Admiralteiskii Pond with H. incongruens

The results of toxicity testing with ostracods were repeated twice with an interval of two months (June and August, 2015) on the sediments from the Admiralteiskii Pond. The repeatability of ostracods survival was very similar in nine samples (figure 3). Only two samples demonstrated difference in ostracods survival in the range of 0-20% and 0-30%, respectively.

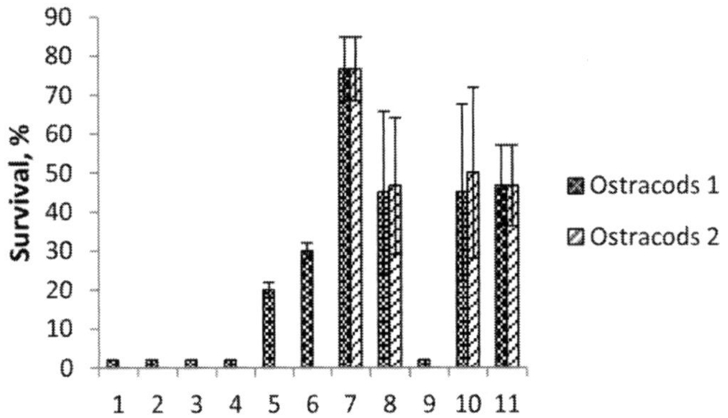

Figure 3. The survival of ostracods exposed to the same sediments with the difference of eight weeks.

3.3. Comparison of toxicity using H. incongruens and H. azteca exposed to sediments from the Admiralteiskii Pond

The comparison of the results of whole sediment toxicity tests with ostracods and amphipods showed good convergence, based on survival, and growth inhibition criteria. The survival of ostracods and amphipods in all samples (except sample no. 7) was less than 80%, thereby characterizing the latter as toxic. The total mortality was observed in samples nos. 1, 2, and 9 (figure 4).

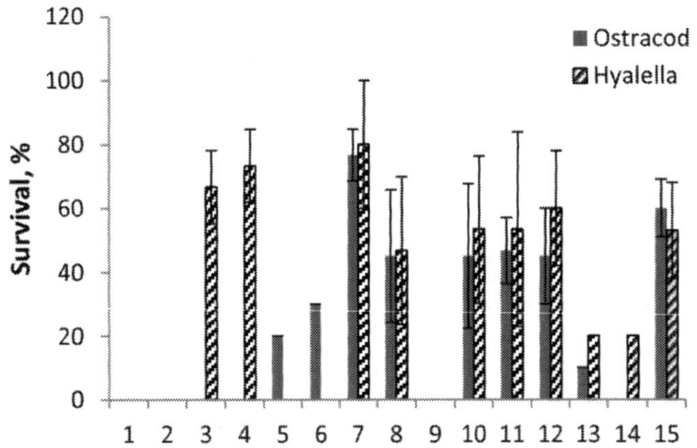

Figure 4. The toxicity of sediments with ostracods and amphipods based on the criterion of "survival".

It was shown that values in 82% of tests results were either identical or very similar. Differences in the survival rate of two species were observed between samples nos. 3-4 where the survival rate of amphipod varied from 20 to 30% and the total mortality of ostracods was observed.

3.4. Physical-chemical characterization of sediments

The level of toxicity effect hardly depends on the content of sediments (table 1). One of the most important roles is played by the factors of sorption and accumulation of pollutants. Most samples were silt mixed with sand in the range of 9.9-55.6%. Only samples nos. 7 and 11 were sandy (71.1-79.5%). The content of clay varied from 0.1 to 2.3%. The content of organic substances was 1.9-42.8%.

Due to the discharge of heavily contaminated wastewaters into the pond during the previous period, sediments accumulated numerous pollutants, especially metals and total petroleum hydrocarbons (TPHs) (table 1). Threshold effect concentrations [13] and TPHs criteria [14] were used for assessment of sediment pollution. The sediment quality criteria were mostly exceeded for TPHs (100%), Cr (100%), Ni (87%), Cu (87%), Zn (73%), Pb (47%), and Cd (34%). Only one sample contained dibenz[a,h]anthracene at the level of 1.9, which is above TEC [13].

4. Discussion

As it was shown in the research with the amphipod *Hyalella 6zteca* [15], the age and, therefore, size of an animal at the start can be essential for its recovery at the end of testing. The recovery of ostracods was 80% or higher in the tests with known age ostracods (1-, 4-, 6- and, 8-d-old). It is well known that the sensitivity increases the smaller the age of test organisms is. If the size (age) of ostracods does not influence their recovery, it is better to use 1-d-old ostracods for testing as recommended by the procedure [8].

It should be also emphasized based on the use of control sediments with different texture that recovery is not dependent on sediment structure (granular size): the obtained recovery was in agreement with the procedure requirements – not less than 80%. The mean loss of organisms in all types of sediments was 15%.

Significant correlation was observed in this study between the survival rate of *Heterocypris incongruens* and *Hyalella 6zteca* (R^2=0.78, $p<0.05$). Chial and Persoone [6] reported about the correlation between the mortality of ostracods in microbiotests and *H. 6zteca* in whole sediment tests based on 26 samples from various watercourses in Flanders, Belgium. The results of this comparative study revealed that the intensity of the toxic effect varied from "nearly identical" to "substantially different" between the amphipod and the ostracod tests. In this regard, it is interesting that some sediments were only slightly toxic to ostracods and very toxic to amphipods, whereas for other samples opposite trends were noted [6]. The pairwise comparison of toxicity data between the two crustacean assays revealed a significant correlation (r = 0.71).

In the comparison of two bioassays (ostracods and the amphipod *H. 6zteca*), De Cooman et al [5] noted that, the mean percentage mortalities for the two bioassays in samples collected immediately after the addition of oil were all extremely toxic to both crustaceans (75-100% mortality). After six weeks, the toxicity of sediments from some plots decreased substantially for the amphipod (mortality around 30%), whereas the ostracod mortality was still between 80% and 90%. The results of the present study concerning the repeatability of ostracod tests (Figure 3) confirm the assumption about retention of ostracod test sensitivity of after eight weeks of sediment store. On the contrary, the survival of ostracods was higher than that of amphipods in two sediment samples (Figure 4).

No significant correlations were observed between the toxicity with both animals and the chemical content of sediments. Despite the high level of sediment pollution by metals (Table 2), no correlation was found between sediment content and toxicity. Based on the chemical interactions between simultaneously extracted metals (SEM) and acid volatile sulfides (AVS), the concentrations of these two components were used to assess the potential toxicity for sediment-dwelling organisms. The content of acid volatile sulfides (AVS) was significantly higher than that of simultaneously extracted metals (SEM). Thus, sulfides were available to bind with metals, thereby rendering them non-bioavailable.

Table 1. The physical-chemical characterization of sediments from the Admiralteiskii Pond (TECQ – quotient of pollution content to sediment quality criteria (TEC) [13], *Russian criteria for TPHs [14]).

Number of sample	1	2	3	4	5	6	7	8	9	10	11	12	13	14	15
Sand, %	37.3	9.9	54.8	39.7	53.5	45.1	79.6	55.6	53.6	23.0	77.1	31.7	24.5	38.8	14.8
Silt, %	62.2	87.8	45.1	59.6	46.2	53.6	20.4	44.2	45.4	75.5	22.9	67.1	74.6	60.5	84.7
Clay, %	0.5	2.3	0.1	0.7	0.3	1.3	<0.1	0.2	1.0	1.5	<0.1	1.2	0.9	0.7	0.5
Corg, %	21.0	33.2	42.8	42.5	41.9	6.9	1.9	8.0	36.6	17.8	5.5	18.1	15.8	31.2	35.8
TECQ Cd	0.5	2.3	11.1	106.1	55.6	0.8	0.6	0.7	0.5	0.7	0.4	5.3	8.2	5.9	2.7
TECQ Cu	22.2	19.0	25.3	45.9	20.6	5.2	3.26	0.9	0.3	2.1	2.0	12.9	28.6	3.8	10.6
TECQ Ni	15.4	20.9	39.7	59.5	39.7	18.7	11.01	2.4	0.9	4.4	0.7	5.9	12.3	25.1	3.8
TECQ Pb	1.1	2.0	1.5	2.7	1.5	0.6	0.4	0.0	0.0	0.0	0.0	1.8	0.5	0.2	1.1
TECQ Zn	5.4	3.5	4.6	12.4	4.3	1.4	0.98	0.6	0.2	1.2	0.8	4.0	2.8	1.9	3.5
TECQ Cr	184.3	83.0	110.6	216.6	80.7	62.2	1.61	43.8	1.2	9.2	3.5	2.1	1.9	1.3	1.1
∑SEM-AVS, mmol kg^{-1}	-74.2	-69.7	-80.9	-49.6	-87.4	-28.4	-44.9	-2.3	0.05	-25.7	-26.7	-52.9	-19.6	-72.3	-53.8
TPHQ*	119.7	117.1	13.9	13.7	13.7	71.2	72.8	53.2	48.7	65.0	73.9	102.5	95.8	23.7	10.4
TECQ Naphthalene	ND[a]	ND	ND	ND	ND	0.04	0.02	0.02	0	0.02	0.02	0.06	0.04	0.02	0.02
TECQ Phenanthrene	ND	ND	ND	ND	ND	0.29	0.17	0.02	0	0	0	0.10	0.06	0	0.01
TECQ Anthracene	ND	ND	ND	ND	ND	0.01	0.01	0.01	0	0	0.02	0.02	0.01	0.02	0.02
TECQ Chrysene	ND	ND	ND	ND	ND	0	0	0	0	0	0	0	0.01	0	0
TECQ Benzo(a)pyrene	ND	ND	ND	ND	ND	0.30	0.07	0.08	0	0.03	0.09	0	0.05	0.07	0.04
TECQ Dibenz(a,h)anthracene	ND	ND	ND	ND	ND	1.88	0.41	0	0	0	0	0.81	0.15	0.20	0

[a]ND – no data

The compliance with the criteria for PAHs indicated fresh oil pollution. The reason for sediments toxicity to ostracods and amphipods was not revealed. No significant correlation between the toxicity effect and chemical content was observed. It seems that toxicity may be related to combined effects of the pollutants or non-detected toxicants.

5. Conclusions

Laboratories should evaluate the ability of their personnel to recover *Heterocypris incongruens* from control and test sediments. Results of ostracod recovery evaluation can be used to separate effects on survival from the inability to recover small organisms from sediment. There were no significant differences between the recovery of ostracods with known age (1, 4, 6, and 8-d-old). The same results were obtained with the exposure of ostracods to sediments with different texture (sand and silt).
Comparison of the survival rates of two crustacean ostracods and amphipods showed good convergence in the testing of heavy contaminated sediments. The sensitivity of ostracods did not change after two month of sediment storage.

Thus, it is possible to conclude that both methods demonstrated sensitivity to sediment pollution, as well as are characterized by good convergence and reproducibility of results. However, the bioassay with *Heterocypris incongruens* ostracods has a number of advantages compared to the routine test with *Hyalella azteca*: no need to maintain laboratory culture of ostracods; shorter time of exposure (6 days against 14 days with amphipods); smaller volume of sample (1 g against 100 g with amphipods). Along with the sensitivity comparable to other toxicity tests, the above-mentioned benefits create favorable conditions for the wide usage of microbiotests with *Heterocypris incongruens* ostracods for toxicity assessment of sediments.

Acknowledgments

We would like to thank Chris Ingersoll for the idea of the present study and providing us kindly with the clean sediment. The work is performed according to the Russian Government Program of Competitive Growth of Kazan Federal University.

References

[1] Kemble N E, Brumbaugh W G, Brunson E L, Dwyer F S, Ingersoll C G, Monda D P and Weodward D F 1994 *Environ. Toxicol. Chem.* **13** 1985 – 97 10.1002/etc.5620131212
[2] Baudo R 2008 *Ann. Ist. Super. Sanita* **44** 233 – 38
[3] Bakaeva E N, Nikanorov A M and Ignatova N A 2009 *Nauka Yuga Ross. Sci. S. Russ.* **5** 84 –93 (In Russian)
[4] Whitacre D M 2012 *Reviews of Environmental Contamination and Toxicology.* (Springer Science & Business Media) p 182
[5] De Cooman W, Blaise C, Janssen C, Detemmerman L, Elst R and Persoone G 2015 *Knowl. Manage. Aquat. Ecosyst.* **416** 10.1051/kmae/2015011
[6] Chial B and Persoone G 2002 *Environ. Toxicol.* **17** 10.1002/tox.10087
[7] Chial B, Persoone G and Blaise C. 2003 *Chemosphere* **52** 10.1016/S0045-6535(03)00186-3
[8] 2012 ISO 14371Water quality – Determination of fresh water sediment toxicity to Heterocypris incongruens (Crustacea, Ostracoda) *ISO Water quality* p 16
[9] 2013 ISO 16303 Water quality – Determination of toxicity of fresh water sediments using Hyalella Azteca *ISO Water quality* p 27
[10] 2015 Standard practice for total digestion of sediment samples for chemical analysis of various metals (D4698-92) *Annual Book of ASTM Standards* **11** (West Conshohocken, PA: ASTM International)
[11] De Deckere E, de Cooman W, Florus M and Devroede-Vanderlinden M P 2000 *A manual for the assessment of sediments in Flanders with the Triad approach.* (Flemish: Ministry of the Flemish Community in collaboration with the Flemish Environmental Agency) p 112.

[12] 2006 ISO 18287 2006 Soil quality – Determination of polycyclic aromatic hydrocarbons (PAH) – Gas chromatographic method with mass spectrometric detection (GC-MS) *ISO Soil quality* p 17
[13] MacDonald D D, Ingersoll C G and Berger T A. 2000 *Arch. Environ. Contam. Toxicol.* **39** 10.1007/s002440010075)
[14] Stepanova N Y, Latypova T R and Latypova V Z. 2015 *Neft. Khoz. Oil Ind.* **3** 106 – 9
[15] Tomacovic M J, Dwyer F J, Greer I E and Ingersoll C G 1995 *Environ. Toxicol. Chem.* **14** 10.1002/etc.5620140708

IOP Conf. Series: Earth and Environmental Science **107** (2017) 012078 doi:10.1088/1755-1315/107/1/012078

The use of expanded clay dust in paint manufacturing

S V Sverguzova, Zh A Sapronova, Yu L Starostina and E A Belovodskiy

Belgorod Shukhov State Tecnological University, Kostyukova str., 46 Belgorod, 308012, Russia

e-mail: starostinairinav@yandex.ru

Abstract. Production increase of useful products is accompanied by the formation and the accumulation of the vast amounts of industrial wastes, the bulk of which is not involved in the recycling processes. An example of such wastes is dust bag filters of ceramsite production. At the large enterprises, the volume of its formation can reach 7-8 tons of dust per day, which is 10-15% of feedstock mass. The studies on the use of ceramsite production dust as filler pigment in the composition of organic mixed primer of red-brown color are carried out in this work. For comparison, red iron oxide pigment (Pg FGM) was used. The results showed that, primer with the use of expanded clay dust is characterized by the short drying time and meets all regulatory requirements.

1. Introduction

Currently in Russia and in the whole world, production volume of useful products increase is accompanied by the formation and the accumulation of vast amounts of industrial wastes. This indicates about irrational use of natural resources, as only a small part becomes the final product but the greatest part of the feedstock withdrawn passes into the heap. In the Russian Federation alone annually, tens of billions tons of industrial waste are produced, the bulk of which is not involved in the recycling processes but takes huge areas of fertile soils and contaminants contained in such wastes in the form of dust particles and dissolved precipitation substances enter the environment [1-3].

The production of building materials is one of the most important factors that have a negative impact on the environment [4-11]. The various production waste of construction materials are: breakage, scrap, non-standard situation, pulps, sludge, dusts [12]. One of these wastes is the bag filters dust of ceramsite production, which is now not in demand, and stored at industrial sites.

According to experts, today in Russia, there are about 160 factories for the ceramsite production. Total annual production is about 3.5 million m^3 [13]. Dust is formed during the preparation of raw materials, ceramsite granules dry, burning, cooling and inoculating of material into fractions with the addition of large pieces [14]. The dust is trapped in factories by dust precipitating chambers, cyclones, bag filters. Utilization of expanded clay dust is a serious problem. In the large ceramsite plants every day 7-8t of dust may be formed that is 10-15% of feedstock mass [14].

Ceramsite production dust, as we can know is widely used in the manufacture of ceramsite gravel (again), fine filler of light weighted concrete (instead of silica sand), ceramic bricks and insulating backfill [15-18].

However, despite many of the described methods, the use of expanded clay dust is mostly stored in the dump. Therefore, the development of new alternative methods of utilization of expanded clay dust is an urgent task. According to a preliminary study of the composition and properties of ceramsite production dust, we propose to use it as a filler pigment.

2. Material and methods

Grain sized composition of raw material was measured by a laser granulometry analyzer at the laser analyzer Microtrac S3500 (US).

The main physico-chemical and mechanical parameters of the considered materials were determined according to the standard fillers pigments procedures [19-21].

X-ray diffraction patterns analyze (XDA) was carried out on a diffractometer DRON-4 using Cu-anode radiation according the method of powder diffraction patterns. The diffraction patterns identification was carried out by catalogue ICDD (International Centre for Diffraction Data (USA)).

Thermal Analysis (DTA) was carried out with the help of derivatograph of model 431Q -1500 and device of synchronous thermal analysis STA 449 F1.

The study of changes in the main indicators of quality of filler pigments were carried out in accordance with GOST[20]. The microstructure of the material was studied by the scanning electron microscope TESCANMIRA 3 LMU (Poland).

Material humidity (%) was determined according to the difference between the initial sample mass (its mass is 1 g) and dried one at 100° C to constant weight. pH of water extract was determined after three minutes of 5g fine-grained dust boiling into 50sm3 of distilled water with the first figure of pH equal to 7.0.

The true density was found by pycnometric way. To determine the bulk density the method ASTM D2854-70 "Standard method of controlling the apparent density of active carbons was used.

The chemical composition of the expanded clay dust sample was determined by X-ray fluorescence analysis (X-ray workstation "ARL 9900 series x-ray work station" with radiation of Co-anode and Kav I is 60 kV and energy dispersive analysis (analyzer meals, combined with ion-electron microscope Quanta 200 3D).

3. Results and discussion

The following materials were used in the work such as

- bag filters dust of ceramsite production of Belgorod plant JBK-1
- red iron oxide pigment (GHG FGM) .

There are a number of specific requirements to the pigment fillers in the paint industry such as: high dispersion, low oil absorption, low density, low cost and availability of raw materials, minimum content of water-soluble impurities (electrolytes). As an example, in table 1 shows some of the properties used in the paint industry of fillers.

Table 1. Filler properties.

Filler	Density, $kg \cdot m^{-3}$	Oil absorption, $g \cdot (100g)^{-1}$	pH of water extract
Barite	4500	6-10	8-9
Dolomite	2850	15-19	10
Chalk	2710	10-17	9-10

The results of physical and chemical properties study of expanded clay dust selected after the bag filters are shown in table 2.

Table 2. Physic chemical properties.

Poured Density, $kg \cdot m^{-3}$	Original Density $kg \cdot m^{-3}$	Oil absorption, $g \cdot (100g)^{-1}$	pH of water extract	Humidity, %
1250	2490	14-17	10.2	1.68

As it can be seen from table 2, such indicators of expanded clay dust as original density, oil absorption, pH of water extract were compared with those ones for traditionally used fillers from table1 Oxide dust composition determined by X-ray fluorescence method is specified in table 3, the elementary composition is given in table 4

Table 3. Oxide composition of expanded clay dust samples, mass. %

№	Oxides													
test	SiO_2	Al_2O_3	Fe_2O_3	CaO	MgO	K_2O	TiO_2	Na_2O	MnO	SO_3	ZrO_2	V_2O_5	ZnO	Cr_2O_3
1	62.4	20.31	6.91	2.76	2.91	2.52	0.93	0.90	0.1	-	0.04	0.03	0.01	0.018
2	61.9	19.17	7.05	4.14	2.79	2.63	0.95	0.97	0.11	0.02	0.05	0.03	0.02	0.017
Medium	62.15	19.74	6.98	3.45	2.85	2.58	0.94	0.93	0.10	0.02	0.05	0.03	0.01	0.018

Table 4. The elementary composition of expanded clay dust samples, [mass. %]

№	Elements														
test	Si	Al	Fe	Ca	Mg	K	Ti	Na	Mn	S	Zr	V	Zn	Cr	O
1	29.21	10.75	4.83	1.97	1.76	2.09	0.56	0.67	0.078	-	0.027	0.018	0.01	0.012	48.02
2	28.98	10.14	4.93	2.96	1.68	2.18	0.57	0.72	0.084	0.009	0.04	0.018	0.013	0.012	47.66
Medium	29.095	10.445	4.88	2.47	1.72	2.14	0.565	0.695	0.081	0.009	0.034	0.018	0.012	0.012	47.84

The analyze results of dispersed dust composition are shown in figure 1.

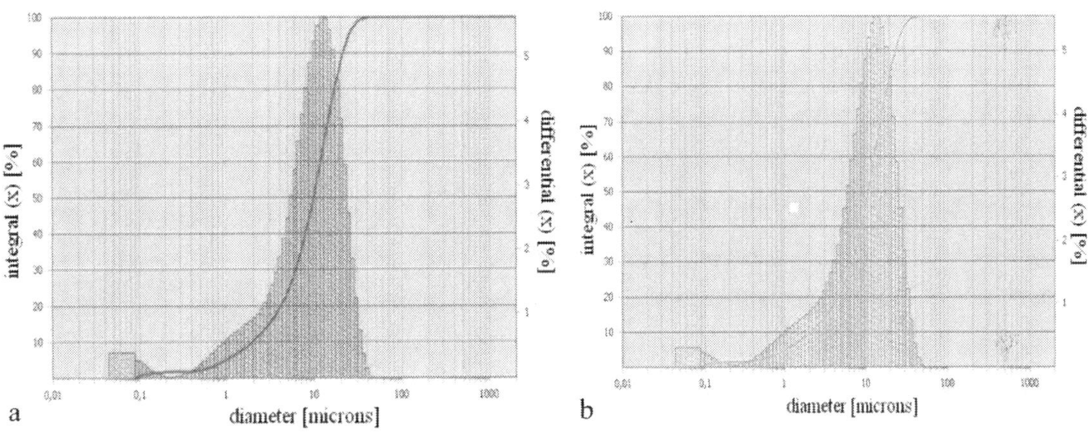

Figure 1. Differential and integral distribution of dust particles, selected after the bag filters (*a* - is the beginning of work, *b* - is the end of work).

Expanded clay dust roentgenograms selected after the bag filter are shown in figure 2 and 3.

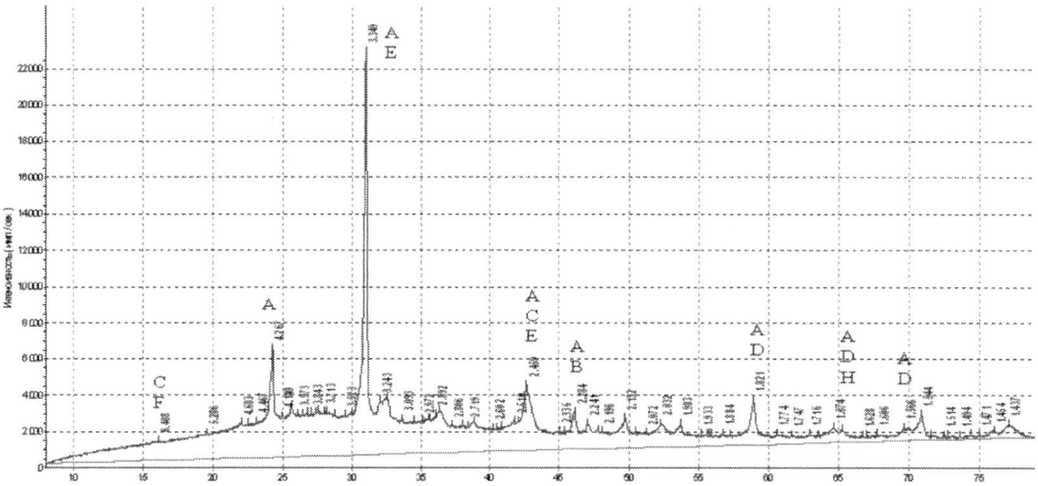

Figure 2. X-ray of expanded clay dust selected after the bag filter at the beginning of work where: *A* - is quartz SiO_2; *B* - is calcite CaCO*3*; *C* - is albite $NaAlSi_3O_8$; *D* - is orthoclase $KAlSi_3O_8$; *E* - is muscovite $KAl_2Si_3AlO_{10}(OH)_2$; *F* - is palygorskite $Mg_5Si_8O_{20}(OH)_2\ 18H_2O$; *H* - is kaolinite $Al_2Si_2O_5(OH)_4$.

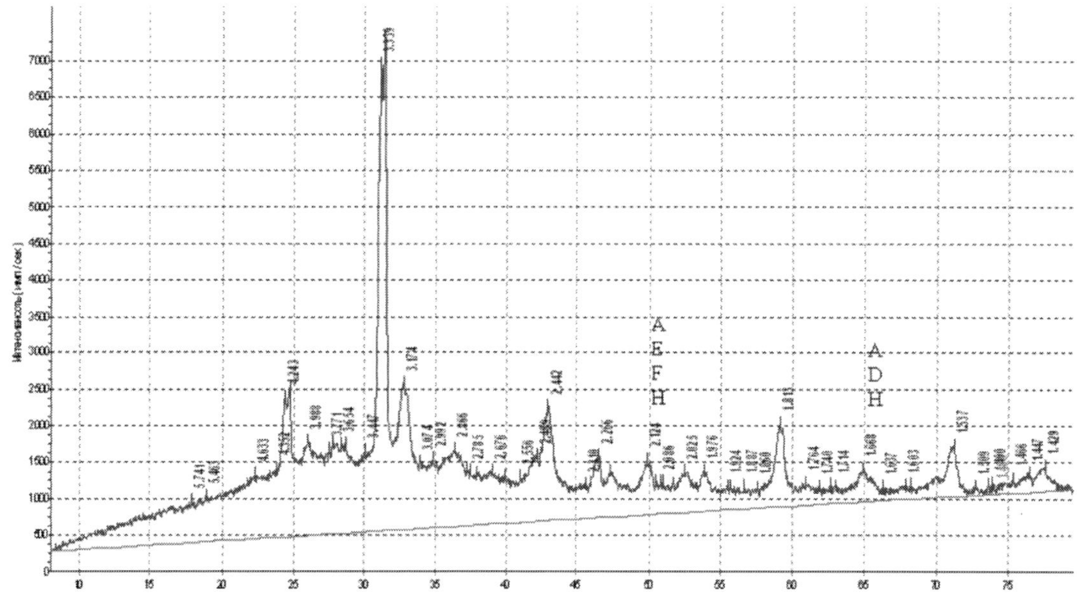

Figure 3. X-ray of expanded clay dust, selected after the bag filter at the end of the work where: *A* - is quartz SiO_2; *D* - is orthoclase $KAlSi_3O_8$; *E* - is muscovite $KAl_2Si_3AlO_{10}(OH)_2$; *F* - is palygorskite $Mg_5Si_8O_{20}(OH)_2 18H_2O$; *H* - is kaolinite $Al_2Si_2O_5(OH)_4$.

For more complete data on the physical and chemical properties of ceramsite production dust, the study investigated the structure of the particle surface with the help of electronic center MIRA3 microscope TESCAN scientific & research center of the higher technologies FGBOU VO "Belgorod State technological University named after V G Shukhov" was considered. Figure 4 shows the photomicrographs of particles dust surface made at different magnifications. In micrographs one can observe a highly developed surface of dust particles with clearly defined interface.

To make the dust much higher, it is dispersed and was further milled with the help of bead mill. For further study, the fraction with a particle size of less than 50 [microns] was selected with the help of a sieve №005. The dust sample prepared in this way was tested for its suitability in paint production in the conditions of the production laboratory of "Paint factory Quil".

Figure 4. Parts of expanded clay dust production surface at different magnification.

Dust tests as a filler pigment were carried out in the following main parameters: dispersibility, the degree of grinding, relative viscosity, hardness, drying time, adhesion, and elasticity. Test specimen of pigment filler was put into the formulation of organic mixed primer of red-brown color for carrying out according to the given data. The red ferrous oxide pigment used in paint industry was applied as a material for compressing. The primer was made on the pigment materials under equal conditions. Samples were put by the method of pneumatic spraying on previously clear and degreased plates (in accordance with GOST 8832-76, Section 3) of grade steel 08 with a size 150 70mm, thickness of 0.8-0.9mm of GOST 16523-70 for determination of stroke indicators, adhesion and appearance of the coating. Film elasticity at bending was determined on plates with the size 90h 120mm, thickness 5mm.

Comparison of the results of the technical characteristics of the primer, prepared using expanded clay dust-based pigments, showed compliance with all regulatory requirements.

4. Conclusion

The test of primer with the use of pigment materials showed similar results. Primer with the use of expanded clay dust corresponds to all regulatory requirements. However, it is necessary to pay attention to the short drying time (drying time indicators at 20°C).

The determination of stroke indicators, adhesion and elasticity were made within 24 hours after material application. The hardness determination was done after 48h after material application.

Acknowledgments
The work has been carried out within the framework of the Flagship University Development Program on the base of BSTU named after VG Shukhov.

References

[1] Petin A N 2004 Geoecological consequences of the iron ore deposits KMA development and the problems of optimal subsoil in the region. *Proc. of the Federal Research Center of Hygiene them. FF Erismann: Regional hygienic problems and strategies to protect public health (Belgorod)* **10** 149 – 158

[2] Kornilov A G, Prisny Yu A, Kolchanov A F and Prisny A V 2008 *Izvestiya Proc. of the Russian Academy of Sciences Geographical Series* **2** 85 – 92

[3] Petin A N 2005 Mineral resources of the Kursk magnetic anomaly and environmental problems of industrial development *Bulletin of Peoples' Friendship University A series of engineering studies* **12** 124 – 25

[4] 2014 On the state and Environmental Protection of the Russian Federation, the environment in 2013 year *State report* (Moscow: The Ministry of Natural Resources of Russia) p 463
[5] 2015 On the state and Environmental Protection of the Russian Federation, the environment in 2014 year *State report* (Moscow: The Ministry of Natural Resources of Russia) p 473
[6] 2016 On the state and Environmental Protection of the Russian Federation, the environment in 2015 year *State report* (Moscow: The Ministry of Natural Resources of Russia) p 640
[7] Lobachyova G K, Zheltobryukhov V F, Prokopov I I and Fomenko A P 2005 *The issue of wastes and modern methods of their processing* (Volgograd: Volgograd State University Press) p 176
[8] Zhenikhov Y N and Ivanova V N 2013 *The waste production and consumption* (Tver: TvGTU) p 164
[9] Chistyakov B Z and Lyamina A I 1984 *The use of mineral industry wastes* (Leningrad: Stroyizdat) p 132
[10] Arbuzova T B, Shabanov V A, Korenkova S F and Chumachenko N G 1993 *Construction of industrial wastes* (Samara: Book Publishing) p 96
[11] 2002 The state of the environment and natural resources of the Belgorod region in 2001 *Annual report* 2002 (Belgorod: Main Department of Natural Resources and Environment Ministry of Russia for the Belgorod region) p 95
[12] 2003 Environment and natural resources of the Belgorod region in 2002 *Annual Report* Ed V S Pashkova (Belgorod: Main Department of Natural Resources and Environment Ministry of Russia for the Belgorod region) p 89
[13] Production of expanded clay as a business http://abcbiznes.ru/biznes-idei/369-proizvodstvo-keramzita-kak-biznes.html
[14] Ivleva I A, Shimanskaya M S and Nemets I I 2011 The technology of gettingceramsite from slightly swelling up clay raw materials *Glass and ceramics* **11** 17 – 8
[15] Alfimova N I and Cherkasov V S 2010 Prospects of using wastes of ceramsite production in building materials. *Bulletin of BSTU named after V G Shukhov* **3** 21 – 4.
[16] Kulikov V A, Abdrakhimov V Z, Kovkov I V 2011 Study of the porous structure of ceramsite on the basis of flotation coal enrichment *Izvestiya vuzov Stroitelstvo* **1** 31 – 7
[17] Alfimova N I, Vishnevskaya Ya Yu, Cherkasov V S and Kalatozi V V 2010 Improving the efficiency of composite astringents using ceramsitewastes production and optimization of the mode hardness. *The Research, nanosystems and energy saving technologies in the industry of buildings materials: a collection of papers of the international scientific-technical conf. (XIX Scientific Conference) (Belgorod)* **1** 36 – 8
[18] Rakhimov R Z, Hamiullin M I and Gaifullin A R 2012 Composite gypsum astringents with the use of ceramsite dust and blast furnace slag *Building materials* **7** 13 – 15
[19] Hodakov G S 1968 *Basic methods of analysis of the dispersion powders* (Moscow: Stroyizdat) p 126
[20] 1999 GOST 10503-71 Oil paints ready for use. Specifications *Russian national standards (GOST)* (Moscow: Publishing House of Standards) p 10
[21] 1999 *GOST* 21119.8-75 The Common test methods for pigments and fillers / definition of oil absorption *Russian national standards (GOST)* (Moscow: Izdatelstvo standarstov) p 5

IOP Conf. Series: Earth and Environmental Science **107** (2017) 012079 doi:10.1088/1755-1315/107/1/012079

Results of adolescent health risk assesment on exposure to habitat water peroral factor in conditions of a large industrial city

E R Valeeva, N V Stepanova, G A Ismagilova, A I Ziyatdinova and D A Semanov

Institute of Fundamental Medicine and Biology, Kazan Federal University, Kremlevskay str, 18, 420008, Russia
e-mail: val_med@mail.ru

Abstract. Results of the non-carcinogenic risk assessment on ingestion of chemical substances with drinking water showed that the risk value corresponded to the allowable level of the non-carcinogenic risk (HQ < 1) for the major part of elements in all zones. The excess of the allowable level is observed only in oil products in the 1st zone (2.05) and the 4th zone (1.04). However, the total hazard index (HI) on combined peroral ingestion of chemical compounds and elements with drinking water in selected zones of the city of Kazan implies a low risk level for adolescents living in the 1st and the 4th zones (3.7 and 3.59) correspondingly, and is dangerous for health. According to the results of analysis carried out in all zones, the following basic critical organs and systems were identified: blood, CNS, kidneys, endocrine system, cardiovascular system, skeletal system and teeth. The total hazard indices in the 1st and the 4th zones deserve particular attention. The following elements: oil products (29.7% - 54.0%), nitrates (in NO3), chloroform and fluorides make a major contribution to the value of risk. In all other zones, irrespective of the value of exposure factors, total hazard quotients indicate alarming and unacceptable risk levels at HI_{Me} = from 4 to 8.67; and at $HI_{95th\ Perc}$ = from 8.7 to 16.8.

1. Introduction

So far, WHO makes global analysis and assessment of drinking-water hygiene (GLAAS) which is the initiative of the United Nations on water resources. Providing access to safe drinking water is one of the most effective instruments in promoting good health. According to WHO estimates, 58% of global burden of disease (DALY/Disability-adjusted life year) or 842 000 deaths per year are due to unsafe water supplies, water sanitation and hygiene [1].

At present, a large body of research was carried out abroad, on the basis of which national and international data bases of the exposure factor values used when assessing the health risk were formed. The most large-scale data on various exposure factors are available in the guidelines of American Environmental Protection Agency (US EPA) which regularly reviews and corrects them in accordance with the new data obtained via public opinion polls [2-4]. The database of the European Commission contains information on exposure factors for 30 countries of the European Union [5]. As for Russia, the studies on exposure factors were carried out in Moscow, Ryazan, Lipetsk, Novodvinsk and several other cities/towns [6]. Currently, the exposure values recommended in foreign guidelines are available for reference and must certainly be corrected with respect to regional peculiarities. The analysis of studies on the risk assessment in our country and the Republic of Tatarstan showed the presence of methodological and toxicometric problems resulting in underestimation of the actual health risk level for the child population associated with uncertainties of the exposure assessment, lack of regional and age differences in the factors exposure and susceptibility to carcinogens [7]. WHO underlines in the Guidelines for drinking-water quality that the approach based on the risk assessment should be applied for justification of management decisions on provision of the drinking-water safety [8]. High level of the adolescent morbidity in disease classes related to habitat markers, in particular, implies the necessity for improving the socio-hygienic monitoring in respect of the territory specificity [9]. As of today, one of the priority factors having an effect on health is the quality of the public drinking water supply [10-15]. Health risk assessment for the adolescent population with selection of priority substances and adverse impacts on the adolescent health is an urgent problem.

Objective: to assess the non-carcinogenic and carcinogenic health risks for the adolescent population (aged 12-16 years) on peroral ingestion of chemical substances with drinking tap water on the basis of the regional and standard exposure factors.

2. Materials and methods

The non-carcinogenic risk assessment on ingestion of chemical substances with drinking water was carried out for the adolescents aged 12-16 years, living in 4 districts (1-Vakhitosky, 2-Sovetsky/Soviet, 3-Kirovsky, 4-Privolzhsky/Volga) of the city of Kazan, and that fact allowed minimizing uncertainties associated with specific regional parameters in exposure assessment. The research areas were selected on the basis of arrangement of permanent stations for monitoring the atmospheric air pollution and the children's polyclinics (No. 1, 2, 3, 4) providing services to these districts with the purpose of subsequent complex assessment of the multi-environmental risk. The risk assessment was carried out according to the data of the Regional Information Fund (RIF) of social and hygienic monitoring and results of the research carried out on the basis of an accredited laboratory of the Federal State-Funded Healthcare Institution "The Center of Hygiene and Epidemiology in the Republic of Tatarstan" in keeping with Guidelines P 2.1.10.1920-04[16].

Non-carcinogenic risk (ingestion route: per os) is assessed by calculating the hazard quotient (HQ). The HI is usually calculated only for the substances having an effect on the same body organs and systems. The approach based on safe (reference) doses and total hazard indices (THI) was used for the non-carcinogenic risk assessment. Life average daily doses (LADD), carcinogenic potential factors (SF) and ADAF were used for assessing the non-carcinogenic risk. Non-carcinogenic and carcinogenic risks were assessed for exposure factors (standard and regional ones) at the median level (Me), (usual exposure range), and the 95th Percentile (P_{95}, maximum rational exposure). The study of the contaminant toxicity was carried out on the basis of chronic daily ingestion of a substance (the peroral route). Characteristics of general toxic effect were identified based on hazard quotients (HQ) of certain substances and hazard indices (HI) for the substances with synergistic effect. Calculation of an average daily dose (ADD) of chemical substances ingested perorally with drinking water was made according to formula 1 [16]:

$$\mathrm{ADD} = \frac{(Cw \times V \times EF \times ED)}{BW \times AT \times 365} \tag{1}$$

where ADD – average daily dose ingested with drinking water ($mg \cdot kg^{-1} \cdot day^{-1}$); CW – the substance concentration in water, ($mg \cdot L^{-1}$); V – the amount of water taken, ($L \cdot day^{-1}$); EF – the exposure frequency, ($days \cdot year^{-1}$); ED – the exposure duration (years); BW – the body weight (kg); AT - the exposure averaging time, years (for non-carcinogens, AT = ED × 365 days);

$$\Sigma HQ = \frac{ADD}{RfD} \tag{2}$$

where RfD is the reference dose (safe effect level) for each of the substances ($\mu g \cdot kg^{-1} \cdot day^{-1}$) [16].

Total hazard quotients (HI) were calculated according to formula 3

$$THI = \Sigma HQ \tag{3}$$

The regional exposure factors investigation was performed in the cross-sectional study. 1560 persons of two age groups: 680 children aged 12-16 years underwent the questionnaire survey. A questionnaire including the assessment parameters of regional exposure factors (EF) was developed by the researchers of the Institute of Fundamental Medicine and Biology under Kazan (Volga Region) Federal University. Owing to the fact that distribution of the quantitative EF was statistically significantly different from normal distribution, the median (Me) and the 95th percentile (Perc) were applied for their presentation. Comparison of the median values of the EF for the child and adult populations was made by means of the Mann –Whitney U criterion (U test). The value of $p < 0.05$ was taken for a critical level of statistical significance.

Statistical analysis of the obtained data was performed with Windows 2007 using standard application program packages Excel 2007 and «STATISTICA».

3. Results

The assessment of the peroral non-carcinogenic risk was carried out in terms of coefficient and hazard indices; critical organs, systems and effects complied with the established reference doses were identified.

We studied the content of 19 priority chemical pollutants including three carcinogens in the drinking water of the city of Kazan for the period from 2010 to 2015. The chlorine residual wasn't mentioned in calculations due to the fact that at present the reference doses are not determined on chronic chloramine ingestion. Owing to the fact that distribution of the values of exposure factor and concentration of pollutants was different from normal distribution, the 95th percentile (Perc) was applied for their presentation. The excess of hygienic regulations was not the major priority criterion. The major criterion was revealing of compounds whose detection rate in drinking water samples exceeded 5% in all the territories under study. The assessment of non-carcinogenic risk on peroral ingestion of chemical substances with drinking water revealed differences in the risk levels at regional values compared with TEF. The results of non-carcinogenic risk assessment on peroral ingestion of chemicals with drinking water showed that the total risk value corresponded to the allowable level only in the 2nd and the 3rd city zones. In all the rest zones, irrespective of the exposure factor values, the total hazard quotients indicate alarming and unacceptable risk levels. An unacceptable total risk level is registered at HI_{Me} = from 4 to 8.67; $HI_{95th\ Perc}$ = from 8,7 to 16.8 is observed in all zones at the median level and the 95-th Percentile (Perc). Total hazard indices calculated at the level of standard EF for the adolescents, apart from the 2nd and the 4th zones (the kidneys – 6.13 and 3.32) and the 1st, 2nd zones (the blood – 3.19 and 3.01), do not exceed the upper borders of the reference level (3.0). The assessment of the total hazard index values calculated on the basis of regional exposure factors (median concentrations and the 95th Percentile (Perc) for the adolescent population of the city of Kazan showed the excess but not of the upper border of the reference level (3.0) for three systems: the blood, the kidneys, the cardiovascular system and the liver (figure 1).

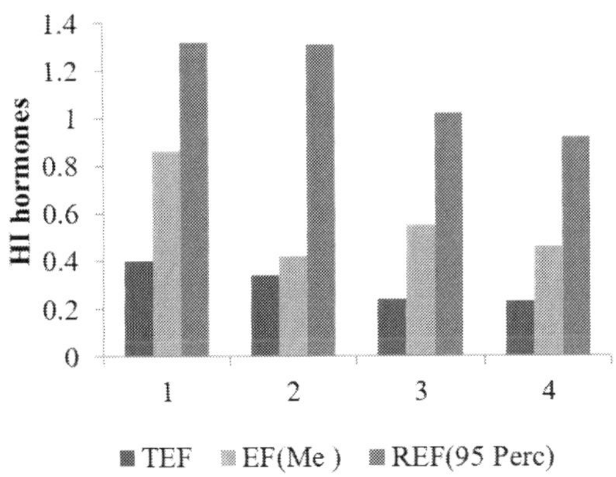

Figure 1. Total hazard indices (HI hormones) for substances with synergistic effect calculated with application of TEF and REF

The values of total hazard indices (HI hormones) of the substances with synergistic effect calculated with application of TEF and REF revealed higher indices in the 1st and the 2nd zones for the adolescent population.

The availability of criteria for subsequent risk assessment – the carcinogenic potential factors on peroral exposure (SFo) was ascertained for chemical carcinogens. Substances belonging to groups 1,2A, 2B according to the IARC and US EPA classification were regarded as potential chemical carcinogens. Carcinogenic risk assessment was carried out for 3 substances contained in drinking water: cadmium, lead and chloroform.

The carcinogenic risk levels for these carcinogenic substances were calculated with application of standard values and regional exposure factors at the level of Me and the 95th perc (table 1).

Table 1. Total carcinogenic risk (TCR) for adolescent health on exposure to carcinogens in drinking-water

TCR	Zones			
	1st	2nd	3rd	4th
REF_{Me}	1. 67 10^{-6}	1.70 10^{-6}	1.05 10^{-6}	1.66 10^{-6}
$REF_{95\ Pers}$	1.49 10^{-5}	1.65 10^{-5}	1.96 10^{-5}	1,46 10^{-5}
TEF	7.44E 10^{-6}	8.23 10^{-6}	9.82 10^{-6}	7.31 10^{-6}
Differences, times:				
REF_{Me} / TEF	0.224	0.207	0.107	0.227
$REF_{95\ Pers}$/ TEF	2.0	2.0	2.0	2.0

Differences in the value of the carcinogenic risk levels made 0.087- 0.323 at EF/TEF, and 2.0 at $EF_{95\ Perc}$/TEF. Chloroform and lead make the major contribution to the value of the carcinogenic risk in all zones. Calculation of the carcinogenic risk levels for these carcinogenic substances was made with the application of correction factors to the carcinogenic potential factor. Chloroform contained in drinking water has no genotoxic effect; therefore, the carcinogenic risk for it was calculated without the application of age coefficients (table 2).

Table 2. Total carcinogenic risk (TCR) for adolescent health on exposure to carcinogens in the drinking water, in respect of correction coefficients (age-dependent adjustment factor - ADAF).

TCR	Zones			
	1st	2nd	3rd	4th
REF_{Me}	2.69 10^{-6}	3.23 10^{-6}	1.53 10^{-6}	3.16 10^{-6}
$REF_{95\ Pers}$	2.54 10^{-5}	3.22 10^{-5}	3.52 10^{-5}	1.96 10^{-5}
TEF	1.27 10^{-5}	1.61 10^{-5}	1.76 10^{-5}	9.80 10^{-6}
Differences, times:				
REF_{Me} / TEF	0.211	0.201	0.087	0.323
REF_{95Pers}/ TEF	2.0	2.0	2.0	2.0

TCR calculated by reference to ADAF exceeded the allowable level from 1.96 in the 2nd zone and by a factor of 1.71 (1.79) in the 1st and the 4th zones at the level of all exposure values. In the 3rd zone, differences are by a factor of 1.34. Comparison of the results of the carcinogenic risk assessment carried out in conformity with the age susceptibility coefficients compared with a traditional approach revealed significant difference in risk levels. TCR calculated with application of the age coefficients exceeds the risk levels obtained without the account of susceptibility of different age groups to carcinogens by a factor of 1.34 – 1.96. The median refers to the 95% quartile upward in the 4th district and downward in the 3rd district. We can suppose that chloroform in the 3rd district and metals in the 4th district make the major contribution in the majority of cases. The highest level of the carcinogenic risk is determined in all zones under study at 95_{Perc} mark (maximum rational exposure). Non-carcinogenic risk on the chloroform inhalation effect made 5.950637, and carcinogenic risk – 0.000156.

Conclusions

Analysis of the risk levels with application of regional factors and age differences in exposure to chemical substances on peroral ingestion with drinking water showed that the application of standard values in the risk assessment methodology resulted in a two-fold underestimation of actual non-carcinogenic and carcinogenic risks for the adolescent health at the level of the 95-th Percentile (Perc) in all zones. The values of the total carcinogenic risk (TCR) on exposure to three carcinogens in drinking water — cadmium, lead and chloroform calculated with regard to ADAF exceeded the risk levels obtained without considering susceptibility to carcinogens in different age groups by a factor of 1.34/1.71 (1.79) and by a factor of 1.96 (in different zones), irrespective of the applied exposure factor. In all other zones, irrespective of the value of exposure factors, total hazard quotients indicate alarming and unacceptable risk levels at HI_{Me} = from 4 to 8.67; an d at $HI_{95th\ Perc}$ = from 8.7 to 16.8.

Acknowledgements

This work was funded by the subsidy allocated to Kazan Federal University for the state assignment in the sphere of scientific activities 19.9777.2017/8.9

References

[1] WHO 2017 *Water sanitation hygiene* www.who.int/water_sanitation_health/diseases-risks/en/

[2] 2008 *Child-specific exposure factors handbook US EPA EPA/600/R-06/096F* (Washington, DC: National Center for Environmental Assessment) p 687

[3] US EPA 1997 *Exposure factors handbook* (Washington, DC: Office of Research and Development) EPA/600/P-95/002 Fa.b.c https://rais.ornl.gov/documents/EFH_Final_1997_EPA600P95002Fa.pdf

[4] US EPA 2011 *Exposure factors handbook* (Washington, DC: Office of Research and Development) EPA/600/P-10/030 p 72

[5] 2007 The European Exposure Factors (ExpoFacts) Sourcebook *Expofacts* http://expofacts.jrc.ec.europa.eu/

[6] Unguryanu T N and Novikov S M 2014 Results of health risk assessment due to exposure to contaminants in drinking water in Russia population (review of literature) *Hygiene and sanitation* **1** 19 – 24

[7] Stepanova N V, Valeeva E R, Ziyatdinova A and Fomina S F 2016 Peculiarities of children's risk assessment on ingestion of chemicals with drinking water *Research Journal of Pharmaceutical, Biological and Chemical Sciences* **7** 1677 – 81

[8] 2011 Guidelines for drinking water quality 4-th edition *Guidelines World Health Organization (WHO)* (Gutenberg: WHO) p 564

[9] Unguryanu T N 2011 Population health risk under comprehensive effect of the drinking water pollutants *Human ecology* **3** 14 – 20

[10] Rakhmanin Yu A, Shashina T A, Unguryanu T N, Novikov S M, Skvortsova N S, Matsyuk A V, Legostaeva T B and Antipanova N A 2012 Characteristics of quantitative values of regional factors exposure in the study area *Hygiene and sanitation* **6** 30 – 3

[11] Jang J-Y, Jo S–N, Kim So-Y and Myung H-N 2014 Overview of the Development of the Korean Exposure Factors Handbook *Journal of Preventive Medicine and Public Health* **47** p 1 – 6

[12] Margot T B and Foos B P 2009 Assessing children's exposures and risks to drinking water contaminants: a manganese case study *Human and Ecological Risk Assessment* **15** 923 –47

[13] WHO 2011 *Summary of principles for evaluating health risks in children associated with exposure to chemicals* (Geneva: WHO) p 56

IOP Conf. Series: Earth and Environmental Science **107** (2017) 012079 doi:10.1088/1755-1315/107/1/012079

[14] US EPA 2011 *Exposure Factors Handbook (Final Report)* (Washington, DC: US EPA) EPA/600/R-09/052F p 1436

[15] Stepanova N V, Valeeva E R, Fomina S F and Ziyatdinova A I 2016 Assessment of non-carcinogenic risk for the health the child population under the consumption of drinking water *Hygiene and sanitation* **95** 1079 – 83

[16] Rakhmanin J A et al 2004 *Guidelines for health risk assessment for the population on exposure to chemical substances polluting the environment (P 2.1.10.1920-04)* (Moscow: Federal Center of the State Committee for Sanitary and Epidemiological Control) p 143

The sustainable utilization of malting industry wastewater biological treatment sludge

T A Vasilenko, A V Svintsov and I V Chernysh

Belgorod Shukhov State Tecnological University, Belgorod, Kostyukova str., 46 308012, Russia

E-mail: land-vna78@list.ru

Abstract. The article deals with the research of using the sludge from malting industry wastewater's biological treatment and the calcium carbonate slurry as organo-mineral fertilizing additives. The sludge, generated as a result of industrial wastewater biological treatment, is subject to dumping at solid domestic waste landfills, which has a negative impact on the environment, though its properties and composition allow using it as an organic fertilizer. The physical and chemical properties of both wastes have been studied; the recommendations concerning the optimum composition of soil mix, containing the above-mentioned components, have been provided. The phytotoxic effect on the germination capacity and sprouts of cress (*Lepidium sativum*), barley (*Hordéum vulgáre*) and oats (*Avena sativa*) in soil mixes has been determined. The heavy metals and arsenic contents in the sludge does not exceed the allowable level; it is also free of pathogenic flora and helminthes.

1. Introduction

The amount of wastewater sludge is constantly growing, as a result of, firstly, the population growth and, secondly, the increase of demand for water resources. The increase of wastewater sludge amount causes environmental problems, as well as the problems of its economically efficient and environmentally sound utilization. There are a number of regions in the Russian Federation where there exists a threat of environmental, sanitary and epidemiological degradation and occurrence of emergency situations due to possible accidents at waste treatment facilities, aeration plants or containment ponds [1].

At present $20 \cdot 109$ tons of wastewater sludge per year is generated in the world [2]. In Russia, the generation of wastewater sludge amounts to 100 million tons per year, with moisture level 98%, i.e. 2 million tons per year in dry matter [3]. In Europe, the amount of wastewater sludge generation is decreasing, despite the high development rates [4]. Of various types of waste disposal, the most common way of wastewater sludge utilization in Russia is its burial or permanent disposal at sludge draining beds. The usage of sludge as a fertilizer is no more than 4–6 % of its total amount [5]. As using sludge as a fertilizer can contaminate soils and plants with various forms of heavy metals, the sludge must be treated and purified from heavy metals before usage [6]. According to the resulting data, the recommended amount of adding the mixed mechanical and biological treatment sludge to the soil for wheat and barley is 30–40% [7]. It was found that, long-term application of sewage sludge may lead to accumulation of Zn, Cu and Ni in soil and plants [8]. A possible method of disposal is the use of sludge or ash in cement and concrete, obtaining from it the use of biogas or as fuel due to the content of organic matter after drying [9–11].

2. Material and methods

The object of research waste is referred to as «The excess dewatered sludge of malting industry wastewater». According to the results of quantitative chemical analysis, this waste contains the following components (table 1). The annual generation of the waste amounts to about 2200 tons per year. It is generated as a result of industrial wastewater biological treatment at one of the Belgorod region malting enterprises (waste water after washing and steeping of barley, removing of float and spraying barley in tanks, cleanup of steeping tanks and malting equipment) and the subsequent dewatering of the excess activated sludge with a flocculation agent on a filter-belt press. At present, it is subject to landfilling.

Table 1. The composition of malting industry wastewater biological treatment sludge.

Component	Content, %
Moisture	81.20
Organic matter	16.98
Iron oxide	0.390
Aluminium oxide	0.003
Sulphates	0.417
Nitrates	0.003
Calcium oxide	0.690
Magnesium oxide	0.021
Phosphates	0.150
Chlorides	0.005
Nickel	0.005
Cadmium	0.0001
Lead	0.001
Ash	0.1349

The heavy metals and arsenic content in the sludge does not exceed the allowable level (table 2). According to GOST R 17.4.3.07-2001 [12] the biological treatment sludge belongs to the first class, which means that it can be used for all species of agricultural crops (except for vegetables, mushrooms, greenstuff and strawberry). It is also suitable for technical and biological recultivation of contaminated soils, by its heavy metals content according to GOST R 54534-2011 [13]. Its microbiological and parasitological indices are within the allowable limits (table 3); as a result of the analysis, there were detected no pathogenic flora or helminthes in the sludge. By the negative impact on the environment, the sludge refers to category IV (low-hazardous substance) [14], which has been determined with two test objects (table 4).

Table 2. Heavy metals content in the biological treatment sludge.

Component	Dry matter concentration, mg kg^{-1}, not above		Sediment under research, mg kg^{-1} of dry matter
	Sediment group I [12]	Sediment group II [12]	
Lead	250	500	3.59
Cadmium	15	30	0.261
Nickel	200	400	19.34
Chrome	500	1000	12.59
Zinc	1750	3500	601.40
Copper	750	1500	51.95
Mercury	7,5	15	0.080
Arsenic	10	20	1.46

Another waste product studied in this work was the waste of chalk production – the calcium carbonate slurry, which has no toxicity properties (as determined with two test objects); its generation amounts to 17000 tons per year. The slurry moisture was 60–70%.

Table 3. Microbiological and parasitological indices.

Index	Measure unit	Content	Guideline value
	Microbiological indices		
General coliform bacteria	Cells g^{-1}	Less than 10	100
Pathogenic enterobacteria	Cells g^{-1}	Not detected	Not allowed
	Parasitological indices		
Viable helminthes eggs and larvae	Specimen kg^{-1}	Not detected	Not allowed
Enteric pathogenic protozoa cysts	Specimen g^{-1}	Not detected	Not allowed

Table 4. The toxicological analysis data of a sample of waste referred to as «The excess dewatered sludge of malting industry wastewater».

Method	Test object	Test length, hours	Target parameter		
			Acute (chronic) toxic effect		Dilution factor, times
Mortality and fertility test	*Ceriodaphnia affinis*	48	presence	$LDF^{a}{}_{50\text{-}48}$	1
			absence	$SDF^{b}{}_{10\text{-}48}$	100
The testing culture optical density measurement	*Chlorella vulgaris* Beijer	22	presence	$IDF^{c}{}_{50\text{-}22}$	1
			absence	$SDF_{20\text{-}22}$	100

[a] LDF – lethal dilution factor, [b] SDF – safe dilution factor, [c] IDF – inhibiting dilution factor

The composition of chalk production waste – the calcium carbonate slurry without regard to moisture is presented in table 5. The waste is diluted with process water and channeled to the industrial sewage, then to the tailing flume of the concentration plant and then by chute to the tailing dump. The water is separated from the slurry by natural filtration and flows to the lower level of the tailing dump, and then is used again in production. The slurry is not used.

Table 5. The slurry composition without regard to moisture

Component	Content, %
Nickel	0.0060
Iron trioxide	0.0420
Calcium carbonate	64.500
Chrome	0.0001
Manganese	0.0210
Silicium dioxide	35.4196
Cadmium	0.0002
Copper	0.0016
Zinc	0.0075
Lead	0.0020

The soil was collected at the territory of BSTU named after V G Shukhov in October 2015. The heavy metals and arsenic content was analyzed according to the methodologies of their detection in agricultural field soils and crop products and by the photometric method of detecting arsenic in soils;

the results are presented in table 6. The pH of salt extract (for potassium chloride) amounted to 6.8 (at the standard rate $pH_{KCl} > 5.5$). The content of components in the soil: humus – 4.9%; hydrolyzable nitrogen – 151 mg/kg; labile phosphorus – 43 mg kg^{-1}; mobile potassium – 530 mg kg^{-1}.

Table 6. Analysis data of heavy metals and arsenic content in the soil

Total forms concentration (mg kg^{-1} of dry matter)							
Component	Pb	Cd	Ni	Cr	Zn	Hg	As
TPC [a] for soil, $pH_{KCl} > 5.5$	130	2.0	80	No	220	2.1	10
The soil under research	15.40	0.332	22.81	21.29	32.43	0.025	3.23

[a] Tentative permissible concentration

3. Results and discussion

The components (slurry, soil and sludge) were mixed at various ratios; the calcium carbonate slurry concentration amounted to 1 %. The biological treatment sludge and the calcium carbonate slurry were dried before use. The biological treatment sludge content amounted to, % of mixture weight: 10.0; 20.0; 30.0; 40.0; 50.0 and 100 %; the pH value of aqueous extract amounted to: 7.56; 7.30; 7.15; 7.12; 7.08 and 7.58 respectively.

In the course of the experiment, 60 seeds of cress *Lepidium sativum* (Cruciferae family) were planted into vegetation vessels and the dependence of germination capacity and sprouts length on the content of biological treatment sludge in the soil mix was researched. All the experiments were carried out in triplicate. As a result of the research there was determined the germination capacity and sprouts length on the 4, 6, 8 and 12 day of the experiment (figure 1 and 2).

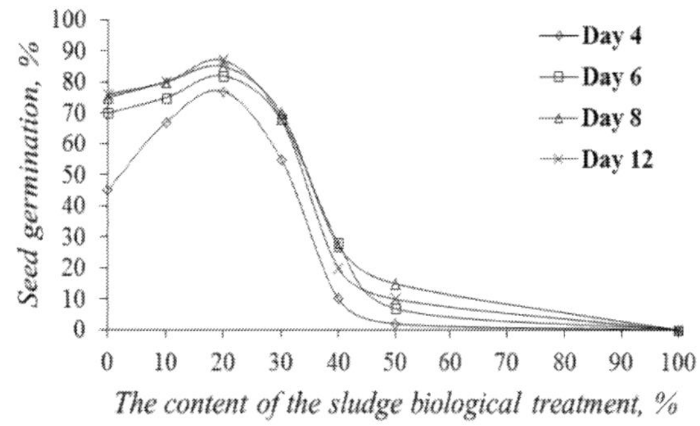

Figure 1. The dependence of cress (*Lepidium sativum*) seeds germination capacity on the biological treatment sludge content in soil mixes.

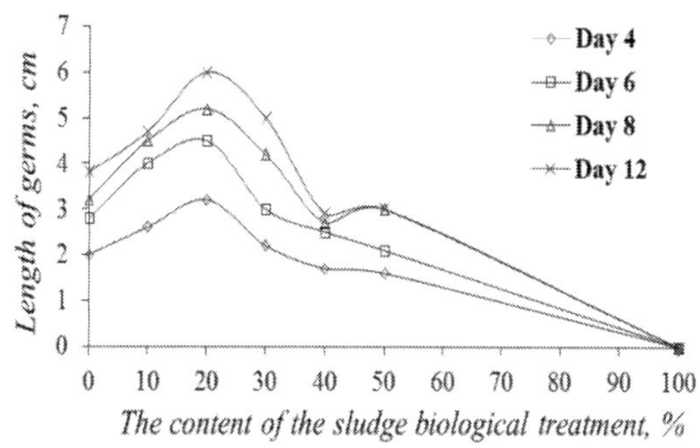

Figure 2. The dependence of cress (*Lepidium sativum*) sprouts length on the biological treatment sludge content in soil mixes.

For growing cress, the sludge content in the soil mix must not exceed 20 %. As depicted from the presented data, the sprouts length becomes smaller with the increase of the sludge concentration. This indicates its phytotoxic effect. At the same time, it belongs to the first class according to GOST R 17.4.3.07-2001 [12]. The possible factor of the biological treatment sludge adverse effect is, that due to its being a waste, the established allowable level of metals content in the sludge was based on sanitation and hygiene standards, and don't take into account the phytotoxic effect on the plants. The nitrate content in the sludge and soil mixes was determined by ionometric technique (table 7). The acceptable level of nitrate content in the soil according to hygienic standards is 130 mg kg^{-1}. The high nitrate concentration in the soil is absolutely safe and not toxic for plants – on the contrary, it fosters the growth of aerial parts of plants, makes the photosynthesis process more active, but in the given concentrations it is a constraining factor.

Table 7. Nitrate concentration in soil mixes and in the biological treatment sludge.

The percentage of sludge in the mix, %	0.0 (the initial soil)	10.0	20.0	30.0	40.0	50.0	100.0 (sludge)
Nitrates, mg kg^{-1}	30.9	372.0	662.0	1626.0	1742.0	2200.0	2468.0

At 10.0 and 20.0 % of biological treatment sludge in the soil mix the germination capacity of *Lepidium sativum* is increased by about 5.0 and 10.0 % respectively. At adding 30.0 % of sludge to the soil the germination capacity index decreases to that of the initial soil. In the soil mix with 30.0 % of sludge the germination capacity is reduced to the level of the initial soil; at 40.0 % of sludge the seeds germination capacity drops considerably, and at 50.0 % approaches zero.

The other crop plants used in the experiment were barley (*Hordéum vulgáre*) and oats (*Avena sativa*). 100 selected seeds of each species were planted to separate vegetation vessels. The experiment was carried out in triplicate. According to the results of germination index fixation, the recommended content of sludge in the soil for oats and barley amounts to 20.0–30.0 %. The germination indices of *Hordéum vulgáre* and *Avena sativa* seeds are presented in figure 3 and 4.

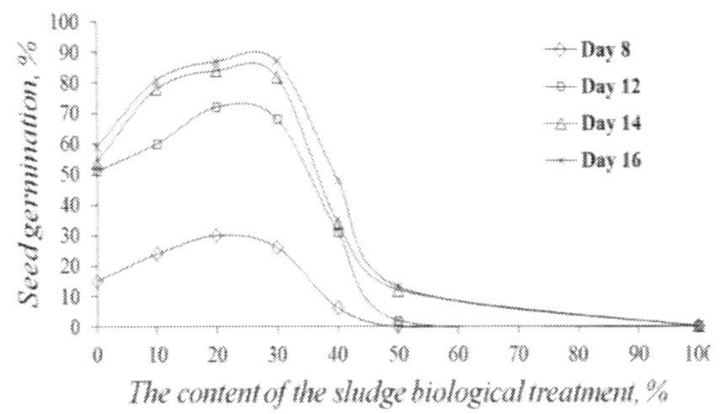

Figure 3. The dependence of *Hordéum vulgáre* seeds germination capacity on the biological treatment sludge content in soil mixes.

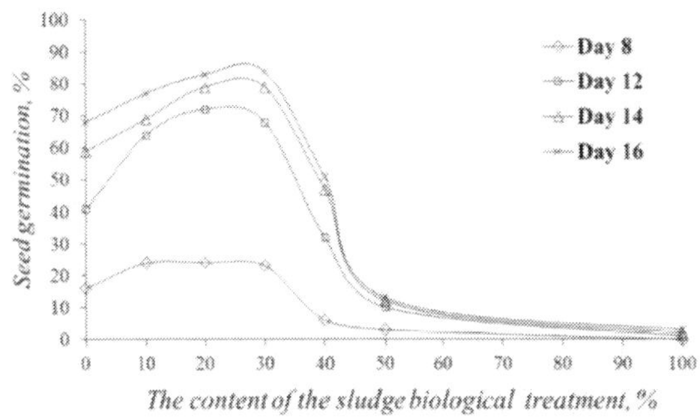

Figure 4. The dependence of *Avena sativa* seeds germination capacity on the biological treatment sludge content in soil mixes.

In figure 5 and 6, the dependence of *Hordéum vulgáre* and *Avena sativa* sprouts length on the biological treatment sludge content in soil mixes is presented.

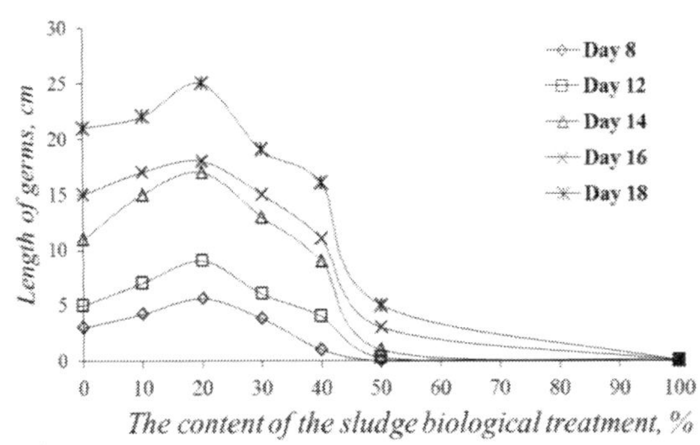

Figure 5. The dependence of *Hordéum vulgáre* sprouts length on the biological treatment sludge content in soil mixes

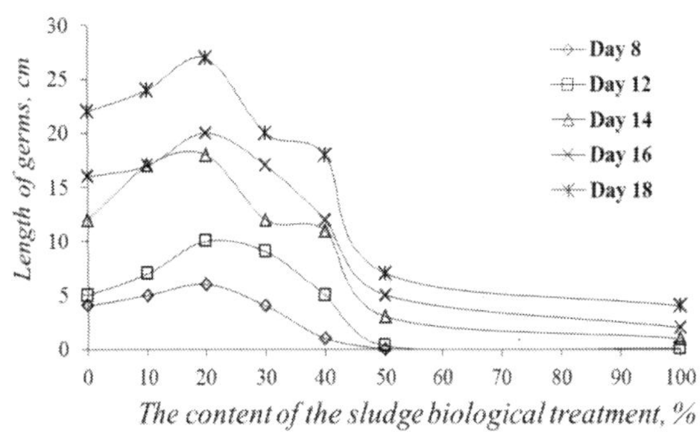

Figure 6. The dependence of *Avena sativa* sprouts length on the biological treatment sludge content in soil mixes

With the increase of sludge content (10.0 and 20.0 %) in the soil mix, the sprouts length increases as well, and then the growth slows down. At the sludge content 40.0 and 50.0 % the growth is inhibited. In the vegetation vessel with 100.0% of sludge the sprouts length of both species amounted to about 5–6 cm on the 18th day (at 20.0 % of sludge the sprouts length on the 18th day was 27 cm at average). So, the largest herbage length was observed in those plants, which were grown in the soil mix with 20.0 % of sludge.

The phytotoxic effect of biological treatment sludge on grain crops, according to its content in the soil by weight for a vegetation vessel was calculated. As we can see in the presented diagrams (figure 7 and 8), the stimulating effect on barley is observed at the sludge content in the soil mix up to 30%, and on oats – at the sludge content up to 40%.

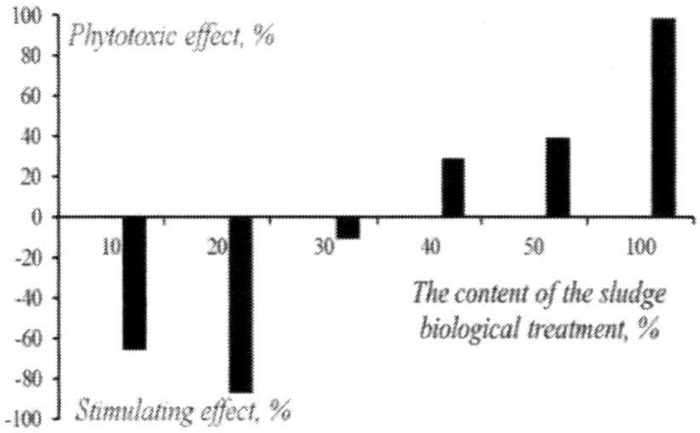

Figure 7. The stimulating and phytotoxic effect of soil mixes with biological treatment sludge and calcium carbonate slurry on the sprouts of *Hordéum vulgáre* (by weight for a vegetation vessel)

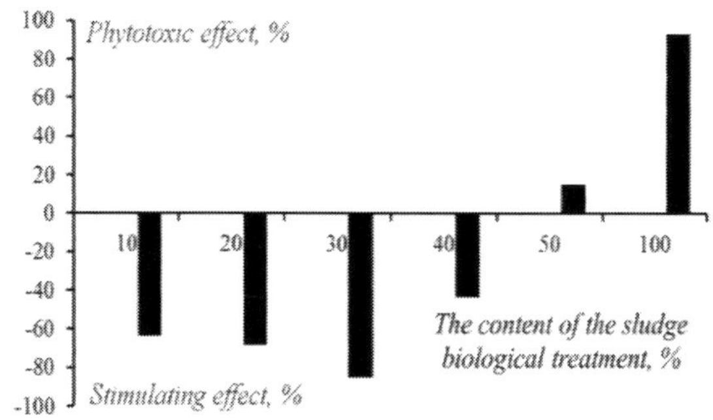

Figure 8. The stimulating and phytotoxic effect of soil mixes with biological treatment sludge and calcium carbonate slurry on *Avena sativa* (by weight for a vegetation vessel)

Hordéum vulgáre is more sensitive to the content of biological treatment sludge. At the sludge content over 40% the phytotoxic effect is observed, which can be calculated by the following formula:

$$PE = \frac{m_c - m_x}{m_c} \cdot 100\%$$

where m_c – the total weight of plants in the check vegetation vessel; m_x – the total weight of plants grown on the presumably phytotoxic medium.

4. Conclusion

As it has been mentioned above, the organo-mineral fertilizer is presented in the form of mixture. Before use, the biological treatment sludge is mixed with the calcium carbonate slurry, and the obtained mixture is pelletized. The initial components are fed from hoppers to the belt conveyer, then through a funnel to the paddle double-shaft mixer, and then to the pelleting mill. Then the pelletized mixture is dried, and after that, the finished product can be sent to storage facilities or to the customers.

The process scheme (figure 9) consists of five process units and appears as consecutive schematic drawings of interconnected machines and equipment, presented on the scheme in form of simplified outlines of the elements, conventional signs, rectangles and other geometrical figures. The biological treatment sludge is generated unevenly throughout the year, so the presented scheme operates periodically. As the biological treatment sludge is accumulated, it can be prone to natural drying, which results in its shrinkage. The line performance is assumed as 8 tons per day. It is recommended to use the double spiral mixer, which provides the continuous organo-mineral fertilizer production process

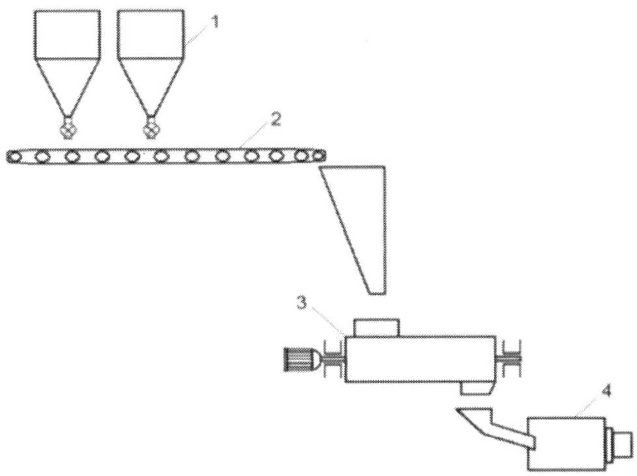

Figure 9. The process scheme of the organo-mineral fertilizer production: *1* – components storage unit; *2* – components transportation unit; *3* – mixing unit; *4* – pelletizing, drying and finished product unit.

The carried-out research has demonstrated the prospects of using the malting industry wastewater biological treatment sludge in amount of 20% of the soil mix, which corresponds to 20 t/ha. The process scheme of the organo-mineral fertilizer production has been designed. The saving of annual environmental charges because of prevention of sludge dumping will amount to 2 million 800 thousand rubles at 2016 values.

References

[1] Pakhnenko E P 2007 *Sewage sludge and other non-traditional organic fertilizers: a training manual* (Moscow: BINOM Laboratory knowledge) p 311
[2] Sergienko L I and Semenova D A 2013 Environmentally safe technology for the disposal of rain water and sewage Volzhsky *Ecological systems and devices* **3** 3 – 7
[3] Galchenko V C and Cermakova A S 2012 Rationale for the use of sewage sludge of municipal wastewater treatment plants as fertilizer *Ecological Herald of Russia* **3** 30 – 4
[4] Bernadiner I M, Stepanova T A, Khoreva P V, Chevychelov D D, Neekolayev D A and Bernadiner M N 2012 Selection of Optimal Trends of Processing and Neutralization of Sewage Sediments *Ecology and Industry of Russia* **6** 44–5
[5] Dovgan' S A 2012 Environmental, Technological and Economic Aspects of Sewage Sediments Utilization *Ecology and Industry of Russia* **5** 28–30
[6] Sinyagina N A and Suldin V B 2014 Effect of sewage sludge on the content of heavy metals in soil and plants *Hygiene and sanitation* **2** 14–5
[7] Vasilenko T A and Hared M A 2016 The application of sludge mechanical and biological treatment domestic and industrial sewage as fertilizer *Bulletin of BSTU named after V G Shukhov* **6** 211–9
[8] Bozkurtl M A, Yarılgaз T and Yazıcı A 2010 The use of sewage sludge as an organic matter source in apple trees *Polish Journal of Environmental Studies* **19** 267–74
[9] Vouk D, Nakic D, Stirmer N and Cheeseman C R 2017 Use of sewage sludge ash in cementitious materials *Reviews on advanced materials science* **49** 158–70
[10] Chin S Ch, Ing D Sh, Kusbiantoro A., Wong Y K, Saffuan W A 2016 Characterization of sewage sludge ash (SSA) in cement mortar *ARPN Journal of Engineering and Applied Sciences* **11** 2242–7

IOP Conf. Series: Earth and Environmental Science **107** (2017) 012080 doi:10.1088/1755-1315/107/1/012080

[11] Rulkens W 2008 Sewage sludge as a biomass resource for the production of energy: overview and assessment of the various options *Energy & Fuels* **22** 9–15
[12] 2008 *GOST R 17.4.3.07-2001 Protection of nature Soil The requirements for the properties of sewage sludge when used as fertilizer* (Moscow: Standartinform) p 5
[13] 2012 *GOST R 54534-2011 Resources Sewage sludge Requirements when used for reclamation of disturbed lands* (Moscow.: Standartinform) p 8
[14] 2014 *On approval of criteria for classification of waste to I–V hazard classes according to the degree of negative impact on the environment* (Order of the Ministry of Russia № 536)

IOP Conf. Series: Earth and Environmental Science **107** (2017) 012081 doi:10.1088/1755-1315/107/1/012081

Anthropogenic impact signatures revealed in the travelling ionospheric disturbances by regional GPS interferometry

V I Zakharov[1,2] and Ya A Ilyushin[1,3]

[1]Atmospheric Physics Department, Physical Faculty, Moscow State University, Moscow, GSP-2, Lengory, 119992, Russia
[2]A.M.Obukhov Institute of Atmospheric Physics Russian Academy of Sciences, Moscow, Pyzhyovskiy pereulok, 3, 119017, Russia
[3]Institute of Radio-engineering and Electronics, Moscow, ulitza Mokhovaya, 11/7, 125009, Russia
e-mail: ilyushin@phys.msu.ru

Abstract. With the technique of the radio interferometry of the navigational satellite signals, we investigate the travelling ionospheric disturbances generated by large urban agglomeration. We resolve detailed structure of the ionospheric disturbances field and attribute disturbances to particular atmospheric wave sources in the agglomeration. Wave generation efficiency revealed from the observational data well agrees with the theoretical prediction derived from energetic considerations.

1. Introduction

Presently, in the whole world big attention is paid to the monitoring of natural environment. Continuously growing population of the whole Earth, increasing anthropogenic load of the nature, intensive exploitation and consumption of natural resources results in degradation of the natural environment, primarily in the most populated regions with well-developed industrial and urban infrastructure. This problem has a lot of issues, some of them are related to a special thermal regime established in large urban areas [1, 2]. Some of them are mentioned in this brief introduction.

Due to high consumption of electrical and thermal energy, organic fuel as well as relatively low albedo of urban landscapes in the solar spectrum, big city is a strong concentrated source of the heat which is deposited in the surrounding atmosphere. Besides, lots of water vapor, soot, urban dust, aerosols etc. emitted in the atmosphere by the city, strongly influence thermal and radiation balance in the atmosphere [3]. All of that causes notable perturbations of natural distributions of the atmospheric air pressure, temperature and humidity in different regions, including the polar ones [4, 5]. These perturbations to a significant extent determine regional atmosphere dynamics, local climate in the city and region and air mass transfer in it. They can also generate atmospheric waves of different types [4, 6] which can reach ionospheric heights [6] under certain circumstances.

Central industrial region of Russia, as one of the most populated and developed regions of the Russian Federation, is characterized by complicated ecological situation and is one of the most problematic from the ecological point of view. Due to that, the city of Moscow where huge population and industry is concentrated within relatively small compact area is a strong source of perturbations of the natural environment in the region.

For that reason, techniques of the instrumental monitoring of the atmospheric dynamics in the city and the region become especially important. These techniques should combine both in situ measurements and remote sensing of the atmospheric parameters, providing characterization of the environment in average at different scales.

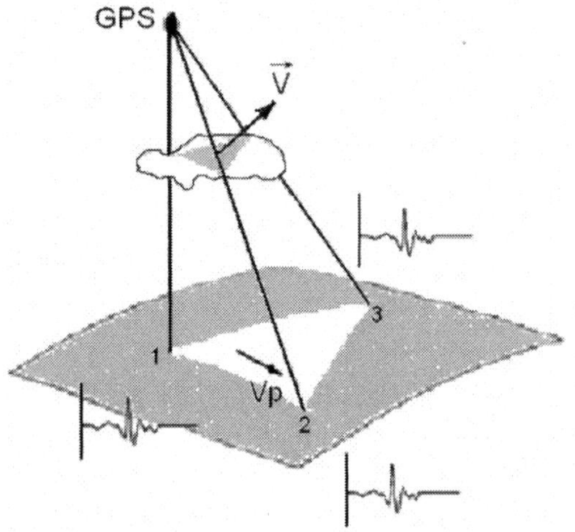

Figure 1. Schematic view of GNSS radio interferometric experiment.

2. Measuring and data processing technique

For detection of the ionospheric inhomogeneity in the GNSS satellite interferometric experiment we register changes of phases of the two navigational signals [7] with the frequencies L1 and L2 with several two-frequency professional GNSS receivers in at least three points (1, 2 and 3, see the Figure 1). Synchronous records of the signals at the observation points are then processed with correlation analysis technique. Similar phase variations in the different records are detected and interpreted as impact of the ionospheric disturbance, traversing the signal trajectory from the satellite to each receiver. Time difference between the appearances of these events at different observing stations is used to estimate the projection of the travel velocity of the disturbances on the ground. Using many GNSS satellites and receivers, one can retrieve all the three components of the travel velocity of the ionospheric object. Any of high orbital GNSS satellites can be used in the experiment, such as GPS, GLONASS and others.

The method of detection of ionospheric disturbances applied in this study is based on the utilization of combination of two registered phases L1 and L2 of the signals of the Global Navigational Satellite System (GNSS) beacons and further signal filtration for the detection of waves with typical periods longer than 5 minutes [7-10] (1)

$$I = \frac{1}{40.308} \frac{f_1^2 f_2^2}{f_1^2 - f_2^2} \left[(L_1\lambda_1 - L_2\lambda_2) + K + \delta L \right], \tag{1}$$

where f_1, f_2 are operating frequencies of the Global Positioning System (GPS), respectively 1575.42 and 1227.6 MHz; L1, L2 are the signal phases at these frequencies; λ_1, λ_1 are the wavelengths, m. K is a constant due to phase definition ambiguity, δL is an instrumental error of phase path difference measurement. The time derivative of the Total Electron Content (TEC) $dI \cdot dt^{-1}$ is a key informative parameter of the method. This allows us to detect quasi-periodical signals without elimination of the phase definition ambiguity.

For detection of the wave structures, the whole receiver network on the ground is structured into cells of three receiving stations each, at which the minimal data set sufficient for the processing is measured. For the analysis, series of measured TEC values (1), derived from phase measurements at each station in the cell are used together with corresponding zenith and azimuth angles known from sp3-files of navigational data. We can determine full vector of the wave structure motion V or its projection onto the Earth surface V_p, as it is shown in the figure 1. Assessment of the accuracy of the GPS interferometry technique [8] and its modification [9-12], applied here, yields for the azimuth angle accuracy not worse ±15 degrees, i.e. the 30 degrees sector in the worst geophysical situation. Important property of the method is its selectivity. This means that only those traveling ionospheric disturbances (TID) are detected which move in the directions not coincident with the observation direction.

In the present work, TID parameters were retrieved from three point registration of GPS signals registered on nine stations of GPS receiving network in the Moscow region. The network included three IGS stations [13] and six stations of navigational and geodetic support system of Moscow region [14] (figure 2). Analysis shows that for the satisfactory results not less than 6-8 measuring cells of three stations each should be used in the given network configuration.

Maximal amplitude of electron concentration variations are observed at the height of maximal electron concentration of the F2 ionospheric layer [8-10], which is assumed to be constant and equal to 250 km during the whole period under investigation [8, 10]. The projection of crossing point of the TID moving direction and level of maximal ionospheric concentration is called the sub-ionospheric point [8-10]. This technique allows determination if the TID position in the geographic coordinate system with the accuracy 0.5 degree or better [8-10].

Ionospheric disturbances revealed with correlation technique are then verified by the cluster analysis [11]. In the work we separated the clusters of the structures, i.e. TID with similar characteristic, observed not less that in three cells simultaneously. In fact, cluster analysis enables effective filtering of the TID not only in the space of their positions in the ground projections, but also in the space of velocities and directions.

As indicators of the ionospheric perturbed state due to solar-terrestrial connections in this study, we considered geomagnetic indices and indices of the solar activity [15]. During the whole observational period, the indices Kp and Dst characterize the ionospheric situation as moderately perturbed and account for it in the interpretation of the obtained data.

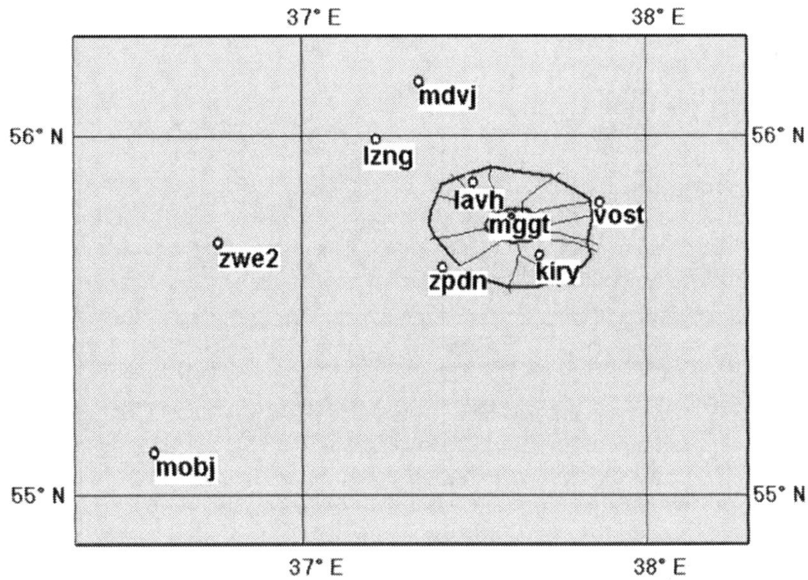

Figure 2. Location of the GPS receiver network in the Moscow region, used in our work. Stations code names are presented.

3. Main results

In this work, we present the results of the continuous radio-interferometric analysis of GPS signals registered during the summer period of 2010 year (97 days) on the Moscow regional GPS receiving network. We show that revealed TIDs group around Moscow and form the ring-shaped pattern (figure 3a). We consider empirical functions of the TID parameters (motion velocity and observed periods). We show that observed distributions of TID are formed by acoustical-gravitational waves and can be connected with a specific thermal regime in the Moscow region during the summer of 2010, in particular the urban heat island (UHI) caused by blocking anticyclone [16-18].

In the present study, we attempted to resolve internal structure of the TID spatial pattern, to retrieve the signatures of the corresponding structure of the heat source. In the Central Economic Region of Russia, the second largest industrial cluster except Moscow city is Tula, the administrative center of the Tula region separated from Moscow by a 200 km distance. Energy consumption of the Tula industrial cluster is roughly 10 times less than Moscow energy consumption. Thus, we expect proportionally less efficiency of the TID generation by Tula city. Both cities are located within the ring of TID, revealed in the Central Economic Region of Russia with our methods. Analysis of the internal structure is not an easy task, because of the limited accuracy of the direction of the TID motion determination.

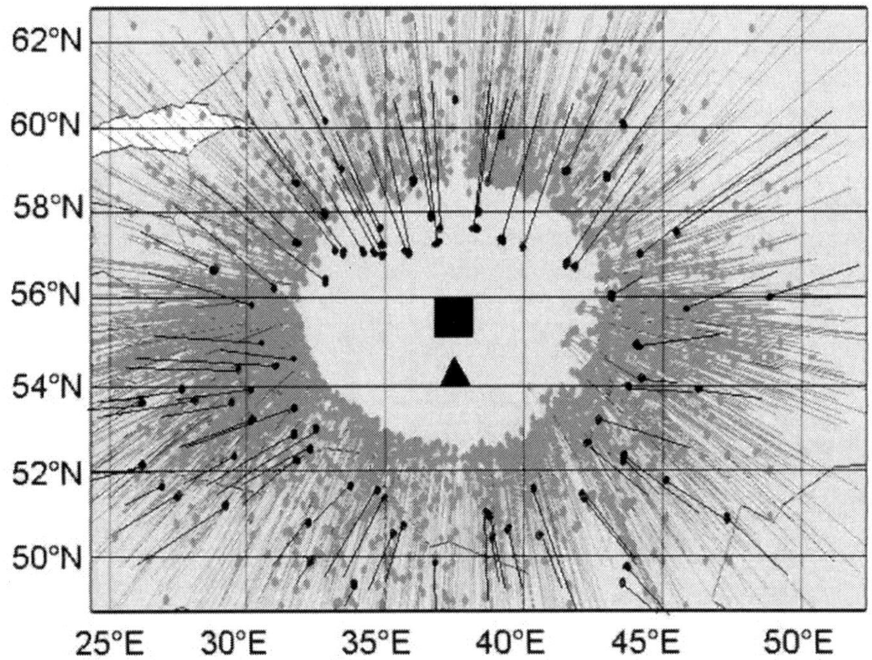

Figure 3. The ring of the travelling ionospheric disturbances (TID) detected in the Moscow region in conditions of summer 2010 [17, 18]. Some of detected waves attributed to the city of Tula as a possible source are marked with darker color. Points denote sub-ionospheric point, vector – direction and proportional value of TIDs speed. Square denotes Moscow center, triangle – the city of Tula.

However, large amount of statistical data collected in the 2010 summer (4000 events, figures 3 a-b) allowed us to filter the raw data with better precision. We searched for the TID which could be related to the acoustical-gravity waves coming from the source in the known location at 45°12' N; 37°37' E (Tula city). After that, we selected the most reliable data with minimal errors that allow to locate the source with the accuracy better than the distance between Moscow and Tula cities.

The results of such precise analysis for 2010 summer period are shown in the figure 3b. The square and the triangle are the marks of Moscow and Tula city centers, respectively. Red vectors denote TID whose source is probably Tula. TID periods and velocities are now indicated in the figure 3b. Points in the figures 3 a-b are the sub-ionospheric points projected on the ground. Our analysis unambiguously shows inhomogeneous structure of the TID ring around Moscow and reveals contributions of neighbouring sources. For Tula, we estimate its TID generation efficiency and contribution to the total TID field at 5-6% level of the whole background. Note that it is a bit less than our expectations based on energetic budget estimate, which predict corresponding efficiency at 10% level.

The TIS identified by us as generated by Tula group in the ring structure, too. Despite not the best configuration of the Moscow GPS network in 2010, it was still capable to provide the statistical data sufficient for localization of different TID sources and assessment of their contributions.

Our analysis shows that local anthropogenically driven atmospheric processes have complicated structure and should not be regarded as a result of action of a single source. Therefore, obtained results are very promising because give us a hope for well-planned and developed experiment for localization and identification of separate TID sources with GPS interferometry methods.

4. Conclusions and remarks

Complicated structure of the TID ring-shaped pattern is established and determined from thorough data analysis. Contributions of regional industrial centers surrounding Moscow were selected and estimated with regard to the common TID field formed mostly by the Moscow city itself. Estimated efficiency of TID generation by local sources is in good agreement with theoretical expectations derived from energy budget considerations.

Obtained results indicate close interaction between atmospheric and ionospheric disturbances in the Moscow region and surrounding regions, which can be related to mesoscale processes of air mass transport in the atmosphere.

5. Acknowledgements

This research was supported by the Russian Science Foundation grant 17-77-20087 "Study of long period perturbations in the ionosphere, lower atmosphere and ocean by means of GNSS sounding" and RFBR grant 16-05-01024 (ZVI, archive GPS-data analysis)

References

[1] Landsberg H E 1981 *The urban climate (*N-Y: Acad Press) p 275
[2] Oke T R 1987 *Boundary Layer Climates (2nd ed)* (London: Routledge) p 435
[3] Blake R, Grimm A, Ichinose T, Horton R, Gaffi S, Jiong S, Bader D and Cecil L D 2011 *Urban climate: Processes, trends, and projections. Climate Change and Cities: First Assessment Report of the Urban Climate Change Research Network* Ed C Rosenzweig et al (Cambridge: University Press) p 81
[4] Hamdi R, Duchene F, Berckmans J, Delcloo A, Vanpoucke C and Termonia P 2016 *Urban Climate* **17** 176 – 95
[5] Varentsov M I, Konstantinov P I, Samsonov T E and Repina I A 2014 *Curr. Problems in remote sensing of the Earth from Space* **1** 329 – 37
[6] Hocke K and Schlegel K 1996 *Ann Geophys.* **14** 917 – 40
[7] Hoffman-Wellenhoft B 1998 *GPS Theory and Practice* (New York: Springer-Verlag Vienna)
[8] Afraimovich E L and Perevalova N P 2006 *GPS-monitoring Earth's upper atmosphere* (Irkutsk: SB RAS) p 480
[9] Hernandez-Pajares M and Juan J M 2006 *J. Geophys. Res* **111** A07S11
[10] Zakharov V I and Kunitsyn V E 2012 *Geomagnetism and Aeronomy* **52** 533 – 45
[11] Zakharov V I and Budnikov P A 2012 *Moscow University Physics Bull.* **67** 26 – 33
[12] Yasyukevich Yu V, Zakharov V I, Kunitsyn V E and Voeikov S V 2015 *Geomagnetism and Aeronomy.* **55** 108 – 117

IOP Conf. Series: Earth and Environmental Science **107** (2017) 012081 doi:10.1088/1755-1315/107/1/012081

[13] SOPAC sopac.ucsd.edu
[14] SNGS of Moskow http://sngo.mggt.ru
[15] Geomagnetic kp and ap Indices https://www.ngdc.noaa.gov/stp/GEOMAG/kp_ap.html
[16] Gorchakov G I, Kadygrov E N, Kunitsyn V E, Zakharov V I, Semutnikova E G, Karpov A V, Kurbatov G A and Satansky S I 2014 *Doklady of the Russian Academy of Science. Earth Science section* **456** 736 – 40
[17] Zakharov V I and Gorchakov G I 2017 *Advances in Space Research* **59** 614 – 18
[18] Zakharov V I, Kunitsyn V E and Gorchakov G I 2017 *Doklady Earth Sci.* **472** 241 – 43

The dynamics of fauna and population of birds in agricultural landscapes on the border of Kazan International Airport

A F Bespalov and A N Belyaev

Institute of Fundamental Medicine and Biology, Kazan Federal University, Kremlevskaya str., 18, 420008, Russia

E-mail: kerwood@mail.ru

Abstract. The dynamics of fauna and population of birds from agricultural lands near Kazan International Airport has been studied for one year. A total of 60 bird species have been registered, of which 26 species must be considered as highly dangerous to flying aircraft. With regard to seasonal activity, the nesting period (April – June) is the most hazardous, since it is characterized by maximum species richness and total population density of birds, as well as their high species diversity. Furthermore, a serious danger is posed by the period of post-nesting nomadic movements and the onset of migrations (July – September) when the species diversity turns out to be the highest and the total population density begins to decrease. It has been recommended based on the obtained results that the current crops should be replaced by cultivars that will be less attractive to birds. In spring and autumn, certain bird hazing and frightening measures must be taken on a more regular basis.

1. Introduction

The need to ensure flight safety, which appeared due to advances in air transportation since the late 20th century, is closely associated with the problem of reducing bird-aircraft collisions. This problem can not be solved efficiently without detailed studying of annual behavioral patterns in birds living near airports and aircraft runways [2, 3, 4], because 70% of all birdstrikes take place in aerodrome areas, especially during takeoff and landing [5].

The aim of this study is to reveal the annual dynamics of fauna and bird populations from agricultural lands near Kazan International Airport, as well as to find out the extent to which their presence on the edge of the airport is dangerous for aircraft flying. The following tasks are set in accordance with the above aim: to assess species richness, total population density, and species diversity of birds inhabiting agricultural lands near the airport territory; to identify the dominant species in the overall bird population; and to describe seasonal aspects in the dynamics of bird fauna and population changes.

Kazan International Airport is a large airport which located 26 km south of Kazan, and has has been operating flights since 1979. Its territory is surrounded by agricultural lands, i.e., fields divided by forest belts and meadows. During the period of study, rye, wheat, medick, potato, and pea plants were recorded in the fields adjacent to the aerodrome.

2. Material and methods

Route counting of birds without limiting transect width was performed from June 2015 to May 2016 (every month) in agricultural lands along the perimeter of the airport borders. The route length was 13 km and remained unchanged throughout the study. All birds, regardless of the distance to them, were registered with subsequent conversion of the obtained data into area units based on average detection ranges [6]. For birds registered as they were flying, corrections were made for the average rate of their movement [7]. Using the materials of route counting, the population density of birds was determined and seasonal aspects in the bird population were assessed. When describing the bird population, only species accounting for more than 10% of the community were considered as either dominants or subdominants, depending on their abundance. In order to assess the population structure of birds, the following parameters were used: species richness (number of species), total population density (density (ind./km^2) of birds in all types of habitats under consideration), species diversity (Shannon-Weaver index). The population of birds was investigated using cluster analysis with the help of the PAST software (version 3.16) based on measuring the Euclidean distance (paired group option).

3. Result and discussions

Sixty bird species were registered during the period of study. Among them, 38 species were passerines. The species richness gradually decreased from June (28 species) (figure 1) to December (2 species), then increased, most sharply from April (21 species) to May (30 species). The largest number of species was found from April to September, thereby indicating the importance of this habitat for bird groups in the nesting period and their further nomadic movements. In winter, birds tend to stay away from agricultural lands.

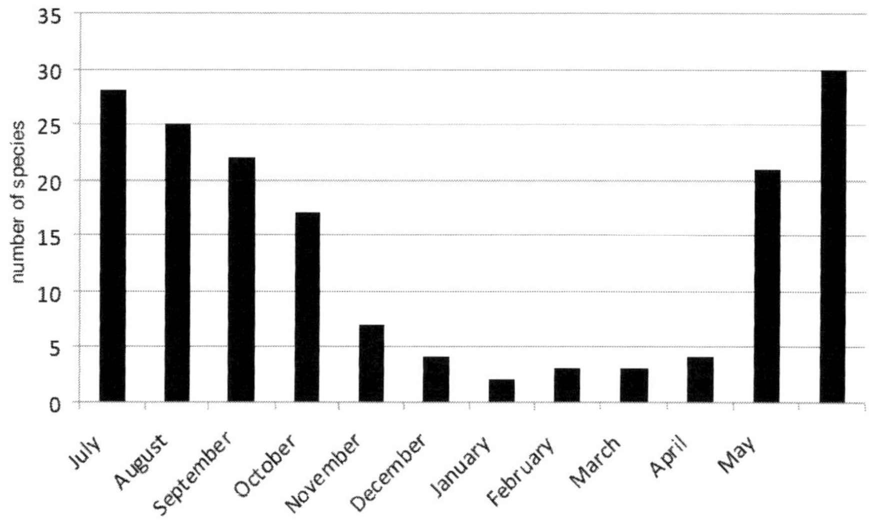

Figure 1. The dynamics of species richness (number of species) of birds in agricultural areas on the border with Kazan International Airport.

The following bird species (relatively large-sized or gregarious) posing a potential risk to aircraft were resident in the territory of study: grey heron (*Ardea cinerea*), bean goose (*Anser fabalis*), mallard (*Anas platyrhynchos*), black kite (*Milvus migrans*), hen harrier (*Circus cyaneus*), pale (*C. macrourus*), Montagu's harrier (*C. pygargus*), Western marsh harrier (*C. aeruginosus*), northern goshawk (*Accipiter gentilis*), common buzzard (*Buteo buteo*), eastern imperial eagle (*Aquila heliacal*), white-tailed eagle (*Haliaeetus albicilla*), common kestrel (*Falco tinnunculus*), grey partridge (*Perdix perdix*), northern lapwing (*Vanellus vanellus*), black-headed gull (*Larus ridibundus*), common wood pigeon (*Columba palumbus*), rock dove (*C. livia*), short-eared owl (*Asio flammeus*), black woodpecker (*Dryocopus martius*), common starling (*Sturnus vulgaris*), rook (*Corvus frugilegus*), hooded crow (*C. cornix*), common raven (*C. corax*), fieldfare (*Turdus pilaris*), and redwing (*T. iliacus*). The occurrence of semiaquatic bird species is due to the relative proximity of the Mesha River. Migratory geese are attracted to large winter crop areas. In these areas predatory species find stable groups of other small-sized birds and mammals. Synanthropic species fly to the territory because of nearby villages and airport facilities, which are favorable for nest building.

The total density of bird population (figure 2) reached a maximum in June (1188.1 ind./km^2) and May (1139.1 ind./km^2) and a minimum during the period from October to March, falling to6.2 ind./km^2 in December. Interestingly, the values of this parameter obtained for the same period turned out to be several times higher in the adjacent small forest habitats [8]. The population density of birds staying in agricultural lands during and, especially, after the nesting period was mainly dependent on the field-grown crop species and the extent of their harvesting, as well as the presence of forest belts adding tree, shrub, and marginal species to the complex of plants [9, 10].

The highest population density was recorded for Eurasian skylark (*Alauda arvensis*) from April to June (up to 443.1 ind./km^2 in May). Moreover, the following bird species were characterized by high population density: common chaffinch (*Fringilla coelebs*) (up to 196.9 ind./km^2 in June) confined to the forest belts; Western yellow wagtail (*Motacilla flava*) (up to 190,8 ind./km^2 in May); and common linnet (*Acanthis cannabina*) (up to 190.8 ind./km^2 from July to September) migrating from one agricultural land to another after the nesting period. The dominance of the former three species in agricultural landscape of the studied territory was reported previously by other researchers [11]. These birds are small-sized and pose little danger to aircraft.

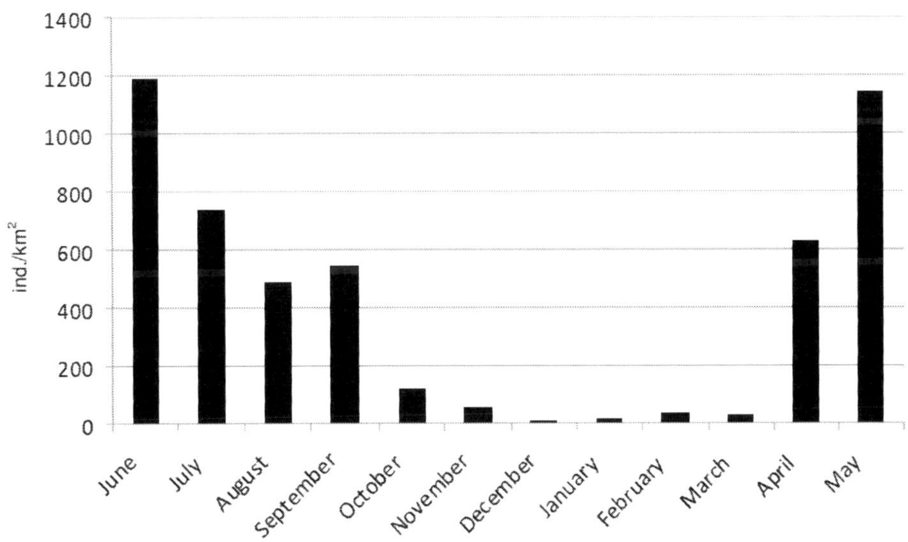

Figure 2. The dynamics of total population density (ind./km^2) of birds in agricultural areas on the border with Kazan International Airport.

Prevailing bird species were *A. arvensis* from April to June (up to 50%), *M. flava* in July (18.9%), *A. cannabina* in August-September (up to 35.4%), great tit (*Parus major*) from October to December (42.1-99.7%), and *P. perdix* from January to March (99.8-37.5%). In different months, the subdominants were tree pipit (*Anthus trivialis*) (up to 13.4%), Eurasian magpie (*Pica pica*) (up to 25%), Eurasian tree sparrow (*Passer montanus*) (up to 31.6%), *F. coelebs* (up to 16.6%), Eurasian siskin (*Spinus spinus*) (up to 13.7%), and European goldfinch (*Carduelis carduelis*) (up to 37.5%).

Therefore, small-sized bird species were generally dominant in the population. They are not very hazardous to aircraft. The flocks of *P. perdix* stayed in the fields and meadows around the aerodrome during the winter period only. These birds may collide with aircraft at low elevations, during takeoff and landing.

The biodiversity index (figure 3) hardly reached 0.02 in winter months, increased from March to July (1.1-2.5), and then decreased from August to November (2.4-1.0). On the whole, these values are not high, and the maximum diversity of birds in July and August indicates the important role of the biotope for post-nesting nomadic movements and autumn migrations.

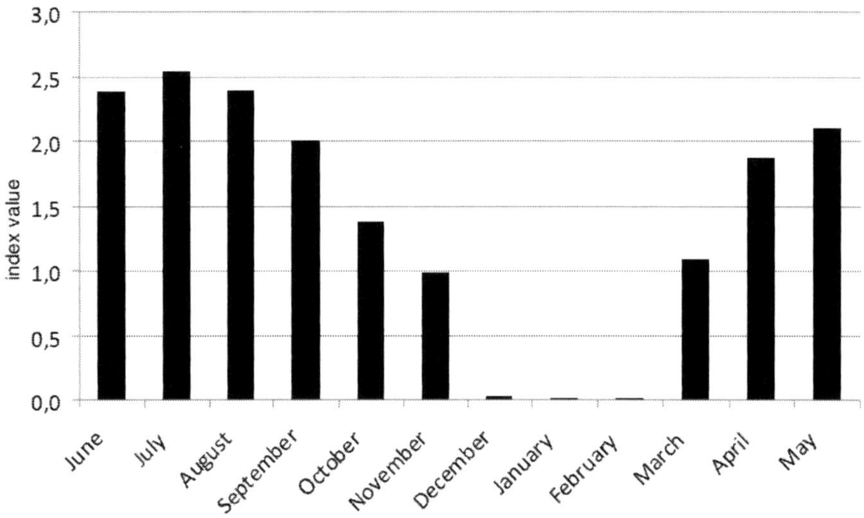

Figure 3. The dynamics of species diversity (Shannon-Weaver index) of birds in agricultural areas on the border of Kazan International Airport.

Cluster analysis revealed seasonal specifics in bird population. The nesting period (April – June) (figure 4) characterized by the maximum values of species richness and total population density of birds, as well as their high species diversity, stands out.

The period of post-nesting nomadic movements and the onset of migrations (July – September) with the maximum species diversity can be distinguished. The species richness and total population density of birds begins to decrease. During this period, the risk of birdstrikes in the airport area is particularly high.

In other months, the fauna and population of birds are generally characterized by the low values of species richness, total population density, and species diversity. Hence, birdstrikes become less likely.

Attention should be paid to the fact that the structure of bird population during the vegetation period differed in some areas of agricultural lands near the airport. The southeastern border with the airport is most dangerous, because the territories adjacent to it are actively used for growing winter crops and the Mesha River is not away from this area. Winter crops attract birds, including predators, as the places for feeding and nesting. The areas around the Mesha River are parts of migratory routes in spring and autumn for many bird species, especially semiaquatic ones, such as geese. In other periods, these areas favor penetration of semiaquatic birds and large-sized predators into the southern border with the airport.

For example, a flock of bean geese was flying northwards along the Mesha River on May 11 (12.24 a.m.), but then changed its course and began to slowly circle at the height of 40 m above the southern border of the airport, heading to its territory for daytime rest and in search for food, because it was attracted by the large green areas of winter crops.

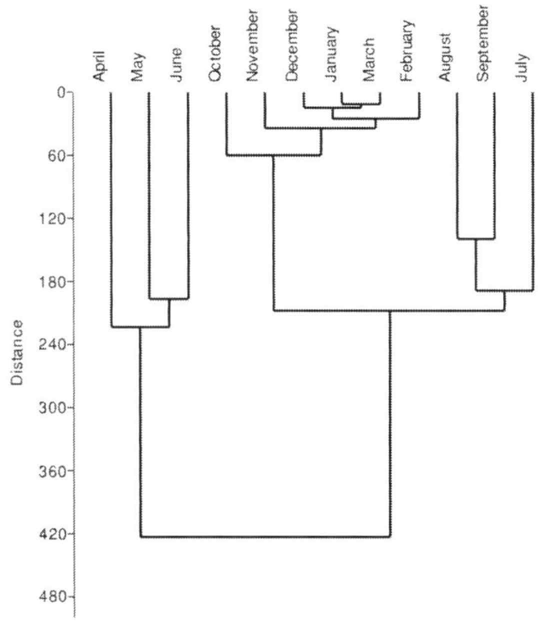

Figure 4. The dendrogram of similarities in bird population in different months.

Furthermore, ten white-tailed eagles from the areas of the Mesha River circled for a long time over this territory. The birds hovered and swooped down on each other with specific squawks, flying below 30 m or above 300 – 400 m. They moved to the airport territory and back to its border, circling above the field with winter crops. In the daytime, black-headed gull flocks (up to 86 ind.) fly above this habitat, including the aerodrome area, from the south. Thus, particular attention should be focused on the southern border of the airport. Winter crops are very attractive for migratory geese and other birds. So, it is appropriate to replace them with cultivars that will be less preferred by birds [12]. In spring and autumn, certain hazing measures should be taken to scare birds flying from the Mesha River.

Therefore, agricultural lands with both open (fields) and forest (forest belts) areas are used by various bird species from open and forest biotopes. Intermittent water bodies which dry out later in spring add even typical semiaquatic birds (mallard) to the bird fauna. After the harvesting period, many seeds are left in fields. They are potential food objects and attract birds during the period of post-nesting nomadic movements (August-September). There are also many sparrow nestlings in the period from June to September, which attract predatory birds, thereby creating the risk of birdstrikes. The studied territory was unattractive to birds from October to March. During this period, only migratory (tits) or passing (*C. cornix*) birds could be seen here in the daytime, and only small flocks of *P. perdix* appear to be more stable, being confined to winter crops.

4. Conclusions

1. A total of 60 bird species were registered, of which 26 species must be considered as highly dangerous for aircraft flying.

2. The species richness of birds decreased from June (28 species) to December (2 species) and then increased, most dramatically from April (21 species) to May (30 species). The number of species was higher in April-September, thereby confirming that the studied habitat plays an important role for bird groups during the nesting period and subsequent post-nesting nomadic movements.

3. The total density of bird population was maximum in May-June and low during the period from October to March, being determined mainly by the field-grown crop species and the extent of their harvesting, as well as the presence of forest belts.

4. The bird population was dominated by small-sized species (*A. arvensis, M. flava, A. cannabina, P. major*) posing little danger to aircraft flying. The flocks of *P. perdix* stayed in the fields and meadows around the aerodrome during the winter period. These birds may collide with aircraft at low elevations, during takeoff and landing.

5. The species diversity index was low during winter months (up to 0.02), then increased from March to July (up to 2.5), and then decreased from August to November (up to 1.0). On the whole, these values are not high, and the maximum diversity of birds in July and August indicates the important role of the biotope for post-nesting nomadic movements and autumn migrations.

6. With regard to seasonal specifics, the most dangerous periods for aircraft flying can be distinguished according to bird population dynamics: the nesting period (April – June), which is characterized by the maximum values of species richness and total population density of birds, as well as the high values of species diversity of birds; the onset of migrations (July – September) with the maximum values of species diversity and the beginning of a decrease in species richness and total population density. In other months, the fauna and population of birds is generally characterized by low values of species richness, total population density, and species diversity.

7. It was recommended based on the obtained results that the current crops must be replaced by cultivars that will not attract birds. In spring and autumn, bird hazing must be carried out on a more regular basis.

References

[1] Dolbeer R A 2013 *Wildlife in Airport Environments: Preventing Animal-Aircraft Collisions through Science-Based Management* (The Johns Hopkins University Press, Baltimore, Maryland, in association with The Wildlife Society) p 1–6
[2] Molodovskii A V and Zaloznykh D V 1999 *Vest. Nizhegorod. Gos. Univ. im. N.I. Lobachevskogo. Ser. Biol.* **1** 39–47
[3] Sodhi N S 2002 *The Auk* **119** 587–595
[4] Blackwell B F, Schmidt P M and Martin J A 2013 *Wildlife in Airport Environments: Preventing Animal-Aircraft Collisions through Science-Based Management* (The Johns Hopkins University Press, Baltimore, Maryland, in association with The Wildlife Society) p 153–165
[5] Borisov V V, Uryadova L P and Shcheblykina L S 2014 *Vestn. Pskov. Gos. Univ. Ser. Estestv. Fiz.-Mat. Nauki* **5** 3–11
[6] Ravkin Yu S 1967 *Nature of Tick-Borne Encephalitis Loci in Altai* (Novosibirsk: Nauka) p 66–75
[7] Ravkin Yu S and Dobrokotov B P 1963 *Organization and Methods for Counting Birds and Hazardous Rodents* (Moscow: Izd. Akad. Nauk SSSR) p 130–137
[8] Bespalov A F and Belyaev A N 2017 *Proc. Lyubishchev Lectures: Current Problems of Ecology and Evolution* (Ulyanovsk: UlGPU im. I. N. Ul'yanova) p 268–272
[9] DeVault T L, Begier M J, Belant J L, Blackwell B F, Dolbeer R A, Martin J A, Seamans T W, and Washburn B E 2013 *Human-Wildlife Interactions* **7(1)** 10–15
[10] Iglay R B, Buckingham B N, Seamans T W, Martin J A, Blackwell B F, Belant J L and DeVault T L. 2017 *Agriculture, Ecosystems and Environment* **242** 34–42
[11] Ivliev V G 2010 *Spatial and Temporal Heterogeneity of Bird Populations in Agro-Landscapes of the Republic of Tatarstan* (Kazan: Novoe Znanie) p 204
[12] DeVault T L, and Washburn B E 2013 *Wildlife in Airport Environments: Preventing Animal-Aircraft Collisions through Science-Based Management* (The Johns Hopkins University Press, Baltimore, Maryland, in association with The Wildlife Society) p 79–90

IOP Conf. Series: Earth and Environmental Science **107** (2017) 012083 doi:10.1088/1755-1315/107/1/012083

Dendroclimatic research on Scots pine growing under the conditions of the raised bog in the Volga-Kama region, Russia

R Chugunov, P Iskandirov and D Tishin

Institute of Environmental Sciences, Kazan Federal University, Kremlevskaya str., 18, 420008, Russia

E-mail: dtishin@kpfu.ru

Abstract. The radial growth dynamics of Scots pine in a large raised bog located in the territory of the Volga-Kama Nature Reserve, Republic of Tatarstan, has been studied. The master chronology correlated positively with the total amount of precipitation in January and negatively with the mean monthly temperature in June. Based on a complex analysis of temperature and precipitation in different months using hierarchical regression trees, three scenarios resulting in increased, average, and decreased radial growth have been suggested.

1. Introduction

Current changes in climatic parameters and subsequent reactions of forest communities at the regional level as a reflection of global environmental transformations are of great importance. Tree plants are the most suitable objects in evaluating these processes, because they are able to capture ecological information with their annual rings [1, 2, 3]. As a result of climate change, it is of special interest to investigate the natural ecosystems under pessimal conditions, i.e., when every minor fluctuation meteorological characteristics (temperature and precipitation) may lead to significant changes in the metabolism of plants [4, 5]. The purpose of this paper is to evaluate the effect of natural and climatic factors on the growth of trees in a large raised bog in the Republic of Tatarstan, the conditions of which can be characterized as pessimal. This bog is located in the territory of the Volga-Kama Nature Reserve, on the southern border of the boreal ecotone. Considerable factual material on the effect of climatic factors on the radial growth of trees has been accumulated in the scientific literature, but available current data on the radial growth of trees in the Volga-Kama region, bog habitats in particular, are insufficient.

The geographical location of the Republic of Tatarstan is not favorable for bog development. Due to the continental climate and the drainage degree of watersheds, along with the well-developed river network, bogs develop mainly in the river valleys. Thus, this territory is dominated by lowland bogs [6]. Sphagnum bogs appear as a result of atmospheric precipitation on depleted sandy soils.

The border between the following two bog zones runs along the territory of Tatarstan: raised pine-sphagnum and lowland grass bogs (East European province), which cover the pre-Kama area, as well as lowland sedge and reed bogs (Central Russian province), which stretch along the pre-Volga and trans-Kama regions [7]. The border between these zones coincides with the isoline of the hydrothermal factor (1.0), which runs from Penza to Kazan and farther eastwards, along the 56th parallel [8]. The main forest-forming species of the Volga-Kama Nature Reserve is Scots pine (*Pinus sylvestris*). It accounts for 70% of the total forest area, thereby offering an attractive object for dendrological studies.

This paper is an attempt to single out the main climatic parameters associated with radial growth, as well as to suggest possible scenarios of radial xylem growth in Scots pine under the conditions of bogs in the Volga-Kama Nature Reserve.

The Volga-Kama Nature Reserve is located on the left bank of the Volga River, 25 km westwards of Kazan. The landscape is dune-hilly with hollows and ancient gullies. The absolute altitudes vary from 65 to 105 m. The soils are sandy, medium podzolic. Peat bogs of limnogenic origin appear in depressions between the dunes.

2. Material and methods

The sample plot was laid on a floating sphagnum bog of Lake Dolgoe (N 66°40' E 34°21'). The forest type is sphagnum-dwarf shrub pine forest with solitary spruce and birch trees. The grass-dwarf shrub layer is dominated by cotton grass (*Eriophorum sp.*), marsh Labrador tea (*Ledum palustre*), leatherleaf (*Chamaedaphne calyculata*), bog-rosemary (*Andromeda polifolia*), cranberry (*Oxycoccus palustris*), and sphagnum moss. Pine is represented by the following two life forms: f. *litwinowii* and f. *uliginosa*; the trees are 125-227 years old and range in height from 3.5 to 15 m, average diameter 19 cm, growth class Va.

The standard methods of dendrochronological studies were used [9]. Material was sampled from 78 trees using a Haglof increment borer at the height of 0.5 m from the root collar. The width of annual rings was measured with the help of a semiautomatic LINTAB-6 system with TSAPWin software package [10]. The wood-ring chronology was constructed and analyzed in the dplR package [11]. Trend was excluded from the individual chronologies of radial growth, because it was suggested that it reflects the age characteristics of trees. The trend was evaluated with the help of spline. The individual chronologies were standardized. In order to construct master chronology, trendless chronologies of individual trees having a low correlation with master chronology were excluded.

The reaction of radial growth of Scots pine to climate was identified based on data on the mean monthly temperature and precipitation amount at the Sadovyi weather station of Sadovyi (1948-2011) 5 km from the bog. The dynamics of correlation between the radial growth of trees and the weather factors was analyzed using sliding correlation.

The complex effect of climatic factors on the radial growth of pine was evaluated using regression trees with a set of rules to determine the radial growth [12]. These models were built with the help of the rpart package [13] in the R statistical programming environment [14]. Since the size of data series on temperature and precipitation is quite limited, it was decided to use as few rules as possible.

3. Result and discussions

Based on the chronologies of radial growth of model trees in the bog area under study, a generalized chronology (series length 196 years) was obtained (figure 1). According to the EPS test, the chronology embraces a sufficient amount of data from 1830 to 2010. Therefore, the subsequent analysis was based on the radial growth index over the last 180 years.

The analysis of the climatic data from the Sadovyi weather station, the nearest one to the studied area, over the last 60 years showed clear trends in rising mean annual temperature and total precipitation. The positive trend for air temperature and total precipitation was 0.02°C/year and 1.5 mm/year, respectively. The observed climate change in the territory of the region takes place in the winter-spring period.

In order to identify the main climatic factors defining the radial growth of Scots pine in the studied area, correlations between the radial growth index and the air temperature and precipitation during the period from September of the previous year to August of the current year were analyzed (Fig. 3). There was a statistically significant response of the radial growth of Scots pine to the precipitation amount in January ($r = 0.5$, significance level <0.01). The negative correlation with the temperature in June was ($r = -0.43$, significance level <0.01).

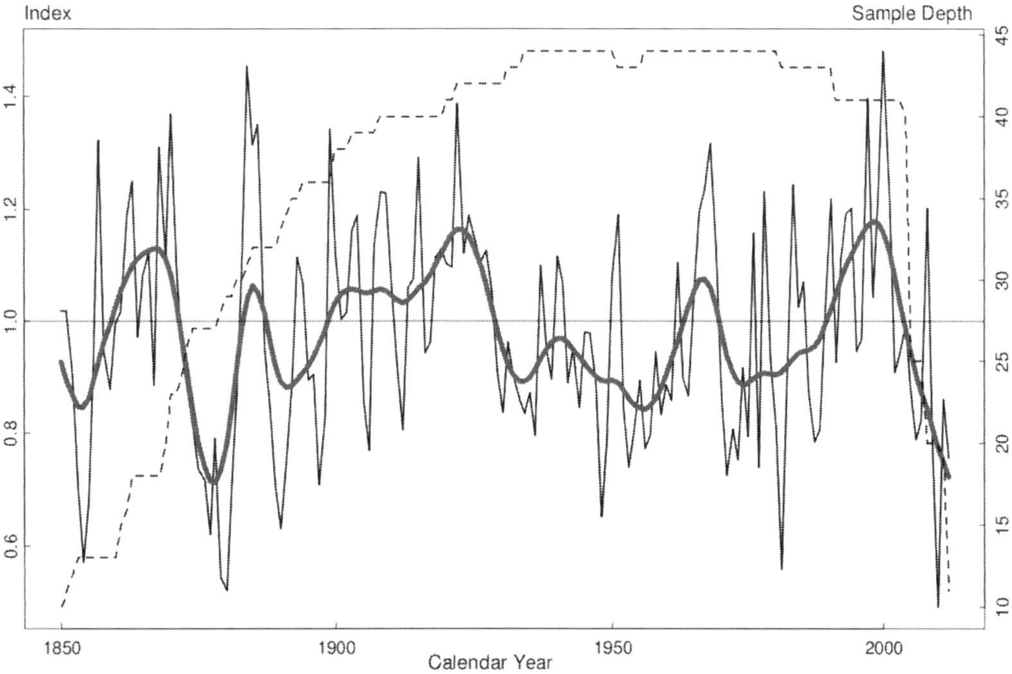

Figure 1. The long-term dynamics of the generalized index chronology DLG of the radial growth in Scots pine. The bold line shows the moving average index of radial growth, with a sliding window diameter of 30 years. The dashed line shows the individual growth chronologies involved in obtaining the master chronology.

The complex effect of climatic factors on the radial growth of Scots pine was evaluated using regression trees with a set of rules to determine the radial growth. Three scenarios resulting in increased, average, and decreased radial growth were elaborated.

The regression models also demonstrated that the key factors influencing the radial growth of Scots pine are the amount of precipitation in January and the average temperature in June, which corresponds to the data of the correlation analysis. The observed radial growth reaction is indicative of the characteristics of the site where the studied tree stand grows. The radial growth of Scots pine in the raised bogs on the southern border with the boreal ecotone depends on the amount of snow during the winter period. As it melts during the spring period, a moisture reserve for trees is formed. Large amounts of snow during the winter period can ease the arid and hot conditions of summer (advantageous scenario). A lack of snow during the winter period can be compensated by cold summer (neutral scenario). January with little snow and June with hot weather are non-advantageous scenarios leading to the formation of narrower annual rings in Scots pine under raised bog conditions.

4. Conclusions

The positive correlation of radial growth and the amount of winter precipitation proves that soil moisture plays an important role at the onset of vegetation. Further physiological activity of Scots pine depends on the summer temperature, which is especially important in the period of maximum radial growth of trees (June-July).

The sliding correlation showed that the climatic response of radial growth to temperature in June and the amount of precipitation in January, which were identified as the key factors as a result of the previous analysis, is not stable in time. This was mostly observed concerning the amount of precipitation in January. The importance of soil moisture reserve associated with the amount of winter precipitation has increased for radial growth during recent decades. The unstable response of radial growth of Scots pine to temperature and precipitation is more likely to be caused by regional climate change over the recent decades as evidenced by the data of weather observations. Interestingly, this reaction may be demonstrated by trees not only because of the climatic fluctuations, but also due to changes in the groundwater regime and the cenogenesis of the raised bog.

References

[1] Vaganov E A, Shiyatov S G and Mazepa V S 1996 *Dendroclimatic Study in the Ural-Siberian Subarctic* (Novosibirsk: Nauka) p 246
[2] Komin G E 1990 *Lesovedenie* **2** 3–11
[3] Shiyatov S G 1986 *Dendrochronology of the Upper Forest Border in the Urals* (Moscow: Nauka) p 137
[4] Mazepa V S 2005 *Can. J. For. Res.* **35** 2082–91
[5] Shiyatov S G 2003 *Pages News* **11** 8–10
[6] Baranov V I 1947 *Bogs and Peatlands of Tataria* (Kazan: Tatgosizdat) p 76
[7] Boch M S 1979 *Ecosystems of Bogs in the USSR* (Moscow: Nauka) p 188
[8] Kolobov N V 1968 *Climate of Middle Volga Region* (Kazan: Izd. Kazan. Univ.) p 253
[9] Shiyatov S G, Vaganov E A, Kirdyanov A V, Kruglov V B, Mazepa V S, Naurzbaev M M and Khantemirov R M 2000 *Methods of Dendrochronology. Part I. Basics of Dendrochronology. Sampling and Obtaining Data from Annual Rings* (Krasnoyarsk: KrasGU) p 80
[10] Rinn F 2005 *TSAPWin – Time Series Analysis and Presentation for Dendrochronology and Related Applications, Version 0.53, User Reference* (Heidelberg) p 91
[11] Bunn A G, Korpela M, Biondi F, Campelo F, Merian P, Qeadan F and Zang Ch 2012 *dplR: Dendrochronology Program Library in R. R Package Version 1.5.6* http://CRAN.R-project.org/package=dplR
[12] Breiman L, Friedman J H, Olshen R A and Stone C J 1984 *Classification and Regression Trees* (Monterey, CA: Wadsworth & Brooks /Cole) p 368
[13] Therneau T M, Atkinson B and Ripley B 2011 *rpart: Recursive Partitioning. R Package Version 3.1-50* http://CRAN.R-project.org/package=rpart
[14] R Development Core Team 2011 *R: A Language and Environment for Statistical Computing* (Vienna, Austria: R Foundation for Statistical Computing) http://www.R-project.org

Cladocera from bottom deposits as an indicator of changes in climate and ecological conditions

L A Frolova

Institute of Geology and Petroleum Technologies, Kazan Federal University, Kremlevskaya str., 4/5, 420008, Russia

E-mail: Larissa.Frolova@kpfu.ru

Abstract. Diatoms, pollen, and remains of higher vegetation are used as indicator groups in paleoecological studies. Using certain groups of zoological indicators such as planktonic and benthic organisms (Ostracoda, Cladocera, Chironomidae) has recently become popular in paleolimnology and paleoecology. This study aims to estimate the possibilities, benefits, problems and prospects of Cladocera use in the composition of zoothanatocoenosis of lakes' sediments as one of the biological indicators in paleoenvironmental studies and paleoreconstructions of abiotic conditions of the past.

1. Introduction

Cladocera is an ancient group of water organisms, which is known to have existed at least since the mid-Mesozoic era [1]. Most of the modern Cladocera were present in the Jurassic period (about 150 million years ago) during the existence of the supercontinent Pangaea [2]. Cladocera can be considered as a whole natural taxonomic unit of superorder or subclass within the Branchiopoda class, and its internal groupings as the rank of orders Ctenopoda, Anomopoda, Onychopoda and Haplopoda, which reflect a significant degree of their representatives' divergence [1]. Difficulties with identifying the taxonomic resolution of Cladocera are associated with an insufficient knowledge of the systematics and its local fauna, the absence of reliable determinants and a very low number of professional experts working on systematics and faunal studies.

2. The method of sampling for paleolimnological or paleoecological analysis of Cladocera communities

It is recommended that sampling for paleolimnological or paleoecological analysis of Cladocera communities of surface sediments or short and long columns of bottom deposits be conducted in the middle (or deepest) part of the lake. Passive movement of Cladocera remains from different littoral substrates of the shore (for example, due to the presence of coastal slope and wind) and their mixing with the remains of pelagic Cladocera has been empirically identified on the basis of numerous studies [3, 4]. Thus, deposits formed in the central part of the lake are supplemented by fossil remains of Cladocera living in different parts and substrates of the lake. Moreover, the analysis of Cladocera communities, conducted by use of paleolimnological methodology, provides more complete information on the species composition in the water than using standard hydrobiological methods with regular numerous sampling during the growing season, since the composition of bottom sediments is

IOP Conf. Series: Earth and Environmental Science **107** (2017) 012084 doi:10.1088/1755-1315/107/1/012084

represented by species living in different periods of the growing season. Using this method for obtaining information about Cladocera communities may be recommended for remote, accessible, unexplored water bodies, or reservoirs with a very short ice-free period, for instance, a relatively unknown Arctic and subarctic lakes [5, 6, 7].

3. Identification of Cladocera fossil remains

Today, systematics of this group is not completely developed, particularly at the level of species, subspecies, and morphs. Not all of Cladocera taxa can be easily identified to the taxonomic rank, required for research. Thus, the taxonomic ambiguity of some taxonomic units obstructs the ecological interpretation of paleoecological data on Cladocera [8].

In addition, there is currently no single, complete and universally recognized "standard" guide, which could be used as a starting point to identify fossil remains of Cladocera. Instead, each researcher collects crumbs of information about fossil Cladocera remains from various literature sources such as scientific articles, reports on research work, faunal summaries and illustrated identification keys.

The chemical composition of the Cladocera body is a determining factor of the fossil remains in deep-water sediments. Chitin, which forms the Cladocera exoskeleton, is a chemically inert material, but the chitin cover of species differs in composition and inclusions and has various thicknesses in different areas of the body; as a result, the quality of preservation of species and their body parts varies.

The holistic chitinous covers of planktonic crustaceans disintegrate after the death (or after the molting) of different parts of the exoskeleton such as carapace, postabdomen, postabdomen claws, mandibles, segments of antenna (only Copepoda) and headshields (only Cladocera), the identification of which is possible up to the different taxonomic ranks [5, 9]. Representatives of *Chydoridae* (species-rich group of benthic Cladocera, mainly living on the bottom or on macrophytes) and *Bosminidae* (planktonic group) families, all exoskeleton components of which are well preserved in samples, are the most valuable for the paleolimnological studies [10]. Their remains are usually preserved in great quantity and tend to reflect quantitative balance and production of species of real-zooplankton community. Not all representatives of 9 existing families of Cladocera [1] are saved equally well. For example, planktonic *Daphniidae*, a key component of aquatic ecosystems is not fully preserved, since their exoskeleton is too fragile to resist decomposition processes by microorganisms and fungi.

4. Cladocera in paleoecological research

Study of Cladocera initiated in Denmark at the end of the XVIII century and then was picked up by other countries of Northern Europe. The earliest references to Cladocera remains findings in bottom sediments of the lakes date to the late nineteenth century [2]. The accumulated scattered information has been systematized by D. G. Frey and that contributed to use of Cladocera as an indicator organism in paleoindication [11].

Classical approaches based the relationship between fossil animal communities and environmental conditions on the presence of certain types of indicators. But it is commonly understood that the ecological indicator value of some Cladocera species is apparently low as confirmed by numerous studies of fossil Cladocera [2, 6]. By identifying individual species we obtain an overview of structure and composition of the Cladocera community. In interpreting the composition of Cladocera communities in lake sediments it is important to understand regional relationships between the community composition and their habitats depending on environmental factors, which are responsible for modern distribution of various taxa, their abundance and the frequency of occurrence. The creation of a calibration database of modern Cladocera from surface sediments of a large number of lakes distributed along the environmental gradient of interest parameter (e.g., pH, trophicity) is one of the best and fastest ways of gathering information about environmental factors determining the composition of Cladocera communities in lakes. Each sample of surface sediment accumulates detailed information concerning the lake and its catchment area. Then the environmental parameters most significantly affecting the Cladocera composition are highlighted by use of multivariate statistical analysis and specialized software.

Cladocera inhabit a variety of biotopes in lakes, mainly occurring on the border between the pelagic and coastal zones, among the rocks, sand, vegetation and on soft soils in the littoral zone. Changes in the proportions between fossil remains of planktonic and littoral species may reflect changes in the relationships between shallow-water and pelagic zones. Littoral Cladocera increase with an expansion of littoral zone, indicating the development of preferred biotopes [2]. But care must be taken in interpreting the ratio of planktonic and littoral species of Cladocera in bottom sediments, as these relationships are influenced not only by changes in the area of the lake and its separate zones, but also by a number of additional abiotic and biotic factors.

Some species of Cladocera prefer a certain type of habitat substrate. Cladocera are found both in the littoral and pelagic zones of the reservoir: the pelagic zone is dominated by representatives of families *Daphniidae* and *Bosminidae*, while in the littoral zone representatives of *Chydoridae* dominate [12]. Phytophilic species, such as *Pleuroxus truncates*, *Syda crystallina*, pelagic such as *Bosmina*, benthic such as *Rhynchotalona*, *Monospilus*, *Chydorus gibbus* and others can be identified according to biotopes [13]. Most species of *Chydoridae* inhabit the thickets, where their biomass sometimes is much greater than in open coastal areas, but some species prefer this type of habitat (*Chydorus gibbis*, *Pleuroxus uncinatus*, *Disparalona rostrata*) [14].

Cladocera is very sensitive to any changes of trophic conditions in the reservoir, therefore they are used to study the eutrophication history of lakes, causes of which can be related to natural and anthropogenic factors [15]. *Daphnia cristata*, *D. longiremus*, *Alonopsis*, *Limnosida*, *Holopedium*, *Bythotrephes* are indicators of oligo- and mesotrophic conditions [16]. Increasing nutrient status in the water body contribute to a decrease in species diversity and significant changes in ratios between species. The decrease of species diversity primarily concerns phytophilous species of the family Chydoridae. At the same time, with the eutrophication in biocenoses the number of species such as *Daphnia galeata*, *D. cuculata*, *Ceriodaphnia reticulata*, *Leptodora*, *Chydorus gibbus*, *Leydigia* appears to greatly increase [13].

The changes in Cladocera community composition in accordance with changing pH have been noted in some paleolimnological studies. The change of acid-base reaction towards an increase in acidity is accompanied by changes in structural and functional relationships in the plankton community, acid-sensitive species extinction, reduction in species richness, changes in total biomass production and Cladocera abundance. The decrease in species richness is noted in Norwegian lakes exposed to acidification, as a result of which acid-sensitive planktonic organisms such as *Daphnia longispina*, *Bythotrephes longimanus*, *Leptodora kindtii* and *Bosmina longirostris* were eliminated [2, 17].

Climate variables, primarily temperature parameters, are the most significant factors affecting quantitative measures — abundance and biomass of zooplankton [9], as well as the composition and structure of Cladocera communities zoothanatocoenosis according to studies in different parts of the world, including Russia [18], Finnish Lapland [19], Canada [20].

Several previous studies, as well as our data, confirm that lake depth is one of the most important abiotic factors affecting the composition of Cladocera communities [3, 18, 21]. For instance, subfossil Cladocera remains are used by paleolimnology to reconstruct changes in lake depth [22]. Small lakes are generally smaller in size, have uniform habitats and have a well-developed littoral zone in comparison with pelagic and profundal parts of the lake [3]. A naturally determined increase in relative abundance of the littoral species in shallow lakes and an increase of pelagic species in deeper lakes were revealed as a result of our studies on a number of Yakutia thermokarst lakes [18].

A number of studies indicate a clear correlation between the main hydrochemical parameters (ionic composition, pH, specific conductivity) and Cladocera species composition. As a rule, we should expect a decrease in the diversity of fauna with rapidly changing values of these parameters [2].

IOP Conf. Series: Earth and Environmental Science **107** (2017) 012084 doi:10.1088/1755-1315/107/1/012084

5. Conclusions

The examples given above demonstrate the value of Cladocera community zoothanatocoenosis as indicators of the impact of various abiotic and biotic environmental factors affecting lake status, such as climate variation, changes in trophic status, water level, ionic composition, etc. Cladocera have been successfully used over the past decade to create transfer models on a line of parameters (temperature, depth, trophicity, etc.). Studies of communities of Cladocera remains from bottom sediments of lakes allow an increase in the use of this group of organisms as bioindicators, particularly for paleolimnological and paleoecological reconstructions, to highlight the theoretical aspects of community ecology and their biogeography [21, 23].

Acknowledgements

The study was performed with financial support of RFBR (project 15-05-04442), as well as at the expense of Russian Government Program of Competitive Growth of Kazan Federal University. Development of the methods for the reconstruction of Holocene conditions using Cladocera as a bioindicator was supported by the RSF grant 16-17-10118.

References

[1] Smirnov N N, Korovchinsky N M, Kotov A A and Sinev A Y 2007 *Proc. Rus. Conf. on Cladocera: systematics and biology* (Nizhny Novgorod: Papanin Institute for Biology of Inland Waters Russian Academy of Sciences, Vector publisher) p 5–73
[2] Korhola A and Rautio M 2001 *Zoological indicators* **4** (Dordrecht: Kluwer Academic Publishers) p 125–165
[3] Frey D G 1988 *J. Paleolimol.* **1** 179–191
[4] Kosareva L R et al. 2017 *ARPN Journal of Engineering and Applied Sciences* **12** (7) 1–15
[5] Ulrich M, Wetterich S, Rudaya N, Frolova L, Schmidt J, Siegert C, Fedorov A N and Zielhofer C 2017 *The Holocene OnlineFirst* doi:10.1177/0959683617708454
[6] Frolova L A 2016 *Proc. Int. Mult. Scient. GeoConf. SGEM 2016, Energy and Clean Technologies,* **4** (2) 601–607
[7] Frolova L A, Ibragimova A G and Fedorova I V *Proc. 16th Int. Mult. Scient. Geo-Conf. SGEM 2016, Energy and Clean Technologies* **4** (2) 579–587
[8] Frey D G 1986 *Handbook of holocene palaeoecology and palaeohydrology* (Great Britain: Whiley & Sons) p 667–701
[9] Rautio M 2001 *Arctic, Antarctic and Alpine Research* **33** 289–298
[10] Hofmann W 1987 *Hydrobiologia* **145** 315–321
[11] Frey D G 1958 *Arch. Hydrobiol.* **54** 209–275
[12] Hann B J 1989 *Geosci. Canada* **16** 17–26
[13] Manuylova E F 1964 *Cladocera of the USSR fauna* (Moscow: Science) p 328
[14] Smirnov N N 1971 *Chydoridae of the world fauna. The fauna of the USSR. Crustaceans vol 1 issue 2* (Leningrad: Science) p 531
[15] Whiteside M C 1970 *Ecological Monographs* **40** 79–118
[16] Szeroczyñska K 2002 *Quaternary International* **95/96** 165–174
[17] Nilssen J P and Sandoy S 1990 *Palaeolimnology and lake acidification* (London: The Royal Society) p 73–83
[18] Frolova L A 2009 *Modern problems of evolution* (Ulyanovsk: Ulyanovsk state pedagogical University) p 416–426
[19] Sarmaja-Korjonen K, Nyman M, Kultti S and Valiranta M 2006 *J. Paleolimnol.* **35** 65–81
[20] Sweetman J N, Rühland K M and Smol J P 2010 *J. Limnol.* **69** 1–12
[21] Frolova L A 2011 *Methodological approaches to the use of biological indicators in paleoecology* (Kazan: Kazan federal university) p 52–87
[22] Korhola A, Tikkanen M and Weckström J 2005 *J. Paleolimnol.* **34** 175–190
[23] Wetterich S, Schirrmeister L, Meyer H, Viehberg F A and Mackensen A 2008 *J. Paleolimnol.* **39** 427–449

IOP Conf. Series: Earth and Environmental Science **107** (2017) 012085 doi:10.1088/1755-1315/107/1/012085

Invasive species in phytocenosis of Sterlitamak town (Republic of Bashkortostan, Russia)

Ya M Golovanov[1], **L M Abramova**[1] **and S S Petrov**[2]

[1] Botanical Garden-Institute Ufa Scientific Centre RAS, Ufa, Russia

[2] Sterlitamak Branch of Bashkir State University, Sterlitamak, Russia

E-mail: jaro1986@mail.ru, abramova.lm@mail.ru

Abstract. There were 69 invasive and potentially invasive species in the flora of Sterlitamak in the Bashkortostan Republic (Russia). Eight are in the most dangerous types of invasive species. The greatest danger is represented by: *Acer negundo*, *Ambrosia trifida*, *Elodea canadensis* and *Xanthium albinum*. Within the boundaries of Sterlitamak, 21 syntaxa (13 associations and 8 derivated communities) are invasive species. This phytocenosis in an urban environment can be prime targets for plant quarantine actions.

1. Introduction

Invasions of alien species have long been a global ecological problem, and therefore interest in the study of invasion processes has increased in the whole world. Papers dedicated to penetration of invasive species are numerous and diverse [1, 2, 3, 4, 5]. The dispersal of alien plants is one of indicators of a drastic change in the ecological situation in the modern period and the result of anthropogenic transformation of the natural vegetation [6]. As Ch. S. Elton [7] stressed, the main cause of mass distribution of invasive species is anthropogenic disturbance of self-regulation processes of ecosystems in the absence of antagonist species. In recent years, research of invasions of alien species in Russia has increased [8, 9, 10, 11], and intensification of invasions of aggressive new species, mainly of North American origin, has been noted.

Currently in the Republic of Bashkortostan (South Urals), expansion and naturalization has been observed of North American invasive species of the family *Asteraceae* Dumort, from genera *Ambrosia* L., *Xanthium* L., *Bidens* L., *Galinsoga* Ruiz & Pav., *Cyclachaena* (Iva) Fresen, and others [8, 12, 13]. Naturalization, i.e. the introduction of new plant species into communities, is considered the highest degree of acclimatization and adaptation to a new habitat.

Cities are the first and largest locations of invasion by new species, which leads to «floristic pollution of the territory». Big cities, such as Sterlitamak in the Republic of Bashkortostan (RB), with their developed transport network and large areas of disturbed territories occupied by ruderal communities, are the locations for introduction and fixing of new synanthropic species, from which they extend further into agricultural ecosystems [14, 15].

2. Material and methods

We studied the flora and vegetation of Sterlitamak during 2014–2016. Sterlitamak is located in the south of Cisurals of the Republic of Bashkortostan in the fluvial plain of the Belaya River. The territory of the town is characterized by a continental climate. The average temperature in January is – 14.7°C, the average temperature of July – +19.2°C, and the average annual amount of precipitation is 410–460 mm. A large branch of a federal highway from Ufa to Orenburg passes through Sterlitamak. The town also possesses a developed industrial complex.

During the research, areas of invasive species were noted, and vegetation plots of these communities were analyzed. Data were sampled according to the methods of Zürich-Montpellier school [16].

Invasive plants of the «black list» of flora of RB, were maintained in a regional «Black book», as recommended, and divided into 4 groups of different invasive status.

Status 1. Species that actively penetrate natural and semi-natural communities, change the shape of ecosystems, disturb syngenetic communications, act as dominants, form homotypical thickets of area, and force out and/or interfere with reproduction of natural flora species.

Status 2. Alien species that actively settle and naturalize in disturbed semi-natural and natural habitats.

Status 3. Alien species that settle and naturalize in disturbed habitats and after further naturalization some of them, apparently, will be able to penetrate semi-natural and natural communities.

Status 4. Potentially invasive species capable of reproduction and demonstrated to drift into adjacent regions as invasive species [17].

3. Result and discussions

In the flora of Sterlitamak, several invasive and potentially invasive species of plants (table 1) are noted. Eight species are placed in the most dangerous types of new invasive status. Mostly similar species penetrate various floodplain habitats, frequently meeting on the shores of reservoirs, and they are characteristic of disturbed habitats. The greatest danger, currently, to ecosystems of the town comes from *Acer negundo, Ambrosia trifida, Elodea canadensis* and *Xanthium albinum.* The first 3 species belong to priority species for research and control [18], and *Ambrosia trifida* for plant quarantine in all territory of the Russian Federation.

The second group (10 species) of invasive species of plants are in Status 2. Similar species commonly occur in various disturbed habitats and are less widely found in natural, undisturbed habitats. The highest invasive potential in Sterlitamak is possessed by *Echinochloa crusgalli* and *Fraxinus pennsylvanica*, which are commonly found in disturbed habitats.

The third group (24 species) contains species commonly found in various ruderal habitats and not naturalized in undisturbed plant communities. Most of the species of this group are the widespread weeds often forming a monodominant phytocenosis. These are such species as: *Atriplex tatarica, Conyza canadensis, Kochia scoparia, Lactuca serriola*, etc.

The most numerous group contains potentially invasive species of plants – 27 species. In the future, these species can occupy Status 1 – 3 in their danger of invasiveness.

Analysis of occurrence of invasive species in the highest levels of disturbed habitats (table 2) has shown that 2 invasive species are the most widespread: *Acer negundo* and *Conyza canadensis* which occur in almost all sinantropic vegetation classes. The most significant fact is presence of juvenile individuals and young growth of *Acer negundo* in practically all investigated coenofloras. That speaks about the very strong invasive potential of this species in anthropogenic transformed coenofloras [19].

Often, invasive species of plants are not only a component of coenofloras, but also form monodominant communities. Currently, 21 syntaxa (13 associations and 8 derivated communities) have been identified in the territory of Sterlitamak, with domination by the above-named species. Similar communities are easily identified by the characteristic invasive species acting in a role of dominant. Similar phytocenosis in the urban environment can be prime targets for phytoquarantine actions.

IOP Conf. Series: Earth and Environmental Science **107** (2017) 012085 doi:10.1088/1755-1315/107/1/012085

Table 1. Invasive and potentially invasive species in Sterlitamak town.

N	Species	Family	Invasive status	Occurrence	Habitat
1.	*Acer negundo*	*Aceraceae*	1	Frequent	Disturbed habitats, afforestation, floodplains
2.	*Ambrosia trifida*	*Asteraceae*	1	Occasional	Disturbed habitats, floodplains
3.	*Bidens frondosa*	*Asteraceae*	1	Occasional	Banks of rivers, disturbed habitats
4.	*Xanthium albinum*	*Asteraceae*	1	Occasional	Banks of rivers, disturbed habitats
5.	*Cyclachaena xanthiifolia*	*Asteraceae*	1	Occasional	Disturbed habitats
6.	*Echinocystis lobata*	*Cucurbitaceae*	1	Frequent	Banks of rivers, osiers
7.	*Elodea canadensis*	*Hydrocharitaceae*	1	Frequent	Rivers, ponds
8.	*Hordeum jubatum*	*Poaceae*	1	Occasional	Road borders, railways
9.	*Carduus acanthoides*	*Asteraceae*	2	Common	Disturbed habitats
10.	*Solidago canadensis*	*Asteraceae*	2	Frequent	Abandoned gardens
11.	*Alyssum turkestanicum*	*Brassicaceae*	2	Very rare	Disturbed steppe
12.	*Cardaria draba*	*Brassicaceae*	2	Occasional	Road borders
13.	*Lupinus polyphyllus*	*Fabaceae*	2	Rare	Abandoned gardens
14.	*Fraxinus pennsylvanica*	*Oleaceae*	2	Frequent	Floodplains, disturbed habitats
15.	*Echinochloa crusgalli*	*Poaceae*	2	Frequent	Banks of rivers, disturbed habitats
16.	*Collomia linearis*	*Polemoniaceae*	2	Very rare	Railways
17.	*Portulaca oleracea*	*Portulacaceae*	2	Occasional	Flower gardens, railways
18.	*Saponaria officinalis*	*Caryophyllaceae*	2	Occasional	Meadows, disturbed habitats
19.	*Amaranthus albus*	*Amaranthaceae*	3	Occasional	Railways, disturbed habitats
20.	*A. blitoides*	*Amaranthaceae*	3	Occasional	Disturbed habitats
21.	*A. retroflexus*	*Amaranthaceae*	3	Common	Disturbed habitats
22.	*Artemisia sieversiana*	*Asteraceae*	3	Occasional	Disturbed habitats
23.	*Conyza canadensis*	*Asteraceae*	3	Frequent	Disturbed habitats
24.	*Galinsoga ciliata*	*Asteraceae*	3	Rare	Flower gardens
25.	*G. parviflora*	*Asteraceae*	3	Rare	Flower gardens, disturbed habitats
26.	*Helianthus tuberosus*	*Asteraceae*	3	Occasional	Abandoned gardens, flower gardens
27.	*Lactuca serriola*	*Asteraceae*	3	Common	Disturbed habitats
28.	*Lepidotheca suaveolens*	*Asteraceae*	3	Occasional	Trampled habitats
29.	*Onopordum acanthium*	*Asteraceae*	3	Rare	Disturbed habitats
30.	*Senecio vernalis*	*Asteraceae*	3	Rare	Road borders
31.	*S. viscosus*	*Asteraceae*	3	Rare	Railways
32.	*Lepidium densiflorum*	*Brassicaceae*	3	Occasional	Trampled habitats
33.	*Atriplex tatarica*	*Chenopodiaceae*	3	Common	Disturbed habitats
34.	*Kochia scoparia*	*Chenopodiaceae*	3	Frequent	Disturbed habitats
35.	*Cuscuta campestris*	*Cuscutaceae*	3	Occasional	Disturbed habitats
36.	*Hippophae rhamnoides*	*Elaeagnaceae*	3	Rare	Abandoned gardens

IOP Conf. Series: Earth and Environmental Science **107** (2017) 012085 doi:10.1088/1755-1315/107/1/012085

N Species	Family	Invasive status	Occurrence	Habitat
37. *Medicago sativa*	*Fabaceae*	3	Occasional	Disturbed habitats
38. *Elsholtzia ciliata*	*Lamiaceae*	3	Very rare	Road borders
39. *Bromus japonicus*	*Poaceae*	3	Occasional	Disturbed habitats
40. *B. squarrosus*	*Poaceae*	3	Frequent	Disturbed habitats
41. *Setaria pumila*	*Poaceae*	3	Common	Disturbed habitats
42. *S. viridis*	*Poaceae*	3	Frequent	Disturbed habitats
43. *Acroptilon repens*	*Asteraceae*	4	Very rare	Disturbed habitats
44. *Phalacroloma septentrionale*	*Asteraceae*	4	Very rare	Road borders
45. *Symphyotrichum novi-belgii*	*Asteraceae*	4	Occasional	Abandoned gardens
46. *S.× salignum*	*Asteraceae*	4	Rare	Abandoned gardens
47. *Impatiens glandulifera*	*Balsaminaceae*	4	Rare	Banks of streams
48. *I. parviflora*	*Balsaminaceae*	4	Very rare	Flower gardens
49. *Symphytum caucasicum*	*Boraginaceae*	4	Rare	Flower gardens
50. *Armoracia rusticana*	*Brassicaceae*	4	Occasional	Abandoned gardens, dumps, meadows
51. *Sisymbrium volgense*	*Brassicaceae*	4	Very rare	Road borders
52. *Sambucus racemosa*	*Caprifoliaceae*	4	Rare	Riparian forests
53. *Bryonia alba*	*Cucurbitaceae*	4	Rare	Abandoned gardens
54. *Caragana arborescens*	*Fabaceae*	4	Occasional	Afforestation, road borders, railways
55. *Galega orientalis*	*Fabaceae*	4	Rare	Road borders, railways, abandoned gardens
56. *Epilobium pseudorubescens*	*Onagraceae*	4	Very rare	Moist habitats
57. *Xanthoxalis stricta*	*Oxalidaceae*	4	Very rare	Lawns
58. *Anisantha tectorum*	*Poaceae*	4	Rare	Railways
59. *Digitaria sanguinalis*	*Poaceae*	4	Rare	Railways, flower gardens
60. *Lolium perenne*	*Poaceae*	4	Frequent	Lawns
61. *Aquilegia vulgaris*	*Ranunculaceae*	4	Occasional	Abandoned gardens
62. *Amelanchier alnifolia*	*Rosaceae*	4	Rare	Afforestation, railways
63. *Cerasus vulgaris*	*Rosaceae*	4	Frequent	Abandoned gardens, railways
64. *Malus domestica*	*Rosaceae*	4	Rare	Abandoned gardens
65. *Rosa pimpinellifolia*	*Rosaceae*	4	Occasional	Abandoned gardens, afforestation, railways
66. *Populus balsamifera*	*Salicaceae*	4	Occasional	Afforestation, disturbed habitats
67. *Typha laxmannii*	*Typhaceae*	4	Very rare	Moist habitats
68. *Ulmus pumila*	*Ulmaceae*	4	Occasional	Road borders, railways, afforestation
69. *Parthenocissus quinquefolia*	*Vitaceae*	4	Occasional	Abandoned gardens, riparian forests, disturbed habitats

Table 2. Representation of invasive species in units of vegetation of the disturbed habitats of Sterlitamak town.

Species/Syntaxa	S. m.		Art.			P-P.	G-U.	Bid.
	Sis.	A-C.	A.	On.	Ag.			
Xanthium albinum	I	-	-	I	-	-	-	V
Bidens frondosa	I	-	-	-	-	-	-	IV
Acer negundo	I	-	IV	II	II	I	-	III
Ambrosia trifida	I	-	-	-	-	-	-	II
Conyza canadensis	I	II	II	II	I	I	-	II
Echinocystis lobata	I	I	-	-	-	-	-	II
Cuscuta campestris	I	-	-	I	-	I	-	II
Cyclachaena xanthiifolia	I	-	-	-	-	I	-	-
Lepidotheca suaveolens	I	-	-	-	-	I	-	-
Hordeum jubatum	-	-	-	-	-	I	-	-
Anisantha tectorum	-	-	-	I	-	-	-	-
Galinsoga ciliata	I	-	-	-	-	-	-	-
G. parviflora	I	II	-	-	-	-	-	-
Lepidium densiflorum	I	-	-	-	-	-	-	-
Fraxinus lanceolata	I	-	-	-	-	-	-	-
Portulaca oleracea	-	I	-	-	-	-	-	-
Impatiens glandulifera	-	-	-	-	-	-	I	-
Total	13	4	2	5	2	6	-	7

Note. I–V — Classes of constancy.
Classes of vegetation: S.m. – *Stellarietea mediae* (Orders: Sis. – *Sisymbrietalia*, A-C. –*Atriplici–Chenopodietalia albi*), Art. – *Artemisietea vulgaris* (Orders: A. – *Artemisietalia vulgaris*, On. – *Onopordetalia acanthii*, Ag. – *Agropyretalia repentis*), P-P. – *Polygono arenastri–Poëtea annuae*, P-A. – *Polygono-Artemisietea austriacae*, G-U. – *Galio–Urticetea*, Bid. – *Bidentetea tripartitae*.

Table 3. Sytaxa with dominating invasive and potentially invasive species in Sterlitamak town.

Syntaxa	Occurrence
Sytaxa with dominating of 1 group species	
Association ***Elodeetum canadensis*** Nedelcu 1967	Frequent
Association ***Chelidonio-Aceretum negundi*** Ishbirdina et Ishbirdin 1989	Frequent
Association ***Polygonetum hydropiperis*** Passarge 1965 variant ***Bidens frondosa***	Occasional
Association ***Salici-Populetum*** (R. Tx. 1931) Meijer Drees 1936 variant ***Acer negundo***	Occasional
Derivated community ***Xanthium albinum*** [***Potentillion anserinae/Bidentetea tripartitae***]	Occasional
Derivated community ***Ambrosia trifida*** [***Potentillion anserinae***]	Occasional
Association ***Ivaetum xanthiifoliae*** Fijałkowski 1967	Rare
Association ***Polygono avicularis–Hordeetum jubati*** Abramova, Golovanov 2016	Rare
Sytaxa with dominating of 2 group species	
Association ***Carduetum acanthoidis*** Felföldy 1942	Frequent
Derivated community ***Cardaria draba*** [***Artemisietea vulgaris***]	Occasional
Association ***Setario pumilae–Echinochloëtum cruris-galli*** Felföldy 1942 corr. Mucina in Mucina et al. 1993	Rare
Sytaxa with dominating of 3 group species	
Association ***Conyzo canadensis–Lactucetum serriolae*** Lohmeyer in Oberdorfer 1957	Frequent
Association ***Convolvulo arvensis–Amaranthetum retroflexi*** Abramova et Sakhapov in Ishbirdin et al. 1988	Frequent

Syntaxa	Occurrence
Association ***Atriplicetum tataricae*** Ubryszy 1949	Frequent
Association ***Kochietum densiflorae*** Gutte et Klotz 1985	Occasional
Derivated community ***Cuscuta campestris*** [***Polygono arenastri–Poëtea annuae***]	Rare
Derivated community ***Bromus squarrosus*** [***Sisymbrietalia/Onopordetalia acanthii***]	Rare
Derivated community ***Galinsoga ciliata*** [***Stellarietea mediae/Molinio-Arrhenatheretea***]	Rare
Derivated community ***Galinsoga parviflora*** [***Stellarietea mediae/Molinio-Arrhenatheretea***]	Rare
Sytaxa with dominating of 4 group species	
Derivated community ***Digitaria sanguinalis*** [***Stellarietea mediae***]	Rare
Association ***Calystegio sepium–Impatientetum glanduliferae***	Rare

4. Conclusions

The intensification of anthropogenic impact within the urban area leads to the formation of a greater number of habitats, favorable for invasions, and active penetration and naturalization of aggressive invasive species of plants. The most vulnerable to this process are areas near railway stations and, in general, railways, highways with adjacent habitats, elevators and banks of water bodies (especially the banks of the Ashkadar and Belaya rivers).

But practically all phytocenoses within city lines contain up to 74% of the alien species [19] which can become further invasive. Therefore, territories of the large cities have to be objects of continuous monitoring of penetration and formation of invasion centers of aggressive, new plant introductions, and these areas demand measures to decrease their numbers by using all suppression methods possible.

Acknowledgment

The reported study was funded by RFBR according to the research project № 17-04-0037.

References

[1] Lonsdale M 1999 *Ecology* **80** (5) 1522–36
[2] Richardson D M and Pysek P 2006 *Progr. Phis. Geogr* **30** (3) 409–431
[3] Chytry M, Maskell L, Pino J, Pyšek P, Vila M, Font X and Smart S 2008 *J. Appl. Ecol.* **45** (2) 448–458
[4] Lambdon P W et al. 2008 *Preslia* **80** 101–149
[5] Pyšek P et al. 2009 *Diversity and Distributions* **15** 891–903
[6] Burda R I 1991 *Anthropogenic Transformation of Flora* (Kiev: Naukova Dumka)
[7] Elton C 1958 *The Ecology of Invasions by Animals and Plants* (London: Methuen)
[8] Abramova L M 2012 *Russian Journal of Ecology* **43** (5) 352–357
[9] Abramova L M 2014 *Bull. Acad. Sci. Rep. Bashk.* **19** (4) 16–27
[10] Vinogradova Yu K, Mayorov S R and Horun L V 2010 *Black book of flora of Central Russia* (Moscow: GEOS)
[11] Panasenko N N 2014 *Ross. Zhurn. Biol. Invas.* **2** 127–132
[12] Abramova L M 1997 *Bot. Journ.* **82** (1) 66–74
[13] Abramova L M 2003 *Bot. Journ.* **88** (4) 67–76
[14] Golovanov Ya M and Abramova L M 2011 *Bull. Voronezh State Univ.* **1** 173–176
[15] Golovanov Ya M and Abramova L M 2013 *Bull. Altai State Univ.* **1** (3) 27–30
[16] Braun-Blanquet J 1964 *Pflanzensociologie* (Wien)
[17] Abramova L M and Golovanov Ya M 2016 *Proc. Ufa Sci. Center RAS* **2** 54–61
[18] Dgebuadze Y Y 2014 *Ross. Zhurn. Biol. Invas.* **1** 2–8
[19] Golovanov Ya M and Abramova L M 2016 *Vegetation of Russia* **28** 28–36

Rare species of the genus *Oxytropis* DC. (Fabaceae) from the flora of the Republic of Bashkortostan under cultivation within the city of Ufa

N V Maslova, O A Elizaryeva, G M Galikeeva and N M Tyutyunova

Ufa Institute of Biology, RAS, prospekt Oktyabrya, 69, 450054, Russia

E-mail: maslovanv-ib-ufa@mail.ru

Abstract. This paper presents the results of investigations introducing eight rare species of the genus *Oxytropis* DC. from the flora of the Republic of Bashkortostan into the city of Ufa. Seven species are characterized as promising ornamental plants. According to the integrated evaluation scale of introduction effectiveness based on seven indices, two species were considered as very promising (*O. spicata* (Pall.) O. et B. Fedtsch., *O. knjazevii* Vasjukov), five species as promising (*O. approximata* Less., *O. baschkirensis* Knjasev, *O. hippolyti* Boriss., *O. gmelinii* Boriss., *O. kungurensis* Knjasev) and one species as non-promising (*O. sordida* (Willd.) Pers.). The ornamental quality assessment of rare species of the genus *Oxytropis* based on ten indices showed that these species can be used to create ornamental compositions in the landscape design style.

1. Introduction

The existing assortment of useful plants employed in green building is the result of a long-term introduction process. An introduction survey is an important element in the development of recommendations for management and protection of useful plants [1, 2]. The introduction of rare and endangered plants into cultivation is an important and effective measure to save their gene pool. Introduction tests showed the possibility of using rare species as ornamental plants [3, 4, 5, 6, 7].

Special programs are under development to protect the biodiversity of rare species of the genus *Oxytropis* DC. from the flora of the Republic of Bashkortostan, including their protection *in situ* and *ex situ*, as well as reintroduction techniques. Introduction is one of the ways to preserve these specie [8, 9, 10]. Species of the genus *Oxytropis* DC. can be recommended as promising ornamental plants [8, 9, 11, 12, 13, 14, 15].

2. Material and methods

This paper presents the results of investigations on introducing eight rare species of the genus *Oxytropis* DC. from the flora of the Republic of Bashkortostan and gives their characteristics as ornamental plants (table 1). Six of them are included in the Red Data Book of the Republic of Bashkortostan (2011) [8] in group 2 (decreasing species): *O. kungurensis* Knjasev (figure 1), *O. approximata* Less. and *O. sordida* (Willd.) Pers. and in group 3 (rare species): *O. baschkirensis* Knjasev (figure 2), *O. gmelinii* Fisch. ex Boriss. (figure 3) and *O. hippolyti* Boriss. (figure 4). Two rare species: *O. spicata* (Pall.) O. et B. Fedtsch. (figure 5) and *O. knjazevii* Vasjukov (*O. knjazevii*

IOP Conf. Series: Earth and Environmental Science **107** (2017) 012086 doi:10.1088/1755-1315/107/1/012086

Vasjukov, nom. nov. – *O. tatarica* Knjaz. 2001 [16] (figure 6) are included in «The List of Plants and Fungi Not Included in the Red Data Book of the Republic of Bashkortostan That Should Be Paid Special Attention to Their Environmental Conditions within the Republic and Monitored» [10]. One (*O. hippolyti*) out of eight rare species is included in the Red Data Book of the Russian Federation (2008) [14]. There are seven endemics and one relic plant among them [10] (table 1).

These species occurred in stony steppes. Regarding the moisture content of their habitats these species are classified as mesophytes (*O. sordida*), xeromesophytes (*O. spicata, O. knjazevii*), mesoxerophytes (*O. baschkirensis, O. kungurensis, O. gmelinii*) and xerophytes (*O. approximata, O. hippolyti*) [10].

Investigations on the species of the genus *Oxytropis* DC. under introduction were launched in 1997 (introduction nursery for rare plants of the South Urals, Ufa Institute of Biology, RAS). The nursery is situated in the Botanical Garden of Ufa (forest-steppe zone of the Bashkir Cis-Urals) [8, 9, 10]. It is located outdoors under full insolation and defended against winds with trees and shrubs. The soil is grey forest clayey loam. The species are introduced with seeds collected in natural habitats.

Figure 1. *Oxytropis kungurensis.*

Figure 2. *Oxytropis baschkirensis.*

Figure 3. *Oxytropis gmelinii.*

Figure 4. *Oxytropis hippolyti.*

Figure 5. *Oxytropis spicata.*

Figure 6. *Oxytropis knjazevii.*

Table 1. Some biological and introduction characteristics of rare species of the genus *Oxytropis* DC. under introduction.

Year of introduction	Number of samples	Flowering time	Flowering season	Corolla colour	Group of promising potential
O. kungurensis; 2 – decreasing species, endemic plant of the South Urals and Central Cis-Urals					
1997	3	Fast-flowering	Spring and summer-flowering	Pink violet	P
O. baschkirensis; 3 – rare species, endemic plant of the South Urals and Trans-Volga Region					
2000	3	Long-flowering	Summer and autumn-flowering	Purple	P
O. approximata; 2 – decreasing species, endemic plant of the South Urals					
2000	2	Medium-flowering	Summer-flowering	Pale yellow	P
O. gmelinii; 3 – rare species, endemic plant of the South Urals					
1997	4	Medium-flowering	Summer-flowering	Pale yellow	P
O. hippolyti; 3 – rare species, endemic plant of the Trans-Volga Region					
2000	4	Medium-flowering	Summer-flowering	Pale yellow	P
O. sordida; 2 – decreasing species, periglacial Pleistocenic relic plant					
2009	1	Non-flowering	Non-flowering	Pale yellow	NP
O. spicata; species under observation, endemic plant					
1997	3	Long-flowering	Spring, summer and autumn-flowering	Pale pink or bluish pink	VP
O. knjazevii; species under observation, endemic plant of the Trans-Volga Region and South Cis-Urals					
1997	2	Long-flowering	Spring, summer and autumn-flowering	Pinkish blue	VP

Note. The conservation status of the species is specified according to the Red Data Book of the Republic of Bashkortostan [10]. Groups of promising potential are given according to Karpisonova [3]: non-promising species (NP) – less than 10 points, not very promising species (NVP) – 10–13 points, promising species (P) – 14–17 points, very promising species (VP) – 18–21 points.

This paper applies conventional research methods [3, 5]. The assessment of the introduction success was performed using the integrated scale developed in the Main Botanical Garden of the Russian Academy of Sciences (Moscow) [3].

The assessment of the ornamental quality was performed using a standard method with our own modifications (table 2) [17] with regard to such assessment indices as winter hardiness and ornamental quality of fruit-bearing plants. Also, changes were made with regard to such indices as corolla colour, its intensity and inflorescence length and density. During the development of the scale we used the point-based assessment card of the ornamental quality of a variety (species) with using a conversion factor for each index so that the maximum index expression would eventually be 100 points [17].

3. Result and discussions

Long-term phenological observations showed that according to their phenorhythmic type all *Oxytropis* species are spring, summer and autumn-flowering plants with a long vegetative period and winter dormancy. Under cultivation they have a steady-state type of phenological development. According to their flowering season, the species are classified in the following way: *O. kungurensis* as spring and summer-flowering (May – June); *O. approximata, O. gmelinii* and *O. hippolyti* as summer-flowering (June – July); *O. baschkirensis* as summer and autumn-flowering (July – September); *O. spicata* and *O. knjazevii* as spring-summer and autum-flowering (May – September) (table 1). According to their flowering time, they are classified in the following way: *O. kungurensis* as fast-flowering; *O. approximata, O. gmelinii* and *O. hippolyti* as medium-flowering; *O. baschkirensis, O. spicata* and *O. knjazevii* as long-flowering.

Secondary flowering in August and September is characteristic of fast and medium-flowering species providing there is a wet and warm autumn. Phenological stages of the *Oxytropis* species overlap each other. Thus, summer-flowering species may have buds, flowers and fruits on one and the same plant from July to August (figure 2). In this connection various *Oxytropis* species can be used for compositions in gardens of continuous bloom.

According to the corolla colour the introduced *Oxytropis* species are subdivided into two groups: "pink flowering" and "yellow flowering." The "pink flowering" group includes *O. baschkirensis, O. kungurensis* and *O. spicata* (corolla colour is from pale pink to light blue or violet). The "yellow flowering" group contains *O. approximata, O. gmelinii* and *O. hippolyti* (corolla colour ranges from almost white to yellowish) (table 1). Intra- and inter-population corolla-colour polymorphism is noted under introduction in *O. gmelinii, O. kungurensis* and *O. spicata*. These species fall within the group of ornamental flowers, with the period of the highest ornamental quality during their flowering stage.

The *Oxytropis* species are propagated only by seeds formed every year. Freshly collected seeds are of good quality. Their germination capacity is 80–100%. Under storage conditions the hardness of the seed coat tends to increase up to 70–90 % and is broken by scarification (germination capacity shows an increase up to 80–90 %). The germination capacity is maintained for five years in *O. kungurensis*, eight years in *O. gmelinii* and *O. baschkirensis*, twelve years in *O. hippolyti* and thirteen years in *O. spicata*.

All species were propagated using the seedling technique. The adaptability of seedlings was 60–100 % depending on the year of observation and time of planting. Self-seeding was also observed in all species. The most frequent and viable self-seeding occurred in *O. spicata* and *O. knjazevii*.

The preservation of plants after transplanting the seedlings may vary over several years depending on the origin of the samples. Thus, for example, the preservation of *O. hippolyti* in the third year of its life was as follows for different populations: 33.3 % of seedlings from the vicinity of the village of Starye Kandry (Tuymazinsky District), 23.5 % from Gurovskaya Mountain (Kushnarenkovsky District), 12.5 % from the eastern bank of Aslykul Lake (Davlekanovsky District) and 0 % from Izmailka Mountain (Bizhbulyksky District) and from the vicinity of the village of Kanly-Turkeevo (Buzdyaksky District).

The rate of ontogenesis in the *Oxytropis* species under cultivation is faster compared to natural populations. Its duration is short (2–5 years) in *O. approximata, O. baschkirensis, O. gmelinii, O.*

hippolyti, *O. spicata* and *O. kungurensis* and longer (5–15 years) in *O. gmelinii*, *O. kungurensis* and *O. spicata*. The flowering stage begins after two to five years of development.

The vital status of the *Oxytropis* species under cultivation is good. Annual observations show that their biometric indices are equal to or even exceed those found in nature. An increase is seen in their height, number of inflorescences, number of flowers in an inflorescence and so on. For example, in *O. gmelinii* under cultivation the height was 40.5 cm, the number of leaves was 91.5 and the number of inflorescences averaged 33.4versus 28.4, 67.7 and 17.4 under natural conditions, respectively. In *O. baschkirensis* under cultivation the height was 33.9 cm, the number of leaves was 54.5 and the number of inflorescences was 30.8 versus 25.3, 21.5 and 6.5 under natural conditions, respectively. The vital form of the plants remains the same (taproot caudiciform herbaceous perennials with polycarpic rosette-type shoots). The indices of seed productivity in the *Oxytropis* species under cultivation are higher than those in nature. For example, in *O. gmelinii* the fruit formation under cultivation was 41.8 % and the productivity factor was 10.2 %, whereas in nature these indices turned out to be equal to 28.3 % and 4.6 %, respectively. In *O. baschkirensis* the same indices are equal to 55.9 % and 7.6 % under cultivation and 35.6 % and 4.0 % in nature.

According to the integrated assessment scale of introduction success (according to assessment indices: fruit bearing intensity, seed germination capacity, seed propagation capability under cultivation through self-seeding, vegetative propagation capability, plant status under cultivation, pest and disease resistance, winter hardiness [3] and also due to their ornamental quality) *O. spicata* and *O. knjazevii* are characterized as very promising; *O. approximata*, *O. baschkirensis*, *O. hippolyti*, *O. gmelinii* and *O. kungurensis* as promising; *O. sordida* as non-promising (the plants are developed until the immature and virginal states and preserve on the trial plot up to three years).

The plants are winter-hardy in introduction nursery conditions; (they are able to survive winter months without covering) and are frost-resistant. No frost damages were observed in spring or autumn. Vegetative parts of the plants were not injured by pests and diseases, although fruits and seeds were affected by insect pests.

The assessment of the ornamental quality of seven species of the genus *Oxytropis* DC. using ten assessment indices is given in table 2. The most intensely coloured corolla is seen in *O. baschkirensis*. This species is also characteristic of the tightest inflorescences with few flowers. The most profusely flowering species with a great many of inflorescences are *O. hippolyti*, *O. approximata* and *O. gmelinii*. These species are also noted for the length and strength of their flowering shoots. *O. kungurensis* forms second-generation grey-green densely pubescent leaves that make its vegetative sphere attractive throughout the entire vegetation season. *O. baschkirensis* does not become less attractive during its fruit-bearing stage. *O. spicata* and *O. knjazevii* are noted for their winter hardiness. They also have the longest flowering period. The total number of points for each species was as follows: *O. baschkirensis* – 81, *O. kungurensis* – 76, *O. spicata* and *O. knjazevii* – 74, *O. hippolyti* – 73, *O. approximata* and *O. gmelinii* – 72. The maximum number of points was received by *O. baschkirensis*.

All the species in the nursery were grown in monoculture. In addition,, some of them were tested in Alpine rock gardens and mixed plantings. For example, *O. gmelinii* was tested in an artificial phytocoenosis together with such rare species as *Allium nutans* L., *Globularia punctata* Lapeyr., *Iris pumila* L., *Dianthus uralensis* Korsh., *Helichrysum arenaria* (L.) Moench, *Heliantemum bashkirorum* (Juz. ex Kupatadze) Tzvel., etc. and also in the Alpine rock garden (figure 3). Such conditions do not hinder the species from blooming, bearing fruits and producing full-fledged seeds.

With due consideration for ornamental properties, environmental features and development rhythms, it is possible to create multi-component compositions of rare plants for landscaping. Based on our research experience we think it reasonable to combine the species of the genus *Oxytropis* DC. with those of the genera *Allium* L. (*A. flavescens* Bess., *A. nutans* L.) *Iris* L. (*I. pumila* L., *I. scariosa* Willd. ex Link.), *Hedysarum* L. (*H. grandiflorum* Pall. and *H. razoumovianum* Fisch. et Helm.).

Table 2. Assessment of the ornamental quality of rare species of the genus *Oxytropis* DC. under introduction.

O. kungurensis	*O. baschkirensis*	*O. approximata*	*O. gmelinii*	*O. hippolyti*	*O. spicata*	*O. knjazevii*
Corolla colour, intensity (20)						
17	18	12	12	12	16	16
Flower size (5)						
5	5	4	4	4	4	4
Floriferous shoot (length and strength) (10)						
8	8	9	9	9	7	7
Inflorescence (number of flowers) (10)						
7	7	8	8	8	10	10
Inflorescence density and length (10)						
9	10	7	7	7	5	5
Abundance of flowering (10)						
5	6	8	8	8	7	7
Ornamental quality of the vegetative sphere (10)						
9	8	7	7	7	6	6
Winter hardiness (5)						
3	3	3	3	4	5	5
Duration of the flowering period (10)						
5	8	7	7	7	9	9
Ornamental quality of fruit-bearing plants (10)						
8	8	7	7	7	5	5
In total (100)						
76	81	72	72	73	74	74

Note. The maximum number of points is given in brackets for each assessment index.

All the assessment data show that the species of the genus *Oxytropis* DC. in question are of practical interest as promising ornamental plants and can be applied in such phytocompositions as rock gardens. These compositions may not only make wonderful park and garden decorations in urban areas, but also fulfill educational and nature-protection functions. The urban environment can give shelter to rare species that will contribute to their conservation.

Thus, according to a set of biological and economic indices seven rare species of the genus *Oxytropis* from the flora of the Republic of Bashkortostan hold promise for growth in the Botanical Garden of Ufa and cultivation under urban conditions of the forest-steppe zone of the southern Cis-Urals. This makes it possible to preserve the species *ex situ*, obtain seeds and planting materials for reintroduction and use them to create ornamental compositions in the landscape design style.

Acknowledgements

The research was performed in 2017–2019 with partial financial support from the Russian Foundation for Basic Research and Government of the Republic of Bashkortostan, Project 17-44-020506 p_a.

References

[1] *Botanicheskie sady v sovremennom mire: teoreticheskie i prikladnye issledovaniya [Botanical gardens in the modern world: theoretical and applied investigations] Proc. of the Rus. Scien. Conf. with International Participans* 2011 (Moscow: KMK Scientific Press ltd.) p 765

[2] Karpinosova R A 1985 *Travyanistye rasteniya shirokolistvennykh lesov SSSR. Ekologo-floristicheskaya i introduktsionnaya kharakteristika [Herbaceous plants of broadleaf forests of the USSR. Eco-floral and introduction characteristics]* (Moscow: Nauka) p 205

[3] Kondratyuk E N and Ostapko V M 1990 *Redkie, endemichnye i reliktovye rasteniya yugo-*

vostoka Ukrainy v prirode i kulture [Rare, endemic and relic plants of south-eastern Ukraine in nature and under cultivation] (Kiev: Naukova dumka) p 152

[4] Trulevich N V 1991 *Ekologo-fitotsenoticheskie osnovy introduktsii [Eco-phytocoenotic basics of plant introduction]* (Moscow: Nauka) p 216

[5] Danilova N S 1993 *Introduktsiya travyanistykh rasteniy flory Yakutii [Introduction of herbaceous perennials from the flora of Yakutia]* (Yakutsk: Yakut SC of the SB of RAS) p 164

[6] Maslova N V, Elizaryeva O A, Kuvatova D N and Asadullina S F 2006 *Izuchenie zapovednoy prirody Yuzhnogo Urala [Wildlife research of the South Urals]* issue 2 (Ufa: Willy Oxler ltd.) p 166–178

[7] Maslova N V, Karimova O A and Abramova L A 2009 *Bioraznoobrazie rasteniy na Yuzhnom Urale v prirode i pri introduktsii [Plant biodiversity of the South Urals in nature and under introduction]* (Ufa: Gilem) p 65–80

[8] *Krasnaya kniga Respubliki Bashkortostan (rasteniya i griby) [Red Data Book of the Republic of Bashkortostan (plants and fungi)]* 2008 vol 1 (Ufa: MediaPrint) p 384

[9] Klobukova-Alisova E N 1960 *Dikorastushchie poleznye i vrednye rasteniya Bashkirii [Wild useful and harmful plants of Baskiria]* vol 2 (Moscow, Leningrad: Publishing House of the USSR Academy of Sciences) p 247

[10] Plennik R Ya and Kuznetsova G V 1972 *Rasteniya prirodnoy flory Sibiri dlya zelenogo stroitelstva [Plants of the Siberian natural flora for green building]* (Novosibirsk) p 98–105

[11] Nicholls G, Lupp R and Ward J B 2002 *Alpine plants of North America. An encyclopedia of mountain flowers from the Rockies to Alaska* (Portland: Timber Press) p 344

[12] *Krasnaya kniga Rossiyskoy Federatsii (rasteniya i griby) [Red Data Book of the Russian Federation (plants and fungi)]* 2008 (Moscow: KMK Scientific Press ltd.) p 855

[13] *Vse o polevykh tsvetakh Rossii: atlas opredelitel [Everything about wildflowers of Russia: Atlas Index]* 2011 (St. Petersburg: SZKEO) p 120

[14] *Metodika gosudarstvennogo sortoispytaniya dekorativnykh kultur [Methods for official variety testing]* 1960 (Moscow: Publishing House of the Ministry of Agriculture of the USSR) p 182

The Dilemmas of Energy: Essential energy services and potentially fatal risks

J H Perkins

The Evergreen State College, 236 Cambridge Avenue, Kensington, California 94708 USA

Email: perkinsj@evergreen.edu

Abstract. During their evolution, humans have made three energy transitions, each marked by the adoption of new ways of procuring energy with attendant changes in lifestyle. Modern civilization arose in the Third Energy Transition, and its major sources of energy come from coal, oil, gas, uranium, and hydropower. Unfortunately, despite its incalculable benefits, the Third Transition can't provide sustainable energy services for the indefinite future. Climate change is the most serious problem. Criteria and standards for each of the currently available, nine primary energy sources indicate the potential feasibility of replacing most or all uses of coal, oil, gas, and uranium with hydropower, solar, wind, biomass, and geothermal. This is the Fourth Energy Transition, promotion of which is strongly supported by considerations of sustainability.

Human life as we know it today is dependent on energy services. If these services were lost, most people would suffer and many would die. To establish a firm base of knowledge about energy, new methods of analysis are needed compared to methods used in the past. I will briefly indicate some of these methods.

1. Energy Transitions

During human evolution, humans have embraced four energy transitions, events based on mastering a new form of energy supply. The First Energy Transition was based on fire and firewood to cook food and provide light, heat, and protection from predators. This Transition probably happened about one million years ago during the evolution of *Homo habilis* to *Homo erectus*. Modern humans, *Homo sapiens*, are dependent upon cooked food. The Second Energy Transition occurred when *H. sapiens* domesticated plants and animals to stabilize and increase the supply of food energy. Agrarian civilization developed, about 10,000 years ago. The Third Energy Transition stemmed from substituting coal for firewood as a source of heat. This Transition began about 1550, and has since moved around the world. Modern civilization replaced agrarian civilization. The Fourth Energy Transition has barely begun, and it arose because of severe problems caused by the energy sources developed during the Third Transition, especially by coal, oil, gas, and uranium. The Fourth Transition has the potential to replace these four energy sources — possibly completely — with renewable energy and energy efficiency.

The Third Transition began in England in the 1500s when England, like all other countries, relied on firewood for its major external source of energy for cooking, heating, lighting, making metals and ceramics, and many other purposes. By the 1500s, however, the English population had risen to over three million people, and the increased population had cleared forests for agriculture and reduced firewood supply by extensive harvests of woodland.

Coal was readily available at shallow depths, and people began to substitute coal for shrinking supplies of firewood. By 1700, the population had reached five million, and coal was commonly used for brewing, distilling, bricks, tiles, glass, pottery, nails, cutlery, salt, sugar, soap, brass, and copper. Moreover, the amount of heat available from coal far surpassed the amount available from firewood, and England began a spurt of unprecedented economic growth. This was the first step of the Third Energy Transition.

Step 2 of the Transition began in 1712 with the successful development of Thomas Newcomen's "atmospheric engine." This machine turned the heat of burning coal into physical motion that could pump water from coal mines. By the early 1700s, coal mines were deep enough to collect water that had to be removed. Newcomen's engine accelerated production of more coal.

Abraham Darby successfully used coal to make coke to replace charcoal for smelting iron ore in 1708 – 1709, and, combined with Newcomen's engine, created a mutually reinforcing cycle: more coal → more iron → more steam engines → more coal, and so forth. England's economy and military power grew rapidly.

Step 3 of the Third Transition began in the late 1700s and early 1800s. Inventors developed steam engines that could power ships and railroad trains. Thus, the heat of coal made mechanical motion, and the motion in turn increased mobility on the seas and the land. No longer was the speed of transport dependent upon fickle winds or the speed of a horse. People began to move themselves and their goods over landscapes and oceans far faster and more cheaply than had ever been done before. The pace of economic growth increased, and in the 1800s the inventions that had begun in England began to spread to other parts of the world.

In addition to coal, people also began to use deposits of oil and gas in addition to coal. Together, coal, oil, and gas comprised what are now called the "fossil fuels," and the combustion of these materials supply 81 percent of the heat energy used worldwide. In the 1940s and 1950s, scientists identified heat from fission of uranium, and nuclear power joined fossil fuels to make the "mineral fuels," which supply 86 percent of the world's heat energy (2015) [2]. I have called these mineral fuels the "big-four fuels" of the Third Energy Transition.

Step 4 of the Third Transition came in 1882, when Thomas Edison and Henry Rogers successfully turned the heat of burning coal and the energy of falling water, respectively, into motion to turn an electric generator. "Thermal electricity" and "hydroelectricity," both initially developed in the United States, augmented the Third Energy Transition by making energy that was easily transported long distances and could produce heat, light, motion, mobility, and telecommunications. Uranium supplemented thermal electricity and hydroelectricity after 1954, when the Union of Soviet Socialist Republics first turned the heat of fission into electricity.

Not only did the big-four fuels, plus hydroelectricity, spread, the amounts of energy deployed to provide energy services rose exponentially during the 20th century. In 1900, all of humanity used the energy equivalent to about 1000 million tons of oil per year, and by 2000, this amount had exceeded 14,000 million tons of oil per year. During the same time, the human population grew from about 1.6 billion to 6.1 billion and today stands at about 7.6 billion [3]. Support of today's population would be impossible without energy services.

Benefits of the Third Energy Transition have been incalculable. It is more accurate to say the Third Transition created "modern civilization," which replaced the "agrarian civilization" that arose from the Second Energy Transition. Agriculture did not disappear, but machines and energy sources replaced the muscle power of people and animals. By 2000, modern civilization occupied most of the globe, and only tiny, isolated pockets of people in remote places lived lives of the First and Second Energy Transitions.

To put it bluntly, almost all people now alive depend upon the energy services of the Third Energy Transition for their comfort, safety, prosperity, and survival. No other options exist: humanity must have the benefits of energy sources or literally billions of people will suffer and die. Only sustainable energy services can preserve modern civilization.

If no problems with the Third Energy Transition existed, then energy would not be a subject of interest. Unfortunately, multiple, serious problems have arisen, and these problems threaten the sustainability of energy services. We turn next to these problems.

2. The Third Energy Transition is not Sustainable

Four challenges have arisen in the Third Energy Transition: climate change, geopolitical tensions, and health and environmental consequences of using the big-four energy sources, plus depletion of the mineral deposits of the big-four. Climate change is by far the most serious and imminent of these four challenges, and it is a global problem. The other three tend to be local or regional problems, and I won't discuss them here.

Briefly summarized, climate change results primarily from the production and release to the atmosphere of carbon dioxide (CO_2) a combustion product of coal, oil, and gas. CO_2 is a greenhouse gas that alters the radiation balance of the earth by absorbing and retaining on earth heat that would otherwise leave the earth as infrared radiation. Historically for the past 800,000 years, the concentration of CO_2 in the atmosphere has varied between about 280 parts per million (ppm) and 300 ppm. The Third Energy Transition has released enough CO_2 to increase its concentration to over 400 ppm. This extra CO_2 has warmed the atmosphere and the oceans by about 1°C in 2016 compared to 1951–1980 [4].

Levels of CO_2 are now at unprecedented, high levels compared to the entire time that modern humans, *Homo sapiens*, have existed (about 300,000 years) and to the time that humans have engaged in agriculture (about 10,000 years). CO_2 levels are still increasing, and the Intergovernmental Panel on Climate Change estimates that humans have injected about 500 billion tons of CO_2 into the atmosphere. If future injections put an additional 500 billion tons of CO_2 into the atmosphere, the predicted temperature increases will be about 2°C, an increase that most scientists believe may disrupt agriculture, cause extreme weather, and disrupt other processes [5].

Future damages from climate change risk being horrendous, but already evidence has accumulated showing damages of USD 1.1×10^{12} from extreme weather events in the United States since 1980 [6]. These include heat waves, droughts, and severe rain storms. Sea-levels have risen threatening coastal areas world-wide, and the ocean has become more acidic due to dissolved CO_2, a threat to aquatic species. Arctic ice has declined, which is likely to produce a positive feedback to further temperature increases. In addition, a wide range of theories and models produces results highly consistent with the already observed events, strongly suggesting that climate change poses existential risks to modern civilization across the earth.

3. Analysis of Sustainability of Energy Services

To meet the needs of the present without compromising the ability of the future to meet their needs. This is the guiding principle of sustainability identified in *Our Common Future*, published in 1986 by the World Commission on Environment and Development. It directs those who are alive now not to use the earth's resources in ways that diminish the ability of future generations to meet their own needs. Often the principle of sustainability is shortened to the "3 E's:" economy, environment, and equity. Managing earth's resources must pay attention to each factor to find sustainable ways of supporting the lives of present and future generation.

Further guiding principles of sustainability analysis comes from the following questions, with my brief answers. Issues of science and technology must be considered in sustainability analysis, but they are not sufficient by themselves to complete the analysis. Ethics, morals, and values also must be included, which makes sustainability analysis a difficult — but necessary — process.

- What is to be sustained? (Energy services)
- Over what area? (Every country)
- Over what time? (Forever)
- At what cost? (Reasonable)
- For whose benefit? (Present and future generations)
- Who pays? (Fairness and equity guide division of costs)

Democracy and rule of law must play a central role in decision-making about energy to find sustainable ways forward. For universities, this means that students must have opportunities to develop general energy literacy so that they understand why the Third Energy Transition, despite its benefits, is not sustainable and must give way to the Fourth Energy Transition and sustainable energy services. Universities must also educate and train the scientists and engineers as well as social-cultural-economic-political experts who have deep knowledge of energy technologies.

The most important task in building sustainable energy services focuses on which of nine primary energy sources are best suited to provide sustainable energy services. Once the sources compatible with sustainable services have been selected, then the technological and social infrastructure needed can be constructed. The big-four energy sources (coal, oil, gas, and uranium) cannot provide sustainable energy services. The five sources that can — if used with proper safeguards — are hydropower, solar, wind, biomass, and geothermal energy.

The conclusion that renewable energy sources can provide sustainable energy services stemmed from developing criteria and standards to assess each of the nine primary energy sources. The results of the analysis divide the criteria into "Traditional Criteria" used in building the Third Energy Transition, and "New Criteria for Sustainability" that assess the ability of each of the nine sources to provide sustainable energy services.

Each of the nine sources is assessed by each criterion with an ordinal, qualitative variable in which +++ means exceptional strength on a criterion and – – – means exceptional and fatal weakness on a criterion. Intermediate values, ranked from strong to weak are ++, +, – , and – –. The details of the analysis are published elsewhere, but the general pattern is clear: the big-four fuels have their greatest strength based on traditional criteria, and the five renewable energy sources are strongest on the criteria for sustainability [7]. Renewable energy has mixed but adequate strength on the traditional criteria, but the big-four fuels have predominantly fatal weaknesses on the criteria for sustainability. It's important to note that energy efficiency scores strongly on both traditional and sustainability criteria.

Sustainability assessment indicates that achieving sustainable energy services requires directing new investments in energy to favour — as much as possible — the Fourth Energy Transition, in which renewable energy and energy efficiency replace the big-four fuels of the Third Transition. Barriers and challenges to this change will arise, but the problems identified with the Third Transition, especially climate change, make the changes imperative for the sake of current and future generations.

Acknowledgement

Perkins was a Fulbright Scholar at Kazan Federal University in Kazan, Russia, during Fall, 2017, and the financial support of the Fulbright Commission, US Department of State, is gratefully acknowledged.

References

[1] Perkins J H 2017 *Changing Energy: The Transition to a Sustainable Future* (Berkeley, University of California Press) p 343
[2] International Energy Agency 2017 *Key World Energy Statistics* p 6
[3] Worldometers, World Population by Year, http://www.worldometers.info/world-population/world-population-by-year/ 29 September 2017
[4] US National Aeronautics and Space Administration, Global Temperature, https://climate.nasa.gov/vital-signs/global-temperature/ 29 September 2017
[5] Intergovernmental Panel on Climate Change 2013 *Summary for Policy Makers* Assessment Report 5 Working Group 1 p 25
[6] US Global Change Research Program 2017 *Climate Science Special Report* June 2017 p 12
[7] Perkins J H 2017 *Changing Energy: The Transition to a Sustainable Future* (Berkeley, University of California Press) p 219 – 268

IOP Conf. Series: Earth and Environmental Science **107** (2017) 012088 doi:10.1088/1755-1315/107/1/012088

Migratory birds are the source of highly toxic organic pollutants for indigenous people in the Russian Arctic

A A Pesiakova, E V Gusakova, A N Trofimova and T Yu Sorokina

Northern (Arctic) Federal University, Severnaya Dvina Emb. 17, Arkhangelsk, 163002, Russia

E-mail: helengusakova@gmail.com

Abstract. Polychlorinated biphenyls are highly toxic organic contaminants. Due to their chemical properties they had wide application in industry and agriculture in the 20th century. In 2001 the production of PCBs has been prohibited almost worldwide. Environmental contamination has been found in soils, water, and air where there were PCB production sites. They have been detected in fish, birds and animals of migratory species, retaining transboarding transfer. Several migratory species of birds (Taiga bean goose, greater white-fronted goose, lesser white fronted goose and barnacle goose) are a diet for indigenous people. PCBs accumulating in the human body affect all systems and organs. This article reviews the contribution of migratory bird species in transboarding transfer of highly toxic contaminants in the Nenets Autonomous Area, Kolguev island (Russian Arctic).

1. Introduction

PCBs (polychlorinated biphenyls) are highly toxic organic contaminants, accumulating in the fat of animals and humans, affecting skin, immune systems, and body organs (liver, respiratory tract). Consequences from PCBs exposure are headaches, depression, tiredness, amnesia, reproduction failure, transformation of microbiota, liver and kidney impairments and cancer [1, 2, 3, 4]. Serious genetic disruptions occur through cytochrome P450 mediated transformation [5].

Polychlorinated biphenyls are chemically stable, low water soluble, non-flammable, electrically insulating [6]. Due to their properties PCBs have a versatile use in industry: electrical capacitors and transformers, rubber/resin production, hydraulic fluids, lubricants, plasticizers, and surface coatings [7].

Severe environmental contamination has been established in areas located in neighborhoods with former PCB production. Major producers were the US, UK, Germany, France, Japan, Italy, USSR and Czechoslovakia. In 2001 the majority of countries in the world signed the Stockholm Convention on Persistent Organic Pollutants, which prohibited the production and utilization of persistent organic pollutants. It is over 10 years since PCB production was stopped, however its accumulation has been found in samples of air, soil, ice, water, sediments, plankton, birds, fish, and mammals [8, 9]. Biphenyls pass through biomagnification in trophic chains and tend to retain transboarding transfer with migratory species (birds, fishes, mammals). For the indigenous people living in the Russian Arctic, migratory animals are a basic part of their diet. For example, in Bugrino (Kolguev island, Nenets Autonomous Area) local people eat approximately 65.3 geese per capita in a year. Increased amounts of PCBs in their blood is a result of their eating habits. Humans, as top trophic members, tend to accumulate highly toxic organic pollutants over the life span. The authors' goal is to analyze risks for indigenous people from PCB transboarding transfer with migratory birds

2. A pathway of PCB transfer to the Russian Arctic

Generally, the diet of the indigenous people from Bugrino village (Kolguev, NAA) consists of migratory birds. Bugrino inhabitants were asked about the most valuable migratory birds species that are caught in the territory and used as a food. Based on the talk with the Department of Natural Resources, Ecology and Agroindustrial Complex of Nenents Autonomous Area, it was found that in the spring and autumn period, the local population catches mainly members of the *Anatidae* family, such species as the Taiga Bean-Goose (*Anser fabalis*), Greater White-fronted Goose (*Anser albifrons*), Lesser White-fronted Goose (*Anser erythropus*), Barnacle Goose (*Branta leucopsis*) as well as the eggs of these species and eggs of the Tundra Swan (*Cygnus columbianus columbianus*) and Long-tailed Duck (*Clangula hyemalis*), all of which could be potential sources of the various PCBs for humans in the Russian north. In avian species, PCB bioaccumulation is related to many factors, such as sex and age of the bird, but the most important factors are diet and residence time in PCB-contaminated areas.

2.1. Migratory pathways of commercially valuable birds

The Taiga Bean-Goose (*Anser fabalis*) and Greater White-fronted goose (*Anser albifrons*) have similarities in many behavior parameters. They are both fully migratory species that have breeding grounds in the high Arctic region. They breed in the spring on areas near lakes, pools, rivers and streams [10]. Their wintering grounds in Europe (mostly in Germany and the Netherlands) [10, 11]. Both of them are herbivorous species. Their diet consists of herbs, grasses, mosses and sedges, and is complemented during the breeding season by berries (e.g. from *Empetrum* or *Vaccinium spp.*). During winter the Taiga Bean-Goose feeds predominantly on agricultural lands taking grain, beans, potatoes and sprouting winter cereal crops. The Greater White-fronted Goose prefers agricultural grains (e.g. corn, oats, wheat, rice and barley, potatoes and sprouting cereals) [10, 11, 12].

The Lesser white-fronted goose (*Anser erythropus*) is also a strongly migratory species; it breeds in the northern Russian region, on the Taimyr Peninsula, to the west and east of the Pechora river and Yamal and Gydan peninsulas. But this species has a different migratory pattern because it winters mainly around the Black and Caspian Seas, in Azerbaijan, Iraq and Uzbekistan [11]. This species is herbivorous, feeding on roots, stems, and leaves and completes its diet with agricultural grains during the winter [12].

The Barnacle goose (*Branta leucopsis*) breeds in the Arctic tundra and winters in the Netherlands and the North of Germany [11, 13]. The diet of this species consists of stems, grasses, sedges, aquatic plants, herbs and shrubs, but in winter it is also completed with agricultural grains and vegetables [10, 11, 12].

Such species as the Tundra swan (*Cygnus columbianus columbianus*) and the Long-tailed Duck (*Clangula hyemalis*) are not included in the indigenous peoples' rations, but the eggs of these species mainly contribute to the diet of the people during the spring. It is known that by laying eggs birds remove some PCBs from their bodies. Therefore, by consuming this product indigenous peoples also can get a large amount of the highly toxic substances. The Tundra Swan and Long-tailed Duck are also migratory species that can transport the PCBs from the contaminated areas to the Arctic region during seasonal migrations. Compared to the previously discussed members from the *Anatidae* family, these two species show a preference for marine foods as well as for plant feedstock. In this case we suggest there is evidence of PCBs bioaccumulation, along with the biomagnification of PCBs via the trophic chain. Due to the consumption of marine and fresh water invertebrates and fish, these birds have a higher chance of obtaining larger amounts of PCBs compared to herbivorous birds.

The Long-tailed Duck (*Clangula hyemalis*) is a species that has a circumpolar range and breeds in the North of Russia. It winters mainly in Germany, Lithuania and Sweden. This species' diet consists

IOP Conf. Series: Earth and Environmental Science **107** (2017) 012088 doi:10.1088/1755-1315/107/1/012088

of animal matter such as crustaceans, molluscs, and some other marine invertebrates and fish, but it also takes freshwater insects and their larvae and some plant materials such as algae and grasses.

The Tundra Swan (*Cygnus columbianus columbianus*) migrates on a narrow front between its Arctic breeding and European wintering grounds. Its wintering occurs mostly in the Netherlands and the United Kingdom [10, 11, 13]. Food preferences of this species include estuarine invertebrates such as molluscs, amphipods and polycheate worms, as well as herbaceous tundra vegetation; but during the winter its diet is completed with agricultural grain and vegetables (e.g. potatoes and sugar beets) [10, 11].

Migratory pathways and PCB contaminated areas are represented on figure 1.

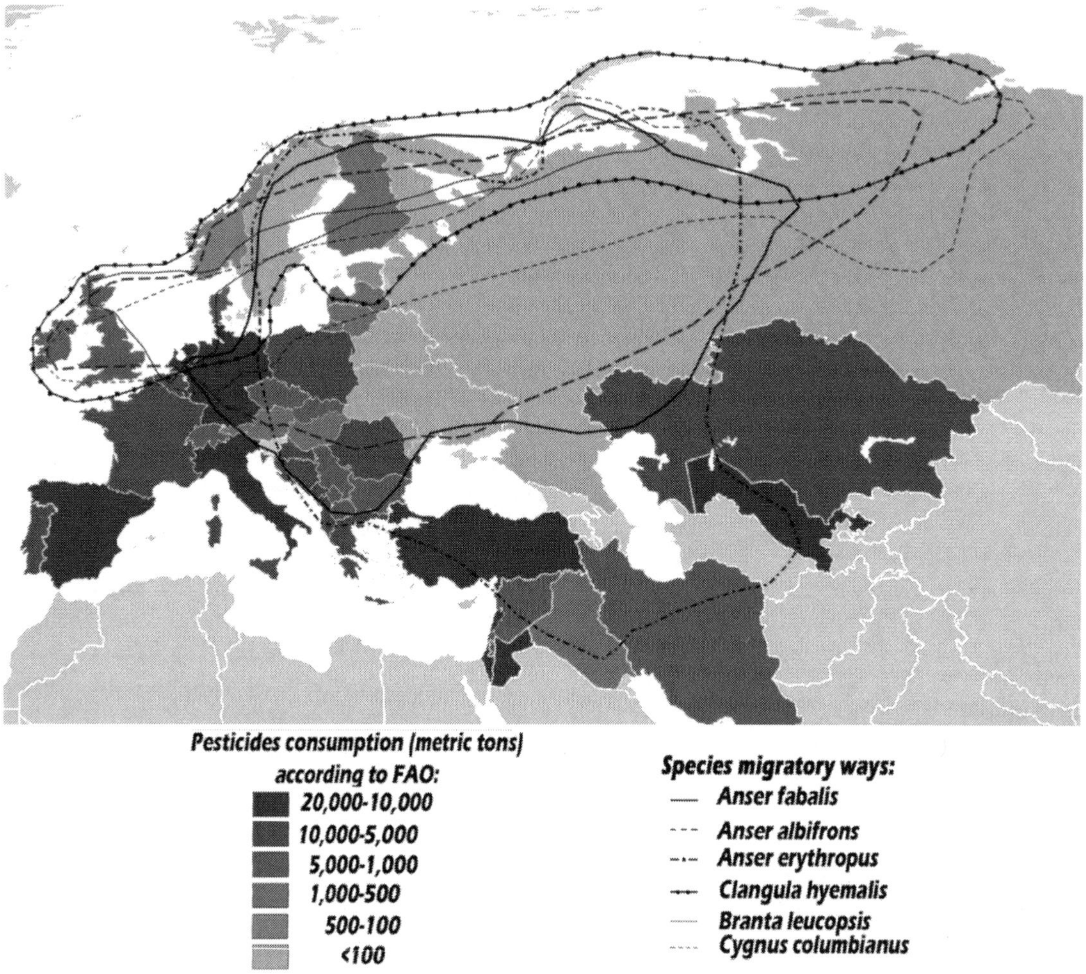

Figure 1. Migratory birds' pathways and pesticides consumption according to the Food and Agriculture Organization of the United Nations.

Based on the study of migratory ways and diet of the bird species that contribute a lot to the diet of the indigenous people of the Russian north we identified some key regions that can be potentially associated with the spread of PCBs.

2.2. *Potential sources of PCBs: agriculture*

Industrial application of PCBs is worldwide, particularly in electrical industry . There is a long history of PCBs use in agriculture. Many of these highly toxic substances were used as pesticides and insecticides in the USA, Europe and the territory of the Soviet Union. The major producers of the PCBs were Monsanto (USA, UK) and Mitsubisi-Monsanto (Japan), Bayer (Germany), Prodelec (France), Caffaro (Italy), Sovol (USSR) and Chemko (Czechoslavakia).

Annually, approximately 2 million tons of pesticides were used worldwide in agriculture, water disinfection and wood preservation until 2001. The chlorine organic chemical heptachlor was used as an insecticide and for etching of corn and sugar beet seeds [14]. Toxaphene was used to control sugar beet and pea pests and to treat cotton and soybean crops. Hexachlorobenzene (HCB) was used as a seed dresser to combat diseases of wheat, rye, buckwheat, soybean and other cereal crops [15]. According to Eurostast data, at the peak of pesticide use (fungicides, herbicides, insecticides and other pesticides) the largest users of these substances in Europe were the Netherlands (9.4 kg/ha), Portugal (5.3 kg/ha), France (4.6 kg/ha),Italy (4 kg/ha), Greece (2.8 kg/ha), Germany (2.5 kg/ha), Jordan (1.4 kg/ha), Czech Republic (1.3 kg/ha) and Iraq (0.1kg/ha).

Despite of the fact that most PCB-containing pesticides were prohibited due to their high toxicity to living organisms and low rate of biodegradation, in some countries the production and use of PCBs (e.g. Dichlorodiphenyltrichloroethane or DDT) still exists (for instance in China, India and Australia and there is a reserve of these substances in many countries) and the contamination of the environment by these substances is still very high. In the Soviet Union even after its prohibition, DDT was used in Uzbekistan as a remedy against the spread of malaria. The soil in Uzbekistan is still full of chlorine and other chemicals from the DDT used in cotton production in the Soviet period, which continues to contaminate farmland [15, 16].

The most dangerous aspect of using pesticides is that only 0.1 to 1% of these substances reach the target area of plants, while 99% of them transfer to the soil, water, air and to agricultural products. The PCBs due to their physical and chemical features are actively accumulating in sediments and biota. They actively accumulate in organisms and multiply their concentration during bird migration via the trophic chain, and as a result, concentration in the final members of the chain increases dramatically. For instance DDT is known to have the following pattern of biomagnification: sediments – 0.014 mg/kg, invertebrates – 0.41 mg/kg, fish – 3 to 6 mg/kg depending on the size, sex, physiological features and specifics of metabolism of different species and their food preferences, lipids of marine fish-eating birds – up to 200 mg/kg [15, 16].

3. Conclusions

Taking into accountmigration routes and diet preferences of the commercial bird species of the NAA, we can suggest that it is a possible way of PCB transport from PCB-contaminated areas of Europe and East Asia. Most of the studied species are herbivorous and winter in the territory of the Netherlands, Germany and Uzbekistan, predominantly on agricultural land (pastures, arable fields, rice-paddies, damp steppe grasslands). Previously these countries actively produced and used various chlorine organic pesticides in agriculture and due to the PCB's slow degradation rate and its ability to be preserved in the environment, these areas still can be a potential source of persistent organic pollution. Biomonitoring analysis will help assess and minimize risks for the population from pollution. We suggest developing a National biomonitoring system of indigenous people exposed to highly toxic organic pollutants, conducted by official authorities in Russia.

Acknowledgements

The work was supported by the Contract from 14th of March 2017 No. 14.Y26.31.0009 "About the Governmental Grant of Russian Federation for state support of scientific research, conducted under the guidance of leading scientists, in Russian educational organizations of higher education, scientific institutions, coordinated by Federal Agency of Scientific Organizations, and State Scientific Centers of Russian Federation, in terms of the subprogram "Institutional development of the research sector" from the state program of Russian Federation "Development of Science and Technology" for 2013-2020.

References

[1] *Environment Canada. Meeting Background Report* 1985 vol IV Prepared for International Experts Meeting on Persistent Organic Pollutants Towards Global Action, Vancouver, Canada, June
[2] Agency for Toxic Substances and Disease Registry (ATSDR) 1997 *"ToxFAQs: Polychlorinated Biphenyls"* URL: http://atsdr1.atsdr.cdc.gov:8080/tfacts17.html, Atlanta, Gerogia, USA,
[3] International Agency for Research on Cancer (IARC). *IARC Monographs*, Supplement 7. URL: http://193.51.164.11/htdocs/Monographs/Suppl7/PolychlorinatedBiphenyls.html, Lyon, France, 1987
[4] *Memorandum of State Committee for Environmental Protection of RF N165*, 13th April, 1999.
[5] Mills R A, Mills C D, Dannan G A, Guengerich F P and Aust S D 1985 *Toxicol. Appi. Pharmat* **78** 96–104
[6] Tanabe S 1988 *Environ. Po. Nut.* **50** 5–28
[7] *Guidelines for Identification of PCBs and Materials Containing PCBs* United Nations Environment Programme. UNEP First Issue Geneva, Switzerland, August 1999
[8] Waid J S 1987 *PCBs and the Environment* vol Z-III (FL:CRC Press, Boca Raton)
[9] Grimm F, Hu D, Kania-Korwel I, Lehmler H, Ludewig G, Hornbuckle K and Robertson L. 2015 *Critical Reviews in Toxicology* **45** 245–272
[10] del Hoyo J, Elliot A and Sargatal J 1992 *Handbook of the Birds of the World* 1: Ostrich to Ducks (Spain: Lynx Edicions)
[11] Kear J 2005 *Ducks, geese and swans 1: general chapters; species accounts (Anhima to Salvadorina)* (Oxford: Oxford University Press)
[12] Johnsgard P A 1978 *Ducks, geese and swans of the World* (Lincoln: University of Nebraska Press)
[13] Madge S and Burn H 1988 *Wildfowl* (London: Christopher Helm)
[14] Borlakoglu J T and Haegele K D 1991 *Comparative Biochemistry and Physiology Part C: Comparative Pharmacology* **100**(3) 327–338
[15] Moiseenko T I 2009 *Aquatic ecotoxicology: fundamental and applied aspects* (Moscow: Science) p 400
[16] Barron M G, Galbraith H and Beltman D 1995 *Comparative Biochemistry and Physiology Part C: Pharmacology, Toxicology and Endocrinology* **112**(1) 1–14

Morphological and ecological preadaptations as the basis of bird synanthropization under transformed environment conditions

I I Rakhimov and K K Ibragimova

Institute of Fundamental Medicine and Biology, Kazan Federal University, Kremlevskaya str., 18, 420008, Russia

E-mail: rakhim56@mail.ru

Abstract. Bird synanthropization is connected with a thorough and serious reconstruction of their biology and is a demonstration of changes currently occurring in the biosphere due to human influence. Nutritional and nesting conditions as well as protection due to urban characteristics are advantage factors that affect their populations. Under these conditions, the adaptive potential of species can be realized. Adaptations to a new and in-distinctive environment appear due to preadaptations. The synanthropization process of species happens without speciation by expression of existing genetic variation of morphological and ecological characteristics.

Preadaptation, as A.Georgievsky notes [1], is an independent evolutionary phenomenon that involves readiness to express neutral or even relatively destructive characteristics; it is not a result, but it is one of the ways to develop new phylogenetic adaptations. Most of the statements about value of preadaptations are considered from the perspective of influence on species formation in micro- and macro-evolution processes. Due to the dynamics of urban conditions and their constant variability, the processes occurring in urbanized ecosystems and in populations of synanthropic birds do not allow considering synanthropization as speciation and the formation of exclusively urban species. But the synanthropization process itself is undoubtedly based on morphological, ecological, and behavioral preadaptations of individual species.

Bird synanthropization is connected with multiple features and seriously restructures the ecology of birds found in areas affected by anthropogenic factors, especially urbanized environments. At the same time, only a small portion of bird species become synanthropic, not all of them. Thus, for most cities of Tatarstan, the Middle Volga region and Russia in general, synanthropic birds are the rock pigeon (*Columba livia*), the black swift (*Apus apus*), the jackdaw (*Corvus monedula*), the house martin (*Delichon urbica*), the house sparrow (*Passer domesticus*) and the tree sparrow (*Passer montanus*). In some regions, three or four more species inhabit urban ecosystems and form synanthropic populations. Ecological characteristics of these species are associated with transformed territories, and their existence is mainly dependent on human influences. Nutritional and nesting conditions as well as protection by certain urban characteristics are some of the advantage factors affecting their populations. Under these conditions, adaptive potentials of species can be expressed [2].

Numerous researchers of birds in cities state that adaptations to unusual environments and realization of adaptive potential occur on the basis of preadaptations and synanthropization processes .

Preadaptations are widespread in nature and appear to be one of the most important ecological mechanisms of evolutionary processes [3]. Living organisms develop features which, under the control of natural selection, lead to formation of qualitatively new adaptations. Properties that have an adaptive value for a previously unrealized interaction between an organism and the environment or for functions not yet acquired by an organism are called preadaptive; the process of expressing a preadaptive trait is called preadaptation [4]. Antropogenic environments present special demands for bird species inhabiting them and always depend on human activity; it is a habitat for species and populations whose reaction to changing conditions are consistent with potential opportunities of their biology. This is especially evident for the synanthropic birds of the cities of Western and Central Europe.

It is not entirely correct to say that bird synanthropization in West and Central Europe began earlier than in Eastern Europe. There are numerous known historical studies about of jackdaws, rooks, sparrows inhabiting cities even in the early historical periods of the formation of states and the construction of urban settlements in all Eastern Europe. But in the past 100–150 years, birds have penetrated and begun living in the cities of Western Europe.

The bird populations of Western and Eastern Europe have similar species but appear to be in different stages of synanthropization. The European jay (*Garrulus glandarius*), the ring dove (*Columba palumbus*), the European partridge (*Pedrix pedrix*) and many other representatives of western avifauna differ markedly in their behavior in anthropogenically transformed territories compared to populations of the same species from Eastern Europe. This indicates many potential opportunities of birds to exhibit adaptations in the process of their synanthropization. As noticed by S. S. Schwarz [3], a certain type of adaptation arises when the adaptations promote prosperity of a species when they are useful but not necessary. Development of such an adaptation enables a species to enter an environment where the adaptation is an essential condition for existence. For these reasons the synanthropization of eurybionts, with wider adaptive opportunities, occurs more actively. It is highly important for birds to be eurythropic and habitat versatile. Feeding with various types of food promotes colonization of anthropogenic environments by birds which are polyphages; they are attracted into the cities by available food resources. The main places of synanthropic birds concentration are dumps, food waste collection places, and agricultural ecosystems [5].

Before mastering the urbanized environment, birds were meant to have a minimum of adaptive features to the later appeared urban environment. Most of the typical synanthropic birds are characterized by wide potential in feeding and nesting, in other words they are preadapted to mastering the urbanized environment. Morphological preadaptations include: features of the structure of limbs and beaks, body size, and flying abilities. For birds, the structure of beak is an essential peculiarity for feeding. It is clearly demonstrated by a well-known example with Galapagos finches [6]. Such a specialization limits ability of a species to thrive on various types of food. Under anthropogenic environmental conditions, nibs specialized in taking food particles from the ground or other solid surfaces are better adapted. For example, crows, thrushes, starlings, pigeons, sparrows share this trait in common. For comparison, the incurved beaks of birds of prey or long beaks of sandpipers and herons are not such a universal hunting tool. Long legs, necks, and tail-feathers acting as rudders are not so useful for living in urban landscapes. The size of a bird also limits penetration into anthropogenic landscape. However, it should be noted that morphological preadaptations can limit, but do not exclude, the appearance of some bird species, even in urbanized areas. The variety of conditions found near human activity might be acceptable in exceptional cases for highly specialized species, especially in response to unstable conditions. Populations of big birds of prey (kites and vultures) in the cities of India are well known [7].

A group of Swedish scientists came to an unusual conclusion; they studied more than 82 bird species in cities and outside them. Having surveyed more than 12 cities, they tried to figure out why some bird species are more adapted to live in cities. It turned out that urban birds have a brain which is larger in relation to the body than the brain of their conspecifics from outside the city. Studies have confirmed that there is a close relationship between behavioral innovations and enlarged brain size in mammals and birds. The results strongly suggest that brain size is one of the critical factors contributing to the survival of animals in the city.

Ecological preadaptations involve ways of hunting and searching for food, nesting level, feeding level, aggregating behavior, migrating, and others. Behavioral preadaptations are associated with mutual tolerance of a species and humans, potential enemies, and birds from different ecological groups.

The method of obtaining food, in other words feeding behavior, undoubtedly plays an important role in the preadaptive capabilities of species. As noted, picking from solid surfaces is the most successful option for synanthropization of bird. But many observations indicate a wide range of ways of obtaining food. So, urban dumps are actively used by storks, marabous, and various gulls. During the plowing of the land are followed not only by rooks and jackdaws but also by sea gulls. A unique case was described in St. Petersburg. Seagulls were snatching pieces of bread in the air from woman on the balcony of a 5-story building. How can one explain such an uncharacteristic, at first sight, behavior of seagulls? The potential of preadaptive opportunities is big enough and, when needed, this preadaptation can be manifested. In a natural environment, most seagulls collect food on shore while ability to catch a food directly in flight can be observed when they take food from other birds [5].

During processes of urbanization, birds enter an urban territory and occupy specific ecological niches within a city. Often they are analogues of their natural biotopes, or niches that are completely unknown in natural ecosystems (urban dumps). As a biotic category, an ecological niche is characterized as an assembly of factors which support organisms' survival and their functional place in an urbancenosis. The variety of ecological niches determines the zoocenosis- stability of a city. The mosaic structure of urban biotopes creates extra opportunities for inhabiting a city by different bird species, and it also creates ecological niches for different groups of birds which are occupied by species belonging to different systematic groups. The penetration of a new environment happens by using already existing habitats with a set of conditions acceptable for a species' survival. Thus, inhabiting anthropogenic landscapes depends on biotopes as the analogues of natural habitats. According to the opinions of several authors, birds inhabit micro-environments of cities that resemble their native niches [8, 9]. Conservation of natural habitats within the city accelerates entrance of many species into anthropogenic landscapes. They serve as buffer and transitional zones for most of potential synanthropes. Thus, the currently observed, wide-spread range of mallard duck (*Anas platyrhynchos*) is due to preadaptive capabilities of these species and lack of chasing by people.

The most important stage of bird species synanthropization is nesting in anthropogenic landscapes. Nesting in anthropogenic landscape is an indicator of the favorable conditions for species within parts of anthropogenic landscape. There is no species that could begin to reproduce in anthropogenic landscape before it acquired a complex of trophic connections. Reproduction in a transformed territory has its own special aspects, because it occurs with changes of some standard nesting characteristics [10]. The changes in reproductive biology are shown in the elongation of a reproductive period, in the change in the number of clutches, and increased fertility. Nesting techniques and some behavioral aspects during the reproductive period are subject to change. In anthropogenic landscapes, birds exhibit unusual nest-building, uncommon location, the use of anthropogenic materials to build a nest, and other features. Due to environmental change in urban areas prolongs feed availability, suitable places for nesting, and the reproductive period of some species, sometimes by large amounts. This, however, is not a complete list of preadaptive features of birds which are on their way to synantropization [11].

The most favorable nesting and natural conditions of birds belonging to the rupicolous complex are close to that of urban environment. This complex is wide spread and mostly specific for cities (rock pigeons, swallows and swifts). The historical process of city colonization by these species has existed for the longest in time, and it is represented in all adaptations of a given group through readiness to nest. Birds of the rupicolous complex easily find suitable places for nesting under urbanized conditions. The range of species of urban birds associated with the tree and shrub layer is less defined. For the Middle Volga cities, they are the chaffinch, the goldfinch, the linnet, the great tit, the collared turtledove and other species, which nest in tree plantations. They actively populate park and garden biotopes and invade available ecological niches, even disturbed ones, which are close to their natural analogues. In cities of Central and Western Europe, the niches of birds involving tree and shrub layers are inhabited by the blackbird, the ring dove, the collared turtle dove, the common jay and other species that are not characteristic of urbanized territories in the east of the European part of Russia and that rarely fly into cities. Thus, different cities of various geographical regions have similar ecological niches which are occupied in each specific case with a different set of species. Semi-aquatic birds need special conditions, however. For example the mallard duck, the seagull, the blue-throated robin and several other species often nest in urban areas with lakes, rivers, or other water bodies. Terrestrial birds that inhabit open biotopes have fewer advantages in urbanized ecosystems. The characteristics of urban localities and urban planning practices exclude large spaces of waste grounds and undeveloped lands. The ecological niches of open biotopes remain unoccupied due to strong anthropogenic pressure.

In the urban environment, over time one can expect a special life form (synanthropic birds) with a complex of new morphological and functional and behavioral adaptations to emerge. The ecological niches of each synantropic species will be represented by its functional role in the community, for example, by its trophic status. D. V. Vladishevsky [8] proposes to introduce a term "trophic convergence" the main idea of which is feeding of several bird species on similar food. In natural conditions, the species after the nesting period that abounds in food (mainly various invertebrates) switch to feed on specific types of food like berries, seeds and others. Meanwhile, in anthropogenic ecosystems single-type food is present all year round. Several authors [12] confirm the emergence of typical feeding behavior, including longer duration of the feeding period, feeding rhythm, and other traits. At the same time, ecological niches of distantly related taxa can strongly converge, for example, dove and jackdaw feeding on dumps; they are of different systematic groups, but their trophic niche is the same.

As a result, we agree with the statement that the formation of urban populations occurs through non-hereditary behavior changes and the adaptations take place not at the level of individuals but at the level of population structure. However, in the long term, the pressure of natural selection under new environmental conditions must lead to evolutionary changes which are inheritable. But there are no facts confirming this statement yet.

References

[1] George A B 1974 *Problem preadaptation* (Leningrad:Nauka) p 148
[2] Marzluff J, Bowman R and Donnelly R 2001 *Avian conservation and ecology in an urbanizing world* (Boston: Kluwer Acad. Publ.) p 1–17
[3] Shvarts S S 1980 *Ecological regularities of evolution* (Moscow:Nauka) p 278
[4] Cuenot L 1911 *La genese des especes animals* (Paris) p 1932
[5] Rakhimov I I 2010 *Birds of anthropogenous landscapes* (Saarbrucke, Germany) p 274
[6] Farrington H L, Lawson L P, Clark C M and Petren K 2014 *Evolution* **68** 2932–44
[7] Galushin V M 2009 *Russian Ornithological Journal* **18** (505) 1434-45
[8] Vladyshevsky D V 1975 *The birds in the anthropogenic landscape* (Novosibirsk: Nauka) p 198
[9] Jokimäki J and Huhta E 2000 *Condor* **102** 838–847
[10] Geyer H and Kontuly T 1993 *International Regional Science Review* **15** (3) 157–177
[11] Luniak M 2004 Proc. 4th Int. Urban Wildlife Symp. p 50–55
[12] Voronov L N 1999 *Bull. Chuvash state ped. Univ.* **7** p 113–116

Chorological classification approach for species and ecosystem conservation practice

T V Rogova, M V Kozevnikova, V E Prokhorov and N O Timofeeva

Institute of Environmental Sciences, Kazan Federal University, 18 Kremlyovskaya str, Kazan 420008, Russia

E-mail: tatiana.rogova@kpfu.ru

Abstract. The habitat type allocation approach based on the EUNIS Habitat Classification and the JUICE version 7 software is used for the conservation of species and ecosystem biodiversity. Using the vegetation plots of the Vegetation Database of Tatarstan, included in the EVA (European Vegetation Archive) and GIVD (Global Index of Vegetation-plots Databases) types of habitats of dry meadows and steppes are distinguished by differing compositions of the leading families composing their flora – *Asteraceae*, *Fabaceae*, *Poaceae* and *Rosaceae*. E12a – Semi-dry perennial calcareous grassland, and E12b - Perennial calcareous grassland and basic steppes were identified. The selected group of relevés that do not correspond to any of the EUNIS types can be considered specific for ecotone forest-steppe landscapes of the southeast of the Republic of Tatarstan. In all types of studied habitats, rare and protected plant species are noted, most of which are South-East-European-Asian species.

1. Introduction

The provisions of the Sustainable Development Concept and the role of biological diversity in providing ecosystem services require new methodological approaches in solving modern problems of biodiversity conservation. In the context of intensive use of landscapes, the successful conservation of species and ecosystems depends not only on knowledge of the distribution of rare and endangered species of plants, animals and fungi in certain areas, but also understanding the relationship of biological species to the ecological conditions of their habitats [1]. Traditionally used species conservation through publication and management of Red Data Books for specific territories is already successful in protecting rare and endangered species. It is much more difficult to identify and ensure the preservation of the entire cenotic and ecosystem diversity. Fragmentation of vegetation cover and successive disruption of communities by land use make it difficult to assess the extent of these changes and preserve natural features [2].

The presence of modern electronic databases of relevé, accumulating information on both rare species and all species forming the community, greatly facilitate preserving cenotic diversity. The inclusion of such databases in the structure of regional and global Databases and in GIS layers, facilitates conservation of ecosystem diversity [3, 4]. Different classification approaches used to identify plant communities and build classification systems helps solve the problem. However, even the ecological approach using the Braun-Blanquet method for floristic classification of vegetation, which is widely used in world practice, does not give a clear picture of the association of individual syntaxonomic units of the system, especially of high ranks. In addition, such ecological approaches do not capture ecotopological conditions, and accordingly they do not facilitate the practical task of preserving the biological diversity of a particular landscape [5]. On the other hand, there are methods of landscape-ecological analysis and identification of territorial units, based primarily on knowledge of the abiotic environment and its influence on the biotic communities [6]. The landscape approach is most convenient for the purposes of ecological mapping and land use optimization based on it, but it does not imply any concern about conservation of the entire biological diversity of the landscape and, especially, its structural units of low rank. In addition, the species of plants, animals and fungi "read" the ecological conditions of landscape in their own way and scale.

In the practice of biodiversity conservation in recent decades, a new direction has successfully developed in European countries, based on the classification of habitat types, both natural and anthropogenic [7, 8, 9]. For the territory of Europe, EUNIS (European Nature Information System) Habitat Classification has been developed, which is successfully applied now not only in EU countries, but also in the Russian regions participating in the European Emerald Network project (Emerald Book of Russian Federation) [10].

2. Material and methods

For the Republic of Tatarstan, an attempt was made to identify habitat types of vascular plant species, based on the EUNIS Habitat Classification approach using the vegetation plots of the Vegetation Database of Tatarstan, included in the EVA (European Vegetation Archive) and GIVD (Global Index of Vegetation-plots Databases, ID: EU-RU-011). Relevés of herbaceous dry grassland communities were noted, mainly in the forest-steppe landscapes of Zavolzhje, Republic of Tatarstan in the southeast of the Russian Plain [11]. Selection of plots was based on the presence of indicator species, determined expertly. The sample, including 92 plots, was subjected to the clusterization procedure by the similarity of species composition. Using the JUICE software version 7.0 [12] the selected relevés were compared and assessed with the types of the EUNIS Habitat Classification and syntaxons of the Braun-Blanquet floristic classification. The floristic structure of the clusters was analyzed based on the leading families rate, life forms rate, and a list of rare and protected plant species.

3. Result and discussions

From the sample of 92 plots, we identified 7 clusters. According to EUNIS Habitat Classification, the most significant relevés correspond to the type of ecotope *E1 – Dry grassland.* Among them most represented habitat was *E12 – Perennial calcareous grassland and basic steppes* (29 plots): *E12a – Semi-dry perennial calcareous grassland* (6 plots) and *E12b – Perennial calcareous grassland and basic steppes* (23 plots). From the group of Dry grassland habitats, one relevé is assigned to type *E11f – Continental-dry-rocky-steppic-grassland-and-dwarf-scrub-on-chalk-outcrops-specialists* and one relevé to type *F31f – Low steppic scrub.* It's very important to note that 56 plots were not assigned to any type of habitats, and a few were assigned to *X06 – Artemisietea vulgaris.* According to the classification of Braun-Blanquet, most of the relevés refer to the class *Festuco-Brometea.*

The dependence of these indices on the specificity of the ecotope can be traced to the structure of the species composition of communities in different habitats, which in turn is based on the range of leading families and by proportions of life forms [13]. In general, for the landscapes of the eastern regions of the European part of Russia (the Republic of Tatarstan, the Ulyanovsk region, the Volga-Ural region), the flora spectrum is characterized by the predominance of the first triad of families - *Asteraceae, Poaceae, Fabaceae* (table 1). For the more western regions (for example, the Moscow region) flora spectrum is characterized *Asteraceae, Poaceae, Rosaceae.* The high position of the *Rosaceae* family testifies to the similarity of the flora composition with the western floras of forest landscapes; the high position of the *Fabaceae* family characterizes the species composition of the more southeastern areas. Based on these general patterns, it can be argued that the habitat *E12b– Perennial calcareous grassland and basic steppes* exhibit features of a more xerophilous southeastern flora, in which the *Fabaceae* occupy the second position, and the *Rosaceae* move to the 4th place. Habitat *E12a - Semi-dry perennial calcareous grassland* reflect a more mesophytic character of species composition, in which *Rosaceae* go to second place and *Fabaceae* to the fourth. The flora composition of a large group of plots (56), not identified by JUICE classification to any of the types of habitats, is more consistent with western forest-steppe landscapes, with a more xerophytic character of the flora in comparison with the western floras, as evidenced by the position of *Rosaceae* in the first triad of the family spectrum and *Fabaceae* in the third position following the *Rosaceae*. A fairly high participation of the phanerophytes (up to 10%) and chamaephytes (up to 11%) in their flora also indicate the forest-steppe features of these vegetation plots and the habitat as a whole. In habitat *E12b – Perennial calcareous grassland and basic steppes*, the share of phanerophytes is not more than 1%, with a large participation of chamaephytes, which indicates more dry habitat conditions.

Table 1. Floristic composition (first five families) of studied habitats and regions of the European part of Russia.

Tatarstan	Tatarstan *E12a*	Tatarstan *E12b*	Tatarstan *Not identified plots*	Moscow region	East of European part of Russia	Ulyanovskiy region	Volga-Ural region
Asteraceae	Asteraceae	Asteraceae	Asteraceae	Asteraceae	Asteraceae	Asteraceae	Asteraceae
Poaceae	Rosaceae	Fabaceae	Rosaceae	Poaceae	Fabaceae	Poaceae	Poaceae
Fabaceae	Poaceae	Poaceae	Fabaceae	Rosaceae	Poaceae	Fabaceae	Fabaceae
Rosaceae	Fabaceae	Rosaceae	Poaceae	Cyperaceae	Lamiaceae	Rosaceae	Brassicaceae
Cyperaceae	Brassicaceae	Brassicaceae	Lamiaceae	Brassicac.	Rosaceae	Brassicaceae	Cyperaceae

Table 2. The proportion of Eastern European, East-European-Asian and endemic species of flora (%)

Distribution area	Not identified plots	E12a	E12b
Eastern European	2.1	1.8	1.7
Eastern-European-Asian	17.8	14.2	16.3
Endemic of the Volga-Don interfluve and Zavolzhje	0.5	0	0.4
Endemic of Volga-Ural Region	0.2	0	0
Endemic of the Middle Volga Region	0.5	0	0.8
Endemic of the Southern Urals and Zavolzhje	0.7	0	0.8

E12a – Semi-dry perennial calcareous grassland
E12b – Perennial calcareous grassland and basic steppes

The presence of relevés not attributed to any of the habitat types of the EUNIS Habitat Classification is most likely due to the vegetation cover ecotone in the conditions of the forest-steppe south-east landscapes of the Republic of Tatarstan, as well as the presence of plant species of the Eastern European and East-European-Asian ranges. Rare and protected plant species included in the Red Book of the Republic of Tatarstan [14] are identified in all types of studied habitats. The largest number of them was recorded in the habitat *E12b*. It is important to note that in the relevés of unidentified group of habitats there are also rare species of dry habitats, many of which have East European-Asian range and endemics (table 2). Among them are: *Astragalus henningii* (Stev.) Klok. (Endemic of the Volga-Don interfluve and Zavolzhje). *Astragalus zingeri* Korsh., *Asperula exasperata* V. Krecz. ex Klok. (Endemic of the Middle Volga Region). *Oxytropis hippolyti* Boriss., *Oxytropis spicata* (Pall.) O. & B. Fedtsch. (Endemic of the Southern Urals and Zavolzhje).

4. Conclusions

It is important to mention that ecological habitats in the south-eastern part of Republic of Tatarstan have distinct features because of the presence of eastern species; it is evident that more detailed investigations have to be done for new types habitat identification and determination of characteristic species for them. The questions of determining the flora composition of syntaxa of low ranks, unions, orders and associations remain open.

References

[1] Dengler J, Boch S, Filibeck G, Charucci A, Dembicz I, Guarino R, Henneberg B, Janisova M, Marceno C, Naginezhad A, Polchaninova N Y, Vassiliev K and Biurrun I 2016 *Bulletin of the Eurasian Dry Grassland Group* **32** 13–30

[2] Rogova T V, Saveliev A A and Mukharamova S S 2005 *Botanicheskiy Zhurnal* **90**(3) 450–460

[3] Schaminee J H J, Janssen J A M, Hennekens S M and Ozinga W A 2011 *Plant Biosystems* **145** 85–90

[4] Dengler J, Oldeland J, Jansen F, Chytry M, Ewald J, Finckh M, Glockler F, Lopez-Gonzalez G, Peet R K and Schaminee J H J 2012 *Biodyversity & Ecology* **4** 1–447

[5] Bruelheide H and Chytry M 2000 *Journal of Vegetation Science* **11** 295–306

[6] Mucher C A, Klijn J A, Wascher D M, Schaminee J H J 2010 *Ecological Indicators* **10**(1) 87–103

[7] Choimaa D, Welk E, Jager E J and Hauck M 2005 *Flora* **200** 376–397

[8] Angelini P, Bianco P, Cardillo A, Francescato C and Oriolo G 2009 *Schede descrittive degli habitat per la cartografia alla scala 1:50 000* (Roma) p 334

[9] Schaminee J H J, Chytry M, Hennekens S M, Mucina L, Rodwell J S, Tichy L 2012 *Development of vegetation syntaxa crosswalks to EUNIS habitata classification and related data sets. Final report EEA/NSV/12/001*

[10] Emerald Book of Russian Federation. The territory of special conservation value of European Russia. Proposals for the identification P.1. 2011-2013. (Moscow) p 1–308

[11] Prokhorov V, Rogova T and Kozhevnikova M 2017 *Phytocoenologia* **47**(3) 309–313

[12] Tichý L 2002 *Journal of Vegetation Science* **13** 451–453

[13] Bakin O V, Rogova T V and Sitnikov A P 2000 *Vascular Plants of Tatarstan* (Kazan: Izd. Kazan. Univ.) p 496

[14] *Red Book of Republic of Tatarstan (animals, plants, fungi), 3nd ed* 2016 ed. Nazirov A A (Kazan) p 760

IOP Conf. Series: Earth and Environmental Science **107** (2017) 012091 doi:10.1088/1755-1315/107/1/012091

Functional approach in estimation of cultural ecosystem services of recreational areas

I S Sautkin and T V Rogova

Institute of Environmental Sciences, Kazan Federal University, Kremlevskaya str.,18, 420008, Russia

E-mail: sautkin.ilia@gmail.com

Abstract. The article is devoted to the identification and analysis of cultural ecosystem services of recreational areas from the different forest plant functional groups in the suburbs of Kazan. The study explored two cultural ecosystem services supplied by forest plants by linking these services to different plant functional traits. Information on the functional traits of 76 plants occurring in the forest ecosystems of the investigated area was collected from reference books on the biological characteristics of plant species. Analysis of these species and traits with the Ward clustering method yielded four functional groups with different potentials for delivering ecosystem services. The results show that the contribution of species diversity to services can be characterized through the functional traits of plants. This proves that there is a stable relationship between biodiversity and the quality and quantity of ecosystem services. The proposed method can be extended to other types of services (regulating and supporting). The analysis can be used in the socio-economic assessment of natural ecosystems for recreation and other uses.

1. Introduction

The concept of ecosystem services (ES) accepted under the UN program, the Millennium Ecosystem Assessment (MEA) [1] has become a formal approach for describing and classifying possible interactions between ecosystems and society [2]. The concept of ES has found wide application at the international level in ecology, economics and politics [3, 4, 5]. Ecosystem services are defined as conditions and processes that have value for people. MEA has defined cultural ES as "the intangible benefits that people derive from ecosystems through spiritual enrichment, cognitive development, reflection, recreation and aesthetic experience, including, for example, knowledge systems, social relations and aesthetic values" [1], [6]. In most studies on the assessment of ES, environmental processes that are not directly related to human well-being are considered [7]. In addition, in research, cultural services are regularly referred to as a category of ES and are therefore recognized as important, but in part because of the focus on economic evaluation, many cultural ES are subsequently ignored [8]. There are several studies on ecosystem-based recreation [9] and the attractiveness of landscapes [10, 11], but other cultural services are usually not characterized.

Modern research, focusing on ecosystem services as an instrument for assessing ecosystems and promoting their sustainable use [1], also draws attention to the ways in which different organisms contribute to the delivery of ecosystem services [12, 13, 14, 15]. The ability of the ecosystem to provide several services involves distribution of the relevant ecosystem properties to determine which organisms or groups of organisms control these properties [12, 16]. Then, it is necessary to identify the key characteristics and mechanisms by which these organisms influence ecosystem properties [17, 18, 19]. This method has been used in a growing number of studies, e.g. [20, 21, 22]. In addition, studies have used the identification of functional traits as a key mechanism by which individual species [15] and species groups [23, 24], affect ecosystem properties. Thus, there is growing recognition that the abundance of functional traits in biotic communities can greatly contribute to the assessment and management of ecosystem services [25, 13].

The objective of this work was an attempt to identify cultural ES and functional traits of plants that supply these services. A technique was tested that allows (a) linking individual plant characteristics and services, (b) their mutual relations, and (c) the contribution of each species to the supply of cultural ES. The study was conducted in pine forests on the territory of the forest park "Lebyazhye" in suburbs of Kazan.

2. Material and methods

Forest park "Lebyazhye" occupies an area of 3729 hectares in the Kirov district of Kazan. About 16% of the area is occupied by buildings, roads, garbage, and wasteland; approximately 70% of the total area is forest. The forest park has an important recreational significance for the population of Kazan and surrounding villages. There are children's recreation camps, health and holiday facilities, and areas for sports competitions. Good transport links cause high traffic volumes to this forest most of the year.

In the forest park, 399 species of higher vascular plants are currently found. Native types of forest phytocoenoses account for no more than 6% of the forest area. Among them are associations of the classes *Vaccinio-Piceetea* Br.-Bl. in Br.-Bl., Siss. Et Vlieger 1939 and *Carpino-Fagetea sylvaticae* Jakucs ex Passarge 1968. Dominants in the forest ground cover of *Pinus sylvestris*-derived communities include *Pteridium aquilinum, Convallaria majalis, Calamagrostis epigeous, Agrostis tenuis, Poa angustifolia*, and other graminoids and ruderal plant species. In broad-leaved forests, *Betula pendula* and *Populus tremula* dominate, and graminoids and ruderal plant species dominate in the ground cover. Small fragments of area are represented by meadows and coastal vegetation around the lakes.

The attractiveness of forests for recreation is largely determined by their aesthetic properties, which depend on various characteristics of the stand, including its composition, because wood species are characterized by different decorative qualities. The composition of the stand influences the stability of forest ecosystems devoted to recreational uses, because woody species differ in their susceptibilities to the effects of recreation. The most attractive are pine forests with a share of other species of 30-50%. Variable tree distributions and low stand densities of pine forests, clustered in groups with a rare juvenile and / or undergrowth, significantly increase the attractiveness of forests for recreation [26].

Nine test plots of 20×20 m^2 were established on four quarters (83, 82, 101, and 102) of the "Lebyazhye" forest park. For each plot a geobotanical description was compiled. The plot laid in quarter 83 belonged to the *Pinus sylvestris* forest at the 2nd and 3rd stages of recreational uses with the dominance of *Convallaria majalis* in the ground cover. Mosses were found only near tree trunks (5-10%). Grass stands cover 80-90%, of which 10-20% are meadow grasses. The plots on 82, 101, and 102 belonged to *Pinus sylvestris* forests at the 3nd and 4rd stages of the recreational use with the dominance of grasses and ruderal herbs. All plots were outside the road network and trampled and dead-cover sites. A total of 76 plant species were identified at all sites.

Based on the methods of Garcia-Llorente [27], the functional features of plants and socio-cultural ES of interest were identified. Information for the search and identification of traits and socio-cultural services was from [28, 29, 30, 31, 32].

For the analysis, the following functional traits were selected:

- Life form: herbs, shrubs, dwarf shrubs and trees;
- Plant height;
- Attractiveness of flowers, fruits and leaves: For fruit, attractiveness is determined by bright color; for flowers, by size, bright color, and pleasant fragrance; and for leaves by size, architecture, presence of pubescence, and changed colors in autumn;

- Early phenology: active growth and flowering until June;
- Lifespan: annual / biennial and perennial;
- Not poisonous.

The classification of functional plant groups was carried out by cluster analysis. Data on 76 described species and 11 selected traits were used. The cluster analysis was carried out according to Ward's algorithm, and the Euclidean distance was used as a metric of similarity [33].

Based on the classification of MEA services, two types of cultural services were selected for research: aesthetic and recreational services. For each type of service, an influential set of functional plant traits was determined.

For aesthetic services, the following traits were selected: life form (herbs, shrubs and dwarf shrubs); plant height; attractiveness of flowers, fruits and leaves; early phenology; and lifespan.

For recreational services, the following traits were selected: life form (trees and herbs); plant height; attractiveness of leaves; lifespan, and not poisonous.

The classification of functional groups and data on the sets of characteristics affecting the selected types of services were used to assess the contributions of each species to the provision of services.

3. Result and discussions

The results of cluster analysis showed the presence of 4 functional groups (figures 1 and 2).

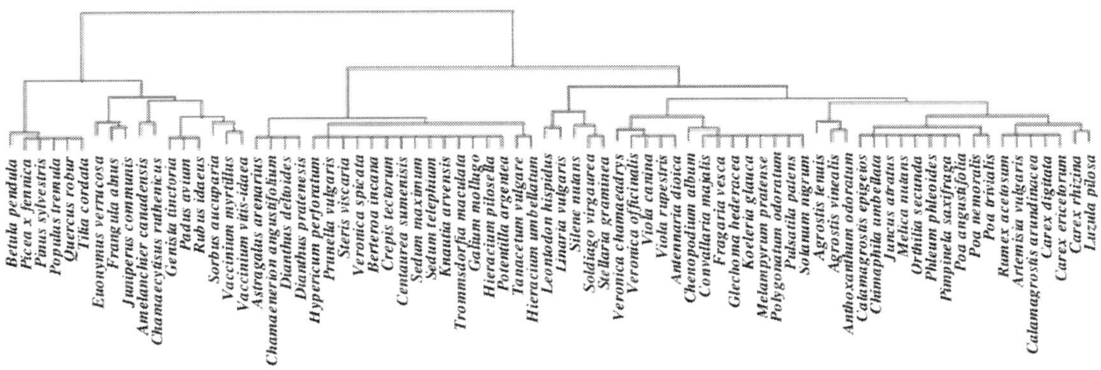

Figure 1. Cluster dendrogram based on 11 functional traits.

Group 1. It includes 6 species (8%) belonging to 5 families (*Pinaceae, Betulaceae, Fagaceae, Tiliaceae* and *Salicaceae*). Species in this group (*Pinus sylvestris, Picea fennica, Betula pendula, Quercus robur, Tilia cordata, Populus tremula*) are tree forms and are characterized as perennials with high growth (more than 20 m) and high leaf attractiveness. This set of indicators indicates the high (100%) contribution of the group to recreational services. In aesthetic services, the contribution of the group is insignificant.

Group 2. It includes 11 species (14%) belonging to 6 families (*Rosaceae, Fabaceae, Ericaceae, Cupressaceae, Celastraceae* and *Rhamnaceae*). Species in this group are predominantly shrubs, and 4 species are dwarf shrubs. Representatives of all six families are characterized by the high contribution to aesthetic services (100%). The leading species are the following representatives of the *Rosaceae* (*Amelanchier canadensis, Padus avium, Rubus idaeus, Sorbus aucuparia*), *Fabaceae* (*Chamaecytisus ruthenicus, Genista tinctoria*) and *Ericaceae* (*Vaccinium myrtillus, Vaccinium vitis-idaea*). Only 3 species (*Padus avium, Sorbus aucuparia, Vaccinium vitis-idaea*) are characterized by a contribution to recreational services (27%).

Group 3. The group includes the largest number of species – 39 (51%) belonging to 16 families (*Asteraceae, Caryophyllaceae, Lamiaceae, Rosaceae, Convallariaceae, Violaceae, Crassulaceae, Scrophulariaceae, Hypericaceae, Onagraceae, Poaceae, Chenopodiaceae, Dipsacaceae, Fabaceae, Rubiaceae, Ranunculaceae* and *Solanaceae*). The group is represented by herbaceous plants.

Representatives of the group have a high contribution to both aesthetic (66%) and recreational services (79%). The greatest effect on services is rendered by the representatives of the *Asteraceae* (*Centaurea sumensis, Trommsdorfia maculata, Hieracium pilosella, Tanacetum vulgare, Hieracium umbellatum, Leontodon hispidus, Solidago virgaurea, Antennaria dioica*), *Scrophulariaceae* (*Veronica spicata, Veronica chamaedrys, Veronica officinalis, Linaria vulgaris*), *Caryophyllaceae* (*Dianthus deltoides, Dianthus pratensis, Steris viscaria, Silene nutans*), *Lamiaceae* (*Prunella vulgaris, Glechoma hederacea*), *Rosaceae* (*Fragaria vesca, Potentilla argentea*), *Convallariaceae* (*Convallaria majalis, Polygonatum odoratum*), *Violaceae* (*Viola canina, Viola rupestris*), and *Crassulaceae* (*Sedum maximum, Sedum telephium*).

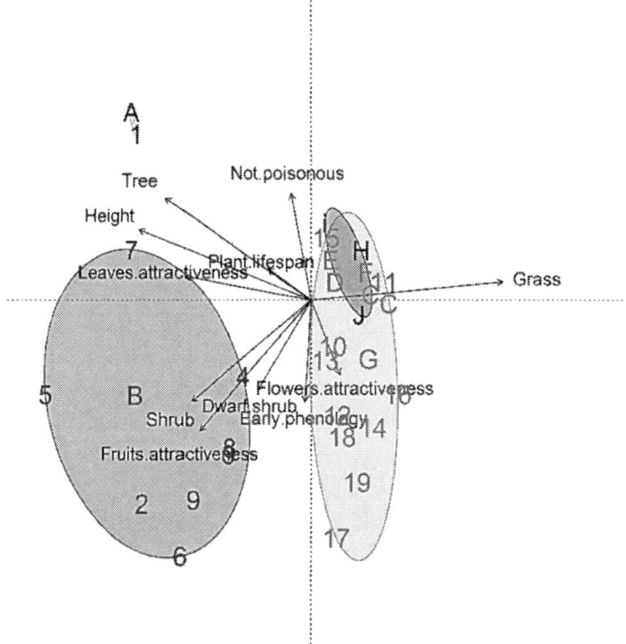

Figure 2. Ordination of species on 11 functional traits.

Note. Group 1. (A, 1). Group 2. (B, 2, 3, 4, 5, 6, 7, 8, 9). Group 3. (C, D, E, F, G, 10, 11, 12, 13, 14, 15, 16, 17, 18, 19). Group 4. (H, J).

Group 4. The group includes mostly graminoid herbs *(Poaceae, Juncaceae* and *Cyperaceae),* but there are representatives of other families also *(Pyrolaceae, Polygonaceae, Asteraceae* and *Apiaceae).* In total, the group has 20 species (27%). A high contribution of the group to recreational services was found (90%), provided by the representatives of the *Poaceae (Agrostis tenuis, Agrostis vinealis, Anthoxanthum odoratum, Calamagrostis arundinacea, Calamagrostis epigeios, Melica nutans, Phleum phleoides, Poa angustifloia, Poa nemoralis, Poa trivialis), Cyperaceae (Carex digitata, Carex ericetorum, Carex rhizina), Juncaceae (Luzula pilosa, Juncus atratus)* and *Pyrolaceae (Chimaphila umbellata, Orthilia secunda).* The contribution of the group to aesthetic services is negligible.

Analysis of the data made possible the following conclusions. Among the 76 species, 41 species (54%) contribute to aesthetic services, and 59 species (77%) to recreational services. The greatest contribution to aesthetic services is provided by beautifully flowering and ornamental shrubs, dwarf shrubs and herbs, as well as plants with early vegetation. They include representatives of the families *Rosaceae* (6 species), *Asteraceae* (5 species), *Caryophyllaceae* (3 species), *Cyperaceae* (3 species), *Fabaceae* (3 species), *Convallariaceae* (2 species), *Crassulaceae* (2 species), *Ericaceae* (2 species), *Lamiaceae* (2 species), *Violaceae* (2 species). The high contribution of sedge to aesthetic services is due to the early vegetation of the species encountered. A lower contribution to aesthetic services is provided by trees, herbs, and ruderal species. Recreational services were more influenced by trees and herbs, and the ruderal species have a low impact. The largest contribution made by the following families: *Poaceae* (10 species), *Asteraceae* (8 species), *Caryophyllaceae* (5 species), *Rosaceae* (4 species), *Scrophulariaceae* (4 species), *Cyperaceae* (3 species), *Crassulaceae* (2 species), *Juncaceae* (2 species), *Pinaceae* (2 species s), *Pyrolaceae* (2 species), *Violaceae* (2 species).

The results obtained show that the contribution of species diversity to services can be characterized through the functional traits of plants. This shows a relationship between biodiversity and the quality and quantity of ecosystem services. The proposed method, in the future, can be extended to other types of services (regulating and supporting); the data can also be used in the socio-economic assessment of natural ecosystems for recreational and other uses. Using more species and functional traits will improve the accuracy of estimates.

References

[1] Millennium Ecosystem Assessment 2005 *Ecosystems and human wellbeing: a framework for assessment* (Washington, DC: Island Press)
[2] Daily G 1997 Nature's services: societal dependence on natural ecosystems (Washington, DC: Island Press)
[3] Costanza R et al. 1997 *Nature* **387** (6630) 253–260
[4] Daily G C et al. 2009 *Front Ecol Environ* **7** (1) 21–28
[5] Daily G C et al. 2000 *Science* **289** (5478) 395–396
[6] Rudolf de Groot P S et al. 2005 Ecosystems and Human Well-Being: Current State and Trends: Findings of the Condition and Trends Working Group ed R Hassan et al. (Washington, DC: Island Press) vol 1 p 455
[7] Egoh B et al. 2007 *Ecol. Econ.* **63** (4) 714–721
[8] Chan K M et al. 2012 *Bioscience* **62** (8) 744–756
[9] Cisneros-Montemayor A M and Sumaila U R 2010 *J. Bioecon.* **12** (3) 245–268
[10] Grêt-Regamey A et al. 2008 *J. Environ. Manage* **89** (3) 197–208
[11] Daniel T C et al. 2012 *P. Natl. A. Sci. USA* **109** (23) 8812–19
[12] Kremen C and Ostfeld R S 2005 *Front. Ecol. Environ.* **3** (10) 540–548
[13] Díaz S et al. 2007 *P. Natl. A. Sci. USA* **104** (52) 20684–89
[14] Suding K N et al. 2008 *Glob. Chang. Biol.* **14** (5) 1125–40
[15] Luck G W et al. 2009 *Bioscience* **59** (3) 223–235
[16] Bengtsson J 1998 *Appl. Soil. Ecol.* **10** (3) 191–199
[17] Grime J P 2001 Plant Strategies, vegetation processes, and ecosystem properties (Hoboken, New Jersey, USA: Wiley)
[18] Eviner V T and Chapin F S 2003 *Annu. Rev. Ecol. Evol. Syst.* **34** (1) 455–485
[19] Hooper D U et al. 2005 *Ecol. Monogr.* **75** (1) 3–35
[20] Kremen C et al. 2007 *Ecol. Lett.* **10** (4) 299–314
[21] Wardle D A et al. 2004 *Science* **304** (5677) 1629–33
[22] Boyero L et al. 2007 *Ecol. Res.* **22** (4) 551–558
[23] Lavorel S and Garnier E 2002 *Funct. Ecol.* **16** (5) 545–556
[24] Díaz S et al. 2004 *J. Veg. Sci.* **15** (3) 295–304
[25] Petchey O L and Gaston K J 2006 *Ecol. Lett.* **9** (6) 741–758
[26] Yushkevich M V 2015 *Proc. BSTU* **1** (174) 111–114
[27] García-Llorente M et al. 2011 *Ecol. Appl.* **21** (8) 3083–103
[28] de Bello F et al. 2010 *Biodivers. Conserv.* **19** (10) 2873–93
[29] Lavorel S and Grigulis K 2012 *J. Ecol.* **100** (1) 128–140
[30] Garnier E et al. 2016 Plant functional diversity: Organism traits, community structure, and ecosystem properties (Oxford, UK: Oxford University Press)
[31] Perez-Harguindeguy N et al. 2013 *Aust. J. Bot.* **61** (3) 167–234
[32] Cornelissen J H C et al. 2003 *Aust. J. Bot.* **51** (4) 335–380
[33] Murtagh F and Legendre P 2014 *J. Classif.* **31** (3) 274–295

Productivity of *Polytrichum juniperinum* Hedw. in forest ecosystem of Tatarstan

N R Shafigullina and E N Karzhavkina

Institute of Environmental Sciences, Kazan Federal University, Kremlevskaya str., 18, 420008, Russia

E-mail: nadiashafigullina@gmail.com

Abstract. The values of linear and phytomass increments for the period May - October 2016 in forests of Tatarstan (central part of European Russia) are studied. The average values of increments for *Polytrichum juniperinum* Hedw. are provided. The relation between habitat characteristics and the productivity of *Polytrichum juniperinum* is discussed.

1. Introduction

It is possible to assess the role of productivity of mosses in conserving biodiversity and contributing to the productivity of the vegetative cover of the Republic of Tatarstan. Despite the apparent importance of the moss communities in the vegetation cover of forest ecosystems, they have not yet been sufficiently studied. The evaluation of forest production does not consider the contributions of the annual production of ground cover. However, it is the ground cover, including moss-lichen, which makes a significant contribution to the annual production of forest and forest-steppe ecosystems [1]. The study of moss growth is important for assessing the state of ecosystems in which bryophytes constitute a significant proportion of the components of the total biomass [2, 3]. In these regions, the amount of annual increments was measured to identify annual production. Such ecosystems include sphagnum bogs, forest complexes of Siberia, as well as tundra and northern coniferous taiga forests [4, 5, 6].

However, at present, the values of linear increments are not known for most of the mosses [7]. This situation is largely due to the fact that the boundaries between annual increments in many mosses are not visually determined and painstaking work is required to obtain the necessary data.

Growth is one of the indicators of productivity, but it has two components: biomass increment and linear increment (including branches). As Schwinning [8] pointed out, unequal growth rates within one species can be the result of the influence of environmental and other factors that are not dependent on productivity. An interesting phenomenon, an increase in biomass and elongation, respectively, can occur at different times. Rincòn and Grime [9] have very clearly shown that the linear increase and increase in the biomass of *Brachythecium rutabulum*, *Thuidium tamariscinum* and *Lophocolea bidentata* can be practically inverse. When the production of dry matter decreased, there was an increase in length, which led to negative biomass production [10]. Some authors [9, 11] concluded that measuring the elongation of the stem can only lead to an inaccurate picture of true productivity. In fact, biomass accumulation and linear growth are events that are not related to each other, and biomass is a better indicator of productivity than linear growth [12].

2. Materials and methods

Polytrichum juniperinum Hedw. is a large plant, and it grows in loose or dense turf, essentially in open habitats: on wastelands, after cutting of timber or burning, in sparse forests and forest belts, on various types of soil outcrops, both natural and man-made, in stony habitats, sometimes on fallen trees, and at the base of trunks.

To study the linear growth and growth of mosses by mass in spring (April 2016), test sites were laid in the territory of Narat-Astinsky Bor (Muslumovsky district, Republic of Tatarstan) and the national park "Nizhnyaya Kama" (Yelabuga district, Republic of Tatarstan).

The sites were chosen with different types of illumination, microrelief and substrate. Sites were in full light, in the opening between trees (penumbra), and under the trees (shadow). The substrata were chosen in a variety of ways: forest soil, stumps, and upturned roots on fallen trees. The microrelief was divided into 4 types: a flat area, a small slope, a slope, and a micro depression. To determine the dynamics of growth and the values of linear increments, the dressing method was used. To identify the mosses, decorative ribbons used for decoration of flower bouquets were split into thin long strips, and the strapping was attached at a distance of 1 cm from the tip of the shoot. In the fall (October 2016), the stalks of the banded shoots were harvested, cut at the dressing site, and collected in envelopes for further processing in the laboratory. The increments, without 1 cm, were measured and recorded and weighed on high-precision scales.

Correlation analysis was performed with the Pearson correlation coefficient. The values of increments were displayed using the ordihull function of the statistical programming environment R 3.2.2 [13]. Statistical data processing was conducted using vegan packages [14] of statistical analysis environment R. A correlation analysis of linear growth and biomass increment was carried out. Graphs of box plot with a spread of growth values with average values were constructed.

3. Results and discussion

In total 117 shoots of *Polytrichum juniperinum* were treated. The increase was measured in autumn during the growing season in April-October. Since *Polytrichum juniperinum* was found in the form of loose and small turf, we marked 12-15 shoots from which an average of 10 pieces were found. The lengths of linear increase vary from 4 to 22 mm (figure 1). There is a wide range of the magnitudes of the increment. The average increment is 12.52 mm. The increase in biomass varies from 2.6 to 19.1 mg. The average increment is 8.21 mg. The differences between the increments of *Polytrichum* samples grown in various forest areas (Nizhnyaya Kama and Narat-Astinsky Bor) are not observed.

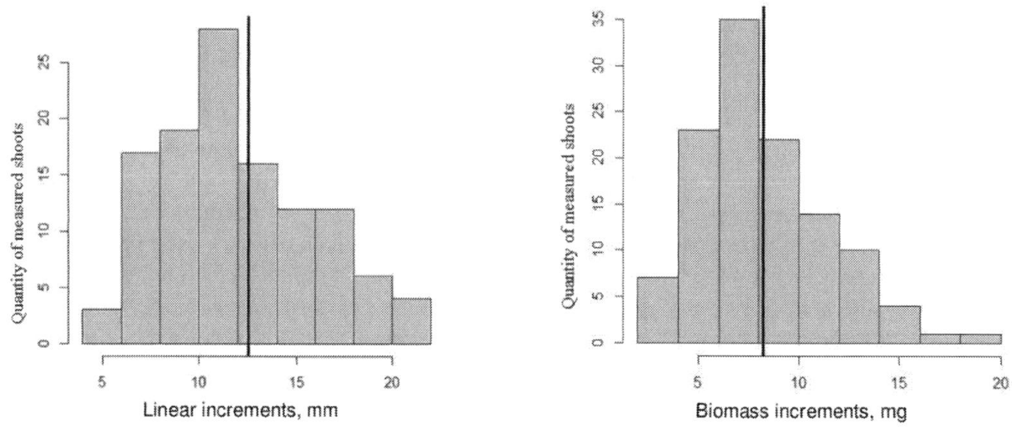

Figure 1. The distribution of increments of *Polytrichum juniperinum.*

Based on our data, the increase in biomass and linear increment react differently to the environmental factors of habitats. The Pearson correlation coefficient between linear growth and biomass increment was 0.676 (95% confidence interval 0.564-0.764, p-value <2.2e-16). Since *Polytrichum juniperinum* possesses monopodial growth and does not have lateral branches, the larger the linear increase, the greater the biomass increment.

3.1. Light

Polytrihum juniperinum is a heliophyte, and its growth in fully illuminated areas is higher than openings between trees and under the trees (figure 2).

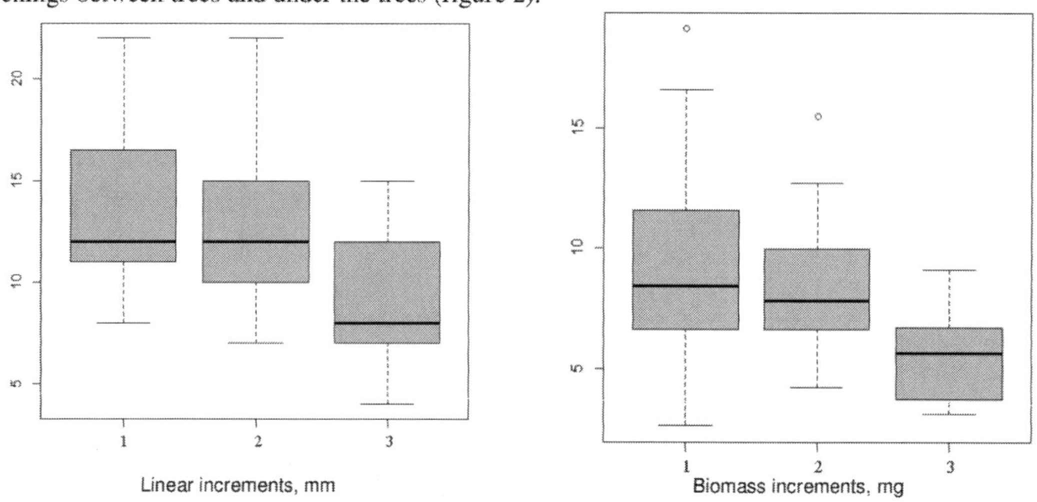

Figure 2. Influence of light on growth of *Polytrichum juniperinum*, 1– open areas in forest meadows; 2 – openings between trees; 3– under trees.

Reduction of light content also affects the growth of biomass and linear growth. Reducing linear growth in penumbra and shadow is statistically significant (the level of significance in the penumbra is 0.0001, in the shadow = 0.004), in contrast to the increase in biomass.

3.2. Substrate

In the forest massifs studied by us, *Polytrichum* occupied such substrates as forest soil, dry stumps, and upturned roots on fallen trees (figure 3).

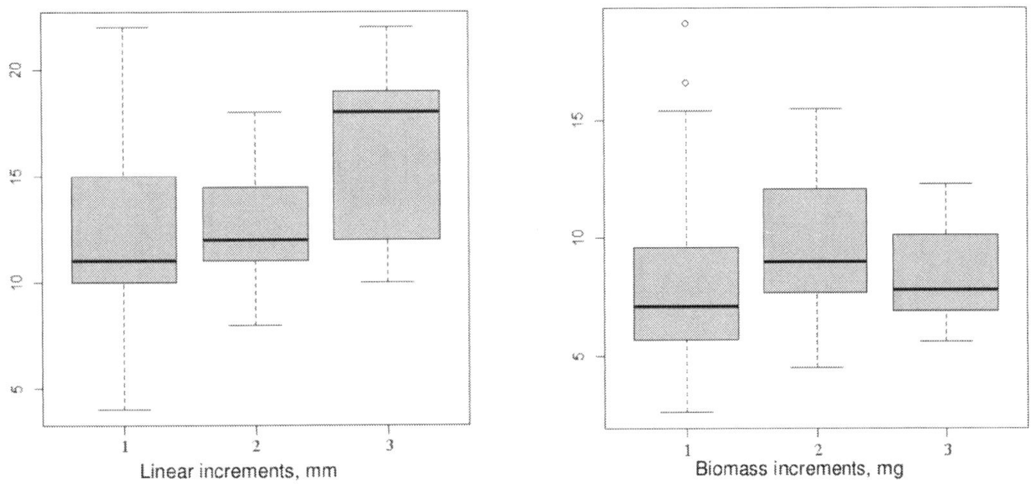

Figure 3. Effect of substrate on growth of *Polytrichum juniperinum.* 1– forest soil; 2 – stumps; 3 – upturned roots on fallen trees.

Polytrichum juniperinum prefers mostly well-drained sandy soils and dry habitats. The dry stump as a substrate has a negative effect on the increment of biomass -1, and the upturned roots is positive +1.5 relative to the forest soil, but these values are not reliable. Linear gain sensitively reacts to the substrate. The largest linear increase is observed on the upturned roots. If we take the values of linear increment in the forest soil for 0, then the gain on the upturned roots will be +7.3 which is statistically significant.

3.3. Microrelief

Microrelief affects the distribution of heat, moisture and light in the microhabitat. *Polytrichum* is found on four types of microrelief: a flat area, a small slope, a slope and a micro depression (figure 4). A small slope and slope negatively affect the biomass of moss, but this is not statistically significant (0.7 and 0.9) with respect to the flat area. The longest moss shoots were observed in a decrease of +8 relative to the flat area (significance level 5.50e-05). Also long shoots characterize the samples collected by a small slope +3.9 and slope +3 (significance level 0.02 and 0.37). Thus, the greatest increase is observed in microdepression.

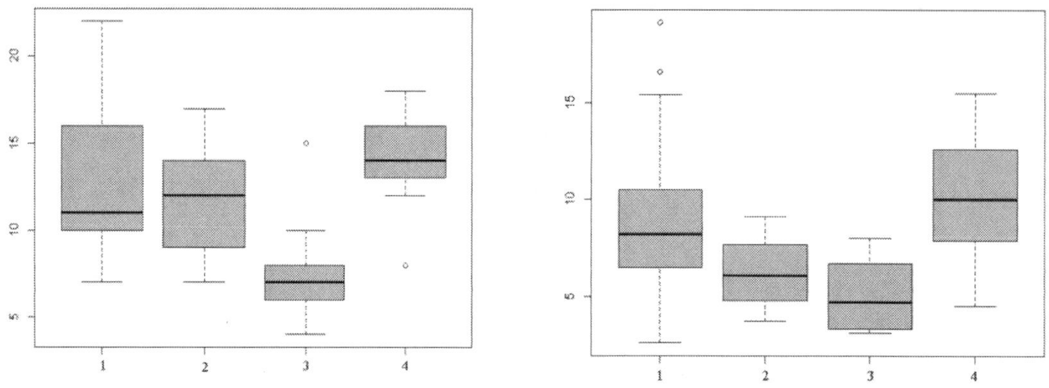

Figure 4. Influence of the microrelief on growth of *Polytrihum juniperinum*. 1– a flat area, 2 – a small slope; 3 – a slope; 4 – microdepression.

4. Conclusions

Polytrichum juniperinum prefers microdepression, open lighted areas and upturned roots as a substrate. Growth on the slope adversely affects the biomass of moss. Reduction of light content also affects the growth of biomass and linear growth. There were no significant differences between the growth rates in various forest areas.

Acknowledgements

We thank Nelli Chizhikova for help in statistical data processing.

References

[1] Goncharova I A and Benkov A V 2015 *Sibirskiy lesnoy zhurnal* **6** 54–61
[2] Yermolayeva O V and Shmakova N Y 2016 *Uchenyye zapiski Petrozavodskogo gosudarstvennogo universiteta, obshchaya biologiya* **8** (161) 40–45
[3] Kostina M V, Safronova G A and Agapov P A 2013 Arctoa **22** 15–22
[4] Kostina M V and Safronova G A 2012 *Botanicheskiy zhurnal* (Moskva KMK) p 88-100
[5] Grabovik S I and Nazarova L E 2013 *Arctoa* **22** 23–26
[6] Ermolaeva O V, Shmakova N Y and Lukyanova L M *Arctoa* **22** 7–14.
[7] Kostina M V, Safronova G A and Barabanshchikova N S 2016 *Bull. MOIP* **121** (1) 53–64
[8] Schwinning S 1993 *Bull. Ecol. Soc. Amer. Program and Abstracts 78th Ann. ESA Meeting* **74**(2) 432
[9] Rincòn E and Grime J P 1989 *J. Ecol.* **77** 447–455
[10] Glime J M 2012 *Bryophyte Ecology* (Electronic Book Vol 1)
[11] Stark L R and McLetchie D N, Mishler B D 2001 *Plant Ecol.* **157** 183–196
[12] Stark L R 2002 *Bryologist* **105** 204-218
[13] R: A language and environment for statistical computing. R Foundation for Statistical Computing, Vienna, Austria.
[14] Oksanen J et al. 2012. vegan: Community Ecology Package. R package version 2.0

IOP Conf. Series: Earth and Environmental Science **107** (2017) 012093 doi:10.1088/1755-1315/107/1/012093

The role of vertical structure in maintaining the stable functioning of meadow phytocenoses of the Middle Ob floodplain

L F Shepeleva

Research Institute of Biology and Biophysics, Tomsk State University, Lenin Ave. 36, 634050, Russia

E-mail: shepelevalf@mail.ru

Abstract. The paper analyses the vertical structure of floodplain meadow communities of the Middle Ob and studies its association with yearly variations of weather conditions, soil regimes, and changes in the species composition of the grass stand. The vertical structure reflects the average annual conditions of floodplain ecotopes and is an adaptive mechanism for the stability of meadow phytocenoses during environmental dynamics.

1. Introduction

Sustainable functioning of biogeocenotic systems is complicated and comprises many aspects [1, 2, 3, 4]. Studying the regularities of the interaction between the vegetative community and its subsystems (species populations, ecological groups and life forms of plants) with environmental conditions and human impacts makes it possible to identify the mechanisms of sustainability and is necessary to solve the practical problems of protection, rational use, restoration of natural communities, and creation of artificial ones [5, 6, 7]. Herbal communities, and especially floodplain meadows, due to their ability to annually renew the above-ground organs [8], quickly respond to the change in various factors and are therefore convenient for studying adaptive mechanisms.

2. Material and methods

Stationary observations of meadow vegetation dynamics and alluvial-soil regimes were carried out in the Podobinsk key area of the Middle Ob floodplain (Tomsk Oblast) in 3 trial sites exhibiting various relief elements, altitude of floodplains, and types of alluvial soils. The mixed-fescue, sedge-bluegrass, soddy-sedge phytocenoses on the sod-meadow, meadow-marsh and marsh peat soils were studied. Synchronously, layer-by-layer temperature measurements and sampling of soils were done to determine the moisture content of nitrogen compounds available for plants, mobile phosphorus, and exchange potassium. The groundwater level was fixed. The flooding regime and weather conditions were taken into account using weather data and gauging stations. This allowed us to perform a complex analysis of the influence of flooding conditions, groundwater dynamics, and weather conditions on the hydration, food regimes of soils, and dynamics of grass stand [9]. The observation time was 7 years, which included extremely dry, extremely wet, and transitional years between them (table 1). The layered indices of soil regimes that were most contrasting years in terms of soil moisture (1983 and 1987), are given in table 2.

Analysis of fluctuations in phytocoenoses was done in accordance with herbage productivity indicators; the composition and ratio of the dominants; changes in the abundance and presence of other species; and the composition and relationship of ecological, biological, and biomorphological groups of species. The ecological groups were distinguished based on bioindication scales by L.G. Ramensky. The ecobiomorphs were characterized by the type of the morphostructure of their underground organs, vegetative mobility, life span, and life strategy. To determine the phytocoenotic indicators the method of accounting squares was used (accounting areas of 1 m^2 in a fivefold repetition). The selection of cuts was carried out during the flowering of most species.

Table 1. The alluvial soils' regimes in the Podobinsk key area of the Middle Ob floodplain [10].

Soil*	Flooding regime**	GWL, cm	RPM, mm	Soil temperature, °C	NH^4	NO^3	P^2O^5	K^2O
					mg/kg of soil		mg/100 g of soil	
ABP	1978-79-AH	40	78.4	12.9	110.8	5.8	8.0	10.9
	1980-83-PI	60	70.5	14.7	52.3	4.7	6.2	13.1
	1984-85-HH	25	99.5	16.8	88.6	5.5	8.6	13.9
	1986-88-HH	20	126.8	13.6	112.2	1.2	8.4	14.5
	1989-90-AH	65	91.2	15.5	23	4.6	38	10.6
	1991-HH	30	207.9	13.8	97.8	2.2	9.7	7.7
	1992-94-AH	20	77.8	15	54.9	4.3	8	9.1
AMB	1978-79-AH	60	49	15.6	51.7	3.2	13.1	12.9
	1980-83-PI	65	45	13	14.5	7.6	6.7	9.7
	1984-85-HH	50	88	15.8	39.4	5.4	7.2	10.5
	1986-88-HH	45	113.3	13.6	40.3	4.2	8.1	10.8
	1989-90-AH	90	47.5	14.9	13	6.3	18.5	8.7
	1991-92-AH	70	43.8	16.4	48.7	2.9	12.6	11.2
	1993-94-AH	75	34	17.9	59	5.4	7.5	9.6
ATM	1978-79-AH	120	37.3	15.8	21.1	8	13.6	6.7
	1980-83-PI	140	12.2	14.8	12.3	2.1	10.7	7.2
	1984-85-HH	120	52.4	14.6	30.3	5.9	7.4	11.4
	1986-88-HH	110	58.8	13.6	28.1	4.7	8.4	12.2
	1989-90-AH	140	15.2	15.8	21.5	1.5	19.5	7.2
	1991-92-AH	100	24.7	17.2	9.5	1.2	11	5
	1993-94-PI	100	11.7	17.3	8.2	0.9	8.2	5.7

*ABP–alluvial bogpeat soil, AMB – alluvial meadow-bog soil, ATM – alluvial turf-meadow soil, GWL – ground water level, RPM – reserves of productive soil moisture
**PI – period of inundation, AH – average humidification period, HH – period of high humidity.

To study the vertical structure of the meadows excavations of the soil profile were done to the depth of propagation of the plants' root systems, the roots and rhizomes were cleared from the soil so that their spatial disposition was not disturbed, then the structure of the root systems was fixed on millimeter paper.

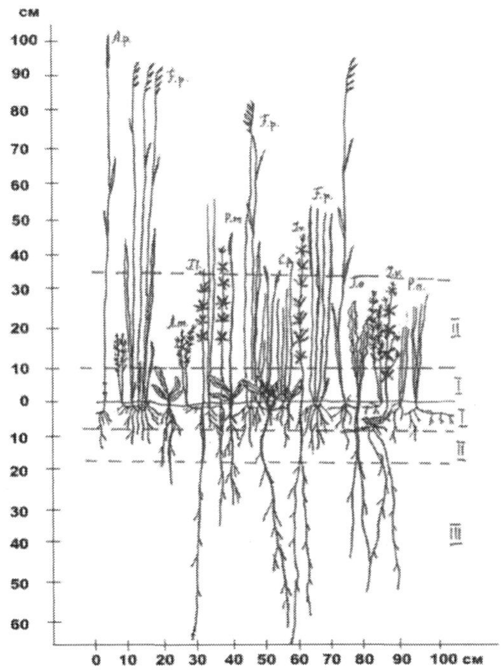

A

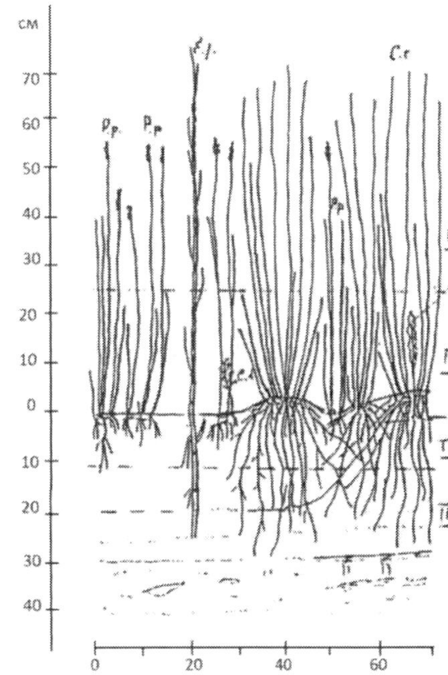

B

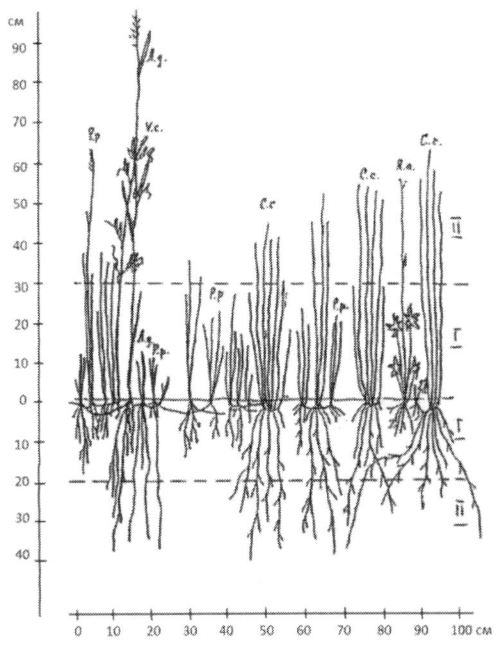

C

Figure1. The vertical structure of the meadow communities in the Podobinsk area of the Middle Ob floodplain:

A) mixed-fescue phytocenosis on alluvial turf-meadow soil

I, II, III – phytocoenogens; *A.m. – Achillea millefolium, P.m. – Plantago media, T.o. – Taraxacum officinale, A.p. – Alopecurus pratensis, F.p. – Festuca pratensis, T.l. – Trifolium lupinaster, C.p. – Carex praecox, G.v. – Galium verum*

B) soddy-sedge phytocenosis on alluvial bog peat soil

I, II–phytocoenogens; *C.c. – Carex cespitosa, C.d. – Cnidium dubium, E.f. – Equisetum fluviatile, P.p. – Poa pratensis, T.o. – Taraxacum officinale*

C) sedge-bluegrass phytocenosis on alluvial meadow-bog soil

I, II – phytocoenogens; *C.c. – Carex cespitosa, Poa pratensis, A.g. – Agrostis gigantea, R.a. – Ranunculus acris, L.p. – Lathyrus palustris*

Soil profile stratification was carried out on the root systems of plants, soil phytocoenogens (rhizohorizons) were distinguished - layers of soil, characterized by the specificity of the roots of certain plant species – figure 1A, B, C.

Table 2. Thermal and water availability in the soil profile of the Podobinsky section of the Middle Ob floodplain in1983 and 1987 [10].

Layer, cm	Soil	Duration of the temperature period		Sum of temperatures		RPM, mm	AMR
		10°C	15°C	10°C	15°C		
			1983 (dry)year				
0-10	ABP	100	80	1,503	1,248	36.9	0.8
	AMB	92	67	1,527	1,050	25.6	0.7
	ATM	100	74	1,591	1,267	16.4	0.5
0-20	ABP	94	74	1,532	1,201	83.9	1.0
	AMB	96	58	1,422	857	54.0	0.6
	ATM	92	73	1,490	1,105	27.5	0.5
0-30	ABP	101	76	1,495	1,193	129.7	0.5
	AMB	93	21	1,326	340	80.0	1.2
	ATM	97	52	1,431	826	41.5	
0-60	ABP					272.6	
	AMB					185.2	
	ATM					112.6	
			1987 (humid) year				
0-10	ABP	105	55	1,745	1,095	58.4	1.3
	AMB	103	58	1,577	1,018	110.1	2.0
	ATM	94	60	1,449	1,002	33.3	
0-20	ABP	106	53	1,570	902	119.3	1.5
	AMB	103	52	1,521	896	158.7	2.4
	ATM	90	40	1,346	650	54.7	1.1
0-30	ABP	103	51	1,533	857	187.1	
	AMB	101	53	1,444	846	222.4	
	ATM	95	32	1,406	563	94.9	
0-60	ABP					297.5	
	AMB					484.1	
	ATM					162.5	

Note. ABP– alluvial bogpeat soil, AMB – alluvial meadow-bog soil, ATM – alluvial turf-meadow soil, RPM – reserves of productive soil moisture, AMR – active moisture range.

3. Result and discussions

It was found that in different types of floodplain soils the active layer occupied by plant roots ranges from 80 cm in soddy-meadow light loamy soils to 30-40 cm in marsh peat and meadow-bog soils of heavy loamy soils. The number of rhizohorizons is also different: in the underground part of the herb-fescue community, three rhizohorizons are distinguished, and in the subterranean part of the sedge-bluegrass and soddy-sedge community there are two of them (figure 1A, B, C). The vertical profile of the soddy-sedge cenosis is characterized by mosaicism, due to the dominance of *Carex cespitosa*.

The vertical underground and above-ground layerage of the mixed herb-fescue phytocenosis on ATM is clearly seen. In the underground part the following rhizohorizons are distinguished: 1– *Achillea millefolium* L. (0-5 cm), 2 – *Festuca pratensis* Huds.(5-20 cm), and 3 – *Galium verum* L. (20-80 cm).It is typical that the root systems of all phytocenosis cereals are confined to the upper rhizohorizons (figure 1A).

Sod-meadow soil in mixed dynamics is characterized by optimal and insufficient moistening, respectively, in communities of in which mesophytes and xeromesophytes predominate. With changes in the weather and flood conditions over the years, the moisture, temperature, and amount of plant nutrients in all soil horizons varies, particularly sharply in the surface ones (table 1, 2).The community's adaptability to these fluctuations is expressed by the fact that in all the rhizogrizones there are the roots of ecologically different plants. Horizon *F. pratensis* (5-20 cm) has the regime of optimal moisture more often than the upper and lower parts of the root-inhabited stratum and is mainly formed by the mesophytes' roots.

In community dynamics, the significance of biological groups of species (cereals, legumes, herbage) is clearly seen, which, in our opinion, contributes to the stability of the biogeochemical exchange of biogeocoenosis. The determining biogeochemical role of cereals is obvious, since this group dominates in the herbal phytocenoses, creating the main part of the above ground and underground phytomass. The herbage yields the maximum phytomass in the arid years, while a group of legumes does so in the transitional periods, when the content of nitrates increases in soils (table 1). Therefore, the plants from groups of herbs and legumes are the important components of the meadow community, as they contribute to its stability.

Smaller differences in species at the ecobiomorph level are also significant. During dry years, xeromesophyll herbs predominate in the grass stand, either with both superficial and deep rooting or only with deep rooting. The reason is that in different years the conditions optimal for absorbing moisture and nutrient elements are formed at particular depths. In this case there is no increase in the mass of more hygrophilous species, the roots of which are developed in the same horizon. This fact shows the reality of the physiological differences between the species belonging to one biomorph but to different ecological groups [11, 12].

The underground structure of the soddy-sedge phytocenosis on ABP is represented by rhizohorizons: bluegrass and sedge soddy (figure 1B). In the horizon of bluegrass (0-10 cm), the roots of mesophilous plants are predominantly distributed, but there are also rhizomes and roots of hygrophilous species (*Lisimachia vulgaris* L., *Caltha palustris* L., *Alisma plantago-aquatica* L.), which do not grow annually in the herbage. In the soddy sedge horizon (10-30 cm) the roots of other dominants of the community, *Agrostis gigantea* Roth., *Poa palustris* L., *Equisetum fluviatile* L., are located.

When the observations began the soil was moist and cold (table 1), and *C. cespitosa, A. gigantea,* and *E. fluviatile* dominated. Without flooding (1981–1983) the upper horizon was well heated, and its moisture content was low (table 2). Flooding years (1984–1985) provided an excess of moisture in all horizons. In subsequent years, the soil profile was excessively moistened, and there was a decrease in the mass of all species, which led to a decrease in the total grass mass.

In this dynamic, the most stable species was *C. cespitosa*, whose roots developed in all horizons, and the rhizomes form tufts on the soil surface. The mesophilous species, whose roots are distributed in the 0-10 cm layer, reached a high abundance in the years without flooding under optimal moisture conditions. The hydrophilic species, whose roots developed in the same horizon, were either not present in the herbage in those years, or had sharply reduced phytomass. They developed in the inundation years with excessive moisture. For the deep-rooted *E. fluviatile*, the years with optimum moisture in the lower soil layers were favorable.

If we consider the structure of the underground part of the soddy-sedge community to indicate the hydrothermal regimes of soils, we can note the following: 1. The confinement of the root systems of mesophilous plants exclusively to the 0-10 cm layer indicates that in ABP the optimum conditions for moistening, aeration and heat supply are possible only in the upper horizon, and possibly include its considerable desiccation. 2. The penetration of the roots of the more moisture-loving mesophytes into the 0-20-cm layer reflects a different hydrothermal situation, when the oscillation of conditions is less significant and the regime of mean, suboptimal moistening prevails. 3. The lower rhizohorizon is constantly excessively moistened and characterized by low heat supply; here the roots of hydrophilic species with a universal type of root system penetrate several horizons. 4. The maximum contrast of regimes is characteristic of an upper rhizohorizon, a high contrast – of a layer of 10-20 cm, the regime stability – of a layer of 20-40 cm.

The vertical structure of the sedge-bluegrass phytocenosis on AMB (figure 1C) is more uniform, since *C. cespitosa* here forms not tussocks, but turfs, 5-10 cm high and about 25 cm in diameter. Soddy sedge and bluegrass dominate, but in some years *Poa palustris* L., *Alopecurus pratensis* L. and

Carex acuta L. dominate. The root systems of mesophilous species are concentrated in the soil layer of 0-15 cm (figure 1C); other plants' roots penetrate into deep soil layers (up to 40 cm), but the main mass of the roots are in the layer up to 25 cm deep.

The difference between the environmental regimes of AMB and ABP soil consists, first, in the absence of clearly pronounced stratification of temperature and moisture changes. Changes in the hydrothermal parameters of the soil, connected with the dynamics of flooding and weather conditions, usually covers the entire profile, although at a depth of about 20 cm there is a limit to the change in regimes. Above this level, a sharp fluctuation of moisture can occur – from excessively humid to arid conditions. In the underlying horizon, the range of changes in the humidification state is from excess to optimum. Lack of moisture is observed in the absence of soil flooding for several years.

This may be the reason why the structure of the rhizohorizons of the sedge-bluegrass community is not clearly expressed; only the roots of mesophytes and poorly-established species from the group of xeromesophytes are definitely confined to the upper soil layer. The more moisture-loving eumesophytes and hydromesophytes in the cenosis maximally spread their underground organs in the upper rhizohorizon; however, their root systems happen to penetrate deeper. The inactive stems-root and short-stemmed herbs are not numerous in the community. Long-stemmed vegetatively mobile cereals and sedges of the most diverse ecological requirements (from xeromesophytes to aerohydrophytes) dominate.

Regimes of meadow-bog soils vary greatly from year to year, which probably determines the spread of plant species with different ecological and biological properties. Thus, in one rhizohorizon, the roots of *P. pratensis* and *A. pratensis* develop, but if the former predominates in dry years, when optimal and insufficient moisture are observed in the surface layer of soils, then the second one occurs during flooding, when the high humidity conditions in this horizon are combined with high temperatures. The abundance of *A. gigantea* decreases with lack of moisture, both in the upper and lower horizons. To develop sedges, excessive moisture and lower temperatures are necessary.

Since ammonification and nitrification processes intensify in the years without flooding, various mesophilic, nitratephylic grasses are widely spread in the herbage – *Cirsium setosum* Bess., *Ranunculus acris* L., *Taraxacum officinale* Wigg. Deterioration of soil aeration, with prolonged flooding of the cenosis for several consecutive years, leads to a high (toxic) ammonia nitrogen concentration (table 1). The hydrophilic sedges and marsh herbs become dominant, which confirms the idea of better adaptive ability of swamp sedges to use ammonium nitrogen for their development [13, 14].

Thus, the nature of constitutional and spatial (vertical) structure helps to understand the mechanisms of adaptation of plant communities to unfavorable environmental conditions. The composition and underground vertical layers of the meadow community are determined by the mean annual fluctuations of the hydrothermal and food conditions in certain soil layers and represent an adaptation to the range of their variation. The soil horizons are heterogeneous in their physical and biological properties. The phytocenosis responds to these conditions by selecting ecologically and biologically different plant species, as well as by the natural placement of their root systems in space. Vertical distribution of root systems along the soil profile carries the information about soil regimes and can be used for indication purposes.

The structural differences between the meadow phytocoenosis cover the functional differences. Morphologically distinct are the swamp and meadow-bog types of communities according to the presence of a coarse and turf form of sedge soddy, which reacts to excessive moistening and weak aeration of bog soils. From the meadow type of phytocenosis, the absence of a longline of stem-root and short-stemmed herbage and the relatively small thickness of the root layer are no less distinct. The basic similarity between the dynamic processes of the swamp and meadow-bog types is reflected in the structure of soil phytocoenogens, when the root systems of mesophytes and aerohydrophytes are combined in the upper soil layer, but these groups of species develop in different years. Increased dynamics of the meadow-bog community compared to the marsh phytocenosis is expressed in the predominance of vegetatively mobile forms.

The meadow type of phytocenosis differs more significantly. The communities of this type are characterized by a considerable thickness of the root-inhabited soil stratum and the presence of 3 or more rhizohorizons in the underground part. The maximum occupancy by the root systems and rhizomes of the upper rhizohorizons indicates the prevailing source of moisture and nutrients with precipitation.

Acknowledgements
The work was published as part of the state task of the Ministry of Education and Science of the Russian Federation, project № 5.4004.2017/4.6.

References

[1] Silvertown I, Doodd K, McConway I and Potts M 1994 *Ecology* **75** (8) 2430–7
[2] Nekrosiene R and Skuodiene R 2012 *Polish journal of Environmental Studies* **21**(3) 703–711
[3] Carr I, Odorico P, McGlathery K and Wiberg P 2012 *Marine Ecology Progress series* **448** 289–301
[4] Wang Z, Hou J, Qu Z, Guo J, Li J 2017 *Applied Ecology and Environmental Research* **15**(3) 1625–51
[5] Gowing D, Youngs E, Gilbert J and Spoor G 1998 *Hydrology in a changing environment* **1** 473–483
[6] Kuzmina Z V, Treshkin S E and Karimova T Y 2015 *Arid Ecosystems* **5** (4) 230–242
[7] Blume-Werry G, Jansson R and Milbau A 2017 *Functional Ecology* **31** 1493–502
[8] Rabotnov T A 1985 *Ecology of meadow grasses* (Moscow) p 176
[9] Pashneva G E and Shepeleva L F 1993 *Ecology* **5** 23–32
[10] Pashneva G E, Tsytsareva L K and Abramova M D 1996 *Biological resources of the Middle Ob floodplain: dynamics and forecast* (Tomsk) p 49–65
[11] Alekseenko L N 1967 *Meadow plants fertility depending on the environmental conditions* p 168
[12] Sudnitsin I I 1979 *Movement of soil moisture and water consumption of plants* p 255
[13] Ianiesch P 1986 *Abb. Westfall. Mus. Naturk.* **2-3** 341–345
[14] Aerts R and. de Caluwe H 1994 *Ecology* **75** (8) 2362–72

Acclimation of *Juglans mandshurica* Maxim. and *Phellodendron amurense* Rupr. in the Middle Volga region

D Tishin, M Fardeeva, N Chizhikova, R Rizatdinov

Institute of Environmental Sciences, Kazan Federal University, Kremlevskaya str., 18, 420008, Russia

E-mail: Denis.Tishin@kpfu.ru

Abstract. This research is the first attempt to analyze the results of acclimation of *J. mandshurica* and *P. amurense* in coniferous-deciduous forests under the conditions of the temperate continental climate of the Middle Volga Region. The study has been performed in the Volga-Kama Nature Reserve (Republic of Tatarstan, Russia) and demonstrated that *J. mandshurica* is a successfully acclimated species. This species naturalized in the forests of the Reserve, being distinguished by a rapid biomass production, high germination capacity of seeds and high number of pre-generative specimens. *P. amurense* can be characterized by the opposite features.

1. Introduction

The adaptation mechanisms of trees outside their native ranges and the abilities of invaders to tolerate the conditions of temperate continental climate of the Middle Volga region are currently of particular research interest. Due to the growing number of studies focused on possible use of various plant species in pharmacology, medicine, food and light industry, landscaping, the problem of searching for new sources of plant materials and their application has become increasingly important. The Far Eastern exotic genera *Juglans* and *Phellodendron* are in the focus of attention with this regard, because they are marked by high edibility characteristics and specific medicinal properties [1, 2, 3]. The purpose of current study is to investigate the ecological and physiological characteristics of *J. mandshurica* and *P. amurense* growing in the forest habitats of the Middle Volga region.

The ecological and biological characteristics of the two Far Eastern tree species under study are similar. *J. mandshurica* is deep rooted species and rarely suffers from a lack of moisture. *J. mandshurica* is a hygromesophyte and is sensitive to drought [2]. This species is light demanding. It prefers fertile, well-moistened and drained soils. *J. mandshurica* is the most frost-resistant representative of the genus *Juglans*, which was confirmed by the previous experiments on its introduction into Central Siberia, St. Petersburg, and Petrozavodsk [4, 5]. The tree blossoms in April – May. *J. mandshurica* is pollinated by wind, propagates by seeds, and regenerates by sprouting. *P. amurense* is heliophilous as is *J. mandshurica*. *P. amurense* is demanding for fertile soil as well and is wind resistant. Its root system is deep and aggressive similarly to *J. mandshurica*. This hygromesophyte prefers moistened and well-drained soils. The tree is winter hardy and regenerates from seeds and root shoots. *P. amurense* is an entomophilous species, blossoms in the first half of summer [6].

The first mass planting of exotic trees and shrubs in the Raifa forest (part of Volga-Kama Nature Reserve) began as early as 1924; the first sites of planting were forest management blocks 41, 64 and the Arboretum of the Reserve. These plantings in blocks 41 and 64 died, the only *J. mandshurica* represented by trees which developed from the seeds brought from the Russian Far East can be found merely in the Arboretum. According to the current annotated list of tree species in the Raifa Arboretum, *Juglans mandshurica* Maxim. and *Phellodendron amurense* Rupr. are located in the "Asia" section of the botanical garden. Nowadays the biggest (up to 19 m in height) and oldest tree of *Juglans mandshurica* is about 85 years. The Arboretum has 14 exotic *Juglans mandshurica* trees aged 60 years and 17-18 m in height, which developed from the seeds of the first planted trees. Currently, *P. amurense* species is represented in the Raifa Arboretum as a single tree with a curved and hollow trunk. The estimated age of this tree is around 80 years, and height of the trunk is 6 m. On the winter hardiness scale, the *P. amurense* tree is given a score II (meaning that no more than 50% of the annual shoot length dies from frost) [4].

During the period from 1932 to 1966, the studies on acclimation and introduction of exotic species were continued in different compartments of the Raifa forest.

J. mandshurica was planted both in pure culture and in mixed stands combined with green ash (*Fraxinus pennsylvanica*), ash-leaved maple (*Acer negundo*), pine (*Pinus sylvetsris*), and siberian pine (*Pinus sibirica*) in Raifa's forest management blocks 87, 75, 76, 68, 88. Vast stands of this species were planted in 1949 (0.5 ha; block 76), as well as in 1941 and 1946 (0.4 ha; block 75). The last investigations on the state of exotic species and their naturalization were carried out in 1972-1973 by E. Deryuga and A. Murzov [7]. They showed that *J. mandshurica* in the Arboretum was in a worse state than in the forest: 1790 individuals planted in pure culture in block 76 were attributed to bonitet class II; the part of 1430 individuals planted in block 75 were classified as stand quality class I if grown in mixture with *F. pennsylvanica* and as quality class II and III if found in mixture with *P. sylvetsris* and *P. sibirica*. The term "bonitet", as used herein, can be defined as an estimate of tree productivity, so the quality class of bonitet I corresponds to tree stands having good access to water and nutrients.

P. amurense was mainly planted in Raifa's forest management blocks 85, 75, 76. Most of *P. amurense* trees found in block 75 were dead, as discovered by [4] in 1973. In a *P. amurense* stand of 0.1 ha (block 75) mixed with *F. pennsylvanica*, the authors [4] described 1350 surviving tree individuals (small trees) aged 24 years and classified as quality class III bonitet. Analyzing the pure culture stand of 600 m2 in block 76, 2779 trees aged 24 years were attributed to stand quality class II [4].

The massive planting of exotic species in the Raifa forest stopped in 1964 when the Volga-Kama Nature Reserve was founded. Since then naturalization of the introduced species proceeded without any economic influence of humans, which is a unique situation. The planted forest stands were inhabited by natural forest tree species *Tillia cordata* and *Acer platanoides*.

The current article summarizes results of investigations of *J. mandshurica* and *P. amurense* that resumed in 2012. The study represents an analysis of the invasion of exotic species into natural communities, assessment of their naturalization and regeneration, and mapping of their distribution in the nature reserve.

2. Material and methods

The investigation was performed in the Raifa area of the Volga-Kama State Nature Biosphere Reserve. The territory under consideration is unique in the diversity of forests and represents the main variants of taiga, mixed, and deciduous communities. The nature reserve is geographically located in the northwest of the Tatarstan Republic. According to the natural zoning [8], its territory lies within the borders of the western Kazan terrace-valley region of Eastern European pine and deciduous-pine forests on high terraces above the floodplain of the Volga River, where sod-podzolic sandy loam soils prevail. The climate is temperate continental. The average temperature is –13.8°C in January and +19.1°C in July. The annual precipitation amount is 509 mm. The Raifa forest district was a scientific experimental station of the higher forest school in the Volga region since 1919. In the years 1922-1964 in Raifa, various forestry studies were carried out, including forest planting and acclimation of exotic tree species [9].

The study was done in Raifa blocks 76 and 75. Living individuals of *J. mandshurica* and *P. amurense* were exposed to an extensive analysis including (i) assessment of the morphometry of vegetative and generative organs to describe ontogenesis; (ii) dendrochronological analysis to specify the calendar age and tree-ring widths by the method described in [9], (iii) dendroclimatic analysis of tree-ring data to reveal the influence of climate on wood growth; (iv) analysis of the survival rate of tree specimens in different age groups; (v) assessment of the population growth and state.

3. Result and discussions

Ontogenetic groups (stages of plant development) and calendar age of examined individuals of *J. mandshurica* and *P. amurense* were determined using the morphological, dendrochronological and statistical methods. This allowed identifying of ontogenetic structure of their modern populations, as well as tree abundance and density of *J. mandshurica* and *P. amurense* generations that appeared in certain years [11].

Table 1. Average height and diameter of *J. mandshurica* and *P. amurense* trunks under various growth conditions.

Location	Calendar age (ontogenetic group)	Trunk height, m	Trunk diameter, cm
	Juglans mandshurica		
Far East [2], native habitat	8 years (im)	2	4
	20 years (v)	6	8
	50 years (g)	14	20
Raifa, block 76 [7], data obtained in 1973	24 years (g1), generation 1	13	11
Raifa, block 76, pure culture [11], data obtained in 2015	7 years (im), generation 3	2	2
	14 years (v), generation 3	8	9
	24 years (g1), generation 2	14	26
	31 years (g1), generation 2	17	40
	37 years (g1), generation 2	20	49
	60 years (g2), generation 1	22	55
Raifa, block 75, mixed planted stand [7], data obtained in 1973	28 years (g1), generation 1	11	10
	35 years (g1), generation 1	14	30
	32 years (g1), generation 1	17	17
Raifa, block 75, mixed planted stand [11], data obtained in 2015-2016	7 years (im), generation 3	2	2
	30 years (g1), generation 2	16	22
	35 years (g1), generation 2	19	27
Raifa, Dendrological Garden, block 31 [4], data obtained in 2015	86 years (g2) generation 1	20	55
	60 years (g1) generation 1	17	38
	Phellodendron amurense		
Far East [6], native habitat	100 years (g2-g3)	20-25	70
Raifa, block 85, mixed planted stand	45 years (g2), generation 1	16	19
Raifa, block 75, mixed planted stand	24 years (g1), generation 1	8	8
Raifa, block 76, pure culture [6] data obtained in 1973	24 years (g1), generation 1	11	8
Raifa, block 76, pure culture [12], data obtained in 2015	67 years (g2), generation 1	23	28
	50 years (g2), generation 2	18	23
Raifa, Dendrological Garden, block 31 [4], data obtained in 2016	80 years (g3), generation 1	6	22

As a result of the dendrochronological analysis of 26 model trees of *J. mandshurica*, we found first-generation trees aged 60-65 years, second-generation trees aged 30-37 years, and third-generation trees aged 1-14 years (table 1). Examining 10 model trees of *P. amurense*, we found first-generation trees aged 65-67 years (forest plantation of 1949) and trees aged about 50 years (seed germination in 1964, possibly the second generation of trees) [12].

Although the ecological demands of the two Far Eastern species are similar, their acclimation was significantly dependent on the biological characteristics. *J. mandshurica* blossoms in May and has wind pollination, these characteristics are favorable for fruit setting: May is a leafless period for most of the native species, this is the time when leaves of the main forest-forming species (linden and maple) just beginning to open, which considerably reduces competition for light and wind. On the contrary, *P. amurense* blossoms in June when deciduous trees (linden and maple) have developed leafy crowns, which limits the amount of light and affects blossoming. *P. amurense* has no natural pollinators in the Middle Volga; this in fact reduces the success of fruit setting, because pollination occurs with the help of occasional insects.

According to the current measurements and to the analysis of published data, generative individuals of *J. mandshurica* undergoing acclimation to the temperate conditions of the Middle Volga region are higher and have bigger diameter of trunks than individuals studied in the natural habitats of the Russian Far East. The third-generation immature trees seem less adapted to the temperate climate: their average height is 1.8 m (maximum height is 2.5 m, registered in block 76), while the diameter varies from 1.7 to 2.3 cm, which often leads to breaking of trunks during the autumn-winter period.

Several sample plots of 200-400 m^2 covered the area studied (both forest blocks 75 and 76). The abundance and density of tree specimens belonging to different plant developmental groups were assessed. The population of *J. mandshurica* (block 75 and 76) has trees of all stages of development: germinating seedlings and juvenile individuals account for 45-55% of the total population (calculated in average over 5 years data), immature and virginile trees account for 10-15%, young generative individuals for 20-30%, and mature generative specimens for about 5%. On the whole, to date, the density of *J. mandshurica* trees representing the first generation in the pure culture plantings (block 76) decreased by 30 times, from 0.36 trees/m² to 0.012 trees/m². In the mixed stand (block 75), the density of *J. mandshurica* was 0.71 trees/m² in 1973, no mature generative trees survived until the present time.

The population of *P. amurense* is declining. There are almost no young trees. In 2016, two shoots were registered at the base of a fallen and decaying tree. In 2015, the density of *P. amurense* was 0.005 trees/m², although the density of this species in 1973 [7] reached 0.21 trees/m², which means that the population has considerably decreased. Most first-generation trees of *J. mandshurica* and *P. amurense* died in the cold winter of 1978-1979 when the temperature did not rise above –40°C during 7-10 days. As a result, natural forest-forming species began to expand, which led to the forest regeneration process finalized by forming of linden forest with goutweed (*Aegopodium podagraria*) and hazelwort (*Asarum europaeum*). This type of forests is characterized by intense competition and poor lightning conditions, which is crucial for successful development of immature and virginile specimens of *J. mandshurica* and germinating seedlings of *P. amurense*.

Table 2. Survival rate and population growth* of pre-generative *J. mandshurica* individuals.

Ontogenetic groups	Survival rate / population growth			
	Plot 1		Plot 2	
	b/t 2015 & 2013	b/t 2016 & 2015	b/t 2015 & 2013	b/t 2016 & 2015
Germinating seedlings	100 % / 96 %	97 % / **–70 %**	100 % / 40 %	100 % / **–86 %**
Juvenile	79 % / 54 %	100 % / 4 %	94 % / 28 %	100 % / 22 %
Immature	88 % / 25 %	100 % / 63 %	81 % / **–19 %**	71 % / **–29 %**

* positive population growth corresponds to increase, negative growth corresponds to decrease; in percent of population increased or decreased

Growth rate and the dependence of germinating seedlings and juvenile trees on light are the important biological characteristics of *J. mandshurica* and *P. amurense*. The survival rate and population growth over a number of years contribute to an understanding of the stability of the population and its development. The calculated survival rate of young trees in the *J. mandshurica* population is shown in table 2.

The survival rate of pre-generative groups of *J. mandshurica* is positive in general, but depends on climatic conditions. Elimination of young trees begins at the immature and virginile plant stages. Due to the lack of incoming light the young trees grow up to 2-6 m in height and develop large leaves and crowns. However, the tree radial growth is very small, thus the trunk diameter is only 2-4 cm. This leads to a situation when the trunk can not withstand the weight of a tree and breaks. Another mechanism of elimination of immature trees aged 4-5 years is death of the apical regeneration bud during the winter period. In this situation, the process of plant development resumes when a new lateral shoot is formed from the dormant lateral buds on the trunk, which prolongs their ontogenesis up to 8-10 years.

The statistical analysis of individual tree-ring chronologies of *J. mandshurica* and *P. amurense* revealed that the rates of the radial growth of *J. mandshurica* are higher than for *P. amurense* (table 3).

Table 3. Statistics of individual tree-ring chronologies for *J. mandshurica* and *P. amurense*

Species	Mean correlation with master chronology	Average tree-ring width, mm	Standard error	Average sensitivity	Autocorrelation
J. mandshurica	0.35	3.3	1.4	0.29	0.62
P. amurense	0.43	2.0	0.7	0.19	0.66

The published studies on *J. mandshurica* report that grow is most active during the first 20 years before the onset of reproduction. The current analysis done for Raifa's *J. mandshurica* shows that the radial growth of the first-generation trees over the first 20 years is 52 mm on average, as well as 89 mm for the second-generation trees, which is significantly higher ($T_{Student} = 3.48$; p-value = 0.002). This fact indicates that the first-generation individuals of *J. mandshurica* acclimated in the Raifa forest to produce the second-generation specimens which are characterized by a more rapid biomass development.

4. Conclusions

The research undertaken allows description of *J. mandshurica* as a successfully acclimated species to the temperate continental climate conditions of the coniferous-deciduous forests of the Middle Volga region: it has naturalized in forests of the nature reserve, has a high biomass productivity and high germination capacity of seeds, and the pre-generative specimens have a high abundance.

P. amurense species do not demonstrate successful acclimation to the conditions of forest habitats of temperate climate. Under the current circumstances, *P. amurense* does not propagate due to the intensive spreading of *Tilia cordata* and due to the development of natural secondary permanent linden forests. The radial growth, despite being stable every year, produces subtle, narrow rings. Thus, the abundance and viability of *P. amurense* decrease due to occasional droughts and cold winters typical for the temperate continental climate of the Middle Volga, due to the habitat shading, due to the low competitiveness of the species and its inability to reproduce under the conditions of the natural forest.

References

[1] Vasil'ev N P and Vasin E A *Biological Diversity. Introduction of Plants. Proc. III Int. Sci. Conf* p 179–180

[2] Reif O Yu 2013 *The State of Forests and Urgent Problems of Forest Management. Proc. All-*

Russ. Conf. with Int. Pariticipation (Khabarovsk: Izd. FBU Dal'NIILKh) p 402–405
[3] Cheng Z, Yao G, Guo R, Huang X and Song S 2017 *Bioorganic & Medicinal Chemistry Letters* **27** 597-601
[4] Ivanov V B 2005 *Trudy VKGZ* **6** 343–375
[5] Gurskii A V 1957 *Main Results of Introduction of Tree Plants in the USSR* (Moscow, Leningrad: Izd. Akad. Nauk SSR) p 304
[6] Abaimov V F 2009 *Dendrology* (Moscow) p 280
[7] Deryuga E S and Murzov A M 1977 *Trudy VKGZ* **3** 61–79
[8] Bakin O V, Rogova T V, Sitnikov A P 2000 *Vascular Plants of Tatarstan* (Kazan: Izd. Kazan. Univ.) p 496
[9] Bakin O V 2001 *Raifa-Sviyazhsk* (Kazan) p 66–70
[10] Shuyatov S G 2000 *Methods of Dendrochronology. Part I: Basics of Dendrochronology. Sampling and Obtaining Data from Tree Rings* (Krasnoyarsk: KrasGU) p 80
[11] Fardeeva M B, Tishin D V and Safiulina N I 2016 *VI Int. Sci. Conf. Biological Diversity. Introduction of Plants* (St. Petersburg) p 380–383
[12] Tishin D V and Fardeeva M B 2017 *Proc. Int. Sci. Conf. Role of Botanical Gardens and Arboreta in Preserving, Studying, and Stable Use of the Diversity of Plant Community* (Minsk) p 139–141

IOP Conf. Series: Earth and Environmental Science **107** (2017) 012095 doi:10.1088/1755-1315/107/1/012095

Mathematical modeling of a dynamic thin plate deformation in acoustoelasticity problems

I B Badriev[1] and V N Paimuhin[1, 2]

[1] Institute of Computational Mathematics and Information Technologies, Kazan Federal University, Kremlyovskaya str.,18, Kazan 420008 Russian Federation
[2] Institute of Aviation, Land Vehicles and Energetics, Kazan National Research Technical University, 10 K.Marks str., Kazan, 420111 Russian Federation

E-mail: ildar.badriev1@mail.ru

Abstract. The coupled problem of planar acoustic wave propagation through a composite plate covered with a second damping layer with a large logarithmic decrement of oscillations is formulated. The aerohydrodynamic interaction of a plate with external acoustic environment is described by three-dimensional wave equations and the mechanical behavior of a two-layer plate by the classical Kirchhoff-Love model. An exact analytic solution of the problem is found for the case of hinged support of the edges of a plate. On the basis of this, the parameters of the covering damping layer were found, under which it is possible to achieve a practically complete damping of the plate vibration under resonant modes of its acoustic loading.

1. Introduction

At present, noise pollution of the environment is one of the actual problems for Russia and a lot of technologically advanced countries of the world. The magnitude of the permissible vibration of any design is determined by its influence on the strength characteristics of the structure and its elements, on the performance, people's health, the work of the installed equipment, etc. In terms of providing strength characteristics, one of the most dangerous modes of dynamic deformation of structures is resonance. It is realized in the construction when the frequencies of its natural oscillations coincide with the frequency of external cyclic action. As it is well known, under this loading regime the amplitude values of the parameters of the dynamic stress-strain state are multiplied. A correct and reliable theoretical definition with the accuracy necessary for practical purposes requires proper consideration in the calculated ratios of the damping properties of structural materials caused by internal friction. The magnitude of the permissible vibration of the structure must be limited by the amount of permissible noise that is formed in the surrounding environment of the acoustic environment as a result of its dynamic interaction with the deforming structure. Vibration problems of mechanical systems are mainly handled by specialists in the mechanics of the deformed body, the dynamics and strength of machinery, instruments and equipment, the strength of aircraft, ships, etc., without paying due attention to the problems created by the construction of noise when deforming them. But questions of formation and propagation of noise are studied by experts in the field of acoustics. The literature devoted to the study of these questions is quite extensive (see, e.g., [1–4]). Nevertheless, studies in the field of the acoustoelasticity of thin-walled structural elements are currently among the topical and priority scientific research. Note that the described problems are very multifaceted, have a complex character and for their solution further experimental and theoretical research is required. In order to reduce the dynamic tension and noise level, many structural elements are made in the form of multilayer structures [5–11]) consisting of a thickness of rigid carriers and soft damping layers. Composite materials based on high-strength carbon or glass fibers have become widely used construction materials for the manufacture of products of a particular purpose by now. They have a sufficiently low level of impact strength, internal friction parameters and shock energy absorption. In connection with this, one of the directions of increasing these properties of structures is their production by gluing from various composite materials. The basis of such materials are traditional composites with carbon or glass fibers, and the protective layer are various types of reinforced or unreinforced elastomers, which have a high impact strength and a high level of absorption of impact energy. In particular, as a protective layer, it is expedient to use composites made of high-strength high-modulus polyethylene fibers (HHMP) with an elastomer matrix. They differ from other high-strength fibers not only by a higher level of specific mechanical characteristics, but also by a minimum coefficient of friction, a positive effect of the strain rate on strength, a sharp increase in strength in the negative temperature region, and other properties. It should be noted that due to poor adhesion with almost many plastics, HHMP were mainly used for making cords and ropes. However, technologies have recently appeared that allow irradiation with cold plasma to increase substantially the adhesion properties of HHMP fibers and to create composites based on them [12–14]. This paper is devoted to the further development of the described direction of research. The main goal is to construct a mathematical model of dynamic deformation of thin plates for small displacements and deformations and allow for internal damping of the layer materials by the Thompson-Kelvin-Voigt model, and also to use this model to solve the problem of sound transmission Wave through a rectangular two-layer plate. As an energy-absorbing coating, elastomeric materials reinforced with rectilinear fibers from HHMP are considered, which have increased values of elastic characteristics while maintaining high values of internal friction parameters. It will be shown that the pasted layers, having high damping properties, allow significantly reducing the amplitude values of deformations and displacements of structural elements of structures under vibration loading conditions, thus creating a low level of sound pressure in the salons. Moreover, the use of such special coatings with large damping properties leads to a multiple decrease in the level of the cyclic stresses formed in the elements and, as a consequence, to a multiple increase in the life (resource) of the structures.

2. The problem statement

We consider a plate consisting of two thin layers with thicknesses t_1, t_2 and made of orthotropic materials. We assign it to an orthogonal Cartesian coordinate system Ox_1x_2z in which the coordinate plane z=0 coincides with the middle plane of the first layer, and the axes x_1, x_2 coincide with the orthotropy axes of the layers materials. Let u_1, u_2, w be the displacements of the points of the plane σ in the directions of the axes x_1, x_2, z through which, according to the Kirchhoff-Love model, at a middle bending of the plate, the components of displacements and deformations of its arbitrary point are determined by the relations

$$U_i = u_i - zw_{,i}, \quad \varepsilon_{ij} = \varepsilon_{ij}^0 - zw_{,ij}, \quad -t_1/2 \le (t_1/2 + t_2); \quad 2\varepsilon_{ij}^0 = u_{i,j} + u_{j,i} + w_{,i}w_{,i}, \tag{1}$$

where the conventional notations for partial derivatives with respect to coordinates x_i were used. If we assume that the materials of the layers have viscoelastic properties, then to describe dynamic deformation processes that vary in time τ according to a harmonic law with circular frequency ω the components of the stresses of the k-th layer with deformation components (1) can be related by defining relations

$$\sigma_{11}^{(k)}=\frac{E_1^{(k)}}{1-\nu_{12}^{(k)}\nu_{21}^{(k)}}\left(1+\frac{\delta_1^{(k)}}{\pi\omega}\frac{\partial}{\partial\tau}\right)\left[\varepsilon_{11}^0+\nu_{21}^{(k)}\varepsilon_{22}^0-z(w_{,11}-\nu_{12}^{(k)}w_{,22})\right]\ \underset{\leftarrow}{\overset{\rightarrow}{1,2}};$$
$$\sigma_{12}^{(k)}=2G_{12}^{(k)}\left(1+\frac{\delta_{12}^{(k)}}{\pi\omega}\frac{\partial}{\partial\tau}\right)(\varepsilon_{12}^0-zw_{,12}). \tag{2}$$

These equalities show a generalization of the relations in [15, 16] used in the articles [17–21] for statement of some acoustoelasticity problems of the plate of isotropic material. Here and below $-t_1/2\le t_1/2$ for $k=1$; $-t_1/2\le(t_1/2+t_2)$ for $k=2$; $\delta_i^{(k)}$, $\delta_{12}^{(k)}$ are the logarithmic decrements of oscillations for axial and shear deformations of the material of the k-th layer, the determination of which is possible on the basis of the technique [22–24]; $E_1^{(k)}$, $E_1^{(k)}$, $E_1^{(k)}$ $\nu_{12}^{(k)}$ are the elastic modules of the first and second kinds, and also the Poisson coefficients. With respect to the internal forces and moments brought to the plane, we have

$$T_{11}=B_1\varepsilon_{11}^0+\widetilde{B}\varepsilon_{22}^0-B_1^{(2)}H(w_{,11}+\nu_{21}^{(2)}w_{,22})/2,\ T_{12}=2B_{12}\varepsilon_{12}^0-B_{12}^{(2)}Hw_{,12};\ \underset{\leftarrow}{\overset{\rightarrow}{1,2}};$$
$$M_{11}=-B_1^{(2)}H(\varepsilon_{11}^0+\nu_{21}^{(2)}\varepsilon_{22}^0)/2+D_1w_{,11}+\widetilde{D}w_{,22},\ M_{12}=-B_{12}^{(2)}H\varepsilon_{12}^0+2D_{12}w_{,12};\ \underset{\leftarrow}{\overset{\rightarrow}{1,2}}. \tag{3}$$

where

$$H=t_1+t_2,\ B_1^{(k)}=\frac{E_1^{(k)}t_k}{1-\nu_{12}^{(k)}\nu_{21}^{(k)}}\left(1+\frac{\delta_1^{(k)}}{\pi\omega}\frac{\partial}{\partial\tau}\right),\ B_{12}^{(k)}=2G_{12}^{(k)}t_k\left(1+\frac{\delta_{12}^{(k)}}{\pi\omega}\frac{\partial}{\partial\tau}\right),$$
$$B_1=B_1^{(k)}+B_2^{(k)},\ \widetilde{B}=B_1^{(1)}\nu_{21}^{(1)}+B_1^{(2)}\nu_{21}^{(2)},\ B_{12}=B_{12}^{(1)}+B_{12}^{(2)};\ \underset{\leftarrow}{\overset{\rightarrow}{1,2}};\ D_s^{(1)}=B_s^{(1)}t_1^2/12,\ D_s=D_s^{(1)}+D_s^{(2)},$$
$$D_s^{(2)}=B_s^{(2)}t_2^2\kappa/12,\ s=1,2,\ D_{12}^{(1)}=B_{12}^{(1)}t_1^2/12,\ D_{12}^{(2)}=B_{12}^{(2)}t_2^2\kappa/12,\ \widetilde{D}=D_1^{(1)}\nu_{21}^{(1)}+D_1^{(2)}\nu_{21}^{(2)},$$
$$D_{12}=D_{12}^{(1)}+D_{12}^{(2)},\ \kappa=4+3t_1^2/t_2^2+6t_1/t_2. \tag{4}$$

For the element selected from the plate, we write the motion equations

$$T_{s1,1}+T_{s2,2}-q\ddot{u}_s=0,\ q=r_1t_1+r_2t_2;\ s=1,2; \tag{5}$$
$$M_{11,11}+2M_{12,12}+M_{22,22}+(T_{11}w_{,1}+T_{12}w_{,2})_{,1}+(T_{12}w_{,1}+T_{22}w_{,2})_{,2}+q\ddot{w}-p=0. \tag{6}$$

where the dots over the letters denote the partial derivatives with respect to τ, r_k is the density of the material of the k-th layer, p is the external aerohydrodynamic load acting on the plate, to be determined. Accounting for this load, along with taking into account the internal friction of the material, is necessary for correct and reliable determination of the parameters of the stress-strain state of structural elements under dynamic loading processes. Below we restrict ourselves to the investigation of two successive stages of deformation of the plate. In the first stage, which is static, a field of stresses in the plate is formed in coordinates x_1, x_2 and equivalent to the forces N_{sj}. It is inhomogeneous, in general. At the second stage, in the vicinity of the stress-strain state of the first stage, a cyclic deformation process is realized with the formation in the plate of such flexural stresses and deformations that have little effect on the generated forces N_{sj} of the first stage. By virtue of the assumptions made, from the conditions $T_{sj}=N_{sj}$, which are valid for small displacements, the last relations (3) can be transformed to the form

$$M_{11}=d_1w_{,11}+\widetilde{d}_1w_{,22}+g_1N_{11}+\widetilde{g}_1N_{22},\ M_{12}=2d_{12}w_{,12}-g_{12}N_{12};\ \underset{\leftarrow}{\overset{\rightarrow}{1,2}}, \tag{7}$$

where

$$d_1 = D_1 - B_1^{(2)} H^2 (G_1 - v_{21}^{(2)} \widetilde{G}_2)/(4B_d),\ \widetilde{d}_1 = \widetilde{D} + B_1^{(2)} H^2 (\widetilde{G}_1 - v_{21}^{(2)} G_2)/(4B_d);\ B_d = B_1 B_2 - \widetilde{B}^2,$$
$$G_1 = B_1^{(2)} B_2 - v_{12}^{(2)} B_1^{(2)} \widetilde{B},\ \widetilde{G}_1 = B_2^{(2)} \widetilde{B} - v_{21}^{(2)} B_1^{(2)} B_2,\ g_1 = -B_1^{(2)} H^2 (B_2 - v_{21}^{(2)} \widetilde{B})/(2B_d);$$
$$\widetilde{g}_1 = B_1^{(2)} H (\widetilde{B} - v_{21}^{(2)} B_1)/(2B_d);\ \ \underset{\leftarrow}{\overset{\rightarrow}{1,2}};\ \ d_{12} = D_{12} - (B_{12}^{(2)} H)^2/(4B_{12});\ g_{12} = B_{12}^{(2)} H/(2B_{12}). \tag{8}$$

Using relations (7), equation (6) is reduced to the resolving equation, expressed in terms of a function w, which in the case $T_{sj} = N_{sj} = const$ is written in the form

$$d_1 w_{,1111} + (\widetilde{d}_1 + \widetilde{d}_2 + 4d_{12}) w_{,1122} + d_2 w_{,2222} + N_{11} w_{,11} + 2N_{12} w_{,12} + N_{22} w_{,22} - p + q\ddot{w} = 0. \tag{9}$$

The boundary and initial conditions for the obtained equation (9) are formulated in the same form as in the classical theory of plates. In accordance with the assumptions made, the total displacements of the second (dynamic) stage of deformation will be equal $\overset{*}{w} = w + \Delta w,\ \ \overset{*}{u_i} = u_i + \Delta u_i$, where Δw, Δu_i are small in comparison with w and u increments of displacements. In this case for the deformation components we have the following relations $\overset{*}{\kappa_{ij}} = w_{ij} + \Delta w_{ij}$, $2\overset{*}{\varepsilon_{ij}} = u_{i,j} + u_{j,i} + w_{,i} w_{,j} + \Delta u_{i,j} + \Delta u_{j,i} = 2\varepsilon_{ij}^0 + \Delta\varepsilon_{ij}$, to which the moments and efforts are corresponded to $\overset{*}{M_{ij}} = M_{ij} + \Delta M_{ij}$, $\overset{*}{T_{ij}} = T_{ij} + \Delta T_{ij}$. The values of ΔT_{ij} have to satisfy the conditions

$$\Delta T_{ij} = B_1 \Delta\varepsilon_{11} + \widetilde{B} \Delta\varepsilon_{22} - B_1^{(2)} H (\Delta w_{,11} + v_{21}^{(2)} \Delta w_{,22})/2 = 0,\quad \Delta T_{12} = 2B_{12} \Delta\varepsilon_{12} + B_{21}^{(2)} H \Delta w_{,12} = 0 \tag{10}$$

and ΔM_{ij} are equal

$$\Delta M_{11} = -B_1^{(2)} H (\Delta\varepsilon_{12}^0 + v_{21}^{(2)} \Delta\varepsilon_{22}^0)/2 + D_1 \Delta w_{,11} + \widetilde{D} \Delta w_{,22},\quad \Delta M_{12} = -B_{12}^{(2)} H \Delta\varepsilon_{12}^0 + 2D_{12} \Delta w_{,12}. \tag{11}$$

Determining from the conditions (10) the quantities $\Delta\varepsilon_{11}^0$, $\Delta\varepsilon_{22}^0$, $2\Delta\varepsilon_{12}^0$ and inserting them into relations (11), we arrive at the relations $\Delta M_{11} = d_1 \Delta w_{,11} + \widetilde{d}_1 \Delta w_{,22};\ \underset{\leftarrow}{\overset{\rightarrow}{1,2}},\ M_{12} = 2d_{12} \Delta w_{,12}$, in which the coefficients d_1, $\widetilde{d}_1$, d_{12} are determined by the formulas (8). It can be shown that in order to determine the function Δw in the approximation under consideration it is necessary that the variational equation of the form

$$\int_{\tau_0}^{\tau_1} \iint_{\sigma} [N_{ij} \Delta w_{,j} \sigma(\Delta w_{,i}) + (d_1 \Delta w_{,11} + \widetilde{d}_1 \Delta w_{,22}) \delta(\Delta w_{,11}) + (\widetilde{d}_2 \Delta w_{,11} + d_2 \Delta w_{,22}) \delta(\Delta w_{,22}) +$$
$$+ 4d_{12} \Delta w_{,12}) \delta(\Delta w_{,12}) - (\rho * q \Delta\ddot{w}) \delta(\Delta w)] dx_1\, dx_2\, d\tau = 0,$$

is satisfied. From the last equation, under assumption $N_{sj} = const$, the motion equation (9) follows, if Δw is replaced by w in it.

3. Construction of solution

Suppose that the plate is rectangular in plan with sizes a, b in the axes $x = x_1$, $y = y_1$ directions, and is surrounded on both sides by the acoustic media "1" and "2" occupying the half-spaces V_1 and V_2 bounded by the plane z=0. A plane harmonic wave, characterized by pressure p_* and frequency ω, is incident on the plate. As a result of its interaction with the plate in the surrounding half-spaces V_1 and

V_2, acoustic waves are excited, which are reflected and radiated in the first medium and radiated in the second medium. These waves with respect to the velocity potentials Φ_*, Φ_1, Φ_2 are described by the wave equations, which we write in the approximation (here and below k=1,2)

$$\Phi_{*,zz} - c_1^{-2}\ddot{\Phi}_* = 0, \quad \Phi_{k,xx} + \Phi_{k,yy} + \Phi_{k,zz} - c_k^{-2}\ddot{\Phi}_k = 0, \tag{12}$$

where c_k is the sound velocities in media "1" and "2". Through functions Φ_*, Φ_k, the pressures p_*, p_k, and velocity components $\overset{*}{v_z}$, $\overset{*}{v_k}$ in half-spaces V_1 and V_2 are determined by the relations (ρ_k are the densities of media "1" and "2")

$$p_* = -\rho_1\dot{\Phi}_*, \quad \overset{*}{v_z} = \Phi_{*,z}, \quad p_k = -\rho_k\dot{\Phi}_* \quad \overset{k}{v_z} = \Phi_{k,z}. \tag{13}$$

In this case, the aerohydrodynamic load p acting on the plate will be equal to

$$p = (p_* + p_1 - p_2)\big|_{z=0}, \tag{14}$$

and the velocity components $\overset{*}{v_z}$, $\overset{*}{v_k}$ for the uninterrupted interaction of acoustic media with the plate, have to satisfy the conditions

$$\dot{w} = (v_z^1 + v_z^2)\big|_{z=0}, \quad \dot{w} = v_z^2\big|_{z=0}. \tag{15}$$

For harmonic waves, the following representations (i is the imaginary unit) are valid

$$\Phi_* = \tilde{\Phi}_* e^{i\omega\tau}, \quad \Phi_k = \tilde{\Phi}_k e^{i\omega\tau}, \quad w = \tilde{w}e^{i\omega\tau}, \tag{16}$$

and for a plate hingedly supported at the edges $x=0$, $x=a$, $y=0$, $y=b$ the solution of equation (9) will have the form

$$w = \sum_{m,n=1,3,\ldots}^{M,N} \tilde{w}_{mn} \sin\lambda_m x \sin\lambda_n y e^{i\omega\tau}, \tag{17}$$

where $\lambda_m = m\pi/a$, $\lambda_n = n\pi/b$; $m,n = 1,3,\ldots,M,N$. By virtue of (15), the solutions of the last two equations in (12) must be sought in the form

$$\Phi_k = \sum_{m,n=1,3,\ldots}^{M,N} \tilde{\Phi}_k^{mn} \sin\lambda_m x \sin\lambda_n y e^{i\omega\tau}. \tag{18}$$

Taking into account the representations (16)–(18), equations (12) are transformed to the form

$$\Phi_{*,zz} + k_1^2\tilde{\Phi}_* = 0, \quad \tilde{\Phi}_{k,zz}^{mn} - (\kappa_k^{mn})^2\tilde{\Phi}_k^{mn} = 0, \tag{19}$$

where

$$\kappa_k^{mn} = \sqrt{\lambda_m^2 + \lambda_n^2 - k_k^2}, \quad k_k^2 = \omega^2 / c_k^2. \tag{20}$$

In view of (16), the first equation for the incident wave has a solution $\tilde{\Phi}_* = A_* e^{-i(k_k - \omega\tau)}$, using which, in accordance with relations (13), we arrive at the dependences

$$p_*\big|_{z=0} = \tilde{p}_*\big|_{z=0} e^{i\omega\tau}, \ \tilde{p}_*\big|_{z=0} = -i\rho_1 A_*, \quad \overset{*}{v_z} = \tilde{v}_z^*\big|_{z=0} e^{i\omega\tau}, \ \tilde{v}_z^*\big|_{z=0} = -ik_1 A_*, \tag{21}$$

where A_* is an integration constant, uniquely determined for given frequency ω and sound pressure level $p_*\big|_{z=0}$. Later, when performing the calculations, we shall consider a parameter A_* to be given, the value of which determines the pressure and the velocity of the incoming wave. The solution of the remaining equations (19) depends on the sign of $(\kappa_k^{mn})^2$. In the case $(\kappa_k^{mn})^2 > 0$ we take them in the form [18–20]

$$\tilde{\Phi}_1^{mn} = B_1^{mn} e^{\kappa_1^{mn} z}, \tilde{\Phi}_2^{mn} = A_2^{mn} e^{-\kappa_2^{mn} z} \text{ if } (\kappa_k^{mn})^2 > 0, \tilde{\Phi}_1^{mn} = B_1^{mn} e^{i\tilde{\kappa}_1^{mn} z}, \tilde{\Phi}_2^{mn} = A_2^{mn} e^{-i\tilde{\kappa}_2^{mn} z} \text{ if } (\kappa_k^{mn})^2 < 0, \tag{22}$$

where $(\widetilde{\kappa}_k^{mn})^2 = -(\kappa_k^{mn})^2$, B_1^{mn}, A_2^{mn} are the integration constants. We expand A_* in a Fourier series

$$A_* = \sum_{m,n=1,3,\ldots}^{M,N} f_{mn} \sin\lambda_m x \sin\lambda_n y e^{i\omega\tau}, \qquad f_{mn} = 16/(\pi^2 mn) \tag{23}$$

and the resulting solutions (22) and equation (17), (21) we subject the conditions (15). As a result, to determine the integration constants, we obtain the dependences

$$B_1^{mn} = i\omega(f_{mn}/c_1 + \widetilde{w}_{mn})/\kappa_1^{mn}, \quad A_2^{mn} = -i\omega\widetilde{w}_{mn}/\kappa_2^{mn}. \tag{24}$$

Here and in the future, the quantities κ_1^{mn} are calculated from formulae (20), if $\lambda_m^2 + \lambda_n^2 - k_k^2 > 0$, and by the formulae $\kappa_k^{mn} = \sqrt{k_k^2 - \lambda_m^2 - \lambda_n^2}$ for $\lambda_m^2 + \lambda_n^2 - k_k^2 < 0$. Using the dependences (24) to determine p, in accordance with relations (13), (14), (16), (18), (21)–(23), one can obtain relations

$$p = \sum_{m,n=1,3,\ldots}^{M,N} \widetilde{p}_{mn} \sin\lambda_m x \sin\lambda_n y e^{i\omega\tau}, \tag{25}$$

where $\widetilde{p}_{mn} = -i\rho_1\omega f_{mn}[1 + t\omega/(c_1\kappa_1^{mn})]A_* + \varphi_{mn}w_{mn}$, $\varphi_{mn} = \rho_1\omega^2/\kappa_1^{mn} + \rho_2\omega^2/\kappa_2^{mn}$. If $N_{12} = 0$, then the substitution of relations (17) and (25) into the motion equation (9) leads to the solution

$$\widetilde{w}_{mn} = -R_{mn}A_*/L_{mn}, \quad R_{mn} = i\rho_1\omega f_{mn}[1 + t\omega/(c_1\kappa_1^{mn})], \qquad L_{mn} = d_{mn}, \tag{26}$$

where $d_{mn} = d_1^*\lambda_m^4 + (\widetilde{d}_1^* + \widetilde{d}_2^* + 4d_{12}^*)\lambda_m^2\lambda_n^2 + d_2^*\lambda_n^4 - N_{11}\lambda_m^2 - N_{22}\lambda_n^2 - q\omega^2$, d_1^*, d_2^*, $\widetilde{d}_1^*$, $\widetilde{d}_2^*$, d_{12}^* are calculated by (4), (8) with the replacement of $B_s^{(k)}$, $B_{12}^{(k)}$ by $B_s^{(k)*} = E_s^{(k)}t_k(1 + i\delta_s^{(k)}/\pi)/(1 - \nu_{12}^{(k)}\nu_{21}^{(k)})$, $B_{12}^{(k)*} = G_{12}^{(k)}t_k(1 + i\delta_{12}^{(k)}/\pi)$. In accordance with the obtained solution (26), the amplitude values of the sound pressures at the points of the boundary planes $z = -t_1/2$ and $z = -t_1/2 + t_2$ of the plate will be equal

$$\widetilde{P}_1 = -i\rho_1\omega \sum_{m,n=1,3,\ldots}^{M,N} [f_{mn}[1 + t\omega/(c_1\kappa_1^{mn}) - i\omega R_{mn}/(\kappa_1^{mn}L_{mn})]A_*, \quad \widetilde{P}_2 = \rho_2\omega \sum_{m,n=1,3,\ldots}^{M,N} R_{mn}A_*/(\kappa_2^{mn}L_{mn}).$$

The given value of A_* and the values of $\widetilde{P}_1$, $\widetilde{P}_2$ found, the sound-insulating properties of the plate and the sound pressure levels at the points of the boundary planes will be characterized by the following parameters ($\widetilde{p}_0$ is the amplitude value of the sound pressure corresponding to the audibility threshold):

$$R_p^0 = -20\lg|\widetilde{P}_2/\widetilde{P}_1|, \quad R_p^- = -20\lg|\widetilde{P}_1/\widetilde{p}_0|, \quad R_p^+ = -20\lg|\widetilde{P}_{21}/\widetilde{p}_0|, \tag{27}$$

and the amplitude values of the dynamic components of the internal bending moments formed in the plate from the action of the sound wave incident on it will be determined from formulae of the form

$M_{11}^{\partial} = \sum_{m,n=1,3,\ldots}^{M,N} [(d_1\lambda_m^2 + \widetilde{d}_1\lambda_n^2)R_{mn}/L_{mn}]A_*, \ldots$, which follow from relations (7), (17) (26).

4. Results and discussion

On the basis of the constructed solutions for a plate made of fiberglass and having characteristics $a = 0.48$ m, $b = 0.56$ m, $t_1 = 2$ mm, $E_1^{(1)} = 18$ GPa, $E_2^{(1)} = 14$ GPa, $\nu_{12}^{(1)} = 0.3$ GPa, $G_{12}^{(1)} = 2.9$ GPa, $\delta_1^{(1)} = 0.03$, $\delta_2^{(1)} = 0.035$, $\delta_{12}^{(1)} = 0.03$, $r_1 = 3000$ kg/m³, at $\widetilde{p}_0 = 2\cdot10^{-5}$ Pa, $A_* = 1$, $N_{ij} = 0$ and at different frequencies f of the sound wave, calculations were performed to determine the amplitude values of the acoustic insulation parameters (27), deflection w_0, and the dynamic component of the internal bending moment M_{11}^{∂}. The results of calculations for a single-layer plate ($t_2 = 0$) and a double-layer plate ($t_2 = 5$ mm) with a damping layer of rubber having the characteristics $E_1^{(2)} = E_2^{(2)} = 500$ MPa, $\nu_{12}^{(2)} = \nu_{21}^{(2)} = 0.4$, $\delta_1^{(2)} = \delta_2^{(2)} = 1.2$, $\delta_{12}^{(2)} = 0.9$, $r_2 = 1500$ kg/m^3 are shown in figure 1, 2. The solid lines correspond to the $t_2 = 0$, dashed lines correspond to the $t_2 = 5$ mm. Analyzing the obtained results, we can conclude that the presence of a damping layer on the plate, which has the considerable logarithmic decrements of oscillations values, practically does not affect the characteristic R_p^-. Depending on the frequency of the incident sound wave, the existence of a layer can lead both to a multiple increase, and to a certain decrease (at resonant frequencies) of the sound insulation parameter R_p^0 (figure 1a). In wide ranges of frequency changes, on the average, it leads to a significant decrease in the characteristic R_p^+ (and, consequently, the sound pressure level in the half-space V_2 (figure 1b)). Due to the introduction of a damping layer, it is possible to reduce significantly the values of the vibro-displacement (figure 2a) and the bending moment (and, consequently, the stresses) (figure 2b).

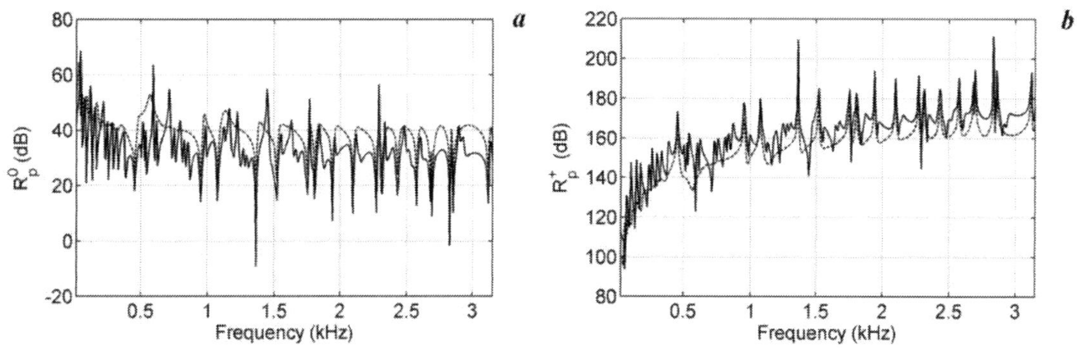

Figure 1. Dependences of the soundproofing parameter (a) and sound pressure parameter (b) on the frequency of oscillation

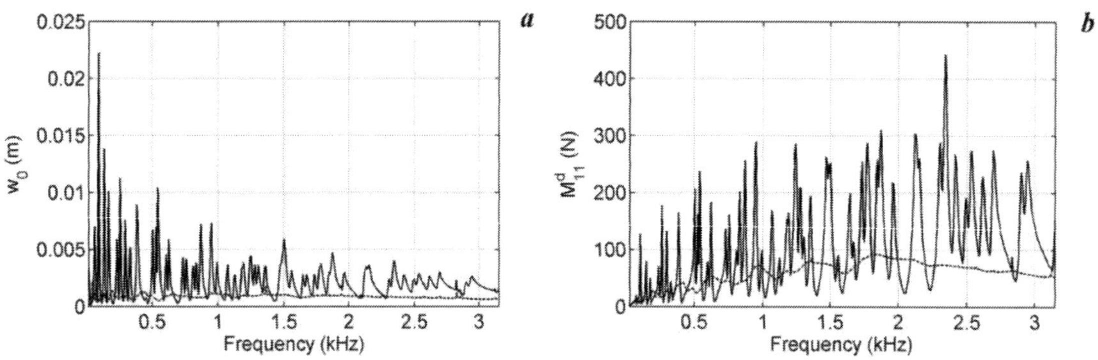

Figure 2. Dependences of amplitude values of deflection (a) and amplitude values of the bending moment (b) on the frequency of oscillations.

The analysis of the obtained results let us formulate the conclusion that functional rubber-like materials, pasted on power thin-walled structural elements in the form of a second thin layer and possessing high internal damping properties (and, consequently, energy absorbing properties) allow to reduce significantly the amplitude values of deformations and displacements at vibrational (even resonant) modes of their loading, thus forming a lowered level of radiated sound pressure. The use of such special coatings with large values of the logarithmic decrement of vibrations leads to a multiple decrease in the level of the cyclic stresses formed in the elements and, as a result, to a multiple increase in the life (resource) of the structures.

In conclusion, we note that the obtained results can be used to establish and study practical problems of acousto-elasticity. They are important from environmental safety point of view, and connected, in particular, with the design of noise-shielding screens.

Acknowledgement
The research results obtained in the framework of the state order Ministry of Education and Science of Russian Federation (project № 9.1395.2017/PP) and with financial support of Russian Science Foundation (project 14-19-00667).

References

[1] Aureli M, Basaran M E and Porfiri M 2012 Nonlinear finite amplitude vibrations of sharp-edged beams in viscous fluids *Journal of Sound and Vibration* **331** 1624-54
[2] Badriev I B, Makarov M V and Paimushin V N 2015 On the interaction of composite plate having a vibration-absorbing covering with incident acoustic wave *Russian Mathematics* **59**(3) 66-71.
[3] Gorshkov A G, Morozov V I, Ponomarev A T and Shklyarchuk F N 2000 *Aerohydroelasticity of structures* (Moscow, Fizmatlit) (in Russian).
[4] Paimushin V N 2014 Dynamic instability of a compressible plate surrounded by an acoustic medium on free surfaces *Doklady Physics* **59**(3) 158-62
[5] Badriev I B, Makarov M V and Paimushin V N 2015 Solvability of a physically and geometrically nonlinear problem of the theory of sandwich plates with transversal-soft core *Russian Mathematics* **59**(10) 57-60
[6] Badriev I B, Makarov M V and Paimushin V N 2016 Mathematical Simulation of Nonlinear Problem of Three-point Composite Sample Bending Test *Procedia Engineering* **150** 1056-62
[7] Sultanov L U and Davydov R L 2013 Numerical analysis of large deformation by finite element method *Magazine of Civil Engineering* **44**(9) 64-8
[8] Badriev I B, Makarov M V and Paimushin V N 2016 Numerical Investigation of Physically Nonlinear Problem of Sandwich Plate Bending *Procedia Engineering* **150** 1050-5
[9] Sultanov L U 2016 Analysis of finite elasto-plastic strains. Medium kinematics and constitutive equations *Lobachevskii Journal of Mathematics* **37**(6) 787-93
[10] Badriev I B, Makarov M V and Paimushin V N 2017 Contact statement of mechanical problems of reinforced on a contour sandwich plates with transversally-soft core *Russian Mathematics* **61**(1) 69-75
[11] Badriev I B, Garipova G Z, Makarov M V, Paimushin V N and Khabibullin R F 2015 Solving Physically Nonlinear Equilibrium Problems for Sandwich Plates with a Transversally Soft Core *Lobachevskii Journal of Mathematics* **36**(4) 474-81
[12] Moon S I and Jang J J 1999 The effect of polybutadiene interlayer on interfacial adhesion and impact properties in oxygen-plasma-treated UHMPE fiber/epoxy composites *Composites. Part A.* **30** 1039–44.
[13] Kudinov V V, Shaekhov M F and Korneeva N V 2004 Effect of Plasma Treatment and Impregnation Technology on the Joint Strength of Epoxy Matrix with a Polyethylene Fiber during the the Production of Composite Materials *Fiz. Khim. Obrab. Mater.* **3** 18–24
[14] Sergeeva E A 2010 Effect of plasma treatment on the structure and properties of high-modulus polyethylene fibers *Voprosyi materialovedeniya* **2** 51-7 (in Russian)
[15] Panovko Ya G 1960 *Internal friction in the case of vibrations of elastic systems* (Moscow: Fizmatgiz) (in Russian)

[16] Khilchevsky V V and Dubenets V G 1977 *Scattering of energy in vibrations of thin-walled structural elements* (Kiev: Vischa shkola) (in Russian)
[17] Gazizullin R K and Paimushin V N 2016 The transmission of an acoustic wave through a rectangular plate between barriers *Journal of applied mathematics and mechanics* **80**(5) 421-32.
[18] Paimushin V N 2015 Sound-wave passage through a composite plate with a vibration-absorbing surface *Doklady physics* **60**(6) 263-8
[19] Paimushin V N 2014 On the Problems of Emission of Sound Waves in the Dynamic Process of Deformation of Plates with Regard for External and Internal Damping *Journal of Mathematical Sciences* **203**(1) 87-103
[20] Paimushin V N, Tarlakovskii D V, Gazizullin R K and Lukashevich 2016 A Investigation of Different Versions of Formulation of the Problem of Soundproofing of Rectangular Plates Surrounded with Acoustic Media *Journal of Mathematical Sciences* **203**(1) 87-103
[21] Badriev I B, Makarov M V and Paimushin V N 2015 On the interaction of composite plate having a vibration-absorbing covering with incident acoustic wave *Russian Mathematics* **59**(3) 66-71
[22] Paimushin V N, Firsov V A, Gyunal I and Egorov A G 2014 Theoretical-Experimental Method for Determining the Parameters of Damping Based on the Study of Damped Flexural Vibrations of Test Specimens 1. Experimental Basis *Mechanics of composite materials* **50**(2) 127-36.
[23] Egorov A G, Kamalutdinov A M, Nuriev A N and Paimushin V N 2014 Theoretical-Experimental Method for Determining the Parameters of Damping Based on the Study of Damped Flexural Vibrations of Test Specimens 2. Aerodynamic Component of Damping *Mechanics of composite materials* **50**(3) 267-78
[24] Paimushin V N, Firsov V A, Gyunal I, Egorov A G and Kayumov R A 2014 Theoretical-experimental method for determining the parameters of damping based on the study of damped flexural vibrations of test specimens. 3. Identification of the characteristics of internal damping *Mechanics of composite materials* **50**(5) 633-46

IOP Conf. Series: Earth and Environmental Science **107** (2017) 012096 doi:10.1088/1755-1315/107/1/012096

MODELING OF HYDRAULIC FRACTURING ON THE BASIS OF THE PARTICLE METHOD

D V Berezhnoi[1], N F Gabsalikova[1], V G Izotov[2] and V V Miheev[1]

[1]N.I. Lobachevsky Institute of Mathematics and Mechanics, Kazan Federal University, Kremlyovskaya str., 18, 420008, Russia
[2]Tatarstan Academy of Sciences, Kazan, Russia

E-mail: berezhnoi.dmitri@mail.ru

Abstract. A technique of calculating the deformation of the soil environment when it interacts with a liquid on the basis of the particle method a is realized. To describe the behavior of the solid and liquid phases of the soil, a classical two-parameter Lennard-Jones interaction potential and its modified version proposed by the authors were chosen. The model problem of deformation and partial destruction of a soil massif under strong pressure from the liquid pumped into it is solved. Analysis of the results shows that the use of the modified Lennard-Jones potential for describing the solid phase of the soil environment makes it possible to describe the process of formation of cracks in the soil during hydraulic fracturing of the formation.

1. Introduction

One of the two basic technologies for the production of shale gas is the technology of hydraulic fracturing. Hydraulic fracturing of a formation is a process that involves the introduction of a mixture of water, sand and chemicals into the gas-bearing rocks under extremely high pressure (500-1500 atm). However, during hydraulic fracturing of the formation, consequences arise which pose a greater threat to the environment. There is a contamination of groundwater chemical reagents. There are destructive processes in the soil and even up to seismic activity. It is also worth mentioning the contamination of the soil from the drainage of waste water and a lot of other technological factors, as well as subsidence of the soil in the areas of hydraulic fracturing. It should be noted that it is very difficult to safely model and understand the danger of hydraulic fracturing in a given area, since the soil is a complex heterogeneous medium. It is possible that we will lose more from the consequences of hydraulic fracturing of the formation. However, it is also necessary to extract resources, especially in the current needs of mankind.

Nevertheless, modern methods of mathematical modeling allow us to investigate heterogeneous media. One of the methods that are capable of modeling destruction, mixing of different types of particles is the particle method. Thus, we will be able to predict the consequences of hydraulic fracturing of the formation for the environment and in advance, if possible, exclude them by reducing the pressure of hydraulic fracturing of the formation, by introducing any substances into the soil, which will increase its connectivity, or completely eliminate hydraulic fracturing in this place, etc.

At present, most of the computational methods are based on the discretization of the domain, and most often finite element and finite difference methods are used [1-12]. Such an approach has perfectly proved itself in solving problems of deformation of complex heterogeneous media of various natures. However, there is a wide class of problems in which loaded media are accompanied by multiple destruction and slippage of fragments, medium cracking, intense mass transfer, including mass mixing effects, etc. When solving such problems, the use of the grid approach is very difficult. A promising class of numerical methods for mechanics of a deformable solid body, adapted for modeling fracture processes, and in principle for the modeling of heterogeneous media, such as soil, is particle methods [13-18]. The theoretical basis for calculating the resulting crack, as well as known analytical models of the hydraulic fracturing process, are given in [19-24]. It should be noted that in the modern sense the term "particle methods" is collective and includes very diverse numerical methods, both relating to the classical representatives of the discrete approach in mechanics, and the meshless algorithms for numerical solution of the equations of continuum mechanics. Moreover, at present, some modern implementations of traditional numerical methods, such as the particle finite element method, are often referred to as particle methods. It should be noted that the particle methods were formed on the basis of the well-known molecular dynamics method used to study the response of the medium at the atomic level. At the same time, the possibilities of atomic description for modeling the behavior of a solid body on the spatial and temporal scales interesting from the point of view of engineering applications are extremely limited, and this description was used for modeling in most of the liquid and gaseous media. At present, the potentials of interatomic interaction for most materials are known. This can not be said about the potentials for describing the behavior of deformed continuous media, especially shear ones. Therefore, it is very important to model soil masses and interacting fluids based on the particle method with different pair interaction potentials for different phases. To describe the behavior of the liquid phase, the Lennard-Jones potential is used in this paper [17], and to describe the deformation of the soil massif, the "modified" Lennard-Jones potential introduced by the authors [25-27].

2. The problem statement

The fundamentals of the particle method, based on the use of pair interaction potentials, are as follows. Let's consider a collection of N material particles interacting with each other in a potential field. The motion of the particles is described on the basis of the following equation of motion:

$$m\ddot{\vec{r}}_k = \sum_{n=1}^{N} \Phi(r_{kn})\vec{r}_{kn} + \vec{\varphi}(\vec{r}_{kn}),$$

where $\vec{r}_k$ and $\vec{v}_k$ are the position and velocity vectors of the k -th particle

$$\vec{r}_{kn} \overset{def}{=} \vec{r}_k - \vec{r}_n,\ \vec{v}_{kn} \overset{def}{=} \vec{v}_k - \vec{v}_n,\ r_{kn} \overset{def}{=} |\vec{r}_{kn}|,\ v_{kn} \overset{def}{=} |\vec{v}_{kn}|,$$

m is the mass of the particle, $\Phi(r)$ describes the conservative component of the interaction between the particles, $\varphi(r)$ describes the external conservative force field. Conservative component of the interaction $\Phi(r)$ can be found by the formula:

$$\Phi(r) \overset{def}{=} \frac{1}{r} f(r),\ f(r) \overset{def}{=} -\dot{\Pi}(r),$$

where $f(r)$ is the scalar potential force, $\Pi(r)$ is the interaction potential.

Two potentials for pair interaction are used to describe continuous media. In the first case, this is the Lennard-Jones potential [17]

$$a = 0.4r,\ \Pi(r) \equiv D\left[(a/r)^{12} - 2(a/r)^{6}\right].$$

In the second case, the modified Lennard-Jones potential [29-31] was used

$$\Pi(r) = D\left[48.97(a/r)^{12} - 93.55(a/r)^{9} + 35.11(a/r)^{8} + 9.93(a/r)^{6}\right],\quad a = 0.4*1.28r.$$

To solve the problems of mechanics by the particle method, one has to deal with a large amount of data and computations. Therefore the most promising technology for solving these problems is heterogeneous computing clusters with the use of graphic accelerators (GPU). One GPU-device is equivalent to solving the problems of seismic wave propagation to about 50 central Xeon processors. However, this technology has certain limitations on the amount of available memory, since the internal memory of GPU devices is much smaller than the RAM of traditional cluster nodes. To implement calculations on graphics accelerators, the open source library VexCL is used. This library simplifies the development of applications using OpenCL technology and allows using intuitive notation for writing basic operations of linear algebra. The calculations were performed on a heterogeneous computing cluster [28-31]. Each of the seven cluster nodes has three AMD Radeon HD 7970 GPUs installed.

3. Numerical results and discussion

In the work in two-dimensional formulation, two model problems of hydraulic fracturing of the reservoir were solved. The calculation area is shown in figure 1. Yellow is a soil environment in which a fluid enters the vertical channel, which is shown in lilac color. Both the soil and the fluid are a collection of material particles, the number of which is horizontally and vertically indicated in figure 1 in thousands. In the first, the Lennard-Jones potential was used to describe the motion of soil and fluid particles. In the second problem, the Lennard-Jones potential was used only to describe the fluid, and to describe the motion of the soil, the "modified" Lennard-Jones potential was used.

At the initial moment of time, the soil is in a position of equilibrium in the potential field of gravity and is stationary. The fluid is immobile and weightless. Further, the fluid is also placed in the gravitational potential field and it starts moving (as a set of material particles) down the channel and, interacting with the soil, leaving the channel. On the interaction front between the particles representing the soil and the fluid, they are partially mixed, and in the soil, cracks can appear.

Figures 2 and 3 show the configurations that the soil mass becomes in the process of fracturing and further mixing of the media, partial fracture and cracking of the soil for the first and second model tasks, respectively.

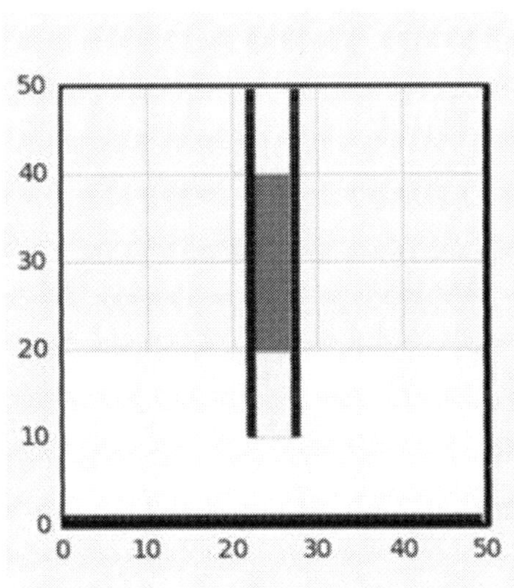

Figure 1. Initial calculation area.

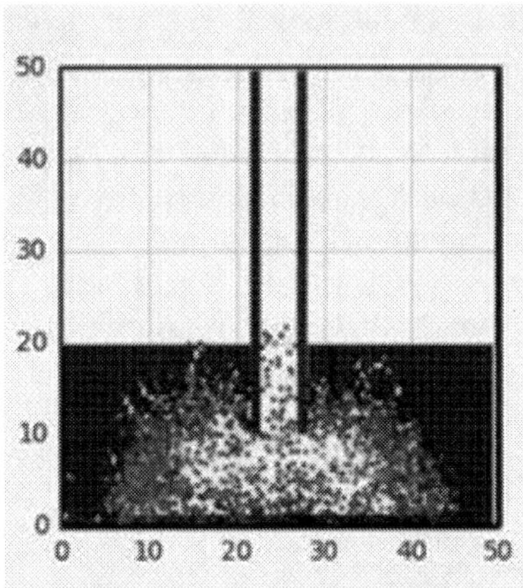

Figure 2. The final state of the formation after soil fracturing with its modeling using the Lennard-Jones potential.

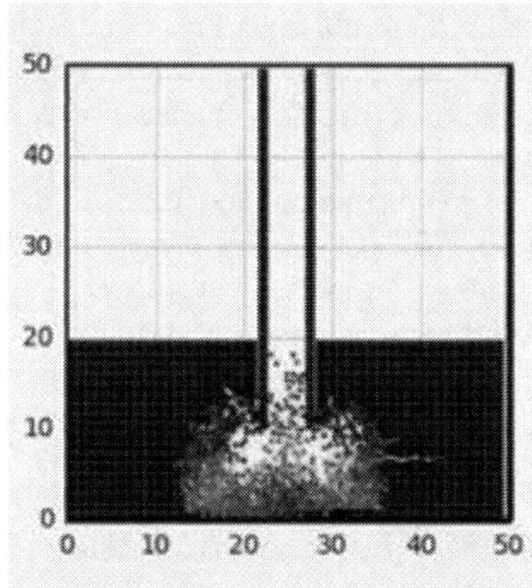

Figure 3. The final state of the formation after soil fracturing with its modeling using the "modified" Lennard-Jones potential.

It should be noted that when modeling a layer of soil using the Lennard-Jones potential, the mixing of fluid particles and soil occurred in the front of their interaction. The use of the "modified" Lennard-Jones potential for soil modeling allows obtaining cracks in the soil, and in general allows obtaining more realistic picture of the fracturing.

4. Conclusions

The method of particles was realized in the work, which makes it possible to investigate the interaction of continuous media of different physical nature (fluid and soil). Based on the Lennard-Jones potential and the previously introduced "modified" Lennard-Jones potential, modelling the problem of deformation and partial destruction of a soil massif under strong pressure from the fluid side was solved. Analysis of the results shows that the use of the "modified" Lennard-Jones potential for describing the soil environment makes it possible to realize the process of brittle fracture in the soil during fracturing, whereas the Lennard-Jones potential allows reproducing only viscous destruction of the soil environment.

Acknowledgement

The reported study was supported by Government of the Republic of Tatarstan and RFBR research projects No. 15-41-02555 and 15-07-05380.

References

[1] Badriev I B, Makarov M V and Paimushin V N 2016 *Procedia Engineering* **150** 1050–55
[2] Badriev I B, Makarov M V and Paimushin V N 2016 *Procedia Engineering* **150** 1056–62
[3] Shigapova F A, Mustakimova R F, Saleeva G T and Sachenkov O A 2015 *International Journal of Applied Engineering Research* **10**(24) 44711–14
[4] Sachenkov O A, Mitryaikin V I, Zaitseva T A and Konoplev Yu G 2014 *Applied Mathematical Sciences* **8**(159) 7889–7897
[5] Sultanov L U 2015 *Procedia Earth and Planetary Science* **15** 119–24
[6] Golovanov A I and Sultanov L U 2005 *International Applied Mechanics* **41**(6) 614–20
[7] Sultanov L U 2014 *Applied Mathematical Sciences* **8**(143) 7117–24
[8] Zienkiewicz O 1977 *The finite element method* (Mc Graw-Hill) 787
[9] Wilkins M L 1967 *Vychislitelnye metody v gidrodinamice* (Mir) 212-63

[10] Evans M W and Harlow F H 1957 *Los Alamos Scientific Laboratory Report No. LA-2139* (Los Alamos)
[11] Berezhnoi D V, Balafendieva I S, Sachenkov A A and Sekaeva L R 2016 *IOP Conference Series: Materials Science and Engineering* **158** 012018
[12] Mokshin E V and Berezhnoi D V 2016 *IOP Conference Series: Materials Science and Engineering* **158** 012069
[13] Hockney R W and Eastwood J W 1981 *Computer simulation using particles* (McGraw-Hill)
[14] Grigorjev Yu N and Vshivkov V A 2000 *Numerical methods "particle-in-cell"* (Nauka)
[15] Bogomolov S V, Zamaraeva A A, Karabelly H and Kuznetsov K V 1998 *Computational Mathematics and Mathematical Physics* **38**(9) 1602
[16] Monaghan J J 2005 *Reports on Progress in Physics* **68**(8) 1703-59
[17] Krivtsov A M and Krivtsova N V 2002 *Far Eastern Mathematical Journal* **3**(2) 254-76
[18] Psahje S G, Smolin A Yu, Stefanov Yu P, Makarov P V and Chertov M A 2004 *Letters Technical Physics* **30**(17) 7-13
[19] Zheltov Y P and Hristianovich S A 1955 *Izvestia AN USSR* **5** 3-41
[20] Perkins T K and Kern L R 1961 *J. Petrol. Technol.* **9** 937–49
[21] Maxwell S C, Rutledge J, Jones R and Fehler M 2010 *Geophysics* **75**(5) 75A129–75A137
[22] Duncan P and Eisner L 2010 *Geophysics* **75**(5) 75A139–75A146.
[23] Qi Zhao, VitaeAndrea Lisjak, Omid Mahabadi, Qinya Liu and Giovanni Grasselli 2014 *Journal of Rock Mechanics and Geotechnical Engineering* **6**(6) 574–81
[24] Soong C Y, Yen T H and Tzeng P Y 2007 *Phys. Rev. E.* **76** 036303
[25] Galimov M R and Birjaltsev E V 2010 *Numerical methods and programming* **11** 77-93
[26] Demidov D *VexCL: a C++ vector expression template library for OpenCL/CUDA* https://github.com/ddemidov/vexcl Acessed 01.09.2017
[27] Demidov D, Karsten A, Rupp K and Gottschling P 2013 *SIAM Journal on Scientific Computing* **35**(5) 453-72
[28] Karsten A, Demidov D and Mulansky M 2014 *Numerical Computations with GPUs* (Springer International Publishing) 125-57
[29] Berezhnoy D V, Gabsalikova N F and Balafendieva I S 2015 *International Journal of Applied Engineering Research* **10**(24) 44677-81
[30] Berezhnoi D V, Gabsalikova N F and Miheev V V 2016 *IOP Conference Series: Materials Science and Engineering* **158** 012016
[31] Berezhnoi D V, Gabsalikova N F, Izotov V G and Miheev V V 2017 *IOP Conference Series: Materials Science and Engineering* **204** 012028

Mathematical simulation of the low-temperature plazma at the interaction with oil products

V Ju Chebakova[1] and I B Badriev[1]

[1]Kazan Federal University, 18 Kremlevskaya Street, Kazan, 420008, Russia

E-mail: vchebakova@mail.ru

Abstract. Self-consistent mathematical models of a non-equilibrium low-temperature plasma of a radio frequency capacitive discharge in argon in different pressure ranges have been created. These mathematical models allow us to calculate the internal structure and discharge parameters for controlling parameters of the plasma torches at solving the target problems in order to optimize their functioning. The methods for numerical realization of the proposed models are developed. The results of numerical calculations for test problems are analyzed.

1. Introduction

At present, the methods for studying the environmental safety problems with oil pollution are rapidly developing. The significant scale of oil inflow into the environment when developing oil and gas resources leads to the fact that this type of pollution is the main one for a lot of oil production areas. Currently, the most promising is the technology of joint processing of oil-containing wastes and/or heavy oil fractions using plasma processing methods applying a low-temperature plasma [1-4]. Plasma-chemical methods provide a higher degree of processing (conversion of raw materials is 96-98% of mass), increase the depth of processing into unsaturated hydrocarbons (more than 75%), allow using heavy oil fractions (kerosene and diesel) as raw materials. These methods also reduce the number of stages and branching of chemical processes.

Plasma technologies have also been found wide application in other areas of environmental protection. For example, in [5] a method is given for controlling algae cyanobacteria *Microcystis aeruginosa* in treatment plants. In [6], a breakthrough system is developed by assisting the acid hydrolysis with two-step microwave treatment to improve the more efficient saccharification process of starch and fiber from sago pith. In [7], the safe use of materials in contact with food in microwave discharges is investigated. When creating plasma plants are used in the petrochemical and chemical industries, it is usually difficult to carry out field experiments, since their carrying out requires considerable resource costs, and besides, experimental methods do not allow obtaining detailed information on internal plasma-chemical processes [8, 9]. Therefore, experimental and computational methods were used for the research. They mutually complement each other, allow solving a lot of problems of physics and chemistry of low-temperature plasma. This approach allows us to relate the internal and external parameters of the discharges. This problem is especially important for optimizing the operation of plasma plants [10-16].

The aim of this work is to create self-consistent mathematical models of the non equilibrium low-temperature plasma of a radio frequency capacitive coupled (RFCC) discharge in argon at various pressures. They allow calculating the structure and internal parameters of the RFCC discharges for controlling the parameters of the RFCC discharge when solving target problems. The corresponding effective numerical algorithms implemented in the form of applied software have been developed to analyze the change in the components of the ionized inert gas as a function of pressure ranges and to identify the main factors affecting internal processes in RFCC discharges. The constructed self-consistent mathematical model describes a capacitive RF discharge between two plane-parallel electrodes, one of which is grounded and the other is connected to an RF generator (the frequency of the RF generator f = 13.56 MHz) with the inter-electrode distance which is smaller, than the dimensions of the electrodes themselves. Under these conditions, the electric field is close to the potential field and the discharge is homogeneous along the electrodes. It allows one-dimensional approximation to be used. Estimates of the time and distance over which electrons lose energy acquired from the electric field show that when simulating a low-pressure RF discharge we can use a non local approximation in which the parameters of the electronic component of the plasma depend on the electron temperature, whereas at higher pressures from the local value of the reduced electric field.

2. Statement of the problem

When constructing the model, we denoted b as the distance between the electrodes. For the coordinate x=0 we take the grounded electrode, for the coordinate x=b we take the loaded one (the x axis is directed perpendicular to the surface of the electrode). When setting the boundary conditions, it is assumed that at x=0 the field is directed to the electrode, if E<0, and from the electrode, if $E \geq 0$, and at x=b (the loaded electrode) the field is directed to the electrode, if from the electrode, if E<0. A self-consistent model of the RFCC discharge at low pressures in the non local approximation contains the following equations with boundary conditions.

- Ion-gas balance equation

$$\frac{dn_+}{dt} + \frac{d}{dx}\left(-D_+\frac{dn_+}{dx} + \mu_+ n_+ E\right) = R_1 n_e N + R_2 n_m^2 + R_3 n_m n_e - R_4 n_e n_+ - R_5 n_e^2 n_+, \quad 0<x<b, t>0,$$

$$G_+(0,t) = \begin{cases} (-n_+\upsilon_+ + \mu_+ n_+ E)\big|_{(0+,t)}, E<0 \\ (-n_+\upsilon_+)\big|_{(0+,t)}, E\geq 0 \end{cases}, \quad G_+(b,t) = \begin{cases} (n_+\upsilon_+ + \mu_+ n_+ E)\big|_{(b-,t)}, E\geq 0 \\ (n_+\upsilon_+)\big|_{(b-,t)}, E<0 \end{cases}.$$

Here n_e, n_+, n_m are the densities of electrons, ions and metastables, respectively, μ_+ is the mobility of ions, $G_+ = -D_+\, dn_+/dx + \mu_+ n_+ E$ is the ion flow, D_+ is the ion diffusion, factor E is an electric field intensity, $(E = -\partial\varphi/\partial x)$, φ is the electric field potential, $\upsilon_+ = \sqrt{8kT_a/(\pi m_+)}/4$ is the average thermal velocity of ions, m_+ is atomic ion mass, k is the Boltzmann's constant, T_a is a temperature of atoms. Hereinafter R_i, i=1…9 are the factors of plasma chemical reaction velocities ($Ar+e \rightarrow Ar^+ + 2e$; $Ar^* + Ar^* \rightarrow Ar + Ar^+ + e$; $Ar^* + e \rightarrow Ar^+ + 2e$; $Ar^+ + e \rightarrow Ar + h_\nu$; $Ar^+ + 2e \rightarrow Ar + e$; $Ar + e \rightarrow Ar^* + e$; $Ar^* \rightarrow Ar + h_\nu$; $Ar^* + Ar \rightarrow 2Ar$; $Ar^* + e \rightarrow Ar + e$) [17–21].

- Electron gas balance equation:

$$\frac{dn_e}{dt} + \frac{d}{dx}\left(-D_e\frac{dn_e}{dx} - \mu_e n_e E\right) = R_1 n_e N + R_2 n_m^2 + R_3 n_m n_e - R_4 n_e n_+ - R_5 n_e^2 n_+, \quad 0<x<b, \quad t>0$$

$$G_e(0,t) = \begin{cases} (-n_e\upsilon_e - \gamma\mu_+ n_+ E)\big|_{(0+,t)}, E<0 \\ (-n_e\upsilon_e - \mu_e n_e E)\big|_{(0+,t)}, E\geq 0 \end{cases} \quad G_+(b,t) = \begin{cases} (n_e\upsilon_e - \gamma\mu_+ n_+ E)\big|_{(b-,t)}, E\geq 0 \\ (n_e\upsilon_e - \mu_e n_e E)\big|_{(b-,t)}, E<0 \end{cases}$$

Here $G_e = -D_e\, dn_e/dx - \mu_e n_e E$ is the electrons flow, μ_e is the mobility of electrons, D_e is the electrons diffusion, γ is the secondary electron emission factor, $\upsilon_e = \sqrt{8kT_e/(\pi m_e)}/4$ is the average thermal velocity electrons, T_e is temperature of electrons, m_e is mass of an electron.

- Poisson's equation for the electric field:

$$-\frac{d^2\varphi}{dx^2} = \frac{q_e}{\varepsilon_0}(n_+ - n_e), \quad a(r-1)<x<b, \quad t>0, \ \varphi(0,t)=0, \quad \varphi(b,t)=V_a\sin(\omega t)$$

where q_e is the electron charge, ε_0 is the electrical constant, ω is the angular frequency of electric field, V_a is voltage peak-to-peak amplitude on charged electrode.

- The balance equation of metastable atoms:

$$\frac{dn_m}{dt}+\frac{d}{dx}\left(-D_m\frac{dn_m}{dx}\right)=R_6n_eN-R_2n_m^2-R_3n_mn_e-R_7n_m-R_8Nn_m-R_9n_mn_e,\quad 0<x<b, t>0$$

$$G_m(0,t)=(-n_m\sqrt{8kT_a/(\pi m_a)}/4)\Big|_{(0+,t)},\quad G_m(b,t)=(-n_m\sqrt{8kT_a/(\pi m_a)}/4)\Big|_{(b-,t)},$$

where $G_m=-D_m dn_m/dx$ is density of flow of metastable atoms, D_m is diffusion factor of metastable argon atoms.

- Balance equation for the electron temperature:

$$\frac{3}{2}k\frac{dn_eT_e}{dt}+\frac{d}{dx}\left(\frac{5}{2}kT_eG_e-\lambda_e\frac{dT_e}{dx}\right)=q_en_e\left\langle\mu_eE^2\right\rangle-Q_{el}Nn_e-IR_1n_eN-I_1R_3n_mn_e,$$

$$T_e(0,t)=T_e(b,t)=T_{e,w}$$

where λ_e is heat transfer ration of electron gas, Here I is the ionization energy of 15.76 eV, I_1 is the step ionization energy, the term Q_{el} is the exchange energy in collisions of elastic scattering.

- Balance equation for the atomic ion gas temperature:

$$\frac{d}{dx}\left(-\lambda_a\frac{dTa}{dx}\right)=j_iE+Q_{el}Nn_e,\ T_a(0,t)=T_a(b,t)=T_w.$$

Here λ_a is the thermal conductivity of atoms, $j_i=q_e\mu_+n_+E$ is the temperature on borders and is equal to temperature of electrodes.

A self-consistent model of the RFCC discharge at low pressures in the non local approximation contains the following equations with boundary conditions.

- The diffusion-drift equation for atomic ions:

$$\frac{\partial n_+}{\partial t}+\frac{\partial G_+}{\partial x}=R_1n_eN+R_2n_m^2+R_3n_mn_e-R_4n_en_+-R_5n_e^2n_++R_{12}n_en_{2+}-R_{11}n_+N^2+R_{20}Nn_{2+},0\le x\le b,t>0,$$

$$G_+(0,t)=\begin{cases}(-n_+\upsilon_++\mu_+n_+E)\big|_{(0+,t)},E<0\\(-n_+\upsilon_+)\big|_{(0+,t)},E\ge 0\end{cases},\ G_+(b,t)=\begin{cases}(n_+\upsilon_++\mu_+n_+E)\big|_{(b-,t)},E\ge 0\\(n_+\upsilon_+)\big|_{(b-,t)},E<0\end{cases}$$

where n_{2+} are densities of molecular positive ions, hereinafter R_i, i=10…20 are velocity factors of plasma-chemical reactions: $2Ar+Ar^+\rightarrow Ar_2^++Ar$, $Ar^*+Ar^*\rightarrow Ar_2^++e$, $Ar_2^++e\rightarrow Ar+Ar^++e$, $Ar_2^++e\rightarrow Ar+Ar$, $Ar_2^*\rightarrow 2Ar+h\nu$, $Ar^*+2Ar\rightarrow Ar_2^*+Ar$, $Ar_2^*+Ar_2^*\rightarrow e+2Ar+Ar_2^+$, $Ar_2^*+Ar^*\rightarrow e+Ar+Ar_2^+$, e+ $Ar_2^*\rightarrow 2Ar^*+e$, $Ar^*+2Ar\rightarrow 3Ar+h\nu$, $Ar_2^++Ar\rightarrow 2Ar+Ar^+$ (see [22–24]).

- The diffusion-drift equation for molecular ions:

$$\frac{\partial n_{2+}}{\partial t}+\frac{\partial}{\partial x}(G_{2+})=R_{10}n_m^2+R_{11}n_+N^2-R_{12}n_en_{2+}-R_{13}n_en_{2+}+R_{16}n_{2*}^{\ 2}+R_{17}n_mn_{2*}-R_{20}Nn_{2+},$$

$$\begin{cases}G_{2+}(0,t)=(-n_{2+}\upsilon_{2+}+\mu_{2+}n_{2+}E)|_{(0+,t)},E<0,\\G_{2+}(0,t)=(-n_{2+}\upsilon_{2+})|_{(0+,t)},E\ge 0\end{cases}\quad\begin{cases}G_{2+}(0,t)=(-n_{2+}\upsilon_{2+}+\mu_{2+}n_{2+}E)|_{(0+,t)},E<0,\\G_{2+}(0,t)=(-n_{2+}\upsilon_{2+})|_{(0+,t)},E\ge 0\end{cases}$$

where $G_{2+}=-D_{2+}\,\partial n_{2+}/\partial x+\mu_{2+}n_{2+}E$ is the density of molecular ion flow, μ_{2+} is the mobility of molecular ions, D_{2+} is the diffusion coefficient of molecular ions, $\upsilon_{2+}=\sqrt{8kT_a/\pi m_+}/4$ is the average thermal velocity of molecular ions, m_{2+} is the mass of atomic ion.

- The diffusion-drift equation for electrons:

$$\frac{\partial n_e}{\partial t}+\frac{\partial G_e}{\partial x}=R_1n_eN+R_2n_m^2+R_3n_mn_e-R_4n_en_+-R_5n_e^2n_++R_{10}n_m^2-R_{13}n_en_{2+}+R_{16}n_{2*}^2+R_{17}n_mn_{2*},$$

$$G_e(0,t)=\begin{cases}(-n_e\upsilon_e-\mu_e n_e E)\,|_{(0+,t)},\, E\geq 0\\ -n_e\upsilon_e-\gamma(\mu_+ n_+ E+\mu_{2+}n_{2+}E)\,|_{(0+,t)},\, E<0\end{cases},$$

$$G_e(b,t)=\begin{cases}(n_e\upsilon_e-\mu_e n_e E)\,|_{(b-,t)},\quad E<0\\ (n_e\upsilon_e-\gamma(\mu_+ n_+ E+\mu_{2+}n_{2+}E))\,|_{(b-,t)},\, E\geq 0\end{cases}$$

where n_{2*} is the density of argon dimmers.

- Poisson's equation for the electric field:

$$-\partial^2\varphi/\partial x^2=q_e(n_+ + n_{2+}-n_e)/\varepsilon_0\quad 0\leq x\leq b, t>0,\ \varphi(0,t)=0\quad ,\varphi(b,t)=V_a\sin(\omega t),$$

- Balance equation of metastable atoms:

$$\partial n_m/\partial t-\partial(G_m)/\partial x=R_6 N n_e+R_{18}n_e n_{2*}-R_2 n_m^2-R_3 n_m n_e-R_7 n_m-R_8 N n_m-$$
$$-R_9 n_m n_e-R_{10}n_m^2-R_{15}N^2 n_m-R_{17}n_m n_{2*}-R_{19}n_m N^2$$
$$G_m(0,t)=(-n_m\sqrt{8kT_a/(\pi m_a)}/4)\Big|_{(0+,t)},\quad G_m(b,t)=(-n_m\sqrt{8kT_a/(\pi m_a)}/4)\Big|_{(b-,t)},$$

- Kinetic equation for argon dimmers:

$$\frac{\partial n_{2*}}{\partial t}=R_{19}n_m N^2+R_{15}N^2 n_m-R_{14}n_{2*}-R_{16}n_{2*}^2-R_{17}n_m n_{2*}-R_{18}n_e n_{2*}+R_{19}n_m N^2,$$

- Kinetic equation for neutral argon atoms:

$$\frac{\partial N}{\partial t}=-R_1 n_e N+R_2 n_m^2+R_4 n_e n_+ +R_5 n_e^2 n_+ -R_6 N n_e+R_7 n_m+R_8 N n_m+R_9 n_m n_e-R_{11}n_+ N^2+$$
$$+R_{12}n_e n_{2+}+2R_{13}n_e n_{2+}+R_{14}n_{2*}-R_{15}N^2 n_m+2R_{16}n_{2*}+R_{17}n_m n_{2*}+R_{19}n_m N^2,$$

the initial conditions are based on ideal gas equation: $P/(kT_a(x,0))=N(x,0)$, $n_{2*}(x,0)=0$.

- Balance equation atomic ion gas temperature:

$$\frac{\mathrm{d}}{\mathrm{d}x}\left(-\lambda_a\frac{\mathrm{d}T_a}{\mathrm{d}x}\right)=j_i E+Q_{el}Nn_e$$

$$,-\lambda_a\,\partial T_a/\partial x|_0=-\chi(T_a(0,t)-T_w),\ -\lambda_a\,\partial T_a/\partial x|_b=\chi(T_a(b,t)-T_w),$$

where λ_a is the heat transfer ration of atom-ion gas.

3. Approximate method

The systems of boundary and initial boundary value problems are characterized by several features (nonlinearity, rapid growth of coefficients in narrow domains, different scales of processes, etc.) complicating its numerical solution and making it impossible to use traditional methods [25-28].

To solve the nonlinear systems proposed in the present paper, we used an approximate method based on preliminary finite-dimensional approximation of the problem by means of difference schemes with subsequent application of an iterative process for its implementation [29–32]. The finite-difference approximation for convection-diffusion equations for charged particles is constructed by the integro-interpolation method using the method of directed differences.

To solve the Cauchy problem, an implicit Eulerian scheme was used. When the quadratic nonlinearity is approximated, the Newtonian linearization is used in the right-hand sides of the equations. The flux densities for ion and electron gas were calculated using the Hummel-type method based on the solution found by the implicit scheme.

The numerical algorithm is based on the demolition of nonlinearity with respect to incoming coefficients (diffusion coefficients, mobility and rates of plasma-chemical reactions) on the lower layer. The linearization of the system was carried out using the Seidel-type method.

4. Results of calculations

The results of the numerical solution of the model problem revealed changes in the composition of the gas during the development of the discharge. If molecular ions first predominate at a small gas temperature and the concentrations of excited atoms and dimers are approximately equal, then when the gas is heated, the concentration of molecular ions and dimers decreases with increasing concentration of atomic ions and dimer reduction. The concentration of molecular ions is greater in the near-electrode layers, where the gas temperature is lower. In this case, a quasineutral region is conserved in the discharge. It was shown that in the case of significant gas heating, the change in the gas temperature in the interelectrode space begins to affect significantly the ratio of contributions to the formation and death of particles of various plasmochemical processes, and, consequently, affects the distribution as well as the fraction of charged (electrons, Molecular ions), and excited particles in the discharge gap, thereby determining the development of the discharge. In conclusion, we note that the results obtained in this paper can be used to develop models for the interaction of low-temperature plasma with various substances. These models play an important role in solving environmental safety problems, which arise, in particular, when processing pipelines for oil and gas and engineering infrastructure from composite materials in order to improve the operational reliability of oilfield pipeline systems. In addition, these models can be used to develop technologies for the joint processing of oily waste and/or heavy oil fractions using low-plasma systems.

Acknowledgments

The work is performed according to the Russian Government Program of Competitive Growth of Kazan Federal University. The publication was carried out with the financial support of the Russian Science Foundation (project No 16-11-10299).

References

[1] Peter Bruggeman and Christophe Leys, 2009 Non-thermal plasmas in and in contact with liquids *J. Phys. D: Appl. Phys.* **42** 053001

[2] Bruggeman P J, Kushner M J et al 2016 Plasma–liquid interactions: a review and roadmap *Plasma Sources Sci. Technol.* **25** 053002

[3] Zhao Li, Kunyao Wu, Jing Cao, Yongfeng Wang 2017 Study on microwave induced pyrolysis of low metamorphic coal and liquefaction residue *IOP Conf. Ser.: Earth Environ. Sci.* **64** 012038

[4] Bingnan Yan, Zhou Jing and Zhao Liang 2017 Study on the influence of the electrode model on discharge characteristics in High-voltage Pulsed Deplugging Technology *IOP Conf. Ser.: Earth Environ. Sci.* **64** 012001

[5] Lapsongpon T, Leungprasert S and Yoshimura C 2017 Pre-chlorination contact time and the removal and control of Microcystis aeroginosa in coagulation *IOP Conf. Ser.: Earth Environ. Sci.* **67** 012011

[6] Sunarti T C, Yanti S D and Ruriani E 2017 Two-steps microwave-assisted treatment on acid hydrolysis of sago pith for bioethanol production *IOP Conf. Ser.: Earth Environ. Sci.* **65** 012052

[7] Zhu X, Li F, Qiu Z Z and Huang J W 2017Migration of DEHP from plastic to food simulants under microwave heating *IOP Conf. Ser.: Earth Environ. Sci.* **64** 012050

[8] Fridman A 2008 *Plasma chemistry* (Cambridge: Cambridge Univ. Press)

[9] Chebert Pascal, Braithwaite Nicholast 2011 *Physics of radio-frequency plasmas*(Cambridge: Cambridge Univ. Press)

[10] Kashapov N F, Saifutdinov A I and Fadeev S A 2014 The model of the positive column of a glow discharge with the influence of the acoustic oscillations *J. Phys.: Conf. Ser.* **567** 012004

[11] Ganea I and Morar R 2017 Factors that increase the electric field of the dielectric barrierozone generators *IOP Conf. Ser.: Mater. Sci. Eng.* **200** 012057

[12] Lebedev Yu A, Tatarinov A V 2004 Electrodynamics of microwaves in a coaxial non-regular waveguide partly filled with plasma *Plasma Sources Science and Technology* **13**(1) 1-7

[13] Lebedev Yu A, Mokeev M V 2001 Characteristics of the electrode plasma of an electrode microwave discharge in hydrogen *Plasma Physics Reports* **27**(5) 418-23
[14] Dautov G, Dautov I, Fayrushin I and Kashapov N 2014 Calculation of distribution of potential near the surface of metal particle in the dust-electron thermal plasma *J. Phys.: Conf. Ser.* **56** 7012005
[15] Zheltukhin V S, Solov'Ev S I, Solov'Ev P S, Chebakova V Yu and Sidorov A M 2016 Third type boundary conditions for steady state ambipolar diffusion equation *IOP Conf. Ser.: Mater. Sci. Eng.* **158** 012102
[16] Chebakova V Ju, Gaisin A F and Zheltukhin V S 2016 Solution of the problem of interaction between capacitive coupled radio-frequency discharge and a sample *IOP Conf. Ser.: Mater. Sci. Eng.* **158** 012024
[17] Diatko N A, Ionokh Yu Z, Meshanov A B and Napartovich A P 2005 Study of dark study of development of positive column of glow discharge in argon *Plasma Physics Reports* **31**(10) 939-53
[18] Lymberopoulos D P, Economou D J 1993 Fluid simulations of glow discharge & Effect of metastable atoms in argon *J. Appl. Phys.* **73**(8) 3668–79
[19] Xi-Ming Zhu and Yi-Kang Pu 2010 A simple collisional–radiative model for low-temperature argon discharges with pressure ranging from 1 Pa to atmospheric pressure: kinetics of Paschen 1s and 2p levels *J. Phys. D: Appl. Phys.* **43** 015204
[20] Martinez E Castonos, Kabouzi Y, Makasheva K and Moisan M 2004 Modeling of microwave-sustained plasmas at atmospheric pressure with application to discharge contraction *Physical review E70* 066405
[21] Hagelaar G J M, Pitchford L C 2005 Solving the Boltzmann equation to obtain electron transport coefficients and rate coefficients for fluid models *Plasma Sources Sci. Techn.* **14** 722-33
[22] Balcon N P, Aanesland A, Hagelaar G J M, Boswell R and Boeuf J P 2007 Atmospheric pressure RF discharge in argon: optical diagnostic, fluid model and applications *28th ICPIG* 957–60
[23] Moravej M, Yang X, Hicks R F, Penelon J and Babayan S E 2006 A radio-frequency nonequilibrium atmospheric pressure plasma operating with argon and oxygen *J. Appl. Phys.* **99** 093305
[24] Bora B, Bhuyan H, Favre M, Wyndham E and Chuaqui H 2011 Diagnostic of Capacitively Coupled Low Pressure Radio Frequency Plasma: An Approach through Electrical Discharge Characteristic *International Journal of Applied Physics and Mathematics* **1**(2) 124–8.
[25] Badriev I B and Karchevskii M M 1989 Convergence of the iterative Uzawa method for the solution of the stationary problem of seepage theory with a limit gradient *Journal of Soviet Mathematics* **45** (4) 1302-9
[26] Badriev I B, Makarov M V and Paimushin V N 2016 Numerical Investigation of Physically Nonlinear Problem of Sandwich Plate Bending *Procedia Engineering* **150** 1050-5
[27] Badriev I B, Makarov M V and Paimushin V N 2016 Mathematical Simulation of Nonlinear Problem of Three-point Composite Sample Bending Test *Procedia Engineering* **150** 1056-62.
[28] Badriev I B, Zadvornov O A and Lyashko A D 2004 A study of variable step iterative methods for variational inequalities of the second kind *Differential Equations* **40**(7) 971-83
[29] Chebakova V J, Zheltukhin V S and Dubrovin V T 2016 Problem of radio-frequency discharge at atmospheric pressure in local statement *Applied Mathematical Sciences* **10** (21-24) 1013-22
[30] Zheltukhin V S, Chebakova V J, Abaidullin R N and Kadyrov R F 2015 On solving of some nonlinear problems of plasma physics *Applied Mathematical Sciences* **9**(85-88) 4351-59
[31] Chebakova V J, Gerasimov A V and Kirpichnikov A P 2016 On the solving of one type of problems of mathematical physics *IOP Conf. Ser.: Mater. Sci. Eng.* **158** 012023
[32] Dubrovin V T, Chebakova V Ju and Zheltukhin V S 2016 Radio-Frequency Discharge at Low Pressure: a Non-Local Problem Statement Approach *Procedia Engineering* **150** 1041–5

Dynamics of intense rainfalls in the southern half of European Russia for the period 1960-2015

N Chizhikova

Institute of Environmental Sciences, Kazan Federal University, Kremlyovskaya str., 18, 420008, Russia

E-mail: Nelly.Chizhikova@kpfu.ru

Abstract. Two time periods (1960-1986 and 1986-2015) were compared in terms of mean annual frequency and mean annual sum of warm season rainfalls to quantify general trends in the changing regime of heavy precipitation over the southern part of European Russia, which is the area of the most intensive agricultural activity. The identified trends were compared with the published assessment of the intensity trends of soil erosion processes. The prevalence of the increasing tendency in the frequency and amount of precipitation is demonstrated, which undergoes against a decrease in the rate of eroded sediment accumulation and against a decrease of linear growth rate of gullies. This result rather proves the crucial contribution of the snowmelt to the soil erosion over the studied area.

1. Introduction

The current century is characterized by global intensification and changes in the structure of the water cycle [1]. In accordance with the fifth assessment report of the Intergovernmental Panel on Climate Change [2], the coming decades are expected to have more intense rains. This theory is already being justified in regional studies in Europe and on the European territory of Russia in particular [3-7]. In general, the greater amount of precipitation leads to the greater soil erosion, which is one of the most important factors of arable land degradation.

This paper presents an analysis of trends in heavy precipitation of the warm season on the southern half of the East European Plain, which was chosen as the territory with the most intensive agricultural activity. The study is bounded to the period 1960-2015 to be comparable with evaluation of soil erosion rate trends based on the dating of sediments deposited in the bottoms of the first-order valleys using the bomb-derived and Chernobyl-derived ^{137}Cs. Such estimates for small catchments of steppe and forest-steppe zones can be found in [8, 9, 10].

2. Materials and methods

2.1. Site and meteorological data

The territory covered by the current study refers to the most cultivated land. This is the southern part of the forest zone, the forest-steppe and steppe zones (figure 1). Here the arable land accounts for 20-80% of the total area, and the role of soil surface runoff induced with rainfall is substantial.

The rainfall daily amount exceeding 10 mm threshold is conventionally indicated as potentially erosive [11], but, as reported by [12], the most destructive runoff power in soil losses is associated with extreme rainfalls featured by more than 40 mm of daily amount. Erosive potential of a rainfall is defined as the product of the kinetic energy and its maximum 30 minute intensity, and is shown to have correlation with daily amount of precipitation [13]. The latter fact makes it possible to use the frequency of erosion-hazardous precipitation, characterized by different daily quantities, as a surrogate to assess the possible trends of soil loss intensity caused by rainfall runoff.

The analysis is based on daily precipitation records freely provided by the Russian Institute of Hydrometeorological Information – World Data Centre (RIHMI-WDC; http://meteo.ru/english/climate/) [14, 15]. Each weather station falling into the study area was characterized by warm season (May-September) time series of frequency and amount of precipitation for six groups of daily sum of precipitation: >10 mm, 10-20 mm, 20-30 mm, 30-40 mm, 40-50 mm and >50 mm (twelve time series per weather station). Season (year) was excluded from the time series if more than 30% of daily data was absent. The stations selected for the analysis have no more than 20% of missing data in time series. 70 stations were chosen in total.

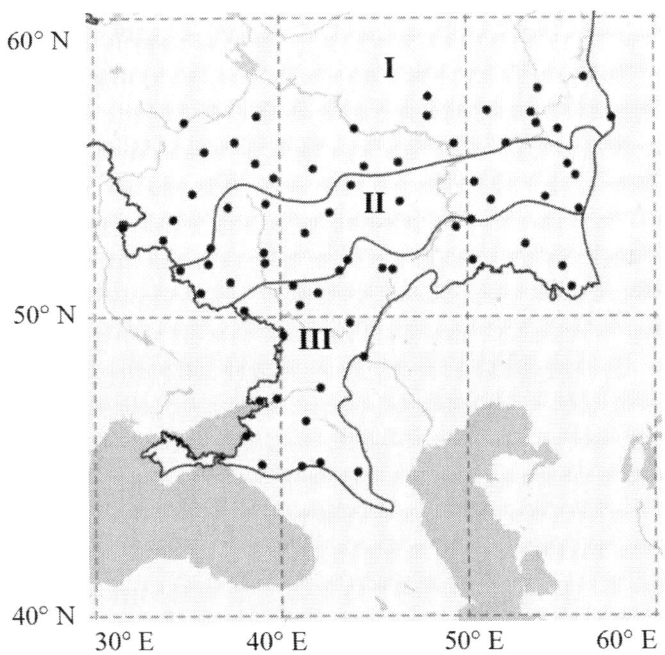

Figure 1. Scheme of landscape zones and weather stations. Zone numbering: (I) forest, (II) forest-steppe, (III) steppe zone. Forest zone is shown partly, because only its southern part is under study.

2.2. Statistical analysis

The aim of the study was to assess precipitation trends assuming its potential to change soil erosion rates and to compare observed rainfall trends (if any found) with available estimates of the trend of erosion processes on arable land in European Russia. Several published estimates of soil erosion rates used in this investigation are based on dating of soil deposition with use of ^{137}Cs which peak concentrations are due to testing of nuclear weapons in the atmosphere in 1954-1963 and the Chernobyl nuclear disaster in 1986 [8, 9, 10]. Because of this timing, all precipitation time series were divided into two time slices: from 1960 to 1986, and from 1986 (inclusive) to 2015. Depending on the changes of precipitation statistics between these two periods, two types of the precipitation dynamics can be articulated: (i) increasing and (ii) zero change or decreasing.

To quantify the change of warm season rainfall frequency the mean annual count of rainfalls during the warm season was calculated for each time period. The type of precipitation dynamics was inferred from the difference between mean annual counts: positive difference corresponds to increasing frequency, and negative difference corresponds to the decreasing rainfall frequency. The statistical significance of the ratio of mean annual counts was assessed using the Poisson test. It should be noted that this test assumes the count of rainfalls to be Poisson distributed, which might not be the case for very rare rains with high daily precipitation sums. So the estimates of statistical significance should be be taken only as additional illustrative evidence of substantial changes. Duration of time base for event count was adjusted at each weather station in accordance with recommendation of [7]: years with more than 30% of missing daily data were excluded from time series.

To quantify the change of amount of warm season precipitation, the difference in the medians of the annual sums of warm season rainfalls was calculated. The statistical significance of the difference was assessed using the Mann-Whitney test.

3. Results

At the beginning period under study (1960-1986), the forest, forest-steppe and steppe zones had similar features in terms of frequency and amount of precipitation, which is summarized in the table 1. On average, during a warm period, the meteorological stations have 3-9 rainfalls exceeding daily amount of 10 mm, which corresponds to 60-180 mm per season. The most erosive rainfall events with daily amount exceeding 40 mm per day of the warm season occur no more often than one event per 10 years.

Table 1. Precipitation characteristics of the warm season for the periods of time.

Daily amount of precipitation	Mean annual frequency of warm season rainfalls (minimum / median / maximum)		Mean summary precipitation of the warm season, mm (minimum / median / maximum)	
	1960-1986	1986-2015	1960-1986	1986-2015
>10 mm	3.4 / 6.7 / 9.4	2.7 / 7.3 / 9.9	61 / 126 / 181	47 / 140 / 226
10-20 mm	2.4 / 4.8 / 6.9	2.1 / 5.0 / 6.7	34 / 69 / 100	29 / 72 / 97
20-30 mm	0.6 / 1.2 / 2.2	0.4 / 1.5 / 2.3	15 / 30 / 55	11 / 36 / 56
30-40 mm	0.1 / 0.4 / 0.8	0.1 / 0.4 / 0.9	2 / 13 / 27	2 / 14 / 32
40-50 mm	0.0 / 0.1 / 0.3	0.0 / 0.2 / 0.6	0 / 5 / 15	0 / 8 / 28
>50 mm	0.0 / 0.1 / 0.4	0.0 / 0.1 / 0.4	0 / 5 / 31	0 / 6 / 29

Comparing the annual characteristics of the precipitation, a slight increase in the frequency and amount of warm season precipitation can be noted, irrespective of their daily amount (table 1). However, it should be emphasized, that the rate of change is tiny. The amplitude of changes is small, as well as counts of stations demonstrating statistically significant changes (tables 2, 3).

Table 2. Statistics of frequency shifts across the studied area. Shifts are computed as the difference between the mean annual warm season rainfall frequencies for the period of 1986-2015 and the 1960-1986.

Daily amount of precipitation	Minimum difference (decrease)	Median difference	Maximum difference (increase)	% of stations with increased frequency (& p-value < 0.05)	% of stations with zero changes or decreased frequency (& p-value < 0.05)
>10 mm	-1.3	0.4	2.7	79 (20)	21 (1)
10-20 mm	-1.0	0.1	1.6	60 (9)	40 (0)
20-30 mm	-0.6	0.2	0.9	70 (11)	30 (3)
30-40 mm	-0.3	0.1	0.5	60 (6)	40 (1)
40-50 mm	-0.3	0.1	0.4	73 (6)	27 (3)
>50 mm	-0.2	0.0	0.2	54 (1)	46 (4)

Analyzing the distribution of the frequency shifts (table 2) and the precipitation sum shifts (table 3), the prevalence of increasing tendencies over the downward ones is evident. The percent of station with upward changes is higher than for stations with zero or negative temporal changes, as well as 50% quantile (median) is above zero almost in all cases of daily amount of precipitation (tables 2, 3). This pattern is the most prominent for the broadest group of precipitation events with daily amount ">10 mm".

Table 3. Statistics of summary precipitation shifts across the studied area. Shifts are computed as the difference between the mean annual warm season rainfall sums for the periods of 1986-2015 and 1960-1986.

Daily amount of precipitation	Minimum difference (decrease)	Median difference	Maximum difference (increase)	% of stations with increased precipitation (& p-value < 0.05)	% of stations with zero changes or decreased precipitation (& p-value < 0.05)
>10 mm	-24	11	68	77 (17)	23 (1)
10-20 mm	-16	2	23	59 (9)	41 (0)
20-30 mm	-16	5	23	69 (15)	31 (3)
30-40 mm	-10	2	18	61 (7)	39 (1)
40-50 mm	-15	4	17	73 (13)	27 (3)
>50 mm	-21	1	15	53 (1)	47 (5)

The spatial patterns of frequency trends and the summary precipitation trends are similar. Despite diverse and heterogeneous pattern, when positive trends alternate with zero and negative ones, the positive trends for precipitation of daily amount >10 mm, 10-20 mm, 20-30 mm, 30-40 mm and 40-50 mm seem to dominate over the northern part of the territory under study (south border of the forest landscape zone) (figure 1a-b, figure 2a-b). The growth is either small or negative in south-eastern part of steppe zone and around Volga-Kama region. The group of rainfalls of daily amount of >50 mm exhibits a distinct increasing tendency around the Central Russian Upland (figure 2c-d).

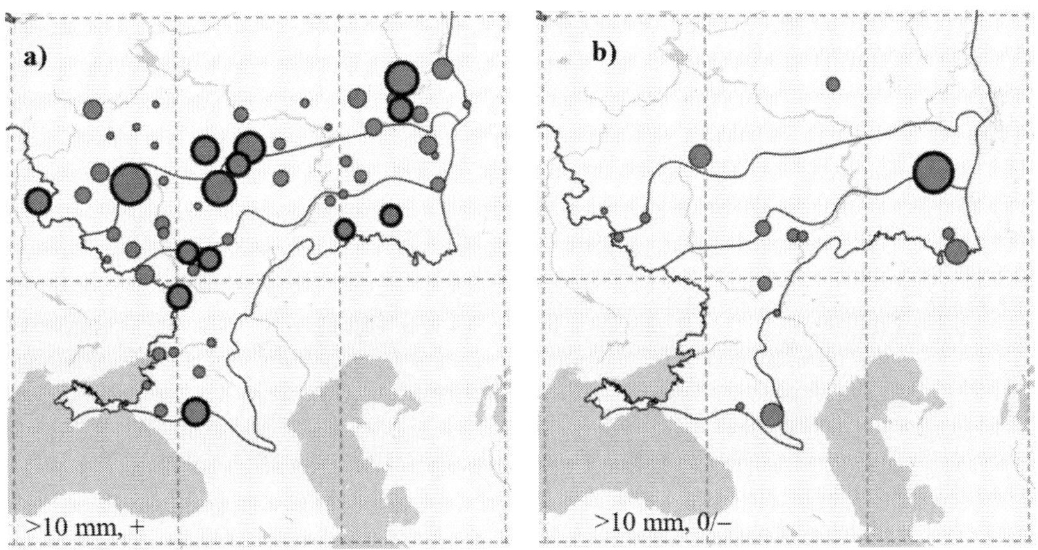

Figure 2. Precipitation frequency shifts between the periods of 1986-2015 and 1960-1986, general group of rainfall daily amount: >10 mm. Size of the points is proportional to the absolute difference of mean annual warm season rainfall frequencies (see statistics in table 2). A: Stations exhibiting increase of warm season rainfall frequency. B: Stations exhibiting zero changes or decrease of warm season rainfall frequency. Points with thicker outline denotes statistically significant rate of seasonal frequencies by 0.05 confidence level by the Poisson test of rates.

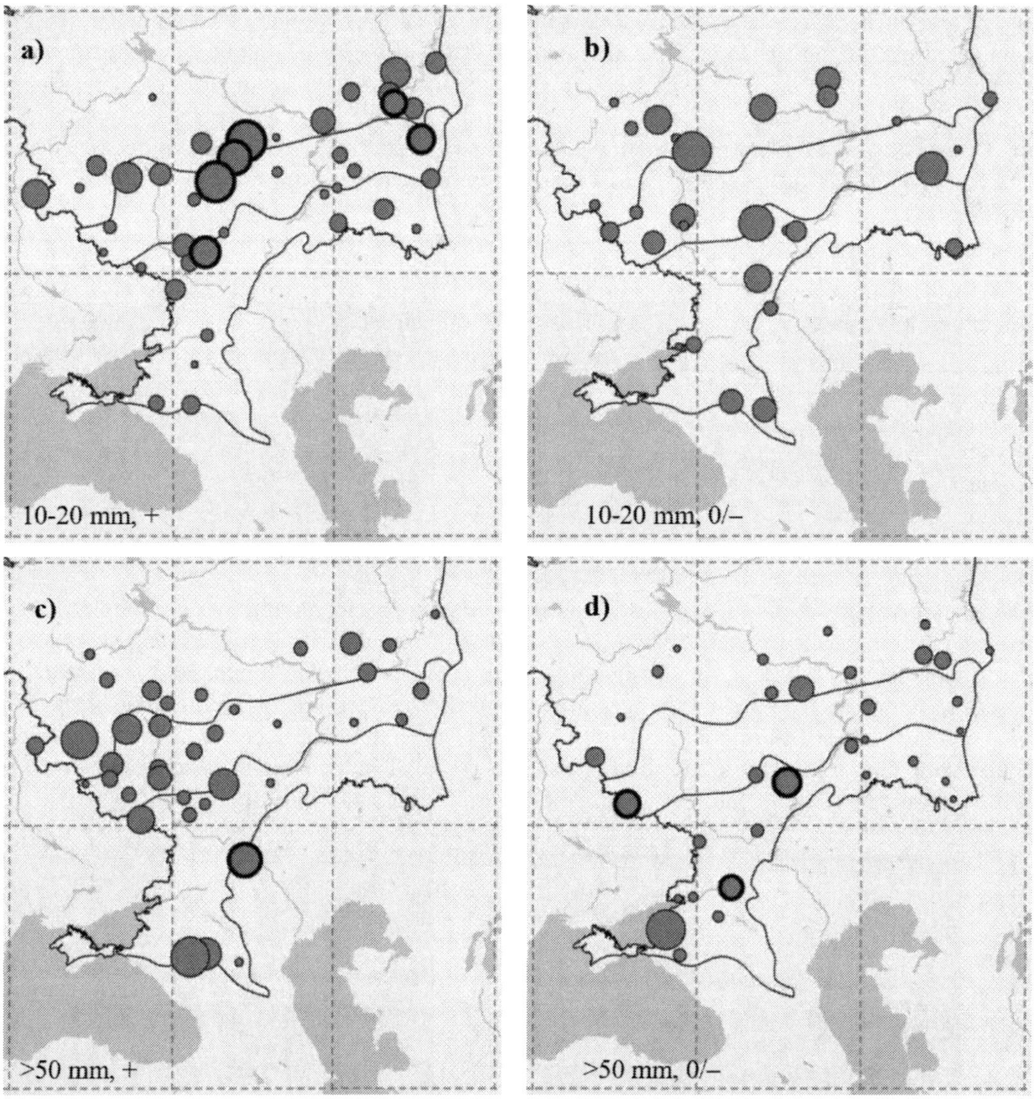

Figure 3. Precipitation seasonal summary precipitation shifts between the periods of 1986-2015 and 1960-1986. Size of the points is proportional to the absolute difference of the mean annual warm season rainfall sums (see statistics in table 3). A, B: Group of rainfall daily amount: 10-20 mm. C, D: Group of rainfall daily amount >50 mm. A, C: Stations exhibiting increase of warm season rainfall sum. B, D: Stations exhibiting zero changes or decrease of warm season rainfall sum. Points with thicker outline denote statistically significant difference of the medians by 0.05 confidence level by the Mann-Whitney test.

4. Discussion

According to the estimates of the volumes of accumulated eroded deposits in the catchments of the forest-steppe zone [8], during the period 1986-2015, the rate of erosion of soils from arable land decreased by half in comparison with the previous period of 1954-1986. The estimates gained in the research cited cover the Central Russian Upland, which meanwhile undergoes the increase of the precipitation of daily amount exceeding 50 mm (figure 3a).

The work of [16] uses the estimates of the linear growth of gullies to characterize the soil erosion rates in the south part of the forest landscape zones (Udmurt Republic). The rate of linear growth of gullies is decreasing [16], which contradicts to dynamics of potential erosive precipitation. The precipitation time series of weather stations around Udmurtia demonstrate upward trends.

This contradiction is evidence that slight changes of the precipitation regime in forest, forest-steppe and steppe zones did not change the general tendency of erosion processes. This outcome argues into the key role of the surface runoff from the slopes during the snowmelt for erosion processes. According to the [17-19], the intensity of surface runoff during snow melting is decreasing because of reduction of frozen soil depth during the winter. If the tendency to an increase in erosion-hazardous precipitation persists as well as the decrease of the intensity of surface runoff during the spring snow melt, an increase in the contribution of showers to soil erosion can be expected.

5. Conclusion

Precipitation regime of the warm season shows upward trends of frequency and total sum at the southern boundary of the forest landscape zone, in forest-steppe and steppe zones. The rate of this increase is moderate or rather small and did not affect the directionality of the erosive processes towards decline.

Acknowledgments
The work was funded by the Russian Science Foundation, project no. 15-17-20006.

References

[1] Durack PJ, Wijffels SE and Matear RJ 2012 Science **336**(6080) 455–8
[2] IPCC 2013 *Climate change 2013: the physical science basis. Contribution of working group I to the fifth assessment report of the intergovernmental panel on climate change* ed T F Stocker, D Qin et al (Cambridge: Cambridge University Press)
[3] Roshydromet 2014 *Second Roshydromet assessment report on climate change and its consequences in Russian Federation* (Moscow: Roshydromet) (in Russian)
[4] Jones RJ, Le Bissonnais Y, Bazzoffi P, Sanchez DJ, Düwel O, Loj G, Øygarden L, Prasuhn V, Rydell B, Strauss P et al 2004 Nature and Extent of Soil Erosion in Europe *EU Reports of the Technical Working Groups Established Under the Thematic Strategy for Soil Protection* ed L Van-Camp, B Bujarrabal et al (Luxembourg: Office for Official Publications of the European Communities) pp 145-85
[5] Bardin MYu and Platova TV 2013 *J. Problems of Ecological Monitoring and Ecosystem Modelling* **25** 71-93 (in Russian)
[6] Groisman PY, Knight RW and Zolina OG 2013 Recent Trends in Regional and Global Intense Precipitation Patterns *Climate Vulnerability: Understanding and Addressing Threats to Essential Resources* (Oxford: Academic Press) pp 25-55
[7] Zolina O 2012 Change in intense precipitation in Europe *Changes in Flood Risk in Europe* (*Special Publication* No. 10) ed ZW Kundzewicz (Wallingford, Oxfordshire: IAHS Press)
[8] Golosov VN, Belyaev VR and Markelov MV 2013 *Hydrol Process* **27** (6) 781–94
[9] Golosov VN, Gennadiev AN, Olson KR, Markelov MV, Zhidkin AP, Chendev YuG and Kovach RG 2011 *Eurasian Soil Sci* **44**(7) 794-801
[10] Golosov VN, Walling DE, Konoplev AV, Ivanov MM, Sharifullin AG 2017 Application of bomb- and Chernobyl-derived radiocaesium for reconstructing changes in erosion rates and sediment fluxes from croplands in areas of European Russia with different levels of Chernobyl fallout. *J Environ Radioactiv* doi.org/10.1016/j.jenvrad.2017.06.022 (in press)
[11] Larionov GA 1993 *Erosion and deflation of soils* (Moscow: Moscow State University Press) (in Russian)
[12] Edwards WM and Owens LB 1991 *J. Soil Water Conserv* **46** 75–8
[13] Larionov GA 1987 Erosion potential of precipitation *Work of water flow* ed RS Chalov (Moscow: Moscow State University Press) pp 17-21 (in Russian)
[14] Veselov VM 2002 *Proc. RIHMI-WDC* (RIHMI-WDC) **170** pp 16–30 (in Russian)

[15] Razuvayev VN, Apasova EG, Martuganov RA, Steurer P and Vose R 1993 *Daily Temperature and Precipitation Data for 223 U.S.S.R. Stations* (Oak Ridge: Oak Ridge National laboratory)
[16] Rysin II, Grigoryev II, Zaitceva MYu and Golosov VN 2017 *MSU Vestnik, Ser. 5: Geography* **1** 63–72 (in Russian)
[17] Rysin II, Golosov VN, Grigoryev II and Zaitceva MYu 2017 *Geomorphology RAS* **1** 90-102 (in Russian)
[18] Park H, Sherstiukov AB, Fedorov AN, Polyakov IV and Walsh JE 2014 *Environ Res Lett* **9** 1–7
[19] Golosov VN, Ivanova NN, Gusarov AV and Sharifullin AG 2017 Assessment of trends of arable soil degradation on the base of the investigation of layered soils by using ^{137}Cs as a chronomarker *Eurasian Soil Science* **10** (in Russian, in press)

IOP Conf. Series: Earth and Environmental Science **107** (2017) 012099 doi:10.1088/1755-1315/107/1/012099

Landscape fire in East Siberia: medical, ecological and economic aspects

N V Efimova[1], V S Rukavishnikov[1], G A Zabuga[2] and T A Elfimova[1]

[1]East-Siberian institute Medical and Ecological Research, 665827, a/b1170, Angarsk, Russia

[2]Irkutsk State University of Railway Transport 664074, Chernyshevsky st, 15, Irkutsk, Russia

Abstract. More than 40 % of the forests in Siberia region are known to have a fire danger of high classes and high burning degrees. This paper describes air pollutants emission (PM10, nitrogen oxides, sulfur dioxide and others) in East Siberian region during a 10-year period in the forests fires focus. A total of 500 to 2000 fires occurred in Irkutsk oblast during the last ten years. At an average annual forest fires cover an area of 1 109 hectares on the model territory (Bratsk city). The plane pollutant emission source with a high productivity is formed on the significant forest fire area occurred in a relatively short-term time periods. The increase in hazard ratios was registered for the ingredients of emission-specific industrial enterprises and capable of accumulating in vegetation: carbon disulphide 1.9 times, fluorine-containing substances 1.8 times during the fire. The economic loss of energy resources resulting from reduced production of firewood was estimated at $ 56.6 million in Irkutsk oblast. The potential risk of negative effects for the respiratory system and cardiovascular system stipulated for the acute inhalation exposure was found to increase on the days, of the fires, as evidenced by the growth of the daily mortality and morbidity rates among the population.

1. Introduction

There are 150 thousand forest fires on average every year in Russia, the US, Canada, and Brazil, and about 1.5 million hectares of land are burned on average in each case. Other countries also experience an annual average of 150 thousand fires, though the average area that catches fire is not large, except for Argentina. Ignition conditions, physicochemical characteristics of flora (tree species, age, forest types, etc.), the weather, climate, season, time of day, and other factors influence a fire's parameters. All these factors can cause a fire to spread quickly, which means that their characteristics and intensity influence the natural ecosystem [2]. Radiation from a forest fire as well as the vast amounts of harmful substances and other elements emitted influence the environment globally, going beyond the region [3, 5, 7, 13]. For example, it is clear that the fire risk is higher when the air temperature is high and the air is dry (when the humidity is relatively low). Some have positive view that a fire modifies natural components of a forest ecosystem, but firstly we would like to have a look at key negative factors of a forest fire, i.e., a forest fire is a strong factor in air pollution; a forest fire generates noxious substances (sulphur dioxide, carbon monoxide, dust, etc.) and destroys the forest ecosystem. It triggers changes in the local ecology and climate, as well as the oxygen balance in the atmosphere; a forest fire destroys the surface of the land more widely than erosion by a rainstorm. It leads to the risk of the ground sinking and collapsing. It changes freshwater flow of rivers, swamps, and lakes, and triggers thawing in permafrost areas.

The many substances that are generated by a forest fire are harmful to people and animals, as well as to vegetation. A large forest fire is threatening. It frequently destroys houses, industrial complexes, and entire villages. Eighty percent of fires that occur in all countries is caused by men, and another 5-7% is caused by nature. The cause of the other fires has not been identified. In Russia since the 1990s every 2-4 years forest and peat fires had catastrophic scales, threatening environmental population of certain subjects and regions of the country [2, 6]. The consequences of the natural fires may be divided into ecological and socio-economic ones. The structure destruction and the function disorders of the natural ecosystems, the alteration of the biogeochemical cycles, the soil humification processes on the territories exposed to pyrogenic factors are known to concern with a number of serious ecological consequences [4, 5, 11]. The potential impact of smoke from prescribed burning on occupational and community smoke exposures and related health effects are of growing concern. A number of studies have also demonstrated or assumed adverse health effects in individuals from communities exposed to smoke from wildland fires [1, 8]. People and material losses were revealed to be more significant social-economic consequences of natural fires, prevalence of which continues to be high in many planet regions in the recent years [13]. So, as evidenced by the preliminary assessment, about a half of the summary economic loss because of the natural fires in 2010 in the European part of Russia has been stipulated for health disorders (people's death, higher applying for medical care) [9].

2. Methods

The Siberia region belongs to a number of the territories with a high degree of woodiness where the forest proportion, mainly light-coniferous ones, is known to amount more than 60% of a common area of its territory. More than 40% of the forests in this region are known to have the fire danger of high classes and high burning degrees. In Russia, areas with the highest fire risks are Irkutsk Oblast, Zabaikal Krai, and Khabarovsk Krai. The more fire dangerous period in this region is known to be correlated with the spring and autumn months when the high air dryness with the prevalence of windy weather may lead to strong drying the soil surface up and form favorable conditions for occurrence and development of forest fires. This work is aimed to calculate the concentrations of CO, NO_x, total PM (TPM) in the distance from the fire focus to the points according to the observation posts for concentration of air pollutants on the residential part on the example of the forests fires in the vicinity of Bratsk (Irkutsk oblast). To characterize the forests fires, the data of the territorial Department of State Forest Service in Bratsk of Irkutsk oblast have been used. The data of the observation posts have been presented by the Centre of Hydrometeorology and environmental monitoring. The values of multiple monitor's data for TPM and gaseous pollutants (SO_2, NO_2, NO, H_2S, formaldehyde and CO) are included. Weather variables from Bratsk were also distributed in this data set, including daily mean temperature, relative humidity, precipitation, pressure, resultant wind speed and direction. To assess the pollutant volumes, the calculation algorithm of summary emissions of pollutants to the atmosphere for the low-based forest fires, as well as the relative values of the emission coefficients have been used:

$$M_{\alpha i} = S_j \cdot K_j \cdot K_{\alpha i} \cdot m_{30j} \tag{1}$$

Here S_j is the forest territory area being exposed to fires, m^2; K_j is the burning fullness coefficient (0.5); $K_{\alpha j}$ is the alpha-pollutant emission coefficient, kg/kg; m_{30j} is the combustible forest material emergence, kg/m^2.

The theoretically-probability mean maximal value C_m (mg/m^3) of the pollutant concentration in the point x during the time moment t was calculated using the equation:

$$C_m = K \frac{P}{\pi \sigma_2 \sigma_3 U} \exp\left[-\frac{1}{2} \left(\frac{x_3}{\sigma_3} \right)^2 \right] \tag{2}$$

Here $P(x', t')$ is the source productivity in the point x' during the moment t'; $\sigma_i(f)$ are the standard deviations (i – 1, 2, 3; f – t-t'); K is the coefficient taking into account emission source type; x_3 is the height of pollutant emission source.

We calculated Hazard coefficient (*HQ*) and Hazard Index (*HI*) by formulae 3 and 4 respectively:

$$HQ = \frac{C_m}{Rfc} \quad (3)$$

$$HI = \sum HQ \quad (4)$$

Here *Rfc* is the reference concentration (mg/m^3) recommended by World Health Organization (WHO) for inhalation exposure.

3. Results and discussion

Analysis of the climatic factors and records on fires in Siberia indicates that the forest fires in this area have cycles of 3-4, 11, and 61-62 years, and shows the humidity distribution in Central and South Siberia from May to June and in North Siberia from July to August. The fire cycle of one of the fire sites was shortened to 11 years and 7 months due to human activity. Each fire in spring generally destroys 45-80 hectares, indicating a level of force that is in the low-mid to high range.

Fires that occur in industrial areas are especially dangerous. Long-term release of industrial wastes exacerbates the destruction of the forest ecosystem, cause an imbalance in air components, and cause harmful substances to accumulate in plant fiber. The amount of smoke generated in the event of a forest fire in an industrial area is 34% in the case of poplar, 51.1% in the case of birch, 37.7% in the case of aspen, and 69.1% in the case of the bark of broadleaf trees [12]. Key air pollutants that are generated in the event of a large forest fire in an industrial area include CO, CO_2, NO_x, SO_2, H_2S, C, HCN, SiO_2, HCHO, organic acid, and benzopyrene. According to a 2015 assessment, only in Irkutsk, the amount of poisonous substances discharged into the air as trees burned down during the fire risk period reached 75 tons.

From among 52 indexes, statistics, such as the number of forest fires, land area burned by fires, number of residents in affected areas, components of the forests that burned (grass, boughs, etc.), and the number of trees that burned down, were used to identify factors that influence the components of poisonous substances that are substantially discharged into the air in the event of a forest fire. Based on such statistics, the research was carried out to discern and forecast the number of fires and risk areas from the discharge of poisonous substances.

Research models were devised in Bratsk City, Irkutsk oblast. A total of 500 to 2000 fires were occurred in Irkutsk oblast in the last ten years. Over a ten-year period in the Irkutsk region, forest fires damaged an average of 61 623 hectares (table1).

Table 1. Average characteristics of the forest fires in the period from 2003 to 2013 in Irkutsk Oblast and Bratsk District.

Territory	Area of forest fires, ha			Mass of pollutants, ton			Average damage from forest fires, thousand rubles
	M±m	Min	Max	M±m	Min	Max	
Bratsky District	1 109±611	40	5 586	1 532±843	55	7 773	68 466
Irkutsk Oblast	61 623±18 312	7 214	193 685	85 088±25 286	9 961	267 439	943 653

In Bratsk district at an average annual forest fires cover an area of 1 109 hectares. On average, during the period under review, 85 088 tons of pollutants were allocated by forest fires in Irkutsk region, and 1 532 tons were allocated in Bratsk district. The economic loss of energy resources resulting from reduced production of firewood for fuel was estimated at 326,000 dollars, and the loss of lumber for industrial use was estimated at 285,000 dollars. The incineration of trees reduced the stores of available lumber, productivity of the forest stemming from the nutritive conditions of the ecosystem's energy and material flow. This is regarded as a loss to the forest ecosystem caused by fire.

Bratsk City has a population of about 300 thousand people. On the territory of Bratsk district, in the period of 17 September and 6 October 2011, forest fires took place on a total area of 6 928 hectares. The period of high altitude of forest territories lasted 20 days and the areas of forest fires were from 10 hectares to thousands of hectares.

As a result of burning forest vegetation in the designated territory, the following pollutants were released into the air: carbon monoxide – 9.1, total particulate matter – 4.5, PM2.5 – 2.2, carbon black – 0.7, acetaldehyde – 0.6, sulfur dioxide – 0.0000648 tons a day. The summary pollutant emission mass was found to be increased according to the fire area. The higher emission masses, independent on the fire area, were characteristic for CO and TPM formed in the smoldering of forest combustible materials stored in the landfills of woodworking enterprises. Traces of smoke and an increase in air pollution were detected at the site of a major forest fire close by a residential area. During the first week after the fire, the hygiene index increased more than four times due to pollutants, carbon monoxide more than five times, nitrogen dioxide at least six times, phenol at least 4.2 times, formaldehyde at least two times, and sulphur dioxide at least 2.5 times.

High productivity of a forest fire over a large area has led to large concentrations of pollutants estimated. The results of calculations are compatible with the CO and TPM at the observation posts well. The real data of the observation post testify the increase in the contents of CO 4 times (1.0 mg/m3), the TPM 1.6 times (0.2 mg/m^3). In the period of massive landscape fires, the ranking of the danger of air pollutants for population of Bratsk can be presented in the form of the following series: PM10 and PM2.5, carbon black, carbon monoxide, nitrogen dioxide, hydrogen sulphide, acrolein, and formaldehyde. Air pollutants have the greatest danger for the respiratory system and cardiovascular system. The increase in hazard ratios was registered for the ingredients of emission-specific industrial enterprises and capable of accumulating in vegetation, namely carbon disulphide 1.9 times, fluorine-containing substances 1.8 times during the fire.

Taking into account the admixture activity trends emitted to the atmospheric air the priority classes of diseases have been determined in the formation of which the air quality may play a significant role (respiratory diseases and circulation organ diseases). Formaldehyde, nitrogen dioxide, TPM, carbonic oxide, sulfur dioxide is known to belong to the ingredients giving the danger level for the classes indicated. The hazard index (HI) stipulated for the short-term exposure was found to be on the average 3.91. The maximal HI = 19.1 was observed to be in September on the day with dangerous wind rate. The analysis of the climatic parameters testified that the higher pollution levels were just observed to be in those days. The prevalence of the daily death cases based on these causes is considered. It was found that, on the average, 0.22 death cases were registered due to the diseases of the cardiovascular system per day and 0.032 were because of the pathology of the respiratory organs. On the day with the maximal HI level 3.0 death cases have been registered due to the cardiovascular diseases (CVD), 1.0 were due to the respiratory organ diseases (ROD). The daily applying for the urgent medical care, based on the causes of studying, did not practically differ and it was 10.5 (CVD) and 10.2 (ROD) of cases in average. During the first two or three days, there was thick smoke, and 1.3% of the city's population went to the hospital. The incidence of deaths due to circulatory system diseases increased 2.5 times. An analysis of the pollution led to statistical materials (r_{sp}=0.46. p=0.002) indicating a more than five-fold increase in the level of danger caused by such short-term impact on health such as deaths from breathing and circulatory system diseases, as well as the use of ambulances. The total number of patients hospitalized increased by 10%. The corresponding figure for children was 17%.

The information assessment of the environmental factors for forming the population mortality rate has been performed using the method by Shepard. The following more significant factors influencing on the population health disorders during the fire periods have been revealed, i.e. the wind rate, the TPM contents, air temperature, the contents of soot and formaldehyde. A number of the significant predictors on the days without the fire may have another order (table 2).

Table 2. Predictor significance for forming the population mortality among the population of Bratsk.

Predictor	Forest fire days	Days without fires
T° C	0.0019	0.0019
TPM	0.0020	0.0014
Wind rate	0.0020	0.0012
Formaldehyde	0.0007	0.0005
Black carbon	0.0011	0.0003
NO_2	0.0003	0.0001
CO	0.0005	0.0001
SO_2	9.5711	6.0679

In addition, the most important factors of harm to the health of residents in the event of a fire were identified. Harm to the health of residents triggers a substantial economic loss of 8.0 million dollars. A rise in deaths also leads to great economic losses. Economic losses stemming from hospital treatment for less than 1% of residents total 120 thousand dollars in seven days, accounting for 0.01% of annual GRDP [7]. The total economic losses calculated based on official materials in Bratsk were 72.1 million dollars. The amount stemming from harm to health (deaths, hospitalization) is presumed to total 3.2 million dollars, costs incurred from extinguishing fires are 0.45 million dollars, and losses to forest resources are 68.5 million dollars.

The forest fires in the neighborhood of the populated areas may contribute to the air pollution and they are known to be one of the causes of their forming smoke-screen. For example, the heaviest smoking has occurred in Moscow due to the wild fires in the central regions of Russia in 2010 that is why the maximal concentration of CO, specific hydrocarbons and the total PM was significantly higher than the allowed level, already in the first decade of August 2010 [9, 10]. The suspended particles with size of 10 micrometers (PM10) or less (0.0004 inches or one-seventh the width of a human hair were found to be more dangerous for human health). Major concerns for human health from exposure to PM10 include the following: effects on breathing and respiratory systems, damage to lung tissue, cancer and premature death [9].

4. Conclusion

Forest fires served to adjust the forest ecosystem during the last century, but now they are regarded as disasters from the perspective of the economy, society, and ecology. There is a periodical occurrence of major fires that require mobilization of tremendous resources and methods to combat and extinguish them. That is why forecasting such disaster is essential. Effective forecasting would minimize losses by improving the ability to respond to such disasters and establishing better ways to respond to them.

References

[1] Analitis A, Georgiadis I and Katsouyanni K 2012 Forest fires are associated with elevated mortality in a dense urban setting *Occup Environ Med* **69** 158–62

[2] Bacciu V, Spano D and Salis M 2015 Emissions from Forest Fires: Methods of Estimation and National Results *The Greenhouse Gas Balance of Italy* 87-102

[3] Chertov O G, Komarov A S, Gryazkin A V, Smirnov A P and Bhatti D S 2013 Simulation modeling of the impact of forest fire on the carbon pool in coniferous forests of European Russia and Central Canada *Contemporary Problems of Ecology* **6**(7) 727-33

[4] Coluzzi R, Masini N, Lanorte A and Lasoponara R 2010 On the Estimation of Fire Severity Using Satellite ASTER Data and Spatial Autocorrelation Statistics *Lecture Notes in Computer Science* **6016** 361-73

[5] Ganteaume A, Camia A, Jappiot M, San-Miguel-Ayanz J and Lampin C A 2013 Review of the main driving factors of forest fire ignition over Europe *Envir Managemen* **51**(3) 651-62.

[6] Isaeva L K, Solovyov S V, Sulimenko V A and Shilin S A 2010 Environmental impacts of forest and peat fires in the Moscow region in 2002-2009 years *Fires and emergencies* **3**(10) 85-91

[7] Mavsar R, Varela E, Corona P, Barbati A and Marsh G 2012 Economic, legal and social aspects of post-fire management Post-Fire Management and Restoration of Southern European Forests *Managing Forest Ecosystems* **24** 45-78

[8] Naeher L, Brauer M, Lipsett M, Simpson C, Koenig J Q, Zelikoff J and Smith K R 2007 Woodsmoke health effects: a review *Inhal Toxicol* **19** 67-106

[9] Shaposhnikov D, Revich B, Bellander T C, Bedada G B, Bottai M, Kharkova T, Kvasha E, Lezina E, Lind T, Semutnikova E and Pershagen G 2014 heat wave and wildfire air pollution related mortality in the summer of 2010 in Moscow *Epidemiol* **25**(3) 359–64

[10] San-Miguel-Ayanz J, Pereira J M C, Boca R, Strobl P, Kucera J and Pekkarinen A 2009 Forest Fires in the European Mediterranean Region: *Mapping and Analysis of Burned Areas, Earth Observation of Wildland Fires in Mediterranean Ecosystem*s Ed. E. Chuvieco (Berlin Heidelberg: Springer-Verlag) pp 189-203

[11] Sherstyukov B G and Sherstyukov A B 2014 Assessment of increase in forest fire risk in Russia till the late 21st century based on scenario experiments with fifth-generation climate models *Rus Meteorology and Hydrology* **39**(5) 292-301

[12] Timofeeva S S, Garmyshev V V and Zyryanov V S 2012 Smoke situation when burning forest combustible materials in urban and suburban areas of cities *Bul of the Irkutsk State Technical University* **3**(62) 50-55

[13] Vadrevu K P, Eaturu A and Badarinath K V S 2010 Fire risk evaluation using multicriteria analysis – a case study *Environ Monitoring and Assessment* **166** 223-39

Mapping of Vegetation with the Geoinformation System and Determining of Carrying Capacity of the Pre-Urals Steppe area for a Newly Establishing Population of the Przewalski Horse *Equus ferus przewalskii* at the Orenburg State Nature Reserve

N I Fedorov[1], **O I Mikhailenko**[2], **T L Zharkikh**[3], **R T Bakirova**[3]

[1] Ufa Institute of Biology RAS, 69 Prospekt Oktyabrya, Ufa, 450054, Russia
[2] Ufa State Petroleum Technological University, 1 Cosmonauts Str, Ufa, 450062, Russia
[3] The Federal Government Funded Institution «The Joint Directorate of State Nature Reserves «Orenburg» and «Shaitan Tau» (FGFI «Orenburg Reserves»), 2/2 Donetskaya Str, Orenburg, 460001, Russia

E-mail: fedorov@anrb.ru, trioksan@mail.ru, russian969@yandex.ru, zapoved@mail.orb.ru

Abstract. Mapping of the vegetation (1:25000) of the Pre-Urals Steppe area at the Orenburg State Nature Reserve was completed in 2016. A map created with the geoinformation system contains 1931 simple and complex polygons for 25 types of vegetation. In a drought year, the average stock of palatable vegetation of the whole area is estimated at 8380 tons dry weight. The estimation is based on the size of areas covered by different types of vegetation, their grass production, the correction coefficients for decreasing of pasture forage stocks in winter and decreasing of production of grass communities in dry years. Based on pasture forage stocks the area could tolerate the maximum population size of 1769 individuals of the Przewalski horse, their average density could be 0.11 horse per ha. Yet, as watering places for animals are limited in Pre-Urals Steppe, grazing pressures on the vegetation next to the water sources may increase in dry years. That is why the above-mentioned calculated maximum population size and density must be reduced at least by half until some additional watering places are established and monitoring of the grazing effect on the vegetation next to the places is carried out regularly. Thus, the maximum size of the population is estimated at 800 to 900 individuals, which is almost 1.5 times more than necessary to establish a self-sustained population of the Przewalski horse.

1. Introduction

At present, the number of living captive and reintroduced Przewalski horses *Equus ferus przewalskii* in the International Studbook is about 2000 individuals [1]. In 2015, the Government of the Russian Federation handed over to the Federal Government Funded Institution (FGFI) «Orenburg Reserves» the territory of the former military training ground of 16538 ha in size to establish a new strictly protected area called Pre-Urals Steppe as the 5th part of the Orenburg State Nature Reserve (hereinafter Orenburg Reserve). The FGFI «Orenburg Reserves» developed the Programme of Establishing of a

Semi-Free Population of the Przewalski Horse in Pre-Urals Steppe. Having been supported by The United Nations Development Programme / Global Environment Facility (GEF) / Ministry of Natural Resources and Ecology of the Russian Federation within the frame of the Project «Improving the Coverage and Management Efficiency of Protected Areas in the Steppe Biome of Russia» FGFI «Orenburg Reserves» transported six Przewalski horse to Pre-Urals Steppe from France in 2015 and 30 horses from Hungary in 2016 and 2017. An evaluation of the carrying capacity of the area for the newly establishing population of the Przewalski horses was necessary for the successful development of the Programme.

The aim of the present study is to create a map of vegetation (1:25000) with the Geoinformation System (hereinafter GIS-map), to analyse the pasture forage stocks of the Pre-Urals Steppe area and to determine the maximum number of the horses the area can tolerate.

2. Materials and methods

A typology of the vegetation of Pre-Urals Steppe was developed by Sergey M. Yamalov, Maria V. Lebedeva, and Yaroslav M. Golovanov from Botanical Garden-Institute, Ufa Scientific Centre RAS in 2016 before the present study. The typology includes 26 types of plant communities with various degrees of anthropogenic disturbance. The analyses of the total productivity of grass available for horses and the production of palatable grass forage were carried out during the period of the maximum grass development. In each type of plant communities, vegetation was cut in 5 sample squares of $1 m^2$ each. The samples were sorted by 6 categories regarding their edibility and palatability for Przewalski horses. They are the following: 1) gramineous plants, 2) legumes, 3) well edible forbs, 4) reluctantly and non-edible forbs, 5) plants with unknown edibility, 6) standing dead plants from the last year. The data on preferred plant species for the Przewalski horse were obtained from an earlier study at the Biosphere Reserve "Askania Nova" [2]. Then the samples were air-dried and weighted to within 0.01 g.

During the main growing season in 2016, the Pre-Urals Steppe area experienced optimum vegetation-growing conditions. The field investigation for the GIS-mapping was carried out in Pre-Urals Steppe in June–August 2016. Geobotanical descriptions of 1354 sample plots of $100 m^2$ each were fulfilled to classify them according to the above-mentioned typology of vegetation. Images obtained from Earth observation satellites Resurs-P, Landsat 7, and Landsat 8 were analysed with software NextGIS Mobile in an Android-based tablet and used as a cartographic base during the field period. Multispectral images obtained from the above-mentioned satellites and analysed with software QGIS were used for GIS- mapping during the desktop period. The resulting outlined polygons were simple and complex contours of vegetation. Simple contours demonstrated vegetation of a single type, complex contours included 2 to 3 types of vegetation. Each dominated type of vegetation was set in an individual layer of the GIS-map. Tables of attributes for each layer contained information about the ratio between the main type and associated types of vegetation of a given plot, bush encroachment, size of the plot, pasture forage stocks, etc.

Calculation of the pasture forage stocks and evaluation of the maximum size of the population of the Przewalski horse for Pre-Urals Steppe were made with the use of the methodological recommendations on optimisation of grazing pressure on natural steppe communities [3]. An estimation of maximum pressure of free-ranging domestic horses *Equus caballus* on lands of the Isle of Vodnyj in the Lake of Manych-Gudilo [4] was used. Data on the pasture food intake by Przewalski horses in winter and summer were used from the study by M K Pozdnyakova and co-authors [5].

3. Results and discussion

There were 1924 simple and complex polygons outlined in pursuing GIS-mapping of the Pre-Urals Steppe area. The most widespread vegetation type was *Stipa capillata* communities. 348 polygons of 2105.8 ha represented it. Recent fallow lands (*Melilotus officinalis*+*Artemisia austriaca* communities) occupied 288.1 ha. There was little woody vegetatione; the size of steppe plots with bush encroachment of ≥10% of their total area were 154.6 ha. Yet, bush communities themselves occupied only 43.2 ha. Basing on the size of vegetation communities and their grass production the palatable grass forage stocks were calculated for the Pre-Urals Steppe area. Totally 30451.7 t of dry weight were estimated (table 1).

Table 1. Characteristics of types of plant communities in the Pre-Urals Steppe area at the Orenburg State Nature Reserve.

Type of a plant community	Number of polygons in the GIS-map	Size, ha	Pasture forage stock, t	
			Totally	Palatable plants
Stipa borysthenica motley-grass species-poor	76	632.7	14223.0	11145.5
Stipa borysthenica motley-grass species-rich	74	292.2	5653.8	3497.1
Agropyron pectinatum	48	1079.9	37830.5	34248.4
Alopecurus arundinaceus	4	35.2	1597.3	1464.5
Elytrigia repens+Poa transbaicalica	151	1049.0	35786.8	27609.0
Stipa zaleskii+Poa transbaicalica	84	927.2	32099.1	19387.4
Poa transbaicalica motley-grass	54	1310.0	55201.7	49265.4
Caragana frutex+Spiraea crenata	71	43.2	822.2	398.6
Stipa lessingiana+Festuca valesiaca	51	1410.4	45869.4	42676.9
Stipa lessingiana+ Serratulla cardunculus	9	384.3	13230.0	7609.2
Galatella villosa+Artemisia lercheana	188	1307.0	29806.2	3981.8
Poa crispa	138	1370.7	26015.3	15216.2
Festuca valesiaca+Artemisia austriaca	194	929.6	20944.1	6423.6
Ceratocarpus arenarius[a]	1	18.2	–	–
Atriplex sagittata[a]	23	144.6	–	–
Artemisia lerchiana	124	908.5	13647.3	7020.3
Serratula cardunculus	20	480.0	10348.8	7052.4
Festuca valesiaca+Falcaria vulgaris+ Malabaila graveolens	17	1245.0	31472.0	26349.1
Festuca valesiaca+Artemisia lerchiana petrophytic	165	240.8	2898.3	1696.2
Festuca valesiaca+Artemisia lerchiana+Galatella villosa	96	154.6	2725.3	1414.0
Artemisia pauciflora	16	73.6	1215.5	542.2
Stipa capillata	291	2105.8	38118.7	35952.6
Puccinellia distans	6	6.4	123.4	84.3
Phragmites australis+Carex acuta[a]		8.6	–	–
Melilotus officinalis+Artemisia austriaca	4	288.1	14758.0	1482.2
Woody vegetation[a]	19	29.0	–	–
Total:	1924	16474.6	434386.6	304516.8

[a] – Reluctantly and non-edible vegetation

The Przewalski horse has the same feeding preferences as the domestic horse [5]. Free-ranging domestic horses consumed 8 to 14 kg dry wt daily during snow-free grazing and 16 kg dry wt daily in winter [4]. The daily forage intake by the Przewalski horse was slightly less; depending on quality of the forage it fluctuated from 7 to 11 kg dry wt per horse [5]. Hence, our approach was based on daily forage intake of 10 kg per horse in summer. In winter, the Przewalski horse consumes gramineous plants preferentially (T L Zharkikh, pers. comm). As gramineous plants averaged 93.2% of the edible pasture forage in Pre-Urals Steppe, the daily food intake was estimated to be 17.2 kg per horse in winter. Consequently, the pasture forage consumption was estimated at 2.14 t per horse during snow-free period lasting 214 days from April 1 to October 30. The forage consumption was estimated at 2.6 t per horse in winter period lasting 151 days from November 1 to March 31. Totally, one Przewalski horse needs 4.74 t dry wt of pasture forage per year.

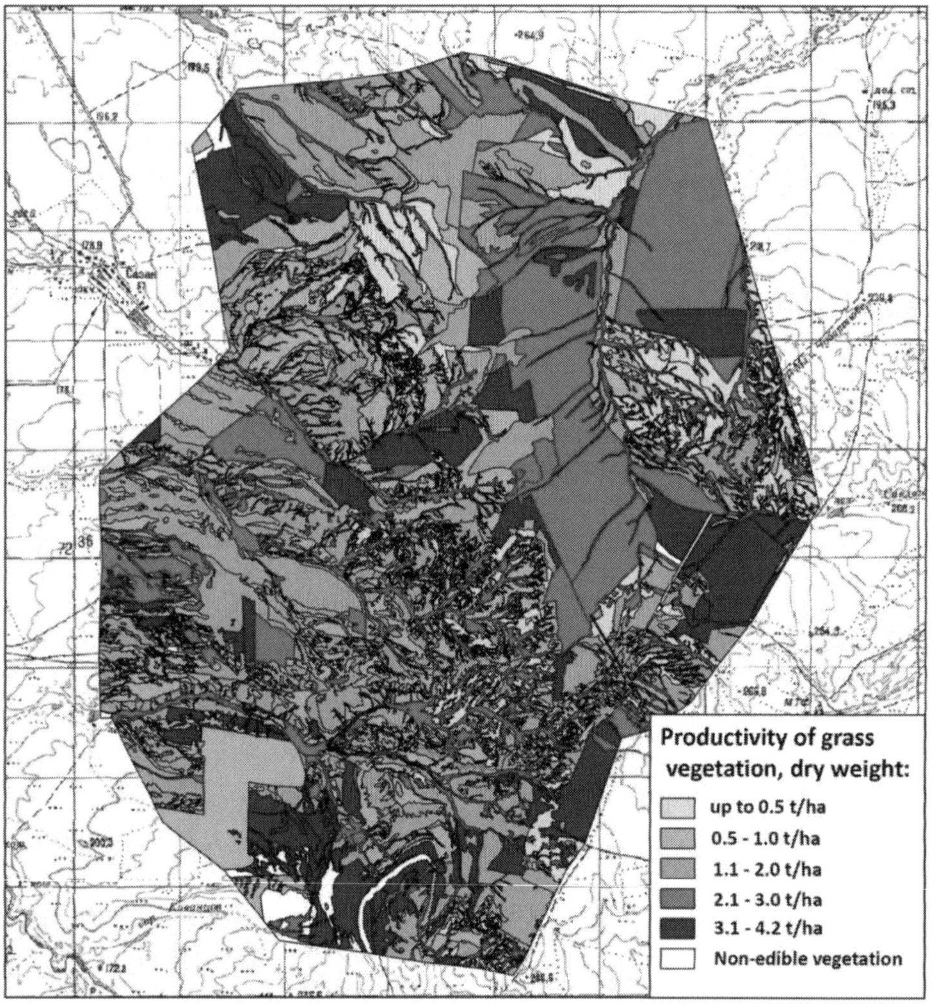

Figure 1. The productivity of palatable grass vegetation in plant communities in the Pre-Urals Steppe area.

Dry steppes are characterized by sharp fluctuations in vegetation production depending of rainfall. Grass productivity grows until the middle of June. In genuine steppe grasslands, half of vegetation dries out in August; petrophytic steppes dry out almost entirely at the same time. Meadow steppes mostly remain green in August. In autumn, the green plant weight of all types of steppe vegetation increases as a result of colder weather and increasing in rainfall. Spring and autumn grazing results in grass regrowth increasing its productivity by 1.5 to 2 times. It compensates for the loss of forage in the second half of summer. [3, 6]

Thus, the productivity of the herb layer of plant communities in the period of their maximum growth could be used as a basis to calculate the maximum number of Przewalski horses the area can tolerate during the snow-free season. Correction coefficient 0.8 should be used to calculate winter edible forage stocks [6]. Extreme drought reduces steppe grass productivity by 2.5 to 3 times [3, 6], so, correction coefficient 0.3 should be used to calculate edible forage stocks. Taking into consideration the correction coefficients, we estimated the average total palatable forage stock at 8379.6 t dry wt in extreme years. Hence, the area could tolerate the maximum population size of 1769 individuals of the Przewalski horse, their average density could be 0.11 horse per ha. However, the estimation was carried out without regard to the shortage of constant sources of water in Pre-Urals Steppe. Grazing pressures on the vegetation next to watering places may be increased in dry years. In this connection, the establishment of additional water sources is essential given the variegated vegetation production in different parts of the area (figure 1). Once new water sources are established the constant monitoring of the state of vegetation in areas with most grazing pressure and its succession trends should be carried out.

Until some additional water sources are provided the above-mentioned calculation of the maximum size of the newly establishing population and the density of Przewalski horses should be halved at least. Thus, the maximum population size should not exceed 800 to 900 individuals. The density of the animals should be about 0.05 to 0.06 horse per ha or 16 to 20 ha per horse. Nevertheless, the estimated maximum number of Przewalski horses for the Pre-Urals Steppe area is 1.5 times more than the minimum required number of animals to establish a self-sustained population [7]. If some new species of large herbivorous are introduced in this area, the present estimation should be reviewed.

Acknowledgments
We greatly acknowledge Luibov G. Linerova, Vladimir Yu. Petrov, and Aleksei A. Kozyr for their kind technical and practical assistance.

References

[1] King S R B, Boyd L, Zimmermann W and Kendall B E 2015 Equus ferus ssp. przewalskii. (errata version published in 2016) *The IUCN Red List of Threatened Species 2015*: e.T7961A97205530 http://dx.doi.org/10.2305/IUCN.UK.2015-2.RLTS.T7961A45172099.en (*Accessed* 08/2017)
[2] Zharkikh T L, Yasynetska N I 2009 A Study of Preferred Feeding Plants of Equids at the Reserve “Askania Nova” *Equus* 194-206
[3] Yunusbaev U B 2001 *Optimisation of grazing pressure on natural steppe communities* (Saratov: Nauchnaya Kniga) (In Russian)
[4] Kaz'min V D, Pozdnyakova M K, Abaturov B D, Kolesnikov M P 2013 Quantitative characterization of the nutrition of the free-living horse (Equus caballus) on Vodnyi Island (Manych-gudilo Lake) *Biology Bulletin* **8** 684–91
[5] Pozdnyakova M K, Zharkikh T L, Yasinetskaya N I, Kolesnikov M P 2011 Quantitative estimation of nutrition of free-ranging group of Przhewalskii horse (Equus przewalskii) in steppe habitat (Askania-Nova Reserve) *Russian Journal of Zoology* **3** 368–76 (In Russian)
[6] Yunusbaev U B 2000 *The steppes of the Bashkir Trans-Urals: pasture digression and possibilities of theirs recovering (on the basis of Baymak District)* PhD Thesis (Ufa) (In Russian)
[7] Franklin I R 1980 Evolutionary change in small populations *Conservation Biology: An Evolutionary–Ecological Perspective* ed M E Soule and B A Wilcox (Sunderland, Mass.: Sinauer Associates) p 135–40

A three-dimensional model of solar radiation transfer in a non-uniform plant canopy

N T Levashova, Yu V Mukhartova

Faculty of Physics, M.V. Lomonosov Moscow State University, 1 - 2, Leninskiye Gory, Moscow, 119991, Russia

E-mail: natasha@npanalytica.ru

Abstract. A three-dimensional (3D) model of solar radiation transfer in a non-uniform plant canopy was developed. It is based on radiative transfer equations and a so-called turbid medium assumption. The model takes into account the multiple scattering contributions of plant elements in radiation fluxes. These enable more accurate descriptions of plant canopy reflectance and transmission in different spectral bands. The model was applied to assess the effects of plant canopy heterogeneity on solar radiation transmission and to quantify the difference in a radiation transfer between photosynthetically active radiation PAR (=0.39-0.72 μm) and near infrared solar radiation NIR ($\Delta\lambda$ = 0.72-3.00 μm). Comparisons of the radiative transfer fluxes simulated by the 3D model within a plant canopy consisted of sparsely planted fruit trees (plant area index, PAI - 0.96 m^2 m^{-2}) with radiation fluxes simulated by a one-dimensional (1D) approach, assumed horizontal homogeneity of plant and leaf area distributions, showed that, for sunny weather conditions with a high solar elevation angle, an application of a simplified 1D approach can result in an underestimation of transmitted solar radiation by about 22% for PAR, and by about 26% for NIR.

1. Introduction

Incoming solar radiation is a key factor influencing all biological processes in Earth's biosphere. It determines the thermal regime of vegetation cover and soil, and affects the plant canopy's photosynthesis and transpiration [1]. Adequate estimation of vegetation, primary production and land surface evapotranspiration requires developing methods and models that enable highly accurate assessment of solar radiation reflection, absorption and transmission in plant canopies with non-uniform structure and diverse optical properties. During the last decades a large number of radiative transfer models based on both relatively simple parameterizations and more sophisticated approaches were developed and applied [1-9]. The most commonly used one-dimensional (1D) approaches consider the vegetation cover as a horizontally homogeneous turbid medium. They are usually based on either the "two-stream approximation" [10, 11] or a more simple approach based on the Beer–*Lambert–Bouguer* law, which assumes an exponential decrease of solar radiation depending on leaf area density [12]. It can be expected that this 1D assumption lead to uncertainties in estimation of radiation reflection, absorption and transmission in a spatially non-uniform vegetation cover. For such heterogeneous plant canopy, it is obviously necessary to use more sophisticated three-dimensional (3D) radiative transfer models that are able to take into account the 3D canopy structure [4, 8].

Within the framework of the study, a 3D radiative transfer model was developed. To quantify the effects of plant canopy heterogeneity on transfer of photosynthetically active radiation (PAR, $\Delta\lambda$ = 0.39-0.72 μm) and near infrared solar radiation (NIR, $\Delta\lambda$ = 0.72-3.00 μm) the radiative transfer parameters simulated by the 3D model were compared with radiation fluxes simulated by a 1D approach.

2. Model description

The 3D radiative transfer model is based on the main equations and assumptions suggested by Y. Ross [1] and further developed and described by Muneni [13, 14] and Kniazikhin et al. [4, 15]. According to this approach, the distribution function of the solar radiation intensity $I_\lambda(\mathbf{r},\Omega)$ of a wavelength λ at each spatial point $\mathbf{r}=\{x,y,z\}$ inside the vegetation canopy is dependent on the solid angle $\Omega=\{\varphi,\theta\}$ and is estimated as a sum of the direct radiation $I_{m,\lambda}(\mathbf{r},\Omega)$ and scattered diffuse radiation $I_{d,\lambda}(\mathbf{r},\Omega)$, where θ is a solar zenith angle. The distribution of the direct radiation intensity inside the vegetation cover is determined from the expression $I_{m,\lambda}(\mathbf{r},\Omega)=Q_{0,\lambda}(\mathbf{r})\delta(\Omega-\Omega_0)$. The probable density of the event that the sunbeam incoming to the upper boundary of the vegetation cover along the direction Ω_0 reaches a certain point $\mathbf{r}$ inside the vegetation without being reflected or scattered by vegetation elements is defined as

$$Q_{0,\lambda}(\mathbf{r})=T_{m,\lambda}(\mathbf{r}-l_{\mathbf{r},\Omega_0}\Omega_0)\exp\left(-\int_0^{l_{\mathbf{r},\Omega_0}}\sigma(\mathbf{r}-s\Omega_0,\Omega_0)ds\right).$$

Here, $T_{m,\lambda}$ is intensity of direct solar radiation at the upper boundary of the vegetation cover, $l_{\mathbf{r},\Omega_0}$ is the distance between the point $\mathbf{r}$ and the upper vegetation boundary along the direction Ω_0 to the sun, $\sigma(\mathbf{r},\Omega)$ is the cross-sectional area of the interaction (scattering and absorption) of solar radiation with vegetation elements. If the uniform distribution of the leaves by the tilt angles is assumed, we can take $\sigma(\mathbf{r},\Omega)=0.5\cdot LAD(\mathbf{r})$, where $LAD(\mathbf{r})$ is the leaf area density at point $\mathbf{r}=\{x,y,z\}$.

The intensity of the scattered radiation is determined from the equation

$$(\Omega\cdot\nabla I_{\lambda,d}(\mathbf{r},\Omega))+\sigma(\mathbf{r},\Omega)I_{\lambda,d}(\mathbf{r},\Omega)=\int_{4\pi}I_{\lambda,d}(\mathbf{r},\Omega')\sigma_{s\lambda}(\mathbf{r},\Omega'\to\Omega)\mathbf{d}\Omega'+\sigma_{s\lambda}(\mathbf{r},\Omega_0\to\Omega)Q_{0,\lambda}(\mathbf{r})$$

with additional conditions on the top and the lateral borders of the vegetation cover $I_{d,\lambda}(\mathbf{r}_t,\Omega)=T_{d,\lambda}$, where $T_{d,\lambda}$ is the incoming scattered radiation into the point $\mathbf{r}_b=\{x,y,z_b\}$. At the bottom border near the soil surface, the value $I_{d,\lambda}(\mathbf{r}_b)$ is determined as $I_{d,\lambda}(\mathbf{r}_b)=\dfrac{\rho_\lambda\mu_0 Q_{0,\lambda}(\mathbf{r}_b)}{1-\rho_\lambda\pi}$ where ρ_λ is a coefficient of solar radiation reflection from the soil surface at the point $\mathbf{r}_b=\{x,y,z_b\}$ at the bottom border. In the equation for scattered radiation, $\sigma_s(\mathbf{r},\Omega',\Omega)$ is the differential cross-section for the scattering of the rays falling in the direction Ω' and scattered in the solid angle $d\Omega$ for respected r and Ω. According to Ross and Myneni [2]

$$\sigma_s(\mathbf{r},\Omega',\Omega)=\frac{1}{2\pi}LAD(\mathbf{r})\int_{2\pi^+}|(\Omega',\Omega_L)|(\gamma_{LD}(\Omega_L,\Omega',\Omega)+\gamma_{LS}(\Omega_L,\Omega',\Omega))d\Omega_L$$

where $\gamma_{LD}(\Omega_L,\Omega'\Omega)=\begin{cases}\pi^{-1}r_{LD}(\alpha')|(\Omega,\Omega_L)|,\ (\Omega,\Omega_L)(\Omega',\Omega_L)<0,\\ \pi^{-1}t_{LD}(\alpha')|(\Omega,\Omega_L)|,\ (\Omega,\Omega_L)(\Omega',\Omega_L)>0,\end{cases}$ is a phase function of leaf scattering: the part of the energy intercepted by photons initially falling in the direction Ω' and scattered after collision of photons with a leaf surface with an external normal Ω_L into a solid angle

Ω; α' is an angle between the incident sunbeams and the perpendicular to the leaf surface, $r_{LD}(\alpha')$ is a hemispherical leaf reflection coefficient, $t_{LD}(\alpha')$ is a hemispherical leaf transmission function; $\gamma_{LS}(\Omega_L,\Omega',\Omega)=F_r(n,\alpha')\delta_2(\Omega,\Omega^*)$ is a mirror reflection component, where $F_r(n,\alpha')=\frac{1}{2}\left(\frac{\sin^2(\alpha'-\theta_s)}{\sin^2(\alpha'+\theta_s)}+\frac{\mathrm{tg}^2(\alpha'-\theta_s)}{\mathrm{tg}^2(\alpha'+\theta_s)}\right)$ is the air refraction index $n=1.5$, $\theta_s=\arcsin\left(\frac{\sin\alpha'}{n}\right)$, δ_2 is a Dirac delta function, Ω^* is the direction of the corresponding leaf normal for the mirror scattering of the incident and reflected rays. The intensity of solar radiation at each point $\mathbf{r}=\{x,y,z\}$ is calculated as the total intensity of the radiation flux in each of the directions determined by all possible Ω.

To solve the system of differential equations, the numerical scheme described by Ross and Myneni [2] is used.

3. Description of experimental plot and model input parameters

For numerical experiments, a one-hectare plot with sparsely planted fruit trees was chosen (figures 1-2). The height of each planted tree is 8 m, maximal diameter of a tree crown is 6 m, the distance from the ground to the crown bottom is 1 m, and diameter of a tree trunk is 0.2 m. The distance between tree trunks is 5.24 m. The total number of trees within selected plot is 181. The plant area density (PAD) includes leaf area densities (LAD) and the density of non-photosynthesizing element of trees (branches and stems). The proportion of LAD in PAD within tree crowns is changed depending on the height above a ground and the distances from the tree stem and crown edge.

The leaf area distribution inside of the tree crown is based on the model of a tree structure described by Wildowski et al. [16] and Olchev et al. [7]. The model assumes that the maximal crown spreads and LAD are observed in the upper part of the trees. Moreover, it is assumed that LAD in the tree crown decreases as the distance from tree crown edge increases. The spectral properties of the leaves and bark are taken from the literature [e.g. 17] (table 1). The modeling experiments were provided for midday hours under maximal sun height (the solar zenith angle, Θ=30° that is equal to solar elevation angle $h_\odot$=60°). It was assumed that direct portion of solar radiation is 75%.

Table 1. Reflection and transmission coefficients of plant elements and soil surface for PAR and NIR.

Spectral bands, $\Delta\lambda$ (μm)	Leaves		Branches and stems		Soil
	r_{LD}	t_{LD}	r_{LD}	t_{LD}	r_S
0.39-0.72	0.06	0.06	0.04	0.00	0.05
0.72-3.00	0.45	0.45	0.40	0.00	0.30

4. Results and discussion

The results of numerical experiments show a strong heterogeneity of radiation patterns within both the entire experimental plot and the crown of individual trees (figures 1 - 2). A strong difference in plant canopy transmission coefficients (K) between PAR and NIR is mainly governed by different reflection and transmission coefficients of leaves, bark and soil for PAR and NIR spectral bands, as well as a higher contribution of the multiple scattering from plant elements for NIR. The estimated difference of K(NIR) - K(PAR), even for relatively sparse canopy, varies between 0.02 to 0.24 (figure 3). Maximal differences are found for sites which are shaded by tree crown leaves and stems.

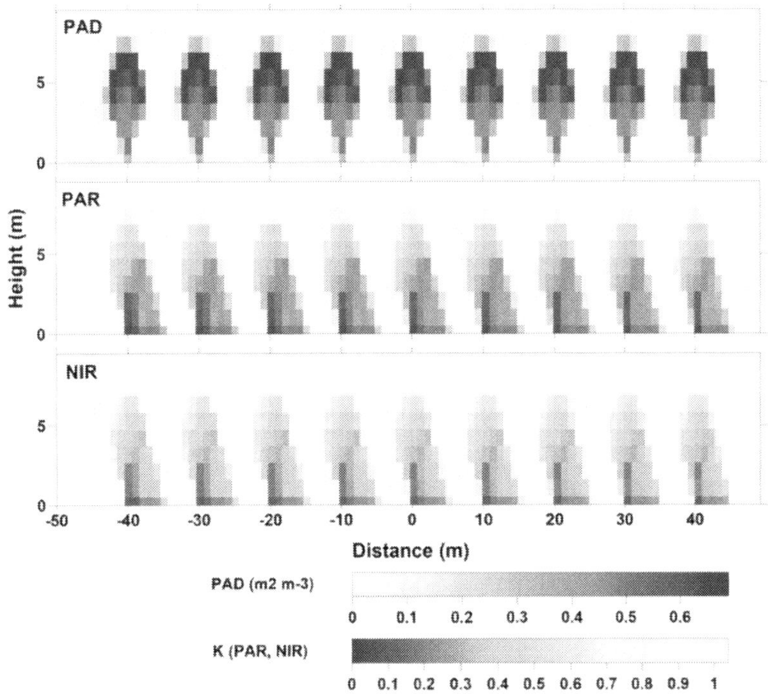

Figure 1. The distribution of plant area density (*PAD*) and the transmission coefficients (*K*) for PAR and NIR along a vertical cross-section through an experimental plot (at $y = 0$, see figure 2).

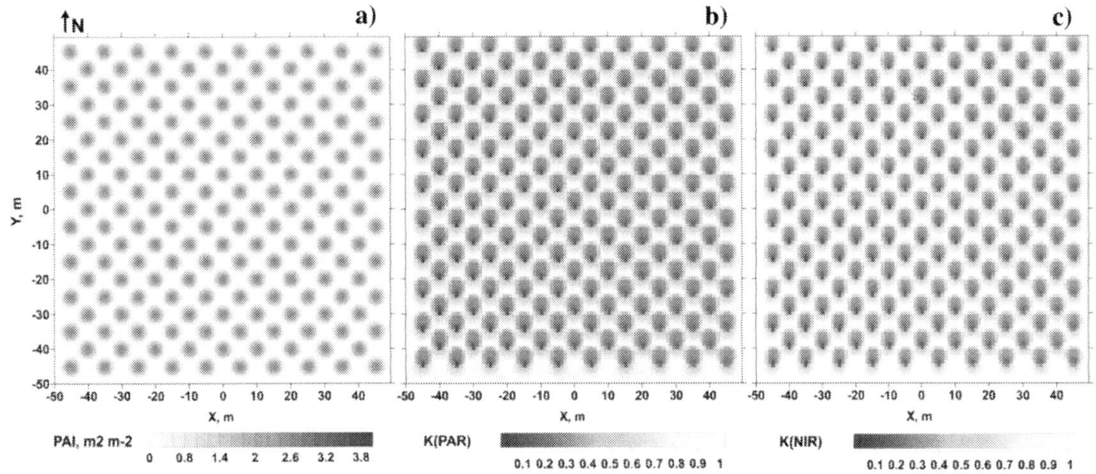

Figure 2. The distribution of the plant area index, *PAI,* (a) and the transmission coefficients (*K*) for PAR (b) and NIR (c) at a ground surface level. The shapes of tree crowns and PAI distribution are shown in the figures for *K*(PAR) and *K*(NIR) as a background.

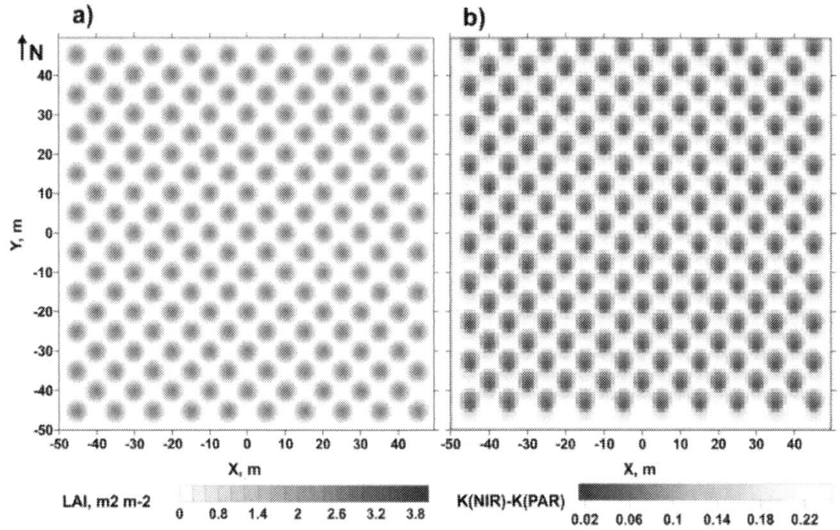

Figure 3. The distribution of *PAI* (a) and the difference between transmission coefficients *K*(NIR) and *K*(PAR) at a ground surface level (b). The shapes of tree crowns and PAI distribution are shown in the right figure for *K*(NIR) - *K*(PAR) as a background.

To estimate the possible uncertainties of solar radiation transfer estimations due to ignoring the spatial vegetation heterogeneity two separate experiments were provided. In the first experiment the solar radiation transfer was simulated for actual 3D plant canopy structure (figure 2). The second experiment imitated 1D plant canopy structure. The plant canopy in this experiment was considered as a horizontally uniform turbid medium. It was also assumed that the PAD of the plants in the experiment is dependent on vertical coordinate only. The values of PAI averaged for entire experimental plot for both modeling experiments were identical.

Comparing the extinction coefficients calculated for PAR and NIR for different levels within a plant canopy showed a significant differences in radiation flux calculations between 3D and 1D plant canopies. Transmission coefficients (K) of solar radiation predicted by the 3D model for both spectral intervals (PAR and NIR) was higher than K values calculated for 1D plant canopy for all canopy layers (tables 1-2). Maximal differences were observed for NIR at a ground surface layer (about 26%). Maximal differences for PAR were observed at 2 m layer above a ground surface. The difference between K calculated for 3D plant canopy in this layer was higher than K calculated for uniform plant canopy by about 22%. The main explanation of such differences could be the free penetration of direct solar beams through the gaps of relatively sparse 3D plant canopy. The difference between transmission of PAR and NIR can be explained by the different scattering coefficients (reflection, transmission) of the leaf and bark for PAR and NIR, as well as by contribution of the multiple scattering processes, which are due to higher leaf reflection and transmission much larger for NIR (table 1).

The modeling experiments provided by Yuan et al. [18], using the 3D radiation transfer model and the two-stream approximation model within the Community Land Model (CLM4.0), showed similar results. The canopy absorption estimated by the 3D model under low zenith angle was lower and canopy transmission was higher than in case of 1D model, due to mainly enhanced transmission through the canopy edges. In these modeling experiments, it was also shown that under the large zenith angle, the canopy absorption calculated using the 3D model can be higher than in case of 1D model, due to increases of the ground and sky shadows as well as the optical path length. Similar effects were described by Huang et al. [19] from simulation results of radiative transfer using a 3D model for a sparse forest canopy.

Table 2. Transmission coefficients (K) for PAR calculated for different levels of a horizontally uniform (1D) and heterogeneous (3D) plant canopy and averaged for entire experimental plot.

Height (m)	Transmission coefficient for vegetation with uniform structure (1D)	Transmission coefficient for vegetation of heterogeneous structure (3D)	The relative difference between K(1D) and K(3D) in %
0	0.47	0.60	-21.8
1	0.53	0.68	-22.2
2	0.52	0.67	-21.6
3	0.54	0.67	-19.0
4	0.59	0.70	-16.3
5	0.68	0.78	-13.2
6	0.80	0.88	-8.9
7	0.89	0.95	-6.8
8	1.00	1.00	0.0

Table 3. Transmission coefficients (K) for NIR calculated for different levels of a horizontally uniform (1D) and heterogeneous (3D) plant canopy and averaged for entire experimental plot.

Height (m)	Transmission coefficient for vegetation with uniform structure (1D)	Transmission coefficient for vegetation of heterogeneous structure (3D)	The relative difference between K(1D) and K(3D) in %
0	0.53	0.72	-26.3
1	0.56	0.74	-24.3
2	0.56	0.73	-23.7
3	0.58	0.73	-21.1
4	0.62	0.75	-17.0
5	0.71	0.82	-13.6
6	0.81	0.90	-10.0
7	0.88	0.96	-8.3
8	1.00	1.00	0.0

Considering the differences in radiative transfer simulations found between 1D and 3D approaches, it should be taken into account the possible uncertainties of applied 3D models arising from the model simplifications and assumptions used to describe canopy structure and radiative transfer. These uncertainties could stem from limitations of the applied numerical methods, etc. In particular, the results of a 3D radiative transfer model comparison experiment (RAMI4PILPS) provided for seven independent 3D models under controlled experimental conditions showed that the averaged deviation from predicted model parameters from some reference amounted to 19% for canopy absorption, 21% for canopy reflection and 26.7% for canopy transmission [8]. Another point that must be taken into account in a comparison of 3D and 1D models is the different complexity of applied 1D models. More sophisticated 1D radiative transfer models based, for example, on the two-stream approximation are able to describe the penetration of direct and diffuse solar radiation through the plant canopy more accurately. In this case, the difference between predicted absorbed and transmitted radiation fluxes by 3D and 1D models can be smaller.

Acknowledgement
This study was supported by a grant from the Russian Science Foundation (14-14-00956).

References

[1] Ross J 1981 *The radiation regime and the architecture of plant stands* (The Netherlands: Dr. W. Junk Publ.)
[2] Ross J and Myneni R B 1991 *Photon-Vegetation Interactions. Applications in Optical Remote Sensing and Plant Ecology* (Heidel: Springer-Verlag)
[3] Gastellu-Etchegorry J P Demarez V Pinel V Zagolski F 1996 *Remote Sens Environ.* **58** 131-156
[4] Knyazikhin Yu, Miessen G, Panferov O and Gravenhorst G 1997 *Agr. Forest Meteorol.* **88** 215–39
[5] Kuusk A and Nilson T 2000 *Remote Sens. Environ.* **72** 244–52
[6] Panferov O, Knyazikhin Y, Myneni RB, Szarzynski J, Engwald S, Schnitzler KG and Gravenhorst G 2001 *IEEE T. Geosci. Remote Sens.* **39** 241–53
[7] Olchev A, Radler K, Sogachev A, Panferov O and Gravenhorst 2009 *Ecol. Model.* **220**(21) 3046-56
[8] Widlowski J-L. *et al.* 2011 *J. Geophys. Res.* **116** G02019
[9] Widlowski J-L, Côté JF and Béland M 2014 *Remote Sens. Environ.* **142** 155-75
[10] Dickinson RE 1983 *Adv. Geophys.* **25** 305–53
[11] Olchev A, Ibrom A., Ross T, Falk U, Rakkibu G, Radler K, Grote S, Kreilein H and Gravenhorst G. 2008 *Ecol Model.* **212** 122–30
[12] Monsi M and Saeki T 1953 *Jpn. J. Bot.* **14** 22–55
[13] Myneni RB, Ross J and Asrar G 1989 *Agr. Forest Meteorol.* **45** 1–153
[14] Myneni RB, Asrar G and Gerstl SAW 1990 *Transport Theor. Stat.* **19** 205-50
[15] Knyazikhin Yu, Marshak A, and Myneni R B 2004 Three-Dimensional Radiative Transfer in Vegetation Canopies *Three-Dimensional Radiative Transfer in the Cloudy Atmosphere* ed A. Davis and A. Marshak (Berlin: Springer-Verlag) pp 617-51
[16] Wildowski J L Verstraete M Pinty B Gobron N 2003 *Allometric Relationships of Selected European Tree Species* (Italy: EC Joint Research Centre)
[17] Gates DM, Keegan HJ, Schleter JC and Weidner VR 1965 *Appl. Optics.* 4 11-20
[18] Yuan H, Dickinson RE, Dai Y, Shaikh MJ, Zhou, Shangguan W and Ji D 2012 *J. Climate* **27**(3) 1168-92
[19] Huang D, Knyazikhin Y, Wang W, Privette J, Deering DW, Stenberg P, Shabanov NV, Konstantinjva B, Myneni RB 2008 *Remote Sens. Environ.* **112**(1) 35-50

Description of bioremediation of soils using the model of a multistep system of microorganisms

A I Lubysheva[1, 2], K A Potashev[1] and O A Sofinskaya[1]

[1]Kazan Federal University, Kremlyovskaya str., 18, Kazan 420008 Russian Federation
[2]This work was funded by the subsidy allocated to Kazan Federal University for the state assignment in the sphere of scientific activities No. 9.9786.2017/8.9

E-mail: nastia.lubysheva@gmail.com

Abstract. The paper deals with the development of a mathematical model describing the interaction of a multi-step system of microorganisms in soil polluted with oil products. Each step in this system uses products of vital activity of the previous step to feed. Six different models of the multi-step system are considered. The equipping of the models with coefficients was carried out from the condition of minimizing the residual of the calculated and experimental data using an original algorithm based on the Levenberg-Marquardt method in combination with the Monte Carlo method for the initial approximation finding.

1. Introduction

Bioremediation is the most environmentally friendly method of soil purification from oil pollutants based on the ability of microorganisms (MO) to oxidize hydrocarbons. To optimize the scenario of recovery measures, a correct mathematical model of the process is necessary. In this case, the main difficulty is the closing of the model of transport of pollutants by kinetic relationships describing the vital activity of microorganisms. This article is devoted to the selection of the most suitable kinetics of vital activity of the MO based on comparison with experimental data.

Unlike most known models [1, 2], the article deals with the case of a multistep MO system. Each step in this system uses products of vital activity of the previous step to feed. The initial substrate was oil.

Several different models of the kinetics of a multistep system are formulated. The equipping of the models with coefficients is carried out from the conditions for minimizing the calculated and experimental data. For this purpose, an original optimization algorithm [3] was implemented, based on the use of the Levenberg – Marquardt method.

2. Basic equations of models

We consider N different biomasses, metabolism products each of which is a substrate for the biomass growth of next stage of consumption. The equations determining the key processes of vital activity of such a multistep MO system can be written in the form

$$\frac{\mathrm{d}M_i}{\mathrm{d}t} = v_i M_i - d_i M_i, \; i = 1..N; \tag{1}$$

where M_i is the concentration of biomass of step i MO in the soil $(\mathrm{mg/g})$; v and d are the growth rate and rate of death of biomass (s^{-1}); t is the process time (s);

$$\frac{\mathrm{d}C_i}{\mathrm{d}t}=\begin{cases}-\alpha_i v_i M_i, i=1,\\ -\alpha_i v_i M_i+\beta_{i-1} v_{i-1} M_{i-1}+\gamma_{i-1} d_{i-1} M_{i-1}, 1<i<N+1,\\ \beta_{i-1} v_{i-1} M_{i-1}+\gamma_{i-1} d_{i-1} M_{i-1}, i=N+1;\end{cases} \tag{2}$$

where C_i is the concentration of substrate $(\mathrm{mg/g})$; α is the economic coefficient; γ and β are the coefficients that determine the quantity of metabolite appearing in the decomposition of the death and the growth of new biomass unit.

Initial conditions

$$\begin{aligned}&M_i(t=0)=M_i^0, \quad i=1..N,\\ &C_i(t=0)=C_i^0, \quad i=1..N+1.\end{aligned} \tag{3}$$

To determine the functions v and d, three models were used. They reflect various options for accounting the effect of metabolites on them.

2.1. Model 1

The rate of production growth is determined by the Monod kinetics [4], and the death rate given by [12]:

$$v_i=\omega_i C_i/(K_i+C_i)\ ,i=1..N, \tag{4}$$

$$d_i=\delta_i(1+C_{i+1}/\varepsilon_i),i=1..N, \tag{5}$$

where ω_i is the maximum growth rate (s^{-1}); K_i is the half saturation constant $(\mathrm{mg/g})$, ε_i is the constant $(\mathrm{mg/g})$.

2.2. Model 2

The growth rate is defined by the formula Monod – Ierusalimskii [5]:

$$v_i=\omega_i \xi_i C_i/[(K_i+C_i)(\xi_i+C_{i+1})]\ ,i=1..N, \tag{6}$$

where ξ_i, K_i are the half saturation constants $(\mathrm{mg/g})$.

The rate of death d is considered as a constant:

$$d_i=\delta_i . i=1..N. \tag{7}$$

2.3. Model 3

In some cases [1] it is possible to improve the description of the production process by the introduction of the inhibition constant into the kinetic equation (4):

$$v_i=\omega_i\left[C_i/(K_i+C_i)\right]\left[K_{1i}/(K_{1i}+C_i)\right]\ , \tag{8}$$

where K_i, K_{1i} are the half saturation constants $(\mathrm{mg/g})$. For the rate of death, we use equation (4).

3. Modification of models

In equation (1), we introduce the control variable K_m $(\mathrm{mg/g})$ [2], which allows us to control the growth rate of the MO

$$\mathrm{d}M_i/\mathrm{d}t=v_i(1-M_i/K_{mi})M_i-d_iM_i . \tag{9}$$

Options for all three models, in which the equation (1) is replaced by equation (9), will be denoted by an additional symbol (a).

Thus, there are 6 models. The models 1 and 2 contain $L=7N$ unknown coefficients, the model 3 contains $8N$, the models 1(a), 2(a) contain $8N$, and the model 3(a) contains $9N$ coefficients.

After normalization using characteristic scales

$$t^* = 1 \text{ day},\ M^* = M_1^0,\ C^* = C_1^0,\ \alpha^*,\ \beta^*,\ \gamma^* = C_1^0/M_1^0$$

the form of the equations remains the same, and the initial conditions take on form

$$M_1(t=0)=1,\ M_i(t=0)=M_i^0,\ \ i=2..N,$$
$$C_1(t=0)=1,\ \ C_i(t=0)=C_i^0,\ \ i=2..N+1.$$

4. Results and discussion

Model coefficients were obtained from the conditions of minimizing the residual

$$R^2 = R_B^2 + R_{C_1}^2 + R_{C_{sum}}^2 \to \min,\quad R_p^2 = \sum_{k=0}^{N_p}\left(\Delta t_p^k/T_p\right)\left(p^k - p\left(t_p^k\right)\right)^2, \tag{10}$$

where N_p, T_p $(p = B, C_1, C_{sum})$ are the numbers of instants of time and the last time moment of the experimental measurements p^k of the indicator p; B is the value of the total respiration of all steps MO; $C_{sum} = \sum_{i=1}^{N} C_i$ is the total concentration of substrates; t_p^k is the time moment k of the indicator p measurement; $p\left(t_p^k\right)$ is the model value of the indicator p at the moment t_p^k; Δt_p^k is the local interval of "time of action" p^k:

$$\Delta t_p^k = 0.5\left(t_n - t_m\right);\ \ n = \min\left(N_p, k+1\right),\ \ m = \max\left(0, k-1\right).$$

The measurements of the total respiration were normalized by the initial value. To calculate the model values $B(t)$, we used equation

$$B(t) = \theta\sum_{i=1}^{N} M_i(t),\quad \theta = 1\Big/\sum_{i=1}^{N} M_i^0. \tag{11}$$

To solve problem (10), an original technique [3] was used. It is based on the use of the Levenberg-Marquardt method [6] with application of the Monte Carlo principle [7]. This technique operates according to the following scheme:

- the set of random points in which the residual (10) is calculated is "pounced" on the region of the sought coefficients;
- L points with the least value of a residual are selected of them;
- the problems of minimizing the functional (10) by the Levenberg-Marquardt method are solved using each selected point as the initial approximation; the model coefficients delivering the best approximation of the experimental data are chosen.

It is shown that the one-step system model is not capable to qualitatively describe the experimental values (figure 1), especially the non-monotonic behavior of the dynamics of the total concentration of substrates C_{sum}, since in the one-step system growth of it is impossible. At the same time its growth is possible in the two-step model, which leads to an improvement in the quality of the approximation of the experimental data (figure 2). Further quality improvement of the model can be achieved due to the control variable. Taking the inhibition constant into account improves the model 1 (a) (table 1). The model 1 (a) and 3 (a) were chosen as the best ones.

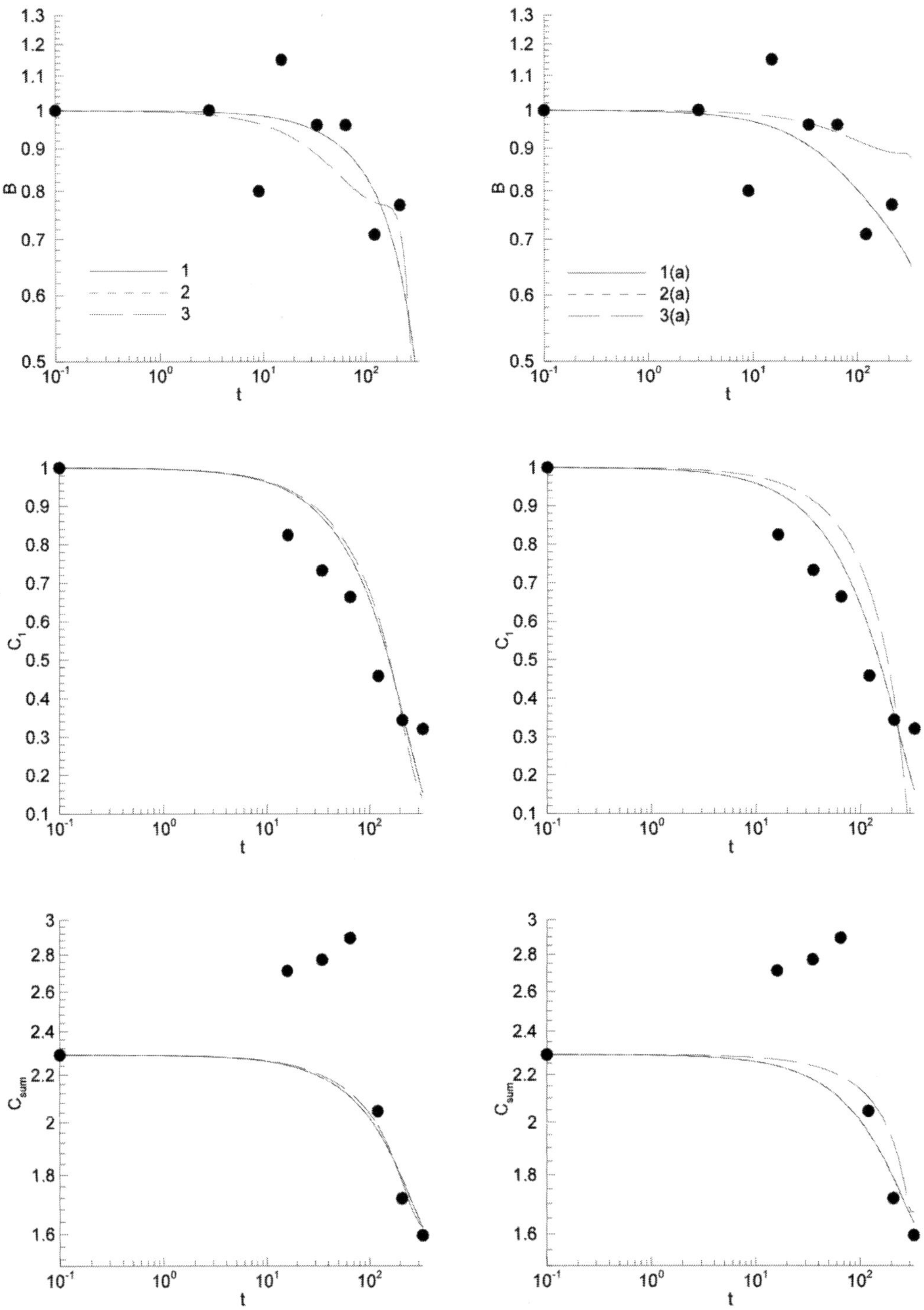

Figure 1. Results for a one-tier system (markers – experimental values, lines – model values).

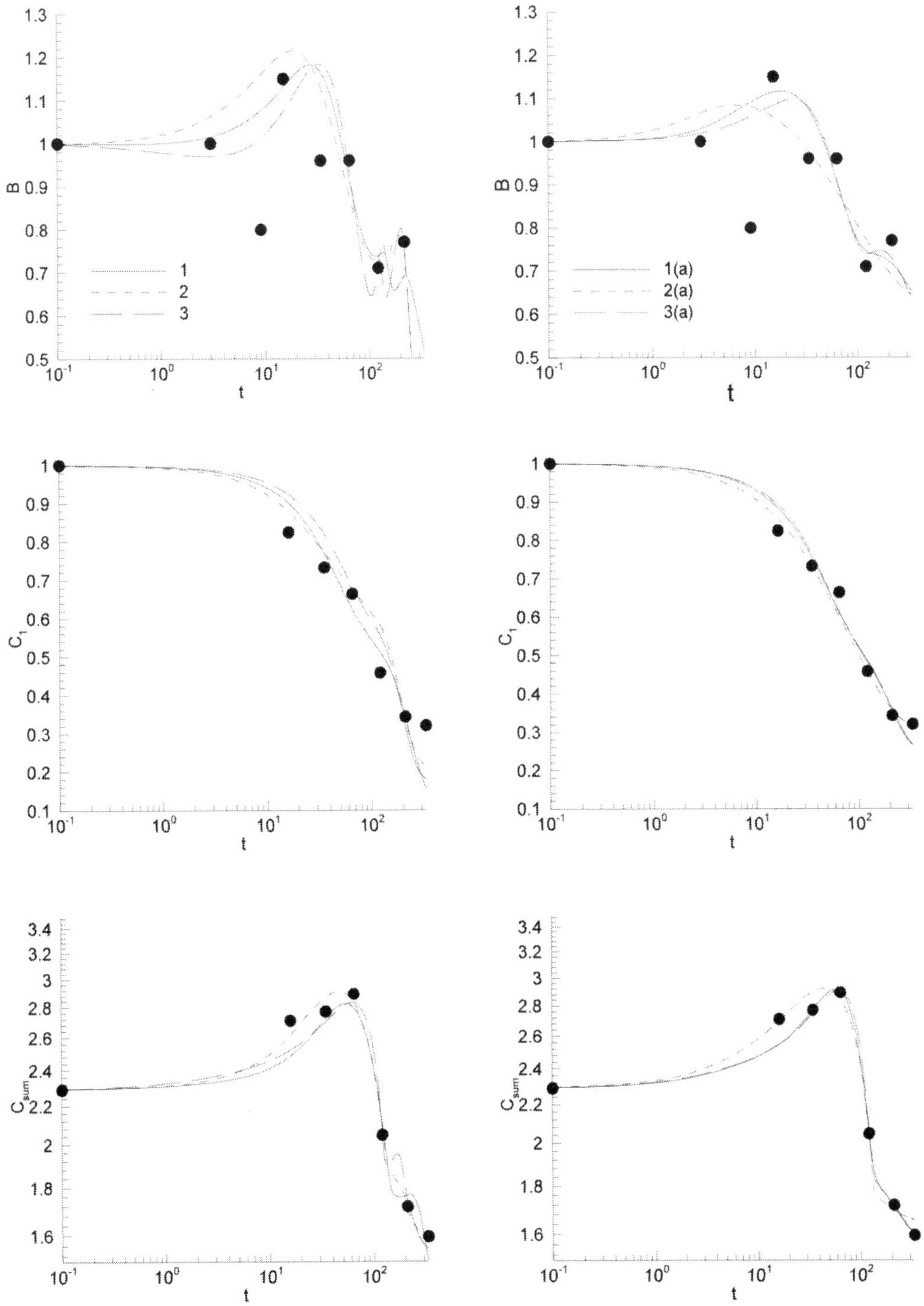

Figure 2. Results for a two-step system (markers – experimental values, lines – model values).

IOP Conf. Series: Earth and Environmental Science **107** (2017) 012102 doi:10.1088/1755-1315/107/1/012102

Table 1. The values of the residuals R.

System	Model					
	1	1(a)	2	2(a)	3	3(a)
One-step	0.13236	0.12906	0.13232	0.12904	0.12947	0.17311
Two-step	0.06466	0.03437	0.05083	0.04057	0.05173	0.03384

Table 2. Optimal values of model coefficients.

Model	Step	α	β	δ	ε	ξ	K	K_1	K_m	ω	γ
1(a)	1	0.1731	0.196	0.0279	1.303	-	0.4895	-	0.0999	0.0609	1.1351
	2	0.1134	0.0005	0.0001	1.7929	-	0.1374	-	0.845	0.3315	0.0001
3(a)	1	0.0587	0.189	0.1	3.2232	-	0.7785	0.7785	0.0001	0.4448	0.2263
	2	0.5	0.0039	0.0001	10	-	0.3802	0.3802	0.8065	0.2243	0.0001

References

[1] Mandli A R and Modak J M 2014 *Mathematical Biosciences* **258** 57–67
[2] Enszer JA, Măceş DA and Stadtherr MA 2015 *Mathematical Biosciences* **267** 97–108
[3] Lubysheva A I and Potashev K A *Proc. of the Lobachevsky math. center* **53** 88–90 (In Russian)
[4] Rubin A B 1977 *The Kinetics of biological processes* (Moscow: MSU) p 330 (In russian)
[5] Varfolomeev S D and Gurevich K G 1999 *Biokinetics: practical course* (Moscow: Fair-press) p 720 (In Russian)
[6] Gill P E, Murray W and Wright M H 1985 *Practical optimization* (Moscow: Mir) p 609 (In Russian)
[7] Ermakov S M 1971 *The Monte Carlo method and related questions* (Moscow: Nauka) p 328 (In Russian)

IOP Conf. Series: Earth and Environmental Science **107** (2017) 012103 doi:10.1088/1755-1315/107/1/012103

A two-dimensional hydrodynamic model of turbulent transfer of CO_2 and H_2O over a heterogeneous land surface

Yu.V. Mukhartova, A.S. Krupenko, P.A. Mangura, N.T. Levashova

Moscow State University, Faculty of Physics, Leninskie Gory, Moscow, 119991, Russia

E-mail: muhartova@yandex.ru

Abstract. A two-dimensional hydrodynamic model was developed and applied to describe turbulent fluxes of CO_2 and H_2O within the atmospheric surface layer over a heterogeneous land surface featuring mosaic vegetation and complex topography. Numerical experiments were carried out with a 4.5-km profile that crosses a hilly region in the central part of European Russia, with the diverse land-use patterns (bare soil, crop areas, grasslands, and forests). The results showed very strong variability of the vertical and horizontal turbulent CO_2 and H_2O fluxes. The standard deviations of the vertical fluxes were estimated for separate profile sections with uniform vegetation cover for daylight conditions in summer, and they were comparable with the mean vertical fluxes for corresponding sections. The highest horizontal turbulent fluxes occurred at the boundaries between different plant communities and at irregularities in surface profile. In some cases, these fluxes reached 10-20% of the absolute values of the mean vertical fluxes for corresponding profile sections. Significant errors in estimating the local and integrated fluxes e.g. when using the eddy covariance technique, can result from ignoring the surface topography, even in the case of relatively large plots with uniform vegetation cover.

1. Introduction

An accurate estimation of turbulent fluxes between heterogeneous land surfaces and the atmosphere is very important scientific task in numerous experimental and modeling studies [1-5]. Modern systems for monitoring greenhouse gas exchange involve a large number of field stations that provide continuous flux measurements in different types of ecosystems and geographical regions. The advantage of such monitoring systems is the use of a unified measuring technique based on the eddy covariance method, which includes a joint system for data analysis and post processing [6-7]. For flux measurements, the method ultimately requires a large and uniform land surface in the upwind direction. Surface heterogeneity may obviously lead to uncertainties in flux estimation, which can result in anomalously high or low values of H_2O and CO_2 fluxes in some cases.

The primary goal of this study is to describe the influence of vegetation heterogeneity and surface topography on CO_2 and H_2O fluxes within the atmospheric surface layer. One of the main objectives are to develop a two-dimensional (2D) model that can describe the turbulent exchange over a non-uniform land surface with complex topography and mosaic vegetation under different thermal stratification of the atmospheric boundary layer (neutral, unstable). Another objective is to quantify the contribution of horizontal advection to the ecosystem's total CO_2 and H_2O exchange as a function of topography and key vegetation properties. A modeling experiment was conducted for a 4.5-km-long profile that crosses a territory with hilly topography and mosaic vegetation in the central part of European Russia.

2. Model description

The developed model is based on solving a system of averaged Navier-Stokes and continuity equations for the wind-speed-averaged components ($\vec{V} = \{V_1, V_2\}$, where V_1 is the horizontal component, and V_2 is the vertical component of the average wind speed) using a 1.5-order closure scheme [8-9]:

$$\frac{\partial V_i}{\partial t} + (\vec{V}, \nabla)V_i = -\frac{1}{\rho_0}\frac{\partial}{\partial x_i}\delta P - \sum_{j=1}^{2}\frac{\partial}{\partial x_j}\left\{\frac{2}{3}\delta_{ij}\bar{e} - K\left(\frac{\partial V_i}{\partial x_j} + \frac{\partial V_j}{\partial x_i}\right)\right\} + g\frac{\delta T_v}{T_0}\delta_{i2} + F_i, \quad i = 1,2,$$

$$\operatorname{div}\vec{V} = 0, \quad \vec{F} = -c_d \cdot LAD \cdot |\vec{V}| \cdot \vec{V}.$$

x_1 and x_2 are the horizontal and vertical coordinates, respectively, ρ_0 is the density of dry air (which is assumed to be constant in this approximation), δP is the deviation of the mean air pressure from the hydrostatic distribution, and $\bar{e}$ is the turbulent kinetic energy (TKE):

$$\bar{e} = \frac{1}{2}\sum_{j=1}^{3}\overline{\left\langle (v'_j)^2 \right\rangle},$$

where the symbol $\overline{\langle\ \rangle}$ denotes the average over a certain time interval and space volume, v'_j is a fluctuating part of the wind speed components ($v_j = V_j + v'_j$), g is the free fall acceleration, T is the temperature of air, δT_v is the deviation of the virtual temperature $\Theta = T \cdot (1 + 0.61q)$ from the adiabatic temperature T_0 for dry air ($\delta T_v = T \cdot (1 + 0.61q) - T_0(x_2)$), q is the specific humidity, $T_0(x_2) = T_{ref} + (h_{ref} - x_2)\gamma_a$, T_{ref} is the given temperature at a certain height h_{ref}, $\gamma_a = g/c_p$, c_p is the specific heat of air at constant atmospheric pressure, δ_{ij} is the Kronecker symbol ($\delta_{ii} = 1$ and $\delta_{ij} = 0$, $i \neq j$), F_i is a component of the viscous drag force induced by the presence of vegetation, LAD is the leaf area density and c_d is the drag coefficient.

The 1.5-order closure scheme assumes that the turbulent fluxes $\overline{\langle v'_i v'_j \rangle}$ can be expressed using the TKE and space derivatives of the averaged wind speed components:

$$\overline{\langle v'_i v'_j \rangle} = \frac{2}{3}\delta_{ij}\bar{e} - K\left(\frac{\partial V_i}{\partial x_j} + \frac{\partial V_j}{\partial x_i}\right),$$

where $K = C_\mu \bar{e}^2 \varepsilon^{-1}$, which represents the turbulent exchange coefficient for momentum, where ε is the dissipation rate for TKE and C_μ is the dimensionless model constant (we used $C_\mu = 0.09$).

The turbulent kinetic energy $\bar{e}$ and the rate of its dissipation ε can be found by solving the following equations [5, 11-15]:

$$\frac{\partial \bar{e}}{\partial t} + (\vec{V}, \nabla)\bar{e} = \operatorname{div}(K_e \nabla \bar{e}) + P_{\bar{e}} - \varepsilon - g\frac{K_T}{T_0}\frac{\partial}{\partial x_2}\delta T_v, \qquad P_{\bar{e}} = -\sum_{i=1}^{2}\sum_{j=1}^{2}\overline{\langle v'_i v'_j \rangle}\frac{\partial V_i}{\partial x_j},$$

$$\frac{\partial \varphi}{\partial t} + (\vec{V}, \nabla)\varphi = \operatorname{div}(K_\varphi \nabla \varphi) + \frac{\varphi}{\bar{e}}\left(C_{\varphi 1}P_{\bar{e}} - C_{\varphi 2}\,\varepsilon - C_{\varphi 3}g\frac{K_T}{T_0}\frac{\partial}{\partial x_2}\delta T_v\right) + \Delta_\varphi, \qquad \varphi = \frac{\varepsilon}{\bar{e}},$$

$$\Delta_\varphi = 12\sqrt{C_\mu}(C_{\varphi 2} - C_{\varphi 1})c_d LAD|\vec{V}|\varphi,$$

where $K_e = K\sigma_e^{-1}$, $K_T = \alpha_a K$, and $K_\varphi = K\sigma_\varphi^{-1}$, which represent the turbulent exchange coefficients for TKE, temperature T, and function φ, respectively. σ_e and σ_φ are the Prandtl number and turbulence Schmidt number, and α_a is the inverse of the Prandtl number for the temperature [14-15]:

$$\alpha_a = \begin{cases} 1.35(1+1.35Ri)^{-1}, & Ri \geq 0, \\ 1.35(1-15Ri)^{1/4}, & Ri < 0, \end{cases} \qquad Ri = \frac{g}{T} \cdot \frac{\partial T/\partial x_2 + \gamma_a}{\sum_{i=1}^{2} (\partial V_i/\partial x_2)^2},$$

Ri is the Richardson number, while $C_{\varphi 1}$, $C_{\varphi 2}$, and $C_{\varphi 3}$ are model constants [10, 14-15].

The main system of equations also includes diffusion and advection equations for turbulent transfer of sensible heat, CO_2 (C_s) and specific humidity (q) at the interface between the soil, vegetation, and atmosphere [5]:

$$\frac{\partial T}{\partial t} + (\vec{V},\nabla)T + \gamma_a \cdot \frac{T_v}{T_0} \cdot V_2 = \mathrm{div}(K_T \cdot \nabla T) + \frac{\gamma_a}{T_0} \cdot K_T \cdot \frac{\partial T}{\partial x_2} + \frac{1}{\rho_0 c_p}(\vec{V},\nabla)\delta P - \frac{H}{\rho_0 c_p},$$

$$\frac{\partial C_s}{\partial t} + (\vec{V},\nabla)C_s = \mathrm{div}(K_C \cdot \nabla C_s) + F_C, \quad \frac{\partial q}{\partial t} + (\vec{V},\nabla)q = \mathrm{div}(K_v \cdot \nabla q) + \frac{E}{\rho}.$$

K_C and K_v are the turbulent exchange coefficients for CO_2 and H_2O respectively, F_C describes the sources/sinks of CO_2 in vegetation, H is the sensible heat flux and E is the evaporation rate. We assumed that K_C and K_v are proportional to the turbulent exchange coefficient for momentum: $K_C = K_v = Sc^{-1}K$, where Sc is the Schmidt number ($Sc = 0.75$).

The vegetation cover is considered as a two-phase medium comprising vegetation and the air. The sensible heat flux H within the vegetation is represented as $H = LAD \cdot \delta H$, where

$$\delta H = \rho_0 c_p \frac{T_l - T}{r_b/2},$$

T_l is the leaf temperature and r_b is the boundary layer resistance, which governs the heat flow rate between the leaf surface and surrounding air. The temperature exchange with the soil surface is described by the lower boundary condition:

$$\rho c_p K_T \sum_{i=1}^{2} n_i \cdot \left(\frac{\partial T}{\partial x_i} + \delta_{i2}\gamma_a\right)\Bigg|_{x_2=h(x_1)} = h_a (T_g - T)\Big|_{x_2=h(x_1)}.$$

The function $h(x_1)$ describes the topography of the study area, $\vec{n} = \{n_1, n_2\}$ is the outer normal vector to the boundary, T_g is the ground surface temperature, and h_a is the exchange coefficient ($h_a = 5.6 + 4|\vec{V}|$ [16]). In this modeling study, we assume that the temperatures T_l and T_g are known.

The evaporation rate E of vegetation is represented as $E = LAD \cdot \delta E$, where

$$\delta E = \frac{\rho_0 c_p}{\lambda \gamma_p} \frac{e_a^*(T_l) - e_a}{r_s + r_b},$$

$e_a^*(T_l)$ is the saturated vapor pressure at temperature T_l, e_a is the vapor pressure near the leaf surface, γ_p is the psychrometric constant, and r_s is the stomatal resistance for H_2O, λ is the specific heat of vaporization. As the lower boundary conditions for q we use the rate of evaporation from the ground surface, which is calculated as:

$$\lambda E = \frac{\Delta(T_g)\cdot R_{soil} + \rho c_p\left(e_a^*(T_g) - e_a\right)\cdot g_a}{\Delta(T_g) + \gamma_p}\cdot w,$$

where $\Delta(T_g)$ is the rate of change of saturation water pressure at a particular air temperature, R_{soil} is the soil radiation balance, g_a is the exchange coefficient for H_2O and w is the soil moisture.

We next describe the photosynthesis of the plant canopy and respiration rates. This was accomplished using an aggregated approach based on a model by Ball et al. [17] using Leuning's modification [18], the Beer-Lambert equation for solar radiation penetration within a plant canopy [19], and an algorithm describing the leaves' stomatal conductance response g_s to incoming photosynthetic active radiation (*PAR*) [20]:

$$F_C = -\frac{LAD}{a_1}(g_s - g_0)(C_s - \Gamma_*)\left(1 + \frac{D_s}{D_0}\right),\quad g_s = g_{max} f(PAR),\quad f(PAR) = 1 - \exp(-\beta_{sl}\cdot PAR),$$

where g_0 is the value of g_s at the light compensation point, Γ_* is the CO_2 compensation point, D_s is the dimensionless water vapor partial pressure deficit at the leaf surface, and a_1 and D_0 are empirical coefficients. The *PAR* profile within the plant canopy is described as:

$$PAR(x_2) = PAR_{hc}\cdot\exp\left(-k\int_{x_2}^{hc} LAD(z)dz\right),$$

where PAR_{hc} is the value of *PAR* at the canopy height hc, and *k* is the extinction coefficient.

The soil CO_2 emission is considered as a function of the soil temperature and moisture. It is also assumed that the soil CO_2 emission rate is dependent on the wind speed and turbulence intensity at the ground surface layer.

3. Modeling experiments

Figure 1 shows the 4.5-km profile used to quantify the influence of surface topography and vegetation heterogeneity on the turbulent CO_2 and H_2O fluxes. The vegetation along the profile is represented by a mosaic of bare soil, agricultural crops, and small woody areas. The vertical and horizontal turbulent fluxes of CO_2 and H_2O were calculated for three main model scenarios. The first one considers the actual surface topography and vegetation heterogeneity with the assumption of neutral thermal stratification of the atmospheric surface layer. The second one considers the vegetation heterogeneity but ignores the surface topography. Neutral atmospheric stratification is also assumed in this scenario. In the third scenario, the ecosystem fluxes are calculated for the actual vegetation heterogeneity and surface topography under unstable atmospheric stratification due to the ground surface and vegetation overheating. All numerical experiments were conducted for different wind directions.

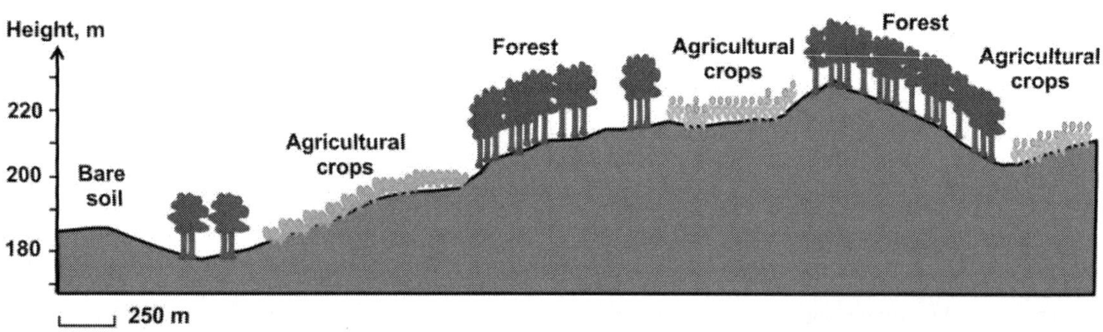

Figure 1. Vegetation and topography along the 4.5-km profile selected for numerical experiments.

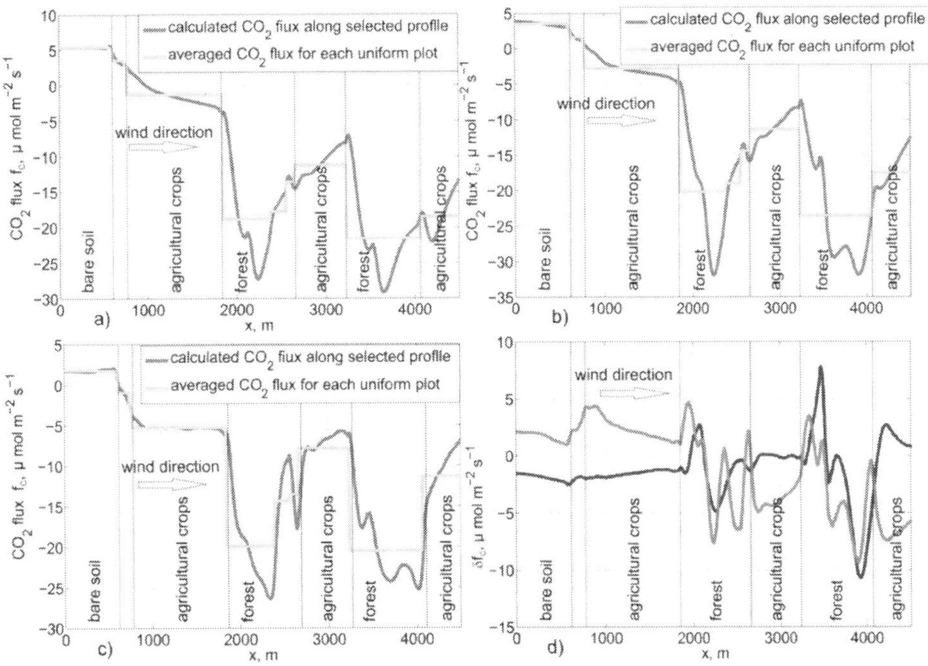

Figure 2. Modeled vertical turbulent CO_2 fluxes at 30 m above a ground surface calculated for different scenarios assuming: a) neutral atmospheric conditions, while ignoring the surface topography; b) neutral atmospheric conditions with the real surface topography; and c) unstable atmospheric conditions with the real surface topography. d) Comparison of fluxes calculated for different model scenarios. The blue line shows the difference between fluxes for the scenarios 1 (b) and 2 (a); the violet line shows the difference between fluxes for the scenarios 1 (b) and 3 (c).

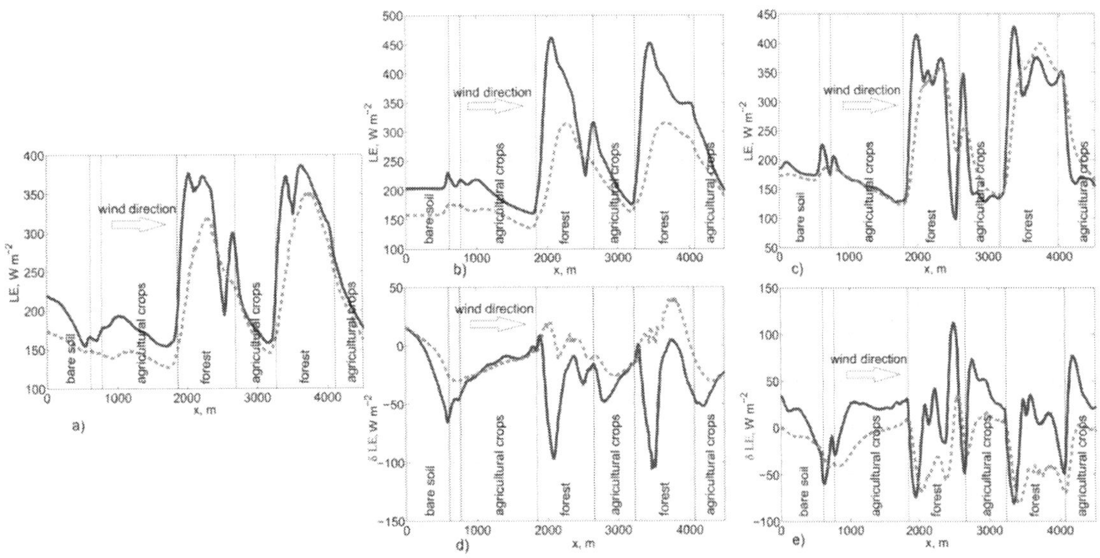

Figure 3. Modeled vertical LE fluxes along the selected profile at 30 m (blue line) and 50 m (dotted violet line) above a ground surface calculated for different scenarios assuming: a) neutral atmospheric conditions with the real surface topography; b) neutral conditions while ignoring the surface topography; and c) unstable conditions with the real surface topography. Comparison of fluxes between different model scenarios: d) difference between fluxes for the scenarios 1 (a) and 2 (b); e) difference between fluxes for the scenarios 1 (a) and 3 (c).

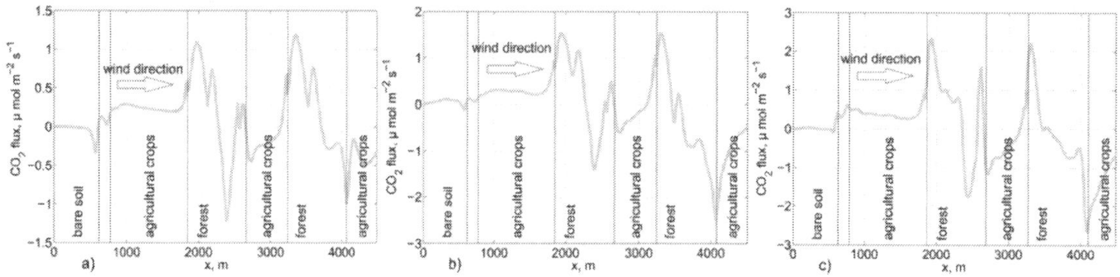

Figure 4. Modeled horizontal turbulent CO_2 fluxes at 30 m above a ground surface along selected profile: a) scenario ignoring the surface topography (neutral stratification); b) scenario considering the real surface topography (neutral stratification); c) scenario considering the real surface topography and unstable stratification of the atmospheric boundary layer.

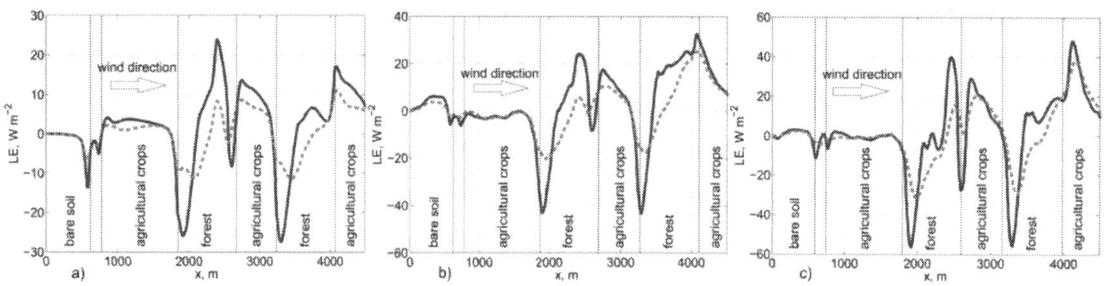

Figure 5. Modeled horizontal LE fluxes along the selected profile at 30 m (blue line) and 50 m (dotted violet line) above a ground surface for scenarios: a) horizontal surface (topography is ignored) and neutral atmospheric stratification; b) neutral atmospheric stratification and real surface topography; c) unstable atmospheric stratification and real surface topography.

4. Results and discussion

The results of the modeling experiments showed that the complex topography and mosaic vegetation cover have a significant influence on the atmospheric CO_2 and H_2O fluxes (figures 2-5). The greatest flux disturbances occurred at the boundaries between different plant communities, especially at windward forest edges and in the profile sections with uniform vegetation cover but with some irregularities in the surface topography (e.g., hilltops, hollows). In particular, uniform forest plots are characterized by significant horizontal heterogeneity of vertical fluxes (figures 2-3) and permanent non-zero horizontal fluxes (figures 4-5), despite the relatively large size of the plots (>500-600 m).

The variability of both the vertical and horizontal fluxes is minimal for scenarios in which the surface topography is ignored (horizontality of the ground surface is suggested). The highest variability occurs in scenarios that assume the real complex topography and unstable thermal stratification of the atmospheric boundary layer. The highest horizontal turbulent fluxes occur at the forest edges, in some cases reaching 10-20% of the absolute values of the mean vertical fluxes for the corresponding forest plot. The standard deviation of the vertical fluxes decreases as the height above the ground surface increases (figure 3).

The flux pattern over agricultural crops is mainly influenced by downwind surface properties. In the case of upwind from the bare soil area, the spatial pattern of vertical fluxes over agricultural crops is relatively uniform, and the horizontal fluxes for all model scenarios varies around zero. In the case of upwind from some forest patches, the air flow disturbances are manifested at a rather long distance from the leeward forest edge (up to 400-500 m), and the steady-state air flow is not achieved, even with model assumptions suggesting a flat topography (scenario 1). There is quite insignificant change in the vertical CO_2 and H_2O fluxes with the height over agricultural fields and bar soil, in contrast to forest sites.

There was very high heterogeneity of atmospheric fluxes in the numerical experiments. This illustrates a great difficulty in finding the optimal position for flux measuring stations for areas with mosaic vegetation structure, even in the case of flat topography. Modern post processing techniques for eddy covariance data use footprint analysis to describe the upwind area where the atmospheric flux measured by instruments is generated [1]. The flux footprint functions estimate the location and relative importance of passive scalar sources that influence flux measurements at a given height of a measuring device. The results are dependent on the device height, atmospheric stability, and surface roughness [21]. Several studies describe the possible effects of surface topography on atmospheric fluxes by estimating the vertical and horizontal advection terms [4]. Further investigation is required to estimate the effects on atmospheric fluxes of the air flow disturbances that arise at the boundaries between different types of vegetation and at topographic irregularities situated at some distance from the measuring systems. Such efforts will require accurate consideration to obtain representative fluxes.

Acknowledgement
This study was supported by grants of the by the Russian Science Foundation (Grant 14-14-00956).

References
[1] Avissar R and Pielke R A 1989 *Mon. Wea. Rev.* **117** 2113–24
[2] Sogachev A, Leclerc M Y, Karipot A, Zhang G and Vesala T 2005 *Agric. For. Meteorol.* **133** 182–15
[3] Frank C and Ruck B 2008 *Forestry* **81**(3) 259–19
[4] Feigenwinter Ch et al. 2008. *Agric. For. Meteorol.* **148**(1) 12–13
[5] Mamkin V *et al* 2016 *Environ. Res. Lett.* **11**(12) 125012
[6] Aubinet M, Vesala T and Papale D (Eds.) 2012 *Eddy Covariance: A Practical Guide to Measurement and Data Analysis*. (Dordrecht, The Netherlands: Springer Atmospheric Sciences, Springer Verlag)
[7] Burba G 2013 *Eddy Covariance Method for Scientific, Industrial, Agricultural and Regulatory Applications: a Field Book on Measuring Ecosystem Gas Exchange and Areal Emission Rates*. (Lincoln, USA: LI-COR Biosciences)
[8] Wyngaard J C 2010 *Turbulence in the Atmosphere* (Cambridge: Cambridge University press)
[9] Garratt J R 1992 *The atmospheric boundary layer* (Cambridge: Cambridge University press)
[10] Wilcox D C 1998 *Turbulence modeling for CFD* (La Cañada: DCW Industries, Inc)
[11] Sogachev A and Panferov O 2006 *Bound. Lay. Meteorol.* **121**(2) 229–38
[12] Mukhartova Yu V, Levashova N T, Olchev A V and Shapkina N E 2015 *Mosc. Univ. Phys. Bull.* **70** (1) 14–8
[13] Ryzhova M S, Mukhartova Yu V, Shapkina N E and Olchev A V 2015 *Mosc. Univ. Phys. Bull.* **70** (6) 490–7
[14] Sogachev A 2009 *Bound. Lay. Meteorol.* **130**(3) 423–35
[15] Sogachev A, Kelly M, Leclerc MY 2012 *Bound. Lay. Meteorol.* **117**(2) 307–21
[16] Kuchling H 2014 Tashenbuch der Physik (München: Carl Hanser Verlag) p 711
[17] Ball J T, Woodrow I E and Berry J A 1987 A model predicting stomatal conductance and its contribution to the control of photosynthesis under different environmental conditions *Progress in Photosynthesis Research* ed I Biggins (Netherlands: Martinus Nijhoff) pp 221–4
[18] Leuning R 1995 *Plant Cell Environ.* **18** 339–17
[19] Monsi M and Saeki T 1953 *Jpn J. Bot.* **14** 22–31
[20] Oltchev A, Ibrom A, Constantin J, Falk M, Richter I, Morgenstern K, Joo Y, Kreilein H and Gravenhorst G 1998 *J. Phys. Chem. Earth* **23**(4) 453–6
[21] Kljun N, Calanca P, Rotach M W and Schmid H P 2004 *Bound. Lay. Meteorol.* **112** 503–21.

Palaeoecological data as a tool to predict possible future vegetation changes in the boreal forest zone of European Russia: a case study from the Central Forest Biosphere Reserve

E Yu Novenko[1, 2], A N Tsyganov[3], A V Olchev[1, 4]

[1] Faculty of Geography, Lomonosov Moscow State University, GSP-1, Leninskie Gory, Moscow, 119991, Russia
[2] Institute of Geography Russian Academy of Science, Staromonetny lane, 29, 119017, Moscow, Russia
[3] Department of Zoology and Ecology, Penza State University, Krasnaya str. 40, 440026, Penza, Russia
[4] A.N. Severtsov Institute of Ecology and Evolution, Russian Academy of Sciences, Moscow, Leninskiy Pr., 33, 119071, Russia

E-mail: aoltche@gmail.com

Abstract. New multi-proxy records (pollen, testate amoebae, and charcoal) were applied to reconstruct the vegetation dynamics in the boreal forest area of the southern part of Valdai Hills (the Central Forest Biosphere Reserve) during the Holocene. The reconstructions of the mean annual temperature and precipitation, the climate moisture index (CMI), peatland surface moisture, and fire activity have shown that climate change has a significant impact on the boreal forests of European Russia. Temperature growth and decreased moistening during the warmest phases of the Holocene Thermal Maximum in 7.0-6.2 ka BP and 6.0-5.5 ka BP and in the relatively warm phase in 3.4-2.5 ka BP led to structural changes in plant communities, specifically an increase in the abundance of broadleaf tree species in forest stands and the suppression of *Picea*. The frequency of forest fires was higher in that period, and it resulted in the replacement of spruce forests by secondary stands with *Betula* and *Pinus*. Despite significant changes in the climatic parameters projected for the 21st century using even the optimistic RCP2.6 scenario, the time lag between climate changes and vegetation responses makes any catastrophic vegetation disturbances (due to natural reasons) in the area in the 21st century unlikely.

1. Introduction

Modern climate changes, which are manifested as an increase in air temperature, changes in precipitation patterns and more prevalent anomalous weather evens such as heat waves and windstorms, have an obviously significant impact on the growth and functioning of boreal forests in Northern Eurasia [1]. To describe future climate variability, it is very important to know how different forest communities will respond to changes in climate conditions in the future [2, 3, 4, 5]. The method of palaeoenvironmental reconstructions is one of the most effective approaches to describe the possible scenarios of the landscape state and vegetation changes in the 21st century [6, 7]. Within the last decades, climate and landscape reconstructions for the Holocene Thermal Maximum (8.0–5.7 ka

BP) and the optimum of the last Interglacial (Eemian, MIS 5e, about 125 ka BP) were applied as palaeo-analogues of the vegetation response considering increases in global temperature of 0.7–1.0°C and 1.7–2.0°C, respectively [7]. Loutre and Berger [8] have suggested that the conditions of the orbital climate forcing during the Holstenian Interglacial (MIS 11, 420–370 ka BP) are the most similar to those of the current interglacial period. Therefore, vegetation dynamics during the Holstenian Interglacial can be used as an analogue to predict the vegetation changes at the end of the 21st century. Multiple Holocene climate reconstruction models for the Eastern and Northern Europe exhibit rapid temperature increases in the Late Glacial and Early Holocene (an analogue of present-day global warming), during the Holocene Thermal Maximum when the mean annual temperatures were 2°C higher than those of the present day [6,9,10,11]. Roughly 5.7–5.5 ka BP, the Holocene Thermal Maximum was followed by gradual climatic cooling that included several warming and cooling phases with temperature fluctuations ranging between 2 and 3°C. The key mechanisms of forest-climate interaction in previous epochs remain unclear, and additional research using aggregated multy-proxy data analysis is necessary.

The present study is focused on reconstructing vegetation dynamics in the southern part of the Valdai Hills at various time intervals during the Holocene. We make use of new multi-proxy palaecological data (pollen, testate amoebae, charcoal, etc.) to interpret past vegetation dynamics. The temporal moistening conditions in the area were derived using the climate moisture index (CMI, [12]). To predict possible regional vegetation changes in the boreal forest zone of central European Russia under climate changes in the 21st century, several scenarios of future vegetation changes are suggested. Available projections of climate change for the 21st century are provided by an ensemble of global climate models such as CMIP5 (IPCC 2013, Coupled Model Intercomparison Project, Phase 5 [1]) for scenarios RCP2.6 and RCP8.5 (Representative Concentration Pathways for possible range of radiative forcing +2.6 W/m^2 and +8.5 W/m^2). These projections show that the mean air temperature for the central part of European Russia may increase at the end of the current century by 2.0–2.5 and 6.0–7.0°C, respectively. This temperature increase will be accompanied by an increase in annual precipitation that may range from 7% (RCP2.6) to 15% (RCP8.5). All of these changes may result in significant alterations in forest cover.

A key region studied in this investigation was the Central Forest State Natural Biosphere Reserve (CFSNBR) situated in the southern part of the Valdai Hills. We opted to focus on the southern boundary of the boreal forest zone. Taking into account the modern trend of increases in global air temperature, forest communities can be very sensitive to environmental changes. A number of very detailed palaeoecological reconstructions of the Holocene climate and vegetation history for the area have been conducted and found to be very useful for this study [6, 13, 14, 15].

For projecting possible vegetation changes in the 21st century, we used the optimistic RCP2.6 climate scenario. The RCP8.5 climatic scenario does not have any analogues in the Holocene, and consequently, it was not used in the present study.

2. Materials and Methods

The CFSNBR is situated roughly 360 km northwest of Moscow (the Tver region, 56°35' N, 32°55' E) in an ecological zone transitioning from taiga to broadleaf forests. The vegetation of the CFSNBR is primary southern taiga forests, and it has been undisturbed by any human activities for at least 86 years. The climate of the study area is temperate and moderately continental with a mean annual temperature is 4.1°C and annual precipitation of roughly 700 mm [4]. The plant cover includes mixed uneven-age spruce (*Picea abies*), birch (*Betula pendula*) and aspen (*Populus tremula*) trees with a small admixture of broadleaf trees (*Tilia cordata, Ulmus laevis, Fraxinus excelsior, Acer platanoides*). Alder (*Alnus glutinosa*) is abundant in the river valleys.

The Holocene vegetation and climate reconstructions are based on palaeoecological data from the large peat bog Staroselsky moch (617 ha), which is located in the southeast part of the CFSNBR. The results of pollen and testate amoebae analysis were published by Novenko et al [14] and Payne et al [15]. The estimation of micro-charcoal concentration in peat core from the small forest peatland (<0.1

ha) located 4 km west of the Staroselsky moch peat bog was used for reconstructing fire frequencies. Available palaeoecological data allowed us to reconstruct environmental changes during the 9000 years, with the exception of the micro-charcoal data, which cover the last 7000 years. The last millennium was excluded from analyses of vegetation-climate interactions due to increasing human impacts and vegetation disturbance.

The mean annual temperature and precipitation during the Holocene (figure 1) were reconstructed with pollen data using the Modern Analogue Technique (MAT). Details of the MAT have been presented in previous publication [13]. Using the MAT, we found a statistically significant analogue in the modern pollen datasets (720 sites in Europe and Siberia) for each fossil pollen assemblage. The climatic characteristics (temperature, precipitation) of the area, where the spectrum analogue was obtained, are accepted as reconstructions of climate and woodland coverage in the past. Climatic information was derived from BRIDGE Earth System Modelling results [16].

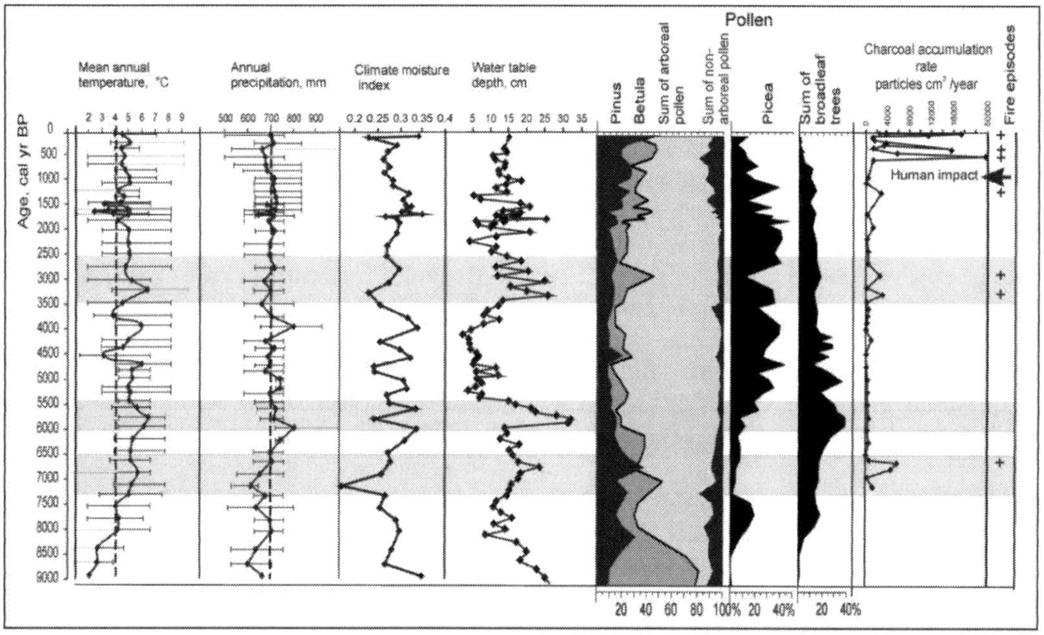

Figure 1. Summary of the results of the multi-proxy palaecological reconstructions for the area of the CFSNBR, including the reconstructions of the mean annual temperature and precipitation (modified after Novenko et al. [14]), climatic moisture index and water table depth inferred by pollen and testate amoebae data from Staroselsky moch peat bog; characteristic of pollen taxa from the pollen records from the Staroselsky moch peat bog; micro-charcoal accumulation rate and fire episodes inferred from charcoal data from the small forest peatland in the area of the CFSNBR. The Sum of broadleaf trees includes pollen of *Quercus, Ulmus, Tilia, Fraxinus, Acer* and *Carpinus*.

Modern analogue calculations were produced using Polygon 2.2.4. A test of the accuracy of the applied method using a database of surface pollen assemblages reveals that the MAT can reproduce the mean annual temperature rather correctly (R^2=0.81; RMSEP=1.5°C). The accuracy of reconstruction of annual precipitation is lower (R^2=0.41, RMSEP=100 mm), and we can determine changes in humidity only as general trends.

To describe the temporal patterns of moistening conditions within the study area during the Holocene, we computed the CMI [12] using the ratio of annual precipitation to annual potential evapotranspiration. The annual potential evaporation was calculated using the Priestley-Taylor equation [17]. This numerical algorithm uses the reconstructed annual air temperature, the forest cover and pollen proportion of coniferous and deciduous tree species as input parameters.

Peatland surface moisture, a robust proxy for climate humidity, was reconstructed as depth to the water table (WTD, cm) using a testate amoeba-based transfer function (reversed weighted averaging regression model) that was specifically developed for the forest zone of European Russia [18]. The transfer function included 80 samples from 18 peatlands located in the taiga, mixed and broadleaf forests and the forest-steppe zones of European Russia. The leave-one-out cross-validation of models constructed for the training dataset revealed a rather high accuracy of the model (R^2=0.74; RMSEP=5.5 cm). Water-table depths inferred by testate amoeba generally reflected the length and severity of the summer moisture deficit, which in raised bogs is primarily controlled by summer precipitation [19].

Micro-charcoal concentration data were transformed into charcoal accumulation rates (CHAR, particles cm^{-2} yr^{-1}) using CharAnalysis software [20], which separates the long-term trends (i.e., background CHAR) from positive deviations (i.e., charcoal peaks) in the CHAR time series. Charcoal peaks represent fire episodes (i.e., one or more fires that occur during the time span of the peak).

3. Results and discussions

The climate conditions of the study area between 9.0 and 8.0 ka BP were relatively cold (figure 1). The mean annual temperature was 2°C lower then at the present time at the beginning of this period and increased to modern values roughly 8.0 ka BP. The mean annual precipitation increased from 600 to 700 mm and a consistent decrease in peatland surface wetness indicated an increase in climate humidity. The CMI decreased from 0.36 (more humid) to 0.26, which corresponds to less humid, drier conditions. During this period, the study area was occupied mainly by birch forest, which has persisted in the region since the early Holocene [13, 21]. The plant macrofossils revealed from the lower part of the peat core suggested an abundance of *Betula pubescens* [14].

Climate warming after 8.0 ka encouraged the expansion of *Picea*, broadleaf trees (*Quercus, Tilia, Ulmus*) and *Corylus* over the study area. The moistening conditions were characterized by high variability in the CMI between 0.17 (low humidity) and 0.30 (moderate humidity). A high WTD in the peatland inferred from testate amoebae data indicated rather humid conditions during the summer periods between 8.0 and 7.0 ka BP (figure 1).

From 7.0 to 5.5 ka BP, the mean annual temperature was roughly 5–7°C, which exceeds modern-day values by 1–3°C (figure 1). These reconstructions agree well with the climatic conditions determined from pollen data for the Holocene Thermal Maximum in the Baltic region and Fennoscandia [10, 11], where the mean annual temperatures were 2°C higher than those at the present time. Pollen-inferred temperature reconstructions from Estonia have shown that the mean annual temperatures reached 8–9°C, which exceeded modern-day ones by 3.0–3.5°C [22].

The annual precipitation in the CFSNBR was roughly 600 mm during the period 7.0–6.5 ka BP. Between 6.5 and 5.5 ka the annual precipitation approached or exceeded modern-day values. The reconstruction of the peatland surface moisture reflected two distinct phases with drier climate conditions (7.0–6.2 ka BP and 6.0–5.5 ka BP). The CMI in the periods decreased to lowest values 0.24 (for a ratio of annual potential evaporation and precipitation of roughly 0.76), and it was lower as the mean values estimate for entire Holocene for the area (CMI=0.285). The WTD during the second phase in the peat bog was extremely low (32 cm), likely due to the reduced summer precipitation and higher summer temperatures.

The proportion of *Picea* pollen reduced significantly between 7.0 and 5.5 ka BP, and it almost disappeared from assemblages during the second warm phase 7.0–6.2 ka BP. At the same time, *Pinus*, *Betula* and broadleaf tree species increased their abundance. The pollen productivity of wind-pollinated plants, such as *Betula, Alnus* and *Pinus*, is much higher than the pollen productivity of *Acer, Ulmus* and *Tilia*. Therefore, the proportions of broadleaf trees in pollen assemblages are often underestimated [23]. In the period 6.0–5.7 ka BP, the fraction of the sum of broadleaf tree pollen reached 40% of the total pollen sum, indicating a broad expansion of temperate deciduous forests.

The warm and relatively dry climatic conditions of the Holocene Thermal Maximum were unfavorable for *Picea* growth. According to data of modern ecological studies of spruce, *Picea abies* prefers moist soils with high seasonal water storage [24] and is highly susceptible to drought [25]. The availability of water is therefore an important factor for the growth of this tree species nowadays.

Climate reconstructions of the study area based on the pollen data from Staroselsky moch peatland revealed climate cooling and an increase in humidity after 5.3 ka BP. These reconstructions coincided with a number of palaeoclimatic reconstructions based on various natural archives in Europe that demonstrated gradual climatic cooling since 5.7 ka BP [6,9,10,11,26], possibly caused by a decrease in summer insolation. Roughly 4.5 ka BP, the mean annual temperature in the study area declined to 3°C and the annual precipitation was close to the modern values. At 4.0 ka BP the annual precipitation increased to 800 mm year. The CMI in the period varied significantly and in wetter years reached 0.35. Higher precipitation levels and lower temperatures promote sufficient soil moisture conditions that resulted in an increasing fraction of *Picea* in forest stands in the study area. The share of *Picea* pollen in assemblages from peat core at Staroselsky moch increased to 30–40% (figure 1). Broadleaf trees remained relatively abundant in the plant cover until 4.0 ka BP and then their fraction gradually decreased. The rise of *Picea* pollen values were traced in pollen diagrams from European Russia and Finland [6, 10], which suggested the expansion of boreal coniferous and mixed coniferous-broadleaf forests.

The next warm and drier phase was detected in climate reconstruction by pollen and testate amoebae data from the Staroselsky moch peat bog between 3.4 and 2.5 ka BP. The mean annual temperature exceeded modern-day values by 1–2°C, the WTD in the peatland dropped to 20–25 cm, and the CMI decreased to 0.23. The proportion of *Picea* fell, and the fraction of *Betula* and broadleaf trees increased. One can expect that *Picea* forests were significantly damaged during this phase by fires.

Charcoal data from the area of the CFSNBR have shown that the Holocene fire activity was low until the last millennium. The conditions of the study area— a flat topography with impeded drainage of soils and an abundance of broadleaf tree species—were replaced by *Picea* in a relatively wet climate. This situation obviously led to hampered burning and low fire activity [27]. Only two distinct fire episodes were detected during the warm and dry phase 3.4–2.5 ka BP, and one fire episode occurred roughly 6.6 ka BP, which corresponded to the oldest dry phase of the Holocene Thermal Maximum. Archeological findings and palaeobotanical indicators are inconclusive in terms of human occupation of the study area during these periods. Consequently, one can assume that the cause of the forest fires was likely summer drought.

Climate reconstructions for the study area have demonstrated relatively humid climate conditions after 2.5 ka BP. The CMI in some years reached 0.33. Our data agree well with European palaeoecological records that highlight climate cooling and an increase in humidity roughly 2.5 ka BP [11, 26]. After 2.5 ka BP, spruce forests recovered in the study area. The *Picea* pollen curve formed a conspicuous maximum between 2.5 and 1.6 ka BP (figure 1). The increase in *Picea* pollen values after 2.5 ka BP was also recorded in a number of sites in Finland, European Russia, Belarus and Estonia [10, 21, 22, 28], findings that suggest a regional expansion of spruce forests.

The last 1000 years have been marked by dramatic changes in vegetation. The proportion of spruce and broadleaf trees in the forest stands decreased abruptly, which coincided with the expansion of birch forests. The micro-charcoal concentration increased by an order of magnitude, and the frequency of fire increased. Pollen records from the area of the CFSNBR have revealed a significant reduction in total forest coverage [13], the presence of cultivated cereals, weeds and ruderal plants [14]. These findings apparently indicate human impacts on plant cover.

4. Projection of future vegetation changes and conclusions

Holocene climatic reconstructions have shown a very high sensitivity of vegetation to climate changes. To predict possible forest vegetation changes during the 21st century, the RCP2.6 scenario was used. For vegetation projections, we can use the warm and dry phases observed during the Holocene Thermal Maximum 7.0–6.2 ka BP and 6.0–5.5 ka BP, as well as the climate warming between 3.4 and 2.5 ka BP. The palaeoecological data that we obtained for these periods have shown that the climate warming and decrease in CMI led to structural changes in plant communities: an increase in the abundance of broadleaf tree species (*Quercus*, *Tilia*, *Ulmus*, *Fraxinus*, *Acer*) in forest stands and the suppression of *Picea* trees. The frequency of forest fires during the aforementioned periods was higher, and it resulted in the replacement of spruce forests by secondary stands containing *Betula* and *Pinus*.

According to our palaeoenvironmental reconstructions, vegetation changes caused by climate warming lasted from 500–1000 years. On the other hand, the observed temperature growth spanned only a few decades. One can expect a lag between climate changes and vegetation response; catastrophic vegetation disturbances are unlikely to occur in the 21st century. Although the palaeoecological data do not fully correspond to the short-term changes expected in the current century, they can be used to assess general trends in vegetation dynamics.

Acknowledgement
This study was supported by grant of the Russian Science Foundation (Grant 16-17-10045).

References

[1] IPCC 2013 *IPCC Climate Change 2013: The Physical Science Basis. Contribution of Working Group I to the Fifth Assessment Report of the IPCC* (Cambridge: Cambridge University Press) p 1535
[2] Bonan GB, Pollard D and Thompson SL 1992 *Nature* **359** 716-8
[3] Brovkin V, Raddatz T, Reick CH, Claussen M and Gayler V 2009 *Geophys. Res. Let.* **36** L07405
[4] Oltchev A *et al* 2002 *Phys. Chem. Earth* **27**(9-10) 675-90
[5] Soja AJ *et al* 2007 *Global Planet Change* **56**(3-4) 274-96
[6] Novenko E, Olchev A, Desherevskaya O and Zuganova I 2009 *Environ Res Let.* **4** 045016
[7] Velichko AA and Borisova OK 2011 *Dokl. Earth Sci.* **438** 681-5
[8] Loutre MF and Berger AL 2003 *Global Planet. Change* **36** 209–17
[9] Davis BAS et al 2003 *Quat. Sci. Rev.* **22** 1701–1716
[10] Heikkilä M and Seppä H 2010 *Boreas* **39** 705–19
[11] Mauri A, Davis BAS, Collins PM and Kaplan JO 2015 *Quat. Sci. Rev.* **112** 109–27
[12] Willmott CJ and Feddema JJ 1992 *Prof. Geogr.* **44**(1) 84–8
[13] Novenko EY, Eremeeva AP and Chepurnaya AA 2014 *Veget Hist Archaeobot.* **23** 109–19
[14] Novenko EYu, Volkova EM, Nosova NB and Zuganova IS 2009 *Quat. Int.* **207** 93–103
[15] Payne RJ, Malysheva E, Tsyganov A, Pampura T, Novenko E, Volkova E, Babeshko K and Mazei Y 2016 *Holocene* **26** 314-26
[16] Gregoire LJ, Valdes PJ, Payne AJ and Kahana R 2011 *Clim. Dyn.* **37** 705–19
[17] Priestley CHB and Taylor RJ 1972 *Monthly Weather Rev.* **100**(2) 81-92
[18] Tsyganov AN, Babeshko KV, Novenko EY, Malysheva EA, Payne RJ and Mazei YA 2017 *Rus J Ecol.* **48**(2) 191–8
[19] Charman DJ 2007 Holocene 17 217–27
[20] Higuera PE, Brubaker LB, Anderson PM, Hu FS and T.A. Brown TA 2009 *Ecol. Monogr.* **79** 201–19
[21] Krementski KV, Borisova OK and Zelikson EM 2000. Paleontolog. J. 34 (1), 67–74
[22] Seppä H and Poska A 2004 *Quat. Res.* **61** 22– 31
[23] Lisitsyna OV, Giesecke T and Hicks S 2011 *Rev. Palaeobot. Palynol.* **166** 311–24
[24] Schmidt-Vogt H 1977 *Die Fichte – Ein Handbuch in zwei Bänden, Band I Taxonomie, Verbreitung, Morphologie, Ökologie, Waldgesellschaften* (Hamburg: Paul Parey).
[25] Cienciala E, Lindroth A, Cermak J, Hallgren JE and Kucera J 1994 *J. Hydrol.* **155** 57–71
[26] Wanner H *et al* 2008 *Quat. Sci. Rev.* **27** 1791–828
[27] Whitlock C, Higuera PE, McWethy DB and Briles CE 2010 *Open Ecol. J.* **3** 6-23
[28] Zernitskaya V, Stančikaitė M, Vlasov B, Šeirienė V, Kisielienė D, Gryguc G and Skipitytė R 2015 *Quat. Int.* **386** 70–82

IOP Conf. Series: Earth and Environmental Science **107** (2017) 012105 doi:10.1088/1755-1315/107/1/012105

Influence of forest cover changes on regional weather conditions: estimations using the mesoscale model COSMO

A V Olchev[1, 2]**, I A Rozinkina**[3]**, E V Kuzmina**[3]**, M A Nikitin**[3]**, G S Rivin**[3, 1]

[1] Faculty of Geography, Lomonosov Moscow State University, GSP-1, Leninskie Gory, Moscow, 119991, Russia
[2] A.N. Severtsov Institute of Ecology and Evolution, Russian Academy of Sciences, Moscow, Leninskiy Pr., 33, 119071, Russia
[3] Hydrometeorological Centre of Russia, B. Predtechensky per., 11-13, 123242, Moscow, Russia

E-mail: aoltche@gmail.com

Abstract. This modeling study intends to estimate the possible influence of forest cover change on regional weather conditions using the non-hydrostatic model COSMO. The central part of the East European Plain was selected as the 'model region' for the study. The results of numerical experiments conducted for the warm period of 2010 for the modeling domain covering almost the whole East European Plain showed that deforestation and afforestation processes within the selected model region of the area about 105 km2 can lead to significant changes in regional weather conditions. The deforestation processes have resulted in an increase of the air temperature and a reduction in the amount of precipitation. The afforestation processes can produce the opposite effects, as manifested in decreased air temperature and increased precipitation. Whereas a change of the air temperature is observed mainly inside of the model region, the changes of the precipitation are evident within the entire East European Plain, even in regions situated far away from the external boundaries of the model region.

1. Introduction

Comprehensive studies of forest – atmosphere interaction at different spatial and temporary scales (from ecosystem to regional and global scales) are important challenges in modern meteorology and ecology nowadays. To examine these scientific problems a wide range of experimental and modeling studies was being provided during the past several decades [1-8]. The main attention in those studies was mainly focused on assessing the response of forest ecosystems to the change of environmental conditions and on determining the influence of deforestation and land-use changes on regional and global weather and climatic conditions in past epochs, under present climate and under different scenarios of projected future climate changes [5, 6, 9-12]. The numerous results obtained, however, do not yet allow a complete assessment of all available relationships and effects arising between atmospheric processes and the land surface. A high degree of uncertainty still remains in the description of vegetation and atmosphere interaction at the regional scale. Reliability of the regional and global models that are usually used in such studies is limited by multiple model simplifications and deficiency of necessary input parameters describing the variety of the processes influencing the vegetation - atmosphere interaction. Furthermore, the accuracy of these simulations is also limited by significant heterogeneity of vegetation cover and surface topography. In this regard, it will be useful to develop and apply more physically reasonable and sophisticated modeling approaches describing the land surface and vegetation properties, vegetation response to the change of ambient conditions and its interaction with the atmosphere. Within the frameworks of the study, numerical experiments using the mesoscale model COSMO to assess the possible influence of forest cover change in the central part of the East European Plain on regional weather conditions have been conducted. The main objective of the study is to quantify the possible effects of forest cover changes on spatial temperature and precipitation patterns within the East European Plain.

2. Material and methods

Within the frameworks of the modeling study to assess the influence of deforestation and afforestation processes on regional weather conditions, a series of numerical experiments using the non-hydrostatic limited-area atmospheric projection model COSMO were instigated [13, 14]. The TERRA model is used within COSMO to describe the land surface-atmosphere interaction [15]. To describe the energy and water transfer at the land surface - atmosphere interface, the model describes: the turbulent exchange within the atmospheric surface layer, stomatal regulation of plant transpiration, and root water uptake from different soil horizons (down to 23 m depth). The parameters characterizing the properties of soil and vegetation are taken from external global land cover data bases (GLC2000 and GLCC) at 30" of arc (about 1 km) resolution and from the digital soil map of the world (DSMW). The results of recent model validation provided using regular meteorological observations from about 150 weather stations located in European part of Russia showed a high quality of COSMO system projections for various parameters and under broad range of weather conditions [14].

The 'model region' selected for the study is situated in the central part of a large modeling domain which covered almost the entire East European Plain and is bounded by geographical coordinates 55° and 59°N and 28° and 37°E. Thus, the area of the model region is about 10^5 km^2. The forests cover about 50% of the area and consist mainly of coniferous tree species in the north and broad-leaved species – in the south. Within the model region the different scenarios of forest cover change were simulated. The modeling study included three main numerical experiments. The first modeling experiment assumed the total deforestation of the model region and replacement of the forest communities by grasslands. The second experiment was a scenario of total afforestation of the model region. It assumed the suspension of logging and active forest regeneration and growth in agricultural and abandoned lands. It also supposes that an increase of forest cover in the scenario is mainly due to pioneer small-leaved tree species (e.g. birch, aspen). In the third 'control' experiment the weather conditions were simulated under the present land-use and vegetation structure of the model region.

The numerical experiments to assess the effect of forest cover changes on regional weather conditions were conducted for a modeling domain located between 50° - 70° N and 15° - 55° E, and covering almost the entire area of the East European plain. The output of the numerical experiments was provided at a 13.2 km grid resolution. The ERA-Interim global atmospheric reanalysis (with 6-hour resolution in time and 0.75°×0.75° – in space) was used to quantify the initial and boundary conditions [16].

For our modeling experiments, the warm period of 2010 (from May to September) was used. It was characterized by various and very contrasting weather conditions. A unique feature of the selected period was extremely hot and dry weather in the middle of summer (July - beginning of August). Such various weather conditions allows us to reflect more clearly on the possible regulating mechanisms of forest vegetation of the energy and water exchange processes under different temperature and soil moisture conditions.

3. Results and discussion

The results of numerical experiments describing the effects of total deforestation and afforestation of the model region on regional weather conditions showed that the modern forest cover and land-use changes can significantly influence the atmospheric parameters, not only within the selected model region, but also in the areas situated within the modeling domain far away from the actual model region (figure 1, tables 1-2). The spatial patterns of the temperature and precipitation changes under deforestation and afforestation scenarios show their very strong spatial heterogeneity. Whereas the significant changes of the air temperature can be seen in internal areas of the model region only, the changes of precipitation amount can be traced within the entire modeling domain (figure 1).

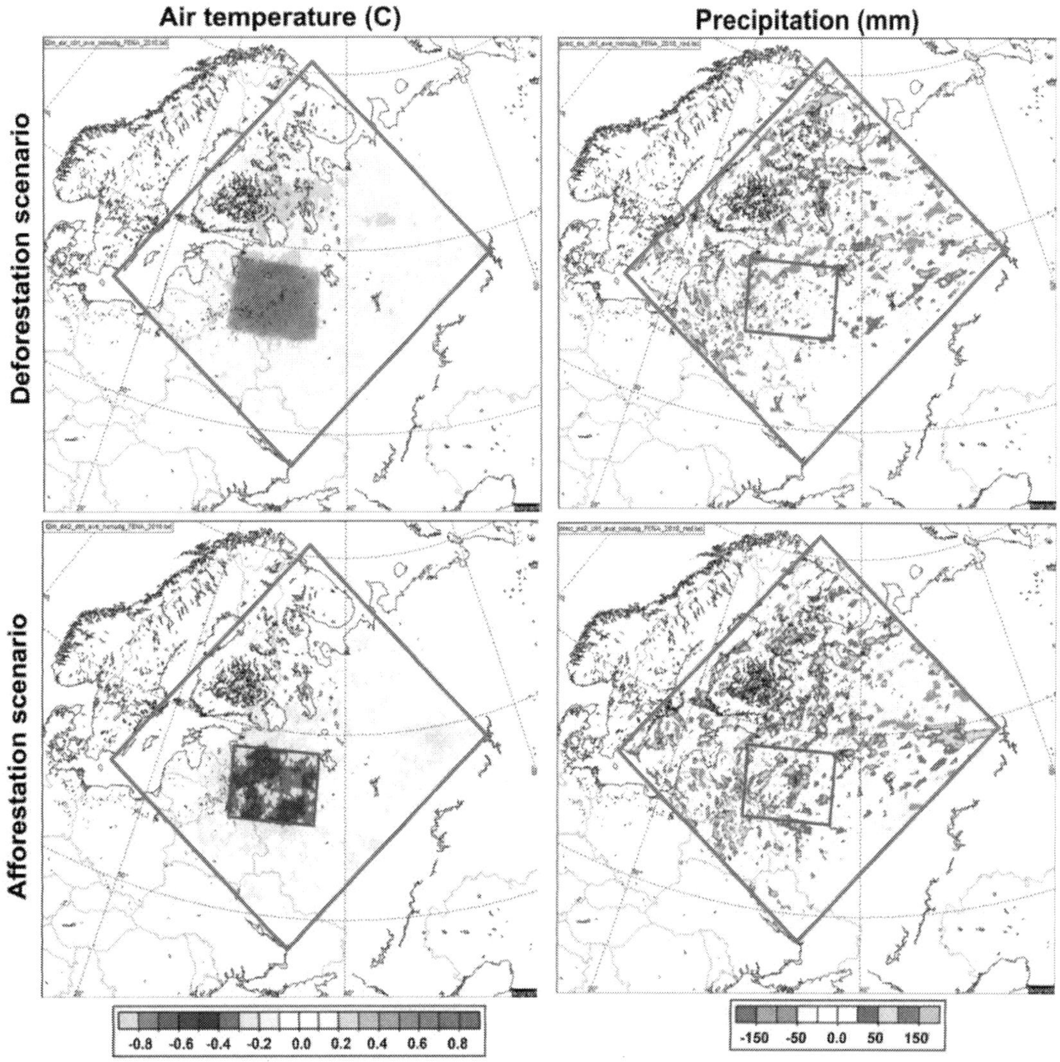

Figure 1. Simulated differences in the air temperature (°C) and precipitation rate (mm) for the selected modeling domain between the deforestation and afforestation scenarios of the model region and the control experiment, assuming present land-use structure for the entire modeling period from May to September 2010. The red square indicates the boundaries of the model region and the grey square shows the boundaries of the selected modeling domain.

Table 1. Simulated air temperatures, T (°C) for the selected model region and entire modeling domain and the temperature differences ΔT (°C) between deforestation and afforestation scenarios in the model region and the control experiment assuming present land-use structure for the modeling period from May to September 2010.

Numerical experiments		Months					
		5	6	7	8	9	Mean
			Model region				
Present land-use	T	9.6	11.9	19.0	15.5	9.2	13.0
Deforestation scenario	T	9.7	12.1	19.6	15.7	9.2	13.3
	ΔT	+0.1	+0.2	+0.6	+0.2	0.0	+0.3
Afforestation scenario	T	9.5	11.8	18.8	15.4	9.1	12.9
	ΔT	-0.1	-0.1	-0.2	-0.1	-0.1	-0.1
			Entire model domain				
Present land-use	T	11.8	14.7	20.9	18.3	10.9	15.3
Deforestation scenario	T	11.9	14.7	21.1	18.4	10.9	15.4
	ΔT	+0.1	0.0	+0.2	+0.1	0.0	+0.1
Afforestation scenario	T	11.8	14.7	20.9	18.3	10.9	15.3
	ΔT	0.0	0.0	0.0	0.0	0.0	0.0

Table 2. Simulated precipitation amount, P (mm), for the selected model region and entire modeling domain and the precipitation differences ΔP (%) between the deforestation and afforestation scenarios in the model region and the control experiment assuming present land-use structure for modeling period from May to September 2010.

Numerical Experiments		Months					
		5	6	7	8	9	Mean
			Model region				
Present land-use	P	75.2	79.4	47.4	52.1	68.4	322.5
Deforestation scenario	P	74.9	77.5	43.6	47.5	67.9	311.4
	ΔP	-0.4	-2.4	-8.0	-8.8	-0.7	-3.4
Afforestation scenario	P	75.6	83.1	49.5	53.4	69.7	331.3
	ΔP	+0.5	+4.8	+4.8	+2.7	+1.9	+2.8
			Entire model domain				
Present land-use	P	50.5	57.8	37.4	49.0	57.6	252.3
Deforestation scenario	P	51.0	56.1	35.9	47.3	57.5	247.8
	ΔP	+1.0	-2.9	-4.0	-3.5	-0.2	-1.8
Afforestation scenario	P	50.6	58.3	37.8	48.9	57.8	253.4
	ΔP	+0.2	+0.9	+1.1	-0.2	+0.3	+0.4

IOP Conf. Series: Earth and Environmental Science **107** (2017) 012105 doi:10.1088/1755-1315/107/1/012105

Under the scenario imitating deforestation processes the mean air temperature within the model region was higher than that was obtained in the control experiment. Under the afforestation scenario the mean air temperature, on the contrary, was lower than under the control conditions (tables 1-2). The maximum temperature differences between both scenarios imitating the forest cover changes and the control experiment were obtained for the summer months (July): +0.6 °C – for the deforestation scenario and -0.1 °C – for the scenario with afforestation, respectively. For the areas situated within the modeling domain but outside of the control region the changes in air temperature are manifested quite poorly due to forest cover change (figure 1). Under the deforestation scenario the temperature difference with respect to the control experiment gradually decreases with an increase of the distance from the external borders of the model region. Any meaningful influence of afforestation of the model region on the average temperature of the entire modeling domain is not revealed. The results of numerical experiments are also showed that under the anomalous weather conditions of 2010 the total deforestation in addition to projected temperature growth also leads to an increase in duration of the periods with abnormally hot weather (the periods with average daily air temperature exceeding the mean climatic values by at least 7 °C). At the same time, for the scenario imitating total afforestation of the model region the duration of the periods with anomalously hot weather, on the contrary, slightly decreased.

Analysis of diurnal variability of the air temperature shows that the maximum differences between the forest cover change scenarios and the control experiment is predicted for daylight hours. Calculations show that for the period of anomalously hot weather in the summer of 2010 the daytime temperatures obtained for the model scenario imitating the total deforestation of the model region could exceed the temperatures of control experiment at modern land-use and forest structure by 2-6 °C. Under the scenario imitating total afforestation the daytime temperatures were slightly lower than the values under the control experiment (by 2-5 °C).

The influence of forest cover change within the model region on precipitation is manifested in an increase of its total amount under the afforestation scenario, and in its decrease under the scenario imitating total deforestation (tables 1-2). The maximum increase of precipitation (+4.8%) for the model region under the afforestation scenario was predicted for June-July, and its maximum decrease (-8.8%) under deforestation scenario for July - August, i.e. in the hottest months of 2010. As already mentioned, the influence of forest cover changes on spatial precipitation pattern is clearly extended beyond the model region to the entire modeling domain of the East European Plain (figure 1). In particular the decrease of precipitation in July for the entire modeling domain under the scenario imitating deforestation processes in relation to precipitation under the control experiment was about 4%. Thus, despite the rather small (about 10% in relation to the size of modeling domain) area of the model region, the modeling experiments show a notable influence of forest cover changes on precipitation pattern within the entire modeling domain of the East European Plain. For the scenario imitating afforestation processes, the total amount of precipitation in July exceeded its amount in the control experiment by 5% for the model region and by about 1% for the entire modeling domain, which also gives evidence for the essential redistribution of precipitation outside the boundaries of the model region.

During the last decades in numerous experimental and modeling studies considerable attention was given to the effects of forest and land-use changes on local and regional meteorological conditions [6, 8, 17 - 19]. Most of the studies show an excess of day and daily mean values of the air temperature in deforested areas over its values in undisturbed forest stands. These results are therefore well in agreement with the estimations obtained in our numerical experiments. The influence of the forest and forest cover change on precipitation is still under-investigated.

The main mechanisms that are responsible for the influence of deforestation processes on weather conditions are: change of surface albedo and surface roughness, decrease of evaporative water loss into the atmosphere in deforested sites, and their additional heating that can lead to the strengthening of local convective instability in the lower troposphere. It is obvious that the influence of all these factors in different seasons, under various synoptic situations and for different land surface types can result in effects that can be quite difficult to predict. The influence of afforestation is manifested in a general increase of water vapor intake into the atmosphere due to the higher transpiration rate of forest vegetation in comparison with evaporation from bare soil or transpiration of grassy vegetation. The water vapor released into the troposphere can be both condensed and immediately returned to the earth as convection rainfall (that occurs quite seldom) or (most often) returned as rain after some delay following its translocation, depending on the current atmospheric conditions.

To parameterize the surface evapotranspiration the COSMO model takes into account aggregated information about land-use structure, forest cover and species composition, surface roughness and albedo, leaf area index (LAI), root depth, and biophysical properties of vegetation cover (e.g. stomatal conductance). Several recent modeling studies [4, 6] showed that, in the case of deforestation of boreal forests, the higher albedo during the winter and spring period can play a key role in regulation of land surface temperature. Higher contribution of snow cover to the albedo of the land surface in deforested areas in comparison with undisturbed forest sites during the winter period and in the early spring can lead to decrease in short-wave radiation balance of ground surface and, as a result, to air temperature reduction. The effects of afforestation can result in the opposite effects. Taking into account a strong influence of surface albedo on temperature regime of land surface, it is also very important to consider a very strong impact of the biophysical properties of vegetation on net radiation and sensible and latent heat fluxes between land surface and the atmosphere. Large differences in the biophysical properties of forest and grassy vegetation have not been completely taken into account in global and regional mathematical models yet. It can obviously result in some uncertainties in model predictions.

4. Conclusion

The results of numerical experiments provided using the COSMO model for warm period of 2010, taken as an example, have shown that deforestation and afforestation processes can result in significant changes of weather conditions in the area of the East European Plain. Numerical experiments indicate the existence of a complex system of relationships between forest vegetation and the atmosphere.

It is obvious that available numerical models are able to predict the possible response of regional weather conditions to changes of land surface and vegetation properties only within the model parameterizations and approximations used for describing the real atmospheric and land surface processes. That requires involving more sophisticated modeling approaches that are allowed to describe the forest - atmosphere interaction in detail and that are validated against experimental data. The numerical experiments provided should be considered as a part of a multidisciplinary study of possible effects of forest cover changes on regional and global weather conditions.

Acknowledgement
This study was supported by a grant from the Russian Science Foundation (14-14-00956).

References

[1] Bonan GB, Pollard D and Thompson SL 1992 *Nature* 359 716-8
[2] Oltchev A, Ibrom A, Morgenstern K, Kreilein H and Gravenhorst G 1999 *J Phys. Chem. Earth* **24**(1-2) 103-10
[3] Amiro BD 2001 *Glob. Change Biol.* **7** 253-68
[4] Brovkin V, Raddatz T, Reick CH, Claussen M and Gayler V 2009 *Geophys. Res. Lett.* 2009 **36** L07405
[5] Brovkin V *et al* 2013 *J Climate* **26**(18) 6859-81
[6] Bathiany S, Claussen M, Brovkin V, Raddatz T and Gayler V *Biogeosciences* **7** 1383–99
[7] Olchev A, Novenko E, Desherevskaya O, Krasnorutskaya K and Kurbatova J 2009 *Environ. Res. Lett.* 4 045007
[8] Mamkin V, Kurbatova J, Avilov V, Mukhartova Yu, Krupenko A, Ivanov D, Levashova N and Olchev A 2016 *Environ. Res. Lett.* **11**(12) 125012

IOP Conf. Series: Earth and Environmental Science **107** (2017) 012105 doi:10.1088/1755-1315/107/1/012105

[9] Oltchev A *et al* 2002 *J Phys Chem Earth* **27**(9-10) 675-90
[10] Lapenis A, Shvidenko A, Schepaschenko D, Nilsson S and Aiyyer A 2005 *Glob. Change Biol.* **11**(12) 2090-102
[11] Olchev A, Radler K, Sogachev A, Panferov O and Gravenhorst G 2009 *J. Ecol. Modell.* **220** 3046–56
[12] Novenko EYu and Olchev AV 2015 *Quat. Int.* **388** 12-22
[13] Baldauf M, Seifert A, Forstner J, Majewski D, Raschendorfer M and Reinhardt T 2011 *Mon Weath Rev.* **139** 3887-905
[14] Rivin G.S *et al.* 2015 *Russ. Meteorol. Hydrol.* **40**(6) 400–10
[15] Doms G *et al* 2011 *A Description of the Nonhydrostatic Regional COSMO Model, part 2: Physical Parameterization* (Offenbach: Deutscher Wetterdienst) p 154
[16] Dee D P *et al* 2011 *Quart. J. Roy. Meteor. Soc.* **553** 553-97
[17] Carlson DW and Groot A 1997 *Agr. Forest Meteorol.* **87** 313-29
[18] Radler K, Oltchev A, Panferov O, Klinck U and Gravenhorst G 2010 *Open Geography J.* **3** 103-14
[19] Kuz'mina E V, Ol'chev A V, Rozinkina I A, Rivin G S, Nikitin M A 2017 *Russ. Meteorol. Hydrol.* **42**(9) 574-81

Recovery of continuous field of ecological-climatic indices based on the meteorological stations data

A A Saveliev, A M Domnina

Institute of Environmental Sciences, Kazan Federal University, Kremlevskay str., 18, 420008, Russia

E-mail: Anatoly.Saveliev.aka.saa@gmail.com

Abstract. The possibility of using INLA (Integrated Nested Laplace Approximations) method for the recovery of continuous field of ecological-climatic indices based on the meteorological stations data has been considered and the accuracy of such recovery assessed. Meteorological data from 13 stations were used to fit INLA model, and independent data from 58 stations were used to test the model. Traditional ecological-climatic indices were modelled (annual sum of effective temperatures, mean annual temperature, mean temperature of the coldest and the warmest month within a year). All indices show good compliance of the model predictions and data test (R2 in range 0.70-0.90).

1. Introduction

Climate is one of the basic complex ecological factors determining conditions for species distribution, especially that of vegetation. The influence of the climate on vegetation is diverse and primarily related to such important conditions of occurrence as heat and humidity. Thus, ecological factor of heat can be estimated by the use of averaged daily temperatures and their derivatives, i.e. annual sum of effective temperatures, mean annual temperature, mean temperature of the coldest and the warmest month within a year. Although these indices are spatially distributed and they can be calculated only for meteorological stations, they can be interpolated for the rest of the area. Recently, Bayesian statistical approaches have been elaborated for spatial-temporal modeling based on determination of the stochastic field indices. This work is devoted to the application of INLA (Integrated Nested Laplace Approximations) method [1, 2] in spatio-temporal modeling of ecological-climatic indices.

2. Input Data

The GSOD data obtained by National Climatic Data Center (NCDC) in Asheville, the USA, in the framework of data exchange program of World Weather Watch (WWW) in accordance with the WMO Resolution 40 (Cg-XII) offering free access for scientific and non-profit institutions have been used as input data. The access was performed via FTP server (ftp://ftp.ncdc.noaa.gov/pub/data/gsod) and involved average, minimal and maximal daily temperatures and information on wind and sediments.

Three derivative ecological-climatic indices have been calculated from average daily temperature, i.e. mean annual temperature (mean.T), mean temperature of July (jul.T), mean temperature of January (jan.T), the number of days with negative temperatures (Sum.NegativeDays), sum of negative temperatures below -10 °C (Sum.Negative10.T) and the Growing degree-day (sum of effective temperatures exceeding 10°C, Sum.Positive10.T).

The area of Middle Volga region limited in the longitude from 43°E to 58°E and in the latitude from 52°N to 58°N was selected for the modelling. This area covers the most part of the Volga region. The area has a distinct latitude climate gradient (boreal ecocline zone). Modeling area is presented by a rectangle of 860x800 km as shown in figure 1 below. Figure 2 shows triangulation net used for the recovery of ecological-meteorological indices.

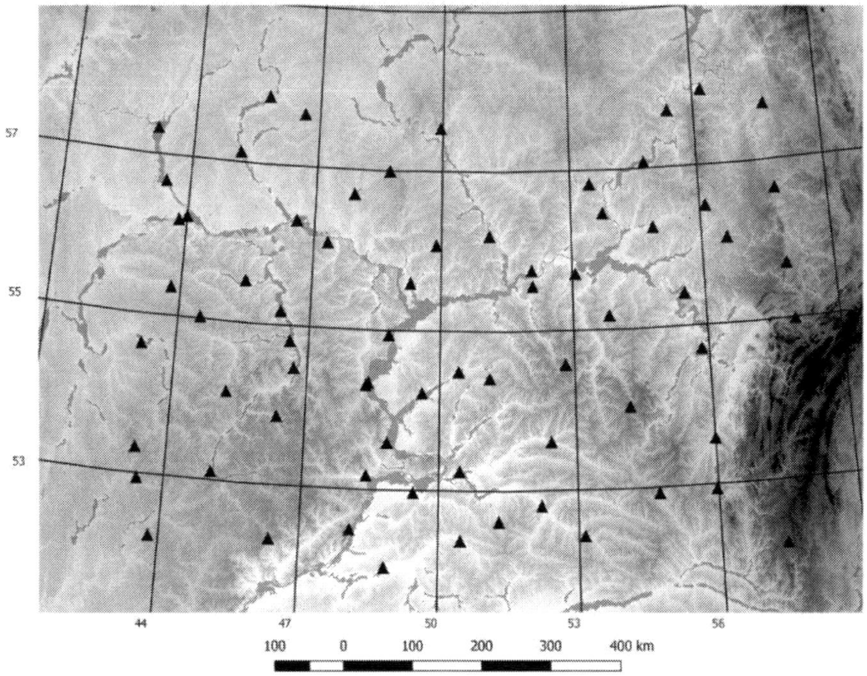

Figure 1. Modeling area and weather observing stations positions.

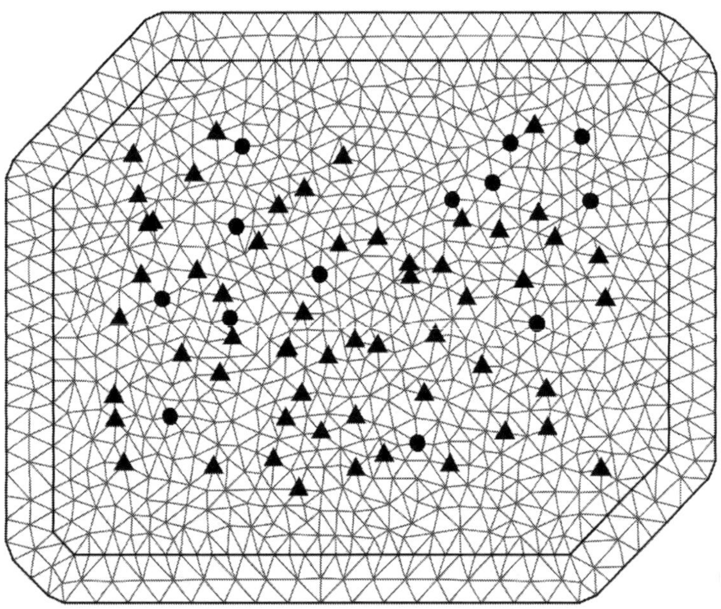

Figure 2. Triangulation net of the modeling area. Round symbols correspond to the stations with input data and triangles to the stations used for the model validation.

Modeling period covers 1980-2016 years reach with input available data. In GSOD database, 71 stations are found with annual information and used as input data source. Other 58 stations with the information covering from 3 to 36 years were applied for modeling results assessment.

3. Methods

The modeling was performed with the statistics system INLA [1] that employs the concept of random fields for the recovery of continuous spatial-temporal field of the distribution of the index to be modeled.

All the parameters modeled indexed as Y were assessed using the same type of model:

$$Y(s, year) = a_0 + a_1 \times Year + a_2 \times H + RF_{mean}(s) + RF_{year}(s, Year) \quad (1)$$

Here $Y(s, year)$ is a modeled index in the spatial location s, in a year $year$, $a_1 \times Year$ is a linear temporal trend (along the years counted from 1980), $a_2 \times H$ is a linear height trend, $RF_{mean}(s)$ is a value of mean annual random field of the index and $RF_{year}(s, Year)$ is a value of random field estimating deviation from mean average value within a $year$.

4. Results and Discussion

The following results have been obtained for linear trends (table 1), Bayesian credible intervals were used to estimate significance.

Table 1. Linear trends.

Index	Years			Height		
	Index	**SD**	**Significance**	**Index**	**SD**	**Significance**
mean.T	0.04	0.01	Yes	-0.010	0.003	Yes
jul.T	0.03	0.03	No	-0.011	0.001	Yes
jan.T	-0.01	0.05	No	-0.011	0.008	No
Sum.NegativeDays	-0.49	0.11	Yes	0.094	0.022	Yes
Sum.Negative10.T	0.13	1.40	No	-0.920	0.510	No
Sum.Positive10.T	7.11	1.59	Yes	-1.305	0.206	Yes

The quality of the modeling of spatial-temporal distribution of the indexes was estimated by comparison of the predicted indices with those observed at 58 test stations. For this purpose, correlation coefficient between the observed and modeled indices was calculated for the test stations within the minimal rectangle bounding 13 stations with the input data (tagged "interpolation" in text), and that for all the stations (tagged "all" in text) providing the comparison of interpolation and extrapolation result. The results obtained are given below in table 2 (correlation).

Table 2. Correlation between observed and modelled indexes.

Index	Correlation (Interpolation)	Correlation (All)	R^2 (Interpolation)
mean.T	0.86	0.87	0.75
jul.T	0.92	0.90	0.81
jan.T	0.89	0.87	0.75
Sum.NegativeDays	0.89	0.88	0.78
Sum.Negative10.T	0.76	0.71	0.51
Sum.Positive10.T	0.88	0.87	0.76

Linear regression results of indices on the corresponding models are given in table 3 (bias, slope and standard deviation).

IOP Conf. Series: Earth and Environmental Science **107** (2017) 012106 doi:10.1088/1755-1315/107/1/012106

Table 3. Linear regression coefficients for models of observed and modelled indices.

Index	Interpolation			All		
	bias	**slope**	**SD**	**bias**	**slope**	**SD**
mean.T	0.91	0.81	0.68	1.54	0.69	0.72
jul.T	1.56	0.93	0.82	3.00	0.86	0.95
jan.T	-0.20	0.95	1.64	-0.81	0.87	1.83
Sum.NegativeDays	3.14	0.97	7.15	21.24	0.83	7.88
Sum.Negative10.T	-82.6	0.72	81.4	-123.8	0.54	91.7
Sum.Positive10.T	155.6	0.85	103.1	235.2	0.78	115.0

Figures 3-5 illustrate some results.

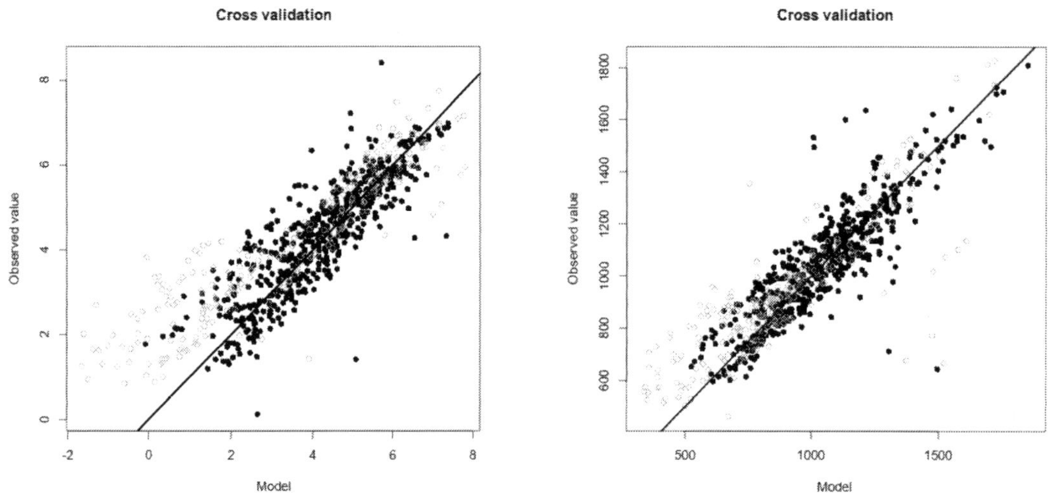

mean.T Sum.Positive10.T

Figure 3. The comparison of model and real indices corresponding to test stations. Filled symbols correspond to interpolation data and open symbols to extrapolation results obtained for the stations outside the initial input data area.

The model obtained makes it possible to estimate continuous distribution of the index along the area used for modeling. Thus, model distribution of effective temperatures for 1980 and 2010 is presented below in figure 3 (obtained with mean height value to provide more smooth contours).

The consideration of the results shows that the applied INLA method is a reliable tool for recovery of continuous field of ecological-climatic indices even if a limited set of data from meteorological stations is available. The method allows operating incomplete temporal data. This makes it convenient for imputation of missed stations data. The spatial distribution of the ecological-climatic indices obtained can be used for territory zoning, especially in agriculture and geobotany. Besides, modeling made it possible to discover statistically significant temporal trend of mean temperature and of the sum of Growing degree-days on the area considered. The applied method (INLA) provides not only the indices of expectation estimate, but also the probability distribution in the form of p.d.f. at each location, as well as uncertainty estimate.

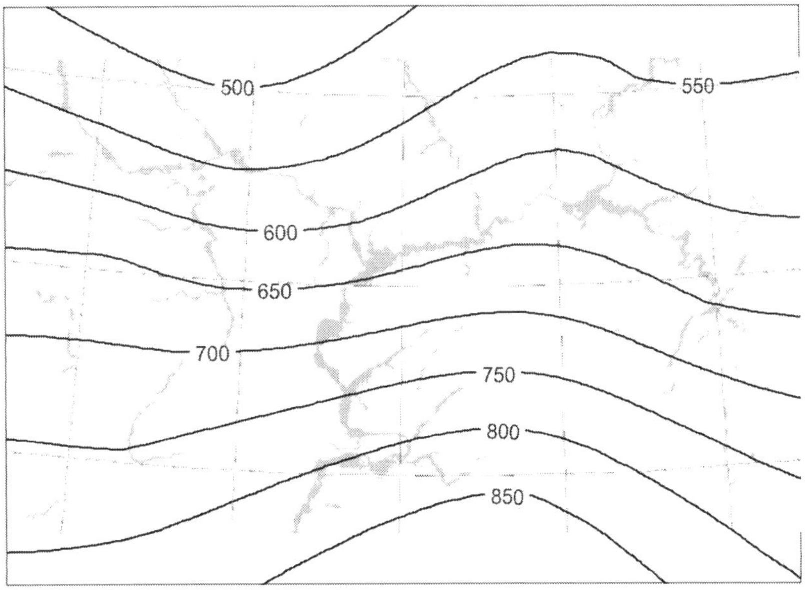

Figure 4. Model distribution of the Growing degree-day (Sum.Positive10.T) values in 1980.

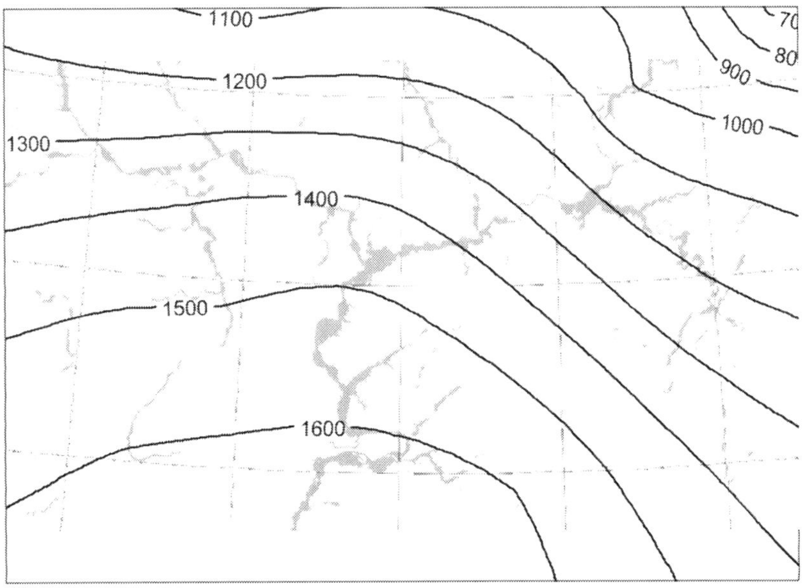

Figure 5. Model distribution of the Growing degree-day (Sum.Positive10.T) values in 2010.

References

[1] Rue H, Martino S and Chopin N 2009 Approximate Bayesian Inference for latent Gaussian models using Integrated Nested Laplace Approximations. *JRSS-series B.* **71**(2) 319-392

[2] Blangiardo M, Cameletti M 2015 *Spatial and Spatio-temporal Bayesian Models with R – INLA.*Wiley, 320 p

IOP Conf. Series: Earth and Environmental Science **107** (2017) 012107 doi:10.1088/1755-1315/107/1/012107

Use of GIS Technologies in Order to Assess the Degree of Transformation of the Land Cover Structure

D A Tretyachenko[1], Y S Greenfieldt[1], O A Klimanova[1], N N Alekseeva[1], M A Arshinova[1], E Y Kolbowsky[1]

[1]Faculty of Geography, Lomonosov Moscow State University, Russia, 119991, Moscow, GSP-1, 1 Leninskiye Gory

E-mail: daria.trt@gmail.com, y.greenfeldt@gmail.com, oxkl@yandex.ru, nalex01@mail.ru, amari_geo@mail.ru, kolbowsky@mail.ru

Abstract. The main purpose of this study was to identify global trends in the transformation of land cover structure for 2001-2012, that have significantly affected the structure and dynamics of modern landscapes.

1. Introduction

In the world practices of scientific research geo-spatial data bases Land Cover Change are wildly used to assess the Degree of Transformation of the Land Cover Structure. They are generated on the basis of spectral radiometer MODIS (Moderate-resolution Imaging Spectroradiometer). Data of the Land Cover Change are the secondary images obtained after processing of MODIS data in certain zones and are distributed in free access in Internet network [4]. These data contain information on spatial distribution of vegetation types of natural and anthropogenic geneses, urban and built-up areas, territories free of vegetation (waste grounds, glaciers, etc.), water bodies, water streams, and over moistened areas [1].

The object of the Land Cover Change analysis is the category of land cover, which in reality is only relatively connected to belt-zone specifics. For instance, the grassland category can be met both in Arctic and sub-Equator latitudes, Mixed Forest category exist within the moderate, subtropical and tropical zones. Therefore, to interpret results on the basis of landscape approach, it is necessary to use independent operational territory landscape units, according which the analysis and assessment of measurements could be conducted. The most applicable for such territory unit seems to be the natural zones with the uniform of hydrothermal, soil-geographic and geo-botanic characteristics.

2. Material and methods

As mentioned above, as a data source on land cover conditions for 2001-2012 is publically available data of Land Cover Change [4]. As a media for analysis the GIS package ArcGIS for Desktop Extension Spatial Analyst is used.

To assess the degree of transformation of the land cover structure, the aggregated global data with horizontal resolution 5'x5' in geographical coordinate system WGS-84 (World Geodetic System – 1984 were used. Considering that the main goal of the analysis was the calculation of area variations in land cover structure over the globe, the data were projected into Cylindrical Equal Area projection. The projection was carried out with the basic complectation of the ArcGIS for Desktop Basic. The cell size of output raster was determined as 1000, firstly for conformity of further calculations of the areas of land cover variations, and secondly, since it provides the optimal raster dimension and does not allow losing the data metallization.

For raster data preprocessed with the method mentioned above the instrument Combine of ArcGIS for Desktop Spatial Analyst was used. Each combination of input data of two rasters was distinguished as a unique zone. For each zone the number of included cells was calculated, which in this case is 1 km ^{2}, since the resolution of analysed rasters is equal to 1000 m. The tables obtained were supplemented with the text description of the land surface (Join Field instrument), also the percent of existed changes within the initial zone (from the zone in 2001) was calculated.

The legend for Land Cover Change includes 17 classes (from 0 to 16), corresponding to classification of International Geosphere Biosphere Programme (IGBP): 0 is for the class of Water,1 is for Evergreen Needleleaf Forest, 2 is for Evergreen Broadleaf Forest, 3 is for Deciduous Needleleaf forest, 4 is for Deciduous Broadleleaf Forest, 5 is for Mixed Forest, 6 is for Closed Shrublands, 7 is for Open Shrublands, 8 is for Woody Savannas, 9 is for Savannas, 10 is for Grasslands, 11 is for Permanent Wetlands, 12 is for Croplands, 13 is for Urban and Built-up, 14 is for Cropland and Natural Vegetation Mosaic, 15 is for Snow and ice, 16 is for Barren or Sparsely Vegetated.

Thus, the data for Land Cover of 2001 and 2012 were compared where the each cell which is characterized with corresponding numerical index from 0 to 16 according to the classification. With the change of the land cover class from 2012 in comparison with 2001 the change of numerical index is occurred. It allows segregating a sell (or the number of sells) as an area of certain types of changes. For example, index 12.5 means that the arable lands covered the cell in 2001 were replaced with mixed forest in 2012.

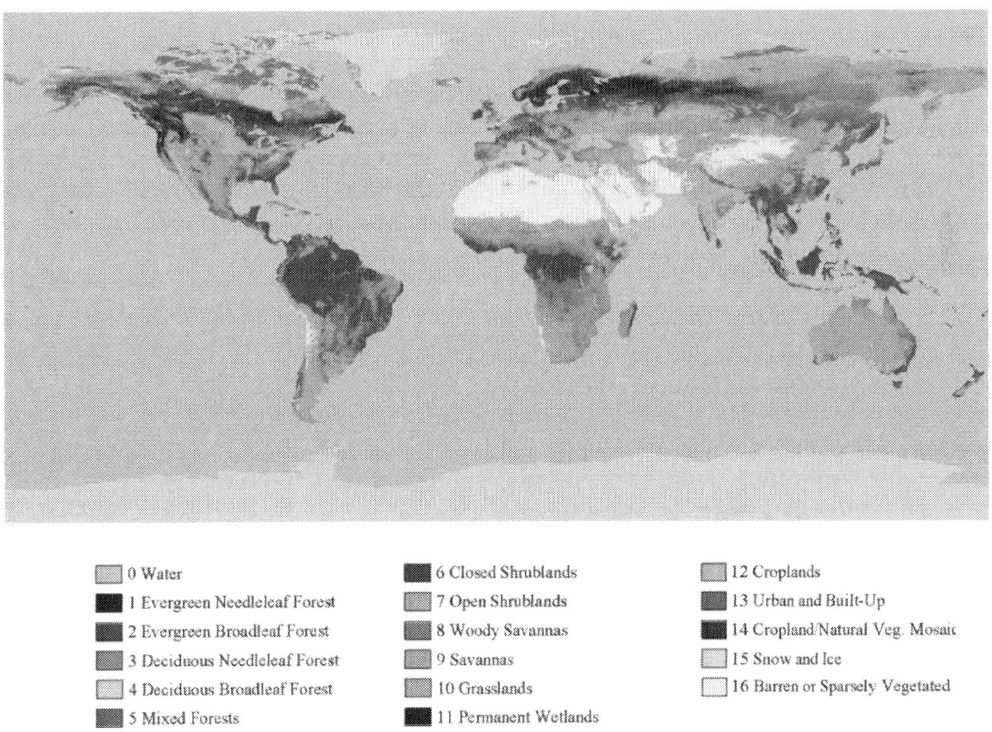

Figure 1. Global Mosaics of the standard MODIS land cover for 2012.

The generated data array allowed calculating the values of relative fractions of the classes of land cover for two time periods and determining the relative decrease/increase of the areas under consequent classes.

Analysis of the data demonstrates a contradictive tendency for different continents and not often corresponding to available statistical data. Among these processes variations of evergreen wide leaves forests in Eurasia and North America could be mentioned, while such forests areas declines in Africa and South America, what contradicts commonly accepted understanding of the process of forest decrease in tropical regions and increase of forested areas in the middle areas. Rarefied bush formations in Africa, South America and Australia demonstrate a tendency to shortening both in absolute and relative measures, which might be related to conditioning of desertification in these areas. Areas covered with permanent snow cover and glaciers all over the World (except for North America) are decreasing. In Africa such class of land cover is not presented at all. Elsewhere the increase of territories without vegetation is observed. The most significant share is observed in Africa. However, we should mention that the temporary changes are generally not high.

Additionally, the transfer between different classes, which let us to suggest that the most common classes and territories occupied by these classes are more or less stable, that in its turn suggests that transaction types for any continent and share of each class are remain stable.

The boundaries of natural zones subjected to analysis of land cover transformation at zonal level are obtained from the electronic atlas of the World “ArcAtlas: Our Earth” (1996) [2]. Along with this the initial map contains the boundaries of geographical bans of the zonal landscape types at the plain and the altitudinal landscape specters in the mountains and intrazonal landscapes (river values, mangroves, saline lands, glaciers, and marshlands). According to explanation note for the map the areas of the natural zones are characterized with the uniformity of hydrothermal, soil-geographic, geo-botanical characteristics, which make them the most appropriate operational-territory unit for conducting spatial-time analysis of land cover. In distinguish of land cover characteristics the natural zones have as an attribute adjunct belongings to a certain geographical belt. Totally 26 natural zones are marked out on the map; they are related to seven geographical belts.

The numerical data were obtained regarding the changes of soil cover categories from 2001 to 2012. All natural zones were divided into three groups. Changes in each group covered: 1) less than 10% zone area (total 6 zones); 2) from 10 to 20% (total 11 zones); 3) over 20% (total 9 zones). Among relatively stable (the first group) are zones occupying huge areas (i.e. tropical deserts) and relatively insignificant by square (i.e. steppes of chilly plateaus). The third zonal group (with maximum variations) includes five zones of subtropical band excluding deserts and semi-deserts, two forest and sparse forests, i.e. moderate climate (taiga) and Arctic tundra bands.

There are the processes of vegetation free land obliteration in the desert zones of moderate and tropical belts. In moderate climate band the square of this area in relation of the general band square is significantly higher (4,81%), than in tropics (1,15%). Semiarid areas of moderate and tropical bans show the similar process specters, i.e. vegetation free land obliteration and bush obliteration. However, semiarid areas of the moderate zone are combined with decrease of bush fraction and obliteration of cultivated lands (probably irrigated in the past). An opposite trend – shortening of projective vegetation cover, i.e. an activation of desertification process – is typical for semiarid areas. In the steppe-forest and steppe zones the dynamics is characterized by increase of plugging but at the same time areas of slinging of cultivated lands is fixed. In the forests of equatorial zone both processes – increase of forest vegetation and its degradation – are observed. In taiga zone and mixed and broad leaf forests the processes of forest vegetation and structural variations determine a role play in increasing of forest vegetation fracture and slinging of previously cultivated lands.

3. Results and discussion

The obtained results demonstrate first of all the inventory of the changes of land cover and need for further deeper analysis that requires additional information. Verification of data obtained in this study based on the key areas supported with the reliable statistical information on the land use and literature data. It is necessary to identify the factors determining the observed variations of the land cover including climatic and anthropogenic (demographic and socio-economic).

The results allowed compiling a preliminary classification of the land cover variations for the considered period and identifying the types of transformation processes at the global scale. Four types of the processes were identified for equatorial and sub-equatorial belts, two types were for tropical belt, six types were for forest and sparse forests of tropical belt, six types were for non-forested zones of subtropics, seven zones were for forested zones of subtropics, seven zones were for forested zones of moderate climate zone, six types were for non-forested areas of the moderate climate zone, three types were for natural zones of sub-polar belt, and three types were identified for polar zone.

The data allow discussing the main trends in land use during 2001–2012 on the global scale, which possibly can impact on the structure and dynamics of existing landscapes. In difference of similar studies [5, 6, 7] the defined transition types and differences cover the entire specter of land cover categories and transformation processes, and give the possibility to investigate the main trajectories of not only the anthropogenic-determined, but also a natural category of the land cover. The use of relatively short (10 years) time interval for changes analysis supplements previously done environmental system changes for 1700-2000 [3].

The identified changes of the certain types, which can be interpreted as a kind of trends of the global processes possible in predictable future, significantly change the state of biosphere. In light of determined regularities it seems reasonable to model certain natural and social factors (climate change, population, density of road network, environmental track, etc.), that impact on the trends and order of reversibility in the determined trends in the land cover transformation.

Acknowledgement
Current study was carried out within the framework of the project of the Russian Foundation of Fundamental Re*search (RFFR)* "Transformation of the World land use structure: analysis and typology of changes" *(grant RFFR P 15-05-06186/15).*

References

[1] Alekseeva N, Klimanova O, Khazieva E 2017 Global land cover database and and prospects of their sage for modern landscapes mapping *Izvestia Russian Academy of Sciences, Geographical series* **1** 126-39
[2] *Atlas ArcAtlas "Our Earth"* 1996
[3] Ellis E C, Goldewijk K K, Siebert S, Lightman D, Ramankutty N 2010 Anthropogenic transformation of the biomes, 1700 to 2000 *Global Ecology and Biogeography* **19** 589–606
[4] *MODIS Land Cover* URL: http://glcf.umd.edu/data/lc/ (Accessed 15.02.2017)
[5] Milleniun Ecosystems Assessment 2005 *Ecosystems and Human Well-being: Synthesis* (Washington, DC, USA: Island Press) 153 p
[6] Mucher C A, Klijn JA, Wascher DM, Schamine'e JHJ 2010 A new European Landscape Classification (LANMAP): A transparent, flexible and user-oriented methodology to distinguish landscapes *Ecological Indicators* **10** 87–103
[7] Sayre R, Dangermond J, Frye C, et al. 2014 *A New Map of Global Ecological Land Units — An Ecophysiographic Stratification Approach* (Washington, DC: Association of American Geographers) 46 p

Geographic Information System and Geoportal «River basins of the European Russia»

O P Yermolaev, S S Mukharamova, K A Maltsev, M A Ivanov, P O Ermolaeva, A I Gayazov, V V Mozzherin, S V Kharchenko, O A Marinina, F N Lisetskii

Institute of Environmental Sciences, Kazan Federal University, Kremlyovskaya str., 18, 420008, Kazan, Russia

E-mail: oyermol@gmail.com

Abstract. Geographic Information System (GIS) and Geoportal with open access «River basins of the European Russia» were implemented. GIS and Geoportal are based on the map of basins of small rivers of the European Russia with information about natural and anthropogenic characteristics, namely geomorphometry of basins relief; climatic parameters, representing averages, variation, seasonal variation, extreme values of temperature and precipitation; land cover types; soil characteristics; type and subtype of landscape; population density. The GIS includes results of spatial analysis and modelling, in particular, assessment of anthropogenic impact on river basins; evaluation of water runoff and sediment runoff; climatic, geomorphological and landscape zoning for the European part of Russia.

1. Introduction

Currently, in Russia, there is no integrated geospatial database or geographic information system tied to the basins of small rivers. The urgency of creating such a complete coverage GIS that is capable to accumulate large volumes of spatial information on natural systems and integrated information on the state of small river basins is in relevancy of studying the results of increasing anthropogenic impact, as well as climatic changes observed in different landscape areas on basin geographic systems.

The large region of Russia – its European part which is some 4 million square kilometres with the most part of population and industrial and agricultural potential of the country – is being fundamentally studied in terms of geographical analysis of minor river basins and river runoff while creating designated geographic information system (GIS) "River basins of the European Russia". The designed GIS is considered to be a base for modern data and knowledge on geographic, hydroclimatic, geoecological and other characteristics of natural resource potential of tens thousands river basins. The thematic information sources are the data from long-term monitoring's, Earth's remote sensing, and accumulated corpus of cartographic materials from state surveys. The GIS includes not only actual information, but also the results of its comprehensive spatial analysis and modelling, assessments of anthropogenic load on river basins, the results of studying the patterns of water runoff formation depending on the landscape and geographical conditions in the European part of Russia.

The information accumulated in the GIS is located on the geoportal "River basins of the European Russia" with public (open) access, which will enable a wide range of representatives of a scientific community and experts in the field of environmental management and protection to obtain thematic and geographic information on natural and anthropogenic characteristics of river basins, as well as it will serve to meet the information needs of the population.

2. Material and methods

Study area is the European part of Russia, except for mountainous areas (figure 1). Scale studies (level of spatial detail) is regional (corresponding to a scale 1: 1 000 000).

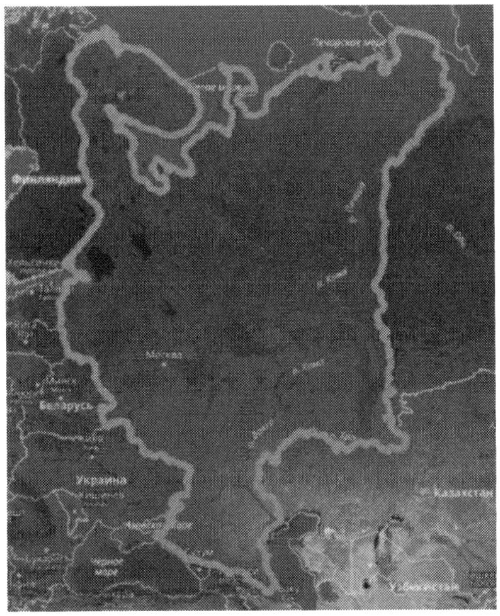

Figure 1. Study area: The European part of Russia.

2.1. Building the layer of river basins – input data and methodology

The basis for making the GIS is an electronic map (vector layer) of river basins in the study area. The river basin acts here as a basic operational and territorial unit for the generalisation of diverse natural and anthropogenic information, as well as for conducting spatial analysis. The basin approach provides, it appears, the most convenient unit of spatial analysis for the Earth's temperate zone, where there is a dense river network, and the main relief-forming role is played by the permanent and temporary streams. An important feature for the use of river basins as territorial units for assessment and analysis is the fact that they are geosystem formation with all inherent properties. Pools also assure the demand for representativeness to the fullest extent. They are easily constructed when the level of generalisation of work changes. A large number of elementary basins covering the study area provide a representative sampling for statistical analysis and modelling.

Currently, there are a number of products obtained through various projects that are models of catchment basins of a particular territorial coverage. Such products differ in their intended purpose, in terms of spatial detail, in the degree of data openness, and etc. Speaking of open products, such projects as HydroSHEDS/HydroBASINS, CCM, and Ecrins [1-3], which cover the territory of the European Russia, should be mentioned first of all. However, the distinctive feature of these global models of basin geosystems is that they have been created for streamflows constructed using digital elevation model (DEM) and are models of catchments of temporary and permanent hydrographic network that is far from the river basins presented on topographic maps. As far as we know, a detailed network of river basins for the entire territory of Russia or at least for large parts of it has not been created yet. Regional geodatabases created by individual researchers or groups of researchers usually within a single constituent entity of the Russian Federation, are characterised by the disunity and lack of a unified methodology for constructing model boundaries. All of this has determined the task of creating our own model of river basins on the territory of the European Russia.

The process of creating a model of river basins (a vector layer of river basins) comprises the following stages [4]:

- initial data selection;
- pre-processing of the relief model;
- construction of pools' boundaries in automatic mode;
- post-processing – automatic and manual correction of pools' boundaries.

As initial data, we have selected DEM GMTED2010 (Global Multi-resolution Terrain Elevation Data 2010) [5] of 250 m spatial resolution and vector map of water objects of the Russian Federation, scale of 1: 1 000 000 (refined DCW). Pre-processing included creation of a single picture from 9 fragments of GMTED2010, correction of DEM using the hydrographic network map [6, 7], elimination of local "swallow holes" that were present in the original model [8, 9]. For the entire study area, river basins' boundaries were allocated in automatic mode according to the algorithm implemented in the Whitebox GAT programme [10].

The vector geometry of the boundaries after the automatic selection with accuracy up to a linear raster cell size was corrected by smoothing the polylines. The topological consistency of the geometry of rivers and basins' boundaries in river mouths (the "convergence" of the basin boundary to the final node of the polyline representing the river) was improved with procedures implemented in ArcGIS and partially (about 5%) in manual way (figure 2).

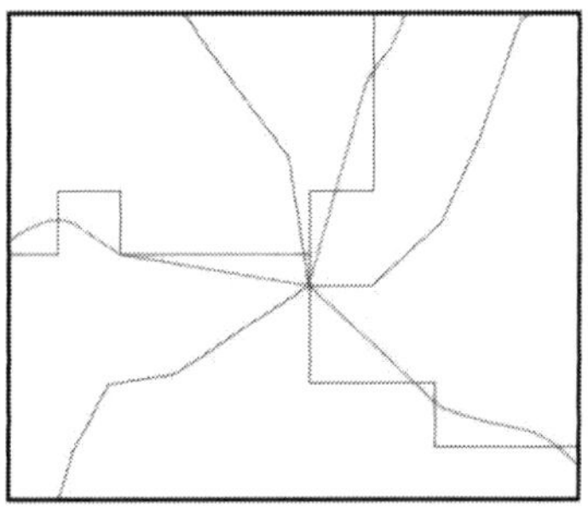

Figure 2. The example of river basins' boundaries before (black lines) and after post-processing (red lines); blue lines are rivers.

The allocation of basins was carried out planarly, that is, not only the basins of small rivers (on the accepted scale they are streamflows of the first order), but also inter-basin (or inter-flow) spaces. In total, layer of river basins comprises 53865 objects.

2.2. Sources of thematic geodata

For the GIS' content, thematic geodata (mainly from open sources), which characterise the natural and resource potential of the territory and correspond to the level of spatial detail of the studies have been processed.

Geomorphometry of the relief. On the basis of the relief model GMTED2010, the derived layers of the morphometric characteristics of study area's relief have been calculated. They are: steepness of slopes, exposure of slopes, erosivity of relief, length of water thread, profile and planned curvature, and order of streamflows. To minimise distortion in the calculation of morphometric indicators, equidistant projection was applied. The steepness (angle of inclination), exposure, curvature, and lengths of water threads were calculated using the ArcGIS software complex, which applies techniques for calculations [11, 12]. The erosivity of the relief and the order of the streamflows were calculated in the WhiteBox GAT software [10]. When constructing the model of water runoff orders, the traditional scheme of the Philosof-Strahler was applied.

Climate. In order to determine the climatic indicators, data of meteorological stations of Roshydromet were used [13]. Sixteen climatic parameters, representing averages, variation, seasonal variation, extreme values of temperature and precipitation, were estimated on the basis of long-term observations of daily air temperature and precipitation at meteorological stations for the period from 1960 to 2014.

Spatial interpolation of these 16 climatic parameters was constructed for the entire territory of the European part of Russia (spatial interpolation method is Multilevel B-spline Approximation).

Land Cover. The source of information on land cover types of the territory was spatial database, gathered based on set of maps of terrestrial ecosystems of Russia *TerraNorte RLC*, elaborated in Space Research Institute (IKI) Russian Academy of Sciences [14]. The main source of the initial information is satellite observation data by spectroradiometer MODIS (MOD09 standard products). The detection method involves multi-annual time-series of spectral reflectance satellite measurements preliminarily cleaned from impact of cloud and other contaminating factors, along with automated land cover types recognition based on a locally-adaptive classification algorithm [15]. The map legend consists of thematic classes, including classes of vegetation cover (forest: evergreen dark needle-leaf, evergreen light needle-leaf, broadleaf, mixed with needle-leaf majority, mixed, mixed with broadleaf majority, deciduous needle-leaf, sparse deciduous needle-leaf; grasslands and shrub lands; tundra; wetlands; recent burns; croplands and non-vegetated areas).

Soils and landscapes. GIS includes coverage of the study area with electronic soil and landscape maps. The electronic soil map of Russia was developed in the Soil Science Institute named after V.V. Dokuchaev and is available as the Unified State Register of Soil Resources of Russia [16, 17]. The polygonal objects of the layer are the digitised contours of the RSFSR soil map at a scale of 1: 2,500,000. Each polygon contains information on the soil cover and soil-forming rock material. The map's explanatory notes include 205 soil varieties, 70 soil complexes, 6 non-soil formations, and 30 variants of soil texture and petrographic composition of soil-forming rock material.

An electronic landscape map of the USSR is available through the link [18] on the website of the International Institute for Applied Systems Analysis (IIASA). This is a digital version of the landscape map of the USSR at a scale of 1: 2,500,000. The section, group, class, type, subtype, sort, sort types, and landscape types are presented as attributes of landscapes. It should be noted that the coordinate system indicated in the map description does not allow us to obtain an exact location (closure error: 15 km). We carried out a refinement of the digital map georeferencing, and selected projection's parameters ensuring its spatial linking with the rest of the geoinformation.

Hydrological geodata. The main source of data on water runoff and runoff of suspended load of rivers in the European Russia is published materials of long-term monitoring observations at hydrological posts – "Surface water resources of the USSR". On the basis of these materials, a database has been created. It presents information on the average annual water discharge and solid discharge for all the years of observation. A part of the data is obtained at the open Global Runoff Data Center (http://www.bafg.de/GRDC/). A part of the information for posts of the Volga-Kama basin has been updated according to data up to 1985 obtained from open sources (http://caspi.ru/), and according to data up to 2013 for posts in the Upper Volga, Kama, and Lower Volga basins, obtained within the framework of work under the contract with FGBU VNIIGMI-MTSD (RIHMI-WDC, http://meteo.ru/data/). The initial information on the localisation of gauging stations is FGBU VNIIGMI-MTSD's open data (http://meteo.ru/data/). To correct the location of posts, information from the volumes of hydrological directories (a name of a river and a settlement, the length of a river from the mouth to the post), layers of topographical map at a scale of 1: 1,000,000, Google Maps, and Yandex.Maps were used. The creation of point objects at gauging stations' layer was carried out in compliance with topological relations with the objects of the hydrographic network map (figure 3).

To study the dependencies of a river runoff on the landscape and geographical conditions of its formation, we have created a vector layer of catchments related to hydrological stations. In order to solve this issue, we used the processed model of the relief GMTED2010 and a layer of point objects representing hydrological stations. The construction of the watersheds' boundaries was performed in an automatic mode using the WhiteBox GAT program [10]. In order to assess the quality of the process of boundary setting, a comparison between the polygon areas of the obtained layer and the values of areas of the gauging stations' basins taken from the hydrological reference books was made.

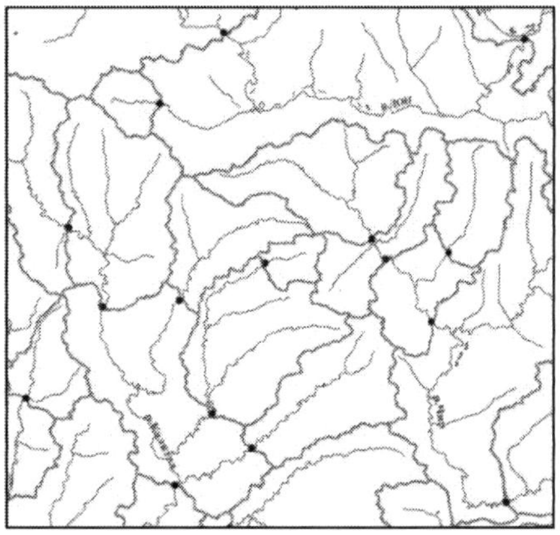

Figure 3. Boundaries of catchments related to hydrological stations: blue lines are rivers, black points are hydrological stations, red lines are boundaries of catchments.

2.3. Generalization of geodata

A generalisation of the thematic information included in the GIS into the operational and territorial units of analysis was done, i.e. river basins and basins related to hydrological stations. For each basin, generalised morphometric characteristics of the relief, average climatic parameters, prevailing soil characteristics, predominant type and subtype of landscape, and percentage of different land cover types were determined. All operations with geodata, processing and analysis (spatial interpolation, generalisation of data pools from thematic raster and vector layers) were performed using programs written in R language (rgdal, raster, MBA packages) [19-24], and with the help of developed SQL-requests in the DBMS PostgreSQL/PostGIS.

3. Results and discussion

3.1. GIS composition

The GIS "River basins of the European Russia" currently includes the following vector and raster cartographic layers: river basins; hydrological stations; catchments related to hydrological stations; digital elevation model; terrain slope; terrain aspect; plan curvature; profile curvature; erosional potential of relief; lengths of the flows; meteorological stations; average annual air temperature; average air temperature in January; average air temperature in July; average maximum of year temperature; average minimum of year temperature; average annual air temperature amplitude; temperature standard deviation for the year; average number of days with temperatures below 8 degrees; frequency of severe frosts (-30 deg. and below); sum of active temperatures; average annual precipitation; average precipitation in May-August; average precipitation for the cold season; average precipitation for the warm season; coefficient of variation of annual precipitation (%); hydrothermal coefficient; bioclimatic layers from WorldClim database; geological map (classes of pre-Quaternary deposits); geomorphological map; soil map; land cover types (vegetation); landscape map; layers of topographic map scale of 1: 1,000,000; boundary of the study area.

Attributes of objects of main layers (in particular layer of river basins (53865 basins) and layer of basins related to hydrological stations (1892 basins) are the following: basin identifier, basin area, predominant soil type in basin, predominant type of soil-forming rock, predominant class of pre-Quaternary deposits, predominant type and subtype of landscape in basin, percentage of woodland, cropland, grassland, shrubland, wetland, lakes in basin, average, minimum and maximum elevation in basin, average slope, average aspect (from "cold" -1 to "warm" 1), erosional potential of relief and flows length in basin, average annual air temperature, average air temperature in January, average air temperature in July, average maximum of year temperature, average minimum of year temperature, average annual air temperature amplitude, mean square temperature deviation for the year, average number of days with temperatures below 8 degrees, frequency of severe frosts, sum of active temperatures, average annual precipitation, average precipitation in May-August, average precipitation

IOP Conf. Series: Earth and Environmental Science **107** (2017) 012108 doi:10.1088/1755-1315/107/1/012108

for the cold season, average precipitation for the warm season, coefficient of variation of annual precipitation, hydrothermal coefficient.

3.2. Spatial analysis based on GIS

The geodata accumulated in the GIS make it possible to study the features of forming and functioning of basin geosystems in the landscape and geographical conditions of the European part of Russia, to conduct thematic mapping of basins with a sufficiently high level of spatial detail, and to solve tasks of comprehensive spatial analysis and modelling. As examples, we present the results of solving two tasks (the solution itself is not discussed in this paper).

Climatic zoning. The method of automated zoning of the study area was developed. It is based on Kohonen Self Organizing Map (SOM) method [25] and uses the authors' software. On the basis of 16 climatic parameters we obtained climatic zoning for the European part of Russia. Basins of small rivers were used as spatial units of analysis. Figure 4 shows the result of climatic zoning and the class ordination in the attribute space. Classes are ordinated by the temperature gradient in the vertical direction on the ordinate plane, and by humidity gradient in the horizontal direction.

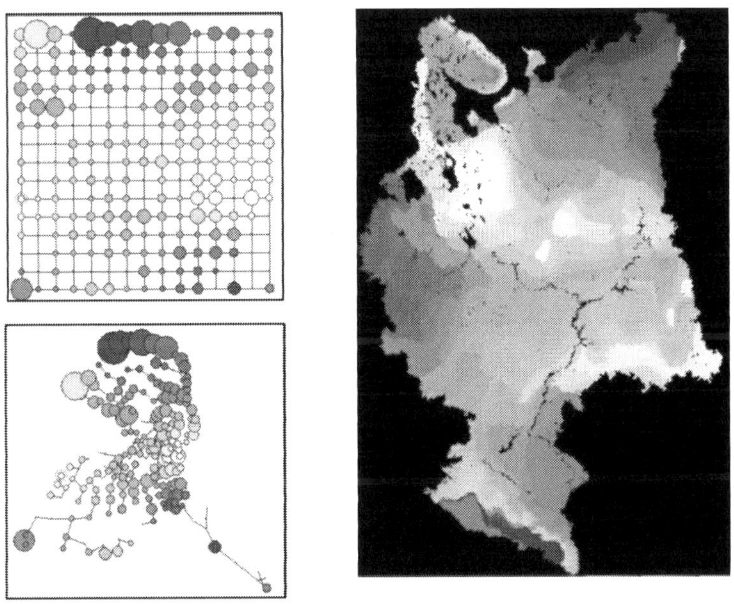

Figure 4. Climatic zoning for the European part of Russia: left are classes in geographical space, right up are neurons (classes) on ordination plane, right bottom is Sammon map of the minimal skeleton of classes (with a thematic palette, the size of the icons is proportional to the number of basins assigned to the corresponding class).

The class of semi-desert dry climate basins (the red color) is clearly distinguished; classes of steppe basins with unstable moisturisation, lower temperature class (the yellow color) and class of steppes with insufficient moisturisation (the bright green color) are close to the previous one in terms of their properties. Then, there are classes of the basins of subboreal and boreal temperate continental climate. They are: forest-steppe with sufficient moisturisation and mixed forests (shades of green, transition to white), and taiga with excessive moisturisation (shades of light-grey and grey-blue). And then, there are basins in the subarctic climate (the blue color). The classes of high mountain climate basins (the purple color) and even the subtropical climate (the pink color) are clearly distinguished. In geographical space (on the map), it can be observed that the obtained classes demonstrate not only the latitudinal (zonal); but also the longitude and sector gradient. In general, the climatic zoning carried out is quite adequate to the published climatic maps of Russia. However, unlike classical climate maps with their clear boundaries between climatic regions or zones, it reflects continual changes in climatic conditions.

Evaluation of water runoff. Dependence of the characteristics of water runoff on the landscape and geographical conditions and anthropogenic stress are studied by statistical methods (GLM, GAM). Statistical models of these relations were constructed. The most significant predictors of specific water discharge model are the following: sum of active temperatures, annual precipitation, average slope, average elevation (for mountain areas), percentage of woodland in basin, percentage of cropland, soil category.

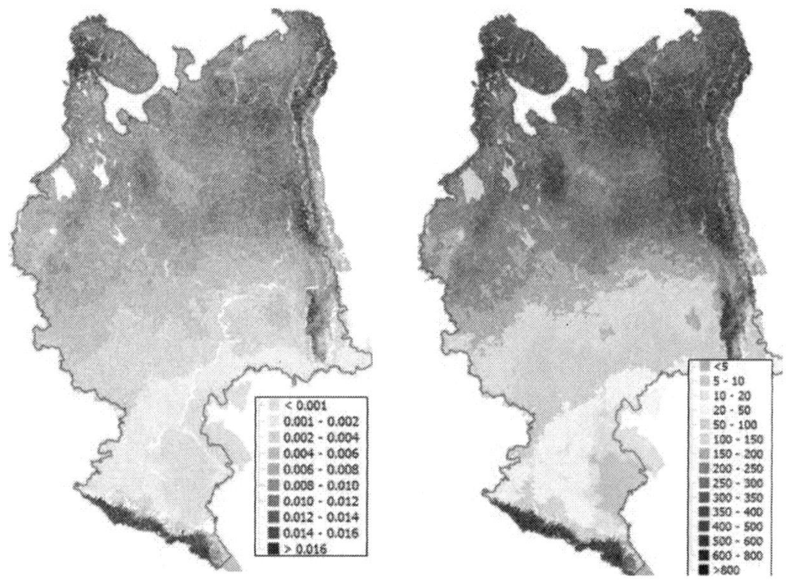

Figure 5. Map of specific water discharge (left) and map of annual water runoff layer (right).

Models explain about 80% of data variability, provide acceptable prediction accuracy and represent the main dependencies between water runoff and its formation conditions at the scale of study.

Spatial extrapolation of the values of water runoff was produced in hydrological unexplored areas of the European Russia on the basis of constructed models. Figure 5 shows the results as a map of specific water discharge and map of annual water runoff layer.

3.3. Geoportal "River basins of the European Russia"

The geoportal "River basins of the European Russia" has been developed and put on the Internet: http://kpfu.ru/ecology/nauchno-issledovatelskaya-rabota/rechnye-bassejny-evropejskoj-rossii; direct link: http://bassepr.kpfu.ru/ .

On the geoportal, with the help of cartographic web services, the major part of the geodata that has been accumulated in the GIS is displayed. It is, first of all, the layer "Pools" that is basins of small rivers and their inter-tributory spaces. On the geoportal, there are also such layers, as "Gauging stations" layer which is hydrological posts of Rosgidromet (Russian Federal Service for Hydrometeorology and Environmental Monitoring) provided with data on average annual water consumption; "Study area" layer that shows the boundary of the studied area, and two layers in the group "Hydrography" that are hydrographic network objects represented on the map M 1:1,000,000. The geoportal enables its visitors to use maps, such as a map of Russia, OpenStreetMaps, Yandex-map, or Google-map, and satellite images (mosaic of synthesised images from Google, Yandex, Roskosmos, ScanEx) as a "background layer" (Figure 6).

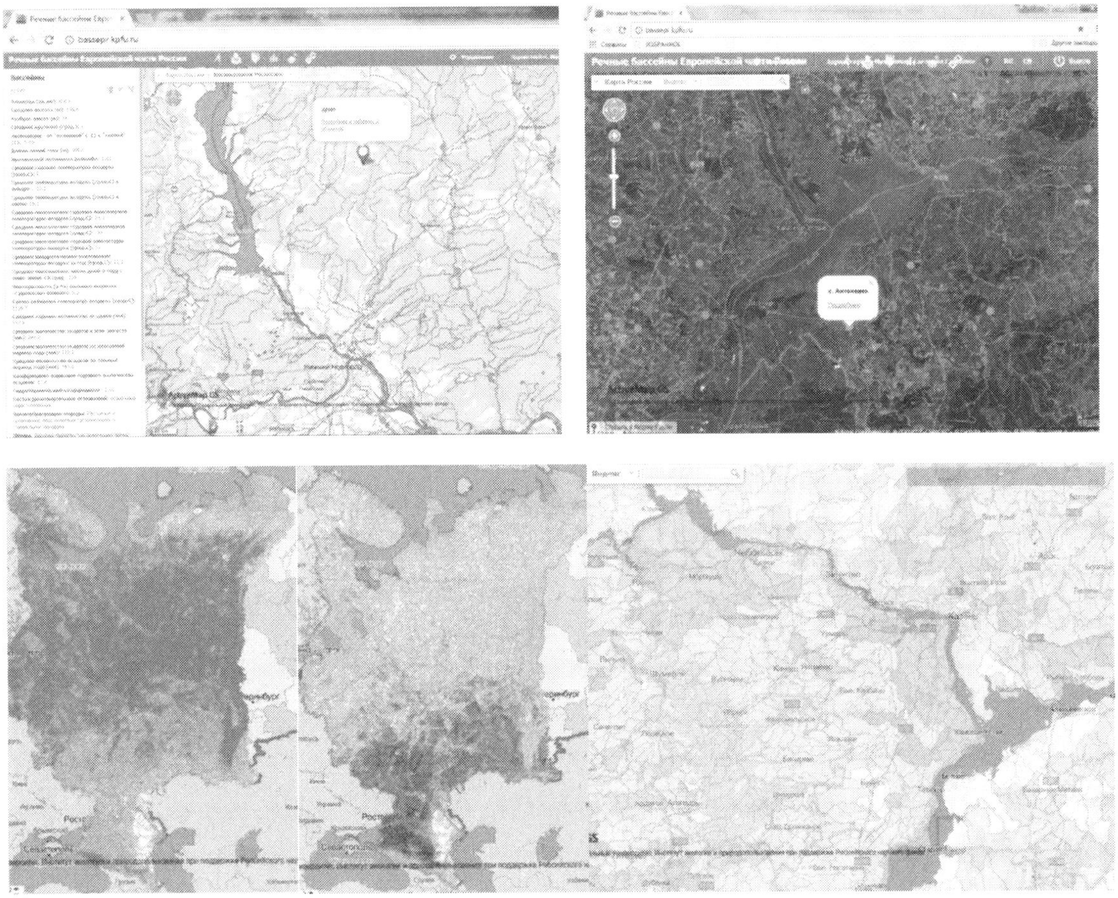

Figure 6. Examples of pages from the geoportal "River basins of the European Russia" on the Internet.

We have provided public (open) access to the developed resource. Any user who visits the geoportal gets the opportunity to familiarise themselves with the map of basins, information on the natural and anthropogenic characteristics of basins, and to save the information they are interested in (in tabular form). One can see the locations of the hydrological observation posts and acquire information on the duration of observations of the water runoff and the flow of suspended load at posts.

We view the developed geoportal as the first version. Expert interviews (via online platform *Testograf*) with the faculty, research scientists (ecologists, geographers, social ecologists, sociologists, etc.), representatives of project and environmental organisations, representatives of the public sector, mass media, and environmental NPOs from different regions of Russia have been conducted to optimise the content of the geoportal. The follow-on revision of the geoportal is carried out to expand the presented geoinformation, improve the functionality, display data on the geoportal in the form of thematic, analytical, and integrated maps, and implement a "download geodata" option.

Acknowledgement
This study was supported by the Russian Science Foundation (project 15-17-10008).

References

[1] Lehner B, Verdin K, Jarvis A 2008 *Eos, Transactions* **89**(10) 93–4
[2] Vogt J et al. 2007 *Report EUR 22920, European Commission – Joint Research Centre* 120p
[3] 2012 *EEA Catchments and Rivers Network System ECRINS v1.1* 111p
[4] Ermolaev O et al. 2017 *Geography and Natural Resources* **38**(2) 131-8
[5] Danielson J, Gesch D 2011 Global multi-resolution terrain elevation data 2010 (GMTED2010) *Open-File Report 2011–1073* (Reston: U.S. Geological Survey) 26p
[6] Mal'tsev K, Yermolayev O 2014 *Geomorfologiya* **1** 45-53 (In Russian)
[7] Yermolaev O, Mal'cev K, Ivanov M 2014 *Geogr. i prirodnye resursy* **3** 32-9 (In Russian)
[8] Pogorelov A, Dumit Zh 2009 *Kuban river basins relief: morphologic analysis* (M.: GEOS) 218p (In Russian)
[9] O'Callaghan J, Mark D 1984 *Comput. Vis. Graph. Image Process.* **28**(3) 323–344
[10] Lindsay J 2014 *Proceedings of the GIS Research* (UK 22nd Annual Conference, The University of Glasgow, UK) 16–18
[11] Burrough P, McDonell R 1998 *Principles of Geographical Information Systems* (Oxford University Press, New York) 190 p
[12] Zeverbergen L, Thorne C 1987 *Earth Surface Processes and Landforms* **12** 47–56
[13] Buligina O, Razuvayev V, Aleksandrova T *Daily Temperature and Precipitation Data for Russia and U.S.S.R. Stations (TTTR)* Patent № 2014620942
[14] Bartalev S et al. 2011 *Current problems in remote sensing of the earth from space* (CPRSES) **8**(4) 285-302 (In Russian)
[15] Bartalev S et al. 2012 *Current problems in remote sensing of the earth from space* (CPRSES) **9**(2) 9-27 (In Russian)
[16] 2014 *Edinyy gosudarstvennyy reestr pochvennyh resursov Rossii* [Unified state register of soil resources of Russia] V. 1.0 Collective monography (M.: V.V. Dokuchaev Soil Inst.) 768 p (In Russian)
[17] Rukhovich D et al. 2013 *Pochvovedenie* **3** 251-67 (In Russian)
[18] *Forests and forestry in Russia. Data and analysis. Landscapes* (http://webarchive.iiasa.ac.at/Research/FOR/forest_cdrom/russian/for_cond_ru.html#landscapes) (Accessed 18.05.2017) (In Russian)
[19] Lee S, Wolberg G, Shin S 1997 *IEEE Transactions on Visualization and Computer Graphics* **3**(3) 229-44
[20] Finley A, Banerjee S 2014 *MBA: Multilevel B-spline Approximation* (R package version 0.0-8)
[21] Wood S 2003 *Journal of the Royal Statistical Society* **65**(1) 95-114
[22] Hijmans R 2014 *raster: Geographic data analysis and modeling* (R package version 2.2-31)
[23] Bivand R, Keitt T, Rowlingson B 2014 *rgdal: Bindings for the Geospatial Data Abstraction Library* (R package version 0.8-16)
[24] R Core Team 2014 R: *A language and environment for statistical computing* (R Foundation for Statistical Computing, Vienna, Austria)
[25] Kohonen T 1997 *Self-organization and Associative Memory* (New York: Springer-Verlag) 428 p

IOP Conf. Series: Earth and Environmental Science **107** (2017) 012109 doi:10.1088/1755-1315/107/1/012109

Calculation of concentration fields of high-inertia aerosol particles in the flow past a cylindrical fibre

T S Zaripov[1], A K Gilfanov[1], S K Zaripov[1], O D Rybdylova[2], S S Sazhin[2]

[1]Kazan Federal University, Kremlevskaya str., 18, Kazan, 420008, Russia
[2]School of Computing, Engineering and Mathematics, University of Brighton, Cockcroft Building, Lewes Road, Brighton BN2 4GJ, United Kingdom

E-mail: zaript@gmail.com

Abstract. The behaviour of high-inertia aerosol particles' concentration fields in stationary gas suspension flows around a cylinder is investigated using a numerical solution to the Navier-Stokes equations and the fully Lagrangian approach for four Stokes numbers (Stk = 0.1, 1, 4, 10) and three Reynolds numbers (Re = 1, 10, 100). It has been shown that the points of maximum particle concentration along each trajectory shift downstream both when Stk and/or Re increase.

1. Introduction

The calculation of concentrations of high-inertia aerosol particles in complex gas-suspension flows is an integral part of the modelling of various engineering processes and natural phenomena, including the processes in internal combustion engines and the transfer of anthropogenic air suspensions (aerosols). The spatial distribution of the concentration of particles is required to determine the intensity of heat and mass transfer between the carrier and dispersed phases and the rates of chemical reactions.

There are two groups of methods for calculating the concentration of suspended aerosol particles: Eulerian-Eulerian and Eulerian-Lagrangian. The motion of the carrier phase in both groups is described by the continuity and Navier-Stokes equations written and solved in Eulerian coordinates. Within the first group of methods, the equations of continuity and motion of a dispersed phase are also solved in Eulerian coordinates [1, 2]. Particular interest has been focused on the method of moments [3], based on finding the moments of the distribution function of particles in several phase variables (dimensions, velocities, temperature, composition). Among the disadvantages of the Eulerian-Eulerian methods are the high computational cost and the need to specify the boundary conditions for the dispersed phase.

Within the framework of the second group of methods, the equations of motion of the dispersed phase are solved for individual particles, and trajectories of particle motion are found. The concentration of the dispersed phase in the Eulerian cell is found by counting the number of particles in it at every time step. It was shown in [4] that in order to obtain reliable results each cell should contain about 10,000 particles even in the two-dimensional case. If the mesh is very fine, then the computational cost of the problem becomes very high.

In [5, 6] alternative methods were suggested, based on the solution to the continuity equation of a dispersed phase along fixed trajectories. These are called fully Lagrangian approaches, and are much cheaper computationally, as the particle concentration is calculated based on the mass conservation equation. In [5], the continuity equation of a dispersed phase along the trajectory is solved using the Eulerian formulation. It was shown in [4] that this approach could not be used in the points/regions with infinitely high particle concentration, which are formed in flows with crossing particle trajectories.

In [6], the equation of continuity for a dispersed phase along a trajectory is written in Lagrangian coordinates, and the problem of finding the concentration is reduced to that of finding the components of the Jacobi matrix of the Eulerian-Lagrangian coordinate transformation. An important advantage of this method is that it can be applied to cases with singularities in particle concentration fields, where the number density becomes infinitely high.

In our previous paper [7] we presented the results of implementation of the fully Lagrangian approach (FLA), proposed in [6], into the ANSYS Fluent CFD package. The results of the investigation of the distribution of the low inertia particle concentrations in a stationary gas-suspension flow around a cylindrical fibre were presented and compared with those of a variant of the Eulerian-Eulerian approach. In this paper our numerical study is focused on the distribution of concentration of high inertia particles.

2. Mathematical model

Let us consider an isothermal, dilute gas-suspension flow around an infinitely long circular cylinder. As a carrier phase, we take the laminar flow of a viscous incompressible fluid (gas), described by the continuity and Navier-Stokes equations. We assume that the volume and mass fractions of the particles are small, which allows us to ignore the interaction between the particles and their influence on the carrier phase. The particles are assumed to be spherical, and the equation of particle motion is written in the framework of the Stokes drag law. The forces of buoyancy and the added masses are ignored remembering that the density of the particle material is much higher than the density of the carrier phase. Also, we assume that the particles are sufficiently large to allow us to ignore the contributions of the Brownian forces. The equations of motion of inertial particles in this case are written as

$$\begin{gathered} \frac{d\boldsymbol{x}}{dt} = \boldsymbol{v}, \\ \frac{d\boldsymbol{v}}{dt} = \frac{1}{\mathrm{Stk}}(\boldsymbol{u} - \boldsymbol{v}), \end{gathered} \tag{1}$$

where $\boldsymbol{x}$ and $\boldsymbol{v}$ are the position and velocity vectors of a particle in the Eulerian coordinates, $\boldsymbol{u}$ is the velocity vector of the carrier phase. The Stokes number Stk is determined as

$$\mathrm{Stk} = \frac{\rho_p d_p^2 U_0}{18\mu L},$$

where d_p are ρ_p are the diameter and density of the particle material, respectively, μ is the viscosity of the carrier phase, U_0 and L are the characteristic velocity and length.

The equation of continuity for a dispersed phase along a chosen trajectory is written in the form

$$n_p = \frac{n_0}{|J|},$$

where n_0 is the concentration of particles at the starting point of the trajectory. To find the components of the Jacobi matrix J, it is necessary to solve the following system of ordinary differential equations along the chosen trajectory [6]

$$\frac{\partial J_{ij}}{\partial t} = \frac{\partial v_i}{\partial x_{i0}} = \omega_{ij}, \tag{2}$$

$$\frac{\partial \omega_{ij}}{\partial t} = \frac{1}{\mathrm{Stk}}\left(\sum_k \left(J_{kj}\frac{\partial u_i}{\partial x_k}\right) - \omega_{ij}\right).$$

The continuity and Navier-Stokes equations for the carrier phase are solved by ANSYS Fluent using the finite volume method. The particle trajectories are calculated by solving Equations (1) using the Runge-Kutta method. Equations (2) for the components of the Jacobi matrix of the Eulerian-Lagrangian coordinate transformation are solved along the trajectories using the Runge-Kutta method.

The geometry of the computational domain for a cylinder of radius d_c is shown in figure 1. The flow of the carrier phase far from the cylinder was assumed to be uniform. The velocity of the carrier phase $u = U_0$ was set at the left, upper and lower boundaries of the domain. On the right boundary, gauge pressure drop was set to zero.

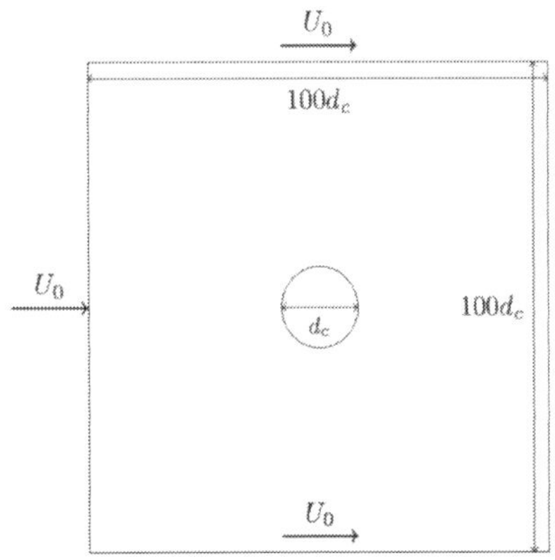

Figure 1. Geometry of the computational domain.

3. Results

Particle concentrations were calculated along 100 trajectories starting at the left boundary of the domain for four Stokes numbers (Stk = 0.1, 1, 4, 10) and three Reynolds numbers of the flow (Re = 1, 10, 100). In figure 2, individual points of the trajectories are shown for four values of Stk at a fixed Re = 10. Points are coloured in shades of grey depending on the concentration of particles: lighter shades correspond to lower concentrations, darker shades correspond to higher concentrations. It can be seen that the region of high concentrations shifts downstream as the Stk increases.

In Figure 3, the curves connecting the points of greatest concentration along each trajectory are shown. It can be seen that these curves shift to the right both with increasing Stk and with increasing Re. The maximum concentration along the trajectories of the deposited particles is always achieved at the point of deposition. Points of the global maximum of the concentration for each Stk are marked by circles. For high-inertia particles (Stk > 1) these points move downstream when Stk and/or Re increase. The global maximum of the concentration of low-inertia particles (Stk < 0.2) is located at the deceleration (stagnation) point in front of the cylinder for all three Re.

IOP Conf. Series: Earth and Environmental Science **107** (2017) 012109 doi:10.1088/1755-1315/107/1/012109

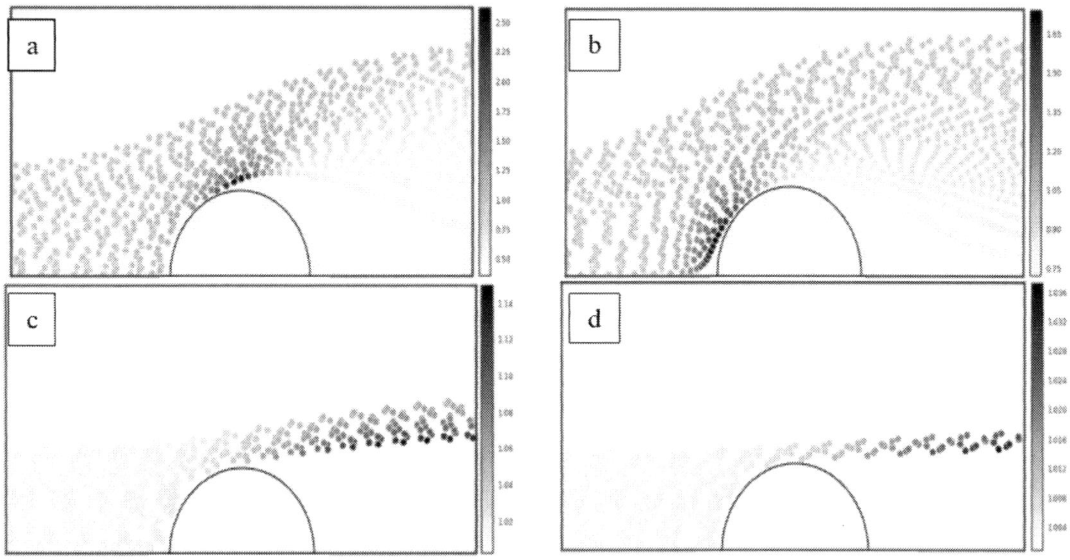

Figure 2. Trajectories and the distribution of particle concentrations at Re = 10 for four Stokes numbers (Stk = 0.1 (a), Stk = 1 (b), Stk = 4 (c), Stk = 10 (d)). The points of the trajectories are coloured depending on the concentration of particles.

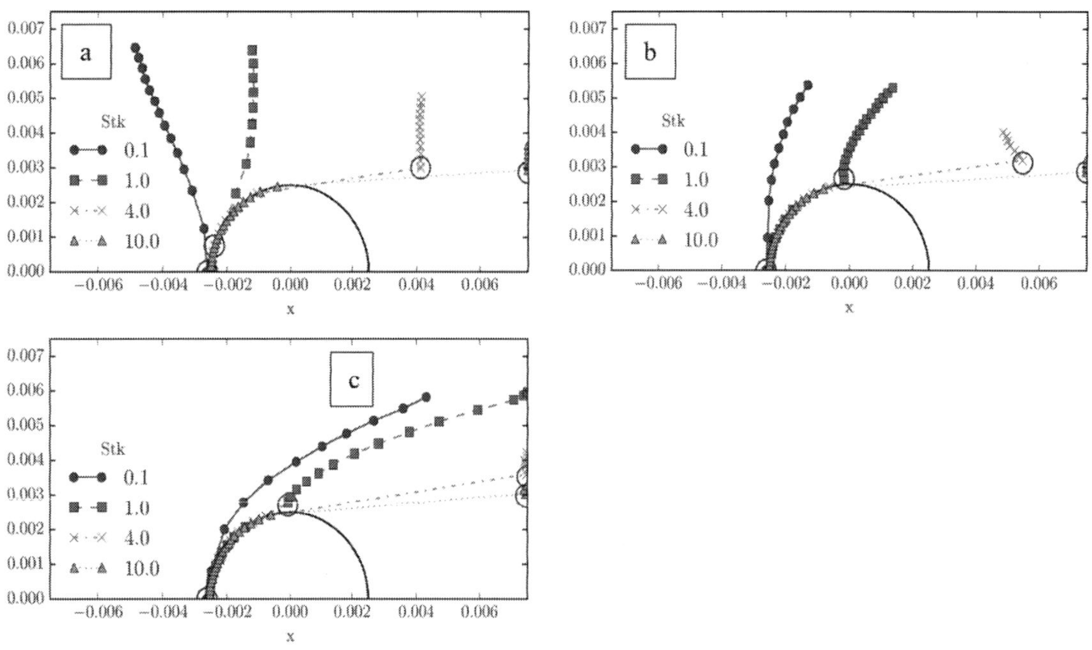

Figure 3. Curves of the greatest concentration along each trajectory for Re = 1 (a), Re = 10 (b), and Re = 100 (c).

IOP Conf. Series: Earth and Environmental Science **107** (2017) 012109 doi:10.1088/1755-1315/107/1/012109

4. Conclusions

The behaviour of the particle concentration in steady-state gas-suspension flows around a cylinder has been investigated for four Stokes numbers (Stk = 0.1, 1, 4, 10) and three Reynolds numbers (Re = 1, 10, 100). It has been shown that the points of the maximum concentration along each trajectory shift downstream when Stk and/or Re increase.

Acknowledgements

This work was supported by RFBR grants (16-51-10024 and 15-01-06135) and the Royal Society (UK) (grant IE 160014).

References

[1] Crowe C T 1982 *Trans. ASME J. Fluids Eng.* **104** 297–303
[2] Ferry J, Balachandar S 2001 *International Journal of Multiphase Flow* **27** 1199–226
[3] Desjardins O, Fox R O, Villedieu P 2008 *Journal of Computational Physics* **227** 2514–39
[4] Healy D P, Young J B 2005 *Proc. R. Soc.* A. **461** 2197–225
[5] Fernandez de la Mora J, Rosner D E 1981 *Physicochem. Hydrodyn.* **2** 1–21
[6] Osiptsov A N 2000 *Astrophysics and Space Science* **274** 377–86
[7] Zaripov T S, Gilfanov A K, Begg S M, Rybdylova O D, Sazhin S S, Heikal M 2017 *Atomization and Sprays* **27** 493–510

Groundwater movement simulation by the software package PM5 for the Sviyaga river adjoining territory in the Republic of Tatarstan

E A Kosterina, Z Sh Isagadzhieva

Institute of Environmental Sciences, Kazan Federal University, 18 Kremlyovskaya str, Kazan 420008, Russia
E-mail: Ekaterina.Kosterina@kpfu.ru

Abstract. Data of the ecological-hydrogeological fieldwork at the Predvolzhye region of the Republic of Tatarstan were analyzed. A geofiltration model of the Buinsk region area near the village of Stary Studenets in the territory of the Republic of Tatarstan was constructed by the PM5 software package. The model can be developed to become the basis for estimation of the groundwater reserves of the territory, modeling the operation of water intake wells, designing the location of water intake wells, and evaluation of their operational capabilities, and constructing sanitary protection zones.

1. Introduction

Mathematical modeling of groundwater filtration is used in calculation of operational water resources, estimation of geofiltration parameters, development of the hydrodynamic model of the operational aquifer, in calculation and justification the boundaries of sanitary protection zones, and also in creation of the forecast of the influence of landscapes on groundwater formation and vice versa [1].

PM5 is a software package, designed for modeling the non-stationary spatial geofiltration of groundwater [2]. Its main component is the ModFlow program for constructing hydrodynamic models of various complexity levels, developed in the US Geological Survey. The PM5 software package and its components are used all over the world to model the reserves and movement of groundwater, as well as the migration of contaminants in addressing the depletion of fresh water supplies, the organization of irrigation in agriculture, and the protection of freshwater from pollution [2-9].

The main problem in the modeling of groundwater filtration is the insufficient hydrogeological study of the territories. This problem is observed both in global models and in regional scale models. The lack of data explains the infrequent analysis of geofiltration processes in hydrogeological models [1].

The purpose of this paper is to analyze and process the ecological-hydrogeological fieldwork data of the Buinsk district near the village of Stary Studenets on the territory of the Republic of Tatarstan and to construct a geofiltration model of the selected territory using the PM5 software package.

2. The problem statement

The territory under the study is located between the rivers: the Karla River in the north, the Sviyaga River in the east and the Chilcha River in the south. The western boundary of the territory is the watershed in the relief. The area under the study is approximately 323 km^2.

Orographically, the selected plot of land is located in the northern part of the Volga Upland. The region has a calm, smooth relief and represents an almost flat lowland ("Buinsky steppes"), dissected by shallow but wide valleys of the left tributaries of the Sviyaga River.

The territory of investigation is located in the zone of a complex geological structure. There is a great variety and variability in the structural-tectonic plan, as well as in the lithologic-facies composition of the water-bearing layers. Underground waters occur in deposits of a wide stratigraphic range. When considering the hydrogeological formation, it is possible to single out a multi-layer system in which alternating permeable and weakly permeable layers are observed. In general, the rock strata in the study area consist of Neogene, Upper Jurassic, Upper Tatar sediments, the Kazan series, Lower Permian (Asselian) sediments. For water supply, the underground waters of the Urzhum and Kazan deposits are mostly often used. The peculiarity of the geological structure of the selected territory is that about half of the area is occupied by the old channel of the Sviyaga River. It is very deep and entirely composed of very good water-conducting rocks. In the rest (western) part of the modeling area, on the contrary, poorly permeable and less abundant rocks with thin (about 1 m thick) aquiferous layers approach the earth's surface.

3. Material and methods

In the selected territory there are 67 engineering-geological and hydro-geological wells and 4 water-intake wells.

The schematization of the hydrogeological conditions of the territory under investigation was carried out mainly on the basis of an analysis of the fund materials for the project "Conducting an ecological-hydrogeological survey of the scale 1:200000 sheets N-38-VI, XII, XVIII, N-39-VII, VIII Predvolzhye, XIII.", Provided by LLC "Scientific and Production Center for Ecological, Geological and Geodetic Research" (Kazan).

To build a numerical geofiltration model, the PM5 software package was implemented. The package is designed to solve stationary and non-stationary problems in conditions of non-pressure, pressure-non-pressure and pressure filtration modes. The software package is represented by the main module and connected to additional tools that can simulate various objects – rivers, wells, drains; boundary conditions of any kind; infiltration recharge and evaporation.

PM5 is a free program created on the basis of ModFlow 2000 [2]. PM5 also includes other components. The model of the geofiltration realized by the ModFlow program in the PM5 structure is based on the equation of the mass balance of the fluid in terms of heads [10, 11]:

$$\frac{\partial}{\partial x}k_x\frac{\partial H}{\partial x}+\frac{\partial}{\partial y}k_y\frac{\partial H}{\partial y}+\frac{\partial}{\partial z}k_z\frac{\partial H}{\partial z}=\eta\frac{\partial H}{\partial t}-\varepsilon,$$

where k_x, k_y, k_z are the hydraulic conductivity coefficients in the directions of the corresponding coordinate axes, ε is the density of external sources; η is the specific storage coefficient; H means head.

The PM5 package contains a finite-difference approximation of the differential problem that bases on the equation mentioned above. The calculation is carried out by an iterative method.

To construct the model of the selected area of the Republic of Tatarstan, a rectangular fragment was selected. It is 27.8 km long from west to east, and 24.3 km from north to south. The simulated area was covered with a regular rectangular grid with the cell size of 100 m x 100 m. The calculation domain was identified by setting approximate boundaries of the territory using the topographical base (figure 1). The position of the wells in the calculation area was marked according to their coordinates.

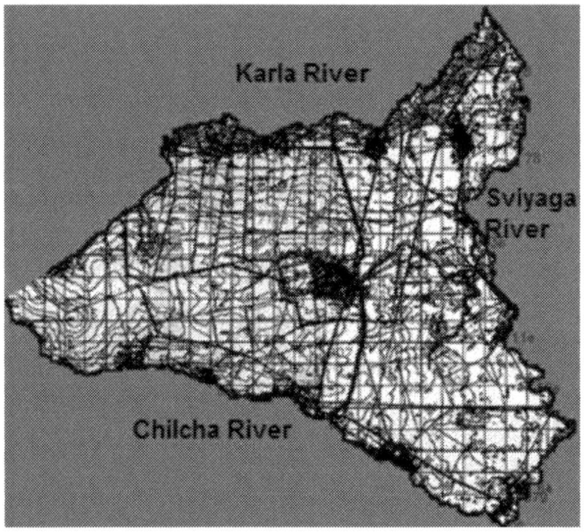

Figure 1. Calculation area in the PM5 package.

Aquifers and water-resistant horizons were combined into three layers of the aquifer complex. For the entire study area, the markings of the top and the bottom of each computational layer were obtained by interpolation with the assignment of interpolation nodes along the contour isolines and the water level values in the rivers, and as well as according to the register of wells from the fund materials (figures 2-5).

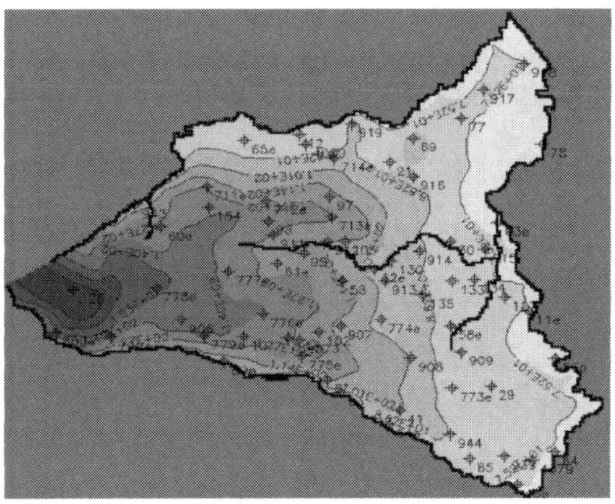

Figure 2. Top of the 1st layer.

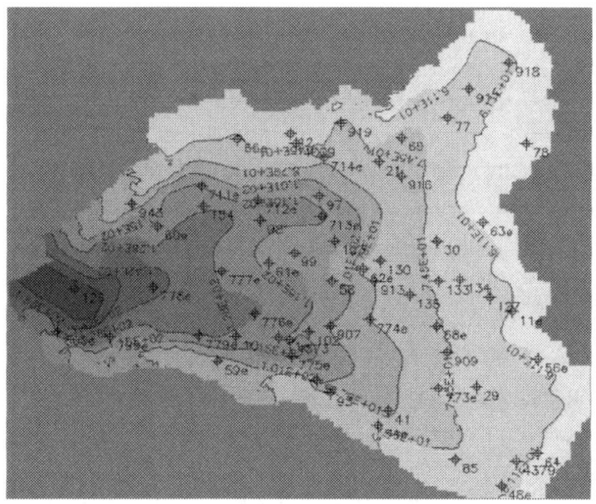

Figure 3. Bottom of the 1st layer (Top of the 2nd layer).

Values of the head were set in the calculation cells in concordance with the water level in the rivers. Head values vary within the following limits: the Sviyaga River – 65.3-70.4 m, the Karla River – 65.3-167 m, the Chilcha River – 70.4-173 m. The water cuts were taken from the topographic base. The values of the water level in each calculation cell were determined between the marks of water cuts by the method of linear interpolation.

The values of the hydraulic conductivity coefficients were determined on the basis of expert estimates, depending on the composition of the aquifer. The porous medium was assumed to be inhomogeneous, isotropic.

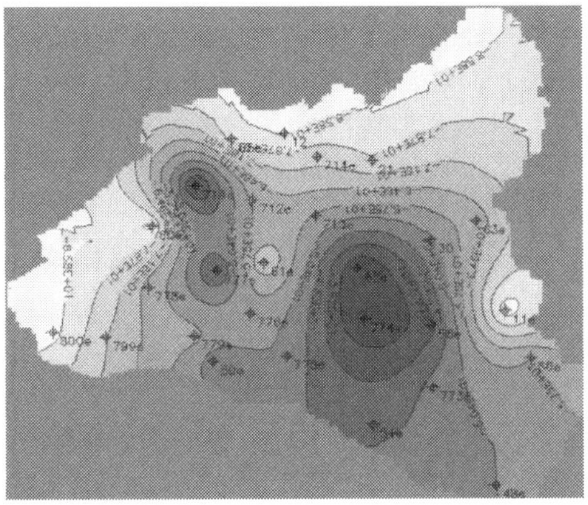

Figure 4. Bottom of the 2nd layer (Top of the 3rd layer).

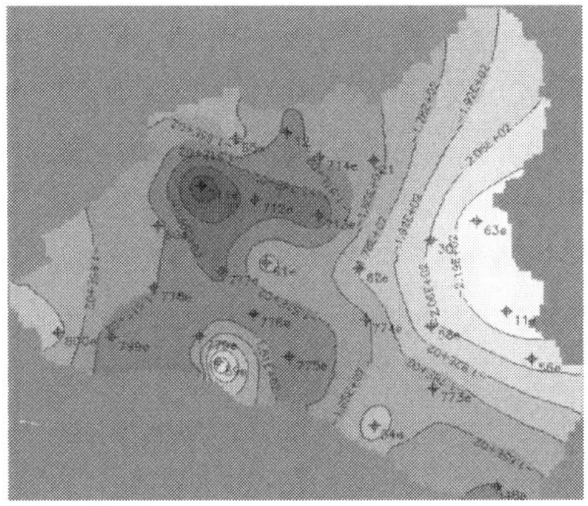

Figure 5. Bottom of the 3rd layer.

The data on mean total annual precipitation were taken at the Geoportal "River basins of European Russia" [12]. Precipitation is 454 mm/year. Infiltration recharge is 20% of this value for well-permeable rocks: 1.86 10^{-4} m/day.

4. Results and discussion

To restore the natural regime on the interfluve, a stationary problem was solved. After specifying all the necessary parameters, calculations were made by the conjugate gradient method. The results were obtained as maps of hydroisogypsum and piezoisogypsum (figures 6-8). The head values do not exceed the absolute relief marks at each point of the simulated area, which is the first sign of the correctness of the solution obtained. Higher head values correspond to higher elevation marks.

A geofiltration model of the Buinsk region area near the village of Stary Studenets in the territory of the Republic of Tatarstan was built. The model can be developed to become the basis for estimation of the groundwater reserves of the territory, modeling the operation of water intake wells, designing the location of water intake wells, estimation of their operational capabilities, and constructing sanitary protection zones.

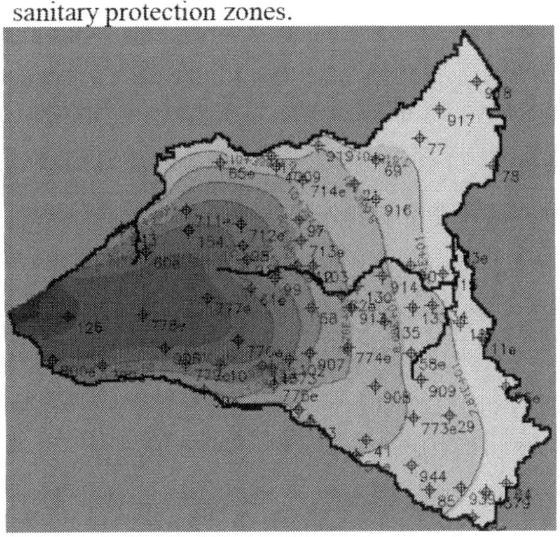

Figure 6. Hydroisogyps map for the first computational layer.

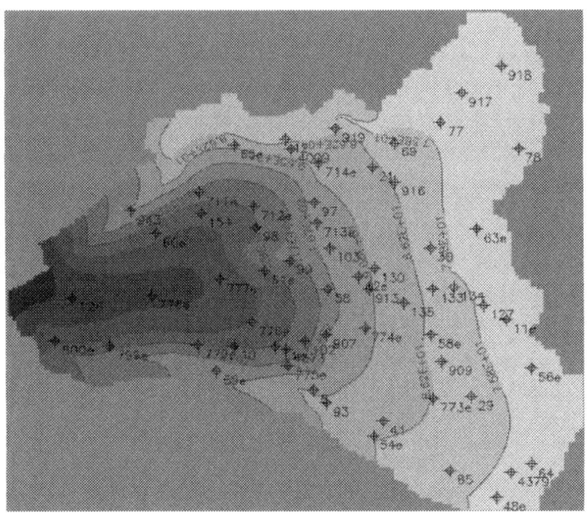

Figure 7. The piezoisogyps map for the second computational layer.

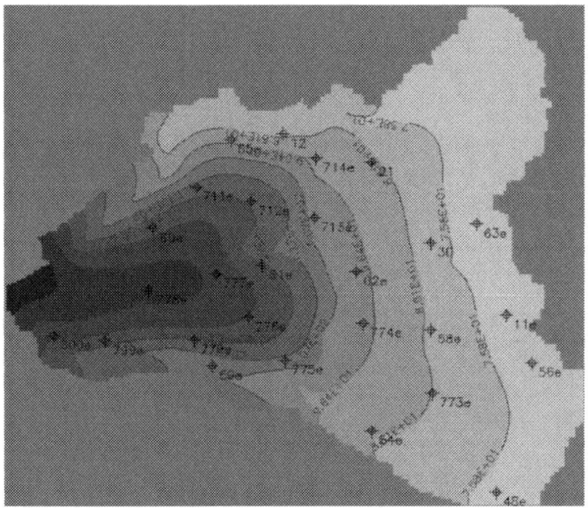

Figure 8. The piezoisogyps map for the third computational layer.

References

[1] Gensel G N, Karachevtsev N F, Konosavsky P K, Kravchuk S V, Mironenko V A, Rumynin V G 1992 *Solution the Problems of Groundwater Protection in Numerical Models* (Moscow: Nedra) p 240 [in Russian]

[2] *Processing Modflow for Windows* URL: http://www.pmwin.net/pmwin5.htm

[3] Chiang W-H, Kinzelbach W 1998 *Processing MODFLOW – a Simulation System for Modelling Groundwater Flow and Pollution* URL: https://www.researchgate.net/publication/238774706_Processing_MODFLOW-a_Simulation_System_for_Modelling_Groundwater_Flow_and_Pollution

[4] Ghodoosipour B 2013 *Three Dimensional Groundwater Modeling in Laxemar-Simepevarp Guaternary Deposits* TRITA-LWR Degree Project 13:36 p 41

[5] de Graaf I E M, Sutanudjaja E H, van Beek L P H, Bierkens M F P 2015 *Hydrol. Earth Syst. Sci.* **19** 823-37 URL: https://www.hydrol-earth-syst-sci.net/19/823/2015/hess-19-823-2015.pdf

[6] Markstrom S L, Niswonger R G, Regan R S, Prudic D E, Barlow P M 2008 *GSFLOW– Coupled Ground-Water and Surface-Water Flow Model Based on the Integration of the Precipitation-Runoff Modeling System (PRMS) and the Modular Ground-Water Flow Model (MODFLOW-2005)* URL: https://pubs.usgs.gov/tm/tm6d1/pdf/tm6d1.pdf

[7] Abu-El-Sha'r W Y, Hatamleh R I 2007 *Jordan Journal of Civil Engineering* **1(2)** 153-72

[8] Alvarez M d P, Trovatto M M, Hernández M A, González N 2012 *Environmental Earth Sciences* **66(7)** 2097-108

[9] Vallner L, Porman A 2016 Hydrology Research **47(4)** 814-34

[10] Bear J, Zaslavsky D, Irmay S 1968 *Physical Principles of Water Percolation and Seepage* (Paris: Unesco) p 465

[11] Mironenko V A, Rumynin V G 1998 *Theoretical Study and Modeling of Geomigration Processes* (Moscow: Moscow State Mining University) p 611 [in Russian]

[12] Ermolaev O P, Mal'tsev K A, Ivanov M A 2014 *Geography and Natural Resources* **35(3)** 222-28

Comparison of two methods of numerical tracking of the soil contamination dynamics during a leak from a pipeline

E A Kosterina

Institute of Environmental Sciences, Kazan Federal University, 18 Kremlyovskaya str, Kazan 420008, Russia
E-mail: Ekaterina.Kosterina@kpfu.ru

Abstract. The situation of leakage of a polluting liquid from a longitudinal crack of the pipeline lying on the ground surface is considered. The two-dimensional nonstationary mathematical model is based on the mass balance equation in terms of pressure, which is satisfied in a domain with an unknown moving boundary. This area corresponds to the area of contaminated zone. A function characterizing the region of action of the equation is introduced, which makes it possible to obtain the formulation of the problem in a fixed domain. Two types of finite-difference approximation of the problem statement are proposed. They differ by approximation of the convective term. Counter-current approximation and approximation along characteristics are used. The results of computational experiments, which are in favor of using the method of characteristics, are presented. The methods application is illustrated by an example of spread of oil pollution.

1. Introduction

Man-caused pollution of soil and water resources, including groundwater, is constantly in the spotlight all over the world. For instance, various aspects of occurrence and propagation of pollution of soil and water resources in various regions of the world (e.g., Europe, China, India, Brazil, Africa) are considered in the works [1-7]. Methods of modeling and numerical simulation of the contaminant transport are studied, for example, in [8-10]. Special attention is paid to modeling of the consequences of catastrophic failures of oil pipelines [3].

This work is devoted to modeling the spread of pollution into the ground from its surface caused by an oil pipeline crack.

The problem under consideration is characterized by the fact that the equation is written in a domain which boundary is mobile and must also be found simultaneously with the solution of the equation inside the domain with this boundary. Such tasks can not be solved by any known software package. The construction of methods for solving free-boundary problems and their investigation is constantly in development [11-23].

2. The problem statement

The situation is considered when a longitudinal crack is formed in the pipe lying on the ground surface and leakage of the polluting liquid occurs through it. As a result, a stain of contamination is formed in the soil (figure 1).

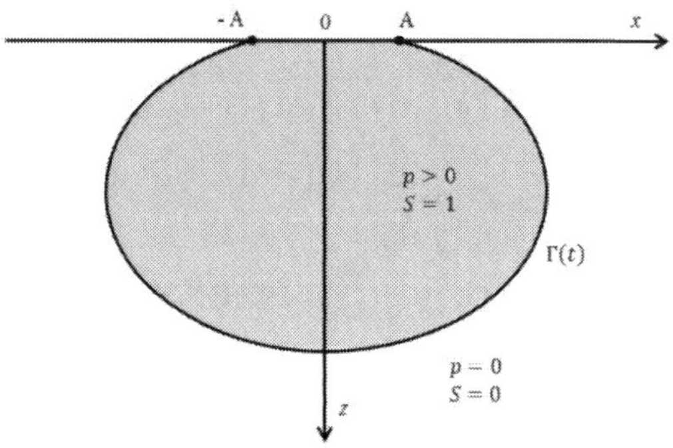

Figure 1. Model situation scheme.

The segment $[-A, A]$ corresponds to the crack in the pipe through which liquid flows. It is assumed that the soil is saturated inside the stain. Outside the contaminated zone pressure and saturation are equal to zero. The boundary of the contaminated zone $\Gamma(t)$ changes its form with time.

Since the contamination zone is symmetrical, one-half of this zone is considered and referred to as Ω.

The mathematical model has the form:

$$\begin{cases} div\,\bar{v} = 0, & (x,z) \in \Omega, t > 0; \\ p = p_0 & x \in [0, A], z = 0, t > 0; \\ \begin{cases} p = 0 \\ u = \frac{v_n}{m} \end{cases} & (x,z) \in \Gamma(t), t > 0; \\ v_x = 0 & x = 0, t > 0; \\ p \geq 0 & (x,z) \in \Omega, t > 0; \\ p = 0 & (x,z) \in \Omega, t = 0, \end{cases} \tag{1}$$

where m is the porosity of the soil, p is the fluid pressure, v is the filtration velocity vector according to the Darcy's law [8]:

$$v_x(x,z,t) = -\frac{k_1^0\,a(z)}{\mu}\frac{\partial p(x,z,t)}{\partial x}, \tag{2}$$

$$v_z(x,z,t) = -\frac{k_2^0\,b(z)}{\mu}\left(\frac{\partial p(x,z,t)}{\partial z} - \rho g\right). \tag{3}$$

Here, k_1^0 and k_2^0 are hydraulic conductivity for x and z directions. To describe the compaction of soil with depth, the model dependence $a(z)$ is introduced in the direction of the x axis, and $b(z)$ is in the z direction. Constants μ, ρ, g mean viscosity, fluid density and acceleration of gravity respectively.

In order to formulate the problem by dimensionless quantities, new variables $\tilde{x} = x \cdot \bar{x}^{-1}$, $\tilde{z} = z \cdot \bar{z}^{-1}$, $\tilde{t} = t \cdot \bar{t}^{-1}$, $\tilde{v} = v \cdot \bar{v}^{-1}$, $\tilde{p} = p \cdot \bar{p}^{-1}$ are introduced, where $\bar{x}$, $\bar{z}$, $\bar{t}$, $\bar{v}$, $\bar{p}$ are the scale of the problem for x, z, t, v, p respectively.

Neglecting the tilde sign above dimensionless variables, a dimensionless formulation of the problem is obtained:

$$\begin{cases} m\frac{\partial S}{\partial t} - k\frac{\partial}{\partial x}\left(a(z)\frac{\partial p}{\partial x}\right) - \frac{\partial}{\partial z}\left(b(z)\frac{\partial p}{\partial z}\right) + \frac{\mathrm{d}b(z)}{\mathrm{d}z} = 0 & (x,z,t) \in \Omega, t > 0; \\ p = p_0 & x \in [0, A], z = 0, t > 0; \\ \begin{cases} p = 0 \\ u = \frac{v_n}{m} \end{cases} & (x,z) \in \Gamma(t), t > 0; \\ v_x = -ka(z)\frac{\partial p}{\partial x} = 0 & x = 0, t > 0; \\ p \geq 0 & (x,z,t) \in \bar{\Omega}, t > 0; \\ S = 0, p = 0 & (x,z,t) \in \Omega, t = 0, \end{cases} \tag{4}$$

where $k = k_1 \cdot k_2^{-1}$, $k_1 = k_1^0\, \bar{t}\,\bar{p} \cdot (\mu\, \bar{x}^2)^{-1}$, $k_2 = k_2^0\, \bar{t}\,\bar{p} \cdot (\mu\, \bar{x}^2)^{-1}$ and $S = 1$ is the saturation of the soil.

3. Material and methods

To formulate the problem in a fixed domain, the Ω is enclosed into the rectangle $Q = [0, L_x] \times [0, L_z]$, the size of which is sufficient to ensure that the contamination never reaches the boundaries of the rectangle (figure 2). The function $\xi(x, z, t)$ is introduced for the fixed domain Q. It is the Heaviside function [24].

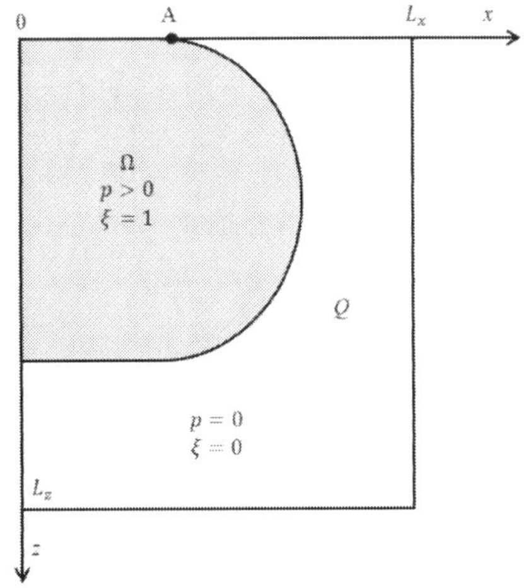

Figure 2. The fixed domain Q.

For the contaminated zone $\xi(x, z, t) = 1$, for the rest of the domain $\xi(x, z, t) = 0$. Further, we assume that the function $\xi\big(p(x, z, t)\big)$ is multi-valued: $\xi \in [0,1]$ on the boundary of the contaminated zone.

Thus, the statement of the problem in a fixed domain is obtained:

$$\begin{cases} m\dfrac{\partial \xi}{\partial t} - k\dfrac{\partial}{\partial x}\left(a(z)\dfrac{\partial p}{\partial x}\right) - \dfrac{\partial}{\partial z}\left(b(z)\left[\dfrac{\partial p}{\partial z} - \xi\right]\right) = 0 & (x, z, t) \in Q,\ t > 0; \\ p = p_0 & x \in [0, \mathrm{A}],\ z = 0,\ t > 0; \\ v_x = -ka(z)\dfrac{\partial p}{\partial x} = 0 & x = 0,\ z \in [0, L_z]; \\ \xi\big(p(x, z, t)\big) \in H(p) \text{ for almost all } (x, z, t) \in Q; & \\ p \geq 0 & (x, z, t) \in \bar{Q}; \\ p = 0, \xi = 0 & x \in [0, L_x],\ z \in [0, L_z],\ t = 0; \\ p = 0, \xi = 0 & x \in [0, L_x],\ z = L_z,\ t > 0; \\ p = 0, \xi = 0 & x = L_x,\ z \in [0, L_z],\ t > 0. \end{cases} \tag{5}$$

Further, the soil is assumed to be homogeneous, i.e. $a(z) = b(z) = 1$ and $k = 1$.

An implicit finite-difference approximation is constructed for the differential problem in a fixed domain. The solution is found by relaxation method [25].

4. Results and discussion

A set of computational experiments is carried out. The pressure decreases with depth for each fixed x and decreases for each z in the direction from the symmetry axis to the contaminated zone boundary.

Comparison of the computational results for the counter-current approximation and approximation along the characteristics is made.

For the calculations, the results of which are presented below as typical, the following values of the parameters are taken: the hydraulic conductivity $k = 1$, the fluid pressure in the pipe $p = 0.1$, $L_x = 0.3$, $L_z = 0.5$, $T = 0.01$.

It should be noted that the grid steps in space for all presented results are $h_x = h_z = 0.0025$. At the same time, the observed point – the maximum depth of contaminant penetration at each time point – moves during one time step to a distance of approximately 0.003 (greater than $h_z = 0.0025$) for the time step $\tau = 0.0005$, and a distance less than h_z for $\tau = 0.00025$. In the latter case, the boundary turns out to be smoother (figure 3).

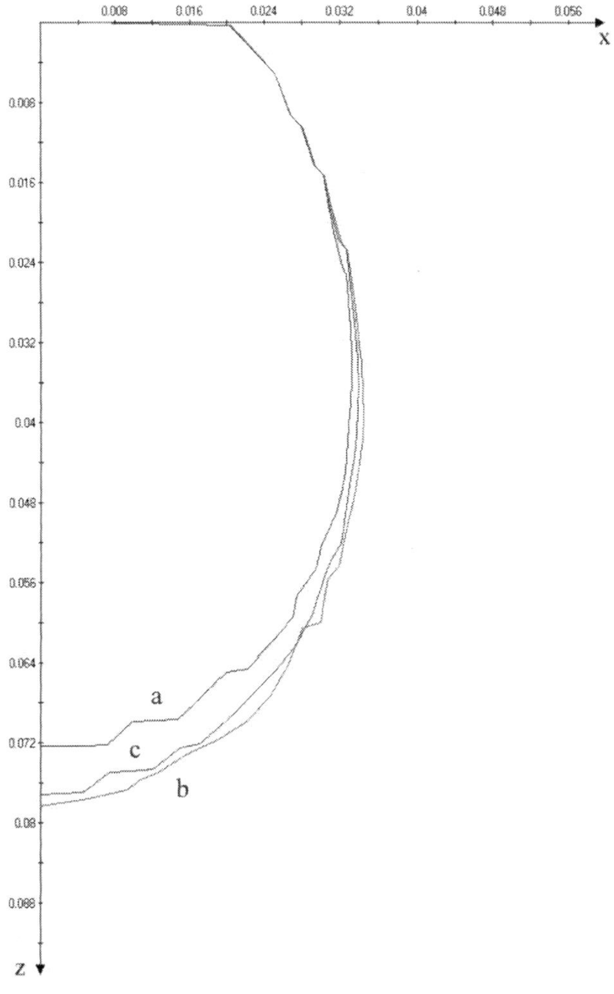

Figure 3. Results of calculations at the last time point ($T = 0.01$): the line (a) corresponds to counter-current approximation; the line (b) corresponds to $\tau = 0.0005$ while the characteristic passes through grid nodes; the line (c) corresponds to $\tau = 0.00025$ while the characteristic passes between the grid nodes.

Let us suppose that the pipeline passes through sandy loam soil. Hydraulic conductivity of the sandy loam soil is $k = 1.1$ Darcy $= 0.11 \cdot 10^{-7}\ \text{cm}^2 = 0.11 \cdot 10^{-11}\ \text{m}^2$. Porosity is $m = 0.2$. The width of the crack let be 3 cm. Groundwater let be at a depth of $\bar{z} = 3$ m. The pollution liquid let be oil at 20°C. Its density is $\rho = 860\ \text{kg} \cdot \text{m}^{-3}$, the viscosity is $\mu = 0.03\ \text{kg} \cdot (\text{m} \cdot \text{sec})^{-1}$. The width of the calculation area with respect to x is $L_x \cdot 3\ \text{m} = 0.3 \cdot 3\ \text{m} = 0.9$ m. The depth of the calculation area $L_z \cdot 3\ \text{m} = 0.5 \cdot 3\ \text{m} = 1.5$ m. Typical filtration rate is

$$\bar{v} = \frac{0.11 \cdot 10^{-11}\text{m}^2 \cdot 860\ \text{kg} \cdot \text{m}^{-3} \cdot 9.8\ \text{m} \cdot \text{sec}^2}{0.03\ \text{kg} \cdot (\text{m} \cdot \text{sec})^{-1}} = 3.09 \cdot 10^{-7} \frac{\text{m}}{\text{sec}} = 0.027 \frac{\text{m}}{\text{day}}.$$

At such an average flow velocity, the contamination will reach the groundwater table in $\bar{t} = \bar{z}/\bar{V} = 9.7 \cdot 10^6 \mathrm{sec} \approx 112.27$ days. It should be noted that the average filtration rate differs from the actual fluid velocity. The time step $\tau = 0.0005 \cdot \bar{t} \approx 0.056$ days ≈ 1 h 20 min. The estimation time in 20 steps will be 1.12 days = 27 hours. The model pressure in the pipe can be estimated by selecting the pressure measurement scale as $\bar{p} = \rho g \bar{z} = 860\ \mathrm{kg} \cdot \mathrm{m}^{-3} \cdot 9.8\ \mathrm{m} \cdot \mathrm{sec}^{-2} \cdot 3\mathrm{m} = 25284\ \mathrm{kg} \cdot \mathrm{m}^{-1}\mathrm{sec}^{-2} = 0.26$ atm (technical atmosphere). For the computational experiment the pressure in the pipe is set 10 times less than $\bar{p}$, which can correspond to an oil storage. In the real pipeline, the liquid moves on a pressure of several atmospheres. Then the calculation area in the program must be set to a much larger size. It can be seen on the figure 3 that during the time in 20 time steps (27 hours) the oil reaches a depth of $z = 0.077$ (line (c)), which is $z = 0.077 \cdot 3\ \mathrm{m} = 0.231\ \mathrm{m}$ (about 23 cm).

Thus, the choice of approximation along the characteristics gives a better result than the counter-current approximation. In this case, there is no scattering of the values of the function ξ, which is responsible for determining the boundary of the contaminated zone, if the grid steps in space and time are chosen so that the characteristics pass through grid nodes.

References

[1] Haile T, Abiye T A 2012 *Environmental Earth Sciences* **67(1)** 71-80
[2] Hundal H S, Singh K, Singh D 2009 *Environmental Monitoring and Assessment* **154(1-4)** 393-400
[3] Vairo T, Magrì S, Quagliati M, Reverberi A P, Fabiano B 2017 *Chemical Engineering Transactions* **57** 373-78
[4] Veado M A R V, Arantes I A, Oliveira A H, Almeida M R M G, Miguel R A, Severo M I, Cabaleiro H L 2006 *Environmental Monitoring and Assessment* **117(1-3)** 157-72
[5] Zaporozec A 1981 *GeoJournal* **5(5)** 457-71
[6] Zhou Y, Khu S-T, Xi B, Su J, Hao F, Wu J, Huo S 2014 *Environmental Earth Sciences* **72(4)** 1243-54
[7] Zou L, Wang S, Liu L, Hashmi M Z, Tang X, Shi J 2015 *Environmental Earth Sciences* **74(4)** 2861-70
[8] Bear J, Zaslavsky D, Irmay S 1968 *Physical Principles of Water Percolation and Seepage* (Paris: Unesco) p 465
[9] Delleur J W 2006 *The Handbook of Groundwater Engineering*, ed J W Delleur (Boca Raton, London, New York: CRC Press) p 1320
[10] Šimůnek J, Jacques D, Langergraber G, Bradford S A, Šejna M, Van Genuchten M Th 2013 *Journal of the Indian Institute of Science* **93(2)** 265-84
[11] Alduncin G 2009 *Applicable Analysis* **88(6)** 921-36
[12] Bazaliy B V, Degtyarev S P 2011 *Journal of Mathematical Physics, Analysis, Geometry* **7(4)** 295-332
[13] Billi L, Farina A 2000 *Quarterly of Applied Mathematics* **58(1)** 85-101
[14] Borsi I, Farina A, Fasano A 2004 *Nonlinear Analysis: Real World Applications* **5(5)** 763-800
[15] Boutet de Monvel A, Loutsenko I, Yermolayeva O 2013 *Analysis and Mathematical Physics* **3** 277-94
[16] Díaz J I, Shmarev S 2009 *Archive for Rational Mechanics and Analysis* **194(1)** 75-103
[17] Farina A, Fasano A 2003 *Mathematical Models and Methods in Applied Sciences* **13(6)** 843-74
[18] Fusi L, Farina A, Rosso F 2015 *International Journal of Engineering Science* **87** 110-8
[19] Kharab A 2009 *International Journal of Pure and Applied Mathematics* **57(2)** 289-97
[20] Kosterina E A, Lapin A V 1995 *Russian Mathematics (Iz. VUZ)* **6** 42-50
[21] Laitinen E, Lapin A V, Pieskä J 2003 *Journal of Computational and Applied Mathematics* **154(2)** 393-413
[22] Othman N L, Suzuki T, Tsuchiya T 2011 *International Journal of Mathematics and Computers in Simulation* **5(1)** 53-60

IOP Conf. Series: Earth and Environmental Science **107** (2017) 012111 doi:10.1088/1755-1315/107/1/012111

[23] Räisänen T 2014 *Efficient numerical methods for simulating continuous casting processes* (Jyväskylä: University of Jyväskylä) p 102
[24] Vladimirov V S 1979 *Generalized functions in mathematical physics* (Moscow: Nauka) p 320 [in Russian]
[25] Samarsky A A, Gulin A V 1989 *Numerical methods* (Moscow: Nauka) p 432 [in Russian]

IOP Conf. Series: Earth and Environmental Science **107** (2017) 012112 doi:10.1088/1755-1315/107/1/012112

Man-induced transformation of mountain meadow soils of Aragats mountain massif (Armenia)

M H Avetisyan

Center for Ecological-Noosphere Studies NAS RA, 0025 Yerevan, Armenia
E-mail: marieta.avetisyan@cens.am

Abstract. The article considers issues of degradation of mountain meadow soils of the Aragats mountain massif of the Republic of Armenia and provides the averaged research results obtained for 2013 and 2014. The present research was initiated in the frames of long-term complex investigations of agroecosystems of Armenia's mountain massifs and covered sod soils of high mountain meadow pasturelands and meadow steppe grasslands lying on southern slope of Mt. Aragats. With a purpose of studying the peculiarities of migration and transformation of flows of major nutrients namely carbon, nitrogen, phosphorus in study mountain meadow and meadow steppe belts of the Aragats massif we investigated water migration of chemical elements and regularities of their leaching depending on different belts. Field measurement data have indicated that organic carbon and humus in a heavily grazed plot are almost twice as low as on a control site. Lysimetric data analysis has demonstrated that heavy grazing and illegal deforestation have brought to an increase in intrasoil water acidity. The results generated from this research support a conclusion that a man's intervention has brought to disturbance of structure and nutrient and water regimes of soils and loss of significant amounts of soil nutrients throughout the studied region.

1. Introduction

It is generally agreed today that human activities interfere with most processes produced by nature and finally lead to disturbances in ecosystem development, soil and vegetation pollution with diverse substances, loss of most essential biogenic substances from soils, decreasing contents of humus, deterioration of soil fertility and productivity of plants and as a consequence to soil degradation and desertification.

One of man-induced factors of transformation of the Aragats massif's soils has been heavy grazing and illegal deforestation. Heavy grazing across the study region has already brought to destruction of soil cover, sod slip and pasture degradation namely replacement of perennials by annual plants, decrease of root penetration depth, soil compaction, deterioration of soil water and air balance that finally leads to soil erosion [1, 2].

In respect of deforestation one should mention that such activities disagree with national forest protection standards and measures and largely contribute to acceleration of eluvial processes and consequent removal of biogenic elements by intrasoil runoff. A change in elemental composition of intrasoil runoff proves that actually there occurs a significant disturbance of balance between major biogenic substances. It is also essential, that in natural conditions the balance is maintained by a dynamic equilibrium of eluvial and illuvial processes running in forest soils [3, 4].

Due to manmade activities in the Aragats massif like other mountain regions of Armenia loss of stored biogenic elements (carbon, nitrogen, phosphorus) from soils is followed by mineralization of a considerable amount of organic matter. Simultaneously biogenic elements change from organic into inorganic form, thus contributing to their leaching from soil to ground and surface waters. One of factors determining loss of biogenic substances is intrasoil runoff [4].

One should mention that lysimetric studies of intrasoil runoff are one of best informative methods when implementing complex landscape-biogeochemical investigations into condition and functioning of mountain ecosystems. Lysimetric solutions characterizing vertical migration of a flow of substances not only partially reflect biogeochemical cyclisity, but also provide direct information about geochemical specificity and functioning of ecosystems [5].

This research was done in the frames of long-term complex investigations of geoecological problems of agroecosystems of Armenia's mountain massifs. We studied high-mountain meadows and pastures found in meadow steppe and alpine belts of the Aragats massif.

With a purpose of studying the peculiarities of migration and transformation of flows of major nutrients namely carbon, nitrogen, phosphorus in mountain meadow and meadow steppe belts of the Aragats massif we studied water migration of chemical elements and regularities of their leaching depending on different belts.

The article considers the averaged research results obtained in 2013 - 2014.

2. Material and methods

The research covered alpine mountain meadow sod soils (2700-3250m a.s.l., pastureland) and meadow steppe (2080-2700m a.s.l., grasslands) belts of southern slope of Mt. Aragats.

Soil sampling and analyses were carried out by accepted methods of landscape-geochemical investigations [6, 7].

Atmospheric deposition and lysimetric waters were sampled from monitoring stations located in the noted belts at 3250m a.s.l. (the Aragats station) and 2080m a.s.l. (the Hamberd station).

The initial stage of analytical treatment of samples included a solid-liquid phase separation of solution.

Vertical soil runoff was studied by lisymetric method that both provide information about the capacity of soil runoff, chemical composition and migration of soil water elements and allows to assess loss of nutrients in the result of leaching [5]. The lysimeters were installed plane into 0-10 and 0-50 cm deep soil layer to minimally damage soil structure and composition.

Macro components in all water samples were determined through accepted hydrochemical methods [8, 9].

3. Results and discussion

According to research results, concentration of solutes in atmospheric precipitation vary within wide limits that is typical of high-mountain regions and is associated with the dynamics of a number of meteorological factors namely temperature, air humidity, wind direction and velocity as well as intensity and amount of atmospheric precipitation [10].

Recent studies of composition of atmospheric precipitation in mountain ecosystems have indicated that over the years the contents of chemical substances in atmospheric precipitation and rates of mineralization (at pH – 7.3-7.4) have been increasing. Meadow steppe vs. alpine belt exhibited high contents of chemical elements which might possibly be connected with high levels of air pollution in the given belt [1].We have also established that in atmospheric precipitation hydrocarbonates and sulfates are present in large quantities. By averaged contents cations were dominated by calcium and magnesium, anions – by sulfates. A descending order of cations is represented by the following descending series: Ca^{2+}>Mg^{2+}>K^{+}>Na^{+}, anions – by SO_4^{2-}>HCO_3^{-}>Cl^{-}.

It is known that the basic ecosystem's equilibrium maintaining mechanism depends on soil cover, where the flow of atmospheric substances is distributed between vegetation and ground waters [3].

The soils of Aragats massif differ by thickness of humus horizon and contents of organic matter. Such differences are mainly determined by the height a.s.l, exposition and steepness of slopes and vegetation cover [10].

Collation between newly obtained and earlier data [11] has indicated that the contents of humus in meadow steppe soils vary 5 to 8%, those of total nitrogen – between 0.20 and 0.33%. The content of total phosphorus is high (0.19-0.26%), and potassium – low (1-1.4%). Those soils are not rich in available nitrogen and phosphorus and are moderately and well supplied with potassium. Mountain meadow sod soils of the alpine belt are characterized by high contents of humus (10-12%) and total nitrogen (0.30-0.80%). As compared with meadow steppe soils, the contents of total phosphorus in mountain meadow sod soils are higher (0.20-0.40%) which is determined by its more intense bioaccumulation in the humus horizon, potassium being significantly lower in mountain meadow sod soils. The latter vs. meadow steppe soils of meadow steppe belt are well lower in total and mobile potassium. This phenomenon is explained by a "lighter" granulometric composition of mountain meadow sod soils of alpine belt. From meadow steppe towards mountain meadow soils the content of humus, total nitrogen, phosphorus and soil acidity increases, whereas that of potassium decreases.

High contents of humus, total nitrogen and phosphorus in mountain meadow sod soils of alpine belt are determined by peculiarities of soil formation processes running under humid climatic conditions and relatively low temperatures. These conditions contribute to the accumulation of organic matter and make it difficult to decompose, which prevents the removal of organic matter from the ecosystem [10].

The important role of soil organic matter is that it improves soil structure when transformed by soil microorganisms. The organic and mineral acids released by microorganisms promote their cementation into water-resistant aggregates. Formation of soil aggregates is supported by microbial and chemical products of organic matter transformation, humine substances, polysaccharides and microorganic cells [3, 12, 13].

Soil organic matter being the major factor of development of a sustainable ecosystem has a double orientation. On the one hand, it serves as a feed source supporting the activities of microorganisms and therefore determining the intensity of redox processes in soil. On the other hand, soil organic matter involved in redox reactions, has a biochemical effect on soil conditions. Humic substances are also exposed to different transformations leading to destruction of water-resistant aggregates [3, 12, 13].

Earlier researches [2, 4] have indicated that organic carbon and humus in a heavily grazed plot are almost twice as low as on a control site. A decrease in soil organic matter is determined primarily by destruction of physical structure of a surface soil horizon, which - accounting for active reaction of excrements in the upper sod layer of soil - consequently leads to violation of stability of biochemical compounds i.e. soil degradation followed by loss of nitrogen and carbon compounds in gaseous form and leaching of elements.

When assessing deforestation-caused consequences as an ecological risk factor two aspects characterizing condition of soils must be taken into consideration. The first aspect includes disturbance of stability of vegetation cover in the result of negative deforestation-caused changes in soils and a possible significant and rapid change in physico-chemical properties of soil under humid climatic conditions and a sharply sloped relief. The second aspect includes assessment not only of changes in ecological conditions, but also associated ecological risks given the use of the natural resources of the study region. These two aspects underlie a full-scale assessment of natural potential and resource power of the study region and ecological changes in it [3, 4]. Hence, the eco-geochemical situation, which emerges on deforested areas leads to loss of nutrients, being determined by disturbance of dynamic equilibrium of soil formation and ecosystem processes.

Lysimetric data obtained from this research (table 1) have indicated that deforestation entails an increase in intrasoil water acidity by 1.4-1.7 units.

The influence of intrasoil water acidity accelerates eluvial processes and therefore contributes to the removal of biogenic macro and micro components by runoff. This in turn leads to a dramatic decrease of soil nutrients at root nutrition [4, 5].

In intrasoil water at a depth of 0-50cm concentration of Ca^{2+}, Mg^{2+}, K^{+}, PO_4^{3-} decreases by 1.5-2.6 times.

It is known that under direct solar exposure humus substances are built up in the uppermost layer, but in mountain conditions the intensity and duration of solar radiation associated with illuvial migration of substances determine the influence of photochemical destruction on humus composition and lower horizons [3, 12], and later on acid solutions influence the mineral composition of soil.

Table 1. Composition of lysimetric waters in mountain steppe zone (before and after deforestation, mg/L).

Study site	*pH*	$C_{org.}$	NO_2^-	NO_3^-	PO_4^{3-}	Ca^{2+}	Mg^{2+}	K^+
				0-10 cm				
Before deforestation	6.5	30.2	1.5	4.2	0.96	92.4	60.0	43.0
(1998)				0-50 cm				
	6.0	22.5	0.3	3.6	0.32	38.9	26.2	28.2
				0-10 cm				
After deforestation	5.7	31.2	1.3	3.8	0.89	79.0	58.1	25.7
(2013-2014)				0-50 cm				
	6.2	25.3	0.2	2.7	0.23	30.2	22.4	19.4

A change in elemental composition of lysimetric waters suggests that actually there occurs a significant disturbance of balance of substances including major components of mineral nutrition of plants: nitrogen, phosphorus, potassium. Under natural conditions the balance of elements is maintained by dynamic equilibrium of eluvial and illuvial processes [2].

Hence, over a relatively short period (15 years) the ecosystem responds to man-induced impacts (heavy grazing, illegal deforestation).

The changes mentioned above caused by an increase in soil acidity have a negative character as migration of substances is followed by disturbance of dynamic equilibrium of eluvial and illuvial processes in soils, changes in condition and physico-chemical properties of soils and intensity of eluvial processes.

The obtained results support a conclusion that in mountain ecosystems, in meadow steppe zone under conditions of intense water exchange on deforested plots peculiarities of transformation of soil and its further development is determined by peculiarities of soil formation processes and a relatively high activity of ecosystems [2, 3, 4].

The results obtained from this research provide exhaustive characteristics of migration and leaching of major nutrients from different soil types of the Aragats mountain massif.

So, man-made activities affect soil nutrient and water regime and trigger changes in mountain forest soils and consequent leaching of significant quantities of nutrients that finally brings to nutrient deficiency and destruction of soil structure, whereas duration and intensity of such activities determine further changes in ecological conditions of the study region.

References

[1] Avetisyan M H and Poghosyan T E 2013 Environmental issues and identification of risk groups in the population (Yerevan: Gitutyun) pp 5–10
[2] Araratyan L, Avetisyan M H and Sakoyan A G 2014 *Geographical bulletin (*Perm: Perm State National Research University) **N1 (28)** pp 94–100
[3] Aleksandrova L N 1980 Soil organic matter and processes of its transformation (Leningrad: Nauka) p 280
[4] Revazyan R H, Sakoyan A G and Avetisyan M H 2011 Interactions between biogenic and abiogenic components in natural and anthropogenic systems (Saint Petersburg: VVM Publishing) pp 500–504
[5] Shilova E I 1972 *Using lysimentric methods in soil science, agrochemistry and landscape science* (Leningrad: Nauka) pp 1–21
[6] Arinushkina E V 1970 Manual on chemical analysis of soils (Moscow: Moscow State University)

p 487
[7] Yudin F A 1972 Agrochemical research methods (Moscow: Kolos) p 272
[8] Alekin O A 1954 Chemical analysis of continental waters (Leningrad: Gidrometizdat) p223
[9] Reznikov A A, Mulikovskaya E P and Sokolov I Yu 1963 Methods of analysis of natural waters (Moscow: Gosgeoltekhizdat) p 334
[10] Babayan G B 1982 An agrochemical characteristic of mountain meadow soils of the Armenian SSR (Yerevan: ArmSSR Academy of Sciences) p136
[11] Avetisyan M H 2016 *Geographical bulletin* (Perm: Perm State National Research University) **N4 (39)** pp 84–91
[12] Eldor A P 2007 Soil microbiology, ecology, and biochemistry (Chennai: Charon Tec Ltd) pp 53–80
[13] Bot A and Benites J 2005 The importance of soil organic matter. Key to drought-resistant soil and sustained food production (Rome: FAO Soils Bulletin 80) pp 5–8, 47–52

IOP Conf. Series: Earth and Environmental Science 107 (2017) 012113 doi:10.1088/1755-1315/107/1/012113

Mathematical models application for mapping soils spatial distribution on the example of the farm from the North of Udmurt Republic of Russia

P M Dokuchaev[1], **J L Meshalkina**[1,2,3], **A M Yaroslavtsev**[2,4,5]

[1]Soil Science Faculty, Lomonosov Moscow State University, Leninskye Gory, GSP-1, bld.12, Moscow, 119991, Russia
[2]Department of Ecology, Russian Timiryazev State Agrarian University, Timiryazevskaya Str. 49, Moscow, 127550, Russia
[3]Dokuchaev Soil Science Institute, Pyzhyovskiy lane 7 building 2, Moscow, 119017, Russia
[4]RUDN University, Miklukho-Maklaya str.6, Moscow, 117198, Russia
[5]School of Natural Sciences, Far Eastern Federal University, Sukhanova St. 8,Vladivostok, 690090, Russia
E-mail: jlmesh@list.ru

Abstract. Comparative analysis of soils geospatial modeling using multinomial logistic regression, decision trees, random forest, regression trees and support vector machines algorithms was conducted. The visual interpretation of the digital maps obtained and their comparison with the existing map, as well as the quantitative assessment of the individual soil groups detection overall accuracy and of the models kappa showed that multiple logistic regression, support vector method, and random forest models application with spatial prediction of the conditional soil groups distribution can be reliably used for mapping of the study area. It has shown the most accurate detection for sod-podzolics soils (Phaeozems Albic) lightly eroded and moderately eroded soils. In second place, according to the mean overall accuracy of the prediction, there are sod-podzolics soils – non-eroded and warp one, as well as sod-gley soils (Umbrisols Gleyic) and alluvial soils (Fluvisols Dystric, Umbric). Heavy eroded sod-podzolics and gray forest soils (Phaeozems Albic) were detected by methods of automatic classification worst of all.

1. Introduction

Soil map is the main scientific document on the basis of which competent evaluation of land funds is possible, as well as the development of a system of practical measures for increasing soil fertility. Renewal of legacy soil maps with the actualization of soil cover state and properties is one of the most important requirements for conducting smart agriculture. At the same time, carrying out large-scale soil mapping by traditional means to the whole territory of Russia is not possible because of the high labor costs for field research and the limited number of specialists. Digital soil mapping (DSM) can be one of solutions to this problem.

At the heart of the DSM is the development of Dokuchaev's ideas, which are expressed in the model SCORPAN [1]. This model considers the classification of soils and soil properties as a function of the soil forming factors and / or indicators, which in the DSM are called covariates. Information on covariates can be obtained from various sources: for example, remote sensing data [2], old soil maps and survey materials [3, 4], digital elevation models [5, 6], data obtained by proximal sensing with measuring electrical, gravitational, magnetic, and electromagnetic fields [7, 8], etc.

Mapping of soils spatial distribution in DSM is realized utilizing algorithms specially developed for these purposes or used for the mathematical solution of many other problems. Two algorithms: multinomial logistic regression (MLR) [9, 10] and support vector machine (SVM) [11, 12], which create the probability vector finding a certain class of soil [9] for each point of the raster, are based on the predictor variables. They differ in functions used to construct those probability vectors. Three other algorithms: decision trees (C5) [10], random forests (RF) [10] and regression trees (RT) [13, 14], are based on so-called decision trees. In each node of such trees, classifying rules are formed, according to which the decision tree assigns a point to a class. The main difference between algorithms is the number of decision trees that are being built: for the decision tree algorithm, only one tree is constructed, and for other two algorithms - numerous.

The mapping process can be described as follows. First, a training set is formed, where for each known point on the Earth's surface soil classification is indicated, as well as the values of soil forming factors and covariates. Based on the training sample, a cartographic model is constructed for the study area. The cartographic model can be expressed in number of decision rules or in the form of coefficients in the equation estimates, depending on the chosen classification algorithm that specifies the spatial structure of a multidimentional space. Then, the diagnostics of soil classification is carried out throughout the study area using the obtained cartographic model, based on data of soil forming factors and indicators.

The main DSM feature is the ability to check sustainability of cartographic models obtained for the training set, and test the accuracy of predictions for an independent random sample. As a result, each map obtained by the DSM is characterized by the degree of uncertainty in the prediction.

The aim of the study was the construction of large-scale digital soil maps using five mathematical models, as well as comparison of digital soil maps with the existing large-scale soil map performed by UralGyprozem staff in 1984 (further called just "1984" map). The research tasks also included: the selection of the optimal pixel size of the image raster for modeling and the verification of stability of a model.

2. Material and methods

The key site is located 8 km to South-West from town Glazov, which is located on the north of Udmurt Republic of Russia. Soil-climatic conditions there are typical for the eastern part of the Eastern European Plain. The climate is moderately continental; the precipitation-evaporation ratio is 1.00-1.33. Vegetation is represented mainly by fir-spruce and pine forests with a significant admixture of birch.

The length of the site from north to south is 4.5 km, and from west to east - 5.2 km (figure 1). The total area is about 19 square km. Maximal elevation difference in the study area is 62 m. The territory is rugged with ravines. Forests occupy 24% of the investigated area and are located along the slopes of deep ravines. There are sod-gley soils (Umbrisols Gleyic) on the ravine slopes. The river Kypka flows from the North-East to the South-West of the study area. Accordingly, floodplain meadows with alluvial soils (Fluvisols Dystric, Umbric) occupy about 5% of the area. Anthropogenic load is significant here: 65% of the territory is occupied by arable land, 5% are hayfields, 6% are household plots of the village of Kypka. Most of the sod-podzolic soils (Albeluvisol Umbric) on arable land are represented by eroded categories. Small flat areas in the lower part of the slopes can be occupied by gray forest soils (Phaeozems Albic). All soils of the area under consideration are with medium- and heavy-loamy soil texture.

On the basis of a topographic map with a scale of 1: 50000, a DEM with a 5 meter resolution was built. Based on the DEM, 35 maps of the morphometric relief variables were constructed, such as slope exposition, steepness of slopes, catchment area, flow power index, topographic index, horizontal curvature, vertical curvature, etc. [6].

During the field survey of 2014-2015, there were 166 survey points laid (figure 1), with the full description of the soil profile. The survey points covered the entire territory and were laid considering mapping of all elementary landscapes positions of the investigated territory.

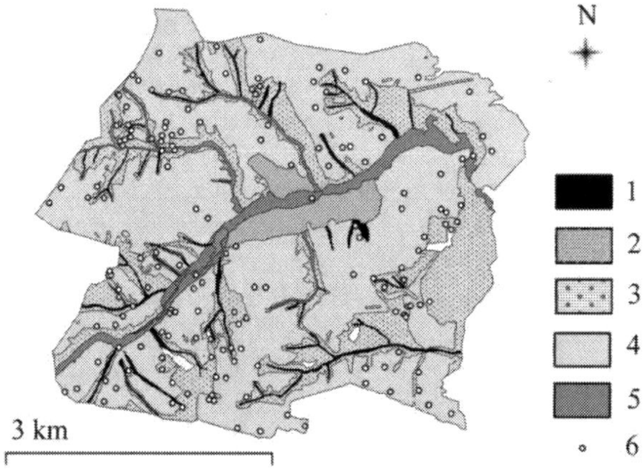

Figure 1. Study area (scale 1:50 000). Legend 1. Ravines. 2. Settlement. 3. Forests. 4. Arable land. 5. Flood plain and water. 6. Sampling points.

The database consisted of 166 profiles, and during simulation procedures it was necessary that for each classification group there should be at least 10 observations, therefore, 8 large soil groups were selected for digital soil mapping: sod-podzolics non-eroded soils, sod-podzolics lightly eroded soils, sod-podzolics moderately eroded soils, sod-podzolics heavy eroded soils, sod-podzolics wrap soils, gray forest soils, sod-gley soils and alluvial soils. Such a set of conditional soil groups allowed to reflect all presented types of soils and distribution of eroded soils in the study area. The latter is important for the proper planning of agricultural activities.

One-way analysis of variances was used for covariates selection: soil groups acted as a factor, and responses were covariates. The input data for modeling were the following covariates: 25 out of 35 morphometric variables, land use, NDVI index.

Modeling of soils spatial distribution was based on five algorithms: multiple logistic regression, decision trees C5, random forest, regression trees and support vector machines.

Models stability was estimated using cross-validation. In this method, a subset of observations is divided randomly into two subsets: the "training" sample and the "test" sample. The model is built using data from training sample, whereas the data from the second sample are used to assess the model quality and stability. For each algorithm, 100 iterations were carried out. As a result, for each algorithm 100 digital maps were constructed. To produce each map, 146 points (88%) of the points were selected randomly from the training sample, and the remaining 20 points (12%) were used as testing one.

For each test sample, a confusion matrix was made between the actual soil groups identified during the field studies and the soil groups predicted for these points as a result of modeling. The sum of the diagonal elements values (X_{ii}) reflects the total number of identically classified points, and the ratio of this quantity to the total number of points (N) is called overall accuracy [15, 16]:

$$A_o = \theta_1 = \frac{\sum_{i=1}^{r} x_{ii}}{N} = \sum_{i=1}^{r} p_{ii}; \qquad p_{kl} = \frac{x_{kl}}{N} \qquad (1)$$

The accuracy of individual classes prediction is calculated similarly. Traditionally, they are divided by the total number of class predictions. This measure of prediction accuracy indicates the likelihood that the observation was classified correctly and called "producer's accuracy", because the model creator is interested in how well a particular class can be predicted.

The kappa coefficient or simply kappa is another statistical measure of the coincidence between the observed and predicted classes. The calculation is based on the difference between how many coincidences are present ("observed" coincidence is the percentage of the overall prediction accuracy), it is possible to expect a coincidence that has turned out just randomly.

We denote this by θ_2:

$$\theta_2 = \sum_{i=1}^{r} \frac{x_{i+}x_{+i}}{N^2}; \tag{2}$$

The index kappa is defined as:

$$\hat{\kappa} = \frac{\theta_1 - \theta_2}{1 - \theta_2} \tag{3}$$

The kappa is equal to 1 when the two maps coincide or when the results of the field survey and the soil map are completely identical. The kappa is zero if θ_1 coincides with θ_2 and a purely coincidental match of the two maps is observed or if the soil map coincides with the results of the field survey. In most cases, the kappa lies from 0 to 1 and shows how similar maps between each other or how close the soil map and survey results.

All indicators produced as a result of modeling iterations were collected in a table in which for cross-validation: kappa coefficient and overall accuracy [15, 16]. For each model, the mean values, medians, standard deviations and the boundaries of the above characteristics were calculated. For each model, two specific implementations of models were selected for further investigation: with average and with maximum value of the kappa. To select the optimal pixel size, curves were plotted showing the ranking of the kappa model depending from pixel size.

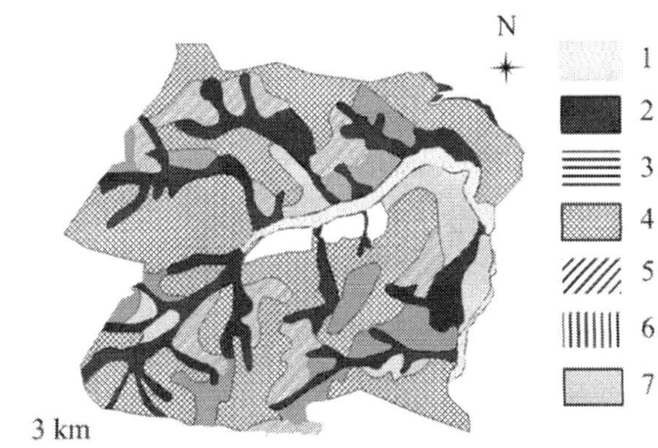

Figure 2. Soil map of 1984 (scale 1:50 000). Conditional soil groups: 1. Alluvial soils (Fluvisols Dystric, Umbric), 2. Sod-gley soils (Umbrisols Gleyic). Sod-podzolics soils (Phaeozems Albic) 3. - non-eroded, 4 - lightly eroded, 5 - moderately eroded, 6 - heavy eroded, 7 gray forest soils (Phaeozems Albic).

A comparison of the obtained soil maps with the UralGiprozem map of 1984 was carried out for 200 random points (figure 2). The following maps were compared: the soil map of 1984, made by the specialists of UralGiprosem, and five digital soil maps obtained on the basis of models with an average kappa value. For each pair of maps being compared, the overall accuracy and kappa were calculated. The following software was used in purpose of the investigation: Quantum GIS 2.8.2, SAGA GIS 2.1.2, Multispec and R software environment 3.3.1 [17].

3. Results and discussion

For the selection of the optimal image raster pixel size, a complete simulation was performed for 5 m, 20 m and 50 m pixels. The simulation results for multiple logistic regression are shown in figure 3. For all five algorithms, decreasing pixel size lead to increasing in model's kappa values. Further results are presented for a pixel size of 5 meter.

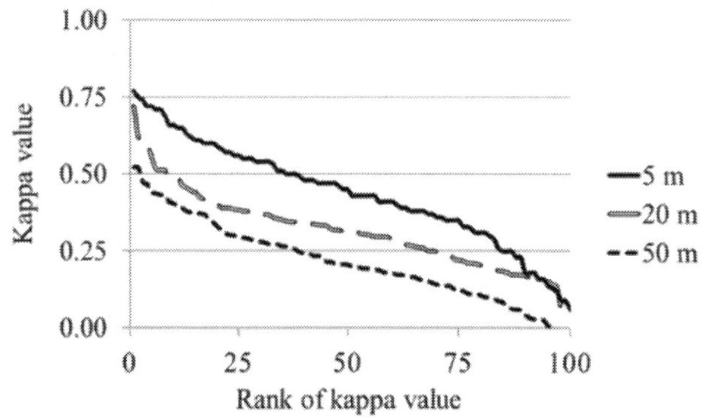

Figure 3. Series of decrease for multiple logistic regression model kappas in cross-validation for different pixel sizes.

Modeling the spatial distribution of soil groups by five algorithms showed (table 1) that the highest average model kappa 0.44 was observed using multiple logistic regression, while the lowest average kappa, equal to 0.26, was detected for model produced with regression trees algorithm. In general, the best prediction was obtained with models produced by multiple logistic regression, support vector machines and random forest methods. Models produced with regression tree algorithm, showed the worst results.

Table 1. Characteristics of the distributions of overall accuracy and kappa for 100 iterations of cross-validation in modeling based on 5 algorithms.

	MNLR	C5	RF	RT	SVM
			overall accuracy		
Mean	61	56	59	49	60
Minimum	30	35	35	30	30
Maximum	85	75	85	70	85
			kappa		
Mean	0.44	0.32	0.40	0.26	0.41
Minimum	0.06	0.00	0.00	0.00	0.00
Maximum	0.77	0.64	0.75	0.52	0.79

Maps for different kappa values and overall accuracy were analyzed. It turned out that in most cases the best results from the expert's point of view were found for maps with the average kappa and overall accuracy values.

From an expert point of view, digital soil maps based on decision tree models are closest to the map of 1984, compared to other digital maps. High level of resemblance was observed with maps produced by multiple logistic regression and the support vector machines models.

Based on visual comparison, the least degree of similarity of the soil map was observed with a map created with regression trees. However, a quantitative comparison with the map of 1984 showed satisfactory coincidence with digital maps: the kappa was 0.27-0.28, and the overall accuracy was 53-54%.

The most accurate detection by models was shown for sod-podzolics lightly eroded and moderately eroded soils (table 2). In second place, according to the mean overall accuracy of the prediction, there were sod-podzolics soils non-eroded and wrap one, as well as sod-gley soils and alluvial soils. Heavy eroded sod-podzolics soils and gray forest soils were detected worst of all. The latter were not determined on the average by any algorithm.

Table 2. Average overall accuracy of soil group predictions in modeling based on 5 algorithms.

Soils	MNLR	C5	RF	RT	SVM
Sod-podzolics non-eroded	71.7	26.0	43.7	68.3	46.2
Sod-podzolics lightly eroded	78.4	86.2	77.6	68.8	75.9
Sod-podzolics moderately eroded	60.9	55.3	67.1	54.7	57.6
Sod-podzolics heavy eroded	19.1	0	0	0	9.1
Sod-podzolics wrap	45.4	5.2	23.5	18.6	42.1
Alluvial soils	95.4	97.3	91.2	11.8	100
Gray forest soils	0	0	0	0	0
Sod-gley soils	43.7	8.2	53.4	0	63.3

In this study, a set of cartographic models for one territory was generated as a result of the DSM algorithms application. The use of mathematical models to describe the spatial distribution of soils is a promising approach in the development of soil cover mapping, which makes it possible to quickly obtain soil maps that are comparable in accuracy to maps constructed by traditional soil survey.

The resulting cartographic materials can form the basis for soil risks assessment and environmental development modeling scenarios.

Acknowledgement
This study has been conducted with support from grant №14-120-14-4266-ScSh and by grant according to the Agreement № 02.A03.21.0008 of the Russian Federation Ministry of Education and Science, grant №ScSh-10347.2016.11 of the Leading Scientific Schools, EU's FP7 grant agreement №603542 (the project LUC4C) and grant № 15-16-30007 of the Russian Science Foundation.

References

[1] McBratney A B, Mendonca Santos M L and Minasny B 2003 *Geoderma* **117** pp 3–52
[2] Mulder V L, de Bruin S, Schaepman M E and Mayr T R 2011 The use of remote sensing in soil and terrain mapping—A review *Geoderma* **162** pp 1–19
[3] Mayr T, Rivas-Casado M, Bellamy P, Palmer R, Zawadzka J and Corstanje R 2010 Digital Soil Mapping (New York: Springer Science + Business Media BV) pp 191–202
[4] Dobos E, Bialkó T, Micheli E and Kobza J 2010 *Digital Soil Mapping* (New York: Springer Science + Business Media BV) pp 309–323
[5] Shary P A, Sharay L S and Mitusov A V 2002 *Geoderma* 107 pp 1–43
[6] Florinsky I V 2011 *Digital Terrain Modeling in Soil Science and Geology* (Amsterdam: Elsevier Academic Press) 432
[7] Huang J, Lark R M, Robinson D A, Lebron I, Keith A M, Rawlins B, Tye A, Kuras O, Raines M and Triantafilis J 2014 *Geoderma* **232–234** pp 69-80
[8] Buchanan S and Triantafilis J 2009 *Ground Water* **47** pp 80–96
[9] Kempen B, Brus D J, Heuvelink G B M and Stoorvogel J J 2009 *Geoderma* **151 (3–4)** 311–326
[10] Brungard CW, Boettinger J, Duniway MC, Wills S A and Edwards J 2015 *Geoderma* **239–240** pp 68–83

IOP Conf. Series: Earth and Environmental Science **107** (2017) 012113 doi:10.1088/1755-1315/107/1/012113

[11] Simin C, Rongqun Z, Liming L and Zhou De 2010 *Math. and Computer Modelling* **51** pp 1319–25
[12] Lamorski K, Pachepsky Y, Slawiński C and Walczak RT 2008 *Soil Sci Soc Am J* **72** pp 1243–47
[13] Lemercier B, Lacoste M, Loum M and Walter C 2012 *Geoderma* **171** pp 75–84
[14] Schmidt K, Behrens T and Scholten T 2008 *Geoderma* **146 (1–2)** pp 138–146
[15] Samsonova V P and Meshalkina YuL 2011 *Moscow University Soil Science Bull* **66 (3)** pp 83–85
[16] Hengl T, Toormanian N, Reuter H I and Malakouti M J 2007 *Geoderma* **140** pp 417–27
[17] R Core Team 2017. R: A language and environment for statistical computing. (Vienna: R Foundation for Statistical Computing) http://www.R-project.org/

The organic matter of the different ages fallow Luvisols

K G Giniyatullin, A A Valeeva, E V Smirnova, R V Okunev, L I Latipova

Institute of Environmental Sciences, Kazan Federal University, Kremlevskay str., 18, 420008, Kazan, Russia

E-mail: ginijatullin@mail.ru

Abstract. The study of the change in the humus state of the fallow Luvisols of different ages under the influence of weeds and meadow vegetation was carried out in dynamics (after 5 years). It is shown that both under weedy and meadow vegetation there is a statistically significant accumulation of organic matter in the upper part of the long-arable horizon. Based on the study of composition and spectral properties of soil organic matter in fallow soils of different ages concluded that the significant qualitative change of the humus state of fallow soils requires significant time, measured at least decades.

1. Introduction

The reduction of arable land is currently a worldwide trend which is most widespread on the territory of modern Russia [1, 2]. The formation of fallow soils there is characterized by the simple abandonment of arable land which is accompanied by spontaneous recovery of the original plant ecosystems. The formation of fallow vegetation in the forest-steppe zone begins with overgrowing of weeds (the pioneer stage of development of fallow soils) and then during a few years formation of the meadow phytocenosis. At the last stage of the development of fallow vegetation in the forest-steppe zone, under favorable conditions, occurs an introduction of woody plants and succession of wood communities [2]. Fallow phytocoenosis at each stage of development should be considered as a modern dynamic state of vegetation which is characterized by a certain biodiversity, productivity, vertical, horizontal structure and mosaic. The stages of development of fallow vegetation will be reflected in the nature of the postagrogenic transformation of soils and, first of all, on the change in their humus state. The purpose of this research work is to study the changes in the humus state of the different-aged fallow Luvisols in dynamics under the influence of weeds and meadow vegetation.

2. Material and methods

Objects of research are the associated plots of different ages of fallow soils which are located in the Volga region of the Republic of Tatarstan (Russia). Fallow plots are confined to one relief element - slightly sloping slope of the south-eastern exposure. Samples were selected in 2010, when the young fallow soil was at the age of 2 years and was in the initial stage of overgrowth with weed vegetation, the old fallow soil of 70-75 years of age was represented by a stable herbage-cereal meadow phytocenosis. The fallow plot of 70-75 years of age is located on the territory of the local airfield (which ceased to exist in the early 90's) on the lands withdrawn from arable land in 1935-1940. This territory was subjected to intensive haymaking and has no signs of overgrowing by woody vegetation. Samples from the old plow horizon were selected layerwise (each 5 cm) on depth to 20 cm by the special drill from 7 sampling points dated for 7 clusters of the hexagonal systematic lattices put separately on each fallow plot. In 2015, repeated sampling was carried out according to the same scheme with an indentation of 1-2 m in an arbitrary direction from the primary points when the fallow soils had the age of 7 and 75-80 years, respectively. Samples to characterize the horizontal spatial inhomogeneity were obtained by mixing of layered samples (0-5, 5-10, 10-15 and 15-20 cm) at sampling points. To characterize the vertical heterogeneity of the old plow horizon there were obtained 4 samples by mixing separately in layers 0-5 cm, 5-10 cm, 10-15 cm, 15-20 cm from the 7 sampling points. After careful selection of the roots, the soil was ground and passed through a 0.1 mm. In the mixed samples the total organic carbon content (C_{org}) and organic carbon soluble in sodium pyrophosphate was determined. For a qualitative characteristic of organic matter (**OM**) of long-arable horizons of fallow soils of different ages (2 years and 70-75 years) IR-Fourier spectrometry was used. The IR spectra of soil OM were obtained by the method of disturbed total internal reflection (DTIR) with the prefix PIKE MIRacleTM in the range 4000-660 cm^{-1}. For correction of soil minerals absorption bands as background used a mineral phase of samples which are selected on depth of 0-5cm, 5-10cm, 10-15cm, 15-20cm. The mineral phase was obtained by burning of soil OM with 30% H_2O_2 [3, 4]. Soil treatment with hydrogen peroxide allows to remove of OM from the surface of clay minerals by 85-95% [5, 6]. After drying, the soil sample without OM was ground and passed through a sieve of 0.1 mm diameter. This procedure allows to obtain IR spectra of OM of soil material with an automatic subtraction of its background mineral phase and provided an opportunity to interpret more informatively absorption bands associated with the OM. Interpretation of IR spectra was carried out on the basis of work Hyeong K., Capuano R.M. [7].

3. Results and discussion

In figure 1 the content of soluble and insoluble organic carbon in sodium pyrophosphate organic carbon in layerwise samples from old-plow horizons of different-aged fallow soils are presented. The total carbon content increases mainly in the upper part of the old plow horizon in accordance with the fallow age. According to the paired t-test, a statistically significant difference in the Corg content in the 0-20 cm layer of the old plow horizon is observed for the samples taken from the 7-point systematic lattices as between 2 and 7-year of fallow soils ($t = 3.72$ at $t_{st} = 2.45$), and between 70-75 and 75-80-year-old ($t=4.56$ при $t_{st}=2.45$). With increasing of the fallow soil age increases the horizontal variability of the Corg content in the old plow horizon. The coefficient of variation in humus content in the 0-20cm layer, according to the analysis of mixed samples, is 7.9% for the fallow soils of 2-year, 10.2% for the 7 years, 20.4% for the 70-75 years and 75-80 years is 21.5%. It can be concluded that, both at the pioneering stage of overgrowing and under meadow vegetation is observed the statistically significant accumulation of OM, which is characterized by a very high horizontal variability significantly exceeding the variability of the initial arable soils.

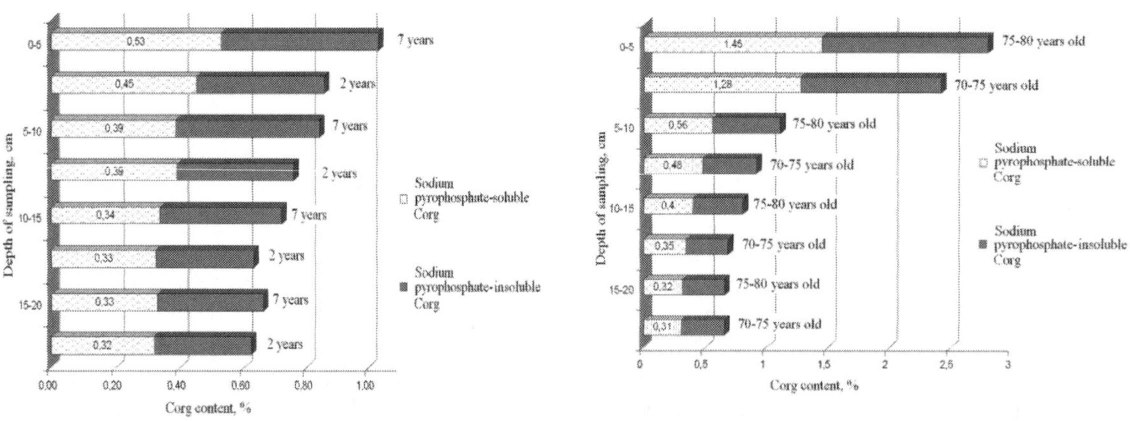

Figure 1. The content of soluble and insoluble in the sodium pyrophosphate of Corg in layered samples from old-arable horizons of uneven-aged fallow soils.

Also it can be concluded that the content of sodium pyrophosphate-soluble OM in the upper part of the old plow horizon increases. However, according to the paired t-test data the difference in the 0-20cm layer of the old-plow horizon of sodium pyrophosphate-soluble OM is statistically insignificant, both between the fallow soils of 2 and 7 years (t = 2.43 at tst = 2.45), and between the fallow soils of 70-75 and 75-80 years (t = 0.59 at tst = 2.45). A comparison of fallow plots with each other using the independent samples t-test shows the statistically significant difference in the content of sodium pyrophosphate-soluble Corg. The value of the two-sample t-test with the same variances for fallow soils of 2 and 70-75 years is 5.40 at tst = 2.18, the value of the two-sample t-test with different variances for fallow soils of 7 and 75-80 years, respectively, 7.01 at tst = 2.20. Thus, it can be concluded that the 5-year period is not sufficient for a significant accumulation of mobile soil OM, which would lead to a noticeable qualitative change of the humus state of postagrogenic soils. Probably, for a significant qualitative change of the humus substances composition the postagrogenic transformation is required for at least decades. Accumulation of OM in shorter intervals proceeds, most likely, due to the accumulation of detrital coarse humus [8]. Therefore, for the spectral estimation of the change of the OM qualitative composition of the old plow horizon of soils were used mixed layered samples from fallow soils which significantly different in age. The absorption frequencies of layered mixed samples from different ages of fallow soils (2 years and 70-75 years) are shown in table 1.

Table 1. Absorption frequencies of humus substances in fallow soils.

Depth of sampling, cm	Wave numbers, cm^{-1} (Group and и vibration)
	Fallow soils 70-75 years old
0 – 5	2960($\nu_{as}CH_3$), 2924($\nu_{as}CH_2$), 2874(ν_sCH_3), 2860(ν_sCH_2), 1732(νC=O), 1451(δCH_2,$\delta_{as}CH_3$), 1414($\delta_{as}CH_3$), 1339(νC-N), 1262(νC(=O)-O), 1160(νO-C-C), 1081(νSi-O-Si), 1020(νSi-O-Si), 875(νO-O), 840(trisubstituted), 731(trisubstituted), 703(ρCH_2)
5 – 10	2952($\nu_{as}CH_3$), 2924($\nu_{as}CH_2$), 2876(ν_sCH_3), 2853(ν_sCH_2), 1723(νC=O), 1460(δCH_2,$\delta_{as}CH_3$), 1378(δ_sCH_3), 1267(νC(=O)-O), 1156(νO-C-C), 1080(νSi-O-Si), 1023(νSi-O-Si), 889(fivesubstituted), 873(νO-O), 730(ρCH_2)
10 – 15	2925($\nu_{as}CH_2$), 2855(ν_sCH_2), 1732(νC=O), 1455(δCH_2), 1263(νC(=O)-O), 1160(νO-C-C), 1018(νSi-O-Si), 777(disubstituted), 700(disubstituted)
15 - 20	1729(νC=O), 1259(νC(=O)-O), 1163(νO-C-C), 1009(νSi-O-Si)
	Fallow soils 2 years old
0 – 5	2927($\nu_{as}CH_2$), 2856(ν_sCH_2), 1731(νC=O), 1450(δCH_2), 1263(νC(=O)-O), 1177(νO-C-C), 1076(νSi-O-Si), 1020(νSi-O-Si), 875(νO-O), 733(ρCH_2), 683(monosubstituted)
5 – 10	2925($\nu_{as}CH_2$), 2854(ν_sCH_2), 1735(νC=O), 1456(δCH_2), 1264(νC(=O)-O), 1162(νO-C-C), 1018(νSi-O-Si), 913($\delta CH=CH_2$), 749(ρCH_2), 682(monosubstituted)
10 – 15	2926($\nu_{as}CH_2$), 2854(ν_sCH_2), 1732(νC=O), 1441(δCH_2), 1264(νC(=O)-O), 1160(νO-C-C), 1072(νSi-O-Si), 1009(νSi-O-Si), 788(disubstituted), 760(ρCH_2), 684(disubstituted)
15 - 20	2924($\nu_{as}CH_2$), 2853(ν_sCH_2), 1732(νC=O), 1457(δCH_2), 1264(νC(=O)-O), 1160(νO-C-C), 1013(νSi-O-Si), 790(disubstituted), 766(ρCH_2), 691(disubstituted)

Application: ρ –wagging vibrations, ν –stretching vibrations, δ –bending vibrations.

In the 70-75-year-old fallow soils from the surface and up to the depth of 15 cm the increase in the absorption intensity in the region of about 3000 $см^{-1}$ corresponding to stretching vibrations of CH_2 and CH_3 groups is observed. The increase in their intensity with depth can be associated with an increase in chain length of saturated hydrocarbons. The frequencies values of the symmetric and asymmetric stretching vibration of methyl ($v_{as}CH_3$, v_sCH_3) and methylene groups ($v_{as}CH_2$ и v_sCH_2) remain constant to the depth of 10 cm. In the upper 10 cm layer of old-fallow soil the presence of the absorption band about 730 cm^{-1} (wagging vibrations δCH_2) indicates the presence of a number of directly related groups - CH_2 - with the amount of carbon atoms more than four. The combined absorption band due to the bending vibrations of the groups CH_2 (δCH_2) and CH_3 ($\delta_{as}CH_3$) in the region of 1460-1451 cm^{-1} and bending vibrations δ_sCH_3 in the region of 1414-1378 cm^{-1} is also observed up to the depth of 10 cm. The change in the symmetrical vibration frequency of the CH_3 group to 1414 cm^{-1} in the 0-5 cm layer of the old-fallow soil is probably due to the addition of the methyl group to the nitrogen atom (CH_3-N), not to the carbon atom. In the underlying horizons (10-15 cm) only absorption bands of stretching and bending vibrations of methylene groups are observed. The frequency value of about 1339 cm^{-1} is observed only at the depth of 5 cm of old - fallow soils (70-75 years), which can be attributed to the stretching vibrations of C-N aromatic amines.

In the region of 875-873 cm^{-1} there are observed the stretching vibrations of the peroxide group (O-O) of aliphatic compounds that are found in the composition of humic substances. The peroxide group is found only in the upper 10 cm layer of 70-75 year old fallow soils and in the 5 cm layer of 2-year fallow soils, which may be due to the autooxidation of organic and organometallic compounds by air oxygen in the upper soil horizons.

In 2-year-old fallow soils the quantity of absorption bands decreases. From the surface and to the depth of 20cm in the region of about 3000 cm_{-1} only the absorption bands of methylene groups (vasCH2, vsCH2, δCH2) are observed with relatively constant frequency. Absorption bands of methyl groups were not detected. The presence in the samples from a depth of 0-5 cm and 5-10 cm of the absorption band near 730 cm^{-1} (wagging vibrations of CH_2) is also indicated by the presence of a number of directly connected groups - CH_2- with more than four carbon atoms. The absorption band of wagging vibrations of methylene groups in samples at depth of 10-20 cm shifts toward higher frequencies that may be due to a decrease in the number of bound methylene groups with each other.

The feature of all studied samples of uneven-aged fallow soils is the presence of common frequencies over the entire depth of the old plow horizon. The typical esters of the absorption band of the stretching vibrations of the carbonyl group (C=O) in the region 1310-1250 cm^{-1}, which corresponds to the stretching vibrations C (= O) -O and in the region of 1180-1130 cm^{-1}, which corresponds to stretching vibrations of O-C-C group were found. Probably, these are esters of aliphatic acids, since absorption characteristic for aromatic compounds is absent. This indicates that these groups are the most s e and permanent structural units of humic substances of the studied soils that have not been mineralized as a result of agricultural use, before the transition of soils to a fallow state.

In the 2-year-old fallow soil inhomogeneous series of intense bands of silicic oxygen valence vibrations are identified, which by layerwise do not differ in intensity and wave numbers. It, probably, due to the fallow soil age in which the secondary accumulation of humus is not significant and changes in the soil OM spectra have to be very weak.

4. Conclusion

A general analysis of the results shows that the presence of Luvisols for 5 years both under fallow vegetation in the pioneer stage of overgrowing, and under developed meadow vegetation leads to the statistically significant quantitative accumulation of humus in the old plow horizon. However, the significant qualitative change in the humus state of fallow soils requires more time, measured at least for decades.

Acknowledgments

This work was supported in part by the Russian Foundation for Basic Research, research project №17-04-00846.

References

[1] Lyuri D I, Goryachkin S V, Karavaeva N A, Denisenko E A and Nefedova T G 2010 *Dynamics of*

Agricultural lands of Russia in XX century and Postagrogenic Restoration of vegetation and soils (Moscow: GEOS) p 416
[2] Kurganova I N, Lopes de Gerenyu V O, Shvidenko A Z and Sapozhnikov P M 2010 Change in the organic carbon pool of abandoned soils in Russia (1990-2004) *Eurasian Soil Science.* **43(3)** pp 333–340
[3] ISO 11277:1998 *Soil quality – Determination of practicle size distribution in mineral soil material – Method by sieving and sedimentation* (Switzerland: Int. Organ. Stand)
[4] Pansu M and Gautheyrou J 2006 *Handbook of soil analysis. Mineralogical, organic and inorganic methods* (Berlin, Heidelberg: Springer-Verlag)
[5] Hyeong K and Capuano R M 2000 The effect of organic matter and the H_2O_2 organic-matter-removal method on the δD of smectite-rich samples *Geochim. Cosmochim. Acta.* **64(22)** pp 3827–3829
[6] Giniyatullin K G, Shinkarev A A (Jr.), Shinkarev A A, Krinari G A, Lygina T Z, Gubaidullina A M, Kornilova A G and Melnikov L V 2012 Irreversible fixation of organic components in labile inter-spaces as a mechanism for the chemical stabilization of clay-organic structures *Eurasian Soil Science* **45(11)** pp1068–1080
[7] Smith A L 1982 *Applied Infrared Spectroscopy* (London: John Wiley and Sons)
[8] Giniyatullin K G, Shinkarev A A, Fazylova A G, Kuzmina K I and Shinkarev A A(Jr) 2012 Spatial Heterogeneity of Secondary Humus-Accumulation in Old-Arable Horizons of Fallow Light-Grey Forest Soils *Uchenye Zapiski Kazanskogo Universiteta. Seriya Estestvennye Nauki.* **154(4)** pp61–70

IOP Conf. Series: Earth and Environmental Science **107** (2017) 012115 doi:10.1088/1755-1315/107/1/012115

Mapping of Soil-Ecological Conditions of a Medium-Size Industrial City (Birobidzhan City, Jewish Autonomous Oblast, FarEast of Russia as an Example)

V B Kalmanova[1], L A Matiushkina[2]

[1] Institute for Complex Analysis of Regional Problems FEB RAS, 4 Sholom-Aleikhem St., 679000, Birobidzhan, Russia
[2] Institute of Water and Ecology Problems FEB RAS, 56 Dikopoltsev St., 680000, Khabarovsk, Russia
E-mail: Kalmanova@yandex.ru; lira@ivep.as.khb.ru

Abstract. The authors analyze soil relations with other elements of the city ecosystem (the position in the landscape, soil-forming rocks and lithology, vegetation and its state) to develop the legend and map of soils in the City of Birobidzhan (scale 1:25 000). The focus of study is the morphological structure of urban soils with different degree of disturbance of these relations under the impact of technical effects, economic and recreational activities of the city population. The soil cover structure is composed of four large ecological groups of soils: natural untransformed, natural with a disturbed surface, anthropogenic soils and technogenic surface formations. Using cartometry of the mapped soil contours the authors created the scheme of soil-ecological city zoning, which in a general way depicts the state of soil ecological functions in the city as well as identified zones of soils with preserved, partially and fully distured ecological functions and zones of local geochemical anomalies at the initial formation stage (environmental risk zones).

1. Introduction

Modern surveys of ecological situation in cities are focused on studies of soil cover and their ability to serve certain ecological functions, e.g. provide a microclimate (moisture, air, and heat exchange) in urban areas, regulate geochemical fluxes which get into surface and groundwaters, support survivability and longevity of plantation and biodiversity in natural urban environments [1-3]. Recently, special attention is also paid to the economic functions of urban soils, which they perform in conditions of various uses and diversity of forms of land ownership. More and more often urban soils act as objects of market relations, in which their value is determined by properties, functions and quality [4, 5].

Various anthropogenically transformed soils and technogenic deposits increasingly become typical to many large cities and industrial centers [6-9]. There are practically no soils as natural formations. In small and medium-size cities natural processes of soil formation are often not much suppressed by technogenic effects. Therefore, development strategies of small towns should include the solution of the problem of "man-environment-soil" and be based on the surveys of conditions and properties of the existing natural soil resource, its careful management and rational development and use [3, 10]. The experience of soils studies in different cities of Russia shows that the term "urban soils" actually implies economical activities in the city. Over 50 papers at symposiums and sections of the VI Congress of the Dokuchaev Soil Science Society in Moscow in 2012 discussed these and other issues of anthropogenic transformation of soils in urban areas.

Modern methodology of urban soils studies in Russia were initiated in the 1990es, when general guidelines and main principles of systematics of urban soils, diagnostics and mapping of soils in big cities were worked out [4, 6, 11, 12]. However, problems in studies of urban soils still remain to great extend unsolved [8, 12, 13].

The goals of our work are to study, classify and map soils in Birobidzhan, a relatively young and developing city of the Amur Region. The obtained results are the first stage of the continuing complex research on monitoring and assessment of the geo-ecological situation in the city.

2. Materials and Methods

2.1. Study area

The authors selected Birobidzhan soils as an object of study because this city is the administrative and industrial center of Jewish Autonomous Region (figure 1) and is ranked as a medium-size city in the southern part of the Far East of Russia. At present Birobidzhan like many cities in this region has the impetus for growth and development. This is largely due to its location in the Middle-Amur basin in the transboundary area between Russia and Heilongjiang Province of China [14]. The geographical coordinates of Birobidzhan are 30 ° 33 '31 "-31 ° 18' 38" N, 118 ° 28 '08 "-119 ° 38' 40" E. Its area is 200 thousand km^2; the population in 2014 reached 74,791 people.

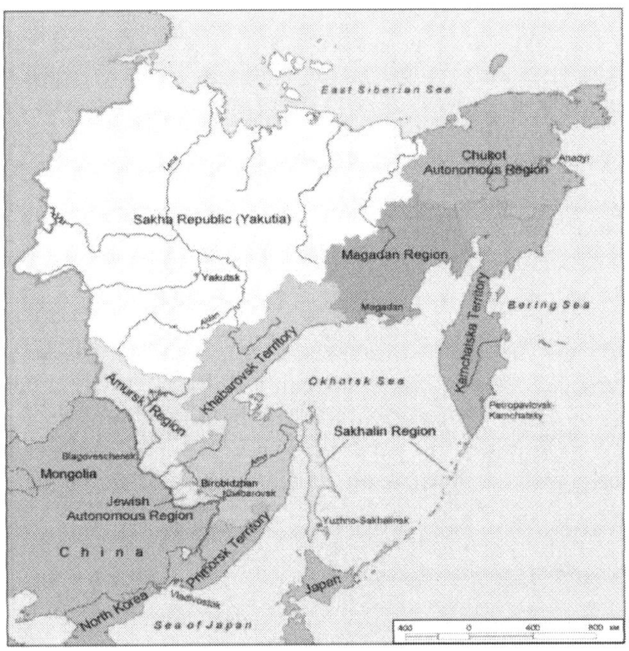

Figure 1. Scheme of Birobidzhan location.

Birobidzhan is situated mostly on low accumulative surfaces (within a high floodplain and 1 st terrace above floodplain of the Bira River). Alluvial deposits are mostly composed of sandy loams, sands, gravel-pebble materials. Sandy and ferruginized clay are less abundant. An elevated relief elements in the form of denudation remnants of low mountains (hills) and erosion-accumulative foothills with slightly undulating surface occupy western and south-western parts of the city. Here, the soil-forming rocks are eluvial-deluvium deposits of bedrock weathering products, mainly effusive (andesite, andesite-basalt, quartz porphyry). The slopes of low mountains are covered with oak-broad-leaved and deciduous-forb woods with numerous representatives of the Manchurian flora. Swampy sedge-reed grass meadows and grass-moss marshes bound the city in the northeast and east, and large massifs of high-floodplain grasslands are located in the city south. Its central part (the city proper) has undergone several stages of construction and redevelopment that greatly altered natural landscapes and soils.

2.2. Data and sources
The authors' field studies of soils and soil-grounds in different functional zones of Birobidzhan provided materials for the present research. The basis for soil mapping was a topographical map (scale 1:25 000). The authors also used unpublished cartographic materials from archives of some scientific and research institutions on Quaternary deposits, geomorphology and hydrological conditions of the city (scale 1:50 000). Data on vegetation are provided in this paper [15].

2.3. Methods
Basic research methods used were comparative geographic studies (geographic interpolation method), field routing and semi-stationary methods of soils and grounds analysis and the method of thematic mapping [16-19]. All the works were based on a systematic approach, i.e. on the analysis of interrelations of soil and other elements of the city ecosystem and taking into account the disturbance of these relations due to technogenic impacts, economic and recreational activities of the city population [7, 20]. Field works included making and describing soil profiles in different functional zones of Birobidzhan (industrial, residential, intercity areas of organized recreation, recreational-landscape and others).

The soil map was compiled using a traditional method [16, 19]. The soil contours on the map were defined by their genetic content with the help of traditional diagnostic characteristics of the soil morphological structure. The authors would like to stress that the main focus of the present study is the analysis of the degree of anthropogenic disturbance of the morphological structure of soils. The map was digitized using ArcView GIS software [17].

3. Results and discussion

3.1. Soil map and Legend
Soil and environmental studies have made it possible to identify four large ecological groups of soils and non-soil substrates in the territory of Birobidzhan: (1) natural relatively untransformed soils; (2) natural soils with a disturbed surface; (3) anthropogenic soil; (4) technogenic surface formations (TSF) [21]. According to these groups, the legend of the map of Birobidzhan soils has four sections and 23 cartographic units. The legend in the full paper version of the map is in the form of a table, which systematizes information about the relief, soil-forming rocks and the vegetation cover. The paper presents an ArcView GIS version of this map (figure. 2).

3.2. Detail description of ecological groups of soils

3.2.1. Natural relatively untransformed soils
Soils of this group are mainly found in the natural environment (outskirts of Birobidzhan) and presented by brown forest soils, sod-brown soils, sod- meadow gley, bog peaty- and peat-gley soils and complexes of floodplain soils. The terms used correspond to the legend of the 1:2.5 M Soil Map of the Russian Federation [22]. The indexation of soil horizons used accordingly [23]. These soils retain specific to natural morphological composition of the soil profile. The brown soils are common within hills under oak-broadleaf forests. By granulometric composition these soils are loam or even clay, gravelly to varying degrees, and depending on the position on the slopes are divided into the brown soils with a small-profile (not fully developed), typical and gleyic. Profile composition is typical brown soils - AY–BM–C, gleic brown soils - AY–BMg–Cg. Within the flat part of the city sod-brown soils and residual alluvial soils with favorable physical and hydro-physical properties (of light

granulometric composition, good permeability) are formed on the elevated bars of the first terrace above the floodplain under the oak-broadleaf forests. Profile composition is AY–[BM–C] al. Sod-meadow gley and peaty-(peat)-gley soils under wet reed-sedge meadows and lowland swamps are most common soils of this group. Profile composition is AY–G–CG and T–G–CG. Complexes of Bira-floodplain soils include sod layered under developed (W–C˜ ˜), sod meadow gleyic (AYg–Cg˜ ˜) and wetlands silty-gley soils (Hmr–G–CG˜ ˜). Soils of the 1st group have undisturbed (or weakly disturbed) ecological functions due to the preserved connection with the lithogenic basis and the vegetation cover and the integrity of the soil profile.

3.2.2 Natural soils with a disturbed surface

The main feature of these soils is the preserved connection with the parent rocks (untransformed middle and lower horizons are always observed in the soil profile). At the same time connections with the vegetation cover are partly or mostly completely disturbed. Changes in the morphological composition and structure in these soils are associated with the upper part of the profile up to 50 cm, where the "urbic" U horizon is formed [6, 8]. These soils are proposed to be called "urbosoils". Profile formular is U (AU)–B–C. Depending on the specifics of anthropogenic impacts the soils may be without upper genetic horizons, called urboabrazems and soils of little farms with a cultivated profile, called urboagrozems. Special varieties of urban soils are formed with the recreational use of natural soils. Under this influence, organogenic horizons of soils valuable in the biogeocenotic terms (forest litter, grass felts humus horizons) are transformed. In general, natural superficially-disturbed soils have in comparison with undisturbed, lower ecological potential, their ecological functions are partially preserved or transformed.

3.2.3 Anthropogenic soils

These soils are typical for built-up Birobidzhan areas. They are diagnosed by the absence of genetic horizons inherent to natural soils. They are primarily formed due to mechanical (and often chemical) soil destructions over 50 cm deep. Urbotechnozems with profile TG–C (TG) are most common in the city. They were formed (and are currently formed in areas of multi-storey housing) from dump, repeatedly mixed on the ground and imported soils. Urbotehnologic profiles consist of soil layers, different in color, thickness, lithology and dumping time and always include fragments of soil horizons, construction, household and industrial wastes in different amounts.

Along with urbotechnozem, created by the technical activity of man, the researchers distinguish in the group of urban anthropogenic soils proper urban soils, the formation of which is associated with the historical accumulation of the cultural layer - a homogeneous dusty-humified substrate with anthropogenic inclusions [6, 8]. However, such soil formations in the surface deposits of Birobidzhan are insignificant because of the youth of the urban settlement and the features of the aerial accumulation of matter. Among the anthropogenic soils on the territory of the city, we should also mention replantozems, the formation of which is associated with the deposition of a relatively fertile layer on bulk (mainly sandy-gravel) soils. Replantozems are mainly soils of lawn and flower beds. Profile formula is U (Utg) –B–C or U–TG–C. Soil fragments (mostly hydromorphic), everywhere buried under asphalt, buildings and accumulations of sand and gravel are classified as ekranozems. In some localities, mainly in industrial and communal areas of the city, there are urbotechnozems significantly contaminated with heavy metals (lead, copper, zinc), petroleum products (oils, fuels, lubricants) and other toxic substances. These variants of urbotechnozems are called intrusions. It was impossible to show all variety of anthropogenic soils on the map of scale 1:25000, that is why, these soils in the old (central) part of the city are shown as urban soil complexes which differ in history and development technologies, a modern economic and functional structure and the percentage ratio of the areas of their natural soils, soils with a disturbed surface and anthropogenic soils (see legend to figure 2).

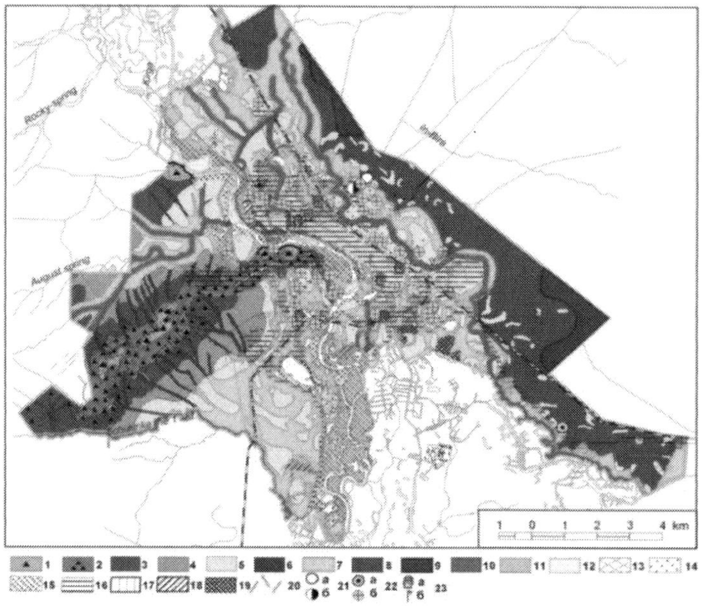

Figure 2. Soil Map of Birobidzhan (original scale 1:25 000).

Legend: **I.** ***Natural relatively untransformed soil***: 1 – *primitive brown earthquakes fragmentary gravelly-stony with outcrops of bedrock to the surface; 2 – small profile (thin) brown soils with lot of or not so much debris; 3 – typical predominantly loamy brown soils with little debris; 4 – gleyic and gley loam and clay brown soils; 5 – brown soils and sod residual alluvial brown soils combined with sod-gley and peaty-gley sod soils; 6 – sod-humus gley soils in the upper parts of bars on deluvium-proluvium sediments; 7 – meadow gley soils in combination with meadow-bog- and peat-sod-gley soils; 8 – peat-bog and peat-gley soils of a lowland type; 9 – bog peat-gley and peat soils of heterotrophic bogs; 10 – floodplain bog peat-gley soils; 11 – floodplain bog silt-gley and gley soils on silty sand and sandy loam; 12 – floodplain weakly-sod layered (gleyic) soils on the sands.* **II.** ***Natural soils with a disturbed surface***: *13 – soils in gardens, greenhouses and farms; 14 – soils of private landholdings; 15 – soils in landscape-recreational zones.* **III.** ***Anthropogenic soils***: *16 – urban soil complex of type I, composed of 75-95% of anthropogenic soils (urbotechnozems, replantozems, ekranozems) and 5-25% of natural soils with a disturbed surface; 17 – urban soil complex of type II, composed of 50-75% of anthropogenic and 25-50 % of natural soils with a disturbed surface; 18 – urban soil complex of type III, composed of less than 50% of anthropogenic and more than 50% of naturally-disturbed soils, 19 – a soil complex of city cemeteries (nekrozems).* **IV.** ***Technogenic surface formations (TSF)***: *20 – spoil grounds in areas of existing and developed (filled with surface waters) quarries; 21 – ash disposals: a – in use, b - closed (reclaimed); 22 – landfills of municipal solid wastes: a – authorized, b – unauthorized; 23 – soil-grounds (intruzems) contaminated with petroleum products, oil, fuel oil, heavy metals, rubber products, etc. in the areas of: a - fuel stations, b - car services (parking lots, car wash, tire repairs, etc.).*

3.2.4. Technogenic surface formations (TSF)

TSF in Birobidzhan spread steadily and increasingly. TSF are open dumps and accumulations of natural and technogenic soils (mining wastes, ash dumps, landfills of domestic and industrial wastes, including unauthorized). TSF systematics and diagnostics are at the development stage. On the map they are shown with extra-scale signs. Ecological functions of TSF are completely lost, their effects on the urban environment are characterized as destabilizing and TSF require measures to reduce their negative effects (recultivation).

3.3 Soil-ecological zoning

The created soil map of Birobidzhan and the analysis of morphological and genetic characteristics of several soil types made possible the zoning of soil-ecological conditions of the city. Four soil-ecological zones are distinguished according to the degree of disturbance of soil ecological functions (figure 3). Zones of soils with fully preserved, partially preserved and completely lost ecological functions occupy 54%, 23% and 14% of the total Birobidzhan area. A potentially environment-hazardous area was also identified in the city. It occupies 8% of the city area and may cause the formation of geochemical anomalies in soils.

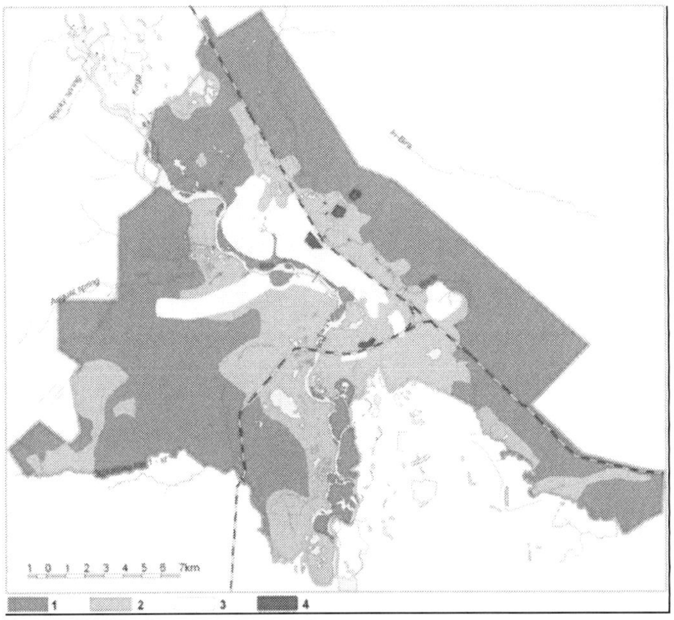

Figure 3. Zoning of Birobidzhan area by the degree of ecological disfunction of the soil cover (original scale 1:25 000).

Legend: *1 – areas with preserved ecological functions of the soil cover; 2 – areas with partially preserved ecological functions of soils; 3 – areas with fully lost ecological functions of soil; 4 – areas with a potentially environment-dangerous anthropogenic soil cover.*

4. Conclusions

The created soil map of Birobidzhan (scale 1:25000) reflects the diversity of soils and non-soil formations in the city. It presents a range of environmental characteristics of the soil cover, sources of potential pollution (landfills, gas stations, car service centers, etc.) and allows to determine the percentage of soils with different levels of disturbance. The schematic map of soil-ecological zoning reflects the summarized state of ecological functions of urban soils, which makes it valuable for the analysis of the overall ecological situation in Birobidzhan. It is important to note that 54% of the city soil cover is made of soils with preserved ecological functions. They are mainly located in the Birobidzhan outskirts, which are open spaces and are potential reserve areas for urban environmental planning and rational land management [3].

References

[1] Dobrovolsky G and Nikitin E 1990 *Soil functions in the biosphere and ecosystems* (Moscow: Nauka) p 261
[2] Kurbatova A and Bashkin V 2004 *Ecological functions of urban soils* (Moscow: Magenta) p 232
[3] Narbut N and Matyushkina L 2005 The ecological role of open spaces in urban area (Khabarovsk as an example) J *Geography and natural resources* **1** pp 46–50.
[4] [Stroganova M, Prokofieva T, Prokhorov A, Lysak L, Sizov A and Yakovlev A 2003 Ecological status of urban soils and economic evaluation of lands *J Soil Science* **7** pp 867–875.
[5] Xue Desheng, Huang Gengzhi, Guan Jingwen and Lin Jiarong 2014 Changing Concepts of City and Urban Planning Practices in Guangzhou (1949–2010) An Approach to Sustainable Urban Development *J Chinese geographical science* **24(5)** pp 607–619.
[6] Gerasimova M et al 2003 *Anthropogenic soils: genesis, geography, restoration Textbook* (Smolensk: Ojkumena) p 268
[7] Likhachev E, Timofeev D, Zhidkov P et al 1997 *City - ecosystem* (Moscow: Media – Press) pp 185–199
[8] Prokofieva T, Martinenko I and Ivannikov F 2011 Systematics of soils and soil-forming rocks of Moscow and the possibility of their inclusion in the general classification *J Soil Science* **5** pp 611–623.
[9] Qu Tao, Zhang Yun, Liu Ran and Dong Mali 2009 Social Effect of Environmental Pollution on Valley-cities in Western China *J Chinese geographical science* **19(1)** pp 008–016.
[10] Narbut N, Antonova L, Matyushkina L and Karavanov K 2002 *Strategy for forming ecological carcass of urban area (Khabarovsk as an example)* (Vladivostok – Khabarovsk: FEB RAS).
[11] Agarkova M, Tselishcheva L and Stroganova M, 1991. Morphological and genetic characteristics of urban soils and their systematics *J Vestnik Moskov. Univ.* Ser. 17, Soil Science. **2**: pp 11–16.
[12] Stroganova M, Myagkova A and Prokofieva T 1997 Classification of urban soils. Problems of anthropogenic soil formation Proc. Int. conf. (M.: V V Dokuchaev Soil Institute RAAS) **2** pp 234–239.
[13] Deeva N and Kerzhentsev A 1998 Methodological problems of soil-ecological mapping. *Soil Science* **9** pp 1112–1118.
[14] Ganzei S 2004 *Transboundary geo-systems in the south of the Russian Far East and in Northeast China* (Vladivostok: Dalnauka).
[15] Kalmanova V 2005 Comprehensive assessment of functional significance and ecological state of Birobidzhan dendroflora *J Regional problems* **6-7** pp 67–72.
[16] Trofimov B 2007 *Ecological-geological maps. Theoretical bases and methods of compolling (ed)* (Moscow: Vysshaya Shkola).
[17] Simakova M 2008 *Guide for medium-scale soil mapping using GIS (ed)* (Moscow: VV Dokuchaev Inst.)
[18] Kalmanova V B 2010 Selection and justification of methods to assess the ecological state of urbanized areas. *Regional problems* **13 (2)** pp 67–71.
[19] Mapping of soils on a large scale (methods, theory, practice) 1971 (Moscow: Nauka)
[20] Doncheva A B 1979 A systematic approach to assessing the impact of an industrial facility on the natural system. In: Problems of geographic forecast (ed) (Moscow: Moscow University Press) pp 93–98.
[21] Matyushkina L and Kalmanova V 2011 Map of soil and environmental conditions of Birobidzhan: structure and content *J Geodesy and Cartography* **9** pp 49–54.
[22] Soil Map of the Russian Federation Scale 1:2.500 000 (1988) In Fridland VM (ed) Dokuchaev Institute of Soil of the Russian Academy of Agricultural Science, GUGK USSR. VASKHNIL, Moscow.
[23] Shishoff L 2004 *Classification and diagnosis of soils of Russia* (Smolensk: Ojkumena).

Lignin in the Organic Matter of the Soils of the Russian Plain as Biomarker of Palaeoenvironment

I V Kovalev, N O Kovaleva

Soil Science Faculty, Lomonosov Moscow State University, Russia, Moscow
E-mail: kovalevmsu@mail.ru

Abstract. It has been shown by the methods of biochemistry, nuclear magnetic resonance, and isotope geochemistry that the proportions of lignin phenols may be used as molecular traces of paleovegetation due to their biochemical and physiological specificity and high resistance to decomposition. Lignin structures have been detected in soils and in iron–manganese concretions. The comparison of the ^{13}C NMR spectra of native lignin preparations isolated from different woody and herbaceous species with those of soil humic acids makes it possible to identify many characteristic shifts of lignin nature in humic acids at 56, 102, 115, 119, 131, 147, 151–152, 160, and 166 ppm. The information role of biomarker has been tested at the reconstruction of paleovegetation in the uplands of the Russian Plain. The representativeness of information has been increased using the isotope analysis ($\delta^{13}C$) and the radiocarbon dating; a new parameter—the composition of lignin phenols—has been introduced to the existing system of biomarkers.

1. Introduction

Biomarkers are organic molecules of the known structure and origin. Separate molecules of lignin phenols, amino acids, amino sugars, lipids can be reliably detected in living organisms, their residues, and waste, as well as in complex associates of humic substances in the soil [8]. Many biomarkers are resistant to degradation and mineralization in soils under specific conditions because of their specific structure and biochemical functions; therefore, they serve as molecular traces of paleobiota and land vegetation and as indicators of the rates of recent and past biochemical processes occurring in the biosphere. Individual organic compounds are well preserved not only in the normal profiles of postlithogenic soils, but also in their redeposition products, soilcolluvial and soilalluvial complexes, pedosediments, pedoliths, derivatives of separate horizons, bottom sediments of water bodies, etc [7].

The information role of biomarkers is important when the pool of molecular products of organic matter decomposition in soils is not identical to the sum of individual waste components because of the masking effect of the mineral matrix. The information role of biomarkers extracted from soils can be increased due to the determination of their structurally specific isotopes [8]. The diagnostics of the structural elements of individual nonspecific compounds in associates of humic substances, e.g., by NMR spectroscopy, in combination with the radiocarbon dating of soil humus increases the representativeness of the obtained information. Although the current database on the contents of similar individual compounds in soils is still poor, there is a significant body of data in the Russian soil science about the nonspecific groups of compounds like lipids, chlorophyll, amino acids, lignin, inositol phosphates, fungal melanins, etc. in soils [7]. A unique methodological basis for the interpretation of biochemical information from soil archives includes the studies of organic matter from buried soils and Fe-Mn nodules.

Lignin is the most prevalent natural phenolic compound of plant origin. The highest contents of lignin are in woody plants: lignin makes up 18–25% of wood biomass in deciduous species and 25–33% in coniferous species, while grasses contain about 4–9% lignin [2]. Chemically, lignin is an irregular threedimensional biopolymer of high molecular weight composed of branched phenylpropane units with colloidal properties [2]. The relative proportions of the component phenols are determined by the phylogenetic origin of plants and thus cause the appearance of many various low- and high-molecular-weight products of lignin decomposition in soils of different ecosystems. Different types of plants and plant communities produce a phylogenetically specific and strictly individual set of phenolic compounds.

The high stability of lignin phenols in buried soils is due to the fact that the phenylpropane moieties after burial enter into carboxylation and fragmentary condensation reactions, and their structural unites become less capable of entering into biochemical reactions, especially at the decreased biological activity [7]. Therefore, the aim of this work was to assess the information role of individual organic compounds of lignin origin and their applicability as molecular traces of paleovegetation, especially in soils of different ages and complex genesis.

2. Materials and methods

The objects of study included: light-gray forest soils (Greyic Phaeozems Albic; Eutric Podzoluvisols), light-gray forest gleyic and gley soils (Greyic Phaeozems Albic; Gleyic Greyzems) and agrogray soils of the southern taiga (Moscow district, 38° 12' E; 55° 02' N,); gray forest soils (Greyic Phaeozems Albic; Haplic Greyzems) under broad-leaved forest of Tul'skie Zaseki Reserve (Tula district, 37° 30' E; 54° 00'N); dark gray forest soils on the microelevations and dark gray forest soils with the second humus horizon (Greyic Phaeozems Albic; Haplic Greyzems) on the microdepressions, as well as agrogray soils and gray forest gleyic and gley soils (Greyic Phaeozems Albic; Gleyic Greyzems) at footslopes of the Bryansk district (33° 38' E; 52° 34' N).

Lignin phenols were isolated using Amelung's version [1] of the Ertel–Hedges procedure [2]. The determination of lignin in the soils included the alkaline oxidation of the sample with copper oxide at 170°C under pressure in a nitrogen atmosphere, the precipitation of HAs, and the preconcentration of phenolic products on compact disposable C18 columns under pressure. Lignin preparations were isolated by the evaporation of ethyl acetate on a rotary evaporator. The phenol components of lignin were derivatized into trimethylsilyl ethers and separated on a gasliquid chromatograph. They were determined on a gas chromatograph–mass spectrometer (Hewlett-Packard, Palo Alto, CA, USA) with a flame ionization detector and a capillary column. Individual reaction products (vanillin, syringic aldehyde (al), syringic acid (ac), p-coumaric acid, and ferulic acid) were identified by comparing the retention times and peaks with those of the known components and amounts used as external standards. The reproducibility of results was 95%. It should be emphasized that the oxidation products of lignin include only methylated lignin structures without any changes in their ring fragments. The alkaline oxidation of vascular pant tissues and their remains by copper oxide in the soil yields 11 phenols [2], which can be grouped in accordance with their chemical nature into three structural families: vanillyl (guaiacyl) (V), syringyl (S), and cinnamyl (C) ones. The two first phenol types are found in mixed oxidation products of plant tissues in the form of aldehydes, ketones, and acids, while the last have only acid forms: ferulic (F) and p-coumaric (K) acids. Thus, the sum of oxidation products reflects the total content of lignin in the sample. ^{13}C NMR spectra were recorded for HA preparations. The preparations were isolated from the studied soils, including iron–manganese nodules, by triple extraction with a 0.1 M NaOH + 0.4 M NaF mixture from the samples decalcified with 0.05 N H_2SO_4. [6]. The dialyzed and frozen-out HA preparations (50 mg) were dissolved in 0.6 mL of 0.3 M NaOD/D2O. Spectra were recorded on a Bruker Avance DRX 500 NMR spectrometer at 25.18 MHz and 290 K.

The isotope composition of carbon in soil organic matter was determined on a Thermo V Plus isotope ratio mass spectrometer and a Thermo Flash 1112 elemental analyzer [8]. Radiocarbon analyses of extractable humic acids were carried out at the Kyiv Radiocarbon laboratory.

3. Results and discussion

Although insufficient data are available about the biochemical compositions of different organisms participating in humification, it is considered established [2] that different types of plant tissues (gymnospermous and angiospermous, woody and nonwoody,cereal and herbaceous) have contrast lignin parameters. Our earlier studies [4, 5, 7] confirmed the revealed regularities and found three known lignin types in different samples of plant tissues. The first type is the lignin of coniferous plants (soft wood lignin), which contains vanillyl (guaiacyl) phenols as the main structural units. The content of cinnamyl (coumaryl and ferulyl) phenols in tissues of gymnospermous plants is low; syringic acids and aldehydes are almost absent, and the S/V ratio is 0. The second type is the lignin of deciduous plants (hard wood lignin) mainly consisting of similar amounts of vanillyl (guaiacyl) and syringyl structures. The lignin in the tissues of deciduous species (beech, chestnut) contains similar amounts of vanillyl and syringyl phenols. The C/V ratios are higher than 0 but lower than 1. The content of cinnamyl (*p*-coumaryl) phenols is close to 0. In the tissues of deciduous plants from the temperate zone (Moscow region), the V: S: C composition ratio is 3:4:1 for smallleaved species, 6:5:1 for broadleaved species (Tula region) [4]. The third type is the lignin of herbaceous plants containing the largest amounts of cinnamyl structural units (coumaric and ferulic acids), which exceeds their content in woody tissues by 4–6 times. The content of syringyl phenols in herbs is similar to their content in the wood of angiospermous plants but exceeds that in leaves by 5–6 times. Ferulic acids are associated with hemicelluloses in the cell walls of grasses; similar amounts of coumaryl and ferulyl phenols are found in herbaceous tissues [5].

Thus, as shown in our earlier series of publications [4, 5, 7], the nature of plant tissues in soils is clearly identified, and the oxidation of lignin biopolymers and their chromatographic separation into simple phenols provide information about the types of plant tissues in soils: the S/V ratio may be used for the separation of gymnospermous and angiospermous plant tissues, the C/V ratio for the separation of organic materials of woody and nonwoody origins, and the K/F ratio for the identification of herbaceous and cereal plant residues.

The ^{13}C NMR spectroscopic study of HA preparations showed that the lignin of higher plants is involved in the formation of specific humus compounds. In general, HAs are similar to soil samples in the content of lignin oxidation products and lignin parameters [7]; they inherit characteristic properties of plant tissues but show a better ordering of structural lignin fragments. Their proportions of vanillyl, syringyl, and cinnamyl phenols strongly vary among the tissue types but are similar within the homotypic ecosystems (1:1:1 for agrocenoses and 3:4:1 for woody cenoses). These facts are embodied in the NMR spectra (figure 1). It is known that the peaks at 147 ppm (region of aromatic bonds) and 56 ppm (region of aliphatic bonds) in the ^{13}C NMR spectra of HAs are due to compounds of lignin origin. The coefficient of correlation between the content of lignin (VSC) in the humus soil horizons of humid landscapes and the ^{13}C NMR peak area of the chemical bonds of lignin origin is 0.94 for the aliphatic region of the HA spectrum (56 ppm) and 0.93 for the aromatic region of the HA spectrum (147 ppm).

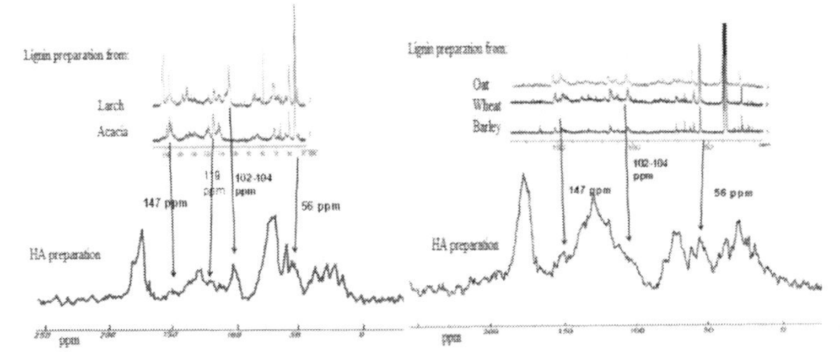

Figure 1. Comparison of the ^{13}C NMR spectra of native lignin preparations and soil humic acids from (a) woody plants and Light- gray forest soil and (b) cereal crops and Dark agrogray soil.

The comparison of the ^{13}C NMR spectra of native lignin preparations from different woody and herbaceous plant species obtained by Karmanov and Kocheva [3], with the spectra of our HA preparations [3], revealed that the signals inherited by HAs from plant tissues are also clearly identified at 102, 115, 119, 131, 152, and 160 ppm (table 1). The sets of peaks vary in the plant species; hence, they should also vary in the HAs from different soils; i.e., the chemical structures of lignin from coniferous and deciduous or cereal and herbaceous species differ not only in the content of the main types of intermonomer bonds β–O–4, β–5, β–β, and C_{arom}–C, but also in the composition of macromolecules, i.e., the proportions of the three types of monomeric units: vanillyl (guaiacyl), syringyl, and pcoumaryl ones (V : S : C). For example, the characteristic signals of the pcoumaryl units from herbs include characteristic peaks at 131.4–131.5 ppm (C-2 and C-6 in phydroxyphenylpropane units) and signals at 160.1 and 166.7 ppm related to the C-4 and C-β atoms in esters of *p*-coumaric structures. The signals at 53.5 and 53.8 ppm indicate the presence of coumaran and pinoresinol structures (in the macromolecules, monomeric units are linked by the β–5 (phenylcoumaran) and β–β (pinoresinol) bonds) [6].

Table 1. Signals of lignin units in the ^{13}C NMR spectra of soil humic acids.

Soil, horizon, cm	53.5-53.8 ppm coumarans pinoresi nols	55.7 ppm vanil lyls	55.9 ppm syrin gyls	102-104 ppm syrin gyls C-2, C-6	115.5 ppm C-2, C-5	119.1 ppm vanil lyls, C-6	131.4-131.5 ppm p-coumaryl C-2, C-6	151.0-152.0 ppm C-3, C-5-OCH_3	160.1 ppm C-4 esters	166.7 ppm C-β couma Ryls
					Moscow region					
Light gray soil, A 0–10	+	+	+	+	-	-	+	-	-	-
Gleyic light gray soil, A g',fs 0-10	+	+	+	+	+	-	+	+	-	+
					Bryansk region					
Gray soil, A 0-30	-	-	+	+	+	+	+	-	-	-
Dark gray soil, A 0-30	+	+	+	+	+	+	+	+	-	-
G 60-70	+	+	+	+	+	+	-	-	-	-

The signals in the region of 100–160 ppm are due to the presence of aromatic structural units in HAs from soils under forest. This region can be subdivided into four intervals: 100–117 ppm, signals of tertiary aromatic carbon atoms located in the ortho position to the C atoms with an oxygen function (C-2 and C-5 in the noncondenced vanillyl (guaiacyl) units or C-2 and C-6 in the syringyl units); 117–125 ppm, signals of tertiary aromatic carbon atoms without C atoms with an oxygen function in the ortho position (C-2 and C-6 in the phydroxyphenylpropane units and C-6 in the vanillyl (guaiacyl) units); 125–142 ppm, signals of quaternary aromatic carbon atoms, mainly C-1 and C-5; and 142–160 ppm, signals due to etherified aromatic carbon atoms. In the ^{13}C NMR spectra of HAs from soils of coniferous forest, characteristic signals are observed at 152.1–152.4 ppm, which are due to the C-3 and C-5 atoms bound to the –OCH_3 groups. The unsubstituted C-2 and C-6 atoms in the syringyl units also give signals at 102–104 ppm, and the signals at 119 ppm indicate the presence of the vanillyl (guaiacyl) units (C-6). In the spectrum of coniferous lignin, there are nopeaks at 152.2 and 152.5 ppm, which are due to the C-3 and C-5 atoms bound to the –OCH_3 groups. However, the uncondensed units of coniferous lignin contain a significant number of the C-5 atoms, which give a characteristic shift at 115.6 ppm. The general view of HA spectra (figure 1) shows that lignin from plants of the southern taiga zone becomes a source of more developed, elongated HA molecules with a well-developed aliphatic moiety (high peaks in the aliphatic region of the spectrum); feruryl and coumaryl phenols from steppe plants form compact structures of chernozem HAs with a well- defined aromatic moiety. Thus, different types of plants and plant communities produce phylogenetically specific and strictly individual set of phenolic acids and aldehydes, which determine the structural features of HAs and may be used as biomarkers in the study of buried soils and the paleoecological reconstructions of past regional environments. We shall present several examples.

It follows from the data presented in table 2 and figure 2 that tissues of angiospermous (woody and herbaceous) plants participate in the formation of humus in the surface and second humus horizons of gray forest soils. Their lignin consists of similar amounts of syringyl and vanillyl structures, which is typical for the tissues of woody species and is embodied in the value of the S/V ratio (about 0.8) (table 2).

Table 2. Lignin parameters in Gray soils of the Tul'skie Zaseki Reserve, mg/g $C_{org.}$.

Soil	Horizon, depth, cm	VSC, мг $г^{-1}$ $C_{орг.}$	(ac/al)v	(ac/al)s	S/V	$\delta^{13}C$, ‰
Gray forest soil	A 0–10	14.60	0.27	0.40	0.80	-26.91
	Bt1 50–60	2.30	0.25	0.24	0.55	-
Gray forest soil	A 0–10	12.20	0.24	0.37	0.79	-26.73
	Bt2 50–60	1.20	0.36	0.48	0.99	-

This ratio is also observed in the soil surface horizons of the Tul'skie Zaseki Reserve (0.7–0.8) (table 2), as well as in the second humus horizons of soils of the Bryansk region (0.7–0.9). The radiocarbon age of the latter soils is 2180 ± 60 years. The vanillyls: syringyls:coumaryls ratios are about 6:5:1 and 5:3:1, respectively, which corresponds to the residues of nonwoody tissues of such broad-leaved species as maple and alder. The cinnamyl (styryl and ferulyl) phenols are mainly extracted from herbaceous tissues; therefore, their complete disappearance from the second humus horizon can also indicate the contribution of woody species to humus formation in the late Holocene. The herbaceous genesis of cultivated cereal residues was revealed in the plow horizons of soils of the Bryansk region (V: S: C = 1:2:1; C/F = 0.79–0.95) and in the humus horizon of agrogray soil on the microelevation. The predominance of herbaceous elements was established from the values C/V = 1.5 and C/F = 1.7 with the domination of coumaryl units in the gley soils of hydromorphic landscape positions., Peak identifiers of cereals at 53.5 and 113.1 ppm are also present in the NMR spectra of HAs from the plow horizons of agrogray soil (Table 1). However, the second humus horizon of soil in the microdepression has no peaks of coumaryl structures (at 113.1 and 166.7 ppm) typical for cereals. On the contrary, the NMR spectra of HAs contain signals of syringyl and vanillyl structures, which confirm the participation of soft wood lignin in the formation of their humus. The isotope mass spectrometry data also reveal contrast isotope profiles for agrogray soils of microdepressions and microelevations (Table 3, figure 2).

The lightening of humus isotope composition in the second humus horizon of the soil in the microdepression to −28.4‰ confirms the conclusion about the participation of deciduous wood tissues in humus formation 2180 ± 60 years ago (Ki-17415) (table 3). A similar isotope composition of humus (−27.6‰) is found in the surface horizons of gray forest soils 330 ± 50 years old (Ki-18777) under the deciduous forest of Bryansk Forest Reserve. At the same time, cynnamyl phenols make heavier the isotope composition of humus carbon to −26.1−26.4‰ in the recent plow horizons and the horizons of the Atlantic age (6690 ± 110, Ki-18776) and to −24.7‰ in the lower part of the humus horizon of agrogray soil on the microelevation. The herbaceous meadow composition of plants is proved by the $\delta^{13}C$ values of −29.49 to −32.69‰ in the second humus horizons of soils in the hydromorphic positions of the landscapes.

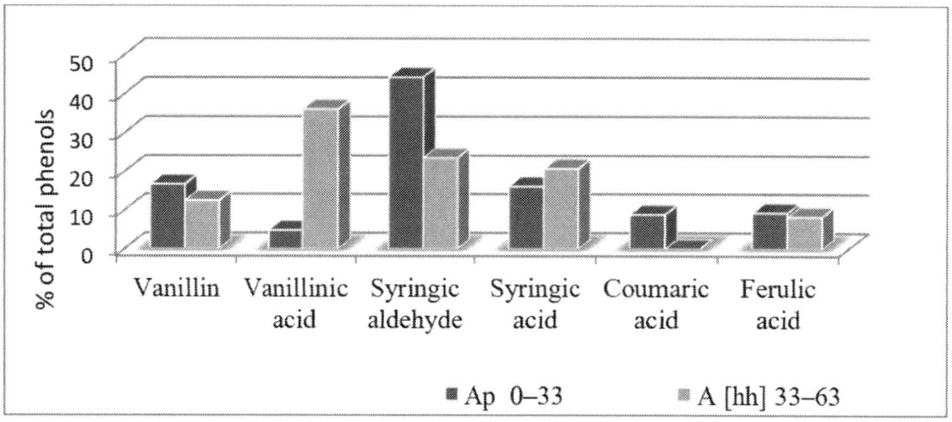

Figure 2. Products of lignin oxidation (% of total phenols) in recent (Ap, 0–33 cm) and second humus ([A], 33–63 cm) horizons of gray forest soils of the Bryansk region.

Table 3. Lignin parameters in Agrogray soil of Bryansk district, mg/g C_{org}.

Horizon, depth, cm	VSC, мг г$^{-1}$С$_{орг.}$	Coumaric acids	Age, ^{14}C	S/V	C/V	C/F	V:S:C	δ^{13}C, ‰
Agrogray soils with a second humus horizon (microdepression)								
Ap 0-33	9.09	0.81	-	2.80	0.85	0.95	1:3:1	-26.41
A[hh] 33-63	0.72	0.00	2180 ± 60 (Ki-17415)	0.91	0.00	0.00	6:5:1	-28.41
Agrogray soils with a second humus horizon (microdepression)								
Ap 0-20	6.77	0.98	-	2.13	1.41	0.88	1:2:1	-26.10
A[hh] 50-80	0.79	0.00	6690 ± 110 (Ki-18776)	0.65	0.00	0.00	5:3:1	-24.10
Gleyic dark gray soil (ravine bottom)								
A 0-32	12.49	1.54	-	1.50	0.94	0.83	1:2:1	-29.49
G 32-54	4.99	0.94	-	2.46	1.49	1.68	1:2:1	-32.69
Agrogray soil (microelevation)								
Ap 0-30	5.80	0.59	-	2.84	1.16	0.79	1:3:1	-26.11
EB 30-55	1.85	0.74	-	3.09	5.43	2.39	1:2:5	-25.39

In the agrogray soils of the southern taiga (plowland in the place of a former forest), the S/V value is slightly higher: about 1 (table 4). This corresponds to the values typical for fresh birch tissues, which are characterized by similar proportions of syringyl and vanillyl units [2]. The development time of these former forest soils was apparently insufficient for the complete transformation of decomposition-resistant lignin compounds in woody tissues. Moreover, the composition proportions of lignin phenols typical for the birch ecosystems of southern taiga remained unchanged not only in the soils, but also in nodules, where the age of humus is from 1600 to 1990 years.

The isotope composition of organic carbon in nodules (–26.7‰) agrees with the paleoecological scenario of their formation under variable moisture conditions 1990 ± 90 years ago, which correspond to the current ecosystem of small-leaved (birch or aspen) forest. However, the isotope compositions of some nodule fractions is lightened because of the presence of coumaric structures from meadow plants to –27.41‰, which indicates more hydromorphic conditions of their formation 1600 ± 80 years ago that replaced the drier period (–26.7‰) of about 2000 years ago. The contents of lignin compounds (VSC) are regularly accumulated in the gley soils and in the second humus horizons and reflect the degree of soil hydromorphism.

Table 4. Lignin parameters in Agrogray soil of Moscow district, mg/g $C_{org.}$.

Soil		VSC, мг г$^{-1}$ $C_{opr.}$	Age, ^{14}C	V:S:C	S/V	δ^{13}C, ‰,
Agrogray soil (micro elevation)	Soil, Ap 0-15	10.92 ± 0.62	-	2:2:1	0,95	-27.11
	Fe-Mn nodules, (1-2 мм)	1.29 ± 0,51	-	2:2:1	0,85	-26.70
Gleyic agrogray in micro depression	Soil, Ap, fs, g'	11.97 ± 0.80	-	2:2:1	1,02	-27.08
	Fe-Mn nodules, (1–2 мм)	1.34 ± 0,02	1990 ± 90 (Ki–17759)	1.6:1.5:1	0,93	-26.73 ± 0.13
	Fe-Mn nodules, (2–3 мм)	1.41 ± 0,30	1600 ± 80 (Ki–17411)	1.6:2:1	1,16	-26.76 ± 0.36
	Fe-Mn nodules, (3–5 мм)	0.81 ± 0,02	1690 ± 110 (Ki–17412)	1.8:3:1	1,51	-26.89 ± 0.36

Thus, analysis of the presented data on the composition of lignin phenols in the recent and buried horizons of different polygenetic soils, reveals that the composition of lignin phenols in soils serves as a molecular trace of land vegetation. In spite of the significant number of lignin peaks in the NMR spectra of HAs, most of them allow only the qualitative identification of biomarker and the fixation of changes in the natural environment during the time intervals determined by radiocarbon dating.

Acknowledgements
We are grateful to Russian Scientific Fund for the financial support of this investigation (grant 17-14-01120) and professor W. Zech (Bayreuth University of FRG)

References
[1] Amelung W 1999 *Zum Klimaeinfluß auf die Organische Substanz Nordamerikanischer Praärieböden*. Bayreuth. p 131
[2] Ertel J R and Hedges J I 1984 The lignin component of humic substances: distribution among the soil and sedimentary humic, fulvic and baseinsoluble fractions *Geochim. Cosmochim.* Acta **48** pp 2065–2074
[3] Karmanov A P, Kocheva I S and Borisenkov M F 2013 Antioxidant properties of lignins *Vestnik of Biology Institute, Komi SC UrDRAS* **10-11** pp 31–38. (in Russian)
[4] Kovalev I V and Kovaleva N O 2008 Biochemistry of lignin in soils of periodic excessive moistening (from the example of agrogray soils) *Eurasian Soil Science* **vol. 41, 10** pp 1066–1076
[5] Kovaleva N O and Kovalev I V 2002 Aromatic Struktures of Lignin in the Organic Matter of Gray Forest Soils *Moscow University Bulletin 17. Soil Science* **2** pp 23–29
[6] Kovaleva N O and Kovalev I V 2003 Peculiarities of organic matter of Fe-Mn nodules in gray forest soils determined by C NMR-spectroscopy *Moscow University Bulletin 17. Soil Science,* **2** pp 23–29
Kovaleva N O and Kovalev I V 2009 Transformation of lignin in surface and buried soils of mountainous landscapes *Eurasian Soil Science* **vol. 42, 11** pp 1270–1281
[7] Schatz A., Zech M., Buggle B., Gulyas S., Hambach U., Markovic S., Sumegi P., and Scholten T. 2011 The Quaternary loess record of Tokaj, Hungary – reconstructing palaeoenviroment, vegetation and climate using stable C and N isotopes and biomarkers *Quaternary International* **240 (1-2)** pp 52–61

Isotopic signature of Tian-Shan mountain soils as a record of climatic changes of the Late Pleistocene and Holocene

N O Kovaleva

Soil Science Faculty, Lomonosov Moscow State University, Russia, Moscow
E-mail: natalia_kovaleva@mail.ru

Abstract Specific features of the polygenetic mountain soils of the Tian-Shan (Kyrgystan) are due to the action of present-day and relict soil processes that vary in age and intensity under the influence of glacier movements and climatic fluctuations. These properties can be used as indicators of paleoclimatic changes. The diagnosis of ancient pedogenesis was based on criteria with the longest response time, namely, soil morphology, characteristics of organic matter, ^{13}C-NMR spectra of soil humic acids, isotope composition of humus and carbonates, and the soil age. The results indicate a glacial climate of the Late Pleistocene, a dry and cold climate during the Early Holocene, warm and dry conditions of soil formation in the Middle Holocene, and humidity climate of the Late Holocene.

1. Introduction

Global warming and impending sea-level rise have attracted an increasing interest in regional climate changes and in the methods of their investigation. The soil cover of the extensive territories of the Tian-Shan is represented by polygenetic soils along with soils buried by the Late Pleistocene and Holocene loess. Specific features of these soils are due to the action of elementary modern and relic soil processes, whose age and intensity varied under the influence of glacier movements and climatic fluctuations. These properties could help to elucidate conditions of soil formation during the Late Pleistocene and Holocene. This paper presents records of ^{13}C/^{12}C ratios of soil organic matter and carbonates, combined with chemical properties of humus and radiocarbon dating, to reconstruct changes in climate and paleoenvironments during the Pleistocene and Holocene in the high mountains of Asia, focusing on the Kirgizskiy Range of the Tian-Shan (Kyrgystan).

2. Materials and methods

A soil chronoseqence was studied in the central part of Kirgizskiy Range (at the longitude of the city Bishkek, 74° 42′ E) at elevations between 1800 and 3400 m above sea level on interfluve of the Alamedin and Ala-Archa Rivers. The interfluve is the highest part of Kirgizskiy Range that was glaciated by the Shopokov Glacier.

The studied valley contains typical glacial features including U-shaped cross-sections, cirques, moraines, and glaciofluvial sediments. Bedrock is comprised of limestone, diorite, porphirite and metamorphic rocks.

This study examined 40 soil profiles, representing the change of soil types from the foothills to the top slope of the Tatyr, Chon-Kurchak, Kurchak-Tor valleys.

The older members of the chronosequence developed on the lowest frontal moraines of the Last Glacial Maximum (LGM) at altitudes of 1900-2100 m. These soils are mountain Chernozem (Chernozems Haplic) profiles with the following horizonation: A-AB-B-BC1-C1-Bb_1-C2-C3-Bb_2-C4-

Bb_3. Vegetation cover is herbage-estragon (*Festuca valesiaca, Artemisia dracunculus, Poa relaxa, Stipa capillata, Phlomis oreophila, Elytrigia repens*) meadow-steppe.

Upvalley, moraines are present at 2100, 2350, 2450 and 2600 m above sea level. These moraines characterize Late Glacial ice advances in Holocene [3]. These soils are polygenetic soils that have buried A and B horizons. These soils can be classified as Chernozem-like soils (Leptosols Mollic) with A-Ab-AB-B-Bt-C1-B_{b1}-C2-B_{b2}-C3 horizonation. Vegetation is a prairie-grass-herbage meadow steppe (*Brachypodium pinnatum, Helictotrichon schellianum, Phleum phleoides, Campanula glomerata, Galium septentrionale, Rananculus polyanthemus*) and dry subalpine meadows (*Phlomis oreophila, Aegopodium alpestre, Alchemilla retropilosa, Dichodon cerastoides, Ligularia thomsonii, Brachypodium pinnatum, Poa angustifolia, Dactylis glomerata, Geranium regelii, Alopecurus pratensis*).

Incomplete soils developed on the frontal moraines with large porphyrite and granite-diorite boulders occur at altitudes of 3100 m. These soils are in the initial stages of their formation and are classified as alpine-meadow soil-rankers (Leptosols Umbric), with A-AC-C horizonation. Vegetation is cobresia meadow from *Kobresia humulis, Alchemilla retropilosa, Allium atrosanguineum, Astragalus alpinus, Potentilla nivea, Carex stenocarpa.*

All samples taken from soil pits were air dried, sieved (<2mm), and ground prior to chemical analysis. In addition, macro-, meso- and micromorphological [1] descriptions were made for all studied profiles. Total C and N, as well as organic C (C_{org}) were analyzed by dry combustion with a Carlo Erba ANA 1500 C/N/S Analyzer. Group and fractional composition of humus was analyzed by the Tyurin method [5]. The optical properties of humic and fulvic acids were measured in the alkaline extract by the spectrophotometer "CF-18" and the coefficient of extinction, color coefficient were determined. These analyses are defined as follows: (1) coefficient of extinction, E, is the optical density of the solution when the λ= 465 nm, the concentration of humic acids is 0.001 %, and the cuvette length is 1 cm; (2) color coefficient $Q = D_{465}/D_{650}$, where D is the optical density at 465 and 620 nm wavelength; (3) Cha/Cfa, where ha – humic acid, fa – fulvic acid.

Radiocarbon analyses of extractable humic acids were carried out at the geochemical laboratory of the Institute of Geography, Russian Academy of Sciences. We also used radiocarbon data, obtained by Maksimov and Pomortsev [6] for soils studied at Leningrad University.

13C NMR spectra were recorded for HA preparations. The preparations were isolated from the studied soils, including iron–manganese nodules, by triple extraction with a 0.1 M NaOH + 0.4 M NaF mixture from the samples decalcified with 0.05 N H2SO4. After the separation from colloids by centrifugation (Beckman Model J-6 centrifuge, 2500 g, 15 min), the HA preparations were purified from mineral salts by electrodialysis (MWCO 12000–14 000) [6]. The dialyzed and frozen-out HA preparations (50 mg) were dissolved in 0.6 mL of 0.3 M NaOD/D2O. Spectra were recorded on a Bruker Avance DRX 500 NMR spectrometer at 25.18 MHz and 290 K.

The $^{13}C/^{12}C$ ($\delta^{13}C$) ratio was measured for all samples on a Finnigan MAT 251 mass spectrometer with PDB as reference and average precision less than 0.1 ‰ at Soil Scince Institute of Bayreuth University (BRD) The isotope composition of carbon is expressed by the $^{13}C/^{12}C$ ratio or the relative density: $\delta^{13}C$, ‰ = 1000[(Rsample/Rstandard – 1)], where R = $^{13}C/^{12}C$. For the PDB standard, $^{13}C/^{12}C$ = 0.01125, and $\delta^{13}C = 0$.

3. Results and discussion

The $^{13}C/^{12}C$ ratio in carbonates and humus has been under examination for loess and soils in order to clarify the type of vegetation and conditions promoting carbonates formation. Based on the radiocarbon chronology, soil morphology, chemical properties and $\delta^{13}C$ data of the recent and buried soils the following major climatic events can be reconstructed.

Mountain Chernozems are not clearly differentiated in terms of humus properties (table 1). Their profile can only be subdivided into upper and lower parts, which are assessed as recent and early Holocene, respectively. The criteria to subdivide the profile are the values of the color coefficient ($Q > 4$ and $3 < Q < 4$), the coefficient of extinction (0.08 and 0.19). This subdivision is in agreement with the climate of the modern meadow-steppe and the effect of the early and middle Holocene xerothermic period in the formation of ancient Chernozems. Soil texture is strongly affected by aeolian silt additions. Loesslike loam contains buried Pleistocene soils (Bb) transitioning to the highly weathered and rubefied metamorphic bedrock. According to Zech et al. [9], the radiocarbon age of such buried soils (Bb_2 and Bb_3 horizons) in loess of North-Western Tian-Shan is 15930±460 and 24300±1160 years BP. Authors assume that the latter paleosol developed during a warmer interstadial which is postulated for oxygen isotope stage 3, between the Sartan (Late Valdai, isotope stage 2) and Zyryanka (Early Valdai, isope stage 4) ice advances. Beginning with the Sartan glaciation about 24 000 years BP, solifluction debris and Shopocov till began to cover the paleosol. $\delta^{13}C$ values of -4 to -5‰ from carbonate in the C3 horizons of the mountain Chernozem (on the LGM moraine) characterized the glacial climate too.

Based on the $\delta^{13}C$ data, -8 to -10‰ in the C1 and C2 horizons the cold and dry climatic period with little alpine desert vegetation and slightly weathered interstadial soils can be reconstructed. A similar horizon is described by Zech in Western Tian-Shan [9] (24300±1160 years BP), as well as by Yan et al. in Tibet (25910±400 years BP) [8]. According to $\delta^{13}C$ in the B_{Ib} horizon (-24.41 ‰), the warm period is reconstructed at about 14030±880 years BP [6] in Chon-Kurchak valley with C-3 type of vegetation. The peak areas of aromatic and aliphatic fragments of the buried soils humic acid molecules are similar - about 28 % in NMR spectra (figure 1).

Cold and dry climate with, probably, CAM-type of photosynthesis ($\delta^{13}C$ in the C1 horizon -7 to -10 ‰) (table 1) predominated, in the Early Holocene at about 10100± 564 and 9130±640 years BP [6]. At the 8-6 kyr BP, the warm and dry period can be reconstructed, when peats spread out in the piedmonts and lowlands due to glacial melting. Meadows with C-3 type of plants developed, based on $\delta^{13}C$ values ranging from –19 to –29 ‰. The age of B horizon at the altitude 2350 and 2450 m in studied valley is 7130 and 6440 years BP respectively [6].

1. The profiles of the Chernozem-like soils, which were formed on the moraines of the Shopocov glacier during the last 10000-13000 years, are clearly differentiated in terms of the morphology and organic matter properties. The upper A horizons have dark-brown color and crumby structure, as well as the values of the color coefficient Q>4. Moreover, the humus type is humic-fulvic in the A horizon. The radiocarbon age of these horizons is 3010±90 years BP (IGAS-1028).
2. Very high values of the humification index, high values of the coefficient of extinction, and the spectral curve slope with the color coefficient 3<Q<4 are intrinsic to the Ab and AB horizons (table 1). The latter being indicative of a more complex structure of humic acids is rich in benzoid compounds typical of mountain Chernozem profiles. Black buried horizon of the humus profile of a previous stage of soil formation has a residual granular structure and provides evidence for a warmer and wetter period than at present, with high plant biomass production. The radiocarbon age of it is 5560±120 years BP (IGAS-1027).

The data in figure 1 and table 1 show that in the buried horizons Ab the peak areas of aromatic part of the humic acid molecules are larger than their halos for humic acids of the surfaces horizons by 1.5 times (30 and 46 % respectively)

Besides that, the Ab horizon has the following relict characteristics: high values of the Cha/Cfa ratio, humic humus with the predominance of humic acids bound to Ca, along with a darker color, the well pronounced granular structure (as distinct from A horizon). Thus, humus composition in the horizons below the A is identical to that of the mountain Chernozems, reflecting the previous stage of Chernozem pattern of the development. This all indicates that dry conditions of soil formation existed in the area of Chernozem before the climatic changes that occured between the Middle and Late Holocene. The Holocene optimum events with a warm and dry climate, air temperatures higher than at the present, a C-3 type of vegetation on the Chernozem soils in the valleys can be reconstructed at 6-4 kyr BP.

At the same time, A horizons forming under modern bioclimatic conditions differ appreciably from the underling horizons with respect to the condensation of aromatic nucleus which are closer to typical

IOP Conf. Series: Earth and Environmental Science **107** (2017) 012117 doi:10.1088/1755-1315/107/1/012117

alpine soils. The peak areas of aromatic part of the mountain meadow alpine soils humic acid molecules are about 26 % (figure 1).

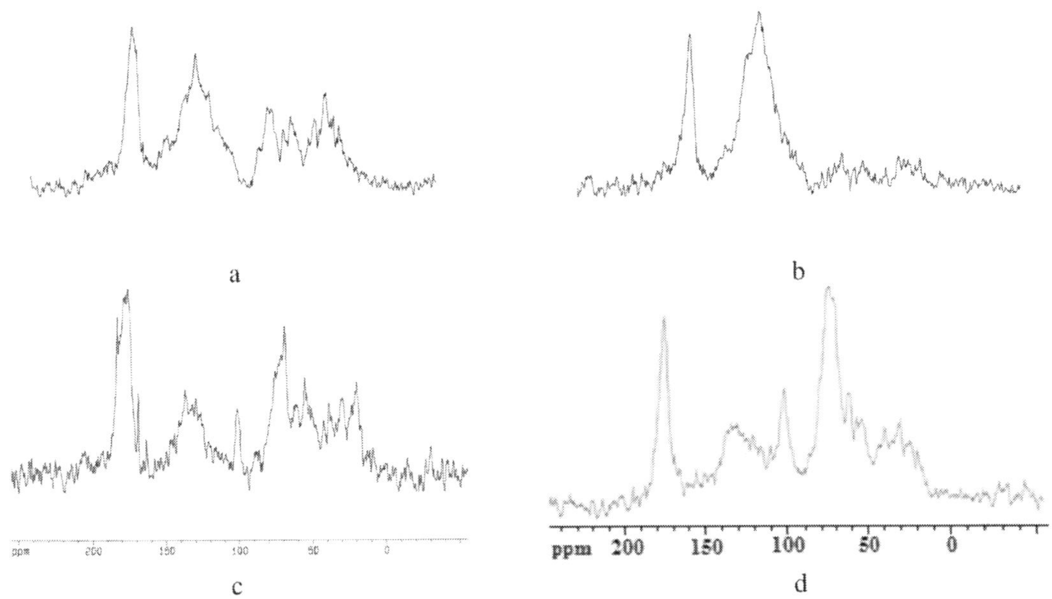

Figure 1. ^{13}C NMR spectra of humic acid from (a) the surface (3010±90 years BP) and (b) buried horizons (5560±120 years BP) of chernozem-like sois, c – from A horizon of mountain-meadow soils (109±1.47 years BP), d – from Bb horizon (14030±880 years BP).

Thus, the Late Holocene climatic changes toward cooling and wet caused the formation of a new subalpine soil type on the profile of mountain Chernozem in the zone from the lower boundary of the LGM (2100 m) to the alpine belt (3000 m). Favorable and relatively stable climatic conditions of Middle Holocene gradually changed to the Late Holocene. Decreasing of $\delta^{13}C$ values in humus, the traces of trees phenolic compaunds in soil humus [5], optical properties of humus acids detect wetter conditions than in Middle Holocene.

Soil in initial stages of formation at the altitude above 3100 m are characterized by the high humus content (up to 15 %), humate-fulvate type of humus, low value of the extinction coefficient (0.06), the considerable green humic acids content, all of which indicate a humid and cold climate of the alpine zone during, at least, the last 100-200 years. The radiocarbon date of the A horizon in the alpine soil is 109±1.47 years BP (IGAS-1029). These soils correspond with glacial advances during the so-called "Little Ace Age" with a maximum advance at about 1850 in the Alpine [9].

Modern subalpine and alpine vegetation and Chernozem-like soils below the 3000 m, and subalpine soils at the altitude above 3000 m began to develop at 4-3 kyr BP, and a trend towards increasing humidity has taken place ever since ($\delta^{13}C$=-28 ‰).

Thus, Late Holocene climatic changes toward cooling and wet caused the formation of a new soil type on the profile of mountain Chernozem. A horizons forming under modern bioclimatic conditions in the subalpine zone of mountains studied differ appreciably from the underling horizons with respect to the condensation of aromatic nucleus which are closer to typical alpine soils (aromaticity of humic acids is 34 and 27% respectively)

Thus, climatic changes in the Late Pleistocene and Holocene in the mountain regions of the Tian-Shan (Kyrgystan) seem to coincide with Tibet and Eastern Europe [7] and correspond to the global climate changes [2, 3, 4, 8, 9]. But, these changes in Central Asia were probably less intensive than those in the other regions.

Table 1. Isotope composition of carbonate and humus carbon and some chemical properties of soils.

Soil, Altitude above sea level, m	Article I. Horizon, depth, cm	Cha Cfk	E	Q	Radiocarbon age, years B.P.	$\delta^{13}C$ humus, ‰	$\delta^{13}C$ carbonates, ‰
	Article II.						
Chernozem	A 12-18	2.2	0.09	4.61	-	- 25.50	-
Haplic	38-44	3.1	0.08	4.20	-	- 25.22	-
1800 m	B 56-76	2.1	0.14	4.15	-	-25.03	-
	76-83	1.4	0.19	3.16	-	-14.91	-
	BC1 83-101	1.6	0.12	3.33	-	-12.30	-10.07
	C1 101-114	1.2	0.15	3.64	-	- 9.00	-7.68
	C2 114-126				-	- 9.59	-8.53
	Bb_1 126-157				-	-9.41	-8.01
	C3 157-187				-	-5.09	-4.36
	Bb_2 187-200				> 24000	-5.59	-4.65
Article III.	A 0-12	1.5	0.08	4.08	-	-25.34	-
eptosols	A_b 37-50	0.4	0.11	4.47	-	-25.39	-
Mollic	B 80-95	1.8	0.17	4.00	-	-19.37	-
2100 m	C_1 100-110	2.1	0.24	3.36	-	-8.32	-7.84
	Bb_1 110-125	1.5	0.23	3.40	14030±880	-24.41	-19.55
	C2 125-135	1.1	0.14	3.34	-	-8.53	-7.80
	C3 135-165	0.4	0.05	3.31	-	-8.62	-7.54
	Bb_2 165-200					-9.11	-8.04
Article IV.	A 4-35	1.0	0.14	4.41	3010±90	-24.91	-
eptosols	A_b 35-50	2.5	0.17	4.87	5560±120	-25.25	-
Mollic	B 87-97	1.7	0.07	3.57	7130± 610	-28.80	-11.94
2350 m	C 97-110	1.2	0.06	3.63	-	-10.58	-8.00
	B_b 110-120	1.3	0.08	3.55	10100±564	-10.36	-9.30
Leptosols	A 0-14	1.6	0.09	4.41	-	-26.01	-
Mollic	23-37	1.3	0.13	4.87	-	-25.43	-
2450 m	A_b 37-50	1.7	0.13	4.07	6440±180	-25.32	-
	BC 73-83	2.8	0.18	3.42	-	-21.01	-11.52
	Bb_1 83-100	0.5	0.01	3.33	9130±640	-11.12	-10.09
	C_1 100-130	0.3	0.03	3.50	-	-10.19	-9.11
	Bb_2 130-150	0.8	0.01	1.50	-	-10.62	-9.27
Leptosols	A 3-10	1.3	0.06	5.03	109±47	-24.18	-
Umbric	AC 10-35	0.9	0.06	5.13	-	-24.21	-
3100 м	C 35-50	0.9	0.09	4.59	-	-23.89	-
		0.8	0.05	5.11	-		

Acknowledgements
We thank Russian Scientific Fund for the financial support of this investigation (grant 17-14-01120) and professor W. Zech (Bayreuth University of FRG)

References

[1] Kovaleva N O 1996 Micromorphological diagnostics of genesis of Northern Tian-Shan soils *Abstracts of the X International Meeting of Soil Micromorphology* Moscow 49
[2] Kovaleva N O and Evdokimova T I 1997 Holocene soils of the Kirgizskii Ridge in the Tian - Shan Mountaines *Eurasian Soil Science* vol.**30** **9** 950–956

[3] Kovaleva N O, Zech W and Haumaier L 2000 Soils as indicators of paleoclimatic changes in the Northern Tian-Shan *Pamir and Tian-Shan International Workshop* Universitet Bayreuth 26

[4] Kovaleva N O 2004 Northern Tian-Shan paleosoil sedimentary sequences as a record of major climatic events in the last 30.000 years. *Revista Mexicana de Ciencias Geologicas.* Vol. **21 1** 71–78

[5] Kovaleva N O and Kovalev I V 2009 Transformation of lignin in surface and buried soils of mountainous landscapes *Eurasian Soil Science* Vol. **42 11** 1270–1281

[6] Maksimov E B 1980 *Lednikovaya istoriaya chrebta Kirgizskii Alatau (The glacial history of the Kirgizskii Alatau range* LGU Leningrad 182 (in Russian)

[7] Schatz A, Zech M, Buggle B, Gulyas S, Hambach U, Markovic S, Sumegi P, and Scholten T 2011 The Quaternary loess record of Tokaj, Hungary – reconstructing palaeoenviroment, vegetation and climate using stable C and N isotopes and biomarkers *Quaternary Int*ernational vol.**240 (1-2)** 52–61

[8] Yan G, Wang F B, Shi G R and Li S F 1999 Palynological and stable isotopic study of palaeoenvironmental changes on the northeastern Tibetan plateau in the last 30,000 years. *Palaeogeography, Palaeoclimatology, Palaeoecology* **153** 147–159

[9] Zech W, Glaser B, Ni A, Petrov M and Lemzin I 2000 Soils as indicators of the Pleistocene and Holocene landscape evolution in the Alay Range (Kyrgystan). *Quaternary International* **65** 147–160

Mercury in the Soil of the Tunka Depression

E E Lyapina[1], A A Cherkashina[2]

[1] Institute of Monitoring of Climatic and Ecological System of SB of RAS
[2] V. B. Sochava Institute of geography of Siberian Branch of the Russian Academy of Science, Irkutsk, Russia
E-mail: eeldv@mail.ru

Abstract. The work evaluates the general distribution of mercury in the soil cover of the Tunkinskaya depression (the Republic of Buryatia, the national park "Tunkinsky"). For this purpose, its gross contents in natural and agrogenically transformed soils were studied: plow lands, fallow lands, hayfields and pastures. The physico-technical characteristics of soils, the content of organic carbon, group composition of humus are determined. The method of processing the results included the calculation of the ecological and geochemical parameters: the concentration coefficient relative to the background, MAC, Clark concentration relative to the Earth's crust, the Earth's soils, the identification of the relationship with the physical and technical characteristics of the soil, and the content of C02, CO2 carbonates, fulvic and humic acids.

1. Introduction

Soil is an important component of biosphere. It has several main functions, such as geological, geoecological, bioecological, bioenergy, hydrological, and meteorological. Soil is an intermedium for regulation of the water and air chemical composition, and for energy and matter exchange in many elements of the global biogeochemical cycles. All these functions determine many interrelations in the global mechanism of geographical environment functioning [6, 7, 10].

Mercury is pervasive, compound, and hydro- and sulfophilous. It is present in all types of ecological environments and has many occurrence forms, which complicate the process of its study. Even in very low concentrations mercury is supertoxic and superpathologic. It is found in the lithosphere in the form of solid compounds, in various gaseous phases and dispersed forms, each of which predominates in concrete physical-chemical conditions, but easily changes from one form to another [1, 2].

2. Methods and research subjects

This study aims to evaluate the general allocation of mercury in the soil mantle of the Tunka Depression (Republic of Buryatia, Tunkinskiy National Park). For that purpose, a total mercury concentration in natural and agrogenically transformed soil samples was examined [1]. Those samples were taken from arable land, set-asides, hay-fields, and pastures. Soil samples were collected during the summer periods of 2015 and 2016, using two methods: envelope method according to [5, 9] and method of study of genetic horizons including soil materials. In total, 26 sites were chosen and 77 soil samples were collected. Under laboratory conditions the samples were air-dried at a room temperature and screened (sieve size is equal to 1 mm).

The Tunka depression is a continuation to the west of the Baikal rift zone. Its length from east to west is 190 km (figure 1). From the south it is bounded by the Khamar-Daban mountains and the spurs of the Khangarul ridge, from the north by the Tunkinsky loaches, from the west by the southeastern chain of the Great Sayan. In the basin there are six separate basins, separated by narrowings of the bed of the Irkut-Bystrinskaya, Torskaya, Tunkinskaya, Turanskaya, Khoytogolskaya and Mondina.

The sample analysis was conducted in the laboratories of International innovative academic center «Uranium Geology», Geoecology and Geochemistry Department, Tomsk Polytechnic University. Mercury content in the earth samples was detected on the Atomic Absorption Spectrometer RA-915+ with pyrolyzer PYRO-915 (pyrolysis method). Precision is equal to 5 ng/g. Mercury concentration was calculated on 1 gram of dry basis. Besides, there were determined physico-technological characteristics of the soil samples, Corg concentration, and humus group composition. Results processing technique included calculation of ecological-geochemical indexes: concentration coefficient with reference to the background, maximum permissible concentration, percentage abundance with reference to the Earth crust and soil, identification of interrelations with physico-technological characteristics of the soil, and also content of Corg, CO_2 carbonates, and fulvic and humic acids.

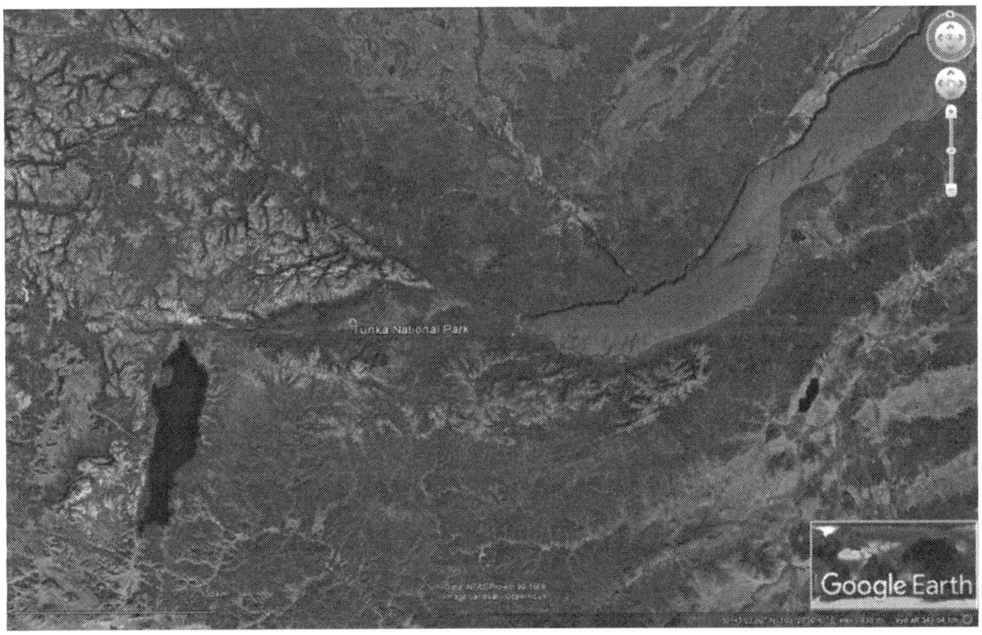

Figure 1. Map of the location of the Tunka depression.

3. Results and discussion

Average mercury concentration in the soil of the Tunka Depression is equal to 23 ng/g, assuming spread from 1 to 124 ng/g (table 1). All depth profiles of mercury concentration in studied soil cross-sections look similar: maximum values appear in the upper part of the profile with subsequent decrease. Acquired data corresponds to principles of mercury allocation in the soil in many other countries [3]. In addition, it is worth mentioning that the highest mercury concentrations are detected in upper horizons in the interval from 0 to 0.2 meters (figure 2). However, the depth of detection of the highest mercury concentrations is inhomogeneous: 26% of the highest mercury concentrations are found at a depth of 0-5 cm, 21% - at a depth of 10-15 cm, and 19% - at a depth of 5-10 cm.

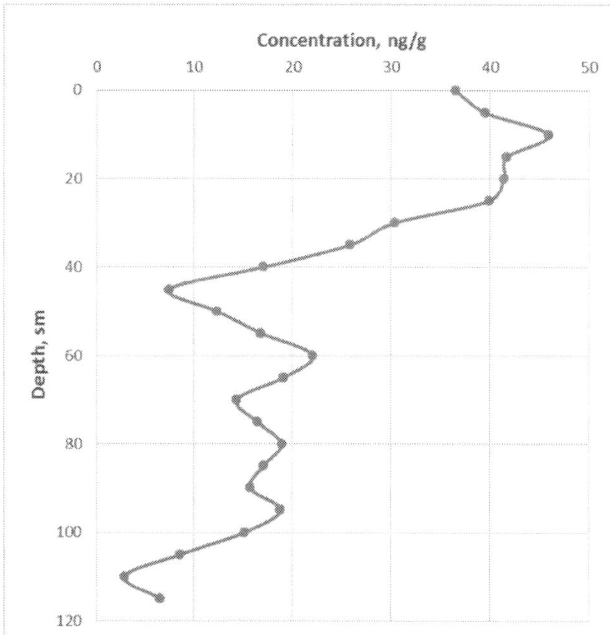

Figure 2. Distribution of mercury with depth in the soils of the Tunka depression.

It is found that upper horizons of the soil hold higher total mercury concentrations than lower horizons. Maximum values are detected in the detritus, peat, humus, and topsoil horizons. Relatively lower mercury concentrations are found in the vegetable and agrohumus horizons. In the soils of alfehumus (podbur and sod podbur soil) and texture-differentiated (sod brown podzolic soil) sections, which are widespread on the studied territory, it is clearly observed that there is a certain increase of mercury concentration in accumulation-illuvial horizons (BHF and BT). It can occur due to a downward migration of mercury in the form of organo-mineral complexes from overlying humus and eluvial horizons [9].

In the studied soil profiles, two kinds of relation between mercury concentration and soil pH are revealed: positive one ($r=1$ when $P=0.02$) in transformed soil and negative one ($r=-0.9$ when $P=0.04$) in natural soil. There is a negative relation between mercury, Corg and CO_2 of carbonates in natural soil ($r=-0.76$ when $P=0.05$ for Corg; $r=-0.88$ when $P=0.04$ for CO_2 of carbonates). Also there is a positive relation between mercury, Corg and CO_2 of carbonates in transformed soil ($r=0.82$ when $P=0.05$ for Corg; $r=0.99$ when $P=0.03$ for CO_2 of carbonates). Interrelation between mercury concentration, and fulvic and humic acids on the studied territory is not identified. However, there is a steady interrelation between mercury and undissolved residue ($r=0.5$ when $P=0.02$).

Average mercury concentration in natural and agrogenically transformed soils varies significantly from 20 ng/g in transformed soil to 29 ng/g in natural soil. It is worth mentioning here that mercury concentration in transformed soil is lower and conforms to regional geochemical background (20 ng/g) [4]. It can be linked to blending of soil layers when plowing: thin humus horizons and detritus containing the highest mercury concentration are blended with underlying mineral horizons with considerably lower mercury concentrations [2].

Total mercury concentrations in the studied soil samples do not exceed maximum permissible concentration, but are higher than those of the regional background. Average rate of this excess over regional background is equal to 1.16 (0.05-6.2). Calculated geoecological parameters of the mercury load on the soil of Tunka Depression reveal the rate of excessive mercury concentrations in the upper layer of the soil. It exceeds abundance of mercury in Earth mantle (up to 0.29 in average (0.01-1.55)) and in Earth soils (up to 2.32 (0.10-12.40)).

Table 1. The content of mercury, depending on the physical and chemical characteristics of the soils of the Tunka depression.

Soil	Soil horizons [11]	Capacity, sm	CHg, ng/g	pH	C_{org}, %
Anthropogenic	P, rr, W, Pw	0-37	20	8.0	1.21
Anthropogenic ledge	AU, W, A0	0-30	16	7.3	1.56
Natural	AH, AU	0-13	29	6.6	4.31
Natural slope	AY, Aye	0-17	41	6.2	6.58
Forest	T, Th, mr	0-14	43	6.4	15.64
Bog	T1, T2h	0-12	23	7.7	53.80

4. Conclusions

As a result of the study, the total content of mercury in various types of both natural and anthropogenically disturbed soils was determined. At the same time, the average content in natural soils exceeds the concentration in soils experiencing anthropogenic impact. In this case, it is worth noting that an increase in the content of mercury in natural soils is associated with more acidic soil conditions. In disturbed soils, the concentration of mercury increases in more alkaline conditions. In addition, natural soils are characterized by an increase in mercury content at lower Corg values and CO_2 of carbonates. In disturbed soils the picture is the opposite. Calculations of geoecological indicators revealed an excess of mercury in the upper horizons above the background and Clarks.

Acknowledgement

This work was partly finacially supported by RFBR 14-05-00183 A (2014-2016) "Technogenesis and forecast of landscape transformation in conditions of global environmental changes, scientific basis for the regulation of nature management".

5. Reference

[1] Grebenshchikova et al. 2008 Geochemistry of the environment of the Baikal region (Baikal geoecological polygon) *Geo* p 234
[2] Guidelines for conducting of field and laboratory researches of soil and plants by the control of environmental pollution with metals 1981 *Gidrometeoizdat* p 108
[3] Kabata-Pendias A and Pendias G 1989 Microelements in soils and plants. Trans. With the English. (Moscow: Mir) p 439
[4] Motuzova G V 2009 Connections of microelements in soils: System organization, ecological significance, monitoring , LIBROKOM p 168
[5] Ivanov V V 1997 Ecological geochemistry of elements (Moscow: Ecology) **5** 576
[6] Yanin E P, Artamonova V G and Polkanova E K 1999 Integrated Security System (Saint Petersburg Mercury) p 26
[7] Belousov V M, I.Yu. Bude and Ya.B. Radzimovich 2000 Physical and Geographical Characteristics and Environmental Issues of the Southwestern Branch of the Baikal Rift Zone Irkutsk p 160
[8] Cherkashina A A and Belozertseva I A 2013 Current state of the landscapes of the Tunka system of Basins. *Proc. Int. Conf. Soils in Space and Time",* (Germany: Ulm) pp 125-126
[9] Fridland V M 1984 *Structure of the soil cover of the world* (Moscow: Mysl) p 235
[10] Sochava V B 1978 Introduction to the theory of geosystems (Novosibirsk: Science) 319
[11] Soil combinations and their genesis 1972 (Moscow: Science) p 215

IOP Conf. Series: Earth and Environmental Science **107** (2017) 012119 doi:10.1088/1755-1315/107/1/012119

Carbon balance assessment by eddy covariance method for agroecosystems with potato plants and oats & vetch mixture on sod-podzolic soils of Russia

J L Meshalkina[1,2,3], A M Yaroslavtsev[2,4,5], I I Vasenev[2,4,5], I V Andreeva[2], M V Tihonova[2]

[1]Soil Science Faculty, Lomonosov Moscow State University, Leninskye Gory, GSP-1, bld.12, 119991, Moscow, Russia
[2]Department of Ecology, Russian Timiryazev State Agrarian University, Timiryazevskaya Str. 49, 127550, Moscow, Russia
[3]Dokuchaev Soil Science Institute, Pyzhyovskiy lane 7 building 2, Moscow, 119017, Russia
[4]RUDN University, Miklukho-Maklaya str.6, 117198, Moscow, Russia
[5]School of Natural Sciences, Far Eastern Federal University, Sukhanova St. 8, 690090, Vladivostok, Russia
E-mail: jlmesh@list.ru

Abstract. The carbon balance for the agroecosystems with potato plants and oats & vetch mixture on sod-podzolics soils was evaluated using the eddy covariance approach. Absorption of carbon was recorded only during the growing season; maximum values were detected for all crops in July. The number of days during the vegetation period, when the carbon stocked in the fields with potatoes and oats & vetch mixture was about the same and accounted for 53-55 days. During this period, the increase in gross primary production (GPP) is well correlated with the crop yields. The curve of the gross primary productivity is closely linked to the phases of development of plants; for potatoes, this graph differs significantly for all phases. Form of oats & vetch mixture biomass curve shown linear increases. Carbon losses were observed for all the studied agroecosystems: for fields with an oats & vetch mixture they were 254 g C m^{-2} y^{-1}, while for fields with potato plants they were 307 g C m^{-2} y^{-1}. Values about 250-300 g C m^{-2} per year may be considered as estimated values for the total carbon uptake for agroecosystems with potato plants and oats & vetch mixture on sod-podzolic soils.

1. Introduction

Most of the papers on the carbon balance assessment are devoted to forestry and natural ecosystems, whereas agroecosystems are not given sufficient attention to. However, in Central Russia, carbon losses in arable lands exceed 9 times the losses observed for natural forest ecosystems [1]. Therefore, the task is to monitor carbon losses for typical agroecosystems with different crops in different regions.

The application of the eddy covariance (EC) method for estimating the carbon balance in the agroecosystem is an example of the introduction of modern technologies into agricultural sciences. The EC technique is a statistical method to measure and calculate vertical turbulent fluxes of greenhouse gases within atmospheric boundary layers; today it is a widely used and accurate method [2]. This method can provide continuous, long-term flux information integrated into the ecosystem scale [3]. Net ecosystem exchange of carbon (NEE) which expresses the potential of each agro-ecosystem to fix C or to release C by both autotrophic and heterotrophic respiration is particularly important when a comprehensive C balance for agro-ecosystem is performed. The traditional way of addressing net carbon exchange of an ecosystem over multiple years involves quantifying temporal changes in biomass and soil carbon. However, changes in soil organic matter become apparent after decades rather than years and thus soil sampling techniques do not allow to get insight changes in the short term. The eddy covariance technique has been used widely to determine NEE [3] because it is scale appropriate and provides a method to assess net CO_2 exchange at ecosystem level, it produces a direct measure of NEE, it is able to measure ecosystem CO_2 exchange across a spectrum of timescales and the area sampled by this technique (footprint) range from hundred meters to several kilometers.

The balance of organic matter in the soil consists of income and the loss of carbon. An equilibrium is established over the time in natural ecosystems in which the annual carbon input is equal to its output. Agroecosystems feature is that a portion of the carbon is taken out with harvest. Carbon status in agroecosystems depends on climate, land use and the level of agricultural technology.

The objective of this work was to evaluate the carbon balance for the agroecosystems with potato plants and oats & vetch mixture on sod-podzolic soils with eddy covariance approach.

2. Material and methods

The research has been carried out at the Precision Farming Experimental Field of the Russian Timiryazev State Agricultural University (RTSAU: 55°55'14'N, 37°33'56'E) situated in Moscow, Russia. The experimental site has a temperate and continental climate and is situated in south taiga zone with Arable Sod-Podzoluvisols (Albeluvisols Umbric) that had around 1% of SOC, pH 5.4 (KCl) and NPK medium-enhanced contents in sandy loam topsoil. Crop rotation included winter wheat (*Triticum sativum L.*), barley (*Hordeum vulgare L.*), potato crop (*Solanum tuberosum L.)* and cereal-legume mixture (*Vicia sativa L.* and *Avena sativa L.*). Two fields of the same crop rotation were studied in 2013-2016 with potato plants and oats & vetch mixture.

Planting of crops was carried out in late April-early May, depending on the weather conditions. Oats & vetch mixture was harvested in July - early August. Potato harvesting was in early September.

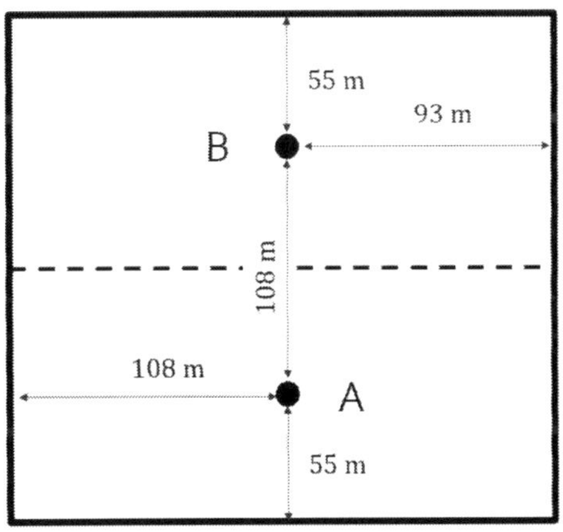

Figure 1. The layout of two EC stations (A and B) located on two adjacent fields, named according to stations.

The agroecosystem CO_2 flux seasonal monitoring was done by two eddy covariance stations located on fields A and B at the distance of 108 m (figure 1). The LI-COR instrumental equipment was the same for both stations. Agroecosystems CO_2 fluxes were measured using the EC system mounted 1.4 m above the soil surface for both stations. The EC system included a three-axis sonic anemometer (CSAT-3, Campbell Scientific Inc., USA) and enclosed path infrared gas analyzer

(IRGA, LI-7200, Li-COR Inc., USA). The flux data were recorded at 20 Hz by a data logger (CR1000, Campbell Scientific Inc., USA) at 30 min intervals. Meteorological parameters were measured simultaneously with the same array of sensors: including net radiation (NR01, Hukseflux Thermal Sensors B.V., the Netherlands), air temperature and relative humidity (HC2S3, Campbell Scientific, Inc., USA) and heat flux at the depths of 8 cm (HFP01, Hukseflux Thermal Sensors B.V., the Netherlands). Soil temperature and water content were measured at 3 depths (5, 20 and 50 cm) with multi-parameter sensor (CS650, Campbell Scientific Inc., USA). Photosynthetic active radiation (PAR) (LI-190SB, Li-Cor Inc., USA) and precipitation (TE525 MM tipping bucket gauge, Texas Electronics, Texas, USA) were measured only on the field A. It was assumed that temporal patterns of PAR and precipitation were similar for the both agroecosystems, because the two towers were situated side by side, at the negligible distance regarding to PAR and precipitation differences. All meteorological data were measured every 10 s and then averaged half-hourly. Raw data were processed using the eddy covariance processing software EddyPro, version 4.1 (LI-COR Inc., USA) to determine NEE with an averaged half-hourly period with the following settings. Data processing followed standard methods and included coordinate rotation with tilt corrections, linear detrending, despiking, time lag corrections, correction of low-pass filtering effects with Moncrieff et al. [4] and Webb–Pearman–Leuning (WPL) correction [5]. Also the dynamic metafile was used to take into account the model the canopy growth and snow melting. The surface roughness was neglected. The flux footprint estimation was made using the model of Kljun et all. [6]. Quality control tests for fluxes (1 to 9 system) were performed according to Foken [7]. Special algorithm was developed for complex footprints filtering because the fields are located close to each other. The data gap filling by Reichstein M, Falge E, Baldocchi D et al. [8] was used.

Based on the assumption that daytime respiration was of similar magnitude and responsiveness as nighttime ecosystem respiration, the daily ecosystem respiration (Reco) was determined as an average of nighttime ecosystem respiration summarized for a 24-hours period. Gross primary production (GPP) was calculated as the balance between daytime NEE and Reco: GPP = Reco– NEE. The calculations were made for daily data.

The difference between micrometeorological conditions on two fields was negligible during the same year. The correlation coefficients between the respective biometric data differed from 0.7 to 0.9. Thus, climatic features and crop type were main drivers of CO_2 flux dynamics on.

3. Results and discussion

Climatic conditions were different in 2013 and 2014: the spring started earlier in 2014; the summer of 2014 was warmer and drier. So, the year 2013 was less favorable for oats & vetch mixture than 2014 due to climatic conditions. As a result, the yield of oats & vetch mixture was two times higher in 2014 compared with 2013 (Table 1).

Table 1. Carbon output (g C m^{-2} per year) with the yield within the agroecosystems during the year.

	Field B 2013	Field A 2014	Field B 2015	Field A 2016
	oats & vetch mixture		potato plants	
Yield (t/ha)	13.4	24.8	28.6	28.8
Water content	80%	80%	-	-
Total:	121	223	210	211

Climatic conditions were quite close in 2015 and 2016; it is possible to consider those years as favorable for potato plants. The harvest of potatoes was approximately the same. The yields for potato plants and oats & vetch mixture in 2014 were close to average yields for the crop rotation.

The cumulative NEE in the agroecosystem with oats & vetch mixture in 2014 was consistently higher than that in 2013 throughout from the mid-February to the beginning of the growing season; after that the curve for 2014 was lower on the graph compared to the curve for 2013 (figure 2).

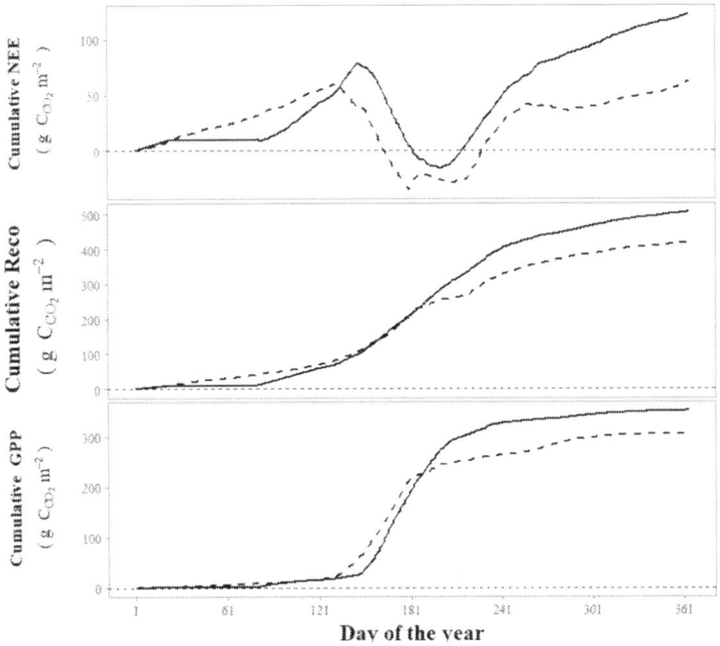

Figure 2. Cumulative net ecosystem exchange of carbon (NEE), ecosystem's respiration (Reco) and gross primary production (GPP) estimated by EC method for agroecosystems with oats & vetch mixture in 2014 (the dotted line) and 2013 (solid line).

The considerable uptake of CO_2 in 2014 comparing to 2013 was observed during the growing season. The agroecosystems with oats & vetch mixture released CO_2 into the atmosphere, but in 2014 the emission was much less. For potato plants the graph of cumulative NEE, Reco and GPP did not show differences in 2015 and 2016.

Absorption of carbon was recorded only during the growing season; maximum values were detected for all crops in July. The number of days during the vegetation period, when the carbon stocked in the fields with potatoes and oats & vetch mixture was about the same and accounted for 53-55 days. During this period, the increase in GPP is well correlated with the crop yields. The curve of the gross primary productivity was closely linked to the phases of development of plants; for potatoes, this figure differed significantly for all phases. Form of oats & vetch mixture biomass increase curve was linear. However, in 2014, with more favorable conditions, the slope of the curve was steeper.

The ecosystem respiration for potato and oats & vetch mixture fields fluctuated from 420 to 580 g C m^{-2} per year and was closely related to weather conditions. Carbon balance assessment is shown in Table 2. The carbon entered to ecosystem by seeding was only 9.5-17% from the carbon taken out with the yield.

The resulting difference in cumulative values of NEE of the CO_2 fluxes between the two agroecosystems during the two years was quite similar (Table 2). Carbon uptake (NEE negative values) was not registered for studied crops. All cases showed CO_2 emission. So, the soil and vegetation respiration during the year exceeded the carbon uptake by the photosynthesis process. The difference of 82 g C m^{-2} per year in NEE for fields with oats & vetch mixture was related with the two times difference in grass yields. NEE for potato plants was positive all the year; taking into consideration only the growing season, NEE for potato plants was about 0 g C m^{-2}.

Table 2. Carbon input and output (g C m^{-2} per year) within the agroecosystems during the year.

	Field B 2013	Field A 2014	Field B 2015	Field A 2016
	oats & vetch mixture		potato plants	
Input in soil with seeds	-22*	-22	-24	-24
Output with the yield	121	223	210	211
NEE of the CO_2 fluxes	145	63	118	123
Total:	**244**	**264**	**304**	**310**

* Negative values correspond to carbon uptake as it is accepted in micrometeorology

Closing balance for the year showed that carbon losses were observed for all the studied agroecosystems: for fields with an oats & vetch mixture they were 254 g C m^{-2} y^{-1}, while for fields with potato plants – 307 g C m^{-2} y^{-1}. Values about 250-300 g C m^{-2} per year may be considered as estimated values for the total carbon uptake for agroecosystems with potato plants and oats & vetch mixture on sod-podzolic soils.

Acknowledgement
This study has been performed with support of grant №14-120-14-4266-ScSh and of grant according to the Agreement № 02.A03.21.0008 of the Russian Federation Ministry of Education and Science, grant №ScSh-10347.2016.11 of the Leading Scientific Schools, EU's FP7 grant agreement №603542 (the project LUC4C) and grant № 15-16-30007 of the Russian Science Foundation.

References
[1] Stolbovoi V S, Nilsson S, Shvidenko A Z and McCallum I 2004 *Ecology* **3** pp 179–184
[2] Burba G 2013 Eddy Covariance Method for Scientific, Industrial, Agricultural and Regulatory Applications: a Field Book on Measuring Ecosystem Gas Exchange and Areal Emission Rates (Lincoln: LI-COR Biosciences)
[3] Baldocchi D D 2003 *Global Change Biol* **9** pp 479–492
[4] Moncrieff J B, Massheder J M, Verhoef A, Elbers J, Heutsunkveld B H, Scott S, de Bruin H, Kabat P, Soegaard H and Jarvis P G 1997 *J Hydrology* **188–189** pp 589–611
[5] Webb E K, Pearman G I and Leuning R 1980 Quart J R Meteorol Soc 106 pp 85–100
[6] Kljun N, Calanca P, Rotach M W and Schmid H P 2004 *Boundary-Layer Meteorology* **112** pp 503–523
[7] Foken T 2003 Angewandte Meteorologie, Mikrometeorologische Methoden (Cham: Springer)
[8] Reichstein M, Falge E and Baldocchi D 2005 *Global Change Biology* **11 (9)** pp 1424–1439

IOP Conf. Series: Earth and Environmental Science **107** (2017) 012120 doi:10.1088/1755-1315/107/1/012120

Assessment of environmental stability of agroserous soil according to indicator of energy potential of organic substances

S G Murtazina, L G Gaffarova, MG Murtazin

Kazan State Agrarian University, Kazan, Karl Marks Str., 65, 420015, Russia
E-mail: gaffarovalylya@mail.ru

Abstract. Studies of the group and fractional composition of humus have determineded that the long-term use of soil (for 20 years) without the use of fertilizers (control) leads to a decrease in the content of humic acids and fulvic acids relative to the initial soil, which indicates an increase in mineralization of the soil humus. Under the influence of a long application of high doses of mineral fertilizers, the content of mobile fractions of humic and fulvic acids in the field rotation increases in the humus content. In systems of agriculture that are not balanced by organic matter, which are predominant in most farms of the Republic of Tatarstan, the use of very high doses of potassium fertilizers is not justified energetically. To compensate for losses of humus and its energy potential in calculating organic fertilizers on backgrounds with high doses of mineral fertilizers, the humification coefficients of organic residues should be increased by 30-40% during the rotational period of 5-6 years, which will reduce the loss of energy reserves and thereby improve the ecological stability of soils and the stability of agricultural landscapes

1. Introduction

The characteristics and direction of the matter flows and energy in the ecosystem, which take part in forming of soils fertility, stability and its productivity level, is determined by the qualitative and quantitative composition of the soil organic matter [1]. The energy function of the soil is related to the energy of its organic and mineral parts. Humus also retains a free form of energy, which can perform work without reducing the entropy of soils [2].

The organic matter of the soil is a source of energy for the organisms inhabiting the soil and it supplies plants with basic nutrients. The energy potential of the organic matter of the soil determines its fertility and ecological stability. To assess the potential of soil resources, to study the ecological capacity and bioenergy potential of the territory, it is necessary to determine the energy potential of the organic matter of the soil.

Evaluation of the humus state of agrogenic soils and the energy potential of humus and organic substances in the ones which are used in systems of agriculture that are not balanced by organic matter with intensive application of mineral fertilizers presents both scientific and practical interest.
Scientifically substantiated solution of these issues is also significant for developing fertility models of agrogenic soils and developing techniques for increasing their environmental sustainability [3].

In modern agriculture, agro-forest soils of arable land are subjected to great anthropogenic pressure and, according to a number of researchers, the intensive use of mineral fertilizers contributes to the destabilization of the humus state of soils [4, 5].

The assessment of the ecological stability of the agro-forest soil in terms of the energy potential of its organic substances during the long-term intensive application of mineral fertilizers in the conditions of stationary field experience in the Prekamje region of the Republic of Tatarstan is carried out for the first time.

2. Materials and methods

For the statistical evaluation of humus content in arable agroseric soils of the Prekamje region of the Republic of Tatarstan, soil and agrochemical survey materials were used. Field experience included studying efficiency of various doses of potassium (40, 80, 120, 160, 200 kg/ha) in nitrogen supply at three levels of soil (60, 120, 180 kg/ha).

Field experiment was organized in 1992 on agroserous soil of the stationary experimental field of the Department of Agrochemistry and Soil Science of the Kazan State Agrarian University in the conditions of field crop rotation with grain crops. Alternation of crops in crop rotation was as follows: 1. Spring wheat; 2. Barley; 3. Oats; 4.Fallow land; 5. Winter rye.

The soil of the experimental plot is agros and medium loamy, before the beginning of the experiment it was characterized by the following agrochemical indicators: the content of humus - 2.8-3.5%; the amount of exchange bases 19-25 mg/eq; hydrolytic acidity i 3-4 mg/eq; the degree of saturation with bases 77-85%; pH of sol. 5.5-6.0; the content of mobile forms of phosphorus and potassium - 100-120 and 110-130 mg per kg of soil, respectively, i.e. it is medium provided by these elements of nutrition.

In order to assess the ecological stability of the soil in intensive farming 20 years after the completion of the 4th rotation of the crop rotation, we studied the humus condition of soils under control (without fertilizers) and in variants of the experiment with fertilizers, and the calculation of the energy potential of organic substances was carried out. The content of humus in the soil was determined by I.V. Tyurin in the modification of Simakov, the group and fractional composition of it - according to the method of B, B. Ponomareva and Plotnikova. Evaluation of the energy potential of organic substances in agroserous soil was conducted by the method of Volodin V.M. and Vasyutenko N.P. (2005), using the data on humus content, soil density and crop yields.

3. Results and discussion

The generalization and statistical evaluation of soil and agrochemical survey data on the Predkamje region of the Republic of Tatarstan showed a decrease of humus content and reserves in gray forest soils in recent years. The obtained factual material in the field experiment indicates significant changes in the humus state of the soil as a result of prolonged application of fertilizers [6,7].

Calculations of the humus balance in the soil of various variants of the experiment with fertilizers have shown that over 20 years of use, a negative balance of humus is formed in it, both while using fertilizers and without them.

Mineral fertilizers contributed to the enhancement of mineralization of humus. The change in the parameters of the qualitative composition of humus is shown in table 1, from which it can be seen that the soil before the experiment was characterized by the predominance of humic acids over fulvic acids, while the ratio of GK to FK is 1.2, which is typical for arable gray forest soils in the republic [8].

Long-term use of soil (for 20 years) without using fertilizers (control) leads to a decrease in the content of humic acids and fulvic acids relative to the initial soil, which indicates an increase in mineralization of soil humus. The type of humus at the same time remains the same (fulvate-humate), but with depth the content of fulvic acids increases, and humic - decreases.

In the experiment with fertilizers, 1000 kg of nitrogen, 1000 kg of phosphorus and 2000 kg of potassium were applied to the soil over 20 years of experience. Such a high saturation of the crop rotation with mineral fertilizers contributed to a decrease in the content of GK in humus and an increase in the content of FA.

- **Table 1.** Dynamics of group and fractional composition of humus in agroserous soil under the influence of intensive application of fertilizers (% of C_{org}.).

Horizon	Humic acids				Fulvic acids					The sum of acids	CGK/ CFK	Humin
	GK*-1	GK -2	GK -3	Total	FK* - 1a	FK -1	FK -2	FK -3	Total			
Initial Soil												
PY	3.4	29.5	7.1	40.0	4.2	10.4	11.7	7.2	33.5	73.5	1.2	26.5
ELB	2.2	27.5	5.2	34.9	6.9	8.5	10.3	9.8	35.5	70.4	1.0	29.6
B1	1.3	24.7	8.7	34.7	9.5	9.9	9.0	11.2	39.6	74.3	0.9	25.7
Control (without fertilizer)												
PY	1.4	26.5	10.1	38.0	4.0	8.4	10.9	7.0	31.3	69.3	1.2	30.7
ELB	1.2	24.6	4.2	32.0	8.0	13.0	9.4	5.6	35.0	67.0	0.9	33.0
B1	1.7	23.0	6.0	31.7	7.2	10.9	9.0	11.5	38.6	70.3	0.8	29.7
Long-fertilized soil (Variant $N_{60}P_{60}K_{120}$)												
PY	8.0	22.5	5.0	35.5	9.0	15.0	7.0	4.5	35.5	71.0	1.0	29.0
ELB	9.0	16.0	3.0	28.0	10.0	16.0	6.0	3.0	35.0	63.0	0.8	37.0
B1	11.0	12.0	3.0	26.0	11.0	10.0	5.0	12.0	38.0	64.0	0.7	36.0

Note. * GK -humic acids, FK -fulvic acids

The ratio of GK to FK was reduced from 1.2 in the initial soil to 1.0, which indicates that under the influence of fertilizers more acidic humus is formed in the plow layer. Changes in the quality of humus are associated with an increase in mobile first fractions of GK and FK, as well as a decrease in the content of the most agronomically valuable fraction associated with calcium (GK-2).

Our studies show a decrease in the content and reserves of humus in the soil under the effect of prolonged use of mineral fertilizers in the crop rotation, which undoubtedly leads to structural changes in energy and mass content.

The energy potential of soil organic matter is represented by the energy potential of humus, which is characterized by the parameters of the energy potential of inert and labile humus, the energy reserve of ungulified organic substances and an additional energy reserve in the form of microbial bodies and the energy of soil reactions.

The calculation of the dynamics of the energy potential of organic substances in gray forest soils used in grain-crop rotation for 20 years shows (table 2) that both in control and in experiments with fertilizers, as humus stocks decrease, structural changes occur in the energy potential of organic soil substances.

Table 2. Structure of the loss of the energy potential of organic substances in the agro-soil of agrocenoses under the influence of fertilizers over 20 years, GJ/ha.

Options	The loss of energy reserves in humic substances	The loss of energy reserves in inert humus	Loss of energy reserve labile humus	The loss of energy in non-humified organic substances	Additional energy losses	Total loss of energy potential of soil organic matter
Control	54.1	50.9	3.24	1.3	4.8	60.3
$N_{60}P_{60} + K_{40}$	90.9	85.5	5.45	2.3	6.8	100.3
$N_{60}P_{60} + K_{120}$	174.8	163.7	10.5	4.3	13.1	191.5
$N_{60}P_{60} + K_{200}$	193.0	181.5	11.6	4.8	19.3	217.7

On the control, the total energy potential of all structural constituents of organic substances decreased by 60.3 GJ / ha over 20 years, and in the variants with mineral fertilizers - by 100-218 GJ /

ha. In this case, increasing doses of potassium (K120-200) against the background of $N_{60}P_{60}$ contributed to an increase in the loss of energy potential of organic substances almost twice.

Acknowledgement

Generalization and statistical evaluation of soil and agrochemical survey data on the Predkamje region of the Republic of Tatarstan showed a decrease in humus content and reserves in agroforest soils.

Calculations of the humus balance in the soil of various variants of the experiment with fertilizers have shown that over 20 years of use, a negative balance of humus is formed in it, both with fertilizers and without them. Mineral fertilizers contributed to the enhancement of mineralization of humus.

Due to the long application of fertilizers, a more acidic humus is formed in the plow layer. Changes in the quality of humus are associated with an increase in mobile first fractions of GK and FK, as well as a decrease in the content of the most agronomically valuable fraction associated with calcium (GK-2).

In systems of agriculture that are not balanced by organic matter, which are predominant in most farms of the Republic of Tatarstan, the use of very high doses of potassium fertilizers is not justified energetically.

To compensate for losses of humus and its energy potential in calculating organic fertilizers on backgrounds with high doses of mineral fertilizers, the humification coefficients of organic residues should be increased by 30-40% during the rotational rotation period of 5-6 years, which will reduce the loss of energy reserves and thereby improve the ecological stability of soils and stability of agrological landscapes.

References

[1] Fokin A D 1994 On the role of organic matter in soils and the functioning of natural and agricultural crops Ecosystems *Soil Science* **4** pp 40–45
[2] Volodin V M and Vasyutenko N 2000 *Methodology for determining the ecological capacity and bioenergy potential of the territory of the agrolandscape* (Kursk: Publishing center UMKS)
[3] Gaffarova L G, Murtazina S G and Murtazin M G 2017 Dynamics of the humus state of gray forest soils of the Predkamje region of Republic of Tatarstan and productivity of crop rotation with long-term fertilization *Grain economy* **2** pp 57–60
[4] Koloskova A V 1985 *Humus condition of soils of the Volga-Kama forest-steppe* Publishing house of Kazan University p 137
[5] Zavyalova N E 2005 Influence of prolonged use of organic and mineral fertilizers on the transformation of organic matter of sod-podzolic soil *Agrochemistry* **6** pp 5–10
[6] Murtazina S G 2005 Assessment of the effect of fertilizers on the fertility of gray forest soils *Modern science-intensive technologies* **10** pp 106–107
[7] Murtazina S G 2014 The effect of intensive application of mineral fertilizers in field crop rotation on its productivity and soil properties *Grain economy of Russia* **2** pp 14–17
[8] Orlov D 2005 *Humus acids of soils* (Moscow: MSU Publishing House) p 420

IOP Conf. Series: Earth and Environmental Science **107** (2017) 012121 doi:10.1088/1755-1315/107/1/012121

Investigation of biological destruction of benzo[a]pyrene andpolycyclic aromatic hydrocarbons of biochar in soil

R V Okunev, E V Smirnova, A R Sharipova, I M Gilmutdinova, K G Giniyatullin

Institute of Environmental Sciences, Kazan Federal University, Kremlevskaya str., 18, 420008, Kazan, Russia

E-mail: tutinkaz@yandex.ru

Abstract. The biological decomposition of benzo[a]pyrene in the concentrations exceeding the MAC (maximum permissible concentration) level in soils by 2, 5 and 10 times was studied in laboratory conditions. The gray forest soil samples were contaminated with benzo[a]pyrene and incubated in optimum for bacterial growth soil moisture for 30 and 60 days. The residual amount of contaminant was monitored by HPLC after extraction with acetone-cyclohexane (2:1). Soil microbial activity was evaluated by measuring basal respiration (BR) and substrate-induced respiration (SID) rates of the soil by gas chromatography. The results of the experiment showed that in 60 days the amount of benzo[a]pyrene in contaminated soils decreased; however, this time was not enough for complete decomposition of pollutant. In this case, benzo[a]pyrene has a negative effect on the BR and SIR rates. Soil contamination affected the BR rate only at high doses (10 MPC), whereas the SIR was a more sensitive indicator of the toxic effect of the pollutant and significantly reacts already at concentrations at the level of 2 MPC. The combination of PAHs isolated from biochar has a strong negative effect on the values of BR and SIR.

1. Introduction

Polycyclic aromatic hydrocarbons (PAHs) have pronounced mutagenic and carcinogenic properties and are dangerous for humans life [1]. Their occurrence in the environment is conditioned by both anthropogenic and natural factors. Anthropogenic sources present various types of industry, including the oil industry, incomplete combustion of coal and waste. Recently, promising soil amendment - biochar is considered as sources of PAHs. In the process of formation of biochar, during the pyrolysis of plant material, various toxic compounds, including PAHs, that affect the activity of the soil microbes, can be formed [2]. In soils, PAHs can undergo chemical transformations, photochemical degradation in the surface layers, can be removed as a result of leaching, and can also undergo biodegradation [3]. Among these substances, benzo[a]pyrene is the most dangerous [4]. The aim of this study was to investigate the interaction of benzo[a]pyrene and PAHs extracted from biochar with soil microbiota by measuring the BR, SIR and estimating the residual amount of benzo[a]pyrene in laboratory conditions.

2. Materials and methods

In the present work the samples of the humus horizon of gray forest soil obtained from a depth of 5-20 cm were used. The soil was formed under broadleaf species (linden, maple, ash) in continental bioclimatic conditions. Physical and chemical soil characteristics are presented in table 1.

Table 1. Physical and chemical soil characteristics.

Soil properties	Values
Humus, %	6,52
Total N, %	0,48
Clay particles with size <0,01мм, %	56,8
pH_{water}	5,7

Benzo[a]pyrene and PAHs isolated from biochar with a mixture of acetone-cyclohexane (linden, pyrolysis temperature - 250-300°C) were chosen as pollutants. In the biochar obtained by the low-temperature pyrolysis of linden wood, a large number of PAHs were detected (the chromatogram is shown in Figure 1). The concentration of PAHs varied from 4.8 to 77.1 μg/kg of biochar. An aliquot of the extract from biochar before introducing into the soil was evaporated in vacuum, followed by dissolution in acetone.

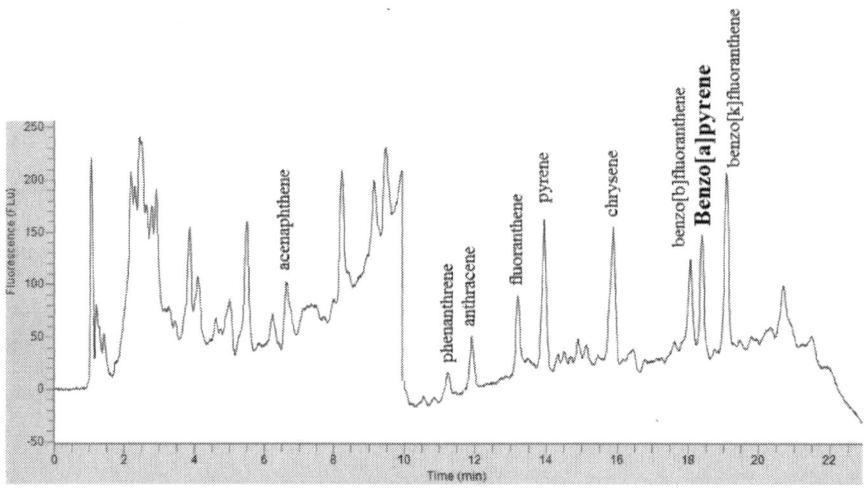

Figure 1. Chromatogram of polynuclear aromatic hydrocarbons extracted from biochar (linden, pyrolysis temperature 250-300 ° C).

The incubation experiment was carried out as follows. Benzo[a]pyrene dissolved in acetone was added to the fresh soil samples (240 g) at concentrations corresponding to 0 (control), 2, 5 and 10 MPC [5]. The PAHs isolated from biochar were added to adjust the concentration of benzo[a]pyrene in the soil to 2 MPC and 1 MPC. In the variant with 1 MPC of PAHs, additional solution with benzo[a]pyrene corresponding to 1 MPC was added. In Russia, the MPC of benzo[a]pyrene in soil is 0.02 mg kg^{-1} [6]. Samples of soils were adjusted to 60-65% of water-holding capacity and incubated at 20°C. Respiratory activity of soil microbiota and the residual concentration of benzo[a]pyrene was measured at 30 and 60 days after the start of the experiment. BR and SIR rates were determined by gas chromatography using a Clarus-580 (Perkin Elmer, USA) chromatograph [2]. Determination of the residual amount of benzo[a]pyrene was carried out after extraction with a mixture of acetone-cyclohexane (2:1) in Soxhlet extractor [7] by HPLC on a Flexar liquid chromatograph, Perkin Elmer, USA [8]. The achieved data were processed statistically with "Statistica-6" program.

3. Results and discussion

On the 30th day of incubation, benzo[a]pyrene introduced into the soil was partially degraded, and its content decreased in the variant with a concentration of 2 MPC by 7.4%, 5 MPC by 13.2%, 10 MPC by 2.4% in comparison with the level of its content at the beginning of the experiment. After 60 days of incubation, the level of the pollutant decreased at 2 MPC by 15.3%, 5 MPC by 19.4%, 10 MPC by 3.9% from its initial concentration (figure 2).

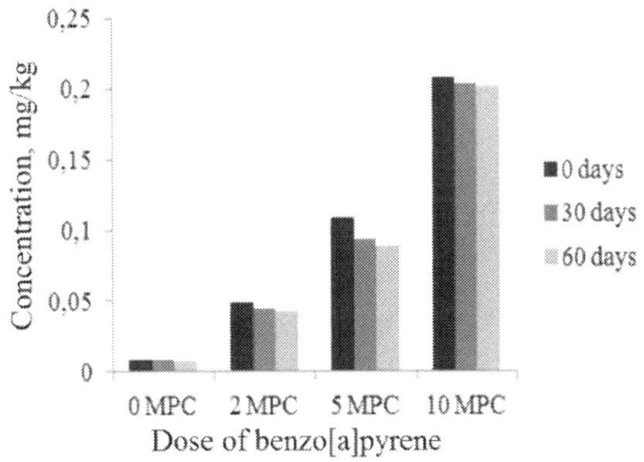

Figure 2. The concentrations of benzo[a]pyrene an the beginning of the experiment (0 days), on the 30th and 60th day of incubation of the soil (in mg/kg soil).

The experiment showed that a 60-day incubation is not sufficient to completely decompose of beno[a]pyrene in soil. Such slow biodegradation of PAHs in soils may be due to their low solubility in the aquatic environment and active sorption on the surfaces of soil minerals, that reduce their bioavailability [9]. Moreover, with increasing contact time, their bioavailability also decreases [10].

Figure 3 shows the values of BR and SIR at different concentrations of benzo[a]pyrene. The intensity of the BR after 30 days of incubation in the control variant (soil without benzo[a]pyrene) was 2.4 ± 0.4 mg C-CO2 kg^{-1} dry soil h^{-1}. A significant reduction of the BR rate to 23.5 ± 5.7 mg C-CO2 kg^{-1} dry soil h^{-1} compared to the control was observed only at a level of benzo[a]pyrene of 10 MPC. A statistically significant difference in BR rate between 30 and 60 days of incubation was not observed.

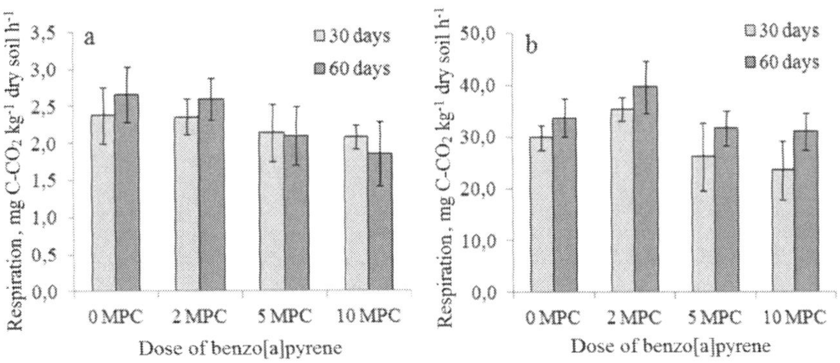

Figure 3. Basal (a) and substrate-induced (b) soil respiration rates at various concentrations of benzo[a]pyrene, mg C-CO2 kg^{-1} dry soil h^{-1}.

The effect of benzo[a]pyrene on SIR is ambiguous. Its application at the concentration of 2 MPC leads to an increase of the intensity of CO_2 release compared with the control from 29.9 ± 2.4 to 35.4 ± 2.3 mg C-CO_2 kg^{-1} dry soil h^{-1} on the 30th day of the experiment and from 33.7 ± 3.6 to 39.7 ± 5.0 mg C-CO_2 kg^{-1} dry soil h^{-1} on the 60th day of the experiment. This is probably due to the part of the microbial community inhibited under the influence of pollutant, whereas tolerant to benzo[a]pyrene part receives an advantage in development. The same effect is observed when soil contaminated with heavy metals or arsenic [11], [12]. Higher dose of benzo[a]pyrene (5 and 10 MPC) reduces SIR rates by suppressing the activity of all microorganisms.

Figure 4 shows the values of BR and SIR in the variant with PAHs contamination extracted from biochar at 1 MPC level (with the addition of 1 MPC of benzo[a]pyrene) and 2 MPC. The intensity of the BR after 30 days of incubation in the control variant was 2.5±0.2 mg C-CO2 kg^{-1} dry soil h^{-1}. A significant decrease in BR compared with the control ones was observed only in the variant with the contamination with PAHs at doze 2 MPC, whereas the application of PAHs at the concentration of 1 MPC with adding to the soil an additional 1 MPC of benzo[a] pyrene, does not have significant effect on the BR rate. Similar results were obtained after 60 days of incubation.

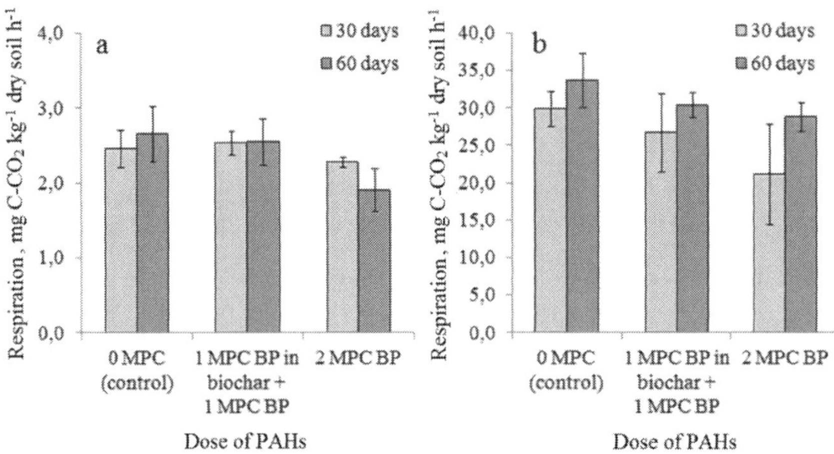

Figure 4. Basal (a) and substrate-induced (b) soil respiration rates at concentrations of PAHs extracted from biochar at the level of 2 MPC of benzo[a]pyrene (2 MPC BP) and 1 MPC with addition of 1 MPC of benzo[a]pyrene (1 MPC BP in biochar+1 MPC BP).

The amount of PAHs obtained from biochar in the concentration providing the benzo[a]pyrene content of 2 MPC exerted a stronger influence on the SIR rates than the concentration of 1 MPC of PAHs with addition of 1 MPC of benzo[a]pyrene.

Thus, benzo[a]pyrene in soils undergoes partial biodegradation within 60 days, however, this time is not enough for its complete decomposition. Pollutant in toxic concentrations (above the MPC), in conditions favorable for the life of soil microbes, has a negative impact on the BR and SIR. The influence of the contaminant on the BR can be traced only at high doses (10 MPC), the SIR responds to soil contamination with benzo[a]pyrene more sensitively. Its difference with the control ones was observed at a dose of 2 MPC.

The use of PAHs isolated from biochars in a concentration that provides the content of benzo[a]pyrene at the level of 2 MPC leads to a more significant inhibition of soil respiration (BR and SIR) than the application of single benzo[a]pyrene at a concentration of 2 MPC or 1 MPC of biochar PAHs with addition of 1 MPC of benzo[a]pyrene.

Acknowledgements
This work was supported in part by the Russian Foundation for Basic Research, research project №17-04-00869 and is performed according to the Russian Government Program of Competitive Growth of Kazan Federal University.

References

[1] Crawford R L and Crawford D L 1996 *Bioremediation: Principles and applications* (Cambridge: Cambridge University Press) 195–207

[2] Smirnova E V, Giniyatullin K G, Okunev R V, Valeeva A A and Guseva I A 2016 *Res J of Pharm, Biol and Chem Sci.* **7(5)** 1360–6

[3] Wilson S C and Jones K C 1993 *Environ Pollut.* **88** 229–49

[4] Hadibarata T 2009 *Interdisciplinary Studies on Environmental Chemistry - Environmental Research in Asia* (Tokyo: Terrapub) 309–16

[5] Smith M J, Flowers T H, Duncan H J and Saito H 2011 *J Hazard Mater.* **192** 1219–25

[6] Hygienic Norms (GN 2.1.7.2041-06) 2006 *The maximum permissible concentration (MPC) of chemical substances in soil* (Moscow)

[7] Fabbri D, Rombola A G, Torri C and Spokas K A 2013 J Anal Appl Pyrol. **103** 60–7

[8] Kayali-Sayadi M N, Rubio-Barroso S, Díaz-Díaz C A and Polo-Díez L M 2000 *Fresenius J Anal Chem.* **368** 697–701

[9] Semple K T, Morriss W J and Paton G I 2003 *Eur J Soil Sci.* **54** 809–18

[10] Hatzinger P B and Alexander M 1995 *Environ Sci Technol.* **29** 537–45

[11] Khan M 2000 *Pakistan J Biol Sci* **3**(7) 1113–8

[12] Okunev R V, Sungatullina L M and Grigoryan B R 2015 *Int J Appl Eng Res.* **10**(24) 44934–8

Digital soil mapping as a basis for climatically oriented agriculture a thematic on the territory of the national crop testing fields of the Republic of Tatarstan, Russia

I A Sahabiev[1], K G Giniyatullin[1], S S Ryazanov[2]

[1]Institute of Environmental Sciences, Kazan Federal University, Kremlevskay str., 18, 420008, Kazan, Russia
[2]Institute for Problems of Ecology and Mineral Wealth Use, Tatarstan Academy of Sciences, Kazan, Russia
E-mail: ilnassoil@yandex.ru

Abstract. The concept of climate-optimized agriculture (COA) of the UN FAO implies the transformation of agriculture techniques in conditions of changing climate. It is important to implement a timely transition to the concept of COA and sustainable development of soil resources, accurate digital maps of spatial distribution of soils and soil properties are needed. Digital mapping of soil humus content was carried out on the territory of the national crop testing fields (NCTF) of the Republic of Tatarstan (Russian Federation) and the accuracy of the maps obtained was estimated.

1. Introduction

The future of the world's agrifood system is linked to the concept of climate-smart agriculture that is based on an integrated approach to transforming agricultural systems in order to effectively support sustainable development and ensure food security in a changing climate [1]. To achieve the goals of ensuring food security and agricultural development, adaptation to climate change and reduction of emission intensity for each yield output will be required. Such a transformation must be carried out without depletion of the natural resource base, which in turn is possible with the improvement of approaches and agrotechnologies. Such technologies aimed at implementing the concept of CSA are called smart technologies.

The Food and Agriculture Organization of the United Nations (FAO) developed the concept of CSA in 2013 [2]. Since then, smart technologies have been gaining increasing popularity in agroecology, agriculture and soil science around the world. The role of "smart" technologies in the ecologization of farming systems, agroecological monitoring, improving the quality of products and utilization of agricultural waste is emphasized in Russia [3].

In the concept of CSA, the management of soil resources is presented as strategies that conserve carbon in the soil, reduce greenhouse gas emissions and promote intensification of production, while strengthening the natural resource base [1]. In this regard, knowledge of the state of soils is fundamental for making decisions about management practices that contribute to the rational land use in view of climate change. The FAO concept emphasizes that "... ideally, soil information should be available in the form of continuous maps ..." [3]. Nevertheless, obtaining information about soils and their properties in the form of maps is a non-trivial task, since it involves a number of difficulties, among which the key one is the high cost of soil surveys. At the same time, it should be understood that with the help of single data not realized in the form of cartographic material, it is practically impossible to make managerial decisions, both in traditional agriculture and in CSA.

Along with traditional soil maps, digital soil maps based on the concept of digital soil mapping (DSM) can be used for CSA purposes. In the last decade, the DSM has received rapid development in soil science and has been singled out as a separate sub-discipline [4]. The essence of the DSM comprises the prediction of soils and their properties on the basis of spatially distributed quantitative characteristics of soil formation factors [5]. Maps based on the DSM concept allow to estimate the accuracy of mapping, which can be improved by using various statistical methods (variants of regression analysis, regression trees, random forests, etc.) and approaches (geostatistics, hybrid methods, linear mixed modeling, etc.).

2. Material and methods

Taking into account the foregoing, digital maps of humus content were created for the territory of two national crop testing fields (Arsky and Zainsky NCTF) of the Republic of Tatarstan (Russian Federation). The territory of the NCTF was chosen as an object of research for several reasons: 1) for a long time the territory of the NCTFs was subjected to intensive agricultural use; 2) NCTFs use scientifically grounded agrotechniques, thanks to which sorts and hybrids of crops studies, and evaluates for their cultivation stability.

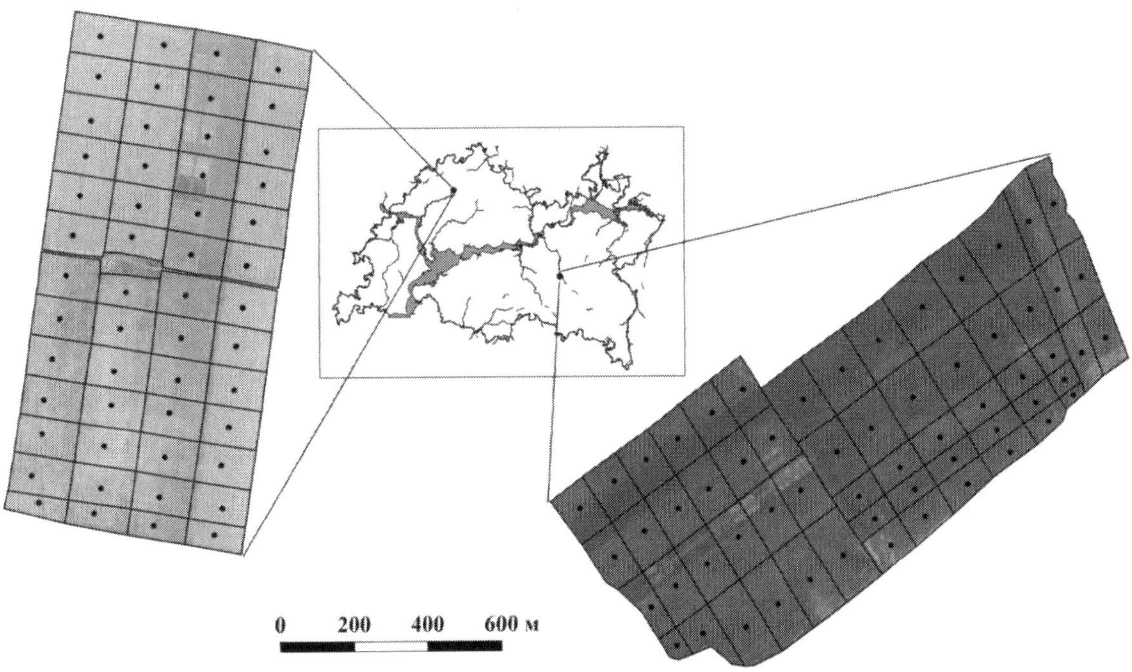

Figure 1. Location of the NCTFs and sampling design.

Zainsky NCTF is located in the south-eastern part of the Republic of Tatarstan (figure. 1). The territory of the test plot occupies an area of 80 hectares. The relief of the field is flat in the northern part that changes to the gentle slope in the southern and south-eastern parts. The soil cover is represented by leached, silt loamy chernozems (Luvic Chernozems (Pachic)) with varying rates of erosion. The most eroded chernozem is located in the eastern and south-eastern parts of the field. The soft eroded soils are located in the northern part of the CTF. Parent rocks are represented by heavy loamy and clayey calcareous deluvium, underlined by the ancient alluvial deposits in the eastern part. Texture is heavy loamy.

Arsky NCTF is located in the northwestern part of the Republic of Tatarstan (figure 1). The testing plot occupies an area of 83 hectares. The soil cover consists of sod-podzolic (Glossic Albic Dystric Retisols (Loamic, Cutanic, Ochric)) and light gray forest soils (Eutric Retisols (Loamic, Cutanic, Ochric)). The granulometric composition of the soils is also heavy loamy. The relief of the testing field is represented by a weak gentle slope in a watershed area. The northern part of the NCTF is on an elevated part of relief, which gradually decreases in the southern direction.

Digital maps of the humus content were created using geostatistical approach. For the purpose of spatial interpolation the regression kriging (RK) was used. The deterministic part of spatial variation was modeled using LASSO- and ridge-regression. The variogram analysis was used for specification of the stochastic part of spatial variability. The ordinary kriging results were used as a baseline to assess the performance of other interpolation techniques.

The relief attributes that have potential impact to spatial humus distribution were used as explanatory variables. The relief attributes were calculated from SRTM1 (spatial resolution – 30 m. (1 arc second)) after preliminary smoothing filtration. The averaged values of the relief attributes within elementary sampling section were assigned to the middle point of the corresponding section.

The use of the ridge and lasso regression was due to the fact that a large set of multicollinear auxiliary variables in the form of relief attributes (32 parameters) was used to construct a digital maps of humus content.

Ridge regression is the implementation of the classical L2-regularization, where into the RSS loss function, that minimizes using the least squares to estimate the model coefficients, the additional term λ is added [6]. This term depends on the estimated coefficients. Thus, instead of RSS, the following expression is minimized:

$$L_2 = RSS + \lambda \sum_{i=1}^{n} \beta_i^2 \qquad (1)$$

This introduces a penalty for excessive values of β_i, and the value of this penalty is proportional to the value of the λ. The λ term is used to adjust the model fitting algorithm. If $\lambda \to 0$, the regularized solution tends to the solution of least squares, that is a simple linear regression. If $\lambda \to \infty$, excessive regularization leads to a degenerate solution of $\beta \to 0$. To find the optimal value of the λ the generalized cross-validation (GCV) is used:

$$GCV = \frac{1}{n}\sum\left(\frac{y_i - \hat{y}_i}{1 - \frac{tr(H)}{n}}\right)^2 \qquad (2)$$

where tr(H) is the effective degrees of freedom.

LASSO stands for least absolute shrinkage and selection operator. The method is similar to ridge regression. The model coefficients is also estimated on the condition of minimizing the loss function taking into account the penalty, but the penalty parameter is calculated somewhat differently:

$$L_1 = RSS + \lambda \sum_{i=1}^{n} |\beta_i| \qquad (3)$$

This leads to qualitatively different estimation of the coefficient vector during the optimization. In addition to the uniform adjustment of all set of β, some of the coefficients become equal to 0, which allows to perform a selection of significant predictors [7].

The spatial interpolation models were compared using leave-one-out cross-validation (LOOCV). In LOOCV each sampling point o_i is removed sequentially. The spatial interpolation model is fit on n-1 observation, and a prediction p_i is made for the excluded observation, using its X values. Several error measurements were calculated using the difference o_i-p_i:

Mean error is given by

$$ME = \frac{1}{n}\sum_{i=1}^{n}(p_i - o_i) \tag{4}$$

Root mean square error is given by

$$RMSE = \left[\frac{1}{n}\sum_{i=1}^{n}(p_i - o_i)^2\right]^{\frac{1}{2}} \tag{5}$$

Ratio of the observed and the predicted variances is given by

$$RVar = \frac{Var[p]}{Var[o]} \tag{6}$$

The model is better if ME is closer to zero and RMSE is smaller. RMSSE should be close to 1. The closer RVar is to 1, the better is the ability of a spatial interpolation method to preserve the observed variance.

The calculation and analysis of the relief attributes were conducted using SAGA GIS [8]. All statistical analyses presented in the article were performed within the statistical environment R [9]. Variogram analysis and spatial interpolation were made using the "gstat" package [10]. The final maps were formed using QGIS [11].

3. Results and discussion

Variogram analyses shows that the crop testing fields have strong differences in the spatial structures of humus content. The presence of spatial anisotropy was assessed using the variogram maps and directional variograms with a horizontal tolerance of ±20°. Parameters of the fitted variogram models are presented in table 1. The spatial structure of the humus content shows the presence of a geometric anisotropy, the direction of which corresponds to the sampling grid orientation (45°). The autocorrelation range in a minor direction (315°) is 0.6 times less.

The spatial structure of humus values of the Arsky NCTF is described by linear isotropic variogram. The autocorrelation range of the values is larger than the size of the field and the semivariogram graph does not reached "plateau" within the field (figure 2). According to Cambardella et al. the ratio nugget/sill makes it possible to characterize the power of the spatial dependence of the studied soil values [12]. Proceeding from this, the values of humus content of the Zainsky and Arsky NCTFs have a strong spatial dependence (table 1).

Table 1. Parameters of the fitted variogram models.

Model	Range	Nugget (C_0)	Partial sill (C_1)	Sill (C_0+C_1)	$C_0/(C_0+C_1)$	Anisotropy main dir.	Anisotropy coef.	**Method**
				Zainsky NCTF				
Sph	800.5	0.03	0.59	0.62	0.05	45^0	0.6	OK
Sph	333.7	0.16	0.13	0.28	0.55	-	-	RK+Ridge
Sph	285.5	0.16	0.09	0.25	0.64	-	-	RK+Lasso
				Arsky NCTF				
Lin	1246.14	0.06	0.22	0.3	0.20	-	-	OK
Lin	1187.689	0.07	0.01	0.09	0.86	-	-	RK+Ridge
Nug	-	0.078	0	0.08	1	-	-	RK+Lasso

Sph – spherical model, Lin – linear model, Nug – nugget effect.

Ridge regression requires an accurate selection of the tuning parameter λ. The leave-one-out cross validation was used to tackle this problem. The ridge regression was fitted with the tuning parameter value for which the cross-validation error was the smallest. In the case of humus, the λ was relatively large (λ=1.43), indicating that the optimal fit involves a high amount of shrinkage relative to the least squares solution.

Inclusion of the terrain variables into the model of regionalization explained the part of spatial variation of the response, which was reflected on the experimental variograms of the model residuals (figure 2). The ridge drift model eliminated the geometric anisotropy and reduced the autocorrelation range and the overall spatial dependence of the humus values (table 1).

The performance of the LASSO regression also depends on the selection of the parameter λ. The method of the parameter selection was the same as for the ridge regression. As opposed to the ridge regression, the LASSO acts as the feature selection method. Because of the small shrinkage, the coefficients of the humus model are very similar to those we can obtain with OLS regression. The results of LASSO modeling indicate the terrain indices that influence the spatial distribution of humus.

In the case of the LASSO regression in the Arsky NCTF, the variogram has the form of a pure nugget effect - the spatial structure of humus is completely described by the regression model (table 1, figure 2). In the case of Zainsky NCTF, the spatial structure of humus is preserved after applying regression (figure2), although the spatial dependence weakens (table 1).

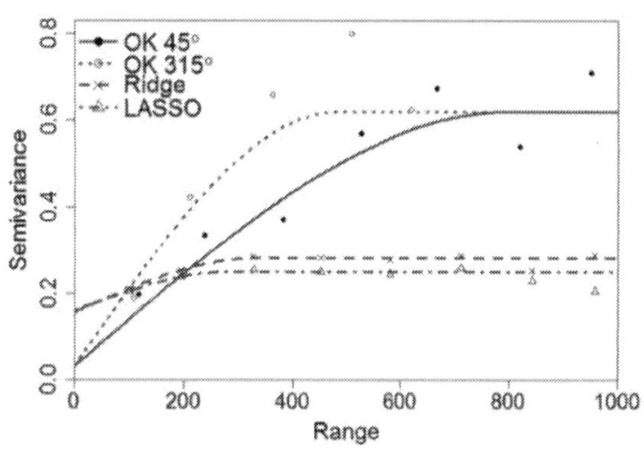

Zainsky NCTF (a)

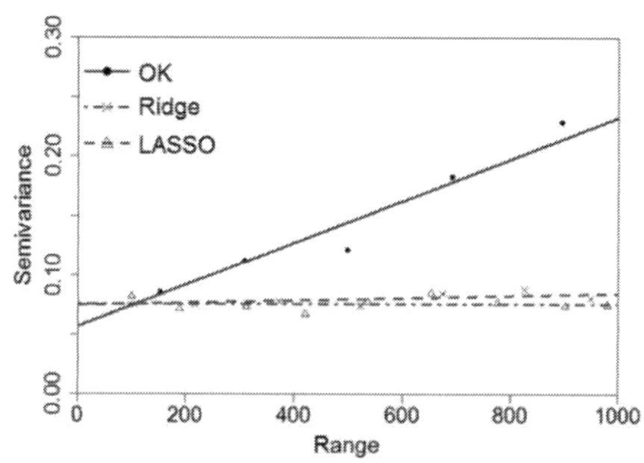

Arsky NCTF (b)

Figure 2. Sampling variograms with the fitted variogram models.

The final maps of the interpolated humus content are shown in figure 3. Digital maps based on the DSM concept have a continuous smoothed outlook, unlike traditional maps and electronic maps created in various GIS systems in which the soil areas are expressed as discrete units, and the values of soil properties are enclosed within the limits of the elementary sampling sites [13].

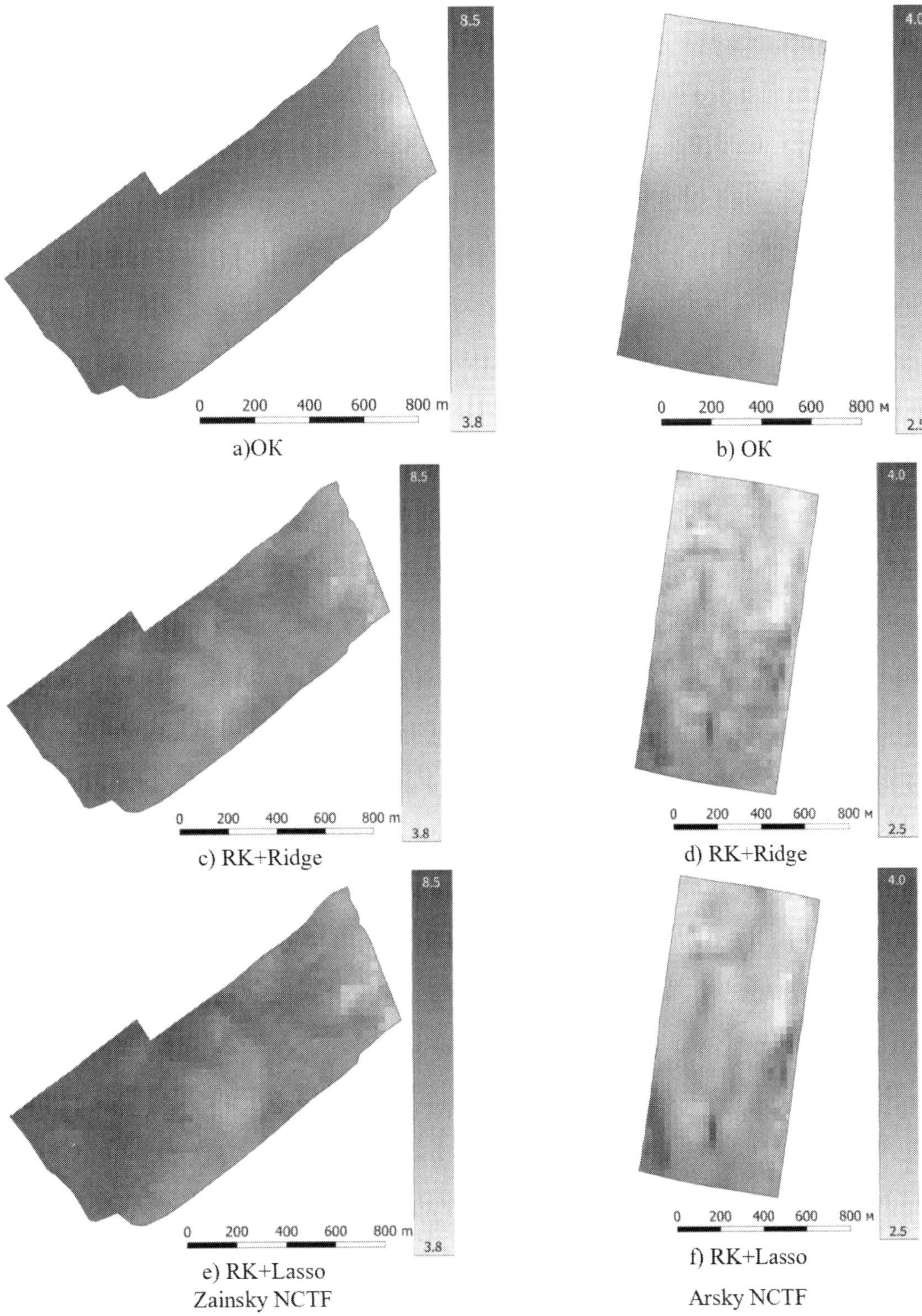

Figure 3. Interpolated maps of the humus content in the Arsky and Zainsky national crop testing fields.

The geomorphological features of the fields, as well as the inhomogeneity of the humus distribution, have caused the differences in the accuracy of the spatial interpolation methods. The results of cross-validation are shown in table 2. The method with the highest performance is the one with the lowest ME and RMSE values. The ability of spatial interpolation to keep the variance of the target variable is estimated from the value of RVar [14].

Table 2. Accuracy assessment.

Leave-one-out cross-validation			Method
ME	RMSE	RVar	
Zainsky NCTF			
-0.015	0.476	0.626	OK
-0.0008	0.536	0.546	RK+Ridge
-0.001	0.583	0.694	RK+Lasso
Arsky NCTF			
-0.0008	0.333	0.391	OK
0.00214	0.343	0.313	RK+Ridge
-0.00004	0.329	0.333	RK+Lasso

Ordinary kriging in the case of Zainsky NCTF shows the lowest RMSE value. A similar picture is consistent with the previous work of the authors, when the OK showed the greatest accuracy in the spatial prediction of the humus content [15]. In the case of the Arsky NCTF, regression kriging with the lasso regression turned out to be the most accurate (table 2). Nevertheless, in some works it is noted that not always lasso and ridge regression have the greatest accuracy of prediction in comparison with other prediction methods [16, 17]. In this case, much depends on the choice of the regularizer and there must be a trade-off with respect to regularization.

In the digital soil mapping, the use of various statistical methods does not guarantee that with the complexity of the method, the productivity and accuracy of the spatial prediction increases. To obtain adequate estimates, several methods must be used. Nevertheless, in comparison with traditional soil mapping, the DSM allows more rapid creation of soil maps. Maps of soil properties, which in the traditional version were not created and were practically replaced by cartograms, can be created with high accuracy in the DSM. At the same time, the concept of the DSM does not violate the principle of the landscape approach in the CSA, since the DSM uses a large set of available environmental variables.

Acknowledgement
This work was supported in part by the Russian Foundation for Basic Research, research project №17-04-00846.

References

[1] *Climate-Smart Agriculture* http://www.fao.org/climate-smart-agriculture/ru/
[2] *Climate smart agriculture* 2013 (Food and Agriculture Organization of the United Nations) p 570
[3] *Russia looks to smart technologies for the future of agroecology and farming.* http://www.fao.org/partnerships/container/news-article/ru/c/881418/
[4] Minasny B and McBratney A B 2016 Digital soil mapping: A brief history and some lessons *Geoderma* **(264)** pp 301–311
[5] Mc Bratney A B, Mendonça S M L and Minasny B 2003 On digital soil mapping *Geoderma* **(117)** pp 3–52

[6] Masticki S E and Shitikov V K 2014 *Statistical analysis and data visualization using R* (Moscow: DMK-Press) p 496 (In Russian)
[7] James G, Witten D, Hastie T and Tibshirani R 2013 *An introduction to Statistical Learning with Applications in R* (Springer New York Heidelberg Dordrecht London) p 440
[8] Conrad O, Bechtel B, Bock M, Dietrich H, Fischer E, Gerlitz L, Wehberg J, Wichmann V and Boehner J 2015 System for Automated Geoscientific Analyses (SAGA) **v. 2.1.4** *Geosci. Model Dev.* (8) pp 1991–2007
[9] *R Core Team. R: A Language and Environment for Statistical Computing* 2016 (Austria: R Foundation for Statistical Computing) http://www.R-project.org/
[10] Pebesma E J 2004 Multivariable geostatistics in S: the gstat package *Computers & Geosciences* (30) pp 683–691
[11] *QGIS Development Team QGIS Geographic Information System 2016 Open Source Geospatial Foundation Project* http://qgis.osgeo.org
[12] Cambardella C, Moorman T, Novak J, Parkin T, Karlen D, Turco R and Konopka A 1994 Field-Scale Variability of Soil Properties in Central Iowa Soils *Soil Sci. Soc. Am. J.* **(58)** pp 1501–1511
[13] Sahabiev I A and Ryazanov S S 2015 Investigation of the spatial variability of soil properties using the geostatistical approach *Russian Journal of applied ecology* **(2)** pp 32–37 (In Russian)
[14] Li J and Heap A D 2011 A review of comparative studies of spatial interpolation methods in environmental sciences: Performance and impact factors *Ecological Informatics* **(6)** pp 228–241.
[15] Ryazanov S S and Sahabiev I A 2016 Comparison of terrain-based drift models to improve the quality of soil predictive mapping at a field scale *Tomsk State University Journal of Biology* **4(36)** pp 21–33.
[16] Nussbaum M, Spiess K, Baltensweiler A, Grob U, Keller A, Greiner L, Schaepman M E and Papritz A 2017 Evaluation of digital soil mapping approaches with large sets of environmental covariates *SOIL Discuss.*
[17] Song X D, Brus D J, Liu F, Li D C, Zhao Y G, Yang J L and Zhang C L 2016 Mapping soil organic carbon content by geographically weighted regression: A case study in the Heihe River Basin, China *Geoderma* **(261)** pp 1–22

The Carbon Isotope Composition of Organic Matter and the Age of Paleosols from Wurm Glaciations Interstadials to Holocene (Bryansk region, Russia)

E M Stolpnikova, N O Kovaleva, I V Kovalev

Lomonosov Moscow State University, Moscow, Russia,
E-mail: natalia_kovaleva@mail.ru

Abstract. Paleosols of Trubchevsk district of the Bryansk region (Russia) lie in landscape with its own characteristic micro-relief, called Trubchevsk Opolye. The radiocarbon data have obtained for two interstadial soils: 16500 ± 230 Ki-17 414, 12930 ± 170 Ki-17 413 years ago. The ratio δ^{13}Corg for underlying sandy pedosediments varies in range -26.5-27.2 ‰ characterizes relatively humid climate. The most lightweight carbon isotope composition ($\delta^{13}C$ = -28.4-29.5 ‰) measured for the Holocene second humus horizon, discovered in micro-depressions of Trubchevsk opolye and dating (in its upper part) 2180 ± 60 Ki-17 415 BP, 1650±60 Ki-18775. It is characterized by a high content of phosphorus, including its strong accumulation of organic compounds (635.8 mg/kg P_2O_5).

1. Introduction

The more studied and widespread Late Pleistocene soil in the East European Plain refers to the climate warming 32-24 kyr BP called Bryansk interval (Paudorf, Shtillfrid "B", etc.). Often this soil is covered by loess and modern (Holocene) soils and has well-defined horizons. However, recent research [1, 6, 11] has discovered that these Late Valdai (Wurm) loess are heterogeneous and are separated by humic or gley horizons (undeveloped soils) in many stratigraphic sections. These horizons determine regional characteristics of climatic fluctuations between 17-10.2 kyr BP. The existence of paleosols levels with different characteristics is evidence of periods with relatively mild climatic conditions or interstadials of the final stage of the Late Valdai glaciation. Another interesting feature of these deposits is the presence of thick sandy loess strata, which recorded the cryogenic deformations. The genesis of these sequences can be attribute to the water-glacial streams activities on right bank of the river Desna.

Thus, the aim of this study is to clarify the regional climate conditions on Late Glacial-Holocene boundary for the Trubchevsk district (Bryansk region), based on the studying of buried soil properties in the loess-sandy-soil series. We have previously shown [5, 6, 7, 8] that the most informative features for paleoecological recovery of the environment are such biomarkers as carbon is isotope composition of humus, different characteristics of the organic matter, organic and inorganic phosphorus content and also the magnetic susceptibility.

2. Materials and methods

The studied sites are located in the Bryansk region on the territory of Trubchevskoe Opolye, where the agrogrey soils (WRB: Albic Luvisols / Grey Luvic Phaeozems) are formed on the Late Wurm loess. Trubchevskoe Opolye is situated on the high right bank of the middle course of the river Desna (on the watershed of the rivers Desna and Sudost) in the middle latitudes of the Russian plain (52°34′80″ N, 33°38′52″ E). The average annual temperature in the study area is +5.4°C. Amount of the precipitation ranges from 550 to 590 mm [11]. Different types of agrogrey soil are formed of different elements of the landscape: their varieties with the second humus horizon are formed in saucer shaped micro-depressions, gleyic variants formed in the middle parts of slopes and gley ones at footslopes, deeper depressions; there are also agrogrey soils without second humus horizon on the micro-elevations. The Late Pleistocene and Holocene loess-paleosoil series are presented in the sediments of quarries in villages Krasnoe (52^{0}33΄70΄΄N, 33^{0}42΄62΄΄E) and Telec (52^{0}34΄51΄΄N, 33^{0}44΄71΄΄E). These quarries expose the thickness of sand and loess deposits.

All samples taken from soil pits were air dried and sieved (<1 mm). Total C and N, as well as organic C (C_{org}) were analyzed by dry combustion with an elemental analyzer Thermo Flash1112. The content of humic and fulvic acids in organic matter was analyzed by the Tyurin method [9]. The group phosphorus analysis allows us to divide organic and inorganic compounds of phosphorus [5]. The inorganic phosphorus content is determined by photometry after 1n H_2SO_4 treatment. Total phosphorus is determined by photometry after the calcination at 500°C (for conversion of phosphorus-organic compounds to soluble forms) and 1n H_2SO_4 acid treatment.

Radiocarbon analyses of extractable humic acids were carried out at the Kyiv Radiocarbon laboratory (Institute of Enviromental Geochemistry, Ukraine). The magnetic susceptibility of soils was measured by kappameter KT-5 [3]. The $^{13}C/^{12}C$ ($\delta^{13}C$) ratio was measured on a mass spectrometer Thermo-Finnigan Delta V Plus IRMS and elemental analyzer Thermo Flash1112. Removal of carbonates from the soil was made by the method of soil fumigation [2]: by decomposing the carbonate under the concentrated hydrochloric acid. Sample of calcareous soil pulverized into powder (30 mg) was placed in open glass bottles and moistened with 50 mkl of distilled water. Then weighing bottles with the sample were placed into a closed desiccator with the glass of concentrated HCl (100ml). After fumigation (24 h) the samples were dried for 4 hours at 60°C or under a hood. For the measuring of the isotopic composition of soil carbonates sample was calcined at 500°C to remove the organic substance.

3. Results and discussion

Last glaciation deposits on the terraces of the Desna river (middle reaches), opened in quarries near villages Krasnoe (first terrace above the floodplain) and Telec (second terrace above the floodplain), are a part of a hollow formed by melted glacial waters flowing into the river. Therefore, these sediments are presented by sandy-loam, sand deposits and loess with sand inclusions [5, 6, 7]. According to Velichko [11], second and first terraces of the river Desna were formed respectively in the first and second half of the Valdai period. Investigated deposits were formed during the Late Valdai 24-10.2 kyr BP (~Wurm IV) and correspond to isotope stage MIS 2 (25-10 kyr BP, by Svendsen et al. [10]). This phase began after Bryansk (Paudorf, Shtilfrid B) Interstadial - 32-24 kyr BP characterized by well-studied and well-expressed soils in many loess strata of the East European Plain [11].

In the studied sediments, we did not find well-expressed Bryansk soil (it is present only in the outcrops at a higher elevation on the bluff right bank of the Desna River). Investigated Last Glacial soils and pedolithosediments have traces of active freezing processes and thawing of deposits. At the same time there was an accumulation of thick sandy deposits on the study territory, which have a layered texture, solifluction traces (soil deformation processes), ferruginizated horizons.

Studied sandy-loess sediments contain two sinlithogenic soils. The radiocarbon dates for these soils are 16500 ± 230 BP (Ki-17414) and 12930 ± 170 BP (Ki-17413), which correspond to the last stage of Valdai glaciation and reflect two interstadials corresponding to Lascaux and Belling interstadials (Wurm IV) and MIS2. Both soils have a number of common traits: gleying profile, iron accumulation, the presence of iron-manganese concretions, lower values of the magnetic susceptibility and increase in humus, organic phosphorus and calcium carbonate content (table 1).

Among the interstadial soils after Bryansk interval (younger than 24 kyr BP) there is well known Trubchevsk humus horizon (gleying level) highlighted by Velichko [11] near our research area (village Golubcha, Bryansk region, depth of the layer 3.8-4.0 m) and dated 16.5-15 kyr BP (Lascaux

IOP Conf. Series: Earth and Environmental Science **107** (2017) 012123 doi:10.1088/1755-1315/107/1/012123

interstadial, Wurm IV). The horizon is described by the authors as a weak brown level with silty loam composition, carbonates and organic stains along the pores, inclusions of sandy material. There are also signs of waterlogging: ferrugination spots and grayish tint in this horizon.

Thus, the study horizon with data 16500 BP from quarry by Telec village is close to the Trubchevsk horizon and Lascaux interstadial. It lies at the depth of 2.5-2.8 m and is represented by light loam with thin layers of sand, carbonates through the pores and Fe-Mn-concretions. Below it the clayey carbonate BCb horizon lies (2.8-3.0 m) which is underlain by ferruginate and carbonate sandy horizon (3.5-4.2 m) - BCfe,b. Perhaps the study horizon crowned permafrost level with processes of carbonates and iron pulling. Dated horizon was covered with noncarbonate loess (C1 1.0-2,5m) with humus streaks and Fe-Mn-concretions. Also the paleosol with profile Bb-BCb-BCfe,b-Gb has good spatial expression. The upper horizon Bb has slightly alkaline pH, depleted humus content (0.5%), contains 6.0% $CaCO_3$ and does not contain P_2O_{5org} (table 1). Humic acids prevail in the humus composition. Horizon BCb has relatively high content of humus (0.7%), organic phosphorus (68.0 mg/kg $P_2O_{5\ org}$), carbonates (5.6% $CaCO_3$) and higher values of the magnetic susceptibility. Underlying sandy horizon BCfe,b enriched with calcium carbonate (10.6%), has low humus content and 44.5 mg/kg of organic phosphorus (P_2O_{5org}).

The results of carbon isotopic composition (figure 1) of organic matter indicate a moderately humid conditions of paleosol and underlying sediments formation ($\delta^{13}C$= -25.6-26.7‰). It also points to the arid climate ($\delta^{13}C$= -23.1-24.5‰) during the subsequent accumulation of loess loam horizon C1 overlying the paleosol. This loess layer was probably formed during the Yaroslavl cryogenic stage of the Valdai glaciation (altynovsky loess by Velichko et. al, [11]) 13-14 kyr BP.

Buried soil with age 12930 ± 170 BP (interstadial Belling) lies in the sediments in a quarry near the village Krasnoe. The study paleosol unlike previously described Trubchevsk soil has a deformed character and stands out in the lenses of humified loam material between ferruginate sandy layers. It has a profile Bca b-BCca,fe b-Gb. Horizon Bb contains a relatively high amount of humus (0.7%) and total phosphorus (409.7 mg / kg, P_2O_5), which corresponds to the humus horizon of the present agrogrey arable soil. It has relatively high values of magnetic susceptibility judging by the increase in humus content. Perhaps it was formed during periods of surface stabilization in drier than Lascaux interstadial conditions. Drier conditions are also illustrated by the results of the carbon isotopic composition of organic matter (heavier values of $\delta^{13}C$= -24.9‰, figure1a) and the high content of calcium carbonate (39.1% CaCO3) (table 1). The horizons lying above and below this soil are not enriched with calcium carbonate. Organic matter from buried soil is saturated with humic acids and do not contain soluble fulvic acids. High amount of organic phosphorus compounds is probably due to greater biological productivity of the landscape. Thus, the interstadial Belling is characterized by drier climate than Lascaux Interstadial (Trubchevsk Interstadial).

However, the beginning of the Holocene is featured by a catastrophic change of climate, vegetation and atmosphere chemical composition. The concentration of carbon dioxide in the atmosphere increased sharply and C-3 plants with strong root systems developed on the waterlogged soil. The low values of the isotopic ratio $\delta^{13}C$ of organic carbon (to $\delta^{13}C$= -28.4 ‰), high content of humus, phosphorus in the second humus horizons (Ab, 50-85 cm) of agrogrey soils (table1) that are common for the investigated territory characterize humid and warm climatic conditions. Second humus horizons distinguished by black color, slightly acidic pH, relatively high humus content (2.6%), are more than in the modern plow horizon. Second humus horizon is characterized by high phosphorus content, including organic phosphorus accumulation (635.8 mg/kg P_2O_5), while the amount of organic phosphorus compounds in modern humus horizon reaches only 178.7 mg/kg of P_2O_5 . Among the low magnetic susceptibility values, a small maximum allocated is caused by the humus accumulation. $\delta^{13}C$ ratio have the lowest values in the profile ($\delta^{13}C$ = -28.4 -29.5‰). Thus, the climate was more humid and had more favorable conditions for the accumulation of organic matter than the modern one. About 2 thousand years ago (2180±60 BP Ki-17415, 1650±60 Ki-18775) chernozem under meadow vegetation with C-3 type of photosynthesis on this aria of modern gray forest soils was formed.

Accumulation of stable highly condensed molecules of organic matter in the second humus horizons of modern soils, apparently, is a relic of the chernozem soil formation stage [4, 8].

Atlantic Chernozem soil is separated from modern gray soil by the Late Holocene podzolic horizon (AE 30-50 cm) with date 2180+60 cal BP. Colder conditions are indicated by lower values of humus and organic phosphorus content as indicators of reduction in biological activity. There is increase in weight values of the isotopic ratios $\delta^{13}C$ to -26 ‰. The lignin phenols ratio for the observed horizons also typical for small-leaved species of wood have been described in our previous studies [6, 7].

Table 1. Paleoecological indicators of soils.

Location	Age, years BP	Horizon, depth, cm	$C_{org.}$, %	$P_2O_{5\ org.}$ mg/kg	$\delta^{13}C_{org.}$ ‰	$CaCO_3$, %	Cha/ Cfa*	χ x10^{-6}, SGSM	$P_2O_{5\ inorg.}$ mg/kg
Quarry, v. Telec with micro-elevation soil		Ap 0-30	1.2	115.0	-26.1	0	0.69	31.5	325.7
		AE 30-35	0.4	17.8	-25.4	0	0.95	31.5	253.4
		EB 35-46	0.2	18.4	-24.7	0		21.1	205.2
		BC 46-65	0.2	10.6	-24.7	0		19.5	186.4
		C 65-85	0.2	0.0	-24.1	0		16.8	232.0
	6690±110	Ab 85-99	0.4	97.2	-25.4	16.7		17.1	258.7
		C1 99-120	0.3	83.9	-25.3	46.0		13.2	261.4
		C2 120-150	0.2	38.9	-25.7	41.3		9.6	258.7
		C3 150-250	0.5	29.9	-23.1	42.0	0.14	12.0	118.03
	16500 ± 230	Bb 250-275	0.3	0.0	-26.5	6.0	4.00	2.5	24.48
		BCb 275-300	0.4	68.0	-25.6	5.6	2.00	23.6	65.08
		ORTZ1 350-420	0.3	44.5	-26.5	10.6		15.3	42.76
		G 770-970	0.1	0.0	-26.6	11.1		2.2	141.74
		C4 970-1000	0.1	0.0	-26.7	10.7		0.0	11.25
		C5 1000-1300	0.2	0.0	-25.7	9.9		6.9	123.96
Quarry, v.Krasnoe with microdepression soil		Ap 0-30	1.8	178.7	-25.0	0	0.92	16.1	334.2
	2180±60, 1650±60	AE 30-50	1.5	311.9	-26.4	0	0.82	11.3	246.8
		Ab 50-85	2.3	635.8	-28.4	0	1.08	19.4	421.6
		AEb 85-100	1.3	389.8	-26.2	0	0.54	13.2	596.4
		Eb 100-125	0.4	125.4	-25.6	0		13.9	591.1
		EBb 124-140	0.6	0.0	-26.1	0		8.7	712.0
		Bb 140-175	0.3	0.0	-26.1	0		11	383.5
		C 175-270	0.2	0.0	-25.3	0		1.1	265.4
		C3,fe,1 445-470	0.2	0.0	-26.6	0		8.6	132.26
	12930 ± 170	Bb 470-495	0.4	106.3	-24.8	39.1	4.00	13.1	303.42
		BCfe,b2 495-500	0.1	0.0	-25.3	35.6		5.4	101.14
		G, 500-510	0.1	0.0	-27.0	0		3.1	129.67
		BCfe,b3 510-520	0.1	7.8	-27.2	0		5.2	62.24
		C4 520-570	0.1	2.6	-26.7	7.6		2.8	77.80

* - Cha - C from humic acids. Cfa - C from fulvic acids

Thus, loess-sandy-soil series are an archive of paleoecological information about the planet global climate system restructuring, which occurred on the investigated territory at the Late Glacial-Holocene boundary. This restructuring had a pulsating character, accompanied by sharp fluctuations in climate, vegetation composition and appearance of new landscapes. There are at least two different interstadials by latest wave of the Valdai (Wurm) glaciation, corresponding to the two buried soils: dry arid episode about 12930 years BP and temperate - humid period - about 16500 years BP for the territory of the middle reaches of the river Desna. Despite rather harsh climate characteristic of this area there is a large number of Paleolithic sites (Borschevo, Khotylevo, Suponevo, Timonovka etc. [11]) which acted both in warmer (interstadials) and cooling (stadials) periods. A sharp increase in the humidity of landscapes in the first half of the Holocene and cooling - in the second half were unidirectional and global.

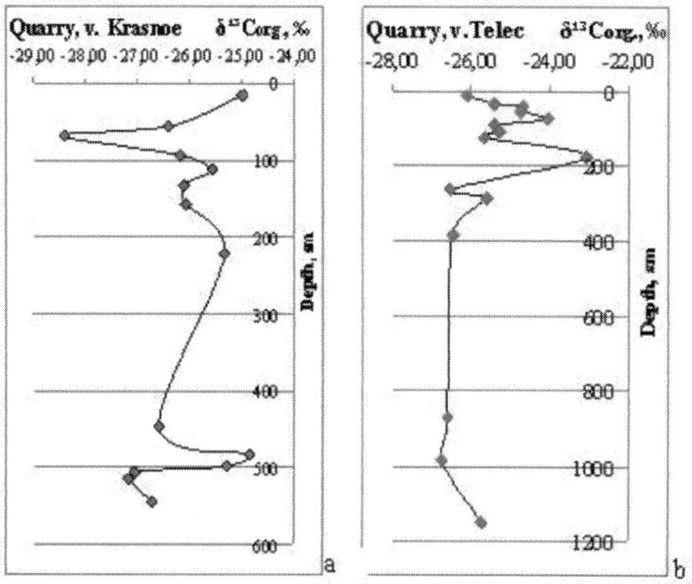

Figure 1. Carbon isotopic composition of organic matter: a - sequence of quarry v. Krasnoe, b - sequence of quarry v. Telec.

Acknowledgements
We thank Russian Scientific Fond for the financial support of this investigation (grant 17-14-01120)

References

[1] Alifanov et. al. 2010 Paleocriogenesis and a variety of soils on the center of East European Moscow p 160 (in Russian)
[2] Harris D et. al. 2001 Acid fumigation of soils to remove carbonates prior to total organic carbon or carbon-13 isotopic analysis *Soil Sci. Am. J.* **65** pp 1853–1856
[3] Kellerman D G 2008 "Magnetometry" Tutorial (Ekaterinburg: Ural State University) 8-9.
[4] Khokhlova O S et al. 2016 Carbonate features in the uppervost layers of Quaternary deposists Northern Armenia, and their significance for paleoenvaronmental reconstruction *Quaternary international* **418** pp 94–104
[5] Kovalev I V and Kovaleva N O 2011 Organophosphates in agrogrey soils with periodic water logging according to the data of ^{13}P NMR-spectroscopy *Eurasian Soil Science* **1** pp 29–37.
[6] Kovaleva N O and Kovalev I V 2002 Aromatic Struktures of Linin in the Organic Matter of Gray Forest Soils *Moscow University Bulletin,* 17 Soil Science, **2**, pp 23–29.
[7] Kovaleva N O and Kovalev I V 2015 Lignin phenols in soils as biomarkers of paleovegetation. *Eurasian Soil Science,* **9** pp 1073–1086.

[8] Kovaleva N O et. al. 2013 Carbon isotope composition in of soil organic matter in the diagnosis of climate change *Int. Conf. Bioindication in Ecologial Assessment of Soils and Related Habitatas* (Russia: Moscow BINOM) pp 98–99.
[9] Orlov D S 1995 Humic Substances of Soils and General Theory of Humification (Oxfordshire: Taylor & Frances) p 266.
[10] Svedsen J I et. al. 2004 Late Quaternary ice sheet history of Northen Eurasia. *Quaternary Science Reviews* **23** pp 1229–1271.
[11] Velichko A A et. al. 1996 Paleocryogenesis, soil cover and agriculture (Moscow: Nauka) p 150

IOP Conf. Series: Earth and Environmental Science **107** (2017) 012124 doi:10.1088/1755-1315/107/1/012124

Characteristics of the surface chemistry of linden pyrochar after removal of labile organic matter

A A Valeeva[1], E V Smirnova[1], K G Giniyatullin[1], V V Vorobev[2], L R Biktasheva[1] A N Grachev[3]

[1]Institute of Environmental Sciences, Kazan Federal University, Kremlevskaya str., 18, 420008, Kazan, Russia
[2]Interdisciplinary Centre of Analytical Microscopy, Kazan Federal University, 420008, Kazan Russia
[3] Kazan National Research Technological University, K. Marksa Str. 68, 420015, Kazan Russia
E - mail: valeyabc@mail.ru

Abstract. The changes of chemical properties of the pyrochar surface were studied in the laboratory experiment that simulated pedogenic transformation of pyrochar under the influence of soil biota. The native pyrochar samples were obtained by pyrolysis of linden wood residues at the temperature of 250°C, 450°C and 650°C. Their modified samples were obtained by removing an easily degradable pool of organic substances that can be used by microorganisms during the first months after application to the soil. In low-temperature linden pyrochar (250°C and 450°C) dominated carboxylic and phenolic surface groups, in high-temperature (650°C) - lactonic groups. After removal of readily decomposable organic substances the acidity of the phenolic and lactonic groups in pyrochar of low-temperature pyrolysis sharply decreased. Characteristic feature of all studied samples is the presence in IR spectra of absorption bands of gyroxyl, carbonyl, methylene groups and organosilicon polymers. The feature of IR spectra of linden pyrochar (250°C and 450°C) is the presence of absorption bands of the stretching vibrations of the tertiary alcohols and phenols C-O group.

1. Introduction

Eco-pyrochar is derived from available agricultural and forestry wastes that do not pose a threat to the environment and human health [1]. They are mainly used to increase the soil fertility and reduce the greenhouse gas emissions [2-4]. The main factors that determine the physical and chemical properties of pyrochar (**PC**), as well as the surface chemistry are plant residues and the technology of their production [5-7], produce and use PCs as sorbents of heavy metals, pesticides, various ions, organic pollutants, as well as to reduce the leaching of nutrients from the soil [8-13]. Their practical application is based on the ability to interact with atoms, ions and molecules of other substances through of chemical groups on the particles surface [14]. The scientific literature describes research works of the study of pyrogenic organic matter (**OM**) surface chemistry [15-18] and there are practically no publications about the study of PC chemical properties changes in soil conditions. The main difficulty of studying the change of PC chemical properties in soils is the physical separation of pyrogenic particles from the soil matrix. The purpose of this work is to study in the model experiment of the pedogenic transformation of PC chemical properties.

2. Material and methods

To study the changes of properties and surface chemistry of the pyrogenic OM the native and modified PCs samples were studied in the model experiment. Native samples were obtained from the stem wood linden of Laishevsky region of RT. The pre-cut samples were initially crushed to a chip size by a knife crusher and then subjected to pyrolysis in a laboratory retort d = 32 mm in a muffle furnace with a vapor-gas mixture withdrawn through a condenser under various temperature conditions of 250°C (**PC**^{250}), 450°C (**PC**^{450}) and 650°C (**PC**^{650}). The weight of the sample was 50g, and the rate of heating with pyrolysis to the regime temperature was 10°C/min. After heating to the regime temperature, the retort was held in the isothermal regime until the gas formation ceased, followed by cooling. The temperature fixation was carried out by chromel alumel thermocouples inside and outside of the retort. The cooled and stabilized PC was re-ground and subjected to separation on sieves. The modified PC were obtained by boiling native samples in distilled water in the flasks under reflux for 1 hour. After removal of OM extracted by hot water the PC was dried to constant weight. Boiling was carried out for the decomposition of the readily decomposable fractions of the PC organic substance which can be used by microorganisms for several months after their introduction into the soil. Carbon extracted by hot water is the indicator of readily decomposable carbon of organic compounds in the soil, i.e. active pool of soil OM [19].

The study of the PC samples morphology has been performed with field emission scanning electron microscope (**SEM**) Merlin (Carl Zeiss). Surface morphology was surveyed with an accelerating voltage of primary electrons of 5 kV and probe current of 300 pA for the minimal impact on the object of investigation. Elemental microanalysis was carried out using energy dispersive spectroscopy using an X-Max (Oxford Instruments) detector mounted on an SEM at an accelerating voltage of 20 kV and probe current of 600 pA. The determination of the functional groups of native and modified PC samples was carried out using the Boehm method [20], taking into account recommendations for standardizing of the experimental conditions [21-23]. The titration was carried out on an automatic ATP titrator with a drop volume of 0.07 ml / 2 minutes to pH = 7. The IR spectra of the samples was obtained by the method of disturbed total internal reflection on a Perkin-Elmer spectrometer with a PIKE MIRacleTM attachment in the range 4000-660 cm^{-1}. Interpretation of IR spectra was carried out on the basis of work [24].

3. Results and discussion

Figure 1 shows the SEM images of microstructure of linden PC samples. These images show that the PC is a highly porous carbon material with micropores which repeats the picture of natural "nutrient" channels of the original wood.

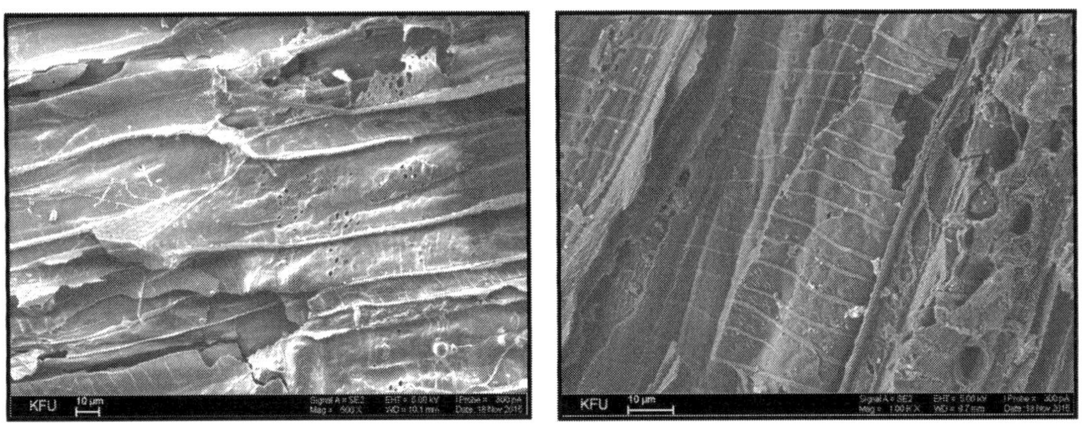

a b

Figure 1. SEM images of native linden PC (a) and modified linden PC after boiling of water (b).

Most of the internal space of the sample is pores with a diameter of 10 - 30 μm. These pores are large for the penetration of root hairs of plants which mainly are in the macroporous space filled with water [25]. However, they are most suitable for growth and development of microorganisms.

Figure 2 shows the areas within the elemental composition of native and modified PC samples surface was quantified.

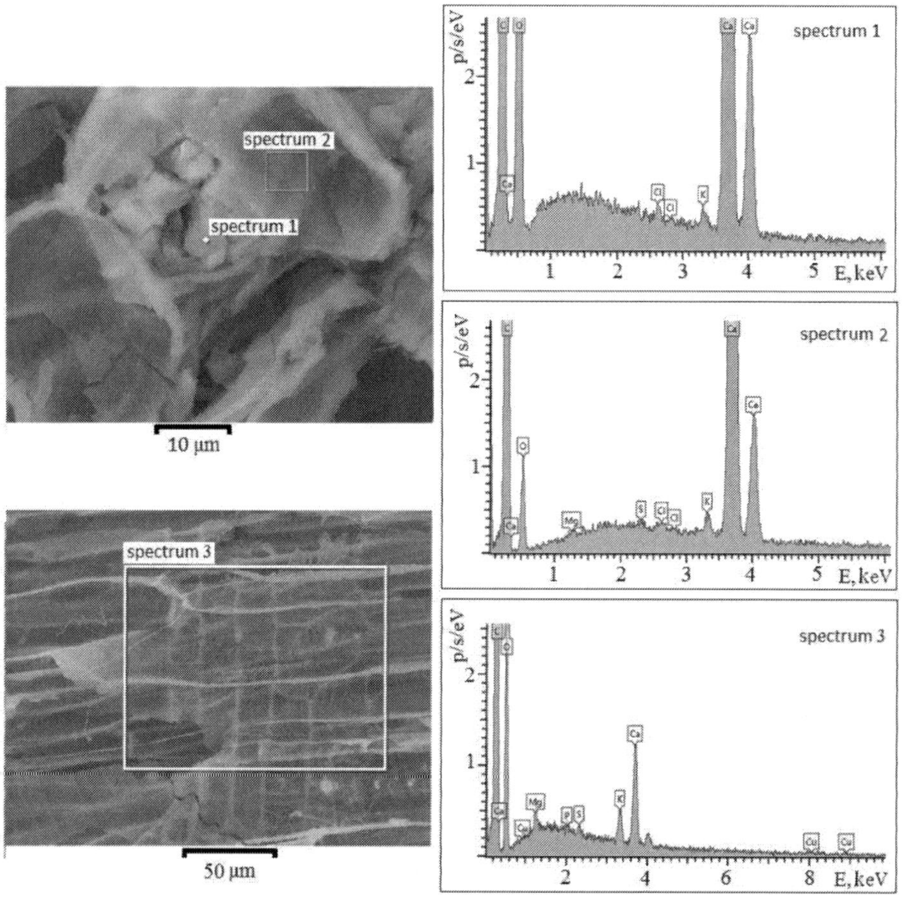

Figure 2. SEM images and elemental composition of native linden PC.

The main part of the elemental composition of the internal space of the PC sample is carbon and oxygen. Among these mineral elements were found macro and microelements of plant nutrition (figure 2, 3). Within the pores with the diameter of about 30 μm were discovered mineral crystals of about 5 μm in size (figure 2; spectrum 1). The increased oxygen content in spectra 1 and 3 indicates that oxygen is not only adsorbed on the surface of the pyrogenic OM, but is also chemically bonded to the mineral elements.

In the modified samples the mineral crystals were scattered over the surface of the PC slice (figure 3; spectrum 4 and 5). In the chemical composition iron and silicon absent in native samples were also found there.

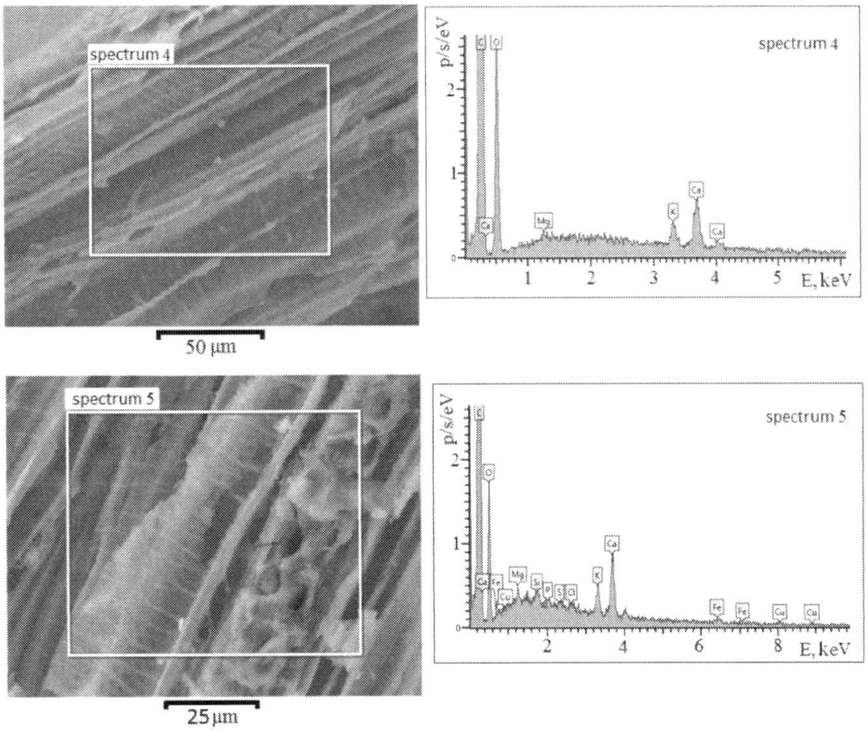

Figure 3. SEM images and elemental composition modified linden PC.

The total acidity of functional groups in native samples was from 297 to 1560 µmol/g, in modified from 367 to 1451 µmol/g (table 1). Carbonyl groups predominate in native samples PC^{250} and their amount increases after removal of organic substances extracted by hot water. The acidity of the lactonic group in the modified samples is on the contrary reduced. Thus after application of linden PC^{250} into the soil the acidity of the phenolic and carboxyl groups will increase.

Table 1. Acidity of PC surface functional groups.

Functional groups	Acidity, µmol/g					
	PC^{250}		PC^{450}		PC^{650}	
	Native	Modified	Native	Modified	Native	Modified
Phenolic	397	464	1167	492	63	85
Lactonic	265	71	322	340	227	224
Carboxylic	852	916	71	33	7	58
Total	1514	1451	1560	865	297	367

The acidity of the PC^{450} functional groups is mainly due to the presence of phenolic and lactonic groups which after the removal of readily decomposable organic substances are sharply reduced. The acidity caused by carboxylic groups is insignificant and after the removal of easily decomposable fractions of OM the acidity carboxylic groups decreases 2 times. The acidity of the PC^{650} functional groups is determined by the presence of lactonic groups. After the removal of the readily decomposable OM pool the slight increase in the acidity of phenolic and carboxylic groups is observed.

Thus, with the increase in the pyrolysis temperature the acidity of the functional groups in PCs decreases. In PC^{250} and PC^{450} predominate carboxylic and phenolic functional groups, in PC^{650} - the basic acidity of the functional groups is formed by lactonic groups.

The PC absorption bands are presented in table 2. A broad absorption band is observed in all samples in the region of stretching vibrations of hydroxy groups (~ 3400 cm^{-1}), but in PC^{650} its intensity is lower and indicates a decrease in the number of hydroxy groups. In native and modified samples the frequencies are characteristic of symmetric and asymmetric valence and deformation vibrations of methylene groups. Also all considered spectra have strong absorption in the 1700-1742 cm^{-1} region which corresponds to the carbonyl group of organic compounds.

In PC samples of high-temperature pyrolysis the absorption band is shifted toward the lower boundary of the usual interval which is probably related with the conjugation of aromatic rings. Characteristic feature of IR spectra is the presence of bands group between 1650 and 1450 cm^{-1} which is typical for stretching vibrations of the aromatic ring.

A broad and very intense band in the range from 1100 cm^{-1} to 1000 cm^{-1} is probably related to the presence of organosilicon polymers in which silicon and oxygen atoms alternate in the molecule.

Table 2. Absorption frequencies of linden PC.

Sample	Wave numbers, cm^{-1} (group and и vibration)
	Pyrolysis temperature of 250°C
Native PC	3340(νOH), 2930($\nu_{as}CH_2$), 2857(ν_sCH_2), 1738 (νC=O), 1606 (νC=C), 1515(νC=C), 1455(δCH_2, $\delta_{as}CH_3$), 1427 (δOH), 1373 (δ_sCH_3), 1321 (νC-O), 1235 (νC(=O)-O), 1159 (νO-C-C), 1031 (cyclo $(R_2SiO_3)_3$), 892 (δOH)
Modified PC	3355(νOH), 2927($\nu_{as}CH_2$), 2856(ν_sCH_2), 1727 (νC=O), 1602(νC=C), 1513(νC=C), 1453(δCH_2, $\delta_{as}CH_3$), 1423 (δOH), 1373 (δ_sCH_3), 1320 (νC-O), 1240 (νC(=O)-O), 1160 (νO-C-C), 1033 (cyclo $(R_2SiO_3)_3$), 884 (νO-O)
	Pyrolysis temperature of 450°C
Native PC	3343 (νOH), 2928 ($\nu_{as}CH_2$), 2856 (ν_sCH_2), 1742 (νC=O), 1607 (νC=C), 1515(νC=C), 1456 (δCH_2, $\delta_{as}CH_3$), 1426 (δOH), 1368 (δ_sCH_3), 1316 (νC-O), 1209 (νC-O), 1159 (νC-O), 1033 (cyclo $(R_2SiO_3)_3$)
Modified PC	3338 (νOH), 2928 ($\nu_{as}CH_2$), 2857 (ν_sCH_2), 1740 (νC=O), 1607(νC=C), 1515(νC=C), 1460 (δCH_2, $\delta_{as}CH_3$), 1428 (δOH), 1368 (δ_sCH_3), 1316 (νC-O), 1218 (νC-O), 1160 (νC-O), 1033 (cyclo $(R_2SiO_3)_3$)
	Pyrolysis temperature of 650°C
Native PC	3341 (νOH), 2926 ($\nu_{as}CH_2$), 2853 (ν_sCH_2), 1700 (νC=O), 1599 (νC=C), 1516(νC=C), 1451 (δCH_2, $\delta_{as}CH_3$), 1428 (δOH), 1377 (δ_sCH_3), 1037 (cyclo $(R_2SiO_3)_3$), 875 (νO-O)
Modified PC	3316 (νOH), 2925 ($\nu_{as}CH_2$), 2855 (ν_sCH_2), 1700 (νC=O), 1598 (νC=C), 1513(νC=C), 1451 (δCH_2, $\delta_{as}CH_3$), 1440 (δOH), 1377(δ_sCH_3), 1036 (cyclo $(R_2SiO_3)_3$), 882 (νO-O)

A feature of the IR spectra of linden PC of the pyrolysis temperature of 250°C and 450°C is the presence of absorption bands of the valence vibrations of the C-O group of tertiary alcohols and phenols near 1150 cm^{-1}, 1410-1310 cm^{-1} and 1200 cm^{-1}, 1410-1310 cm^{-1} respectively.

4. Conclusion

Analysis of the study of results showed that in the linden PC^{250} and PC^{450} dominated carboxylic and phenolic functionalities groups, in the linden PC^{650} dominated lactone groups. After removal of readily decomposable OM the acidity of the phenolic and lantonic groups in the PC of low-temperature pyrolysis considerably decreases. The acidity of functional groups in the PC^{650} is determined by the presence of lactonic groups. After the removal of the readily decomposable OM fractions there is a slight increase in the acidity of phenolic and carboxyl groups. Characteristic feature of all studied samples is the presence in the IR spectra of absorption bands of gyroxyl, carbonyl, methylene groups and organosilicon polymers. A feature of the IR spectra of linden PC^{250} and PC^{450} is the presence of absorption bands of the stretching vibrations of the tertiary alcohols and phenols C-O group.

Acknowledgments
This work was supported in part by the Russian Foundation for Basic Research, research project №17-04-00869 and is performed according to the Russian Government Program of Competitive Growth of Kazan Federal University

References:
[1] Duwiejuah A B, Cobbina S J and Bakobie N 2017 Review of Eco-Friendly Biochar Used in the Removal of Trace Metals on Aqueous Phases *International Journal of Environmental Bioremediation & Biodegradation*. **5(2)** pp 27–40
[2] Whitman T, Nicholson C F, Torres D and Lehmann J 2011 Climate change impact of biochar cook stoves in western Kenyan farm households: system dynamics model analysis *Environ. Sci. Technol.* **45** pp 3687–3694
[3] Kloss S, Zehetner F, Wimmer B, Buecker J, Rempt F and Soja G 2014 Biochar application to temperate soils: Effects on soil fertility and crop growth under greenhouse conditions. J. *Plant Nutr. Soil Sci.* **177(1)** pp 3–15
[4] Matovic D 2011 Biochar as a viable carbon sequestration option: global and Canadian perspective *Energy* **36** pp 2011–2016.
[5] Ling Zhao, Xinde Cao, Ondrej Masek and Andrew Zimmerman 2013 Heterogeneity of biochar properties as a function of feedstock sources and production temperatures *Journal of Hazardous Materials* **256–257** pp 1–9
[6] Giniyatullin K G, Smirnova E V, Grigoryan B R and Valeeva A A 2015 The Possibility of Use Research Methods of Soil Organic Matter for Assess the Biochar Properties Research *Journal of Pharmaceutical, Biological and Chemical Sciences.* **vol 6(4)** pp 194–201
[7] Jindo K, Mizumoto H. Sawada Y, Sanchez-Monedero M A and Sonoki T 2014 Physical and chemical characterization of biochars derived from different agricultural residues *Biogeosciences* **11** pp. 6613–6621
[8] [8] Moreno-Castilla C 2004 Adsorption of organic molecules from aqueous solutions on carbon materials. *Carbon* **42** pp 83–94
[9] Cao X D, Ma L N, Gao B and Harris W 2009 Dairy-Manure Derived Biochar Effectively Sorbs Lead and Atrazine *Environmental Science & Technology*. **43(9)** pp 3285–3291
[10] Mohan D, Jr. Pittman C U, Bricka M, Smith F, Yancey B, Mohammad J, Steele P H, Alexandre-Franco M F, Gomez-Serrano V and Gong H 2007 Sorption of arsenic, cadmium, and lead by chars produced from fast pyrolysis of wood and bark during bio-oil production *Journal of Colloid and Interface Science* **310(1)** pp 57–73
[11] Lehmann J, da Silva J P, Steiner C, Nehls T, Zech W and Glaser B 2003 Nutrient availability and leaching in an archaeological Anthrosol and a Ferralsol of the Central Amazon basin: fertilizer, manure and charcoal amendments *Plant and Soil*. **249** pp 343–357
[12] Novak J M, Busscher W J, Laird D L, Ahmedna M, Watts D W and Niandou M A S 2009 Impact of Biochar Amendment on Fertility of a Southeastern Coastal Plain Soil *Soil Science* **174** pp 105–112
[13] Valeeva A A, Grigoryan B R, Bayan M R, Giniyatullin K G, Vandyukov A and E Evtygin V G 2015 Adsorption of Methylene Blue by Biochar Produced Through Torrefaction and Slow Pyrolysis from Switchgrass *Research Journal of Pharmaceutical, Biological and Chemical Sciences* **vol 6(4)** pp8–17
[14] Radovic L R, Moreno-Castilla C and Rivera-Utrilla J 2000 Carbon materials as adsorbents in aqueous solutions *Chemistry and physics of carbon* **vol 27**. (New York: Marcel Dekker) pp 227–405

[15] Gorgulho H F, Mesquita J P, Pereira M F R and Figueiredo J L 2008 Characterization of the surface chemistry of carbon materials by potentiometric titrations and temperature-programmed desorption *Carbon.* **46** pp 1544–1555
[16] Boehm H P 1994 Some aspects of the surface chemistry of carbon blacks and other carbons *Carbon* **32(5)** pp 759–769
[17] Chen J P and Wu S 2004 Acid/Base-Treated Activated Carbons: Characterization of Functional Groups and Metal Adsorptive Properties *Langmuir* **20** pp 2233–2242
[18] Salame I and Teresa J 2001 Bandosz Surface Chemistry of Activated Carbons: Combining the Results of Temperature-Programmed Desorption, Boehm, and Potentiometric Titrations *Journal of Colloid and Interface Science* **240** pp 252–258
[19] Korschens M, Schulz E and Behm R 1990 Heißwasserloslicher C und N im Boden als Kriterium fur das N-Nachlieferungsvermogen *Zentralbl. Mikrobiol.* **145** pp 305–311
[20] Boehm H P 1966 Chemical identification of surface groups Advances in Catalysis (New York: Academic Press) ed Eley DD, Herman Pines, Paul B Weisz pp 179–274
[21] Goertzen S L, Theriault K D, Oickle A M, Tarasuk A C and Andreas H A 2010 Standardization of the Boehm titration. Part I. CO_2 expulsion and endpoint determination *Carbon.* **48** pp 1252–1261
[22] Oickle A M, Goertzen S L, Hopper K R, Abdalla Y O and Andreas H A 2010 Standardization of the Boehm titration: Part II. Method of agitation, effect of filtering and dilute titrant *Carbon* **48** pp 3313–3322
[23] Kim Y S, Yang S J, Lim H J, Kim T and Park C R 2012 A simple method for determining the neutralization point in Boehm titration regardless of the CO_2 effect *Carbon* **50** pp 3315–3323
[24] Smith A L 1982 *Applied Infrared Spectroscopy* (London: John Wiley and Sons)
[25] Joseph S, et al 2010 An investigation into the reactions of biochar in soil *Aust. J. Soil Res.* **48** pp 501–515

IOP Conf. Series: Earth and Environmental Science **107** (2017) 012125 doi:10.1088/1755-1315/107/1/012125

Comparative recreational assessment of Karaganda city public green spaces

I S Akylbekova and T Yu Zengina

Faculty of Geography, Department of Environmental Management, Lomonosov Moscow State University, Leninskie Gory 1, Moscow 119991, Russia

E-mail: indira_msu@mail.ru

Abstract. This article represents evaluation of recreation environment on the territory of the large industrial city of Karaganda, located in the dry steppe zone of Central Kazakhstan. A comparison of quantitative and qualitative indicators, level of recreational attractiveness and providing the citizens with public green spaces, allowed to make a more complete characterization the urban recreation places and to identify the city districts, which require prioritized fundraising for development of existing parks and public gardens, and for creation of new territories of recreational purpose. Based on the results of conducted expert assessment and sociological survey of visitors, the main problems of urban green areas were identified and also the most high-demand trends and practical recommendations for their improvement and further use were proposed.

1. Introduction

Currently, with the increasing of functional and planning structure complexity of the cities, the importance of urban planning, as well as the role of environmental and social open public spaces significantly increased [1, 2]. In this regard, one of the main instruments of urban development is the creation and development of parks and other green areas, which can simultaneously perform several functions: improve outer appearance of the cities, improve environmental conditions and promote development of urban public life.

So, *public green spaces* (hereafter PGS) in addition to their primary purpose – organization of various forms of mass leisure activities for population – also provide several other functions like planning, sanitary-hygienic and aesthetic roles. PGS are multifunctional and specialized parks, public gardens, boulevards, short-time recreation areas by water, citywide green places and district public centers [3]. Performing the role of major recreation facilities in the city, these areas must respond the current needs of society and have all necessary recreational conditions for different social groups. Therefore, their effective organization and development is one of the main priorities in creating a comfortable urban environment and provision the population with PGS, and the degree of their recreational attraction is considered as important social and environmental indicator that determine the quality of city life [4, 5].

Provision of population with PGS is estimated as square area of parks, public gardens, parkways and other city- and district level recreation facilities per person (m^2/person) [6]. Estimation of this index is carried out in accordance with the standards of urban land improvement, which provide hygienic conditions of population life. Generally, the rules of landscaping vary depending on the characteristics of an area, in which the city is located, and on the number of its inhabitants.

Assessment of recreational attractiveness of PGS includes calculation of a number of indicators of transport accessibility, safety, cleanness and maintenance, level of improvement, as well as a set of objects and services for a variety of recreational activities. This integrated assessment is intended for analysis of the PGS current state and identification of existing deficiencies and the most high-demand directions for their further use and development.

The comparison of these two parameters (provision of PGS and their recreational attractiveness) allows to describe the level of satisfaction the public needs in a short-time daily recreation in the natural complex territories of the city more fully.

2. Study area

In Kazakhstan, the issues of improving the efficiency social and ecological functions of the PGS are most acute in the city of Karaganda - a large industrial center of the country, where the lands disturbed by industrial activities, which not only pose a serious threat to the population health and lives, but also considerably complicate a creation of an integrated green spaces system [7]. Sparseness of forest stands is also caused by adverse soil and climatic conditions of dry steppe zone, where the city is located.

Currently, the area of mining badlands, formed as a result of long-term intensive development of coal deposits, is 28% of the city total area [8]. Moreover, most of them are located in the central part of Karaganda that was the main reason for formation of non-compact urban planning structure and isolated location of residential areas from the city centre.

As a result, Karaganda is distinguished by disjointed and uneven location of recreation areas that defines considerable difference in areal and qualitative provision the people with urban districts, including public green spaces, which have various design. In this regard, this study was aimed to evaluate and compare the certain city districts according to level of provision the population with public green spaces and with recreation space within them (i.e. in terms of their recreational attractiveness).

All parks and the largest public gardens of study area and boulevards located within the five districts of Karaganda were selected as research targets. These are Noviy Gorod, Yugo-Vostok, Maikuduk, Prishakhtinsk and Sortirovka. In total, 19 public green spaces were studied: 5 parks, 10 public gardens and 4 boulevards.

3. Material and Methods

Assessment of recreational attractiveness of PGS was conducted by means of individual expert scoring, which was supplemented with information obtained in the course of the sociological survey.

As such surveys in Kazakhstan are implemented very recently, the expert scoring was based on the criteria, which was used to build the social rating of Moscow and Moscow region parks [9,10]. Evaluation criteria were at first adapted to the climatic and social conditions of Karaganda city. For example, due to dry and hot summers, particular attention was paid to the arrangement of gazebos, canopies and other sun shelters, and presence of water bodies, fountains and shaded leisure areas.

Altogether there were offered 6 key evaluation criteria, amounting 41 positions for parks evaluation and 33 for public gardens and boulevards (Table 1). Unequal number of items is explained by different functional purposes of objects, their sizes and specific character of location in the city planning structure.

For each estimated item, 3-digit rating scale was used, where a point 0 corresponded to the lowest value, and point 2 – the highest value. Thus, maximum amount of value to be gained by park was 82 points and for public garden and boulevard – 66 points.

Table 1. Key criteria and evaluation of Karaganda public green spaces recreational attractiveness.
** items that have not been used for evaluation of public gardens and boulevards*

Key criterion	Evaluation items
I. Transport accessibility	1. distance from the main entrance to the nearest public transport 2. availability of parking for personal transport* 3. number of parking stalls at the parking space*
II. Quality of parking infrastructure	***A) marking of entrance zone and availability of indicators for orientation*** 4. quality of equipped entrances 5. presence of pedestrian crossings next to entrances 6. visual identification of entrances* 7. availability of park layout, indicators and navigation stands* ***B) arrangement of road and walkway network*** 8. presence of cracks, potholes and other deformations of the roadway surfacing 9. the possibility of tracks using by people with limited mobility 10. furnishing ladders with ramps ***C) availability and general condition of main land improvement facilities:*** 11. public lavatories* 12. lavatories for wheelchair users* 13. benches 14. gazebos, canopies and sun and rain shelters 15. garden furniture
III. Sufficiency of specially equipped spaces and services for various recreation	***A) availability of specially equipped spaces for recreation:*** 16. playgrounds 17. sports grounds 18. fairground attractions* 19. sites for mass events and stages 20. bicycle lanes 21. lawns, picnic grass plots 22. table game places 23. water pond rest areas 24. fountain rest areas 25. shadow rest areas 26. quiet rest areas ***B) availability of services and opportunities for comfortable rest and entertainment:*** 27. fast-food restaurants 28. water selling points 29. game and sports equipment rent 30. mass cultural events
IV. Rest security	31. police posts 32. availability of street lamps throughout PGS territory 33. availability of first aid posts* 34. presence of homeless and other marginal groups representatives 35. presence of people drinking alcoholic beverages 36. presence of stray dogs
V. Well-keeping and cleanness of territory	37. presence of rubbish and level of PGS littering 38. availability of litter bins throughout PGS territory 39. availability of dog runs
VI. Presence of identification objects distinguishing PGS from the others	40. memorials, sculptural complexes and other show places structures 41. landscaped architecture sites (specially designed water ponds, parkways, groves and etc.)

A social survey of visitors was conducted at the same time with an expert assessment in the 4 most visited parks (Central Park, Park Pobedy (Victory Park), Park Pyatidesyatiletiya Nezavisimosti Respublici Kazakhstan (Kazakhstan 50th Independence Anniversary park), Park Zheleznodorzhnikov (Railroad men cultural center park) and 2 city public gardens (public garden "Vechniy ogon"" (Eternal flame) and public garden on the Respublica Avenue).

The respondents were asked to complete the questionnaire and to evaluate PGS on 5-digit scoring scale on the key criteria applied for the expert scoring assessment. Altogether, 120 visitors were interviewed. Most of them were pupils/students (40%), parents with children (34%) and pensioners (20%).

The responses enabled us to clarify and specify the results of expert assessment and thereafter to evaluate the recreational conditions within each of recreation area surveyed in most exactly manner. For this aim, at the final stage of study, total amount of points were estimated on each PGS, including points of all key criteria. Thus, the results of the expert assessment actually contained elements of collective expertise.

To be able to compare the final results, the amount of points gained by each object of research was converted into relative value - percentage of the maximum value possible for the number of points for objects of this type. Further, based on this variation indicator (%), the explored territories were evaluated (ranked) according to the degree of recreational attractiveness for each of the city districts.

4. Results and discussion

The assessment showed that the integrated evaluation indicators of recreational attractiveness of Karaganda city PGS are much lower than the maximum possible total sum of evaluation points and vary within 20 to 60%.

Out of 19 researched territories, 7 objects (2 parks, 4 public gardens, 1 boulevard) have values above 50%, 9 objects (2 parks, 5 public gardens, 2 boulevards) - 30-50%, 3 objects (1 park, 1 public garden, 1 boulevard) - less than 30%. Most attractive leisure areas are Park Imeni Tridsatiletiya VLKSM (Central Park) in Noviy Gorod district (59.7%) and the public garden of Molodezhniy cultural center in Prishakhtinsk (59%) and the least attractive ones are Park of Zheleznodorzhnikov cultural center in Sortirovka (20,7%) and the boulevard on Satpayev street in Noviy Gorod (19,6%). Consequently, there are no parks, public gardens or boulevards in Karaganda with a high degree of recreational attractiveness that would completely satisfy the needs of population.

In addition, the ratings differ not only between different city districts, but also have significant intraregional differences (Table 2). Thus, the greatest difference in the rating of PGS (up to 40%) was fixed in Noviy Gorod, where 3 of 8 explored objects had values above 50%, 4 objects – 30-50%, and 1 object - 30%. A small rating difference equal to 1-3% was fixed in Yugo-Vostok district leisure areas. Within the other districts, indicators of recreational attractiveness vary from 12.6 to 33.3 percents.

Comparison of the results of a comprehensive assessment of recreational attractiveness with PGS provision levels is highly interesting.

According to the construction norms and regulations of the Republic of Kazakhstan (ROK SNiP standard), public green space area standard in the cities with high population size, including Karaganda, should be 10 m^2/person [11]. With a population of 499,6 million people (2016) and public green space with total area of 3164,3 m^2, on the average, there is 6,33 m2 per one Karaganda resident, which is below the standard by 1.5 times. However, due to the existing city planning structure, this average figure is not representative.

Level of PGS provision calculated for different city regions (Table 3) varies considerably. So, only for one urban district – Noviy Gorod – it exceeds the standard nearly 1.5 times; for the other districts it varies from 50% of normal value to a value 5 times less than the standard.

Thus, the most favorable conditions for recreation are created in the existing public green spaces of administrative and business district Yugo-Vostok. Parks and gardens recently designed are highly appreciated by the local population (integrated indicator of their recreational attractiveness is more than 50%), but they are very few in number to provide the leisure areas for all residents of this intensely built-up area. Calculated indicator of PGS provision is more than twice below the standard and in total is 4,16 m^2/person.

Table 2. A comprehensive assessment of Karaganda public green spaces recreational attractiveness.

Urban district	Public green spaces	*Total score, %*	*Average score by district, %*
Noviy gorod	Park Imeni Tridsatiletiya VLKSM (Central Park)	59,7	44,3
	Boulevard on Mir street	54,5	
	Public garden "Vechniy ogon" (Eternal flame)	51,5	
	Public garden "Shalkyma"	50,0	
	Boulevard on Tolepova street	43,9	
	Boulevard on Abdirova street	39,3	
	Public garden near Lenin cinema	36,3	
	Boulevard on Satpayev street	19,6	
Yugo-Vostok	Public gardens on the Respublica Avenue	53,0	51,4
	Kazakhstan 10th Independence Anniversary park	51,2	
	Park Pobedy (Victory Park)	50,0	
Maikuduk	Public garden of Noviy Maikuduk cultural center	54,5	47,8
	Public gardens on Kuzembaeva street	50,0	
	Kazakhstan 50th Independence Anniversary park	39,0	
Prishakhtinsk	Public garden of Molodezhniy cultural center	59,0	42,8
	Public garden of 21 microregion	43,9	
	Public garden in Uzenka	25,7	
Sortirovka	Public garden on Drujby street	33,3	27
	Park of Zheleznodorzhnikov	20,7	

Table 3. Provision of Karaganda city population with public green spaces (PGS).

Urban district	**PGS area, m^2**	**Number of population, ths. persons**	**Specific provision with PGS (m^2 per man)***	**Level of provision with PGS (% of SNiP standard)**
Noviy gorod	2 044,8	140,6	14,50	**145,0**
Sortirovka	148,1	27,7	5,34	**53,4**
Yugo-Vostok	536,8	128,8	4,16	**41,6**
Maikuduk	312,7	138,7	2,25	**22,5**
Prishakhtinsk *(including neighboring villages)*	121,9	63,8	1,91	**19,1**

* *under the ROK SNiP 3.01-01-2008 standard for PGS area– 10 m^2/man*

Reverse situation is observed in residential-administrative and cultural-historical centre Noviy Gorod, which PGS provision level is 1.5 times above the standard (14.5 m^2/person), but average level of their comprehensive recreational attractiveness is lower than in Yugo-Vostok district and amounts to 44,3 % of the possible value. This suggests that, despite the large number of parks, public gardens and boulevards, there are not so many attractive leisure areas.

In other city districts, level of population provision with PGS is considerably below the standard at quite low level of their recreational appeal. This means that most of Karaganda residents are not provided sufficiently with leisure areas that meet the requirements on quantitative indicator (PGS area per resident) or qualitative (level of improvement, their transport accessibility, etc.). The most challenging situation is observed in *Prishakhtinsk*, connecting several towns with small industrial enterprises, which PGS provision amounts to 1.91 m^2/person and in Maikuduk - a large industrial district - which rate is 2.25 m^2/person. Recreation environment within these districts also differs. In the first case, general rate of recreation attractiveness of PGS range from 59,0% to 25.7%, in the second one - from 54.5% to 39.0%. In Sortirovka, the rail transportation maintenance center, level of population PGS provision is slightly higher (5,34 m^2/person), but recreational facilities are the worst in the city and level of recreational attractiveness is less than 30% of the possible rate.

The comparison of study results with PGS provision indicators allows to split the Karaganda districts into 3 groups, where:

1 – indicators of PGS provision above the standard at a relatively low level of their recreational attractiveness (below 50%) (Noviy Gorod);

2 – indicators of PGS provision below the standard at a relatively high level of their recreational attractiveness (above 50%) (Yugo-Vostok);

3 – indicators of PGS provision below the standard at a relatively low level of their recreational attractiveness (below 50%) (Maikuduk, Prishakhtinsk, Sortirovka).

The analysis allowed not only to identify the city districts requiring prioritized funds to provide the residents with public green spaces, but also to determine the most popular areas for further development. So, in Noviy Gorod, leisure facilities within the existing parks and gardens should be improved; in Yugo-Vostok district, it is important to create new green areas of recreational purpose; in the other city districts it is necessary to carry out both things.

In addition, assessment of recreation attractiveness has allowed identifying the *main problems of Karaganda city* public green spaces and to develop appropriate *practical recommendations*.

As it turned out, many of the city leisure areas are heavily littered that requires location of additional litterbins. Considering almost widespread presence of damaged park furniture, it is necessary to enhance control over the public tranquility in the parks. Poor maintenance of public green spaces is also proved by weaker condition of most green plants. To improve their environmental and recreational functions, it is necessary, first of all, to ensure regular watering and timely sanitary pruning and felling [12]. The main problem of transport accessibility is the absence of specially designated transport parking areas in the most of parks' territories, while they must be placed not farther than 400 m from the park entrance [13].

In estimating of the organization quality of public green areas, a number of problems were identified. These are first of all: 1) insufficient number or absence of gazebos, canopies and other sun and rain shelters; 2) deficiency of facilities and services both for passive and active recreation. In this regard, it is necessary to locate extra park gazebos and sheltered benches at all city recreational areas and install more playgrounds and sports grounds, to expand assortment of sports and game equipment available for rent and hold a variety of cultural and entertainment events.

In general, Karaganda city may become more comfortable and interesting for living if each district leisure areas will differ with unique features, such as spectacular landscape designs, memorials, sculptural complexes, as well as various city events such as music festivals, concerts and etc.

5. Conclusions

The assessment showed that exactly comparison of quantitative and qualitative indicators of public green spaces provision and recreational attractiveness of PGS allows to make a more exactly evaluation of the recreation environment in each city district.

So, it was revealed that none of Karaganda districts was sufficiently provided with landscaped areas of recreational attractiveness high level. The main problems of public green spaces surveyed are associated with their poor maintenance and littering, paucity of facilities and services for various recreations, as well as lack of unique objects defining their distinctness and attractiveness for visitors. In addition, expert assessment showed that there are no necessary conditions for people with limited mobility in the city. This can be considered a significant disadvantage, since nowadays ensuring recreational needs of all population categories, including disabled persons, is one of the primary tasks of urban planning.

Summarizing the results of expert assessment, including sociological survey, we can conclude that in Karaganda it is necessary, first of all, to pay attention to development of public green spaces with district importance and which are not municipal. It is extremely important to create the multifunctional parks and public gardens in each city district, meeting diverse needs of visitors.

To create new territories of recreational purpose and for efficient development and improvement of existing parks and public gardens, it is necessary to use modern methods of architectural-planning management and landscape design of public green spaces, and also it is very important to consider public opinion solving the issues related to improvement of urban recreation environment [14, 15].

It should be noted that foreign experience of recreational use of disturbed lands is most actual for Karaganda city. Impressive examples of mining badlands recovery and turning into common recreation places are presented in many countries. The most widespread of them are creation of water ponds and common recreation places for various recreational uses replacing the worked out quarries (swimming, fishing, water sports, etc.). In addition, man-made landscapes are being actively transformed to urban gardens and parks. At the same time, green areas are formed in such a way to constitute a single recreation zone, covering all parts of the old industrial center [16]. So, current experience confirms the possibility of involvement the planning structure in Karaganda city not only the local disturbed areas, but also large territories of mining badlands.

Therefore, recommendations and suggestions for improvement and development of Karaganda city public green spaces are consistent with the applicable standards of urban design, meet the modern approaches to the creation of public spaces and sensible to the wishes of the interviewed visitors. In the future, the study results may be included in urban development programs and also used at developmental projects for the city recreational zones development.

References

[1] Gehl J 2010 Cities for People (Washington - Covelo - London: Island Press) 128 p
[2] Kochurov B I and Ivashkina I V 2014 The role of open spaces in maintaining the ecological balance in the large city *Ecology of urban areas* **1** pp 6-13
[3] *All-Union State Standard 28329-89* 2006 Urban planting. Terms and definitions (Moscow: Standartinform)
[4] Jacobs J 1961 *The Death and Life of Great American Cities* (New York: Random House) 474 p.
[5] Nefedov V A 2015 *How to return the city back to people* (Moscow: Iskusstvo-XXI vek)
[6] Kravchuk L A 2011 *Structurally functional organization of landscape-recreational territories of cities in Belarus* (Minsk: Belarus nauka)
[7] Akylbekova I S and Zengina T Yu 2015 Karaganda city modern urban land use research and mapping *Geodesy and Cartography* **6** pp 34-40
[8] Akylbekova I S and Tzengina T Yu 2014 Mining badlands in the urban land use structure (the case study of the Karaganda city, Kazakhstan). *Engineering for Green Development: Proceedings of the 1st Russian-Japanese Collaboration Seminar for Sustainable Environment* (MSU Publishers) 154-159
[9] *Methods and sociological tools for social rating composing of parks in Moscow for a family holiday* 2012 (Moscow: RIA Novosti)
[10] *Rating of urban parks in Moscow region* 2014 (Moscow: HSE Vysokovsky Graduate School of Urbanism)
[11] *Construction norms and regulations of the Republic of Kazakhstan* 3.01-01-2008 Urban planning. Layout and development of urban and rural communities (Astana: Agency for Construction, Housing and Utilities of the Republic of Kazakhstan)
[12] Akylbekova I S, Zengina T Yu and Mustafina A B 2014 Update problems of Karaganda city

urban greening and ecosystem restoration *Science and Management III International Research and Practice Conference "Urgent Issues in Science and Management: New Challenges and Solutions* **3** pp15-17
[13] *Construction norms and regulations of the Republic of Kazakhstan* 3.01-03-2010 Regulations for land improvement of settlements (Almaty: KazNIISSA)
[14] Best Management Practices Used at Urban Parks in National and International Locations: *A Background Report for the National Mall Plan* 2007 (Washington, D.C.: National Park Service U.S. Department of the Interior)
[15] Garvin A *What Makes a Great City* 2016 (Washington, D.C.: Island Press)
[16] Olenkov V D 2002 *Disturbed areas in urban planning: the restoration, use, aeration mode* (Chelyabinsk: South Ural State University Publ.)

Wetlands 'Zaymische' as a promising protected natural territories in the republic of Tatarstan

N Yu Assanova and N M Mingasova

Kazan (Volga) Federal University, Institute of Management, Economics and Finance, Department of Environmental Engineering and Water Management,
Kremlevskay str.,18, Kazan, 420008, Russia
E-mail: assanova-nadezhda@mail.ru

Abstract: The article reviews the data of a comprehensive survey of wetlands of the Kuibyshev reservoir (Zelenodolsk district of Republic of Tatarstan). The study discusses wetlands as one of the key elements of the ecological frame of the city of Kazan and Republic of Tatarstan. Change of the status to reserve or national park is recommended for the conservation of wetlands.

1. Introduction

The Republic of Tatarstan is rich in water resources that could potentially be considered as objects of international importance. For instance, system of islands and shoals in the vicinity of Bulgars is present in the Perspective list of the Ramsar Convention. Many experts noted that taking into account the 'year of the water protection zones in Tatarstan' (2016), first and foremost there is a need to create protected areas of local and regional significance [1]. As protected areas were not created, the same task is on 'year of the environment and open spaces in Tatarstan Republic' (2017).

One of the areas that can be considered as potential protected area of Federal and regional importance is wetlands of the Kuibyshev reservoir in Zelenodolsk district of Tatarstan near Kazan. Territory includes Islands, peninsulas and shallow water; has a high natural value, social significance and could be included in the list of wetlands of international importance. The studied area is located in the left Bank of the Kuybyshev reservoir, on the border of Kazan (village Zaymische) and Zelenodolsky district (village Oktyabrsky). Is contains deep water area, Islands with shallow waters, inland lakes, canals and marshes.

Nowadays the water area is partially transformed as a result of reclamation of the artificial land. Reclamation work was carried out on an area of about 600 hectares. It still retains high natural value, social significance and could potentially be included in the list of wetlands of international importance.

It should be noted that right now shoal was formed 50 years ago with the creation of the Kuybyshev reservoir of the flooded territories of the Volga river basin. Currently, the area is covered with numerous Islands in the confluence of the Volga and Sviyaga rivers and it is one of the key centers of distribution of waterfowl and water birds, important fish species for the whole Middle Volga region.

During the course of hydraulic monitoring of the status of the territory (2014-2017) is was noted that these wetlands have a great potential for recovery in the absence of intense economic activity.

IOP Conf. Series: Earth and Environmental Science **107** (2017) 012126 doi:10.1088/1755-1315/107/1/012126

This area can be recognized as one of the most promising protected areas in the Republic of Tatarstan, and measures to status change should be included to the 'year of ecology in Russia' program.

2. Results and discussion

Significance to form ecological frame of the city of Kazan and Republic of Tatarstan

The creation of a comfortable and supportive environment of vital functions of population is the main objective of spatial planning and urban development. Therefore, the adoption of economically feasible, environmentally and socially acceptable management decisions is impossible without the development of ecological-natural frame, guarantor of sustainable development of the territory [2].

For a city with population of over 1 million people there is a minimum of 25 % green space of the total area of the city. This is necessary to ensure the self-cleaning the air and creating a favorable environment for the life of citizens. At the same time, experts estimate the situation with the formation of ecological-natural frame in Kazan in recent years as unfavorable.

A natural complex of islands and peninsulas in the Zaymishche area is located on the border of the Kirov district of Kazan and the Zelenodolsk district of Tatarstan. In the immediate vicinity, at distance in 5 to 8 km there are boundaries of the Volga-Kama state natural biosphere reserve, state natural reserve "Sviyazhsky", urban forest Park "Lebyazhye" (protected areas of local importance), as well as buffer zone of the Volga-Kama biosphere reserve.

The mutual location of protected and green areas in this region clearly shows the value of the territory as a buffer and transit component of the ecological frame of Kazan and Republic of Tatarstan. In addition, the area is located at the intersection of the "green ring" of the city of Kazan and the "water arc". From this point of view it is not only a transit element, but also one of the key components of the ecological frame of the city.

Thus, the area of islands and shoals of the Kuibyshev reservoir near the Zaymishche and Oktyabrsky villages is one of the key components of the environmental frame of the Republic of Tatarstan linking urban ecological framework of city and national environmental frame.

Assessment of the current ecological status of wetlands

The comprehensive ecological survey of the territory was undertaken in 2014 with the participation of a group of researchers of Kazan (Volga) Federal University [3,4]. In addition, monitoring studies was conducted in 2015 – 2016. These studies identified more than 450 species of animals and plants. More than 50 species of plants and animals included in the Red Book (list of endangered species) of Russia, Red Book of the Tatarstan Republic, Red Books of Russian regions and the European list (Table 1).

The vegetation cover of the floodplain represented forest ecosystem types, meadow ecosystem types (grass, forb grass) as well as aquatic, riparian, lake and wetland ecosystems [4]. Botanical research in 2014 identified a list of 122 species of vascular plants, including 11 species of plants included in the Red book of Tatarstan: *Galatella rossica*, *Carex bohemica*, *Cladium mariscus*, *Scirpus radicans*, *Gentiana pneumonanthe*, *Senecio tataricus*, *Iris sibirica*, *Nyphoides peltata*, *Nympaea candida*, *Sparganium minimum*, *Salvinia natans*. They all have the 'Category 3' status of protection: rare species.

Study of the fauna of the region identified 3 species of mammals listed in the Red book of Tatarstan: *Myotis daubentoni*, *Vespertilio marinus*, *Mustera erminae*. In our studies, 17 species of mammals were observed; the mammal fauna is characterized by a mixture of wetland and forest component with the presence of some commensal species. During the course of avifauna studies we noted 92 species of birds. The following species are listed in the Red Book of Tatarstan: *Botaurus stellaris*, *Cygnus olor*, *Circus cyaneus*, *Circus pygargus*, *Haliaeetus albicilla*, *Columba oenas*, *Asio otus*, *Asio flammeus*, *Strix uralensis*, *Alce doatthis*, *Upupa epops*, *Caprimulgus europaeus*. Also, according to the data of fauna, the following rare species identified in the period 2012-2015: *Cygnus cygnus*, *Anser anser*, *Tringa nebularia*, *Tringa tetanus*, *Tringa stagnatilis Haematopus ostralegus*;

Pandion haliaetus, *Limosa limosa*, *Sterna albifrons*. Thus, the total number of rare species of birds recorded on site is 21.

Haliaeetus albicilla is a rare bird for the entire territory of Russia (Red Book of Russia, 3rd category).

It is necessary to consider the historical abundance on migration of grey herons, gull, waders, ducks and importance of the territory as a span for many species.

The complex of amphibians and reptiles for coastal and island territories includes 6 types. A single *Bombina bombina* was observed in 2014 while more than 200 of them were observed in 2015. *Bufo buf* is a rare species observed among reptiles.

Course of our research in 2014 revealed following list of rare species of insects: *Bryodema tuberculatum*, *Cycindela germanica*, *Calosoma sycophanta*, *Carabus clathratus*, *Trypocopris vernalis*, *Polyommatus daphnis*, *Callimorpha dominula*, *Bombus muscorum*. *Osmoderma barnabita*, *Ephialtes manifestator* were noted earlier from available data of Observatory (Oktyabrsky) [5]. There is number of species excluded from red book of Tatarstan in 2015 due to recovery of populations in the region, but they also need attention (*Myrmeleon formicarius, Calopteryx virgo, Oryctes nasicornis, Copris lunaris, Vanessa atalanta*).

Following arachnids are marked rare in Tatarstan: *Argyroneta aquatic*, *Lycosa singoriensis*, *Lepidurus arus*. Among large aquatic crustaceans as a regular form *Astacus leptodactylus* could be marked, which, in addition to their role in maintaining biodiversity, an indicator of water purity.

It is especially necessary to note the presence of *Calosoma sycophanta* on the coastal areas, tree species included in the Red book of the Russian Federation (category 2) and the European red list [6].

The ichthyofauna is very diverse and includes more than 40 species of fish, including such valuable species as *Acipenser ruthenus*, 41 species of phytoplankton 30 zooplankton species and 13 species belonging to the group of zoobenthos were revealed in the study of aquatic communities in the summer of 2014.

Table 1. Biodiversity and rare species of animals and plants of the islands and shoals of the Kuibyshev reservoir near Zaymishche, Oktyabrsky.

	Total	Red Book Tatarstan	Red Book Russia
Plants	122	11	-
Arthropods	91	13	1
Fish	40	-	-
Amphibians, reptiles	6	2	-
Birds	101	21	1
Mammals	20	3	-

Insects of the family of ground beetles (Carabidae) as one of the largest and most ecologically diverse families of the Coleoptera order and they are are a good model object for studying changes in ecosystems under the influence of anthropogenic factors. Spectrum of life forms of ground beetles as indicator of soil and plant conditions considered in the works of I.Kh. Sharova, S.Y. Gryuntal [7-9].

Collection of ground beetles were conducted in habitats located on the peninsular part of the study area, including tall grass and meadows, pine forests, riparian willow areas.

Tables 2 - 4 presents the characteristics of species of carabid life forms (imago) and ecological groups (habitat preferences).

In terms of structures of life forms zoophagous is dominating (78.2% of numerical abundance and 79.1% of the species); litter-soil stratobionts are the most abundant (30,4 species and 26.8% of numerical abundance), followed by epigeobionts (17,4; 13,7 %, respectively) and flying epigeobionts (13 and 24,7 %, respectively). This spectrum characterizes ecosystems as intact, despite intensive human impact on this area.

Table 2. The species composition of Carabidae in coastal areas.

	Vital form	Environmental group
Zoophagous		
Cycindela germanica L.	Epigeobiont flying	Meadow
Cycindela hybrida L.	Epigeobiont flying	Meadow
Cycindela maritima L.	Epigeobiont flying	Steppe mesophilous
Calosoma sycophanta L.	Epigeobiont walking, dendroepigeobiont	Forest
Carabus clathratus L.	Epigeobiont walking	Forest
Carabus granulatus L.	Epigeobiont walking	Forest-wetland
Carabus nemoralis L.	Epigeobiont walking	Forest
Carabus hortensis L.	Epigeobiont walking	Forest
Broscus cephalotes L.	Geobiont running-digging	Field
Broscus semistriatus Dej.	Geobiont running-digging	Steppe-field
Licinus depressus Pk.	Stratobiont litter-soil	Meadow
Badister unipustulatus Bon.	Stratobiont litter	Forest-steppe
Pterostichus (Poecilus) lepidus Leske.	Stratobiont litter-soil	Meadow-field
Pterostichus (Poecilus) cupreus L.	Stratobiont litter-soil	Meadow-field
Pterostichus niger Sch.	Stratobiont litter-soil	Forest mesophilous
Pterostichus oblongopunctatus F.	Stratobiont litter-soil	Forest
Pterostichus melanarius Ill.	Stratobiont litter-soil	Forest
Pterostichus mannerheimi Dej.	Stratobiont litter-soil	Forest-wetland
Myxophytophagous		
Harpalus flavescens Pill.	Geochortobiont	Forest mesophilous
Harpalus hospes Sturm.	Geochortobiont	Meadow mesophilous
Ophonus rufipes Dej.	Stratechortobiont	Meadow-field
Ophonus stictus Steph.	Stratechortobiont	Meadow-steppe
Amara aenea Dej.	Geochortobiont	Meadow-field

Table 3. Spectrum of life forms of imago Carabidae in coastal habitats.

Vital form	Abundance	
	Species	Numerical
Class Zoophagous		
1. Epigeobionts flying	13,0	24,7
2. Epigeobionts walking	17,4	13,7
3. Epigeobionts walking, subgroup dendroepigeobionts	4,3	4,6
4. Geobionts running-digging	8,8	7,9
5. Stratobionts litter-soil	30,4	26,8
6. Stratobionts litter	4,3	1,4
Class Myxophytophagus		
1. Stratechortobionts	8,8	9,4
2. Geochortobionts	13,0	11,7

Among the environmental groups forest and meadow species are dominatimg (of 33,4 and 26.6% of numerical abundance), followed by species are confined to various habitats (15,2 %).

There are species that are indicator of forest preservation and wetland habitats: *Carabus clathratus, Calosoma sycophanta, Pterostichus sp.*

Table 4. Spectrum of ecological groups of Carabidae in the coastal habitats.

Environmental group	Abundance	
	Species	Numerical
Forest	34,7	33,4
Meadow	17,4	26,6
Meadow-field	17,4	16,3
Forest-wetland	8,7	2,8
Meadow-steppe	4,3	5,6
Forest-steppe	4,3	1,1
Steppe	8,7	8,5
Field	4,3	5,6

This information about the state of the vegetation and fauna and information about the habitat of this rare species, convince of the need to preserve habitats to successfully maintain the reproduction of the species of flora and fauna. To solve this problem it is necessary to provide normal flow of hydrological regime of habitats of riparian ecosystems, promising to eliminate the capital building to strengthen control over observance of environmental legislation.

Proposals for the establishment of protected areas 'Wetlands Zaymishche'

The creation of the protected area 'Wetlands Zaymische' pursues the following main objectives:

- preservation and restoration of the unique natural complex of islands, peninsulas and shallow water wetlands in the Kuybyshev reservoir;
- the inclusion of the territory of the natural complex to the ecological frame of the city of Kazan and national ecological frame, ensuring their interconnection;
- the inclusion of the territory as a valuable "natural core" to the system of territorial development of Oktyabrsky rural settlement; the creation of conditions for regulated ecotourism and recreation in the natural environment.

Given these factors, a Natural Park (protected areas of regional significance) status change is highly reccomended. In accordance with article 15 "the Regime of special protection of territories of national parks" of the Russian Federation Law № 33-FFederal Law "About protected natural territories" it is recommended to differentiate the regime of special protection and allocate 5 functional areas: 1) reserved area; 2) specially protected area; 3) zone intended for the development of environmental education, excursions, rest and recreation of visitors; 4) zone of heritage nature use; 5) recovery zone with natural boundaries of the shoreline and islands.

Also, taking into account the anthropogenic disturbance of individual sites affected by hydraulic sustainable functioning, areas should include optimization of the hydrological regime (flow blurring), planting higher aquatic vegetation, planting herbaceous meadow plants and other events.

3. Conclusion

'Zaymische' Wetlands, containing of shallow waters, canals, pools and open waters, are one of the most valuable areas of the Kuibyshev reservoir in the Republic of Tatarstan in terms of biodiversity conservation, purification and reproduction of fish. Currently, there is the process of overgrowing vegetation of sandy mounds formed as a result of the backfill area in 2011-2013.

Given the location next to Kazan city, the area has significant recreational impact that must be considered when granting the status of protected areas. The most suitable status is a Natural Park status of national value, or National Park of Federal significance.

Acknowledgment
Authors are deeply grateful to prof. T. Rogova, assoc. V. Prokhorov, assoc. N. Zaripova and other colleagues who participated in conducting complex studies and whose materials are referred to in this paper.

References
[1] Fayzrakhmanova Yu I and Assanova N Yu 2016 Year of water protection zones: from the sources to the great water *Kazan* **1** (Kazan Idel-Press) p 12-23
[2] Grydnev D Z 2011 The natural and ecological framework of the territory is the basis for the adoption of town-planning decisions as part of the documents for the territorial planning of municipalities *Territory and Planning* **1** p 96-103
[3] Integrated ecological survey of shallow water in the Kuibyshev reservoir in the area of the village October Zelenodolsky district of the Republic of Tatarstan for the purposes of environmental certification *Report on the implementation of works* 2015 Federal University Kazan 150 p
[4] Assanova N Yu, Mingazova N M, Rogova T V, Prokhorov V E, Pavlov Yu I, Zaripova N R, Palagushkina O V, Derevenskaya O Yu, Ivanova V M and Mukhachev S G Biodiversity of shallow water and islands of the Kuibyshev reservoir in the area of the village October Zelenodolsky district of the RT *Abstracts of III International Congress "Clear Water"* Kazan p 31-35
[5] *The Red Book of the Republic of Tatarstan (animals, plants, mushrooms*) 2016 The third edition Kazan Idel-Press 760 p
[6] *The Red Data Book of the Russian Federation (animals)* 2001 RAS (Moscow: Astrel) 862 p
[7] Sharova I Kh 1978 Life forms of ground beetles as indicators of soil-plant conditions *Problems of soil zoology Abstracts of the VI All-Union Conference* p 273-274
[8] Sharova I Kh 1981 *Life forms of beetles (Coleoptera, Carabidae)* (Moscow: Nauka) 283 p
[9] Grüntal S and Butovsky R 1997 Beetles (Coleoptera, Carabidae) as indicators of recreational impact on forest ecosystems *Entomological Review* **LXXVI** charter 3 p 547-554

IOP Conf. Series: Earth and Environmental Science **107** (2017) 012127 doi:10.1088/1755-1315/107/1/012127

Methodology of ecooriented assessment of constructive schemes of cast in-situ RC framework in civil engineering*

I P Avilova[1] and M O Krutilova[2]

[1] Professor, Belgorod State Technological University named after V.G. Shoukhov, Belgorod, Russia
[2] PhD student, Belgorod State Technological University named after V.G. Shoukhov, Belgorod, Russia
E-mail: kafeun@mail.ru

Abstract. Economic growth is the main determinant of the trend to increased greenhouse gas (GHG) emission. Therefore, the reduction of emission and stabilization of GHG levels in the atmosphere become an urgent task to avoid the worst predicted consequences of climate change. GHG emissions in construction industry take a significant part of industrial GHG emission and are expected to consistently increase. The problem could be successfully solved with a help of both economical and organizational restrictions, based on enhanced algorithms of calculation and amercement of environmental harm in building industry. This study aims to quantify of GHG emission caused by different constructive schemes of RC framework in concrete casting. The result shows that proposed methodology allows to make a comparative analysis of alternative projects in residential housing, taking into account an environmental damage, caused by construction process.

1. Introduction

Russia has emerged as a major player in international climate politics due to its decisive role into force of the Kyoto Protocol, its position as a major global fossil-fuel supplier, and its significant rate of global emissions (fourth-largest, with a rate about 5%) [1]. The building sector should be prioritized in order to be able to reach a sustainable society because, as of 2010, the building sector constituted almost 30% of global CO_2 emission (IEA) [2]. In Russia, the emission rates are 7% of total emission in addition to the transport services sector accounts for about 10%, which allows estimating the total emission in construction industry, including related logistics in the range of 12-15%. It makes the task of reducing the environmental harm caused by construction industry quite significant. The 2013 Prognosis of Socio-economic Development until 2030 (developed by the Ministry of economic development of the Russian Federation) forecasts that greenhouse gas (GHG) emissions will reach 75% of the 1990 level by 2020, thereafter declining to 70% by 2030 as a result of energy-efficiency measures, increased labor productivity, and renewable energy policies [3]. In this context, the issues of enhancement of practical algorithms of calculation and penalization of environmental harm in building activities will be a part of basis for successful solution of the problem.

*The study was carried out in the framework of the Program of flagship university development on the base of Belgorod State Technological University named after V.G. Shoukhov

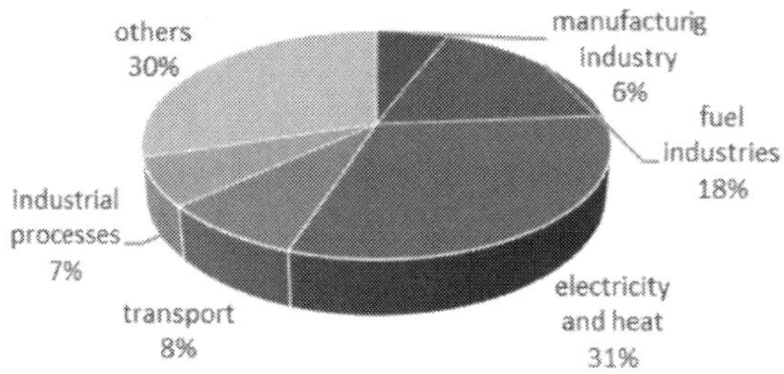

Figure 1. Russia's GHG emission rate according to economic sector, 2012 [1].

2. The problem statement and methods

As environmental issues continue to become increasingly significant, the energy required for construction industry and for the material production is getting of greater importance. GHG emissions at the construction phase are not usually considered in life cycle assessment of GHG emission and which called indirect emission. It should be noted that building materials (including manufacturing), construction works (including building equipment) and transport – all of these processes consume energy and therefore emit GHG [4]. Despite the fact that during the life cycle of the building a number of studies on energy consumption and GHG emission was carried out, practical methods for GHG emission quantifying at the construction stage have not been still implemented in construction design.

The building's total consumed energy depends on the amount of the material and installation work used in construction stage and their previously consumed energy [5]. In this case the environmental safety grade of used materials and the energy safety grade of used construction technology are important. Thus, it is necessary to consider environmental pollution and try to decrease emissions at the pre-project and project stages, thereby to mitigate the environmental consequences of the construction of buildings and structures. Firstly the problem deals with erecting of reinforced concrete (RC) structures as the most common and energy consumed constructions.

Constructive schemes active used in RC frameworks in Russian civil engineering in last 7 years presented in Figure 2. Average share of cast-in-situ frameworks sustainably increased by 14% per year, nowadays totally including more than 18% of newly constructed residential houses [6].

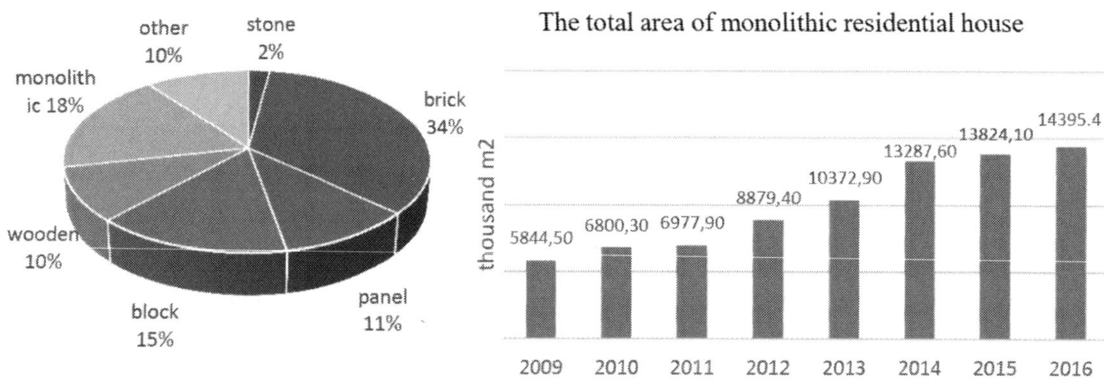

Figure 2. Arrangement of residential houses according to constructive materials in the Russian Federation in 2009-2017.

At pre-project stage in civil RC constructions design, in most cases, several constructive schemes are allowable, but till nowadays comparative analysis of alternative schemes is carrying out according to such economic issues as construction cost, duration of construction and consumed manpower.

Meanwhile alternative schemes often drastically differ by energy and resource consumption. This important issue has almost no reflection in construction prices and could not be taken into account according actual construction code, because of lack of practical methods of environmental harm quantifying. Thus, improved algorithms of economic environmental analysis should be updated by the calculation and penalization (CP) method of GHG emission, caused by alternative constructive schemes. Proposed CP consists of four main stages [4]:

1. Estimating of basic expenditures of building the multistory RC frames by alternative constructive schemes.
2. Determination of GHG emission caused by building materials and construction works based on actual estimating codes.
3. Monetary penalization of revealed GHG emission
4. Comparative estimation of construction cost incl. eco- penal fines for alternative constructive schemes.

3. Results and discussion

Let's take a closer look to the importance and significance of CP at pre-project stage in a comparative analysis of possible constructive schemes of RC frame forming residential house (MRH).

Stage 1. According to ABC analysis **Ошибка! Источник ссылки не найден.** conducted for typical MRH RC frames one can highlight the most significant and expensive structural materials and installation works. It allows to preliminary assess the effectiveness of "green" building technologies in various constructive schemes and select the safest one for project stage. Based on the authors' data, the most costly and resource-intensive lines in estimates of MRH RC frames are: 1. Concrete mix's materials. 2. Preparing of concrete mix. 3. Pouring of concrete mix. 4. Steel reinforcements. 5. Formwork construction. 6. Reinforcements and concrete mix transportation.

According to preliminary analysis three mostly preferred constructive scheme of MRH RC frame are(figure 3):

1. Frequently placed square-shaped columns and separate rigid walls (scheme 1);
2. Distantly placed rectangular-shaped pylons and rigid cores (scheme 2);
3. System of longitudinal and transverse walls (scheme 3).

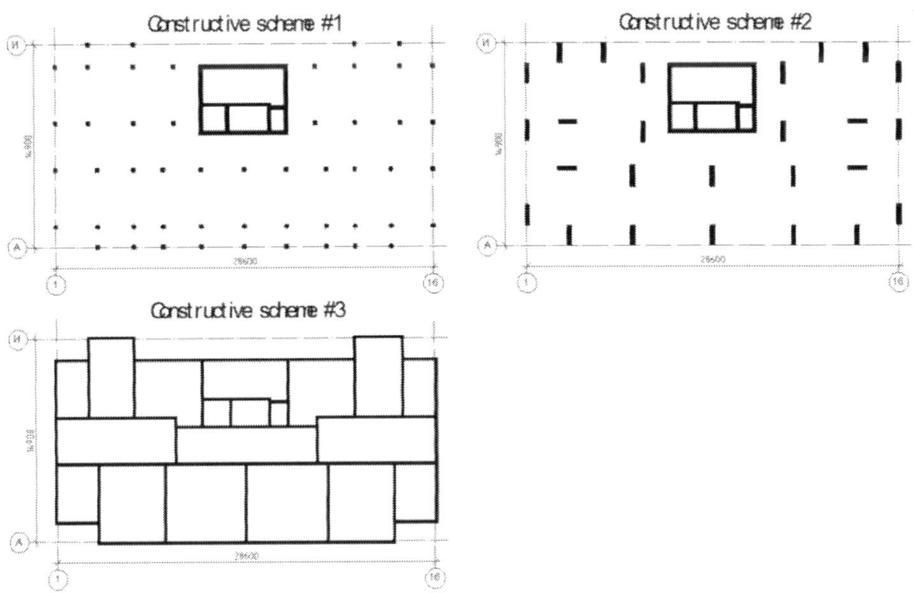

Figure 3. Constructive schemes of MRH RC frame.

For estimating of basic expenditures in building of MRH RC frames by constructive schemes shown above the quantity and cost of necessary materials, machines and equipment should be determined, also $1m^2$ of constructed framework. Table 1 shows list of quantities and estimated costs for three constructive schemes determined by using of Federal Unit Rates for construction works in Russia [8].

Table 1. Estimates of basic expenditures forconstructive schemes of MRH RC frame.

	Scheme 1	Scheme 2	Scheme #3
Material expenditures			
1. Concrete, m^3	1590	2031	2704
2. Steel reinforcements, t	215	155	180
Machines expenditures			
Cranes, hrs	79	105	184
Trucks, hrs	851	945	1227
Mixers, hrs	34	43	71
Welding machines, hrs	101	155	218
Total construction cost before penalties, mlnRUR	**16.31**	**24.13**	**41.22**

Stage 2. At this stage, GHG emission caused by main construction materials (concrete and steel) and works is quantified. For certain, all the embodied GHG emission, caused by producing and installation of constructions, using engines and equipment, wasting of energy to provide all kinds of necessary in-situ management need to be estimated separately. All the building materials during its producing, transportation and installation require energy consumption and cause GHG emissionas well [9]. ABC analysis at previous stage allowed to highlight the basic construction materials and works, which GHG emission's determination should be enough for comparative purposes.

Based on [10, 11] GHG emission from the production of materials shown in tables 2-3.

Table 2. GHGemission from the production of $1m^3$ of typical concrete.

GHG emission	Unit	
Total	kg	**1,50848**
CO_2	kg	1,5
CO	g	0,86
NO_x	g	2,3
SO_x	g	3,3
CH_4	g	1,7
HC	g	0,32

Table 3. GHG emission from steel production processes.

Production process	Unit	GHG Emission
Integrated steel mill	Tons per ton product	1,6-2,2
Scrap based production in a mini-mill		0,6-0,9
Scrap substitutes (DRI) in a mini-mill		1,4-2,0

Main equipment for RC frame's construction (trucks, cranes, mixers) were selected by specific machines' models. Each machine has its engine power and performance specifications, affecting the GHG emission. Based on [12] GHG emission from the specified machines shown in table 4.

Table 4. GHG emission from the specified machines

Equipment type	Manufacturer	Engine power, kW	GHG emission
Trucks	Kamaz 5501	176	69 g/s
Cranes	Potain MDT-180	120	46 g/s
Mixer	Putzmeister BSA 1407 D	115	77 g/s
Welding machine	Stavr SAI-180 BTL	6	33g/kg

According to list of quantities GHG emission were determined for construction materials and works (including machines and equipment) in table 5, 6.

Table 5. GHG emission calculator: construction processes(scheme1).

Machines	Volume
1. Trucks	
Truck's engine power, kW	176
Lifting capacity, t	10
Estimated working time, hr	1227
Estimated GHG emission, g/sec	24.5
Estimated GHG emission total, t	**108**
2. Cranes	
Crane's engine power, kW	120
Lifting capacity, t	8
Estimated working time, hr	184
Estimated GHG emission, g/sec	46
Estimated GHG emission total, t	**18.9**
3. Mixer	
Mixer's engine power, kW	115
Mixing capacity, m3/hr	71
Estimated working time, hr	57
Estimated GHG emission, g/sec	77
Estimated GHG emission total, t	**15.8**
4. Welding machines	
WM's power, kW	6
Estimated working time, hr	218
Estimated GHG emission, g/kg	33
Estimated GHG emission total, t	**5.9**

Materials	Volume
1. Reinforced concrete (RC)	
Estimated GHG for cement production, g/cub.m	1508
Estimated concrete volume, cub.m	2704
Estimated GHG emission total, t	**4.1**
2. Construction Steel	
Estimated GHG emission for steel production, t/t	0.6
Estimated GHG emission total, t	**108**

Total GHG emission caused by materials, t	**112**
Total GHG emission caused by machines, t	**149**
Total GHG emission caused by materials, t	**116**
Total GHG emission caused by construction process, t	**265**

Table 6. GHG emission calculator: results(scheme 1, 2, 3).

Constructive schemes	Scheme 1	Scheme 2	Scheme 3
Total GHG emission caused by machines, t	105	115	149
Total GHG emission caused by materials, t	134	99	116
Total GHG emission caused by construction process, t	238	214	265

Stage 3. At this stage the revealed GHG emission should be monetary penalized. All actual damage, determined in GHG-emission volume for compared constructive schemes at the previous stages should be expressed in monetary equivalent (penalties) according proposed method of estimating: eco-penalty is equal to the Cost Of Prevention (COP) before or Cost Of Sanitation (COS) after harm caused [13].

$$\text{Eco-penalty} = COS + COP = RCT \times k5 + (COP_1 \times k_1 + COP_2 \times k_2 + COP_3 \times k_3) \times k_4 \ (1),$$

As the basis for the assignment of eco-penalties is GHG emissions per unit of material/work/services, COP is defined as:

$$COP = (COP_1 \times k_1 + COP_2 \times k_2 + COP_3 \times k_3) \times k_4 \ (2),$$

COP_1—cost of Research and Development (R&D), aimed to environmental sustainability of the material/technology over the past year divided by GHG emission in the country's production;

COP_2—difference in cost of materials/services produced by the most eco-friendly technology for the current year to the cost of the same materials/services produced by the technology specified by manufacturer, which is penalized.

COP_3—difference in amount of material/services, produced in the penultimate and preceding year multiplied by the average across the region the cost of material/services.

$k_1...k_3$— weight indexes, k_4— activity index.

COP_1encourages research funding, COP_2—implementation of green technologies, COP_3— using of energy-efficient constructive schemes. Indexes $k_1...k_3$ are set by the authorities, giving importance to a particular direction of greening construction, and, accordingly, using a certain management politics at the construction market. Index k_4 allows to level the peaks and valleys of activity in the market, reducing and increasing financial responsibility for environmental project's deficiencies, depending on whether authorities fosters or hinders the local construction industry according with the current trends of the regional's ecology.

The Kyoto Protocol determines total number of permitted emissions of GHG on the planet establishing regional carbon taxes (RCT) at the market [2]. Thus, cost of sanitation (COS) is defined as:

$$COS = RCT \times k_5 \quad (3),$$

k_5 — accumulated cost of regional constructions divided by current year's GDP.

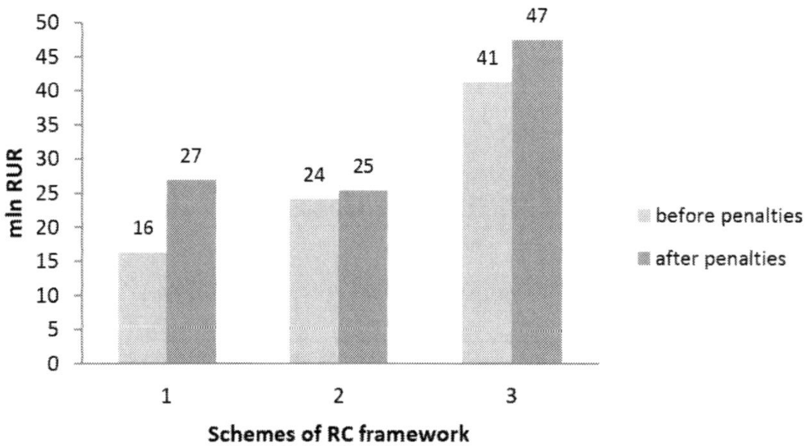

Figure 4. Construction cost of different types of RC frameworks before and after penalization (specimen of ecooriented assessment).

4. Conclusion

Summarizing the above, we can say that conducted method of eco-oriented assessment of constructive schemes of monolithic RC frame in civil engineering allows to carry out comparative analysis of alternative designs in residential housing according to taking into account of embodied environmental harm, caused by construction process. This engineering technique based on selecting the number of basic construction materials, works, machines help to assess construction costs after eco-penalization and could be efficient additional tool at the pre-project eco-environmental analysis of civil structures. The same method could be implemented in comparative analysis and looking for environmentally safest possible design in widespread civil constructions. Observed penalty method is the effective way to decrease environmental harm caused by construction sector itself by means of enhancement of construction legislation.

Thus, according to the figure 4 alternative constructive schemes are ranked by economic efficiency as: 1, 2, 3 (before penalties). After implementing CP-procedure the alternative constructive schemes are ranked by eco-oriented economic efficiency as: 2, 1, 3. That means constructive scheme #2 is environmentally fittest and should be considered in further at the project stage.

References

[1] Korppoo A, Kokorin A 2017 *Russia's 2020 GHG emission target: Emission trends and implementation, Climate Policy* **17 (2)** 113-130
[2] *Energy Technology Perspectives* 2010 International Energy Agency - OECD Style Guide: Rules and Conventions
[3] *Prognosis of Socio-economic Development until 2030* Available at: http://www.consultant.ru/document/cons_doc_LAW_144190/ (accessed 14 August 2017)
[4] Avilova I, Naumov A and Krutilova M 2017 *International Multidisciplinary Scientific GeoConference SGEM*, Methodology of cost-effective eco-directed structural design **53** 255-261
[5] DimoudiA, Tompa C 2008 *Resources, Conservation and Recycling*, Energy and environmental indicators related to construction of office buildings **53** 86–95
[6] *Russia's Federalstatestatisticsservice* Available at: http://www.gks.ru (accessed 14 August 2017)
[7] Dickie H F 1951 *ABC inventory analysis shoots for dollars not pennies. Factory Management and Maintenance* **109 (7)** 92-94
[8] Avilova I, Naumov A, Shchenyatskaya M 2016*Journal of Fundamental and Applied Sciences* Improving the economic efficiency of construction investments by means of technological risks management **8(2)** 1502-1518
[9] Grabovy P, Naumov A, Avilova I 2016 *International Business Management* Scientific aspects of productivity management in the investment and construction sector **10**(7) 1354-1364
[10] Lindahl M, Rydh C, Tingström J 2002 *Livscykelanalys – en metod för miljöbedömning av varor och tjänster (in Swedish)* Studentlitteratur
[11] Greenhouse Gas Emission From Major Industrial Sources - III Iron and steel production 2000 IEA Greenhouse Gas R&D programme (IEA GHG) PH3/30
[12] Lewis M 2009 *Estimating fuel use and emission rates of nonroad diesel construction equipment performing representative duty cycles* North Carolina State University 3357743
[13] Avilova I, Krutilova M and Peresypkina E 2017 Economic incentives of green standards in civil and municipal engineering *International Multidisciplinary Scientific GeoConference SGEM* **53** 551-557

State and business co-operation in settling socio-economic issues: forward to sustainable development of ecologically unfavorable regions

N N Bashkirova[1] and S N Lessovaia[2]

[1]IHigher School of the State Audit, Lomonosov Moscow State University, 119234, Leninskie Gori, 1 build. 4/13, Moscow, Russia.
[2] St. Petersburg State University, Institute of Earth Sciences, 199034, Universitetskaya nab, 7/9, St. Petersburg, Russia
E-mail: nadezhda.bashkirova@gmail.com

Abstract. The complexity of socio-economic issues of mono-cities located in the ecologically unfavorable regions of Eurasia was disclosed. The economically strategic role of city-forming mining enterprises and their impact on ecological situation was revealed. The general conception of settling the socio-economic problems of mono-cities located in ecologically unfavorable regions was worked out. Various approached to the concentration of financial resources for economic and ecological sustainable development of the regions located on the north of Eurasia holding nature protection actions were submitted. Based on performed critical analysis of the positive international experience of ecological taxation some approaches to reforming current Russian system of ecological taxation were suggested. It was revealed that increasing the social responsibilities of business in the field of waste recycling, environmental protection and monitoring of ecological conditions of territories and state and business co-operation are the most efficient opportunities in settling socio-economic issues of ecologically unfavorable regions.

1. Introduction

Nowadays the problem of industrial impact and influence on the surrounded areas remains one of the most important environmental problems. It is well known that the ecological sustainability of regions stays in conflict with the industrial strategy of the regional development, which determined the level and life stability of the population.

Not only each component but also the whole ecosystem is affected by the industrial impact. This impact destroys the links of the components interdependency and as consequence the sustainability of whole biosphere on the global level. It is impossible to take cardinal measures in the form of closing of industrial objects because most of them are strategic enterprises of the national or regional level. Very often those enterprises are city-forming (mono-cities/ one company town) and they are the only opportunity for employment and financial stability of citizens. City-forming enterprises generally provide activity and life conditions in the city and surrounding territories. Labor market situation of mono-cities strongly affects on the socio- political stability in general in the region. The economy of Russia is substantially represented by the city – forming enterprises whose contribution to the Gross Domestic Product (GDP) is approximately 20-40% depending on the methodology of assessment [1]. City-forming enterprises should be treated as a unique Russian economic-geographical phenomenon. Those enterprises provide in natural indicators about 64 % of oil extraction, 83% of gas extraction, 53

% of coal extraction and almost 90% of nickel. [2]. A significance of problem mentioned above is underlined by the fact that list of mono-cities with risk of socio-economic situation deterioration is approved by the Government of Russian Federation, Order № 1398-r as of 29.07.2014. The following fact is geographically determined: in Russia mining companies are generally concentrated in the "cold" regions of Eurasia, so the ecosystems are characterized by the cold climate conditions.

The population of mono-cities located in the "cold" regions of Eurasia is especially unprotected in social and economic aspects, which is totally explainable by not only the single option employment of population, which has no alternatives, but also by the fact, that the ecosystems of these regions are the most sensitive to anthropogenic influence, especially industrial impact [3]. Extraction and processing of minerals determinate the technogenic initiated influence on the ecosystems and the health of population of the mono-cities. The specificity of the technologies in present time does not allow using the direct methods, which can protect the industrial induced contaminations.

The example of mono-city, which could be taken under consideration, is town Nikel (Murmansk region, Kola Peninsula). The city-forming enterprise is Mining and Metallurgical Factory "Pechenganikel", which was founded in 1945 and so far is one of the most important factory of mining and processing of sulfide cooper-nickel ores in Russia. Despite the factory retooling [4] the negative impacts on the population health have been found: increase of the frequency and numbers of diseases of reproductive system, respiratory, cancer and childhood diseases; compromised immune system; and decrease in life expectancy [5]. Evaluation of the industrial impact on the ecosystem has been also reported [6]. It was demonstrated that the contamination traces was found within 7-10 km distance from the factory. Soils and vegetation here are characterized by the features and characteristics, which are common for the ecosystems in this zone. Nevertheless a permanent monitoring of pollutants concentrations in the soils and vegetation including mushrooms and berries collected in these areas and used for human consumption is absent.

2. The problem statement

Thus, the investigation of a list of problems of mono-city given above on an example of town Nickel has revealed that besides the most considered and deeply studied block of problems connected with (a) contamination of not only each component but the whole ecosystem, and (b) health of citizens, one more aspect of problems is reasonable to disclose. Socio-economic aspects of ecological sustainability of unfavorable regions usually escape from the consideration.

Based on the lack of such approach in the investigations the aim of current research is to work out the general conception of settling the socio-economic problems of mono-cities located in ecologically unfavorable regions. According to the main target the following issues were under consideration:

1. Analysis of possibilities of destruction affected by industrial impact on the ecosystem by the Mining enterprise was undertaken.
2. Analysis of current ecological taxation system in Russia was performed with the aim of working out the possible ways of its optimization.
3. Positive international experience of ecological taxation was generalized in order to disclose the abilities of its implementation in Russia.
4. Ways of increasing the social responsibilities of business in the field of waste recycling; environmental protection and monitoring of ecological conditions of territories were presented.

3. Results and discussion

The following attempts concerning the improvement of the ecological situation in the areas, which are under the permanent impact of the Factory contamination(s) were undertaken in 2004–2008 years. The large area including technological barrens was covered by the remediation. Two variants of the remediation were used: (i) ecological approach (i.e. planting of trees and perennial plants, the use of fertilizers and meliorant substances) and (ii) the most expensive and money consuming approach, in which the soil cover is changed. Taking into account the amount of expenses and the significant work intense of the remediation results [7] it was disclosed that even huge monetary investments were not successful to return the considered ecosystem back to its initial uncontaminated level. One of the most significant reasons of unsuccessfulness of such approaches is extremely high sensitivity of ecosystems of Eurasian cold regions to the external impact. As it was indicated above closing of the city-forming enterprises is absolutely impossible because they are the only opportunity for employment and financial stability of citizens. And closing of such enterprises will evidently destroy the socio-economic balance of the region. It is logically to find legal and economic mechanism of providing relevantly comfortable life conditions for the population of mentioned region.

Taking into account the specificity of the mono-city, the research of approaches to create the financial resources for environmental protection on the municipal level is of interest. According to the Budget Code of the Russian Federation (subpart 3 part 3 of the Article 21) budgets of the budgetary system of the Russian Federation, including municipal budgets, are meant to cover expenses on environmental protection, which include the following items: 1) ecological control; 2) waste collecting and disposal, sewage water treatment; 3) protection of biological objects (vegetation, animals etc.) and their habitats; 4) development of applied scientific research in the field of environmental protection.

In this case, it seems to be logically to assign a duty of holding nature protection actions to municipal unit, on which territory there are objects, carrying out negative impact on the environment due to their functioning specifics. However, actually there is a question of insufficiency and / or the actual absence in the municipal budget of funds for holding the specified actions.

At the same time, the Budget Code of the Russian Federation hasn't provided existence of the trust funds, in which the funds allocated for environmental protection at various levels of the budgetary system would be accumulated. It is logically clear that nature-resource payments have to be a source of such funds. Due to absence of the trust funds for holding nature protection actions, a sources of their financing actually are the local budgets in general filled at the expense of tax (a land tax, the individual property tax and also a share of the corporate profit tax enlisted in the local budget), non-tax (income from use of property and financial aid of higher budgets) and other (free aid, etc.) income.

Payments for utilization of natural resources represent set of payments of two types introduced by Russian legislation: for environmental management and for negative impact on the environment. The complex character of nature-resource payments is shown in variety of their functions, the most important of which are compensatory, stimulating, fiscal and ecological [8]. According to the Budget Code of the Russian Federation, 55% of the amount charged as the payment for negative impact on the environment is allocated to the municipal budgets, whereas the Mineral Extraction tax is not enlisted in the municipal budget in general, in full coming to the federal budget. Insufficiency of own funds of municipal budgets results in impossibility of carrying out effective of full-scale nature protection actions on the territory of municipal unit including environmental control and applied scientific research in the field of environmental protection.

In this case, it is absolutely essential to reconsider approaches to creation of funds and generating resources on performing nature protection actions. Increase of payment for environmental pollution will lead to unreasonable increase of fiscal burden to enterprises. In the situation when the economy is overcoming consequences of a world economic crisis with great difficulties, it will negatively affect the investment and tax climate in general. For the company like «Pechenganikel'», substantial increase of fiscal burden inevitably will happen due to reduction of funds which can be potentially directed to expanded reproduction, updating of technological base, expenses on research and development, stimulation of activity of personnel.

Therefore, increase of the extended prices paid by the enterprises for environmental pollution is not an effective way of accumulating resources for providing nature protection actions. One of the efficient ways of settling this problem without overloading tax burden of the city-forming enterprises, is reconsidering the current model of tax federalism in Russia. Excessive concentration of tax income on the federal level without corresponding changes of expenses authorities does not stimulate regions the expansion of own tax base, reduces interest to increase in the budgetary income and rationalization of the budgetary expenses. However now the emphasis is placed on subsidized forms of regulation of the inter-budgetary relations; problems of depressed areas not only remain, but also are aggravated;

problems of regulation of financial flows on all vertical of the budgetary system are not solved. Orientation on more tax self-sufficiency is reasonable to consider as a powerful stimulus of activities of regional authorities and local governments for development of financial and economic capacity of territories.

For this purpose it seems to be reasonable to reconsider the proportion of contributions to regional and municipal budgets as on resource payments (like Mineral Extraction tax and a payment for negative impact on the environment), and on the key taxes (say, value added tax) at the moment coming in full to the federal budget of the Russian Federation. This measure will positively affect to significant growth of self-sufficiency of municipal units and will base a possibility of creation their own resources for carrying out at the local level nature protection actions. But the fiscal burden of the enterprises of the considered region will not be increased.

In search of approaches to settle the problem of creation funds for holding nature protection actions on the local (municipal) level it is reasonable to consider the international experience of ecological taxation. Distinctive feature of the European tax system is a significant amount of such taxes, which are in general neutral for the state budget. Historically at the moment of ecological taxes introduction there was a decrease of contributions to social funds for the corresponding amounts and / or adjustments in the income taxation were made.

The analysis of the ecological legislation of Great Britain is of the utmost interest, the nature protection legislation of the Great Britain was actively developing in the 1990th and the beginning of the 2000th. On the basis of the analysis and creative application of experience of the ecological taxation in Great Britain it is possible to make a conclusion about reasonability of consideration as the perspective direction of reforming of system of taxes and fees in the Russian Federation, the mechanism of tax deductions (e.g., for the corporate profit tax) [9]. Such tax deduction can be offered for the enterprises, which are carrying out adverse impact on the environment due to specifics of the activity makes investments in the project of environmental protection and control, in particular, in holding independently financed actions for environmental contamination monitoring and applied scientific research in the field of environmental protection. We note that this measure will not lead to direct increase of income of local and regional budgets, but will allow to use effectively mechanisms of the state – private partnership in the solution of environmental problems.

It is hard to underestimate the role of state and business co-operation in solving problems of ecologically adverse territories. Identification and implementation of additional opportunities provided by private sector of economy is critical necessary in those cases, when the government is not capable to cope with a problem effectively. Not only nature-protection issues, but all complex of socio-economic problems is finance-consuming and there is shortage of state/ regional/ local budget for financing those activity. In current situation, business could provide the relevant financial recourses for covering such expenses. On the one hand, business helps to find funds for solving problems for which there is not enough money on the state or regional level. On the other hand, there is an outstanding opportunity for business to strengthen the image demonstrating the policy of social responsibility.

According to the open documentation Kolsk mining company in 2015 undertook the following measures: 1) clearing of a technical sludge pond on the industrial site in Monchegorsk; 2) preparation in reconstruction of the treatment plant of mine water (mine "Severnyi"); 3) building of treatment plant for a salt drain of nickel refinement is continued; 4) the subsequent work on cleaning the outflow of the «Severniy» mine continues.

The above-mentioned company develops ecological partnership and co-operation with Finland and Norway in order to share positive international experience of nature-protection activities organization. In solving ecological problem state and business co-operation can be applied in various ways. However, the most widespread way is environmental monitoring provided by the enterprise itself. This activity is quite traditional way of environmental care, but following development of this work is essential for further efficiency. The idea of expansion of responsibility of the organizations, which activity is inevitably connected with environmental pollution, consists in improvement of the organization of waste utilization management at the enterprise. Such idea assumes that beside those problems, which are solved by the producer inside the organization, tasks existing outside the enterprise also have to be implemented. Creation of joint infrastructure for performing nature protection actions by the organizations (including environmental monitoring), which are carrying out environmental pollution, allows to accumulate financial resources in volumes, when each separate enterprise is not able to afford, solving environmental problems in an individual order.

In our opinion, the most effective and productive mechanism of expansion of responsibility is co-operation of producers. This way is generally focused on the consumer and allows the state to reduce administrative burden in this area. At the same time the lack of the governmental support sometimes leads to serious negative consequences. Also should not skip the fact that unconscientious producers can enter the association with the purpose of getting personally both financial and reputational benefit.

Such policy of expansion of social responsibility by the organizations whose activity is unavoidably connected with environmental pollution, in general is favorable to the state, because allows to reach the solution of tasks in the sphere of environmental control and holding nature protection actions, without providing additional financial resources from budgets of the budgetary system of the country. However, without the state support achievement of the target becomes unreal. One of the most significant problems is insufficiently developed legislation. To eliminate this defect it is necessary to define accurately a role and responsibilities of each member on such co-operation:

1. producers undertake financing of all projects running by the co-operation;
2. the government establishes standards and rules of recycling;
3. regional authorities can update requirements and also have to provide control;
4. municipal authorities support realization of this policy at the correspondent level;
5. merging of producers operates this activity on behalf of their members in partnership with authorities and also inform the citizens and the state on results of the carried-out work.

One of the directions of realization this state-private partnership in the solving of problems of ecologically adverse territories, is the co-operation of the state and the enterprises polluting the environment for monitoring of zones with the increased level of air and the soil pollution, and also creation of the research laboratories in joint maintaining the state and enterprises.

As the result, the above mentioned approaches disclosed in this research is the development of the statement concerning of socio-economic sustainability of mono-cities located in ecologically unfavorable regions [10].

4. Conclusion

Based on the analysis of the specific features of the mining enterprises impact on the environment we can conclude that the ecological sustainability of regions is keeping in conflict with the industrial strategy of the regional development, which determined the level and life stability of the population.

As it was revealed, just only ecological approaches and mechanisms including expensive and work intense remediation are not successful to return the considered ecosystem back to its initial uncontaminated level and cannot provide the sustainable development of the ecologically unfavorable regions. It is impossible to take cardinal measures in the form of closing of industrial objects because most of them are strategical enterprises on the national or regional level and being city-forming ones they are the only opportunity for employment and financial stability of citizens. In this case, it is essential to provide social support and financial stability for population of ecologically unfavorable region. At least actions for environmental contamination monitoring and applied scientific research in the field of environmental protection should be performed. Due to the insufficiency and / or the actual absence of funds for holding the specified actions in the local budget, it is necessary to search out resources for undertaking these activities. Increase of payment for environmental pollution will lead to unreasonable increase of fiscal burden of enterprises. According to our opinion, the most effective and productive way of solving socio-economic problems of ecologically unsustainable region is strong co-operation between state and business and increasing the social responsibilities of business. Creation of joint infrastructure for performing nature protection actions by the organizations (including, environmental monitoring) which are carrying out environmental pollution allows to accumulate financial resources in volumes which each separate enterprise is not able to afford, solving environmental problems in an individual order.

Acknowledgement.
The research was supported by St. Petersburg State University (project 18.38.418.2015) and the Informational Legal System ConsultantPlus.

References

[1] Maslova A 2011 Mono-cities in Russia: Problems and Solutions (Monogoroda v Rossii: problemy i resheniya). *J. Kontury globalnyh transformacij: politika, ehkonomika, pravo.* **4** (5) pp 16–28

[2] Sulakshin S 2010 *Criteria and Basics of the Modernization of Russia* (Kriterii I Osnovy Modernizacii Rossii) (Moscow: Scientific expert) pp 27-29

[3] Sumina O and Lessovaia S 2016 Clay minerals in the loose substrate of quarries affected by vegetation in the cold environment (Siberia, Russia) *In Biogenic-Abiogenic Interactions in Natural and Anthropogenic Systems, Lecture Notes in Earth System Sciences* (Editors V. Frank-Kamenetskaya et al.) (Switzerland: Springer International Publishing) pp 249 – 59

[4] Khlebosolova O (Editor) 2012 *Kola Mining Company (Industrial Sites "Nikel" and "Zapolyarnyj"): Influence on Terrestrial Ecosystems (*Kolskaya gorno metallurgicheskaya kompaniya (promyshlennye ploshchadki "Nikel" i "Zapolyarnyj"): vliyanie na nazemnye ehkosistemy (Ryazan: NP "Golos Gubernii") p 92

[5] Mining Company "Norilskij Nikel": impact on the environmental and health of population (Gorno-metallurgicheskaya kompaniya "Norilskij nikel": vliyanie na okruzhayushchuyu sredu i zdorove lyudej) 2010 Report of Bellona (doklad obedineniya Bellona) (http://bellona.org/assets/sites/4/fil_nikel-report-bellona-2010-ru.pdf)

[6] Koptsik S, Koptsik G, Strand L, Venn K and Amlid D 1999 Multiparametric analysis of the acidity and cation exchange capacity in forest soils affected by acid atmospheric precipitation. *J. Eurasian Soil Science* **32** (8) pp 902–910

[7] Koptsik G, Koptsik S and Smirnova I 2016 Alternative technologies for remediation of technogenic barrens in the Kola subarctic *J. Eurasian Soil Science* **49** (11) pp 1294–1309

[8] Ponkratov V 2014. Tax maneuver in Russian oil production industry. *J. Neftyanoe khozyaystvo – Oil Industry.* **9** pp 58-61

[9] Bashkirova N 2017 Possibilities of application of the Great Britain experience in ecological taxation in Russia developing (Vozmozhnosti primeneniya opyta Velikobritanii v razvitii ehkologicheskogo nalogooblozheniya Rossii) *Proc. Ensuring Economic Security of Russia in Modern Conditions (Obespechenie Ehkonomicheskoj Bezopasnosti Rrossii v Sovremennyh Usloviyah)* (Moscow: Belyi Veter) pp 16-22

[10] Bashkirova N and Lessovaia S 2017 Approaches to the resources creation for nature protection actions on the ecologically adverse territories (Podhody k formirovaniyu istochnikov dlya provedeniya prirodoohrannyh meropriyatij na ehkologicheski neblagopriyatnyh territoriyah) *J. Vestnik Universiteta. Series 26 "State Audit".* **3** pp 14-23

IOP Conf. Series: Earth and Environmental Science **107** (2017) 012129 doi:10.1088/1755-1315/107/1/012129

Evaluation of the lake Lyabiazhie (Kazan, Russia) state by indicators of communities of hydrobionts

O Derevenskaya and N Urazaeva

Kazan Federal University, Kazan, Kremlevskay str.,18, 420008, Russia
E-mail: oderevenskaya@mail.ru

Abstract. The article presents the results of studies of zooplankton and zoobenthos communities of Lake Lebyazhye (Kazan, Russia) performed in 2015-2016. Over the past decades, under the influence of anthropogenic and natural factors, the level of water decreased and the area of Lake Lebyazhye decreased. Now the water level in the lake is artificially maintained by the supply of groundwater, which led to a change in the type of water, an increase in mineralization. As a result, species composition of zooplankton decreased, a complex of dominant species changed, and quantitative indicators of zooplankton and zoobenthos communities decreased. The results of studies of the zooplankton and zoobenthos communities of Lake Lebyazhye can be used to justify measures for the lake's eco-rehabilitation.

1. Introdution

The problem of preserving water bodies located in urban areas is becoming increasingly urgent. Lake Lebyazhie is located in the urban forest park "Lebyazhye" (Kazan, Russia). Earlier the lake was a system of four reservoirs: Great, Small, Light, and Dry Swan, connected by ducts. Now there is only one lake left, and the others have dried up. Talking of the origin of the lake, it is supposedly dune or deflation-karst [1]. The water level is artificially maintained by the supply of groundwater from wells in Small Lebyazhye Lake. According to estimations in 2015, the area of Lake. M. Lebyazhye is 3.38 hectares, the volume of the water mass is 31 thousand m3, the maximum depth is 2.9 m, the average depth is 0.92 m, the water volume is 31 thousand m^3, the length of the coastline is 1630 m, the catchment area is 73.7 hectares.

Previously, the water in the Lebyazhye Lake was a hydrocarbonate-calcium type, it was "very soft", with "medium" mineralization. Submission of groundwater to the lake led to an increase in mineralization of water (from 151.5 mg/l to 975.2 mg/l). Indirectly, the content of salts in water can be recognized from the value of the electrical conductivity of water. Its magnitude in different years was 121-410 μS/cm in Small Lebyazhye Lake. In 2015, this indicator was 1030-1280 μS/cm, and in 2016 it was 730-1500 μS/cm, which characterizes the mineralization as "medium-high". The sulfate content was 21.8 mg/l earlier; 513.7 mg/l in 2015, then sulfates prevailed from the anions in 2015. The water in the lakes was observed previously as hydrocarbonate-sulphate-calcium type, medium mineralization, 'very soft'. At the present time it is 'very hard', with high mineralization.

The lake is a valuable recreational facility, therefore it is necessary to take measures to preserve the lake and maintain high water quality. Currently, the project of eco-rehabilitation of Lake Lebyazhye has been developed and is being implemented. It is planned to restore the Great and Light Lebyazhye lakes within their former borders. It is planned to deepen the lakes up to 3-4 m, to restore the waterproof layer and also to build a pumping station and a pipeline to supply water from the Emerald lake. Conservation of the Lake Lebyazhye includes an assessment of the state of the zooplankton and zoobenthos communities in the complex of actions. It is an important and necessary element, since it will allow selecting the most optimal strategy for the conservation and restoration of the reservoir.

The purpose of the work was to assess the state of Lake Lebyazhye (Kazan, Russia) according to the indicators of the communities of hydrobionts.

There are many methods to assess the water quality, each of them has its own advantages and disadvantages. The abundance of species of living creatures that live in water bodies and the complexity of their interaction with the environment and other organisms made it possible for emergence of methods for biological assessment of the state of natural waters. Hydrobiological analysis is an important element and allows determining the ecological state of water bodies and the consequences of their contamination [2]. The driver to use bioindication as a method for assessing water quality is close relationship between environmental quality and the structure of benthic and plankton communities of aquatic ecosystems. Biological indicators are a promising element of the monitoring system for surface water pollution and allow to determine the ecological status and trophic status of water bodies; to assess the quality of waters as habitats of organisms; determine the combined effect of the combined effect of pollutants; localize the source of pollution; to establish the type of pollutants and the occurrence of secondary water pollution [3, 4].

Zooplankton in aquatic ecosystems that carry out a huge process of transferring the energy of primary production to subsequent trophic levels, participate in the processes of natural self-purification and creation of secondary products [5]. Some characteristics of zooplankton are recommended for bioindication, such as the index of species diversity, the average individual mass of zooplankton, the ratio of crustacean and rotifer biomass, biomass of filter feeders and predators, etc. [6-11]. All these characteristics of zooplankton vary due to contamination and eutrophication. The dynamics of changes in the qualitative and quantitative composition of zooplankton and data on the dominance of various species will allow us to conclude the change of water quality over a certain period of time. Therefore, the greatest interest in environmental studies is the long-term and repeated observation of zooplankton [12].

According to many experts [13], zoobenthos is one of the most informative and reliable bioindicators of the state of the environment and its anthropogenic changes. Benthic invertebrates and their communities are sensitive indicators of pollution with nutrients and toxic substances, acidification and eutrophication of water bodies [14, 15]. The duration of life cycles of zoobenthos organisms is much higher in comparison with planktonic organisms. Consequently, the state of zoobenthos communities can reflect the level of chronic pollution. In addition, generally benthic invertebrates lead a sedentary lifestyle, therefore the state of zoobenthos communities clearly characterizes not only the ecological state of the reservoir as a whole, but also its individual areas [14, 15].

2. Material and methods

Samples of zooplankton and zoobenthos were collected at Lake Lebyazhye during the period from May to September of 2015 and 2016. Samples were collected every 10-14 days from 1-4 stations.
Zooplankton samples were collected by filtering 50 liters of water through an Apstein network with a mesh size of 100 μm. Cameral processing was conducted according to generally accepted methods [12]. The values of the abundance of zooplankton were calculated for each station. Biomass was calculated from the equations connecting the length of organisms with their mass. Assessment of water quality by zooplankton was carried out using saprobity indices.

Features of the community structure were identified using the Shannon and Simpson indices. Sampling of zoobenthos was carried out at stations measuring 20x20 cm using a scraper. Fixation and cameral processing were carried out using conventional techniques. Individual weights of organisms were found by weighing on scales. Assessment of water quality by zoobenthos was performed using the Woodiviss and Mayer indexes.

3. Results and discussion

Zooplankton. Zooplankton of Lake Maloe Lebyazhye in 2015-2016 was characterized as follows: 68 species of zooplankton were identified, including 43 species of rotifers- (63%), 17 species of cladocera (25%), 8 species of copepods (12%). 44 species of zooplankton were identified in 2015, and 43 in 2016. The index of similarity of Serensen-Chekanovsky was 64%. Thus, there is certain variability in species composition. Partly the reason for this could be the low number of individual species and zooplankton in general, which may cause short-term species representation in the community. Another reason could be the fact that their presence, due to low numbers, is not always detectable.

The composition of the dominant species reflects the state of the community. In 2015, *Asplanchna priodonta* (Gosse, 1850), *Chydorus sphaericus* (O.F. Muller, 1785), *Keratella cochlearis* (Gosse, 1851) dominated most of the population. The composition of the dominant complex in different periods formed from 2 to 4 species of zooplankton. Biomass was dominated by 1 to 6 species, constantly, including *A. priodonta*, often *C. sphaericus* and *Thermocyclops oithonoides* (Sars GO, 1863). Average number of dominant species was 2.8.

In 2016, *Brachionus calyciflorus* Pallas, 1766, *Polyarthra dolichoptera* Idelson, 1925, *C. sphaericus* and *A. ptiodonta* dominated most. The rotifer *A. ptiodonta* was predominant in biomass in most cases, *C. sphaericus* and *T. oithonoides* were less frequently included in the dominant complex. The average number of dominant species, both in number and in biomass, was 3.5.

The abundance and biomass of the community are important characteristics of its condition. They differed slightly in Lake Lebyazhye (Table 1). Rotifers predominated in numbers and biomass. In terms of zooplankton biomass, the reservoir is α-oligotrophic (according to the classification of S.P. Kitaev (1984).

Table 1. The abundance (N, thousand ind./m^3) and biomass (B, g/m^3) of the zooplankton of the lake Lebyazhye.

Group	2015 (n=12)		2016 (n=11)	
	N	B	N	B
Rotifera	22,71±9,75	0,048±0,031	15,31±5,26	0,017±0,008
Cladocera	3,71±0,94	0,012±0,003	0,90±0,16	0,003±0,001
Copepoda	16,26±4,29	0,050±0,017	6,73±1,78	0,032±0,011
Total	42,68±13,14	0,111±0,038	22,95±5,60	0,054±0,014

Due to unfavorable conditions there were low quantitative indicators of zooplankton. The most important for zooplankton, in our opinion, is the high mineralization of water due to the artificial supply of groundwater to the lake in order to maintain the water level.
The peak of zooplankton numbers in 2015 and 2016 was observed in the second half of May and was mainly caused by Rotifera. The second largest peak was observed in June in 2015 and in the second half of August in 2016. The biomass peak was observed in May and in the second half of August in 2015, and in late August and early July in 2016 (dominated by Copepoda) (Figure 1).

The index of saprobity, characterizing the level of contamination with organic substances, corresponds to the β-mesosaprobic zone, moderately polluted waters, III class of water quality. The index of the species diversity of Shannon characterizes the alignment of the community structure. The index values were relatively high and amounted to 2.65 ± 0.13. The community is relatively aligned, i.e. species are represented by approximately the same number of individuals, however, with the small number. The absence of a concentration of dominance is also indicated by the values of the Simpson index (0.74 ± 0.05 in 2015 and 0.77 ± 0.03 in 2016).

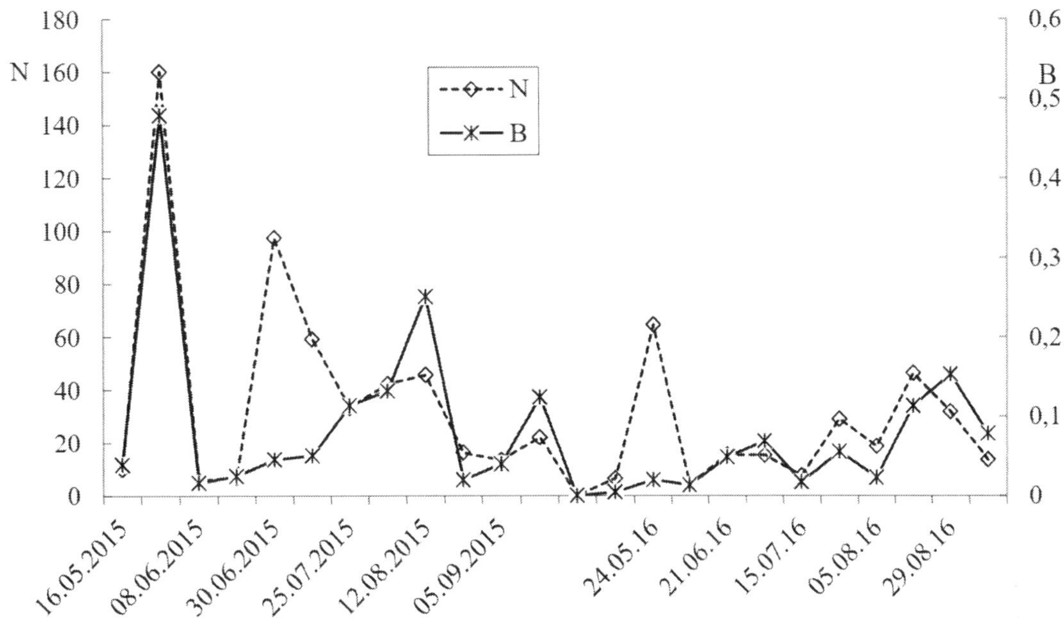

Figure 1. Number (N, thousand ind./m^3) and biomass (B, g / m^3) of zooplankton in 2015-2016.

Zoobenthos. Benthic organisms in 2015-2016 were represented by seven taxa: Lymnea, Oligochaeta, larvae fam. Chironomidae, Agabus (fam. Corixidae), dragonfly larvae (Odonata), Diptera larvae, Ephemeroptera larvae. In 2015, there were 6 taxa, in 2016 - 2.

Larvae fam. Chironomidae (in 92% of samples) and worms (Oligochaeta) (in 30% of samples) were most frequently encountered in the samples in 2015. The remaining groups of organisms were found in no more than two samples. In 2016, all samples contained larvae of family Chironomidae and only in July there were Ephemeroptera larvae. The species richness of zoobenthos is estimated as "low".

Reduction of species diversity in 2016 may be associated with a decrease in the water level in the lake. From May to July 2016, the lake was not replenished with groundwater due to pump failure. Exposure of the bottom areas previously covered by water could lead to the death of inactive benthic organisms, especially the inhabitants of macrophyte thickets.

Very low quantitative indicators of zoobenthos were observed in 2015. The maximum values of abundance and biomass were recorded in May and amounted to 175 ind/m^2 and 116.6 mg/m^2 respectively; the minimum values were observed in September (5 ind/m^2 and 6.8 mg/m^2). The average abundance was 39 ± 16 ind/m^2, and average biomass was 45.1 ± 10.1 mg/m^2 (Figure 2).

In 2016 the abundance and biomass of the zoobenthos were higher; the maximum values of abundance and biomass were 1225 ind/m^2 and 1837 mg/m^2 respectively. The average abundance and biomass were 234 ± 116 ind/m^2 and 351 ± 173 mg/m^2. The greatest contribution to the total abundance and biomass of zoobenthos was made by the larvae of mosquitoes (fam. Chironomidae), especially in 2016.

Some features of the structure of zoobenthos can be caused by a change in the type of water and the value of mineralization, which is associated with the artificial replenishment of the lake with groundwater. So, for example, earlier in the lake there was an abundance of mollusks-vivipar (*Vivivarus contectus* (Millet, 1813)), who prefers soft water. At present, the zoobenthos of these mollusks are not found.

According to zoobenthos, the water in the Lebyazhye Lake is characterized as "dirty" to "very dirty": in 2015 the average value of the Mayer index was 1.91 ± 0.27, and the Woodiviss index had values of 1.33 ± 0.21 (Figure 3). In 2016, the values of the indices were higher and amounted to 1.6 ± 0.4 and 1.8 ± 0.5 respectively. In general, the benthic community was represented by a small number of species, which may be due to changes in the chemical composition of the water, the unstable water level in the lake, the use of insecticides in the recreational area and active recreational loads on the lake.

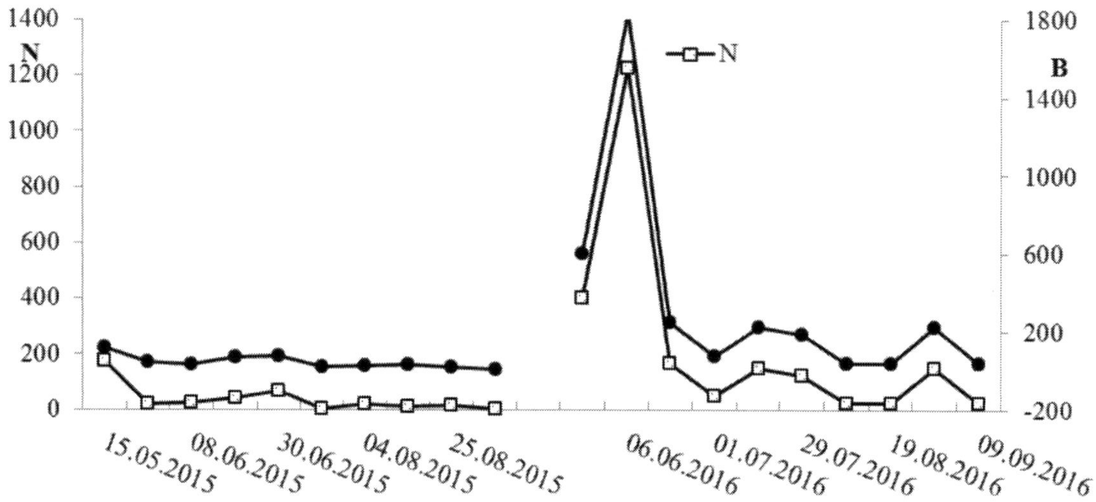

Figure 2. Dynamics of abundance (N, ind./m^2) and biomass (B, mg/m^2) of the zoobenthos of Lake Lebyazhye.

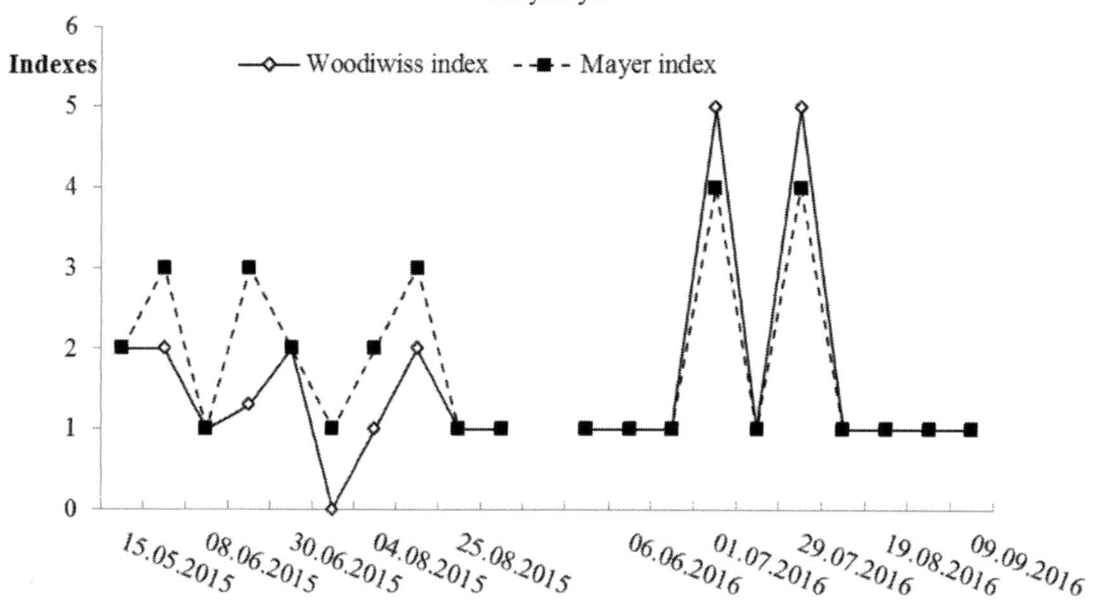

Figure 3. The values of the Mayer and Woodywiss indices in 2015-2016.

Thus, as a result of the studies, 68 zooplankton species were found. Zoobenthos was represented by seven taxa. The composition of zooplankton was dominated by *A. priodonta, B. longirostris, T. oithonoides, P. vulgaris, P. dolichoptera, C sphaericus.* Larvae of family Chironomidae predominated in most cases in the zoobenthos. Quantitative indicators of zooplankton and zoobenthos were low. Indicators of zooplankton estimated water of the Lebyazhye lake as "moderately polluted", which corresponds to the third class of water quality. The state of the zoobenthos communities characterizes the reservoir as "very dirty".

References

[1] *Essays on the geography of Tataria* 1957 (Kazan: Tatkhnigoizdat) p 357
[2] *Manual on methods of hydrobiological analysis of surface waters and bottom sediments* Ed. VA Abakumova 1983 (Leningrad: Gidrometeoizdat) p 240
[3] Shitikov V K, Rosenberg G S and Zinchenko T D 2003 *Quantitative hydroecology: methods of system identification* (Togliatti: IEEB RAS) p 463
[4] *Ecology of the city of Kazan* 2005 (Kazan: FEN Publishing House) p 300.
[5] Kuznetsova M A 2002 Changes in the structural and functional characteristics of zooplankton in the course of eutrophication of diverse lakes in the aspect of the succession concept on the example of the lakes of the East European Plain (Nizhny Novgorod) p 380.
[6] Andronikova I N 1996 *Structural and functional organization of zooplankton of lake ecosystems of different trophic types* (Sankt-Petersburg: Science) p 189
[7] Ejsmont-Karabin J and Karabin A 2013 The suitability of zooplankton as lake ecosystem indicators: crustacean trophic state index *Polish Journal of Ecology* **61(3)** pp 561-573
[8] Haberman J, Haldna M 2014 Indices of zooplankton community as valuable tools in assessing the trophic state and water quality of eutrophic lakes: long term study of Lake Võrtsjärv *J. Limnol.* **73 (2)** pp 263-273 ORIGINAL ARTICLEDOI: 10.4081/jlimnol.2014.828J.
[9] Ochocka A, Pasztaleniec A 2016 Sensitivity of plankton indices to lake trophic conditions *Environ Monit Assess* **188** p 622
[10] Derevenskaya O Y, Umyarova R M 2016 Zooplankton as an indicator of river ecological condition *International Journal Of Pharmacy & Technology* **8 (2)** pp 14567-14574.
[11] Derevenskaya O Y, Mingazova N M and Pavlova L R 2015 Lake water quality of Kazan city (Russia) Kaban lake in the anthropogenic pollution conditions and improving actions implementation *International Journal of Applied Engineering Research* **10(24)** pp 44682-44687.
[12] *Methodical recommendations for the collection and processing of materials in hydrobiological research on freshwater reservoirs. Zooplankton and its products* 1982 (Leningrad, ZIN USSR-GOSNIORCH) p 33
[13] Bakanov A I 1999 Use of combined indices for monitoring freshwater reservoirs by zoobenthos *Water resources* **26(1)** pp 108-111
[14] Bezmaternykh D M 2007 *Zoobenthos as an indicator of the ecological state of aquatic ecosystems in western Siberia: analytic overview* (Novosibirsk: Gos. Pub. Scientific and Technical Library of the Siberian Branch of the Russian Academy of Sciences) p 87.
[15] Bezmaternykh D M, Miseiko G N and Tushkova G I 2001 *Biological analysis of the quality of fresh water* (Barnaul: AltSU) p 201.

IOP Conf. Series: Earth and Environmental Science **107** (2017) 012130 doi:10.1088/1755-1315/107/1/012130

Ecological planning of urbanized areas in the south of the Far East (Birobidzhan city as an example)

V B Kalmanova

Institute for Complex Analysis of Regional Problems FEB RAS, 4 Sholom-Aleikhem St., Birobidzhan, 679000, Russia
E-mail: Kalmanova@yandex.ru

Abstract. Ecological planning of urbanized areas is an urgent demand of the time, because more than 70% of Russia's population lives in cities. The article describes that the city's ecological planning is an important part of the area's organization in its development strategy. The principles and features of the urban area's ecological organization are proposed. The basis for environmental planning is the ecological and functional zoning of urban areas. The algorithm of ecological-functional zoning is developed to optimize the quality of the urban environment. Based on it, it is possible to identify the planning structure's features, justify anthropogenic pressure on the natural components of the urban environment, etc. The article briefly presents the possibility of using the main conditions of the ecological framework in the planning of urban areas. Considering the perspective trends of the formation and development of cities in the south of the Far East, the ecological problems caused by regional natural and anthropogenic causes (features of relief, climate, functional-planning structure) are considered. The need for environmental planning of cities in the south of the Far East is shown. The results of the ecological framework's formation of Birobidzhan city based on its ecological and functional zoning are described. The total area of open unreformed spaces in the city is calculated to be 60.8%, which can serve as the main elements of the ecological framework and perspective reserve areas for ecological planning. The cartographic model of Birobidzhan's ecological framework is presented, which is the result and model of this type of planning. The practical use of the proposed model will facilitate the adoption of effective management decisions aimed at stabilized development of the city.

1. Introduction

Currently, 73% of the Russian population is concentrated in cities. In some countries, this proportion is even higher. Generally, the growth of cities leads to environmental problems, which are the result of numerous related problems. According to environmental indicators, approximately 15% of Russian territory is in a critical or near-critical condition. 60% of Russia's population lives in unfavorable environmental conditions (eg, 76% of the Northwestern economic region, 52% - Volga region , 54% of the West-Siberian region live in environmentally dangerous conditions (I category of danger); 46% of the population of the south of the Far East live in the region of the 2nd danger category, etc.) [1]. A certain "input" to the creation of these problems, together with the state of the natural environment and the development of natural processes, is also contributed by the territorial organization of the city (features of the planning structure, location of functional zones).

The issues of territorial planning in the spatial development of the city acquire a key importance, because they allow an approach to different types of urban areas and in each individual case to determine qualitatively different directions of environmental and town-planning policies and practices.

Ecological planning of urban areas allows not only to create favorable conditions for living and recreation of the population in residential areas, but also to protect ecologically valuable areas (recreational, coastal, water protection and sanitary protection zones) from illegal and chaotic building.

Recently, the planning of territories is increasingly focused on the inclusion of environmental functions, which are inherent in individual components of nature - landscapes, soils, vegetation, etc. Thereby, the functional zoning of urban territories in its various aspects acquires significance, one of which is the ecological and functional zoning. In contrast to the traditional segmentation of urban areas, which are divided into residential, industrial, communal, warehouse, sanitary-protective, landscape-recreational and other zones by type of use, this zoning takes into account the value and priority of ecological functions performed by urban ecosystems and their individual components and which have a environment-forming, environment-stabilizing (environmental control) or environmental destabilizing value [2]. Zones with a dominance of environment-forming functions create a basis for the ecological framework of area. In turn, ecologic -functional zoning and the ecological framework of the urban area serve as the basis for the development of a local ecological regulatory framework [3].

For cities in the south of the Far East, the experience of environmental planning is practically non-existent, in general, planning of the territory is subordinated to town-planning goals.

The Far East is one of the most urbanized regions of the Russian Federation, in connection with the natural features of the territory - 80% of the population is concentrated in cities. Technogenic pollution of the Far East's cities cannot determine an overwhelming part of their territory as favorable for human habitation. Many cities in the south of the Far East were formed as "power centers" with military and political goals, without taking into account the natural features of the territory to the detriment of the ecological state of the environment. A special feature of many Far Eastern cities is the gradual build-up of industrial potential, which was carried out at a time when economic growth was a priority to the detriment of environmental requirements. As a result, there is no demarcation zone (buffer zone) between the residential and industrial sectors. Some elements of infrastructure and social life are often located in the city. In the formation of the planning structure, the wind rose, geomorphological indicators, landscape and climatic differences of individual regions were not taken into account. The main results of negative impact are: pollution of ground and surface water, disruption of the hydrological regime; air pollution; land disturbance, etc. It is also important to note the specific features of the natural backgrounds, most often not quite comfortable, in which our cities "live". High vulnerability of natural systems, the character of the dynamics of natural processes do not "forgive the misses" in planning and lack of monitoring. This is exhibited not only in obvious and acute environmental problems, but also in higher economic costs for maintaining urban infrastructure (for example, road construction, the construction of sewage collectors, etc.) [4]. The acuteness of problems and the combination of causes leading to discomfort in different parts of the city and in the different towns of the region are changing.

The development of any city requires a complex geo-ecological approach to solve the issues of territorial planning and improving the state of environment. In the new social-economic conditions, it is necessary to develop approaches to geoecological research and integrated assessment of the territory, which allows determine the limitations of urban building, as well as to plan and adjust the prior urban and environmental measures [2, 3, 5].

2. Materials and Methods

Among the cities of the Far East, Birobidzhan was chosen as an object for the research, because it belongs to the category of medium-sized cities of the Far East, where the ecological situation is complicated by an indigested planning structure (chaotic location of functional zones, lack of sanitary protection zones). [6]. Unlike other cities of the Far East, Birobidzhan possesses a sufficient number of plantations (21%), open spaces (60.8%) and free zones (51%), related to potential reserve areas for ecological planning. Considering that, it is possible to develop a model for the optimal organization of the urban environment in order to improve the ecological situation in the city.

To optimize the quality of the urban environment, an algorithm for the ecological planning of the territory is proposed, which provides a clearly order of actions (figure 1). Ecological planning requires an objective and comprehensive ecological assessment of the state of the urban environment. Since a single integral indicator of ecological condition does not exist, the criteria for assessing natural components are a number of bio-indicators, spatial and dynamic indicators.

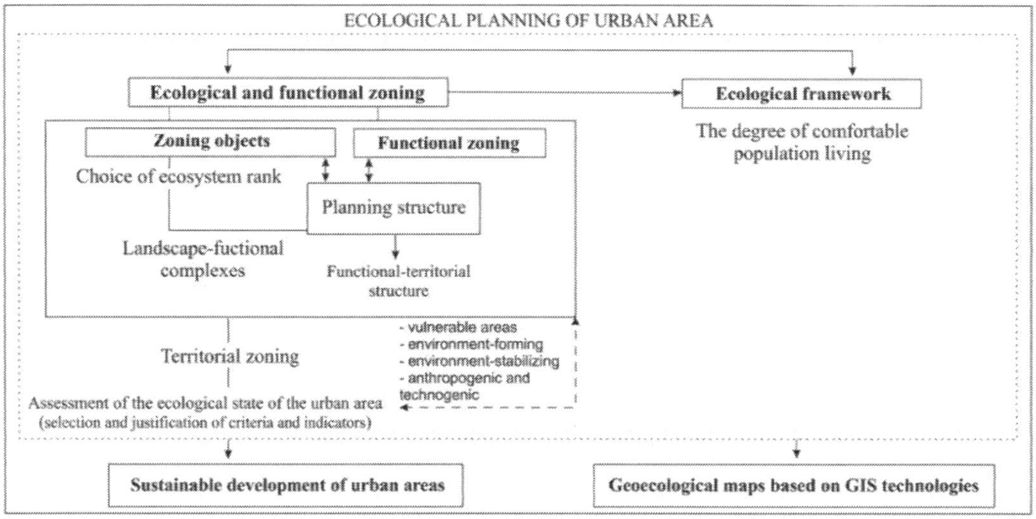

Figure 1. Algorithm of urban area's ecological planning.

At the regional level, each city should develop its own specific set of criteria for assessing the geoecological state of the natural environment, taking into account natural and anthropogenic factors, types of technogenic pressure. But the assessment should be made according to the same principles [1, 5, 7]: 1) the basic geoecological problems of the region should be identified; a systematic analysis of natural-anthropogenic and technogenic factors should be carried out; 2) a set of evaluative criteria and indicators should be developed; 3) classify the ecological state of the urban environment; 4) evaluative taxa- sectors, should be allocated; 5) a scale of expert evaluation should be developed; 6) calculations of evaluation points for all components of the urbanized environment and indicators for each site should be carried out and zoning of the territory should be made according to the geoecological state.

The result and model of this type of planning is the formation of the ecological framework of the urban area (EFUA), which should be preceded by the ecological and functional zoning of the territory, meaning the ranking of urban landscapes from the standpoint of their sustainability, combined with the performed ecological functions.

The published statistical data, literature, cartographic and stock materials of ICARP FEB RAS (Institute for complex analysis of regional problems of the Russian Academy of Sciences, Far Eastern branch), Federal State Statistics Service of Jewish Autonomous Region (Evrstat), The Federal Service for Controling in the Sphere of Nature Management of Jewish Autonomous Region were used in the working process. The main methods of research were comparative-geographic, semi-stationary methods of studying natural components, as well as a method of thematic mapping.

The methodology and principles for constructing the map "Ecological framework of Khabarovsk city", created by the Institute of Water and Environmental Problems (IWEP) of the Far East Branch of Russian Academy of Sciences [2]. The map of the ecological framework of Birobidzhan is made on a scale of 1: 25000 by the traditional method and digitized using the ArcView GIS software.

3. Results and discussion

Formation of the ecological framework of Birobidzhan was preceded by geoecological analysis of the territory and its ecological and functional zoning.

Based on the results of the geoecological assessment, zoning of the Birobidzhan territory was carried out according to the pollution level, which showed that 4% of the city area has a very high, 8% to high, 10% to above average, 47% to average pollution levels, 31% to relatively clean areas cities.

Ecological and functional zoning of the territory assumes the ranging of urban landscapes from the standpoint of their sustainability, united by the performed ecological functions. According to the ecological and functional zoning of the territory, a structure was developed for Birobidzhan: the core, first order elements of the urban area's ecological framework (area structure), second and third order elements of the urban area's ecological framework (linear structure) were identified (table 1, figure 2)

Table 1. Structure of ecological and functional zones of the Birobidzhan territory

Ecological and functional zone	Element of ecological-functional zone	Main functions
Vulnerable (7.1% of the total area of urban land)	Ravines, ash dumps, dumps, wastelands, industrial wastelands, quarries, plots of debris and landslides.	Destruction of natural and natural-anthropogenic complexes as a result of erosion and geological processes; Dust formation.
Environment-forming (46%)	1. Monuments of nature of regional importance (farm, dendropark). 2. Natural territories of local importance (Park of Culture and Recreation). 3. Landscaping and recreational areas. 4. Swamps.	Genetic conservation of biodiversity; microclimate formation; conservation of ecosystems of natural and anthropogenic value, as well as natural habitats of plants, terrestrial and soil fauna.
Environment-stabilizing (16.4%)	1. Natural territories of local importance (Square of Victory, Veterans, Teatralny; "Friendship of Peoples" Square, etc.). 2. Public gardens. 3.Boulevards, squares, promenades. 4. Cemeteries, memorial parks. 5. Water protection zone along the Bira river and other small rivers. 6. Protective zones along motorways, railways.	Erosion-stabilizing; ensuring water quality and normal hydrological regime; stabilization of the ratio of oxygen and carbon dioxide in the air; wind regime regulation.
Anthropogenic and technogenic (30.5%)	1. Residential, industrial, agricultural areas, 2. Linear and core infrastructure systems.	Ensuring the citizens life.

In the formation of an ecological framework of Birobidzhan, special attention was paid to the fact, that the main elements of the EFUA were connected by linear elements and adjoined to watercourses and water protection zones. The last ones, extending beyond the boundaries of the urban area, joined the landscape complexes of local and regional levels. Ash dumps and wastelands are perspective for the ecological framework. Dumps, ash dumps and quarries account 1.8% of the total area of Birobidzhan, while this number should be less than 1%. Wastelands (3.2%), including technogenic ones, are found in every district of the city. Depending on the plans for the future development of the urban area, they could be used to form EFUA elements.

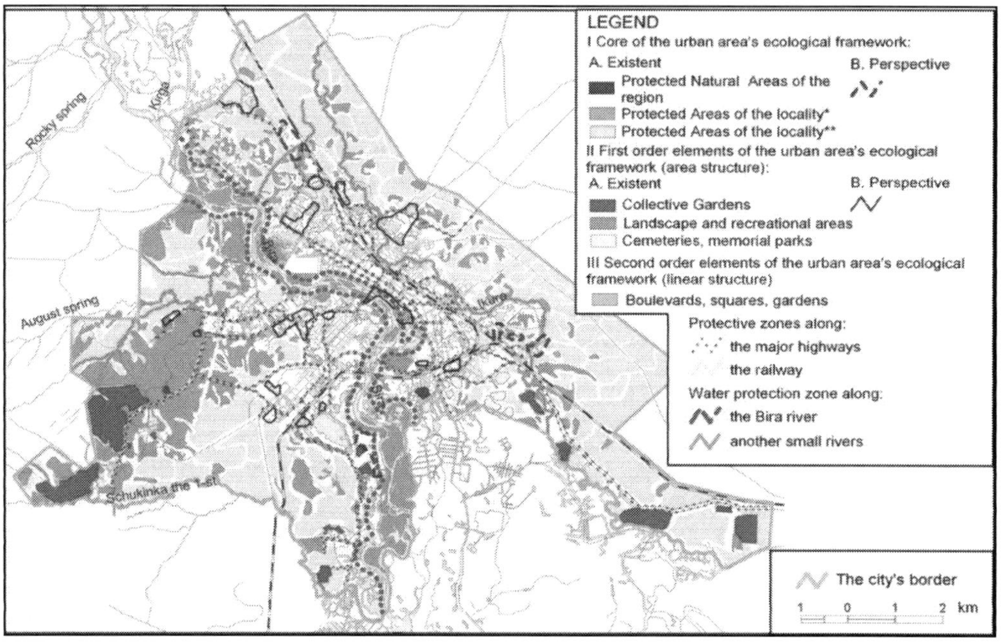

Note: the dominant function (* environment-forming, ** environment-stabilizing)

Figure 2. Ecological framework of the Birobidzhan territory.

In the formation of EFUA, a special attention is given to a system of relative green areas of the city that perform environment-forming and environment-stabilizing functions. Determining the functional significance of green plantations, they should be considered as an integral part of general system – EFUA; and considering the structure and state - as an independent system - the "green" framework of the city. In Birobidzhan public green spaces comprise 18% of the city's area, restricted - 0.5%, special use - 2.8%. For 1 person is 561.6 m^2, taking into account the urban forests located in the environs of the city. Nevertheless, if we consider the territory within the urban development, there is a lack of green spaces: per person account for 4 m^2, at a rate of 21 m^2. It is very difficult to reach the norm, as the city is being built exactly within the boundaries of development and, accordingly, the green zones are being destroyed. The number of plantings of all categories has declined and continues to decline; their quality is deteriorating. Over the past 5 years there has been a general reduction in vegetation by 30%. It is possible to achieve the norm due to a sufficient number of open spaces (60.8%).

The main component of the EFUA is open spaces of the city, which include territories covered with greenery. Due to them, it is possible: 1) to minimize unfavourable climate and other environmental conditions; 2) to bring the number of green spaces to the regulatory parameters, which can perform protective, health and other functions; 3) to preserve and improve the quality of basin that can perform recreational functions; 4) to increase the ability of air to self-purify .

The obtained results can be used as a basis for optimizing the urban environment. For recommendations, the following options for planting the urban environment are proposed:

1) creation of one continuous green frame;
2) renovation and preservation of the connection between urban planting and the suburban area;
3) presence of large green areas, which are the centers of high biodiversity concentration;
4) maximum equability and accessibility of green areas of general use for city residents;
5) rational planning of planting, in accordance with the general plan for the development of the city;

6) creation of plantations, ecologically adapted to local climatic conditions and anthropogenic factors; protection of green spaces located on the territory of the city, regardless of the form of property, where these plantations are located (reconstruction of green spaces, if possible);

7) creation of a plan for the planting's regularization on the territory of the city, taking into account their functional significance;

8) increasing the level of workers' skills associated with the greening of the city.

4. Conclusions

An analysis of the complex of social and environmental problems, which are existed in the cities of the Far East, makes possible to come to conclusions about the causes and acuity of problems (especially architectural and planning). This is the result of natural and climatic conditions, but to a large extent - the effect of inefficient or poor-quality town-planning and engineering activities since the period of the formation and development of cities.

Birobidzhan is unique city that is different from other cities in the south of the Far East by the presence of a natural component. It belongs to the category of medium-sized cities in the south of the Far East, many of which receive an impulse for growth and development in recent years. Taking into account the intensification of economic activity within the Birobidzhan industrial hub in the short term, the most important component of environmental policy is the organization of the territory, at which its ecological part should be proactive and focused on carrying out the territorial structure of the economy in accordance with its resource base, orientation of strategic development plans, geosystem organization of natural complexes.

Consequently, environmental planning of the territory, taking into account the algorithm of its implementation and certain research methods, represents an important block of the integrated analysis of urbanized territories, which is necessary for making effective decisions with a view to formulating a city development strategy. The basis of its formation is ecological framework. Without legal status, the ecological framework of the urban area or its main provisions (developments) can be included in the general plan of the city. The mutual understanding of the developers of the ecological framework and the general plan can contribute to this. However, the general plan can effectively "work", if there are legal mechanisms for its implementation. Therefore, the creation of a local environmental regulatory framework becomes extremely important.

References

[1] Zaikanov V and Minakova T 2005 *Geoecological assessment of territories* (Moscow: Science) 319 p

[2] Narbut N, Antonova L, Matyushkina L, Klimina E and Karavanov K 2002 *Strategy for the formation of the ecological framework of urban territory (Khabarovsk, for example)* (Vladivostok-Khabarovsk: FEB RAS) 129 p

[3] Mirzekhanova Z and Narbut N 2013 Ecological foundations of organization of urban territories (on the example of Khabarovsk) *J. Pacific Geology* **32 (4)** pp 111-120

[4] Kalmanova V 2015 Ecological and hygienic condition of cities in the south of the Far East as a consequence of the development of the region *J. Regional problems* **18 (2)** pp 37-43

[5] Azpeitia A and Azkarate A 2016 Urban Planning and Sustainable Development in The 21st Century, Conceptual and Management Issues *J. IOP Conf. Series: Earth and Environmental Science* **44** pp 1-6

[6] Kalmanova V 2016 Ecological features of the functional-planning structure of medium and small cities in the south of the Far East (based on the example of Birobidzhan) InterCarto / *InterGIS* **22 (2)** pp 273-286

[7] Kalmanova V and Suhoveeva A 2014 Selection and justification of urban environment quality indicators *J. News of the Samara Scientific Center of the Russian Academy of Sciences. Social, humanitarian, biomedical sciences* **16 (5-2)** pp 878-882

Inventory of water objects for purposes of development objectives and design of natural frame of Kazan (Russia)

N M Mingazova

Kazan (Volga) Federal University,
Institute of Management, Economics and Finance,
Department of Environmental Engineering and Water Management,
Kazan, 420008, Russia
E-mail: nmingas@mail.ru

Abstract
The article is devoted to the results of the inventory and certification of water objects in the city of Kazan in 2007-2017. Inventory of water objects contributes to the formation of the natural framework of the city, improvement of the quality of the environment, preservation of the local gene pool of flora and fauna in urban conditions. During the research, 236 small lakes, rivers and their tributaries, as well as wetlands, were identified. The environmental passports were developed for each of them. Registries of water bodies were developed for administrative regions. The inventory and certification of water objects are an effective measure in the conservation of lakes and rivers in urban conditions.

1. Introduction

Small lakes of world are subject to human impact, including industrial, agricultural, communal and recreational [1]. This is especially the case within cities and settlements. Silting, pollution and development of the banks influence reduction in catchment area and water area, deterioration of water quality, loss of biological diversity [2,3].

According to the cadastral records of the Institute of Ecology of Natural Systems, records of the Republic of Tatarstan (Russia) display about 10 thousand small lake [4], however only a little more than 8100 lakes have been preserved as per records in the 2000s. About 20% of the republic's lake fund has disappeared due to formation of new water reservoirs and as a result of agricultural impact (siltation due to removal of suspended solids from arable land and due to erosion of shores and pollution from watering and grazing).

A bright example is the anthropogenic transformation of the lakes of Kazan (Russia), an industrial city with a population of more than 1 million people, where in the 19-20 century about 50 small lakes with an area less than 1 ha have disappeared.

Meanwhile, small lakes and rivers serve as sites for preserving the fauna and flora, providing biological diversity and sustainable development of the territories. Lakes, being accumulating systems, in fact, serve as indicators of the state of the territories. Small lakes and rivers are important links in the "green corridors" of the natural (ecological) frame of territories; they often play an important recreational role.

Therefore, it is extremely important to provide regional and federal measures for the conservation of small water bodies (lakes and rivers), to adopt special programs for the revival and restoration of small lakes and rivers in Russia, and to provide accounting and inventory work.

2. Inventory of water bodies in Kazan

A distinctive feature of the territory of Kazan is diverse water bodies (about 250), that are widely used for recreational purposes, water supply and sanitation. Large water bodies, such as the Kuibyshev Reservoir, the Kazanka River, Kaban and Lebyazhye lakes are present on maps of various sizes and are monitored by environmental authorities. Small objects mostly were not studied, there was no information about them, and they were mostly ignored by city structures and environmental services. Lack of information on the number of small water bodies, their characteristics and ecological state is noticed in the 2000s. Thus, cases of destruction of water bodies due to the trade of water fund as the land for housing and communal construction increased sharply.

All this caused serious concern among the population, public environmental organizations, environmental authorities, deputies of the republic. "Improvement of the Environment of Kazan in 2004-2008" program and certification of water objects was suggested as an alternative to the uncontrolled use of water resources the inventory.

3. Methodology

Large-scale works on inventory and certification of water objects were carried out, considering the experience of the Nizhny Novgorod University [5,6], in Kazan in 2002-2008. During the inventory all available lakes, rivers, ponds and wetlands were identified and investigated in each administrative district. Cartographic and topographic materials were preliminarily analyzed, space images were analyzed. Clarification of the boundaries of the reservoir directly at the site of the water body was conducted later.

Methodological approaches to the inventory and certification of water bodies were developed in 2007-2008, the work continues to the present [7].

Environmental passports were developed within the course of environmental certification, filled with data for each water body. In addition, 7 registers of water bodies were compiled for each district of the city.

The following information on water bodies was available: the type of water body; administrative location; land user; GPS coordinates; explication of a water body in the region; origin; morphometric characteristics (area, length, width, depth); bathymetric chart; character of water exchange; type of water (ionic composition); assessment of water quality by physicochemical indicators; the results of microbiological and hydrobiological studies; composition of higher aquatic vegetation; composition of aquatic and near-water vertebrates; anthropogenic impact on the water body; environmental assessment; the value of the water body [6]. This information was covered fully in environmental passports and briefly in registers.

4. Results and discussion

Data on the number of water bodies per administrative district

Aviastroitelny administrative district of Kazan at the time of the research (2007-2008) had 32 water objects were identified from 42 known for 2002. These included 26 lakes, 4 marshs and 2 rivers. The most valuable natural sites are the Solonka River, Sukhaya River, Lesnoye Lake, Borisoglebskoe Lake, wetlands in the Sukhaya river valley and in the Borisoglebsk village.

6 water bodies were identified in the Vakhitovsky administrative district: the Nizhniy Kaban lake (area 47.5 ha); Kazanka river, the Bulak channel and 2 water basins with a total area of 1.44 ha. The reservoirs of the Vakhitovsky district have a large historical and recreational significance for the residents of Kazan.

54 water bodies were identified on the territory of the Kirov region: 34 lakes (with a total area of 142.5 ha), including valuable recreational lakes such as Bolshoe Glubokoe, Maloe Glubokoe, Izumrudnoe, Lebyazhye and others; 7 small (0.5 ha) artificial reservoirs; 7 wetlands with the area of 38.2 ha; 1 channel (0.6 ha). The southern part of the area adjoins the Kuibyshev reservoir, there are 4 gulfs (39.3 ha).

7 water bodies were identified and inspected in detail within the Moscow administrative district. In addition, inventory revealed that 3 previously existing reservoirs were filled up (Tolstovsky pond, underground channel of the Komarovka River, the second lake along Serova Street). Inventory discovered 1 river (part of the mouth of the Kazanka River), 5 lakes with a total area of 2.55 ha, and 1 channel (0.036 ha).

34 water bodies were identified in 2007-2008 out of 66 known in 2002 within the Novo-Savinovsky administrative district. This included river Kazanka; 1 duct; 27 lakes with a total area of 34.83 ha; 3 wetlands (total area 51.53 ha) and 2 drainage ditches. During the inventory it was revealed that 32 water bodies ceased to exist due to backfilling for the construction of residential buildings, parking lots and gas stations, some reservoirs are covered with garbage. Objects of high species richness are the wetlands in the floodplain of the Kazanka River and the Kazanka river channel, the wetlands in the Victory Park, the Small Chaykovy Lake and the Big Chaykovy Lake (there is a massive nesting of colonial bird species - gull and tern seagulls, high diversity duck birds).

66 water bodies were identified for the Privolzhsky administrative region, including 1 reservoir - Kuibyshev reservoir (area of 4375 ha in the city), 2 gulfs, 6 streams, 38 lakes (total area 264.4 ha), 1 river -

IOP Conf. Series: Earth and Environmental Science **107** (2017) 012131 doi:10.1088/1755-1315/107/1/012131

Noksa River, 1 pond (area 0.508 ha), 2 swimming pools (1,2 ha) and 5 drainage ditches. During the inventory, it was also revealed that 10 water bodies no longer exist due to backfilling for the construction of residential buildings, parking lots, gas stations, lowering the water level and drying out. The most valuable water objects in the Privolzhsky district include: Lake Middle Kaban - the largest lake in Tatarstan (119 ha), Lake Upper Kaban and the Poduválje Lake-Bay. Those are favorite recreational areas for residents of Kazan. Other important water bodies include Boriskovo village wetland complex and Lake Shanghai as they are the habitat of rare species of birds; Pobedilovsky Bay and the Volga River and Kuibyshev reservoir itself.

51 water bodies were identified in the Soviet administrative district and on the border with the Vysokogorsky district. Through the territory of the district there are 3 rivers (the Kazanka river and its tributaries - rivers Noksa and Kinderka); 1 stream (Vertelevka); 3 ponds (with a total area of 0.24 ha); 4 basins (3 of them are sedimentation tanks of 0.2 ha); 1 wetland (next to the Torfyanoy village) with a total area of 55.1 ha; 1 bay and 38 lakes (a total area of 49.18 ha). The lacustrine marsh Torfyanoy includes oxbow lake of origin, a number of lakes and a channel, wetlands. During the inventory it was also revealed that 3 water bodies do not exist due to backfilling with debris, lowering of the water level and drying out (Pioneer basin Lake, Malye Klyki village reservoir and the Karyer settlement reservoir). The most valuable water objects of the Soviet district are the waterways (rivers Kazanka, Kinderka and Noksa, the Torfyanoy village lacustrine marsh, the old Kazanka River, Karasinoye lake, lakes along Kazanka river, Derbyshky village cemetery lakes.

Due to the high natural value of many water bodies, it is possible to recommend assigning them the status of protected natural areas as conservation sites of rare plant and animal species.

Comparative analysis and identified trends

During the 2007-2008 inventory it was revealed that Volga region (66), the Soviet (51) and Kirov (54) administrative districts have bigger number of water objects, although Novo-Savinovsky District was the richest one in terms of water bodies in 2002 (66 objects in 2002, 34 in 2007, with a tendency to further reduce the number and area of water bodies). Aviastroitelniy and Novo-Savinovsky districts are now referred as "medium-rich" of water bodies with 32 and 34 objects respectively. The smallest number of water bodies was found in Vakhitovsky and Moscow districts, 6 and 7 water objects respectively.

Given that a number of watercourses in the city pass through several regions (the Kazanka River, Noksa River), as well as the reservoir, as well as the connection of the bays with the reservoir, 236 water bodies were identified in Kazan, according to the inventory for 2007-2008 [7].

58 water bodies were not found in Kazan. In comparison with the previously known information (10 in Privolzhsky district, 3 in Soviet district, 10 in Aviastroitelniy district, 32 in Novo-Savinovsky and 3 in Moscow administrative district).

Consequently, by 2008, due to the construction of the urban area, about 20% of water bodies (1/5 part) ceased to exist. 236 water bodies were noted by 2007-2008, but by 2017 fife water bodies had completely ceased to exist or were flooded to a considerable extent in the Novo-Savinovsky area, 1- in the Moskovsky District, 1- in the Soviet district. For the rest of the areas monitoring is needed to clarify the information.

Biodiversity of water objects of Kazan

The problem of destruction and degradation of small water bodies is a very acute environmental problem for the city, especially given the fact that water bodies within the city serve as conservation sites for biological diversity, ensuring sustainable development of the territories.

This is confirmed by data on biological diversity obtained during the inventory. It is revealed that the water objects of Kazan are distinguished by a significant variety of aquatic and coastal plants and animals. The specious composition varied significantly, with 190 species of phytoplankton composition, 185 species of aquatic and coastal vegetation, 204 species of zooplankton, 163 species of zoobenthos. In affition, 30 fish species, 11 amphibian species, 4 reptiles, 104 bird species and 14 mammal species were revealed. Up to 95% of the species found live in small lakes, and floodplain lakes are the most valuable for the species diversity of plants and animals.

5 species of vascular plants, 2 species of zoobenthos and 16 species of vertebrates, listed in the Red Book (recording endangered species) of the Republic of Tatarstan [8] were also revealed. The most valuable in respect of the species diversity of aquatic and coastal plants and animals are the water bodies located on the first above-flood terrace of the rivers Kazanka and Volga (Novo-Savinovsky and Privolzhsky districts).

The collected materials can serve as a basis for creating sections on rare aquatic plants and invertebrates in the regional "Red Book". Water bodies are an essential link in the conservation of biological diversity of aquatic plants and animals and need protection and improvement policies.

Practical implementation of inventory results

The results of the inventory of water bodies in Kazan have become a serious database for studying and further monitoring of the state of urban lakes, rivers and wetlands. The results of the analysis of the state of water bodies on physicochemical and hydrobiological indicators were included in the databases of the Laboratory for Optimization of Aquatic Ecosystems of the Kazan (Volga Region) Federal University.

On the basis of these results, a number of hydroecological questions were studied, including biological diversity, limnological classifications, monitoring of the state of hydrological and hydrochemical parameters [9]. This is important for controlling of anthropogenic eutrophication of lakes [10].

As a result, 173 small and very small lakes with the area from 0.01 to 129 hectares were revealed in Kazan.

The conducted inventory of water objects in Kazan ended with the development of ecological passports, registers and lists of water bodies. Due to the lack of a mechanism for the approval of environmental passports in Russia in general, environmental passports have not been approved as normative documents in Kazan so far. They exist, in fact, in the status of reporting materials with inventory results. This circumstance does not prevent, however, the protection of water bodies.

So, the results of the inventory and the presence of environmental passports helped protect 5 small lakes from development, as law enforcement agencies took into account the presence of an environmental passport as a document confirming the presence of a water body.

5. Conclusion

To conclude, the experience of the inventory of water bodies, which began from Nizhny Novgorod and Kazan, gradually spreads to other regions.

Inventory and certification of water bodies undoubtedly serve as an effective way to save small lakes and rivers. The works serve the purposes of territorial development. Particularly, they allow developing elements of the ecological framework of the territory (the river as a "core" or "green corridors" of the frame, the lake as "nuclei" and "nodes" of the natural frame) on their basis. Materials are used in the design and operation of environmental authorities.

For greater efficiency of work on inventory and environmental passportization, it is necessary to develop the procedure for approving environmental passports as normative documents at the level of the Ministry for Ecology of the Russian Federation.

References

[1] Servos M R, Munkittrick K R, Constantin G, Mngodo R, Aladin N, Choowaew S, Hap N, Kidd K A, Odada E, Parra O, Phillips G, Ryanzhin S, Urrutia R 2013 Science and management of transboundary lakes: Lessons learned from the global environment facility program *Environmental Development* **(7)** pp 17-31

[2] Gelashvili D B, Okhapkin A G, Doronina A I, Kolkutin V I and Ivanova E F 2005 *Ecological condition of water bodies in Nizhny Novgorod* (N. Novgorod: UNN) 270 p

[3] Mingazova N M, Derevenskaya O Yu 2000 The concept and methodology of the restoration of small lakes *Hydrobiological Journal* **36 (2)** p 30-42

[4] *Lakes of the Middle Volga Region* / Ed. Sorokin I N, Petrova R S 1976 (Leningrad: Nauka) 236 p

[5] Gelashvili D B, Zinchenko T D and Rozenberg G M 2007 Certification of urban water bodies (methodical recommendations) // *Exchange of intellectual property* **6(7)** P 17-21

[6] Rozenberg G S, Gelashvili D B, Zinchenko T D and Pereshibaylov L A 2001 On the environmental passportization of urban water objects *J. News of the Samara Scientific Center of the Russian Academy of Sciences* **3(2)** P 254-264

[7] *Water objects of the Kazan city. Register of water objects of Novo-Savinovsky district* 2015 /Ed. Mingazova N M (Kazan: Foliant) 116 p

[8] *The Red Data Book of the Russian Federation (animals)* RAS 2001 (Moscow: Astrel) 862 p

[9] Derevenskaya O Y, Mingazova N M, Pavlova L R 2015 Lake water quality of Kazan city (Russia) Kaban lake in the anthropogenic pollution conditions and improving actions

implementation *International Journal of Applied Engineering Research* **10(24)** pp 44682-44687
[10] Ansari AA, Gill S S, Lanza G R, Rast W *2011 Eutrophication: Causes, consequences and control* (Publisher: Springer) XIII p 394

Role of the state in solving the environmental problems of the industrial monoprofile cities

L M Musina[1] and M U Neucheva[2]

[1]The Bashkir Academy of Public Administration and Management, the Republic of Bashkortostan, Ufa, 40 Zaki Validi Street, 450008, Russia
E-mail: teacherlilia@yandex.ru
[2]The Bashkir Academy of Public Administration and Management, the Republic of Bashkortostan, Ufa, 40 Zaki Validi Street, 450008, Russia
E-mail: neucheva@mail.ru

Abstract. Nowadays the problem of sustainable socio-economic development of monotowns refers to one of the priority issues of the state policy. The author analyzes monotowns state policy support in Russia with main focus on programs aimed at ecological restoration of industrial monoprofile cities. The processes of program control in monotowns within the state economic policy are analyzed. In order to evaluate the results of programs (of city-forming enterprises and monotowns level) the principles of development of criteria development system have been substantiated. The environmental situation of monotowns depends on a complex system of interaction between the city (represented by its people and municipal authorities), private capital and the state. Long-term sustainable development of monotowns requires the interests of all three parties to be in balance. This is possible to achieve by increasing the social responsibility of businesses, increasing the development of local government and urban identity and active influence of local communities on the activities of the municipal authorities.

1. Introduction

Mono-profile towns were created in Soviet times with the command economy, when everything was predictable and scheduled. In market conditions the system of "one enterprise — one city" proved to be unstable. The cities were built for industry, not for people, this is poorly suitable for life. In order to improve the quality of life in the industrial centers the combination of social responsibility of businesses with the development of local self-government and increase of citizens' participation may be needed. This is unlikely to be effective unless the level of industrial pollution is decreased.

Every ninth Russian, a total of about 15.6 million people, lives in a monotown. There are 319 such cities in the country. At least a quarter of their population works in the city-forming enterprise. Most of the monotowns in Russia are in Kemerovo region (19), followed by Sverdlovsk (15), Chelyabinsk and Nizhny Novgorod region (12).

The main problems of monotowns is their dependence on market conditions, unemployment, environmental pollution, lack of infrastructure, including education and health, the outflow of the young population.

Modern state of the greater part of the Russian monotowns of the Urals and Siberia is characterized by extremely high level of industrialization, a predominance of nature intensive industries and obsolete technologies (raw- and semi-raw materials for export). The ecological conditions of those cities are critical and to overcome the environmental crisis is a crucial issue. Environmental problems of the regions are a consequence of the previous unsustainable policies of productive forces and desire to create the country's largest industrial enterprises at the expense of the excessive concentration of production [1].

Monotowns overcoming environmental crisis is possible under the condition of the implementation of sustainable development strategy that combines the economic, environmental and social principles. The most important issue in the transition to the sustainable development is the mechanisms of realization of ecologically oriented strategy [2]. The priority one is the economic mechanism of local administration activities, providing sufficient funding to environmental protection measures and effective influence on the improvement of the ecological situation in the cities.

2. The problem statement

The provisions on environmental protection, developed at different times by G. Motz, R. Wettmann, E. Gerelli, M. Eland, R. Cellerino, G. Panella and others [3]. Consideration of issues of economic management in the municipalities is based on the works of such researchers as B. Burkhard, F. Müller, and others [4]. The issues of improving the economic mechanism in the field of environmental protection are considered in the works of Y. Vasileva, O. Shalina, G. Tokareva, E. Baykova, and others [5]. The information base is statistical materials, regulatory documents, reporting of business entities.

3. Results and discussion

Considerable attention to the modernization of monotowns is paid by the Government of the Russian Federation and development institutions. According to the approved by the government of the Russian Federation (Order No. 1398-R) on 29 July 2014 updated list of mono-profile municipal formations in Russia include 313 monotowns in 62 constituent entities of the Russian Federation.

In order to develop proposals for the provision of monotowns of Russia with state support under the Government Commission on economic development and integration the working group on modernization of monotowns has been working since 2010; it comprises authorized representatives of Federal Executive authorities, development institutions, public organizations and other interested parties.

Aiming the economy diversification and creation of alternative work places in new and modernized industries in 2014-2016, 49 monotowns of Russia received state support for a total amount of 24 billion rubles, including the creation of objects of engineering and municipal infrastructure, stipulated by the comprehensive investment plans for the modernization of 17 billion rubles. Support was provided to 35 monotowns in 2014 and in 2015-2016.

As per assessment of "Vnesheconombank" it allowed to reduce the average unemployment rate from 2.6% to 1.8 % for the period from 31.12.2010 till 30.06.2017 and created 100 279 additional permanent work places [6].

To assess the socio-economic situation in monotowns analysis of all monotowns of the Russian Federation was carried out in 2016 with consideration of negative forecasts for the suspension (closing), the transfer of employees to part-time in a number of industries in such sectors as nonferrous and ferrous metallurgy, car industry, forestry, light industryIt involved participation of the Ministry of industrial trade of Russia, the Ministry of economic development of Russia and Ministry of labor of Russia.

The list approved in 2014, provides for the gradation of monotowns into 3 categories, including category 1, representing towns with the most difficult socio-economic situation (75 monotowns of Russia). The subparagraph "b" of clause1 of the List of Orders of the President of the Russian Federation V. Putin dated May 21, 2014 № PR-1146 provides for the provision of possible 2015-2017 state support to monotowns included in category 1 of the List.

There are 6 single-industry municipalities (towns) in the Republic of Bashkortostan in accordance with the Decree of the Government of the Russian Federation dated July 29, 2014 No. 1398-R. At 1st

January 2017 there are more than 400 thousand people living there, which represents almost 10% of the total population of the Republic:

Belebey and Kumertau, with the most difficult socio-economic situation (the first category); Neftekamsk Beloretsk with the risk of deterioration of the socio-economic status (second category); Uchaly and Blagoveshchensk with a stable socio-economic situation (the third category) [7].

The objectiveness of the analysis of socio-economic situation of the above-listed municipalities raises doubts about the conducted classification and justification for the exclusion of certain municipalities from the list of monotowns.

So, the inclusion of the city of Uchaly in the number of monotowns with a stable socio-economic situation is very conditional. In the industry structure 50,1% of the total volume of production is accounted for the chemical industry represented by OJSC "Uchalinskiy Azot" and JSC "Arnas". This indicates that the city's economy continuing to be mono-profile. Sectoral structure of the economy of the city along with the passage of the P316 route through the city causes the unfavorable situation in the city. 1,204 tons of solids 0.037 tons sulfur dioxide, 3,976 tons of carbon monoxide and 4,705 tons of nitric oxide were released into the atmosphere in 2016 in the city of Uchaly. The general level of atmospheric pollution is assessed as high and is determined by the value of the standard index equal to 2,775.

The result is an incidence rate. The level of general child incidence of disease in the city of Uchaly in the last 6 years has grown by 48.6 %, among teenagers - by 60.2 %. Moreover, among the environmentally caused diseases there are respiratory, oncological pathology, neoplasms, congenital anomalies, diseases of the skin, subcutaneous tissue and diseases of the digestive system. The presence of these diseases exceeds regional indicators by 1.7 and 1.3 times in Uchaly.

Another point indicating the instability of the situation in the city is the decline over the last 10 years in the population of the city by 11 %. Thus, according to the social indicators, the city belongs to the category of high social risk and needs government support. This means that when considering issues on state support of monotowns attention should be paid not only to the development of relevant industries, but also to the social sphere.

The city of Blagoveshchensk of the Republic of Bashkortostan was excluded from the new list of monotowns; the share of the petroleum and petrochemical industries in the economy exceeds 50 % [8].

Solving many problems of monotowns of Russia is seen through the activities of the Fund development of monotowns by the Government. It will operate under the auspices of "Vnesheconombank". About 25 billion rubles from the Federal budget will be allocated for the implementation of investment projects in monotowns of the Russian Federation in 2015-2017. Official media reported that the Fund will be established as a nonprofit organization, in which "Vnesheconombank" will make a starter asset contribution of 16.4 million RUR.

To help the problem of single-industry territories funds will be granted on the following conditions: up to 90% of financing is ready to be provided by the Fund with the condition that the remaining 10% will be allocated by the regional or municipal budgets. In addition, the Fund of development of monotowns is going to prepare the investment projects documentation and will also provide training of management teams [9].

It is not really clear how to find the funds in the conditions of high subsidization of a number of regions and municipalities. The actual volume of expenditures of regional budgets per capita in rubles in the Republic of Bashkortostan 1,5–2 times lower than the national average level. Expenses on national economy needs, in 2016 were, respectively, 16.8 and 17.9% of costs of the regional budget. It should be noted that this subsection includes the costs on transportation and a number of areas necessary for proper functioning of the regional economy.

Municipal educational organizations are even more limited in spending their budgets.

The conducted analysis gives the grounds to formulate recommendations for improving state regulation policy of development of monotowns. Special attention should be paid to attracting private investment. This should expand the responsibility area of regions regarding this issue. Regional lists of monotowns taking into account the specifics of the socio-economic situation in situ must be created to develop the legal framework outlining the status of monotowns in the regions. For example, the regional law on the status of monotowns was already developed and adopted in the Kemerovo region. However, this document poorly spells out specific measures of state support of monotowns. It is necessary to develop regional programs promoting the development of monotowns in the regions of their presence.

It is expedient to consider foreign practices in this field for the rational organization of Russian monotowns development. The experience of Germany is of significant interest in reorganizing monotowns. It should be noted that the development of comprehensive plans of modernization of monotowns in Germany often involves large consulting firms, with significant analytical and organizational capabilities, as well as authority in state bodies and financial circles. AS&P (albert Speer & Partners) dealing with issues of restructuring of monotowns for several decades can be referred to such large structures. The main stages of restructuring of monotowns used by the Corporation AS&P are as follows [9]: the study of a particular situation (the spectrum of issues of a monotown); comprehensive analysis and assessment of all factors; study of international experience in development and modernization of similar monotowns; the development strategy of the city; the determination of a set of short-, medium - and long-term measures on modernization of a monotown; the determination of the amounts and directions of targeted subsidies and grants; development of a system of state-private partnership in the field of modernization of a monotown. After substantiation of such a comprehensive reform program of a monotown, it is further transferred to the State. It goes through wide public discussion and after its adoption, AS&P is responsible for its implementation at all stages of realization. The most successful project of the company AS&P in the field in Germany is Ruhr, transformation from a depressed coal-mining city into diversified post-industrial city.

In accordance with the developed plan for 1980-2002 the number of employed in the service sector increased by 300 thousand people. At the same time new mobile and competitive businesses in the car industry, precision mechanics industry, and electrical engineering industry were created. Conditions for accelerated development of telecommunications and information technology were created in this direction, this reoriented educational programs of the Ruhr Region. As a result, by 2009 the Ruhr was not a single-industry town anymore. Another successful project is Frankfurt — Gerst. Until the end of 1990 it was a depressed industrial area. In the 2000s it was a small town and a national industrial park was established, focused on the development and commercialization of advanced technologies, development of technical and technological basic and R&D Projects. As a result, 22 thousand jobs were established within the industrial Park. The scientific town of Agneshof (a suburb of Berlin) is also an example. On its territory there was previously located a military base and a military airfield. At the end of the conversion of the territory science and technology park, tele- and film- studio were established, and also the buildings of the Humboldt University were placed. Now there are about 800 firms, including 14200 employed on the territory of the town.

The problem of functioning and development of monotowns substantially escalated in the early 1980s in the UK, in the areas heavily dependent on coal and steel industry. Within the 1981-1997 period, 124 coal mining were closed, resulting in a 193 000 workers losing their jobs and thus, 25% of men out of work [6]. The main problems that existed in single-industry UK cities were: significant levels of environmental pollution and, consequently, less satisfactory state of health of people living in these towns comparing to the average for the country; internal migration. A comprehensive solution of the problem of monotowns in the UK included the construction of new houses (including social housing). It also included the formation of efficient transport infrastructure with the aim of increasing mobility of the population, giving them a chance to work in neighboring towns territories; long-term consistent modernization of coal mines and steel plants; encouraging the establishment of new forms of economic activity (for example, organization of the famous music festival). There is a form of public-private partnership in the field of modernization of single-industry towns, as the formation of "urban development corporations". As an example of a single-industry city we use Cardiff Bay, Wales [6]. After the Second World War the coal industry gradually degraded, and the area went into decline. In 1987 the Cardiff Bay Development Corporation was founded. The aim was to create the structure to attract private investment in the development of a monotown, by 1997 Cardiff Bay became a rapidly developing leading center of tourism and leisure of the province of Wales.

We believe that with regard to the Republic of Bashkortostan, the creation of regional industrial parks may be the effective mechanism of support of monotowns at the regional level. Those considered promising institutions for attracting private investment. A key direction of improvement of state policy of support of monotowns should be the extension of the degree of involvement of regions in these processes in order to maximize the multiplier effect of the investment of public funds. The number of industrial parks as institutions for attracting private investment at the regional level should be increased. Residents of industrial parks should have the rights to have tax incentives and administrative preferences, the subsidization of interest on loans, lease incentives, etc.

Another important area of economic diversification in monotowns is the development of small businesses. The low level of entrepreneurial activity in monotowns of the Republic of Bashkortostan can be explained by low entrepreneurial competence. Most of the implemented programs of support of businesses are formal. It is necessary that these programs act as a platform for interaction between employers, governments and financial institutions.

4. Conclusion

Suggestions and recommendations based on the analysis of the results of implementation of state support measures of monotowns in the Republic of Bashkortostan and can be universally applicable in Russia:

- intensification of support not only of economic development of monotowns, but also of the social sphere;
- active environmental policies in monotowns;
- the possibility of replacing the mandatory co-financing by the regions or municipalities in the amount of 10% of the Fund;
- development of regional programs of monotowns, including further dissemination and development of the practice of creating industrial parks;
- implementation of practice-oriented educational programs on training of personnel for small and medium businesses on the design approach basis.

References

[1] Kryukova E M, Razumovskiy S L and Vetrova E A 2013 Mono-Town in the System of Economic Notions of the Russian Federation *World Applied Sciences Journal (Education, Law, Economics, Language and Communication)* **27** pp 162-166

[2] ESAWADI 2013 Utilizing the ecosystem services approach for water framework directive implementation synthesis report work package 5: synthesis and policy recommendations. ESAWADI project. http: //www.esawadi.eu/IMG/pdf/ESAWADI_European_Synthesis_ENG_vf.pdf (Accessed 27 April 2017)

[3] Motz G, Wettmann R, Gerelli E, Eland M, Cellerino R and Panella G 1988 Environment and regional development in less favoured/declining industrial regions. Preparatory study for an ERDF-environment programme. *Final report to the Commission of the European Communities. Directorate General for Regional Policy* (Basle) p 419

[4] Burkhard B, Müller F 2008 Indicating ecosystem health and integrity *Sustainability In River Basins. Oekom-Verlag* (München) pp 35-58

[5] Vasileva Y, Shalina O, Tokareva G and Baykova E 2017 Economic Growth in Russia: Influence of Oil Shock and Macro-Economic Factors *Integration and Clustering for Sustainable Economic* Growth (Switzerland: Springer) p 553-557

[6] De Groot R S, Alkemade R, Braat L, Hein L, & Willemen L. 2010 Challenges in integrating the concept of ecosystem services and values in landscape planning, management and decision making *Ecological complexity* **7(3)** pp 260-272.

[7] Díaz S, Fargione J, Chapin F and Tilman D 2006 Biodiversity loss threatens human well-being *PLoS biology* ***4*(8)** p 277.
[8] Arrow K J, Cropper M L, Eads G C, Hahn R W, Lave L B, Noll R G ... and Stavins R N 1997 Is there a role for benefit-cost analysis in environmental, health, and safety regulation? *Environment and Development Economics* pp 195-221
[9] García-Llorente M, Iniesta-Arandia I, Willaarts B A, Harrison P A, Berry P, Castro A J ... and Del Mar Bayo M 2015 Biophysical and sociocultural factors underlying spatial trade-offs of ecosystem services in semiarid watersheds *Ecology and Society* **20(3)**:39

Influence of unreasoned economic activity on the condition of macrophytes of the Bol'shoye Goluboye Lake

O V Palagushkina, N R Zaripova, N M Mingazova and T O Yarutkin

Kazan Federal University, Institute of management, economics and finance, Kremlevskay str.,18, 420008, Russia
E-mail: opalagushkina@mail.ru

Abstract. The ecosystem of Lake Bolshoye Goluboe had undergone a strong anthropogenic impact in 2013 as a result of the implementation of the dam reconstruction project. Studies in 2014 have shown that the implementation of the project for the reconstruction of the Bolshoye Goluboe dam has negatively affected on the species richness of macrophytes. The total species composition of the lake and species richness of the water core decreased twofold, *Hippuris vulgaris* L., *Zannichellia palustris* L, *Ceratophyllum demersum* L., and the species listed in the Red Book of the Republic of Tatarstan - *Batrachium circinatum* (Sibth.) Spach disappeared from the species composition. The area occupied by macrophyte communities has decreased by 55%.

1. Introduction

Unique natural objects require careful treatment in the implementation of measures for their improvement and verified environmentally decisions. Bolshoye Goluboe Lake, which is characterized by a special blue color of water, refers to the type of "blue", karst cold-water sulphate lakes are extremely rare for the Middle Volga Region. Special features are also the properties of the Great Blue Lake therapeutic sapropel silt. Silts of the lake used to Kazan sanatoriums [1,2].

2. Study site and the problem statement

Lake Bolshoye Goluboe is located in the Vysokogorsky district of the Republic of Tatarstan and is part of the State Nature Reserve of the regional importance "Blue Lakes". The name of the lake is associated with the blue color of water combined with high transparency. By origin lake is oxbow-karstic with complex elongated shape, is located in the rear part of the floodplain terrace of the River Kazanka. The lake's area is 4.60 hectares, the maximum depth is 15.7 m, and the water volume is 59.8 thousand m^3. Underground power of Bolshoye Goluboe Lake is due to the discharge of deep aquifers Permian two sinkholes located in its northern part. The chemical composition of the water in the lake is neutral, very hard, sulfate-calcium with a mineralization about 2.2-2.5 g / l (brackish).

The water level in the lake is 5 m above the water's edge in the Kazanka River to the low water, the width of the jumper separating the lake and the river in the narrowest part is 20 m. The lake does not have permanent tributaries, but from it flows a stream that flows into the Kazanka River 10.6 km from the mouth. The water flow in the stream varies from 0.6 to 0.9 m^3 / s by seasons. The main type of modern anthropogenic impact on Lake Bolshoye Goluboye is recreation. Also in the coastal zone of the lake, there is an irregular grazing of livestock, which introduces about 126 kg / year of nitrogen and 13 kg / year of phosphorus [2,3,4].

In November 2013, the ecosystem of the Lake Bolshoye Goluboe had undergone strong anthropogenic impacts resulting from the project for the reconstruction of the dam that separates the lake from the river Kazanka and strengthen the banks drain the water from the lake. At the time of the implementation of the project, old dam, that maintained the water level in the lake destroyed, in the place below the drain (Crystal Waterfall), the excavator sank. The flow of water from the lake was reoriented into a new ditch with unsettled shores and without a retaining dam. As a result of a temporary change in the hydrological regime, the level of water in the lake dropped sharply, with water retained only in the karst volcanoes (Bolshaya and Malaya Puchini), through which groundwater is discharged. Sapropel silt in winter conditions froze and dried up. After the reconstruction of the dam, the water level in the lake restored, but the water drain led to the disruption and disappearance of the habitats of many species of aquatic plants and animals, to the loss of large amounts of sapropel silt. In connection with this, the main goal of the research was to identify the current state of macrophytes in Lake Bolshoye Goluboe in comparison with investigations of previous years.

3. Results and discussion

Detailed studies of the vegetation of Lake Bolshoye Goluboye, conducted in 1998-2000, attributed it to the macrophyte type [5,6].

Production of organic matter of such reservoirs is created mainly due to the photosynthetic activity of immersed plants characterized by rapid growth, intensive metabolism, increased requirements for the content of minerals in water and soils and represented by charofities and green algae, green mosses, and Hippúris vulgáris and other species. The formation of vegetation of brackish karst lakes affects not only natural factors, such as the transparency and salinity of the waters, but also anthropogenic eutrophication. Moderate anthropogenic eutrophication and high transparency of water favored the development of submerged vegetation of the lake Bolshoye Goluboye to 2013, the lake characterized by underwater meadows of green water plants and mosses and charophytes [2,7].

Research in 1998-2000 allowed identifying in the Bolshoye Goluboye Lake 29 species of macrophytes of 18 families. Hydrobotanical studies were carried out using conventional methods. The investigated plants were determined at the site of collection and laid in the herbarium for further clarification. Only dominant species were considered. In the course of the road surveys, the boundaries of the location of the associations were visually marked, and they were recorded on the map charts. The evaluation of the degree of overgrowth was given by the classification of V.M. Katanskaya [5]. Such special hydrophysical properties of the lake as cold water, high transparency, fluidity of waters, increased mineralization caused the presence of communities of stenobiontic species in it. Among the higher vascular plants, the communities of submerged attached, tall and small-leaved plants with dominants *Myriophylum verticulatum*, *Potamogeton pectinatus*, *Hippuris vulgaris*, *Utricularia vulgaris* were most common. Were widely distributed, but occupied small areas of the community, formed by representatives of the family *Lemnaceae*. The important role-played charophytes - in autumn happening death of most of the vascular plants and charophytes continued to actively vegetate and bear fruit, often replacing edificator species. Floating and submerged algal clusters formed by macroscopic green filamentous algae from the genus *Cladophora* played a large role. As a species-edificator, forming almost monodominant communities along the slopes of karst funnels of Lake Bolshoye Goluboye, the water moss *Fontinalis antipyretica* appeared. In the lake, there was a lack of a vegetation belt with floating leaves. The belt of wetland vegetation, located along the coast, was weakly expressed. 12 species of the "water core" were identified for Lake Bolshoye Goluboe. In the reservoir, there was an uneven type of overgrowth with a degree - excessive, more than 70% [2].

Geographical and ecological analysis of the flora showed that the flora of brackish-water karst lakes is composed of various geographic elements with a predominance of species with a wide range wide of distribution area, which indicates the relative youth of this flora and its immigration character.

The geographical structure of the flora of the "water core" of Lake Bolshoye Goluboe is typical for cold-water lakes in the central part of Russia. Among the higher vascular plants, communities of immersed, tall and small-leaved plants with dominants *Myrophyllum spicatum, Potamogeton pectinatus, Hippuris vulgaris, Urticularia vulgaris*. Quite widespread, but occupy small areas of the community, formed by representatives of the family *Lemnaceae*.

Great is the role of floating and submerged algal clusters, which are often the edificators of communities. In autumn, when the majority of vascular plants die, the cows continue to actively vegetate and bear fruit, often replacing the edifying species. In many brackish water mineralized reservoirs, the role of edificator belongs to water moss Fontinalis antipyretica, having a wide ecological amplitude and form monodominant or mixed communities on the slopes of karst holes and craters. Especially noted the absence in such lakes of the vegetation belt with leaves floating on the surface. Belt of wetland vegetation along the coasts is often poorly expressed [8,9].

Species of the "water core" in 1998-2000: *Lemna minor* L., *Lemna trisulca* L., *Potamogeton pectinatus* L., *Zannichellia palustris* L., *Ceratophyllum demersum* L., *Myrophyllum spicatum* L., *Hippuris vulgaris* L., *Batrachium circinatum* (Sibth.) Spach., *Cinclidotus fontinaloides* (Hedw.) P.beauv., *Fontinalis antipyretica* Hedw., *Cladophora glomerata* (L.) Kutz., *Chara contraria* A.Br.

Species that exist only in brackish-water karst lakes in the Middle Volga Region probably do not exist, but they form a plant complex with species with intensive growth, metabolism, and a high demand for mineral content in water and soils [1].

Summer studies in 2014, conducted according to the standard procedure [10], found only 14 species of macrophytes belonging to 10 families. The vascular plants represented by the families *Alismataceae, Lemnaceae, Potamogetonaceae, Halogaraceae, Ranunculaceae, Solanaceae, Umbelliferae* and were mainly confined to the belt of wetland vegetation along the shores. Charophytes algae were represented by the species *Chara contraria* A.Br. and formed a small area community in the southern part of the lake. Green filamentous algae of the genus *Cladophora* formed clusters in the coastal shallow water zone. Amblystegiaceae and Fontinaliaceae mosses were noted on the slopes of funnels.

According to the 2014 "water core" species of the lake composed of 6 species: *Lemna minor* L., *Lemna trisulca* L., *Potamogeton pectinatus* L., *Myrophyllum spicatum* L., *Fontinalis antipyretica* Hedw., *Chara contraria* A.Br.

The degree of overgrowth of the lake bottom by macrophytes was no more than 15%.

The similarity index of the macrophyte species composition of the two study periods was 65.1% [11,12]. Thus, the conducted studies showed that the implementation of the project for the reconstruction of the dam of Lake Bolshoye Goluboye negatively affected on the species richness of macrophytes.

4. Conglusion

The total species composition of the lake and species of the water core decreased twofold, *Hippuris vulgaris* L., *Zannichellia palustris* L, *Ceratophyllum demersum* L., and the species listed in the Red Book of the Republic of Tatarstan [13] *Batrachium circinatum* (Sibth.) Spach disappeared from the species composition. The area occupied by macrophyte communities has decreased by 55%.

Taking into account that the project of dam and drain reconstruction passed state ecological expertise, it is obvious that the environmental consequences during the examination were not taken into account.

Acknowledgement

Research in 1998-2000 were conducted with the support of the federal target program “Integration”.

References

[1] Aristov V, Marikhin V 1981 The Blue Lake *Soviet Tataria*
[2] *Unique ecosystems of brackish-water karst lakes of the Middle Volga* region Ed. Alimov A F, Mingazova N M 2001 (Kazan: Kazan University) p 256

[3] Mingazova N M and Monasypov M A 2003 Abiotic characteristics of the brackish water karstic lakes of middle volga region and their connection with limnogenesis *Environmental radioecology and applied ecology Kazan* **9(3)** pp 11–17
[4] Mingazova N M, Derevenskaya O Y, Palagushkina O V, Pavlova L R, Barieva F F and Monosipov M A 2005 The typology and biological diversity of karst lakes in the Middle Volga region *Mamagement of Lake Basins for their Sustainable Use: Global Experience and African Issues of 11th World Lakes Conference. Abstracts Volume* Nairobi Kenya p 95
[5] Pokrovskaya T N, Mironova N Ya and Shilkroth G S 1983 *Macrophytic lakes and their eutrophication* (Moscow: Nauka) p 152
[6] Mingazova N, Aladin N, Ivanova M, Golubkov S, Derevanskaya D, Palagushkina O, Unkovskaya E, Plotnikov I, Piriulin D, Dgakova L, Smurov F, Pavlova L, Barieva F, Monasipov M and Nazarova L 2005 Features of the structuralle-functional organization and evolution ecosystems of unique brackish karstic lakes (Russia) *Proceedings of the 11th World Lake Conference* (Nairobi Kenia) **2** pp 363-367
[7] Mingazova N M 2003 Middle Volga Lakes of Russia: Typology, Biodiversity, Ecological problems and Restoration Opportunities *Global Threats of Large Lakes: Managing in an Environment of Instability and Uppredisstability. Abstracts of 46th Conference on Great Lake Research, 10th World Lake Conference* (DePaul University Chicago Illinois USA) p 241
[8] Katanskaya V M 1981 *Higher aquatic vegetation of continental reservoirs of the USSR* (Leningrad: Nauka) p 187
[9] Khomyakova I M 1941 Vegetation of the Lake Goluboe *Uchenye zapiski Kazanskogo universiteta* **101 (3)** pp 5-6
[10] *Methods of studying of biogeocenoses of inland water bodies* 1975 Moscow: Nauka 240 p
[11] Papchenkov V G and Solovyova V A 1995 Analysis of the flora of the ponds of the Middle Volga region *Botanichesky zhurnal* **80(7)** pp 59-67
[12] Sorensen T 1948 A new method of establishimg groups of equal amplitude in plant sociology based on similarity of species content and its application to analysis of vegetation on Danish common *Kgl. Dan. videnskab. selskab. Boil .skr* **5(4)** pp 1-34
[13] *Red Book of the Republic of Tatarstan (animals, plants, mushrooms)* 2006 Second edition (Kazan: Idel-Press) p 832

IOP Conf. Series: Earth and Environmental Science **107** (2017) 012134 doi:10.1088/1755-1315/107/1/012134

Economic regionalization and choice of strategic development directions of municipalities of the Republic of Tatarstan

M V Panasyuk

Kazan (Volga region) Federal University,
18 Kremlevskaya Str., Kazan 420008 Russia
E-mail: mp371@yandex.ru

Abstract. This paper shows the results of economic regionalization and zoning of the Republic of Tatarstan, conducted in 2017. The latest experience of economic regionalization and zoning of the Republic of Tatarstan in 2007 – 2015 is exposed. The economic regionalization problem is solved on the basis of new method and algorithm that uses quantitative measures which characterize spatial and economic features of generated economic regions including their internal and average connectivity, homogeneity, compactness, socio-economic development level and life quality of the population. Three nodal and one homogeneous economic region in the Republic of Tatarstan were identified. The results of economic zoning within homogeneous economic region led to the conclusion about two existing economic zones. They have the potential for developing new economic growth pole and three economic centers – growth points with specialization on agro-industrial sector.

1. Introduction

Economic regionalization and economic zoning are designed to evaluate the structure and organization of regional economic space. Economic region (ER) is determined as the territorial part of the economy, characterized by unity of territory and economics, internal economic interconnectivity and existing economic specialization [1, 2]. Economic zone (EZ) is an area within which there is a similar intensity of performance of economic characteristics [3]. The economic region differs from the economic zone, first of all, by the fact that it is formed as a result of the development of socio-economic relations between economic entities. For this reason, when economic regionalization is performed, special attention is paid to the analysis of the commodity, human, information and energy flows, etc. that requires development of special approaches and methods for solving problems of regionalizing.

In economic geography types of nodal and homogeneous economic regions are determined. Nodal region has one or more centers (nodes) which integrate the part of regional economic space. The territory of homogeneous region is characterized by internal economic homogeneity and absence of significant differences between the indicators of economic subjects' activity. Homogeneous region does not have conspicuous centers of territorial development. However, they can be in the initial phase of their genesis, not manifesting themselves at the present time.

2. Experience in economic regionalization of the Republic of Tatarstan

Problems of economic regionalization and zoning of the Republic of Tatarstan (RT) have repeatedly been in focus of the regional authorities and scientists since the late 1920-ies. Among the most important studies in the field of economic regionalization and zoning the RT territory which were obtained in recent years, there must be marked:

I. The Concept of Territorial Economic Policy of the Republic of Tatarstan, established by Department of Territorial Development of the Economics Ministry of the Republic of Tatarstan in 2009. It has identified six economic zones [4]:

1. Kazan agglomeration. It includes the urban district of Kazan and six municipal districts: Verkhneuslonsky, Vysokogorsky, Zelenodolsky, Laishevsky, Pestrechinsky and Atninsky.
2. Predvolzhskaya economic zone. It includes six municipal districts – Apastovsky, Buinsky, Drozhzhanovsky, Kaybitsky, Kamsko-Ustyinsky and Tetyushsky.
3. Predkamskaya economic zone which includes seven municipal districts – Arsky, Baltasinsky, Kukmorsky, Mamadyshsky, Rybno-Slobodsky, Sabinsky and Tyulyachinsky.
4. Zakamskaya economic zone. It includes seven municipal districts – Alkeevsky, Alekseevsky, Aksubaevsky, Novosheshminsky, Nurlatsky, Spassky and Chistopolsky.
5. Naberezhnochelninskaya agglomeration which includes Naberezhnye Chelny urban district and nine municipal districts - Agryzsky, Aktanyshsky, Menzelinsky, Muslyumovsky, Mendeleevsky, Elabuzhsky, Nizhnekamsky, Tukaevsky and Zainsky.
6. South-East economic zone. It includes 8 municipal districts - Almetyevsky, Aznakaevsky, Bugulminsky, Bavlinsky, Leninogorsky, Sarmanovsky, Cheremshansky and Yutazinsky.

II. The "Development and Allocation of Labor Forces of the Republic of Tatarstan on Basis of Cluster Approach till 2020 and for the Period till 2030" Program, which was developed in 2007. Seven economic regions were identified in this program (Figure 1):

1. Metropolitan region (Kazan, Zelenodolsk cities, and Atninsky, Verkhneuslonsky, Vysokogorsky, Zelenodolsky and Laishevsky Pestrechinsky municipal districts).
2. Kamsky region (Naberezhnye Chelny city, and Nizhnekamsky, Elabuzhsky, Mendeleevsky, Zainsky and Tukayevsky municipal districts).
3. Western region (Apastovsky, Buinsky, Drozhzhanovsky, Kaibitsky, Kamsko-Ustyinsky and Tetyushsky municipal districts).
4. Northern region (Arsky, Baltasinsky, Kukmorsky, Mamadyshsky, Sabynsky, Rybno-Slobodsky and Tyulyachinsky municipal districts).
5. Southern region (Alkeevsky, Alekseevsky, Aksubaevsky, Chistopolsky, Novosheshminsky, Nurlatsky and Spassky municipal districts).
6. Oil region (Almetyevskи, Bugulminsky, Leninogorsky, Sarmanovsky and Cheremshansky municipal districts).
7. Eastern region (Agryzsky, Aktanyshsky, Menzelinsky, Muslyumovsky, Aznakaevsky, Yutazinsky and Bavlinsky municipal districts).

III. The Strategy of Socio-Economic Development of the Republic of Tatarstan till 2030 (the Strategy 2030) [5]. Economic zones and economic regions do not differ in this Strategy. Economic zoning was undertaken and there were identified three economic zones – Kazanskaya, Kamskaya and Almetyevskay (Figure 2, Table 1).

Analysis of the existing economic regionalization experience of the Republic of Tatarstan territory highlighted its imperfections caused by improper level of validity. The necessity of organization of multilevel strategic territorial development system in the Republic of Tatarstan, which must include different territorial levels from individual settlements, municipal districts, cities, agglomerations and up to the regional level, caused the development of the quantitative approach based on the systems of economic and spatial indices and criteria. It let to formulate the quantitative framework for finding optimal variant of economic regionalization.

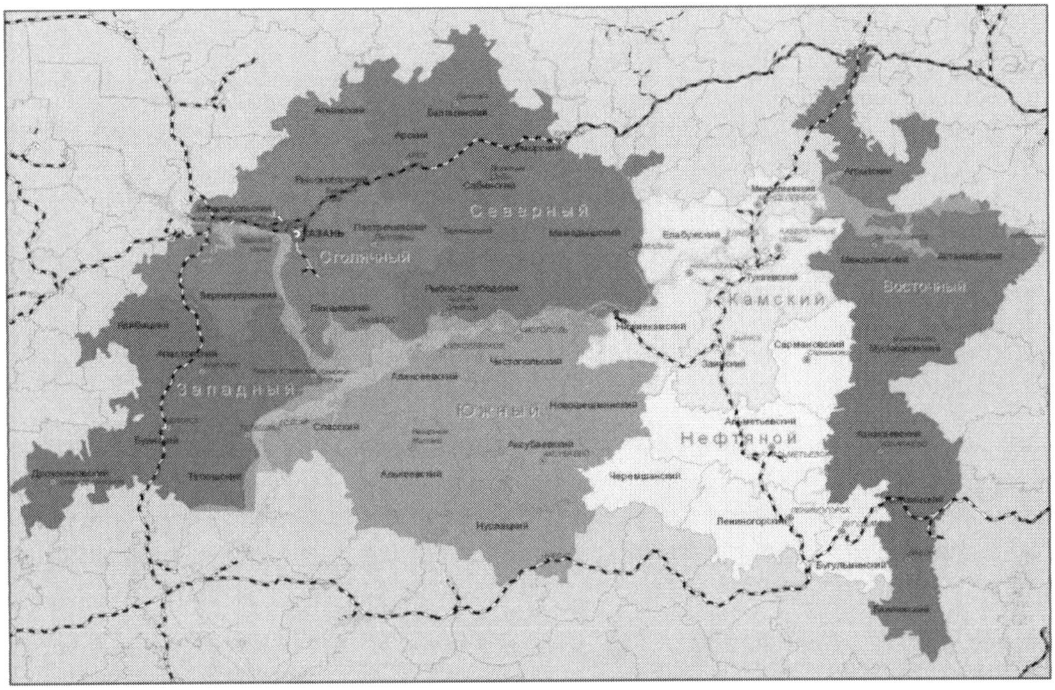

Figure 1. Economic regions of the Republic of Tatarstan in the ""Development and Allocation of Labor Forces of the Republic of Tatarstan on Basis of Cluster Approach till 2020 and for the Period till 2030" Program.

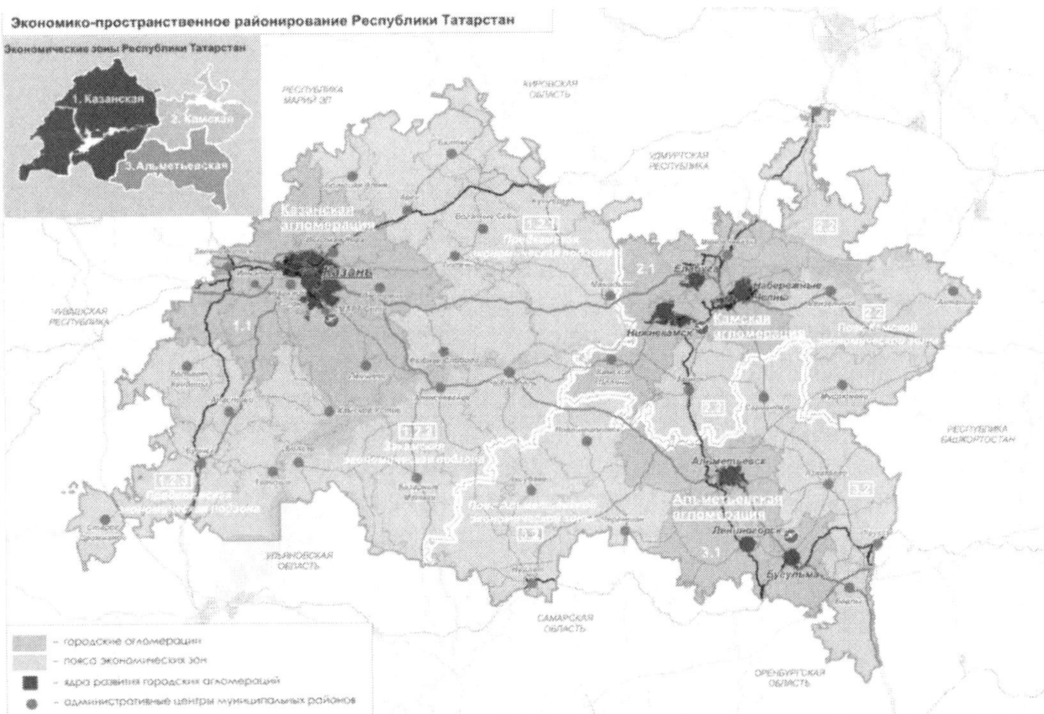

Figure 2. Economic zoning of the Republic of Tatarstan in its Strategy 2030.

IOP Conf. Series: Earth and Environmental Science **107** (2017) 012134 doi:10.1088/1755-1315/107/1/012134

Table 1. Economic zones of the Republic of Tatarstan in the Strategy 2030.

1. Kazan Economic Zone			
1.1. Kazan urban agglomeration	*Bets of \Kazan Economic Zone*		
	1.2.1 Predkamskaya economic subzone	*1.2.2. Zakamskaya economic subzone*	*1.2.3. Predvolzhskaya economic subzone*
Kazan	Arsky municipal district	Alekseevsky municipal district	Apastovsky municipal district
Zelenodolsky municipal district	Rybno-Slobodsky municipal district	Chistopolsky municipal district	Kamsko-Ustyinsky municipal district
Vysokogorsky municipal district	Tyulyachinsky municipal district	Spassky municipal district	Buinsky municipal district
Pestrechinsky municipal district	Sabinsky municipal district	Alkeevsky municipal district	Kaibitsky municipal district
Laishevsky municipal district	Kukmorsky municipal district		Tetyushsky municipal district
Verhneuslonsky municipal district	Baltasinsky municipal district		Drozhzhanovsky municipal district
	Mamadyshsky municipal district		
	Atninsky municipal district		

2. Kama Economic Zone	
2.1. Kama urban agglomeration	*2.2. Belt of emerging impact of Kama Economic Zone*
Naberezhnye Chelny city	Agryzsky municipal district
Nizhnekamsky municipal district	Aktanyshsky municipal district
Elabuzhsky municipal district	Muslyumovsky municipal district
Tukaevsky municipal district	Menzelinsky municipal district
Mendeleevsky municipal district	Zainsky municipal district

3. Almetyevsk Economic Zone	
3.1. Almetyevsk urban agglomeration	*3.2. Belt of Almetyevsk Economic Zone*
Almetyevsky municipal district	Aznakaevsky municipal district
Bugulminsky municipal district	Aksubaevsky municipal district
Leninogorsky municipal district	Bavlinsky municipal district
	Novsheshminsky municipal district
	Nurlatsky municipal district
	Sarmanovsky municipal district
	Cheremshansky municipal district
	Yutazinsky municipal district

3. Methodology

The solution to the problems of economic regionalization of the Republic of Tatarstan territory was based on the author's approach and method [6], as well as on the results of research in the field of identification and analysis "economic center – economic periphery" system of the Republic of Tatarstan.

By reference to the definition of economic region as integrated spatial formation characterized by close internal economic interrelatedness, economic regionalization in the first place should be based on the analysis of the degree of development of socio-economic relations between territorial units included in the region. Therefore, the first stage of economic regionalization is identification of the nodal regions system. For this, a search of territories with high level of interrelation (cohesion) between regional economic centers (nodes) and their peripheral parts [7, 8] is undertaken.

The next stage of economic regionalization is appropriate if there are territorial units that were not included in nodal regions system, i.e. if they are economically weakly associated with their centers. On the basis of their totality, a system of homogeneous regions is formed. The crucial factor and criterion of economic regionalization in this case is the similarity/difference of macroeconomic indicators of socio-economic development level and life quality, spatial differentiation of economic specialization, etc.

Proceeding from this, the method and algorithm of regionalization includes two main stages:

1) Identification of nodal economic regions system. It is based on the analysis of development of socio-economic relations, i.e. the internal economic connectivity of territories of economic regions, as well as their relative economic homogeneity and compactness.

2) Identification of homogeneous economic regions system. It is based on the analysis of degree of their internal economic homogeneity, i.e. on similarities in values of macroeconomic indicators, as well as on compactness of identified homogeneous regions.

To identify the nodal regions system the quantitative measure (coefficient) of connectivity that evaluates the ratio (structure) of economic relations of the municipal district with economic centers was developed. To identify the composition and the borders of nodal region there was used a threshold value of the measure of connectivity, which specifies the lower bound of connectivity, from which municipal district could be included in the economic region. Increasing threshold value during economic region forming specifies the conditions of reduction of composition and size of an economic region.

In the same way a threshold of inner homogeneity of economic region is given. It defines the value below which the quantitative measure of homogeneity, characterizing the degree of macroeconomic indicators similarity of municipal districts that form a part of emerging economic region, could not be decreased. To evaluate homogeneity of groups of municipal regions the mathematical methods of grouping/classification are used.

The measure of compactness allows estimating relative compactness of generated nodal and homogeneous economic regions, avoiding generation variants of regionalization with complex contour boundaries and not enough compact territory.

The combination of threshold values of the measures of connectivity, homogeneity and compactness of economic regions leads to formation of individual variants of economic regionalization. Average connectivity, average homogeneity and average compactness of the nodal economic regions, average homogeneity and average compactness of homogeneous economic regions are calculated for each variant of regionalization.

A comparison of these indicators for which a higher value describes better regionalization variant specifies objective basis for quality assessment of regionalization and zoning, and allows choosing optimal variant of economic regions system.

4. Results

Economic regionalization problem solution was based on the expert estimation system for analysis of commodity and human flows, and given main macroeconomic indicators characterizing behavior of the socio-economic systems of municipal districts in the Republic of Tatarstan in the analyzed period.

Analysis of economic relations was carried out on the basis of the expert estimation system of commodity and human flows between the municipal districts and centers of forming nodal economic regions, including push-pull labor migration. Determining the values of points was carried out in view of volume of economic activity, population, distances between settlements and the quality of connecting transport routes.

Three main economic centers of the Republic of Tatarstan were selected as centers of nodal regions – Kazansky (with cities of Kazan and Zelenodolsk), Kamsky (cities of Naberezhnye Chelny, Nizhnekamsk, Elabuga), South-East (cities of Almetyevsk, Bugulma and Leninogorsk).

Analysis of economic homogeneity/heterogeneity of economic regions and zones was conducted using 35 main macroeconomic indicators, including gross regional product, value added, investment in fixed capital, gross agricultural product, turnover of small and medium enterprises, number of employed in economy, population incomes, level of registered unemployment, etc. All indicators were analyzed in dynamics from 2014 to 2016 and on the forecast period in 2017-2019.

For solving problems of economic regionalization and zoning of the Republic of Tatarstan, R language and its statistical packages were used, with the help of which the software package that provides the implementation of relevant calculations and output the graphic results was developed.

The results of the optimal variant of economic regionalization of the Republic of Tatarstan were chosen by calculations of average connectivity, homogeneity and compactness of economic regions. They are presented in figure 3 (graphic output of the regionalization and classification software package). It is characterized by high value of average connectivity (8.0) along with highest average of homogeneity of the nodal and homogeneous regions (0.945, with a 0.19 and 1.70, respectively, for nodal and homogeneous economic regions) and relatively high degree of compactness of regional territories among all calculated variants.

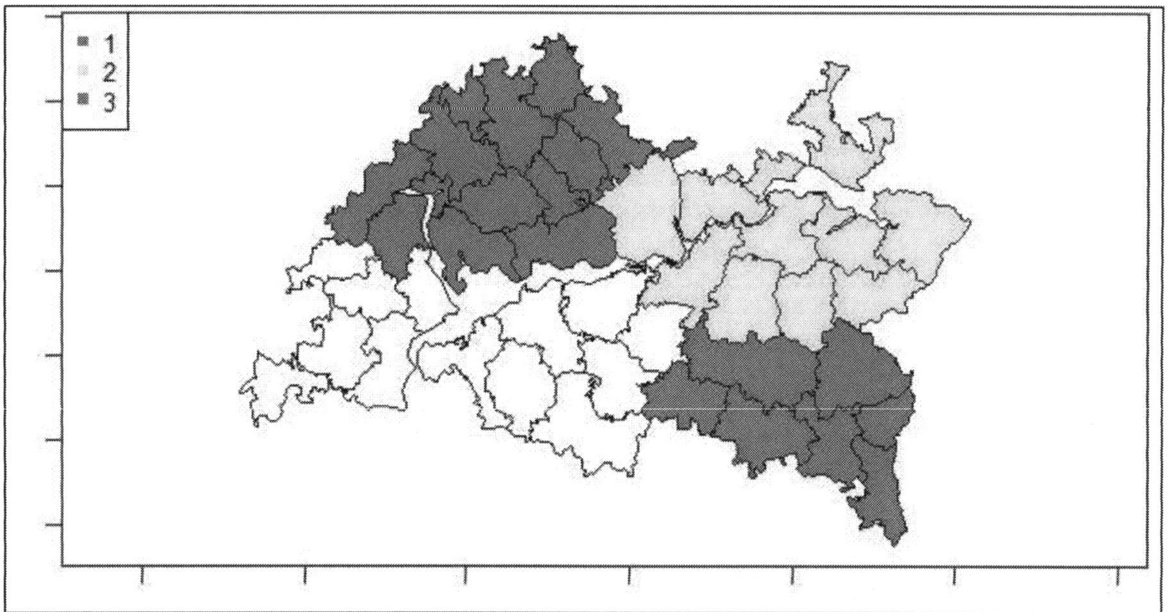

Figure 3. The optimal variant of the economic regionalization of the Republic of Tatarstan (white color highlights the South-West homogeneous economic region).

IOP Conf. Series: Earth and Environmental Science **107** (2017) 012134 doi:10.1088/1755-1315/107/1/012134

The results of the economic regionalization allowed identifying three nodal economic regions of the Republic of Tatarstan with strong economic centers, each of which has particular strategic directions of development:

- Kazansky economic region with the economic centers of Kazan and Zelenodolsk cities.
- Kamsky economic region with the economic centers of Naberezhnye Chelny, Nizhnekamsk and Elabuga cities.
- South-East economic region with economic centers of Almetyevsk, Bugulma and Leninogorsk cities.

Analysis of the set of elaborated variants of economic regionalization shows that the optimal variant, obtained according to the parameters of internal connectivity, homogeneity and compactness, includes a special territorial economic formation - South-West homogeneous economic region. It includes Apastovsky, Buinsky, Drozhzhanovsky, Kaybitsky, Kamsko-Ustyinsky, Tetyushsky, Alekseevsky, Alkeevsky, Aksubaevsky, Novosheshminsky, Nurlatsky, Spassky and Chistopolsky municipal districts.

Research in internal homogeneity and revealing homogeneous groups [9] of municipal districts within the territory of South-West homogeneous economic region also shows that within it there are two economic zones – Privolzhskaya and Zakamskaya.

Privolzhskaya economic zone includes Apastovsky, Buinsky, Drozhzhanovsky, Kaybitsky, Kamsko-Ustyinsky and Tetyushsky municipal districts.

Zakamskaya economic zone includes Alekseevsky, Alkeevsky, Aksubaevsky, Novosheshminsky, Nurlatsky, Spassky and Chistopolsky municipal districts.

These economic zones of the South-West economic region in general are characterized by internal economic homogeneity, internal similar level of economic development and life quality of their population. They differ to some extent from each other by natural and economic environment, economic specialization (agricultural in the Privolzhskaya and agro-industrial in the Zakamskaya economic zone).

5. Conclusions

The economic regions and zones were identified with the purpose of forming basis for defining and developing new economic growth poles and economic centers - growth points of the Republic of Tatarstan.

Analysis of main economic parameters of economic dynamics and levels of socio-economic development of municipal districts which were included in Privolzhskaya and Zakamskaya economic zones of the South-West economic region shows certain internal dissimilarity of these economic zones. This led to the identification of future-oriented municipal regions and economic centers – economic growth points and development points of the Republic of Tatarstan economy. These include the Buinsky, Nurlatsky and Chistopolsky municipal districts and Buinsk, Chistopol and Nurlat towns.

Conclusions on special economic status of Chistopolsky, Buinsksky and Nurlatsky municipal districts within the South-West homogeneous economic region are confirmed by the results of classification/grouping on 35 main macroeconomic indicators of these municipal areas.

The analysis of the main macroeconomic indicators dynamics of Buinsksky, Nurlatsky and Chistopolsky municipal regions in the period of 2014 – 2016 and according to the forecast on 2017-2019 shows that their economy is growing more rapidly than on the average for municipal regions of the South-West economic region.

Conducted research shows that the existing agricultural and industrial potential, favorable conditions and resources for agricultural activities, favorable geographical situation of Privolzhskaya and Zakamskaya economic zones determine the creation of a new agro-industrial centers and agro-industrial clusters in the Buinsky, Nurlatsky and Chistopolsky municipal districts.

Economic growth of agro-industrial branches of these economic semiperipheral centers will have stimulating effect on the development of peripheral areas of their economic zones. It would increase their levels of socio-economic development and life quality of the population, and would reduce the rate of demographic degradation in rural areas of Predvolzhskaya and Zakamskaya economic zones and would preclude, therefore, the depletion of their periphery.

References

[1] Knowles R and Wareing J 1990 Economic and social geography Amsterdam Elsevier 352 p
[2] Duque J C, Ramos R and Suriñach J 2006 Supervised Regionalization Methods: A Survey, International Regional Science Review **30(3)** p 195 - 220
[3] Alaev E B 1983 Socialno-ehkonomicheskaya geografiya: ponyatijno terminologicheskij slovar Moscow Mysl 350 p
[4] *Koncepciya territorialnoj ekonomicheskoj politiki Respubliki Tatarstan.* Available at http://mert.tatarstan.ru/rus/file/pub/pub_25254.doc (accessed date - July 21 2017)
[5] *Strategiya socialno-ehkonomicheskogo razvitiya Respubliki Tatarstan do 2030 goda.* Available at http://mert.tatarstan.ru/strategiya-sotsialno-ekonomicheskogo-razvitiya.htm (accessed date - July 21 2017)
[6] Panasyuk M V 2014 Method of Integrated Assessment of Regional System *Mediterranean Journal of Social Science* **5 (8)** p 149-153
[7] Parr J B 2012 Spatial-structure differences between urban and regional systems *Annals of Regional Science* **49 (2)** p 293 - 303
[8] Sohn J 2014 Industry classification considering spatial distribution of manufacturing activities *Area* **46(1)** p 101-110
[9] Jain A, Murty M and Flynn P 1999 Data clustering: A review *ACM Computing Surveys* **31(3)** p 264 –323

IOP Conf. Series: Earth and Environmental Science **107** (2017) 012135 doi:10.1088/1755-1315/107/1/012135

Analysis of natural prerequisites for the development of ecological tourism in the Belgorod region

Eu A Pendyurin, S N Glamazda, O N Genenko and S A Ryadnova

Belgorod state Institute of arts and culture, Korolyova str., 7
308033, Russia
E-mail: svetlana-glamazda@yandex.ru

Abstract. The tourism industry is related to entertainment, leisure, new impressions, emotions and pleasure. Tourism liberates people from a sense of fatigue by a specific change of environment and activity. Eco-tourism today is becoming one of the most promising developing sector of tourism business. Ecotourism is something average between the measured leisurely walks and extreme sports. It is contemplative and moderately informative at the same time, usually sporty and mobile, quite impressive. Analysis of natural prerequisite for the development of ecological tourism is one of the stages of site assessment as natural resources are one of the important determinants of its use. The Belgorod region has high recreational and touristic potential. On its territory there are unique natural and recreational resources. Exposing, analyzing tourist recreation resources to tourism development in the Belgorod region, want to pay attention to the large potential of this field in the region.

1. Introduction

Many people connect the tourist industry with the entertainment, leisure, new impressions, emotions and pleasure. Tourism exempts people from a sense of fatigue by a specific change of atmosphere and activity. The tourism industry firmly is established in human life with its natural tendency to discover and understand the unexplored countries, monuments of nature, history and culture, customs and traditions of different countries. The last decade was marked by intensive development of tourism and recreation sphere because of the economic, environmental and social needs of the society. The economic component determines the high profitability of this activity, the possibility of obtaining investment for starting a business, small costs while opening new work places; ecological – the need to restore the health and vitality of people in environmentally clear conditions through the reduction of threat of the living space; social – population growth of the planet, its cultural level, prosperity, expansion of the range of needs that include the desire for a healthy lifestyle, active recreation [1, 2].

Appearance and development of ecological tourism was preceded by different kinds of background. Among them, a significant role increases due to the large number of tourists, the anthropogenic workload on natural, cultural and historical tourism objects [3, 4].

Ecotourism today is one of the most promising developing sector of tourism business. Environmental journey makes to follow certain rules and obligations of travel behavior and the voluntary acceptance of certain obligations towards the tourist area. Ecotourism is something average between the measured leisurely walks and extreme sports. It is contemplative and moderately informative at the same time, usually sporty and mobile, quite impressive [5].

2. The research problem

One of the problems of ecological tourism is to teach people to love, to honor and adore nature, to understand its principles, to handle it economically and with care. Tourist hotels, campgrounds, resorts, which are located between the plots of undisturbed wilderness, where pay homage to the lessons of ecology, preservation of natural and cultural heritage, are becoming more fashionable and entice even the experienced, trained and environmentally conscious travelers.

Ecotourism appeared on the basis of the demand for organic, untouched corners of nature. It is a journey in to places with relatively untouched, pristine nature. The main purpose of ecotourism is understanding the natural, cultural and ethnographic features of the territory, without disturbing the integrity of natural ecosystems. Such economic conditions, in which protection of nature is profitable to the local population, must be created.

Ecotourism includes several key criteria that are the most enticing for travelers. Most important for travelers in the survey area are the natuive plants, animals, natural landscape, geological rarity; and the next largest components are the features of the cultural environment, where the emphasis is made on the study and understanding of resources, the activities of tourists and other participants of the company have a sparing effect on the physical and cultural environment in the attended region [6].

Ecotourism should be linked with the concept of sustainable tourism, should not exceed the recreational capacity of the visited territories, to be acceptable to local communities and to support them. The difference of ecological tourism from other leisure activities is that it satisfies a desire and encourages communication with nature, prevents and reduces the adverse impact on natural objects and culture, and encourages tour operators and tourists to promote the protection of nature and socio-economic development [7].

Ecotourism has long been separated and became an independent kind of tourist industry. This component has natural, historical and cultural potential for the territory or region, including natural features, socio-cultural component to the traditions and customs, peculiarities of the household and economic activities [8, 9].

Analysis of natural prerequisite for the development of ecological tourism is one of the stages of site assessment as natural resources are one of the important determinants of its use.

Since resources that can be used to meet the population's needs for recreation and tourism, are allocated on the planet surface unevenly, a growing number of people travels with recreational purposes and motives. As a rule, recreational resources determine the formation of tourism potential in a particular region. Nature and traditional culture of the Russian Federation has great potential for development of recreational tourism.

3. Results and discussion

The Belgorod region is not exception and refers to Chernozem recreation region. Climatic conditions are relatively fair and are characterized as moderately comfortable. The region is rich in recreational potential, stable political environment, diversity of historical and cultural heritage, which contributes to the development of ecological tourism [10].

The Belgorod region has high recreational and tourist potential. There are unique natural and recreational resources, objects of national and cultural and historical heritage on its territory [11]. Important economic, cultural, social and sporting events are given here. Subjecting the analysis of tourist recreation resources and tourism development in the Belgorod region we want to pay attention to the large potential of this field in the region. The Belgorod region has rich origins of spiritual, moral, aesthetic and artistic culture. The cultural heritage of the Belgorod region was created over several centuries: accumulating, enriched, passed down from generation to generation. The region occupies the sixth place in ten of the best Russian regions with high living standards, and is an attractive Russian region in terms of tourism development. The tourist base on the territory of Belgorod region counts more than 2 000 monuments of culture and history, including 908 monuments of archeology, 245 monuments of architecture, 745 monuments of military glory, 35 of them are monuments of history and culture of Federal significance. Unique natural areas of the Belgorod region are united in the wilderness area [12].

According to health and climate characteristics the Belgorod region is belonged to the developing the health effect.

The rocks, composing the surface, exposure of slopes, cross-country territories for tourism routes and tourist transport, mostly do not prevent the installation of transmission and in practice have no effect on the permeability of the territory. Composition of rocks and terrain are often the sources of natural attractions that are in the tourist industry are used as objects of natural excursions, and they are a powerful factor that controls land use.

The same amount of territory has a considerable weight in the formation of the tourist environment and largely determines its quality. Ecological and landscape conditions of the Belgorod region provide an opportunity for development in this area these types of basic recreational classes of permanent and temporary activities, such as cycling, hiking, skiing and sledding with the use of natural topography, equestrian walks, amateur and sport fishing, tours on inland waters. Along with the natural landscapes, which area takes about 9% of the region territory, agricultural and mining landscapes, which are recently recommended for use as objects of ecological tourism.

Belgorod is a region of water-shorts, but inland waters are widely used in tourist activities and contribute to the development of the recreational structures near the water reservoir. Rivers and large reservoirs are applicable for recreational purposes associated with the use of swimming facilities in a greater degree. Currently they are used for organizing and conducting various types of basic recreational activities, from swimming and relaxing in the warm season, relaxing by the water during the cool period associated with the contemplation, riding on boats, pleasure boats, jet skis, motor boats, water bicycles, water skis. There are the mineral waters in the region, which are the hydropathical resource. Hydrochemical composition of common mineral deposits makes the region promising for the development of territorial recreation systems.

The flora and fauna of the region determine the diversity of types of tourist and recreational classes and crafts. Native vegetation determines the aesthetic value of the territory. In the Belgorod region are growing the most attractive russian birch and oak forests, pine forests, feather-grass steppe and mixed grass meadows. Vegetation cover is an important factor in tourism, providing a huge impact on the permeability of the area. In the Belgorod region, road and path network is in a satisfactory condition. Representatives of the local flora and fauna, causing huge scientific and educational interest relating to rare plants, animals, and plant communities represent a resource base for eco-tourism. There are primarily protected areas. Maintaining biological diversity is an important aspect of sustainable tourism development in the region.

In order to preserve the specific natural systems and objects, to maintain the biological diversity and general ecological balance, in accordance with the Federal law "On specially protected natural territories" and the decree of the President of the Russian Federation and the decree of the head of administration of the region created a network of specially protected natural territories. It included the earliest existing reserves as well as the newly created protected areas. The largest number of protected areas was declared in Central, Alexeyevsky, Volokonovsky, Yakovlevsky, Valuysky, Starooskolsky areas. Today, protected areas are natural complexes that have environmental, cultural, and aesthetic and recreation importance and a special protection regime. The system of specially protected natural territories of the Belgorod region was formed mainly in 1991-1995. The initial point in creation of the special protected natural territories of region can be considered the establishing of the state natural reserve "Forest on the Vorskla" in 1924 on the territory of the former Sheremetev's "Zakaznaya small woods".

There is a network of specially protected natural territories in Belgorod region. It includes a state nature reserve "Belogorie" and regional protected areas.

Natural areas with the possibility of establishing conservation territories include: tracts "Borky", "Gorodishche" and "Snowy Alps" in Valuyskiy regon, tracts "Gniloye", "Kamennoye", "Pozharishche" in Veydelevsky region, steppe plots and chalky rock exposure along the river Aydar of

Rovensky region, etc. Prepare the materials for the creation of new nature reserves and nature parks ("Prokhorovsky", "Gubkinsky"), the protected area "Aydarsky", and also protected areas around health-improving areas and mounds.

Currently, the area of specially protected natural territories in Belgorod region reached 51074 hectares, which is 1.9 % of its territory. Protected areas of regional importance 210 nature reserves with a total area 34498 hectares, 13 therapeutic areas with a total area of 328 hectares, 138 natural monuments with a total area of 96 hectares, 2 parks, nature area of 12,000 hectares, 1 dendrological Park and a Botanical garden with an area of 82 hectares, 2131 hectares are areas of the reserve "Belogorie", which has a Federal status [13].

The objects are placed in the region unevenly.

Table 1. Specially protected natural territories of Belgorod region

No.	Region	Protected natural areas of the region		
		Quantity	Area, ha	The share of the relevant area,%
1	Alekseevsky	18	802,24	0,45
2	Belgorod	13	15815,5	10,72
3	Belgorodsky	5	1502,5	9,81
4	Borisovsky	14	2457,83	3,78
5	Valuysky	29	2467,97	1,31
6	Veydelevsky	5	75,01	0,06
7	Volokonovsky	20	468,15	0,36
8	Graivoronsky	8	268,74	0,31
9	Gubkinsky	12	1150,03	0,75
10	Ivnyansky	9	535,04	0,61
11	Korochansky	8	55,08	0,04
12	Krasnensky	5	115,03	0,14
13	Krasnogvardeysky	82	2692,22	1,53
14	Krasnoyaruzhsky	5	184,1	0,38
15	Novooskolsky	6	1017,04	0,73
16	Prokhorovsky	12	72,08	0.05
17	Rakityansky	11	296,02	0,33
18	Rovensky	9	1338,1	0,98
19	Starooskolsky	17	1588,01	0,94
20	Chernyansky	13	1341,06	1,09
21	Shebekinsky	29	1242,96	0,67
22	Yakovlevsky	22	763,14	0,70
Belgorod region		352	51074	1,9

After analyzing these tables we can conclude that the leaders in the number of objects of protected areas are Krasnogvardeysky (82), Valuysky and Shebekinsky (for 29 units). But among the specially protected natural territories the leaders are Belgorod (about 15800 ha), Krasnogvardeysky (about 2700 ha) and Borisovsky (about 2500 ha) areas.

Analysis of landscape and natural recreational potential for development of ecological tourism in the area allows to highlight areas of high and medium potential to meet the needs of the population in natural environment.

The region can highlight areas of high environmental and potential for recreation and tourism. The first category – the Borisovsky region, the second – Belgorod, Yakovlevsky, Prokhorovsky, Gubkinsky, Chernyavsky, Novooskolsky, Krasnogvardeysky, Valuysky and Rovensky regions, the remaining territory of the region according to the degree of satisfaction of population needs in the environmental relaxation can be attributed to the third category.

Complex (landscape) reserves are aimed to protect both flora and fauna of the territory; they can also include separate monuments of nature. Among the most unique we can list the following: tracts "Borky", "Gorodishche" (Valuysky region) – marmots colony, chalky rock exposure and outcrops, relict communities; "Bolshoy log" (Krasnensky region) - relict plants and spring.

Botanical reserves reflect a representative floristic composition of the region and are the most studied ones. The greatest interest is Botanical reserves with calciphyte flora and relict species: "Becarukovsky Bor" (Shebekinsky region), the number of areas in Valuysky, Veydelevsky areas; areas of forb-meadow and forb-feather grass steppes: "Hvoshchevatoye" (Krasnensky region), "Gniloye" (Veydelevsky region), etc.; the groves: "Cherepennoye" (Gubkinsky region), "Korabelnaya oak-wood" genetic reserve of pedunculate oak (Alexeevsky).

There are 352 of specially protected natural territories in Belgorod region. There are no specialized entomological sanctuaries in the area. There are various groups of invertebrates living on the specially protected natural territories: steppe, forest, oak-wood, the number of species of relict and endemic. From this point of view, any of the categories of especially protected natural territories can be attributed to the category of complex protected territories. As an example: "Korabelnaya oak-wood " is only known in the region as the habitat of wingless locust; "Solomenskaya oak-wood" – a number of species which are protected at the federal and regional levels.

It is important to note the existence of monuments of nature. There are the smallest protected areas. The vast number of monuments of nature is taken by the springs, karst springs, and long-living oaks (400 years). These objects can be named as symbols of Belgorod region [14].

Represented in the network of specially protected natural territories, objects reflect fairly representative group of unique areas: the karst funnel "Gornyashka" (Stary Oskol); geological profile of the Chalky career of Belgorod Plant of building materials; "Monastic cavern" (Valuysky region); old estate (Stankevich, Boldyrev, etc.) and the Park "Arkhangelsky" (Gubkinsky region); burial mound "Sciffskoye gorodishche" (Borisovsky region); underground monastery "Cholkinskiye caves" (Chernyansky region). This kind of protected areas has significant development potential.

We should also note the importance of parks and green areas. They are concentrated mainly around the regional centers. These include the artificial planting of pine and oak forests – the largest specially protected natural areas of region.

4. Conclusions

After analyzing the territory of the Belgorod region, we can assume that tourists suggest a visit of ecological routes and paths. For the design and justification of ecological routes, modern techniques are currently used to create training and educational ecological paths and routes. In the region there are already a number of popular ecotourism routes, which, unfortunately, have a lack of information. As a rule, these tours are a mix of educational, popular-scientific and thematic tours, which run on a specially equipped ecological trails.

For the design and justification of ecological routes, modern techniques are currently used to create attractive tourism products that meet the preferences of target consumers (coordinated solutions in terms of assortment, quantity, completeness and level of service of the tourist product). In the development and design of environmental routes various factors are taken in account: physical-geographical and socio-historical properties of the area, within the tourism product is developed, age, professional, ethnic and psychological criteria of the group, which is focused on the tourism product, qualification of guides, investment opportunities of ecotourism development. In the region, there are already a number of popular ecotourism routes, which, unfortunately, have little demand and a lack of information.

Evaluation of objects and phenomena of cultural and historical heritage suggests the possibility of development of cultural-educational, religious, pilgrimage, rural, ethnographic, ecological and scientific tourism on the territory of Belgorod region. Thus, the Belgorod region has a rich and diverse recreation potential conducive to development of ecological tourism.

Now travel agencies offer a variety of routes, including an all inclusive environmental scientific-popular weekend tours that cover all the splendor of nature Belgorod region, where the journey will be momentous trip – a discovery which will stimulate the development of tourism in the region and to solve the problem of unilateral approach of travel companies.

References

[1] Birzhakov M B 2007 *Introduction in tourism* (St. Petersburg: Publishing House Gerda) p 576
[2] Holloway J Christophe 1994 The Business of Tourism (London)
[3] Borisoglebskaya L N, Poishina A A and Bogomazova I V 2008 *The tourism industry is a promising component of the socio-economic district of the Belgorod region* (Science list of BelSU **5** chapter 7 pp 268-276
[4] *On the development of tourism in the Belgorod region* 2015 (Belgorod: publisher Belgorod the regional statistics Committee) p 20
[5] Bernecker P 1997 *Fundamentals of tourism* Geneva
[6] Pendyurin Eu A, Istomina V V and Glamazda S N 2013 Assessment of natural resource potential of the Belgorod region for development of ecological tourism *Science. Art. Culture* **2** pp 228-232
[7] Violante M Q 2011 *Turismo y ocio en lassoiedades: dinamismo, desarolloturistico y riesgos* (Espana Ediciones Academicas) 564 p
[8] *Australian National Ecotourism Strategy* 1992(Canberra: Commonwealth Department of Tourism) p 682
[9] Ceballos-Lascurian H 1991 Tourism, ecotourism and protected *areas Ecotourism and Resource Conservation: A Collection of Papers Madison* **1** p 24
[10] Wight P 1993 *Ecotourism: Ethic or Eco-sell?* Journ. of Travel Res. **31** p 3
[11] Parchomenko I G 2010 *Belgorod land on the map* (Belgorod) p 436
[12] Honey M 2008 *Ecotourism and Sustainable Development* (Washington: Island Press) 549 p
[13] Zateev A A 2011 Actual problems of ecological tourism in contemporary Russia *Geography in school* **3** p 75
[14] Dorofeev A A and Chochlova E P 2010 Ecological tourism as a factor of nature protection and conservation of biodiversity *Bulletin of Tver State University* **32** pp 122-134

IOP Conf. Series: Earth and Environmental Science **107** (2017) 012136 doi:10.1088/1755-1315/107/1/012136

The peculiarity of small lakes formation in the urban territory of Kazan, Russia

I S Shigapov

Department of Environmental Engineering and Water Resources Management, Institute of Management, Economics and Finance, Kazan Federal University, 1 Mezhlauka Str., Kazan, Russia.
shigapov.irshat@gmail.com

Abstract. An analysis was made for the origin of 190 small lakes in the territory of Kazan, Russia. The obtained data made it possible to identify the main mechanisms for the formation of small lakes. Eight factors determining the formation of lakes were identified: seven natural factors and human impact. The natural factors of lakes formation are precipitation, air temperature fluctuations, wind influence, mineral composition, water runoff influence, groundwater, water currents and waves. These factors are responsible for five processes: human activity, fluvial, suffusion, karst erosion, and aeolian processes. These processes through seven mechanisms form different types of lake origin in the territory of Kazan.

1. Introduction

Nowadays, the task of revealing the regularities of the lakes formation depending on the landscape conditions, as well as the problems of the development of water ecosystems under the influence of human impact acquires great scientific importance. Human impact is expressed in the chemical pollution of lakes by substances such as heavy metals [1, 2], organic [3, 4] and oil products [5, 6]. But its influence also express in changing of morphometric parameters of lakes [7], which determining their influence to urban ecosystems [8] and the opportunity for using this lakes for recreation [9]. There are many publications about city lake restoration and water quality management [10, 11, 12], however for this we should have the knowledge about evolution of lakes, which include information about origin and transformation of lake basin.

The account of external factors of lakes formation allows to estimate their current state reasonably and predict the course of their further evolution taking into account changes in external conditions and processes occurring in the lake itself. At present, there is much experience in the study of large lakes, their basic morphometric and ecological characteristics have been established [13]. However, the data of small lakes (with an area of 1-10 km^2) is still insufficient, due to the large number of lakes and variety of their characteristics.

We analyzed the origin of 190 lakes in the territory of Kazan. All investigated lakes are in the catchment basin of the Volga River, and are small shallow lakes (up to 1 km^2 and maximum depths of up to 2 m). In this work we used our own regional classification of lakes origin. The obtained data make it possible to present the evolution of Lake Morpholitogenesis in the following form: conditions - factors - processes - mechanism of formation of lake basins - genetic types of lakes.

The main concepts of Lake Morpholitogenesis, first of all, include the concept of "process", which determines the successive change of states and natural links of following one after another stages of evolution of Lake Morpholitigenesis. Concept “factor” is understood in this instance as propulsion of making process or one of necessary conditions. Concept "mechanism" determines sequence and interaction of processes and events, which conditioning the direction of process (result of action). At the same time we used the assumption that the lake formation in the territory of Kazan occurs in the relatively stable natural conditions of the plain on the Middle Volga, when morpholithological, climatic and anthropogenic conditions have become the main conditions, because their variations cause a significant change in the processes lake basins formation, intensity of sedimentation and the degradation of the lake

2. Factors in the lakes formation on the territory of Kazan city

For the Kazan area, we identified five processes that determine the lake basin formation: anthropogenic process, suffusion, karst, fluvial, and aeolian (figure 1). These processes are caused by 8 factors: anthropogenic factor, which can both form new lakes and modify existing [14, 15] and 7 factors which have natural origin. Three of the factors are climatic: atmospheric precipitation, air temperature, and wind, which affects both evaporation and detalization of coastline.

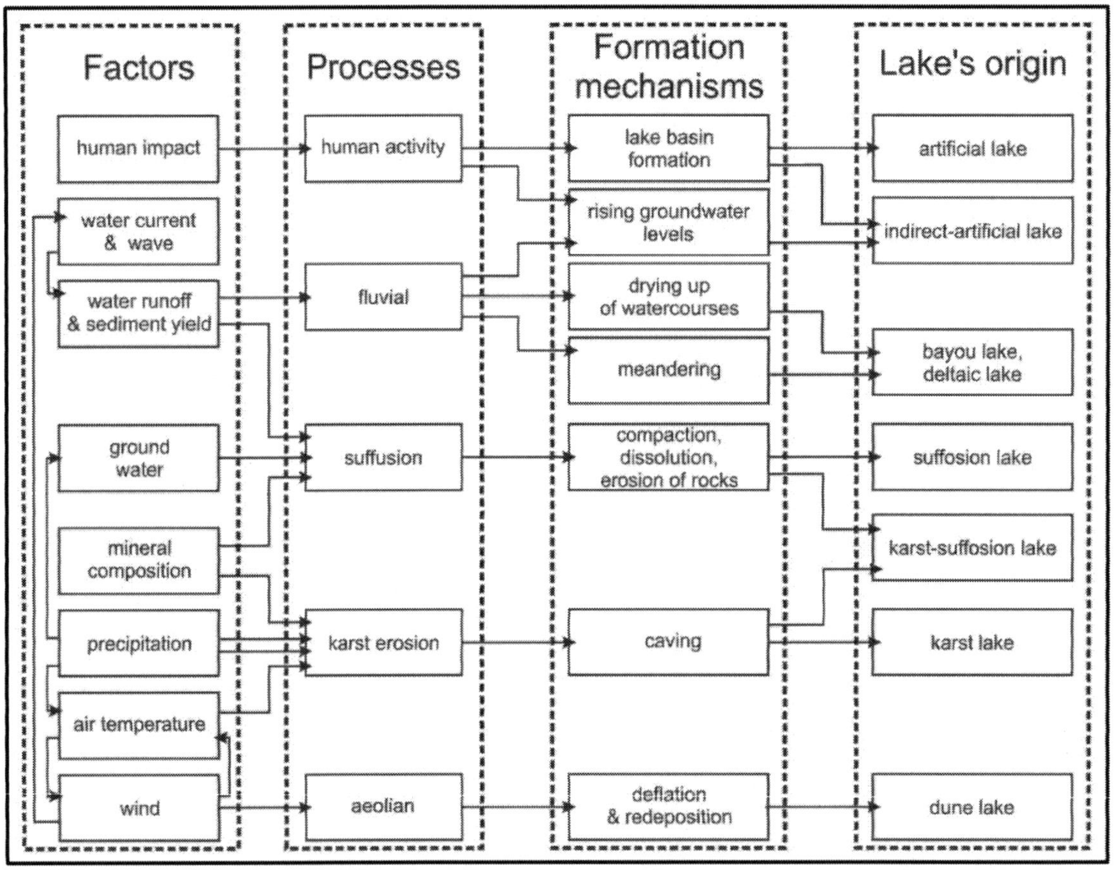

Figure 1. The structural-functional scheme of the lake hollows formation in the Kazan City.

Other factors are geological and hydrogeological: the composition of rocks and groundwater, the impact of surface water, this should include impact of water flow and formation of proluvial deposits.

For example, the Basin of Bolshoe Glubokoe Lake and Maloe Glubokoe Lake, according to the old maps were a single lake, which was divided by the alluvial cone of a large ravine. Currently in the north-western coast of Bolshoe Glubokoe lake is the beach, formed by alluvial cone of several ravines. According to our data, during 10 years in the northwestern coast of the lake the coastline has receded by more than 7 m. The presence of the above processes allows us to identify 6 mechanisms for the lake basin formation in the territory of Kazan.

The anthropogenic impact mechanism is expressed both in the formation of basin and in raising the level of groundwater. Due to aeolian processes, the impact of the deflation and redeposition mechanisms is the cause of dune lakes formation. According to one of the hypotheses, Lebyazhye lake system was formed in this way. As a result of mechanisms of meandering and the drying up of watercourses, were formed the bayou and deltaic lakes.

Karst processes by dissolving rocks and forming cavities form karst funnels in which karst lakes are formed. In compaction and dissolution processes, as a result of the compaction of rocks, suffusion lakes are formed. On the territory of Kazan often occurs the result of combined processes: suffusion-karst lakes.

3. The development of lakes on the territory of Kazan city

After the formation, lakes are developing, changing their size and physico-chemical parameters of water. Eventually lakes are filling with sediment, overgrown and dries. On the territory of Kazan we found out the examples both of gradually or sudden changes of lakes condition. The results of the analysis of this process is shown in figure 2.

In the transgression stage the lake's hollow is flooding (transgression), the water level and water volume in the lake is increasing, and the area and depth of the lake basin is increasing too.

For lakes of the Kazan city we found out note a strong anthropogenic influence, as about a third of the lakes in the city of Kazan have an artificial origin (54 lakes). The human activity can be expressed as an indirect influence on the course of natural processes and purposeful changes in the lake watershed.

Indirect human impact determines the mechanism of raising the level of ground water in hydraulic flooding due to changes in the level of the Kuybyshev reservoir because in the Kazan city groundwater usually is at a depth of 0.2-5 m. Due to hydraulic flooding, intensified fluvial processes have formed bayou and deltaic lakes. The increase of groundwater level and seasonal changes in the level of the Kuybyshev reservoir, undoubtedly, have an impact on the hydrological regime of lakes, especially during the spring floods, but this effect is not yet known.

One of the consequence of construction is underflooding causes the formation of lakes. For example, in the Novo-Savinovsky district of Kazan city, the construction in residential neighborhoods was the cause of formation of many lakes: Kroshka Lake, Ugolok Lake etc. Some kind of indirect human impacts (e.g., mining) determine the mechanism of the formation of basins with subsequent filling them with water (Emerald Lake near Yudino village formed on the site of a former sand quarry).

Direct human impact affects the processes of formation and development of lakes through the mechanism of the formation of the basins of artificial lakes due to dams construction (School Lake, Farmer Lake) or digging depressions (Central Lake in the Park of Victory). In addition, this mechanism can be expressed through the purification and aeration of water bodies overgrown (Large Chuikova Lake and Small Chujkova Lake, Nizhny Kaban Lake, Mar'ino Lake) which can be called the restored lakes.

Natural processes of formation of lakes on the territory of the Kazan city is represented by the following types. Karst processes through the mechanism of the collapse of vaults of caves lead to the formation of karst lakes (Large Glubokoe Lake). Suffusion processes due to the action of the compaction mechanism and erosion of rocks form the suffusion lakes (Chishmale Lake).

Despite the fact that the waves on the lakes of the Kazan city rarely exceed 0.3 m in height, their abrasive activity is of great importance in the formation of coastline. For example, the North-Western shore of the Central Lake in the Victory Park for two years has retreated by about 0.7 m.

You also need to consider the complexity of the process of formation and development of lakes, which proceeding with the participation of several mechanisms. For example, in the case of combined effects of the mechanisms of the collapse of vaults of caves and compaction, erosion and dissolution of rocks are formed by suffusion-karst lakes (Krugloe Lake).

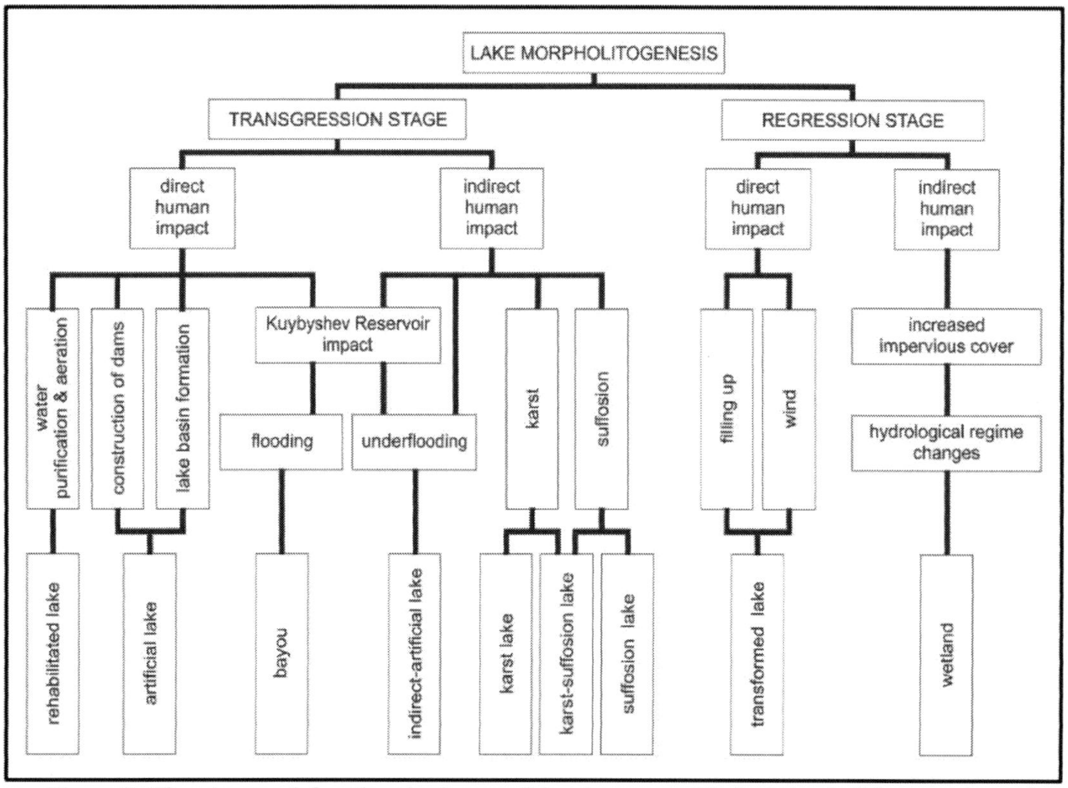

Figure 2. The structural-functional scheme of development of Lake Morpholithogenesis on the territory of Kazan City.

The regression stage of the lakes development is characterized by the degradation of lakes, as a result, the size of the lake basin and the volume of water mass is reduced and water quality deteriorates.

Features of sedimentation and eutrophication are determined by the peculiarities of the lake hollow, but in urban areas the special role played by the human factor.

The intensity of lakes degradation is different. The degradation rate depends primarily on the size of lakes; because of the greater volume of water mass and, consequently, the dissolving ability large lakes are degraded more slowly (in the Central part of the Kazan city remained only relatively large lakes). However, long-term chronic pollution in them is still there are irreversible changes (in the 1980 on the lakes Kaban had to carry out special cleaning events).

The evolutionary path of the regression stage involves reducing the volume of water mass and depth of the lake basin as a result of gradual accumulation of sediments and decreasing of an income part of water balance. It is associated with the degradation mechanisms of lakes: accumulation, eutrophication, drying-up, caused by both natural and anthropogenic processes.

4. Conclusion

Altogether, the study has shown that in the Kazan City the formation of lake hollows due to the 7 factors of natural origin (wind, water temperature, precipitation, water currents and waves, water runoff and sediment yield, underground water, and composition of rocks) and anthropogenic factor.

Anthropogenic process can be expressed as an indirect influence on the course of natural processes and targeted changes in the lake basin.

References

[1] Shang S, Zhong W, Wei Z, Zhu C, Ye S, Tang X, Chen Y, Tian L and Chen B 2017 Heavy Metals in Surface Sediments of Lakes in Guangzhou Public Parks in China and Their Relations with Anthropogenic Activities and Urbanization *Human and Ecological Risk Assessment* **23(8)** pp 2002-2016

[2] Rabajczyk A, Jóźwiak M A, Jóźwiak M and Kozłowski R 2011 Heavy Metals (Cd, Pb, Cu, Zn, Cr) in Bottom Sediments and the Recultivation of Kielce Lake *Polish Journal of Environmental Studies* **20 (4)** pp 1013–19

[3] Honkonen O and Rantalainen A L 2016 Transport of Urban-Derived Organic Contaminants into a Boreal Lake: A Case Study with Passive Samplers *Boreal Environment Research* **6095** pp 87–100

[4] Andersson M and Eggen O A 2015 Urban contamination sources reflected in inorganic pollution in urban lake deposits, Bergen, Norway *Environmental Science Processes & Impacts* **17(4)** pp 854–67

[5] Yan B, Bopp R, Abrajano T, Chaky D and Chillrud S 2014 Source Apportionment of Polycyclic Aromatic Hydrocarbons (PAHs) into Central Park Lake, New York City, over a Century of Deposition *Environmental Toxicology and Chemistry* **33 (5)** pp 985–92

[6] López L, Mogollón J L, Aponte A and Bifano C 2000 Identification of Anthropogenic Organic Contamination Associated with the Sediments of a Hypereutropic Tropical Lake, Venezuela *Environmental Geochemistry and Health* **22 (1)** pp 55–74

[7] Steele M K and Heffernan J B 2014 Morphological Characteristics of Urban Water Bodies: Mechanisms of Change and Implications for Ecosystem Function *Ecological Applications* **24 (5)** pp 1070–84

[8] Chen W 2012 Study on Urban Lake Ecosystem Health Assessment *Applied Mechanics and Materials* **178-181** pp 304-308

[9] Tandyrak R, Grochowska J and Łopata M 2017 Hydrochemical Parameters and Trophic State of an Urban Lake Used for Recreation *Journal of Elementology* **22 (4)** pp 1347-58

[10] Dunalska J, Grochowska J, Wiśniewski G and Napiórkowska-Krzebietke A 2015 Can We Restore Badly Degraded Urban Lakes? *Ecological Engineering* **82** pp 432–41

[11] Zhong J, Huang T, Xie Y, Wang X, Peng D 2006 Study of Water Quality Improvement and the Protective Technique of Urban Rivers and Lakes *Journal of Xi'an University of Architecture and Technology* **38 (6)** pp 771–76

[12] Grochowska J K and Brzozowska R 2013 The Influence of Different Recultivation Methods on the Water Buffer Capacity in a Degraded Urban Lake *Knowledge and Management of Aquatic Ecosystems* **01(410)** pp 1–13

[13] Mingazova N M, Zamaletdinov R I, Derevenskaya O Yu, Palagushkina O V, Nabeeva E G, Pavlova L R, Shigapov I S, Mingaliev R R, Nazarov N G and Zaripova N R 2015 The Impact of XXVII Summer Universiade on the Environment in Kazan *Mediterranean Journal of Social Sciences* **6 (1S3)** pp 470–74

[14] Nabeeva E G, Shalyamova R P and Shigapov I S 2015 Detecting Parameters of Impact Assessment of SDW Landfill Sites on Water Bodies (on the Example of the Samosyrovsk Land Fill, Kazan, Russia) *Research Journal of Pharmaceutical, Biological and Chemical Sciences* **6 (6)** pp 1395–1403

[15] Shaliamova R P, Nabeeva E G and Shigapov I S 2015 The Contaminating Impact of Surface Water Runoff from the MSW Landfill on the River Krutovka (through the Example of Samosyrovskaya Landfill, Kazan, Russia) *International Journal of Applied Engineering Research* **10 (24)** pp 44801–7

IOP Conf. Series: Earth and Environmental Science **107** (2017) 012137 doi:10.1088/1755-1315/107/1/012137

A study of temporal dynamics and spatial variability of power frequency electromagnetic fields in Saint-Petersburg

V I Sturman

Federal State Budget-Financed Educational Institution of Higher Education The Bonch-Bruevich Saint-Petersburg State University of Telecommunications,
22 Prospekt Bolshevikov, St. Petersburg, 193232 Russia
E-mail: st@izh.com

Abstract. This paper studies spatial distribution and temporal dynamics of power frequency electric and magnetic fields in Saint-Petersburg. It was determined that sanitary-protection and exclusion zones of the standard size high-voltage transmission lines (HVTL) do not always ensure maximum allowable limits of the electrical field depression. A dependence of the electric field strength on meteorological factors was defined. A series of sources create a city-wide background for magnetic fields. That said, the heavier the man-caused load is, the higher the mean values of magnetic induction are. Abnormally high values of magnetic induction are explained by the influence of underground electric cables.

1. Introduction.

Electromagnetic fields are still an insufficiently studied and poorly controlled environmental factor. Influence of electromagnetic fields on a human body is associated with the induction by a field of electrical currents of various strength and frequency inside the body, which may lead to both positive and negative consequences. People exposed to electromagnetic fields of increased intensity suffer from cardiovascular system malfunctions, metabolic disorders, as well as disorders of endocrine, immune, and reproductive systems, which is widely discussed in numerous sources ([1-3] et al).

In Russia, the protection of people from the exposure to electromagnetic fields is assured by sanitary-protection and exclusion zones created along HTVL, as well as by hygiene certification of household and commercial appliances and devices. However, Sanitary Norms and Regulations SanPiN 2971-84 [4] and SanPiN 2.2.1/2.1.1.1200-03 [5] define sanitary-protection zones only for HVTL 330 kV (20 meters wide), HVTL 500 kV (30 meters wide), HVTL 750 kV (40 meters wide), and HVTL 1150 kV (55 meters wide). Sanitary-protection zones shall be differentiated from exclusion zones, which are established not for the protection of people from the exposure to electromagnetic fields, but for the protection of high-voltage transmission lines. The width of exclusion zones is 20 meters for HVTL 110, 25 meters for HVTL 220, and 30 meters for HVTL 330 and HVTL 500 calculated from the projection of outermost phase conductors. That said, maximum exposure limits are 1 kV/m for residential zones and 5 kV/m for populated areas outside residential zones. Procedure [6] outlines the rules applicable to the exclusion zones of HVTL. A number of restrictions is applied to the territory of exclusion zones, for example, it is prohibited to arrange any events that involve mass gatherings. The paper does not consider the question of a possible dependence of electric fields of HVTL on weather conditions, and the question of mix and interaction of fields from various sources, a phenomenon known as electromagnetic smog in the scientific literature.

National health organizations in most countries of the world are considered negative influence of low-frequency electromagnetic fields (including power frequencies 50 and 60 Hz) to be not proved. So, in most part of the world there is no permissible exposure limits for low-frequency electric and magnetic fields. Now it is a subject of investigations, in particular by Scientific committee on the new and again revealed risks for health (SCENIHR) – advisory organ at the European commission on safety issues of consumers, public health and environment. Risks of a leukemia, damages of DNA and infringement of immune system are considered as possible effects of low-frequency electromagnetic fields [7]. Approximate safe limit of magnetic induction [7] is 0,4 μT (400 nT), that 10 times less most rigid among hygienic norms operating in Russia (5 μT for premises, children's, preschool, school, general educational and medical institutions according to Hygienic norms GN 2.1.8/2.2.4.2262-07 [8]).

2. Materials and Methods

The study of the power frequency electromagnetic fields in Saint-Petersburg was carried out from October 2016 through June 2017 by means of Gigahertz Solutions ME 3830 B M/E Analyser.

This meter can measure the strength of an electric field between 1 and 2000 V/m and the intensity of a magnetic field (magnetic induction) between 1 and 2000 nT. In several cases, the upper measurable value of 2000 V/m was not enough and it was possible only to state that this value was exceeded and to determine dimensions of a related zone. The below three types of measurements were taken:

- profile monitoring measurements, taken under wires in places of maximum sag and at a distance of 10, 15, 20, 25, and 30 meters from these places (from 19 to 32 series of measurements for 4 profiles);
- point monitoring measurements, taken to determine the stability of characteristics of electric fields with regard to the series of sources for these fields in residential areas (from 32 to 61 measurements in each of 4 points);
- point single measurements, taken at separate points of the city (approximately 300 measurements).

The measurements were taken in the places of maximum wire sag 1,8 meters above the ground. Information about weather conditions was taken at the Internet-service http://rp5.ru/Weather in Saint-Petersburg.

3. Results and Discussion

Strength of electric fields

Table 1 contains statistical characteristics of a series of observations over strength. Strength values exceeded 2 kV/m in all cases when measurements were taken directly under wires. Going away from the wires, the repeatability of exceedance of the given value was gradually diminishing, but it was not enough to determine statistical characteristics.

The study results show that electromagnetic fields with strength more than 1-2 V/m (maybe regarded as a city-wide background or a measurement error) were observed only in the vicinity of HVTL, not farther than 100-120 meters. However, according to data contained in Table 2, the width of zones with high values of the strength of the HVTL electric fields varied and did not always fall within the limits of the above-mentioned dimensions of sanitary-protection and exclusion zones.

In case of HVTL 110 kV near the crossing of Blyukher and Kondratyevskiy avenues, the values higher than maximum allowable limits of the strength at the boundary of the exclusion zone (20 m) were observed in 11 instances out of 24 (45.8%); in case of HVTL 330 kV near the Devyatkino subway station, the values higher than maximum allowable limits of the strength at the boundary of the sanitary-protection zone (20 m) were observed in 27 instances out of 32 (84.4%), and at the boundary of the exclusion zone (30 m) – in 20 instances out of 32 (62.5%). In case of two remaining HVTL, the values higher than maximum allowable limits of the strength were not observed at the boundaries of sanitary-protection and exclusion zones. Strength values in the nearby residential houses were lower than the maximum allowable levels in all instances.

Table 1. Statistical characteristics of intensity measurement results in case of profile monitoring

Monitored locations	Number of measurements	Mean value, kV/m	Root mean square deviation (δ)	Coefficient of variation, %
HVTL 330 kV near the crossing of Blyukher and Kondratyevskiy avenues, 20 meters from wires	30	525	261.6	49.8
HVTL 330 kV near the crossing of Blyukher and Kondratyevskiy avenues, 25 meters from wires	30	194	124.2	64.0
HVTL 330 kV near the crossing of Blyukher and Kondratyevskiy avenues, 30 meters from wires	30	104	33.5	32.2
HVTL 110 kV near the crossing of Blyukher and Kondratyevskiy avenues, 15 meters from wires	30	1257	489.4	38.9
HVTL 110 kV near the crossing of Blyukher and Kondratyevskiy avenues, 20 meters from wires	30	808	334.6	41.4
HVTL 330 kV near the Devyatkino subway station, 30 meters from wires	32	1221	426	34.9
HVTL 330 kV near the Devyatkino subway station, 50 meters from wires	32	688	189.9	27.6
HVTL 330 kV near the Devyatkino subway station, 70 meters from wires (the nearest building)	32	68	114.4	168.2
HVTL 220 kV near the Devyatkino subway station, 20 meters from wires	32	819	257.2	31.4
HVTL 220 kV near the Devyatkino subway station, 25 meters from wires	32	685	199.5	29.1

Table 2. Width of zones where values of the strength of the HVTL electric fields exceed 1 kV/m (maximum exposure limits for residential zones) and 2 kV/m

Monitored locations	Number of measurements	Over 1 kV/m, m			Over 2 kV/m, m		
		min	max	average	min	max	average
HVTL 330 kV near the crossing of Blyukher and Kondratyevskiy avenues	21	18	20	19	14	18	16.5
HVTL 110 kV near the crossing of Blyukher and Kondratyevskiy avenues	21	0	23	17.5	0	14.5	9.1
HVTL 330 kV near the Devyatkino subway station	19	26	50	41.6	13	30	20.9
HVTL 220 kV near the Devyatkino subway station	19	14	22	18.9	11	17	14.8

The major factor influencing the strength of electric fields is a load of HVTL. However, characteristics of the same lines vary depending on meteorological conditions. Dielectric properties of the air depend on its humidity, and not so much on relative humidity, as on the absolute one. Table 3 provides correlation dependences.

Table 3. Coefficients of correlation between strength measures in the HVTL monitoring range and meteorological characteristics

Monitored locations	Coefficients of correlation considering measures: Absolute humidity	Relative humidity	Air temperature	Atmospheric pressure
HVTL 330 kV near the crossing of Blyukher and Kondratyevskiy avenues, 20 meters from wires	0.681	-0.145	0.652	-0.573
HVTL 330 kV near the crossing of Blyukher and Kondratyevskiy avenues, 25 meters from wires	0.737	-0.061	0.617	-0.564
HVTL 110 kV near the crossing of Blyukher and Kondratyevskiy avenues, 15 meters from wires	0.762	0.579	0.353	-0.602
HVTL 110 kV near the crossing of Blyukher and Kondratyevskiy avenues, 20 meters from wires	0.732	0.370	0.453	-0.526
HVTL 330 kV near the Devyatkino subway station, 50 meters from wires	0.629	-0.149	0.511	-0.272
HVTL 220 kV near the Devyatkino subway station, 20 meters from wires	0.726	0.518	0.760	-0.565
HVTL 220 kV near the Devyatkino subway station, 25 meters from wires	0.736	0.461	0.829	-0.503

As it was technically impossible to measure strength values higher than 2 kV/m, we managed to calculate correlations with meteorological characteristics only in the monitoring points relatively remote from the wires. Therefore, an additional calculation was done to determine the correlation ratio between the width of zones with values exceeding 1 kV/m and 2 kV/m and meteorological characteristics. Table 4 provides the calculation results.

Table 4. Coefficient of correlation between the width of zones with values exceeding 1 kV/m and 2 kV/m and meteorological characteristics

Monitored locations	Coefficients of correlation considering measures: Absolute humidity		Relative humidity		Air temperature		Atmospheric pressure	
	< 1 kV/m	< 2 kV/m	< 1 kV/m	< 2 kV/m	< 1 kV/m	< 2 kV/m	< 1 kV/m	< 2 kV/m
HVTL 330 kV near the crossing of Blyukher and Kondratyevskiy avenues	0.620	0.749	0.537	0.473	0.313	0.578	-0.382	-0.573
HVTL 110 kV near the crossing of Blyukher and Kondratyevskiy avenues	0.688	0.761	0.501	0.502	0.310	0.483	-0.504	-0.631
HVTL 330 kV near the Devyatkino subway station	0.687	0.461	0.522	0.244	0.740	0.551	-0.539	-0.411
HVTL 220 kV near the Devyatkino subway station	0.741	0.631	0.596	0.508	0.693	0.638	-0.638	-0.652

Correlation coefficients shown in Tables 3 and 4 are indicative of the average and strong dependencies existing between the strength of an electric field and meteorological characteristics. An absolute humidity is the major meteorological characteristic, which directly influences the result. Moreover, the closest correlations, as a rule, are obtained for the same factor. The influence of remaining characteristics is indirect and is manifested as factors forming the absolute humidity. Atmospheric pressure and temperature hardly influence the strength of an electric field directly; however, the increase of temperature leads to higher evaporation, which, in its turn, results in higher absolute humidity (direct relationship), while the pressure increase (anticyclonic weather) is associated with lower humidity (inverse relationship). The correlation sharply decreases along the borders of zones of HVTL influence, up to insignificant values.

Magnetic Fields

Magnetic fields, characterized by measures of magnetic induction, do not produce such sharp maximums under high-voltage transmission lines as electric fields do, and can be described by a less rapid decrease with the increase of distance from HVTL (Figure 1), as well as by higher and omnipresent city-wide background.

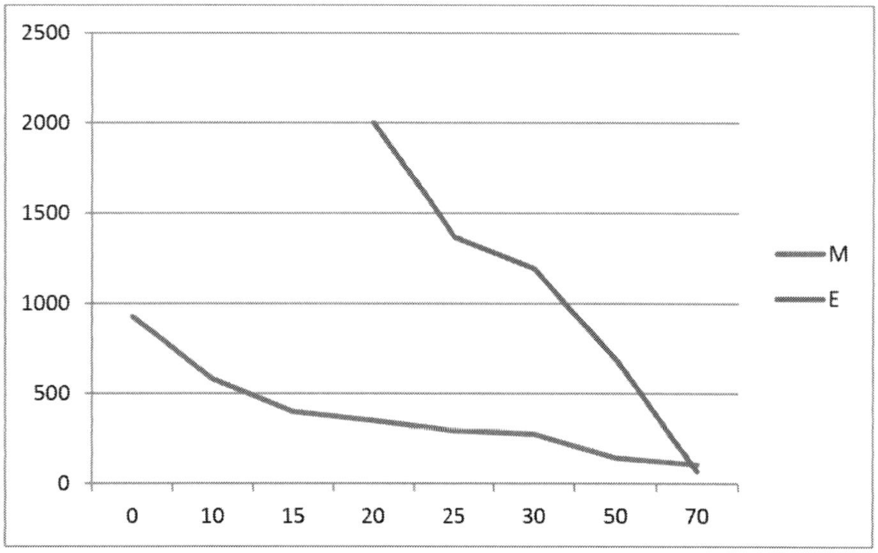

Figure 1. Mean values of strength and magnetic induction at various distances from HVTL 330 kV near the Devyatkino subway station.
Legend: E – strength of an electric field, V/m; M – magnetic induction, nT; along the horizontal axis which shows the distance in meters from the outermost wire.

Values of magnetic induction equaling 1-2 nT, i.e. those within the measurement error, were observed only in the central part of Sosnovka park, at a distance more than 0.5 kilometers from the nearest possible sources. Measures of magnetic induction both HVTL and at the monitoring points of residential areas are less variable (Table 5) and substantially less dependent or independent altogether upon meteorological characteristics (Table 6).

As relatively stable characteristics, the measures of magnetic induction demonstrate a fairly clear spatial differentiation depending on the territory use pattern and housing type (Table 7).

Abnormal values are related to the influence of cables and can be determined not only by statistical data, but also by sharp increase of values next to the ground or to the cable sheath. Comparing the characteristics of territories with different use patterns and housing types, it becomes obvious that the intensity of the power frequency magnetic fields depends on an overall man-caused environmental load.

Table 5. Characteristics of a set of measurements of magnetic induction taken at monitoring points in residential areas

Monitored locations	Number of measure-ments	Mean value, nT	Root mean square deviation (δ)	Coefficient of variation, %
15 Ushinskiy st., building 1	60	179.5	34.4	19.15
114 Grazhdanskiy avenue, building 6	57	84.6	20.4	24.1
near the Grazhdanskiy prospekt subway station	61	36.2	8.2	24.5
near the Ulitsa Dybenko subway station	34	18.3	5.9	32.1

Table 6. Coefficients of correlation between the measures of magnetic induction observed at the monitoring points in residential areas and meteorological characteristics

Monitored locations	Coefficients of correlation considering measures:			
	Absolute humidity	Relative humidity	Air tempe-rature	Atmospheric pressure
15 Ushinskiy st., building 1	-0.368	0.233	-0.413	0.125
114 Grazhdanskiy avenue, building 6	-0.321	0.267	-0.403	0.192
near the Grazhdanskiy prospekt subway station	0.355	-0.281	0.399	-0.187
near the Ulitsa Dybenko subway station	-0.502	0.292	-0.588	0.332

Table 7. Basic results of magnetic induction measurements at Saint-Petersburg's territories with different use patterns and housing types

Use pattern, housing type	Number of measure ments	Mean value, nT	Root mean square deviation (δ)	Coeffici-ent of variation, %	Number of abnormal values above 3δ limit	Mean value excluding abnormal values, nT
Densely built-up historical center	136	279				
Including streets	93	325	310	95.4	4	**271**
Including yards	30	228	211	92.4	1	**195**
Including squares and other spare areas for housing development	13	56				**56**
Mixed housing	18	145				**145**
Modern housing	94	90				
Including streets	61	68	67	99.2	1	**63**
Including space inside city blocks	33	134	205	153.2	1	**97**
Parks, gardens	30	9	10	109.2	1	**7**

4. Conclusions

Results of the study suggest that standard dimensions of sanitary-protection and exclusion zones are not sufficient to ensure the electromagnetic safety of people. The revealed dependence of the strength of electric fields of high-voltage transmission lines on meteorological conditions indicates that the dimensions of sanitary-protection zones shall be differentiated based on climatic parameters. The dimensions of sanitary-protection and exclusion zones of high-voltage transmission lines shall consider climatic factors, which influence dielectric properties of the ambient air. These zones shall be

widened in areas with a humid climate and narrowed in dry regions. There are grounds to believe that most problems with the strength of electric fields will be discovered in the territories with warm and humid climate; in case of Russia, this is, in the first place, the Black Sea coast of the Caucasus.
Measures of magnetic induction can be regarded as indirect indicators of the challenging ecological situation (geoindicators). Advantages of this bioindicator include simple and quick measurements and complete absence of dependence on previous conditions (inertia), while disadvantages include the need to take into account specific sources of fields leading to local anomalies.

References

[1] Rudakov M L 1998 *Electromagnetic fields and safety of people* (St.-Petersburg: Russkoye geograficheskoye obshchestvo publishing house) p 32
[2] Tyasto M I, Ptitsyina N G, Kopytenko Yu A et al. 1995 Influence of electromagnetic fields of natural and anthropogenic origin on the frequency of various pathologies in St.-Petersburg *Biophysics* (Moscow: Nauka) **40(4)** pp 839–847
[3] Yakovleva M I 1973 *Physiological mechanisms of action of electromagnetic fields* (St.-Petersburg: Meditsina publishing house) p 175
[4] SanPiN 2971-84. Sanitary norms and regulations for the protection of people from the influence of an electric field created by AC overhead power lines of power frequency (Moscow) 1984. 8 p. (approved by the chief public health official of the USSR on February 23 1984 № 2971-84) Available at http://www.consultant.ru/document/cons_doc_LAW_102214/ (accessed date – July 16 2017)
[5] SanPiN 2.2.1/2.1.1.1200-03. Sanitary-protection zones and sanitary classification of enterprises, buildings and other facilities (new revision) / The resolution of the chief public health official of the Russian Federation № 74 dated September 25 2007 Available at http://base.garant.ru/12158477/ (accessed date – July 16 2017)
[6] On the procedure of determining exclusion zones of power grid facilities and establishing special conditions of use of land located within such zones / The resolution of the Russia's Government № 160 dated February 24 2009 Available at http://base.garant.ru/12165555/ (accessed date – July 16 2017)
[7] Opinion on Possible effects of Electromagnetic Fields (EMF), Radio Frequency Fields (RF) and Microwave Radiation on human health expressed at the 27th CSTEE plenary meeting Brussels October 30 2001 Available at http://ec.europa.eu/health/ph_risk/committees/sct/documents/out128_en.pdf (accessed date – July 16, 2017).
[8] Hygienic Norms GN 2.1.8/2.2.4.2262-07 Maximum allowable levels of magnetic fields of 50 Hz frequency inside residential and public buildings and in residential areas / The resolution of the chief public health official of the Russian Federation № 60 dated August 21 2007 Available at http://base.garant.ru/12156058/ accessed date – July 16 2017)

IOP Conf. Series: Earth and Environmental Science **107** (2017) 012138 doi:10.1088/1755-1315/107/1/012138

THE ROLE OF FINANCIAL SECTOR IN PROVIDING SUSTAINABLE DEVELOPMENT GOALS

G F Tokareva[1], O I Shalina[2] and E E Barkova[3]

[1]Department of economy Ufa State Aviation Technical University, Ufa, Karl Marx Street, 12, 450008, Russia
E-mail: sapfo79@mail.ru
[2]Department of economy Ufa State Aviation Technical University, Ufa, Karl Marx Street, 12, 450008, Russia
E-mail: olenkash@bk.ru
[3]Department of economy Ufa State Aviation Technical University, Ufa, Karl Marx Street, 12, 450008, Russia
E-mail: k_barkova@mail.ru

Abstract. The article contains the analysis of the current ecological situation in the Republic of Bashkortostan as an oil production region of Russia. It was emerged that the decrease in emissions from means of transport caused by change in the calculation methodology. The statistical data on environmental control costs and payments for the use of natural resources in the Russian Federation for the period from 2008 to 2016 years have been analyzed. The measures aimed to create favorable conditions for the sustainable social and economic development of the region have been determined. The authors proposed alternative sources of financing the environmental protection: the issuance of ecological («green») bonds and the attraction of ecological deposits to finance environment-oriented investment projects. The main significant investment directions of the funds, obtained due to the emission of domestic ecological loan bonds, corporate ecological bonds and concessional environment-oriented bonds have been identified.

1. Introduction

The current ecological situation in the Republic of Bashkortostan (hereinafter, RB) is mainly determined by its resource-industrial potential. For many years Bashkortostan is one of the main oil-producing regions of Russia. The energetics, oil refining, chemical, gas and petrochemical industries, and oil engineering were developed on the basis of oil production.

From an ecological point of view, the Republic of Bashkortostan is determined as a complex region, since a significant concentration of harmful industries in machine engineering, chemistry, petrochemistry contributes to air and water pollution more than in neighboring regions of Russia. As in most regions of Russia, the ecological situation in RB becomes worse, since the process of harmful substances accumulation in the environment takes place, new substances are synthesized, the effect of which on humans has not been studied yet. The self-purification limits of biocoenosis near the industrialized territories are often exceeded.

The man-caused factors of the ecological disaster zone appear in the development of old oil and gas fields and around oil refineries (fresh ground waters in the south, west and north-west of the republic are unsuitable for drinking and household). There are problems of lands and water pollution with heavy metals in the mining industry of the Trans-Urals, almost permanent smog over major cities (sources - mainly motor transport and oil refineries), expanding of fresh water contamination areas (including springs) within boundaries of cities and large settlements without sewage networks and treatment facilities, the drying of forests due to the decrease of the groundwater level, the reduction of biodiversity, the locally increased radiation background on the oil industry, development of cities without taking into account environmental factors, etc.

2. The problem statement or Material and methods

In civilized countries, the main tool forcing the enterprises that harm the environment of the urban territory, to modernize production and comply with environmental legislation is the system of fines and administrative measures. In Russia offenders of the atmospheric air protection rules are threatened only by the imposition of administrative fine from 180 thousand to 250 thousand rubles, or the suspension of production activities for a period up to 90 days. In 2016, the Ministry of ecology of RB for air pollution imposed in the Republic 26 administrative fines for the sum of more than 1 million rubles. Also the activities of 7 legal entities and individual entrepreneurs were suspended for the harmful substances emission into the atmosphere without special permit.

The amount of fines imposed on polluting industries is simply ridiculous considering their multi-billion dollar turnover. It is easier for the companies to pay the fine than to spend money on replacing expensive equipment or installing filters.

In Russia the existing financial mechanism of environmental protection is disorganized and consists of separate structural units. It can be conditionally divided to the following subsystems: financing of measures for the protection of the natural environment; forecasting and preparation of environmental programs; environmental pricing and taxation; the payment for natural resources and environmental pollution; environmental insurance. Despite the importance of the named subsystems in environmental financing, the degree of development and practical use of them is different, some of them have not yet received due development.

In recent years, the economic community paid a lot of attention to the problem of studying the financing of environmental activities in Russia. A lot of scientific work has been devoted to the development of the environmental taxation directions in Russia, the calculation procedure and environmental taxes payment, among which it is necessary to mention the works of Maiburov I A, Gorsky I V, Tischenko A N, Devereux M P, Bashkirova N N, etc. [1-3]

Informational and empirical base of the research are: the Constitution of the Russian Federation, federal laws, Decrees of the President of the Russian Federation.

The statistical and comparative historical analysis was based on the data of the State Statistics Committee of Russia, the Ministry of Finance of the Russian Federation, and the Federal Tax Service of Russia; the regulatory and legislative base of the Russian Federation on incomes and expenditures of the Russian budget system, periodicals, expert and analytical centers and institutes, scientific and practical conferences.

3. Results and discussion

There are a number of serious environmental problems in the Republic of Bashkortostan.

First, the air pollution, which is directly linked to increased levels of productivity in the economy of the Republic of Bashkortostan. The main emissions about 70% from mobile sources account for enterprises of the fuel and energy complex (oil refining, petrochemical, oil mining and electric power industries).Moreover for the last six years, the emissions from industrial enterprises in the Republic have not practically reduced. So, if in 2008 the emissions of polluting substances into the atmosphere amounted to 417.4 thousand tons, in 2016 they reached 448.9 thousand tons. This means that the existing production capacity in the Republic of Bashkortostan must constantly upgrade and the best available technologies must be used in the field of environmental protection. The decrease in emissions from mobile sources from 713.4 million tons in 2008 to 334.7 thousand tons in 2016 caused by change in the calculation methodology.

Second, it is the pollution of water basin. In RB the main influence on surface water bodies provide industry and housing and utilities sector. The share of enterprises of fuel and energy, chemical and petrochemical complexes in 2016 accounted for more than 57.7% from the total volume of sewage waters discharge into surface water bodies in the Republic, the share of housing and utilities sector is 34.0%.

The share of polluted sewage waters in 2016 was about 63% (305.1 million m^3) from the total volume of sewage waters discharged to the water bodies of the Republic. The share of sewage waters classified as «insufficiently purified» in 2016 was more than 99% (304.89 million m^3) from the volume of sewage waters, classified as «contaminated» [4].

The water bodies contamination in the Republic is primarily connected with the inefficiency or lack of treatment facilities. The main reasons for their inefficient work, as in previous years, are the use of obsolete technologies and the deterioration of basic production assets. The existing treatment facilities, even if they are operating in the design mode, do not reach the established quality standards for water bodies, and the entire amount of sewage waters that have been treated cannot be classified as «normatively purified».

At the same time for the last 6 years there has been a gradual reduction in the volume of polluted sewage waters discharge into the water bodies. So, in 2010 the sewage waters volume in this category was 380.85 million m^3. Moreover, the share of contaminated sewage waters in 2010 amounted to about 70% from the total volume of waste waters discharged in water objects of the Republic.

This is achieved due to the implementation of environmental protection measures, mainly through the introduction of new or reconstruction of existing treatment facilities.

Third, it is the contamination of urban territories by household waste due to:
- imperfection of environmental protection legislation;
- high deterioration of the process equipment;
- the use at the enterprises of the Republic of high-waste technologies;
- low rate of introduction of modern resource and energy-saving technologies.

The amount of waste generation from 2008 to 2016 increased from 42.82 to 44.95 million tons. The share of used and neutralized waste in total volume of generated waste in 2016 remained on the level of 2008 and equal to 17%. The amount of waste generation of I - IV hazard classes for the last 9 years has decreased by 25%, and the share of used and neutralized waste in total volume of generated waste of I - IV hazard classes has increased from 55.4 to 61.3%.

The development of the waste management system of production and consumption in the territory of the Republic of Bashkortostan will allow by 2020 to increase to 72%the share of used and neutralized waste in total generated waste volume of I - IV hazard classes.

Former industrial development of the Republic led to the fact that at the moment, on its territory there are objects of accumulated environmental damage of previous years, which puts the population health at risk. The most important tasks at present are the neutralization and the subsequent involvement of these facilities in the economic development of the region. Implementation of the state program «Ecology and natural resources of the Republic of Bashkortostan» (hereinafter - state program) aims to create favorable conditions for sustainable social-economic development of the region and ensuring the implementation of the powers transferred to the Federal laws to subjects of the Russian Federation.

The state program provides a complex of interrelated technical, organizational, technological, economic and environmental protection (ecological) measures with appropriate financial, regulatory, scientific and informational support in accordance with the legislation.

In accordance with the Principles of the state policy in the field of ecological development of the Russian Federation for the period up to 2030, the strategic goal of the state policy in the field of ecological development is the solution of social-economic problems, providing ecologically oriented economic growth, preservation of favorable environment, biological diversity and natural resources to meet needs of present and future generations, realization of the right of every person to favorable environment, strengthening the law in the field of environmental protection and ecological safety.

Taking into account the priorities of the state policy in the field of environmental protection, the objectives of the state program are defined:

- to develop the water sector in the Republic of Bashkortostan;
- to conserve and sustainably use the natural resources of the Republic of Bashkortostan;
- to develop the waste management system of production and consumption in the territory of the Republic of Bashkortostan;
- to maintain a favorable ecological situation in the Republic of Bashkortostan.

To achieve the objectives of the Program envisages the following tasks:

- to increase the level of water supply for the population and its protection from floods and other negative water impacts;
- to conserve natural resources and biodiversity;
- to create the infrastructure for ecologically and sanitary-epidemiologically safe collection, recycling (utilization), neutralization and disposal of waste;
- to reduce the total anthropogenic load on the environment.

The implementation of these complex events requires substantial financial resources.

The volume of resource provision for the state program includes the budgetary appropriations of the Republic of Bashkortostan, the federal budget (subsidies and subventions), local budgets of municipal areas and city districts of the Republic of Bashkortostan, as well as funds from non-budgetary sources.

The total amount of financial support of the national program in 2014-2020 will amount 14048293.0 thousand rubles.

It should be noted that in the present time, for the enterprises as users of natural resources and local government institutions there are no material incentives for the construction of sewage treatment plants, gas-purifying equipment, land reclamation, etc., Complaints for environmental damage filed very seldom. The principle of «polluter pays» is not effective due to the insignificance of tariffs and penalties, absence of the plant operation interruption in case of environmental violations. Modernization of the economy requires a lot of money that can be allocated to a private owner only under government pressure and public opinion.

The creation of a sustainable system of the environmental sphere financing in Russia actualizes the importance of alternative sources of financing for environmental protection, such as: formation of nature protection funds, innovative environmental funds, nature conservation funds of enterprises; formation of ecological storages; ecological insurance; system of ecological concessional investment loans, etc.

All listed sources of financing for environmental protection are summarized under a common concept of «green financing».

The world practice is already familiar with this experience [5, 6]. For example, ecological or «green» bonds (pollution control revenue bonds) are issued by the state, municipal authorities, corporations, international financial institutions for financing projects on environmental protection:

- the World Bank Treasury, together with the Environmental Department, has been issuing «green» bonds since 2008 to support the World Bank credit activities for environmental projects to mitigate the effects of climate change and adaptation to its change. For more than 9 years «green» bonds were issued at amount over 6.4 billion dollars (more than 67 transactions in 17 currencies) [7];
- american corporations are issuing «green» bonds to finance the construction of treatment plants since the 1980s [8, 9];
- in November 2013, «green» bonds were released at market by three large global corporations: the French energy group EDF, the Swedish real estate company Group Vasakronan, and one of the largest American banks Bank of America Merill Lynch. Securities were in good demand from investors, despite the fact that, for example, if the rate of income was 1.359%, than Bank of America Merill Lynch placed bonds for $ 500 million [10]. Funds received from the «green» bonds issue, were aimed at financing different scale of investment projects, in particular the construction of an ecological farm in China and hydroelectric power plants in Chile, etc. [11]

The issuance of debt securities, which link the future of the green economy in 2016 grew by 120%, covering 1.4% of the world financial market and making $93.4 billion, largely thanks to an influx of Chinese borrowers on to the market. They account for a third of all bonds that helped to raise $32.9 billion for environmental projects. According to the Moody's Agency, the environmental bonds in 2017 will be released twice more - at $206 billion.

Green bonds are the debt instruments used to raise funds for projects related to renewable energy, energy efficiency, environmentally friendly transport or low-carbon economy. Such bonds worldwide are issued not only by banks and private companies but also by governments, municipal and local authorities. The first regional bonds were issued by the American state of Massachusetts, and municipal ones - by Gothenburg (Sweden). They were followed by the Canadian provinces Ontario, New York, Johannesburg (South Africa) and others. The major event in early 2017 was the release of environmental bonds by France government: their total value amounted to $7.5 billion, which will focus on investment in clean energy and projects to combat global warming.

Green bonds until 2016 were considered obsolete financial instrument for Russia. However, the prices decline for the hydrocarbon fuel and the global trend in the development of clean energy affected the trends in the Russian economy.

In Russia, the concepts of «green bond» and «trust bond» are missing in the regulatory framework. The Bank of Russia is considering implementation of green financing scheme, taking into account the existing international experience.

If the Central Bank of Russia will change instructions and will set a low reserve requirements for green bonds, a new class of financial assets can appear in Russia that will significantly enhance the development of infrastructure and debt capital market.

Despite the fact that Russia has still not ratified the Paris climate agreement, for further modernization and maintaining competitiveness it will still have to increase energy efficiency and develop clean technologies that are in the initial stages of development and their implementation is capital-intensive. In this regard, «green» bonds can be successfully used to raise capital in new businesses and projects.

The issuance of corporate ecological bonds will allow companies to fulfill their social obligations to the society to support environmental safety activities at the expense of the profits to invest in ecological and energy-efficient production and the creation of ecological products (goods, works).

To ensure attractiveness for investors, the interest rates on corporate environmental bonds should be at the level of ordinary corporate bonds.

Examples of the use of corporate ecological bonds in Russia are also rare. For example, the project realization for the enhancement of the river water area and the gulf of Saint-Petersburg allows providing treatment of wastewater at the 95% level. The project was carried out by the company «Vodokanal of Saint- Petersburg» on the conditions of co-financing, as well as accounting for ecological bonds for a total amount of about 5 billion rubles for a period of five years, which provided funding for about one third of the project.

Thus, in connection with the increase of the worldwide interest in environmental issues, financial institutions are able to structure and invest in the protection of the environment, become more and more popular. According to the authors opinion, financial support for the transition to a sustainable economy is intended to ensure socially responsible investment and mainly specialized bonds («green» bonds) issued by corporations, governments and international financial institutions.

References

[1] Bashkirova N 2017 Possibilities of application of the Great Britain experience in developing of ecological taxation in Russia *Proc. Ensuring Economic Security of Russia in Modern Conditions (Obespechenie Ehkonomicheskoj Bezopasnosti Rrossii v Sovremennyh Usloviyah)* (Moscow: Belyj Veter) pp 16-22

[2] Foster J, Lowe A and Winkelman S 2011 The value of green infrastructure for urban climate adaptation *Center for Clean Air Policy* p 750

[3] VasilevaY, Shalina O, Tokareva G and Baykova E 2017 Economic Growth in Russia: Influence of Oil Shock and Macro-Economic Factors *Integration and Clustering for Sustainable Economic* Growth (Switzerland: Springer) pp 553-557

[4] Burkhard B and Müller F. 2008 Indicating ecosystem health and integrity. *Sustainability In River Basins. Oekom-Verlag* (München) pp 35-58

[5] Díaz S, Fargione J, Chapin F and Tilman D 2006 Biodiversity loss threatens human well-being *PLoS biology 4***(8)** p 277

[6] García-Llorente M, Iniesta-Arandia I, Willaarts B A, Harrison P A, Berry P, Castro A J ... and Del Mar Bayo M 2015 Biophysical and sociocultural factors underlying spatial trade-offs of ecosystem services in semiarid watersheds *Ecology and Society* **20(3)**:39

[7] The site of World Bank (http://treasury.worldbank.org/greenbondsM)

[8] Arrow K J, Cropper M L, Eads G C, Hahn R W, Lave L B, Noll R G. ... and Stavins R N 1997 Is there a role for benefit-cost analysis in environmental, health and safety regulation *Environment and Development Economics* pp 195-221

[9] Motz G, Wettmann R, Gerelli E, Eland M, Cellerino R and Panella G. 1988 Environment and regional development in less favoured / declining industrial regions. Preparatory study for an ERDF-environment programme. *Final report to the Commission of the European Communities. Directorate General for Regional Policy* (Basle) p 419

[10] ESAWADI 2013 Utilizing the ecosystem services approach for water framework directive implementation synthesis report work package 5: synthesis and policy recommendations. ESAWADI project. (http: //www.esawadi.eu/IMG/pdf/ESAWADI_European_Synthesis_ENG_vf.pdf. Accessed 27 April 2017)

[11] De Groot R S, Alkemade R, Braat L, Hein L and Willemen L 2010 Challenges in integrating the concept of ecosystem services and values in landscape planning, management and decision making *Ecological complexity* 7(3) pp 260-272

The problems of formation and conservation of the green frame (green carcass) of the satellite city (on the example of Zelenodolsk)

J Zakirova[1] and S Khusnutdinova[2]

[1]Associate professor, Kazan state university of architecture and engineering, Kazan, RF
[2]Associate professor, Kazan federal university, Kazan, RF
E-mail: jzakirova@gmail.com

Abstract. The article is devoted to the study of the problems of formation and conservation of the green frame (green carcass) of a satellite city of a monocentric city agglomeration. Nowadays the green spaces fulfill not only ecological, but also social and economic functions. This is especially important for mono-industrial and satellite cities. Zelenodolsk is satellite city of Kazan agglomeration. This city has significant natural, geographical, industrial aspects. The article shows the possibilities of forming the green frame of the city and its socio-economic use.

1. Introduction

Urban landscaping is considered not only in the context of environmental problems of cities, but also closely penetrate into the social and economic challenges of modern times.

Geographically, the urban green areas are made up of urban parks or forests, parks of historical buildings, natural heritage, yards of schools and kindergartens, hospital gardens, community sport parks and stadiums, green areas of residential territories, landscaping of beaches, graveyards, linear plantings of streets, alleys, paths, along rivers, roads and other objects, green buffer zones.

Green spaces, parks and banks of a rivers and green frame (green carcass) at whole play significant role in formation of comfort city life [1-3].

At the same time, a significant proportion of urban green spaces carry or accompany social functions; therefore, the terms natural-recreational or ecological framework of urbanized territories are the closest and fully reflective of the concept of "green frame" ("green carcass").

2. The problem statement

At present, for the city administration the problems of the formation and preservation of the green frame come to the fore in the conditions of constant competition of municipalities for human resources. As practice shows, today it is largely true for small towns, where the industrial nature of the socio-economic development still remains in priority and consequently environmental issues are very acute.

The formation of a green frame plays a special role in creating comfortable conditions for satellite cities, especially in monocentric agglomerations. The topic of agglomeration has been actively discussed in the scientific literature in recent years, including the case-study [4-7].

The city of Zelenodolsk belongs to the category of medium cities. The population currently stands at 98,763 people, the area of the city is 37.73 square km. The city of Zelenodolsk has been performing the functions of the administrative center of Zelenodolsk district since 1958. According to its socio-demographic characteristics, the city is young.

The key characteristics of the city, which describe, have an influence and determine its socio-economic and, as a consequence, a spatial development, are two aspects. The first - Zelenodolsk is a classic satellite city of the monocentric Kazan agglomeration, what imposes certain risks and at the same time the opportunities for the development of the city.

The second aspect – Zelenodolsk included in the list of monoindustry cities of the Russian Federation with the most difficult socio-economic situation.

The Zelenodolsky Municipal District occupies a landscape and geographical position unique for the Republic of Tatarstan - on two banks of the Volga river (total area - 1396 square km) and has significant potential from the point of view of natural and ecological resources - the Volga-Kamsky Biosphere Reserve, the historical monument of nature "The Raifa Ensemble Desert", Sviyazhsky historical-architectural and natural-landscape complex. According to the Territorial Development Plan ("Tatinvestgrazhdanproekt", 2010) in the Zelenodolsky Municipal District, the area of forests performing forestry and forest park functions is 25048 hectares, of which 13152 hectares are forest parks, which is 27% of all forest park forests of the suburban zone Kazan. The entire forest-park part of the green zones of cities is concentrated in the northern part of the district, around the city of Zelenodolsk, the village of Vasilyevo and the city of Kazan, forming a kind of green belt.

The undeniable advantage of the city itself is a combination of a coastal position on the Volga river (according to the "Strategy RT-2030" a significant part of the Zelenodolsky municipal district and the city of Zelenodolsk itself are included in the eco-zone "Volga-Kama Stream") and the preserved green frame and significant areas of greenery. According to the General plan of the city, the current supply of green spaces in the territory of Zelenodolsk is 329.64 hectares, of which:

- green areas of common use - 129.98 hectares;
- green territory for special purposes - 173.88 hectares;
- green area of limited use - 25.78 hectares.

Currently, the area of green common areas is 13.1 m2 / person, which corresponds to the established requirements of at least 7 m2 / person for medium-sized cities. After the implementation of the activities of the general plan of Zelenodolsk woodland and park areas in the city should be more than 340 hectares.

However, there are some limitations in terms of urban planning structure: exits to the banks of the Volga largely blocked by city-forming and related industrial enterprises as well as communal and warehouse areas.

Just like in other Volga cities, river navigation is practically absent in Zelenodolsk compared to the period of the USSR, when active river navigation of both cargo and passenger traffic allowed the city to maintain active communication with the territory on the opposite shore region. Also a small shipping is at a very low level, yachting is not developed. Currently, despite the fact that the Zelenodolsky District is the only one in Tatarstan located on the two banks of the Volga, a connection with the opposite shore by means of navigation is not carried out. All these factors are potential areas of active development of the city, both economical and social, which are currently unrevealed. With an involvement of young people to the water sports (nowadays the "School of Young Sailors" functions in the city, figure 1), yachting can become a highlight of the city and an attractive touristic and recreational area.

Figure 1. View of the Volga (photo by Yu.A. Zakirova).

Actual aspects of the formation of the green framework are the needs of the city's population in the conservation of green spaces, a clear imbalance in the provision of recreational and public spaces of central and eastern residential areas, as well as recreational opportunities in natural conditions.

Zelenodolsk can become a city "host" for the weekend residents of the agglomeration center - Kazan, experiencing a significant limitation of green areas, subject to the appropriate organization of leisure and services for leisure time.

Tourists can become another segment of consumers of recreational and natural services of the green frame of Zelenodolsk, because one of the most popular destinations for travel can be called "tourist ray" from Kazan to Raifa, Vasilyevo, town-island Sviyazhsk. The green frame (green carcass) of the city is formed on the basis of the existing structural and planning organization of the city as well as natural and landscape factors. The city is planned to consist of two residential areas (mainly medium and multi-storey buildings), an array of individual residential buildings and a production zone linearly along the bank of the Volga river.

With respect to the water area, the city is located on two terraces, the difference in the marks is 56 m, with a rather steep forested slope (pine, mixed forests). On the lower terrace (the first terrace above the floodplain, 1-2 km wide) there is an extensive production area, the railway station "Zeleni Dol", a railway, a small area of individual residential development, gardens and a children's recreation camp "Chaika". On the upper terrace (the second terrace above the floodplain, located 2 km from the shore of the Volga) is the main array of residential buildings.

Forests adjacent to the city boundary belong to the Zelenodolsky, Vasilievsky, and Aishinsky forest ranges - these are forests of a forest-park and green zones. Surface water bodies are represented by the river Volga (Kuibyshev reservoir), small watercourses and several lakes (a city lake in the central part of the city, Lake Nadezhda, a system of lakes on the western and northern outskirts of the city in a forest massif, etc.).

Residential areas of the city - Central ("old" city) and Eastern (generally accepted among the population name "Mirny"), significantly differ both in the territorial planning plan, and in their natural and social saturation.

The central residential area was actively formed and built in the period from the 30th to the 90s. XX century in accordance with the norms and principles of Soviet urban development. The territory is a quarter building with a complete "set" of social and cultural objects, recreation and leisure facilities, which currently have different degrees of depreciation and improvement. Due to the fact that the area is "clamped" in planning communications (the railroad from the east side), the adjacent territory of the state forest fund (from the north and the west side), steep terrain and industrial zone (from the south), its territory did not "spread out", which allowed preserving the territorial and social balance. The eastern residential area began to be built in the late 80's. XX century. Here, as a planning unit, is adopted a microdistrict. Currently, there are 4 microdistricts built in the period of the 80's - the 90's XX century (Sectional 5-10 storey building) and the territory built up already in the XXI century. It continues to actively build up now (housing construction).

From the point of view of sustainable development the East part of city is currently not in the best condition. There is a lack of social facilities - schools, polyclinics, well-maintained green spaces, recreational and leisure open public spaces.

This difference in the planning organization of the territory and the nature of landscaping today also determine the difference in approaches and techniques in the formation of elements of the green framework and their linkage in the residential areas, as well as in the structure of individual residential areas and industrial zones.

On the territory of the Central Residential District, the elements of the green framework are formed and linked, so here the direction of activity can be defined as the reconstruction of the city nucleus (the urban forest park and the city lake, the sports and recreation center “Mayak”, the central city park), the local green spaces and linear elements of the green frame, for the restoration of "green logistics" by instruments of tactical urbanism and landscape design.

The project is proposed the formation of Large and Small boulevards, on which major public objects, green cores and local green zones are being piled. Green links go from the boulevards to the north towards the forest range and southwards to the forest zone, forming outcrops to the coastal zone.

On the territory of the Eastern residential area, there is no green frame, no green cores, no linear connections, no local green spaces, therefore, reconstruction and purposeful urban policy on the formation of a green frame are required, taking into account existing natural landscape features and the existing planning structure. Now this is the most urgent issue for the city administration, since the quality of the environment, the quality of the urban social infrastructure, is one of the determining factors in the competition of cities for human capital and the retention of the population from migration to the nearest large city. Here the main "green" directions are the Stolichnaya street and Komarov str., on which squares are formed. These connections lead to the main green core - the sports park "Maidan".

It should also be noted that the forests of the green and forest park zones outside the city are regularly and intensively used as recreation for residents of the city (both in the Central residential area and the Northeast). The purposeful formation of the green framework of the city, in particular the creation of discrete buffer recreational zones on the border of the residential zone will reduce the anthropogenic load on the natural complex in the suburban area.

Spatial solutions (compactness, planning structure laid in the era of Soviet urban planning) and a high degree of gardening allow its inhabitants to consider their city as convenient space for children life. The driving culture and the features of intraurban traffic allow people to feel safe and comfortable in the streets of the city, unlike in Kazan.

However, in the course of discussions with residents, a rather significant problem is emerging: the city does not provide opportunities for free time and self-expression for teens. At the same time, landscape features allow the city to create conditions for fashionable and popular sports, even for so-called extreme sports. Moreover, in the abandoned territories, young people independently created similar sports grounds (Figure 2), which can be the basis for creating a city of an extreme park, attractive for teenagers and young people from Kazan. The creation of sports grounds will allow attract tourists and help the townspeople to lead an active lifestyle [8-10].

Figure 2. Species points on the ravine "Maidan", the place of Sabantuy, as well as the place of spontaneous recreation of residents (author of photographs by Yu.A. Zakirova).

Special attention should be given to the youth environment, its requests and opportunities to the city, because Kazan is a magnet for young, ambitious, active people, both because of their university environment, and the diversity of places for young people - coffee shops, time-cafes, parks, bike parkings, etc. Zelenodolsk practically does not have branches of universities (the Branch of the Kazan State Technical University named after A.N. Tupolev, Zelenodolsk branch of Kazan Innovational University named after V.G. Timiryasov), which also has a negative effect on its attractiveness for senior school pupils and students.

Another very important task for the general successful development of the city, and in particular for the formation of a full-fledged green frame is the formation of exits to a large water area and the creation of recreational infrastructure in coastal areas.

The provision of access to water (the Volga) was one of the main objectives of the master plan of 1975, which was solved by organizing a city center at the site of removal from the coastal territories of a old manufacturing plant. This project solution has not been implemented. Full-scale analysis of the coastal zone (conducted in the period June-August 2017) showed that there are territorial resources for the organization of various recreational areas (wastelands, wetlands, outlets to municipal land). The existing planning structure makes it possible to ensure the availability of these potential recreational areas by public transport, bicycle traffic, and the development of residential and public (entertainment, sports) functions by redeveloping the territories of some industrial enterprises in the coastal zone, which will also lead to the development of river navigation.

3. Results and discussion

Thus, in the course of current study and project work we made a research on the topic "Green Frame Zelenodolsk". On the basis of a full-scale survey of the territory and analysis of city-planning, strategic documentation, problems and resources of the formation of the natural and recreational frame in the city were revealed the two different territorial part of the city. A model of the green frame (green carcass) of Zelenodolsk was formed, with the allocation of the order of development, the reconstruction and the creation of its elements.

In general, the formation of a green frame in the city will help to save and partially restore lost components of the natural environment, increase the biodiversity of the urban environment, improve the quality of the environment, attract private investment for the development of urban infrastructure (sports facilities, recreation and recreation areas), increase a tourist flow.

Based on full-scale conducting of the territory, analysis of town-planning, strategic documentation, problems and resources of forming a natural and recreational framework on the territory of the city were revealed. A sociological study was realized by using the method of mental maps. A model of the green frame of the city of Zelenodolsk has been formed, with the allocation of the order of development, the reconstruction and the creation of its elements.

As the design measures of citywide importance, the following is identified: the formation of natural and recreational zones in the river side and linear exits to the Volga River.

As a project for the implementation of the concept of the Green Frame formation in Zelenodolsk Central District, the following is highlighted: the reconstruction of existing natural and recreational cores, reconstruction and improvement of the central city park "Avangard", reconstruction and development of the recreational and sports base "Mayak", the formation of linear links between recreational cores, the project "Urban boulevards", the formation of buffer zones on the border between urban development and forest area in order to reduce the anthropogenic load on the natural complex.

As the project activities on the territory of the Eastern region, the following was identified: the formation of local recreational spaces in the existing building, the formation of a sports and recreational core and linear links-boulevards leading to it, the project of the sports park "Maidan", the boulevard on the Stolichnaya str. and Komarova str..

The developed concept or model of a green framework of a city gives the basis for actualization of the General plan of a city, updating of Rules of land use and building, becomes the basis for development of Rules of complex accomplishment of territories of a city and their acceptance as the local normative certificate.

Thus, the project will be used a) by the management bodies of municipal entities (cities) in the planning, development of urban projects; b) will be in demand by public organizations and municipal services, whose goals are related to the current state and improvement of the urban environment, the comfort of citizens living; c) will be used by commercial organizations for long-term planning of their activities and placement of enterprises, taking into account the specific urban environment of a particular city territory.

References

[1] Gehl J 2010 *Cities for People* (Washington - Covelo - London: Island Press) 128 p

[2] Lynch K 1960 *The Image of the City* (Cambridge, MA: MIT Press)
[3] Day C. Places of the Soul: Architecture and Environmental Design as Healing Art 2004 (Oxford, Architectural Press) p 323
[4] Fang C, Yu D 2017 Urban agglomeration: An evolving concept of an emerging phenomenon, *Landscape and Urban Planning* **162** p 126-136
[5] Sweet M N, Bullivant B and Kanaroglou P S 2017 Are major canadian city-regions monocentric, polycentric, or dispersed? *Urban Geography* **38(3)** pp 445-471
[6] Bagan H and Yamagata Y 2012 Landsat analysis of urban growth: How Tokyo became the world's largest megacity during the last 40 years *Remote Sensing of Environment* **127** pp 210-222
[7] Zakirova J A, Khusnutdinova S R and Urinova A O 2016 Formation of functional-spatial model of the sintering belt of Kazan *News of KSUAE* **1(35)** pp 87-94
[8] Hinch T and Holt N 2017 Sustaining places and participatory sport tourism events *Journal of Sustainable Tourism* **25(8)** pp 1084-1099
[9] Hodeck A and Hovemann G 2016 Motivation of active sport tourist in a German highland destination – a cross-seasonal comparision *Journal of Sport and Tourism* **20(3-4)** pp 335-348
[10] Tomik R, Kosmala G and Ardenska A 2017 Active sport tourism in Poland: environmental condition and motivational aspect *European Journal of Geography* **8(1)** pp129–138

REPRODUCTIVE INDICATORS OF LEGUMINOUS PLANTS AS A CHARACTERISTIC OF THE ECOLOGICAL STATE OF URBAN AREAS

R I Zamaletdinov[1], S M Okulova[2], E A Gavrilova[1] and A A Zakhvatova[1]

[1]Kazan (Volga) Federal University, Institute of Management, Economics and Finance, Department of Environmental Engineering and Water Management, Kazan, 420008, Russia
E-mail: i.ricinus@rambler.ru
[2]Institute of Ecology of Natural Systems of the Academy of Sciences of the Republic of Tatarstan, Kazan, 420087, Russia
E-mail: dilich1@yandex.ru

Abstract. This article examines the results of many years of research on the reproductive performance of six species of leguminous plants (FabaceaeLind., 1836) under conditions of urbanization of habitat (Kazan). The range of variability of the main reproductive indices in six species is illustrated: the potential productivity, the actual productivity of the six main types of leguminous plants. The features of variability of seed death at different stages of development are shown depending on habitat conditions. It is established that the main regularities of changes in reproductive parameters depending on habitat conditions are manifested both in native species and in the introduced species Caraganaarborescens Lam., 1785. Based on the results of the study we made conclusion about the advisability of monitoring the reproductive parameters of leguminous plants for indicating the state of the environment in a large city.

1. Introduction

At present, it is considered that the anthropogenic factor by the type of influence on biological systems does not fundamentally differ from the natural ones. However, in terms of intensity and combination of various human activities, the anthropogenic factor should be regarded as unique. The fact is that the economic activity of man in terms of the volume and depth of the transformations is now comparable to geological processes that took millions of years [1]. Modern cities represent a quintessence of the diversity of natural and anthropogenic environmental factors on living organisms. In this regard, the study of the reaction of living organisms to changes in habitat should be considered as adaptive mechanisms and apply them as a characteristic of the ecological state of urban areas.

The organism and its environment are in a dynamic equilibrium. The physical and chemical characteristics of the environment change under the influence of anthropogenic influences, which leads to a disturbance of the dynamic equilibrium of natural ecosystems. Such a situation makes studies that determine the quality of the environment and the degree of its comfort for living organisms particularly important. Basic information is required on any global and local environmental changes based on the response of various ecosystem components. However, the diversity and intensity of anthropogenic impact on the natural environment continues to grow. Evaluation of the cumulative impact of the entire variety of combinations of different effects becomes generally impossible.

The assessment of the quality of the natural environment in cities where most of the world's population lives is especially important. One of the modern and most promising methods for assessing the quality of the environment is the bioindication analysis, which provides an integral assessment of the situation, since living organisms summarize all biologically important environmental data and, without exception, reflect its state as a whole.

The process of urbanization leads to changes in the biology of each species. For a more detailed study of the ongoing processes, systematic monitoring studies are necessary. Reproduction is one of the central functions of a living organism. The probability of leaving offspring can be considered as a property of adaptability [2]. In most studies, the deterioration in the parameters of the generative sphere under the influence of anthropogenic transformation of the habitat is displayed.

This report contains results of monitoring the reproductive parameters of leguminous species in urbanized areas, using the example of Kazan.

2. Material and methods

The material for this article was collected in the territory of Kazan in the period 2000-2015. The objects of research were the following types of legumes:

- Chinales *Lathirus sylvestris* L., 1753
- Chinalug *Lathyrus pratensis* L., 1753
- Spring's spring (soviet grass) *Lathyrus vernus* L., 1753
- Caragana tree (acacia yellow) *Caragana arborescens* Lam., 1785
- Mouse *Vicia cracca* L., 1753
- Pickled vegetables *Vicia sepium* L., 1753

The choice of these species was due, firstly, to their widespread prevalence both in conditions of intense anthropogenic press, and, secondly, by the convenience of carrying out the assessment (a relatively small number of seeds in the fetus with a clear differentiation of the stages of seed development).

Assessment of reproductive parameters was carried out according to the following indicators:

1. Potential fertility.
2. The ovules.
3. Underdeveloped seeds.
4. Actual fecundity.

The indicator of potential fertility includes the sum of indicators 2-4. The ovules (seedlets) are the seeds that have stopped their development at the early stages. Undeveloped seeds are seeds that stopped their development at later stages. Actual fecundity is fully developed seeds. In other words, actual fecundity can be represented in the form of an expression of potential fertility - ovules - underdeveloped seeds = actual fertility.

Actual fecundity is fully developed seeds. In other words, actual fecundity can be represented in the form of an expression of potential fertility - ovules - underdeveloped seeds = actual fertility.

For an adequate assessment of the state of the environment, we conducted research; we used the approach of dividing city territory into anthropogenic impact gradient, which was tested earlier in similar works at other sites [3, 4].

The gradient of anthropogenic impact in city conditions was compiled on the basis of functional zoning of the territory. Within the city of Kazan, we identified four main zones [5, 6]:

- I - the industrial zone of the city (zone of the greatest anthropogenic impact);
- II zone - zone of multi-storey building;
- III - low-rise building zone;
- IV - zone outside of the city (zone of least human impact).

100-120 ture fruits were harvested for samples of each habitat.

3. Results and discussion

Habitats in urban areas for most organisms can be considered as an analogy of island populations. The presence of insurmountable reproductive barriers actually gives grounds to consider these isolated populations as independent units of the microevolutionary process [4].

Habitats in an urbanized area for most plant species can be considered as an analog of island populations, where the reproduction mechanism is well traced, which allows the implementation of microevolutionary processes [7, 8]. Thus, we can consider the dynamics of reproductive parameters as a mechanism. Currently, it is considered that the stressful conditions of the habitat lead to an increase in potential fertility [9]. This gives us grounds to consider fertility indicators as an indicator of the body's response to environmental changes.

The state of the maternal individual primarily acts as a factor limiting the indicators of seed production. The influence of pollution, weather conditions and other external factors in this case has indirect significance.

Great importance has the amount of developed seeds in the fetus or, so-called in the literature, the seeds that have been made, since they are the ones that are involved in the resumption of the population. The realization of potential seed productivity depends on the number of underdeveloped seeds and unfertilized ovules. In most of the studied species, a large percentage of death is made from unfertilized ovules, in some - underdeveloped seeds and ovules and rarely underdeveloped seeds.

According to the obtained data, with increasing intensity of anthropogenic impact (from zone IV to I), an increase in potential fertility is observed. This corresponds to the general trend, which was previously described in the literature. In fact, this means that in the conditions of urbanization, beans have an adaptive mechanism for ensuring the effectiveness of seed reproduction. Figure 1 shows the average dynamics of reproductive indicators of leguminous plants in Kazan.

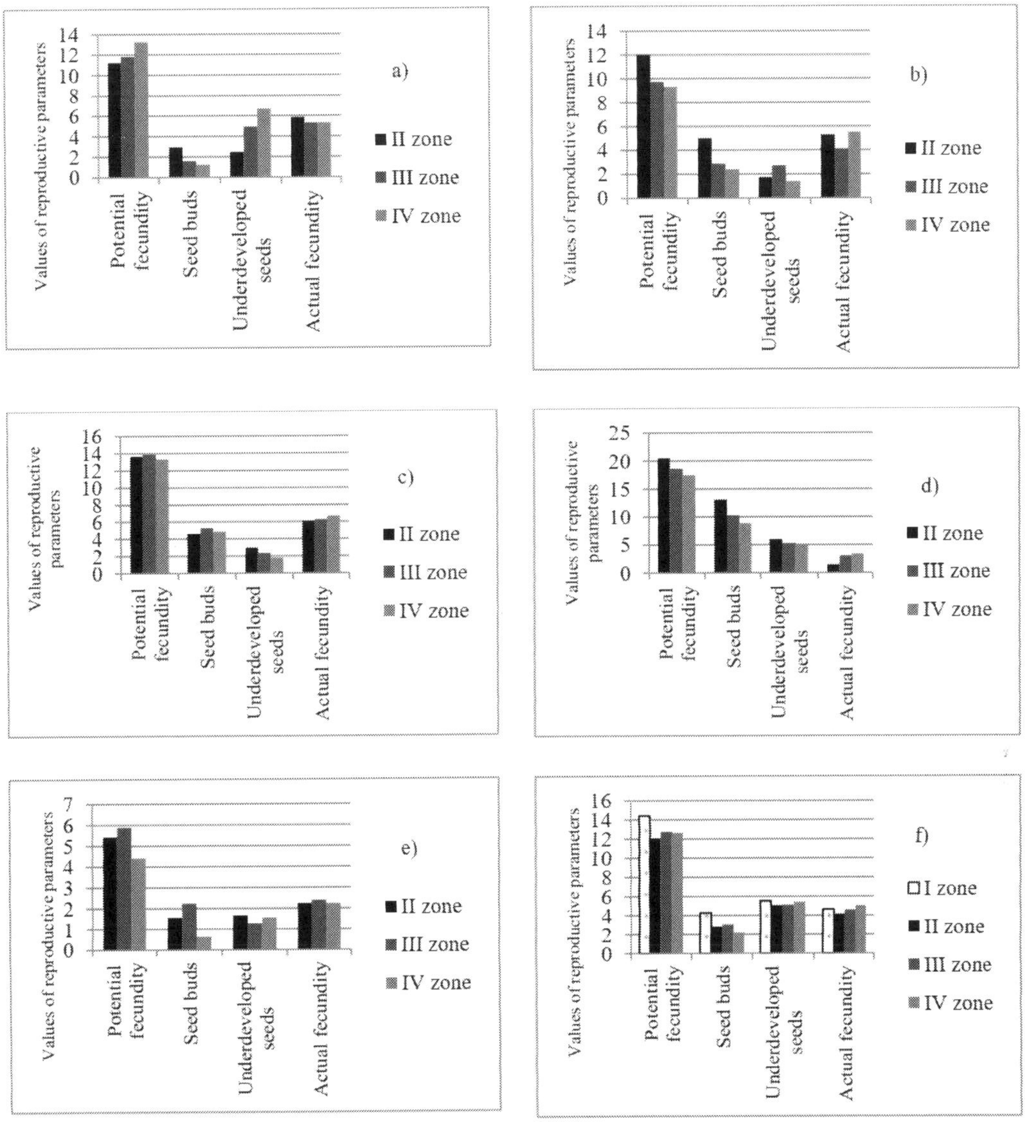

Figure 1. Dynamics of reproductive parameters of leguminous plants in the gradient of urbanization: a) *Lathirus sylvestris*; b) *Lathyrus pratensis*; c) *Lathyrus vernus*; d) *Caragana arborescens*; e) *Vicia cracca*; f) *Vicia sepium*

We found out that in different habitats the character of dynamics of reproductive parameters for each species in different habitats can vary. However, for each of the studied species, the same patterns are observed. These are the following patterns.

1. High values of potential fertility are identified for populations that have a low actual fertility.

Accordingly, low values of potential fertility, as a rule, indicate a high level of actual fertility. The increase in potential fertility should be considered in this case, as a response to the deterioration of living conditions.

2. The high mortality rate at the stage of ovules in the population is associated with a lower mortality rate at the stage of underdeveloped seeds. This picture may be the evidence of similar results that are typical for other organisms, particularly for amphibians [3]. Obviously, in this case a large role is played by a high level of stabilizing selection in conditions of increased stress from the environment [10].

3. Fertility of native species is approximately at the same level under similar growing conditions. The introduced species of *Caragana arborescens* has significantly higher values of potential fertility, as well as a high level of death at all levels of development. However, the patterns described above are common for this species. This circumstance should be considered for the possibility of using this species as an indicator of the state of the environment under urbanization conditions. According to the data received, the overall trend is a decrease in actual fertility with an increase in the urbanization gradient. A similar trend has been identified for all species.

In general, there is a tendency for an increase in mortality in the early stages of development (ovule) for populations growing under conditions of intense anthropogenic press. Large death in later stages of development (underdeveloped seeds) is more typical of populations growing in less transformed habitats.

Thus, comparative studies of changes occurring in populations of living organisms from biotopes that are subject to anthropogenic influence to varying degrees are promising. Conditions with different degrees of anthropogenic impact are important both in theoretical and applied aspects.

On the one hand, the study of reproductive parameters is an object for studying microevolutionary processes in conditions of diverse anthropogenic impact. This method of bioindication is quite informative considering all the stress factors, both chemical and physical, for estimating and making predictions of the suitability of the surrounding environment of the studied city for the urban population.

By the state of the reproductive parameters of beans, it is possible to judge the quality of the natural environment in urbanized areas; bioindication integrates all chemical and physical stress factors and is most informative for social and hygienic assessments of the suitability of the urban natural environment for humans. On the other hand, the registered features of organisms can be used for bioindication and monitoring of ecosystems [11, 12].

Acknowledgment
The authors express their sincere gratitude to Professor N M Mingazova for supporting this study.

References

[1] Vernadsky V I 1978 *A living substance* (Moscow: Science) p 358
[2] Sapunov V B 1986 Quantitative estimation of the limits of intraspecific variability *Journal common. biology* **47(6)** P 790-798
[3] Egorov Yu E 1973 *Mechanisms of divergence* (Moscow: Nauka) p 173
[4] Vershinin V L 2002 Ecological specific and microevolution of amphibian populations of urbanized territories *Advances in amphibian research in the former Soviet Union* **7** Pensoft Publishers (Moscow-Sophia) pp 1-161
[5] Zamaletdinov R I 2003 Ecology of amphibians in the conditions of a large city (on the example of Kazan) *Author's abstract dissertation candidate. Biol. Sciences (*Kazan) p 24
[6] Khayrutdinov I Z 2010 Ecology of reptiles of urbanized territories: (on the example of Kazan) *The abstract of the dissertation candidate Biological sciences* (Kazan) p 24
[7] Laurance W F 2008 Theory meets reality: How habitat fragmentation research has transcended island biogeographic theory *Biological conservation* **141** pp 1731-1744
[8] Jacquemyn H, De Meester L, Jongejans E and Honnay O 2012 Evolutionary changes in plant reproductive traits, the following habitat fragmentation and their consequences for population fitness *Journal of Ecology* **100** pp 76-87
[9] Artemiev Yu T 1980 *Biology of reproduction (microevolutionary aspects)* (Kazan: KSU) p 87
[10] Shmalgauzen I I 1946 *Factors of evolution (the theory of stabilizing selection)* (M-L: USSR) p 396
[11] Okulova S M 2005 Reproductive characteristics of urban plant populations (by the example of legumes) *Ecology of the city of Kazan (*Kazan: Fen) pp171-176
[12] Schwartz S S 1989 *Ecological patterns of evolution* (Moscow: Nauka) p 153

Institute of Physics
Dirac House, Temple Back
Bristol BS1 6BE UK

ISSN: 1755-1307
ISBN 978-1-5108-5773-5

CURRAN ASSOCIATES INC.
proceedings.com